SMITHSONIAN MISCELLANEOUS COLLECTIONS
VOLUME 71, NUMBER 1

SMITHSONIAN PHYSICAL TABLES

FOURTH REPRINT OF SEVENTH REVISED EDITION

PREPARED BY
FREDERICK E. FOWLE
AID, SMITHSONIAN ASTROPHYSICAL OBSERVATORY

(PUBLICATION 2539)

CITY OF WASHINGTON
PUBLISHED BY THE SMITHSONIAN INSTITUTION
1929

ADVERTISEMENT.

In connection with the system of meteorological observations established by the Smithsonian Institution about 1850, a series of meteorological tables was compiled by Dr. Arnold Guyot, at the request of Secretary Henry, and the first edition was published in 1852. Though primarily designed for meteorological observers reporting to the Smithsonian Institution, the tables were so widely used by physicists that it seemed desirable to recast the work entirely. It was decided to publish three sets of tables, each representative of the latest knowledge in its field, and independent of one another, but forming a homogeneous series. The first of the new series, Meteorological Tables, was published in 1893, the second, Geographical Tables, in 1894, and the third, Physical Tables, in 1896. In 1909 yet another volume was added, so that the series now comprises: Smithsonian Meteorological Tables, Smithsonian Geographical Tables, Smithsonian Physical Tables, and Smithsonian Mathematical Tables.

The fourteen years which had elapsed in 1910 since the publication of the first edition of the Physical Tables, prepared by Professor Thomas Gray, had brought such changes in the material upon which the tables must be based that it became necessary to make a radical revision for the fifth and sixth revised editions published in 1910 and 1914. The latter edition was reprinted thrice. For the present seventh revision extended changes have been made with the inclusion of new data on old and new topics.

CHARLES D. WALCOTT,
Secretary of the Smithsonian Institution.

June, 1919.

PREFACE TO 7TH REVISED EDITION.

The present edition of the Smithsonian Physical Tables entails a considerable enlargement. Besides the insertion of new data in the older tables, about 170 new tables have been added. The scope of the tables has been broadened to include tables on astrophysics, meteorology, geochemistry, atomic and molecular data, colloids, photography, etc. In the earlier revisions the insertion of new matter in a way to avoid renumbering the pages resulted in a somewhat illogical sequence of tables. This we have tried to remedy in the present edition by radically rearranging the tables; the sequence is now, — mathematical, mechanical, acoustical, thermal, optical, electrical, etc.

Many suggestions and data have been received: from the Bureau of Standards, — including the revision of the magnetic, mechanical, and X-ray tables, — from the Coast and Geodetic Survey (magnetic data), the Naval Observatory, the Geophysical Laboratory, Department of Terrestrial Magnetism, etc.; from Messrs. Adams of the Mount Wilson Observatory, Adams of the Geophysical Laboratory (compressibility tables), Anderson (mechanical tables), Dellinger, Hackh, Humphreys, Mees and Lovejoy of the Eastman Kodak Co. (photographic data), Miller (acoustical data), Van Orstrand, Russell of Princeton (astronomical tables), Saunders, Wherry and Larsen (crystal indices of refraction), White, Worthing and Forsythe and others of the Nela Research Laboratory, Zahm (aeronautical tables). To all these and others we are indebted for valuable criticisms and data. We will ever be grateful for further criticisms, the notification of errors, and new data.

FREDERICK E. FOWLE.

ASTROPHYSICAL OBSERVATORY,
SMITHSONIAN INSTITUTION,
May, 1919.

NOTE TO REPRINT OF 7TH REVISED EDITION.

Opportunity comes with this reprint to insert in the plates a number of corrections as well as some newer data. Gratitude is especially due to Messrs. Wherry and Smith of the Bureau of Chemistry, Department of Agriculture, for suggestions.

FREDERICK E. FOWLE.

ASTROPHYSICAL OBSERVATORY,
SMITHSONIAN INSTITUTION,
March, 1921.

NOTE TO SECOND REPRINT OF 7TH REVISED EDITION.

Again opportunity is taken to alter the plates for a few corrections. Several tables, especially those connected with molecular physics, have been considerably changed to allow for the rapid advances of the last few years. The data for spectrum series has been revised consistently with the now generally accepted nomenclature. Certain standard values adopted for the International Critical Tables, prepared under the auspices of the International Research Council, have been inserted. These will be found included in the table on page 408. Some of the added data has been inserted in the appendix to avoid renumbering the pages of the body of the Tables for this reprint; it will be inserted in proper sequence in the next edition. Gratitude is especially due to the members of the Bureau of Standards, of the Nela Research Laboratory, to Dr. White of the Geophysical Laboratory, Dr. Washburn of the National Research Council, and others for data and suggestions.

Frederick E. Fowle.

Astrophysical Observatory,
Smithsonian Institution,
February, 1923.

TABLE OF CONTENTS.

Introduction: units of measurement, dimensional and conversion formulae, standards: xxiii

General discussion, xxiii; Fundamental units, xxiii; Derived units, xxiv; Conversion factors and dimensional formulae, xxv; Dimensional reasoning, xxv.

Dimensional formulae: xxvi

Geometrical and mechanical units, xxvi; Heat units, xxviii; Electric and magnetic units, xxix; Electrostatic system, xxx; Electromagnetic system, xxxi.

Fundamental standards: xxxiii

Standards of length, xxxiv; Standards of mass, xxxiv; Standards of time, xxxiv; Standards of temperature, xxxiv.

Numerically different systems of units: xxxv

Proposed systems of units (table I), xxxv; Gaussian systems, xxxv; Practical electromagnetic system, xxxvi; International electric units, xxxvi.

The standards of the International Electric Units: xxxviii

Resistance, xxxviii: Mercury standards, xxxviii; Secondary standards, xxxix; Resistance standards in practice, xxxix; Absolute ohm, xxxix.

Current, xl: Silver voltameter, xl; Resistance standards used in current measurements, xli; Absolute ampere, xli.

Electromotive force, xli: International volt, xli; Weston normal cell, xli; Portable Weston cell, xliii; Absolute and semi-absolute volts, xliii.

Quantity of electricity, xliv: Standards, xliv.

Capacity, xliv.

Inductance, xliv: Inductance standards, xliv.

Power and energy, xlv: Watt, xlv; Standards and measurement, xlv.

Magnetic units, xlv: Table II. — The ordinary and ampere-turn units xlvi.

TABLE		PAGE
1.	Spelling and abbreviations of common units of weight and measure . .	2
2.	Fundamental and derived units, conversion factors	3
	(*a*) Fundamental units	3
	(*b*) Derived units	3
3.	Tables for converting U. S. weights and measures:	
	(1) Customary to metric	5
	(2) Metric to customary	6
4.	Miscellaneous equivalents U. S. and metric weights and measures . .	7

5. Equivalents of metric and British imperial weights and measures:
(1) Metric to imperial 8
(2) Multiples, metric to imperial 9
(3) Imperial to metric 10
(4) Multiples, imperial to metric 11

MATHEMATICAL TABLES

6. Derivatives and integrals 12
7. Series 13
8. Mathematical constants 14
9. Reciprocals, squares, cubes and square roots of natural numbers . . . 15
10. Logarithms, 4-place, 1000–2000 24
11. Logarithms, 4-place. 26
12. Antilogarithms, 4-place 28
13. Antilogarithms, 4-place, 0.9000–1.0000 30
14. Circular (trigonometric) functions, arguments (°, ′) 32
15. " " " " (radians) 37
16. Logarithmic factorials, n!, n = 1 to 100 40
17. Hyperbolic functions 41
18. Factorials, 1 to 20 47
19. Exponential functions 48
20. Values of e^{x^2} and e^{-x^2} and their logarithms 54
21. " " $e^{\frac{\pi}{4}x}$ " $e^{-\frac{\pi}{4}x}$ " " " 55
22. " " $e^{\frac{\sqrt{\pi}}{4}x}$ " $e^{-\frac{\sqrt{\pi}}{4}x}$ " " 55
23. " " e^{x} " e^{-x} " " " , x fractional 56
24. Least squares: probability integral, argument hx 56
25. " " " x/r 57
26. values of $0.6745\sqrt{1/(n-1)}$ 57
27. " " $0.6745\sqrt{1/n(n-1)}$ 58
28. " " $0.8453\sqrt{1/n(n-1)}$. 58
29. " " $0.8453\,\{1/n\sqrt{n-1}\}$ 58
30. formulae 59
31. Inverse probability integral, diffusion integral 60
32. Logarithms of gamma function, n between 1 and 2 62
33. Values for the first seven zonal harmonics, $\Theta = 0°$ to $\Theta = 90°$ 64
34. Cylindrical harmonics, 0th and 1st orders, $x = 0$ to 3.5, 6-place. . . . 66
35. " " " " " " $x = 4$ to 15, 4-place 68
36. (a) 1st ten roots cylindrical harmonic of zeroth order = 0 68
(b) " fifteen " " " " first " = 0 68
Notes, general formulae of Bessel's functions 68
37. Values for $\int_0^{\frac{\pi}{2}} (1 - \sin^2\Theta \sin^2\Phi)^{\pm\frac{1}{2}}\, d\Phi$; argument Θ; also logs 69
38. Moments of inertia, radii of gyration, corresponding weights 70
39. International atomic weights, valencies 71

40. Volume of glass vessel from weight of its volume of H_2O or Hg 72
41. Reductions of weighings in air to vacuo 73
42. Reductions of densities in air to vacuo 73

MECHANICAL PROPERTIES

43. Introduction and definitions 74
44. Ferrous metals and alloys: Iron and iron alloys 75
45. " " " " carbon steels 76
46. " " " " heat treatments 76
47. " " " " alloy steels 77
48. " " " " steel wire, specification values 78
49. " " " " " " experimental values 78
50. " " " " semi-steel 78
51. " " " " steel wire rope, specification values . . . 79
52. " " " " plow-steel rope, " " 79
53. " " " " steel wire rope, experimental values . . . 79
54. Aluminum, miscellaneous 80
55. Aluminum: (*a*) sheet, experimental values 80
" (*b*) " specification values 81
56. Aluminum alloys 81
57. Copper: miscellaneous experimental values 82
58. " rolled, experimental values 82
59. " wire, specification values, hard-drawn 82
60. " " medium hard-drawn 83
61. " " soft or annealed 83
62. " plates 83
63. Copper alloys: nomenclature 83
64. " " copper-zinc alloys or brasses 84
" " copper-tin " " bronzes 84
65. " " three or more metals 85
66. Miscellaneous alloys 88
67. " metals: tungsten; zinc; white metal 89
68. Cement and concrete: (*a*) cement 90
" " " (*b*) cement and cement mortars 90
" " " (*c*) concrete 91
69. Stone and clay products: (*a*) American building stones 92
" " " " (*b*) Bavarian building stones 92
" " " " (*c*) American building bricks 93
" " " " (*d*) brick piers, terra-cotta piers 93
" " " " (*e*) various bricks 93
70. (*a*) Sheet rubber 94
(*b*) Leather belting 94
71. Manilla rope 95
72. Woods: hardwoods, metric units 96
73. " conifers, metric units 97

74. Woods: hardwoods, English units 98
75. " conifers, English units 99
76. Rigidity Modulus 100
77. Variation of moduli of rigidity with the temperature 100
78. Interior friction, variation with the temperature 101
79. Hardness 101
80. Relative hardness of the elements 101
81. Poisson's ratio 101
82. Elastic moduli of crystals, formulae 102
83. " " " " numerical results 103

Compressibility of Gases

84. Compressibility of O, air, N, H, different pressures and temperatures . 104
85. " " ethylene at " " " " . 104
86. Relative gas volumes at various pressures, H, N, air, O, CO_2 104
87. Compressibility of carbon dioxide, pressure-temperature variation . . 105
88. " " gases, values of 105
89. " " air and oxygen between 18° and 22° C 105
90. Relation between pressure, temperature and volume of sulphur dioxide 106
91. " " " " " " " ammonia . . . 106

92. Compressibility of liquids 107
93. " " solids 108

Densities

94. Specific gravities corresponding to the Baumé scale 109
95. Densities of the solid and liquid elements 110
96. " " various woods 112
97. " " " solids 113
98. " " " alloys 114
99. " " " natural and artificial minerals 115
100. " " molten tin and tin-lead eutectic 115
101. Weight in grams per square meter of sheet metal 116
102. " " various common units of sheet metal 116
103. Densities of various liquids 117
104. Density of air-free water between 0° and 41° C 118
105. Relative volume of water between 0° and 36° C 119
106. Density and volume of water, −10° to 250° C 120
107. " " " " mercury, −10° to 360° C 121
108. Density of aqueous solutions of salts, bases and acids 122
109. " " " ethyl alcohol, temperature variation 124
110. " " " methyl alcohol, cane-sugar, sulphuric acid. . . . 126
111. " " various gases 127
112. Volume of gases, — values of $1 + 0.00367\,t$:
(*a*) for values of *t* between 0° and 10° C by 0.1° steps 128
(*b*) " " " " " −90° and +1990° C by 10° steps . 129

(c) logarithms for t between $-49°$ and $399°$ C by $1°$ steps . . 130
(d) " " " " $+400°$ and $1990°$ C by $10°$ steps . 132
113. Density of moist air: $h/760$, h from 1 to 9 133
114. " " " " log $h/760$, h from 80 to 800 133
115. " " " " $0.378e$ in equation $h = B - 0.378e$ 135

116. Maintenance of air at definite humidities 135
117. Pressure of mercury and water columns 136

BAROMETRIC TABLES

118. Reduction of barometer to standard temperature 137
119. " " " " " gravity, in. and mm, altitude ter n 138
120. " " " " latitude 45°, 0° to 45°, mm 139
121. " " " " " " 45° to 90°, mm 140
122. " " " " " " 0° to 45°, inches 141
123. " " " " " " 45° to 90°, inches 142
124. Correction to barometer for capillarity, mm and inches 143
125. Volume of mercury meniscus in mm^3 143
126. Barometric pressure corresponding to the boiling point of water:
(a) metric scale 144
(b) inch scale 144
127. Determination of heights by the barometer 145

ACOUSTICS

128. Velocity of sound in solids 146
129. " " " " liquids and gases 147
130. Musical scales 148
131. " " 148
132. Fundamental tone, its harmonics and equal tempered scale 149
133. Relative strength of the partials of musical instruments 149
134. Characteristics of the vowels 149
135. Miscellaneous sound data 149

AERODYNAMICS

136. Kinetics of bodies in resisting medium, Stokes law 150
137. Flow of gas through tubes 150
138. Air pressure, large square normal planes, various speeds 151
139. Correction factor for small square normal planes 151
140. Effect of aspect ratio 152
141. Ratios of pressures on inclined and normal planes 152
142. Skin friction 152
143. Variation of air resistance with aspect and angle 153
144. " " " " " shape and size 153
145. " " " " " " " " and speed 153

146. Friction 154

147. Lubricants 154
148. Lubricants for cutting tools 154

VISCOSITY

149. Viscosity of fluids and solids, general considerations 155
150. " " water in centipoises, temperature variation 155
151. " " ethyl-alcohol-water mixtures, temperature variation . . 155
152. " and density of sucrose aqueous solutions, temp. variation . 156
153. " " " " glycerol " " at 20° C 156
154. " " " " castor oil, temperature variation 156
155. " of miscellaneous liquids 157
156. " " organic liquids 158
157. Specific viscosity of solutions, density and temperature variation . . 159
158. " " " " atomic concentrations, 25° C 163
159. Viscosity of gases and vapors 164
160. " " " " " : temperature and pressure variation . 165

161. Diffusion of an aqueous solution into pure water 166
162. " " vapors 167
163. " " gases and vapors 168
164. " " metals into metals 168
165. Solubility of inorganic salts in water, temperature variation 169
166. " " a few organic salts in water, temperature variation . . 170
167. " " gases in water 170
168. " " change of, produced by uniform pressure 171
169. Absorption of gases by liquids 172
170. Capillarity and surface tension, water and alcohol in air 173
171. " " " " miscellaneous liquids in air 173
172. " " " " aqueous solutions of salts 173
173. " " " " liquids-air, -water, -mercury 174
174. " " " " liquids at solidifying point 174
175. " " " " thickness of soap films 174

VAPOR PRESSURES

176. Vapor pressures of elements 175
177. " " and rates of evaporation, Mo, W, Pt 175
178. " " organic liquids 176
179. " " of ethyl alcohol 178
180. " " " methyl alcohol 178
181. " " " (*a*) carbon disulphide 179
(*b*) chlorobenzene 179
(*c*) bromobenzene 179
(*d*) aniline 179
(*e*) methyl salicylate 180
(*f*) bromonaphthalene 180
(*g*) mercury 180

182. Vapor pressure of solutions of salts in water 181
183. Pressure of saturated water vapor over ice, low temperatures 183
184. " " " " " " water, low temperatures . . 183
185. " " " " " " " 0° to 374° C 183
186. Weight in g per m^3 of saturated water vapor 185
187. Weight in grains per ft^3 of saturated water vapor 185
188. Pressure of aqueous vapor in atmosphere, various altitudes 185
189. " " " " " " sea-level 186
190. Relative humidity, arguments, vapor pressure and dry temperature . 187
191. " " " wet and dry thermometers 189

THERMOMETRY

192. Stem correction for thermometers, centigrade 190
193. " " " " Jena glass, 0° to 360° C 190
194. " " " " " " 0° " " " 191
195. " " " " " " normal, 0° to 100° C . 191
196. Gas and mercury thermometers, formulae 192
197. Comparison of hydrogen and 16^{III} thermometers, 0° to 100° C 192
198. " " " " 59^{III} " 0° " 100° C. . . . 192
199. " " " " 16^{III} and 59^{III} thermometers, −5° to −35° C. 192
200. Comparison of air and 16^{III} thermometers, 0° to 300° C 193
201. " " " " 59^{III} " 100° to 200° C 193
202. " " hydrogen and various mercury thermometers 194
203. " " air and high temperature (59^{III}) mercury thermometer 194
204. " " H, toluol, alcohol, petrol ether, pentane thermometers 194
205. Platinum resistance thermometry 195
206. Thermodynamic scale; temperature of ice point, Kelvin scale 195
207. Standard points for the calibration of thermometers 195
208. Calibration of thermo-element, Pt−Pt·Rh 196
209. " " " Cu–constantan. 196

210. Mechanical equivalent of heat, summary to 1900 (Ames) 197
211. " " " " best value 197
212. Conversion factors, work units 197
213. English and American horse power, altitude and latitude variation . 197

MELTING AND BOILING POINTS

214. Melting points of the chemical elements 198
215. Boiling points of the chemical elements 199
216. Melting points, effect of pressure 200
217. Freezing point of water, effect of pressure 200
218. Boiling point, effect of pressure 200
219. Inorganic compounds, melting and boiling points, densities 201

220. Organic compounds, melting and boiling points, densities: 203
(a) Paraffin series 203
(b) Olefine series 203
(c) Acetylene series 204
(d) Monatomic alcohols 204
(e) Alcoholic ethers 204
(f) Ethyl ethers 204
(g) Miscellaneous 205
221. Melting points of various mixtures of metals 206
222. " " " " " " " 206
223. Low-melting-point alloys 206
224. Transformation and melting points, minerals and eutectics 207
225. Lowering of freezing points by salts in solution 208
226. Raising of boiling points by salts in solution 210
227. Freezing mixtures 211
228. Critical temperatures, pressures, volumes, densities of gases 212

THERMAL CONDUCTIVITY

229. Thermal conductivity of metals and alloys 213
230. " " insulators, high temperatures 214
231. " " various substances 214
232. " " building materials 215
233. " " various insulators 216
234. " " water and salt solutions 216
235. " " organic liquids 217
236. " " gases 217
237. Diffusivities 217

EXPANSION COEFFICIENTS

238. Linear expansion of the elements 218
239. " " " miscellaneous substances 219
240. Cubical expansion of solids 220
241. " " " liquids 221
242. " " " gases 222

SPECIFIC HEATS

243. Specific heats of elements 223
244. Heat capacities, true and mean specific heats, and latent heats of fusion of the metallic elements, 0° to 1600° C 225
245. Atomic heats, atomic volumes, specific heats at 50° K, elements . . 226
246. Specific heats of various solids 227
247. " " " water and mercury 227
248. " " " various liquids 228
249. " heat of saturated liquid ammonia, −50° to +50° C 228
250. Heat contents of saturated liquid ammonia, −50° to +50° C 228

251. Specific heats of minerals and rocks 229
252. " " (true and mean) of silicates, 0° to 1400° C 229
253. " " of gases and vapors, also c_p/c_v 230

LATENT HEATS

254. Latent heats of vaporization 231
255. " " " " formulae 232
256. " " " " ammonia 232
257. "Latent heat of pressure variation" of liquid ammonia. 232
258. Latent and total heats of vaporization of elements, — theoretical . . 233
259. Properties of saturated steam 234
260. Latent heats of fusion 240

HEATS OF COMBUSTION, FORMATION, ETC.

261. Heats of combustion of some carbon compounds 241
262. " " " " miscellaneous compounds 241
263. Heat values and analyses of various fuels: (*a*) coals and coke 242
(*b*) peats and woods . . . 242
(*c*) liquid fuels 242
(*d*) gases 242
264. Chemical and physical properties of explosives 243
265. Additional data on explosives 244
266. Ignition temperatures of gaseous mixtures 244
267. Explosive decomposition, ignition temperatures 244
268. Flame temperatures 244
269. Thermochemical data: heats of formation from elements 245
270. " " " " " of ions 246
271. " " " " neutralization. 246
272. " " " " dilution of sulphuric acid 246

RADIATION

273. Radiation formulae and constants for perfect (black-body) radiator . 247
274. " in calories for perfect radiator, various temperatures . . . 247
275. " distribution in spectrum for various temperatures 247
276. Black-body spectrum intensities, 50° to 20000° K 248
277. Relative emissive powers of various bodies for total radiation 249
278. Emissivities of metals and oxides 249
279. " " " " " 249
280. Temperature scale for tungsten, — color, black-body and true temperatures 250
281. Color minus brightness temperature for carbon 250

COOLING BY RADIATION, CONDUCTION, AND CONVECTION

282. Cooling by radiation and convection: ordinary pressures 251
283. " " " " " different pressures 251

284. Cooling by radiation and convection: very small pressures 252
285. " " " " " temperature and pressure effect 252
286. Conduction of heat across air spaces, ordinary temperatures 253
287. Convection of heat in air at ordinary temperatures 253
288. " and conduction of heat by gases at high temperatures: . 254
(a) s as function of a/B 254
(b) ϕ in watts per cm. as $f(T° K)$ 254
289. Heat losses from incandescent filaments: 255
(a) Wires of platinum sponge 255
(b) " " bright platinum 255

THE EYE AND RADIATION

290. Spectral variation of sensitiveness as function of intensity (Lumination intensities under various circumstances) 256
291. Threshold sensibility as related to field brightness 256
292. Heterochromatic threshold sensibility 257
293. Contrast or photometric sensibility 257
294. Glare sensibility 257
295. Rate of adaptation of sensibility 257
296. Apparent diameter of pupil and flux density at retina 258
297. Relative visibility of radiation of different wave-lengths 258
298. Miscellaneous eye data: physiological; persistence of vision; sensibility to small differences of color; flicker 258

PHOTOMETRIC TABLES

299. Photometric definitions and units 259
300. " standards 260
301. Intrinsic brightness of various light sources 260
302. Visibility of white lights 260
303. Brightness, Crova wave-length, mechanical equivalent of light . . . 261
304. Luminous and total intensity and radiant luminous efficiency of a black-body; minimum energy necessary for light sensation . . . 261
305. Color of light emitted by various sources 261
306. Efficiency of various electric lights 262

PHOTOGRAPHIC DATA

307. Numerical constants characteristic of a photographic plate 263
308. Relative speeds of various photographic materials 263
309. Variation of resolving power with plate and developer 263
310. Photographic efficiencies of various lights 264
311. Relative intensification of various intensifiers 264

SPECTRUM WAVE-LENGTHS

312. Wave-lengths of the Fraunhofer lines 265
313. Standards: Red cadmium line, 76 cm, 15° C, Angstroms 266

314. Standards: International secondary Fe arc standards, Angstroms . . 266
315. " International secondary Fe arc standards, Angstroms . . 266
316. " Neon wave-lengths 266
317. " International tertiary Fe arc standards, Angstroms . . . 267
318. Reduction of wave-lengths in air to standard conditions: 268
(a) $(d - d_0)/d_0 \times 1000$; B, 60 to 78 cm, t, 9° to 35° C 268
(b) $\delta = \lambda_0(n_0 - n_0')(d - d_0)/d_0$ 268
319. Spectra of the elements 269
320. Spectrum lines of the elements (Kayser) 270
321. Standard solar wave-lengths (Rowland) 272
322. Spectrum series, general discussion 275
323. " " limits of some of the series 276
324. " " first terms and vibration differences 276

Indices of Refraction

325. Indices of refraction of glass (American) 277
326. Dispersion of glasses of Table 325 277
327. Indices of refraction of glasses made by Schott and Gen, Jena. . . . 278
328. Dispersion of Jena glasses 278
329. Changes of indices for 1° C change for some Jena glasses 278
330. Index of refraction of rock-salt in air 279
331. Change of indices for 1° C change for rock-salt 279
332. Index of refraction of silvine (potassium chloride) in air 279
333. " " " " fluorite in air 280
334. Change of indices for 1° C change for fluorite 280
335. Index of refraction of iceland spar ($CaCO_3$) in air 280
336. " " " " nitroso-dimethyl-aniline. 280
337. " " " " quartz (SiO_2) 281
338. Indices of refraction for various alums 281
339. " " refraction: Selected isotropic minerals 282
340. " " " Miscellaneous isotropic solids 283
341. " " " Selected uniaxial minerals (positive) 284
" " " " " " (negative) 284
342. " " " Miscellaneous uniaxial crystals. 285
343. " " " Selected biaxial minerals (a) positive 286
" " " " " " (b) negative 287
344. " " " Miscellaneous biaxial crystals 289
345. " " " Liquefied gases, oils, fats, waxes 289
346. " " " Liquids relative to air 290
347. " " " Solutions of salts and acids relative to air . . 291
348. " " " Gases and vapors 292
349. " " " Air, 15° C, 76 cm; also corrections for reducing wave-lengths and frequencies in air to vacuo (see Table 318) . 293
350. Media for microscopic determinations of refractive indices: liquids with $n_D(0.589\mu) = 1.74$ to 1.87 294

351. Media for microscopic determinations of refractive indices: resin-like substances, $n_D(0.589\mu) = 1.88$ to 2.10 294
352. Media for microscopic determinations of refractive indices: permanent standard resinous media, $n_D = 1.546$ to 1.682 294

353. Optical constants of metals 295
354. " " " " 296

Reflecting Power

355. Reflecting power of metals (see Table 359) 296
356. Light reflected when incident light is normal to transparent medium . 297
357. " " " n is near unity or equals $1 + dn$, $i = 0°$ to $90°$ . . 297
358. " " " $n = 1.55$, $i = 0°$ to $90°$, polarization percentages. 297
359. Reflecting power of metals (see Table 355) 298
360. Percentage diffuse reflection from various substances 298
361. Reflecting power of pigments, $\lambda = 0.44\mu$ to 0.70μ 299
362. Infra-red diffuse reflecting power of dry pigments 299
363. Reflecting power of powders (white light) 300
364. Variation of reflecting power of matt and silvered surfaces with angle 300
365. Infra-red reflectivity of tungsten, temperature variation 300

Transmissive Powers

366. Transmissibility of radiation by dyes, $\lambda = 0.44\mu$ to 0.70μ 301
367. " " " " Jena glasses, 0.375 to 3.1μ 302
368. " " " " " " 0.280 to 0.644μ 302
369. " by Jena ultra-violet glasses, 0.280 to 0.397μ 302
370. " of radiation by glasses (American) 0.5 to 5.0μ . . . 303
371. " by same glasses for various lights 304
372. " of radiation by substances of Tables 330 to 338. . . 305
373. Color screens (Landolt) 306
374. " " (Wood) 306
375. " " (Jena glasses) 307
376. Transmissibility of radiation by water, $\lambda = 0.186\mu$ to 0.945μ 307
377. Transmission percentages of radiation by moist air, 0.75μ to 20μ . . 308
378. Long-wave absorption by gases, 23μ to 314μ 309
379. Properties with wave-lengths $108 \pm \mu$:
(*a*) Percentage reflection 309
(*b*) Percentage transparency 309
(*c*) Transparency of black absorbers, 2μ to 108μ 309
380. Rotation of plane of polarized light by solutions 310
381. " " " " " " " sodium chlorate and quartz . 310

382. Electrical equivalents 311

Electromotive Powers

383. Data for voltaic cells: (*a*) double-fluid cells 312
(*b*) single-fluid cells 313
(*c*) standard cells 313
(*d*) secondary cells 313
384. Contact potential differences, solids with liquids and liquids with liquids in air 314
385. Contact potential differences between metals in salt solutions 316
386. Thermoelectric power of metals 317
387. " " " alloys 318
388. " " against platinum 319
389. " " " " of platinum-rhodium alloys . 319
390. " " pressure effect 320
391. Peltier and Thomson heats, pressure effects 320
392. Peltier effect 321
393. " " Fe–constantan, Ni–Cu, 0° to 560° C 321
394. " electromotive force in millivolts 322
395. The tribo-electric series 322

Electrical Resistance

396. Auxiliary table for computing wire resistances 322
397. Resistivity of metals and some alloys, temperature coefficients . . . 323
398. Resistance of metals under pressure, temperature coefficients 326
399. Resistance of mercury and manganin under pressure 326
400. Conductivity and resistivity of miscellaneous alloys 327
401. Conducting power of alloys, temperature coefficients 328
402. Allowable carrying capacity of rubber-covered copper wires 329
403. Resistivities at high and low temperatures 330
404. Volume and surface resistivity of solid dielectrics 331
405. Variation of resistance of glass and porcelain with temperature . . . 332
(*a*) Temperature coefficients for glass, porcelain and quartz . . 332

Wire Tables

406. Tabular comparison of wire gages 333
407. Introduction; mass and volume resistivity of copper and aluminum . 334
408. Temperature coefficients of copper 335
409. Reduction to standard temperature, copper 335
410. Annealed copper wire table, English units, B. & S. gage 336
411. " " " " Metric units, B. & S. gage 339
412. Hard-drawn aluminum wire table, English units, B. & S. gage . . . 342
413. " " " " Metric units, B. & S. gage 343

414. Ratio of alternating to direct current resistances for copper wire . . 344
415. Maximum diameter of wires for high-frequency-alternating-to-direct-current ratio of 1.01 344

Electrolysis

416. Electrochemical equivalents 345
417. Conductivity of a few dilute solutions 346
418. Electrochemical equivalents and densities of nearly normal solutions 346
419. Specific molecular conductivity of solutions 347
420. " " " " " limiting values 348
421. " " " " " temperature coefficient . 348
422. Equivalent conductivity of salts, acids, bases in solution 349
423. " " " some additional salts in solution 351
424. " conductance of separate ions 352
425. Hydrolysis of ammonium acetate and ionization of water 352

Dielectric Strength

426. Steady potential for spark in air, ball electrodes 353
427. Alternating potential for spark in air, ball electrodes 353
428. Steady and alternating potential for longer sparks in air 354
429. Effect of pressure of the gas on the dielectric strength 354
430. Dielectric strength of various materials. 355
431. Potential in volts for spark in kerosene 355

Dielectric Constants

432. Specific inductive capacity of gases, atmospheric pressure 356
433. Variation of dielectric constant with the temperature (gases) 356
434. " " " " " " pressure (gases) 357
435. Dielectric constant of liquids 357
436. " " " liquids, temperature coefficients 359
437. " " " liquefied gases 359
438. " " " standard solutions for calibration 360
439. " " " solids. 360
440. " " " crystals 361

Wireless Telegraphy

441. Wave-lengths, frequencies and oscillation constants 362
442. Antennae resistances for various wave-lengths and heights 364

443. Dielectric properties of non-conductors 364

Magnetic Properties

444. Definitions and general discussion 365
445. Composition and magnetic properties of iron and steel (old data) . . 367
446. Magnetic properties of iron and steel 368
447. Cast iron in intense fields 368
448. Corrections for ring specimens 368

449. Magnetic properties of various types of iron and steel 369
450. " " " a specimen of very pure iron (0.017 % C) . . 369
451. " " " electrical sheets 369
452. " " " American magnet steel 370
453. " " " a ferro-cobalt alloy 370
454. " " " a ring sample transformer steel, weak field . . 370
455. " " " iron in very weak fields 370
456. Permeability of some specimens of Table 445 371
457. Magnetic properties of soft iron at 0° and 100° C 371
458. " " " steel at 0° and 100° C 371
459. Magnetism and temperature, critical temperature 372
460. Temperature variation for paramagnetic substances 372
461. " effect on susceptibility of diamagnetic elements 372
462. " " " " " paramagnetic elements . . . 372
463. Magnetic properties of cobalt at 0° and 100° C 373
464. " " " nickel " " " " " 373
465. " " " magnetite 373
466. " " " Lowmoor wrought iron 373
467. " " " Vicker's tool steel 373
468. " " " Hadfield's manganese steel 373
469. " " " saturation values for steels 373
470. Demagnetizing factors for rods 374
471. " " " " 374
472. Dissipation of energy in cyclic mangetization, Steinmetz constant . 375
473. Energy losses in transformer steels. 376
474. Magnetic susceptibility 377

Magneto-optic Rotation

475. Magneto-optic rotation, general discussion 378
476. " " solids, Verdet's constant 379
477. " " liquids, Verdet's constant 380
478. " " salt and acid water solutions 381
479. " " gases, Verdet's constant 382
480. Verdet's and Kundt's constants 382
481. Values of Kerr's constant 383
482. Dispersion of Kerr effect 383
483. " " " " 383

Various Magnetic Effects

484. Resistance of metals, variation in transverse magnetic field (Bi) . . 384
485. Increase of resistance in transverse magnetic field (Ni) 384
486. Change of resistance of various metals in magnetic field 384
487. Transverse galvanomagnetic and thermomagnetic effects 385
488. Variation of Hall constant with temperature 385

Cathode and Canal Rays

489. Cathode and canal rays 386
490. Speed of cathode rays 386
491. Cathodic sputtering 386

Röntgen (X-rays) Rays

492. X-rays, general properties 387
493. Röntgen secondary rays 387
494. Corpuscular rays 388
495. Intensity of X-rays; ionization 388
496. Mass absorption coefficients, λ/d 389
497. Absorption coefficients of characteristic radiations in gases 389
498. X-ray spectra and atomic numbers 390
(*a*) K-series 390
(*b*) L-series 391
(*c*) M-series 392
(*d*) Tungsten X-ray spectrum 392
499. X-ray absorption spectra and atomic numbers 393

Radioactivity

500. Relative phosphorescence excited by radium 394
501. The production of α particles (Helium) 394
502. Heating effect of radium and its emanation 394
503. Stopping powers of various substances for α rays 395
504. Absorption of β rays by various substances 395
505. " " γ rays by various substances 395
506. Table of miscellaneous properties 396
507. Total number of ions produced by the α, β, and γ rays 398
508. Amount of radium emanation 398
509. Vapor pressure of the radium emanation in cm of Hg 398
510. References to spectra of radioactive substances 398

Molecular, Atomic and Ionic Data

511. Molecular velocities 399
512. " free paths, collision frequencies and diameters 399
513. Cross sections and lengths of some organic molecules 400
514. Size of diffracting units in crystals 400
515. Electrons; Rutherford atom; Bohr atom; Magnetic field of atom . . 401
516. Electron emission from hot metals 403
517. Photo-electric effect 403
518. Ionizing potentials, resonance potentials, single line spectra 403
519. Contact (Volta) potentials 404
(*a*) Electron affinity of the elements 404
(*b*) Voltage of electrolytic cells 404
520. Ionic mobilities and diffusions, — ionic mobilities 405
521. Diffusion coefficients 405

Colloids

522. General properties of colloids 406
523. Molecular weights of colloids 406
524. Brownian movement 406
525. Adsorption of gas by finely divided particles 407
526. Heats of adsorption 407
527. Molecular heats of adsorption and liquefaction 407

528. Miscellaneous constants, atomic, molecular, etc. 408
529. Radiation wave-length limits 408
530. Periodic system of the elements (Mendelejeff) 409
531. Atomic numbers 409
532. Periodic system of the elements and radioactive isotopes (Hackh) . . 410

Astronomical Data

533. Stellar spectra and related characteristics 411
534. The Harvard spectrum classification 411
535. Apex and velocity of solar motion 411
536. Motion of the stars 412
537. Distances of the stars 412
538. Brightness of the stars 413
539. Masses and densities 413
540. Miscellaneous astronomical data 414
541. The first-magnitude stars 415
542. Wolf's observed sun-spot numbers, 1750 to 1917 415
543. Length of degrees on the earth's surface 416
544. Equation of time 416
545. Planetary data 416
546. Numbers and equivalent light of the stars 417
547. Albedos 417
548. Duration of sunshine 417
549. The solar constant 418
550. Solar spectrum energy and its transmission through the atmosphere . 418
551. Intensity of solar energy in various sections of spectrum 418
552. Distribution of brightness (radiation) over the solar disk 418
553. Transmission of radiation through moist and dry air (see Table 376) . 419
554. Brightness of sky at altitudes of 1730 m and sea-level 419
555. Relative distribution in normal spectrum of sun and sky light . . . 419
556. Air masses 419
557. Relative intensity of solar radiation for various months 420

Meteorological Data

558. Mean monthly and yearly temperatures for representative stations . 420
559. The earth's atmosphere, variation with latitude, miscellaneous . . . 421

560. Variation of percentage composition with altitude, pressure 421
561. " " temperature, pressure and density with altitude 421
562. Temperature variation over the earth's surface 422
563. " " with depth, land and ocean 422

564. Geochemical data. Meteorites 423

GEODETICAL DATA

565. Acceleration of gravity, sea-level and different latitudes and altitudes 424
566. Gravity as observed at various stations and reduced to sea-level. . . 425
567. " corrected to sea-level and for topography and compensation. 426
568. Length of seconds pendulum, sea-level and various latitudes 427
569. Miscellaneous geodetic data 427

TERRESTRIAL MAGNETISM

570. Secular change of declination, 1810 to 1920 428
571. Dip or inclination, various longitudes and latitudes, 1915 430
572. " " " secular change, 1855 to 1915 430
573. Horizontal intensity, various longitudes and latitudes, 1915 431
574. " " secular change, 1860 to 1915 431
575. Total intensity, various longitudes and latitudes 432
576. " " secular change, 1855 to 1915 432
577. Agonic line, 1800 to 1915 433
578. Mean magnetic character by months and years, 1906 to 1917 . . . 433
579. Recent values of the magnetic elements at magnetic observatories . . 434

APPENDIX, DEFINITIONS 435
Table 580. Temperature measurements 438
" 581. Adsorption of H and He by cocoanut charcoal 439
" 582. Adsorption by charcoal at low pressures and temperatures 439
" 583. Adsorption of hydrogen by palladium black 439
" 584. Audibility of sound ; f (pressure and frequency) 440
" 322. Spectrum series (concluded from p. 275) 441
" 518. Ionizing and resonance potentials (concl'd from p. 403) . 442
" 585. List of stars known to be within five parsecs 443
INDEX 445

The following tables appearing in the 6th revised edition of these tables have been omitted:
Table 222. Newton's rings; thicknesses corresponding to various colors.
" 292. Ratio of the electrostatic to the electromagnetic unit of electricity = v (historical).
" 293. Absolute measurement of currents and the E.M.F. of standard cells (historical).
" 295. Contact difference of potential in volts; solids with liquids and liquids with liquids.
" 305. Various determinations of the ohm (historical).

INTRODUCTION.

UNITS OF MEASUREMENT. DIMENSIONAL AND CONVERSION FORMULAE.

The quantitative measure of anything is expressed by two factors, — one, a certain definite amount of the kind of physical quantity measured, called the unit, the other, the number of times this unit is taken. A distance is stated as 5 meters. The purpose in such a statement is to convey an idea of this distance in terms of some familiar or standard unit distance. Similarly quantity of matter is referred to as so many grams; of time, as so many seconds, or minutes, or hours.

The numerical factor definitive of the magnitude of any quantity must depend on the size of the unit in terms of which the quantity is measured. For example, let the magnitude factor be 5 for a certain distance when the mile is used as the unit of measurement. A mile equals 1760 yards or 5280 feet. The numerical factor evidently becomes 8800 and 26400, respectively, when the yard or the foot is used as the unit. Hence, to obtain the magnitude factor for a quantity in terms of a new unit, multiply the old magnitude factor by the ratio of the magnitudes of the old and new units; that is, by the number of the new units required to make one of the old.

The different kinds of quantities measured by physicists fall fairly definitely into two classes. In one class the magnitudes may be called extensive, — in the other, intensive. To decide to which class a quantity belongs, it is often helpful to note the effect of the addition of two equal quantities of the kind in question. If twice the quantity results, then the quantity has extensive (additive) magnitude. For instance, two pieces of platinum, each weighing 5 grams, added together, weigh 10 grams; on the other hand, the addition of one piece of platinum at 100° C to another at 100° C does not result in a system at 200° C. Volume, entropy, energy may be taken as typical of extensive, — density, temperature and magnetic permeability, of intensive magnitudes.

The measurement of quantities having extensive magnitude is a comparatively direct process. Those having intensive magnitude must be correlated with phenomena which may be measured extensively. In the case of temperature, a typical quantity with intensive magnitude, various methods of measurement have been devised, such as the correlation of magnitudes of temperature with the varying lengths of a thread of mercury.

Fundamental Units. — It is desirable that the fewest possible fundamental unit quantities should be chosen. Simplicity should regulate the choice, — simplicity 1st, psychologically, in that they should be easy to grasp mentally, and 2nd, physically, in permitting as straightforward and simple definition as possible of the complex relationships involving them. Further it seems desirable that the units should be extensive in nature. It has been found possible to express all measurable physical quantities in terms of five such units: 1st, geometrical considerations — length, surface, etc., — lead to the need of a length; 2nd, kinematical considerations — velocity, acceleration, etc., — introduce time; 3rd, mechanics — treating of masses instead of immaterial points — introduces matter with the need of a fundamental unit of mass; 4th, electrical, and 5th, thermal considerations require two more such quantities. The discovery of new classes of phenomena may require further additions.

As to the first three fundamental quantities, simplicity and good use sanction the choice of a length, L, a time interval, T, and a mass, M. For the measurement of electrical quantities, good use has sanctioned two fundamental quantities, — the dielectric constant, K, the basis of the "electrostatic" system and the magnetic permeability, μ, the basis of the "electromagnetic" system. Besides these two systems involving electrical considerations, there is in common use a third one called the "international" system which will be referred to later. For the fifth, or thermal fundamental unit, temperature is generally chosen.[1]

Derived Units. — Having selected the fundamental or basic units, — namely, a measure of length, of time, of mass, of permeability or of the dielectric constant, and of temperature, — it remains to express all other units for physical quantities in terms of these. Units depending on powers greater than unity of the basic units are called "derived units." Thus, the unit volume is the volume of a cube having each edge a unit of length. Suppose that the capacity of some volume is expressed in terms of the foot as fundamental unit and the volume number is wished when the yard is taken as the unit. The yard is three times as long as the foot and therefore the volume of a cube whose edge is a yard is $3 \times 3 \times 3$ times as great as that whose edge is a foot. Thus the given volume will contain only 1/27 as many units of volume when the yard is the unit of length as it will contain when the foot is the unit. To transform from the foot as old unit to the yard as new unit, the old volume number must be multiplied by 1/27, or by the ratio of the magnitude of the old to that of the new unit of volume. This is the same rule as already given, but it is usually more convenient to express the transformations in terms of the fundamental units directly. In the present case, since, with the method of measurement here adopted, a volume number is the cube of a length-number, the ratio of two units of volume is the cube of the ratio of the intrinsic values of the two units of length. Hence, if l is the ratio of the magnitude of the old to that of the new unit of length, the ratio of the corresponding units of volume is l^3. Similarly the ratio of two units of area would be l^2, and so on for other quantities.

[1] Because of its greater psychological and physical simplicity, and the desirability that the unit chosen should have extensive magnitude, it has been proposed to choose as the fourth fundamental quantity, a quantity of electrical charge, e. The standard unit of electrical charge would then be the electronic charge. For thermal needs, entropy has been proposed. While not generally so psychologically easy to grasp as temperature, entropy is of fundamental importance in thermodynamics and has extensive magnitude. (R. C. Tolman, The Measurable Quantities of Physics, Physical Review, 9, p. 237, 1917.)

Conversion Factors and Dimensional Formulae. — For the ratios of length, mass, time, temperature, dielectric constant and permeability units the small bracketed letters, $[l]$, $[m]$, $[t]$, $[\theta]$, $[k]$, and $[\mu]$ will be adopted. These symbols will always represent simple numbers, but the magnitude of the number will depend on the relative magnitudes of the units the ratios of which they represent. When the values of the numbers represented by these small bracketed letters as well as the powers of them involved in any particular unit are known, the factor for the transformation is at once obtained. Thus, in the above example, the value of l was $1/3$, and the power involved in the expression for volume was 3; hence the factor for transforming from cubic feet to cubic yards was l^3 or $1/3^3$ or $1/27$. These factors will be called *conversion factors.*

To find the symbolic expression for the conversion factor for any physical quantity, it is sufficient to determine the degree to which the quantities length, mass, time, etc., are involved. Thus a velocity is expressed by the ratio of the number representing a length to that representing an interval of time, or $[L/T]$, and acceleration by a velocity number divided by an interval-of-time number, or $[L/T^2]$, and so on, and the corresponding ratios of units must therefore enter in precisely the same degree. The factors would thus be for the just stated cases, $[l/t]$ and $[l/t^2]$. Equations of the form above given for velocity and acceleration which show the dimensions of the quantity in terms of the fundamental units are called *dimensional equations.* Thus $[E] = [ML^2T^{-2}]$ will be found to be the dimensional equation for energy, and $[ML^2T^{-2}]$ the dimensional formula for it. These expressions will be distinguished from the conversion factors by the use of bracketed capital letters.

In general, if we have an equation for a physical quantity,

$$Q = CL^aM^bT^c,$$

where C is a constant and L, M, T represent length, mass, and time in terms of one set of units, and it is desired to transform to another set of units in terms of which the length, mass, and time are $L_{\prime}$, $M_{\prime}$, $T_{\prime}$, we have to find the value of $L_{\prime}/L$, $M_{\prime}/M$, $T_{\prime}/T$, which, in accordance with the convention adopted above, will be l, m, t, or the ratios of the magnitudes of the old to those of the new units.

Thus $L_{\prime} = Ll$, $M_{\prime} = Mm$, $T_{\prime} = Tt$, and if $Q_{\prime}$ be the new quantity number,

$$\begin{aligned} Q_{\prime} &= CL_{\prime}^aM_{\prime}^bT_{\prime}^c, \\ &= CL^al^aM^bm^bT^ct^c = Ql^am^bt^c, \end{aligned}$$

or the conversion factor is $[l^am^bt^c]$, a quantity precisely of the same form as the dimension formula $[L^aM^bT^c]$.

Dimensional equations are useful for checking the validity of physical equations. Since physical equations must be homogeneous, each term appearing in them must be dimensionally equivalent. For example, the distance moved by a uniformly accelerated body is $s = v_0t + \frac{1}{2}at^2$. The corresponding dimensional equation is $[L] = [(L/T)T] + [(L/T^2)T^2]$, each term reducing to $[L]$.

Dimensional considerations may often give insight into the laws regulating physical phenomena.[1] For instance Lord Rayleigh, in discussing the intensity

[1] See "On Physically Similar Systems; Illustrations of the Use of Dimensional Equations," E. Buckingham, Physical Review, (2) 4, 345, 1914; also Phil. Mag. 42, 696, 1921.

of light scattered from small particles, in so far as it depends upon the wave-length, reasons as follows:[1]

"The object is to compare the intensities of the incident and scattered ray; for these will clearly be proportional. The number (i) expressing the ratio of the two amplitudes is a function of the following quantities: — T, the volume of the disturbing particle; r, the distance of the point under consideration from it; λ, the wave-length; b, the velocity of propagation of light; D and D', the original and altered densities: of which the first three depend only on space, the fourth on space and time, while the fifth and sixth introduce the consideration of mass. Other elements of the problem there are none, except mere numbers and angles, which do not depend upon the fundamental measurements of space, time, and mass. Since the ratio i, whose expression we seek, is of no dimensions in mass, it follows at once that D and D' occur only under the form $D : D'$, which is a simple number and may therefore be omitted. It remains to find how i varies with T, r, λ, b.

"Now, of these quantities, b is the only one depending on time; and therefore, as i is of no dimensions in time, b cannot occur in its expression. We are left, then, with T, r, and λ; and from what we know of the dynamics of the question, we may be sure that i varies directly as T and inversely as r, and must therefore be proportional to $T \div \lambda^2 r$, T being of three dimensions in space. In passing from one part of the spectrum to another λ is the only quantity which varies, and we have the important law:

"When light is scattered by particles which are very small compared with any of the wave-lengths, the ratio of the amplitudes of the vibrations of the scattered and incident light varies inversely as the square of the wave-length, and the intensity of the lights themselves as the inverse fourth power."

The dimensional and conversion-factor formulae for the more commonly occurring derived units will now be developed.

Area is referred to a unit square whose side is the unit of length. The area of a surface is expressed as

$$S = CL^2,$$

where the constant C depends on the contour of the surface and L is a linear dimension. If the surface is a square and L the length of a side, C is unity; if a circle and L its diameter, C is $\pi/4$. The dimensional formula is therefore $[L^2]$ and the conversion factor $[l^2]$. (Since the conversion factors are always of the same dimensions as the dimensional formulae they will be omitted in the subsequent discussions. A table of them will be found on page 3.)

Volume is referred to a unit cube whose edge is the unit of length. The volume of a body is expressed as

$$V = CL^3.$$

The constant C depends on the shape of the bounding surfaces. The dimensional formula is $[L^3]$.

Density is the quantity of matter per unit volume. The dimensional formula is $[M/V]$ or $[ML^{-3}]$.

Ex. — The density of a body is 150 pd. per cu. ft.: required the density in grains per cu. in. Here m, the number of grains in a pd., = 7000; l, the number of in. in a ft., = 12; $ml^{-3} = 7000/12^3 = 4.051$. The density is $150 \times 4.051 = 607.6$ grains/cu. in.

The specific gravity of a body is the ratio of a density to the density of a standard substance. The dimensional formula and conversion factor are both unity.

[1] Philosophical Magazine, (4) 41, p. 107, 1871.

Velocity, v, of a body is dL/dt, or the ratio of a length to a time. The dimensional formula is $[LT^{-1}]$.

Angle is measured by the ratio of the length of an arc to its radius. The dimensional formula is unity.

Angular Velocity is the ratio of the angle described in a given time to that time. The dimensional formula is $[T^{-1}]$.

Linear Acceleration is the rate of change of velocity or $a = dv/dt$. The dimensional formula is $[VT^{-1}]$ or $[LT^{-2}]$.

Ex. — A body acquires velocity at a uniform rate and at the end of one minute moves at the rate of 20 kilometers per hour: what is the acceleration in centimeters per second per second? Since the velocity gained was 20 km per hour in one minute, the acceleration was 1200 km per hour per hour. $l = 100000$, $t = 3600$, $lt^{-2} = 100000/3600^2 = 0.00771$; the acceleration = $.00771 \times 1200 = 9.26$ cm/sec.

Angular Acceleration is rate of change of angular velocity. The dimensional formula is [(angular velocity)/T] or $[T^{-2}]$.

Momentum, the quantity of motion in the Newtonian sense, is measured by the product of the mass and velocity of the body. The dimensional formula is $[MV]$ or $[MLT^{-1}]$.

Moment of Momentum of a body with reference to a point is the product of its momentum by the distance of its line of motion from the point. The dimensional formula is $[ML^2T^{-1}]$.

Moment of Inertia of a body round an axis is expressed by the formula Σmr^2, where m is the mass of any particle of the body and r its distance from the axis. The dimensional formula for the sum is the same as for each element and is $[ML^2]$.

Angular Momentum of a body is the product of its moment of inertia and angular velocity. The dimensional formula is $[ML^2T^{-1}]$.

Force is measured by the rate of change of momentum it can produce. The dimensional formulae for force and "time rate of change of momentum" are therefore the same, the ratio of a momentum to a time $[MLT^{-2}]$.

Ex. — When mass is expressed in lbs., length in ft., and time in secs., the unit force is called the poundal. When grams, cms, and secs. are the corresponding units, the unit of force is called the dyne. Find the number of dynes in 25 poundals. Here $m = 453.59$, $l = 30.48$, $t = 1$; $mlt^{-2} = 453.59 \times 30.48 = 13825$ nearly. The number of dynes is $13825 \times 25 = 345625$ approximately.

Moment of Couple, Torque, or **Twisting Motive** can be expressed as the product of a force and a length. The dimensional formula is $[FL]$ or $[ML^2T^{-2}]$.

Intensity of Stress is the ratio of the total stress to the area over which the stress is distributed. The dimensional formula is $[FL^{-2}]$ or $[ML^{-1}T^{-2}]$.

Intensity of Attraction, or **"Force at a Point,"** is the force of attraction per unit mass on a body placed at the point. The dimensional formula is $[FM^{-1}]$ or $[LT^{-2}]$, the same as acceleration.

Absolute Force of a Center of Attraction, or "**Strength of a Center,**" is the intensity of force at unit distance from the center, and is the force per unit mass at any point multiplied by the square of the distance from the center. The dimensional formula is $[FL^2M^{-1}]$ or $[L^3T^{-2}]$.

Modulus of Elasticity is the ratio of stress intensity to percentage strain. The dimensional of percentage strain, a length divided by a length, is unity. Hence the dimensional formula of a modulus of elasticity is that of stress intensity $[ML^{-1}T^{-2}]$.

Work is done by a force when the point of application of the force, acting on a body, moves in the direction of the force. It is measured by the product of the force and the displacement. The dimensional formula is $[FL]$ or $[ML^2T^{-2}]$.

Energy. — The work done by the force produces either a change in the velocity of the body or a change of its shape or configuration, or both. In the first case it produces a change of kinetic energy, in the second, of potential energy. The dimensional formulae of energy and work, representing quantities of the same kind, are identical $[ML^2T^{-2}]$.

Resilience is the work done per unit volume of a body in distorting it to the elastic limit or in producing rupture. The dimensional formula is $[ML^2T^{-2}L^{-3}]$ or $[ML^{-1}T^{-2}]$.

Power or Activity is the time rate of doing work, or if W represents work and P power, $P = dw/dt$. The dimensional formula is $[WT^{-1}]$ or $[ML^2T^{-3}]$, or for problems in gravitation units more conveniently $[FLT^{-1}]$, where F stands for the force factor.

Exs. — Find the number of gram-cms in one ft.-pd. Here the units of force are the attraction of the earth on the pound and the gram of matter. (In problems like this the terms "grams" and "pd." refer to force and not to mass.) The conversion factor is $[fl]$, where f is 453.59 and l is 30.48. The answer is $453.59 \times 30.48 = 13825$.

Find the number of ft.-poundals in 1000000 cm-dynes. Here $m = 1/453.59$, $l = 1/30.48$, $t = 1$; $ml^2t^{-2} = 1/453.59 \times 30.48^2$, and $10^6ml^2t^{-2} = 10^6/453.59 \times 30.48^2 = 2.373$.

If gravity produces an acceleration of 32.2 ft./sec./sec., how many watts are required to make one horse-power? One horse-power is 550 ft.-pds. per sec., or $550 \times 32.2 = 17710$ ft.-poundals per second. One watt is 10^7 ergs per sec., that is, 10^7 dyne-cms per sec. The conversion factor is $[ml^2t^{-3}]$, where m is 453.59, l is 30.48, and t is 1, and the result has to be divided by 10^7, the number of dyne-cms per sec. in the watt. $17710\, ml^2t^{-3}/10^7 = 17710 \times 453.59 \times 30.48^2/10^7 = 746.3$.

HEAT UNITS.

Quantity of Heat, measured in dynamical units, has the same dimensions as energy $[ML^2T^{-2}]$. Ordinary measurements, however, are made in *thermal units*, that is, in terms of the amount of heat required to raise the temperature of a unit mass of water one degree of temperature at some stated temperature. This involves the unit of mass and some unit of temperature. If we denote temperature numbers by Θ, the dimensional formula for quantity of heat, H, will be $[M\Theta]$. Unit volume is sometimes used instead of unit mass in the measurement of heat, the units being called *thermometric units*. The dimensional formula now changed by the substitution of volume for mass is $[L^3\Theta]$.

Specific Heat is the relative amount of heat, compared with water as standard substance, required to raise unit mass of different substances one degree in temperature and is a simple number.

Coefficient of Thermal Expansion of a substance is the ratio of the change of length per unit length (linear), or change of volume per unit volume (voluminal), to the change of temperature. These ratios are simple numbers, and the change of temperature varies inversely as the magnitude of the unit of temperature. The dimensional formula is $[\Theta^{-1}]$.

Thermal Conductivity, or **Specific Conductance,** is the quantity of heat, H, transmitted per unit of time per unit of surface per unit of temperature gradient. The equation for conductivity is therefore $K = H/L^2T\Theta/L$, and the dimensional formula $[H/\Theta LT] = [ML^{-1}T^{-1}]$ in thermal units. In thermometric units the formula becomes $[L^2T^{-1}]$, which properly represents diffusivity, and in dynamical units $[MLT^{-3}\Theta^{-1}]$.

Thermal Capacity is mass times the specific heat. The dimensional formula is $[M]$.

Latent Heat is the quantity of heat required to change the state of a body divided by the quantity of matter. The dimensional formula is $[M\Theta/M]$ or $[\Theta]$; in dynamical units it is $[L^2T^{-2}]$.

NOTE. — When Θ is given the dimensional formula $[L^2T^{-2}]$, the formulae in thermal and dynamical units are identical.

Joule's Equivalent, J, is connected with the quantity of heat by the equation $ML^2T^{-2} = JH$ or $JM\Theta$. The dimensional formula of J is $[L^2T^{-2}\Theta^{-1}]$. In dynamical units J is a simple number.

Entropy of a body is directly proportional to the quantity of heat it contains and inversely proportional to its temperature. The dimensional formula is $[M\Theta/\Theta]$ or $[M]$. In dynamical units the formula is $[ML^2T^{-2}\Theta^{-1}]$.

EXS. — Find the relation between the British thermal unit, the large or kilogram-calorie and the small or gram-calorie, sometimes called the "therm." Referring all the units to the same temperature of the standard substance, the *British thermal unit* is the amount of heat required to warm one pound of water 1° C, the *large calorie*, 1 kilogram of water, 1° C, the *small calorie* or *therm*, 1 gram, 1° C. (1) To find the number of kg-cals. in one British thermal unit. $m = .45359$, $\theta = 5/9$; $m\theta = .45359 \times 5/9 = .25199$. (2) To find the number therms in one kg-cal. $m = 1000$, and $\theta = 1$; $m\theta = 1000$. (3) Hence the number of small calories or therms in one British thermal unit is $1000 \times .25199 = 251.99$.

ELECTRIC AND MAGNETIC UNITS.

A system of units of electric and magnetic quantities requires four fundamental quantities. A system in which length, mass, and time constitute three of the fundamental quantities is known as an "absolute" system. There are two absolute systems of electric and magnetic units. One is called the electrostatic, in which the fourth fundamental quantity is the dielectric constant, and one is called the electromagnetic, in which the fourth fundamental quantity is magnetic permeability. Besides these two systems there will be described a third in common use called the "international" system.

In the electrostatic system, unit quantity of electricity, Q, is the quantity which exerts unit mechanical force upon an equal quantity a unit distance from it in a vacuum. From this definition the dimensions and the units of all the other electric and magnetic quantities follow through the equations of the mathematical theory of electromagnetism. The mechanical force between two quantities of electricity in any medium is

$$F = \frac{QQ'}{Kr^2},$$

where K is the dielectric constant, characteristic of the medium, and r the distance between the two points at which the quantities Q and Q' are located. K is the fourth quantity entering into dimensional expressions in the electrostatic system. Since the dimensional formula for force is $[MLT^{-2}]$, that for Q is $[M^{\frac{1}{2}}L^{\frac{3}{2}}T^{-1}K^{\frac{1}{2}}]$.

The electromagnetic system is based upon the unit of the magnetic pole strength. The dimensions and the units of the other quantities are built up from this in the same manner as for the electrostatic system. The mechanical force between two magnetic poles in any medium is

$$F = \frac{mm'}{\mu r^2},$$

in which μ is the permeability of the medium and r is the distance between two poles having the strengths m and m'. μ is the fourth quantity entering into dimensional expressions in the electromagnetic system. It follows that the dimensional expression for magnetic pole strength is $[M^{\frac{1}{2}}L^{\frac{3}{2}}T^{-1}\mu^{\frac{1}{2}}]$.

The symbols K and μ are sometimes omitted in the dimensional formulae so that only three fundamental quantities appear. There are a number of objections to this. Such formulae give no information as to the relative magnitudes of the units in the two systems. The omission is equivalent to assuming some relation between mechanical and electrical quantities, or to a mechanical explanation of electricity. Such a relation or explanation is not known.

The properties K and μ are connected by the equation $1/\sqrt{K\mu} = v$, where v is the velocity of an electromagnetic wave. For empty space or for air, K and μ being measured in the same units, $1/\sqrt{K\mu} = c$, where c is the velocity of light in vacuo, 3×10^{10} cm per sec. It is sometimes forgotten that the omission of the dimensions of K or μ is merely conventional. For instance, magnetic field intensity and magnetic induction apparently have the same dimensions when μ is omitted. This results in confusion and difficulty in understanding the theory of magnetism. The suppression of μ has also led to the use of the "centimeter" as a unit of capacity and of inductance; neither is physically the same as length.

ELECTROSTATIC SYSTEM.

Quantity of Electricity has the dimensional formula $[M^{\frac{1}{2}}L^{\frac{3}{2}}T^{-1}K^{\frac{1}{2}}]$, as shown above.

Electric Surface Density of an electrical distribution at any point on a surface is measured by the quantity per unit area. The dimensional formula is the ratio of the formulae for quantity of electricity and for area or $[M^{\frac{1}{2}}L^{-\frac{1}{2}}T^{-1}K^{\frac{1}{2}}]$.

Electric Field Intensity is measured by the ratio of the force on a quantity of electricity at a point to the quantity of electricity. The dimensional formula is therefore the ratio of the formulae for force and electric quantity or $[MLT^{-2}/M^{\frac{1}{2}}L^{\frac{3}{2}}T^{-1}K^{\frac{1}{2}}]$ or $[M^{\frac{1}{2}}L^{-\frac{1}{2}}T^{-1}K^{-\frac{1}{2}}]$.

Electric Potential and Electromotive Force. — Change of potential is proportional to the work done per unit of electricity in producing the change. The dimensional formula is the ratio of the formulae for work and electrical quantity or $[ML^2T^{-2}/M^{\frac{1}{2}}L^{\frac{3}{2}}T^{-1}K^{\frac{1}{2}}]$ or $[M^{\frac{1}{2}}L^{\frac{1}{2}}T^{-1}K^{-\frac{1}{2}}]$.

Capacity of an Insulated Conductor is proportional to the ratio of the quantity of electricity in a charge to the potential of the charge. The dimensional formula is the ratio of the two formulae for electric quantity and potential or $[M^{\frac{1}{2}}L^{\frac{3}{2}}T^{-1}K^{\frac{1}{2}}/M^{\frac{1}{2}}L^{\frac{1}{2}}T^{-1}K^{-\frac{1}{2}}]$ or $[LK]$.

Specific Inductive Capacity is the ratio of the inductive capacity of the substance to that of a standard substance and therefore is a number.

Electric Current is quantity of electricity flowing past a point per unit of time. The dimensional formula is the ratio of the formulae for electric quantity and for time or $[M^{\frac{1}{2}}L^{\frac{3}{2}}T^{-1}K^{\frac{1}{2}}/T]$ or $[M^{\frac{1}{2}}L^{\frac{3}{2}}T^{-2}K^{\frac{1}{2}}]$.

Electrical Conductivity, like the corresponding term for heat, is quantity per unit area per unit potential gradient per unit of time. The dimensional formula is $[M^{\frac{1}{2}}L^{\frac{3}{2}}T^{-1}K^{\frac{1}{2}}/L^2(M^{\frac{1}{2}}L^{\frac{1}{2}}T^{-1}K^{-\frac{1}{2}}/L)T]$ or $[T^{-1}K]$.

Resistivity is the reciprocal of conductivity. The dimensional formula is $[TK^{-1}]$.

Conductance of any part of an electric circuit, not containing a source of electromotive force, is the ratio of the current flowing through it to the difference of potential between its ends. The dimensional formula is the ratio of the formulae for current and potential or $[M^{\frac{1}{2}}L^{\frac{3}{2}}T^{-2}K^{\frac{1}{2}}/M^{\frac{1}{2}}L^{\frac{1}{2}}T^{-1}K^{-\frac{1}{2}}]$ or $[LT^{-1}K]$

Resistance is the reciprocal of conductance. The dimensional formula is $[L^{-1}TK^{-1}]$.

Exs. — Find the factor for converting quantity of electricity expressed in ft.-grain-sec. units to the same expressed in c.g.s. units. The formula is $[m^{\frac{1}{2}}l^{\frac{3}{2}}t^{-1}k^{\frac{1}{2}}]$, in which $m = 0.0648$, $l = 30.48$, $t = 1$, $k = 1$; the factor is $0.0648^{\frac{1}{2}} \times 30.48^{\frac{3}{2}}$, or 42.8.

Find the factor required to convert electric potential from mm-mg-sec. units to c.g.s. units. The formula is $[m^{\frac{1}{2}}l^{\frac{1}{2}}t^{-1}k^{-\frac{1}{2}}]$, in which $m = 0.001$, $l = 0.1$, $t = 1$, $k = 1$; the factor is $0.001^{\frac{1}{2}} \times 0.1^{\frac{1}{2}}$, or 0.01.

Find the factor required to convert electrostatic capacity from ft.-grain-sec. and specific-inductive capacity 6 units to c.g.s. units. The formula is $[lk]$ in which $l = 30.48$, $k = 6$; the factor is 30.48×6, or 182.88.

ELECTROMAGNETIC SYSTEM.

Many of the magnetic quantities are analogues of certain electric quantities. The dimensions of such quantities in the electromagnetic system differ from those of the corresponding electrostatic quantities in the electrostatic system only in the substitution of permeability μ for K.

Magnetic Pole Strength or **Quantity of Magnetism** has already been shown to have the dimensional formula $[M^{\frac{1}{2}}L^{\frac{3}{2}}T^{-1}\mu^{\frac{1}{2}}]$.

Magnetic Flux characterizes the magnetized state of a magnetic circuit. Through a surface inclosing a magnetic pole it is proportional to the magnetic pole strength. The dimensional formula is that for magnetic pole strength.

Magnetic Field Intensity or **Magnetizing Force** is the ratio of the force on a magnetic pole placed at the point and the magnetic pole strength. The dimensional formula is therefore the ratio of the formulae for a force and magnetic quantity, or $[MLT^{-2}/M^{\frac{1}{2}}L^{\frac{3}{2}}T^{-1}\mu^{\frac{1}{2}}]$ or $[M^{\frac{1}{2}}L^{-\frac{1}{2}}T^{-1}\mu^{-\frac{1}{2}}]$.

Magnetic Potential or **Magnetomotive Force** at a point is measured by the work which is required to bring unit quantity of positive magnetism from zero potential to the point. The dimensional formula is the ratio of the formulae for work and magnetic quantity, $[ML^2T^{-2}/M^{\frac{1}{2}}L^{\frac{3}{2}}T^{-1}\mu^{\frac{1}{2}}]$ or $[M^{\frac{1}{2}}L^{\frac{1}{2}}T^{-1}\mu^{-\frac{1}{2}}]$.

Magnetic Moment is the product of the pole strength by the length of the magnet. The dimensional formula is $[M^{\frac{1}{2}}L^{\frac{5}{2}}T^{-1}\mu^{\frac{1}{2}}]$.

Intensity of Magnetization of any portion of a magnetized body is the ratio of the magnetic moment of that portion and its volume. The dimensional formula is $[M^{\frac{1}{2}}L^{\frac{5}{2}}T^{-1}\mu^{\frac{1}{2}}/L^3]$ or $[M^{\frac{1}{2}}L^{-\frac{1}{2}}T^{-1}\mu^{\frac{1}{2}}]$.

Magnetic Induction is the magnetic flux per unit of area taken perpendicular to the direction of the magnetic flux. The dimensional formula is $[M^{\frac{1}{2}}L^{\frac{3}{2}}T^{-1}\mu^{\frac{1}{2}}/L^2]$ or $[M^{\frac{1}{2}}L^{-\frac{1}{2}}T^{-1}\mu^{\frac{1}{2}}]$.

Magnetic Susceptibility is the ratio of intensity of magnetization produced and the intensity of the magnetic field producing it. The dimensional formula is $[M^{\frac{1}{2}}L^{-\frac{1}{2}}T^{-1}\mu^{\frac{1}{2}}/M^{\frac{1}{2}}L^{-\frac{1}{2}}T^{-1}\mu^{-\frac{1}{2}}]$ or $[\mu]$.

Current, I, flowing in circle, radius r, creates magnetic field at its center, $2\pi I/r$. Dimensional formula is product of formulae for magnetic field intensity and length or $[M^{\frac{1}{2}}L^{\frac{1}{2}}T^{-1}\mu^{-\frac{1}{2}}]$.

Quantity of Electricity is the product of the current and time. The dimensional formula is $[M^{\frac{1}{2}}L^{\frac{1}{2}}\mu^{-\frac{1}{2}}]$.

Electric Potential, or **Electromotive Force,** as in the electrostatic system, is the ratio of work to quantity of electricity. The dimensional formula is $[ML^2T^{-2}/M^{\frac{1}{2}}L^{\frac{1}{2}}\mu^{-\frac{1}{2}}]$ or $[M^{\frac{1}{2}}L^{\frac{3}{2}}T^{-2}\mu^{\frac{1}{2}}]$.

Electrostatic Capacity is the ratio of quantity of electricity to difference of potential. The dimensional formula is $[L^{-1}T^2\mu^{-1}]$.

Resistance of a Conductor is the ratio of the difference of potential between its ends and the constant current flowing. The dimensional formula is $[M^{\frac{1}{2}}L^{\frac{3}{2}}T^{-2}\mu^{\frac{1}{2}}/M^{\frac{1}{2}}L^{\frac{1}{2}}T^{-1}\mu^{-\frac{1}{2}}]$ or $[LT^{-1}\mu]$.

Conductance is the reciprocal of resistance, and the dimensional formula is $[L^{-1}T\mu^{-1}]$.

Conductivity is the quantity of electricity transmitted per unit area per unit potential gradient per unit of time. The dimensional formula is $[M^{\frac{1}{2}}L^{\frac{1}{2}}\mu^{-\frac{1}{2}}/L^2(M^{\frac{1}{2}}L^{\frac{3}{2}}T^{-2}\mu^{\frac{1}{2}}/L)\,T]$ or $[L^{-2}T\mu^{-1}]$.

Resistivity is the reciprocal of conductivity as just defined. The dimensional formula is $[L^2T^{-1}\mu]$.

Self-inductance is for any circuit the electromotive force produced in it by unit rate of variation of the current through it. The dimensional formula is the product of the formulae for electromotive force and time divided by that for current or $[M^{\frac{1}{2}}L^{\frac{3}{2}}T^{-2}\mu^{\frac{1}{2}} \times T \div M^{\frac{1}{2}}L^{\frac{1}{2}}T^{-1}\mu^{-\frac{1}{2}}]$ or $[L\mu]$.

Mutual Inductance of two circuits is the electromotive force produced in one per unit rate of variation of the current in the other. The dimensional formula is the same as for self-inductance.

Electric Field Intensity is the ratio of electric potential or electromotive force and length. The dimensional formula is $[M^{\frac{1}{2}}L^{\frac{1}{2}}T^{-2}\mu^{\frac{1}{2}}]$.

Magnetic Reluctance is the ratio of magnetic potential difference to magnetic flux. The dimensional formula is $[L^{-1}\mu^{-1}]$.

Thermoelectric Power is measured by the ratio of electromotive force and temperature. The dimensional formula is $[M^{\frac{1}{2}}L^{\frac{3}{2}}T^{-2}\mu^{\frac{1}{2}}\theta^{-1}]$.

Coefficient of Peltier Effect is measured by the ratio of the quantity of heat and quantity of electricity. The dimensional formula is $[ML^2T^{-2}/M^{\frac{1}{2}}L^{\frac{1}{2}}\mu^{-\frac{1}{2}}]$ or $[M^{\frac{1}{2}}L^{\frac{3}{2}}T^{-2}\mu^{\frac{1}{2}}]$, the same as for electromotive force.

Exs. — Find the factor required to convert intensity of magnetic field from ft.-grain-min. units to c.g.s. units. The formula is $[m^{\frac{1}{2}}l^{-\frac{1}{2}}t^{-1}\mu^{-\frac{1}{2}}]$; $m = 0.0648$, $l = 30.48$, $t = 60$, and $\mu = 1$; the factor is $0.0648^{\frac{1}{2}} \times 30.48^{-\frac{1}{2}}$, or 0.046108.

How many c.g.s. units of magnetic moment make one ft.-grain-sec. unit of the same quantity? The formula is $[m^{\frac{1}{2}}l^{\frac{5}{2}}t^{-1}\mu^{\frac{1}{2}}]$; $m = 0.0648$, $l = 30.48$, $t = 1$, and $\mu = 1$; the number is $0.0648^{\frac{1}{2}} \times 30.48^{\frac{5}{2}}$, or 1305.6.

If the intensity of magnetization of a steel bar is 700 in c.g.s. units, what will it be in mm-mg-sec. units? The formula is $[m^{\frac{1}{2}}l^{\frac{1}{2}}t^{-1}\mu^{\frac{1}{2}}]$; $m = 1000$, $l = 10$, $t = 1$, $\mu = 1$; the intensity is $700 \times 1000^{\frac{1}{2}} \times 10^{\frac{1}{2}}$, or 70000.

Find the factor required to convert current from c.g.s. units to earth-quadrant-10^{-11} gram-sec. units. The formula is $[m^{\frac{1}{2}}l^{\frac{1}{2}}t^{-1}\mu^{-\frac{1}{2}}]$; $m = 10^{11}$, $l = 10^{-9}$, $\mu = 1$; the factor is $10^{\frac{11}{2}} \times 10^{-\frac{9}{2}}$, or 10.

Find the factor required to convert resistance expressed in c.g.s. units into the same expressed in earth-quadrant-10^{-11} gram-sec. units. The formula is $[lt^{-1}\mu]$; $l = 10^{-9}$, $t = 1$, $\mu = 1$; the factor is 10^{-9}.

FUNDAMENTAL STANDARDS.

The choice of the nature of the fundamental quantities already made does not sufficiently define the system for measurements. Some definite unit or arbitrarily chosen standard must next be taken for each of the fundamental quantities. This fundamental standard should have the qualities of permanence, reproducibility and availability and be suitable for accurate measures. Once chosen and made it is called the primary standard and is generally kept at some central bureau, — for instance, the International Bureau of Weights and Measures at Sèvres, France. A primary standard may also be chosen and made for derived units (e.g., the international ohm standard), when it is simply a standard closely representing the unit and accepted for practical purposes, its value having been fixed by certain measuring processes. Secondary or reference standards are accurately compared copies, not necessarily duplicates, of the primaries for use in the work of standardizing laboratories and the production of working standards for everyday use.

Standard of Length. — The primary standard of length which now almost universally serves as the basis for physical measurements is the meter. It is defined as the distance between two lines at 0° C on a platinum-iridium bar deposited at the International Bureau of Weights and Measures. This bar is known as the International Prototype Meter, and its length was derived from the "métre des Archives," which was made by Borda. Borda, Delambre, Laplace, and others, acting as a committee of the French Academy, recommended that the standard unit of length should be the ten-millionth part of the length, from the equator to the pole, of the meridian passing through Paris. In 1795 the French Republic passed a decree making this the legal standard of length, and an arc of the meridian extending from Dunkirk to Barcelona was measured by Delambre and Mechain for the purpose of realizing the standard. From the results of that measurement the meter bar was made by Borda. The meter is now defined as above and not in terms of the meridian length; hence subsequent measures of the length of the meridian have not affected the length of the meter.

Standard of Mass. — The primary standard of mass now almost universally used as the basis for physical measurements is the kilogram. It is defined as the mass of a certain piece of platinum-iridium deposited at the International Bureau of Weights and Measures. This standard is known as the International Prototype Kilogram. Its mass is equal to that of the older standard, the "kilogram des Archives," made by Borda and intended to have the same mass as a cubic decimeter of distilled water at the temperature of 4° C.

Copies of the International Prototype Meter and Kilogram are possessed by the various governments and are called National Prototypes.

Standard of Time. — The unit of time universally used is the mean solar second, or the 86400th part of the mean solar day. It is based on the average time of one rotation of the earth on its axis relatively to the sun as a point of reference = 1.002 737 91 sidereal second.

Standard of Temperature. — The standard scale of temperature as adopted by the International Committee of Weights and Measures (1887) depends on the constant-volume hydrogen thermometer. The hydrogen is taken at an initial pressure at 0° C of one meter of mercury, 0° C, sea-level at latitude 45°. The scale is defined by designating the temperature of melting ice as 0° and of condensing steam as 100° under standard atmospheric pressure. This is known as the Centigrade scale (abbreviated C).

A scale independent of the properties of any particular substance, and called the thermodynamic, or absolute scale, was proposed in 1848 by Lord Kelvin. In it the temperature is proportional to the average kinetic energy per molecule of a perfect gas. The temperature of melting ice is taken as 273.13°, that of the boiling point, 373.13°. The scale of the hydrogen thermometer varies from it only in the sense that the behavior of hydrogen departs from that of a perfect gas. It is customary to refer to this scale as the Kelvin scale (abbreviated K).

NUMERICALLY DIFFERENT SYSTEMS OF UNITS.

The fundamental physical quantities which form the basis of a system for measurements have been chosen and the fundamental standards selected and made. Custom has not however generally used these standards for the measurement of the magnitudes of quantities but rather multiples or submultiples of them. For instance, for very small quantities the micron (μ) or one-millionth of a meter is often used. The following table [1] gives some of the systems proposed, all built upon the fundamental standards already described. The centimeter-gram-second (cm-g-sec. or c.g.s.) system proposed by Kelvin is the only one generally accepted.

TABLE I.

PROPOSED SYSTEMS OF UNITS.

	Weber and Gauss	Kelvin c.g.s.	Moon 1891	Giorgi MKS (Prim. Stds.)	France 1914	B. A. Com., 1863	Practical (B. A. Com., 1873)	Strout 1891
Length	mm	cm	dm	m	m	m	10^9 cm	10^9 cm
Mass	mg	g	Kg	Kg	10^6 g	g	10^{-11} g	10^{-9} g
Time	sec.	sec.	$\frac{\text{sec.}}{10}$	sec.	sec.	sec.	sec.	sec.

Further the choice of a set of fundamental physical quantities to form the basis of a system does not necessarily determine how that system shall be used in measurements. In fact, upon any sufficient set of fundamental quantities, a great many different systems of units may be built. The electrostatic and electromagnetic systems are really systems of electric quantities rather than units. They were based upon the relationships $F = QQ'/Kr^2$ and $mm'/\mu r^2$, respectively. Systems of units built upon a chosen set of fundamental physical quantities may differ in two ways: (1) the units chosen for the fundamental quantities may be different; (2) the defining equations by which the system is built may be different.

The electrostatic system generally used is based on the centimeter, gram, second, and dielectric constant of a vacuum. Other systems have appeared, differing from this in the first way, — for instance using the foot, grain and second in place of the centimeter, gram and second. A system differing from it in the second way is that of Heaviside which introduces the factor 4π at different places than is usual in the equations. There are similarly several systems of electromagnetic units in use.

Gaussian Systems. — "The complexity of the interrelations of the units is increased by the fact that not one of the systems is used as a whole, consistently for all electromagnetic quantities. The 'systems' at present used are therefore combinations of certain of the systems of units.

[1] Circular 60 of the Bureau of Standards, Electric Units and Standards, 1916. The subsequent matter in this introduction is based upon this circular.

"Some writers [1] on the theory of electricity prefer to use what is called a Gaussian system, a combination of electrostatic units for purely electrical quantities and electromagnetic units for magnetic quantities. There are two such Gaussian systems in vogue, — one a combination of c.g.s. electrostatic and c.g.s electromagnetic systems, and the other a combination of the two corresponding Heaviside systems.

"When a Gaussian system is used, caution is necessary when an equation contains both electric and magnetic quantities. A factor expressing the ratio between the electrostatic and electromagnetic units of one of the quantities has to be introduced. This factor is the first or second power of c, the number of electrostatic units of electric charge in one electromagnetic unit of the same. There is sometimes a question as to whether electric current is to be expressed in electrostatic or electromagnetic units, since it has both electric and magnetic attributes. It is usually expressed in electrostatic units in the Gaussian system."

It may be observed from the dimensions of K given in Table 1 that $[1/K\mu] = [L^2/T^2]$ which has the dimensions of a square of a velocity. This velocity was found experimentally to be equal to that of light, when K and μ were expressed in the same system of units. Maxwell proved theoretically that $1/\sqrt{K\mu}$ is the velocity of any electromagnetic wave. This was subsequently proved experimentally. When a Gaussian system is used, this equation becomes $c/\sqrt{K\mu} = v$. For the ether $K = 1$ in electrostatic units and $\mu = 1$ in electromagnetic units. Hence $c = v$ for the ether, or the velocity of an electromagnetic wave in the ether is equal to the ratio of the c.g.s. electromagnetic to the c.g.s. electrostatic unit of electric charge. This constant c is of primary importance in electrical theory. Its most probable value is 2.9986×10^{10} centimeters per second.

"Practical" Electromagnetic System. — This electromagnetic system is based upon the units of 10^9 cm, 10^{-11} gram, the sec. and μ of the ether. It is never used as a complete system of units but is of interest as the historical basis of the present International System. The principal quantities are the resistance unit, the ohm $= 10^9$ c.g.s. units; the current unit, the ampere $= 10^{-1}$ c.g.s. units; and the electromotive force unit, the volt $= 10^8$ c.g.s. units.

The International Electric Units. — The units used in practical measurements, however, are the "International Units." They were derived from the "practical" system just described, or as the latter is sometimes called, the "absolute" system. These international units are based upon certain concrete standards presently to be defined and described. With such standards electrical comparisons can be more accurately and readily made than could absolute measurements in terms of the fundamental units. Two electric units, the international ohm and the international ampere, were chosen and made as nearly equal as possible to the ohm and ampere of the "practical" or "absolute" system.

[1] For example, A. G. Webster, "Theory of Electricity and Magnetism," 1897; J. H. Jeans, "Electricity and magnetism," 1911; H. A. Lorentz, "The Theory of Electrons," 1909; and O. W. Richardson, "The Electron Theory of Matter," 1914.

This system of units, sufficiently near to the "absolute" system for the purpose of electrical measurements and as a basis for legislation, was defined as follows:

"1. The *International Ohm* is the resistance offered to an unvarying electric current by a column of mercury at the temperature of melting ice, 14.4521 grams in mass, of a constant cross-sectional area and of a length of 106.300 centimeters.

"2. The *International Ampere* is the unvarying electric current which, when passed through a solution of nitrate of silver in water, in accordance with specification II attached to these Resolutions, deposits silver at the rate of 0.00111800 of a gram per second.

"3. The *International Volt* is the electrical pressure which, when steadily applied to a conductor the resistance of which is one international ohm will produce a current of one international ampere.

"4. The *International Watt* is the energy expended per second by an unvarying electric current of one international ampere under the pressure of one international volt."

In accordance with these definitions, a value was established for the electromotive force of the recognized standard of electromotive force, the Weston normal cell, as the result of international coöperative experiments in 1910. The value was 1.0183 international volts at 20° C.

The definitions by the 1908 International Conference supersede certain definitions adopted by the International Electrical Congress at Chicago in 1893. Certain of the units retain their Chicago definitions, however. They are as follows:

"*Coulomb*. As a unit of quantity, the *International Coulomb*, which is the quantity of electricity transferred by a current of one international ampere in one second.

"*Farad*. As a unit of capacity, the *International Farad*, which is the capacity of a condenser, charged to be a potential of one international volt by one international coulomb of electricity.

"*Joule*. As a unit of work, the *Joule*, which is equal to 10^7 units of work in the c.g.s. system, and which is represented sufficiently well for practical use by the energy expended in one second by an international ampere in an international ohm.

"*Henry*. As the unit of induction, the *Henry*, which is the induction in a circuit when the electromotive force induced in this circuit is one international volt, while the inducing current varies at the rate of one ampere per second."

"The choice of the ohm and ampere as fundamental was purely arbitrary. These are the two quantities directly measured in absolute electrical measurements. The ohm and volt have been urged as more suitable for definition in terms of arbitrary standards, because the primary standard of electromotive force (standard cell) has greater simplicity than the primary standard of current (silver voltameter). The standard cell is in fact used, together with resistance standards, for the actual maintenance of the units, rather than the silver voltameter and resistance standards. Again, the volt and ampere have some claim for consideration for fundamental definition, both being units of quantities more fundamental in electrical theory than resistance."

For all practical purposes the "international" and the "practical" or "absolute" units are the same. Experimental determination of the ratios of the corresponding units in the two systems have been made and the mean results are given in Table 382. These ratios represent the accuracy with which it was possible to fix the values of the international ohm and ampere at the time they were defined (London Conference of 1908). It is unlikely that the definitions of the international units will be changed in the near future to make the agreement any closer. An act approved July 12, 1894, makes the International units as above defined the legal units in the United States of America.

THE STANDARDS OF THE INTERNATIONAL ELECTRICAL UNITS.

RESISTANCE

Resistance. — The definition of the international ohm adopted by the London Conference in 1908 is accepted practically everywhere.

Mercury Standards. — Mercury standards conforming to the definition were constructed in England, France, Germany, Japan, Russia and the United States. Their mean resistances agree to about two parts in 100,000. To attain this accuracy, elaborate and painstaking experiments were necessary. Tubes are never quite uniform in cross-section; the accurate measurement of the mass of mercury filling the tube is difficult, partly because of a surface film on the walls of the tube; the greatest refinements are necessary in determining the length of the tube. In the electrical comparison of the resistance with wire standards, the largest source of error is in the filling of the tube. These and other sources of error necessitated a certain uniformity in the setting up of mercury standards and at the London Conference the following specifications were drawn up:

SPECIFICATION RELATING TO MERCURY STANDARDS OF RESISTANCE.

The glass tubes used for mercury standards of resistance must be made of a glass such that the dimensions may remain as constant as possible. The tubes must be well annealed and straight. The bore must be as nearly as possible uniform and circular, and the area of cross-section of the bore must be approximately one square millimeter. The mercury must have a resistance of approximately one ohm.

Each of the tubes must be accurately calibrated. The correction to be applied to allow for the area of the cross-section of the bore not being exactly the same at all parts of the tube must not exceed 5 parts in 10,000.

The mercury filling the tube must be considered as bounded by plane surfaces placed in contact with the ends of the tube.

The length of the axis of the tube, the mass of mercury the tube contains, and the electrical resistance of the mercury are to be determined at a temperature as near to 0° C as possible. The measurements are to be corrected to 0° C.

For the purpose of the electrical measurements, end vessels carrying connections for the current and potential terminals are to be fitted on to the tube. These end vessels are to be spherical in shape (of a diameter of approximately four centimeters) and should have cylindrical pieces attached to make connections with the tubes. The outside edge of each end of the tube is to be coincident with the inner surface of the corresponding end vessel. The leads which make contact with the mercury are to be of thin platinum wire fused into glass. The point of entry of the current lead and the end of the tube are to be at opposite ends of a diameter of the bulb; the potential lead is to be midway between these two points. All the leads must be so thin that no error in the resistance is introduced through conduction of heat to the mercury. The filling of the tube with mercury for the purpose of the resistance measurements must be carried out under the same conditions as the filling for the determination of the mass.

The resistance which has to be added to the resistance of the tube to allow for the effect of the end vessels is to be calculated by the formula

$$A = \frac{0.80}{1063\pi}\left(\frac{1}{r_1} + \frac{1}{r_2}\right) \text{ ohm},$$

where r_1 and r_2 are the radii in millimeters of the end sections of the bore of the tube.

The mean of the calculated resistances of at least five tubes shall be taken to determine the value of the unit of resistance.

For the purpose of the comparison of resistances with a mercury tube the measurements shall be made with at least three separate fillings of the tube.

Secondary Standards. — Secondary standards, derived from the mercury standards and used to give values to working standards, are certain coils of manganin wire kept in the national laboratories. Their resistances are adjusted to correspond to the unit or its decimal multiples or submultiples. The values assigned to these coils are checked from time to time with the similar coils of the other countries. The value now in use is based on the comparison made at the U. S. Bureau of Standards in 1910 and may be called the "1910 ohm." Later measurements on various mercury standards checked the value then used within 2 parts in 100,000. Thus the basis of resistance measurement is maintained not by the mercury standards of a single laboratory, but by all the mercury standards of the various national laboratories; it is furthermore the same in all countries, except for very slight outstanding discrepancies due to the errors of measurement and variations of the standards with time.

Resistance Standards in Practice. — In ordinary measurements, working standards of resistance are usually coils of manganin wire (approximately 84 per cent Cu + 12 per cent Mn + 4 per cent Ni). They are generally used in oil which carries away the heat developed by the current and facilitates regulation and measurement of the temperature. The best type is inclosed in a sealed case for protection against atmospheric humidity. Varying humidity changes the resistance of open coils often to several parts in 10,000 higher in summer than in winter. While sealed 1 ohm and 0.1 ohm coils may remain constant to about 1 part in 100,000.

Absolute Ohm. — The absolute measurement of resistance involves the precise determination of a length and a time (usually an angular velocity) in a medium of unit permeability. Since the dimensional formula of resistance in the electromagnetic system is $[L\mu/T]$, such an absolute measurement gives R not in cm/sec. but in cm $\times\ \mu$/sec. The definitions of the ohm, ampere and volt by the 1908 London conference tacitly assume a permeability equal to unity. The relation of the international ohm to the absolute ohm has been measured in different ways involving revolving coil, revolving disk, and alternate current methods. Probably the most accurate determination was made in 1913 by F. E. Smith of the National Physical Laboratory of England, using a modification of the Lorentz revolving disk method. His result was

$$1 \text{ international ohm} = 1.00052 \pm 0.00004 \text{ absolute ohms},$$

or, in other words, while one international ohm is represented by a mercury column 106.300 cm long as specified above, one absolute ohm requires a similar column 106.245 cm long. Table 305 of the 6th revised edition of these tables contains data relative to the various determinations of the ohm.

CURRENT.

The Silver Voltameter. — The silver voltameter is a concrete means of measuring current in accordance with the definition of the international ampere. As used for the realization of the international ampere "it consists of a platinum cathode in the form of a cup holding the silver nitrate solution, a silver anode partly or wholly immersed in the solution, and some means to prevent anode slime and particles of silver mechanically detached from the anode from reaching the cathode. As a standard representing the international ampere, the silver voltameter includes also the chronometer used to measure time. The degree of purity and the mode of preparation of the various parts of the voltameter affect the mass of the deposit. There are numerous sources of error, and the suitability of the silver voltameter as a primary standard of current has been under investigation since 1893. Differences of as much as 0.1 per cent or more may be obtained by different procedures, the larger differences being mainly due to impurities produced in the electrolyte (by filter paper, for instance). Hence, in order that the definition of current be precise, it must be accompanied by *specifications* for using the voltameter."

The original specifications were recognized to be inadequate and an international committee on electrical units and standards was appointed to complete the specifications. It was also recognized that in practice standard cells would replace secondary current standards so that a value must be fixed for the electromotive force of the Weston normal cell. This was attempted in 1910 at the Bureau of Standards by representatives of that institution together with one delegate each from the Physikalische-Technische Reichanstalt, The National Physical Laboratory and the Laboratoire Central d'Electricité. Voltameters from all four institutions were put in series under a variety of experimental conditions. Standard Weston cells and resistance standards of the four laboratories were also intercompared. From the joint comparison of standard cells and silver voltameters particular values were assigned to the standard cells from each laboratory. The different countries thus have a common basis of measurement maintained by the aid of standard cells and resistance standards derived from the international voltameter investigation of 1910.

It was not found possible to draw up satisfactory and final specifications for the silver voltameter. Provisional specifications were submitted by the U. S. Bureau of Standards and more complete specifications have been proposed in correspondence between the national laboratories and members of the international committee since 1910, but no agreement upon final specifications has yet been reached.

Resistance Standards Used in Current Measurements. — Precise measurements of currents require a potentiometer, a standard cell and a resistance standard. The resistance must be so designed as to carry the maximum current without undue heating and consequent change of resistance. Accordingly the resistance metal must have a small temperature resistance coefficient and a sufficient area in contact with the air, oil, or other cooling fluid. It must have a small thermal electromotive force against copper. Manganin satisfies these conditions and is usually used. The terminals of the standard must have sufficient contact area so that there shall be no undue heating at contacts.[1] It must be so designed that the current distribution does not depend upon the mode of connection to the circuit.

Absolute Ampere. — The absolute ampere (10^{-1}c.g.s. electromagnetic units) differs by a negligible amount from the international ampere. Since the dimensional formula of the current in the electromagnetic system is $[L^{\frac{1}{2}}M^{\frac{1}{2}}/T\mu^{\frac{1}{2}}]$ which is equivalent to $[F^{\frac{1}{2}}/\mu^{\frac{1}{2}}]$, the absolute measurement of current involves fundamentally the measurement of a force in a medium of unit permeability. In most measurements of high precision an electrodynamometer has been used of the form known as a current balance. A summary of the various determinations will be found in Table 293 of the 6th Revised Edition of these tables.

The best value is probably the mean of the determinations made at the U. S. Bureau of Standards, the National Physical Laboratory and at the University of Gröningen, which gives

1 international ampere = 0.99991 absolute ampere.

The separate values were 0.99992, 0.99988 and 0.99994, respectively. "The result may also be expressed in terms of the electrochemical equivalent of silver, which, based on the '1910 mean voltameter,' thus equals 0.00111810 g per absolute coulomb. By the definition of the international ampere, the value is 0.00111800 g per international coulomb."

ELECTROMOTIVE FORCE.

International Volt. — "The international volt is derived from the international ohm and ampere by Ohm's law. Its value is maintained by the aid of the Weston normal cell. The national standardizing laboratories have groups of such cells, to which values in terms of the international ohm and ampere have been assigned by international experiments, and thus form a basis of reference for the standardization of the standard cells used in practical measurements."

Weston Normal Cell. — The Weston normal cell is the standard used to maintain the international volt and, in conjunction with resistance standards, to maintain the international ampere. The cell is a simple voltaic combination

[1] See "Report to the International Committee on Electrical Units and Standards," 1912, p. 199. For the Bureau of Standards investigations see Bull. Bureau of Standards, 9, pp. 209, 493; 10, p. 475, 1912-14; 13, p. 147, 1915; 9, p. 151, 1912; 13, pp. 447, 479, 1916.

having its anode or negative electrode of cadmium amalgam, consisting of 10 per cent by weight of cadmium and 90 per cent mercury. The cathode, or positive electrode, is pure mercury covered with a paste consisting of mercurous sulphate, cadmium-sulphate crystals, and solution. The electrolyte is cadmium-sulphate solution in contact with an excess of cadmium-sulphate crystals. The containing vessel is of glass, usually in the H form. Connection is made to the electrodes by platinum wires sealed into the glass. The cells are sealed, preferably hermetically, and in use are submerged in a constant-temperature oil bath. The resistance of a cell is about 600 to 1000 ohms. The Weston cell used with potentiometers is not the Weston normal cell, but differs from it only slightly, the cadmium-sulphate solution not being saturated. It is described in the next section below.

One of the great advantages of the Weston normal cell is its small change of electromotive force with change of temperature. At any temperature, t (centigrade), between 0° and 40°, $E_t = E_{20} - 0.0000406\ (t - 20) - 0.00000095\ (t - 20)^2 + 0.00000001\ (t - 20)^3$. This temperature formula was adopted by the London conference of 1908. That this formula may apply, the cell must be of a strictly uniform temperature throughout. One leg of the cell has a large positive and the other leg a large negative temperature coefficient. If the temperature of one leg changes faster than the other, the formula does not hold.

When the best of care is taken as to purity of materials and mode of procedure, Weston normal cells are reproducible within 1 part in 100,000. The source of the greatest variations has probably been in the mercurous sulphate. Cells using the best samples of this material have an electromotive force the constancy of which over a period of one year is about 1 part in 100,000. Only very meager specifications for the cell have as yet been agreed upon internationally, however, and the procedures in various laboratories differ in some respects.[1]

The basis of measurements of electromotive force is the same in all countries as the result of the joint international experiments of 1910. As already stated, a large number of observations were made at that time with the silver voltameter, and a considerable number of Weston normal cells from the national laboratories of England, France, Germany and the United States were compared. From the results of these voltameter experiments and from resistance measurements, the value

1.0183 international volts at 20° C

was assigned to the Weston normal cell. A mean of the groups of cells from the four laboratories was taken as most accurately representing the Weston normal

[1] For the preliminary specifications which have been issued and the reports of the various investigations on the standard cells see the following references: Preliminary specifications, Wolff and Waters, Bull. B. of S. 3, p. 623, 1907; Clark and Weston Standard Cells, Wolff and Waters, ditto, 4, p. 1, 1907; Temperature formula of Weston Standard Cell, ditto, 5, p. 309, 1908; The materials, reproducibility, etc., of the Weston Cell, Helett, Phys. Rev. 22, p. 321, 1906; 23, p. 166, 1906; 27, pp. 33, 337, 1908; Mercurous sulphate, etc., Steinwehr, Zs. für Electroch. 12, p. 578, 1906; German value of cell, Jaeger and Steinwehr, ditto, 28, p. 367, 1908; National Physical Laboratory researches, Smith, Phil. Trans. 207, p. 393, 1908; On the Weston Cell, Haga and Boerema, Arch. Neerland, des Sci. Exactes, 3, p. 324, 1913.

cell. Each laboratory has means of preserving the unit. Any discrepancies between the bases of the different countries at the present time would be due only to possible variations in the reference cells of the national laboratories. Such discrepancies are probably less than 2 parts in 100,000.

The figure 1.0183 has been in use since January 1, 1911. The value used in the United States before 1911, 1.019126 at 20° C or 1.0189 at 25° C, was assigned to a certain group of cells maintained as the standard of electromotive force at the Bureau of Standards. The high value is partly due to the use of commercial mercurous sulphate in the cells. The old and the new values, 1.01926 and 1.0183, thus apply to different groups of cells. The group of cells to which the value 1.019126 was assigned before 1910 differed by 26 microvolts from the mean of the international group, such that the international group to which the value 1.0183 is now assigned had the value 1.019126 + 0.000026, or 1.019152, in terms of the old United States basis. The difference between 1.019152 and 1.0183 is 0.000852.

The electromotive force of any Weston cell as now given is therefore 0.000852 volt smaller than on the old United States basis, i.e., the present international volt is 84 parts in 100,000 larger than the old international volt of the United States.

Upon the new international basis the Clark cell set up according to the old United States legal specifications has an emf of 1.4328_0 international volts at 15° C. The Clark cell set up (with specially purified mercurous sulphate) according to improved specifications used at the Bureau of Standards has an emf of 1.4325_0 international volts at 15° C or 1.4263_7 at 20° C.

Portable Weston Cells. — The standard cell used in practice is the Weston portable cell. It is like the Weston normal cell except that the cadmium-sulphate solution at ordinary temperatures is unsaturated. As usually made, the cadmium-sulphate solution is saturated at about 4° C; at higher temperatures the crystals are dissolved. Plugs of asbestos or other material hold the chemicals in place. Its resistance is usually about 200 to 311 ohms. The change of emf, wholly negligible in most electrical measurements, is less than 0.00001 volt per degree C. The two legs of the cell have large and opposite temperature coefficients so that care must be taken that the temperature of the cell is kept uniform and the cell must be protected from draughts or large changes of temperature. The electromotive force of a portable cell ranges from 1.0181 to 1.0191 international volts and must be determined by comparison with standards. It decreases very slightly with time, usually less than 0.0001 volt per year.

Absolute and Semi-absolute Volt. — Since the direct determination of the volt in absolute measure presents great difficulties, it is derived by Ohm's law from the absolute measures of the ohm and ampere. From the absolute values of these already given,

$$1 \text{ international volt} = 1.00043 \text{ absolute volts.}$$

The electromotive force of the Weston normal cell at 20° C is 1.0183_0 international volts and 1.0187_4 absolute volts. A semi-absolute volt is that potential difference which exists between the terminals of a resistance of one *international* ohm when the latter carries a current of one *absolute* ampere. The emf of the Weston normal cell may be taken as 1.01821 semi-absolute volts at 20° C.

QUANTITY OF ELECTRICITY.

The international unit of quantity of electricity is the coulomb. The faraday is the quantity of electricity necessary to liberate 1 gram equivalent in electrolysis. It is equivalent to 96,500 coulombs.

Standards. — There are no standards of electric quantity. The silver voltameter may be used for its measurement since under ideal conditions the mass of metal deposited is proportional to the amount of electricity which has flowed.

CAPACITY.

The unit generally used for capacity is the international microfarad or the one-millionth of the international farad. Capacities are commonly measured by comparison with standard capacities. The values of the standards are determined by measurement in terms of resistance and time. The standard is some form of condenser consisting of two sets of metal plates separated by a dielectric. The condenser should be surrounded by a metal shield connected to one set of plates rendering the capacity independent of the surroundings. An ideal condenser would have a constant capacity under all circumstances, with zero resistance in its leads and plates, and no absorption in the dielectric. Actual condensers vary with the temperature, atmospheric pressure, and the voltage, frequency, and time of charge and discharge. A well-constructed air condenser with heavy metal plates and suitable insulating supports is practically free from these effects and is used as a standard of capacity.

Practically air condenser plates must be separated by 1 mm or more and so cannot be of great capacity. The more the capacity is increased by approaching the plates, the less the mechanical stability and the less constant the capacity. Condensers of great capacity use solid dielectrics, preferably mica sheets with conducting plates of tinfoil. At constant temperature the best mica condensers are excellent standards. The dielectric absorption is small but not quite zero, so that the capacity of these standards with different methods of measurement must be carefully determined.

INDUCTANCE.

The henry, the unit of self-inductance, is also the unit of mutual inductance. The henry has been known as the "quadrant" and the "secohm." The length of a quadrant or quarter of the earth's circumference is approximately 10^9 cms. and a henry is 10^9 cms. of inductance. Secohm is a contraction of second and ohm; the dimensions of inductance are $[TR]$ and this unit is based on the second and ohm.

Inductance Standards. — Inductance standards are measured in international units in terms of resistance and time or resistance and capacity by alternate-current bridge methods. Inductances calculated from dimensions are in absolute electromagnetic units. The ratio of the international to the absolute henry is the same as the ratio of the corresponding ohms.

Since inductance is measured in terms of capacity and resistance by the bridge method about as simply and as conveniently as by comparison with standard inductances, it is not necessary to maintain standard inductances. They are however of value in magnetic, alternating-current, and absolute electrical measurements. A standard inductance is a circuit so wound that when used in a circuit it adds a definite amount of inductance. It must have either such a form or so great an inductance that the mutual inductance of the rest of the circuit upon it may be negligible. It usually is a wire coil wound all in the same direction to make self-induction a maximum. A standard, the inductance of which may be calculated from its dimensions, should be a single layer coil of very simple geometrical form. Standards of very small inductance, calculable from their dimensions, are of some simple device, such as a pair of parallel wires or a single turn of wire. With such standards great care must be used that the mutual inductance upon them of the leads and other parts of the circuit is negligible. Any inductance standard should be separated by long leads from the measuring bridge or other apparatus. It must be wound so that the distributed capacity between its turns is negligible; otherwise the apparent inductance will vary with the frequency.

POWER AND ENERGY.

Power and energy, although mechanical and not primarily electrical quantities, are measurable with greater precision by electrical methods than in any other way. The watt and the electric units were so chosen in terms of the c.g.s. units that the product of the current in amperes by the electromotive force in volts gives the power in watts (for continuous or instantaneous values). The international watt, defined as "the energy expended per second by an unvarying electric current of one international ampere under an electric pressure of one international volt," differs but little from the absolute watt.

Standards and Measurements. — No standard is maintained for power or energy. Measurements are always made in electrical practice in terms of some of the purely electrical quantities represented by standards.

MAGNETIC UNITS.

C.G.S. units are generally used for magnetic quantities. American practice is fairly uniform in names for these units: the c.g.s. unit of magnetomotive force is called the "gilbert," of reluctance, the "oersted," following the provisional definitions of the American Institute of Electrical Engineers (1894). The c.g.s. unit of flux is called the "maxwell" as defined by the 1900 Paris conference. The name "gauss" is used unfortunately both for the unit of induction (A.I.E.E. 1894) and for the unit of magnetic field intensity or magnetizing force. "This double usage, recently sanctioned by engineering societies, is based upon the mathematical convenience of defining both induction and magnetizing force as the force on a unit magnetic pole in a narrow cavity in the material, the cavity being in one case perpendicular, in the other parallel, to the direction of the magnetization: this definition however applies only in the ordinary electromagnetic units. There are a number of reasons for considering induction and magnetizing force as two physically distinct quantities, just as electromotive force and current are physically different."

In the United States "gauss" has been used much more for the c.g.s. unit of induction than for the unit of magnetizing force. The longer name of "maxwell per cm^2" is also sometimes used for this unit when it is desired to distinguish clearly between the two quantities. The c.g.s. unit of magnetizing force is usually called the "gilbert per cm."

A unit frequently used is the ampere-turn. It is a convenient unit since it eliminates 4π in certain calculations. It is derived from the "ampere turn per cm." The following table shows the relations between a system built on the ampere-turn and the ordinary magnetic units.[1]

TABLE II.

THE ORDINARY AND THE AMPERE-TURN MAGNETIC UNITS.

Quantity		Ordinary magnetic units.	Ampere-turn units.	Ordinary units in 1 ampere-turn unit
Magnetomotive force	$\mathfrak{F}$	Gilbert	Ampere-turn	$4\pi/10$
Magnetizing force	H	Gilbert per cm.	Ampere-turn per cm.	$4\pi/10$
Magnetic flux	Φ	Maxwell	Maxwell	1
Magnetic induction	B	Maxwell per cm.^{2} Gauss	Maxwell per cm.^{2} Gauss	1
Permeability	μ			1
Reluctance	R	Oersted	Ampere-turn per Maxwell	$4\pi/10$
Magnetization intensity	J		Maxwell per cm.^{2}	$1/4\pi$
Magnetic susceptibility	κ			$1/4\pi$
Magnetic pole strength	m		Maxwell	$1/4\pi$

[1] Dellinger, International System of Electric and Magnetic Units, Bull. Bureau of Standards, 13. p. 599, 1916.

PHYSICAL TABLES

SPELLING AND ABBREVIATIONS OF THE COMMON UNITS OF WEIGHT AND MEASURE.

The spelling of the metric units is that adopted by the International Committee on Weights and Measures and given in the law legalizing the metric system in the United States (1866). The period is omitted after the metric abbreviations but not after those of the customary system. The exponents "2" and "3" are used to signify area and volume respectively in the metric units. The use of the same abbreviation for singular and plural is recommended. It is also suggested that only small letters be used for abbreviations except in the case of A. for acre, where the use of the capital letter is general. The following list is taken from circular 47 of the U. S. Bureau of Standards.

Unit.	Abbreviation.	Unit.	Abbreviation.
acre	A	kilogram	kg
are	a	kiloliter	kl
avoirdupois	av.	kilometer	km
barrel	bbl.	link	li.
board foot	bd. ft.	liquid	liq.
bushel	bu.	liter	l
carat, metric	c	meter	m
centare	ca	metric ton	t
centigram	cg	micron	μ
centiliter	cl	mile	mi.
centimeter	cm	milligram	mg
chain	ch.	milliliter	ml
cubic centimeter	cm^3	millimeter	mm
cubic decimeter	dm^3	millimicron	mμ
cubic dekameter	dkm^3	minim	min. *or* ♏
cubic foot	cu. ft.	ounce	oz.
cubic hectometer	hm^3	ounce, apothecaries'	oz. ap. *or* ℥
cubic inch	cu. in.	ounce, avoirdupois	oz. av.
cubic kilometer	km^3	ounce, fluid	fl. oz.
cubic meter	m^3	ounce, troy	oz. t.
cubic mile	cu. mi.	peck	pk.
cubic millimeter	mm^3	pennyweight	dwt.
cubic yard	cu. yd.	pint	pt.
decigram	dg	pound	lb.
deciliter	dl	pound, apothecaries'	lb. ap.
decimeter	dm	pound, avoirdupois	lb. av.
decistere	ds	pound, troy	lb. t.
dekagram	dkg	quart	qt.
dekaliter	dkl	rod	rd.
dekameter	dkm	scruple, apothecaries'	s. ap. *or* ℈
dekastere	dks	square centimeter	cm^2
dram	dr.	square chain	sq. ch.
dram, apothecaries'	dr. ap. *or* ʒ	square decimeter	dm^2
dram, avoirdupois	dr. av.	square dekameter	dkm^2
dram, fluid	fl. dr.	square foot	sq. ft.
fathom	fath.	square hectometer	hm^2
foot	ft.	square inch	sq. in.
firkin	fir.	square kilometer	km^2
furlong	fur.	square meter	m^2
gallon	gal.	square mile	sq. mi.
grain	gr.	square millimeter	mm^2
gram	g	square rod	sq. rd.
hectare	ha	square yard	sq. yd.
hectogram	hg	stere	s
hectoliter	hl	ton	tn.
hectometer	hm	ton, metric	t
hogshead	hhd.	troy	t.
hundredweight	cwt.	yard	yd.
inch	in.		

Table 2.

FUNDAMENTAL AND DERIVED UNITS.

Conversion Factors.

To change a quantity from one system of units to another: substitute in the corresponding conversion factor from the following table the ratios of the magnitudes of the *old* units to the *new* and multiply the old quantity by the resulting number. For example: to reduce velocity in miles per hour to feet per second, the conversion factor is lt^{-1}; $l = 5280/1$, $t = 3600/1$, and the factor is 5280/3600 or 1.467. Or we may proceed as follows: e. g., to find the equivalent of 1 c.g.s. unit of angular momentum in the pd.ft.m. unit, from the Table 1 g cm²/sec.$=x$ lb. ft.²/min, where x is the factor sought. Solving, $x=$1g/lb. × cm²/ft.² × min./sec.$=1 \times .002205 \times .001076 \times 60 = .0001425$.

The dimensional formulæ lack one quality which is needed for completeness, an indication of their vector characteristics; such characteristics distinguish plane and solid angle, torque and energy, illumination and brightness.

(*a*) Fundamental Units.

The fundamental units and conversion factors in the systems of units most commonly used are: Length [l]; Mass [m]; Time [t]; Temperature [θ]; and for the electrostatic system, Dielectric Constant [k]; for the electromagnetic system, Permeability [μ]. The formulae will also be given for the International System of electric and magnetic units based on the units length, resistance [r], current [i], and time.

(*b*) Derived Units.

Name of unit. (Geometrical and dynamical.)	Conversion factor. $[m^x l^y t^z]$ x	y	z
Area, surface	0	2	0
Volume	0	3	0
Angle	0	0	0
Solid angle	0	0	0
Curvature	0	−1	0
Angular velocity	0	0	−1
Linear velocity	0	1	−1
Angular acceleration	0	0	−2
Linear acceleration	0	1	−2
Density	1	−3	0
Moment of inertia	1	2	0
Intensity of attraction	0	1	−2
Momentum	1	1	−1
Moment of momentum	1	2	−1
Angular momentum	1	2	−1
Force	1	1	−2
Moment of couple, torque	1	2	−2
Work, energy	1	2	−2
Power, activity	1	2	−3
Intensity of stress	1	−1	−2
Modulus of elasticity	1	−1	−2
Compressibility	−1	1	2
Resilience	1	−1	−2
Viscosity	1	−1	−1

Name of units. (Heat and light.)	Conversion factor. $[m^x l^y t^z \theta^v]$ x	y	z	v
Quantity of heat:				
thermal units	1	0	0	1
thermometric units	0	3	0	1
dynamical units	1	2	−2	0
Coefficient of thermal expansion	0	0	0	−1
Thermal conductivity:				
thermal units	1	−1	−1	0
thermometric units or diffusivity	0	2	−1	0
dynamical units	1	1	−3	−1
Thermal capacity	1	0	0	0
Latent heat:				
thermal units	0	0	0	1
dynamical units	0	2	−2	0
Joule's equivalent	0	2	−2	1
Entropy:				
heat in thermal units	1	0	0	0
heat in dynamical units	1	2	−2	1
Luminous intensity	0	0	0	1*
Illumination	0	−2	0	1*
Brightness	0	−2	0	1*
Visibility	−1	−2	3	1*
Luminous efficiency	−1	−2	3	1*

* For these formulæ the numbers in the last column are the exponents of F where F refers to the luminous flux. For definitions of these quantities see Table 299, page 259.

FUNDAMENTAL AND DERIVED UNITS.

Conversion Factors.

(*b*) DERIVED UNITS.

NAME OF UNIT. (Electric and magnetic.)	Sym-bol.*	CONVERSION FACTOR. Electrostatic system. $m^x l^y t^z k^v$				Electromagnetic system. $m^x l^y t^z \mu^v$				emu / esu †	International system. $r^x i^y l^z t^v$			
		x	*y*	*z*	*v*	*x*	*y*	*z*	*v*		*x*	*y*	*z*	*v*
Quantity of electricity	Q	½	$\frac{3}{2}$	−1	½	½	½	0	−½	c	0	1	0	1
Electric displacement	D	½	−½	−1	½	½	$-\frac{3}{2}$	0	−½	c	0	1	−2	1
Electric surface density	D	½	−½	−1	½	½	$-\frac{3}{2}$	0	−½	c	0	1	−2	1
Electric field intensity	ℰ	½	−½	−1	−½	½	½	−2	½	1/c	1	1	1	0
Electric potential	V	½	½	−1	−½	½	$\frac{3}{2}$	−2	½	1/c	1	1	0	0
Electromotive force	E	½	½	−1	−½	½	$\frac{3}{2}$	−2	½	1/c	1	1	0	0
Electrostatic capacity	C	0	1	0	1	0	−1	2	−1	c^2	−1	0	0	0
Dielectric constant	K	0	0	0	1	0	−2	2	−1	c^2	−1	0	−1	1
Specific inductive capacity	—	0	0	0	0	0	0	0	0		0	0	0	0
Current	I	½	$\frac{3}{2}$	−2	½	½	½	−1	−½	c	0	1	0	0
Electric conductivity	γ	0	0	−1	1	0	−2	1	−1	c^2	−1	0	−1	0
Resistivity	ρ	0	0	1	−1	0	2	−1	1	$1/c^2$	1	0	1	0
Conductance	g	0	1	−1	1	0	−1	1	−1	c^2	−1	0	0	0
Resistance	R	0	−1	1	−1	0	1	−1	1	$1/c^2$	1	0	0	0
Magnetic pole strength	m	½	½	0	−½	½	$\frac{3}{2}$	−1	½	1/c	1	1	0	1
Quantity of magnetism	m	½	½	0	−½	½	$\frac{3}{2}$	−1	½	1/c	1	1	0	1
Magnetic flux	Φ	½	½	0	−½	½	$\frac{3}{2}$	−1	½	1/c	1	1	0	1
Magnetic field intensity	H	½	½	−2	½	½	−½	−1	−½	c	0	0	−1	0
Magnetizing force	H	½	½	−2	½	½	−½	−1	−½	c	0	0	−1	0
Magnetic potential	Ω	½	$\frac{3}{2}$	−2	½	½	½	−1	−½	c	0	1	0	0
Magnetomotive force	ℱ	½	$\frac{3}{2}$	−2	½	½	½	−1	−½	c	0	1	0	0
Magnetic moment	—	½	$\frac{3}{2}$	0	−½	½	$\frac{5}{2}$	−1	½	1/c	1	1	1	1
Intensity magnetization	J	½	$-\frac{3}{2}$	0	−½	½	−½	−1	½	1/c	1	1	−2	1
Magnetic induction	B	½	$-\frac{3}{2}$	0	−½	½	−½	−1	½	1/c	1	1	−2	1
Magnetic susceptibility	κ	0	−2	2	−1	0	0	0	1	$1/c^2$	1	0	−1	1
Magnetic permeability	μ	0	−2	2	−1	0	0	0	1	$1/c^2$	1	0	−1	1
Current density	—	½	−½	−2	½	½	$-\frac{3}{2}$	−1	−½	c	0	1	−2	0
Self-inductance	ℒ	0	−1	2	−1	0	1	0	1	$1/c^2$	1	0	0	1
Mutual inductance	ℳ	0	−1	2	−1	0	1	0	1	$1/c^2$	1	0	0	1
Magnetic reluctance	ℛ	0	1	−2	1	0	−1	0	−1	c^2	−1	0	0	−1
Thermoelectric power‡	—	½	½	−1	−½‡	½	$\frac{3}{2}$	−2	½‡	1/c	1	1	0	0‡
Peltier coefficient‡	—	½	½	−1	−½‡	½	$\frac{3}{2}$	−2	½‡	1/c	1	1	0	0‡

* As adopted by American Institute of Electrical Engineers, 1915.
† *c* is the velocity of an electromagnetic wave in the ether = 3×10^{10} approximately.
‡ This conversion factor should include $[\theta^{-1}]$.

TABLE 3.

TABLES FOR CONVERTING U. S. WEIGHTS AND MEASURES.*

(1) CUSTOMARY TO METRIC.

LINEAR.

	Inches to millimeters.	Feet to meters.	Yards to meters.	Miles to kilometers.
1	25.4001	0.304801	0.914402	1.60935
2	50.8001	0.609601	1.828804	3.21869
3	76.2002	0.914402	2.743205	4.82804
4	101.6002	1.219202	3.657607	6.43739
5	127.0003	1.524003	4.572009	8.04674
6	152.4003	1.828804	5.486411	9.65608
7	177.8004	2.133604	6.400813	11.26543
8	203.2004	2.438405	7.315215	12.87478
9	228.6005	2.743205	8.229616	14.48412

CAPACITY.

	Fluid drams to milliliters or cubic centimeters.	Fluid ounces to milliliters.	Liquid quarts to liters.	Gallons to liters.
1	3.70	29.57	0.94633	3.78533
2	7.39	59.15	1.89267	7.57066
3	11.09	88.72	2.83900	11.35600
4	14.79	118.29	3.78533	15.14133
5	18.48	147.87	4.73167	18.92666
6	22.18	177.44	5.67800	22.71199
7	25.88	207.01	6.62433	26.49733
8	29.57	236.58	7.57066	30.28266
9	33.27	266.16	8.51700	34.06799

SQUARE.

	Square inches to square centimeters.	Square feet to square decimeters.	Square yards to square meters.	Acres to hectares.
1	6.452	9.290	0.836	0.4047
2	12.903	18.581	1.672	0.8094
3	19.355	27.871	2.508	1.2141
4	25.807	37.161	3.345	1.6187
5	32.258	46.452	4.181	2.0234
6	38.710	55.742	5.017	2.4281
7	45.161	65.032	5.853	2.8328
8	51.613	74.323	6.689	3.2375
9	58.065	83.613	7.525	3.6422

WEIGHT.

	Grains to milligrams.	Avoirdupois ounces to grams.	Avoirdupois pounds to kilograms.	Troy ounces to grams.
1	64.7989	28.3495	0.45359	31.10348
2	129.5978	56.6991	0.90718	62.20696
3	194.3968	85.0486	1.36078	93.31044
4	259.1957	113.3981	1.81437	124.41392
5	323.9946	141.7476	2.26796	155.51740
6	388.7935	170.0972	2.72155	186.62088
7	453.5924	198.4467	3.17515	217.72437
8	518.3913	226.7962	3.62874	248.82785
9	583.1903	255.1457	4.08233	279.93133

CUBIC.

	Cubic inches to cubic centimeters.	Cubic feet to cubic meters.	Cubic yards to cubic meters.	Bushels to hectoliters.
1	16.387	0.02832	0.765	0.35239
2	32.774	0.05663	1.529	0.70479
3	49.161	0.08495	2.294	1.05718
4	65.549	0.11327	3.058	1.40957
5	81.936	0.14159	3.823	1.76196
6	98.323	0.16990	4.587	2.11436
7	114.710	0.19822	5.352	2.46675
8	131.097	0.22654	6.116	2.81914
9	147.484	0.25485	6.881	3.17154

1 Gunter's chain	=	20.1168	meters.
1 sq. statute mile	=	259.000	hectares.
1 fathom	=	1.829	meters.
1 nautical mile	=	1853.25	meters.
1 foot	=	0.304801	meter.
1 avoir. pound	=	453.5924277	grams.
15432.35639 grains	=	1.000	kilogram.

According to an executive order dated April 15, 1893, the United States yard is defined as 3600/3937 meter, and the avoirdupois pound as 1/2.20462 kilogram.

1 meter (international prototype) = 1553164.13 times the wave-length of the red Cd. line. Benoit, Fabry and Perot. C. R. 144, 1907 differs only in the decimal portion from the measure of Michelson and Benoit 14 years earlier.

The length of the nautical mile given above and adopted by the U. S. Coast and Geodetic Survey many years ago, is defined as that of a minute of arc of a great circle of a sphere whose surface equals that of the earth (Clarke's Spheroid of 1866).

* Quoted from sheets issued by the United States Bureau of Standards.

TABLES FOR CONVERTING U. S. WEIGHTS AND MEASURES.

(2) METRIC TO CUSTOMARY.

LINEAR.

	Meters to inches.	Meters to feet.	Meters to yards.	Kilometers to miles.
1	39.3700	3.28083	1.093611	0.62137
2	78.7400	6.56167	2.187222	1.24274
3	118.1100	9.84250	3.280833	1.86411
4	157.4800	13.12333	4.374444	2.48548
5	196.8500	16.40417	5.468056	3.10685
6	236.2200	19.68500	6.561667	3.72822
7	275.5900	22.96583	7.655278	4.34959
8	314.9600	26.24667	8.748889	4.97096
9	354.3300	29.52750	9.842500	5.59233

CAPACITY.

	Milliliters or cubic centimeters to fluid drams.	Centiliters to fluid ounces.	Liters to quarts.	Decaliters to gallons.	Hectoliters to bushels.
1	0.27	0.338	1.0567	2.6418	2.8378
2	0.54	0.676	2.1134	5.2836	5.6756
3	0.81	1.014	3.1701	7.9253	8.5135
4	1.08	1.353	4.2268	10.5671	11.3513
5	1.35	1.691	5.2836	13.2089	14.1891
6	1.62	2.029	6.3403	15.8507	17.0269
7	1.89	2.367	7.3970	18.4924	19.8647
8	2.16	2.705	8.4537	21.1342	22.7026
9	2.43	3.043	9.5104	23.7760	25.5404

SQUARE.

	Square centimeters to square inches.	Square meters to square feet.	Square meters to square yards.	Hectares to acres.
1	0.1550	10.764	1.196	2.471
2	0.3100	21.528	2.392	4.942
3	0.4650	32.292	3.588	7.413
4	0.6200	43.055	4.784	9.884
5	0.7750	53.819	5.980	12.355
6	0.9300	64.583	7.176	14.826
7	1.0850	75.347	8.372	17.297
8	1.2400	86.111	9.568	19.768
9	1.3950	96.875	10.764	22.239

WEIGHT.

	Milligrams to grains.	Kilograms to grains.	Hectograms to ounces avoirdupois.	Kilograms to pounds avoirdupois.
1	0.01543	15432.36	3.5274	2.20462
2	0.03086	30864.71	7.0548	4.40924
3	0.04630	46297.07	10.5822	6.61387
4	0.06173	61729.43	14.1096	8.81849
5	0.07716	77161.78	17.6370	11.02311
6	0.09259	92594.14	21.1644	13.22773
7	0.10803	108026.49	24.6918	15.43236
8	0.12346	123458.85	28.2192	17.63698
9	0.13889	138891.21	31.7466	19.84160

CUBIC.

	Cubic centimeters to cubic inches.	Cubic decimeters to cubic inches.	Cubic meters to cubic feet.	Cubic meters to cubic yards.
1	0.0610	61.023	35.314	1.308
2	0.1220	122.047	70.269	2.616
3	0.1831	183.070	105.943	3.924
4	0.2441	244.094	141.258	5.232
5	0.3051	305.117	176.572	6.540
6	0.3661	366.140	211.887	7.848
7	0.4272	427.164	247.201	9.156
8	0.4882	488.187	282.516	10.464
9	0.5492	549.210	317.830	11.771

WEIGHT.

	Quintals to pounds av.	Milliers or tonnes to pounds av.	Kilograms to ounces Troy.
1	220.46	2204.6	32.1507
2	440.92	4409.2	64.3015
3	661.39	6613.9	96.4522
4	881.85	8818.5	128.6030
5	1102.31	11023.1	160.7537
6	1322.77	13227.7	192.9045
7	1543.24	15432.4	225.0552
8	1763.70	17637.0	257.2059
9	1984.16	19841.6	289.3567

By the concurrent action of the principal governments of the world an International Bureau of Weights and Measures has been established near Paris. Under the direction of the International Committee, two ingots were cast of pure platinum-iridium in the proportion of 9 parts of the former to 1 of the latter metal. From one of these a certain number of kilograms were prepared, from the other a definite number of meter bars. These standards of weight and length were intercompared, without preference, and certain ones were selected as International prototype standards. The others were distributed by lot, in September, 1889, to the different governments, and are called National prototype standards. Those apportioned to the United States were received in 1890, and are kept at the Bureau of Standards in Washington, D. C.

The metric system was legalized in the United States in 1866.

The International Standard Meter is derived from the Mètre des Archives, and its length is defined by the distance between two lines at 0° Centigrade, on a platinum-iridium bar deposited at the International Bureau of Weights and Measures.

The International Standard Kilogram is a mass of platinum-iridium deposited at the same place, and its weight in vacuo is the same as that of the Kilogram des Archives.

The liter is equal to the quantity of pure water at 4° C (760 mm. Hg. pressure) which weighs 1 kilogram and = 1.000027 cu. dm. (Trav. et Mem. Bureau Intern. des P. et M. 14, 1910, Benoit.)

MISCELLANEOUS EQUIVALENTS OF U. S. AND METRIC WEIGHTS AND MEASURES.*

(For other equivalents than those below, see Table 3.)

LINEAR MEASURES.

1 mil (.001 in.) = 25.4001 μ
1 in. = .000015783 mile
1 hand (4 in.) = 10.16002 cm
1 link (.66 ft.) = 20.11684 cm
1 span (9 in.) = 22.86005 cm
1 fathom (6 ft.) = 1.828804 m
1 rod (25 links) = 5.029210 m
1 chain (4 rods) = 20.11684 m
1 light year (9.5×10^{12} km) = 5.9×10^{12} miles
1 par sec (31×10^{12} km) = 19×10^{12} miles
$\frac{1}{64}$ in. = .397 mm
$\frac{1}{32}$ in. = .794 mm
$\frac{1}{16}$ in. = 1.588 mm
$\frac{1}{8}$ in. = 3.175 mm
$\frac{1}{4}$ in. = 6.350 mm
$\frac{1}{2}$ in. = 12.700 mm
1 Ångström unit = .0000000001 m
1 micron (μ) = .000001 m = .00003937 in.
1 millimicron (mμ) = .000000001 m
1 m = 4.970960 links = 1.093611 yds.
= .198838 rod = .0497096 chain

SQUARE MEASURES.

1 sq. link (62.7264 sq. in.) = 404.6873 cm^2
1 sq. rod (625 sq. links) = 25.29295 m^2
1 sq. chain (16 sq. rods) = 404.6873 m^2
1 acre (10 sq. chains) = 4046.873 m^2
1 sq. mile (640 acres) = 2.589998 km^2
1 km^2 = .3861006 sq. mile
1 m^2 = 24.7104 sq. links = 10.76387 sq. ft.
= .039537 sq. rod. = .00247104 sq. chain

CUBIC MEASURES.

1 board foot (144 cu. in) = 2359.8 cm^3
1 cord (128 cu. ft.) = 3.625 m^3

CAPACITY MEASURES.

1 minim (♏) = .0616102 ml
1 fl. dram (60♏) = 3.69661 ml
1 fl. oz. (8 fl. dr.) = 1.80469 cu. in.
= 29.5729 ml
1 gill (4 fl. oz.) = 7.21875 cu. in. = 118.292 ml
1 liq. pt. (28.875 cu. in.) = .473167 l
1 liq. qt. (57.75 cu. in.) = .946333 l
1 gallon (4 qt., 231 cu. in.) = 3.785332 l
1 dry pt. (33.6003125 cu. in.) = .550599 l
1 dry qt. (67.200625 cu. in.) = 1.101198 l
1 pk. (8 dry qt., 537.605 cu. in.) = 8.80958 l
1 bu. (4 pk., 2150.42 cu. in.) = 35.2383 l
1 firkin (9 gallons) = 34.06799 l
1 liter = .264178 gal. = 1.05671 liq. qt.
= 33.8147 fl. oz. = 270.518 fl. dr.
1 ml = 16.2311 minims.
1 dkl = 18.620 dry pt. = 9.08102 dry qt.
= 1.13513 pk. = .28378 bu.

MASS MEASURES.

Avoirdupois weights.

1 grain = .064798918 g
1 dram av. (27.34375 gr.) = 1.771845 g
1 oz. av. (16 dr. av.) = 28.349527 g
1 lb. av. (16 oz. av. or 7000 gr.)
= 14.583333 oz. ap. (℥) or oz. t.
= 1.2152778 or 7000/5760 lb. ap. or t.
= 453.5924277 g
1 kg = 2.204622341 lb. av.
1 g = 15.432356 gr. = .5643833 dr. av.
= .03527396 oz. av.
1 short hundred weight (100 lb.)
= 45.359243 kg
1 long hundred weight (112 lb.)
= 50.802352 kg
1 short ton (2000 lb.)
= 907.18486 kg
1 long ton (2240 lb.)
= 1016.04704 kg
1 metric ton = 0.98420640 long ton
= 1.1023112 short tons

Troy weights.

1 pennyweight (dwt., 24 gr.) = 1.555174 g; gr., oz., pd. are same as apothecary

Apothecaries' weights.

1 gr. = 64.798918 mg
1 scruple (℈, 20 gr.) = 1.2959784 g
1 dram (ʒ, 3 ℈) = 3.8879351 g
1 oz. (℥, 8 ʒ) = 31.103481 g
1 lb. (12℥, 5760 gr.) = 373.24177 g
1 g = 15.432356 gr. = 0.771618 ℈
= 0.2572059 ʒ = .03215074 ℥
1 kg = 32.150742 ℥ = 2.6792285 lb.

1 metric carat = 200 mg = 3.0864712 gr.

U. S. $\frac{1}{2}$ dollar should weigh 12.5 g and the smaller silver coins in proportion.

* Taken from Circular 47 of the U. S. Bureau of Standards, 1915, which see for more complete tables.

TABLE 5.

EQUIVALENTS OF METRIC AND BRITISH IMPERIAL WEIGHTS AND MEASURES.*

(1) METRIC TO IMPERIAL. (For U.S. Weights and Measures, see Table 3.)

LINEAR MEASURE.

1 millimeter (mm.) (.001 m.)	= 0.03937 in.
1 centimeter (.01 m.)	= 0.39370 "
1 decimeter (.1 m)	= 3.93701 "
1 METER (m.)	= 39.370113 " ; 3.280843 ft. ; 1.09361425 yds.
1 dekameter (10 m.)	= 10.93614 "
1 hectometer (100 m.)	= 109.361425 "
1 kilometer (1,000 m.)	= 0.62137 mile.
1 myriameter (10,000 m.)	= 6.21372 miles.
1 micron	= 0.001 mm.

SQUARE MEASURE.

1 sq. centimeter	= 0.1550 sq. in.
1 sq. decimeter (100 sq. centm.)	= 15.500 sq. in.
1 sq. meter or centiare (100 sq. dcm.)	= 10.7639 sq. ft. ; 1.1960 sq. yds.
1 ARE (100 sq. m.)	= 119.60 sq. yds.
1 hectare (100 ares or 10,000 sq. m.)	= 2.4711 acres.

CUBIC MEASURE.

1 cub. centimeter (c.c.) (1,000 cubic millimeters)	= 0.0610 cub. in.
1 cub. decimeter (c.d.) (1,000 cubic centimeters)	= 61.024 " "
1 CUB. METER or stere (1,000 c.d.)	= 35.3148 cub. ft. ; 1.307954 cub. yds.

MEASURE OF CAPACITY.

1 milliliter (ml.) (.001 liter)	= 0.0610 cub. in.
1 centiliter (.01 liter)	= 0.61024 " " ; 0.070 gill.
1 deciliter (.1 liter)	= 0.176 pint.
1 LITER (1,000 cub. centimeters or 1 cub. decimeter)	= 1.75980 pints.
1 dekaliter (10 liters)	= 2.200 gallons.
1 hectoliter (100 ")	= 2.75 bushels.
1 kiloliter (1,000 ")	= 3.437 quarters.

APOTHECARIES' MEASURE.

1 cubic centimeter (1 gram w't)	= 0.03520 fluid ounce. ; 0.28157 fluid drachm. ; 15.43236 grains weight.
1 cub. millimeter	= 0.01693 minim.

AVOIRDUPOIS WEIGHT.

1 milligram (mgr.)	= 0.01543 grain.
1 centigram (.01 gram.)	= 0.15432 "
1 decigram (.1 ")	= 1.54324 grains.
1 GRAM	= 15.43236 "
1 dekagram (10 gram.)	= 5.64383 drams.
1 hectogram (100 ")	= 3.52739 oz.
1 KILOGRAM (1,000 ")	= 2.2046223 lb. ; 15432.3564 grains.
1 myriagram (10 kilog.)	= 22.04622 lbs.
1 quintal (100 ")	= 1.96841 cwt.
1 millier or tonne (1,000 kilog.)	= 0.9842 ton.

TROY WEIGHT.

1 GRAM	= 0.03215 oz. Troy. ; 0.64301 pennyweight. ; 15.43236 grains.

APOTHECARIES' WEIGHT.

1 GRAM	= 0.25721 drachm. ; 0.77162 scruple. ; 15.43236 grains.

NOTE.—The METER is the length, at the temperature of 0° C., of the platinum-iridium bar deposited at the International Bureau of Weights and Measures at Sèvres, near Paris, France.

The present legal equivalent of the meter is 39.370113 inches, as above stated.

The KILOGRAM is the mass of a platinum-iridium weight deposited at the same place.

The LITER contains one kilogram weight of distilled water at its maximum density (4° C.), the barometer being at 760 millimeters.

*In accordance with the schedule adopted under the Weights and Measures (metric system) Act, 1897.

TABLE 5.

EQUIVALENTS OF METRIC AND BRITISH IMPERIAL WEIGHTS AND MEASURES.

(2) METRIC TO IMPERIAL. (For U.S. Weights and Measures, see Table 3.)

LINEAR MEASURE.

	Millimeters to inches.	Meters to feet.	Meters to yards.	Kilometers to miles.
1	0.03937011	3.28084	1.09361	0.62137
2	0.07874023	6.56169	2.18723	1.24274
3	0.11811034	9.84253	3.28084	1.86412
4	0.15748045	13.12337	4.37446	2.48549
5	0.19685056	16.40421	5.46807	3.10686
6	0.23622068	19.68506	6.56169	3.72823
7	0.27559079	22.96590	7.65530	4.34960
8	0.31496090	26.24674	8.74891	4.97097
9	0.35433102	29.52758	9.84253	5.59235

MEASURE OF CAPACITY.

	Liters to pints.	Dekaliters to gallons.	Hectoliters to bushels.	Kiloliters to quarters.
1	1.75980	2.19975	2.74969	3.43712
2	3.51961	4.39951	5.49938	6.87423
3	5.27941	6.59926	8.24908	10.31135
4	7.03921	8.79902	10.99877	13.74846
5	8.79902	10.99877	13.74846	17.18558
6	10.55882	13.19852	16.49815	20.62269
7	12.31862	15.39828	19.24785	24.05981
8	14.07842	17.59803	21.99754	27.49692
9	15.83823	19.79778	24.74723	30.93404

SQUARE MEASURE.

	Square centimeters to square inches.	Square meters to square feet.	Square meters to square yards.	Hectares to acres.
1	0.15500	10.76393	1.19599	2.4711
2	0.31000	21.52786	2.39198	4.9421
3	0.46500	32.29179	3.58798	7.4132
4	0.62000	43.05572	4.78397	9.8842
5	0.77500	53.81965	5.97996	12.3553
6	0.93000	64.58357	7.17595	14.8263
7	1.08500	75.34750	8.37194	17.2974
8	1.24000	86.11143	9.56794	19.7685
9	1.39501	96.87536	10.76393	22.2395

WEIGHT (AVOIRDUPOIS).

	Milligrams to grains.	Kilograms to grains.	Kilograms to pounds.	Quintals to hundred-weights.
1	0.01543	15432.356	2.20462	1.96841
2	0.03086	30864.713	4.40924	3.93683
3	0.04630	46297.069	6.61387	5.90524
4	0.06173	61729.426	8.81849	7.87365
5	0.07716	77161.782	11.02311	9.84206
6	0.09259	92594.138	13.22773	11.81048
7	0.10803	108026.495	15.43236	13.77889
8	0.12346	123458.851	17.63698	15.74730
9	0.13889	138891.208	19.84160	17.71572

CUBIC MEASURE. / APOTHECARIES' MEASURE.

	Cubic decimeters to cubic inches.	Cubic meters to cubic feet.	Cubic meters to cubic yards.	Apothecaries' Measure. Cub. centimeters to fluid drachms.
1	61.02390	35.31476	1.30795	0.28157
2	122.04781	70.62952	2.61591	0.56314
3	183.07171	105.94428	3.92386	0.84471
4	244.09561	141.25904	5.23182	1.12627
5	305.11952	176.57379	6.53977	1.40784
6	366.14342	211.88855	7.84772	1.68941
7	427.16732	247.20331	9.15568	1.97098
8	488.19123	282.51807	10.46363	2.25255
9	549.21513	317.83283	11.77159	2.53412

AVOIRDUPOIS (cont.) / TROY WEIGHT. / APOTHECARIES' WEIGHT.

	Avoirdupois (*cont.*) Milliers or tonnes to tons.	Troy Weight. Grams to ounces Troy.	Troy Weight. Grams to penny-weights.	Apothecaries' Weight. Grams to scruples.
1	0.98421	0.03215	0.64301	0.77162
2	1.96841	0.06430	1.28603	1.54324
3	2.95262	0.09645	1.92904	2.31485
4	3.93683	0.12860	2.57206	3.08647
5	4.92103	0.16075	3.21507	3.85809
6	5.90524	0.19290	3.85809	4.62971
7	6.88944	0.22506	4.50110	5.40132
8	7.87365	0.25721	5.14412	6.17294
9	8.85786	0.28936	5.78713	6.94456

TABLE 5.

EQUIVALENTS OF BRITISH IMPERIAL AND METRIC WEIGHTS AND MEASURES.

(3) IMPERIAL TO METRIC. (For U.S. Weights and Measures, see Table 3.)

LINEAR MEASURE.

1 inch	=	25.400 millimeters.
1 foot (12 in.) . .	=	0.30480 meter.
1 YARD (3 ft.) . .	=	0.914399 "
1 pole (5½ yd.) . .	=	5.0292 meters.
1 chain (22 yd. or 100 links)	=	20.1168 "
1 furlong (220 yd.)	=	201.168 "
1 mile (1,760 yd.) .	=	1.6093 kilometers.

SQUARE MEASURE.

1 square inch . .	=	6.4516 sq. centimeters.
1 sq. ft. (144 sq. in.)	=	9.2903 sq. decimeters.
1 SQ. YARD (9 sq. ft.)	=	0.836126 sq. meters.
1 perch (30¼ sq. yd.)	=	25.293 sq. meters.
1 rood (40 perches)	=	10.117 ares.
1 ACRE (4840 sq. yd.)	=	0.40468 hectare.
1 sq. mile (640 acres)	=	259.00 hectares.

CUBIC MEASURE.

1 cub. inch	=	16.387 cub. centimeters.
1 cub. foot (1728 cub. in.)	=	0.028317 cub. meter, or 28.317 cub. decimeters.
1 CUB. YARD (27 cub. ft.)	=	0.76455 cub. meter.

APOTHECARIES' MEASURE.

1 gallon (8 pints or 160 fluid ounces)	=	4.5459631 liters.
1 fluid ounce, f ℥ (8 drachms)	=	28.4123 cubic centimeters.
1 fluid drachm, f ʒ (60 minims)	=	3.5515 cubic centimeters.
1 minim, ♏ (0.91146 grain weight)	=	0.05919 cubic centimeters.

NOTE. — The Apothecaries' gallon is of the same capacity as the Imperial gallon.

MEASURE OF CAPACITY.

1 gill	=	1.42 deciliters.
1 pint (4 gills) . . .	=	0.568 liter.
1 quart (2 pints) . .	=	1.136 liters.
1 GALLON (4 quarts)	=	4.5459631 "
1 peck (2 galls.) . .	=	9.092 "
1 bushel (8 galls.) .	=	3.637 dekaliters.
1 quarter (8 bushels)	=	2.909 hectoliters.

AVOIRDUPOIS WEIGHT.

1 grain	=	64.8 milligrams.
1 dram	=	1.772 grams.
1 ounce (16 dr.) . .	=	28.350 "
1 POUND (16 oz. or 7,000 grains)	=	0.45359243 kilogr.
1 stone (14 lb.) . .	=	6.350 "
1 quarter (28 lb.) .	=	12.70 "
1 hundredweight (112 lb.)	=	50.80 " 0.5080 quintal.
1 ton (20 cwt.) .	=	1.0160 tonnes or 1016 kilograms.

TROY WEIGHT.

1 Troy OUNCE (480 grains avoir.)	=	31.1035 grams.
1 pennyweight (24 grains)	=	1.5552 "

NOTE. — The Troy grain is of the same weight as the Avoirdupois grain.

APOTHECARIES' WEIGHT.

1 ounce (8 drachms)	=	31.1035 grams.
1 drachm, ʒ i (3 scruples)	=	3.888 "
1 scruple, ℈ i (20 grains)	=	1.296 "

NOTE. — The Apothecaries' ounce is of the same weight as the Troy ounce. The Apothecaries' grain is also of the same weight as the Avoirdupois grain.

NOTE. — The YARD is the length at 62° Fahr., marked on a bronze bar deposited with the Board of Trade.
The POUND is the weight of a piece of platinum weighed in vacuo at the temperature of 0° C., and which is also deposited with the Board of Trade.
The GALLON contains 10 lb. weight of distilled water at the temperature of 62° Fahr., the barometer being at 30 inches.

Table 5.

EQUIVALENTS OF BRITISH IMPERIAL AND METRIC WEIGHTS AND MEASURES.

(4) IMPERIAL TO METRIC. (For U.S. Weights and Measures, see Table 3.)

LINEAR MEASURE.

	Inches to centimeters.	Feet to meters.	Yards to meters.	Miles to kilometers.
1	2.539998	0.30480	0.91440	1.60934
2	5.079996	0.60960	1.82880	3.21869
3	7.619993	0.91440	2.74320	4.82803
4	10.159991	1.21920	3.65760	6.43737
5	12.699989	1.52400	4.57200	8.04671
6	15.239987	1.82880	5.48640	9.65606
7	17.779984	2.13360	6.40080	11.26540
8	20.319982	2.43840	7.31519	12.87474
9	22.859980	2.74320	8.22959	14.48408

MEASURE OF CAPACITY.

	Quarts to liters.	Gallons to liters.	Bushels to dekaliters.	Quarters to hectoliters.
1	1.13649	4.54596	3.63677	2.90942
2	2.27298	9.09193	7.27354	5.81883
3	3.40947	13.63789	10.91031	8.72825
4	4.54596	18.18385	14.54708	11.63767
5	5.68245	22.72982	18.18385	14.54708
6	6.81894	27.27578	21.82062	17.45650
7	7.95544	31.82174	25.45739	20.36591
8	9.09193	36.36770	29.09416	23.27533
9	10.22842	40.91367	32.73093	26.18475

SQUARE MEASURE.

	Square inches to square centimeters.	Square feet to square decimeters.	Square yards to square meters.	Acres to hectares.
1	6.45159	9.29029	0.83613	0.40468
2	12.90318	18.58058	1.67225	0.80937
3	19.35477	27.87086	2.50838	1.21405
4	25.80636	37.16115	3.34450	1.61874
5	32.25794	46.45144	4.18063	2.02342
6	38.70953	55.74173	5.01676	2.42811
7	45.16112	65.03201	5.85288	2.83279
8	51.61271	74.32230	6.68901	3.23748
9	58.06430	83.61259	7.52513	3.64216

WEIGHT (Avoirdupois).

	Grains to milligrams.	Ounces to grams.	Pounds to kilograms.	Hundredweights to quintals.
1	64.79892	28.34953	0.45359	0.50802
2	129.59784	56.69905	0.90718	1.01605
3	194.39675	85.04858	1.36078	1.52407
4	259.19567	113.39811	1.81437	2.03209
5	323.99459	141.74763	2.26796	2.54012
6	388.79351	170.09716	2.72155	3.04814
7	453.59243	198.44669	3.17515	3.55616
8	518.39135	226.79621	3.62874	4.06419
9	583.19026	255.14574	4.08233	4.57221

	CUBIC MEASURE.			Apothecaries' Measure.
	Cubic inches to cubic centimeters.	Cubic feet to cubic meters.	Cubic yards to cubic meters.	Fluid drachms to cubic centimeters.
1	16.38702	0.02832	0.76455	3.55153
2	32.77404	0.05663	1.52911	7.10307
3	49.16106	0.08495	2.29366	10.65460
4	65.54808	0.11327	3.05821	14.20613
5	81.93511	0.14158	3.82276	17.75767
6	98.32213	0.16990	4.58732	21.30920
7	114.70915	0.19822	5.35187	24.86074
8	131.09617	0.22653	6.11642	28.41227
9	147.48319	0.25485	6.88098	31.96380

	Avoirdupois (*cont.*).	Troy Weight.		Apothecaries' Weight.
	Tons to milliers or tonnes.	Ounces to grams.	Pennyweights to grams.	Scruples to grams.
1	1.01605	31.10348	1.55517	1.29598
2	2.03209	62.20696	3.11035	2.59196
3	3.04814	93.31044	4.66552	3.88794
4	4.06419	124.41392	6.22070	5.18391
5	5.08024	155.51740	7.77587	6.47989
6	6.09628	186.62088	9.33104	7.77587
7	7.11233	217.72437	10.88622	9.07185
8	8.12838	248.82785	12.44139	10.36783
9	9.14442	279.93133	13.99657	11.66381

TABLE 6.

DERIVATIVES AND INTEGRALS.*

$d\,ax$	$= a\,dx$	$\int x^n\,dx$	$= \frac{x^{n+1}}{n+1}$, unless $n = -1$
$d\,uv$	$= \left(u\frac{dv}{dx} + v\frac{du}{dx}\right)dx$	$\int \frac{dx}{x}$	$= \log x$
$d\,\frac{u}{v}$	$= \left(\frac{v\frac{du}{dx} - u\frac{dv}{dx}}{v^2}\right)dx$	$\int e^x\,dx$	$= e^x$
$d\,x^n$	$= nx^{n-1}\,dx$	$\int e^{ax}dx$	$= \frac{1}{a}\,e^{ax}$
$d\,f(u)$	$= d\,\frac{f(u)}{du}.\,\frac{du}{dx}.\,dx$	$\int x\,e^{ax}\,dx$	$= \frac{e^{ax}}{a^2}\,(ax-1)$
$d\,e^x$	$= e^x\,dx$	$\int \log x\,dx$	$= x\log x - x$
$d\,e^{ax}$	$= a\,e^{ax}\,dx$	$\int u\,dv$	$= uv - \int v\,du$
$d\log_e x$	$= \frac{1}{x}\,dx$	$\int (a+bx)^n\,dx$	$= \frac{(a+bx)^{n+1}}{(n+1)b}$
$d\,x^x$	$= x^x\,(1+\log_e x)$		
$d\sin x$	$= \cos x\,dx$	$\int (a^2+x^2)^{-1}\,dx$	$= \frac{1}{a}\tan^{-1}\frac{x}{a} = \frac{1}{a}\sin^{-1}\frac{x}{\sqrt{x^2+a^2}}$
$d\cos x$	$= -\sin x\,dx$	$\int (a^2-x^2)^{-1}\,dx$	$= \frac{1}{2a}\log\frac{a+x}{a-x}$
$d\tan x$	$= \sec^2 x\,dx$	$\int (a^2-x^2)^{-\frac{1}{2}}\,dx$	$= \sin^{-1}\frac{x}{a}$, or $-\cos^{-1}\frac{x}{a}$
$d\cot x$	$= -\csc^2 x\,dx$	$\int x(a^2 \pm x^2)^{-\frac{1}{2}}\,dx$	$= \pm(a^2 \pm x^2)^{\frac{1}{2}}$
$d\sec x$	$= \tan x\sec x\,dx$	$\int \sin^2 x\,dx$	$= -\frac{1}{2}\cos x\sin x + \frac{1}{2}x$
$d\csc x$	$= -\cot x\,.\,\csc x\,dx$	$\int \cos^2 x\,dx$	$= \frac{1}{2}\sin x\cos x + \frac{1}{2}x$
$d\sin^{-1} x$	$= (1-x^2)^{-\frac{1}{2}}\,dx$	$\int \sin x\cos x\,dx$	$= \frac{1}{2}\sin^2 x$
$d\cos^{-1} x$	$= -(1-x^2)^{-\frac{1}{2}}\,dx$	$\int (\sin x\cos x)^{-1}\,dx$	$= \log\tan x$
$d\tan^{-1} x$	$= (1+x^2)^{-1}\,dx$	$\int \tan x\,dx$	$= -\log\cos x$
$d\cot^{-1} x$	$= -(1+x^2)^{-1}\,dx$	$\int \tan^2 x\,dx$	$= \tan x - x$
$d\sec^{-1} x$	$= x^{-1}(x^2-1)^{-\frac{1}{2}}\,dx$	$\int \cot x\,dx$	$= \log\sin x$
$d\csc^{-1} x]$	$= -x^{-1}(x^2-1)^{-\frac{1}{2}}\,dx$	$\int \cot^2 x\,dx$	$= -\cot x - x$
$d\sinh x$	$= \cosh x\,dx$	$\int \csc x\,dx$	$= \log\tan\frac{1}{2}x$
$d\cosh x$	$= \sinh x\,dx$	$\int x\sin x\,dx$	$= \sin x - x\cos x$
$d\tanh x$	$= \mathrm{sech}^2 x\,dx$	$\int x\cos x\,dx$	$= \cos x + x\sin x$
$d\coth x$	$= -\mathrm{csch}^2 x\,dx$	$\int \tanh x\,dx$	$= \log\cosh x$
$d\,\mathrm{sech}\,x$	$= -\mathrm{sech}\,x\tanh dx$	$\int \coth x\,dx$	$= \log\sinh x$
$d\,\mathrm{csch}\,x$	$= -\mathrm{csch}\,x\,.\,\coth x\,dx$	$\int \mathrm{sech}\,x\,dx$	$= 2\tan^{-1} e^x = \mathrm{gd}\,u$
$d\sinh^{-1} x$	$= (x^2+1)^{-\frac{1}{2}}\,dx$	$\int \mathrm{csch}\,x\,dx$	$= \log\tanh\frac{x}{2}$
$d\cosh^{-1} x$	$= (x^2-1)^{-\frac{1}{2}}\,dx$	$\int x\sinh x\,dx$	$= x\cosh x - \sinh x$
$d\tanh^{-1} x$	$= (1-x^2)^{-1}\,dx$	$\int x\cosh x\,dx$	$= x\sinh x - \cosh x$
$d\coth^{-1} x$	$= (1-x^2)^{-1}\,dx$	$\int \sinh^2 x\,dx$	$= \frac{1}{2}(\sinh x\cosh x - x)$
$d\,\mathrm{sech}^{-1} x$	$= -x^{-1}(1-x^2)^{-\frac{1}{2}}\,dx$	$\int \cosh^2 x\,dx$	$= \frac{1}{2}(\sinh x\cosh x + x)$
$d\,\mathrm{csch}^{-1} x$	$= -x^{-1}(x^2+1)^{-\frac{1}{2}}$	$\int \sinh x\cosh x\,dx$	$= \frac{1}{4}\cosh(2x)$

* See also accompanying table of derivatives. For example: $\int \cos.\,x\,dx = \sin.\,x +$ constant.

SERIES.

$$(x+y)^n = x^n + \frac{n}{1} x^{n-1} y + \frac{n(n-1)}{2!} x^{n-2} y^2 + \ldots$$

$$\frac{n(n-1)\ldots(n-m+1)}{m!} x^{n-m} y^m + \ldots \qquad (y^2 < x^2)$$

$$(1 \pm x)^n = 1 \pm nx + \frac{n(n-1)x^2}{2!} \pm \frac{n(n-1)(n-2)x^2}{3!} + \ldots + \frac{(\pm 1)^k n!\, x^k}{(n-k)!\, k!} + \ldots \qquad (x^2 < 1)$$

$$(1 \pm x)^{-n} = 1 \mp nx + \frac{n(n+1)}{2!} x^2 \mp \frac{n(n+1)(n+2)x^3}{3!} + \ldots$$

$$(\mp 1)^k \frac{(n+k-1)\, x^k}{(n-1)!\, k!} + \ldots \qquad (x^2 < 1)$$

$$(1 \pm x)^{-1} = 1 \mp x + x^2 \mp x^3 + x^4 \mp x^5 + \ldots \qquad (x^2 < 1)$$

$$(1 \pm x)^{-2} = 1 \mp 2x + 3x^2 \mp 4x^3 + 5x^4 \mp 6x^5 + \ldots \qquad (x^2 < 1)$$

$$f(x+h) = f(x) + hf'(x) + \frac{h^2}{2!} f''(x) + \ldots + \frac{h^n}{n!} f^{(n)}(x) + \ldots \qquad \text{Taylor's series.}$$

$$f(x) = f(0) + \frac{x}{1} f'(0) + \frac{x^2}{2!} f''(0) + \ldots \frac{x^n}{n!} f^{(n)}(0) + \ldots \qquad \text{Maclaurin's series.}$$

$$e = \lim\left(1 + \frac{1}{n}\right)^n = 1 + \frac{1}{1!} + \frac{1}{2!} + \frac{1}{3!} + \frac{1}{4!} + \ldots$$

$$e^x = 1 + x + \frac{x^2}{2!} + \frac{x^3}{3!} + \frac{x^4}{4!} + \ldots \qquad (x^2 < \infty)$$

$$a^x = 1 + x \log a + \frac{(x \log a)^2}{2!} + \frac{(x \log a)^3}{3!} + \ldots. \qquad (x^2 < \infty)$$

$$\log x = \frac{x-1}{x} + \frac{1}{2}\left(\frac{x-1}{x}\right)^2 + \frac{1}{3}\left(\frac{x-1}{x}\right)^3 + \ldots \qquad (x > \tfrac{1}{2})$$

$$= (x-1) - \tfrac{1}{2}(x-1)^2 + \tfrac{1}{3}(x-1)^3 - \ldots \qquad (2 > x > 0)$$

$$= 2\left[\frac{x-1}{x+1} + \frac{1}{3}\left(\frac{x-1}{x+1}\right)^3 + \frac{1}{5}\left(\frac{x-1}{x+1}\right)^5 + \ldots.\right] \qquad (x > 0)$$

$$\log(1+x) = x - \tfrac{1}{2}x^2 + \tfrac{1}{3}x^3 - \tfrac{1}{4}x^4 + \ldots. \qquad (x^2 < 1)$$

$$\sin x = \frac{1}{2i}(e^{ix} - e^{-ix}) = x - \frac{x^3}{3!} + \frac{x^5}{5!} - \frac{x^7}{7!} + \ldots \qquad (x^2 < \infty)$$

$$\cos x = \frac{1}{2}(e^{ix} + e^{-ix}) = 1 - \frac{x^2}{2!} + \frac{x^4}{4!} - \frac{x^6}{6!} + \ldots = 1 - \text{versin}\, x \qquad (x^2 < \infty)$$

$$\tan x = x + \frac{x^3}{3} + \frac{2x^5}{15} + \frac{17x^7}{315} + \frac{62}{2835} x^9 + \ldots \qquad \left(x^2 < \frac{\pi^2}{4}\right)$$

$$\sin^{-1} x = \frac{\pi}{2} - \cos.^{-1} x = x + \frac{x^3}{6} + \frac{1}{2}\cdot\frac{3}{4}\cdot\frac{x^5}{5} + \frac{1}{2}\cdot\frac{3}{4}\cdot\frac{5}{6}\cdot\frac{x^7}{7} + \ldots \qquad (x^2 < 1)$$

$$\tan^{-1} x = \frac{\pi}{2} - \cot.^{-1} x = x - \frac{1}{3}x^3 + \frac{1}{5}x^5 - \frac{1}{7}x^7 + \ldots \qquad (x^2 < 1)$$

$$= \frac{\pi}{2} - \frac{1}{x} + \frac{1}{3x^3} - \frac{1}{5x^5} + \ldots \qquad (x^2 > 1)$$

$$\sinh x = \frac{1}{2}(e^x - e^{-x}) = x + \frac{x^3}{3!} + \frac{x^5}{5!} + \frac{x^7}{7!} + \ldots \qquad (x^2 < \infty)$$

SERIES.

$$\cosh x = \frac{1}{2}\left(e^{x} + e^{-x}\right) = 1 + \frac{x^2}{2!} + \frac{x^4}{4!} + \frac{x^6}{6!} + \ldots \qquad (x^2 < \infty)$$

$$\tanh x = x - \frac{1}{3}x^3 + \frac{2}{15}x^5 - \frac{17}{315}x^7 + \ldots \qquad (x^2 < \tfrac{1}{4}\pi^2)$$

$$\sinh^{-1} x = x - \frac{1}{2}\frac{x^3}{3} + \frac{1}{2}\cdot\frac{3}{4}\cdot\frac{x^5}{5} - \frac{1}{2}\cdot\frac{3}{4}\cdot\frac{5}{6}\frac{x^7}{7} + \ldots \qquad (x^2 < 1)$$

$$= \log 2x + \frac{1}{2}\frac{1}{2x^2} - \frac{1}{2}\frac{3}{4}\frac{1}{4x^4} + \frac{1}{2}\frac{3}{4}\frac{5}{6}\frac{1}{6x^6} - \ldots \qquad (x^2 > 1)$$

$$\cosh^{-1} x = \log 2x - \frac{1}{2}\frac{1}{2x^2} - \frac{1}{2}\frac{3}{4}\frac{1}{4x^4} - \frac{1}{2}\frac{3}{4}\frac{5}{6}\frac{1}{6x^6} - \ldots \qquad (x^2 > 1)$$

$$\tanh^{-1} x = x + \frac{1}{3}x^3 + \frac{1}{5}x^5 + \frac{1}{7}x^7 + \ldots \qquad (x^2 < 1)$$

$$\mathrm{gd}\, x = \phi = x - \frac{1}{6}x^3 + \frac{1}{24}x^5 - \frac{61}{5040}x^7 + \ldots \qquad (x \text{ small})$$

$$= \frac{\pi}{2} - \mathrm{sech}.\, x - \frac{1}{2}\frac{\mathrm{sech}^3 x}{3} - \frac{1}{2}\frac{3}{4}\frac{\mathrm{sech}^5 x}{5} - \ldots \qquad (x \text{ large})$$

$$x = \mathrm{gd}^{-1}\phi = \phi + \frac{1}{6}\phi^3 + \frac{1}{24}\phi^5 + \frac{61}{5040}\phi^7 + \ldots \qquad \left(\phi < \frac{\pi}{2}\right)$$

$$f(x) = \frac{1}{2}b_0 + b_1\cos\frac{\pi x}{c} + b_2\cos\frac{2\pi x}{c} + \ldots$$
$$+ a_1 \sin\frac{\pi x}{c} + a_2\cos\frac{2\pi x}{c} + \ldots \qquad (-c < x < c)$$

$$a_m = \frac{1}{c}\int_{-c}^{+c} f(x)\sin\frac{m\pi x}{c}\,dx$$

$$b_m = \frac{1}{c}\int_{-c}^{+c} f(x)\cos\frac{m\pi x}{c}\,dx$$

TABLE 8.—MATHEMATICAL CONSTANTS.

	Numbers.	Logarithms.
$e = 2.71828\ 18285$	$\pi = 3.14159\ 26536$	0.49714 98727
$e^{-1} = 0.36787\ 94412$	$\pi^2 = 9.86960\ 44011$	0.99429 97454
$M = \log_{10} e = 0.43429\ 44819$	$\frac{1}{\pi} = 0.31830\ 98862$	9.50285 01273
$(M)^{-1} = \log_e 10 = 2.30258\ 50930$	$\sqrt{\pi} = 1.77245\ 38509$	0.24857 49363
$\log_{10}\log_{10} e = 9.63778\ 43113$	$\frac{\sqrt{\pi}}{2} = 0.88622\ 69255$	9.94754 49407
$\log_{10} 2 = 0.30102\ 99957$	$\frac{1}{\sqrt{\pi}} = 0.56418\ 95835$	9.75142 50637
$\log_e 2 = 0.69314\ 71806$	$\frac{2}{\sqrt{\pi}} = 1.12837\ 91671$	0.05245 50593
$\log_{10} x = M.\log_e x$	$\sqrt{\frac{\pi}{2}} = 1.25331\ 41373$	0.09805 99385
$\log_B x = \log_e x.\ \log_B e$	$\sqrt{\frac{2}{\pi}} = 0.79788\ 45608$	9.90194 00615
$= \log_e x \div \log_e B$	$\frac{\pi}{4} = 0.78539\ 81634$	9.89508 98814
$\log_e \pi = 1.14472\ 98858$	$\frac{\sqrt{\pi}}{4} = 0.44311\ 34627$	9.64651 49450
$\rho = 0.47693\ 62762$	$\frac{4}{3}\pi = 4.18879\ 02048$	0.62208 86093
$\log \rho = 9.67846\ 03565$	$\frac{e}{\sqrt{2\pi}} = 1.08443\ 75514$	0.03520 45477

VALUES OF RECIPROCALS, SQUARES, CUBES, SQUARE ROOTS, OF NATURAL NUMBERS.

n	$1000.\frac{1}{n}$	n^2	n^3	$\sqrt{n}$	n	$1000.\frac{1}{n}$	n^2	n^3	$\sqrt{n}$
10	100.000	100	1000	3.1623	**65**	15.3846	4225	274625	8.0623
11	90.9091	121	1331	3.3166	66	15.1515	4356	287496	8.1240
12	83.3333	144	1728	3.4641	67	14.9254	4489	300763	8.1854
13	76.9231	169	2197	3.6056	68	14.7059	4624	314432	8.2462
14	71.4286	196	2744	3.7417	69	14.4928	4761	328509	8.3066
15	66.6667	225	3375	3.8730	**70**	14.2857	4900	343000	8.3666
16	62.5000	256	4096	4.0000	71	14.0845	5041	357911	8.4261
17	58.8235	289	4913	4.1231	72	13.8889	5184	373248	8.4853
18	55.5556	324	5832	4.2426	73	13.6986	5329	389017	8.5440
19	52.6316	361	6859	4.3589	74	13.5135	5476	405224	8.6023
20	50.0000	400	8000	4.4721	**75**	13.3333	5625	421875	8.6603
21	47.6190	441	9261	4.5826	76	13.1579	5776	438976	8.7178
22	45.4545	484	10648	4.6904	77	12.9870	5929	456533	8.7750
23	43.4783	529	12167	4.7958	78	12.8205	6084	474552	8.8318
24	41.6667	576	13824	4.8990	79	12.6582	6241	493039	8.8882
25	40.0000	625	15625	5.0000	**80**	12.5000	6400	512000	8.9443
26	38.4615	676	17576	5.0990	81	12.3457	6561	531441	9.0000
27	37.0370	729	19683	5.1962	82	12.1951	6724	551368	9.0554
28	35.7143	784	21952	5.2915	83	12.0482	6889	571787	9.1104
29	34.4828	841	24389	5.3852	84	11.9048	7056	592704	9.1652
30	33.3333	900	27000	5.4772	**85**	11.7647	7225	614125	9.2195
31	32.2581	961	29791	5.5678	86	11.6279	7396	636056	9.2736
32	31.2500	1024	32768	5.6569	87	11.4943	7569	658503	9.3274
33	30.3030	1089	35937	5.7446	88	11.3636	7744	681472	9.3808
34	29.4118	1156	39304	5.8310	89	11.2360	7921	704969	9.4340
35	28.5714	1225	42875	5.9161	**90**	11.1111	8100	729000	9.4868
36	27.7778	1296	46656	6.0000	91	10.9890	8281	753571	9.5394
37	27.0270	1369	50653	6.0828	92	10.8696	8464	778688	9.5917
38	26.3158	1444	54872	6.1644	93	10.7527	8649	804357	9.6437
39	25.6410	1521	59319	6.2450	94	10.6383	8836	830584	9.6954
40	25.0000	1600	64000	6.3246	**95**	10.5263	9025	857375	9.7468
41	24.3902	1681	68921	6.4031	96	10.4167	9216	884736	9.7980
42	23.8095	1764	74088	6.4807	97	10.3093	9409	912673	9.8489
43	23.2558	1849	79507	6.5574	98	10.2041	9604	941192	9.8995
44	22.7273	1936	85184	6.6332	99	10.1010	9801	970299	9.9499
45	22.2222	2025	91125	6.7082	**100**	10.0000	10000	1000000	10.0000
46	21.7391	2116	97336	6.7823	101	9.90099	10201	1030301	10.0499
47	21.2766	2209	103823	6.8557	102	9.80392	10404	1061208	10.0995
48	20.8333	2304	110592	6.9282	103	9.70874	10609	1092727	10.1489
49	20.4082	2401	117649	7.0000	104	9.61538	10816	1124864	10.1980
50	20.0000	2500	125000	7.0711	**105**	9.52381	11025	1157625	10.2470
51	19.6078	2601	132651	7.1414	106	9.43396	11236	1191016	10.2956
52	19.2308	2704	140608	7.2111	107	9.34579	11449	1225043	10.3441
53	18.8679	2809	148877	7.2801	108	9.25926	11664	1259712	10.3923
54	18.5185	2916	157464	7.3485	109	9.17431	11881	1295029	10.4403
55	18.1818	3025	166375	7.4162	**110**	9.09091	12100	1331000	10.4881
56	17.8571	3136	175616	7.4833	111	9.00901	12321	1367631	10.5357
57	17.5439	3249	185193	7.5498	112	8.92857	12544	1404928	10.5830
58	17.2414	3364	195112	7.6158	113	8.84956	12769	1442897	10.6301
59	16.9492	3481	205379	7.6811	114	8.77193	12996	1481544	10.6771
60	16.6667	3600	216000	7.7460	**115**	8.69565	13225	1520875	10.7238
61	16.3934	3721	226981	7.8102	116	8.62069	13456	1560896	10.7703
62	16.1290	3844	238328	7.8740	117	8.54701	13689	1601613	10.8167
63	15.8730	3969	250047	7.9373	118	8.47458	13924	1643032	10.8628
64	15.6250	4096	262144	8.0000	119	8.40336	14161	1685159	10.9087

VALUES OF RECIPROCALS, SQUARES, CUBES, SQUARE ROOTS, OF NATURAL NUMBERS.

n	$1000.\frac{1}{n}$	n^2	n^3	$\sqrt{n}$	n	$1000.\frac{1}{n}$	n^2	n^3	$\sqrt{n}$
120	8.33333	14400	1728000	10.9545	**175**	5.71429	30625	5359375	13.2288
121	8.26446	14641	1771561	11.0000	176	5.68182	30976	5451776	13.2665
122	8.19672	14884	1815848	11.0454	177	5.64972	31329	5545233	13.3041
123	8.13008	15129	1860867	11.0905	178	5.61798	31684	5639752	13.3417
124	8.06452	15376	1906624	11.1355	179	5.58659	32041	5735339	13.3791
125	8.00000	15625	1953125	11.1803	**180**	5.55556	32400	5832000	13.4164
126	7.93651	15876	2000376	11.2250	181	5.52486	32761	5929741	13.4536
127	7.87402	16129	2048383	11.2694	182	5.49451	33124	6028568	13.4907
128	7.81250	16384	2097152	11.3137	183	5.46448	33489	6128487	13.5277
129	7.75194	16641	2146689	11.3578	184	5.43478	33856	6229504	13.5647
130	7.69231	16900	2197000	11.4018	**185**	5.40541	34225	6331625	13.6015
131	7.63359	17161	2248091	11.4455	186	5.37634	34596	6434856	13.6382
132	7.57576	17424	2299968	11.4891	187	5.34759	34969	6539203	13.6748
133	7.51880	17689	2352637	11.5326	188	5.31915	35344	6644672	13.7113
134	7.46269	17956	2406104	11.5758	189	5.29101	35721	6751269	13.7477
135	7.40741	18225	2460375	11.6190	**190**	5.26316	36100	6859000	13.7840
136	7.35294	18496	2515456	11.6619	191	5.23560	36481	6967871	13.8203
137	7.29927	18769	2571353	11.7047	192	5.20833	36864	7077888	13.8564
138	7.24638	19044	2628072	11.7473	193	5.18135	37249	7189057	13.8924
139	7.19424	19321	2685619	11.7898	194	5.15464	37636	7301384	13.9284
140	7.14286	19600	2744000	11.8322	**195**	5.12821	38025	7414875	13.9642
141	7.09220	19881	2803221	11.8743	196	5.10204	38416	7529536	14.0000
142	7.04225	20164	2863288	11.9164	197	5.07614	38809	7645373	14.0357
143	6.99301	20449	2924207	11.9583	198	5.05051	39204	7762392	14.0712
144	6.94444	20736	2985984	12.0000	199	5.02513	39601	7880599	14.1067
145	6.89655	21025	3048625	12.0416	**200**	5.00000	40000	8000000	14.1421
146	6.84932	21316	3112136	12.0830	201	4.97512	40401	8120601	14.1774
147	6.80272	21609	3176523	12.1244	202	4.95050	40804	8242408	14.2127
148	6.75676	21904	3241792	12.1655	203	4.92611	41209	8365427	14.2478
149	6.71141	22201	3307949	12.2066	204	4.90196	41616	8489664	14.2829
150	6.66667	22500	3375000	12.2474	**205**	4.87805	42025	8615125	14.3178
151	6.62252	22801	3442951	12.2882	206	4.85437	42436	8741816	14.3527
152	6.57895	23104	3511808	12.3288	207	4.83092	42849	8869743	14.3875
153	6.53595	23409	3581577	12.3693	208	4.80769	43264	8998912	14.4222
154	6.49351	23716	3652264	12.4097	209	4.78469	43681	9129329	14.4568
155	6.45161	24025	3723875	12.4499	**210**	4.76190	44100	9261000	14.4914
156	6.41026	24336	3796416	12.4900	211	4.73934	44521	9393931	14.5258
157	6.36943	24649	3869893	12.5300	212	4.71698	44944	9528128	14.5602
158	6.32911	24964	3944312	12.5698	213	4.69484	45369	9663597	14.5945
159	6.28931	25281	4019679	12.6095	214	4.67290	45796	9800344	14.6287
160	6.25000	25600	4096000	12.6491	**215**	4.65116	46225	9938375	14.6629
161	6.21118	25921	4173281	12.6886	216	4.62963	46656	10077696	14.6969
162	6.17284	26244	4251528	12.7279	217	4.60829	47089	10218313	14.7309
163	6.13497	26569	4330747	12.7671	218	4.58716	47524	10360232	14.7648
164	6.09756	26896	4410944	12.8062	219	4.56621	47961	10503459	14.7986
165	6.06061	27225	4492125	12.8452	**220**	4.54545	48400	10648000	14.8324
166	6.02410	27556	4574296	12.8841	221	4.52489	48841	10793861	14.8661
167	5.98802	27889	4657463	12.9228	222	4.50450	49284	10941048	14.8997
168	5.95238	28224	4741632	12.9615	223	4.48430	49729	11089567	14.9332
169	5.91716	28561	4826809	13.0000	224	4.46429	50176	11239424	14.9666
170	5.88235	28900	4913000	13.0384	**225**	4.44444	50625	11390625	15.0000
171	5.84795	29241	5000211	13.0767	226	4.42478	51076	11543176	15.0333
172	5.81395	29584	5088448	13.1149	227	4.40529	51529	11697083	15.0665
173	5.78035	29929	5177717	13.1529	228	4.38596	51984	11852352	15.0997
174	5.74713	30276	5268024	13.1909	229	4.36681	52441	12008989	15.1327

VALUES OF RECIPROCALS, SQUARES, CUBES, AND SQUARE ROOTS, OF NATURAL NUMBERS.

n	$1000.\frac{1}{n}$	n^2	n^3	$\sqrt{n}$	n	$1000.\frac{1}{n}$	n^2	n^3	$\sqrt{n}$
230	4.34783	52900	12167000	15.1658	**285**	3.50877	81225	23149125	16.8819
231	4.32900	53361	12326391	15.1987	286	3.49650	81796	23393656	16.9115
232	4.31034	53824	12487168	15.2315	287	3.48432	82369	23639903	16.9411
233	4.29185	54289	12649337	15.2643	288	3.47222	82944	23887872	16.9706
234	4.27350	54756	12812904	15.2971	289	3.46021	83521	24137569	17.0000
235	4.25532	55225	12977875	15.3297	**290**	3.44828	84100	24389000	17.0294
236	4.23729	55696	13144256	15.3623	291	3.43643	84681	24642171	17.0587
237	4.21941	56169	13312053	15.3948	292	3.42466	85264	24897088	17.0880
238	4.20168	56644	13481272	15.4272	293	3.41297	85849	25153757	17.1172
239	4.18410	57121	13651919	15.4596	294	3.40136	86436	25412184	17.1464
240	4.16667	57600	13824000	15.4919	**295**	3.38983	87025	25672375	17.1756
241	4.14938	58081	13997521	15.5242	296	3.37838	87616	25934336	17.2047
242	4.13223	58564	14172488	15.5563	297	3.36700	88209	26198073	17.2337
243	4.11523	59049	14348907	15.5885	298	3.35570	88804	26463592	17.2627
244	4.09836	59536	14526784	15.6205	299	3.34448	89401	26730899	17.2916
245	4.08163	60025	14706125	15.6525	**300**	3.33333	90000	27000000	17.3205
246	4.06504	60516	14886936	15.6844	301	3.32226	90601	27270901	17.3494
247	4.04858	61009	15069223	15.7162	302	3.31126	91204	27543608	17.3781
248	4.03226	61504	15252992	15.7480	303	3.30033	91809	27818127	17.4069
249	4.01606	62001	15438249	15.7797	304	3.28947	92416	28094464	17.4356
250	4.00000	62500	15625000	15.8114	**305**	3.27869	93025	28372625	17.4642
251	3.98406	63001	15813251	15.8430	306	3.26797	93636	28652616	17.4929
252	3.96825	63504	16003008	15.8745	307	3.25733	94249	28934443	17.5214
253	3.95257	64009	16194277	15.9060	308	3.24675	94864	29218112	17.5499
254	3.93701	64516	16387064	15.9374	309	3.23625	95481	29503629	17.5784
255	3.92157	65025	16581375	15.9687	**310**	3.22581	96100	29791000	17.6068
256	3.90625	65536	16777216	16.0000	311	3.21543	96721	30080231	17.6352
257	3.89105	66049	16974593	16.0312	312	3.20513	97344	30371328	17.6635
258	3.87597	66564	17173512	16.0624	313	3.19489	97969	30664297	17.6918
259	3.86100	67081	17373979	16.0935	314	3.18471	98596	30959144	17.7200
260	3.84615	67600	17576000	16.1245	**315**	3.17460	99225	31255875	17.7482
261	3.83142	68121	17779581	16.1555	316	3.16456	99856	31554496	17.7764
262	3.81679	68644	17984728	16.1864	317	3.15457	100489	31855013	17.8045
263	3.80228	69169	18191447	16.2173	318	3.14465	101124	32157432	17.8326
264	3.78788	69696	18399744	16.2481	319	3.13480	101761	32461759	17.8606
265	3.77358	70225	18609625	16.2788	**320**	3.12500	102400	32768000	17.8885
266	3.75940	70756	18821096	16.3095	321	3.115 6	103041	33076161	17.9165
267	3.74532	71289	19034163	16.3401	322	3.10559	103684	33386248	17.9444
268	3.73134	71824	19248832	16.3707	323	3.09598	104329	33698267	17.9722
269	3.71747	72361	19465109	16.4012	324	3.08642	104976	34012224	18.0000
270	3.70370	72900	19683000	16.4317	**325**	3.07692	105625	34328125	18.0278
271	3.69004	73441	19902511	16.4621	326	3.06748	106276	34645976	18.0555
272	3.67647	73984	20123648	16.4924	327	3.05810	106929	34965783	18.0831
273	3.66300	74529	20346417	16.5227	328	3.04878	107584	35287552	18.1108
274	3.64964	75076	20570824	16.5529	329	3.03951	108241	35611289	18.1384
275	3.63636	75625	20796875	16.5831	**330**	3.03030	108900	35937000	18.1659
276	3.62319	76176	21024576	16.6132	331	3.02115	109561	36264691	18.1934
277	3.61011	76729	21253933	16.6433	332	3.01205	110224	36594368	18.2209
278	3.59712	77284	21484952	16.6733	333	3.00300	110889	36926037	18.2483
279	3.58423	77841	21717639	16.7033	334	2.99401	111556	37259704	18.2757
280	3.57143	78400	21952000	16.7332	**335**	2.98507	112225	37595375	18.3030
281	3.55872	78961	22188041	16.7631	336	2.97619	112896	37933056	18.3303
282	3.54610	79524	22425768	16.7929	337	2.96736	113569	38272753	18.3576
283	3.53357	80089	22665187	16.8226	338	2.95858	114244	38614472	18.3848
284	3.52113	80656	22906304	16.8523	339	2.94985	114921	38958219	18.4120

VALUES OF RECIPROCALS, SQUARES, CUBES, AND SQUARE ROOTS OF NATURAL NUMBERS.

n	$1000.\frac{1}{n}$	n^2	n^3	$\sqrt{n}$	n	$1000.\frac{1}{n}$	n^2	n^3	$\sqrt{n}$
340	2.94118	115600	39304000	18.4391	**395**	2.53165	156025	61629875	19.8746
341	2.93255	116281	39651821	18.4662	396	2.52525	156816	62099136	19.8997
342	2.92398	116964	40001688	18.4932	397	2.51889	157609	62570773	19.9249
343	2.91545	117649	40353607	18.5203	398	2.51256	158404	63044792	19.9499
344	2.90698	118336	40707584	18.5472	399	2.50627	159201	63521199	19.9750
345	2.89855	119025	41063625	18.5742	**400**	2.50000	160000	64000000	20.0000
346	2.89017	119716	41421736	18.6011	401	2.49377	160801	64481201	20.0250
347	2.88184	120409	41781923	18.6279	402	2.48756	161604	64964808	20.0499
348	2.87356	121104	42144192	18.6548	403	2.48139	162409	65450827	20.0749
349	2.86533	121801	42508549	18.6815	404	2.47525	163216	65939264	20.0998
350	2.85714	122500	42875000	18.7083	**405**	2.46914	164025	66430125	20.1246
351	2.84900	123201	43243551	18.7350	406	2.46305	164836	66923416	20.1494
352	2.84091	123904	43614208	18.7617	407	2.45700	165649	67419143	20.1742
353	2.83286	124609	43986977	18.7883	408	2.45098	166464	67917312	20.1990
354	2.82486	125316	44361864	18.8149	409	2.44499	167281	68417929	20.2237
355	2.81690	126025	44738875	18.8414	**410**	2.43902	168100	68921000	20.2485
356	2.80899	126736	45118016	18.8680	411	2.43309	168921	69426531	20.2731
357	2.80112	127449	45499293	18.8944	412	2.42718	169744	69934528	20.2978
358	2.79330	128164	45882712	18.9209	413	2.42131	170569	70444997	20.3224
359	2.78552	128881	46268279	18.9473	414	2.41546	171396	70957944	20.3470
360	2.77778	129600	46656000	18.9737	**415**	2.40964	172225	71473375	20.3715
361	2.77008	130321	47045881	19.0000	416	2.40385	173056	71991296	20.3961
362	2.76243	131044	47437928	19.0263	417	2.39808	173889	72511713	20.4206
363	2.75482	131769	47832147	19.0526	418	2.39234	174724	73034632	20.4450
364	2.74725	132496	48228544	19.0788	419	2.38663	175561	73560059	20.4695
365	2.73973	133225	48627125	19.1050	**420**	2.38095	176400	74088000	20.4939
366	2.73224	133956	49027896	19.1311	421	2.37530	177241	74618461	20.5183
367	2.72480	134689	49430863	19.1572	422	2.36967	178084	75151448	20.5426
368	2.71739	135424	49836032	19.1833	423	2.36407	178929	75686967	20.5670
369	2.71003	136161	50243409	19.2094	424	2.35849	179776	76225024	20.5913
370	2.70270	136900	50653000	19.2354	**425**	2.35294	180625	76765625	20.6155
371	2.69542	137641	51064811	19.2614	426	2.34742	181476	77308776	20.6398
372	2.68817	138384	51478848	19.2873	427	2.34192	182329	77854483	20.6640
373	2.68097	139129	51895117	19.3132	428	2.33645	183184	78402752	20.6882
374	2.67380	139876	52313624	19.3391	429	2.33100	184041	78953589	20.7123
375	2.66667	140625	52734375	19.3649	**430**	2.32558	184900	79507000	20.7364
376	2.65957	141376	53157376	19.3907	431	2.32019	185761	80062991	20.7605
377	2.65252	142129	53582633	19.4165	432	2.31481	186624	80621568	20.7846
378	2.64550	142884	54010152	19.4422	433	2.30947	187489	81182737	20.8087
379	2.63852	143641	54439939	19.4679	434	2.30415	188356	81746504	20.8327
380	2.63158	144400	54872000	19.4936	**435**	2.29885	189225	82312875	20.8567
381	2.62467	145161	55306341	19.5192	436	2.29358	190096	82881856	20.8806
382	2.61780	145924	55742968	19.5448	437	2.28833	190969	83453453	20.9045
383	2.61097	146689	56181887	19.5704	438	2.28311	191844	84027672	20.9284
384	2.60417	147456	56623104	19.5959	439	2.27790	192721	84604519	20.9523
385	2.59740	148225	57066625	19.6214	**440**	2.27273	193600	85184000	20.9762
386	2.59067	148996	57512456	19.6469	441	2.26757	194481	85766121	21.0000
387	2.58398	149769	57960603	19.6723	442	2.26244	195364	86350888	21.0238
388	2.57732	150544	58411072	19.6977	443	2.25734	196249	86938307	21.0476
389	2.57069	151321	58863869	19.7231	444	2.25225	197136	87528384	21.0713
390	2.56410	152100	59319000	19.7484	**445**	2.24719	198025	88121125	21.0950
391	2.55754	152881	59776471	19.7737	446	2.24215	198916	88716536	21.1187
392	2.55102	153664	60236288	19.7990	447	2.23714	199809	89314623	21.1424
393	2.54453	154449	60698457	19.8242	448	2.23214	200704	89915392	21.1660
394	2.53807	155236	61162984	19.8494	449	2.22717	201601	90518849	21.1896

VALUES OF RECIPROCALS, SQUARES, CUBES, AND SQUARE ROOTS OF NATURAL NUMBERS.

n	$1000.\frac{1}{n}$	n^2	n^3	$\sqrt{n}$	n	$1000.\frac{1}{n}$	n^2	n^3	$\sqrt{n}$
450	2.22222	202500	91125000	21.2132	**505**	1.98020	255025	128787625	22.4722
451	2.21729	203401	91733851	21.2368	506	1.97628	256036	129554216	22.4944
452	2.21239	204304	92345408	21.2603	507	1.97239	257049	130323843	22.5167
453	2.20751	205209	92959677	21.2838	508	1.96850	258064	131096512	22.5389
454	2.20264	206116	93576664	21.3073	509	1.96464	259081	131872229	22.5610
455	2.19780	207025	94196375	21.3307	**510**	1.96078	260100	132651000	22.5832
456	2.19298	207936	94818816	21.3542	511	1.95695	261121	133432831	22.6053
457	2.18818	208849	95443993	21.3776	512	1.95312	262144	134217728	22.6274
458	2.18341	209764	96071912	21.4009	513	1.94932	263169	135005697	22.6495
459	2.17865	210681	96702579	21.4243	514	1.94553	264196	135796744	22.6716
460	2.17391	211600	97336000	21.4476	**515**	1.94175	265225	136590875	22.6936
461	2.16920	212521	97972181	21.4709	516	1.93798	266256	137388096	22.7156
462	2.16450	213444	98611128	21.4942	517	1.93424	267289	138188413	22.7376
463	2.15983	214369	99252847	21.5174	518	1.93050	268324	138991832	22.7596
464	2.15517	215296	99897344	21.5407	519	1.92678	269361	139798359	22.7816
465	2.15054	216225	100544625	21.5639	**520**	1.92308	270400	140608000	22.8035
466	2.14592	217156	101194696	21.5870	521	1.91939	271441	141420761	22.8254
467	2.14133	218089	101847563	21.6102	522	1.91571	272484	142236648	22.8473
468	2.13675	219024	102503232	21.6333	523	1.91205	273529	143055667	22.8692
469	2.13220	219961	103161709	21.6564	524	1.90840	274576	143877824	22.8910
470	2.12766	220900	103823000	21.6795	**525**	1.90476	275625	144703125	22.9129
471	2.12314	221841	104487111	21.7025	526	1.90114	276676	145531576	22.9347
472	2.11864	222784	105154048	21.7256	527	1.89753	277729	146363183	22.9565
473	2.11416	223729	105823817	21.7486	528	1.89394	278784	147197952	22.9783
474	2.10970	224676	106496424	21.7715	529	1.89036	279841	148035889	23.0000
475	2.10526	225625	107171875	21.7945	**530**	1.88679	280900	148877000	23.0217
476	2.10084	226576	107850176	21.8174	531	1.88324	281961	149721291	23.0434
477	2.09644	227529	108531333	21.8403	532	1.87970	283024	150568768	23.0651
478	2.09205	228484	109215352	21.8632	533	1.87617	284089	151419437	23.0868
479	2.08768	229441	109902239	21.8861	534	1.87266	285156	152273304	23.1084
480	2.08333	230400	110592000	21.9089	**535**	1.86916	286225	153130375	23.1301
481	2.07900	231361	111284641	21.9317	536	1.86567	287296	153990656	23.1517
482	2.07469	232324	111980168	21.9545	537	1.86220	288369	154854153	23.1733
483	2.07039	233289	112678587	21.9773	538	1.85874	289444	155720872	23.1948
484	2.06612	234256	113379904	22.0000	539	1.85529	290521	156590819	23.2164
485	2.06186	235225	114084125	22.0227	**540**	1.85185	291600	157464000	23.2379
486	2.05761	236196	114791256	22.0454	541	1.84843	292681	158340421	23.2594
487	2.05339	237169	115501303	22.0681	542	1.84502	293764	159220088	23.2809
488	2.04918	238144	116214272	22.0907	543	1.84162	294849	160103007	23.3024
489	2.04499	239121	116930169	22.1133	544	1.83824	295936	160989184	23.3238
490	2.04082	240100	117649000	22.1359	**545**	1.83486	297025	161878625	23.3452
491	2.03666	241081	118370771	22.1585	546	1.83150	298116	162771336	23.3666
492	2.03252	242064	119095488	22.1811	547	1.82815	299209	163667323	23.3880
493	2.02840	243049	119823157	22.2036	548	1.82482	300304	164566592	23.4094
494	2.02429	244036	120553784	22.2261	549	1.82149	301401	165469149	23.4307
495	2.02020	245025	121287375	22.2486	**550**	1.81818	302500	166375000	23.4521
496	2.01613	246016	122023936	22.2711	551	1.81488	303601	167284151	23.4734
497	2.01207	247009	122763473	22.2935	552	1.81159	304704	168196608	23.4947
498	2.00803	248004	123505992	22.3159	553	1.80832	305809	169112377	23.5160
499	2.00401	249001	124251499	22.3383	554	1.80505	306916	170031464	23.5372
500	2.00000	250000	125000000	22.3607	**555**	1.80180	308025	170953875	23.5584
501	1.99601	251001	125751501	22.3830	556	1.79856	309136	171879616	23.5797
502	1.99203	252004	126506008	22.4054	557	1.79533	310249	172808693	23.6008
503	1.98807	253009	127263527	22.4277	558	1.79211	311364	173741112	23.6220
504	1.98413	254016	128024064	22.4499	559	1.78891	312481	174676879	23.6432

VALUES OF RECIPROCALS, SQUARES, CUBES, AND SQUARE ROOTS OF NATURAL NUMBERS.

n	$1000.\frac{1}{n}$	n^2	n^3	$\sqrt{n}$	n	$1000.\frac{1}{n}$	n^2	n^3	$\sqrt{n}$
560	1.78571	313600	175616000	23.6643	**615**	1.62602	378225	232608375	24.7992
561	1.78253	314721	176558481	23.6854	616	1.62338	379456	233744896	24.8193
562	1.77936	315844	177504328	23.7065	617	1.62075	380689	234885113	24.8395
563	1.77620	316969	178453547	23.7276	618	1.61812	381924	236029032	24.8596
564	1.77305	318096	179406144	23.7487	619	1.61551	383161	237176659	24.8797
565	1.76991	319225	180362125	23.7697	**620**	1.61290	384400	238328000	24.8998
566	1.76678	320356	181321496	23.7908	621	1.61031	385641	239483061	24.9199
567	1.76367	321489	182284263	23.8118	622	1.60772	386884	240641848	24.9399
568	1.76056	322624	183250432	23.8328	623	1.60514	388129	241804367	24.9600
569	1.75747	323761	184220009	23.8537	624	1.60256	389376	242970624	24.9800
570	1.75439	324900	185193000	23.8747	**625**	1.60000	390625	244140625	25.0000
571	1.75131	326041	186169411	23.8956	626	1.59744	391876	245314376	25.0200
572	1.74825	327184	187149248	23.9165	627	1.59490	393129	246491883	25.0400
573	1.74520	328329	188132517	23.9374	628	1.59236	394384	247673152	25.0599
574	1.74216	329476	189119224	23.9583	629	1.58983	395641	248858189	25.0799
575	1.73913	330625	190109375	23.9792	**630**	1.58730	396900	250047000	25.0998
576	1.73611	331776	191102976	24.0000	631	1.58479	398161	251239591	25.1197
577	1.73310	332929	192100033	24.0208	632	1.58228	399424	252435968	25.1396
578	1.73010	334084	193100552	24.0416	633	1.57978	400689	253636137	25.1595
579	1.72712	335241	194104539	24.0624	634	1.57729	401956	254840104	25.1794
580	1.72414	336400	195112000	24.0832	**635**	1.57480	403225	256047875	25.1992
581	1.72117	337561	196122941	24.1039	636	1.57233	404496	257259456	25.2190
582	1.71821	338724	197137368	24.1247	637	1.56986	405769	258474853	25.2389
583	1.71527	339889	198155287	24.1454	638	1.56740	407044	259694072	25.2587
584	1.71233	341056	199176704	24.1661	639	1.56495	408321	260917119	25.2784
585	1.70940	342225	200201625	24.1868	**640**	1.56250	409600	262144000	25.2982
586	1.70648	343396	201230056	24.2074	641	1.56006	410881	263374721	25.3180
587	1.70358	344569	202262003	24.2281	642	1.55763	412164	264609288	25.3377
588	1.70068	345744	203297472	24.2487	643	1.55521	413449	265847707	25.3574
589	1.69779	346921	204336469	24.2693	644	1.55280	414736	267089984	25.3772
590	1.69492	348100	205379000	24.2899	**645**	1.55039	416025	268336125	25.3969
591	1.69205	349281	206425071	24.3105	646	1.54799	417316	269586136	25.4165
592	1.68919	350464	207474688	24.3311	647	1.54560	418609	270840023	25.4362
593	1.68634	351649	208527857	24.3516	648	1.54321	419904	272097792	25.4558
594	1.68350	352836	209584584	24.3721	649	1.54083	421201	273359449	25.4755
595	1.68067	354025	210644875	24.3926	**650**	1.53846	422500	274625000	25.4951
596	1.67785	355216	211708736	24.4131	651	1.53610	423801	275894451	25.5147
597	1.67504	356409	212776173	24.4336	652	1.53374	425104	277167808	25.5343
598	1.67224	357604	213847192	24.4540	653	1.53139	426409	278445077	25.5539
599	1.66945	358801	214921799	24.4745	654	1.52905	427716	279726264	25.5734
600	1.66667	360000	216000000	24.4949	**655**	1.52672	429025	281011375	25.5930
601	1.66389	361201	217081801	24.5153	656	1.52439	430336	282300416	25.6125
602	1.66113	362404	218167208	24.5357	657	1.52207	431649	283593393	25.6320
603	1.65837	363609	219256227	24.5561	658	1.51976	432964	284890312	25.6515
604	1.65563	364816	220348864	24.5764	659	1.51745	434281	286191179	25.6710
605	1.65289	366025	221445125	24.5967	**660**	1.51515	435600	287496000	25.6905
606	1.65017	367236	222545016	24.6171	661	1.51286	436921	288804781	25.7099
607	1.64745	368449	223648543	24.6374	662	1.51057	438244	290117528	25.7294
608	1.64474	369664	224755712	24.6577	663	1.50830	439569	291434247	25.7488
609	1.64204	370881	225866529	24.6779	664	1.50602	440896	292754944	25.7682
610	1.63934	372100	226981000	24.6982	**665**	1.50376	442225	294079625	25.7876
611	1.63666	373321	228099131	24.7184	666	1.50150	443556	295408296	25.8070
612	1.63399	374544	229220928	24.7386	667	1.49925	444889	296740963	25.8263
613	1.63132	375769	230346397	24.7588	668	1.49701	446224	298077632	25.8457
614	1.62866	376996	231475544	24.7790	669	1.49477	447561	299418309	25.8650

VALUES OF RECIPROCALS, SQUARES, CUBES, AND SQUARE ROOTS OF NATURAL NUMBERS.

n	$1000.\frac{1}{n}$	n^2	n^3	$\sqrt{n}$	n	$1000.\frac{1}{n}$	n^2	n^3	$\sqrt{n}$
670	1.49254	448900	300763000	25.8844	**725**	1.37931	525625	381078125	26.9258
671	1.49031	450241	302111711	25.9037	726	1.37741	527076	382657176	26.9444
672	1.48810	451584	303464448	25.9230	727	1.37552	528529	384240583	26.9629
673	1.48588	452929	304821217	25.9422	728	1.37363	529984	385828352	26.9815
674	1.48368	454276	306182024	25.9615	729	1.37174	531441	387420489	27.0000
675	1.48148	455625	307546875	25.9808	**730**	1.36986	532900	389017000	27.0185
676	1.47929	456976	308915776	26.0000	731	1.36799	534361	390617891	27.0370
677	1.47710	458329	310288733	26.0192	732	1.36612	535824	392223168	27.0555
678	1.47493	459684	311665752	26.0384	733	1.36426	537289	393832837	27.0740
679	1.47275	461041	313046839	26.0576	734	1.36240	538756	395446904	27.0924
680	1.47059	462400	314432000	26.0768	**735**	1.36054	540225	397065375	27.1109
681	1.46843	463761	315821241	26.0960	736	1.35870	541696	398688256	27.1293
682	1.46628	465124	317214568	26.1151	737	1.35685	543169	400315553	27.1477
683	1.46413	466489	318611987	26.1343	738	1.35501	544644	401947272	27.1662
684	1.46199	467856	320013504	26.1534	739	1.35318	546121	403583419	27.1846
685	1.45985	469225	321419125	26.1725	**740**	1.35135	547600	405224000	27.2029
686	1.45773	470596	322828856	26.1916	741	1.34953	549081	406869021	27.2213
687	1.45560	471969	324242703	26.2107	742	1.34771	550564	408518488	27.2397
688	1.45349	473344	325660672	26.2298	743	1.34590	552049	410172407	27.2580
689	1.45138	474721	327082769	26.2488	744	1.34409	553536	411830784	27.2764
690	1.44928	476100	328509000	26.2679	**745**	1.34228	555025	413493625	27.2947
691	1.44718	477481	329939371	26.2869	746	1.34048	556516	415160936	27.3130
692	1.44509	478864	331373888	26.3059	747	1.33869	558009	416832723	27.3313
693	1.44300	480249	332812557	26.3249	748	1.33690	559504	418508992	27.3496
694	1.44092	481636	334255384	26.3439	749	1.33511	561001	420189749	27.3679
695	1.43885	483025	335702375	26.3629	**750**	1.33333	562500	421875000	27.3861
696	1.43678	484416	337153536	26.3818	751	1.33156	564001	423564751	27.4044
697	1.43472	485809	338608873	26.4008	752	1.32979	565504	425259008	27.4226
698	1.43266	487204	340368392	26.4197	753	1.32802	567009	426957777	27.4408
699	1.43062	488601	341532099	26.4386	754	1.32626	568516	428661064	27.4591
700	1.42857	490000	343000000	26.4575	**755**	1.32450	570025	430368875	27.4773
701	1.42653	491401	344472101	26.4764	756	1.32275	571536	432081216	27.4955
702	1.42450	492804	345948408	26.4953	757	1.32100	573049	433798093	27.5136
703	1.42248	494209	347428927	26.5141	758	1.31926	574564	435519512	27.5318
704	1.42045	495616	348913664	26.5330	759	1.31752	576081	437245479	27.5500
705	1.41844	497025	350402625	26.5518	**760**	1.31579	577600	438976000	27.5681
706	1.41643	498436	351895816	26.5707	761	1.31406	579121	440711081	27.5862
707	1.41443	499849	353393243	26.5895	762	1.31234	580644	442450728	27.6043
708	1.41243	501264	354894912	26.6083	763	1.31062	582169	444194947	27.6225
709	1.41044	502681	356400829	26.6271	764	1.30890	583696	445943744	27.6405
710	1.40845	504100	357911000	26.6458	**765**	1.30719	585225	447697125	27.6586
711	1.40647	505521	359425431	26.6646	766	1.30548	586756	449455096	27.6767
712	1.40449	506944	360944128	26.6833	767	1.30378	588289	451217663	27.6948
713	1.40252	508369	362467097	26.7021	768	1.30208	589824	452984832	27.7128
714	1.40056	509796	363994344	26.7208	769	1.30039	591361	454756609	27.7308
715	1.39860	511225	365525875	26.7395	**770**	1.29870	592900	456533000	27.7489
716	1.39665	512656	367061696	26.7582	771	1.29702	594441	458314011	27.7669
717	1.39470	514089	368601813	26.7769	772	1.29534	595984	460099648	27.7849
718	1.39276	515524	370146232	26.7955	773	1.29366	597529	461889917	27.8029
719	1.39082	516961	371694959	26.8142	774	1.29199	599076	463684824	27.8209
720	1.38889	518400	373248000	26.8328	**775**	1.29032	600625	465484375	27.8388
721	1.38696	519841	374805361	26.8514	776	1.28866	602176	467288576	27.8568
722	1.38504	521284	376367048	26.8701	777	1.28700	603729	469097433	27.8747
723	1.38313	522729	377933067	26.8887	778	1.28535	605284	470910952	27.8927
724	1.38122	524176	379503424	26.9072	779	1.28370	606841	472729139	27.9106

VALUES OF RECIPROCALS, SQUARES, CUBES, AND SQUARE ROOTS OF NATURAL NUMBERS.

n	$1000.\frac{1}{n}$	n^2	n^3	$\sqrt{n}$	n	$1000.\frac{1}{n}$	n^2	n^3	$\sqrt{n}$
780	1.28205	608400	474552000	27.9285	**835**	1.19760	697225	582182875	28.8964
781	1.28041	609961	476379541	27.9464	836	1.19617	698896	584277056	28.9137
782	1.27877	611524	478211768	27.9643	837	1.19474	700569	586376253	28.9310
783	1.27714	613089	480048687	27.9821	838	1.19332	702244	588480472	28.9482
784	1.27551	614656	481890304	28.0000	839	1.19190	703921	590589719	28.9655
785	1.27389	616225	483736625	28.0179	**840**	1.19048	705600	592704000	28.9828
786	1.27226	617796	485587656	28.0357	841	1.18906	707281	594823321	29.0000
787	1.27065	619369	487443403	28.0535	842	1.18765	708964	596947688	29.0172
788	1.26904	620944	489303872	28.0713	843	1.18624	710649	599077107	29.0345
789	1.26743	622521	491169069	28.0891	844	1.18483	712336	601211584	29.0517
790	1.26582	624100	493039000	28.1069	**845**	1.18343	714025	603351125	29.0689
791	1.26422	625681	494913671	28.1247	846	1.18203	715716	605495736	29.0861
792	1.26263	627264	496793088	28.1425	847	1.18064	717409	607645423	29.1033
793	1.26103	628849	498677257	28.1603	848	1.17925	719104	609800192	29.1204
794	1.25945	630436	500566184	28.1780	849	1.17786	720801	611960049	29.1376
795	1.25786	632025	502459875	28.1957	**850**	1.17647	722500	614125000	29.1548
796	1.25628	633616	504358336	28.2135	851	1.17509	724201	616295051	29.1719
797	1.25471	635209	506261573	28.2312	852	1.17371	725904	618470208	29.1890
798	1.25313	636804	508169592	28.2489	853	1.17233	727609	620650477	29.2062
799	1.25156	638401	510082399	28.2666	854	1.17096	729316	622835864	29.2233
800	1.25000	640000	512000000	28.2843	**855**	1.16959	731025	625026375	29.2404
801	1.24844	641601	513922401	28.3019	856	1.16822	732736	627222016	29.2575
802	1.24688	643204	515849608	28.3196	857	1.16686	734449	629422793	29.2746
803	1.24533	644809	517781627	28.3373	858	1.16550	736164	631628712	29.2916
804	1.24378	646416	519718464	28.3549	859	1.16414	737881	633839779	29.3087
805	1.24224	648025	521660125	28.3725	**860**	1.16279	739600	636056000	29.3258
806	1.24069	649636	523606616	28.3901	861	1.16144	741321	638277381	29.3428
807	1.23916	651249	525557943	28.4077	862	1.16009	743044	640503928	29.3598
808	1.23762	652864	527514112	28.4253	863	1.15875	744769	642735647	29.3769
809	1.23609	654481	529475129	28.4429	864	1.15741	746496	644972544	29.3939
810	1.23457	656100	531441000	28.4605	**865**	1.15607	748225	647214625	29.4109
811	1.23305	657721	533411731	28.4781	866	1.15473	749956	649461896	29.4279
812	1.23153	659344	535387328	28.4956	867	1.15340	751689	651714363	29.4449
813	1.23001	660969	537367797	28.5132	868	1.15207	753424	653972032	29.4618
814	1.22850	662596	539353144	28.5307	869	1.15075	755161	656234909	29.4788
815	1.22699	664225	541343375	28.5482	**870**	1.14943	756900	658503000	29.4958
816	1.22549	665856	543338496	28.5657	871	1.14811	758641	660776311	29.5127
817	1.22399	667489	545338513	28.5832	872	1.14679	760384	663054848	29.5296
818	1.22249	669124	547343432	28.6007	873	1.14548	762129	665338617	29.5466
819	1.22100	670761	549353259	28.6182	874	1.14416	763876	667627624	29.5635
820	1.21951	672400	551368000	28.6356	**875**	1.14286	765625	669921875	29.5804
821	1.21803	674041	553387661	28.6531	876	1.14155	767376	672221376	29.5973
822	1.21655	675684	555412248	28.6705	877	1.14025	769129	674526133	29.6142
823	1.21507	677329	557441767	28.6880	878	1.13895	770884	676836152	29.6311
824	1.21359	678976	559476224	28.7054	879	1.13766	772641	679151439	29.6479
825	1.21212	680625	561515625	28.7228	**880**	1.13636	774400	681472000	29.6648
826	1.21065	682276	563559976	28.7402	881	1.13507	776161	683797841	29.6816
827	1.20919	683929	565609283	28.7576	882	1.13379	777924	686128968	29.6985
828	1.20773	685584	567663552	28.7750	883	1.13250	779689	688465387	29.7153
829	1.20627	687241	569722789	28.7924	884	1.13122	781456	690807104	29.7321
830	1.20482	688900	571787000	28.8097	**885**	1.12994	783225	693154125	29.7489
831	1.20337	690561	573856191	28.8271	886	1.12867	784996	695506456	29.7658
832	1.20192	692224	575930368	28.8444	887	1.12740	786769	697864103	29.7825
833	1.20048	693889	578009537	28.8617	888	1.12613	788544	700227072	29.7993
834	1.19904	695556	580093704	28.8791	889	1.12486	790321	702595369	29.8161

VALUES OF RECIPROCALS, SQUARES, CUBES, AND SQUARE ROOTS OF NATURAL NUMBERS.

n	$1000.\frac{1}{n}$	n^2	n^3	$\sqrt{n}$	n	$1000.\frac{1}{n}$	n^2	n^3	$\sqrt{n}$
890	1.12360	792100	704969000	29.8329	**945**	1.05820	893025	843908625	30.7409
891	1.12233	793881	707347971	29.8496	946	1.05708	894916	846590536	30.7571
892	1.12108	795664	709732288	29.8664	947	1.05597	896809	849278123	30.7734
893	1.11982	797449	712121957	29.8831	948	1.05485	898704	851971392	30.7896
894	1.11857	799236	714516984	29.8998	949	1.05374	900601	854670349	30.8058
895	1.11732	801025	716917375	29.9166	**950**	1.05263	902500	857375000	30.8221
896	1.11607	802816	719323136	29.9333	951	1.05152	904401	860085351	30.8383
897	1.11483	804609	721734273	29.9500	952	1.05042	906304	862801408	30.8545
898	1.11359	806404	724150792	29.9666	953	1.04932	908209	865523177	30.8707
899	1.11235	808201	726572699	29.9833	954	1.04822	910116	868250664	30.8869
900	1.11111	810000	729000000	30.0000	**955**	1.04712	912025	870983875	30.9031
901	1.10988	811801	731432701	30.0167	956	1.04603	913936	873722816	30.9192
902	1.10865	813604	733870808	30.0333	957	1.04493	915849	876467493	30.9354
903	1.10742	815409	736314327	30.0500	958	1.04384	917764	879217912	30.9516
904	1.10619	817216	738763264	30.0666	959	1.04275	919681	881974079	30.9677
905	1.10497	819025	741217625	30.0832	**960**	1.04167	921600	884736000	30.9839
906	1.10375	820836	743677416	30.0998	961	1.04058	923521	887503681	31.0000
907	1.10254	822649	746142643	30.1164	962	1.03950	925444	890277128	31.0161
908	1.10132	824464	748613312	30.1330	963	1.03842	927369	893056347	31.0322
909	1.10011	826281	751089429	30.1496	964	1.03734	929296	895841344	31.0483
910	1.09890	828100	753571000	30.1662	**965**	1.03627	931225	898632125	31.0644
911	1.09769	829921	756058031	30.1828	966	1.03520	933156	901428696	31.0805
912	1.09649	831744	758550528	30.1993	967	1.03413	935089	904231063	31.0966
913	1.09529	833569	761048497	30.2159	968	1.03306	937024	907039232	31.1127
914	1.09409	835396	763551944	30.2324	969	1.03199	938961	909853209	31.1288
915	1.09290	837225	766060875	30.2490	**970**	1.03093	940900	912673000	31.1448
916	1.09170	839056	768575296	30.2655	971	1.02987	942841	915498611	31.1609
917	1.09051	840889	771095213	30.2820	972	1.02881	944784	918330048	31.1769
918	1.08932	842724	773620632	30.2985	973	1.02775	946729	921167317	31.1929
919	1.08814	844561	776151559	30.3150	974	1.02669	948676	924010424	31.2090
920	1.08696	846400	778688000	30.3315	**975**	1.02564	950625	926859375	31.2250
921	1.08578	848241	781229961	30.3480	976	1.02459	952576	929714176	31.2410
922	1.08460	850084	783777448	30.3645	977	1.02354	954529	932574833	31.2570
923	1.08342	851929	786330467	30.3809	978	1.02249	956484	935441352	31.2730
924	1.08225	853776	788889024	30.3974	979	1.02145	958441	938313739	31.2890
925	1.08108	855625	791453125	30.4138	**980**	1.02041	960400	941192000	31.3050
926	1.07991	857476	794022776	30.4302	981	1.01937	962361	944076141	31.3209
927	1.07875	859329	796597983	30.4467	982	1.01833	964324	946966168	31.3369
928	1.07759	861184	799178752	30.4631	983	1.01729	966289	949862087	31.3528
929	1.07643	863041	801765089	30.4795	984	1.01626	968256	952763904	31.3688
930	1.07527	864900	804357000	30.4959	**985**	1.01523	970225	955671625	31.3847
931	1.07411	866761	806954491	30.5123	986	1.01420	972196	958585256	31.4006
932	1.07296	868624	809557568	30.5287	987	1.01317	974169	961504803	31.4166
933	1.07181	870489	812166237	30.5450	988	1.01215	976144	964430272	31.4325
934	1.07066	872356	814780504	30.5614	989	1.01112	978121	967361669	31.4484
935	1.06952	874225	817400375	30.5778	**990**	1.01010	980100	970299000	31.4643
936	1.06838	876096	820025856	30.5941	991	1.00908	982081	973242271	31.4802
937	1.06724	877969	822656953	30.6105	992	1.00806	984064	976191488	31.4960
938	1.06610	879844	825293672	30.6268	993	1.00705	986049	979146657	31.5119
939	1.06496	881721	827936019	30.6431	994	1.00604	988036	982107784	31.5278
940	1.06383	883600	830584000	30.6594	**995**	1.00503	990025	985074875	31.5436
941	1.06270	885481	833237621	30.6757	996	1.00402	992016	988047936	31.5595
942	1.06157	887364	835896888	30.6920	997	1.00301	994009	991026973	31.5753
943	1.06045	889249	838561807	30.7083	998	1.00200	996004	994011992	31.5911
944	1.05932	891136	841232384	30.7246	999	1.00100	998001	997002999	31.6070

Table 10.

LOGARITHMS.

N.	0	1	2	3	4	5	6	7	8	9	10
100	0000	0004	0009	0013	0017	0022	0026	0030	0035	0039	0043
101	0043	0048	0052	0056	0060	0065	0069	0073	0077	0082	0086
102	0086	0090	0095	0099	0103	0107	0111	0116	0120	0124	0128
103	0128	0133	0137	0141	0145	0149	0154	0158	0162	0166	0170
104	0170	0175	0179	0183	0187	0191	0195	0199	0204	0208	0212
105	0212	0216	0220	0224	0228	0233	0237	0241	0245	0249	0253
106	0253	0257	0261	0265	0269	0273	0278	0282	0286	0290	0294
107	0294	0298	0302	0306	0310	0314	0318	0322	0326	0330	0334
108	0334	0338	0342	0346	0350	0354	0358	0362	0366	0370	0374
109	0374	0378	0382	0386	0390	0394	0398	0402	0406	0410	0414
110	0414	0418	0422	0426	0430	0434	0438	0441	0445	0449	0453
111	0453	0457	0461	0465	0469	0473	0477	0481	0484	0488	0492
112	0492	0496	0500	0504	0508	0512	0515	0519	0523	0527	0531
113	0531	0535	0538	0542	0546	0550	0554	0558	0561	0565	0569
114	0569	0573	0577	0580	0584	0588	0592	0596	0599	0603	0607
115	0607	0611	0615	0618	0622	0626	0630	0633	0637	0641	0645
116	0645	0648	0652	0656	0660	0663	0667	0671	0674	0678	0682
117	0682	0686	0689	0693	0697	0700	0704	0708	0711	0715	0719
118	0719	0722	0726	0730	0734	0737	0741	0745	0748	0752	0755
119	0755	0759	0763	0766	0770	0774	0777	0781	0785	0788	0792
120	0792	0795	0799	0803	0806	0810	0813	0817	0821	0824	0828
121	0828	0831	0835	0839	0842	0846	0849	0853	0856	0860	0864
122	0864	0867	0871	0874	0878	0881	0885	0888	0892	0896	0899
123	0899	0903	0906	0910	0913	0917	0920	0924	0927	0931	0934
124	0934	0938	0941	0945	0948	0952	0955	0959	0962	0966	0969
125	0969	0973	0976	0980	0983	0986	0990	0993	0997	1000	1004
126	1004	1007	1011	1014	1017	1021	1024	1028	1031	1035	1038
127	1038	1041	1045	1048	1052	1055	1059	1062	1065	1069	1072
128	1072	1075	1079	1082	1086	1089	1092	1096	1099	1103	1106
129	1106	1109	1113	1116	1119	1123	1126	1129	1133	1136	1139
130	1139	1143	1146	1149	1153	1156	1159	1163	1166	1169	1173
131	1173	1176	1179	1183	1186	1189	1193	1196	1199	1202	1206
132	1206	1209	1212	1216	1219	1222	1225	1229	1232	1235	1239
133	1239	1242	1245	1248	1252	1255	1258	1261	1265	1268	1271
134	1271	1274	1278	1281	1284	1287	1290	1294	1297	1300	1303
135	1303	1307	1310	1313	1316	1319	1323	1326	1329	1332	1335
136	1335	1339	1342	1345	1348	1351	1355	1358	1361	1364	1367
137	1367	1370	1374	1377	1380	1383	1386	1389	1392	1396	1399
138	1399	1402	1405	1408	1411	1414	1418	1421	1424	1427	1430
139	1430	1433	1436	1440	1443	1446	1449	1452	1455	1458	1461
140	1461	1464	1467	1471	1474	1477	1480	1483	1486	1489	1492
141	1492	1495	1498	1501	1504	1508	1511	1514	1517	1520	1523
142	1523	1526	1529	1532	1535	1538	1541	1544	1547	1550	1553
143	1553	1556	1559	1562	1565	1569	1572	1575	1578	1581	1584
144	1584	1587	1590	1593	1596	1599	1602	1605	1608	1611	1614
145	1614	1617	1620	1623	1626	1629	1632	1635	1638	1641	1644
146	1644	1647	1649	1652	1655	1658	1661	1664	1667	1670	1673
147	1673	1676	1679	1682	1685	1688	1691	1694	1697	1700	1703
148	1703	1706	1708	1711	1714	1717	1720	1723	1726	1729	1732
149	1732	1735	1738	1741	1744	1746	1749	1752	1755	1758	1761

LOGARITHMS.

N.	0	1	2	3	4	5	6	7	8	9	10
150	1761	1764	1767	1770	1772	1775	1778	1781	1784	1787	1790
151	1790	1793	1796	1798	1801	1804	1807	1810	1813	1816	1818
152	1818	1821	1824	1827	1830	1833	1836	1838	1841	1844	1847
153	1847	1850	1853	1855	1858	1861	1864	1867	1870	1872	1875
154	1875	1878	1881	1884	1886	1889	1892	1895	1898	1901	1903
155	1903	1906	1909	1912	1915	1917	1920	1923	1926	1928	1931
156	1931	1934	1937	1940	1942	1945	1948	1951	1953	1956	1959
157	1959	1962	1965	1967	1970	1973	1976	1978	1981	1984	1987
158	1987	1989	1992	1995	1998	2000	2003	2006	2009	2011	2014
159	2014	2017	2019	2022	2025	2028	2030	2033	2036	2038	2041
160	2041	2044	2047	2049	2052	2055	2057	2060	2063	2066	2068
161	2068	2071	2074	2076	2079	2082	2084	2087	2090	2092	2095
162	2095	2098	2101	2103	2106	2109	2111	2114	2117	2119	2122
163	2122	2125	2127	2130	2133	2135	2138	2140	2143	2146	2148
164	2148	2151	2154	2156	2159	2162	2164	2167	2170	2172	2175
165	2175	2177	2180	2183	2185	2188	2191	2193	2196	2198	2201
166	2201	2204	2206	2209	2212	2214	2217	2219	2222	2225	2227
167	2227	2230	2232	2235	2238	2240	2243	2245	2248	2251	2253
168	2253	2256	2258	2261	2263	2266	2269	2271	2274	2276	2279
169	2279	2281	2284	2287	2289	2292	2294	2297	2299	2302	2304
170	2304	2307	2310	2312	2315	2317	2320	2322	2325	2327	2330
171	2330	2333	2335	2338	2340	2343	2345	2348	2350	2353	2355
172	2355	2358	2360	2363	2365	2368	2370	2373	2375	2378	2380
173	2380	2383	2385	2388	2390	2393	2395	2398	2400	2403	2405
174	2405	2408	2410	2413	2415	2418	2420	2423	2425	2428	2430
175	2430	2433	2435	2438	2440	2443	2445	2448	2450	2453	2455
176	2455	2458	2460	2463	2465	2467	2470	2472	2475	2477	2480
177	2480	2482	2485	2487	2490	2492	2494	2497	2499	2502	2504
178	2504	2507	2509	2512	2514	2516	2519	2521	2524	2526	2529
179	2529	2531	2533	2536	2538	2541	2543	2545	2548	2550	2553
180	2553	2555	2558	2560	2562	2565	2567	2570	2572	2574	2577
181	2577	2579	2582	2584	2586	2589	2591	2594	2596	2598	2601
182	2601	2603	2605	2608	2610	2613	2615	2617	2620	2622	2625
183	2625	2627	2629	2632	2634	2636	2639	2641	2643	2646	2648
184	2648	2651	2653	2655	2658	2660	2662	2665	2667	2669	2672
185	2672	2674	2676	2679	2681	2683	2686	2688	2690	2693	2695
186	2695	2697	2700	2702	2704	2707	2709	2711	2714	2716	2718
187	2718	2721	2723	2725	2728	2730	2732	2735	2737	2739	2742
188	2742	2744	2746	2749	2751	2753	2755	2758	2760	2762	2765
189	2765	2767	2769	2772	2774	2776	2778	2781	2783	2785	2788
190	2788	2790	2792	2794	2797	2799	2801	2804	2806	2808	2810
191	2810	2813	2815	2817	2819	2822	2824	2826	2828	2831	2833
192	2833	2835	2838	2840	2842	2844	2847	2849	2851	2853	2856
193	2856	2858	2860	2862	2865	2867	2869	2871	2874	2876	2878
194	2878	2880	2882	2885	2887	2889	2891	2894	2896	2898	2900
195	2900	2903	2905	2907	2909	2911	2914	2916	2918	2920	2923
196	2923	2925	2927	2929	2931	2934	2936	2938	2940	2942	2945
197	2945	2947	2949	2951	2953	2956	2958	2960	2962	2964	2967
198	2967	2969	2971	2973	2975	2978	2980	2982	2984	2986	2989
199	2989	2991	2993	2995	2997	2999	3002	3004	3006	3008	3010

TABLE 11.

LOGARITHMS.

N	0	1	2	3	4	5	6	7	8	9	P.P. 1	2	3	4	5
10	0000	0043	0086	0128	0170	0212	0253	0294	0334	0374	4	8	12	17	21
11	0414	0453	0492	0531	0569	0607	0645	0682	0719	0755	4	8	11	15	19
12	0792	0828	0864	0899	0934	0969	1004	1038	1072	1106	3	7	10	14	17
13	1139	1173	1206	1239	1271	1303	1335	1367	1399	1430	3	6	10	13	16
14	1461	1492	1523	1553	1584	1614	1644	1673	1703	1732	3	6	9	12	15
15	1761	1790	1818	1847	1875	1903	1931	1959	1987	2014	3	6	8	11	14
16	2041	2068	2095	2122	2148	2175	2201	2227	2253	2279	3	5	8	11	13
17	2304	2330	2355	2380	2405	2430	2455	2480	2504	2529	2	5	7	10	12
18	2553	2577	2601	2625	2648	2672	2695	2718	2742	2765	2	5	7	9	12
19	2788	2810	2833	2856	2878	2900	2923	2945	2967	2989	2	4	7	9	11
20	3010	3032	3054	3075	3096	3118	3139	3160	3181	3201	2	4	6	8	11
21	3222	3243	3263	3284	3304	3324	3345	3365	3385	3404	2	4	6	8	10
22	3424	3444	3464	3483	3502	3522	3541	3560	3579	3598	2	4	6	8	10
23	3617	3636	3655	3674	3692	3711	3729	3747	3766	3784	2	4	5	7	9
24	3802	3820	3838	3856	3874	3892	3909	3927	3945	3962	2	4	5	7	9
25	3979	3997	4014	4031	4048	4065	4082	4099	4116	4133	2	3	5	7	9
26	4150	4166	4183	4200	4216	4232	4249	4265	4281	4298	2	3	5	7	8
27	4314	4330	4346	4362	4378	4393	4409	4425	4440	4456	2	3	5	6	8
28	4472	4487	4502	4518	4533	4548	4564	4579	4594	4609	2	3	5	6	8
29	4624	4639	4654	4669	4683	4698	4713	4728	4742	4757	1	3	4	6	7
30	4771	4786	4800	4814	4829	4843	4857	4871	4886	4900	1	3	4	6	7
31	4914	4928	4942	4955	4969	4983	4997	5011	5024	5038	1	3	4	6	7
32	5051	5065	5079	5092	5105	5119	5132	5145	5159	5172	1	3	4	5	7
33	5185	5198	5211	5224	5237	5250	5263	5276	5289	5302	1	3	4	5	6
34	5315	5328	5340	5353	5366	5378	5391	5403	5416	5428	1	3	4	5	6
35	5441	5453	5465	5478	5490	5502	5514	5527	5539	5551	1	2	4	5	6
36	5563	5575	5587	5599	5611	5623	5635	5647	5658	5670	1	2	4	5	6
37	5682	5694	5705	5717	5729	5740	5752	5763	5775	5786	1	2	3	5	6
38	5798	5809	5821	5832	5843	5855	5866	5877	5888	5899	1	2	3	5	6
39	5911	5922	5933	5944	5955	5966	5977	5988	5999	6010	1	2	3	4	6
40	6021	6031	6042	6053	6064	6075	6085	6096	6107	6117	1	2	3	4	5
41	6128	6138	6149	6160	6170	6180	6191	6201	6212	6222	1	2	3	4	5
42	6232	6243	6253	6263	6274	6284	6294	6304	6314	6325	1	2	3	4	5
43	6335	6345	6355	6365	6375	6385	6395	6405	6415	6425	1	2	3	4	5
44	6435	6444	6454	6464	6474	6484	6493	6503	6513	6522	1	2	3	4	5
45	6532	6542	6551	6561	6571	6580	6590	6599	6609	6618	1	2	3	4	5
46	6628	6637	6646	6656	6665	6675	6684	6693	6702	6712	1	2	3	4	5
47	6721	6730	6739	6749	6758	6767	6776	6785	6794	6803	1	2	3	4	5
48	6812	6821	6830	6839	6848	6857	6866	6875	6884	6893	1	2	3	4	4
49	6902	6911	6920	6928	6937	6946	6955	6964	6972	6981	1	2	3	4	4
50	6990	6998	7007	7016	7024	7033	7042	7050	7059	7067	1	2	3	3	4
51	7076	7084	7093	7101	7110	7118	7126	7135	7143	7152	1	2	3	3	4
52	7160	7168	7177	7185	7193	7202	7210	7218	7226	7235	1	2	2	3	4
53	7243	7251	7259	7267	7275	7284	7292	7300	7308	7316	1	2	2	3	4
54	7324	7332	7340	7348	7356	7364	7372	7380	7388	7396	1	2	2	3	4

LOGARITHMS.

N.	0	1	2	3	4	5	6	7	8	9	P. P. 1	2	3	4	5
55	7404	7412	7419	7427	7435	7443	7451	7459	7466	7474	1	2	2	3	4
56	7482	7490	7497	7505	7513	7520	7528	7536	7543	7551	1	2	2	3	4
57	7559	7566	7574	7582	7589	7597	7604	7612	7619	7627	1	2	2	3	4
58	7634	7642	7649	7657	7664	7672	7679	7686	7694	7701	1	1	2	3	4
59	7709	7716	7723	7731	7738	7745	7752	7760	7767	7774	1	1	2	3	4
60	7782	7789	7796	7803	7810	7818	7825	7832	7839	7846	1	1	2	3	4
61	7853	7860	7868	7875	7882	7889	7896	7903	7910	7917	1	1	2	3	4
62	7924	7931	7938	7945	7952	7959	7966	7973	7980	7987	1	1	2	3	3
63	7993	8000	8007	8014	8021	8028	8035	8041	8048	8055	1	1	2	3	3
64	8062	8069	8075	8082	8089	8096	8102	8109	8116	8122	1	1	2	3	3
65	8129	8136	8142	8149	8156	8162	8169	8176	8182	8189	1	1	2	3	3
66	8195	8202	8209	8215	8222	8228	8235	8241	8248	8254	1	1	2	3	3
67	8261	8267	8274	8280	8287	8293	8299	8306	8312	8319	1	1	2	3	3
68	8325	8331	8338	8344	8351	8357	8363	8370	8376	8382	1	1	2	3	3
69	8388	8395	8401	8407	8414	8420	8426	8432	8439	8445	1	1	2	3	3
70	8451	8457	8463	8470	8476	8482	8488	8494	8500	8506	1	1	2	2	3
71	8513	8519	8525	8531	8537	8543	8549	8555	8561	8567	1	1	2	2	3
72	8573	8579	8585	8591	8597	8603	8609	8615	8621	8627	1	1	2	2	3
73	8633	8639	8645	8651	8657	8663	8669	8675	8681	8686	1	1	2	2	3
74	8692	8698	8704	8710	8716	8722	8727	8733	8739	8745	1	1	2	2	3
75	8751	8756	8762	8768	8774	8779	8785	8791	8797	8802	1	1	2	2	3
76	8808	8814	8820	8825	8831	8837	8842	8848	8854	8859	1	1	2	2	3
77	8865	8871	8876	8882	8887	8893	8899	8904	8910	8915	1	1	2	2	3
78	8921	8927	8932	8938	8943	8949	8954	8960	8965	8971	1	1	2	2	3
79	8976	8982	8987	8993	8998	9004	9009	9015	9020	9025	1	1	2	2	3
80	9031	9036	9042	9047	9053	9058	9063	9069	9074	9079	1	1	2	2	3
81	9085	9090	9096	9101	9106	9112	9117	9122	9128	9133	1	1	2	2	3
82	9138	9143	9149	9154	9159	9165	9170	9175	9180	9186	1	1	2	2	3
83	9191	9196	9201	9206	9212	9217	9222	9227	9232	9238	1	1	2	2	3
84	9243	9248	9253	9258	9263	9269	9274	9279	9284	9289	1	1	2	2	3
85	9294	9299	9304	9309	9315	9320	9325	9330	9335	9340	1	1	2	2	3
86	9345	9350	9355	9360	9365	9370	9375	9380	9385	9390	1	1	2	2	3
87	9395	9400	9405	9410	9415	9420	9425	9430	9435	9440	0	1	1	2	2
88	9445	9450	9455	9460	9465	9469	9474	9479	9484	9489	0	1	1	2	2
89	9494	9499	9504	9509	9513	9518	9523	9528	9533	9538	0	1	1	2	2
90	9542	9547	9552	9557	9562	9566	9571	9576	9581	9586	0	1	1	2	2
91	9590	9595	9600	9605	9609	9614	9619	9624	9628	9633	0	1	1	2	2
92	9638	9643	9647	9652	9657	9661	9666	9671	9675	9680	0	1	1	2	2
93	9685	9689	9694	9699	9703	9708	9713	9717	9722	9727	0	1	1	2	2
94	9731	9736	9741	9745	9750	9754	9759	9763	9768	9773	0	1	1	2	2
95	9777	9782	9786	9791	9795	9800	9805	9809	9814	9818	0	1	1	2	2
96	9823	9827	9832	9836	9841	9845	9850	9854	9859	9863	0	1	1	2	2
97	9868	9872	9877	9881	9886	9890	9894	9899	9903	9908	0	1	1	2	2
98	9912	9917	9921	9926	9930	9934	9939	9943	9948	9952	0	1	1	2	2
99	9956	9961	9965	9969	9974	9978	9983	9987	9991	9996	0	1	1	2	2

ANTILOGARITHMS.

	0	1	2	3	4	5	6	7	8	9	P. P. 1	2	3	4	5
.00	1000	1002	1005	1007	1009	1012	1014	1016	1019	1021	0	0	1	1	1
.01	1023	1026	1028	1030	1033	1035	1038	1040	1042	1045	0	0	1	1	1
.02	1047	1050	1052	1054	1057	1059	1062	1064	1067	1069	0	0	1	1	1
.03	1072	1074	1076	1079	1081	1084	1086	1089	1091	1094	0	0	1	1	1
.04	1096	1099	1102	1104	1107	1109	1112	1114	1117	1119	0	1	1	1	1
.05	1122	1125	1127	1130	1132	1135	1138	1140	1143	1146	0	1	1	1	1
.06	1148	1151	1153	1156	1159	1161	1164	1167	1169	1172	0	1	1	1	1
.07	1175	1178	1180	1183	1186	1189	1191	1194	1197	1199	0	1	1	1	1
.08	1202	1205	1208	1211	1213	1216	1219	1222	1225	1227	0	1	1	1	1
.09	1230	1233	1236	1239	1242	1245	1247	1250	1253	1256	0	1	1	1	1
.10	1259	1262	1265	1268	1271	1274	1276	1279	1282	1285	0	1	1	1	1
.11	1288	1291	1294	1297	1300	1303	1306	1309	1312	1315	0	1	1	1	2
.12	1318	1321	1324	1327	1330	1334	1337	1340	1343	1346	0	1	1	1	2
.13	1349	1352	1355	1358	1361	1365	1368	1371	1374	1377	0	1	1	1	2
.14	1380	1384	1387	1390	1393	1396	1400	1403	1406	1409	0	1	1	1	2
.15	1413	1416	1419	1422	1426	1429	1432	1435	1439	1442	0	1	1	1	2
.16	1445	1449	1452	1455	1459	1462	1466	1469	1472	1476	0	1	1	1	2
.17	1479	1483	1486	1489	1493	1496	1500	1503	1507	1510	0	1	1	1	2
.18	1514	1517	1521	1524	1528	1531	1535	1538	1542	1545	0	1	1	1	2
.19	1549	1552	1556	1560	1563	1567	1570	1574	1578	1581	0	1	1	1	2
.20	1585	1589	1592	1596	1600	1603	1607	1611	1614	1618	0	1	1	1	2
.21	1622	1626	1629	1633	1637	1641	1644	1648	1652	1656	0	1	1	2	2
.22	1660	1663	1667	1671	1675	1679	1683	1687	1690	1694	0	1	1	2	2
.23	1698	1702	1706	1710	1714	1718	1722	1726	1730	1734	0	1	1	2	2
.24	1738	1742	1746	1750	1754	1758	1762	1766	1770	1774	0	1	1	2	2
.25	1778	1782	1786	1791	1795	1799	1803	1807	1811	1816	0	1	1	2	2
.26	1820	1824	1828	1832	1837	1841	1845	1849	1854	1858	0	1	1	2	2
.27	1862	1866	1871	1875	1879	1884	1888	1892	1897	1901	0	1	1	2	2
.28	1905	1910	1914	1919	1923	1928	1932	1936	1941	1945	0	1	1	2	2
.29	1950	1954	1959	1963	1968	1972	1977	1982	1986	1991	0	1	1	2	2
.30	1995	2000	2004	2009	2014	2018	2023	2028	2032	2037	0	1	1	2	2
.31	2042	2046	2051	2056	2061	2065	2070	2075	2080	2084	0	1	1	2	2
.32	2089	2094	2099	2104	2109	2113	2118	2123	2128	2133	0	1	1	2	2
.33	2138	2143	2148	2153	2158	2163	2168	2173	2178	2183	0	1	1	2	2
.34	2188	2193	2198	2203	2208	2213	2218	2223	2228	2234	1	1	2	2	3
.35	2239	2244	2249	2254	2259	2265	2270	2275	2280	2286	1	1	2	2	3
.36	2291	2296	2301	2307	2312	2317	2323	2328	2333	2339	1	1	2	2	3
.37	2344	2350	2355	2360	2366	2371	2377	2382	2388	2393	1	1	2	2	3
.38	2399	2404	2410	2415	2421	2427	2432	2438	2443	2449	1	1	2	2	3
.39	2455	2460	2466	2472	2477	2483	2489	2495	2500	2506	1	1	2	2	3
.40	2512	2518	2523	2529	2535	2541	2547	2553	2559	2564	1	1	2	2	3
.41	2570	2576	2582	2588	2594	2600	2606	2612	2618	2624	1	1	2	2	3
.42	2630	2636	2642	2649	2655	2661	2667	2673	2679	2685	1	1	2	2	3
.43	2692	2698	2704	2710	2716	2723	2729	2735	2742	2748	1	1	2	3	3
.44	2754	2761	2767	2773	2780	2786	2793	2799	2805	2812	1	1	2	3	3
.45	2818	2825	2831	2838	2844	2851	2858	2864	2871	2877	1	1	2	3	3
.46	2884	2891	2897	2904	2911	2917	2924	2931	2938	2944	1	1	2	3	3
.47	2951	2958	2965	2972	2979	2985	2992	2999	3006	3013	1	1	2	3	3
.48	3020	3027	3034	3041	3048	3055	3062	3069	3076	3083	1	1	2	3	4
.49	3090	3097	3105	3112	3119	3126	3133	3141	3148	3155	1	1	2	3	4

ANTILOGARITHMS.

	0	1	2	3	4	5	6	7	8	9	P. P.				
											1	2	3	4	5
.50	3162	3170	3177	3184	3192	3199	3206	3214	3221	3228	1	1	2	3	4
.51	3236	3243	3251	3258	3266	3273	3281	3289	3296	3304	1	2	2	3	4
.52	3311	3319	3327	3334	3342	3350	3357	3365	3373	3381	1	2	2	3	4
.53	3388	3396	3404	3412	3420	3428	3436	3443	3451	3459	1	2	2	3	4
.54	3467	3475	3483	3491	3499	3508	3516	3524	3532	3540	1	2	2	3	4
.55	3548	3556	3565	3573	3581	3589	3597	3606	3614	3622	1	2	2	3	4
.56	3631	3639	3648	3656	3664	3673	3681	3690	3698	3707	1	2	3	3	4
.57	3715	3724	3733	3741	3750	3758	3767	3776	3784	3793	1	2	3	3	4
.58	3802	3811	3819	3828	3837	3846	3855	3864	3873	3882	1	2	3	4	4
.59	3890	3899	3908	3917	3926	3936	3945	3954	3963	3972	1	2	3	4	5
.60	3981	3990	3999	4009	4018	4027	4036	4046	4055	4064	1	2	3	4	5
.61	4074	4083	4093	4102	4111	4121	4130	4140	4150	4159	1	2	3	4	5
.62	4169	4178	4188	4198	4207	4217	4227	4236	4246	4256	1	2	3	4	5
.63	4266	4276	4285	4295	4305	4315	4325	4335	4345	4355	1	2	3	4	5
.64	4365	4375	4385	4395	4406	4416	4426	4436	4446	4457	1	2	3	4	5
.65	4467	4477	4487	4498	4508	4519	4529	4539	4550	4560	1	2	3	4	5
.66	4571	4581	4592	4603	4613	4624	4634	4645	4656	4667	1	2	3	4	5
.67	4677	4688	4699	4710	4721	4732	4742	4753	4764	4775	1	2	3	4	5
.68	4786	4797	4808	4819	4831	4842	4853	4864	4875	4887	1	2	3	4	6
.69	4898	4909	4920	4932	4943	4955	4966	4977	4989	5000	1	2	3	5	6
.70	5012	5023	5035	5047	5058	5070	5082	5093	5105	5117	1	2	4	5	6
.71	5129	5140	5152	5164	5176	5188	5200	5212	5224	5236	1	2	4	5	6
.72	5248	5260	5272	5284	5297	5309	5321	5333	5346	5358	1	2	4	5	6
.73	5370	5383	5395	5408	5420	5433	5445	5458	5470	5483	1	3	4	5	6
.74	5495	5508	5521	5534	5546	5559	5572	5585	5598	5610	1	3	4	5	6
.75	5623	5636	5649	5662	5675	5689	5702	5715	5728	5741	1	3	4	5	7
.76	5754	5768	5781	5794	5808	5821	5834	5848	5861	5875	1	3	4	5	7
.77	5888	5902	5916	5929	5943	5957	5970	5984	5998	6012	1	3	4	5	7
.78	6026	6039	6053	6067	6081	6095	6109	6124	6138	6152	1	3	4	6	7
.79	6166	6180	6194	6209	6223	6237	6252	6266	6281	6295	1	3	4	6	7
.80	6310	6324	6339	6353	6368	6383	6397	6412	6427	6442	1	3	4	6	7
.81	6457	6471	6486	6501	6516	6531	6546	6561	6577	6592	2	3	5	6	8
.82	6607	6622	6637	6653	6668	6683	6699	6714	6730	6745	2	3	5	6	8
.83	6761	6776	6792	6808	6823	6839	6855	6871	6887	6902	2	3	5	6	8
.84	6918	6934	6950	6966	6982	6998	7015	7031	7047	7063	2	3	5	6	8
.85	7079	7096	7112	7129	7145	7161	7178	7194	7211	7228	2	3	5	7	8
.86	7244	7261	7278	7295	7311	7328	7345	7362	7379	7396	2	3	5	7	8
.87	7413	7430	7447	7464	7482	7499	7516	7534	7551	7568	2	3	5	7	9
.88	7586	7603	7621	7638	7656	7674	7691	7709	7727	7745	2	4	5	7	9
.89	7762	7780	7798	7816	7834	7852	7870	7889	7907	7925	2	4	5	7	9
.90	7943	7962	7980	7998	8017	8035	8054	8072	8091	8110	2	4	6	7	9
.91	8128	8147	8166	8185	8204	8222	8241	8260	8279	8299	2	4	6	8	9
.92	8318	8337	8356	8375	8395	8414	8433	8453	8472	8492	2	4	6	8	10
.93	8511	8531	8551	8570	8590	8610	8630	8650	8670	8690	2	4	6	8	10
.94	8710	8730	8750	8770	8790	8810	8831	8851	8872	8892	2	4	6	8	10
.95	8913	8933	8954	8974	8995	9016	9036	9057	9078	9099	2	4	6	8	10
.96	9120	9141	9162	9183	9204	9226	9247	9268	9290	9311	2	4	6	8	11
.97	9333	9354	9376	9397	9419	9441	9462	9484	9506	9528	2	4	7	9	11
.98	9550	9572	9594	9616	9638	9661	9683	9705	9727	9750	2	4	7	9	11
.99	9772	9795	9817	9840	9863	9886	9908	9931	9954	9977	2	5	7	9	11

Table 13.

ANTILOGARITHMS.

	0	1	2	3	4	5	6	7	8	9	10
.900	7943	7945	7947	7949	7951	7952	7954	7956	7958	7960	7962
.901	7962	7963	7965	7967	7969	7971	7973	7974	7976	7978	7980
.902	7980	7982	7984	7985	7987	7989	7991	7993	7995	7997	7998
.903	7998	8000	8002	8004	8006	8008	8009	8011	8013	8015	8017
.904	8017	8019	8020	8022	8024	8026	8028	8030	8032	8033	8035
.905	8035	8037	8039	8041	8043	8045	8046	8048	8050	8052	8054
.906	8054	8056	8057	8059	8061	8063	8065	8067	8069	8070	8072
.907	8072	8074	8076	8078	8080	8082	8084	8085	8087	8089	8091
.908	8091	8093	8095	8097	8098	8100	8102	8104	8106	8108	8110
.909	8110	8111	8113	8115	8117	8119	8121	8123	8125	8126	8128
.910	8128	8130	8132	8134	8136	8138	8140	8141	8143	8145	8147
.911	8147	8149	8151	8153	8155	8156	8158	8160	8162	8164	8166
.912	8166	8168	8170	8171	8173	8175	8177	8179	8181	8183	8185
.913	8185	8187	8188	8190	8192	8194	8196	8198	8200	8202	8204
.914	8204	8205	8207	8209	8211	8213	8215	8217	8219	8221	8222
.915	8222	8224	8226	8228	8230	8232	8234	8236	8238	8239	8241
.916	8241	8243	8245	8247	8249	8251	8253	8255	8257	8258	8260
.917	8260	8262	8264	8266	8268	8270	8272	8274	8276	8278	8279
.918	8279	8281	8283	8285	8287	8289	8291	8293	8295	8297	8299
.919	8299	8300	8302	8304	8306	8308	8310	8312	8314	8316	8318
.920	8318	8320	8321	8323	8325	8327	8329	8331	8333	8335	8337
.921	8337	8339	8341	8343	8344	8346	8348	8350	8352	8354	8356
.922	8356	8358	8360	8362	8364	8366	8368	8370	8371	8373	8375
.923	8375	8377	8379	8381	8383	8385	8387	8389	8391	8393	8395
.924	8395	8397	8398	8400	8402	8404	8406	8408	8410	8412	8414
.925	8414	8416	8418	8420	8422	8424	8426	8428	8429	8431	8433
.926	8433	8435	8437	8439	8441	8443	8445	8447	8449	8451	8453
.927	8453	8455	8457	8459	8461	8463	8464	8466	8468	8470	8472
.928	8472	8474	8476	8478	8480	8482	8484	8486	8488	8490	8492
.929	8492	8494	8496	8498	8500	8502	8504	8506	8507	8509	8511
.930	8511	8513	8515	8517	8519	8521	8523	8525	8527	8529	8531
.931	8531	8533	8535	8537	8539	8541	8543	8545	8547	8549	8551
.932	8551	8553	8555	8557	8559	8561	8562	8564	8566	8568	8570
.933	8570	8572	8574	8576	8578	8580	8582	8584	8586	8588	8590
.934	8590	8592	8594	8596	8598	8600	8602	8604	8606	8608	8610
.935	8610	8612	8614	8616	8618	8620	8622	8624	8626	8628	8630
.936	8630	8632	8634	8636	8638	8640	8642	8644	8646	8648	8650
.937	8650	8652	8654	8656	8658	8660	8662	8664	8666	8668	8670
.938	8670	8672	8674	8676	8678	8680	8682	8684	8686	8688	8690
.939	8690	8692	8694	8696	8698	8700	8702	8704	8706	8708	8710
.940	8710	8712	8714	8716	8718	8720	8722	8724	8726	8728	8730
.941	8730	8732	8734	8736	8738	8740	8742	8744	8746	8748	8750
.942	8750	8752	8754	8756	8758	8760	8762	8764	8766	8768	8770
.943	8770	8772	8774	8776	8778	8780	8782	8784	8786	8788	8790
.944	8790	8792	8794	8796	8798	8800	8802	8804	8806	8808	8810
.945	8810	8813	8815	8817	8819	8821	8823	8825	8827	8829	8831
.946	8831	8833	8835	8837	8839	8841	8843	8845	8847	8849	8851
.947	8851	8853	8855	8857	8859	8861	8863	8865	8867	8870	8872
.948	8872	8874	8876	8878	8880	8882	8884	8886	8888	8890	8892
.949	8892	8894	8896	8898	8900	8902	8904	8906	8908	8910	8913

ANTILOGARITHMS.

	0	1	2	3	4	5	6	7	8	9	10
.950	8913	8915	8917	8919	8921	8923	8925	8927	8929	8931	8933
.951	8933	8935	8937	8939	8941	8943	8945	8947	8950	8952	8954
.952	8954	8956	8958	8960	8962	8964	8966	8968	8970	8972	8974
.953	8974	8976	8978	8980	8983	8985	8987	8989	8991	8993	8995
.954	8995	8997	8999	9001	9003	9005	9007	9009	9012	9014	9016
.955	9016	9018	9020	9022	9024	9026	9028	9030	9032	9034	9036
.956	9036	9039	9041	9043	9045	9047	9049	9051	9053	9055	9057
.957	9057	9059	9061	9064	9066	9068	9070	9072	9074	9076	9078
.958	9078	9080	9082	9084	9087	9089	9091	9093	9095	9097	9099
.959	9099	9101	9103	9105	9108	9110	9112	9114	9116	9118	9120
.960	9120	9122	9124	9126	9129	9131	9133	9135	9137	9139	9141
.961	9141	9143	9145	9147	9150	9152	9154	9156	9158	9160	9162
.962	9162	9164	9166	9169	9171	9173	9175	9177	9179	9181	9183
.963	9183	9185	9188	9190	9192	9194	9196	9198	9200	9202	9204
.964	9204	9207	9209	9211	9213	9215	9217	9219	9221	9224	9226
.965	9226	9228	9230	9232	9234	9236	9238	9241	9243	9245	9247
.966	9247	9249	9251	9253	9256	9258	9260	9262	9264	9266	9268
.967	9268	9270	9273	9275	9277	9279	9281	9283	9285	9288	9290
.968	9290	9292	9294	9296	9298	9300	9303	9305	9307	9309	9311
.969	9311	9313	9315	9318	9320	9322	9324	9326	9328	9330	9333
.970	9333	9335	9337	9339	9341	9343	9345	9348	9350	9352	9354
.971	9354	9356	9358	9361	9363	9365	9367	9369	9371	9373	9376
.972	9376	9378	9380	9382	9384	9386	9389	9391	9393	9395	9397
.973	9397	9399	9402	9404	9406	9408	9410	9412	9415	9417	9419
.974	9419	9421	9423	9425	9428	9430	9432	9434	9436	9438	9441
.975	9441	9443	9445	9447	9449	9451	9454	9456	9458	9460	9462
.976	9462	9465	9467	9469	9471	9473	9475	9478	9480	9482	9484
.977	9484	9486	9489	9491	9493	9495	9497	9499	9502	9504	9506
.978	9506	9508	9510	9513	9515	9517	9519	9521	9524	9526	9528
.979	9528	9530	9532	9535	9537	9539	9541	9543	9546	9548	9550
.980	9550	9552	9554	9557	9559	9561	9563	9565	9568	9570	9572
.981	9572	9574	9576	9579	9581	9583	9585	9587	9590	9592	9594
.982	9594	9596	9598	9601	9603	9605	9607	9609	9612	9614	9616
.983	9616	9618	9621	9623	9625	9627	9629	9632	9634	9636	9638
.984	9638	9641	9643	9645	9647	9649	9652	9654	9656	9658	9661
.985	9661	9663	9665	9667	9669	9672	9674	9676	9678	9681	9683
.986	9683	9685	9687	9689	9692	9694	9696	9698	9701	9703	9705
.987	9705	9707	9710	9712	9714	9716	9719	9721	9723	9725	9727
.988	9727	9730	9732	9734	9736	9739	9741	9743	9745	9748	9750
.989	9750	9752	9754	9757	9759	9761	9763	9766	9768	9770	9772
.990	9772	9775	9777	9779	9781	9784	9786	9788	9790	9793	9795
.991	9795	9797	9799	9802	9804	9806	9808	9811	9813	9815	9817
.992	9817	9820	9822	9824	9827	9829	9831	9833	9836	9838	9840
.993	9840	9842	9845	9847	9849	9851	9854	9856	9858	9861	9863
.994	9863	9865	9867	9870	9872	9874	9876	9879	9881	9883	9886
.995	9886	9888	9890	9892	9895	9897	9899	9901	9904	9906	9908
.996	9908	9911	9913	9915	9917	9920	9922	9924	9927	9929	9931
.997	9931	9933	9936	9938	9940	9943	9945	9947	9949	9952	9954
.998	9954	9956	9959	9961	9963	9966	9968	9970	9972	9975	9977
.999	9977	9979	9982	9984	9986	9988	9991	9993	9995	9998	0000

CIRCULAR (TRIGONOMETRIC) FUNCTIONS.

(Taken from B. O. Peirce's "Short Table of Integrals," Ginn & Co.)

RADIANS.	DEGREES.	SINES.		COSINES.		TANGENTS.		COTANGENTS.			
		Nat.	Log.	Nat.	Log.	Nat.	Log.	Nat.	Log.		
0.0000	0°00′	.0000	∞	1.0000	0.0000	.0000	∞	∞	∞	90°00′	1.5708
0.0029	10	.0029	7.4637	1.0000	.0000	.0029	7.4637	343.77	2.5363	50	1.5679
0.0058	20	.0058	.7648	1.0000	.0000	.0058	.7648	171.89	.2352	40	1.5650
0.0087	30	.0087	.9408	1.0000	.0000	.0087	.9409	114.59	.0591	30	1.5621
0.0116	40	.0116	8.0658	.9999	.0000	.0116	8.0658	85.940	1.9342	20	1.5592
0.0145	50	.0145	.1627	.9999	.0000	.0145	.1627	68.750	.8373	10	1.5563
0.0175	1°00′	.0175	8.2419	.9998	9.9999	.0175	8.2419	57.290	1.7581	89°00′	1.5533
0.0204	10	.0204	.3088	.9998	.9999	.0204	.3089	49.104	.6911	50	1.5504
0.0233	20	.0233	.3668	.9997	.9999	.0233	.3669	42.964	.6331	40	1.5475
0.0262	30	.0262	.4179	.9997	.9999	.0262	.4181	38.188	.5819	30	1.5446
0.0291	40	.0291	.4637	.9996	.9998	.0291	.4638	34.368	.5362	20	1.5417
0.0320	50	.0320	.5050	.9995	.9998	.0320	.5053	31.242	.4947	10	1.5388
0.0349	2°00′	.0349	8.5428	.9994	9.9997	.0349	8.5431	28.636	1.4569	88°00′	1.5359
0.0378	10	.0378	.5776	.9993	.9997	.0378	.5779	26.432	.4221	50	1.5330
0.0407	20	.0407	.6097	.9992	.9996	.0407	.6101	24.542	.3899	40	1.5301
0.0436	30	.0436	.6397	.9990	.9996	.0437	.6401	22.904	.3599	30	1.5272
0.0465	40	.0465	.6677	.9989	.9995	.0466	.6682	21.470	.3318	20	1.5243
0.0495	50	.0494	.6940	.9988	.9995	.0495	.6945	20.206	.3055	10	1.5213
0.0524	3°00′	.0523	8.7188	.9986	9.9994	.0524	8.7194	19.081	1.2806	87°00′	1.5184
0.0553	10	.0552	.7423	.9985	.9993	.0553	.7429	18.075	.2571	50	1.5155
0.0582	20	.0581	.7645	.9983	.9993	.0582	.7652	17.169	.2348	40	1.5126
0.0611	30	.0610	.7857	.9981	.9992	.0612	.7865	16.350	.2135	30	1.5097
0.0640	40	.0640	.8059	.9980	.9991	.0641	.8067	15.605	.1933	20	1.5068
0.0669	50	.0669	.8251	.9978	.9990	.0670	.8261	14.924	.1739	10	1.5039
0.0698	4°00′	.0698	8.8436	.9976	9.9989	.0699	8.8446	14.301	1.1554	86°00′	1.5010
0.0727	10	.0727	.8613	.9974	.9989	.0729	.8624	13.727	.1376	50	1.4981
0.0756	20	.0756	.8783	.9971	.9988	.0758	.8795	13.197	.1205	40	1.4952
0.0785	30	.0785	.8946	.9969	.9987	.0787	.8960	12.706	.1040	30	1.4923
0.0814	40	.0814	.9104	.9967	.9986	.0816	.9118	12.251	.0882	20	1.4893
0.0844	50	.0843	.9256	.9964	.9985	.0846	.9272	11.826	.0728	10	1.4864
0.0873	5°00′	.0872	8.9403	.9962	9.9983	.0875	8.9420	11.430	1.0580	85°00′	1.4835
0.0902	10	.0901	.9545	.9959	.9982	.0904	.9563	11.059	.0437	50	1.4806
0.0931	20	.0929	.9682	.9957	.9981	.0934	.9701	10.712	.0299	40	1.4777
0.0960	30	.0958	.9816	.9954	.9980	.0963	.9836	10.385	.0164	30	1.4748
0.0989	40	.0987	.9945	.9951	.9979	.0992	.9966	10.078	.0034	20	1.4719
0.1018	50	.1016	9.0070	.9948	.9977	.1022	9.0093	9.7882	0.9907	10	1.4690
0.1047	6°00	.1045	9.0192	.9945	9.9976	.1051	9.0216	9.5144	0.9784	84°00′	1.4661
0.1076	10	.1074	.0311	.9942	.9975	.1080	.0336	9.2553	.9664	50	1.4632
0.1105	20	.1103	.0426	.9939	.9973	.1110	.0453	9.0098	.9547	40	1.4603
0.1134	30	.1132	.0539	.9936	.9972	.1139	.0567	8.7769	.9433	30	1.4574
0.1164	40	.1161	.0648	.9932	.9971	.1169	.0678	8.5555	.9322	20	1.4544
0.1193	50	.1190	.0755	.9929	.9969	.1198	.0786	8.3450	.9214	10	1.4515
0.1222	7°00′	.1219	9.0859	.9925	9.9968	.1228	9.0891	8.1443	0.9109	83°00′	1.4486
0.1251	10	.1248	.0961	.9922	.9966	.1257	.0995	7.9530	.9005	50	1.4457
0.1280	20	.1276	.1060	.9918	.9964	.1287	.1096	7.7704	.8904	40	1.4428
0.1309	30	.1305	.1157	.9914	.9963	.1317	.1194	7.5958	.8806	30	1.4399
0.1338	40	.1334	.1252	.9911	.9961	.1346	.1291	7.4287	.8709	20	1.4370
0.1367	50	.1363	.1345	.9907	.9959	.1376	.1385	7.2687	.8615	10	1.4341
0.1396	8°00′	.1392	9.1436	.9903	9.9958	.1405	9.1478	7.1154	0.8522	82°00′	1.4312
0.1425	10	.1421	.1525	.9899	.9956	.1435	.1569	6.9682	.8431	50	1.4283
0.1454	20	.1449	.1612	.9894	.9954	.1465	.1658	6.8269	.8342	40	1.4254
0.1484	30	.1478	.1697	.9890	.9952	.1495	.1745	6.6912	.8255	30	1.4224
0.1513	40	.1507	.1781	.9886	.9950	.1524	.1831	6.5606	.8169	20	1.4195
0.1542	50	.1536	.1863	.9881	.9948	.1554	.1915	6.4348	.8085	10	1.4166
0.1571	9°00′	.1564	9.1943	.9877	9.9946	.1584	9.1997	6.3138	0.8003	81°00′	1.4137
		Nat.	Log.	Nat.	Log.	Nat.	Log.	Nat.	Log.		
		COSINES.		SINES.		COTANGENTS.		TANGENTS.		DEGREES.	RADIANS.

CIRCULAR (TRIGONOMETRIC) FUNCTIONS.

RADIANS.	DEGREES.	SINES. Nat.	SINES. Log.	COSINES. Nat.	COSINES. Log.	TANGENTS. Nat.	TANGENTS. Log.	COTANGENTS. Nat.	COTANGENTS. Log.		
0.1571	9°00′	.1564	9.1943	.9877	9.9946	.1584	9.1997	6.3138	0.8003	81°00′	1.4137
0.1600	10	.1593	.2022	.9872	.9944	.1614	.2078	6.1970	.7922	50	1.4108
0.1629	20	.1622	.2100	.9868	.9942	.1644	.2158	6.0844	.7842	40	1.4079
0.1658	30	.1650	.2176	.9863	.9940	.1673	.2236	5.9758	.7764	30	1.4050
0.1687	40	.1679	.2251	.9858	.9938	.1703	.2313	5.8708	.7687	20	1.4021
0.1716	50	.1708	.2324	.9853	.9936	.1733	.2389	5.7694	.7611	10	1.3992
0.1745	10°00′	.1736	9.2397	.9848	9.9934	.1763	9.2463	5.6713	0.7537	80°00′	1.3963
0.1774	10	.1765	.2468	.9843	.9931	.1793	.2536	5.5764	.7464	50	1.3934
0.1804	20	.1794	.2538	.9838	.9929	.1823	.2609	5.4845	.7391	40	1.3904
0.1833	30	.1822	.2606	.9833	.9927	.1853	.2680	5.3955	.7320	30	1.3875
0.1862	40	.1851	.2674	.9827	.9924	.1883	.2750	5.3093	.7250	20	1.3846
0.1891	50	.1880	.2740	.9822	.9922	.1914	.2819	5.2257	.7181	10	1.3817
0.1920	11°00′	.1908	9.2806	.9816	9.9919	.1944	9.2887	5.1446	0.7113	79°00′	1.3788
0.1949	10	.1937	.2870	.9811	.9917	.1974	.2953	5.0658	.7047	50	1.3759
0.1978	20	.1965	.2934	.9805	.9914	.2004	.3020	4.9894	.6980	40	1.3730
0.2007	30	.1994	.2997	.9799	.9912	.2035	.3085	4.9152	.6915	30	1.3701
0.2036	40	.2022	.3058	.9793	.9909	.2065	.3149	4.8430	.6851	20	1.3672
0.2065	50	.2051	.3119	.9787	.9907	.2095	.3212	4.7729	.6788	10	1.3643
0.2094	12°00′	.2079	9.3179	.9781	9.9904	.2126	9.3275	4.7046	0.6725	78°00′	1.3614
0.2123	10	.2108	.3238	.9775	.9901	.2156	.3336	4.6382	.6664	50	1.3584
0.2153	20	.2136	.3296	.9769	.9899	.2186	.3397	4.5736	.6603	40	1.3555
0.2182	30	.2164	.3353	.9763	.9896	.2217	.3458	4.5107	.6542	30	1.3526
0.2211	40	.2193	.3410	.9757	.9893	.2247	.3517	4.4494	.6483	20	1.3497
0.2240	50	.2221	.3466	.9750	.9890	.2278	.3576	4.3897	.6424	10	1.3468
0.2269	13°00′	.2250	9.3521	.9744	9.9887	.2309	9.3634	4.3315	0.6366	77°00′	1.3439
0.2298	10	.2278	.3575	.9737	.9884	.2339	.3691	4.2747	.6309	50	1.3410
0.2327	20	.2306	.3629	.9730	.9881	.2370	.3748	4.2193	.6252	40	1.3381
0.2356	30	.2334	.3682	.9724	.9878	.2401	.3804	4.1653	.6196	30	1.3352
0.2385	40	.2363	.3734	.9717	.9875	.2432	.3859	4.1126	.6141	20	1.3323
0.2414	50	.2391	.3786	.9710	.9872	.2462	.3914	4.0611	.6086	10	1.3294
0.2443	14°00′	.2419	9.3837	.9703	9.9869	.2493	9.3968	4.0108	0.6032	76°00′	1.3265
0.2473	10	.2447	.3887	.9696	.9866	.2524	.4021	3.9617	.5979	50	1.3235
0.2502	20	.2476	.3937	.9689	.9863	.2555	.4074	3.9136	.5926	40	1.3206
0.2531	30	.2504	.3986	.9681	.9859	.2586	.4127	3.8667	.5873	30	1.3177
0.2560	40	.2532	.4035	.9674	.9856	.2617	.4178	3.8208	.5822	20	1.3148
0.2589	50	.2560	.4083	.9667	.9853	.2648	.4230	3.7760	.5770	10	1.3119
0.2618	15°00′	.2588	9.4130	.9659	9.9849	.2679	9.4281	3.7321	0.5719	75°00′	1.3090
0.2647	10	.2616	.4177	.9652	.9846	.2711	.4331	3.6891	.5669	50	1.3061
0.2676	20	.2644	.4223	.9644	.9843	.2742	.4381	3.6470	.5619	40	1.3032
0.2705	30	.2672	.4269	.9636	.9839	.2773	.4430	3.6059	.5570	30	1.3003
0.2734	40	.2700	.4314	.9628	.9836	.2805	.4479	3.5656	.5521	20	1.2974
0.2763	50	.2728	.4359	.9621	.9832	.2836	.4527	3.5261	.5473	10	1.2945
0.2793	16°00′	.2756	9.4403	.9613	9.9828	.2867	9.4575	3.4874	0.5425	74°00′	1.2915
0.2822	10	.2784	.4447	.9605	.9825	.2899	.4622	3.4495	.5378	50	1.2886
0.2851	20	.2812	.4491	.9596	.9821	.2931	.4669	3.4124	.5331	40	1.2857
0.2880	30	.2840	.4533	.9588	.9817	.2962	.4716	3.3759	.5284	30	1.2828
0.2909	40	.2868	.4576	.9580	.9814	.2994	.4762	3.3402	.5238	20	1.2799
0.2938	50	.2896	.4618	.9572	.9810	.3026	.4808	3.3052	.5192	10	1.2770
0.2967	17°00′	.2924	9.4659	.9563	9.9806	.3057	9.4853	3.2709	0.5147	73°00′	1.2741
0.2996	10	.2952	.4700	.9555	.9802	.3089	.4898	3.2371	.5102	50	1.2712
0.3025	20	.2979	.4741	.9546	.9798	.3121	.4943	3.2041	.5057	40	1.2683
0.3054	30	.3007	.4781	.9537	.9794	.3153	.4987	3.1716	.5013	30	1.2654
0.3083	40	.3035	.4821	.9528	.9790	.3185	.5031	3.1397	.4969	20	1.2625
0.3113	50	.3062	.4861	.9520	.9786	.3217	.5075	3.1084	.4925	10	1.2595
0.3142	18°00′	.3090	9.4900	.9511	9.9782	.3249	9.5118	3.0777	0.4882	72°00′	1.2566
		Nat.	Log.	Nat.	Log.	Nat.	Log.	Nat.	Log.	DEGREES.	RADIANS.
		COSINES.		SINES.		COTANGENTS.		TANGENTS.			

CIRCULAR (TRIGONOMETRIC) FUNCTIONS.

RADIANS.	DEGREES.	SINES. Nat.	SINES. Log.	COSINES. Nat.	COSINES. Log.	TANGENTS. Nat.	TANGENTS. Log.	COTANGENTS. Nat.	COTANGENTS. Log.		
0.3142	18°00′	.3090	9.4900	.9511	9.9782	.3249	9.5118	3.0777	0.4882	72°00′	1.2566
0.3171	10	.3118	.4939	.9502	.9778	.3281	.5161	3.0475	.4839	50	1.2537
0.3200	20	.3145	.4977	.9492	.9774	.3314	.5203	3.0178	.4797	40	1.2508
0.3229	30	.3173	.5015	.9483	.9770	.3346	.5245	2.9887	.4755	30	1.2479
0.3258	40	.3201	.5052	.9474	.9765	.3378	.5287	2.9600	.4713	20	1.2450
0.3287	50	.3228	.5090	.9465	.9761	.3411	.5329	2.9319	.4671	10	1.2421
0.3316	19°00′	.3256	9.5126	.9455	9.9757	.3443	9.5370	2.9042	0.4630	71°00′	1.2392
0.3345	10	.3283	.5163	.9446	.9752	.3476	.5411	2.8770	.4589	50	1.2363
0.3374	20	.3311	.5199	.9436	.9748	.3508	.5451	2.8502	.4549	40	1.2334
0.3403	30	.3338	.5235	.9426	.9743	.3541	.5491	2.8239	.4509	30	1.2305
0.3432	40	.3365	.5270	.9417	.9739	.3574	.5531	2.7980	.4469	20	1.2275
0.3462	50	.3393	.5306	.9407	.9734	.3607	.5571	2.7725	.4429	10	1.2246
0.3491	20°00′	.3420	9.5341	.9397	9.9730	.3640	9.5611	2.7475	0.4389	70°00′	1.2217
0.3520	10	.3448	.5375	.9387	.9725	.3673	.5650	2.7228	.4350	50	1.2188
0.3549	20	.3475	.5409	.9377	.9721	.3706	.5689	2.6985	.4311	40	1.2159
0.3578	30	.3502	.5443	.9367	.9716	.3739	.5727	2.6746	.4273	30	1.2130
0.3607	40	.3529	.5477	.9356	.9711	.3772	.5766	2.6511	.4234	20	1.2101
0.3636	50	.3557	.5510	.9346	.9706	.3805	.5804	2.6279	.4196	10	1.2072
0.3665	21°00′	.3584	9.5543	.9336	9.9702	.3839	9.5842	2.6051	0.4158	69°00′	1.2043
0.3694	10	.3611	.5576	.9325	.9697	.3872	.5879	2.5826	.4121	50	1.2014
0.3723	20	.3638	.5609	.9315	.9692	.3906	.5917	2.5605	.4083	40	1.1985
0.3752	30	.3665	.5641	.9304	.9687	.3939	.5954	2.5386	.4046	30	1.1956
0.3782	40	.3692	.5673	.9293	.9682	.3973	.5991	2.5172	.4009	20	1.1926
0.3811	50	.3719	.5704	.9283	.9677	.4006	.6028	2.4960	.3972	10	1.1897
0.3840	22°00′	.3746	9.5736	.9272	9.9672	.4040	9.6064	2.4751	0.3936	68°00′	1.1868
0.3869	10	.3773	.5767	.9261	.9667	.4074	.6100	2.4545	.3900	50	1.1839
0.3898	20	.3800	.5798	.9250	.9661	.4108	.6136	2.4342	.3864	40	1.1810
0.3927	30	.3827	.5828	.9239	.9656	.4142	.6172	2.4142	.3828	30	1.1781
0.3956	40	.3854	.5859	.9228	.9651	.4176	.6208	2.3945	.3792	20	1.1752
0.3985	50	.3881	.5889	.9216	.9646	.4210	.6243	2.3750	.3757	10	1.1723
0.4014	23°00′	.3907	9.5919	.9205	9.9640	.4245	9.6279	2.3559	0.3721	67°00′	1.1694
0.4043	10	.3934	.5948	.9194	.9635	.4279	.6314	2.3369	.3686	50	1.1665
0.4072	20	.3961	.5978	.9182	.9629	.4314	.6348	2.3183	.3652	40	1.1636
0.4102	30	.3987	.6007	.9171	.9624	.4348	.6383	2.2998	.3617	30	1.1606
0.4131	40	.4014	.6036	.9159	.9618	.4383	.6417	2.2817	.3583	20	1.1577
0.4160	50	.4041	.6065	.9147	.9613	.4417	.6452	2.2637	.3548	10	1.1548
0.4189	24°00′	.4067	9.6093	.9135	9.9607	.4452	9.6486	2.2460	0.3514	66°00′	1.1519
0.4218	10	.4094	.6121	.9124	.9602	.4487	.6520	2.2286	.3480	50	1.1490
0.4247	20	.4120	.6149	.9112	.9596	.4522	.6553	2.2113	.3447	40	1.1461
0.4276	30	.4147	.6177	.9100	.9590	.4557	.6587	2.1943	.3413	30	1.1432
0.4305	40	.4173	.6205	.9088	.9584	.4592	.6620	2.1775	.3380	20	1.1403
0.4334	50	.4200	.6232	.9075	.9579	.4628	.6654	2.1609	.3346	10	1.1374
0.4363	25°00′	.4226	9.6259	.9063	9.9573	.4663	9.6687	2.1445	0.3313	65°00′	1.1345
0.4392	10	.4253	.6286	.9051	.9567	.4699	.6720	2.1283	.3280	50	1.1316
0.4422	20	.4279	.6313	.9038	.9561	.4734	.6752	2.1123	.3248	40	1.1286
0.4451	30	.4305	.6340	.9026	.9555	.4770	.6785	2.0965	.3215	30	1.1257
0.4480	40	.4331	.6366	.9013	.9549	.4806	.6817	2.0809	.3183	20	1.1228
0.4509	50	.4358	.6392	.9001	.9543	.4841	.6850	2.0655	.3150	10	1.1199
0.4538	26°00′	.4384	9.6418	.8988	9.9537	.4877	9.6882	2.0503	0.3118	64°00′	1.1170
0.4567	10	.4410	.6444	.8975	.9530	.4913	.6914	2.0353	.3086	50	1.1141
0.4596	20	.4436	.6470	.8962	.9524	.4950	.6946	2.0204	.3054	40	1.1112
0.4625	30	.4462	.6495	.8949	.9518	.4986	.6977	2.0057	.3023	30	1.1083
0.4654	40	.4488	.6521	.8936	.9512	.5022	.7009	1.9912	.2991	20	1.1054
0.4683	50	.4514	.6546	.8923	.9505	.5059	.7040	1.9768	.2960	10	1.1025
0.4712	27°00′	.4540	9.6570	.8910	9.9499	.5095	9.7072	1.9626	0.2928	63°00′	1.0996
		Nat.	Log.	Nat.	Log.	Nat.	Log.	Nat.	Log.		
		COSINES.		SINES.		COTANGENTS.		TANGENTS.		DEGREES.	RADIANS.

CIRCULAR (TRIGONOMETRIC) FUNCTIONS.

RADIANS.	DEGREES.	SINES.		COSINES.		TANGENTS.		COTANGENTS.			
		Nat.	Log.	Nat.	Log.	Nat.	Log.	Nat.	Log.		
0.4712	27°00′	.4540	9.6570	.8910	9.9499	.5095	9.7072	1.9626	0.2928	63°00′	1.0996
0.4741	10	.4566	.6595	.8897	.9492	.5132	.7103	1.9486	.2897	50	1.0966
0.4771	20	.4592	.6620	.8884	.9486	.5169	.7134	1.9347	.2866	40	1.0937
0.4800	30	.4617	.6644	.8870	.9479	.5206	.7165	1.9210	.2835	30	1.0908
0.4829	40	.4643	.6668	.8857	.9473	.5243	.7196	1.9074	.2804	20	1.0879
0.4858	50	.4669	.6692	.8843	.9466	.5280	.7226	1.8940	.2774	10	1.0850
0.4887	28°00′	.4695	9.6716	.8829	9.9459	.5317	9.7257	1.8807	0.2743	62°00′	1.0821
0.4916	10	.4720	.6740	.8816	.9453	.5354	.7287	1.8676	.2713	50	1.0792
0.4945	20	.4746	.6763	.8802	.9446	.5392	.7317	1.8546	.2683	40	1.0763
0.4974	30	.4772	.6787	.8788	.9439	.5430	.7348	1.8418	.2652	30	1.0734
0.5003	40	.4797	.6810	.8774	.9432	.5467	.7378	1.8291	.2622	20	1.0705
0.5032	50	.4823	.6833	.8760	.9425	.5505	.7408	1.8165	.2592	10	1.0676
0.5061	29°00′	.4848	9.6856	.8746	9.9418	.5543	9.7438	1.8040	0.2562	61°00′	1.0647
0.5091	10	.4874	.6878	.8732	.9411	.5581	.7467	1.7917	.2533	50	1.0617
0.5120	20	.4899	.6901	.8718	.9404	.5619	.7497	1.7796	.2503	40	1.0588
0.5149	30	.4924	.6923	.8704	.9397	.5658	.7526	1.7675	.2474	30	1.0559
0.5178	40	.4950	.6946	.8689	.9390	.5696	.7556	1.7556	.2444	20	1.0530
0.5207	50	.4975	.6968	.8675	.9383	.5735	.7585	1.7437	.2415	10	1.0501
0.5236	30°00′	.5000	9.6990	.8660	9.9375	.5774	9.7614	1.7321	0.2386	60°00′	1.0472
0.5265	10	.5025	.7012	.8646	.9368	.5812	.7644	1.7205	.2356	50	1.0443
0.5294	20	.5050	.7033	.8631	.9361	.5851	.7673	1.7090	.2327	40	1.0414
0.5323	30	.5075	.7055	.8616	.9353	.5890	.7701	1.6977	.2299	30	1.0385
0.5352	40	.5100	.7076	8601	.9346	.5930	.7730	1.6864	.2270	20	1.0356
0.5381	50	.5125	.7097	.8587	.9338	.5969	.7759	1.6753	.2241	10	1.0327
0.5411	31°00′	.5150	9.7118	.8572	9.9331	.6009	9.7788	1.6643	0.2212	59°00′	1.0297
0.5440	10	.5175	.7139	.8557	.9323	.6048	.7816	1.6534	.2184	50	1.0268
0.5469	20	.5200	.7160	.8542	.9315	.6088	.7845	1.6426	.2155	40	1.0239
0.5498	30	.5225	.7181	.8526	.9308	.6128	.7873	1.6319	.2127	30	1.0210
0.5527	40	.5250	.7201	.8511	.9300	.6168	.7902	1.6212	.2098	20	1.0181
0.5556	50	.5275	.7222	.8496	.9292	.6208	.7930	1.6107	.2070	10	1.0152
0.5585	32°00′	.5299	9.7242	.8480	9.9284	.6249	9.7958	1.6003	0.2042	58°00′	1.0123
0.5614	10	.5324	.7262	.8465	.9276	.6289	.7986	1.5900	.2014	50	1.0094
0.5643	20	.5348	.7282	.8450	.9268	.6330	.8014	1.5798	.1986	40	1.0065
0.5672	30	.5373	.7302	.8434	.9260	.6371	.8042	1.5697	.1958	30	1.0036
0.5701	40	.5398	.7322	.8418	.9252	.6412	.8070	1.5597	.1930	20	1.0007
0.5730	50	.5422	.7342	.8403	.9244	.6453	.8097	1.5497	.1903	10	0.9977
0.5760	33°00′	.5446	9.7361	.8387	9.9236	.6494	9.8125	1.5399	0.1875	57°00′	0.9948
0.5789	10	.5471	.7380	.8371	.9228	.6536	.8153	1.5301	.1847	50	0.9919
0.5818	20	.5495	.7400	.8355	.9219	.6577	.8180	1.5204	.1820	40	0.9890
0.5847	30	.5519	.7419	.8339	.9211	.6619	.8208	1.5108	.1792	30	0.9861
0.5876	40	.5544	.7438	.8323	.9203	.6661	.8235	1.5013	.1765	20	0.9832
0.5905	50	.5568	.7457	.8307	.9194	.6703	.8263	1.4919	.1737	10	0.9803
0.5934	34°00′	.5592	9.7476	.8290	9.9186	.6745	9.8290	1.4826	0.1710	56°00′	0.9774
0.5963	10	.5616	.7494	.8274	.9177	.6787	.8317	1.4733	.1683	50	0.9745
0.5992	20	.5640	.7513	.8258	.9169	.6830	.8344	1.4641	.1656	40	0.9716
0.6021	30	.5664	.7531	.8241	.9160	.6873	.8371	1.4550	.1629	30	0.9687
0.6050	40	.5688	.7550	.8225	.9151	.6916	.8398	1.4460	.1602	20	0.9657
0.6080	50	.5712	.7568	.8208	.9142	.6959	.8425	1.4370	.1575	10	0.9628
0.6109	35°00′	.5736	9.7586	.8192	9.9134	.7002	9.8452	1.4281	0.1548	55°00′	0.9599
0.6138	10	.5760	.7604	.8175	.9125	.7046	.8479	1.4193	.1521	50	0.9570
0.6167	20	.5783	.7622	.8158	.9116	.7089	.8506	1.4106	.1494	40	0.9541
0.6196	30	.5807	.7640	.8141	.9107	.7133	.8533	1.4019	.1467	30	0.9512
0.6225	40	.5831	.7657	.8124	.9098	.7177	.8559	1.3934	.1441	20	0.9483
0.6254	50	.5854	.7675	.8107	.9089	.7221	.8586	1.3848	.1414	10	0.9454
0.6283	36°00′	.5878	9.7692	8090	9.9080	.7265	9.8613	1.3764	0.1387	54°00′	0.9425
		Nat.	Log.	Nat.	Log.	Nat.	Log.	Nat.	Log.		
		COSINES.		SINES.		COTANGENTS.		TANGENTS.		DEGREES.	RADIANS.

CIRCULAR (TRIGONOMETRIC) FUNCTIONS.

RADIANS.	DEGREES.	SINES.		COSINES.		TANGENTS.		COTANGENTS.			
		Nat.	Log.	Nat.	Log.	Nat.	Log.	Nat.	Log.		
0.6283	36°00′	.5878	9.7692	.8090	9.9080	.7265	9.8613	1.3764	0.1387	54°00′	0.9425
0.6312	10	.5901	.7710	.8073	.9070	.7310	.8639	1.3680	.1361	50	0.9396
0.6341	20	.5925	.7727	.8056	.9061	.7355	.8666	1.3597	.1334	40	0.9367
0.6370	30	.5948	.7744	.8039	.9052	.7400	.8692	1.3514	.1308	30	0.9338
0.6400	40	.5972	.7761	.8021	.9042	.7445	.8718	1.3432	.1282	20	0.9308
0.6429	50	.5995	.7778	.8004	.9033	.7490	.8745	1.3351	.1255	10	0.9279
0.6458	37°00′	.6018	9.7795	.7986	9.9023	.7536	9.8771	1.3270	0.1229	53°00′	0.9250
0.6487	10	.6041	.7811	.7969	.9014	.7581	.8797	1.3190	.1203	50	0.9221
0.6516	20	.6065	.7828	.7951	.9004	.7627	.8824	1.3111	.1176	40	0.9192
0.6545	30	.6088	.7844	.7934	.8995	.7673	.8850	1.3032	.1150	30	0.9163
0.6574	40	.6111	.7861	.7916	.8985	.7720	.8876	1.2954	.1124	20	0.9134
0.6603	50	.6134	.7877	.7898	.8975	.7766	.8902	1.2876	.1098	10	0.9105
0.6632	38°00′	.6157	9.7893	.7880	9.8965	.7813	9.8928	1.2799	0.1072	52°00′	0.9076
0.6661	10	.6180	.7910	.7862	.8955	.7860	.8954	1.2723	.1046	50	0.9047
0.6690	20	.6202	.7926	.7844	.8945	.7907	.8980	1.2647	.1020	40	0.9018
0.6720	30	.6225	.7941	.7826	.8935	.7954	.9006	1.2572	.0994	30	0.8988
0.6749	40	.6248	.7957	.7808	.8925	.8002	.9032	1.2497	.0968	20	0.8959
0.6778	50	.6271	.7973	.7790	.8915	.8050	.9058	1.2423	.0942	10	0.8930
0.6807	39°00′	.6293	9.7989	.7771	9.8905	.8098	9.9084	1.2349	0.0916	51°00′	0.8901
0.6836	10	.6316	.8004	.7753	.8895	.8146	.9110	1.2276	.0890	50	0.8872
0.6865	20	.6338	.8020	.7735	.8884	.8195	.9135	1.2203	.0865	40	0.8843
0.6894	30	.6361	.8035	.7716	.8874	.8243	.9161	1.2131	.0839	30	0.8814
0.6923	40	.6383	.8050	.7698	.8864	.8292	.9187	1.2059	.0813	20	0.8785
0.6952	50	.6406	.8066	.7679	.8853	.8342	.9212	1.1988	.0788	10	0.8756
0.6981	40°00′	.6428	9.8081	.7660	9.8843	.8391	9.9238	1.1918	0.0762	50°00′	0.8727
0.7010	10	.6450	.8096	.7642	.8832	.8441	.9264	1.1847	.0736	50	0.8698
0.7039	20	.6472	.8111	.7623	.8821	.8491	.9289	1.1778	.0711	40	0.8668
0.7069	30	.6494	.8125	.7604	.8810	.8541	.9315	1.1708	.0685	30	0.8639
0.7098	40	.6517	.8140	.7585	.8800	.8591	.9341	1.1640	.0659	20	0.8610
0.7127	50	.6539	.8155	.7566	.8789	.8642	.9366	1.1571	.0634	10	0.8581
0.7156	41°00′	.6561	9.8169	.7547	9.8778	.8693	9.9392	1.1504	0.0608	49°00′	0.8552
0.7185	10	.6583	.8184	.7528	.8767	.8744	.9417	1.1436	.0583	50	0.8523
0.7214	20	.6604	.8198	.7509	.8756	.8796	.9443	1.1369	.0557	40	0.8494
0.7243	30	.6626	.8213	.7490	.8745	.8847	.9468	1.1303	.0532	30	0.8465
0.7272	40	.6648	.8227	.7470	.8733	.8899	.9494	1.1237	.0506	20	0.8436
0.7301	50	.6670	.8241	.7451	.8722	.8952	.9519	1.1171	.0481	10	0.8407
0.7330	42°00′	.6691	9.8255	.7431	9.8711	.9004	9.9544	1.1106	0.0456	48°00′	0.8378
0.7359	10	.6713	.8269	.7412	.8699	.9057	.9570	1.1041	.0430	50	0.8348
0.7389	20	.6734	.8283	.7392	.8688	.9110	.9595	1.0977	.0405	40	0.8319
0.7418	30	.6756	.8297	.7373	.8676	.9163	.9621	1.0913	.0379	30	0.8290
0.7447	40	.6777	.8311	.7353	.8665	.9217	.9646	1.0850	.0354	20	0.8261
0.7476	50	.6799	.8324	.7333	.8653	.9271	.9671	1.0786	.0329	10	0.8232
0.7505	43°00′	.6820	9.8338	.7314	9.8641	.9325	9.9697	1.0724	0.0303	47°00′	0.8203
0.7534	10	.6841	.8351	.7294	.8629	.9380	.9722	1.0661	.0278	50	0.8174
0.7563	20	.6862	.8365	.7274	.8618	.9435	.9747	1.0599	.0253	40	0.8145
0.7592	30	.6884	.8378	.7254	.8606	.9490	.9772	1.0538	.0228	30	0.8116
0.7621	40	.6905	.8391	.7234	.8594	.9545	.9798	1.0477	.0202	20	0.8087
0.7650	50	.6926	.8405	.7214	.8582	.9601	.9823	1.0416	.0177	10	0.8058
0.7679	44°00′	.6947	9.8418	.7193	9.8569	.9657	9.9848	1.0355	0.0152	46°00′	0.8029
0.7709	10	.6967	.8431	.7173	.8557	.9713	.9874	1.0295	.0126	50	0.7999
0.7738	20	.6988	.8444	.7153	.8545	.9770	.9899	1.0235	.0101	40	0.7970
0.7767	30	.7009	.8457	.7133	.8532	.9827	.9924	1.0176	.0076	30	0.7941
0.7796	40	.7030	.8469	.7112	.8520	.9884	.9949	1.0117	.0051	20	0.7912
0.7825	50	.7050	.8482	.7092	.8507	.9942	.9975	1.0058	.0025	10	0.7883
0.7854	45°00′	.7071	9.8495	.7071	9.8495	1.0000	0.0000	1.0000	0.0000	45°00′	0.7854
		Nat.	Log.	Nat	Log.	Nat.	Log.	Nat.	Log.	DEGREES.	RADIANS.
		COSINES.		SINES.		COTANGENTS.		TANGENTS.			

Table 15.

CIRCULAR (TRIGONOMETRIC) FUNCTIONS.

RADIANS.	SINES.		COSINES.		TANGENTS.		COTANGENTS.		DEGREES.
	Nat.	Log.	Nat.	Log.	Nat.	Log.	Nat.	Log.	
0.00	0.00000	— ∞	1.00000	0.00000	— ∞	— ∞	∞	∞	00°00′
.01	.01000	7.99999	0.99995	9.99998	0.01000	8.00001	99.997	1.99999	00 34
.02	.02000	8.30100	.99980	.99991	.02000	.30109	49.993	.69891	01 09
.03	.03000	.47706	.99955	.99980	.03001	.47725	33.323	.52275	01 43
.04	.03999	.60194	.99920	.99965	.04002	.60229	24.987	.39771	02 18
0.05	0.04998	8.69879	0.99875	9.99946	0.05004	8.69933	19.983	1.30067	02°52′
.06	.05996	.77789	.99820	.99922	.06007	.77867	16.647	.22133	03 26
.07	.06994	.84474	.99755	.99894	.07011	.84581	14.262	.15419	04 01
.08	.07991	.90263	.99680	.99861	.08017	.90402	12.473	.09598	04 35
.09	.08988	.95366	.99595	.99824	.09024	.95542	11.081	.04458	05 09
0.10	0.09983	8.99928	0.99500	9.99782	0.10033	9.00145	9.9666	0.99855	05°44′
.11	.10978	9.04052	.99396	.99737	.11045	.04315	9.0542	.95685	06 18
.12	.11971	.07814	.99281	.99687	.12058	.08127	8.2933	.91873	06 53
.13	.12963	.11272	.99156	.99632	.13074	.11640	7.6489	.88360	07 27
.14	.13954	.14471	.99022	.99573	.14092	.14898	7.0961	.85102	08 01
0.15	0.14944	9.17446	0.98877	9.99510	0.15114	9.17937	6.6166	0.82063	08°36′
.16	.15932	.20227	.98723	.99442	.16138	.20785	6.1966	.79215	09 10
.17	.16918	.22836	.98558	.99369	.17166	.23466	5.8256	.76534	09 44
.18	.17903	.25292	.98384	.99293	.18197	.26000	5.4954	.74000	10 19
.19	.18886	.27614	.98200	.99211	.19232	.28402	5.1997	.71598	10 53
0.20	0.19867	9.29813	0.98007	9.99126	0.20271	9.30688	4.9332	0.69312	11°28′
.21	.20846	.31902	.97803	.99035	.21314	.32867	4.6917	.67133	12 02
.22	.21823	.33891	.97590	.98940	.22362	.34951	4.4719	.65049	12 36
.23	.22798	.35789	.97367	.98841	.23414	.36948	4.2709	.63052	13 11
.24	.23770	.37603	.97134	.98737	.24472	.38866	4.0864	.61134	13 45
0.25	0.24740	9.39341	0.96891	9.98628	0.25534	9.40712	3.9163	0.59288	14°19′
.26	.25708	.41007	.96639	.98515	.26602	.42491	3.7592	.57509	14 54
.27	.26673	.42607	.96377	.98397	.27676	.44210	3.6133	.55790	15 28
.28	.27636	.44147	.96106	.98275	.28755	.45872	3.4776	.54128	16 03
.29	.28595	.45629	.95824	.98148	.29841	.47482	3.3511	.52518	16 37
0.30	0.29552	9.47059	0.95534	9.98016	0.30934	9.49043	3.2327	0.50957	17°11′
.31	.30506	.48438	.95233	.97879	.32033	.50559	3.1218	.49441	17 46
.32	.31457	.49771	.94924	.97737	.33139	.52034	3.0176	.47966	18 20
.33	.32404	.51060	.94604	.97591	.34252	.53469	2.9195	.46531	18 54
.34	.33349	.52308	.94275	.97440	.35374	.54868	2.8270	.45132	19 29
0.35	0.34290	9.53516	0.93937	9.97284	0.36503	9.56233	2.7395	0.43767	20°03′
.36	.35227	.54688	.93590	.97123	.37640	.57565	2.6567	.42435	20 38
.37	.36162	.55825	.93233	.96957	.38786	.58868	2.5782	.41132	21 12
.38	.37092	.56928	.92866	.96786	.39941	.60142	2.5037	.39858	21 46
.39	.38019	.58000	.92491	.96610	.41105	.61390	2.4328	.38610	22 21
0.40	0.38942	9.59042	0.92106	9.96429	0.42279	9.62613	2.3652	0.37387	22°55′
.41	.39861	.60055	.91712	.96243	.43463	.63812	2.3008	.36188	23 29
.42	.40776	.61041	.91309	.96051	.44657	.64989	2.2393	.35011	24 04
.43	.41687	.62000	.90897	.95855	.45862	.66145	2.1804	.33855	24 38
.44	.42594	.62935	.90475	.95653	.47078	.67282	2.1241	.32718	25 13
0.45	0.43497	9.63845	0.90045	9.95446	0.48306	9.68400	2.0702	0.31600	25°47′
.46	.44395	.64733	.89605	.95233	.49545	.69500	2.0184	.30500	26 21
.47	.45289	.65599	.89157	.95015	.50797	.70583	1.9686	.29417	26 56
.48	.46178	.66443	.88699	.94792	.52061	.71651	1.9208	.28349	27 30
.49	.47063	.67268	.88233	.94563	.53339	.72704	1.8748	.27296	28 04
0 50	0.47943	9.68072	0.87758	9.94329	0.54630	9.73743	1.8305	0.26257	28°39′

CIRCULAR (TRIGONOMETRIC) FUNCTIONS.

Radians.	Sines.		Cosines.		Tangents		Cotangents.		Degrees.
	Nat.	Log.	Nat.	Log.	Nat.	Log.	Nat.	Log.	
0.50	0.47943	9.68072	0.87758	9.94329	0.54630	9.73743	1.8305	0.26257	28°39′
.51	.48818	.68858	.87274	.94089	.55936	.74769	.7878	.25231	29 13
.52	.49688	.69625	.86782	.93843	.57256	.75782	.7465	.24218	29 48
.53	.50553	.70375	.86281	.93591	.58592	.76784	.7067	.23216	30 22
.54	.51414	.71108	.85771	.93334	.59943	.77774	.6683	.22226	30 56
0.55	0.52269	9.71824	0.85252	9.93071	0.61311	9.78754	1.6310	0.21246	31°31′
.56	.53119	.72525	.84726	.92801	.62695	.79723	.5950	.20277	32 05
.57	.53963	.73210	.84190	.92526	.64097	.80684	.5601	.19316	32 40
.58	.54802	.73880	.83646	.92245	.65517	.81635	.5263	.18365	33 14
.59	.55636	.74536	.83094	.91957	.66956	.82579	.4935	.17421	33 48
0.60	0.56464	9.75177	0.82534	9.91663	0.68414	9.83514	1.4617	0.16486	34°23′
.61	.57287	.75805	.81965	.91363	.69892	.84443	.4308	.15557	34 57
.62	.58104	.76420	.81388	.91056	.71391	.85364	.4007	.14636	35 31
.63	.58914	.77022	.80803	.90743	.72911	.86280	.3715	.13720	36 06
.64	.59720	.77612	.80210	.90423	.74454	.87189	.3431	.12811	36 40
0.65	0.60519	9.78189	0.79608	9.90096	0.76020	9.88093	1.3154	0.11907	37°15′
.66	.61312	.78754	.78999	.89762	.77610	.88992	.2885	.11008	37 49
.67	.62099	.79308	.78382	.89422	.79225	.89886	.2622	.10114	38 23
.68	.62879	.79851	.77757	.89074	.80866	.90777	.2366	.09223	38 58
.69	.63654	.80382	.77125	.88719	.82534	.91663	.2116	.08337	39 32
0.70	0.64422	9.80903	0.76484	9.88357	0.84229	9.92546	1.1872	0.07454	40°06′
.71	.65183	.81414	.75836	.87988	.85953	.93426	.1634	.06574	40 41
.72	.65938	.81914	.75181	.87611	.87707	.94303	.1402	.05697	41 15
.73	.66687	.82404	.74517	.87226	.89492	.95178	.1174	.04822	41 50
.74	.67429	.82885	.73847	.86833	.91309	.96051	.0952	.03949	42 24
0.75	0.68164	9.83355	0.73169	9.86433	0.93160	9.96923	1.0734	0.03077	42°58′
.76	.68892	.83817	.72484	.86024	.95045	.97793	.0521	.02207	43 33
.77	.69614	.84269	.71791	.85607	.96967	.98662	.0313	.01338	44 07
.78	.70328	.84713	.71091	.85182	.98926	9.99531	1.0109	.00469	44 41
.79	.71035	.85147	.70385	.84748	1.0092	0.00400	0.99084	9.99600	45 16
0.80	0.71736	9.85573	0.69671	9.84305	1.0296	0.01268	0.97121	9.98732	45°50′
.81	.72429	.85991	.68950	.83853	.0505	.02138	.95197	.97862	46 25
.82	.73115	.86400	.68222	.83393	.0717	.03008	.93309	.96992	46 59
.83	.73793	.86802	.67488	.82922	.0934	.03879	.91455	.96121	47 33
.84	.74464	.87195	.66746	.82443	.1156	.04752	.89635	.95248	48 08
0.85	0.75128	9.87580	0.65998	9.81953	1.1383	0.05627	0.87848	9.94373	48°42′
.86	.75784	.87958	.65244	.81454	.1616	.06504	.86091	.93496	49 16
.87	.76433	.88328	.64483	.80944	.1853	.07384	.84365	.92616	49 51
.88	.77074	.88691	.63715	.80424	.2097	.08266	.82668	.91734	50 25
.89	.77707	.89046	.62941	.79894	.2346	.09153	.80998	.90847	51 00
0.90	0.78333	9.89394	0.62161	9.79352	1.2602	0.10043	0.79355	9.89957	51°34′
.91	.78950	.89735	.61375	.78799	.2864	.10937	.77738	.89063	52 08
.92	.79560	.90070	.60582	.78234	.3133	.11835	.76146	.88165	52 43
.93	.80162	.90397	.59783	.77658	.3409	.12739	.74578	.87261	53 17
.94	.80756	.90717	.58979	.77070	.3692	.13648	.73034	.86352	53 51
0.95	0.81342	9.91031	0.58168	9.76469	1.3984	0.14563	0.71511	9.85437	54°26′
.96	.81919	.91339	.57352	.75855	.4284	.15484	.70010	.84516	55 00
.97	.82489	.91639	.56530	.75228	.4592	.16412	.68531	.83588	55 35
.98	.83050	.91934	.55702	.74587	.4910	.17347	.67071	.82653	56 09
.99	.83603	.92222	.54869	.73933	.5237	.18289	.65631	.81711	56 43
1.00	0.84147	9.92504	0.54030	9.73264	1.5574	0.19240	0.64209	9.80760	57°18′

CIRCULAR (TRIGONOMETRIC) FUNCTIONS.

RADIANS.	SINES.		COSINES.		TANGENTS.		COTANGENTS.		DEGREES.
	Nat.	Log.	Nat.	Log.	Nat.	Log.	Nat.	Log.	
1.00	0.84147	9.92504	0.54030	9.73264	1.5574	0.19240	0.64209	9.80760	57°18′
.01	.84683	.92780	.53186	.72580	.5922	.20200	.62806	.79800	57 52
.02	.85211	.93049	.52337	.71881	.6281	.21169	.61420	.78831	58 27
.03	.85730	.93313	.51482	.71165	.6652	.22148	.60051	.77852	59 01
.04	.86240	.93571	.50622	.70434	.7036	.23137	.58699	.76863	59 35
1.05	0.86742	9.93823	0.49757	9.69686	1.7433	0.24138	0.57362	9.75862	60°10′
.06	.87236	.94069	.48887	.68920	.7844	.25150	.56040	.74850	60 44
.07	.87720	.94310	.48012	.68135	.8270	.26175	.54734	.73825	61 18
.08	.88196	.94545	.47133	.67332	.8712	.27212	.53441	.72788	61 53
.09	.88663	.94774	.46249	.66510	.9171	.28264	.52162	.71736	62 27
1.10	0.89121	9.94998	0.45360	9.65667	1.9648	0.29331	0.50897	9.70669	63°02′
.11	.89570	.95216	.44466	.64803	2.0143	.30413	.49644	.69587	63 36
.12	.90010	.95429	.43568	.63917	.0660	.31512	.48404	.68488	64 10
.13	.90441	.95637	.42666	.63008	.1198	.32628	.47175	.67372	64 45
.14	.90863	.95839	.41759	.62075	.1759	.33763	.45959	.66237	65 19
1.15	0.91276	9.96036	0.40849	9.61118	2.2345	0.34918	0.44753	9.65082	65°53′
.16	.91680	.96228	.39934	.60134	.2958	.36093	.43558	.63907	66 28
.17	.92075	.96414	.39015	.59123	.3600	.37291	.42373	.62709	67 02
.18	.92461	.96596	.38092	.58084	.4273	.38512	.41199	.61488	67 37
.19	.92837	.96772	.37166	.57015	.4979	.39757	.40034	.60243	68 11
1.20	0.93204	9.96943	0.36236	9.55914	2.5722	0.41030	0.38878	9.58970	68°45′
.21	.93562	.97110	.35302	.54780	.6503	.42330	.37731	.57670	69 20
.22	.93910	.97271	.34365	.53611	.7328	.43660	.36593	.56340	69 54
.23	.94249	.97428	.33424	.52406	.8198	.45022	.35463	.54978	70 28
.24	.94578	.97579	.32480	.51161	.9119	.46418	.34341	.53582	71 03
1.25	0.94898	9.97726	0.31532	9.49875	3.0096	0.47850	0.33227	9.52150	71°37′
.26	.95209	.97868	.30582	.48546	.1133	.49322	.32121	.50678	72 12
.27	.95510	.98005	.29628	.47170	.2236	.50835	.31021	.49165	72 46
.28	.95802	.98137	.28672	.45745	.3413	.52392	.29928	.47608	73 20
.29	.96084	.98265	.27712	.44267	.4672	.53998	.28842	.46002	73 55
1.30	0.96356	9.98388	0.26750	9.42732	3.6021	0.55656	0.27762	9.44344	74°29′
.31	.96618	.98506	.25785	.41137	.7471	.57369	.26687	.42631	75 03
.32	.96872	.98620	.24818	.39476	.9033	.59144	.25619	.40856	75 38
.33	.97115	.98729	.23848	.37744	4.0723	.60984	.24556	.39016	76 12
.34	.97348	.98833	.22875	.35937	.2556	.62896	.23498	.37104	76 47
1.35	0.97572	9.98933	0.21901	9.34046	4.4552	0.64887	0.22446	9.35113	77°21′
.36	.97786	.99028	.20924	.32064	.6734	.66964	.21398	.33036	77 55
.37	.97991	.99119	.19945	.29983	.9131	.69135	.20354	.30865	78 30
.38	.98185	.99205	.18964	.27793	5.1774	.71411	.19315	.28589	79 04
.39	.98370	.99286	.17981	.25482	.4707	.73804	.18279	.26196	79 38
1.40	0.98545	9.99363	0.16997	9.23036	5.7979	0.76327	0.17248	9.23673	80°13′
.41	.98710	.99436	.16010	.20440	6.1654	.78996	.16220	.21004	80 47
.42	.98865	.99504	.15023	.17674	6.5811	.81830	.15195	.18170	81 22
.43	.99010	.99568	.14033	.14716	7.0555	.84853	.14173	.15147	81 56
.44	.99146	.99627	.13042	.11536	7.6018	.88092	.13155	.11908	82 30
1.45	0.99271	9.99682	0.12050	9.08100	8.2381	0.91583	0.12139	9.08417	83°05′
.46	.99387	.99733	.11057	.04364	8.9886	.95369	.11125	.04631	83 39
.47	.99492	.99779	.10063	.00271	9.8874	.99508	.10114	.00492	84 13
.48	.99588	.99821	.09067	8.95747	10.983	1.04074	.09105	8.95926	84 48
.49	.99674	.99858	.08071	.90692	12.350	.09166	.08097	.90834	85 22
1.50	0.99749	9.99891	0.07074	8.84965	14.101	1.14926	0.07091	8.85074	85°57′

CIRCULAR FUNCTIONS AND FACTORIALS.

TABLE 15 (*continued*). — Circular (Trigonometric) Functions.

RADIANS.	SINES.		COSINES.		TANGENTS.		COTANGENTS.		DEGREES.
	Nat.	Log	Nat.	Log	Nat.	Log.	Nat.	Log.	
1.50	0.99749	9.99891	0.07074	8.84965	14.101	1.14926	0.07091	8.85074	85°57′
.51	.99815	.99920	.06076	.78361	16.428	.21559	.06087	.78441	86 31
.52	.99871	.99944	.05077	.70565	19.670	.29379	.05084	.70621	87 05
.53	.99917	.99964	.04079	.61050	24.498	.38914	.04082	.61086	87 40
.54	.99953	.99979	.03079	.48843	32.461	.51136	.03081	.48864	88 14
1.55	0.99978	9.99991	0.02079	8.31796	48.078	1.68195	0.02080	8.31805	88°49′
.56	0.99994	9.99997	.01080	8.03327	92.621	1.96671	.01080	8.03329	89 23
.57	1.00000	0.00000	.00080	6.90109	1255.8	3.09891	.00080	6.90109	89 57
.58	0.99996	9.99998	−.00920	7.96396n	108.65	2.03603	−.00920	7.96397n	90 32
.59	0.99982	9.99992	−.01920	8.28336n	52.067	1.71656	−.01921	8.28344n	91 06
1.60	0.99957	9.99981	−0.02920	8.46538n	34.233	1.53444	−0.02921	8.46556n	91°40′

90° = 1.570 7963 radians.

TABLE 16.—Logarithmic Factorials.

Logarithms of the products 1.2.3. ......*n*, *n* from 1 to 100.

See Table 18 for Factorials 1 to 20.

See Table 32 for log. Γ (*n*+1), values of *n* between 1 and 2.

n.	log (*n*!)	*n*.	log (*n*!)	*n*.	log (*n*!)	*n*.	log (*n*!)
1	0.000000	**26**	26.605619	**51**	66.190645	**76**	111.275425
2	0.301030	27	28.036983	52	67.906648	77	113.161916
3	0.778151	28	29.484141	53	69.630924	78	115.054011
4	1.380211	29	30.946539	54	71.363318	79	116.951638
5	2.079181	30	32.423660	55	73.103681	80	118.854728
6	2.857332	**31**	33.915022	**56**	74.851869	**81**	120.763213
7	3.702431	32	35.420172	57	76.607744	82	122.677027
8	4.605521	33	36.938686	58	78.371172	83	124.596105
9	5.559763	34	38.470165	59	80.142024	84	126.520384
10	6.559763	35	40.014233	60	81.920175	85	128.449803
11	7.601156	**36**	41.570535	**61**	83.705505	**86**	130.384301
12	8.680337	37	43.138737	62	85.497896	87	132.323821
13	9.794280	38	44.718520	63	87.297237	88	134.268303
14	10.940408	39	46.309585	64	89.103417	89	136.217693
15	12.116500	40	47.911645	65	90.916330	90	138.171936
16	13.320620	**41**	49.524429	**66**	92.735874	**91**	140.130977
17	14.551069	42	51.147678	67	94.561949	92	142.094765
18	15.806341	43	52.781147	68	96.394458	93	144.063248
19	17.085095	44	54.424599	69	98.233307	94	146.036376
20	18.386125	45	56.077812	70	100.078405	95	148.014099
21	19.708344	**46**	57.740570	**71**	101.929663	**96**	149.996371
22	21.050767	47	59.412668	72	103.786996	97	151.983142
23	22.412494	48	61.093909	73	105.650319	98	153.974368
24	23.792706	49	62.784105	74	107.519550	99	155.970004
25	25.190646	50	64.483075	75	109.394612	100	157.970004

TABLE 17.

HYPERBOLIC FUNCTIONS.

u	sinh. u Nat.	sinh. u Log.	cosh. u Nat.	cosh. u Log.	tanh. u Nat.	tanh. u Log.	coth. u Nat.	coth. u Log.	gd u
0.00	0.00000	— ∞	1.00000	0.00000	0.00000	— ∞	∞	∞	00°00′
.01	.01000	8.00001	.00005	.00002	.01000	7.99999	100.003	2.00001	0 34
.02	.02000	.30106	.00020	.00009	.02000	8.30097	50.007	1.69903	1 09
.03	.03000	.47719	.00045	.00020	.02999	.47699	33.343	1.52301	1 43
.04	.04001	.60218	.00080	.00035	.03998	.60183	25.013	1.39817	2 17
0.05	0.05002	8.69915	1.00125	0.00054	0.04996	8.69861	20.017	1.30139	2 52
.06	.06004	.77841	.00180	.00078	.05993	.77763	16.687	.22237	3 26
.07	.07006	.84545	.00245	.00106	.06989	.84439	14.309	.15561	4 00
.08	.08009	.90355	.00320	.00139	.07983	.90216	12.527	.09784	4 35
.09	.09012	.95483	.00405	.00176	.08976	.95307	11.141	.04693	5 09
0.10	0.10017	9.00072	1.00500	0.00217	0.09967	8.99856	10.0333	1.00144	5 43
.11	.11022	.04227	.00606	.00262	.10956	9.03965	9.1275	0.96035	6 17
.12	.12029	.08022	.00721	.00312	.11943	.07710	8.3733	.92290	6 52
.13	.13037	.11517	.00846	.00366	.12927	.11151	7.7356	.88849	7 26
.14	.14046	.14755	.00982	.00424	.13909	.14330	7.1895	.85670	8 00
0.15	0.15056	9.17772	1.01127	0.00487	0.14889	9.17285	6.7166	0.82715	8 34
.16	.16068	.20597	.01283	.00554	.15865	.20044	6.3032	.79956	9 08
.17	.17082	.23254	.01448	.00625	.16838	.22629	5.9389	.77371	9 42
.18	.18097	.25762	.01624	.00700	.17808	.25062	5.6154	.74938	10 15
.19	.19115	.28136	.01810	.00779	.18775	.27357	5.3263	.72643	10 49
0.20	0.20134	9.30392	1.02007	0.00863	0.19738	9.29529	5.0665	0.70471	11 23
.21	.21155	.32541	.02213	.00951	.20697	.31590	4.8317	.68410	11 57
.22	.22178	.34592	.02430	.01043	.21652	.33549	4.6186	.66451	12 30
.23	.23203	.36555	.02657	.01139	.22603	.35416	4.4242	.64584	13 04
.24	.24231	.38437	.02894	.01239	.23550	.37198	4.2464	.62802	13 37
0.25	0.25261	9.40245	1.03141	0.01343	0.24492	9.38902	4.0830	0.61098	14 11
.26	.26294	.41986	.03399	.01452	.25430	.40534	3.9324	.59466	14 44
.27	.27329	.43663	.03667	.01564	.26362	.42099	3.7933	.57901	15 17
.28	.28367	.45282	.03946	.01681	.27291	.43601	3.6643	.56399	15 50
.29	.29408	.46847	.04235	.01801	.28213	.45046	3.5444	.54954	16 23
0.30	0.30452	9.48362	1.04534	0.01926	0.29131	9.46436	3.4327	0.53564	16 56
.31	.31499	.49830	.04844	.02054	.30044	.47775	.3285	.52225	17 29
.32	.32549	.51254	.05164	.02187	.30951	.49067	.2309	.50933	18 02
.33	.33602	.52637	.05495	.02323	.31852	.50314	.1395	.49686	18 34
.34	.34659	.53981	.05836	.02463	.32748	.51518	.0536	.48482	19 07
0.35	0.35719	9.55290	1.06188	0.02607	0.33638	9.52682	2.9729	0.47318	19 39
.36	.36783	.56564	.06550	.02755	.34521	.53809	.8968	.46191	20 12
.37	.37850	.57807	.06923	.02907	.35399	.54899	.8249	.45101	20 44
.38	.38921	.59019	.07307	.03063	.36271	.55956	.7570	.44044	21 16
.39	.39996	.60202	.07702	.03222	.37136	.56980	.6928	.43020	21 48
0.40	0.41075	9.61358	1.08107	0.03385	0.37995	9.57973	2.6319	0.42027	22 20
.41	.42158	.62488	.08523	.03552	.38847	.58936	.5742	.41064	22 52
.42	.43246	.63594	.08950	.03723	.39693	.59871	.5193	.40129	23 23
.43	.44337	.64677	.09388	.03897	.40532	.60780	.4672	.39220	23 55
.44	.45434	.65738	.09837	.04075	.41364	.61663	.4175	.38337	24 26
0.45	0.46534	9.66777	1.102970	.04256	0.42190	9.62521	2.3702	0.37479	24 57
.46	.47640	.67797	.10768	.04441	.43008	.63355	.3251	.36645	25 28
.47	.48750	.68797	.11250	.04630	.43820	.64167	.2821	.35833	25 59
.48	.49865	.69779	.11743	.04822	.44624	.64957	.2409	.35043	26 30
.49	.50984	.70744	.12247	.05018	.45422	.65726	.2016	.34274	27 01
0.50	0.52110	9.71692	1.12763	0.05217	0.46212	9.66475	2.1640	0.33525	27 31

TABLE 17 (*continued*).

HYBERBOLIC FUNCTIONS.

u	sinh. u Nat.	sinh. u Log.	cosh. u Nat.	cosh. u Log.	tanh. u Nat.	tanh. u Log.	coth. u Nat.	coth. u Log.	gd u
0.50	0.52110	9.71692	1.12763	0.05217	0.46212	9.66475	2.1640	0.33525	27°31′
.51	.53240	.72624	.13289	.05419	.46995	.67205	.1279	.32795	28 02
.52	.54375	.73540	.13827	.05625	.47770	.67916	.0934	.32084	28 32
.53	.55516	.74442	.14377	.05834	.48538	.68608	.0602	.31392	29 02
.54	.56663	.75330	.14938	.06046	.49299	.69284	.0284	.30716	29 32
0.55	0.57815	9.76204	1.15510	0.06262	0.50052	9.69942	1.9979	0.30058	30 02
.56	.58973	.77065	.16094	.06481	.50798	.70584	.9686	.29416	30 32
.57	.60137	.77914	.16690	.06703	.51536	.71211	.9404	.28789	31 01
.58	.61307	.78751	.17297	.06929	.52267	.71822	.9133	.28178	31 31
.59	.62483	.79576	.17916	.07157	.52990	.72419	.8872	.27581	32 00
0.60	0.63665	9.80390	1.18547	0.07389	0.53705	9.73001	1.8620	0.26999	32 29
.61	.64854	.81194	.19189	.07624	.54413	.73570	.8378	.26430	32 58
.62	.66049	.81987	.19844	.07861	.55113	.74125	.8145	.25875	33 27
63	.67251	.82770	.20510	.08102	.55805	.74667	.7919	.25333	33 55
.64	.68459	.83543	.21189	.08346	.56490	.75197	.7702	.24803	34 24
0.65	0.69675	9.84308	1.21879	0.08593	0.57167	9.75715	1.7493	0.24285	34 52
.66	.70897	.85063	.22582	.08843	.57836	.76220	.7290	.23780	35 20
.67	.72126	.85809	.23297	.09095	.58498	.76714	.7095	.23286	35 48
.68	.73363	.86548	.24025	.09351	.59152	.77197	.6906	.22803	36 16
.69	.74607	.87278	.24765	.09609	.59798	.77669	.6723	.22331	36 44
0.70	0.75858	9.88000	1.25517	0.09870	0.60437	9.78130	1.6546	0.21870	37 11
.71	.77117	.88715	.26282	.10134	.61068	.78581	.6375	.21419	37 38
.72	.78384	.89423	.27059	.10401	.61691	.79022	.6210	.20978	38 05
.73	.79659	.90123	.27849	.10670	.62307	.79453	.6050	.20547	38 32
.74	.80941	.90817	.28652	.10942	.62915	.79875	.5895	.20125	38 59
0.75	0.82232	9.91504	1.29468	0.11216	0.63515	9.80288	1.5744	0.19712	39 26
.76	.83530	.92185	.30297	.11493	.64108	.80691	.5599	.19309	39 52
.77	.84838	.92859	.31139	.11773	.64693	.81086	.5458	.18914	40 19
.78	.86153	.93527	.31994	.12055	.65271	.81472	.5321	.18528	40 45
.79	.87478	.94190	.32862	.12340	.65841	.81850	.5188	.18150	41 11
0.80	0.88811	9.94846	1.33743	0.12627	0.66404	9.82219	1.5059	0.17781	41 37
.81	.90152	.95498	.34638	.12917	.66959	.82581	.4935	.17419	42 02
.82	.91503	.96144	.35547	.13209	.67507	.82935	.4813	.17065	42 28
.83	.92863	.96784	.36468	.13503	.68048	.83281	.4696	.16719	42 53
.84	.94233	.97420	.37404	.13800	.68581	.83620	.4581	.16380	43 18
0.85	0.95612	9.98051	1.38353	0.14099	0.69107	9.83952	1.4470	0.16048	43 43
.86	.97000	.98677	.39316	.14400	.69626	.84277	.4362	.15723	44 08
.87	.98398	.99299	.40293	.14704	.70137	.84595	.4258	.15405	44 32
.88	.99806	.99916	.41284	.15009	.70642	.84906	.4156	.15094	44 57
.89	1.01224	0.00528	.42289	.15317	.71139	.85211	.4057	.14789	45 21
0.90	1.02652	0.01137	1.43309	0.15627	0.71630	9.85509	1.3961	0.14491	45 45
.91	.04090	.01741	.44342	.15939	.72113	.85801	.3867	.14199	46 09
.92	.05539	.02341	.45390	.16254	.72590	.86088	.3776	.13912	46 33
.93	.06998	.02937	.46453	.16570	.73059	.86368	.3687	.13632	46 56
.94	.08468	.03530	.47530	.16888	.73522	.86642	.3601	.13358	47 20
0.95	1.09948	0.04119	1.48623	0.17208	0.73978	9.86910	1.3517	0.13090	47 43
.96	.11440	.04704	.49729	.17531	.74428	.87173	.3436	.12827	48 06
.97	.12943	.05286	.50851	.17855	.74870	.87431	.3356	.12569	48 29
.98	.14457	.05864	.51988	.18181	.75307	.87683	.3279	.12317	48 51
.99	.15983	.06439	.53141	.18509	.75736	.87930	.3204	.12070	49 14
1.00	1.17520	0.07011	1.54308	0.18839	0.76159	9.88172	1.3130	0.11828	49 36

TABLE 17 (*continued*).
HYPERBOLIC FUNCTIONS.

u	sinh. u Nat.	sinh. u Log.	cosh. u Nat.	cosh. u Log.	tanh. u Nat.	tanh. u Log.	coth u Nat.	coth u Log.	gd u
1.00	1.17520	0.07011	1.54308	0.18839	0.76159	9.88172	1.3130	0.11828	49°36′
.01	.19069	.07580	.55491	.19171	.76576	.88409	.3059	.11591	49 58
.02	.20630	.08146	.56689	.19504	.76987	.88642	.2989	.11358	50 21
.03	.22203	.08708	.57904	.19839	.77391	.88869	.2921	.11131	50 42
.04	.23788	.09268	.59134	.20176	.77789	.89092	.2855	.10908	51 04
1.05	1.25386	0.09825	1.60379	0.20515	0.78181	9.89310	1.2791	0.10690	51 26
.06	.26996	.10379	.61641	.20855	.78566	.89524	.2728	.10476	51 47
.07	.28619	.10930	.62919	.21197	.78946	.89733	.2667	.10267	52 08
.08	.30254	.11479	.64214	.21541	.79320	.89938	.2607	.10062	52 29
.09	.31903	.12025	.65525	.21886	.79688	.90139	.2549	.09861	52 50
1.10	1.33565	0.12569	1.66852	0.22233	0.80050	9.90336	1.2492	0.09664	53 11
.11	.35240	.13111	.68196	.22582	.80406	.90529	.2437	.09471	53 31
.12	.36929	.13649	.69557	.22931	.80757	.90718	.2383	.09282	53 52
.13	.38631	.14186	.70934	.23283	.81102	.90903	.2330	.09097	54 12
.14	.40347	.14720	.72329	.23636	.81441	.91085	.2279	.08915	54 32
1.15	1.42078	0.15253	1.73741	0.23990	0.81775	9.91262	1.2229	0.08738	54 52
.16	.43822	.15783	.75171	.24346	.82104	.91436	.2180	.08564	55 11
.17	.45581	.16311	.76618	.24703	.82427	.91607	.2132	.08393	55 31
.18	.47355	.16836	.78083	.25062	.82745	.91774	.2085	.08226	55 50
.19	.49143	.17360	.79565	.25422	.83058	.91938	.2040	.08062	56 09
1.20	1.50946	0.17882	1.81066	0.25784	0.83365	9.92099	1.1995	0.07901	56 29
.21	.52764	.18402	.82584	.26146	.83668	.92256	.1952	.07744	56 47
.22	.54598	.18920	.84121	.26510	.83965	.92410	.1910	.07590	57 06
.23	.56447	.19437	.85676	.26876	.84258	.92561	.1868	.07439	57 25
.24	.58311	.19951	.87250	.27242	.84546	.92709	.1828	.07291	57 43
1.25	1.60192	0.20464	1.88842	0.27610	0.84828	9.92854	1.1789	0.07146	58 02
.26	.62088	.20975	.90454	.27979	.85106	.92996	.1750	.07004	58 20
.27	.64001	.21485	.92084	.28349	.85380	.93135	.1712	.06865	58 38
.28	.65930	.21993	.93734	.28721	.85648	.93272	.1676	.06728	58 55
.29	.67876	.22499	.95403	.29093	.85913	.93406	.1640	.06594	59 13
1.30	1.69838	0.23004	1.97091	0.29467	0.86172	9.93537	1.1605	0.06463	59 31
.31	.71818	.23507	.98800	.29842	.86428	.93665	.1570	.06335	59 48
.32	.73814	.24009	2.00528	.30217	.86678	.93791	.1537	.06209	60 05
.33	.75828	.24509	.02276	.30594	.86925	.93914	.1504	.06086	60 22
.34	.77860	.25008	.04044	.30972	.87167	.94035	.1472	.05965	60 39
1.35	1.79909	0.25505	2.05833	0.31352	0.87405	9.94154	1.1441	0.05846	60 56
.36	.81977	.26002	.07643	.31732	.87639	.94270	.1410	.05730	61 13
.37	.84062	.26496	.09473	.32113	.87869	.94384	.1381	.05616	61 29
.38	.86166	.26990	.11324	.32495	.88095	.94495	.1351	.05505	61 45
.39	.88289	.27482	.13196	.32878	.88317	.94604	.1323	.05396	62 02
1.40	1.90430	0.27974	2.15090	0.33262	0.88535	9.94712	1.1295	0.05288	62 18
.41	.92591	.28464	.17005	.33647	.88749	.94817	.1268	.05183	62 34
.42	.94770	.28952	.18942	.34033	.88960	.94919	.1241	.05081	62 49
.43	.96970	.29440	.20900	.34420	.89167	.95020	.1215	.04980	63 05
.44	.99188	.29926	.22881	.34807	.89370	.95119	.1189	.04881	63 20
1.45	2.01427	0.30412	2.24884	0.35196	0.89569	9.95216	1.1165	0.04784	63 36
.46	.03686	.30896	.26910	.35585	.89765	.95311	.1140	.04689	63 51
.47	.05965	.31379	.28958	.35976	.89958	.95404	.1116	.04596	64 06
.48	.08265	.31862	.31029	.36367	.90147	.95495	.1093	.04505	64 21
.49	.10586	.32343	.33123	.36759	.90332	.95584	.1070	.04416	64 36
1.50	2.12928	0.32823	2.35241	0.37151	0.90515	9.95672	1.1048	0.04328	64 51

Table 17 (*continued*).

HYPERBOLIC FUNCTIONS.

u	sinh. u		cosh. u		tanh. u		coth. u		gd. u
	Nat.	Log.	Nat.	Log.	Nat.	Log.	Nat.	Log.	
1.50	2.12928	0.32823	2.35241	0.37151	0.90515	9.95672	1.1048	0.04328	64° 51′
.51	.15291	.33303	.37382	.37545	.90694	.95758	.1026	.04242	65 05
.52	.17676	.33781	.39547	.37939	.90870	.95842	.1005	.04158	65 20
.53	.20082	.34258	.41736	.38334	.91042	.95924	.0984	.04076	65 34
.54	.22510	.34735	.43949	.38730	.91212	.96005	.0963	.03995	65 48
1.55	2.24961	0.35211	2.46186	0.39126	0.91379	9.96084	1.0943	0.03916	66 02
.56	.27434	.35686	.48448	.39524	.91542	.96162	.0924	.03838	66 16
.57	.29930	.36160	.50735	.39921	.91703	.96238	.0905	.03762	66 30
.58	.32449	.36633	.53047	.40320	.91860	.96313	.0886	.03687	66 43
.59	.34991	.37105	.55384	.40719	.92015	.96386	.0868	.03614	66 57
1.60	2.37557	0.37577	2.57746	0.41119	0.92167	9.96457	1.0850	0.03543	67 10
.61	40146	.38048	.60135	.41520	.92316	.96528	.0832	.03472	67 24
.62	.42760	.38518	.62549	.41921	.92462	.96597	.0815	.03403	67 37
.63	.45397	.38987	.64990	.42323	.92606	.96664	.0798	.03336	67 50
.64	.48059	.39456	.67457	.42725	.92747	.96730	.0782	.03270	68 03
1.65	2.50746	0.39923	2.69951	0.43129	0.92886	9.96795	1.0766	0.03205	68 15
.66	.53459	.40391	.72472	.43532	.93022	.96858	.0750	.03142	68 28
.67	.56196	.40857	.75021	.43937	.93155	.96921	.0735	.03079	68 41
.68	.58959	.41323	.77596	.44341	.93286	.96982	.0720	.03018	68 53
.69	.61748	.41788	.80200	.44747	.93415	.97042	.0705	.02958	69 05
1.70	2.64563	0.42253	2.82832	0.45153	0.93541	9.97100	1.0691	0.02900	69 18
.71	.67405	.42717	.85491	.45559	.93665	.97158	.0676	.02842	69 30
.72	.70273	.43180	.88180	.45966	.93786	.97214	.0663	.02786	69 42
.73	.73168	.43643	.90897	.46374	.93906	.97269	.0649	.02731	69 54
.74	.76091	.44105	.93643	.46782	.94023	.97323	.0636	.02677	70 05
1.75	2.79041	0.44567	2.96419	0.47191	0.94138	9.97376	1.0623	0.02624	70 17
.76	.82020	.45028	.99224	.47600	.94250	.97428	.0610	.02572	70 29
.77	.85026	.45488	3.02059	.48009	.94361	.97479	.0598	.02521	70 40
.78	.88061	.45948	.04925	.48419	.94470	.97529	.0585	.02471	70 51
.79	.91125	.46408	.07821	.48830	.94576	.97578	.0574	.02422	71 03
1.80	2.94217	0.46867	3.10747	0.49241	0.94681	9.97626	1.0562	0.02374	71 14
.81	.97340	.47325	.13705	.49652	.94783	.97673	.0550	.02327	71 25
.82	3.00492	.47783	.16694	.50064	.94884	.97719	.0539	.02281	71 36
.83	.03674	.48241	.19715	.50476	.94983	.97764	.0528	.02236	71 46
.84	.06886	.48698	.22768	.50889	.95080	.97809	.0518	.02191	71 57
1.85	3.10129	0.49154	3.25853	0.51302	0.95175	9.97852	1.0507	0.02148	72 08
.86	.13403	.49610	.28970	.51716	.95268	.97895	.0497	.02105	72 18
.87	.16709	.50066	.32121	.52130	.95359	.97936	.0487	.02064	72 29
.88	.20046	.50521	.35305	.52544	.95449	.97977	.0477	.02023	72 39
.89	.23415	.50976	.38522	.52959	.95537	.98017	.0467	.01983	72 49
1.90	3.26816	0.51430	3.41773	0.53374	0.95624	9.98057	1.0458	0.01943	72 59
.91	.30250	.51884	.45058	.53789	.95709	.98095	.0448	.01905	73 09
.92	.33718	.52338	.48378	.54205	.95792	.98133	.0439	.01867	73 19
.93	.37218	.52791	.51733	.54621	.95873	.98170	.0430	.01830	73 29
.94	.40752	.53244	.55123	.55038	.95953	.98206	.0422	.01794	73 39
1.95	3.44321	0.53696	3.58548	0.55455	0.96032	9.98242	1.0413	0.01758	73 48
.96	.47923	.54148	.62009	.55872	.96109	.98276	.0405	.01724	73 58
.97	.51561	.54600	.65507	.56290	.96185	.98311	.0397	.01689	74 07
.98	.55234	.55051	.69041	.56707	.96259	.98344	.0389	.01656	74 17
.99	.58942	.55502	.72611	.57126	.96331	.98377	.0381	.01623	74 26
2.00	3.62686	0.55953	3.76220	0.57544	0.96403	9.98409	1.0373	0.01591	74 35

TABLE 17 (*continued*).

HYPERBOLIC FUNCTIONS.

u	sinh. u		cosh. u		tanh. u		coth. u.		gd. u
	Nat.	Log.	Nat.	Log.	Nat.	Log.	Nat.	Log.	
2.00	3.62686	0.55953	3.76220	0.57544	0.96403	9.98409	1.0373	0.01591	74°35′
.01	.66466	.56403	.79865	.57963	.96473	.98440	.0366	.01560	74 44
.02	.70283	.56853	.83549	.58382	.96541	.98471	.0358	.01529	74 53
.03	.74138	.57303	.87271	.58802	.96609	.98502	.0351	.01498	75 02
.04	.78029	.57753	.91032	.59221	.96675	.98531	.0344	.01469	75 11
2.05	3.81958	0.58202	3.94832	0.59641	0.96740	9.98560	1.0337	0.01440	75 20
.06	.85926	.58650	.98671	.60061	.96803	.98589	.0330	.01411	75 28
.07	.89932	.59099	4.02550	.60482	.96865	.98617	.0324	.01383	75 37
.08	.93977	.59547	.06470	.60903	.96926	.98644	.0317	.01356	75 45
.09	.98061	.59995	.10430	.61324	.96986	.98671	.0311	.01329	75 54
2.10	4.02186	0.60443	4.14431	0.61745	0.97045	9.98697	1.0304	0.01303	76 02
.11	.06350	.60890	.18474	.62167	.97103	.98723	.0298	.01277	76 10
.12	.10555	.61337	.22558	.62589	.97159	.98748	.0292	.01252	76 19
.13	.14801	.61784	.26685	.63011	.97215	.98773	.0286	.01227	76 27
.14	.19089	.62231	.30855	.63433	.97269	.98798	.0281	.01202	76 35
2.15	4.23419	0.62677	4.35067	0.63856	0.97323	9.98821	1.0275	0.01179	76 43
.16	.27791	.63123	.39323	.64278	.97375	.98845	.0270	.01155	76 51
	.32205	.63569	.43623	.64701	.97426	.98868	.0264	.01132	76 58
.18	.36663	.64015	.47967	.65125	.97477	.98890	.0259	.01110	77 06
.19	.41165	.64460	.52356	.65548	.97526	.98912	.0254	.01088	77 14
2.20	4.45711	0.64905	4.56791	0.65972	0.97574	9.98934	1.0249	0.01066	77 21
.21	.50301	.65350	.61271	.66396	.97622	.98955	.0244	.01045	77 29
.22	.54936	.65795	.65797	.66820	.97668	.98975	.0239	.01025	77 36
.23	.59617	.66240	.70370	.67244	.97714	.98996	.0234	.01004	77 44
.24	.64344	.66684	.74989	.67668	.97759	.99016	.0229	.00984	77 51
2.25	4.69117	0.67128	4.79657	0.68093	0.97803	9.99035	1.0225	0.00965	77 58
.26	.73937	.67572	.84372	.68518	.97846	.99054	.0220	.00946	78 05
.27	.78804	.68016	.89136	.68943	.97888	.99073	.0216	.00927	78 12
.28	.83720	.68459	.93948	.69368	.97929	.99091	.0211	.00909	78 19
.29	.88684	.68903	.98810	.69794	.97970	.99109	.0207	.00891	78 26
2.30	4.93696	0.69346	5.03722	0.70219	0.98010	9.99127	1.0203	0.00873	78 33
.31	.98758	.69789	.08684	.70645	.98049	.99144	.0199	.00856	78 40
.32	5.03870	.70232	.13697	.71071	.98087	.99161	.0195	.00839	78 46
.33	.09032	.70675	.18762	.71497	.98124	.99178	.0191	.00822	78 53
.34	.14245	.71117	.23878	.71923	.98161	.99194	.0187	.00806	79 00
2.35	5.19510	0.71559	5.29047	0.72349	0.98197	9.99210	1.0184	0.00790	79 06
.36	.24827	.72002	.34269	.72776	.98233	.99226	.0180	.00774	79 13
.37	.30196	.72444	.39544	.73203	.98267	.99241	.0176	.00759	79 19
.38	.35618	.72885	.44873	.73630	.98301	.99256	.0173	.00744	79 25
.39	.41093	.73327	.50256	.74056	.98335	.99271	.0169	.00729	79 32
2.40	5.46623	0.73769	5.55695	0.74484	0.98367	9.99285	1.0166	0.00715	79 38
.41	.52207	.74210	.61189	.74911	.98400	.99299	.0163	.00701	79 44
.42	.57847	.74652	.66739	.75338	.98431	.99313	.0159	.00687	79 50
.43	.63542	.75093	.72346	.75766	.98462	.99327	.0156	.00673	79 56
.44	.69294	.75534	.78010	.76194	.98492	.99340	.0153	.00660	80 02
2.45	5.75103	0.75975	5.83732	0.76621	0.98522	9.99353	1.0150	0.00647	80 08
.46	.80969	.76415	.89512	.77049	.98551	.99366	.0147	.00634	80 14
.47	.86893	.76856	.95352	.77477	.98579	.99379	.0144	.00621	80 20
.48	.92876	.77296	6.01250	.77906	.98607	.99391	.0141	.00609	80 26
.49	.98918	.77737	.07209	.78334	.98635	.99403	.0138	.00597	80 31
2.50	6.05020	0.78177	6.13229	0.78762	0.98661	9.99415	1.0136	0.00585	80 37

TABLE 17 (*continued*).

HYPERBOLIC FUNCTIONS.

u	sinh. u		cosh. u		tanh. u		coth. u		gd. u
	Nat.	Log.	Nat.	Log.	Nat.	Log.	Nat.	Log.	
2.50	6.05020	0.78177	6.13229	0.78762	0.98661	9.99415	1.0136	0.00585	80° 37′
.51	.11183	.78617	.19310	.79191	.98688	.99426	.0133	.00574	80 42
.52	.17407	.79057	.25453	.79619	.98714	.99438	.0130	.00562	80 48
.53	.23692	.79497	.31658	.80048	.98739	.99449	.0128	.00551	80 53
.54	.30040	.79937	.37927	.80477	.98764	.99460	.0125	.00540	80 59
2.55	6.36451	0.80377	6.44259	0.80906	0.98788	9.99470	1.0123	0.00530	81 04
.56	.42926	.80816	.50656	.81335	.98812	.99481	.0120	.00519	81 10
.57	.49464	.81256	.57118	.81764	.98835	.99491	.0118	.00509	81 15
.58	.56068	.81695	.63646	.82194	.98858	.99501	.0115	.00499	81 20
.59	.62738	.82134	.70240	.82623	.98881	.99511	.0113	.00489	81 25
2.60	6.69473	0.82573	6.76901	0.83052	0.98903	9.99521	1.0111	0.00479	81 30
.61	.76276	.83012	.83629	.83482	.98924	.99530	.0109	.00470	81 35
.62	.83146	.83451	.90426	.83912	.98946	.99540	.0107	.00460	81 40
.63	.90085	.83890	.97292	.84341	.98966	.99549	.0104	.00451	81 45
.64	.97092	.84329	7.04228	.84771	.98987	.99558	.0102	.00442	81 50
2.65	7.04169	0.84768	7.11234	0.85201	0.99007	9.99566	1.0100	0.00434	81 55
.66	.11317	.85206	.18312	.85631	.99026	.99575	.0098	.00425	82 00
.67	.18536	.85645	.25461	.86061	.99045	.99583	.0096	.00417	82 05
.68	.25827	.86083	.32683	.86492	.99064	.99592	.0094	.00408	82 09
.69	.33190	.86522	.39978	.86922	.99083	.99600	.0093	.00400	82 14
2.70	7.40626	0.86960	7.47347	0.87352	0.99101	9.99608	1.0091	0.00392	82 19
.71	.48137	.87398	.54791	.87783	.99118	.99615	.0089	.00385	82 23
.72	.55722	.87836	.62310	.88213	.99136	.99623	.0087	.00377	82 28
.73	.63383	.88274	.69905	.88644	.99153	.99631	.0085	.00369	82 32
.74	.71121	.88712	.77578	.89074	.99170	.99638	.0084	.00362	82 37
2.75	7.78935	0.89150	7.85328	0.89505	0.99186	9.99645	1.0082	0.00355	82 41
.76	.86828	.89588	.93157	.89936	.99202	.99652	.0080	.00348	82 45
.77	.94799	.90026	8.01065	.90367	.99218	.99659	.0079	.00341	82 50
.78	8.02849	.90463	.09053	.90798	.99233	.99666	.0077	.00334	82 54
.79	.10980	.90901	.17122	.91229	.99248	.99672	.0076	.00328	82 58
2.80	8.19192	0.91339	8.25273	0.91660	0.99263	9.99679	1.0074	0.00321	83 02
.81	.27486	.91776	.33506	.92091	.99278	.99685	.0073	.00315	83 07
.82	.35862	.92213	.41823	.92522	.99292	.99691	.0071	.00309	83 11
.83	.44322	.92651	.50224	.92953	.99306	.99698	.0070	.00302	83 15
.84	.52867	.93088	.58710	.93385	.99320	.99704	.0069	.00296	83 19
2.85	8.61497	0.93525	8.67281	0.93816	0.99333	9.99709	1.0067	0.00291	83 23
.86	.70213	.93963	.75940	.94247	.99346	.99715	.0066	.00285	83 27
.87	.79016	.94400	.84686	.94679	.99359	.99721	.0065	.00279	83 31
.88	.87907	.94837	.93520	.95110	.99372	.99726	.0063	.00274	83 34
.89	.96887	.95274	9.02444	.95542	.99384	.99732	.0062	.00268	83 38
2.90	9.05956	0.95711	9.11458	0.95974	0.99396	9.99737	1.0061	0.00263	83 42
.91	.15116	.96148	.20564	.96405	.99408	.99742	.0060	.00258	83 46
.92	.24368	.96584	.29761	.96837	.99420	.99747	.0058	.00253	83 50
.93	.33712	.97021	.39051	.97269	.99531	.99752	.0057	.00248	83 53
.94	.43149	.97458	.48436	.97701	.99443	.99757	.0056	.00243	83 57
2.95	9.52681	0.97895	9.57915	0.98133	0.99454	9.99762	1.0055	0.00238	84 00
.96	.62308	.98331	.67490	.98565	.99464	.99767	.0054	.00233	84 04
.97	.72031	.98768	.77161	.98997	.99475	.99771	.0053	.00229	84 08
.98	.81851	.99205	.86930	.99429	.99485	.99776	.0052	.00224	84 11
.99	.91770	.99641	.96798	.99861	.99496	.99780	.0051	.00220	84 15
3.00	10.01787	1.00078	10.06766	1.00293	0.99505	9.99785	1.0050	0.00215	84 18

HYPERBOLIC FUNCTIONS.

u	sinh. u		cosh. u		tanh. u		coth. u		gd. u
	Nat.	Log.	Nat.	Log.	Nat.	Log.	Nat.	Log.	
3.0	10.0179	1.00078	10.0677	1.00293	0.99505	9.99785	1.0050	0.00215	84°18′
.1	11.0765	.04440	11.1215	.04616	.99595	.99824	.0041	.00176	84 50
.2	12.2459	.08799	12.2866	.08943	.99668	.99856	.0033	.00144	85 20
.3	13.5379	.13155	13.5748	.13273	.99728	.99882	.0027	.00118	85 47
.4	14.9654	.17509	14.9987	.17605	.99777	.99903	.0022	.00097	86 11
3.5	16.5426	1.21860	16.5728	1.21940	0.99818	9.99921	1.0018	0.00079	86 32
.6	18.2855	.26211	18.3128	.26275	.99851	.99935	.0015	.00065	86 52
.7	20.2113	.30559	20.2360	.30612	.99878	.99947	.0012	.00053	87 10
.8	22.3394	.34907	22.3618	.34951	.99900	.99957	.0010	.00043	87 26
.9	24.6911	.39254	24.7113	.39290	.99918	.99964	.0008	.00036	87 41
4.0	27.2899	1.43600	27.3082	1.43629	0.99933	9.99971	1.0007	0.00029	87 54
.1	30.1619	.47946	30.1784	.47970	.99945	.99976	.0005	.00024	88 06
.2	33.3357	.52291	33.3507	.52310	.99955	.99980	.0004	.00020	88 17
.3	36.8431	.56636	36.8567	.56652	.99963	.99984	.0004	.00016	88 27
.4	40.7193	.60980	40.7316	.60993	.99970	.99987	.0003	.00013	88 36
4.5	45.0030	1.65324	45.0141	1.65335	0.99975	9.99989	1.0002	0.00011	88 44
.6	49.7371	.69668	49.7472	.69677	.99980	.99991	.0002	.00009	88 51
.7	54.9690	.74012	54.9781	.74019	.99983	.99993	.0002	.00007	88 57
.8	60.7511	.78355	60.7593	.78361	.99986	.99994	.0001	.00006	89 03
.9	67.1412	.82699	67.1486	.82704	.99989	.99995	.0001	.00005	89 09
5.0	74.2032	1.87042	74.2099	1.87046	0.99991	9.99996	1.0001	0.00004	89 14

TABLE 18.—Factorials.

See Table 16 for logarithms of the products 1.2.3. . . . *n* from 1 to 100.
See Table 32 for log. Γ (*n*+1) for values of *n* between 1.000 and 2.000.

n	$\frac{1}{n:}$	*n*: = 1. 2. 3. 4 . . . *n*	*n*
1	1.	1	1
2	0.5	2	2
3	.16666 66666 66666 66666 66667	6	3
4	.04166 66666 66666 66666 66667	24	4
5	.00833 33333 33333 33333 33333	120	5
6	0.00138 88888 88888 88888 88889	720	6
7	.00019 84126 98412 69841 26984	5040	7
8	.00002 48015 87301 58730 15873	40320	8
9	.00000 27557 31922 39858 90653	3 62880	9
10	.00000 02755 73192 23985 89065	36 28800	10
11	0.00000 00250 52108 38544 17188	399 16800	11
12	.00000 00020 87675 69878 68099	4790 01600	12
13	.00000 00001 60590 43836 82161	62270 20800	13
14	.00000 00000 11470 74559 77297	8 71782 91200	14
15	.00000 00000 00764 71637 31820	130 76743 68000	15
16	0.00000 00000 00047 79477 33239	2092 27898 88000	16
17	.00000 00000 00002 81145 72543	35568 74280 96000	17
18	.00000 00000 00000 15619 20697	6 40237 37057 28000	18
19	.00000 00000 00000 00822 06352	121 64510 04088 32000	19
20	.00000 00000 00000 00041 10318	2432 90200 81766 40000	20

TABLE 19.

EXPONENTIAL FUNCTION.

x	$\log_{10}(e^x)$	e^x	e^{-x}	x	$\log_{10}(e^x)$	e^x	e^{-x}
0.00	0.00000	1.0000	1.000000	0.50	0.21715	1.6487	0.606531
.01	.00434	.0101	0.990050	.51	.22149	.6653	.600496
.02	.00869	.0202	.980199	.52	.22583	.6820	.594521
.03	.01303	.0305	.970446	.53	.23018	.6989	.588605
.04	.01737	.0408	.960789	.54	.23452	.7160	.582748
0.05	0.02171	1.0513	0.951229	0.55	0.23886	1.7333	0.576950
.06	.02606	.0618	.941765	.56	.24320	.7507	.571209
.07	.03040	.0725	.932394	.57	.24755	.7683	.565525
.08	.03474	.0833	.923116	.58	.25189	.7860	.559898
.09	.03909	.0942	.913931	.59	.25623	.8040	.554327
0.10	0.04343	1.1052	0.904837	0.60	0.26058	1.8221	0.548812
.11	.04777	.1163	.895834	.61	.26492	.8404	.543351
.12	.05212	.1275	.886920	.62	.26926	.8589	.537944
.13	.05646	.1388	.878095	.63	.27361	.8776	.532592
.14	.06080	.1503	.869358	.64	.27795	.8965	.527292
0.15	0.06514	1.1618	0.860708	0.65	0.28229	1.9155	0.522046
.16	.06949	.1735	.852144	.66	.28663	.9348	.516851
.17	.07383	.1853	.843665	.67	.29098	.9542	.511709
.18	.07817	.1972	.835270	.68	.29532	.9739	.506617
.19	.08252	.2092	.826959	.69	.29966	.9937	.501576
0.20	0.08686	1.2214	0.818731	0.70	0.30401	2.0138	0.496585
.21	.09120	.2337	.810584	.71	.30835	.0340	.491644
.22	.09554	.2461	.802519	.72	.31269	.0544	.486752
.23	.09989	.2586	.794534	.73	.31703	.0751	.481909
.24	.10423	.2712	.786628	.74	.32138	.0959	.477114
0.25	0.10857	1.2840	0.778801	0.75	0.32572	2.1170	0.472367
.26	.11292	.2969	.771052	.76	.33006	.1383	.467666
.27	.11726	.3100	.763379	.77	.33441	.1598	.463013
.28	.12160	.3231	.755784	.78	.33875	.1815	.458406
.29	.12595	.3364	.748264	.79	.34309	.2034	.453845
0.30	0.13029	1.3499	0.740818	0.80	0.34744	2.2255	0.449329
.31	.13463	.3634	.733447	.81	.35178	.2479	.444858
.32	.13897	.3771	.726149	.82	.35612	.2705	.440432
.33	.14332	.3910	.718924	.83	.36046	.2933	.436049
.34	.14766	.4049	.711770	.84	.36481	.3164	.431711
0.35	0.15200	1.4191	0.704688	0.85	0.36915	2.3396	0.427415
.36	.15635	.4333	.697676	.86	.37349	.3632	.423162
.37	.16069	.4477	.690734	.87	.37784	.3869	.418952
.38	.16503	.4623	.683861	.88	.38218	.4109	.414783
.39	.16937	.4770	.677057	.89	.38652	.4351	.410656
0.40	0.17372	1.4918	0.670320	0.90	0.39087	2.4596	0.406570
.41	.17806	.5068	.663650	.91	.39521	.4843	.402524
.42	.18240	.5220	.657047	.92	.39955	.5093	.398519
.43	.18675	.5373	.650509	.93	.40389	.5345	.394554
.44	.19109	.5527	.644036	.94	.40824	.5600	.390628
0.45	0.19543	1.5683	0.637628	0.95	0.41258	2.5857	0.386741
.46	.19978	.5841	.631284	.96	.41692	.6117	.382893
.47	.20412	.6000	.625002	.97	.42127	.6379	.379083
.48	.20846	.6161	.618783	.98	.42561	.6645	.375311
.49	.21280	.6323	.612626	.99	.42995	.6912	.371577
0.50	0.21715	1.6487	0.606531	1.00	0.43429	2.7183	0.367879

EXPONENTIAL FUNCTION.

x	$\log_{10}(e^x)$	e^x	e^{-x}	x	$\log_{10}(e^x)$	e^x	e^{-x}
1.00	0.43429	2.7183	0.367879	1.50	0.65144	4.4817	0.223130
.01	.43864	.7456	.364219	.51	.65578	.5267	.220910
.02	.44298	.7732	.360595	.52	.66013	.5722	.218712
.03	.44732	.8011	.357007	.53	.66447	.6182	.216536
.04	.45167	.8292	.353455	.54	.66881	.6646	.214381
1.05	0.45601	2.8577	0.349938	1.55	0.67316	4.7115	0.212248
.06	.46035	.8864	.346456	.56	.67750	.7588	.210136
.07	.46470	.9154	.343009	.57	.68184	.8066	.208045
.08	.46904	.9447	.339596	.58	.68619	.8550	.205975
.09	.47338	.9743	.336216	.59	.69053	.9037	.203926
1.10	0.47772	3.0042	0.332871	1.60	0.69487	4.9530	0.201897
.11	.48207	.0344	.329559	.61	.69921	5.0028	.199888
.12	.48641	.0649	.326280	.62	.70356	.0531	.197899
.13	.49075	.0957	.323033	.63	.70790	.1039	.195930
.14	.49510	.1268	.319819	.64	.71224	.1552	.193980
1.15	0.49944	3.1582	0.316637	1.65	0.71659	5.2070	0.192050
.16	.50378	.1899	.313486	.66	.72093	.2593	.190139
.17	.50812	.2220	.310367	.67	.72527	.3122	.188247
.18	.51247	.2544	.307279	.68	.72961	.3656	.186374
.19	.51681	.2871	.304221	.69	.73396	.4195	.184520
1.20	0.52115	3.3201	0.301194	1.70	0.73830	5.4739	0.182684
.21	.52550	.3535	.298197	.71	.74264	.5290	.180866
.22	.52984	.3872	.295230	.72	.74699	.5845	.179066
.23	.53418	.4212	.292293	.73	.75133	.6407	.177284
.24	.53853	.4556	.289384	.74	.75567	.6973	.175520
1.25	0.54287	3.4903	0.286505	1.75	0.76002	5.7546	0.173774
.26	.54721	.5254	.283654	.76	.76436	.8124	.172045
.27	.55155	.5609	.280832	.77	.76870	.8709	.170333
.28	.55590	.5966	.278037	.78	.77304	.9299	.168638
.29	.56024	.6328	.275271	.79	.77739	.9895	.166960
1.30	0.56458	3.6693	0.272532	1.80	0.78173	6.0496	0.165299
.31	.56893	.7062	.269820	.81	.78607	.1104	.163654
.32	.57327	.7434	.267135	.82	.79042	.1719	.162026
.33	.57761	.7810	.264477	.83	.79476	.2339	.160414
.34	.58195	.8190	.261846	.84	.79910	.2965	.158817
1.35	0.58630	3.8574	0.259240	1.85	0.80344	6.3598	0.157237
.36	.59064	.8962	.256661	.86	.80779	.4237	.155673
.37	.59498	.9354	.254107	.87	.81213	.4883	.154124
.38	.59933	.9749	.251579	.88	.81647	.5535	.152590
.39	.60367	4.0149	.249075	.89	.82082	.6194	.151072
1.40	0.60801	4.0552	0.246597	1.90	0.82516	6.6859	0.149569
.41	.61236	.0960	.244143	.91	.82950	.7531	.148080
.42	.61670	.1371	.241714	.92	.83385	.8210	.146607
.43	.62104	.1787	.239309	.93	.83819	.8895	.145148
.44	.62538	.2207	.236928	.94	.84253	.9588	.143704
1.45	0.62973	4.2631	0.234570	1.95	0.84687	7.0287	0.142274
.46	.63407	.3060	.232236	.96	.85122	.0993	.140858
.47	.63841	.3492	.229925	.97	.85556	.1707	.139457
.48	.64276	.3929	.227638	.98	.85990	.2427	.138069
.49	.64710	.4371	.225373	.99	.86425	.3155	.136695
1.50	0.65144	4.4817	0.223130	2.00	0.86859	7.3891	0.135335

EXPONENTIAL FUNCTION.

x	$\log_{10}(e^x)$	e^x	e^{-x}	x	$\log_{10}(e^x)$	e^x	e^{-x}
2.00	0.86859	7.3891	0.135335	2.50	1.08574	12.182	0.082085
.01	.87293	.4633	.133989	.51	.09008	.305	.081268
.02	.87727	.5383	.132655	.52	.09442	.429	.080460
.03	.88162	.6141	.131336	.53	.09877	.554	.079659
.04	.88596	.6906	.130029	.54	.10311	.680	.078866
2.05	0.89030	7.7679	0.128735	2.55	1.10745	12.807	0.078082
.06	.89465	.8460	.127454	.56	.11179	.936	.077305
.07	.89899	.9248	.126186	.57	.11614	13.066	.076536
.08	.90333	8.0045	.124930	.58	.12048	.197	.075774
.09	.90768	.0849	.123687	.59	.12482	.330	.075020
2.10	0.91202	8.1662	0.122456	2.60	1.12917	13.464	0.074274
.11	.91636	.2482	.121238	.61	.13351	.599	.073535
.12	.92070	.3311	.120032	.62	.13785	.736	.072803
.13	.92505	.4149	.118837	.63	.14219	.874	.072078
.14	.92939	.4994	.117655	.64	.14654	14.013	.071361
2.15	0.93373	8.5849	0.116484	2.65	1.15088	14.154	0.070651
.16	.93808	.6711	.115325	.66	.15522	.296	.069948
.17	.94242	.7583	.114178	.67	.15957	.440	.069252
.18	.94676	.8463	.113042	.68	.16391	.585	.068563
.19	.95110	.9352	.111917	.69	.16825	.732	.067881
2.20	0.95545	9.0250	0.110803	2.70	1.17260	14.880	0.067206
.21	.95979	.1157	.109701	.71	.17694	15.029	.066537
.22	.96413	.2073	.108609	.72	.18128	.180	.065875
.23	.96848	.2999	.107528	.73	.18562	.333	.065219
.24	.97282	.3933	.106459	.74	.18997	.487	.064570
2.25	0.97716	9.4877	0.105399	2.75	1.19431	15.643	0.063928
.26	.98151	.5831	.104350	.76	.19865	.800	.063292
.27	.98585	.6794	.103312	.77	.20300	.959	.062662
.28	.99019	.7767	.102284	.78	.20734	16.119	.062039
.29	.99453	.8749	.101266	.79	.21168	.281	.061421
2.30	0.99888	9.9742	0.100259	2.80	1.21602	16.445	0.060810
.31	1.00322	10.074	.099261	.81	.22037	.610	.060205
.32	.00756	.176	.098274	.82	.22471	.777	.059606
.33	.01191	.278	.097296	.83	.22905	.945	.059013
.34	.01625	.381	.096328	.84	.23340	17.116	.058426
2.35	1.02059	10.486	0.095369	2.85	1.23774	17.288	0.057844
.36	.02493	.591	.094420	.86	.24208	.462	.057269
.37	.02928	.697	.093481	.87	.24643	.637	.056699
.38	.03362	.805	.092551	.88	.25077	.814	.056135
.39	.03796	.913	.091630	.89	.25511	.993	.055576
2.40	1.04231	11.023	0.090718	2.90	1.25945	18.174	0.055023
.41	.04665	.134	.089815	.91	.26380	.357	.054476
.42	.05099	.246	.088922	.92	.26814	.541	.053934
.43	.05534	.359	.088037	.93	.27248	.728	.053397
.44	.05968	.473	.087161	.94	.27683	.916	.052866
2.45	1.06402	11.588	0.086294	2.95	1.28117	19.106	0.052340
.46	.06836	.705	.085435	.96	.28551	.298	.051819
.47	.07271	.822	.084585	.97	.28985	.492	.051303
.48	.07705	.941	.083743	.98	.29420	.688	.050793
.49	.08139	12.061	.082910	.99	.29854	.886	.050287
2.50	1.08574	12.182	0.082085	3.00	1.30288	20.086	0.049787

TABLE 19 (*continued*).

EXPONENTIAL FUNCTION.

x	$\log_{10}(e^x)$	e^x	e^{-x}	x	$\log_{10}(e^x)$	e^x	e^{-x}
3.00	1.30288	20.086	0.049787	3.50	1.52003	33.115	0.030197
.01	.30723	.287	.049292	.51	.52437	.448	.029897
.02	.31157	.491	.048801	.52	.52872	.784	.029599
.03	.31591	.697	.048316	.53	.53306	34.124	.029305
.04	.32026	.905	.047835	.54	.53740	.467	.029013
3.05	1.32460	21.115	0.047359	3.55	1.54175	34.813	0.028725
.06	.32894	.328	.046888	.56	.54609	35.163	.028439
.07	.33328	.542	.046421	.57	.55043	.517	.028156
.08	.33763	.758	.045959	.58	.55477	.874	.027876
.09	.34197	.977	.045502	.59	.55912	36.234	.027598
3.10	1.34631	22.198	0.045049	3.60	1.56346	36.598	0.027324
.11	.35066	.421	.044601	.61	.56780	.966	.027052
.12	.35500	.646	.044157	.62	.57215	37.338	.026783
.13	.35934	.874	.043718	.63	.57649	.713	.026516
.14	.36368	23.104	.043283	.64	.58083	38.092	.026252
3.15	1.36803	23.336	0.042852	3.65	1.58517	38.475	0.025991
.16	.37237	.571	.042426	.66	.58952	.861	.025733
.17	.37671	.807	.042004	.67	.59386	39.252	.025476
.18	.38106	24.047	.041586	.68	.59820	.646	.025223
.19	.38540	.288	.041172	.69	.60255	40.045	.024972
3.20	1.38974	24.533	0.040762	3.70	1.60689	40.447	0.024724
.21	.39409	.779	.040357	.71	.61123	.854	.024478
.22	.39843	25.028	.039955	.72	.61558	41.264	.024234
.23	.40277	.280	.039557	.73	.61992	.679	.023993
.24	.40711	.534	.039164	.74	.62426	42.098	.023754
3.25	1.41146	25.790	0.038774	3.75	1.62860	42.521	0.023518
.26	.41580	26.050	.038388	.76	.63295	.948	.023284
.27	.42014	.311	.038006	.77	.63729	43.380	.023052
.28	.42449	.576	.037628	.78	.64163	.816	.022823
.29	.42883	.843	.037254	.79	.64598	44.256	.022596
3.30	1.43317	27.113	0.036883	3.80	1.65032	44.701	0.022371
.31	.43751	.385	.036516	.81	.65466	45.150	.022148
.32	.44186	.660	.036153	.82	.65900	.604	.021928
.33	.44620	.938	.035793	.83	.66335	46.063	.021710
.34	.45054	28.219	.035437	.84	.66769	.525	.021494
3.35	1.45489	28.503	0.035084	3.85	1.67203	46.993	0.021280
.36	.45923	.789	.034735	.86	.67638	47.465	.021068
.37	.46357	29.079	.034390	.87	.68072	.942	.020858
.38	.46792	.371	.034047	.88	.68506	48.424	.020651
.39	.47226	.666	.033709	.89	.68941	.911	.020445
3.40	1.47660	29.964	0.033373	3.90	1.69375	49.402	0.020242
.41	.48094	30.265	.033041	.91	.69809	.899	.020041
.42	.48529	.569	.032712	.92	.70243	50.400	.019841
.43	.48963	.877	.032387	.93	.70678	.907	.019644
.44	.49397	31.187	.032065	.94	.71112	51.419	.019448
3.45	1.49832	31.500	0.031746	3.95	1.71546	51.935	0.019255
.46	.50266	.817	.031430	.96	.71981	52.457	.019063
.47	.50700	32.137	.031117	.97	.72415	.985	.018873
.48	.51134	.460	.030807	.98	.72849	53.517	.018686
.49	.51569	.786	.030501	.99	.73283	54.055	.018500
3.50	1.52003	33.115	0.030197	4.00	1.73718	54.598	0.018316

EXPONENTIAL FUNCTION.

x	log₁₀(eˣ)	eˣ	e⁻ˣ	x	log₁₀(eˣ)	eˣ	e⁻ˣ
4.00	1.73718	54.598	0.018316	4.50	1.95433	90.017	0.011109
.01	.74152	55.147	.018133	.51	.95867	.922	.010998
.02	.74586	.701	.017953	.52	.96301	91.836	.010889
.03	.75021	56.261	.017774	.53	.96735	92.759	.010781
.04	.75455	.826	.017597	.54	.97170	93.691	.010673
4.05	1.75889	57.397	0.017422	4.55	1.97604	94.632	0.010567
.06	.76324	.974	.017249	.56	.98038	95.583	.010462
.07	.76758	58.557	.017077	.57	.98473	96.544	.010358
.08	.77192	59.145	.016907	.58	.98907	97.514	.010255
.09	.77626	.740	.016739	.59	.99341	98.494	.010153
4.10	1.78061	60.340	0.016573	4.60	1.99775	99.484	0.010052
.11	.78495	.947	.016408	.61	2.00210	100.48	.009952
.12	.78929	61.559	.016245	.62	.00644	101.49	.009853
.13	.79364	62.178	.016083	.63	.01078	102.51	.009755
.14	.79798	.803	.015923	.64	.01513	103.54	.009658
4.15	1.80232	63.434	0.015764	4.65	2.01947	104.58	0.009562
.16	.80667	64.072	.015608	.66	.02381	105.64	.009466
.17	.81101	.715	.015452	.67	.02816	106.70	.009372
.18	.81535	65.366	.015299	.68	.03250	107.77	.009279
.19	.81969	66.023	.015146	.69	.03684	108.85	.009187
4.20	1.82404	66.686	0.014996	4.70	2.04118	109.95	0.009095
.21	.82838	67.357	.014846	.71	.04553	111.05	.009005
.22	.83272	68.033	.014699	.72	.04987	112.17	.008915
.23	.83707	.717	.014552	.73	.05421	113.30	.008826
.24	.84141	69.408	.014408	.74	.05856	114.43	.008739
4.25	1.84575	70.105	0.014264	4.75	2.06290	115.58	0.008652
.26	.85009	.810	.014122	.76	.06724	116.75	.008566
.27	.85444	71.522	.013982	.77	.07158	117.92	.008480
.28	.85878	72.240	.013843	.78	.07593	119.10	.008396
.29	.86312	.966	.013705	.79	.08027	120.30	.008312
4.30	1.86747	73.700	0.013569	4.80	2.08461	121.51	0.008230
.31	.87181	74.440	.013434	.81	.08896	122.73	.008148
.32	.87615	75.189	.013300	.82	.09330	123.97	.008067
.33	.88050	.944	.013168	.83	.09764	125.21	.007987
.34	.88484	76.708	.013037	.84	.10199	126.47	.007907
4.35	1.88918	77.478	0.012907	4.85	2.10633	127.74	0.007828
.36	.89352	78.257	.012778	.86	.11067	129.02	.007750
.37	.89787	79.044	.012651	.87	.11501	130.32	.007673
.38	.90221	79.838	.012525	.88	.11936	131.63	.007597
.39	.90655	80.640	.012401	.89	.12370	132.95	.007521
4.40	1.91090	81.451	0.012277	4.90	2.12804	134.29	0.007447
.41	.91524	82.269	.012155	.91	.13239	135.64	.007372
.42	.91958	83.096	.012034	.92	.13673	137.00	.007299
.43	.92392	.931	.011914	.93	.14107	138.38	.007227
.44	.92827	84.775	.011796	.94	.14541	139.77	.007155
4.45	1.93261	85.627	0.011679	4.95	2.14976	141.17	0.007083
.46	.93695	86.488	.011562	.96	.15410	142.59	.007013
.47	.94130	87.357	.011447	.97	.15844	144.03	.006943
.48	.94564	88.235	.011333	.98	.16279	145.47	.006874
.49	.94998	89.121	.011221	.99	.16713	146.94	.006806
4.50	1.95433	90.017	0.011109	5.00	2.17147	148.41	0.006738

EXPONENTIAL FUNCTION.

x	$\log_{10}(e^x)$	e^x	e^{-x}	x	$\log_{10}(e^x)$	e^x	e^{-x}
5.00	2.17147	148.41	0.006738	5.0	2.17147	148.41	0.006738
.01	.17582	149.90	.006671	.1	.21490	164.02	.006097
.02	.18016	151.41	.006605	.2	.25833	181.27	.005517
.03	.18450	152.93	.006539	.3	.30176	200.34	.004992
.04	.18884	154.47	.006474	.4	.34519	221.41	.004517
5.05	2.19319	156.02	0.006409	5.5	2.38862	244.69	0.004087
.06	.19753	157.59	.006346	.6	.43205	270.43	.003698
.07	.20187	159.17	.006282	.7	.47548	298.87	.003346
.08	.20622	160.77	.006220	.8	.51891	330.30	.003028
.09	.21056	162.39	.006158	.9	.56234	365.04	.002739
5.10	2.21490	164.02	0.006097	6.0	2.60577	403.43	0.002479
.11	.21924	165.67	.006036	.1	.64920	445.86	.002243
.12	.22359	167.34	.005976	.2	.69263	492.75	.002029
.13	.22793	169.02	.005917	.3	.73606	544.57	.001836
.14	.23227	170.72	.005858	.4	.77948	601.85	.001662
5.15	2.23662	172.43	0.005799	6.5	2.82291	665.14	0.001503
.16	.24096	174.16	.005742	.6	.86634	735.10	.001360
.17	.24530	175.91	.005685	.7	.90977	812.41	.001231
.18	.24965	177.68	.005628	.8	.95320	897.85	.001114
.19	.25399	179.47	.005572	.9	.99663	992.27	.001008
5.20	2.25833	181.27	0.005517	7.0	3.04006	1096.6	0.000912
.21	.26267	183.09	.005462	.1	.08349	1212.0	.000825
.22	.26702	184.93	.005407	.2	.12692	1339.4	.000747
.23	.27136	186.79	.005354	.3	.17035	1480.3	.000676
.24	.27570	188.67	.005300	.4	.21378	1636.0	.000611
5.25	2.28005	190.57	0.005248	7.5	3.25721	1808.0	0.000553
.26	.28439	192.48	.005195	.6	.30064	1998.2	.000500
.27	.28873	194.42	.005144	.7	.34407	2208.3	.000453
.28	.29307	196.37	.005092	.8	.38750	2440.6	.000410
.29	.29742	198.34	.005042	.9	.43093	2697.3	.000371
5.30	2.30176	200.34	0.004992	8.0	3.47436	2981.0	0.000335
.31	.30610	202.35	.004942	.1	.51779	3294.5	.000304
.32	.31045	204.38	.004893	.2	.56121	3641.0	.000275
.33	.31479	206.44	.004844	.3	.60464	4023.9	.000249
.34	.31913	208.51	.004796	.4	.64807	4447.1	.000225
5.35	2.32348	210.61	0.004748	8.5	3.69150	4914.8	0.000203
.36	.32782	212.72	.004701	.6	.73493	5431.7	.000184
.37	.33216	214.86	.004654	.7	.77836	6002.9	.000167
.38	.33650	217.02	.004608	.8	.82179	6634.2	.000151
.39	.34085	219.20	.004562	.9	.86522	7332.0	.000136
5.40	2.34519	221.41	0.004517	9.0	3.90865	8103.1	0.000123
.41	.34953	223.63	.004472	.1	.95208	8955.3	.000112
.42	.35388	225.88	.004427	.2	.99551	9897.1	.000101
.43	.35822	228.15	.004383	.3	4.03894	10938.	.000091
.44	.36256	230.44	.004339	.4	.08237	12088.	.000083
5.45	2.36690	232.76	0.004296	9.5	4.12580	13360.	0.000075
.46	.37125	235.10	.004254	.6	.16923	14765.	.000068
.47	.37559	237.46	.004211	.7	.21266	16318.	.000061
.48	.37993	239.85	.004169	.8	.25609	18034.	.000055
.49	.38428	242.26	.004128	.9	.29952	19930.	.000050
5.50	2.38862	244.69	0.004087	10.0	4.34294	22026.	0.000045

Table 20.

EXPONENTIAL FUNCTIONS.

Value of e^{x^2} and e^{-x^2} and their logarithms.

x	e^{x^2}	$\log e^{x^2}$	e^{-x^2}	$\log e^{-x^2}$
0.1	1.0101	0.00434	0.99005	$\bar{1}$.99566
2	1.0408	01737	96079	98263
3	1.0942	03909	91393	96091
4	1.1735	06949	85214	93051
5	1.2840	10857	77880	89143
0.6	1.4333	0.15635	0.69768	$\bar{1}$.84365
7	1.6323	21280	61263	78720
8	1.8965	27795	52729	72205
9	2.2479	35178	44486	64822
1.0	2.7183	43429	36788	56571
1.1	3.3535	0.52550	0.29820	$\bar{1}$.47450
2	4.2207	62538	23693	37462
3	5.4195	73396	18452	26604
4	7.0993	85122	14086	14878
5	9.4877	97716	10540	02284
1.6	1.2936 $\times$ 10	1.11179	0.77305 $\times 10^{-1}$	$\bar{2}$.88821
7	1.7993 "	25511	55576 "	74489
8	2.5534 "	40711	39164 "	59289
9	3.6966 "	56780	27052 "	43220
2.0	5.4598 "	73718	18316 "	26282
2.1	8.2269 "	1.91524	0.12155 "	$\bar{2}$.08476
2	1.2647 $\times 10^{2}$	2.10199	79071 $\times 10^{-2}$	$\bar{3}$.89801
3	1.9834 "	29742	50418 "	70258
4	3.1735 "	50154	31511 "	49846
5	5.1801 "	71434	19305 "	28566
2.6	8.6264 "	2.93583	0.11592 "	$\bar{3}$.06417
7	1.4656 $\times 10^{3}$	3.16601	68233 $\times 10^{-3}$	$\bar{4}$.83399
8	2.5402 "	40487	39367 "	59513
9	4.4918 "	65242	22263 "	34758
3.0	8.1031 "	90865	12341 "	09135
3.1	1.4913 $\times 10^{4}$	4.17357	0.67055 $\times 10^{-4}$	$\bar{5}$.82643
2	2.8001 "	44718	35713 "	55282
3	5.3637 "	72947	18644 "	27053
4	1.0482 $\times 10^{5}$	5.02044	95402 $\times 10^{-5}$	$\bar{6}$.97956
5	2.0898 "	32011	47851 "	67989
3.6	4.2507 "	5.62846	0.23526 "	$\bar{6}$.37154
7	8.8205 "	94549	11337 "	05451
8	1.8673 $\times 10^{6}$	6.27121	53553 $\times 10^{-6}$	$\bar{7}$.72879
9	4.0329 "	60562	24796 "	39438
4.0	8.8861 "	94871	11254 "	05129
4.1	1.9975 $\times 10^{7}$	7.30049	0.50062 $\times 10^{-7}$	$\bar{8}$.69951
2	4.5809 "	66095	21830 "	33905
3	1.0718 $\times 10^{8}$	8.03010	93303 $\times 10^{-8}$	$\bar{9}$.96990
4	2.5582 "	40794	39089 "	59206
5	6.2296 "	79446	16052 "	20554
4.6	1.5476 $\times 10^{9}$	9.18967	0.64614 $\times 10^{-9}$	$\overline{10}$.81033
7	3.9225 "	59357	25494 "	40643
8	1.0142 $\times 10^{10}$	10.00614	98595 $\times 10^{-10}$	$\overline{11}$.99386
9	2.6755 "	42741	37376 "	57259
5.0	7.2005 "	85736	13888 "	14264

TABLE 21.

EXPONENTIAL FUNCTIONS.

Values of $e^{\frac{\pi}{4}x}$ and $e^{-\frac{\pi}{4}x}$ and their logarithms.

x	$e^{\frac{\pi}{4}x}$	$\log e^{\frac{\pi}{4}x}$	$e^{-\frac{\pi}{4}x}$	$\log e^{-\frac{\pi}{4}x}$
1	2.1933	0.34109	0.45594	$\bar{1}$.65891
2	4.8105	.68219	.20788	.31781
3	1.0551×10	1.02328	$.94780 \times 10^{-1}$	$\bar{2}$.97672
4	2.3141 "	.36438	.43214 "	.63562
5	5.0754 "	.70547	.19703 "	.29453
6	1.1132×10^{2}	2.04656	0.89833×10^{-2}	$\bar{3}$.95344
7	2.4415 "	.38766	.40958 "	.61234
8	5.3549 "	.72875	.18674 "	.27125
9	1.1745×10^{3}	3.06985	$.85144 \times 10^{-3}$	$\bar{4}$.93015
10	2.5760 "	.41094	.38820 "	.58906
11	5.6498 "	3.75203	0.17700 "	$\bar{4}$.24797
12	1.2392×10^{4}	4.09313	$.80700 \times 10^{-4}$	$\bar{5}$.90687
13	2.7178 "	.43422	.36794 "	.56578
14	5.9610 "	.77532	.16776 "	.22468
15	1.3074×10^{5}	5.11641	$.76487 \times 10^{-5}$	$\bar{6}$.88359
16	2.8675 "	5.45751	0.34873 "	$\bar{6}$.54249
17	6.2893 "	.79860	.15900 "	.20140
18	1.3794×10^{6}	6.13969	$.72495 \times 10^{-6}$	$\bar{7}$.86031
19	3.0254 "	.48079	.33053 "	.51921
20	6.6356 "	.82188	.15070 "	.17812

TABLE 22.

EXPONENTIAL FUNCTIONS.

Values of $e^{\frac{\sqrt{\pi}}{4}x}$ and $e^{-\frac{\sqrt{\pi}}{4}x}$ and their logarithms.

x	$e^{\frac{\sqrt{\pi}}{4}x}$	$\log e^{\frac{\sqrt{\pi}}{4}x}$	$e^{-\frac{\sqrt{\pi}}{4}x}$	$\log e^{-\frac{\sqrt{\pi}}{4}x}$
1	1.5576	0.19244	0.64203	$\bar{1}$.80756
2	2.4260	.38488	.41221	.61512
3	3.7786	.57733	.26465	.42267
4	5.8853	.76977	.16992	.23023
5	9.1666	.96221	.10909	.03779
6	14.277	1.15465	0.070041	$\bar{2}$.84535
7	22.238	.34709	.044968	.65291
8	34.636	.53953	.028871	.46047
9	53.948	.73198	.018536	.26802
10	84.027	.92442	.011901	.07558
11	130.88	2.11686	0.0076408	$\bar{3}$.88314
12	203.85	.30930	.0049057	.69070
13	317.50	.50174	.0031496	.49826
14	494.52	.69418	.0020222	.30582
15	770.24	.88663	.0012983	.11337
16	1199.7	3.07907	0.00083355	$\bar{4}$.92093
17	1868.6	.27151	.00053517	.72849
18	2910.4	.46395	.00034360	.53605
19	4533.1	.65639	.00022060	.34361
20	7060.5	.84883	.00014163	.15117

EXPONENTIAL FUNCTIONS AND LEAST SQUARES.

TABLE 23.—Exponential Functions.

Value of e^x and e^{-x} and their logarithms.

x	e^x	$\log e^x$	e^{-x}	x	e^x	$\log e^x$	e^{-x}
1/64	1.0157	0.00679	0.98450	1/3	1.3956	0.14476	0.71653
1/32	.0317	.01357	.96923	1/2	.6487	.21715	.60653
1/16	.0645	.02714	.93941	3/4	2.1170	.32572	.47237
1/10	.1052	.04343	.90484	1	.7183	.43429	.36788
1/9	.1175	.04825	.89484	5/4	3.4903	.54287	.28650
1/8	1.1331	0.05429	0.88250	3/2	4.4817	0.65144	0.22313
1/7	.1536	.06204	.86688	7/4	5.7546	.76002	.17377
1/6	.1814	.07238	.84648	2	7.3891	.86859	.13534
1/5	.2214	.08686	.81873	9/4	9.4877	.97716	.10540
1/4	.2840	.10857	.77880	5/2	12.1825	1.08574	.08208

TABLE 24.—Least Squares.

Values of $P=\frac{2}{\sqrt{\pi}}\int_0^{hx} e^{-(hx)^2}\, d\,(hx)$.

This table gives the value of P, the probability of an observational error having a value positive or negative equal to or less than x when h is the measure of precision, $P=\frac{2}{\sqrt{\pi}}\int_0^{hx} e^{-(hx)^2} d(hx)$. For values of the inverse function see the table on Diffusion.

hx	0	1	2	3	4	5	6	7	8	9
0.0		.01128	.02256	.03384	.04511	.05637	.06762	.07886	.09008	.10128
.1	.11246	.12362	.13476	.14587	.15695	.16800	.17901	.18999	.20094	.21184
.2	.22270	.23352	.24430	.25502	.26570	.27633	.28690	.29742	.30788	.31828
.3	.32863	.33891	.34913	.35928	.36936	37938	.38933	.39921	.40901	.41874
.4	.42839	.43797	.44747	.45689	.46623	.47548	.48466	.49375	.50275	.51167
0.5	.52050	.52924	.53790	.54646	.55494	.56332	.57162	.57982	.58792	.59594
.6	.60386	.61168	.61941	.62705	.63459	.64203	.64938	.65663	.66378	.67084
.7	.67780	.68467	.69143	.69810	.70468	.71116	.71754	.72382	.73001	.73610
.8	.74210	.74800	.75381	.75952	.76514	.77067	.77610	.78144	.78669	.79184
.9	.79691	.80188	.80677	.81156	.81627	.82089	.82542	.82987	.83423	.83851
1.0	.84270	.84681	.85084	.85478	.85865	.86244	.86614	.86977	.87333	.87680
.1	.88021	.88353	.88679	.88997	.89308	.89612	.89910	.90200	.90484	.90761
.2	.91031	.91296	.91553	.91805	.92051	.92290	.92524	.92751	.92973	.93190
.3	.93401	.93606	.93807	.94002	.94191	.94376	.94556	.94731	.94902	.95067
.4	.95229	.95385	.95538	.95686	.95830	.95970	.96105	.96237	.96365	.96490
1.5	.96611	.96728	.96841	.96952	.97059	.97162	.97263	.97360	.97455	.97546
.6	.97635	.97721	.97804	.97884	.97962	.98038	.98110	.98181	.98249	.98315
.7	.98379	.98441	.98500	.98558	.98613	.98667	.98719	.98769	.98817	.98864
.8	.98909	.98952	.98994	.99035	.99074	.99111	.99147	.99182	.99216	.99248
.9	.99279	.99309	.99338	.99366	.99392	.99418	.99443	.99466	.99489	.99511
2.0	.99532	.99552	.99572	.99591	.99609	.99626	.99642	.99658	.99673	.99688
.1	.99702	.99715	.99728	.99741	.99753	.99764	.99775	.99785	.99795	.99805
.2	.99814	.99822	.99831	.99839	.99846	.99854	.99861	.99867	.99874	.99880
.3	.99886	.99891	.99897	.99902	.99906	.99911	.99915	.99920	.99924	.99928
.4	.99931	.99935	.99938	.99941	.99944	.99947	.99950	.99952	.99955	.99957
2.5	.99959	.99961	.99963	.99965	.99967	.99969	.99971	.99972	.99974	.99975
.6	.99976	.99978	.99979	.99980	.99981	.99982	.99983	.99984	.99985	.99986
.7	.99987	.99987	.99988	.99989	.99989	.99990	.99991	.99991	.99992	.99992
.8	.99992	.99993	.99993	.99994	.99994	.99994	.99995	.99995	.99995	.99996
.9	.99996	.99996	.99996	.99997	.99997	.99997	.99997	.99997	.99997	.99998
3.0	.99998	.99999	.99999	1.00000						

Taken from a paper by Dr. James Burgess 'on the Definite Integral $\frac{2}{\sqrt{\pi}}\int_0^t e^{-t^2} dt$, with Extended Tables of Values.' Trans. Roy. Soc. of Edinburgh, vol. xxxix, 1900, p. 257.

TABLE 25.

LEAST SQUARES.

This table gives the values of the probability P, as defined in last table, corresponding to different values of x/r where r is the "probable error." The probable error r is equal to $0.47694/h$.

$\frac{x}{r}$	0	1	2	3	4	5	6	7	8	9
0.0	.00000	.00538	.01076	.01614	.02152	.02690	.03228	.03766	.04303	.04840
0.1	.05378	.05914	.06451	.06987	.07523	.08059	.08594	.09129	.09663	.10197
0.2	.10731	.11264	.11796	.12328	.12860	13391	.13921	.14451	.14980	.15508
0.3	.16035	.16562	.17088	.17614	.18138	.18662	.19185	.19707	.20229	.20749
0.4	.21268	.21787	.22304	.22821	.23336	.23851	.24364	.24876	.25388	.25898
0.5	.26407	.26915	.27421	.27927	.28431	.28934	.29436	.29936	.30435	.30933
0.6	.31430	.31925	.32419	.32911	.33402	.33892	.34380	.34866	.35352	.35835
0.7	.36317	.36798	.37277	.37755	.38231	.38705	.39178	.39649	.40118	.40586
0.8	.41052	.41517	.41979	.42440	.42899	.43357	.43813	.44267	.44719	.45169
0.9	.45618	.46064	.46509	.46952	.47393	.47832	.48270	.48705	.49139	.49570
1.0	.50000	.50428	.50853	.51277	.51699	.52119	.52537	.52952	.53366	.53778
1.1	.54188	.54595	.55001	.55404	.55806	.56205	.56602	.56998	.57391	.57782
1.2	.58171	.58558	.58942	.59325	.59705	.60083	.60460	.60833	.61205	.61575
1.3	.61942	.62308	.62671	.63032	.63391	.63747	.64102	.64454	.64804	.65152
1.4	.65498	.65841	.66182	.66521	.66858	.67193	.67526	.67856	.68184	.68510
1.5	.68833	.69155	.69474	.69791	.70106	.70419	.70729	.71038	.71344	.71648
1.6	.71949	.72249	.72546	.72841	.73134	.73425	.73714	.74000	.74285	.74567
1.7	.74847	.75124	.75400	.75674	.75945	.76214	.76481	.76746	.77009	.77270
1.8	.77528	.77785	.78039	.78291	.78542	.78790	.79036	.79280	.79522	.79761
1.9	.79999	.80235	.80469	.80700	.80930	.81158	.81383	.81607	.81828	.82048
2.0	.82266	.82481	.82695	.82907	.83117	.83324	.83530	.83734	.83936	.84137
2.1	.84335	.84531	.84726	.84919	.85109	.85298	.85486	.85671	.85854	.86036
2.2	.86216	.86394	.86570	.86745	.86917	.87088	.87258	.87425	.87591	.87755
2.3	.87918	.88078	.88237	.88395	.88550	.88705	.88857	.89008	.89157	.89304
2.4	.89450	.89595	.89738	.89879	.90019	.90157	.90293	.90428	.90562	.90694
2.5	.90825	.90954	.91082	.91208	.91332	.91456	.91578	.91698	.91817	.91935
2.6	.92051	.92166	.92280	.92392	.92503	.92613	.92721	.92828	.92934	.93038
2.7	.93141	.93243	.93344	.93443	.93541	.93638	.93734	.93828	.93922	.94014
2.8	.94105	.94195	.94284	.94371	.94458	.94543	.94627	.94711	.94793	.94874
2.9	.94954	.95033	.95111	.95187	.95263	.95338	.95412	.95484	.95557	.95628
	0	1	2	3	4	5	6	7	8	9
3	.95698	.96346	.96910	.97397	.97817	.98176	.98482	.98743	.98962	.99147
4	.99302	.99431	.99539	99627	.99700	.99760	.99808	.99848	.99879	.99905
5	.99926	.99943	.99956	.99966	.99974	.99980	.99985	.99988	.99991	.99993

TABLE 26.

LEAST SQUARES.

Values of the factor $0.6745\sqrt{\frac{1}{n-1}}$.

This factor occurs in the equation $r_s = 0.6745\sqrt{\frac{\Sigma v^2}{n-1}}$ for the probable error of a single observation, and other similar equations.

n	0	1	2	3	4	5	6	7	8	9
00			0.6745	0.4769	0.3894	0.3372	0.3016	0.2754	0.2549	0.2385
10	0.2248	0.2133	.2034	.1947	.1871	.1803	.1742	.1686	.1636	.1590
20	.1547	.1508	.1472	.1438	.1406	.1377	.1349	.1323	.1298	.1275
30	.1252	.1231	.1211	.1192	.1174	.1157	.1140	.1124	.1109	.1094
40	.1080	.1066	.1053	.1041	.1029	.1017	.1005	.0994	.0984	.0974
50	0.0964	0.0954	0.0944	0.0935	0.0926	0.0918	0.0909	0.0901	0.0893	0.0886
60	.0878	.0871	.0864	.0857	.0850	.0843	.0837	.0830	.0824	.0818
70	.0812	.0806	.0800	.0795	.0789	.0784	.0779	.0774	.0769	.0764
80	.0759	.0754	.0749	.0745	.0740	.0736	.0732	.0727	.0723	.0719
90	.0715	.0711	.0707	.0703	.0699	.0696	.0692	.0688	.0685	.0681

TABLE 27.—LEAST SQUARES.

Values of the factor $0.6745\sqrt{\frac{1}{n(n-1)}}$.

This factor occurs in the equation $r_0 = 0.6745\sqrt{\frac{\Sigma v^2}{n(n-1)}}$ for the probable error of the arithmetic mean.

n =		1	2	3	4	5	6	7	8	9
00			0.4769	0.2754	0.1947	0.1508	0.1231	0.1041	0.0901	0.0795
10	0.0711	0.0643	.0587	.0540	.0500	.0465	.0435	.0409	.0386	.0365
20	.0346	.0329	.0314	.0300	.0287	.0275	.0265	.0255	.0245	.0237
30	.0229	.0221	.0214	.0208	.0201	.0196	.0190	.0185	.0180	.0175
40	.0171	.0167	.0163	.0159	.0155	.0152	.0148	.0145	.0142	.0139
50	0.0136	0.0134	0.0131	0.0128	0.0126	0.0124	0.0122	0.0119	0.0117	0.0115
60	.0113	.0111	.0110	.0108	.0106	.0105	.0103	.0101	.0100	.0098
70	.0097	.0096	.0094	.0093	.0092	.0091	.0089	.0088	.0087	.0086
80	.0085	.0084	.0083	.0082	.0081	.0080	.0079	.0078	.0077	.0076
90	.0075	.0075	.0074	.0073	.0072	.0071	.0071	.0070	.0069	.0068

TABLE 28.—LEAST SQUARES.

Values of the factor $0.8453\sqrt{\frac{1}{n(n-1)}}$.

This factor occurs in the approximate equation $r = 0.8453\frac{\Sigma v}{\sqrt{n(n-1)}}$ for the probable error of a single observation.

n =		1	2	3	4	5	6	7	8	9
00			0.5978	0.3451	0.2440	0.1890	0.1543	0.1304	0.1130	0.0996
10	0.0891	0.0806	.0736	.0677	.0627	.0583	.0546	.0513	.0483	.0457
20	.0434	.0412	.0393	.0376	.0360	.0345	.0332	.0319	.0307	.0297
30	.0287	.0277	.0268	.0260	.0252	.0245	.0238	.0232	.0225	.0220
40	.0214	.0209	.0204	.0199	.0194	.0190	.0186	.0182	.0178	.0174
50	0.0171	0.0167	0.0164	0.0161	0.0158	0.0155	0.0152	0.0150	0.0147	0.0145
60	.0142	.0140	.0137	.0135	.0133	.0131	.0129	.0127	.0125	.0123
70	.0122	.0120	.0118	.0117	.0115	.0113	.0112	.0111	.0109	.0108
80	.0106	.0105	.0104	.0102	.0101	.0100	.0099	.0098	.0097	.0096
90	.0094	.0093	.0092	.0091	.0090	.0089	.0089	.0088	.0087	.0086

TABLE 29.—LEAST SQUARES.

Values of $0.8453\frac{1}{n\sqrt{n-1}}$.

This factor occurs in the approximate equation $r = 0.8453\frac{\Sigma v}{n\sqrt{n-1}}$ for the probable error of the arithmetical mean.

n =		1	2	3	4	5	6	7	8	9
00			0.4227	0.1993	0.1220	0.0845	0.0630	0.0493	0.0399	0.0332
10	0.0282	0.0243	.0212	.0188	.0167	.0151	.0136	.0124	.0114	.0105
20	.0097	.0090	.0084	.0078	.0073	.0069	.0065	.0061	.0058	.0055
30	.0052	.0050	.0047	.0045	.0043	.0041	.0040	.0038	.0037	.0035
40	.0034	.0033	.0031	.0030	.0029	.0028	.0027	.0027	.0026	.0025
50	0.0024	0.0023	0.0023	0.0022	0.0022	0.0021	0.0020	0.0020	0.0019	0.0019
60	.0018	.0018	.0017	.0017	.0017	.0016	.0016	.0016	.0015	.0015
70	.0015	.0014	.0014	.0014	.0013	.0013	.0013	.0013	.0012	.0012
80	.0012	.0012	.0011	.0011	.0011	.0011	.0011	.0010	.0010	.0010
90	.0010	.0010	.0010	.0009	.0009	.0009	.0009	.0009	.0009	.0009

LEAST SQUARES.

Observation equations :

$$a_1z_1 + b_1z_2 + \ldots\ l_1z_q = M_1,\ \text{weight } p_1$$
$$a_2z_1 + b_2z_2 + \ldots\ l_2z_q = M_2.\ \text{weight } p_2$$
$$\cdot\ \cdot\ \cdot\ \cdot\ \cdot\ \cdot\ \cdot\ \cdot\ \cdot\ \cdot$$
$$a_nz_1 + b_nz_2 + \ldots\ l_nz_q = M_n,\ \text{weight } p_n.$$

Auxiliary equations :

$$[paa] = p_1a_1^2 + p_2a_2^2 + \ldots\ p_na_n^2.$$
$$[pab] = p_1a_1b_1 + p_2a_2b_2 + \ldots\ p_na_nb_n.$$
$$\cdot\ \cdot\ \cdot\ \cdot\ \cdot\ \cdot\ \cdot\ \cdot\ \cdot$$
$$[paM] = p_1a_1M_1 + p_2a_2M_2 + \ldots\ p_na_nM_n.$$

Normal equations :

$$[paa]z_1 + [pab]z_2 + \ldots\ [pal]z_q = [paM]$$
$$[pab]z_1 + [pbb]z_2 + \ldots\ [pbl]z_q = [pbM]$$
$$\cdot\ \cdot\ \cdot\ \cdot\ \cdot\ \cdot\ \cdot\ \cdot\ \cdot$$
$$[pla]z_1 + [plb]z_2 + \ldots\ [pll]z_q = [plM].$$

Solution of normal equations in the form,

$$z_1 = A_1[paM] + B_1[pbM] + \ldots\ L_1[plM]$$
$$z_2 = A_2[paM] + B_2[pbM] + \ldots\ L_2[plM]$$
$$\cdot\ \cdot\ \cdot\ \cdot\ \cdot\ \cdot\ \cdot\ \cdot$$
$$z_q = A_n[paM] + B_n[pbM] + \ldots\ L_n[plM],$$

gives :

$$\text{weight of } z_1 = p_{z1} = (A_1)^{-1};\ \text{probable error of } z_1 = \frac{r}{\sqrt{p_{z_1}}}$$
$$\text{weight of } z_2 = pz_2 = (B_2)^{-1};\ \text{probable error of } z_2 = \frac{r}{\sqrt{p_{z_2}}}$$
$$\cdot\ \cdot\ \cdot\ \cdot\ \cdot\ \cdot\ \cdot\ \cdot\ \cdot\ \cdot\ \cdot\ \cdot$$
$$\text{weight of } z_q = p_{z_q} = (L_n)^{-1};\ \text{probable error of } z_q = \frac{r}{\sqrt{p_{z_q}}}$$

wherein

$$r = \text{probable error of observation of weight unity}$$
$$= 0.6745\sqrt{\frac{\Sigma\, pv^2}{n-q}}.\ \text{(q unknowns.)}$$

Arithmetical mean, n observations:

$$r = 0.6745\sqrt{\frac{\Sigma\ v^2}{n-1}} = \frac{0.8453\ \Sigma\ v}{\sqrt{n(n-1)}}.$$ (approx.) = probable error of observation of weight unity.

$$r_o = 0.6745\sqrt{\frac{\Sigma\ v^2}{n\,(n-1)}} = \frac{0.8453\ \Sigma\ v}{n\sqrt{n-1}}.$$ (approx.) = probable error of mean.

Weighted mean, n observations:

$$r = 0.6745\sqrt{\frac{\Sigma\, p\, v^2}{n-1}};\ r_o = \frac{r}{\sqrt{\Sigma p}} = 0.6745\sqrt{\frac{\Sigma\ p\ v^2}{(n-1)\ \Sigma\ p}}$$

Probable error (R) of a function (Z) of several observed quantities $z_1, z_2, \ldots$ whose probable errors are respectively, $r_1, r_2, \ldots$.

$$Z = f\,(z_1, z_2, \ldots)$$
$$R^2 = \left(\frac{\partial Z}{\partial z_1}\right)^2 r_1^2 + \left(\frac{\partial Z}{\partial z_2}\right)^2 r_2^2 + \ldots$$

Examples :

$Z = z_1 \pm z_2 + \ldots$	$R^2 = r_1^2 + r_2^2 + \ldots$
$Z = Az_1 \pm Bz_2 \pm \ldots$	$R^2 = A^2\, r_1^2 + B^2 r_2^2 + \ldots$
$Z = z_1\, z_2.$	$R^2 = z_1^2\, r_2^2 + z_2^2 r_1^2.$

TABLE 31.

DIFFUSION.

Inverse * values of $v/c = 1 - \frac{2}{\sqrt{\pi}} \int_0^q e^{-q^2} dq$.

$\log x = \log(2q) + \log\sqrt{kt}$. t expressed in seconds.
$= \log\delta + \log\sqrt{kt}$. t expressed in days.
$= \log\gamma + \log\sqrt{kt}$. " " years.
k = coefficient of diffusion.†
c = initial concentration.
v = concentration at distance x, time t.

v/c	log $2q$	$2q$	log δ	δ	log γ	γ
0.00	+∞	+∞	+∞	+∞	∞	∞
.01	0.56143	3.6428	3.02970	1070.78	4.31098	20463.
.02	.51719	3.2900	2.98545	967.04	.26674	18481.
.03	.48699	3.0690	.95525	902.90	.23654	17240.
.04	.46306	2.9044	.93132	853.73	.21261	16316.
0.05	0.44276	2.7718	2.91102	814.74	4.19231	15571.
.06	.42486	2.6598	.89311	781.83	.17440	14942.
.07	.40865	2.5624	.87691	753.20	.15820	14395.
.08	.39372	2.4758	.86198	727.75	.14327	13908.
.09	.37979	2.3977	.84804	704.76	.12933	13469.
0.10	0.36664	2.3262	2.83490	683.75	4.11619	13067.
.11	.35414	2.2602	.82240	664.36	.10369	12697.
.12	.34218	2.1988	.81044	646.31	.09173	12352.
.13	.33067	2.1413	.79893	629.40	.08022	12029.
.14	.31954	2.0871	.78780	613.47	.06909	11724.
0.15	0.30874	2.0358	2.77699	598.40	4.05828	11436.
.16	.29821	1.9871	.76647	584.08	.04776	11162.
.17	.28793	1.9406	.75619	570.41	.03748	10901.
.18	.27786	1.8961	.74612	557.34	.02741	10652.
.19	.26798	1.8534	.73624	544.80	.01753	10412.
0.20	0.25825	1.8124	2.72651	532.73	4.00780	10181.
.21	.24866	1.7728	.71692	521.10	3.99821	9958.9
.22	.23919	1.7346	.70745	509.86	.98874	9744.1
.23	.22983	1.6976	.69808	498.98	.97937	9536.2
.24	.22055	1.6617	.68880	488.43	.97010	9334.6
0.25	0.21134	1.6268	2.67960	478.19	3.96089	9138.9
.26	.20220	1.5930	.67046	468.23	.95175	8948.5
.27	.19312	1.5600	.66137	458.53	.94266	8763.2
.28	.18407	1.5278	.65232	449.08	.93361	8582.5
.29	.17505	1.4964	.64331	439.85	.92460	8406.2
0.30	0.16606	1.4657	2.63431	430.84	3.91560	8233.9
.31	.15708	1.4357	.62533	422.02	.90662	8065.4
.32	.14810	1.4064	.61636	413.39	.89765	7900.4
.33	.13912	1.3776	.60738	404.93	.88867	7738.8
.34	.13014	1.3494	.59840	396.64	.87969	7580.3
0.35	0.12114	1.3217	2.58939	388.50	3.87068	7424.8
.36	.11211	1.2945	.58037	380.51	.86166	7272.0
.37	.10305	1.2678	.57131	372.66	.85260	7122.0
.38	.09396	1.2415	.56222	364.93	.84351	6974.4
.39	.08482	1.2157	.55308	357.34	.83437	6829.2
0.40	0.07563	1.1902	2.54389	349.86	3.82518	6686.2
.41	.06639	1.1652	.53464	342.49	.81593	6545.4
.42	.05708	1.1405	.52533	335.22	.80662	6406.6
.43	.04770	1.1161	.51595	328.06	.79724	6269.7
.44	.03824	1.0920	.50650	320.99	.78779	6134.6
0.45	0.02870	1.0683	2.49696	314.02	3.77825	6001.3
.46	.01907	1.0449	.48733	307.13	.76862	5869.7
.47	.00934	1.0217	.47760	300.33	.75889	5739.7
.48	9.99951	0.99886	.46776	293.60	.74905	5611.2
.49	.98956	0.97624	.45782	286.96	.73911	5484.1
0.50	9.97949	0.95387	2.44775	280.38	3.72904	5358.4

† Kelvin, Mathematical and Physical Papers, vol. III. p. 428; Becker, Am. Jour. of Sci. vol. III. 1897, p. 280.

* For direct values see table 24.

DIFFUSION.

v/c	log 2q	2q	log δ	δ	log γ	γ
0.50	9.97949	0.95387	2.44775	280.38	3.72904	5358.4
.51	.96929	.93174	.43755	273.87	.71884	5234.1
.52	.95896	.90983	.42722	267.43	.70851	5111.0
.53	.94848	.88813	.41674	261.06	.69803	4989.1
.54	.93784	.86665	.40610	254.74	.68739	4868.4
0.55	9.92704	0.84536	2.39530	248.48	3.67659	4748.9
.56	.91607	.82426	.38432	242.28	.66561	4630.3
.57	.90490	.80335	.37316	236.13	.65445	4512.8
.58	.89354	.78260	.36180	230.04	.64309	4396.3
.59	.88197	.76203	.35023	223.99	.63152	4280.7
0.60	9.87018	0.74161	2.33843	217.99	3.61973	4166.1
.61	.85815	.72135	.32640	212.03	.60770	4052.2
.62	.84587	.70124	.31412	206.12	.59541	3939.2
.63	.83332	.68126	.30157	200.25	.58286	3827.0
.64	.82048	.66143	.28874	194.42	.57003	3715.6
0.65	9.80734	0.64172	2.27560	188.63	3.55689	3604.9
.66	.79388	.62213	.26214	182.87	.54343	3494.9
.67	.78008	.60266	.24833	177.15	.52962	3385.4
.68	.76590	.58331	.23416	171.46	.51545	3276.8
.69	.75133	.56407	.21959	165.80	.50088	3168.7
0.70	9.73634	0.54493	2.20459	160.17	3.48588	3061.1
.71	.72089	.52588	.18915	154.58	.47044	2954.2
.72	.70495	.50694	.17321	149.01	.45450	2847.7
.73	.68849	.48808	.15675	143.47	.43804	2741.8
.74	.67146	.46931	.13972	137.95	.42101	2636.4
0.75	9.65381	0.45062	2.12207	132.46	3.40336	2531.4
.76	.63550	.43202	.10376	126.99	.38505	2426.9
.77	.61646	.41348	.08471	121.54	.36600	2322.7
.78	.59662	.39502	.06487	116.11	.34616	2219.0
.79	.57590	.37662	.04416	110.70	.32545	2115.7
0.80	9.55423	0.35829	2.02249	105.31	3.30378	2012.7
.81	.53150	.34001	1.99975	99.943	.28104	1910.0
.82	.50758	.32180	.97584	94.589	.25713	1807.7
.83	.48235	.30363	.95061	89.250	.23190	1705.7
.84	.45564	.28552	.92389	83.926	.20518	1603.9
0.85	9.42725	0.26745	1.89551	78.615	3.17680	1502.4
.86	.39695	.24943	.86521	73.317	.14650	1401.2
.87	.36445	.23145	.83271	68.032	.11400	1300.2
.88	.32940	.21350	.79766	62.757	.07895	1199.4
.89	.29135	.19559	.75961	57.492	3.04090	1098.7
0.90	9.24972	0.17771	1.71797	52.236	2.99926	998.31
.91	.20374	.15986	.67200	46.989	.95329	898.03
.92	.15239	.14203	.62065	41.750	.90194	797.89
.93	.09423	.12423	.56249	36.516	.84378	697.88
.94	9.02714	.10645	.49539	31.289	.77668	597.98
0.95	8.94783	0.08868	1.41609	26.067	2.69738	498.17
.96	.85082	.07093	.31907	20.848	.60036	398.44
.97	.72580	.05319	.19406	15.633	.47535	298.78
.98	.54965	.03545	.01791	10.421	.29920	199.16
.99	.24859	.01773	0.71684	5.21007	1.99813	99.571
1.00	$-\infty$	0.00000	$-\infty$	0.00000	$-\infty$	0.000

Table 32.

GAMMA FUNCTION.*

Value of $\log \int_0^\infty e^{-x} x^{n-1} dx + 10.$

Values of the logarithms + 10 of the "Second Eulerian Integral" (Gamma function) $\int_0^\infty e^{-x} x^{n-1} dx$ or $\log \Gamma(n) + 10$ for values of n between 1 and 2. When n has values not lying between 1 and 2 the value of the function can be readily calculated from the equation $\Gamma(n+1) = n\Gamma(n) = n(n-1) \ldots (n-r)\Gamma(n-r)$.

n	0	1	2	3	4	5	6	7	8	9
1.00	9.99——	97497	95001	92512	90030	87555	85087	82627	80173	77727
1.01	75287	72855	70430	68011	65600	63196	60798	58408	56025	53648
1.02	51279	48916	46561	44212	41870	39535	37207	34886	32572	30265
1.03	27964	25671	23384	21104	18831	16564	14305	12052	09806	07567
1.04	05334	03108	00889	98677	96471	94273	92080	89895	87716	85544
1.05	9.9883379	81220	79068	76922	74783	72651	70525	68406	66294	64188
1.06	62089	59996	57910	55830	53757	51690	49630	47577	45530	43489
1.07	41455	39428	37407	35392	33384	31382	29387	27398	25415	23430
1.08	21469	19506	17549	15599	13655	11717	09785	07860	05941	04029
1.09	02123	00223	98329	96442	94561	92686	90818	88956	87100	85250
1.10	9.9783407	81570	79738	77914	76095	74283	72476	70676	68882	67095
1.11	65313	63538	61768	60005	58248	56497	54753	53014	51281	49555
1.12	47834	46120	44411	42709	41013	39323	37638	35960	34288	32622
1.13	30962	29308	27659	26017	24381	22751	21126	19508	17896	16289
1.14	14689	13094	11505	09922	08345	06774	05209	03650	02096	00549
1.15	9.9699007	97471	95941	94417	92898	91386	89879	88378	86883	85393
1.16	83910	82432	80960	79493	78033	76578	75129	73686	72248	70816
1.17	69390	67969	66554	65145	63742	62344	60952	59566	58185	56810
1.18	55440	54076	52718	51366	50019	48677	47341	46011	44687	43368
1.19	42054	40746	39444	38147	36856	35570	34290	33016	31747	30483
1.20	9.9629225	27973	26725	25484	24248	23017	21792	20573	19358	18150
1.21	16946	15748	14556	13369	12188	11011	09841	08675	07515	06361
1.22	05212	04068	02930	01796	00669	99546	98430	97318	96212	95111
1.23	594015	92925	91840	90760	89685	88616	87553	86494	85441	84393
1.24	83350	82313	81280	80253	79232	78215	77204	76198	75197	74201
1.25	9.9573211	72226	71246	70271	69301	68337	67377	66423	65474	64530
1.26	63592	62658	61730	60806	59888	58975	58067	57165	56267	55374
1.27	54487	53604	52727	51855	50988	50126	49268	48416	47570	46728
1.28	45891	45059	44232	43410	42593	41782	40975	40173	39376	38585
1.29	37798	37016	36239	35467	34700	33938	33181	32429	31682	30940
1.30	9.9530203	29470	28743	28021	27303	26590	25883	25180	24482	23789
1.31	23100	22417	21739	21065	20396	19732	19073	18419	17770	17125
1.32	16485	15850	15220	14595	13975	13359	12748	12142	11541	10944
1.33	10353	09766	09184	08606	08034	07466	06903	06344	05791	05242
1.34	04698	04158	03624	03094	02568	02048	01532	01021	00514	00012
1.35	9.9499515	99023	98535	98052	97573	97100	96630	96166	95706	95251
1.36	94800	94355	93913	93477	93044	92617	92194	91776	91362	90953
1.37	90549	90149	89754	89363	88977	88595	88218	87846	87478	87115
1.38	86756	86402	86052	85707	85366	85030	84698	84371	84049	83731
1.39	83417	83108	82803	82503	82208	81916	81630	81348	81070	80797
1.40	9.9480528	80263	80003	79748	79497	79250	79008	78770	78537	78308
1.41	78084	77864	77648	77437	77230	77027	76829	76636	76446	76261
1.42	76081	75905	75733	75565	75402	75243	75089	74939	74793	74652
1.43	74515	74382	74254	74130	74010	73894	73783	73676	73574	73476
1.44	73382	73292	73207	73125	73049	72976	72908	72844	72784	72728

* Legendre's "Exercises de Calcul Intégral," tome ii.

GAMMA FUNCTION.

n	0	1	2	3	4	5	6	7	8	9
1.45	9.9472677	72630	72587	72549	72514	72484	72459	72437	72419	72406
1.46	72397	72393	72392	72396	72404	72416	72432	72452	72477	72506
1.47	72539	72576	72617	72662	72712	72766	72824	72886	72952	73022
1.48	73097	73175	73258	73345	73436	73531	73630	73734	73841	73953
1.49	74068	74188	74312	74440	74572	74708	74848	74992	75141	75293
1.50	9.9475449	75610	75774	75943	76116	76292	76473	76658	76847	77040
1.51	77237	77437	77642	77851	78064	78281	78502	78727	78956	79189
1.52	79426	79667	79912	80161	80414	80671	80932	81196	81465	81738
1.53	82015	82295	82580	82868	83161	83457	83758	84062	84370	84682
1.54	84998	85318	85642	85970	86302	86638	86977	87321	87668	88019
1.55	9.9488374	88733	89096	89463	89834	90208	90587	90969	91355	91745
1.56	92139	92537	92938	93344	93753	94166	94583	95004	95429	95857
1.57	96289	96725	97165	97609	98056	98508	98963	99422	99885	00351
1.58	500822	01296	01774	02255	02741	03230	03723	04220	04720	05225
1.59	05733	06245	06760	07280	07803	08330	08860	09395	09933	10475
1.60	9.9511020	11569	12122	12679	13240	13804	14372	14943	15519	16098
1.61	16680	17267	17857	18451	19048	19649	20254	20862	21475	22091
1.62	22710	23333	23960	24591	25225	25863	26504	27149	27798	28451
1.63	29107	29766	30430	31097	31767	32442	33120	33801	34486	35175
1.64	35867	36563	37263	37966	38673	39383	40097	40815	41536	42260
1.65	9.9542989	43721	44456	45195	45938	46684	47434	48187	48944	49704
1.66	50468	51236	52007	52782	53560	54342	55127	55916	56708	57504
1.67	58303	59106	59913	60723	61536	62353	63174	63998	64825	65656
1.68	66491	67329	68170	69015	69864	70716	71571	72430	73293	74159
1.69	75028	75901	76777	77657	78540	79427	80317	81211	82108	83008
1.70	9.9583912	84820	85731	86645	87563	88484	89409	90337	91268	92203
1.71	93141	94083	95028	95977	96929	97884	98843	99805	00771	01740
1.72	602712	03688	04667	05650	06636	07625	08618	09614	10613	11616
1.73	12622	13632	14645	15661	16681	17704	18730	19760	20793	21830
1.74	22869	23912	24959	26009	27062	28118	29178	30241	31308	32377
1.75	9.9633451	34527	35607	36690	37776	38866	39959	41055	42155	43258
1.76	44364	45473	46586	47702	48821	49944	51070	52199	53331	54467
1.77	55606	56749	57894	59043	60195	61350	62509	63671	64836	66004
1.78	67176	68351	69529	70710	71895	73082	74274	75468	76665	77866
1.79	79070	80277	81488	82701	83918	85138	86361	87588	88818	90051
1.80	9.9691287	92526	93768	95014	96263	97515	98770	00029	01291	02555
1.81	703823	05095	06369	07646	08927	10211	11498	12788	14082	15378
1.82	16678	17981	19287	20596	21908	23224	24542	25864	27189	28517
1.83	29848	31182	32520	33860	35204	36551	37900	39254	40610	41969
1.84	43331	44697	46065	47437	48812	50190	51571	52955	54342	55733
1.85	9.9757126	58522	59922	61325	62730	64139	65551	66966	68384	69805
1.86	71230	72657	74087	75521	76957	78397	79839	81285	82734	84186
1.87	85640	87098	88559	90023	91490	92960	94433	95909	97389	98871
1.88	800356	01844	03335	04830	06327	07827	09331	10837	12346	13859
1.89	15374	16893	18414	19939	21466	22996	24530	26066	27606	29148
1.90	9.9830693	32242	33793	35348	36905	38465	40028	41595	43164	44736
1.91	46311	47890	49471	51055	52642	54232	55825	57421	59020	60621
1.92	62226	63834	65445	67058	68675	70294	71917	73542	75170	76802
1.93	78436	80073	81713	83356	85002	86651	88302	89957	91614	93275
1.94	94938	96605	98274	99946	01621	03299	04980	06663	08350	10039
1.95	9.9911732	13427	15125	16826	18530	20237	21947	23659	25375	27093
1.96	28815	30539	32266	33995	35728	37464	39202	40943	42688	44435
1.97	46185	47937	49693	51451	53213	54977	56744	58513	60286	62062
1.98	63840	65621	67405	69192	70982	72774	74570	76368	78169	79972
1.99	81779	83588	85401	87216	89034	90854	92678	94504	96333	98165

TABLE 33.

ZONAL SPHERICAL HARMONICS.*

Degrees	P_1	P_2	P_3	P_4	P_5	P_6	P_7
0	+ 1.0000	+ 1.0000	+ 1.0000	+ 1.0000	+ 1.0000	+ 1.0000	+ 1.0000
1	.9998	.9995	.9991	.9985	.9977	.9968	.9957
2	.9994	.9982	.9963	.9939	.9909	.9872	.9830
3	.9986	.9959	.9918	.9863	.9795	.9714	.9620
4	.9976	.9927	.9854	.9758	.9638	.9495	.9329
5	+ 0.9962	+ 0.9886	+ 0.9773	+ 0.9623	+ 0.9437	+ 0.9216	+ 0.8962
6	.9945	.9836	.9674	.9459	.9194	.8881	.8522
7	.9925	.9777	.9557	.9267	.8911	.8492	.8016
8	.9903	.9709	.9423	.9048	.8589	.8054	.7449
9	.9877	.9633	.9273	.8803	.8232	.7570	.6830
10	+ 0.9848	+ 0.9548	+ 0.9106	+ 0.8532	+ 0.7840	+ 0.7045	+ 0.6164
11	.9816	.9454	.8923	.8238	.7417	.6483	.5462
12	.9781	.9352	.8724	.7920	.6966	.5891	.4731
13	.9744	.9241	.8511	.7582	.6489	.5273	.3980
14	.9703	.9122	.8283	.7224	.5990	.4635	.3218
15	+ 0.9659	+ 0.8995	+ 0.8042	+ 0.6847	+ 0.5471	+ 0.3983	+ 0.2455
16	.9613	.8860	.7787	.6454	.4937	.3323	+ .1700
17	.9563	.8718	.7519	.6046	.4391	.2661	+ .0961
18	.9511	.8568	.7240	.5624	.3836	.2002	+ .0248
19	.9455	.8410	.6950	.5192	.3276	.1353	− .0433
20	+ 0.9397	+ 0.8245	+ 0.6649	+ 0.4750	+ 0.2715	+ 0.0719	− 0.1072
21	.9336	.8074	.6338	.4300	.2156	+ .0106	.1664
22	.9272	.7895	.6019	.3845	.1602	− .0481	.2202
23	.9205	.7710	.5692	.3386	.1057	− .1038	.2680
24	.9135	.7518	.5357	.2926	.0525	− .1558	.3094
25	+ 0.9063	+ 0.7321	+ 0.5016	+ 0.2465	+ 0.0009	− 0.2040	− 0.3441
26	.8988	.7117	.4670	.2007	− .0489	.2478	.3717
27	.8910	.6908	.4319	.1553	− .0964	.2869	.3922
28	.8829	.6694	.3964	.1105	− .1415	.3212	.4053
29	.8746	.6474	.3607	.0665	− .1839	.3502	.4113
30	+ 0.8660	+ 0.6250	+ 0.3248	+ 0.0234	− 0.2233	− 0.3740	− 0.4102
31	.8572	.6021	.2887	− .0185	.2595	.3924	.4022
32	.8480	.5788	.2527	− .0591	.2923	.4053	.3877
33	.8387	.5551	.2167	− .0982	.3216	.4127	.3671
34	.8290	.5310	.1809	− .1357	.3473	.4147	.3409
35	+ 0.8192	+ 0.5065	+ 0.1454	− 0.1714	− 0.3691	− 0.4114	− 0.3096
36	.8090	.4818	.1102	.2052	.3871	.4031	.2738
37	.7986	.4567	.0755	.2370	.4011	.3898	.2343
38	.7880	.4314	.0413	.2666	.4112	.3719	.1918
39	.7771	.4059	.0077	.2940	.4174	.3497	.1470
40	+ 0.7660	+ 0.3802	− 0.0252	− 0.3190	− 0.4197	− 0.3236	− 0.1006
41	.7547	.3544	.0574	.3416	.4181	.2939	− .0535
42	.7431	.3284	.0887	.3616	.4128	.2610	− .0064
43	.7314	.3023	.1191	.3791	.4038	.2255	+ .0398
44	.7193	.2762	.1485	.3940	.3914	.1878	+ .0846
45	+ 0.7071	+ 0.2500	− 0.1768	− 0.4063	− 0.3757	− 0.1484	+ 0.1271
46	.6947	.2238	.2040	.4158	.3568	− .1078	.1667
47	.6820	.1977	.2300	.4227	.3350	− .0665	.2028
48	.6691	.1716	.2547	.4270	.3105	− .0251	.2350
49	.6561	.1456	.2781	.4286	.2836	+ .0161	.2626
50	+ 0.6428	+ 0.1198	− 0.3002	− 0.4275	− 0.2545	+ 0.0564	+ 0.2854

* Calculated by Mr. C. E. Van Orstrand for this publication.

Table 33 (*continued*).

ZONAL SPHERICAL HARMONICS.

Degrees	P_1	P_2	P_3	P_4	P_5	P_6	P_7
50	+ 0.6428	+ 0.1198	− 0.3002	− 0.4275	− 0.2545	+ 0.0564	+ 0.2854
51	.6293	.0941	.3209	.4239	.2235	.0954	.3031
52	.6157	.0686	.3401	.4178	.1910	.1326	.3154
53	.6018	.0433	.3578	.4093	.1571	.1677	.3221
54	.5878	.0182	.3740	.3984	.1223	.2002	.3234
55	+ 0.5736	− 0.0065	− 0.3886	− 0.3852	− 0.0868	+ 0.2297	+ 0.3191
56	.5592	.0310	.4016	.3698	− .0509	.2560	.3095
57	.5446	.0551	.4131	.3524	− .0150	.2787	.2947
58	.5299	.0788	.4229	.3331	+ .0206	.2976	.2752
59	.5150	.1021	.4310	.3119	+ .0557	.3125	.2512
60	+ 0.5000	− 0.1250	− 0.4375	− 0.2891	+ 0.0898	+ 0.3232	+ 0.2231
61	.4848	.1474	.4423	.2647	.1229	.3298	.1916
62	.4695	.1694	.4455	.2390	.1545	.3321	.1572
63	.4540	.1908	.4471	.2121	.1844	.3302	.1203
64	.4384	.2117	.4470	.1841	.2123	.3240	.0818
65	+ 0.4226	− 0.2321	− 0.4452	− 0.1552	+ 0.2381	+ 0.3138	+ 0.0422
66	.4067	.2518	.4419	.1256	.2615	.2997	+ .0022
67	.3907	.2710	.4370	.0955	.2824	.2819	− .0375
68	.3746	.2895	.4305	.0651	.3005	.2606	− .0763
69	.3584	.3074	.4225	.0344	.3158	.2362	− .1135
70	+ 0.3420	− 0.3245	− 0.4130	− 0.0038	+ 0.3281	+ 0.2089	− 0.1485
71	.3256	.3410	.4021	+ .0267	.3373	.1791	.1808
72	.3090	.3568	.3898	.0568	.3434	.1472	.2099
73	.2924	.3718	.3761	.0864	.3463	.1136	.2352
74	.2756	.3860	.3611	.1153	.3461	.0788	.2563
75	+ 0.2588	− 0.3995	− 0.3449	+ 0.1434	+ 0.3427	+ 0.0431	− 0.2730
76	.2419	.4122	.3275	.1705	.3362	+ .0070	.2850
77	.2250	.4241	.3090	.1964	.3267	− .0290	.2921
78	.2079	.4352	.2894	.2211	.3143	− .0644	.2942
79	.1908	.4454	.2688	.2443	.2990	− .0990	.2913
80	+ 0.1736	− 0.4548	− 0.2474	+ 0.2659	+ 0.2810	− 0.1321	− 0.2835
81	.1564	.4633	.2251	.2859	.2606	.1635	.2708
82	.1392	.4709	.2020	.3040	.2378	.1927	.2536
83	.1219	.4777	.1783	.3203	.2129	.2193	.2321
84	.1045	.4836	.1539	.3345	.1861	.2431	.2067
85	+ 0.0872	− 0.4886	− 0.1291	+ 0.3468	+ 0.1577	− 0.2638	− 0.1778
86	.0698	.4927	.1038	.3569	.1278	.2810	.1460
87	.0523	.4959	.0781	.3648	.0969	.2947	.1117
88	.0349	.4982	.0522	.3704	.0651	.3045	.0755
89	.0175	.4995	.0262	.3739	.0327	.3105	.0381
90	+ 0.0000	− 0.5000	− 0.0000	+ 0.3750	+ 0.0000	− 0.3125	− 0.0000

Table 34.

CYLINDRICAL HARMONICS OF THE 0TH AND 1ST ORDERS

Values when $n = 0$ and 1 of the Bessel function $J_n(x)$

$$= \frac{x^n}{2^n\Gamma(n+1)}\left\{1 - \frac{x^2}{2^2(n+1)} + \frac{x^4}{2^4.2!(n+1)(n+2)} \ldots\right\}. \qquad J_1(x) = -J_0'(x) = \frac{dJ_0(x)}{dx}.$$

x	$J_0(x)$	$J_1(x)$	x	$J_0(x)$	$J_1(x)$	x	$J_0(x)$	$J_1(x)$	x	$J_0(x)$	$J_1(x)$
.00	unity	zero	**.50**	.938470	.242268	**1.00**	.765198	.440051	**1.50**	.511828	.557937
.01	.999975	.005000	.51	.936024	.246799	.01	.760781	.443286	.51	.506241	.559315
.02	.999900	.010000	.52	.933534	.251310	.02	.756332	.446488	.52	.500642	.560653
.03	.999775	.014998	.53	.930998	.255803	.03	.751851	.449658	.53	.495028	.561951
.04	.999600	.019996	.54	.928418	.260277	.04	.747339	.452794	.54	.489403	.563208
.05	.999375	.024992	**.55**	.925793	.264732	**1.05**	.742796	.455897	**1.55**	.483764	.564424
.06	.999100	.029987	.56	.923123	.269166	.06	.738221	.458966	.56	.478114	.565600
.07	.998775	.034979	.57	.920410	.273581	.07	.733616	.462001	.57	.472453	.566735
.08	.998401	.039968	.58	.917652	.277975	.08	.728981	.465003	.58	.466780	.567830
.09	.997976	.044954	.59	.914850	.282349	.09	.724316	.467970	.59	.461096	.568883
.10	.997502	.049938	**.60**	.912005	.286701	**1.10**	.719622	.470902	**1.60**	.455402	.569896
.11	.996977	.054917	.61	.909116	.291032	.11	.714898	.473800	.61	.449698	.570868
.12	.996403	.059892	.62	.905184	.295341	.12	.710146	.476663	.62	.443985	.571798
.13	.995779	.064863	.63	.903209	.299628	.13	.705365	.479491	.63	.438262	.572688
.14	.995106	.069829	.64	.900192	.303893	.14	.700556	.482284	.64	.432531	.573537
.15	.994383	.074789	**.65**	.897132	.308135	**1.15**	.695720	.485041	**1.65**	.426792	.574344
.16	.993610	.079744	.66	.894029	.312355	.16	.690856	.487763	.66	.421045	.575111
.17	.992788	.084693	.67	.890885	.316551	.17	.685965	.490449	.67	.415290	.575836
.18	.991916	.089636	.68	.887698	.320723	.18	.681047	.493098	.68	.409528	.576520
.19	.990995	.094572	.69	.884470	.324871	.19	.676103	.495712	.69	.403760	.577163
.20	.990025	.099501	**.70**	.881201	.328996	**1.20**	.671133	.498289	**1.70**	.397985	.577765
.21	.989005	.104422	.71	.877890	.333096	.21	.666137	.500830	.71	.392204	.578326
.22	.987937	.109336	.72	.874539	.337170	.22	.661116	.503334	.72	.386418	.578845
.23	.986819	.114241	.73	.871147	.341220	.23	.656071	.505801	.73	.380628	.579323
.24	.985652	.119138	.74	.867715	.345245	.24	.651000	.508231	.74	.374832	.579760
.25	.984436	.124026	**.75**	.864242	.349244	**1.25**	.645906	.510623	**1.75**	.369033	.580156
.26	.983171	.128905	.76	.860730	.353216	.26	.640788	.512979	.76	.363229	.580511
.27	.981858	.133774	.77	.857178	.357163	.27	.635647	.515296	.77	.357422	.580824
.28	.980496	.138632	.78	.853587	.361083	.28	.630482	.517577	.78	.351613	.581096
.29	.979085	.143481	.79	.849956	.364976	.29	.625295	.519819	.79	.345801	.581327
.30	.977626	.148319	**.80**	.846287	.368842	**1.30**	.620086	.522023	**1.80**	.339986	.581517
.31	.976119	.153146	.81	.842580	.372681	.31	.614855	.524189	.81	.334170	.581666
.32	.974563	.157961	.82	.838834	.376492	.32	.609602	.526317	.82	.328353	.581773
.33	.972960	.162764	.83	.835050	.380275	.33	.604329	.528407	.83	.322535	.581840
.34	.971308	.167555	.84	.831228	.384029	.34	.599034	.530458	.84	.316717	.581865
.35	.969609	.172334	**.85**	.827369	.387755	**1.35**	.593720	.532470	**1.85**	.310898	.581849
.36	.967861	.177100	.86	.823473	.391453	.36	.588385	.534444	.86	.305080	.581793
.37	.966067	.181852	.87	.819541	.395121	.37	.583031	.536379	.87	.299262	.581695
.38	.964224	.186591	.88	.815571	.398760	.38	.577658	.538274	.88	.293446	.581557
.39	.962335	.191316	.89	.811565	.402370	.39	.572266	.540131	.89	.286631	.581377
.40	.960398	.196027	**.90**	.807524	.405950	**1.40**	.566855	.541948	**1.90**	.281819	.581157
.41	.958414	.200723	.91	.803447	.409499	.41	.561427	.543726	.91	.276008	.580896
.42	.956384	.205403	.92	.799334	.413018	.42	.555981	.545464	.92	.270201	.580595
.43	.954306	.210069	.93	.795186	.416507	.43	.550518	.547162	.93	.264397	.580252
.44	.952183	.214719	.94	.791004	.419965	.44	.545038	.548821	.94	.258596	.579870
.45	.950012	.219353	**.95**	.786787	.423392	**1.45**	.539541	.550441	**1.95**	.252799	.579446
.46	.947796	.223970	.96	.782536	.426787	.46	.534029	.552020	.96	.247007	.578983
.47	.945533	.228571	.97	.778251	.430151	.47	.528501	.553559	.97	.241220	.578478
.48	.943224	.233154	.98	.773933	.433483	.48	.522958	.555059	.98	.235438	.577934
.49	.940870	.237720	.99	.769582	.436783	.49	.517400	.556518	.99	.229661	.577349
.50	.938470	.242268	**1.00**	.765198	.440051	**1.50**	.511828	.557937	**2.00**	.223891	.576725

TABLE 34 (*continued*).

CYLINDRICAL HARMONICS OF THE 0TH AND 1ST ORDERS.

$J_1(x) = -J_0'(x)$. Other orders may be obtained from the relation, $J_{n+1}(x) = \frac{2n}{x} J_n(x) - J_{n-1}(x)$.

$J_{-n}(x) = (-1)^n J_n(x)$.

x	$J_0(x)$	$J_1(x)$	x	$J_0(x)$	$J_1(x)$	x	$J_0(x)$	$J_1(x)$	x	$J_0(x)$	$J_1(x)$
2.00	.223891	.576725	**2.50**	−.048384	.497094	**3.00**	−.260052	.339059	**3.50**	−.380128	.137378
.01	.218127	.576060	.51	−.053342	.494606	.01	−.263424	.335319	.51	−.381481	.133183
.02	.212370	.575355	.52	−.058276	.492086	.02	−.266758	.331563	.52	−.382791	.128989
.03	.206620	.574611	.53	−.063184	.489535	.03	−.270055	.327789	.53	−.384060	.124795
.04	.200878	.573827	.54	−.068066	.486953	.04	−.273314	.323998	.54	−.385287	.120601
2.05	.195143	.573003	**2.55**	−.072923	.484340	**3.05**	−.276535	.320191	**3.55**	−.386472	.116408
.06	.189418	572139	.56	−.077753	.481696	06	−.279718	.316368	.56	−.387615	.112216
.07	.183701	.571236	.57	−.082557	.479021	.07	−.282862	.312529	.57	−.388717	.108025
.08	.177993	.570294	.58	−.087333	.476317	.08	−.285968	.308675	.58	−.389776	.103836
.09	.172295	.569313	.59	−.092083	.473582	.09	−.289036	.304805	.59	−.390793	.099650
2.10	.166607	.568292	**2.60**	−.096805	.470818	**3.10**	−.292064	.300921	**3.60**	−.391769	.095466
.11	.160929	.567233	.61	−.101499	.468025	.11	−.295054	.297023	.61	−.392703	.091284
.12	.155262	.566134	.62	−.106165	.465202	.12	−.298005	.293110	.62	−.393595	.087106
.13	.149607	.564997	.63	−.110803	.462350	.13	−.300916	.289184	.63	−.394445	.082931
.14	.143963	.563821	.64	−.115412	.459470	.14	−.303788	.285244	.64	−.395253	.078760
2.15	.138330	.562607	**2.65**	−.119992	.456561	**3.15**	−.306621	.281291	**3.65**	−.396020	.074593
.16	.132711	.561354	.66	−.124543	.453625	.16	−.309414	.277326	.66	−.396745	.070431
.17	.127104	.560063	.67	−.129065	.450660	.17	−.312168	.273348	.67	−.397429	.066274
.18	.121509	.558735	.68	−.133557	.447668	.18	−.314881	.269358	.68	−.398071	.062122
.19	.115929	.557368	.69	−.138018	.444648	.19	−.317555	.265356	.69	−.398671	.057975
2.20	.110362	.555963	**2.70**	−.142449	.441601	**3.20**	−.320188	.261343	**3.70**	−.399230	.053834
.21	.104810	.554521	.71	−.146850	.438528	.21	−.322781	.257319	.71	−.399748	.049699
.22	.099272	.553041	.72	−.151220	.435428	.22	−.325335	.253284	.72	−.400224	.045571
.23	.093749	.551524	.73	−.155559	.432302	.23	−.327847	.249239	.73	−.400659	.041450
.24	.088242	.549970	.74	−.159866	.429150	.24	−.330319	.245184	.74	−.401053	.037336
2.25	.082750	.548378	**2.75**	−.164141	.425972	**3.25**	−.332751	.241120	**3.75**	−.401406	.033229
.26	.077274	.546750	.76	−.168385	.422769	.26	−.335142	.237046	.76	−.401718	.029131
27	.071815	.545085	.77	−.172597	.419541	.27	−.337492	.232963	.77	−.401989	.025040
.28	.066373	.543384	.78	−.176776	.416288	.28	−.339801	.228871	.78	−.402219	.020958
.29	.060947	.541646	.79	−.180922	.413011	.29	−.342069	.224771	.79	−.402408	.016885
2.30	.055540	.539873	**2.80**	−.185036	.409709	**3.30**	−.344296	.220663	**3.80**	−.402556	.012821
.31	.050150	.538063	.81	−.189117	.406384	.31	−.346482	.216548	.81	−.402664	.008766
.32	.044779	.536217	.82	−.193164	.403035	.32	−.348627	.212425	.82	−.402732	.004722
.33	.039426	.534336	.83	−.197177	.399662	.33	−.350731	.208296	.83	−.402759	.000687
.34	.034092	.532419	.84	−.201157	.396267	.34	−.352793	.204160	.84	−.402746	−.003337
2.35	.028778	.530467	**2.85**	−.205102	.392849	**3.35**	−.354814	.200018	**3.85**	−.402692	−.007350
.36	.023483	.528480	.86	−.209014	.389408	.36	−.356793	.195870	.86	−.402599	−.011352
.37	.018208	.526458	.87	−.212890	.385945	.37	−.358731	.191716	.87	−.402465	−.015343
.38	.012954	.524402	.88	−.216733	.382461	.38	−.360628	.187557	.88	−.402292	−.019322
.39	.007720	.522311	.89	−.220540	.378955	.39	−.362482	.183394	.89	−.402079	−.023289
2.40	.002508	.520185	**2.90**	−.224312	.375427	**3.40**	−.364296	.179226	**3.90**	−.401826	−.027244
.41	−.002683	.518026	.91	−.228048	.371879	.41	−.366067	.175054	.91	−.401534	−.031186
.42	−.007853	.515833	.92	−.231749	.368311	.42	−.367797	.170878	.92	−.401202	−.035115
.43	−.013000	.513606	.93	−.235414	.364722	.43	−.369485	.166699	.93	−.400832	−.039031
.44	−.018125	.511346	.94	−.239043	.361113	.44	−.371131	.162516	.94	−.400422	−.042933
2.45	−.023227	.509052	**2.95**	−.242636	.357485	**3.45**	−.372735	.158331	**3.95**	−.399973	−.046821
.46	−.028306	.506726	.96	−.246193	.353837	.46	−.374297	.154144	.96	−.399485	−.050695
.47	−.033361	.504366	.97	−.249713	.350170	.47	−.375818	.149954	.97	−.398959	−.054555
.48	−.038393	.501974	.98	−.253196	.346484	.48	−.377296	.145763	.98	−.398394	−.058400
.49	−.043401	.499550	.99	−.256643	.342781	.49	−.378733	.141571	.99	−.397791	−.062229
2.50	−.048384	.497094	**3.00**	−.260052	.339059	**3.50**	−.380128	.137378	**4.00**	−.397150	.066043

CYLINDRICAL HARMONICS OF 0TH AND 1ST ORDERS.

TABLE 35. — 4-place Values for $x = 4.0$ to 15.0.

x	$J_0(x)$	$J_1(x)$	x	$J_0(x)$	$J'(x)$
4.0	−.3972	−.0660	9.5	−.1939	+.1613
.1	−.3887	−.1033	.6	−.2090	.1395
.2	−.3766	−.1386	.7	−.2218	.1166
.3	−.3610	−.1719	.8	−.2323	.0928
.4	−.3423	−.2028	.9	−.2403	.0684
4.5	−.3205	−.2311	10.0	−.2459	.0435
.6	−.2961	−.2566	.1	−.2490	+.0184
.7	−.2693	−.2791	.2	−.2496	−.0066
.8	−.2404	−.2985	.3	−.2477	−.0313
.9	−.2097	−.3147	.4	−.2434	−.0555
5.0	−.1776	−.3276	10.5	−.2366	−.0789
.1	−.1443	−.3371	.6	−.2276	−.1012
.2	−.1103	−.3432	.7	−.2164	−.1224
.3	−.0758	−.3460	.8	−.2032	−.1422
.4	−.0412	−.3453	.9	−.1881	−.1603
5.5	−.0068	−.3414	11.0	−.1712	−.1768
.6	+.0270	−.3343	.1	−.1528	−.1913
.7	.0599	−.3241	.2	−.1330	−.2039
.8	.0917	−.3110	.3	−.1121	−.2143
.9	.1220	−.2951	.4	−.0902	−.2225
6.0	.1506	−.2767	11.5	−.0677	−.2284
.1	.1773	−.2559	.6	−.0446	−.2320
.2	.2017	−.2329	.7	−.0213	−.2333
.3	.2238	−.2081	.8	+.0020	−.2323
.4	.2433	−.1816	.9	.0250	−.2290
6.5	.2601	−.1538	12.0	.0477	−.2234
.6	.2740	−.1250	.1	.0697	−.2157
.7	.2851	−.0953	.2	.0908	−.2060
.8	.2931	−.0652	.3	.1108	−.1943
.9	.2981	−.0349	.4	.1296	−.1807
7.0	.3001	−.0047	12.5	.1469	−.1655
.1	.2991	+.0252	.6	.1626	−.1487
.2	.2951	.0543	.7	.1766	−.1307
.3	.2882	.0826	.8	.1887	−.1114
.4	.2786	.1096	.9	.1988	−.0912
7.5	.2663	.1352	13.0	.2069	−.0703
.6	.2516	.1592	.1	.2129	−.0489
.7	.2346	.1813	.2	.2167	−.0271
.8	.2154	.2014	.3	.2183	−.0052
.9	.1944	.2192	.4	.2177	+.0166
8.0	.1717	.2346	13.5	.2150	.0380
.1	.1475	.2476	.6	.2101	.0590
.2	.1222	.2580	.7	.2032	.0791
.3	.0960	.2657	.8	.1943	.0984
.4	.0692	.2708	.9	.1836	.1165
8.5	.0419	.2731	14.0	.1711	.1334
.6	.0146	.2728	.1	.1570	.1488
.7	−.0125	.2697	.2	.1414	.1626
.8	−.0392	.2641	.3	.1245	.1747
.9	−.0653	.2559	.4	.1065	.1850
9.0	−.0903	.2453	14.5	.0875	.1934
.1	−.1142	.2324	.6	.0679	.1999
.2	−.1367	.2174	.7	.0476	.2043
.3	−.1577	.2004	.8	.0271	.2066
.4	−.1768	.1816	.9	.0064	.2069
9.5	−.1939	.1613	15.0	−.0142	.2051

TABLE 36. — Roots.

(*a*) 1st 10 roots of $J_0(x) = 0$

Higher roots may be calculated to better than 1 part in 10,000 by the approximate formula

$$R_m = R_{m-1} + \pi$$

$$R_1 = 2.404826$$
$$R_2 = 5.520078$$
$$R_3 = 8.653728$$
$$R_4 = 11.791534$$
$$R_5 = 14.930918$$
$$R_6 = 18.071064$$
$$R_7 = 21.211637$$
$$R_8 = 24.352472$$
$$R_9 = 27.493479$$
$$R_{10} = 30.634606$$

(*b*) 1st 15 roots of $J_1(x) = \frac{dJ_0(x)}{dx} = 0$ with corresponding values of maximum or or minimum values of $J_0(x)$.

No. of root (n)	Root = x_n.	$J_0(x_n)$.
1	3.831706	−.402759
2	7.015587	+.300116
3	10.173468	−.249705
4	13.323692	+.218359
5	16.470630	−.196465
6	19.615859	+.180063
7	22.760084	−.167185
8	25.903672	+.156725
9	29.046829	−.148011
10	32.189680	+.140606
11	35.332308	−.134211
12	38.474766	+.128617
13	41.617094	−.123668
14	44.759319	+.119250
15	47.901461	−.115274

Higher roots may be obtained as under (*a*).

NOTES. $y = J_n(x)$ is a particular solution of Bessel's equation,

$$x^2 \frac{d^2y}{dx^2} + x\frac{dy}{dx} + (x^2 - n^2)y = 0.$$

The general formula for $J_n(x)$ is

$$J_n(x) = \sum_0^\infty \frac{(-1)^s x^{n+2s}}{2^{n+2s}\pi s\ \pi(n+s)},$$

or

$$= \sum_0^\infty \frac{(-1)^s x^{n+2s}}{2^{n+2s} s!\,(n+s)!}$$

when n is an integer and

$$J_{n+1}(x) = \frac{2n}{x} J_n(x) - J_{n-1}(x),$$

and

$$J_1(x) = \frac{dJ_0(x)}{dx},$$

$$J_{-n}(x) = (-1)^n J_n(x).$$

Tables 35 to 36 are based upon Gray and Matthews' reprints from Dr. Meissel's tables. See also Reports of British Association, 1907–1916.

TABLE 37.

ELLIPTIC INTEGRALS.

Values of $\int_0^{\frac{\pi}{2}}(1-\sin^2\theta\sin^2\phi)^{\pm\frac{1}{2}}\,d\phi$.

This table gives the values of the integrals between 0 and $\pi/2$ of the function $(1-\sin^2\theta\sin^2\phi)^{\pm\frac{1}{2}}\,d\phi$ for different values of the modulus corresponding to each degree of θ between 0 and 90.

θ	$\int_0^{\frac{\pi}{2}}\frac{d\phi}{(1-\sin^2\theta\sin^2\phi)^{\frac{1}{2}}}$		$\int_0^{\frac{\pi}{2}}(1-\sin^2\theta\sin^2\phi)^{\frac{1}{2}}d\phi$		θ	$\int_0^{\frac{\pi}{2}}\frac{d\phi}{(1-\sin^2\theta\sin^2\phi)^{\frac{1}{2}}}$		$\int_0^{\frac{\pi}{2}}(1-\sin^2\theta\sin^2\phi)^{\frac{1}{2}}d\phi$	
	Number.	Log.	Number.	Log.		Number.	Log.	Number.	Log.
0°	1.5708	0.196120	1.5708	0.196120	45°	1.8541	0.268127	1.3506	0.130541
1	5709	196153	5707	196087	6	8691	271644	3418	127690
2	5713	196252	5703	195988	7	8848	275267	3329	124788
3	5719	196418	5697	195822	8	9011	279001	3238	121836
4	5727	196649	5689	195591	9	9180	282848	3147	118836
5°	1.5738	0.196947	1.5678	0.195293	50°	1.9356	0.286811	1.3055	0.115790
6	5751	197312	5665	194930	1	9539	290895	2963	112698
7	5767	197743	5649	194500	2	9729	295101	2870	109563
8	5785	198241	5632	194004	3	9927	299435	2776	106386
9	5805	198806	5611	193442	4	2.0133	303901	2681	103169
10°	1.5828	0.199438	1.5589	0.192815	55°	2.0347	0.308504	1.2587	0.099915
1	5854	200137	5564	192121	6	0571	313247	2492	096626
2	5882	200904	5537	191362	7	0804	318138	2397	093303
3	5913	201740	5507	190537	8	1047	323182	2301	089950
4	5946	202643	5476	189646	9	1300	328384	2206	086569
15°	1.5981	0.203615	1.5442	0.188690	60°	2.1565	0.333753	1.2111	0.083164
6	6020	204657	5405	187668	1	1842	339295	2015	079738
7	6061	205768	5367	186581	2	2132	345020	1920	076293
8	6105	206948	5326	185428	3	2435	350936	1826	072834
9	6151	208200	5283	184210	4	2754	357053	1732	069364
20°	1.6200	0.209522	1.5238	0.182928	65°	2.3088	0.363384	1.1638	0.065889
1	6252	210916	5191	181580	6	3439	369940	1545	062412
2	6307	212382	5141	180168	7	3809	376736	1453	058937
3	6365	213921	5090	178691	8	4198	383787	1362	055472
4	6426	215533	5037	177150	9	4610	391112	1272	052020
25°	1.6490	0.217219	1.4981	0.175545	70°	2.5046	0.398730	1.1184	0.048589
6	6557	218981	4924	173876	1	5507	406665	1096	045183
7	6627	220818	4864	172144	2	5998	414943	1011	041812
8	6701	222732	4803	170348	3	6521	423596	0927	038481
9	6777	224723	4740	168489	4	7081	432660	0844	035200
30°	1.6858	0.226793	1.4675	0.166567	75°	2.7681	0.442176	1.0764	0.031976
1	6941	228943	4608	164583	6	8327	452196	0686	028819
2	7028	231173	4539	162537	7	9026	462782	0611	025740
3	7119	233485	4469	160429	8	9786	474008	0538	022749
4	7214	235880	4397	158261	9	3.0617	485967	0468	019858
35°	1.7312	0.238359	1.4323	0.156031	80°	3.1534	0.498777	1.0401	0.017081
6	7415	240923	4248	153742	1	2553	512591	0338	014432
7	7522	243575	4171	151393	2	3699	527613	0278	011927
8	7633	246315	4092	148985	3	5004	544120	0223	009584
9	7748	249146	4013	146519	4	6519	562514	0172	007422
40°	1.7868	0.252068	1.3931	0.143995	85°	3.8317	0.583396	1.0127	0.005465
1	7992	255085	3849	141414	6	4.0528	607751	0086	003740
2	8122	258197	3765	138778	7	3387	637355	0053	002278
3	8256	261406	3680	136086	8	7427	676027	0026	001121
4	8396	264716	3594	133340	9	5.4349	735192	0008	000326
45°	1.8541	0.268127	1.3506	0.130541	90°	∞	∞	1.0000	——

MOMENTS OF INERTIA, RADII OF GYRATION, AND WEIGHTS.

In each case the axis is supposed to traverse the centre of gravity of the body. The axis is one of symmetry. The mass of a unit of volume is w.

Body.	Axis.	Weight.	Moment of Inertia I_0.	Square of Radius of Gyration ρ_0^2.
Sphere of radius r	Diameter	$\frac{4\pi w r^3}{3}$	$\frac{8\pi w r^5}{15}$	$\frac{2r^2}{5}$
Spheroid of revolution, polar axis $2a$, equatorial diameter $2r$	Polar axis	$\frac{4\pi w a r^2}{3}$	$\frac{8\pi w a r^4}{15}$	$\frac{2r^2}{5}$
Ellipsoid, axes $2a$, $2b$, $2c$	Axis $2a$	$\frac{4\pi w abc}{3}$	$\frac{4\pi w abc(b^2+c^2)}{15}$	$\frac{b^2+c^2}{5}$
Spherical shell, external radius r, internal r'	Diameter	$\frac{4\pi w(r^3-r'^3)}{3}$	$\frac{8\pi w(r^5-r'^5)}{15}$	$\frac{2(r^5-r'^5)}{5(r^3-r'^3)}$
Ditto, insensibly thin, radius r, thickness dr	Diameter	$4\pi w r^2 dr$	$\frac{8\pi w r^4 dr}{3}$	$\frac{2r^2}{3}$
Circular cylinder, length $2a$, radius r	Longitudinal axis $2a$	$2\pi w a r^2$	$\pi w a r^4$	$\frac{r^2}{2}$
Elliptic cylinder, length $2a$, transverse axes $2b$, $2c$	Longitudinal axis $2a$	$2\pi w abc$	$\frac{\pi w abc(b^2+c^2)}{2}$	$\frac{b^2+c^2}{4}$
Hollow circular cylinder, length $2a$, external radius r, internal r'	Longitudinal axis $2a$	$2\pi w a(r^2-r'^2)$	$\pi w a(r^4-r'^4)$	$\frac{r^2+r'^2}{2}$
Ditto, insensibly thin, thickness dr	Longitudinal axis $2a$	$4\pi w a r dr$	$4\pi w a r^3 dr$	r^2
Circular cylinder, length $2a$, radius r	Transverse diameter	$2\pi w a r^2$	$\frac{\pi w a r^2(3r^2+4a^2)}{6}$	$\frac{r^2}{4}+\frac{a^2}{3}$
Elliptic cylinder, length $2a$, transverse axes $2a$, $2b$	Transverse axis $2b$	$2\pi w abc$	$\frac{\pi w abc(3c^2+4a^2)}{6}$	$\frac{c^2}{4}+\frac{a^2}{3}$
Hollow circular cylinder, length $2a$, external radius r, internal r'	Transverse diameter	$2\pi w a(r^2-r'^2)$	$\frac{\pi w a}{6}\left\{\begin{matrix}3(r^4-r'^4)\\+4a^2(r^2-r'^2)\end{matrix}\right\}$	$\frac{r^2+r'^2}{4}+\frac{a^2}{3}$
Ditto, insensibly thin, thickness dr	Transverse diameter	$4\pi w a r dr$	$\pi w a(2r^3+\frac{4}{3}a^2 r)dr$	$\frac{r^2}{2}+\frac{a^2}{3}$
Rectangular prism, dimensions $2a$, $2b$, $2c$	Axis $2a$	$8wabc$	$\frac{8wabc(b^2+c^2)}{3}$	$\frac{b^2+c^2}{3}$
Rhombic prism, length $2a$, diagonals $2b$, $2c$	Axis $2a$	$4wabc$	$\frac{2wabc(b^2+c^2)}{3}$	$\frac{b^2+c^2}{6}$
Ditto	Diagonal $2b$	$4wabc$	$\frac{2wabc(c^2+2a^2)}{3}$	$\frac{c^2}{6}+\frac{a^2}{3}$

(Taken from Rankine.)

For further mathematical data see Smithsonian Mathematical Tables, Becker and Van Orstrand (Hyperbolic, Circular and Exponential Functions); Functionentafeln, Jahnke und Emde (xtgx, x^{-1}tgx, Roots of Transcendental Equations, $a + bi$ and $re^{\vartheta i}$, Exponentials, Hyperbolic Functions, $\int_0^x \frac{\sin u}{u}du$, $\int_x^\infty \frac{\cos u}{u}du$, $\int_\infty^{-x} \frac{e^{-u}}{u}du$, Fresnel Integral, Gamma Function, Gauss Integral $\frac{2}{\sqrt{\pi}}\int_0^x e^{-x^2}dx$, Pearson Function $e^{-\frac{1}{2}\pi\nu}\int_0^\pi \sin^r e^{\nu x}dx$, Elliptic Integrals and Functions, Spherical and Cylindrical Functions, etc.). For further references see under Tables, Mathematical, in the 11th ed. Encyclopædia Britannica. See also Carr's Synopsis of Pure Mathematics and Mellor's Higher Mathematics for Students of Chemistry and Physics.

TABLE 39.

INTERNATIONAL ATOMIC WEIGHTS. VALENCIES.

The International Atomic Weights are quoted from the report of the International Committee on Atomic Weights (Journal American Chemical Society. 44, 427, 1922).

Substance.	Symbol.	Relative atomic wt. Oxygen = 16.	Valency.	Substance.	Symbol.	Relative atomic wt. Oxygen = 16.	Valency.
Aluminum	Al	27.0	3.	Mercury	Hg	200.6	1, 2.
Antimony	Sb	120.2	3, 5.	Molybdenum	Mo	96.0	4, 6.
Argon	A	39.9	0.	Neodymium	Nd	144.3	3.
Arsenic	As	74.96	3, 5.	Neon	Ne	20.2	0.
Barium	Ba	137.37	2.	Nickel	Ni	58.68	2, 3.
Bismuth	Bi	209.0	3, 5.	Niton (Ra emanation)	Nt.	222.4	—
Boron	B	10.9	3.	Nitrogen	N	14.008	3, 5.
Bromine	Br	79.92	1.	Osmium	Os	190.9	6, 8.
Cadmium	Cd	112.40	2.	Oxygen	O	16.00	2.
Cesium	Cs	132.81	1.	Palladium	Pd	106.7	2, 4.
Calcium	Ca	40.07	2.	Phosphorus	P	31.04	3, 5.
Carbon	C	12.005	4.	Platinum	Pt	195.2	2, 4.
Cerium	Ce	140.25	3, 4.	Potassium	K	39.10	1.
Chlorine	Cl	35.46	1.	Praseodymium	Pr	140.9	3.
Chromium	Cr	52.0	2, 3, 6.	Radium	Ra	226.0	2.
Cobalt	Co	58.97	2, 3.	Rhodium	Rh	102.9	3.
Columbium	Cb	93.1	5.	Rubidium	Rb	85.45	1.
Copper	Cu	63.57	1, 2.	Ruthenium	Ru	101.7	6, 8.
Dysprosium	Dy	162.5	3.	Samarium	Sa	150.4	3.
Erbium	Er	167.7	3.	Scandium	Sc	45.1	3.
Europium	Eu	152.0	3.	Selenium	Se	79.2	2, 4, 6.
Fluorine	F	19.0	1.	Silicon	Si	28.1	4.
Gadolinium	Gd	157.3	3.	Silver	Ag	107.88	1.
Gallium	Ga	70.1	3.	Sodium	Na	23.00	1.
Germanium	Ge	72.5	4.	Strontium	Sr	87.63	2.
Glucinum	Gl	9.1	2.	Sulphur	S	32.06	2, 4, 6.
Gold	Au	197.2	1, 3.	Tantalum	Ta	181.5	5.
Helium	He	4.00	0.	Tellurium	Te	127.5	2, 4, 6.
Holmium	Ho	163.5	3.	Terbium	Tb	159.2	3.
Hydrogen	H	1.008	1.	Thallium	Tl	204.0	1, 3.
				Thorium	Th	232.15	4.
Indium	In	114.8	3.				
Iodine	I	126.92	1.	Thulium	Tm	169.9	3.
Iridium	Ir	193.1	4.	Tin	Sn	118.7	2, 4.
Iron	Fe	55.84	2, 3.	Titanium	Ti	48.1	4.
Krypton	Kr	82.92	0.	Tungsten	W	184.0	6.
				Uranium	U	238.2	4, 6.
Lanthanum	La	139.0	3.				
Lead	Pb	207.20	2, 4.	Vanadium	V	51.0	3, 5.
Lithium	Li	6.94	1.	Xenon	Xe	130.2	0.
Lutecium	Lu	175.0	3.	Ytterbium	Yb	173.5	3.
Magnesium	Mg	24.32	2.	Yttrium	Yt	89.33	3.
Manganese	Mn	54.93	2, 3, 7.	Zinc	Zn	65.37	2.
				Zirconium	Zr	90.6	4.

VOLUME OF A GLASS VESSEL FROM THE WEIGHT OF ITS EQUIVALENT VOLUME OF MERCURY OR WATER.

If a glass vessel contains at $t°$ C, P grammes of mercury, weighed with brass weights in air at 760 mm. pressure, then its volume in c. cm.

at the same temperature, t, : $V = PR = P\frac{p}{d}$,

at another temperature, t_1, : $V = PR_1 = Pp/d\,\{1 + \gamma\,(t_1 - t)\}$

p = the weight, reduced to vacuum, of the mass of mercury or water which, weighed with brass weights, equals 1 gram;

d = the density of mercury or water at $t°$ C,

and γ = 0.000 025, is the cubical expansion coefficient of glass.

Temperature t	WATER.			MERCURY.		
	R.	R_1, $t_1 = 10°$.	R_1, $t_1 = 20°$.	R.	R_1, $t_1 = 10°$.	R_1, $t_1 = 20°$.
0°	1.001192	1.001443	1.001693	0.0735499	0.0735683	0.0735867
1	1133	1358	1609	5633	5798	5982
2	1092	1292	1542	5766	5914	6098
3	1068	1243	1493	5900	6029	6213
4	1060	1210	1460	6033	6144	6328
5	1068	1193	1443	6167	6259	6443
6	1.001092	1.001192	1.001442	0.0736301	0.0736374	0.0736558
7	1131	1206	1456	6434	6490	6674
8	1184	1234	1485	6568	6605	6789
9	1252	1277	1527	6702	6720	6904
10	1333	1333	1584	6835	6835	7020
11	1.001428	1.001403	1.001653	0.0736969	0.0736951	0.0737135
12	1536	1486	1736	7103	7066	7250
13	1657	1582	1832	7236	7181	7365
14	1790	1690	1940	7370	7297	7481
15	1935	1810	2060	7504	7412	7596
16	1.002092	1.001942	1.002193	0.0737637	0.0737527	0.0737711
17	2261	2086	2337	7771	7642	7826
18	2441	2241	2491	7905	7757	7941
19	2633	2407	2658	8039	7872	8057
20	2835	2584	2835	8172	7988	8172
21	1.003048	1.002772	1.003023	0.0738306	0.0738103	0.0738288
22	3271	2970	3220	8440	8218	8403
23	3504	3178	3429	8573	8333	8518
24	3748	3396	3647	8707	8449	8633
25	4001	3624	3875	8841	8564	8748
26	1.004264	1.003862	1.004113	0.0738974	0.0738679	0.0738864
27	4537	4110	4361	9108	8794	8979
28	4818	4366	4616	9242	8910	9094
29	5110	4632	4884	9376	9025	9210
30	5410	4908	5159	9510	9140	9325

Taken from Landolt, Börnstein, and Meyerhoffer's Physikalisch-Chemische Tabellen.

REDUCTIONS OF WEIGHINGS IN AIR TO VACUO.

TABLE 41.

When the weight M in grams of a body is determined in air, a correction is necessary for the buoyancy of the air equal to $M\delta(1/d - 1/d_1)$ where δ = the density (wt. of 1 ccm in grams = 0.0012) of the air during the weighing, d the density of the body, d_1 that of the weights. δ for various barometric values and humidities may be determined from Tables 153 to 155. The following table is computed for $\delta = 0.0012$. The corrected weight $= M + kM/1000$.

Density of body weighed d.	Correction factor, k.			Density of body weighed d.	Correction factor, k.		
	Pt. Ir. weights $d_1 = 21.5$.	Brass weights 8.4.	Quartz or Al. weights 2.65.		Pt. Ir. weights $d_1 = 21.5$.	Brass weights 8.4.	Quartz or Al. weights 2.65.
.5	+2.34	+2.26	+1.95	1.6	+0.69	+0.61	+0.30
.6	+1.94	+1.86	+1.55	1.7	+ .65	+ .56	+ .25
.7	+1.66	+1.57	+1.26	1.8	+ .62	+ .52	+ .21
.75	+1.55	+1.46	+1.15	1.9	+ .58	+ .49	+ .18
.80	+1.44	+1.36	+1.05	2.0	+ .54	+ .46	+ .15
.85	+1.36	+1.27	+0.96	2.5	+ .43	+ .34	+ .03
.90	+1.28	+1.19	+ .88	3.0	+ .34	+ .26	— .05
.95	+1.21	+1.12	+ .81	4.0	+ .24	+ .16	— .15
1.00	+1.14	+1.06	+ .75	6.0	+ .14	+ .06	— .25
1.1	+1.04	+0.95	+ .64	8.0	+ .09	+ .01	— .30
1.2	+0.94	+ .86	+ .55	10.0	+ .06	— .02	— .33
1.3	+ .87	+ .78	+ .47	15.0	+ .03	— .06	— .37
1.4	+ .80	+ .71	+ .40	20.0	+ .004	— .08	— .39
1.5	+ .75	+ .66	+ .35	22.0	— .001	— .09	— .40

TABLE 42.—Reductions of Densities in Air to Vacuo.

(This correction may be accomplished through the use of the above table for each separate weighing.)

If s is the density of the substance as calculated from the uncorrected weights, S its true density, and L the true density of the liquid used, then the vacuum correction to be applied to the uncorrected density, s, is $0.0012(1 - s/L)$.

Let W_s = uncorrected weight of substance, W_l = uncorrected weight of the liquid displaced by the substance, then by definition, $s = LW_s/W_l$. Assuming D to be the density of the balance of weights, $W_s\{1 + 0.0012(1/S - 1/D)\}$ and $W_l\{1 + 0.0012(1/L - 1/D)\}$ are the true weights of the substance and liquid respectively (assuming that the weighings are made under normal atmospheric corrections, so that the weight of 1 cc. of air is 0.0012 gram).

Then the true density $S = \dfrac{W_s\{1 + 0.0012(1/S - 1/D)\}}{W_l\{1 + 0.0012(1/L - 1/D)\}}L.$

But from above $W_s/W_l = s/L$, and since L is always large compared with 0.0012,

$$S - s = 0.0012(1 - s/L).$$

The values of $0.0012(1 - s/L)$ for densities up to 20 and for liquids of density 1 (water), 0.852 (xylene) and 13.55 (mercury) follow:

(See reference below for discussion of density determinations).

Density of substance s.	Corrections.			Density of substance s	Corrections.	
	L = 1 Water.	L = 0.852 Xylene.	L = 13.55 Mercury.		L = 1 Water.	L = 13.55 Mercury.
0.8	+0.00024	-	-	11.	—0.0120	+0.0002
0.9	+ .00012	-	-	12.	— .0132	+ .0001
1.	0.0000	—0.0002	+0.0011	13.	— .0144	0.0000
2.	— .0012	— .0016	+ .0010	14.	— .0156	0.0000
3.	— .0024	— .0030	+ .0009	15.	— .0168	— .0001
4.	— .0036	— .0044	+ .0008	16.	— .0180	— .0002
5.	— .0048	— .0058	+ .0008	17.	— .0192	— .0003
6.	— .0060	— .0073	+ .0007	18.	— .0204	— .0004
7.	— .0072	— .0087	+ .0006	19.	— .0216	— .0005
8.	— .0084	— .0101	+ .0005	20.	— .0228	— .0006
9.	— .0096	— .0115	+ .0004			
10.	— .0108	— .0129	+ .0003			

Johnston and Adams, J. Am. Chem. Soc. 34, p. 563, 1912.

TABLE 43.

MECHANICAL PROPERTIES.*

* Compiled from various sources by Harvey A. Anderson, C.E., Assistant Engineer Physicist, U. S. Bureau of Standards.

The mechanical properties of most materials vary between wide limits; the following figures are given as being representative rather than what may be expected from an individual sample. Figures denoting such properties are commonly given either as specification or experimental values. Unless otherwise shown, the values below are experimental. Credit for information included is due the U. S. Bureau of Standards; the Am. Soc. for Testing Materials; the Soc. of Automotive Eng.; the Motor Transport Corps, U. S. War Dept.; the Inst. of Mech. Eng.; the Inst. of Metals; Forest Products Lab.; Dept. of Agriculture (Bull. 556); Moore's Materials of Engineering; Hatfield's Cast Iron; and various other American, English and French authorities.

The specified properties shown are indicated minimums as prescribed by the Am. Soc. for Testing Materials, U. S. Navy Dept., Panama Canal, Soc. of Automotive Eng., or Intern. Aircraft Standards Board. In the majority of cases, specifications show a range for chemical constituents and the average value only of this range is quoted. Corresponding average values are in general given for mechanical properties. In general, tensile test specimens were 12.8 mm (0.505 in.) diameter and 50.8 mm (2 in.) gage length. Sizes of compressive and transverse specimens are generally shown accompanying the data.

All data shown in these tables are as determined at ordinary room temperature, averaging 20° C (68° F.). The properties of most metals and alloys vary considerably from the values shown when the tests are conducted at higher or lower temperatures.

The following definitions govern the more commonly confused terms shown in the tables. In all cases the stress referred to in the definitions is equal to the total load at that stage of the test divided by the original cross-sectional area of the specimen (or the corresponding stress in the extreme fiber as computed from the flexure formula for transverse tests).

Proportional Limit (abbreviated P-limit). — Stress at which the deformation (or deflection) ceases to be proportional to the load (determined with extensometer for tension, compressometer for compression and deflectometer for transverse tests).

Elastic Limit. — Stress which produces a permanent elongation (or shortening) of 0.001 per cent of the gage length, as shown by an instrument capable of this degree of precision (determined from set readings with extensometer or compressometer). In transverse tests the extreme fiber stress at an appreciable permanent deflection.

Yield Point. — Stress at which marked increase in deformation (or deflection) of specimen occurs without increase in load (determined usually by drop of beam or with dividers for tension, compression or transverse tests).

Ultimate Strength in Tension or Compression. — Maximum stress developed in the material during test.

Modulus of Rupture. — Maximum stress in the extreme fiber of a beam tested to rupture, as computed by the empirical application of the flexure formula to stresses above the transverse proportional limit.

Modulus of Elasticity (Young's Modulus). — Ratio of stress within the proportional limit to the corresponding strain, — as determined with an extensometer. Note: All moduli shown are obtained from tensile tests of materials, unless otherwise stated.

Brinell Hardness Numeral (abbreviated B. h. n.). — Ratio of pressure on a sphere used to indent the material to be tested to the area of the spherical indentation produced. The standard sphere used is a 10-mm diameter hardened steel ball. The pressures used are 3000 kg for steel and 500 kg for softer metals, and the time of application of pressure is 30 seconds. Values shown in the tables are based on spherical areas computed in the main from measurements of the diameters of the spherical indentations, by the following formula:

$$\text{B. h. n.} = P \div \pi t D = P \div \pi D(D/2 - \sqrt{D^2/4 - d^2/4}).$$

P = pressure in kg, t = depth of indentation, D = diameter of ball, and d = diameter of indentation, — all lengths being expressed in mm. Brinell hardness values have a direct relation to tensile strength, and hardness determinations may be used to define tensile strengths by employing the proper conversion factor for the material under consideration.

Shore Scleroscope Hardness. — Height of rebound of diamond pointed hammer falling by its own weight on the object. The hardness is measured on an empirical scale on which the average hardness of martensitic high carbon steel equals 100. On very soft metals a "magnifier" hammer is used in place of the commonly used "universal" hammer and values may be converted to the corresponding "universal" value by multiplying the reading by $\frac{4}{7}$. The scleroscope hardness, when accurately determined, is an index of the tensile elastic limit of the metal tested.

Erichsen Value. — Index of forming quality of sheet metal. The test is conducted by supporting the sheet on a circular ring and deforming it at the center of the ring by a spherical pointed tool. The depth of impression (or cup) in mm required to obtain fracture is the Erichsen value for the metal. Erichsen standard values for trade qualities of soft metal sheets are furnished by the manufacturer of the machine corresponding to various sheet thicknesses. (See Proc. A. S. T. M. 17, part 2, p. 200, 1917.)

Alloy steels are commonly used in the heat treated condition, as strength increases are not commensurate with increases in production costs for annealed alloy steels. Corresponding strength values are accordingly shown for annealed alloy steels and for such steels after having been given certain recommended heat treatments of the Society of Automotive Engineers. The heat treatments followed in obtaining the properties shown are outlined on the pages immediately following the tables on steel. It will be noted that considerable latitude is allowed in the indicated drawing temperatures and corresponding wide variations in physical properties may be obtained with each heat treatment. The properties vary also with the size of the specimens heat treated. The drawing temperature is shown with the letter denoting the heat treatment, wherever the information is available.

TABLE 44. — Ferrous Metals and Alloys — Iron and Iron Alloys.

Metal. Grade.	Yield point. Tension kg/mm²	Ultimate strength. Tension kg/mm²	Yield point. Tension lb/in²	Ultimate strength. Tension lb/in²	Elong. in 50.8 mm (2 in.). Per cent.	Reduct. in area. Per cent.	Hardness. Brinell at 3000 kg	Hardness. Scleroscope.
Iron:								
Electrolytic* (remelt): as forged...	34.0	38.5	48,500	55,000	33.0	83.0	95 †	18
annealed 900° C.	12.5	27.0	18,000	38,000	52.0	87.0	75 †	—
Gray cast‡(19 mm diam. bars)	indet.	{17.5	indet.	{25,000	negli	gible	{100	{24
	—	26.5}	—	38,000}	—	—	150}	40}
Malleable cast, American (after	{14.0	{24.5	{20,000	{35,000	{15.0	{15.0	—	—
Hatfield)..................	31.5}	40.0}	45,000}	57,000}	4.5}	4.5}	—	—
European (after Am. Malleable	{19.0	{29.5	{27,000	{42,000	{6.0	{6.0	—	—
Castings Ass.)...............	28.0}	45.5}	40,000}	65,000}	2.0}	2.0}	—	—
(run of 24 successive heats, 1919)§	—	40.8	—	58,000	21.6	—	—	—
Commercial wrought...........	{19.5	{34.0	{28,000	{48,000	{40.0	{45.0	—	{25
	22.5}	37.0}	32,000}	53,000}	30.0}	35.0}	—	30}
Silicon alloys‖ **Si 0.01:** as forged...	29.5	31.5	41,800	45,200	35.0	78.0	—	—
(Melted in vacuo) ann. 970° C	11.0	24.5	16,000	34,900	53.0	81.5	—	—
(Note: C max. 0.01 per cent)								
Si 1.71: as forged...........	48.0	53.5	68,100	76,300	37.0	82.0	—	—
annealed 970° C.......	25.0	38.0	35,800	54,200	50.0	90.6	—	—
Si 4.40: as forged............	66.0	74.0	94,000	105,000	6.0	7.5	—	—
annealed 970° C	51.0	64.5	72,900	91,600	24.0	25.1	—	—
Aluminum alloys¶ **Al 0.00:** as forged	35.5	38.5	50,700	54,700	26.0	84.3	—	—
(Melted in vacuo) ann. 1000° C	12.5	24.5	17,600	34,900	60.0	93.5	—	—
(Note: C max. 0.01 per cent)								
Al 3.08: as forged...........	48.0	54.5	68,200	77,500	21.0	76.4	—	—
annealed 1000° C	22.5	37.5	31,800	53,400	51.0	85.3	—	—
Al 6.24: as forged...........	54.5	60.5	77,700	86,000	28.0	74.7	—	—
annealed 1000° C	37.5	49.0	53,400	69,800	27.0	55.5	—	—

Composition, approximate:
Electrolytic, C 0.0125 per cent; other impurities less than 0.05 per cent.
Cast, gray: Graphitic, C 3.0, Si 1.3 to 2.0, Mn 0.6 to 0.9, S max. 0.1, P max. 1.2.
A. S. T. M. Spec. A48 to 18 allows S max. 0.10, except S max. 0.12 for heavy castings.
Malleable: American "Black Heart," C 2.8 to 3.5, Si 0.6 to 0.8, Mn max. 0.4, S max. 0.07, P max. 0.2.
European "Steely Fracture," C 2.8 to 3.5, Si 0.6 to 0.8, Mn 0.15, S max. 0.35, P max. 0.2.

Compressive Strengths [Specimens tested: 25.4 mm (1 in.) diam. cylinders 76.2 mm (3 in.) long].
Electrolytic iron 56.5 kg/mm² or 80,000 lb/in².
Gray and malleable cast iron 56.5 to 84.5 kg/mm² or 80,000 to 120,000 lb/in².
Wrought iron, approximately equal to tensile yield point (slightly above P-limit).

Density:
Electrolytic iron............ 7.8 g/cm³ or 487 lb/ft³ Malleable iron............... 7.6 g/cm³ or 474 lb/ft³
Cast iron.................. 7.2 g/cm³ or 449 lb/ft³ Wrought iron............... 7.85 g/cm³ or 490 lb/ft³

Ductility: — Normal Erichsen values for good trade quality sheets, 0.4 mm (0.0156 in.)

Thickness, soft annealed.	Depth. mm	Depth. in.
Sheet metal hoop iron, polished	9.5	0.374
Charcoal iron tinned sheet	7.5	0.295
Second quality tinned sheet	6.7	0.264

Modulus of elasticity in tension and compression:
Electrolytic iron.... 17,500 kg/mm² or 25,000,000 lb/in² Malleable iron... 17,500 kg/mm² or 25,000,000 lb/in²
Cast iron.......... 10,500 kg/mm² or 15,000,000 lb/in² Wrought iron.... 17,500 kg/mm² or 25,000,000 lb/in²

Modulus of elasticity in shear:
Electrolytic iron....... 7030 kg/mm² or 10,000,000 lb/in² Cast iron .... 8450 kg/mm² or 12,000,000 lb/in²
Wrought iron...................... 7030 kg/mm² or 10,000,000 lb/in²

Scleroscope hardness values shown are as determined with the Shore Universal hammer.

Strength in Shear:
Electrolytic (remelt)
P-limit.............. 8.4 kg/mm² or 12,000 lb/in²
Ultimate strength...... 21.1 kg/mm² or 30,000 lb/in²
Commercial wrought
P-limit........... 21.1 kg/mm² or 30,000 lb/in²
Ultimate strength.. 35.0 kg/mm² or 50,000 lb/in²

Transverse strength, from flexure formula:
Gray cast iron
Modulus of rupture, 33.0 kg/mm² or 47,000 lb/in²
"Arbitration Bar," 31.8 mm (1¼ in.) diameter, or 304.8 mm (12 in.) span; minimum central load at rupture 1130 to 1500 kg (2500 to 3300 lb.); minimum central deflection at rupture 2.5 mm (0.1 in.), (A. S. T. M. Spec. A 48-18).

* Properties of Swedish iron (impurities less than 1 per cent) approximate those of electrolytic iron.
† These two values of B. h. n. only are as determined at 500 kg pressure.
‡ U. S. Navy specifies minimum tensile strength of 14.1 kg/mm² or 20,000 lb/in².
§ Averages for a U. S. foundry.
‖ From T. D. Yensen, University of Illinois, Engr. Exp. Station, Bulletin No. 83, 1915 (shows Si 4.40 as alloy of maximum strength).
¶ From T. D. Yensen, University of Illinois, Engr. Exp. Station, Bulletin No. 95, 1917.

MECHANICAL PROPERTIES OF MATERIALS.

TABLE 45. — Carbon Steels — Commercial Experimental Values.

S. A. E. (Soc. of Automotive Eng., U. S. A.) classification scheme used as basis for steel groupings. First two digits S. A. E. Spec. No. show steel group number, and last two (or three in case of five figures) show carbon content in hundredths of one per cent.

The first lines of properties for each steel show values for the rolled or forged metal in the annealed or normalized condition. Comparative heat-treated values show properties after receiving modified S. A. E. heat treatment as shown below (Table 46). The P-limit and ductility of cast steel average slightly lower and the ultimate strength 10 to 15 per cent higher than the values shown for the same composition steel in the annealed condition. The properties of rolled steel (raw) are approximately equal to those shown for the annealed condition, which represents the normalized condition of the metal rather than the soft annealed state.

The data for heat-treated strengths are average values for specimens for heat treatment ranging in size from ½ to 1½ in. diameter. The final drawing or quenching temperature for the properties shown is indicated in degrees C with the heat treatment letter, wherever the information is available. In general, specimens were drawn near the lower limit of the indicated temperature range.

Metal.	S.A.E. spec. no.	Nominal contents per cent.	S.A.E. heat treatment.	P-limit. Tension kg/mm²	Ultimate strength. Tension kg/mm²	P-limit. Tension lb/in²	Ultimate strength. Tension lb/in²	Elong. in 50.8 mm (2 in.). Per cent	Reduct. in area. Per cent	Hardness. Brinell @ 3000 kg.	Hardness. Scleroscope.
Steel, carbon	1010	See Spec. No.	Ann.	24.0	32.0	34,500	46,000	37.0	72.0	—	18
	1010		A	27.0	42.0	39,000	60,000	30.0	62.0	120	24
	1020	(Mn 0.45)	Ann.	28.0	38.0	39,500	54,400	32.0	68.0	100	17
	1020		H 230° C	35.0	56.0	49,500	79,500	20.0	59.0	176	35
	1045	(Mn 0.65)	Ann.	40.0	50.0	57,500	71,300	23.0	54.0	168	27
	1045		H 260° C	62.0	86.0	88,000	123,000	13.5	36.0	290	45
	1095	(Mn 0.35)	Ann.	42.0	56.0	59,500	79,000	21.0	51.0	187	29
	1095		F 510° C	84.0	123.0	120,000	175,000	6.0	18.0	551	75

Specification values: **Steel, castings,** Ann. A.S.T.M. A27-16, Class B;* P max. 0.06; S max. 0.05.

Grade.	Yield point.	Ultimate tensile strength kg/mm²	Ultimate tensile strength lb/in²	Per cent elong. 50.8 mm or 2 in.	Per cent reduct. area.
Hard	0.45 ultimate	56.2	80,000	15	20
Medium	0.45 "	49.2	70,000	18	25
Soft	0.45 "	42.2	60,000	22	30

Structural Steel: Rolled: S max. 0.05; P-Bess. max. 0.10; -O-H. max. 0.06.

Tension: Yield Point min. = 0.5 ultimate; ultimate = 38.7 to 45.7 kg/mm² or 55,000 to 65,000 lb/in² with 22% min. elongation in 50.8 mm (2 in.).

* Average carbon contents: steel castings, C 0.30 to 0.40; structural steel, C 0.15 to 0.30 (mild carbon or medium hard steel).

TABLE 46. — Explanation of Heat Treatment Letters used in Table of Steel Data.

Motor Transport Corps Modified S. A. E. Heat Treatments for Steels. (S. A. E. Handbook, Vol. 1, pp. 9d and 9e, 1915, q. v. for alternative treatments.)

Heat Treatment A. — After forging or machining (1) carbonize at a temperature between 870 and 930° C. (1600 and 1700° F.); (2) cool slowly; (3) reheat to 760 to 820° C. (1400 to 1500° F.) and quench in oil.

Heat Treatment D. — After forging or machining: (1) heat to 820 to 840° C. (1500 to 1550° F.); (2) quench; (3) reheat to 790 to 820° C. (1450 to 1500° F.); (4) quench; (5) reheat to 320 to 650° C. (600 to 1200° F.) and cool slowly.

Heat Treatment F. — After shaping or coiling: (1) heat to 775 to 800° C. (1425 to 1475° F.); (2) quench; (3) reheat to 200 to 480° C. (400 to 900° F.) in accordance with degree of temper required and cool slowly.

Heat Treatment H. — After forging or machining: (1) heat to 820 to 840° C. (1500 to 1550° F.); (2) quench; (3) reheat to 230 to 650° C. (450 to 1200° F.) and cool slowly.

Heat Treatment L. — After forging or machining: (1) carbonize at a temperature between 870 and 950° C. (1600 and 1750° F.), preferably between 900 and 930° C. (1650 and 1700 F.); (2) cool slowly in carbonizing material; (3) reheat to 790 to 820° C. (1450 to 1500° F.); (4) quench; (5) reheat to 700 to 760° C. (1300 to 1400° F.); (6) quench; (7) reheat to 120 to 260° C. (250 to 500° F.) and cool slowly.

Heat Treatment M. — After forging or machining: (1) heat to 790 to 820° C. (1450 to 1500° F.); (2) quench; (3) reheat to between 260 and 680° C. (500 and 1250° F.) and cool slowly.

Heat Treatment P. — After forging or machining: (1) heat to 790 to 820° C. (1450 to 1500° F.); (2) quench; (3) reheat to 750 to 770° C. (1375 to 1425° F.); (4) quench; (5) reheat to 260 to 650° C. (500 to 1200° F.) and cool slowly.

Heat Treatment T. — After forging or machining: (1) heat to 900 to 950° C. (1650 to 1750° F.); (2) quench; (3) reheat to 260 to 700° C. (500 to 1300° F.) and cool slowly.

Heat Treatment U. — After forging: (1) heat to 830 to 870° C. (1525 to 1600° F.), hold half an hour; (2) cool slowly; (3) reheat to 900 to 930° C. (1650 to 1700° F.); (4) quench; (5) reheat to 180 to 290° C. (350 to 550° F.) and cool slowly.

Heat Treatment V. — After forging or machining: (1) heat to 900 to 950° C. (1650 to 1750° F.); (2) quench; (3) reheat to between 200 and 650° C. (400 and 1200° F.) and cool slowly.

EDITOR'S NOTE: Oil quenching is recommended wherever the instructions specify "quench," inasmuch as the data in the table are taken from tests of automobile parts which must resist considerable vibration and which are usually small in section. The quenching medium must always be carefully considered.

TABLE 47. — Alloy Steels — Commercial Experimental Values.

Metal.	S. A. E. spec. no.	Nominal contents, per cent.	S. A. E. heat treatment.	P-limit. Tension kg/mm²	Ultimate strength. Tension kg/mm²	P-limit. Tension lb/in²	Ultimate strength. Tension lb/in²	Elong. in 50.8 mm (2 in.). Per cent.	Reduct. in area. Per cent.	Hardness. Brinell @3000 kg.	Hardness. Scleroscope.
Steel, nickel . .	2315	—	Ann.	30.0	38.0	42,500	54,000	32.0	60.0	138	—
	2315	—	H	53.0	76.0	75,000	107,500	18.0	55.0	321	43
	2335	Ni 3.50	Ann.	39.0	48.0	55,000	68,000	24.0	53.0	165	—
	2335		H	106.0	131.0	151,000	186,000	15.0	51.0	465	62
	2345	(Mn 0.65)	Ann.	44.0	55.0	62,500	78,000	21.0	48.0	172	—
	2345		H	136.0	149.0	193,000	212,000	12.0	45.0	570	76
	Invar	Ni 36.0 C 0.40	Ann.	50.0	77.5	71,000	110,000	30.0	50.0	—	—
nickel chrome	3120	Ni 1.25	Ann.	34.0	44.0	49,000	62,000	23.0	53.0	155	22
	3120	Cr 0.60	H 450° C	60.0	82.0	85,000	116,000	23.0	48.0	270	36
	3135		Ann.	40.0	50.0	57,000	71,300	20.0	46.0	182	30
	3135	(Mn 0.65)	H or D	88.0	121.0	125,000	172,000	18.0	43.0	330	44
	3220		Ann.	39.0	49.0	55,000	69,000	21.0	50.0	170	—
	3220	Ni 1.75	H or D	77.0	106.0	110,000	151,000	23.0	48.0	375	50
	3250	Cr 1.10	Ann.	44.0	55.0	62,000	78,000	19.0	42.0	180	—
	3250	(Mn 0.45)	M	134.0	183.0	190,000	260,000	16.0	32.0	480	64
	3320		Ann.	32.0	42.0	46,000	59,500	21.0	50.0	—	—
	3320	Ni 3.50	L	77.0	105.0	110,000	150,000	23.0	48.0	375	50
	3340	Cr 1.50	Ann.	39.0	52.0	56,000	74,000	18.0	45.0	—	—
	3340	(Mn 0.45)	P	120.0	163.0	170,000	232,000	18.0	42.0	479	64
chromium.	51120	Cr 1.00	Ann.	44.0	58.0	62,000	82,000	16.0	31.0	—	—
	51120	(Mn 0.35)	M or P	144.0	193.0	205,000	275,000	7.0	26.0	500	66
	52120	Cr 1.20	Ann.	44.0	58.0	62,000	82,000	13.0	24.0	—	—
	52120	(Mn 0.35)	M or P	141.0	178.0	200,000	253,000	7.0	25.0	524	70
chrome vanadium	6130	(Mn 0.65)	Ann.	43.0	59.0	61,500	84,500	23.0	51.0	152	—
	6130	Cr 0.95 V 0.18	T	84.0	115.0	120,000	163,000	16.0	43.0	432	59
	6195	(Mn 0.35)	Ann.	48.0	63.0	68,200	90,000	16.0	38.0	—	—
	6195		U	176.0	232.0	250,000	330,000	8.c	24.0	562	75
silico-manganese	9250	Si 1.95	Ann.	42.0	54.0	60,000	77,000	16.0	28.0	—	—
	9250	Mn 0.70	V	91.0	122.0	130,000	174,000	14.0	24.0	441	59
	9X30	Si 0.85	Ann.	48.0	61.0	68,000	87,000	13.0	22.0	—	—
	9X30	Mn 1.75	V	113.0	148.0	160,000	211,000	12.0	21.0	470	63
tungsten . .	(C–73)	W 2.4	Ann.	34.0	59.0	48,100	84,200	20.5	31.5	—	—
	(C–70)	W 9.7	Ann.	63.0	89.0	90,000	126,000	14.0	22.1	—	—
	(C–47)	W 15.6	Quench 1065° Draw 205° C	158.5	175.0	225,000	248,000	6.0	43.0	520	64

GENERAL NOTE. — Table on steels after Motor Transport Corps, Metallurgical Branch of Engineering Division, Table No. 88.

Maximum allowable P 0.045 or less, maximum allowable S 0.05 or less.

Silicon contents were not determined by Motor Transport Corps in preparing table, except for silico-manganese steels.

Compressive strengths:
For all steels approx. equal to yield point in tension (slightly above P-limit).

Density:
Steel weighs about 7.85 g/cm³ or 490 lb/ft³

Ductility, Erichsen values:
0.75 mm (0.029 in.) thick, low carbon soft annealed sheet (B. S.), depth of indentation 12.0 mm or 0.472 in.
1.30 mm (0.050 in.) thick, low carbon soft annealed sheet (B. S.), depth of indentation 12.5 mm or 0.492 in.

Modulus of elasticity in tension and compression:
For all steels approx. 21,000 kg/mm² = 30,000,000 lb/in².

Modulus of elasticity in shear:
For all steels approx. 8400 kg/mm² = 12,000,000 lb/in².

Scleroscope hardness values shown are as determined with the Shore Universal hammer.

Strength in shear:
P-limit and ultimate strength each about 70 per cent corresponding tensile values.

MECHANICAL PROPERTIES.

TABLE 48. — Steel Wire — Specification Values.

(After I. A. S. B. Specification 3S12, Sept., 1917, for High-strength Steel Wire.)

S. A. E. Carbon Steel, No. 1050 or higher number specified (see Carbon steels above). Steel used to be manufactured by acid open-hearth process, to be rolled, drawn, and then uniformly coated with pure tin to solder readily.

American or B. and S. wire gage.	Diameter.		Req'd twists in 203.2 mm or 8 in.	Weight.		Req'd bends thru 90°	Spec. minimum tensile strength.			
	mm	in.		kg/100 m	lb/100 ft.		kg	lb.	kg/mm²	lb/in²
6	4.115	0.162	16	10.44	7.01	5	2040	4500	154	219,000
7	3.665	.144	19	8.28	5.56	6	1680	3700	161	229,000
8	3.264	.129	21	6.55	4.40	8	1360	3000	164	233,000
9	2.906	.114	23	5.21	3.50	9	1135	2500	172	244,000
10	2.588	.102	26	4.12	2.77	11	910	2000	172	244,000
11	2.305	.091	30	3.28	2.20	14	735	1620	179	254,000
12	2.053	.081	33	2.60	1.74	17	590	1300	177	252,000
13	1.828	.072	37	2.06	1.38	21	470	1040	179	255,000
14	1.628	.064	42	1.64	1.10	25	375	830	181	258,000
15	1.450	.057	47	1.30	0.87	29	300	660	182	259,000
16	1.291	.051	53	1.03	0.69	34	245	540	186	264,000
17	1.150	.045	60	0.81	0.55	42	195	425	188	267,000
18	1.024	.040	67	0.65	0.43	52	155	340	190	270,000
19	0.912	.036	75	0.51	0.34	70	125	280	193	275,000
20	0.812	.032	85	0.41	0.27	85	100	225	197	280,000
21	0.723	.028	96	0.32	0.22	105	80	175	200	284,000

NOTE. — Number of 90° bends specified above to be obtained by bending sample about 4.76 mm (0.188 in.) radius, alternately, in opposite directions.

(Above specification corresponds to U. S. Navy Department Specification 22W6, Nov. 1, 1916, for tinned, galvanized or bright aeroplane wire.)

TABLE 49. — Steel Wire — Experimental Values.

(Data from tests at General Electric Company laboratories.) "Commercial Steel Music Wire (Hardened)."

Diameter.		Ultimate strength.	
mm	in.	kg/mm² tension lb/in²	
12.95	0.051	226.0	321,500
11.70	.046	249.0	354,000
9.15	.036	253.0	360,000
7.60	.030	260.0	370,000
6.35	.025	262.0	372,500
4.55	.018	265.5	378,000
2.55*	.010	386.5	550,000
1.65*	.0065	527.0	750,000
4.55†	.018	49.2	70,000

* For 4.55 mm wire drawn cold to indicated sizes. † For 4.55 mm (0.018 in.) wire annealed in H_2 at 850° C.

TABLE 50. — Semi-steel.

Test results at Bureau of Standards on 155-mm shell, Jan. 1919.
Microstructure — matrix resembling pearlitic steel, embedded in which are flakes of graphite.
Composition-Comb. C 0.60 to 0.76, Mn 0.88, P 0.42 to 0.43, S 0.077 to 0.088, Si 1.22 to 1.23, graphitic C 2.84 to 2.94.

Metal.	P-limit.	Ultimate strength.	P-limit.	Ultimate strength.	P-limit.	Ultimate strength.	P-limit.	Ultimate strength.	Hardness.	
	Tension kg/mm²		Tension lb/in²		Compression kg/mm²		Compression lb/in²		Brinell @3000 kg	Scleroscope.
Semi-steel:										
Graph. C 2.85, Comb. C 0.76	7.9	19.8	11,200	28,200	24.3	72.6	34,500	103,000	176	—
Graph. C 2.92, Comb. C 0.60	4.2	14.9	6,000	21,200	18.3	61.4	26,000	87,300	170	—

Tension specimens 12.7 mm (0.5 in.) diameter, 50.8 mm (2 in.) gage length; elongation and reduction of area negligible.
Compression specimens 20.3 mm (0.8 in.) diameter, 61.0 mm (2.4 in.) long; failure occurring in shear.
Tension set readings with extensometer showed elastic limit of 2.1 kg/mm² or 3000 lb/in².
Modulus of elasticity in tension — 9560 kg/mm² or 13,600,000 lb/in².

TABLE 51. — Steel-wire Rope — Specification Values.

Cast steel wire to be of hard crucible steel with minimum tensile strength of 155 kg/mm^2 or 220,000 lb/in^2 and minimum elongation of 2 per cent in 254 mm (10 in.).

Plow steel wire to be of hard crucible steel with minimum tensile strength of 183 kg/mm^2 or 260,000 lb/in^2 and minimum elongation of 2 per cent in 254 mm (10 in.).

Annealed steel wire to be of crucible cast steel, annealed, with minimum tensile strength of 77 kg/mm^2 or 110,000 lb/in^2 and minimum elongation of 7 per cent in 254 mm (10 in.).

Type A: 6 strands with hemp core and 19 wires to a strand (= 6 × 19), or 6 strands with hemp core and 18 wires to a strand with jute, cotton or hemp center.

Type B: 6 strands with hemp core, and 12 wires to a strand with hemp center.

Type C: 6 strands with hemp core, and 14 wires to a strand with hemp or jute center.

Type AA: 6 strands with hemp core, and 37 wires to a strand (= 6 × 37) or 6 strands with hemp core and 36 wires to a strand with jute, cotton or hemp center.

Description.	Diameter.		Approx. weight.		Minimum strength.	
	mm	in.	kg/m	lb/ft	kg	lb.
Galv. cast steel, Type A......	9.5	$\frac{3}{8}$	0.31	0.21	3,965	8,740
“ “ “ “ “......	12.7	$\frac{1}{2}$	0.55	0.37	6,910	15,230
“ “ “ “ “......	25.4	1	2.23	1.50	27,650	60,960
“ “ “ “ “......	38.1	$1\frac{1}{2}$	5.06	3.40	63,485	139,960
Galv. cast steel, Type AA.....	9.5	$\frac{3}{8}$	0.35	0.22	3,840	8,460
“ “ “ “ “.....	12.7	$\frac{1}{2}$	0.58	0.39	7,410	16,330
“ “ “ “ “.....	25.4	1	2.23	1.50	27,650	60,960
“ “ “ “ “.....	38.1	$1\frac{1}{2}$	5.28	3.55	59,735	131,690
Galv. cast steel, Type B......	9.5	$\frac{3}{8}$	0.25	0.17	2,995	6,600
“ “ “ “ “......	12.7	$\frac{1}{2}$	0.42	0.28	5,210	11,500
“ “ “ “ “......	25.4	1	1.68	1.13	20,890	46,060
“ “ “ “ “......	38.1	$1\frac{1}{2}$	3.94	2.65	47,965	105,740
Galv. cast steel, Type C......	25.4	1	1.59	1.07	18,825	41,500
“ “ “ “ “......	41.3	$1\frac{5}{8}$	4.35	2.92	51,575	113,700
Galv. plow steel, Type A......	9.5	$\frac{3}{8}$	0.31	0.21	4,690	10,340
“ “ “ “ “......	12.7	$\frac{1}{2}$	0.55	0.37	8,165	18,000
“ “ “ “ “......	25.4	1	2.23	1.50	32,675	72,040
“ “ “ “ “......	36.5	$1\frac{7}{16}$	4.66	3.13	69,140	152,430
Galv. plow steel, Type AA.....	9.5	$\frac{3}{8}$	0.33	0.22	4,540	10,000
“ “ “ “ “.....	12.7	$\frac{1}{2}$	0.58	0.39	8,750	19,300
“ “ “ “ “.....	25.4	1	2.35	1.58	32,250	71,100
“ “ “ “ “.....	41.3	$1\frac{5}{8}$	6.18	4.15	83,010	183,000

TABLE 52. — Plow Steel Hoisting Rope (Bright).

(After Panama Canal Specification No. 302, 1912.)

Wire rope to be of best plow steel grade, and to be composed of 6 strands, 19 wires to the strand, with hemp center. Wires entering into construction of rope to have an elongation in 203.2 mm or 8 in. of about $2\frac{1}{2}$ per cent.

Diameter.		Spec. minimum strength.		Diameter.		Spec. minimum strength.	
mm	in.	kg	lb.	mm	in.	kg	lb.
9.5	$\frac{3}{8}$	5,215	11,500	38.1	$1\frac{1}{2}$	74,390	164,000
12.7	$\frac{1}{2}$	9,070	20,000	50.8	2	127,000	280,000
19.0	$\frac{3}{4}$	20,860	46,000	63.5	$2\frac{1}{2}$	207,740	458,000
25.4	1	34,470	76,000	69.9	$2\frac{3}{4}$	249,350	550,000

TABLE 53. — Steel-wire Rope — Experimental Values.

(Wire rope purchased under Panama Canal Spec. 302 and tested by U. S. Bureau of Standards, Washington, D. C.)

Description and analysis.	Diameter.		Ultimate strength.		Ultimate strength (net area).	
	mm	in.	kg	lb.	kg/mm^2	lb/in^2
Plow Steel, 6 strands × 19 wires C 0.90, S 0.034, P 0.024, Mn 0.48, Si 0.172..............	50.8	2	137,900	304,000	129.5	184,200
Plow Steel, 6 strands × 25 wires C 0.77, S 0.036, P 0.027, Mn 0.46, Si 0.152..............	69.9	$2\frac{3}{4}$	314,800	694,000	151.2	214,900
Plow Steel, 6 × 37 plus 6 × 19 C 0.58, S 0.032, P 0.033, Mn 0.41, Si 0.160..............	82.6	$3\frac{1}{4}$	392,800	866,000	132.2	187,900
Monitor Plow Steel, 6 × 61 plus 6 × 19, C 0.82, S 0.025, P 0.019, Mn 0.23, Si 0.169............	82.6	$3\frac{1}{4}$	425,000	937,000	142.5	202,400

Recommended allowable load for wire rope running over sheave is one fifth of specified min. strength.

TABLE 54. — Aluminum.

Metal, approx. composition, per cent.	Condition.	Density or weight.		P-limit.	Ultimate strength.	P-limit.	Ultimate strength.	Elong. in 50.8 mm (2 in.).	Reduct. of area.	Hardness.	
		gm per cm^3	lb. per ft^3	Tension, kg/mm^2		Tension, $lb./in^2$		Per cent.		Brinell @ 500 kg	Scleroscope.
ALUMINUM: Av. Al 99.3 Imp., Fe and Si. .	Cast, sand at 700° C.	2.57	160.5	6.0 to 7.0	8.0 to 9.8	8,500 to 10,000	12,000 to 14,000	29 to 15	36 to 22	25 to 26	4 to 5
	Cast, sand and heat treated Ann. 500° C, air cooled.	—	—	—	8.9 to 9.6	—	12,600 to 13,600	28 to 18	30 to 22	25 to 27	4 to 5
	Cast, chill.	2.57	160.5	6.0	9.0	9,000	13,000	20.0	—	26	5
	Sheet, ann.	2.69	168.0	6.0	9.0	8,500	13,500	23.0	25.0	—	—
	Sheet, hard.	2.70	168.5	14.0	21.0	20,000	30,000	4.0	25.0	—	14
	Bars, hard.	2.70	168.5	15.0	23.0	22,000	33,000	—	35.0	—	—
	Wire, hard.	2.70	168.5	21.0	28.0	30,000	40,000	6.0	50.0	—	—

Compressive strength: cast, yield point 13.0 kg/mm^2 or 18,000 lb/in^2; ultimate strength 47.0 kg/mm^2 or 67,000 lb/in^2.

Modulus of elasticity: cast, 6900 kg/mm^2 or 9,810,000 lb/in^2 at 17° C.

TABLE 55. — Aluminum Sheet.

(*a*) *Grade A* (*Al min.* 99.0) *Experimental Erichsen and Scleroscope Hardness Values.*

[From tests on No. 18 B. & S. Gage sheet rolled from 6.3 mm (0.25 in.) slab. Iron Age v. 101, page 950].

Heat treatment annealed.	Thickness, mm	Indentation, mm	Scleroscope hardness.
None (as rolled).	1.08	6.83	14.0
@ 200° C, 2 hours	1.09	8.86	8.0
@ 300° C, 2 hours	1.07	10.17	4.5
@ 400° C, 2 hours	1.08	9.40	4.5
@ 200° C, 30 min.	1.07	7.97	11.8
@ 400° C, 30 min.	1.08	9.80	4.5

(*b*) *Specification Values.* — (1) Cast: U. S. Navy 49 Al, July 1, 1915; Al min. 94, Cu max. 6, Fe max. 0.5, Si max. 0.5, Mn max. 3.

Minimum tensile strength 12.5 kg/mm^2 or 18,000 lb/in^2 with minimum elongation of 8 per cent in 50.8 mm (2 in.).

(2) Sheet, Grade A: A. S. T. M. 25 to 18T; Al min. 99.0; minimum strengths and elongations.

Gage, sheet thicknesses.			Temper, No. hardness.	Tensile strength.		Elong. in 50.8 mm or 2 in. per cent.	
(B. & S.)	mm	in.		kg/mm^2	lb/in^2		
12 to 16 incl.	2.052 to 1.293	0.0808 to .0509	1 Soft, Ann.	8.8	12,500	30	Sheets of temper No. 1 to withstand being bent double in any direction and hammered flat; temper No. 2 to bend 180° about radius equal to thickness without cracking.
			2 Half-hard	12.5	18,000	7	
			3 Hard	15.5	22,000	4	
17 to 22 incl.	1.152 to 0.643	.0453 to .0253	1 Soft, Ann.	8.8	12,500	20	
			2 Half-hard	12.5	18,000	5	
			3 Hard	17.5	25,000	2	
23 to 26 incl.	0.574 to 0.404	.0226 to .0159	1 Soft, Ann.	8.8	12,500	10	
			2 Half-hard	12.5	18,000	5	
			3 Hard	21.0	30,000	2	

NOTE. — Tension test specimen to be taken parallel to the direction of cold rolling of the sheet.

ALUMINUM ALLOY.

Alloy, approx. composition per cent.	Condition, per cent reduction.	Density or weight. gm/cm³	lb/ft³	P-limit. Tension, kg/mm²	Ultimate strength. Tension, kg/mm²	P-limit. Tension, lb/in²	Ultimate strength. Tension, lb/in²	Elong. in 50.8 mm (2 in.). per cent.	Reduct. of area. per cent.	Hardness. Brinell @ 500 kg	Hardness. Scleroscope.
Aluminum — Copper.	Cast, chill....	—	—	5.3	10.5	7,500	15,000	24.0	34.0	—	—
Al 98 Cu 1 Imp. max. 1	Rolled, 70%..	—	—	19.0	21.0	27,000	30,000	4.0	—	—	—
Al 96 Cu 3 Imp. max. 1	Cast, chill....	—	—	8.1	13.7	11,500	19,500	12.0	21.0	—	—
	Rolled, 70%..	—	—	25.0	28.8	35,000	41,000	5.5	—	—	—
Al 94 Cu 5 Imp. max. 1	Cast, chill....	—	—	10.0	15.0	14,500	21,500	7.0	14.0	—	—
	Rolled, 70%..	—	—	23.0	27.0	33,000	38,000	6.0	—	—	—
Al 92 Cu 8: Alloy No. 12..............	Cast, sand....	2.88	180	7.7 to	10.5 to	11,000 to	15,000 to	4.0 to	3.5 to	50 to	13 to
		—	—	10.5	16.2	15,000	23,000	None	None	65	18
Al 90–92 Cu 7–8.5 Imp. max. 1.7.....	Cast*........	2.9	181	—	12.7	—	18,000	1.0	—	—	—
Copper, Magnesium..	Cast at 700° C.	—	—	3.2 to	9.6 to	4,500 to	13,600 to	2.0 to	0.5 to	74 to	17 tc
Al 9.52 Cu 4.2 Mg 0.6				4.6	13.3	6,500	18,900	0	0	74	18
	Ann. 500° C...	—	—	4.6	17.3	6,500	24,900	3.0	1.0	80	21
Duralumin or 17S Alloy Al 94 Cu 4 Mg 0.5...............	Ann..........	2.8	174	25.0	42.0	35,100	59,500	21.1	29.5	—	—
	Rolled 70%...	—	—	53.0	56.0	75,400	79,600	4.0	13.2	—	—
	Rolled heat tr'd †......	—	—	23.4	39.0	33,400	55,300	25.5	26.0	—	—
Copper, Manganese..	Cast, chill....	—	—	10.0	14.0	14,300	20,300	5.0	—	—	—
Al 96 Cu 2 Mn 2	Rolled, 20 mm	—	—	19.0	27.0	27,100	38,200	16.0	28.0	—	—
Al 96 Cu 3 Mn 1.....	Cast, chill....	—	—	11.3	19.0	16,200	27,000	14.0	—	—	—
Naval Gun Factory...	Cast, sand....	2.8	175	—	14.0	—	20,000	12.0	—	—	—
Al 97 Cu 1.5 Mn 1....	Forged.......	—	—	14.0	19.0	19,500	27,800	12.0	47.0	—	—
Al 94 Cu max. 6 Mn max. 3............	Minimum ‡...	—	—	—	12.7	—	18,000	8.0	—	—	—
Copper, Nickel, Mg Mn...............	Cast at 700° C.	—	—	3.5 to	17.9 to	5,000 to	25,500 to	6.0 to	8.5 to	54 to	9 to
Al 93.5 Cu 3.5 Ni 1.5 Mg 1 Mn 0.5.....		—	—	9.8	23.2	14,000	33,000	1.5	1.0	86	25
Copper, Nickel Mn...	Cast at 700° C.	—	—	—	14.5 to	—	20,600 to	6.0 to	11.0 to	50 to	9 to
Al 94.2 Cu 3 Ni 2 Mn 0.8...............					21.4		30,500	1.0	2.0	91	27
Magnesium:											
Magnalium Al 95 Mg 5	Cast, sand....	2.5	156	5.6	15.5	8,000	22,000	7.0	8.5	—	—
Al 77–98, Mg 23–2...	Cast, chill....	2.4 to	150 to	—	29.5 to	—	42,000 to	—	—	—	—
		2.57	160		45.0		64,000				
Nickel Al 97 Ni 2.....	Cast, chill....	—	—	4.0	11.0	5,800	14,900	21.0	36.0	—	—
	Drawn, cold..	—	—	14.0	16.0	19,700	22,700	13.0	37.0	—	—
	Rolled, hot...	—	—	8.0	13.0	11,900	18,200	28.0	52.0	—	—
Al 95 Ni 5..........	Cast, chill....	—	—	6.0	15.0	9,000	21,700	9.0	11.0	—	—
	Drawn, cold..	—	—	16.0	20.0	22,900	27,900	8.0	24.0	—	—
	Rolled, hot...	—	—	9.0	16.0	13,500	22,300	22.0	36.0	—	—
Nickel Copper:											
Al 93.5 Ni 5.5 Cu 1..	Cast, chill....	—	—	7.0	17.0	10,700	24,800	6.0	8.0	—	—
Al 91.5 Ni 4.5 Cu 4..	Cast, chill....	—	—	7.0	18.0	9,900	25,200	4.0	5.0	—	—
Al 92 Ni 5.5 Cu 2....	Drawn, cold..	—	—	22.0	27.0	31,700	37,800	8.0	15.0	—	—
	Rolled, hot...	—	—	13.0	22.0	18,200	31,500	16.0	24.0	—	—
Zinc, Copper:											
Al 88.6 Cu 3 Zn 8.4..	Cast at 700° C.	—	—	4.7	18.5	6,700	26,300	8.0	7.5	50	10
	Ann. 500° C.	—	—	4.4	20.2	6,200	28,800	8.0	7.5	50	10
Al 81.1 Cu 3 Zn 15.9.	Cast at 700° C.	3.1	193	9.8	24.7	14,000	35,100	2.0	2.0	74	15
	Ann. 500° C...	—	—	9.8	29.0	14,000	41,200	4.0	4.0	70	15

* Specification Values: Alloy "No. 12": A. S. T. M. B26–18T, tentative specified minimums for aluminum, copper.
† Quenched in water from 475° C. after heating in a salt bath. Modulus of elasticity for Duralumin averages 7000 kg/mm² or 10,000,000 lb/in².
‡ Specification values: Aluminum castings; U. S. Navy 49 Al, July 1, 1915 (Impurities: Fe max. 0.5, Si max. 0.5).

MECHANICAL PROPERTIES.

TABLE 57. — Copper.

Metal and approx. composition. Per cent.	Condition.	Density or weight. gm/cm³	Density or weight. lb/ft³	P-limit. Tension, kg/mm²	Ultimate strength. Tension, kg/mm²	P-limit. Tension, lb/in²	Ultimate strength. Tension, lb/in²	Elong. in 50.8 mm (2 in.) Per cent.	Reduct. in area. Per cent.	Hardness. Brinell @ 500 kg	Hardness. Scleroscope.
Copper:											
99.9: electrolytic	Ann. 200° C......	8.89	555	6.0	27.0	8,500	38,000	50.0	50.0	40	7
Cu 99.6........	Cast............	8.85	552	7.0	18.0	10,000	25,000	20.0	60.0	80	8
Rolled.........	Hard, 40% reduct	8.89	555	14.0	35.0	20,000	50,000	5.0	8.0	94	—
	Ann. at 500° C...	8.90	556	indet.	25.0	indet.	35,000	50.0	60.0	42	6
Cu 99.6........	Drawn cold, 50% reduct.........	—	—	26.0	35.0	37,000	50,000	9.0	—	—	18
Cu 99.9 *.......	No Ann. (96% reduction).......	—	—	—	47.3	—	67,400	0.8	64.5	—	—
	Ann. 750° C after drawing cold...	—	—	—	21.9	—	31,200	24.5	76.0	—	—
Cu 99.9 †.......	Drawn hot (64% reduction).....	—	—	—	33.0	—	46,800	4.3	70.5	—	—

* Wire drawn cold from 3.18 mm (0.125 in.) to 0.64 mm (0.025 in.) Bull. Am. Inst. Min. Eng., Feb., 1919.
† Wire drawn at 150° C from 0.79 mm (0.031 in.) to 0.64 mm (0.025 in.) (Jeffries, *loc. cit.*).
Compression, cast copper, Ann. 15.9 mm (0.625 in.) diam. by 50.8 mm (2 in.) long cylinders.
Shortened 5 per cent at 22.0 kg/mm² or 31,300 lb/in² load.
" 10 " " " 29.0 kg/mm² " 41,200 lb/in² "
" 20 " " " 39.0 kg/mm² " 55,400 lb/in² "
Shearing strength, cast copper 21.0 kg/mm² or 30,000 lb/in²
Modulus of elasticity, electrolytic 12,200 kg/mm² or 17,400,000 lb/in²
" " " cast 7,700 kg/mm² or 11,000,000 lb/in²
" " " drawn, hard 12,400 kg/mm² or 17,600,000 lb/in²

TABLE 58. — Rolled Copper — Specification Value.

Specification values: U. S. Navy Dept., 47C2, minimums for rolled copper, — Cu min. 99.5

Description, temper and thickness.	Tensile strength. kg/mm²	Tensile strength. lb/in²	Elong. in 50.8 or 2 in. — per cent.
Rods, bars, and shapes:			
Soft........................	21.0	30,000	25
Hard: to 9.5 mm (⅜ in.) incl..........	35.0	50,000	10
Hard: 9.5 mm to 25.4 mm (1 in.)..........	31.5	45,000	12
Hard: 25.4 mm to 50.8 mm (2 in.).........	28.0	40,000	15
Hard: over 50.8 mm (2 in.)...............	24.5	35,000	20
Sheets and plates:			
Soft........................	21.0 to 28.0	30,000 to 40,000	25 to 25
Hard........................	24.5	35,000	18

TABLE 59. — Copper Wire — Specification Values.

Specific Gravity 8.89 at 20° C (68° F).

Copper wire: Hard Drawn (and **Hard-rolled flat copper** of thicknesses corresponding to diameters of wire) Specification values. (A. S. T. M. B1-15, and U. S. Navy Dept., 22W3, Mar. 1, 1915.)

Diameter. mm	Diameter. in.	Minimum tensile strength. kg/mm²	Minimum tensile strength. lb/in²	Maximum elongation, per cent in 254 mm (10 in.).
11.68	.460	34.5	49,000	2.75
10.41	.410	35.9	51,000	3.25
9.27	.365	37.1	52,800	2.80
8.25	.325	38.3	54,500	2.40
7.34	.289	39.4	56,100	2.17
6.55	.258	40.5	57,600	1.98
5.82	.229	41.5	59,000	1.79
				in 1524 mm (60 in.)
5.18	.204	42.2	60,100	1.24
4.62	.182	43.0	61,200	1.18
4.12	.162	43.7	62,100	1.14
3.66	.144	44.3	63,000	1.09
3.25	.128	44.8	63,700	1.06
2.90	.114	45.2	64,300	1.02
2.59	.102	45.7	64,900	1.00
2.31	.091	46.0	65,400	0.97
2.06	.081	46.2	65,700	0.95
1.83	.072	46.3	65,900	0.92
1.63	.064	46.5	66,200	0.90
1.45	.057	46.7	66,400	0.89
1.30	.051	46.8	66,600	0.87
1.14	.045	47.0	66,800	0.86
1.02	.040	47.1	67,000	0.85

P-limit of hard-drawn copper wire must average 55 per cent of ultimate tensile strength for four largest sized **wires** in table, and 60 per cent of tensile strength for smaller sizes.

TABLE 60. — Copper Wire — Medium Hard-drawn.

(A. S. T. M. B2–15) Minimum and Maximum Strengths.

Diameter.		Tensile strength.				Elongation, minimum per cent in 254 mm (10 in.).
		Minimum.		Maximum.		
mm	in.	kg/mm²	lb/in²	kg/mm²	lb/in²	
11.70	0.460	29.5	42,000	34.5	49,000	3.75
6.55	.258	33.0	47,000	38.0	54,000	2.50
						in 1524 mm (60 in.)
4.12	.162	34.5	49,000	39.5	56,000	1.15
2.59	.102	35.5	50,330	40.5	57,330	1.04
1.02	.040	37.0	53,000	42.0	60,000	0.88

Representative values only from table in specifications are shown above.
P-limit of medium hard-drawn copper averages 50 per cent of ultimate strength.

TABLE 61. — Copper Wire — Soft or Annealed.

(A. S. T. M. B3–15) Minimum Values.

Diameter.		Minimum tensile strength.		Elongation in 254 mm (10 in.), per cent.
mm	in.	kg/mm²	lb/in²	
11.70 to 7.37	0.460 to 0.290	25.5	36,000	35
7.34 to 2.62	0.289 to 0.103	26.0	37,000	30
2.59 to 0.53	0.102 to 0.021	27.0	38,500	25
0.51 to 0.08	0.020 to 0.003	28.0	40,000	20

NOTE. — Experimental results show tensile strength of concentric-lay copper cable to approximate 90 per cent of combined strengths of wires forming the cable.

TABLE 62. — Copper Plates.

(A. S. T. M. B11–18) for Locomotive Fire Boxes. Specification Values.

Minimum requirements.	Tensile strength.		Elong. in 203.2 mm (8 in.), per cent.
	kg/mm²	lb/in²	
Copper, Arsenical, As 0.25–0.50 Impurities, max. 0.12........	22.0	31,000	35
Copper, Non-arsenical: Impurities, max. 0.12.......	21.0	30,000	30

NOTE. — Copper to be fire-refined or electrolytic, hot-rolled from suitable cakes.

TABLE 63. — Copper Alloys.

The general system of nomenclature employed has been to denominate all simple copper-zinc alloys as **brasses,** copper-tin alloys as **bronzes,** and three or more metals alloys composed primarily of either of these two combinations as alloy brasses or bronzes, e.g., "Zinc bronze" for U. S. Government composition "G" Cu 88 per cent, Sn 10 per cent, Zn 2 per cent. Alloys of the third type noted above, together with other alloys composed mainly of copper, have been called **copper alloys,** with the alloying elements other than minor impurities listed as modifying copper in the order of their relative percentages.

In some instances, the scientific name used to denote an alloy is based upon the deoxidizer used in its preparation, which may appear either as a minor element of its composition or not at all, e.g., phosphor bronze.

Commercial names are shown below the scientific names. Care should be taken to specify the chemical composition of a commercial alloy, as the same name frequently applies to widely varying compositions.

MECHANICAL PROPERTIES OF MATERIALS.

TABLE 64. — Copper-zinc Alloys or Brasses; Tin Alloys or Bronzes.

Metal and approx. composition, per cent.	Condition.	Density or weight. gm/cm³	Density or weight. lb/ft³	P-limit. Tension, kg/mm²	Ultimate strength. Tension, kg/mm²	P-limit. Tension, lb/in²	Ultimate strength. Tension, lb/in²	Elong. in 50.8 mm (2 in.). Per cent.	Reduct. in area. Per cent.	Hardness. Brinell @ 500 kg	Hardness. Sclero-scope.
Brass:											
Cu 90 Zn 10†	Sand cast	—	—	—	20.0	—	29,000	22	—	—	—
	Cold rolled, hard	—	—	—	39.0	—	55,000 *	5 *	—	60	20
	Cold rolled, soft.	8.7	543	—	26.0	—	37,000 *	40 *	70	47	10
Cu 80, Zn 20 ‡	Sand cast	—	—	—	25.0	—	35,000	31	32	—	—
	Cold rolled, hard	—	—	—	53.0	—	75,000 *	5 *	—	75	28
	Cold rolled, soft.	8.6	537	—	29.0	—	42,000 *	50 *	85	46	12
Cu 70, Zn 30	Sand cast	8.4	524	—	28.0	—	40,000	35	—	37	—
Cu 66 Zn 34 Std. sheet	Cold rolled, hard	8.5	530	—	42.0	—	60,000	5 *	—	75§	26
	Cold rolled, soft.	8.4	524	—	34.0	—	48,000 *	50 *	85	45	12
Cu 60, Zn 40	Sand cast	—	—	15.5	32.2	21,800	45,800	15	22	—	—
Muntz metal	Cold rolled, hard	8.4	522	31.5	49.0	45,000	70,000	30	50	—	—
Bronze:											
Cu 97.7, Sn 2.3	Cast	—	—	6.0	19.5	8,500	28,000	20	—	—	—
	Rolled	—	—	7.6	34.0	10,800	48,000	55	75	—	—
Cu 90, Sn 10	Cast or gun bronze or bell metal	8.78	548	7.2	23.0	10,300	33,000	10	—	—	23
Cu 80, Sn 20	Cast	8.81	550	7.1	22.5	10,100	32,000	1.5	—	—	—
Cu 70, Sn 30	Cast	8.84	552	1.4	5.0	2,000	7,000	0.5	—	—	—

Compressive Strengths, **Brasses:**

Cu 90, Zn 10, cast 21.0 kg/mm² or 30,000 lb/in²
Cu 80, Zn 20, cast 27.4 kg/mm² or 39,000 lb/in²
Cu 70, Zn 30, cast 42.0 kg/mm² or 60,000 lb/in²
Cu 60, Zn 40, cast 52.5 kg/mm² or 75,000 lb/in²
Cu 50, Zn 50, cast 77.0 kg/mm² or 110,000 lb/in²

Modulus of elasticity, — cast brass, — average 9100 kg/mm² or 13,000,000 lb/in²
Erichsen values: Soft slab, 1.3 mm (0.05 in.) thick, no rolling, depth of impression 13.8 mm (0.55 in.).
Hard sheet, 1.3 mm, rolled 38% reduction, depth of impression 7.3 mm (0.29 in.).
Hard sheet, 0.5 mm, rolled 60% reduction, depth of impression 3.7 mm (0.15 in.).

Compressive Ultimate Strengths, Cast Bronzes:

Cu 97.7, Sn 2.3 to 24.0 kg/mm² or 34,000 lb/in²
Cu 90, Sn 10 to 39.0 kg/mm² or 56,000 lb/in²
Cu 80, Sn 20 to 83.0 kg/mm² or 118,000 lb/in²
Cu 70, Sn 30 to 105.0 kg/mm² or 150,000 lb/in²

Specification value, A. S. T. M., B 22–18 T, for specimen = cylinder 645 sq. mm (1 sq. in.) area, 25.4 mm (1 in.) long.
Cu 80, Sn 20: minimum compressive elastic limit = 17.0 kg/mm² or 24,000 lb/in²
Modulus of elasticity for bronzes varies from 7000 kg/mm² or 10,000,000 lb/in² to 10,000 kg/mm² or 15,500,000 lb/in²

* Values marked thus are S. A. E. Spec. values. (See S. A. E. Handbook, Vol. I, p. 13a, rev. December, 1913.
† Red metal. ‡ Low brass or bell metal.
§ A. S. T. M. Spec. B19–18T requires B.h.n. of 51–65 kg/mm² @ 5000 kg pressure for 70: 30 annealed sheet brass.

FOOT NOTES TO TABLE 65, PAGE 85.

* Tensilite, Cu 67, Zn 24, Al 4.4, Mn 3.8, P 0.01 compressive P-limit: 42.2 kg/mm² or 60,000 lb/in² and 1.33 per cent set for 70.3 kg/mm² or 100,000 lb/in² load.
† Compressive P-limit 20.0 to 28.2 kg/mm² or 28,500 to 40,000 lb/in²
‡ Compressive ultimate strength 54.5 kg/mm² or 77,500 lb/in²
§ Compressive P-limit 4.2 kg/mm² or 6000 lb/in² and 40 per cent set for 70.3 kg/mm² or 100,000 lb/in²
¶ Modulus of elasticity 9840 kg/mm² or 14,000,000 lb/in²
|| Values are for yield point. ** Minimum values for ingots.
†† Rolled manganese bronze (U. S. N.) Cu 57 to 60, Zn 40 to 37, Fe max. 2.0, Sn 0.5 to 1.5; 2.9 per cent increase for thickness 25.4 mm (1 in.) and under.
‡‡ Ni 9 per cent, B.h.n. = 130 as rolled; B.h.n. = 50 as annealed at 930° C.
U. S. Navy Dept. Spec. 46S 3a, June 1, 1917: German silver Cu 60 to 67, Zn 18 to 22, Ni min. 15, no mechanical requirements.
For list of 30 German silver alloys, see Braunt, "Metallic Alloys," p. 314, — "best" (Hiorns), "hard Sheffield," Cu 46, Zn 20, Ni 34.
§§ Platinoid Cu 60, Zn 24, Ni 14, W 1 to 2; high electric resistance alloy with mechanical properties as nickel brass.
|||| Specification Values, Naval Brass Castings, U. S. Navy, 46B 10b, Dec. 1, 1917 for normal proportions Cu 62, Zn 37, Sn 1, min. tensile strength 17.5 kg/mm² or 25,000 lb/in² with 15 per cent elongation in 50.8 mm (2 in.).

TABLE 65. — Copper Alloys — Three (or more) Components.

Alloy and approx. composition per cent.	Condition.	Density or weight. gm per cm³	Density or weight. lb. per ft³	P-limit. Tension, kg/mm²	Ultimate strength. Tension, kg/mm²	P-limit. Tension, lb/in²	Ultimate strength. Tension, lb/in²	Elong. in 50.8 mm (2 in.). Per cent.	Reduct. of area. Per cent.	Hardness. Brinell @ 500 kg	Hardness. Scleroscope.
Brass, Aluminum...	Cast										
Cu 57, Zn 42, Al 1...		—	—	—	40.0	—	57,000	50.0	—	—	—
Cu 55, Zn 41, Al 4...		—	—	—	60.0	—	85,400	16.5	—	—	—
Cu 62.9, Zn 33.3, Al 3.8		—	—	—	56.2	—	80,000	—	—	—	—
Cu 70.5, Zn 26.4, Al 3.1		—	—	13.4	33.0	19,000	47,000	50.0	—	—	—
Alum., Manganese..	Cast, tensilite*										
Cu 64, Zn 29, Al 3.1, Mn 2.5, Fe 1.2...		—	—	21.1	68.8	30,000	98,000	16.0	17.0	130	—
Alum., Vanadium..											
Cu 58.5, Zn 38.5, Al 1.5, V 0.03.......	Cold drawn...	—	—	35.6	57.0	50,600	81,400	12.0	14.0	—	—
Iron:											
Cu 56, Zn 41.5, Fe 1.	Cast.........	—	—	—	50.7 to 59.2	—	72,000 to 84,000	35.0 to 22.0	35.0 to 25.0	109 to 119	—
Aich's Metal Cu60,Zn38.2,Fe1.8	Cast.........	8.42	526	—	40.3	—	57,300	—	—	—	—
Delta Metal											
Cu 57, Zn 42, Fe 1..	Cast, sand....	—	—	—	31.7	—	45,000	10.0	—	—	—
	Rolled, hard..	—	—	—	42.2	—	60,000	17.0	—	—	—
Cu 65, Zn 30, Fe 5..	Rolled hard...	—	—	—	45.5	—	65,000	—	—	—	—
Iron, Tin:											
Cu56.5,Zn40,Fe1.5, Sn 1.0†.........	Cast.........	—	—	23.2 to 26.0	49.2 to 52.8	33,000 to 37,000	70,000 to 75,000	35.0 to 20.0	35.0 to 22.0	104 to 119	—
Sterro metal:											
Cu 55, Zn 42.4 Fe 1.8, Sn 0.8.....	Cast.........	8.4	525	—	42.5	—	60,500	—	—	—	—
	Forged........	—	—	—	53.6	—	76,200	—	—	—	—
	Hard drawn..	—	—	—	58.5	—	83,100	—	—	—	—
Lead or Yellow brass	Cast.........	8.5	531	—	23.2 to 27.5	—	33,000 to 39,000	30.0 to 26.0	35.0 to 30.0	—	—
Cu 60 to 63.5, Zn 35 to 33.5, Pb 5 to 3.	Sheet ann.....	—	—	—	25.5	—	42,000	50.0	—	—	—
	Sheet hard....	—	—	—	42.9	—	61,000	30.0	—	—	—
Lead, Tin or Red brass......... Cu83,Zn7,Pb6,Sn4	Cast .... ‡	8.6	535	11.0	21.0	16,000	30,000	17.0	19.0	—	7.0
Cu 78, Zn 9.5, Pb 10, Sn 2...........	Cast.........	8.87	554	8.4	18.6	12,000	26,500	22.0	24.9	—	—
Yellow brass: Cu 70, Zn 27, Pb 2, Sn 1...........	Cast §........	8.4	524	7.4	20.7	10,500	29,500	25.0	28.5	53.0	—
Manganese or Manganese bronze											
Cu 58, Zn 39, Mn 0.05	Cast, sand ¶..	8.3	520	21.1 to 24.6	49.2 to 52.7	30,000 to 35,000‖	70,000 to 75,000	30.0 to 22.0	32.0 to 25.0	109 to 119	18 to 19
(Sn, Fe, Al, Pb.)	Cast, chill....	—	—	22.5 to 26.0‖	52.7 to 56.3	32,000 to 37,000‖	75,000 to 80,000	32.0 to 25.0	34.0 to 28.0	119 to 130	18 to 22
Cu 60, Zn 39 Mn, tr	Rolled.......	8.3	520	31.5	52.5	45,000	75,000	25.0	28.0	—	30
Specification values:											
U. S. Navy, 46 B 16a **.........		—	—	—	49.2	—	70,000	20.0	—	—	—
U. S. N., 46 B 15a	Rolled††	—	—	24.6	49.2	35,000	70,000	30.0	—	—	—
Manganese Vanadium:											
Cu 58.6, Zn 38.5, Al 1.5 Mn 0.5, V 0.03.	Cold drawn...	—	—	35.6	57.0	50,600	81,400	12.0	14.0	—	—
Nickel: Nickel silver, Cu 60.4, Zn 31.8, Ni 7.7....	Cast.........	8.5	530	10.8	25.3	15,400	36,000	40.5	42.0	46	—
German silver, Cu 61.6, Zn 17.2, Ni 21.1.......		8.7	544	13.2	28.8	18,800	40,900	28.5	25.1	80	—
Cu 60.6, Zn 11.8, Ni 27.3........		8.8	547	16.7	37.6	23,700	53,500	32.0	31.4	67	—
Fine wire: Cu58,Zn24,Ni18 Nickel silver ‡‡	Drawn hard..	8.5	530	—	105.5	—	150,000	—	—	—	—
Nickel Tungsten: §§											
Tin:											
Cu 61, Zn 38, Sn 1...	Cast, sand....	—	—	11.0	30.0	15,700	42,600	29.6	32.0	—	—
Naval brass, as above	Ann. after rolling........	—	—	26.0	43.5	37,000	62,000	25.0	37.0	—	—
Tobin bronze: as below...........	Cast.........	8.3	51[illegible]	17.6	42.2	25,000	60,000	—	—	—	—
Cu 58.2, Zn 39.5, Sn 2.3.........	Rolled.......	8.4	524	38.0	56.0	54,000	79,000	35.0	40.0	—	—
Cu 55, Zn 43, Sn 2.	Cast‖‖	—	—	—	48.4	—	68,900	48.0	70.0	—	—

For Footnotes see page 84.

MECHANICAL PROPERTIES.

TABLE 65. — Copper Alloys — Three (or more) Components.

Alloy and approx. composition per cent.	Condition.	Density or weight.		P-limit.	Ultimate strength.	P-limit.	Ultimate strength.	Elong. in 50.8 mm (2 in.).	Reduct. of area.	Hardness.	
		gm per cm³	lb. per ft³	Tension, kg/mm²		Tension, lb/in²		Per cent.		Brinell @ 500 kg.	Scleroscope.
Brass, Tin — (continued):											
Rods:* 0 to 12.7 mm (½ in.)		—	—	19.0	42.2	27,000	60,000	35.0	To bend 120°		
12.7 to 25.4 mm (1 in.)		—	—	18.3	40.8	26,000	58,000	40.0	cold about radius equal		
over 25.4 mm (in.) diam..		—	—	17.6	38.0	25,000	54,000	40.0	to diameter.		
Shapes, all................		—	—	15.7	39.4	22,400	56,000	30.0	"		
Plates to 12.7 mm (½ in.)		—	—	19.3	38.7	27,500	55,000	32.0	"		
over 12.7 mm (½ in.) thick		—	—	17.6	39.4	25,000	56,000	35.0	"		
Tubing (wall thickness) 0 to 3.2 mm (⅛ in.)........		—	—	21.1	42.2	30,000	60,000	28.0	—	—	—
3.2 to 6.4 mm (¼ in.).....		—	—	19.7	38.7	28,000	55,000	32.0	—	—	—
over 6.4 mm (¼ in.)......		—	—	18.3	35.1	26,000	50,000	35.0	—	—	—
Vanadium:											
Victor bronze, V 0.03, Cu 58.6, Zn 38.5, Al 1.5, Fe 1.0	Cold drawn	—	—	56.5	64.5	80,000	92,000	11.5	29.0	—	—
U. S. Navy † 49 B 1b....		—	—	15.8	38.7	22,500	55,000	25.0	—	—	—
Bronze, Aluminum.	See Cu. Al										
Lead:											
Cu 89, Sn 10, Pb 1.........	Cast ‡.....	—	—	—	15.5	—	22,000	—	—	—	—
Cu 88, Sn 10, Pb 2.........	Cast §.....	—	—	13.4 to 16.2	21.1 to 24.6	19,000 to 23,000	30,000 to 35,000	20.0 to 15.0	26.0 to 18.0	65 to 70	—
Cu 80, Sn 10, Pb 10........	Cast, sand.	8.8	549	10.9	22.1	15,500	31,400	13.5	12.0	63	—
	Cast, chill..	—	—	12.8	24.7	18,200	35,200	4.5	3.5	85	—
Lead, Phosphor:											
Cu 80, Sn 10, Pb 10, P trace	Cast......	9.1	570	11.0	21.0	16,000	30,000	6.0	3.5	65	12
Lead Zinc, Red brass:	Cast||.....	—	—	13.8	18.8	19,600	26,800	11.0	11.5	—	8.0
Cu 81, Sn 7, Pb 9, Zn 3.....	Cast ¶.....	8.9	555	13.4 to 14.1	21.1 to 24.6	19,000 to 20,000	30,000 to 35,000	18.0 to 15.0	24.0 to 22.0	50 to 55	—
Cu 88, Sn 8, Pb 2, Zn 2......	Cast......	—	—	—	21.8 to 26.0	—	31,000 to 37,000	20.0 to 16.0	—	57 to 59	—
Lead, Zinc Phosphor:											
Cu 73.2, Sn 11.3, Pb 12.0, Zn 2.5, P 1..............	Cast **....	—	—	10.5	21.4	15,000	30,400	4.0	3.3	—	11
Manganese:											
Cu 88, Sn 10, Mn 2.........	Cast......	—	—	9.0	19.1	12,800	27,200	25.0	—	—	—
Nickel, Zinc:											
Cu 88, Sn 5, Ni 5, Zn 2 (1)...	Cast††	—	—	9.2	28.6	13,100	40,700	32.0	28.0	—	—
Cu 89, Sn 4, Ni 4, Zn 3 (2)...	Cast††	—	—	8.1	27.9	11,500	39,700	31.0	31.0	—	—
Phosphor:											
Cu 95, Sn 4.9, P 0.1........	Rolled.....	8.6	535	28.0	46.0	40,000	65,000	30.0	—	—	37
Cu 89, Sn 10.5, P 0.5.......	Cast.......	—	—	11.2 to	21.8 to	16,000 to	31,000 to	6.0 to	—	72 to	—
Cu 80, Sn 20, P max. 1......	Cast ‡‡....			14.1	24.6	20,000	35,000	10.0		77	
Rods and bars §§ up to 12.7 mm (½ in.)...........		—	—	42.2||||	56.2	60,000	80,000	12.0	Required to bend cold		
(minimum) over 12.7 mm to 25.4 mm (1 in.).....		—	—	28.1||||	42.2	40,000||||	60,000	20.0	through 120° about radi-		
over 25.4 mm (1 in.).....		—	—	21.1||||	38.7	30,000||||	55,000	25.0	us equal to		
Sheets and plates §§ spring temper...............		—	—	—	63.2	—	90,000	—	thickness. "		
Medium temper.........		—	—	17.6||||	35.1	25,000||||	50,000	25.0	"		

Bronze, Phosphor: spring wire, hard-drawn or hard-rolled (U. S. Navy Spec. 22 W5, Dec. 1, 1915). Cu 94; Sn min. 4.5, Zn max 0.3, Fe max. 0.1, Pb max. 0.2, P 0.05 to 0.50; max. elong. in 203 mm (8 in.) = 4 per cent.

Diameter (group limits).	Min. tensile strength.		Diameter (group limits).		Min. tensile strength.	
	kg/mm²	lb/in²	mm	in.	kg/mm²	lb/in²
Up to 1.59 mm or 0.0625 in..........	95.0	135,000	to 6.35	to 0.250	77.5	110,000
Over 1.59 mm to 3.17 mm (0.125 in.)..	88.0	125,000	to 9.52	to 0.375	74.0	105,000

* Specification Values, Rolled Brass, Cu 62, Zn 37, Sn 1, min. properties after U. S. Navy Spec., 1918.
† Specification Values: Jan. 3, 1916, Vanadium Bronze Castings, Cu 61, Zn 38, Sn max. 1 (incl. V). Mimima.
‡ Compressive P-limit 15.5 kg/mm² or 22,000 lb/in²
§ Compressive P-limit 10.5 kg/mm² or 15,000 lb/in² and 28 per cent set for 70 kg/mm² or 100,000 lb/in²
|| Ultimate compressive strength, 54.2 kg/mm² or 77,100 lb/in² (Cu 76, Sn 7, Pb 13, Zn 4).
¶ Compressive P-limit 8.8 to 9.1 kg/mm² or 12,500 to 13,000 lb/in², and 34 to 35 per cent set for 70 kg/mm²
** Compression: ultimate strength 49.5 kg/mm² or 70,500 lb/in²
†† Modulus of Elasticity: (1) 12,200 kg/mm² or 17,300,000 lb/in²; (2) 10,500 kg/mm² or 14,900,000 lb/in²
‡‡ Compressive P-limit 17.6 to 28.1 kg/mm² or 25,000 to 40,000 lb/in² and 6 to 10 per cent set for 70 kg/mm² or 100,000 lb/in² load.
Specification Values: U. S. Navy 46 B 5c, Mar. 1, 1917, Cu 85 to 90, Sn 6 to 11, Zn max. 4: Cast, Grade 1. — Impurities max. 0.8; min. tensile strength 31.6 kg/mm² or 45,000 lb/in² with 20 per cent elong. in 50.8 mm (2 in.).
¶ Grade 2. — Impurities max. 1.6; min. tensile strength 21.1 kg/mm² or 30,000 lb/in² with 15 per cent elong. in 50.8 mm (2 in.).
§§ Specification Values: U. S. Navy 46B 14b, Mar. 1, 1916, Cu min. 94, Sn min. 3.5, P 0.50, rolled or drawn.
|||| Minimum yield points specified: for P-limits assume 66 per cent of values shown.

MECHANICAL PROPERTIES.

TABLE 65. — Copper Alloys — Three (or more) Components.

Alloy and approx. composition. per cent.	Condition.	Density or weight. gm per cm³	Density or weight. lb. per in³	P-limit. Tension, kg/mm²	Ultimate strength. Tension, kg/mm²	P-limit. Tension, lb/in²	Ultimate strength. Tension, lb/in²	Elong. in 50.8 mm (2 in.) Per cent.	Reduct. in area. Per cent.	Hardness. Brinell @ 500 kg.	Hardness. Scleroscope.
Bronze:											
Silicon	Cast	—	—	—	46.0	—	65,000	—	—	—	—
Cu 70, Zn 29.5, Si 0.5	Drawn, hard	—	—	—	74.0	—	105,000	—	—	—	—
Zinc * Comp. "G"	Cast	8.6	535	8.6	27.4	12,200	38,900	25.0	21.0	64	13
Admiralty gun metal	Cast†	—	—	5.6 to	22.5 to	8,000 to	32,000 to	25.0 to	25.0 to	65 to	10 to
Comm'c'l range				8.4	26.7	12,000	38,000	10.0	12.0	75	20
Spec. values	Cast (mins.)	—	—	—	21.1	—	30,000	14.0	—	—	—
Cu 88, Sn 8, Zn 4	Cast ‡	8.5	530	7.7	27.5	11,000	39,200	30.5	24.0	58	11
Cu 85, Sn 13, Zn 2	Cast	—	—	—	26.7	—	38,000	2.5	2.5	—	25
Zinc, Lead											
Cu 90, Sn 6.5, Zn 2, Pb 1.5	Cast §	—	—	8.4 to	23.9 to	12,000 to	34,000 to	33.0 to	34.0 to	50 to	—
Rods and bars ‖ up to				11.2	28.1	16,000	40,000	25.0	26.0	60	
12.7 mm (½ in.)		—	—	28.1	56.2	40,000	80,000	30.0	Required to bend cold through 120° about radius equal to thickness.		
over 12.7 mm to 25.4 mm (1 in.)		—	—	26.4	52.7	37,500	75,000	30.0			
over 25.4 mm (1 in.)		—	—	24.6	50.7	35,000	72,000	30.0			
Shapes,‖ all thicknesses		—	—	26.4	52.7	37,500	75,000	30.0			
Sheets and plates,‖ to 12.7 mm (½ in.)		—	—	27.4	54.8	39,000¶	78,000	30.0	"		
over 12.7 mm (½ in.)		—	—	26.4	52.7	37,500	75,000	30.0	"		
Aluminum Tin:											
Cu 88.5, Al 10.4, Sn 1.2	Cast, chill	—	—	26.0	48.0	36,700	68,000	4.5	5.5	189	32
Aluminum Titanium:											
Cu 90, Al 10	Cast **	—	—	13.9	52.0	19,800	74,000	19.5	23.7	100	25
	Quench, 800° C.	—	—	29.0	74.0	40,500	105,200	1.0	0.8	262	—
Cu 89, Al 10, Fe 1	Cast ††	7.58	473	14.1 to	45.7 to	20,000 to	65,000 to	30.0 to	30.0 to	93 to	25 to
				17.6	56.2	25,000	80,000	20.0	20.0	100	26
Lead:											
Cu 71.9, Pb 27.5, Sn 0.5	Cast	—	—	—	4.2 to	—	6,000 to	3.0 to	4.2 to	—	—
					4.6		6,600	3.2	6.7		
Nickel, Aluminum:											
Cu 82.1, Ni 14.6, Al 2.5, Zn 0.7 ‡‡	Forged	—	—	44.5	90.0	63,300	128,000	10.0	12.0	—	—
Cu 85, Sn 5, Zn 5, Pb 5	Cast §§	—	—	10.5 to	19.0 to	15,000 to	27,000 to	20.0 to	20.0 to	50 to	—
				13.4	23.2	19,000	33,000	16.0	15.0	62	
Cu 83, Sn 14, Zn 2, Pb 1	Cast	—	—	10.5 to	16.2 to	15,000 to	23,000 to	4.0 to	4.0 to	—	20
				13.4	19.0	19,000	27,000	0.5	0.5	—	24
Zinc, Phosphor ("Non Gran")											
Cu 86, Sn 11, Zn 3, Ptr.	Cast	—	—	13.0	25.0	19,000	35,000	9.0	—	—	—
Vanadium, See Brass, Vanadium.											
Copper, Aluminum or Aluminum Bronze:											
Cu 90, Al 10	Cast, sand ‖ ‖	7.5–	468–	13.9 to	51.1 to	19,800 to	72,700 to	28.8 to	30.0 to	102 to	25 to
		7.45	465	23.3	60.0	33,200	85,500	21.7	22.4	106	26
Cu 92.5, Al 7.2	Rolled, and ann.	—	—	7.0	37.5	9,600	53,500	91.0	72.9	81	19
Aluminum, Iron or Sillman bronze	Wrought	—	—	9.8	59.3	14,000	84,400	11.5	—	—	—
	Cast	—	—	8.1	55.5	11,500	78,850	14.5	—	—	—
Cu 86.4, Al 9.7, Fe 3.9	Cast, sand	—	—	14.0	54.0	20,000	77,000	24.5	25.0	100	—
Cu 88.5, Al 10.5, Fe 1.0	Quenched 850° C. drawn 700° C.	—	—	28.0	65.0	40,000	92,000	14.0	18.5	140	—

* Gov't. Bronze: Cu 88, Sn 10, Zn 2 (values shown are averages for 30 specimens from five foundries tested at the Bureau of Standards).

† Compressive P-limit 10.5 kg/mm² or 15,000 lb/in² with 29 per cent set for 70 kg/mm² or 100,000 lb/in² load.

‡ Values from same series of tests as first values for "88–10–2," averages for 26 specimens from five foundries tested at Bureau of Standards.

§ Compressive P-limit 9.1 kg/mm² or 13,000 lb/in² with 34 per cent set for 70 kg/mm² or 100,000 lb/in² load.

‖ Specification minimums: U. S. Navy 46B17, Dec. 2, 1918, for hot-rolled aluminum bronze, Cu 85 to 87, Al 7 to 9, Fe 2.5 to 4.5. Specification values under P-limit are for yield point.

¶ Two and six tenths per cent increase in strength up to 762 mm (30 in.) width.

** Compressive P-limit: cast, 14.1 kg/mm² or 20,000 lb/in² with 11.4 per cent set at 70 kg/mm² or 100,000 lb/in² load.

†† Compressive P-limit: cast, 12.7 to 14.1 kg/mm² or 18,000 to 20,000 lb/in² with 13 to 15 per cent set at 700 kg/mm² or 100,000 lb/in² load.

‡‡ Modulus of elasticity 14,800 kg/mm² or 21,150,000 lb/in²

§§ Compressive P-limit 8.4 kg/mm² or 12,000 lb/in² with 36 per cent set for 70.3 kg/mm², or 100,000 lb/in² load.

‖ ‖ High values are after Jean Escard "L'Aluminum dans L'Industrie," Paris, 1918. Compressive P-limit 13.5 kg/mm² or 19,200 lb/in² with 13.5 per cent set for 70.3 kg/mm² or 100,000 lb/in² load.

MECHANICAL PROPERTIES.

TABLE 66. — Miscellaneous Metals and Alloys.

Metal or alloy. Approx. composition, per cent.	Condition.	Density or weight. gm per cm³	Density or weight. lb. per ft.³	P-limit. Tension, kg/mm²	Ultimate strength. Tension, kg/mm²	P-limit. Tension, lb/in²	Ultimate strength. Tension, lb/in²	Elong. in 50.8 mm (2 in.) Per cent.	Reduct. of area. Per cent.	Hardness. Brinell @ 500 kg.	Hardness. Scleroscope.
* **Cobalt**, Co 99.7	Cast	8.8	550	—	23.1	—	33,000	—	—	121	—
	Ann.	8.9	556	—	26.0	—	37,000	—	—	48	—
Gold, Au 100	Cast	19.3	1203	—	18.0	—	25,000	25.0	—	—	20
	Drawn hard	—	—	—	26.0	—	37,000	—	—	—	—
Copper, Au 90, Cu 10	Drawn hard	17.2	1073	—	45.8	—	65,100	—	—	—	—
Copper, Silver, Au 58, Cu 30 Ag 12	Drawn hard	—	—	—	102.0	—	145,000	—	—	—	—
Lead, Pb†	Cast	11.38	710	—	1.3	—	1,780	—	—	8	3
(Comm'c'l.)	Rolled hard	11.40	711	—	2.3	—	3,300	—	—	—	—
	Drawn soft	—	—	—	1.7	—	2,420	—	—	—	—
	Drawn hard	—	—	—	2.2	—	3,130	—	—	—	—
Antimony ‡Pb95.5, Sb4.5	Cast	10.5	655	2.8	4.5	4,000	6,400	—	—	—	—
Magnesium, Mg	Drawn hard	1.7	106	—	21.0	—	30,000	—	—	—	—
		1.74	109	—	23.2	—	33,000	—	—	—	—
Nickel, Ni 98.5	Cast	8.3	518	16.7 **	26.7	23,800 **	38,000	5.7	6.1	76	—
Ni 99.95	Wrought, ann.	8.7	543	12.6	29.9	17,900	42,500	11.0	—	—	—
Ni 98.5	Wrought, com.	—	—	—	46.0	—	65,000	—	—	83	35
Ni	Rolled hard, "	—	—	—	64.7	—	92,000	11.0	—	—	—
Ni	Rolled ann. "	—	—	—	53.4	—	76,000	35.0	—	—	—
Ni	Drawn hard, D = 1.65 mm or 0.065 in.	—	—	—	109.0	—	155,000	—	—	—	—
Copper, iron, manganese or Monel metal:											
Ni 67, Cu 28, Fe 3, Mn 2.	Cast	8.9	555	21.2	49.3	30,100	70,000	18.0	20.0	—	21
	Rolled	—	—	55.1	73.8	78,400	104,900	31.3	61.7	—	27
Ni 66, Cu 28, Fe 3.5, Mn 2.5	Wrought	—	—	28.3	64.8	40,300	92,200	46.3	70.2	—	—
Ni 71, Cu 27, Fe 2 §	Drawn hard	—	—	—	112.5	—	160,000	—	—	—	—
46 M1a ‖	Cast, minimums.	—	—	22.8 **	45.7	32,500 **	65,000	25.0	—	—	—
46 M 7b ‖	Rolled, min., rods and bars ¶	—	—	28.1 **	56.2	40,000 **	80,000	32.0	—	—	—
‖	Rolled, minimum, sheets and plates	—	—	21.1	45.7	30,000	65,000	15.0	—	—	—
Palladium, Pd	Drawn hard	12.1	755	—	27.0	—	39,000	—	—	—	—
Platinum, Pt	Drawn hard	21.5	1342	—	37.3	—	53,000	18.0	—	—	24
	Drawn ann.	—	—	—	24.6	—	35,000	50.0	—	—	13
Silver, Ag 100	Cast	10.5	655	—	28.1	—	40,000	—	—	—	—
	Drawn hard	10.57	660	—	36.0	—	51,200	—	—	59	32
Copper, Ag 75, Cu 25	Drawn hard	—	—	—	77.0	—	109,500	—	—	—	—
Tantalum, Ta	Drawn hard	16.6	1035	—	91.0	—	130,000	—	—	—	—
Tin, Sn 99.8††	Cast	7.3	456	1.1	2.8	1,600	4,000	35.0	—	14	8
	Rolled	—	—	—	3.7	—	5,300	—	—	—	—
	Drawn hard	—	—	—	7.0	—	10,000	—	—	—	—
Antimony, Copper, Zinc (Britannia Metal): Sn 81, Sb 16, Cu 2, Zn 1.											
Zinc, Aluminum, etc. (aluminum solder):											
Sn 63, Zn 18, Al 13, Cu 3, Sb 2, Pb 1	Cast	—	—	—	10.2	—	14,500	1.9	1.5	—	—
Sn 62, Zn 15, Al 11, Pb 8, Cu 3, Sb 1	Cast	—	—	—	9.1	—	13,000	1.6	1.3	—	—
Zinc, aluminum: Sn 86, Zn 9, Al 5	Cast, chill	—	—	—	8.6	—	12,200	41.0	81.0	—	—
Aluminum, zinc, cadmium: Sn 78, Al 9, Zn 8, Cd 5	Cast, chill	—	—	—	10.1	—	14,300	18.0	41.0	—	—

Antimony: Modulus of Elasticity 7960 kg/mm² or 11,320,000 lb/in² (Bridgman).
* Compressive strength: cast and annealed, 86.0 kg/mm² or 122,000 lb/in².
Comm'c'l. comp., C 0.06, cast, tensile, ultimate, 42.8 kg/mm² or 61,000 lb/in², with 20 per cent elongation in 50.8 or 2 in. Compression, ultimate 123.0 kg/mm² or 175,000 lb/in²
Stellite, Co 59.5, Mo 22.5, Cr 10.8, Fe 3.1, Mn 2.0, C 0.9, Si 0.8. Brinell hardness 512 at 3000 kg. density 8.3.
† Modulus of elasticity, cast or rolled, 492 kg/mm² or 700,000 lb/in²; drawn hard 703 kg/mm² or 1,000,000 lb/in²
‡ For compressive test data on lead-base babbitt metal, see table following zinc.
§ Modulus of elasticity 15,800 kg/mm² or 22,500,000 lb/in².
‖ Specification values, U. S. Navy, Monel metal, Ni min. 60, Cu min. 23, Fe max. 3.5, Mn max. 3.5, C + Si max. 0.8, Al max. 0.5.
¶ Values shown are subject to slight modifications dependent on shapes and thicknesses.
** Values are for yield point.
†† Compressive strength: cast, 4.5 kg/mm² or 6,400 lb/in²
Modulus of elasticity: cast av. 2,810 kg/mm² or 4,000,000 lb/in²; rolled av. 401.0 kg/mm² or 5,700,000 lb/in²

TABLE 67. — Miscellaneous Metals and Alloys.

(a) TUNGSTEN AND ZINC.

Metal or alloy approx. comp. per cent.	Condition.	Density or weight. gm per cm^3	Density or weight. lb. per ft^3	P-limit. Tension, kg/mm^2	Ultimate strength. Tension, kg/mm^2	P-limit. Tension, lb/in^2	Ultimate strength. Tension, lb/in^2	Elong. in 50.8 mm (2 in.) Per cent	Reduct. of area. Per cent	Hardness. Brinell @ 500 kg.	Hardness. Sclero-scope.
Tungsten, W 99.2 *	Ingot sintered, D = 5.7 mm or 0.22 in.	18.0	1124	—	12.7	—	18,000	0.0	0.0	—	—
	Swaged rod, D = 0.7 mm or 0.03 in.	—	—	—	151.0	—	215,000	4.0	28.0	—	—
	Drawn hard, D = 0.029 mm or 0.00114 in.	—	—	—	415.0	—	590,000	—	65.0	—	—
	Swaged and drawn hot 97.5% reduction†	—	—	—	164.0	—	233,500	3.2	14.0	—	—
	Same as above and equiaxed at 2000°C in H_2‡	—	—	—	118.0	—	168,000	0.0	0.0	—	—
Zinc, §Zn.	Cast	7.0	437	(Impurities Pb, Fe and Cd)				—	—	—	—
	Coarse crystalline	—	—	—	2.8 to	—	4,000 to	—	—	42 to	8 to
	Fine crystalline	—	—	—	8.4	—	12,000	—	—	48	10
	Rolled (with grain or direction of rolling)	—	—	2.0	19.0	2,900	27,000	—	—	—	—
	Rolled (across grain or direction of rolling)	—	—	4.1	25.3	5,800	36,000	—	—	—	—
	Drawn hard	7.1	443	—	7.0	—	10,000	—	—	—	—

* Commercial composition for incandescent electric lamp filaments containing thoria (ThO_2) approx. 0.75 per cent after Z Jeffries Am. Inst. Min. Eng. Bulletin 138, June, 1918.

† After Z Jeffries Am. Inst. Min. Eng. Bulletin 149, May, 1919.

‡ Ordinary annealing treatment makes W brittle, and severe working, below recrystallization or equiaxing temperature, produces ductility W rods which have been worked and recrystallized are stronger than sintered rods. The equiaxing temperature of worked tungsten, with a 5-min. exposure, varies from 2200° C for a work rod with 24 per cent reduction, to 1350° C for a fine wire with 100 per cent reduction. Tungsten wire, D = 0.635 mm or 0.025 in.

§ Compression on cylinder 25.4 mm (1 in.) by 65.1 mm (2.6 in.), at 20 per cent deformation:
For spelter (cast zinc) free from Cd, av. 17.2 kg/mm^2 or 24,500 lb/in^2.
For spelter with Cd 0.26, av. 27.4 kg/mm^2 or 39,000 lb/in^2. (See Proc. A. S. T. M., Vol. 13, pl. 19.)
Modulus of rupture averages twice the corresponding tensile strength.
Shearing strength: rolled, averages 13.6 kg/mm^2 or 194,000 lb/in^2.
Modulus of elasticity: cast, 7,750 kg/mm^2 or 11,025,000 lb/in^2
Modulus of elasticity. rolled, 8450 kg/mm^2 or 12,000 000 lb/in^2. (Moore, Bulletin 52, Eng. Exp. Sta. Univ. of Ill.)

(b) WHITE METAL BEARING ALLOYS (BABBITT METAL).

A. S. T. M. vol. xviii, I, p. 491.

Experimental permanent deformation values from compression tests on cylinders 31.8 mm (1¼ in.) diam. by 63.5 mm (2½ in.) long, tested at 21° C (70° F.) (Set readings after removing loads.)

Alloy No	Formula, per cent. Sn	Sb	Cu	Pb	Pouring temp. C	F.	Weight. g/cm^3	$lb./ft^3$	Permanent deformation @ 21° C: @ 454 kg = 1000 lb. mm	in.	@ 2268 kg = 5000 lb. mm	in.	@ 4536 kg = 10,000 lb. mm	in.	Hardness. Brinell @ 21° C	@ 500 kg @ 100° C
	Tin Base.															
1	91.0	4.5	4.5	—	440	824	7.34	458	0.000	0.0000	0.025	0.0010	0.380	0.0150	28.6	12.8
2 *	89.0	7.5	3.5	—	432	808	7.39	461	.000	.0000	.038	.0015	.305	.0120	28.3	12.7
3	83.3	8.3	8.3	—	491	916	7.46	465	.025	.0010	.114	.0045	.180	.0070	34.4	15.7
4	75.0	12.0	3.0	10.0	360	680	7.52	469	.013	.0005	.064	.0025	.230	.0090	29.6	12.8
5	65.0	15.0	2.0	18.0	350	661	7.75	484	.025	.0010	.076	.0030	.230	.0090	29.6	11.8
	Lead Base.															
6	20.0	15.0	1.5	63.5	337	638	9.33	582	.038	.0015	.127	.0050	.457	.0180	24.3	11.1
7	10.0	15.0	—	75.0	329	625	9.73	607	.025	.0010	.127	.0050	.583	.0230	24.1	11.7
8	5.0	15.0	—	80.0	329	625	10.04	627	.051	.0020	.229	.0090	1.575	.0620	20.9	10.3
9	5.0	10.0	—	85.0	319	616	10.24	640	.102	.0040	.305	.0120	2.130	.0840	19.5	8.6
10	2.0	15.0	—	83.0	325	625	10.07	629	.025	.0010	.254	.0100	3.910	.1540	17.0	8.9
11	—	15.0	—	85.0	325	625	10.28	642	.025	.0010	.254	.0100	3.020	.1190	17.0	9.9
12	—	10.0	—	90.0	334	634	10.67	666	0.064	0.0025	0.432	0.0170	7.240	0.2850	14.3	6.4

* U S. Navy Spec. 46M2b (Cu 3 to 4.5, Sn 88 to 89.5, Sb 7.0 to 8.0) covers manufacture of anti-friction-metal castings. (Composition W.)

NOTE. — *See also* Brass, Lead (yellow brass), Brass, Lead-Tin (Red Brass); Bronze, Phosphor, etc., under Copper alloys

TABLE 68. — Cement and Concrete.

(a) Cement.

CEMENT: Specification Values (A. S. T. M. C9 to 17, C10 to 09, and C9 to 16T). Minimum strengths based on tests of 645 mm^2 (1 in^2) cross section briquettes for tension, and cylinders 50.8 mm (2 in.) diameter by 101.6 mm (4 in.) length for compression. Mortar composed of 1 part cement to 3 parts Ottawa sand by volume; specimens kept in damp closet for first 24 hours and in water from then on until tested.

Cement (1:3 mortar tested).	Specific gravity.	Age, days.	Tension.		Compression.	
			kg/mm^2	lb/in^2	kg/mm^2	lb/in^2
Std. Portland......	3.10	7	0.16	200	0.85	1,200
White Portland....	3.07	28	.24	300	1.60	2,000
Natural Av........	2.85	7	.03	50	—	—
Natural...........	—	28	0.09	125	—	—

(b) Cement and Cement Mortars.

CEMENT AND CEMENT MORTARS. — Bureau of Standards Experimental Values. Compressive Strengths of Portland cement mortars of uniform plastic consistency. Data from tests on 50.8 mm (2 in.) cubes stored in water. Sand: Potomac River, representative concrete sand.

Cement.	Sand.	Water, per cent.	Age, days.	Compressive strength.	
Proportions by volume.				kg/mm^2	lb/in^2
1	0	30.0	7	4.20	5,970
			28	6.40	9,120
1	1	16.0	7	3.10	4,440
			28	4.75	6,750
1	2	13.6	7	2.05	2,900
			28	3.10	4,440
1	3	13.9	7	1.25	1,780
			28	2.05	2,890
1	9	15.1	7	0.10	120
			28	0.15	200

NOTE. — (From Bureau of Standards Tech. Paper 58.) **Neat cement** briquettes mixed at plastic consistency (water 21 per cent) show 0.52 kg/mm^2 or 740 lb/in^2 tensile strength at 28 days' age;

1 **Cement:** 3 **Ottawa sand-mortar** briquettes, mixed at plastic consistency (water 9 per cent) show 0.28 kg/mm^2 or 400 lb/in^2 tensile strength at 28 days' age.

(*c*) CONCRETE.

CONCRETE: Compressive strengths. Experimental values for various mixtures. Results compiled by Joint Committee on Concrete and Reinforced Concrete. Final Report adopted by the Committee July 1, 1916. Data are based on tests of cylinders 203.2 mm (8 in.) diameter and 406.4 mm (16 in.) long **at 28 days age.**

American Standard Concrete Compressive Strengths.

Aggregate.	Units.	Mix.				
		1:3	1:4½	1:6	1:7½	1:9
Granite, trap rock	kg/mm²	2.3	2.0	1.5	1.3	1.0
	lb/in²	3300	2800	2200	1800	1400
Gravel, hard limestone and hard sandstone	kg/mm²	2.1	1.8	1.4	1.1	0.9
	lb/in²	3000	2500	2000	1600	1300
Soft limestone and soft sandstone	kg/mm²	1.5	1.3	1.1	0.8	0.7
	lb/in²	2200	1800	1500	1200	1000
Cinders	kg/mm²	0.6	0.5	0.4	0.4	0.3
	lb/in²	800	700	600	500	400

NOTE. — Mix shows ratio of cement (Portland) to combined volume of fine and coarse aggregate (latter as shown).

Committee recommends certain fractions of tabular values as safe working stresses in reinforced concrete design, which may be summarized as follows:
Bearing, 35 per cent of compressive strength;
Compression, extreme fiber, 32.5 per cent of compressive strength;
Vertical shearing stress 2 to 6 per cent of compressive strength, depending on reinforcing;
Bond stress, 4 and 5 per cent of compressive strength, for plain and deformed bars, respectively.

Modulus of Elasticity to be assumed as follows:

For concrete with strength.		Assume modulus of elasticity.	
kg/mm²	lb/in²	kg/mm²	lb/in²
up to 0.6	up to 800	530	750,000
0.6 to 1.5	800 to 2200	1400	2,000,000
1.5 to 2.0	2200 to 2900	1750	2,500,000
over 2.0	over 2900	2100	3,000,000

(See Joint Committee Report, Proc. A. S. T. M. v. XVII, 1917, p. 201.)

EDITOR'S NOTE. — The values shown in the table above are probably fair values for the compressive strengths of concretes made with average commercial material, although higher results are usually obtained in laboratory tests of specimens with high grade aggregates. Observed values on 1 : 2 : 4 gravel concrete show moduli of elasticity up to 3160 kg/mm² or 4,500,000 lb/in² and compressive strengths to 4.2 kg/mm² or 6000 lb/in²

Tensile strengths average 10 per cent of values shown from compressive strengths.

Shearing strengths average from 75 to 125 per cent of the compressive strengths; the larger percentage representing the shear of the leaner mixtures (for direct shear, Hatt gives 60 to 80 per cent of crushing strength).

Compressive strengths of natural cement concrete average from 30 to 40 per cent of that of Portland cement concrete of the same proportioned mix.

Transverse strength: modulus of rupture of 1 : 2½ : 5 concrete at 1 and 2 months equal to one sixth crushing strength at same age (Hatt).

Weight of granite, gravel and limestone, 1 : 2 : 4 concretes averages about 2.33 g/cm³ or 145 lb/ft³; that of cinder concrete of same mix is about 1.85 g/cm³ or 115 lb/ft³

Concrete, 1 : 2 : 4 *Mix, Compressive Strengths at Various Ages.*

Experimental Values: one part cement, two parts Ohio River sand and four parts of coarse aggregate as shown. Compressive tests made on 203.2 mm (8 in.) diameter cylinders, 406.4 mm (16 in.) long. (After Pittsburgh Testing Laboratory Results. *See Rwy Age,* vol. 64, Jan. 18, 1918, pp. 165–166.)

Coarse aggregate.	Unit.	Age.			
		14 days.	30 days.	60 days.	180 days.
Gravel	kg/mm²	1.35	1.61	2.06	2.67
	lb/in²	1921	2294	2925	3798
Limestone	kg/mm²	1.24	1.53	2.35	3.11
	lb/in²	1758	2174	3343	4426
Trap rock	kg/mm²	1.45	1.67	2.36	3.39
	lb/in²	2063	2386	3360	4819
Granite	kg/mm²	1.49	1.61	2.14	2.92
	lb/in²	2122	2292	3043	4151
Slag No. 1	kg/mm²	1.75	2.16	2.37	3.38
	lb/in²	2484	3075	3365	4803
Slag No. 2	kg/mm²	1.37	1.78	2.06	2.64
	lb/in²	1941	2525	2930	3753

NOTE. — Maximum and minimum test results varied about 5 per cent above or below average values shown above.

MECHANICAL PROPERTIES.

TABLE 69. — Stone and Clay Products.

(a) Strength and Stiffness of American Building Stones.*

Stone.	Weight, average.		Compression. Ultimate strength.			Flexure. Modulus of rupture.			Shear. Ultimate strength.			Flexure, modulus of elasticity.		
			Average.		Range per cent.	Average.		Range per cent.	Average.		Range per cent.	Average.		Range per cent.
	g/cm³	lb/ft³	kg/mm²	lb/in²		kg/mm²	lb/in²		kg/mm²	lb./in²		kg/mm²	lb/in²	
Granite...	2.6	165	14.20	20,200	25	1.15	1600	30	1.60	2300	20	5300	7,500,000	25
Marble...	2.7	170	8.85	12,600	25	1.05	1500	50	0.90	1300	25	5750	8,200,000	50
Limestone	2.6	160	6.30	9,000	95	0.85	1200	100	1.00	1400	45	5900	8,400,000	65
Sandstone.	2.2	135	8.80	12,500	50	1.05	1500	55	1.20	1700	45	2300	3,300,000	100

* Values based on tests of American building stones from upwards of twenty-five localities, made at Watertown (Mass.) Arsenal (Moore, p. 184). Each value shown under "Range" is one half the difference between maximum and minimum locality averages expressed as a percentage of the average for the stone.

(b) Strength and Stiffness of Bavarian Building Stone.*

Stone.	Weight, average.		Compression. Ultimate strength.			Flexure. Modulus of rupture.			Shear. Ultimate Strength.†			Flexure. Modulus of elasticity.		
			Average.		Range per cent.	Average.		Range per cent.	Average.		Range per cent.	Average.		Range per cent.
	g/cm³	lb/ft³	kg/mm²	lb/in²		kg/mm²	lb/in²		kg/mm²	lb/in²		kg/mm²	lb/in²	
Granite..	2.66	165	13.70	19,500	5	0.90	1300	5	1.00	1420	0	1600	2,300,000	30
Marble ‡.	2.16	135	5.60	8,000	15	0.30	450	—	0.45	620	50	3450	4,900,000	—
Limestone	2.48	155	8.10	11,500	5	1.10	1550	45	0.60	870	20	2350	3,350,000	90
Sandstone	2.30	145	8.10	11,500	75	0.45	650	55	0.50	680	35	2500	3,550,000	35

* Values based on careful tests by Bauschinger, "Communications," Vol. 10.
† Shearing strength determined perpendicular to bed of stone.
‡ Values are for Jurassic limestone.

General Notes. — 1. Later transverse strength (flexure) tests on Wisconsin building stones (Johnson's "Materials of Construction," 1918 ed., p. 255) show moduli of rupture as follows: Granite, 1.90 to 2.75 kg/mm² or 2710 to 3910 lb/in²; limestone, 0.80 to 3.30 kg/mm² or 1160 to 4660 lb/in²; sandstone, 0.25 to 0.95 kg/mm² or 360 to 1320 lb/in².

2. Good slate has a modulus of rupture of 4.90 kg/mm² or 7000 lb/in² (*loc. cit.*, p. 257).

TABLE 69.— Stone and Clay Products.

(*c*) Strengths of American Building Bricks.*

Brick — description.	Absorption average per cent.	Compression. Min. ult. strength.		Flexure. Min. modulus rupture.	
		kg/mm²	lb/in²	kg/mm²	lb/in²
Class A (Vitrified)	5	3.50	5000	0.65	900
Class B (Hard burned)	12	2.45	3500	0.40	600
Class C (Common firsts)	18	1.40	2000	0.30	400
Class D (Common)	—	1.05	1500	0.20	300

* After A. S. T. M. Committee C–3, Report 1913, and University laboratories' tests for Committee C–3 (Johnson, p. 281).

(*d*) Strength in Compression of Brick Piers and of Terra-cotta Block Piers.

Tabular values are based on test data from Watertown Arsenal, Cornell University, U. S. Bureau of Standards, and University of Ill. (Moore, p. 185).

Brick or block used.	Mortar.	Compression.* Av. ult. strength.	
		kg/mm²	lb/in²
Vitrified brick	1 part P.† cement : 3 parts sand	1.95	2800
Pressed (face) brick	1 part P. cement : 3 parts sand	1.40	2000
Pressed (face) brick	1 part lime : 3 parts sand	1.00	1400
Common brick	1 part P. cement : 3 parts sand	0.70	1000
Common brick	1 part lime : 3 parts sand	0.50	700
Terra-cotta brick	1 part P. cement : 3 parts sand	2.10	3000

* Building ordinances of American cities specify allowable working stresses in compression over bearing area of 12.5 per cent (vitrified brick) to 17.5 per cent (common brick) of corresponding ultimate compressive strength shown in table.

† P. denotes Portland.

(*e*) Strength of Compression of Various Bricks.

Reasonable minimum average compressive strengths for other types of brick than building brick are noted by Johnson, "Materials of Construction," pp. 289 ff., as follows:

Brick.	kg/mm²	lb/in²
sand-lime	2.10	3000
sand-lime (*German*)	1.53	2180 (av. 255 tests)
paving	5.60	8000
acid-refractory	0.70	1000
silica-refractory	1.40	2000

The specific gravity of brick ranges from 1.9 to 2.6 (corresponding to 120 to 160 lb/ft³).

Building tile: hollow clay blocks of good quality,— minimum compressive strength: 0.70 kg/mm² or 1000 lb/in². Tests made for A. S. T. M. Committee C–10 (A. S. T. M. Proc. XVII, I, p. 334) show compressive strengths ranging from 0.45 to 8.70 kg/mm² or 640 to 12,360 lb/in² of net section, corresponding to 0.05 to 4.20 kg/mm² or 95 to 6000 lb/in² of gross section. Recommended safe loads (Marks, "Mechanical Engineers' Handbook," p. 625) for effective bearing parts of hollow tile: hard fire-clay tiles 0.06 kg/mm² or 80 lb./in²; ordinary clay tiles 0.04 kg/mm² or 60 lb/in²; porous terra-cotta tiles 0.03 kg/mm² or 40 lb/in.² The specific gravity of tile ranges from 1.9 to 2.5 corresponding to a weight of 120 to 155 lb/ft³.

TABLE 70. — Rubber and Leather.

(a) Rubber, — Sheet.*

Grade.	Ultimate strength.				Ult. elongation.		Set.‡	
	Longitudinal.†		Transverse.		Longit.	Transv.	Longit.	Transv.
	kg/mm²	lb/in²	kg/mm²	lb/in²	per cent.		per cent.	
1	1.92	2730	1.81	2575	630	640	11.2	7.3
2	1.45	2070	1.43	2030	640	670	6.0	5.0
3	0.84	1200	0.89	1260	480	555	22.1	16.3
4	1.30	1850	1.20	1700	410	460	34.0	24.0
5	0.48	690	0.36	510	320	280	27.5	25.0
6	0.62	880	0.48	690	315	315	34.3	25.9

* Data from Bureau of Standards Circular 38.
† Longitudinal indicates direction of rolling through the calendar.
‡ Set measured after 300 per cent elongation for 1 minute with 1 minute rest.

The specific gravity of rubber averages from 0.95 to 1.25, corresponding to an average weight of 60 to 80 lb/ft³.

Four-ply rubber belts show an average ultimate tensile strength of 0.63 to 0.65 kg/mm² or 890 to 930 lb./in² (Benjamin), and a working tensile stress of 0.07 to 0.11 kg/mm² or 100 to 150 lb./in² is recommended (Bach).

(b) Leather, — Belting.

Oak tanned leather from the center or back of the hide:

Minimum tensile strengths of belts (Marks, p. 622) { single 2.8 kg/mm² or 4000 lb./in²; double 2.5 kg/mm² or 3600 lb./in²

Maximum elongation for one hour application of 1.6 kg/mm² or 2250 lb./in² stress { single 13.5 per cent; double 12.5 per cent.

Modulus of elasticity of leather varies from an average value of 12.5 kg/mm² or 17,800 lb/in² (new) to 22.5 kg/mm² or 32,000 lb./in² (old).

Chrome leather has a tensile strength of 6.0 to 9.1 kg/mm² or 8500 to 12,900 lb/in².

The specific gravity of leather varies from 0.86 to 1.02, corresponding to a weight of 53.6 to 63.6 lb./ft³.

TABLE 71. — Manila Rope.

Manila Rope, Weight and Strength — Specification Values. From U. S. Government Standard Specifications adopted April 4, 1918.

Rope to be made of manila or Abaca fiber with no fiber of grade lower than U. S. Government Grade I, to be three-strand,* medium-laid, with maximum weights and minimum strengths shown in the table below, lubricant content to be not less than 8 nor more than 12 per cent of the weight of the rope as sold.

Approximate diameter.		Circumference.		Maximum net weight.		Minimum breaking strength.	
mm	in.	mm	in.	kg/m	lb/ft.	kg	lb.
6.3	1/4	19.1	3/4	0.029	0.0196	320	700
7.9	5/16	25.4	1	0.044	0.0286	540	1,200
9.5	3/8	28.6	1 1/8	0.061	0.0408	660	1,450
11.1	7/16	31.8	1 1/4	0.080	0.0539	790	1,750
11.9	15/32	34.9	1 3/8	0.095	0.0637	950	2,100
12.7	1/2	38.1	1 1/2	0.109	0.0735	1,110	2,450
14.3	9/16	44.5	1 3/4	0.153	0.1029	1,430	3,150
15.9	5/8	50.8	2	0.195	0.1307	1,810	4,000
19.1	3/4	57.2	2 1/4	0.241	0.1617	2,220	4,900
20.6	13/16	63.5	2 1/2	0.284	0.1911	2,680	5,900
22.2	7/8	69.9	2 3/4	0.328	0.2205	3,170	7,000
25.4	1	76.2	3	0.394	0.2645	3,720	8,200
27.0	1 1/16	82.6	3 1/4	0.459	0.3087	4,310	9,500
28.6	1 1/8	88.9	3 1/2	0.525	0.3528	4,990	11,000
31.8	1 1/4	95.2	3 3/4	0.612	0.4115	5,670	12,500
33.3	1 5/16	101.6	4	0.700	0.4703	6,440	14,200
34.9	1 3/8	108.0	4 1/4	0.787	0.5290	7,260	16,000
38.1	1 1/2	114.3	4 1/2	0.875	0.5879	7,940	17,500
39.4	1 9/16	120.7	4 3/4	0.984	0.6615	8,840	19,500
41.2	1 5/8	127.0	5	1.094	0.7348	9,750	21,500
44.5	1 3/4	140.0	5 1/2	1.312	0.8818	11,550	25,500
50.8	2	152.4	6	1.576	1.059	13,610	30,000
52.4	2 1/16	165.1	6 1/2	1.823	1.225	15,420	34,000
57.2	2 1/4	177.8	7	2.144	1.441	17,460	38,500
63.5	2 1/2	190.5	7 1/2	2.450	1.646	19,730	43,500
66.7	2 5/8	203.2	8	2.799	1.881	22,220	49,000
73.0	2 7/8	215.9	8 1/2	3.136	2.107	24,940	55,000
76.2	3	228.6	9	3.543	2.381	27,670	61,000
79.4	3 1/8	241.3	9 1/2	3.936	2.645	30,390	67,000
82.5	3 1/4	254.0	10	4.375	2.940	33,110	73,000

* Four-strand, medium-laid rope when ordered may run up to 7% heavier than three-strand rope of the same size, and must show 95% of the strength required for three-strand rope of the same size.

Common and botanical name.	Specific gravity, oven-dry, based on vol. when green.	Specific gravity, oven-dry, based on vol. oven-dry.	Static bending. P-limit, kg/mm²	Static bending. Modulus of rupture, kg/mm²	Static bending. Modulus of elasticity, kg/mm²	Impact bending. P-limit, kg/mm²	Impact bending. 22.7 kg hammer fall for failure — m.	Compression. Parallel to grain. P-limit kg/mm²	Compression. Parallel to grain. Ultimate. kg/mm²	Compression. Perpendicular to grain P-limit, kg/mm²	Shear. Parallel to grain ult. st. kg/mm²	Tension. Perpendicular to grain ult. st. kg/mm²	Hardness. Load to ½ imbed 11.3 mm d. ball end kg	Hardness. Load to ½ imbed 11.3 mm d. ball side kg
1	**4**	**5**	**6**	**7**	**8**	**9**	**10**	**11**	**12**	**13**	**14**	**15**	**16**	**17**
Alder, red (*Alnus oregona*)	0.37	0.43	2.65	4.55	830	5.60	0.56	1.85	2.10	0.22	0.54	0.27	250	200
Ash, black (*Fraxinus nigra*)	0.46	0.53	1.85	4.20	720	5.10	0.81	1.15	1.60	0.31	0.61	0.35	270	250
Ash, white (forest grown) (*Fraxinus americana*)	0.52	0.60	3.45	6.40	950	8.25	0.91	2.30	2.70	0.57	0.89	0.44	455	401
Ash, white (second growth) (*Fraxinus americana*)	0.58	0.71	4.30	7.60	1150	9.70	1.19	2.70	2.90	0.56	1.13	0.56	515	490
Aspen (*Populus tremuloides*)	0.36	0.42	2.05	3.75	590	4.85	0.71	1.10	1.50	0.14	0.44	0.13	120	145
Basswood (*Tilia americana*)	0.33	0.40	1.90	3.50	725	4.35	0.43	1.20	1.55	0.15	0.43	0.20	125	115
Beech (*Fagus atropunicea*)	0.54	0.66	3.15	5.80	875	7.30	1.02	1.80	2.30	0.43	0.85	0.56	430	370
Birch, paper (*Betula papyrifera*)	0.47	0.60	2.05	4.10	710	5.50	1.14	1.20	1.55	0.21	0.56	0.27	180	220
Birch, yellow (*Betula lutea*)	0.54	0.66	3.25	6.05	1080	8.25	1.02	1.90	2.40	0.32	0.78	0.34	370	340
Butternut (*Juglans cinerea*)	0.36	0.40	2.05	3.80	680	5.15	0.61	1.40	1.70	0.19	0.53	0.30	185	175
Cherry, black (*Prunus serotina*)	0.47	0.53	2.95	5.65	920	7.20	0.84	2.10	2.50	0.31	0.80	0.40	340	300
Chestnut (*Castanea dentata*)	0.40	0.46	2.20	3.95	655	5.55	0.61	1.45	1.75	0.27	0.56	0.30	240	190
Cottonwood (*Populus deltoides*)	0.37	0.43	2.05	3.75	710	5.05	0.53	1.25	1.60	0.17	0.48	0.29	175	155
Cucumber tree (*Magnolia acuminata*)	0.44	0.52	2.95	5.20	1100	6.55	0.76	1.95	2.20	0.29	0.70	0.31	270	235
Dogwood (flowering) (*Cornus florida*)	0.64	0.80	3.40	6.20	830	5.00	1.47	—	2.55	0.73	1.07	—	640	640
Elm, cork (*Ulmus racemosa*)	0.58	0.66	3.25	6.70	840	7.75	1.27	2.00	2.70	0.53	0.89	0.47	445	450
Elm, white (*Ulmus americana*)	0.44	0.54	2.55	4.85	725	5.70	0.86	1.60	2.00	0.23	0.65	0.39	275	250
Gum, blue (*Eucalyptus globulus*)	0.62	0.80	5.35	7.85	1430	10.00	1.02	3.40	3.70	0.72	1.09	0.45	595	610
Gum, cotton (*Nyssa aquatica*)	0.46	0.52	2.95	5.15	740	6.30	0.76	1.95	2.40	0.42	0.84	0.42	365	320
Gum, red (*Liquidambar styraciflua*)	0.44	0.53	2.60	4.80	810	7.05	0.84	1.70	1.95	0.32	0.75	0.36	285	235
Hickory pecan (*Hicoria pecan*)	0.60	0.69	3.65	6.90	960	8.65	1.35	2.15	2.80	0.63	1.04	0.48	575	595
Hickory, shagbark (*Hicoria ovata*)	0.64	—	4.15	7.75	1105	10.10	1.83	2.40	3.20	0.70	0.93	—	—	—
Holly, American (*Ilex opaca*)	0.50	0.61	2.40	4.55	630	6.25	1.30	1.40	1.85	0.43	0.80	0.43	390	360
Laurel, mountain (*Kalmia latifolia*)	0.62	0.74	4.10	5.90	650	7.20	0.81	—	3.00	0.78	1.18	—	635	590
Locust, black (*Robinia pseudacacia*)	0.66	0.71	6.20	9.70	1300	12.90	1.12	4.40	4.80	1.01	1.24	0.54	740	715
Locust, honey (*Gleditsia triacanthos*)	0.60	0.67	3.95	7.20	910	8.30	1.20	2.35	3.10	1.00	1.17	0.66	655	630
Magnolia (evergreen) (*Magnolia foetida*)	0.46	0.53	2.55	4.80	780	6.20	1.37	1.55	1.90	0.40	0.73	0.43	355	335
Maple, silver (*Acer saccharinum*)	0.44	0.51	2.20	4.10	660	4.80	0.74	1.35	1.75	0.32	0.74	0.39	305	270
Maple, sugar (*Acer saccharum*)	0.56	0.66	3.50	6.40	1040	8.50	0.91	2.20	2.80	0.53	0.97	0.54	455	415
Oak, canyon live (*Quercus chrysolepsis*)	0.70	0.84	4.45	7.45	945	7.90	1.20	2.85	3.30	1.04	1.20	0.68	720	715
Oak, red (*Quercus rubra*)	0.56	0.65	2.60	5.40	910	7.30	1.04	1.65	2.25	0.51	0.79	0.52	465	430
Oak, white (*Quercus alba*)	0.60	0.71	3.30	5.85	880	7.55	1.07	2.10	2.50	0.59	0.88	0.54	510	480
Persimmon (*Diospyros virginiana*)	0.64	0.78	3.95	7.05	965	8.50	1.04	2.15	2.95	0.78	1.03	0.54	565	580
Poplar, yellow (*Liriodendron tulipifera*)	0.37	0.42	2.25	3.95	850	5.65	0.43	1.40	1.80	0.22	0.56	0.32	190	155
Sycamore (*Platanus occidentalis*)	0.46	0.54	2.30	4.60	745	6.20	0.84	1.70	2.00	0.32	0.71	0.44	320	275
Walnut, black (*Juglans nigra*)	0.51	0.56	3.80	6.70	1000	8.40	0.94	2.55	3.05	0.42	0.86	0.43	435	410
Willow, black (*Salix nigra*)	0.34	0.41	1.25	2.75	395	3.60	0.91	0.70	1.05	0.15	0.44	0.30	160	165

NOTE. — Results of tests on sixty-eight species; test specimens, small clear pieces, 50.8 by 50.8 mm in section, 762 mm long for bending; others, shorter. Data taken from Bulletin 556, Forest Service, U. S. Dept. of Agriculture, containing data on 130,000 tests. See pages 87 and 99 for explanation of columns.

Common and botanical name.	Specific gravity, oven-dry, based on vol. when green.	Specific gravity, oven-dry, based on vol. oven-dry.	Static bending. P-limit, kg/mm²	Static bending. Modulus of rupture, kg/mm²	Static bending. Modulus of elasticity, kg/mm²	Impact bending. P-limit, kg/mm²	Impact bending. 22.7 kg hammer fall for failure — m.	Compression. Parallel to grain. P-limit. kg/mm²	Compression. Parallel to grain. Ultimate. kg/mm²	Compression. Perpendicular to grain P-limit, kg/mm²	Shear. Parallel to grain ult. st, kg/mm²	Tension. Perpendicular to grain ult. st. kg/mm²	Hardness. Load to ½ imbed 11.3 mm d. ball, end kg	Hardness. Load to ½ imbed 11.3 mm d. ball, side kg
1	4	5	6	7	8	9	10	11	12	13	14	15	16	17
Cedar, incense (*Libocedrus decurrens*)	0.35	0.36	2.75	4.35	590	5.15	0.43	2.00	2.20	0.32	0.58	0.20	260	175
Cedar, Port Orford (*Chamaecyparis lawsoniana*)	0.41	0.47	2.75	4.80	1055	6.55	0.64	2.10	2.30	0.27	0.62	0.17	255	220
Cedar, western red (*Thuja plicata*)	0.31	0.34	2.30	3.65	670	5.05	0.43	1.75	2.00	0.22	0.51	0.15	195	118
Cedar, white (*Thuja occidentalis*)	0.29	0.32	1.85	2.95	450	3.75	0.38	1.00	1.40	0.20	0.44	0.17	145	104
Cypress, bald (*Taxodium distichum*)	0.41	0.47	2.80	4.80	835	5.60	0.61	2.20	2.45	0.33	0.58	0.20	215	175
Fir, amabilis (*Abies amabilis*)	0.37	0.42	2.75	4.45	915	5.50	0.53	1.70	2.00	0.22	0.47	0.17	165	140
Fir, balsam (*Abies balsamea*)	0.34	0.41	2.10	3.45	675	4.85	0.41	1.55	1.70	0.15	0.43	0.23	135	135
Fir, Douglas (1) (*Pseudotsuga taxifolia*)	0.45	0.52	3.50	5.50	1110	6.60	0.63	2.40	2.80	0.37	0.64	0.14	230	215
Fir, Douglas (2) (*Pseudotsuga taxifolia*)	0.40	0.44	2.55	4.50	830	6.40	0.51	1.80	2.10	0.32	0.62	0.25	205	180
Fir, grand (*Abies grandis*)	0.37	0.42	2.55	4.30	915	5.70	0.56	1.90	2.10	0.24	0.53	0.16	190	165
Fir, noble (*Abies nobilis*)	0.35	0.41	2.40	4.00	900	5.55	0.51	1.70	1.90	0.22	0.49	0.13	135	115
Fir, white (*Abies concolor*)	0.35	0.44	2.75	4.20	795	5.05	0.46	1.85	1.95	0.31	0.51	0.18	175	150
Hemlock, eastern (*Tsuga canadensis*)	0.38	0.44	2.95	4.70	790	5.55	0.51	1.90	2.30	0.35	0.62	0.18	230	185
Hemlock, western (*Tsuga heterophylla*)	0.38	0.43	2.40	4.30	835	5.50	0.51	1.60	2.05	0.25	0.57	0.18	245	195
Larch, western (*Larix occidentalis*)	0.48	0.59	3.25	5.25	950	6.60	0.61	2.30	2.70	0.39	0.65	0.16	215	205
Pine, Cuban (*Pinus heterophylla*)	0.58	0.68	3.95	6.20	1150	7.95	0.94	2.80	3.15	0.41	0.72	0.20	260	285
Pine, loblolly (*Pinus taeda*)	0.50	0.59	3.10	5.30	970	6.70	0.81	2.00	2.50	0.39	0.63	0.20	185	205
Pine, lodgepole (*Pinus contorta*)	0.38	0.44	2.10	3.85	760	5.05	0.51	1.50	1.85	0.22	0.49	0.15	145	150
Pine, longleaf (*Pinus palustris*)	0.55	0.64	3.80	6.10	1150	7.60	0.86	2.70	3.10	0.42	0.75	0.20	250	270
Pine, Norway (*Pinus resinosa*)	0.44	0.51	2.60	4.50	970	5.35	0.71	1.75	2.20	0.25	0.55	0.13	165	155
Pine, pitch (*Pinus rigida*)	0.47	0.54	2.60	4.70	790	6.40	0.74	1.50	2.15	0.36	0.67	0.25	210	220
Pine, shortleaf (*Pinus echinata*)	0.50	0.58	3.15	5.65	1020	7.90	0.99	2.50	2.70	0.34	0.63	0.23	220	255
Pine, sugar (*Pinus lambertiana*)	0.36	0.39	2.30	3.75	685	4.70	0.43	1.65	1.85	0.25	0.50	0.19	150	145
Pine, western white (*Pinus monticola*)	0.39	0.45	2.45	4.00	935	5.35	0.58	1.95	2.15	0.21	0.50	0.18	150	150
Pine, western yellow (*Pinus ponderosa*)	0.38	0.42	3.20	3.65	710	4.70	0.48	1.45	1.75	0.24	0.48	0.20	140	145
Pine, white (*Pinus strobus*)	0.36	0.39	2.40	3.75	750	4.55	0.46	1.65	1.90	0.22	0.45	0.18	135	135
Spruce, red (*Picea rubens*)	0.48	0.41	2.40	4.00	830	5.05	0.46	1.65	1.95	0.25	0.54	0.15	190	160
Spruce, Sitka (*Picea sitchensis*)	0.34	0.37	2.10	3.85	830	5.05	0.74	1.60	1.85	0.23	0.55	0.16	195	170
Tamarack (*Larix laricina*)	0.49	0.56	2.95	5.05	875	5.50	0.71	2.20	2.45	0.34	0.66	0.18	180	170
Yew, western (*Taxus brevifolia*)	0.60	0.67	4.55	7.10	695	9.20	0.97	2.40	3.25	0.73	1.14	0.32	610	520

NOTE. — The data above are extracted from tests on one hundred and twenty-six species of wood made at the Forest Products Laboratory, Madison, Wisconsin. Bulletin 556 records results of tests on air-dry timber also, but only data on green timber are shown, as the latter are based on a larger number of tests and on tests which are not influenced by variations in moisture content. The strength of dry material usually exceeds that of green material, but allowable working stresses in design should be based on strengths of green timber, inasmuch as the increase of strength due to drying is a variable, uncertain factor and likely to be offset by defects. All test specimens were two inches square, by lengths as shown.

COLUMN NOTES. — **2**, Locality where grown, — see Tables 74 and 75; **3**, Moisture includes all matter volatile at 100° C expressed as per cent of ordinary weight; **5**, Weight, air dry is for wood with 12 per cent moisture; for density, see metric unit tables 72 and 73; **6–10**, 762 mm (30 in.) long specimen on 711.2 mm (28 in.) span, with load at center.

 TABLE 74. — Hardwoods Grown in U. S. (English Units).

Common and botanical name.	Locality where grown.	Moisture content, green, per cent.	Weight. Green. lb/ft³	Weight. Air-dry. lb/ft³	Static bending. P-limit, lb/in²	Static bending. Modulus of rupture, lb/in²	Static bending. Modulus of elasticity 1000 × lb/in²	Impact bending. P-limit, lb/in²	Compression. Parallel to grain. P-limit. lb/in²	Compression. Perpendicular to grain, P-limit lb/in²	Shear. Parallel to grain, ult. st. lb/in²	Tension. Perpendicular to grain, ult. st. lb/in²
1	2	3	4	5	6	7	8	9	11	13	14	15
Alder, red (*Alnus oregona*)	Wash.	98	46	28	3800	6500	1170	8000	2650	310	770	390
Ash, black (*Fraxinus nigra*)	Mich. and Wis.	83	53	34	2600	6000	1020	7200	1620	430	870	490
Ash, white (forest grown) (*Fraxinus americana*)	Ark. and W. Va.	43	46	40	4900	9100	1350	11700	3230	800	1260	620
Ash, white (2d growth) (*Fraxinus americana*)	N. Y.	40	51	46	6100	10800	1640	13800	3820	790	1600	790
Aspen (*Populus tremuloides*)	Wis.	107	47	27	2900	5300	840	6900	1620	200	620	180
Basswood (*Tilia americana*)	Wis. and Pa.	103	41	26	2700	5000	1030	6200	1710	210	610	280
Beech (*Fagus atropunicea*)	Ind. and Pa.	62	55	44	4500	8200	1240	10400	2550	610	1210	760
Birch, paper (*Betula papyrifera*)	Wis. and Pa.	72	51	38	2900	5800	1010	7800	1650	300	790	380
Birch, yellow (*Betula lutea*)	Wis.	68	58	45	4600	8600	1540	11700	2760	450	1110	480
Butternut (*Juglans cinerea*)	Tenn. and Wis.	104	46	27	2900	5400	970	7300	1960	270	760	430
Cherry, black (*Prunus serotina*)	Pa.	55	46	36	4200	8000	1310	10200	2940	440	1130	570
Chestnut (*Castanea dentata*)	Md. and Tenn.	122	55	30	3100	5600	930	7900	2040	380	800	430
Cottonwood (*Populus deltoides*)	Mo.	111	49	29	2900	5300	1010	7200	1770	240	680	410
Cucumber tree (*Magnolia acuminata*)	Tenn.	80	50	33	4200	7400	1560	9300	2760	410	990	440
Dogwood (flowering) (*Cornus florida*)	Tenn.	62	65	54	4800	8800	1180	7100	—	1030	1520	—
Elm, cork (*Ulmus racemosa*)	Wis.	50	54	45	4600	9500	1190	11000	2870	750	1270	660
Elm, white (*Ulmus americana*)	Wis. and Pa.	88	52	35	3600	6900	1030	8100	2290	390	920	560
Gum, blue (*Eucalyptus globulus*)	Cal.	79	70	54	7600	11200	2010	14200	4870	1020	1550	640
Gum, cotton (*Nyssa aquatica*)	La.	97	56	34	4200	7300	1050	9000	2760	590	1190	600
Gum, red (*Liquidambar styraciflua*)	Mo.	81	50	36	3700	6800	1150	10000	2360	460	1070	510
Hickory, pecan (*Hicoria pecan*)	Mo.	63	61	46	5200	9800	1370	12300	3040	960	1480	680
Hickory, shagbark (*Hicoria ovata*)	O., Miss., Pa. and W. Va.	60	64	51	5900	11000	1570	14400	3430	1000	1320	—
Holly, American (*Ilex opaca*)	Tenn.	82	57	40	3400	6500	900	8900	1970	610	1130	610
Laurel, mountain (*Kalmia latifolia*)	Tenn.	62	62	49	5800	8400	920	10200	—	1110	1670	—
Locust, black (*Robinia pseudacacia*)	Tenn.	40	58	49	8800	13800	1850	18300	6280	1430	1760	770
Locust, honey (*Gleditsia triacanthos*)	Mo. and Ind.	63	61	47	5600	10200	1290	11800	3320	1420	1660	930
Magnolia (evergreen) (*Magnolia foetida*)	La.	117	62	35	3600	6800	1110	8800	2200	570	1040	610
Maple, silver (*Acer saccharinum*)	Wis.	66	46	34	3100	5800	940	6800	1950	460	1050	560
Maple, sugar (*Acer saccharum*)	Ind., Pa. and Wis.	60	56	44	5000	9100	1480	12100	3120	750	1380	770
Oak, canyon live (*Quercus chrysolepsis*)	Cal.	62	71	56	6300	10600	1340	11200	4050	1480	1700	970
Oak, red (*Quercus rubra*)	Ark., La., Ind. and Tenn.	84	64	45	3700	7700	1290	10400	2330	730	1120	740
Oak, white (*Quercus alba*)	Ark., La. and Ind.	68	62	47	4700	8300	1250	10700	2990	830	1250	770
Persimmon (*Diospyros virginiana*)	Mo.	58	63	53	5600	10000	1370	12100	3030	1110	1470	770
Poplar, yellow (*Liriodendron tulipifera*)	Tenn.	64	38	28	3200	5600	1210	8000	2000	310	790	460
Sycamore (*Platanus occidentalis*)	Ind. and Tenn.	83	52	35	3300	6500	1060	8800	2390	450	1000	630
Walnut, black (*Juglans nigra*)	Ky.	81	58	39	5400	9500	1420	11900	3600	600	1220	570

NOTE. — Results of tests on sixty-eight species; test specimens, small clear pieces, 2 by 2 inches in section, 30 inches long for bending; others, shorter. Tested in a green condition. Data taken from Bulletin 556, Forest Service, U. S. Dept. of Agriculture, containing data on 130,000 tests. See pages 97 and 99 for explanation of columns.

Common and botanical name.	Locality where grown.	Moisture content, green, per cent.	Weight. Green. lb/ft³	Weight. Air-dry. lb/ft³	Static bending. P-limit, lb/in²	Static bending. Modulus of rupture, lb/in²	Static bending. Modulus of elasticity 1000 × lb/in²	Impact bending. P-limit, lb/in²	Compression. Parallel to grain P-limit. lb/in²	Compression. Perpendicular to grain, P-limit lb/in²	Shear. Parallel to grain, ult. st. lb/in²	Tension. Perpendicular to grain, ult. st. lb/in²
1	**2**	**3**	**4**	**5**	**6**	**7**	**8**	**9**	**11**	**13**	**14**	**15**
Cedar, incense (*Libocedrus decurrens*)	Cal. and Ore.	108	45	24	3900	6200	840	7300	2870	460	830	280
Cedar, Port Orford (*Chamaecyparis lawsoniana*)	Ore.	52	39	31	3900	6800	1500	9300	3970	380	880	240
Cedar, western red (*Thuja plicata*)	Wash. and Mont.	39	27	23	3300	5200	950	7100	2500	310	720	210
Cedar, white (*Thuja occidentalis*)	Wis.	55	28	21	2600	4200	640	5300	1420	290	620	240
Cypress, bald (*Taxodium distichum*)	La. and Mo.	87	48	30	4000	6800	1190	8000	3100	470	820	280
Fir, amabilis (*Abies amabilis*)	Ore. and Wash.	102	47	27	3900	6300	1300	7800	2380	320	670	240
Fir, balsam (*Abies balsamea*)	Wis.	117	45	25	3000	4900	960	6900	2220	210	610	180
Fir, Douglas (1) (*Pseudotsuga taxifolia*)	Wash. and Ore.	36	38	34	5000	7800	1580	9400	3400	530	910	200
Fir, Douglas (2) (*Pseudotsuga taxifolia*)	Mont. and Wyo.	38	34	32	3600	6400	1180	9100	2520	450	880	350
Fir, grand (*Abies grandis*)	Mont. and Ore.	94	44	27	3600	6100	1300	8100	2680	340	700	230
Fir, noble (*Abies nobilis*)	Ore.	41	31	26	3400	5700	1280	7900	2370	310	700	180
Fir, white (*Abies concolor*)	Cal.	156	56	26	3900	6000	1130	7200	2610	440	730	260
Hemlock (eastern) (*Tsuga canadensia*)	Tenn. and Wis.	105	48	29	4200	6700	1120	7900	2710	500	880	260
Hemlock (western) (*Tsuga heterophylla*)	Wash.	71	41	29	3400	6100	1190	7800	2290	350	810	260
Larch, western (*Larix occidentalis*)	Mont. and Wash.	58	48	37	4600	7500	1350	9400	3250	560	920	230
Pine, Cuban (*Pinus heterophylla*)	Fla.	47	53	45	5600	8800	1630	11300	3950	590	1030	290
Pine, loblolly (*Pinus taeda*)	Fla., N. and S. Car.	70	54	39	4400	7500	1380	9500	2870	550	900	280
Pine, lodgepole (*Pinus contorta*)	Col., Mont. and Wyo.	65	39	28	3000	5500	1080	7200	2100	310	690	220
Pine, longleaf (*Pinus palustris*)	Fla., La. and Miss.	47	50	43	5400	8700	1630	10800	3840	600	1070	290
Pine, Norway (*Pinus resinosa*)	Wis.	54	42	34	3700	6400	1380	7500	2470	360	780	190
Pine, pitch (*Pinus rigida*)	Tenn.	85	54	35	3700	6700	1120	9100	2100	510	950	350
Pine, shortleaf (*Pinus echinata*)	Ark. and La.	64	50	37	4500	8000	1450	11200	3650	480	890	330
Pine, sugar (*Pinus lambertiana*)	Cal.	123	50	26	3300	5300	970	6700	2340	350	710	270
Pine, western white (*Pinus monticola*)	Mont.	58	39	30	3500	5700	1330	7600	2770	300	710	250
Pine, western yellow (*Pinus ponderosa*)	Col., Mont., Ariz., Wash. and Cal.	95	46	28	3100	5200	1010	6700	2080	340	680	280
Pine, white (*Pinus strobus*)	Wis.	74	39	27	3400	5300	1070	6500	2370	310	640	260
Spruce, red (*Picea rubens*)	N. H. and Tenn.	43	34	28	3400	5700	1180	7200	2360	350	770	220
Spruce, Sitka (*Picea sitchensis*)	Wash.	53	33	26	3000	5500	1180	7900	2280	330	780	230
Tamarack (*Larix laricina*)	Wis.	52	47	38	4200	7200	1240	7800	3010	480	860	260
Yew, western (*Taxus brevifolia*)	Wash.	44	54	45	6500	10100	990	13100	3400	1040	1620	450

COLUMN NOTES (*continued*). — (**7**) recommended allowable working stress (interior construction): $\frac{1}{6}$ tabular value; experimental results on tests of air-dry timber in small clear pieces average 50 per cent higher; kiln-dry, double tabular values; (**10**) repeated falls of 50-lb. hammer from increasing heights; **11–12**, 203.2-mm (8 in.) long specimen loaded on ends with deformations measured in a 152.4-mm (6 in.) gage length; (**12**) allowable working stress $\frac{1}{4}$ tabular crushing strength; (**13**) 152.4-mm (6 in.) long block loaded on its side with a central bearing area of 2580.6-mm² (4 in²) allowable working stress, $\frac{2}{3}$ tabular value. (**14**) 50.8-mm by 50.8-mm (2 in.) projecting lip sheared from block; allowable working stress, $\frac{1}{6}$ tabular value; (**15**) 63.5-mm ($2\frac{1}{2}$ in.) specimen with 25.4-mm (1 in.) free loaded length; allowable working stress, $\frac{1}{6}$ tabular value. (**16–17**) for values in lbs. multiply values of metric tables by 2.2.

ELASTIC MODULI.

TABLE 76.—Rigidity Modulus.

If to the four consecutive faces of a cube a tangential stress is applied, opposite in direction on adjacent sides, the modulus of rigidity is obtained by dividing the numerical value of the tangential stress per unit area (kg. per sq. mm.) by the number representing the change of angles on the non-stressed faces, measured in radians.

Substance.	Rigidity Modulus.	Reference.	Substance.	Rigidity Modulus.	Reference.
Aluminum	3350	14	Quartz fibre	2888	20
" cast	2580	5	" "	2380	21
Brass	3550	10	Silver	2960	5
"	3715	11	"	2650	10
" cast, 60 Cu + 12 Sn	3700	5	"	2566	16
Bismuth, slowly cooled	1240	5	" hard-drawn	2816	11
Bronze, cast, 88 Cu + 12 Sn	4060	5	Steel	8290	16
Cadmium, cast	2450	5	" cast	7458	15
Copper, cast	4780	5	" cast, coarse gr.	8070	5
"	4213	18	" silver-	7872	11
"	4450	10	Tin, cast	1730	5
"	4664	19	"	1543	19
Gold	2850	5	Zinc	3880	5
"	3950	14	"	3820	19
Iron, cast	5210	5	Platinum	6630	16
"	6706	15	"	6220	22
"	7975	10	Glass	2350	–
"	6940	7	"	2730	–
"	8108	16	Clay rock	1770	23
"	7505	14	Granite	1280	23
Magnesium, cast	1710	5	Marble	1190	23
Nickel	7820	5	Slate	2290	23
Phosphor bronze	4359	11			

References 1–16, see Table 48.
17 Grätz, Wied. Ann. 28, 1886.
18 Savart, Pogg. Ann. 16, 1829.
19 Kiewiet, Diss. Göttingen, 1886.
20 Threlfall, Philos. Mag. (5) 30, 1890.
21 Boys, Philos. Mag. (5) 30, 1890.
22 Thomson, Lord Kelvin.
23 Gray and Milne.
24 Adams-Coker, Carnegie Publ. No. 46, 1906.

TABLE 77. — Variation of the Rigidity Modulus with the Temperature.

$n_t = n_o\,(1 - at - \beta t^2 - \gamma t^3)$, where t = temperature Centigrade.

Substance.	n_o	$\alpha 10^6$	$\beta 10^8$	$\gamma 10^{10}$	Authority.
Brass	2652	2158	48	32	Pisati, Nuovo Cimento, 5, 34, 1879.
"	3200	455	36	–	Kohlrausch-Loomis, Pogg. Ann. 141.
Copper	3972	2716	—23	47	Pisati, loc. cit.
"	3900	572	28	–	K and L, loc. cit.
Iron	8108	206	19	—11	Pisati, loc. cit.
"	6940	483	12	–	K and L, loc. cit.
Platinum	6632	111	50	—8	Pisati, loc. cit.
Silver	2566	387	38	11	" " "
Steel	8290	187	59	—9	" " "

$n_t{}^* = n_{15}\,[1 - \alpha\,(t - 15)]$; Horton, Philos. Trans. 204 A, 1905.

Copper	4.37*	α =.00039	Platinum	6.46*	α = .00012	Tin	1.50*	α = .00416
Copper (commercial)	3.80	.00038	Gold	2.45	.00031	Lead	0.80	.00164
Iron	8.26	.00029	Silver	2.67	.00048	Cadmium	2.31	.0058
Steel	8.45	.00026	Aluminum	2.55	.00148	Quartz	3.00	.00012

* Modulus of rigidity in 10^{11} dynes per sq. cm.

TABLE 78.—Interior Friction at Low Temperatures.

C is the damping coefficient for infinitely small oscillations; T, the period of oscillation in seconds; N, the second modulus of elasticity. Guye and Schapper, C. R. 150, p. 963, 1910.

Substance		Cu	Ni	Au	Pd	Pt	Ag	Quartz
Length of wire in cm.		22.5	22.2	22.3	22.2	23.0	17.2	17.3
Diameter in mm.		.643	.411	.609	.553	.812	.601	.612
100° C	C	24.1	1.34	27.5	1.67	2.98	55.8	—
	T	2.381s	3.831s	3.010s	2.579	1.143s	1.808s	—
	$N\times10^{-11}$	3.32	7.54	2.55	5.08	5.77	2.71	—
0° C	C	5.88	.417	4.82	1.25	4.60	7.19	4.69
	T	2.336s	3.754s	2.969s	2.571s	1.133s	1.759s	1.408s
	$N\times10^{-11}$	3.45	7.85	2.62	5.12	—	2.87	2.26
—195° C	C	3.64	.556	6.36	.744	3.02	1.64	1.02
	T	2.274s	3.577s	2.902s	2.552s	1.111s	1.694s	1.425s
	$N\times10^{-11}$	3.64	8.65	2.74	5.19	6.10	3.18	2.20

TABLE 79.—Hardness.

Agate	7.	Brass	3–4.	Iridosmium	7.	Sulphur	1.5–2.5
Alabaster	1.7	Calamine	5.	Iron	4–5.	Stibnite	2.
Alum	2–2.5	Calcite	3.	Kaolin	1.	Serpentine	3–4.
Aluminum	2.	Copper	2.5–3.	Loess (0°)	0.3	Silver	2.5–3.
Amber	2–2.5	Corundum	9.	Magnetite	6.	Steel	5–8.5
Andalusite	7.5	Diamond	10.	Marble	3–4.	Talc	1.
Anthracite	2.2	Dolomite	3.5–4.	Meerschaum	2–3.	Tin	1.5
Antimony	3.3	Feldspar	6.	Mica	2.8	Topaz	8.
Apatite	5.	Flint	7.	Opal	4–6.	Tourmaline	7.3
Aragonite	3.5	Fluorite	4.	Orthoclase	6.	Wax (0°)	0.2
Arsenic	3.5	Galena	2.5	Palladium	4.8	Wood's metal	3.
Asbestos	5.	Garnet	7.	Phosphorbronze	4.		
Asphalt	1–2.	Glass	4.5–6.5	Platinum	4.3		
Augite	6.	Gold	2.5–3.	Platin-iridium	6.5		
Barite	3.3	Graphite	0.5–1.	Pyrite	6.3		
Beryl	7.8	Gypsum	1.6–2.	Quartz	7.		
Bell-metal	4.	Hematite	6.	Rock-salt	2.		
Bismuth	2.5	Hornblende	5.5	Ross' metal	2.5–3.0		
Boric acid	3.	Iridium	6.	Silver chloride	1.3		

From Landolt-Börnstein-Meyerhoffer Tables: Auerbachs, Winklemann, Handb. der Phys. 1891.

TABLE 80.—Relative Hardness of the Elements.

C	10.0	Ru	6.5	Cu	3.0	Au	2.5	Sn	1.8	Li	0.6
B	9.5	Mn	5.0	Sb	3.0	Te	2.3	Sr	1.8	P	0.5
Cr	9.0	Pd	4.8	Al	2.9	Cd	2.0	Ca	1.5	K	0.5
Os	7.0	Fe	4.5	Ag	2.7	S	2.0	Ga	1.5	Na	0.4
Si	7.0	Pt	4.3	Bi	2.5	Se	2.0	Pb	1.5	Rb	0.3
Ir	6.5	As	3.5	Zn	2.5	Mg	2.0	In	1.2	Cs	0.2

Rydberg, Zeitschr. Phys. Chem. 33, 1900.

TABLE 81.—Ratio, ρ, of Transverse Contraction to Longitudinal Extension under Tensile Stress. (Poisson's Ratio.)

Metal	Pb	Au	Pd	Pt	Ag	Cu	Al	Bi	Sn	Ni	Cd	Fe
ρ	0.45	0.42	0.39	0.39	0.38	0.35	0.34	0.33	0.33	0.31	0.30	0.28

From data from Physikalisch-Technischen Reichsanstalt, 1907.

ρ for: marbles, 0.27; granites, 0.24; basic-intrusives, 0.26; glass, 0.23. Adams-Coker, 1906.

TABLE 82.

ELASTICITY OF CRYSTALS.*

The formulæ were deduced from experiments made on rectangular prismatic bars cut from the crystal. These bars were subjected to cross bending and twisting and the corresponding Elastic Moduli deduced. The symbols $\alpha\ \beta\ \gamma$, $\alpha_1\ \beta_1\ \gamma_1$ and $\alpha_2\ \beta_2\ \gamma_2$ represent the direction cosines of the length, the greater and the less transverse dimensions of the prism with reference to the principal axis of the crystal. E is the modulus for extension or compression, and T is the modulus for torsional rigidity. The moduli are in grams per square centimeter.

Barite.

$$\frac{10^{10}}{E} = 16.13\alpha^4 + 18.51\beta^4 + 10.42\gamma^4 + 2(38.79\beta^2\gamma^2 + 15.21\gamma^2\alpha^2 + 8.88\alpha^2\beta^2)$$

$$\frac{10^{10}}{T} = 69.52\alpha^4 + 117.66\beta^4 + 116.46\gamma^4 + 2(20.16\beta^2\gamma^2 + 85.29\gamma^2\alpha^2 + 127.35\alpha^2\beta^2)$$

Beryl (Emerald).

$$\frac{10^{10}}{E} = 4.325 \sin^4\phi + 4.619 \cos^4\phi + 13.328 \sin^2\phi \cos^2\phi$$

$$\frac{10^{10}}{T} = 15.00 - 3.675 \cos^4\phi_2 - 17.536 \cos^2\phi \cos^2\phi_1$$

where $\phi\ \phi_1\ \phi_2$ are the angles which the length, breadth, and thickness of the specimen make with the principal axis of the crystal.

Fluorite.

$$\frac{10^{10}}{E} = 13.05 - 6.26\,(\alpha^4 + \beta^4 + \gamma^4)$$

$$\frac{10^{10}}{T} = 58.04 - 50.08\,(\beta^2\gamma^2 + \gamma^2\alpha^2 + \alpha^2\beta^2)$$

Pyrite.

$$\frac{10^{10}}{E} = 5.08 - 2.24\,(\alpha^4 + \beta^4 + \gamma^4)$$

$$\frac{10^{10}}{T} = 18.60 - 17.95\,(\beta^2\gamma^2 + \gamma^2\alpha^2 + \alpha^2\beta^2)$$

Rock salt.

$$\frac{10^{10}}{E} = 33.48 - 9.66\,(\alpha^4 + \beta^4 + \gamma^4)$$

$$\frac{10^{10}}{T} = 154.58 - 77.28\,(\beta^2\gamma^2 + \gamma^2\alpha^2 + \alpha^2\beta^2)$$

Sylvite.

$$\frac{10^{10}}{E} = 75.1 - 48.2\,(\alpha^4 + \beta^4 + \gamma^4)$$

$$\frac{10^{10}}{T} = 306.0 - 192.8\,(\beta^2\gamma^2 + \gamma^2\alpha^2 + \alpha^2\beta^2)$$

Topaz.

$$\frac{10^{10}}{E} = 4.341\alpha^4 + 3.460\beta^4 + 3.771\gamma^4 + 2\,(3.879\beta^2\gamma^2 + 2.856\gamma^2\alpha^2 + 2.39\alpha^2\beta^2)$$

$$\frac{10^{10}}{T} = 14.88\alpha^4 + 16.54\beta^4 + 16.45\gamma^4 + 30.89\beta^2\gamma^2 + 40.89\gamma^2\alpha^2 + 43.51\alpha^2\beta^2$$

Quartz.

$$\frac{10^{10}}{E} = 12.734\,(1 - \gamma^2)^2 + 16.693\,(1 - \gamma^2)\gamma^2 + 9.705\gamma^4 - 8.460\beta\gamma\,(3\alpha^2 - \beta^2)$$

$$\frac{10^{10}}{T} = 19.665 + 9.060\gamma_2^2 + 22.984\gamma^2\gamma_1^2 - 16.920\,[(\gamma\beta_1 + \beta\gamma_1)\,(3\alpha\alpha_1 - \beta\beta_1) - \beta_2\gamma_2)]$$

* These formulæ are taken from Voigt's papers (Wied. Ann. vols. 31, 34, and 35).

TABLE 83.

ELASTICITY OF CRYSTALS.

Some particular values of the Elastic Moduli are here given. Under E are given moduli for extension or compression in the directions indicated by the subscripts and explained in the notes, and under T the moduli for torsional rigidities round the axes similarly indicated. Moduli in grams per sq. cm.

(a) ISOMETRIC SYSTEM.*

Substance.	E_a	E_b	E_c	T_a	Authority.
Fluorite	1473×10^6	1008×10^6	910×10^6	345×10^6	Voigt.†
Pyrite	3530×10^6	2530×10^6	2310×10^6	1075×10^6	"
Rock salt	419×10^6	349×10^6	303×10^6	129×10^6	"
"	403×10^6	339×10^6	—	—	Koch.‡
Sylvite	401×10^6	209×10^6	—	—	"
"	372×10^6	196×10^6	—	655×10^5	Voigt.
Sodium chlorate	405×10^6	319×10^6	—	—	Koch.
Potassium alum	181×10^6	199×10^6	—	—	Beckenkamp.§
Chromium alum	161×10^6	177×10^6	—	—	"
Iron alum	186×10^6	—	—	—	"

(b) ORTHORHOMBIC SYSTEM.‖

Substance.	E_1	E_2	E_3	E_4	E_5	E_6	Authority.
Barite	620×10^6	540×10^6	959×10^6	376×10^6	702×10^6	740×10^6	Voigt.
Topaz	2304×10^6	2890×10^6	2652×10^6	2670×10^6	2893×10^6	3180×10^6	"

Substance.	$T_{12} = T_{21}$	$T_{13} = T_{31}$	$T_{23} = T_{32}$	Authority.
Barite	283×10^6	293×10^6	121×10^6	Voigt.
Topaz	1336×10^6	1353×10^6	1104×10^6	"

In the MONOCLINIC SYSTEM, Coromilas (Zeit. für Kryst. vol. 1) gives

Gypsum $\begin{cases} E_{max} = 887 \times 10^6 \text{ at } 21.9° \text{ to the principal axis.} \\ E_{min} = 313 \times 10^6 \text{ at } 75.4° \text{ " " "} \end{cases}$

Mica $\begin{cases} E_{max} = 2213 \times 10^6 \text{ in the principal axis.} \\ E_{min} = 1554 \times 10^6 \text{ at } 45° \text{ to the principal axis.} \end{cases}$

In the HEXAGONAL SYSTEM, Voigt gives measurements on a beryl crystal (emerald). The subscripts indicate inclination in degrees of the axis of stress to the principal axis of the crystal.

$E_0 = 2165 \times 10^6$, $E_{45} = 1796 \times 10^6$, $E_{90} = 2312 \times 10^6$,
$T_0 = 667 \times 10^6$, $T_{90} = 883 \times 10^6$. The smallest cross dimension of the prism experimented on (see Table 82), was in the principal axis for this last case.

In the RHOMBOHEDRAL SYSTEM, Voigt has measured quartz. The subscripts have the same meaning as in the hexagonal system.

$E_0 = 1030 \times 10^6$, $E_{-45} = 1305 \times 10^6$, $E_{+45} = 850 \times 10^6$, $E_{90} = 785 \times 10^6$,
$T_0 = 508 \times 10^6$, $T_{90} = 348 \times 10^6$.

Baumgarten ¶ gives for calcite

$E_0 = 501 \times 10^6$, $E_{-45} = 441 \times 10^6$, $E_{+45} = 772 \times 10^6$, $E_{90} = 790 \times 10^6$.

* In this system the subscript *a* indicates that compression or extension takes place along the crystalline axis, and distortion round the axis. The subscripts *b* and *c* correspond to directions equally inclined to two and normal to the third and equally inclined to all three axes respectively.
† Voigt, "Wied. Ann." 31, p. 474, p. 701, 1887; 34, p. 981, 1888; 36, p. 642, 1888.
‡ Koch, "Wied. Ann." 18, p. 325, 1882.
§ Beckenkamp, "Zeit. für Kryst." vol. 10.
‖ The subscripts 1, 2, 3 indicate that the three principal axes are the axes of stress; 4, 5, 6 that the axes of stress are in the three principal planes at angles of 45° to the corresponding axes.
¶ Baumgarten, "Pogg. Ann." 152, p. 369, 1879.

COMPRESSIBILITY OF GASES.

TABLE 84.—Relative Volumes at Various Pressures and Temperatures, the volumes at 0° C and at 1 atmosphere being taken as 1 000 000.

Atm.	Oxygen.			Air.			Nitrogen.			Hydrogen.		
	0°	99°.5	199°.5	0°	99°.4	200°.4	0°	99°.5	199°.6	0°	99°.3	200°.5
100	9265	–	–	9730	–	–	9910	–	–	–	–	–
200	4570	7000	9095	5050	7360	9430	5195	7445	9532	5690	7567	9420
300	3208	4843	6283	3658	5170	6622	3786	5301	6715	4030	5286	6520
400	2629	3830	4900	3036	4170	5240	3142	4265	5331	3207	4147	5075
500	2312	3244	4100	2680	3565	4422	2780	3655	4515	2713	3462	4210
600	2115	2867	3570	2450	3180	3883	2543	3258	3973	2387	3006	3627
700	1979	2610	3202	2288	2904	3502	2374	2980	3589	2149	2680	3212
800	1879	2417	2929	2168	2699	3219	2240	2775	3300	1972	2444	2900
900	1800	2268	2718	2070	2544	3000	2149	2616	3085	1832	2244	2657
1000	1735	2151	–	1992	2415	2828	2068	–	–	1720	2093	–

Amagat: C. R. 111, p. 871, 1890; Ann. chim. phys. (6) 29, pp. 68 and 505, 1893.

TABLE 85.—Ethylene.

pv at 0° C and 1 atm. = 1.

Atm.	0°	10°	20°	30°	40°	60°	80°	100°	137°.5	198°.5
46	–	0.562	0.684	–	–	–	–	–	–	–
48	–	0.508	–	–	–	–	–	–	–	–
50	0.176	0.420	0.629	0.731	0.814	0.954	1.077	1.192	1.374	1.652
52	–	0.240	0.598	–	–	–	–	–	–	–
54	–	0.229	0.561	–	–	–	–	–	–	–
56	–	0.227	0.524	–	–	–	–	–	–	–
100	0.310	0.331	0.360	0.403	0.471	0.668	0.847	1.005	1.247	1.580
150	0.441	0.459	0.485	0.515	0.551	0.649	0.776	0.924	1.178	1.540
200	0.565	0.585	0.610	0.638	0.669	0.744	0.838	0.946	1.174	1.537
300	0.806	0.827	0.852	0.878	0.908	0.972	1.048	1.133	1.310	1.628
500	1.256	1.280	1.308	1.337	1.367	1.431	1.500	1.578	1.721	1.985
1000	2.289	2.321	2.354	2.387	2.422	2.493	2.566	2.643	2.798	–

Amagat, C. R. 111, p. 871, 1890; 116, p. 946, 1893.

TABLE 86.—Relative Gas Volumes at Various Pressures.

The following table, deduced by Mr. C. Cochrane, from the *PV* curves of Amagat and other observers, gives the relative volumes occupied by various gases when the pressure is reduced from the value given at the head of the column to 1 atmosphere:

Gas. (Temp. = 16° C.).	Relative volume which the gas will occupy when the pressure is reduced to atmospheric from					
	1 atm.	50 atm.	100 atm.	120 atm.	150 atm.	200 atm.
"Perfect" gas	1	50	100	120	150	200
Hydrogen	1	48.5	93.6	111.3	136.3	176.4
Nitrogen	1	50.5	100.6	120.0	147.6	190.8
Air	1	50.9	101.8	121.9	150.3	194.8
Oxygen	1	—	105.2	—	—	212.6
Oxygen (at 0° C.)	1	52.3	107.9	128.6	161.9	218.8
Carbon dioxide	1	69.0	477*	485*	498*	515*

* Carbon dioxide is liquid at pressures greater than 90 atmospheres.

COMPRESSIBILITY OF GASES.

TABLE 87.—Carbon Dioxide.

Pressure in meters of mercury.	Relative values of pv at —								
	18°.2	35°.1	40°.2	50°.0	60°.0	70°.0	80°.0	90°.0	100°.0
30	liquid	2360	2460	2590	2730	2870	2995	3120	3225
50	–	1725	1900	2145	2330	2525	2685	2845	2980
80	625	750	825	1200	1650	1975	2225	2440	2635
110	825	930	980	1090	1275	1550	1845	2105	2325
140	1020	1120	1175	1250	1360	1525	1715	1950	2160
170	1210	1310	1360	1430	1520	1645	1780	1975	2135
200	1405	1500	1550	1615	1705	1810	1930	2075	2215
230	1590	1690	1730	1800	1890	1990	2090	2210	2340
260	1770	1870	1920	1985	2070	2166	2265	2375	2490
290	1950	2060	2100	2170	2260	2340	2440	2550	2655
320	2135	2240	2280	2360	2440	2525	2620	2725	2830

Atm.	Relative values of pv ; pv at 0° C. and 1 atm. = 1.										
	0°	10°	20°	30°	40°	60°	80°	100°	137°	198°	258°
50	0.105	0.114	0.680	0.775	0.750	0.984	1.096	1.206	1.380	–	–
100	0.202	0.213	0.229	0.255	0.309	0.661	0.873	1.030	1.259	1.582	1.847
150	0.295	0.309	0.326	0.346	0.377	0.485	0.681	0.878	1.159	1.530	1.818
300	0.559	0.578	0.599	0.623	0.649	0.710	0.790	0.890	1.108	1.493	1.820
500	0.891	0.913	0.938	0.963	0.990	1.054	1.124	1.201	1.362	1.678	–
1000	1.656	1.685	1.716	1.748	1.780	1.848	1.921	1.999	–	–	–

Amagat, C. R. 111, p. 871, 1890; Ann. chim. phys. (5) 22, p. 353, 1881; (6) 29, pp. 68 and 405, 1893.

TABLE 88.—Compressibility of Gases.

Gas.	$\frac{p.v.\ (\frac{1}{2}\text{ atm.})}{p_0v_0\ (1\text{ atm.})}$.	$\frac{1}{p.v.}\frac{d(p.v.)}{dp} = a$.	t	a t = 0	Density. O = 32, 0°C P = 76cm	Density. Very small pressure.
O_2	1.00038	—.00076	11.2°	—.00094	32.	32.
H_2	0.99974	+.00052	10.7	+.00053	2.015 (16°)	2.0173
N_2	1.00015	—.00030	14.9	—.00056	28.005	28.016
CO	1.00026	—.00052	13.8	—.00081	28.000	28.003
CO_2	1.00279	—.00558	15.0	—.00668	44.268	44.014
N_2O	1.00327	—.00654	11.0	—.00747	44.285	43.996
Air	1.00026	—.00046	11.4	–	–	–
NH_3	1.00632	–	–	–	–	–

Rayleigh, Zeitschr. Phys. Chem. 52, p. 705, 1905.

TABLE 89.—Compressibility of Air and Oxygen between 18° and 22° C.

Pressures in meters of mercury, pv, relative.

Air	p	24.07	34.90	45.24	55.30	64.00	72.16	84.22	101.47	214.54	304.04
	pv	26968	26908	26791	26789	26778	26792	26840	27041	29585	32488
O_2	p	24.07	34.89	–	55.50	64.07	72.15	84.19	101.06	214.52	303.03
	pv	26843	26614	–	26185	26050	25858	25745	25639	26536	28756

Amagat, C. R. 1879.

RELATION BETWEEN PRESSURE, TEMPERATURE AND VOLUME OF SULPHUR DIOXIDE AND AMMONIA.*

TABLE 90.—Sulphur Dioxide.

Original volume 100000 under one atmosphere of pressure and the temperature of the experiments as indicated at the top of the different columns.

Pressure in Atmos.	Corresponding Volume for Experiments at Temperature —		
	58°.0	99°.6	183°.2
10	8560	9440	–
12	6360	7800	–
14	4040	6420	–
16	–	5310	–
18	–	4405	–
20	–	4030	–
24	–	3345	–
28	–	2780	3180
32	–	2305	2640
36	–	1935	2260
40	–	1450	2040
50	–	–	1640
60	–	–	1375
70	–	–	1130
80	–	–	930
90	–	–	790
100	–	–	680
120	–	–	545
140	–	–	430
160	–	–	325

Volume.	Pressure in Atmospheres for Experiments at Temperature —		
	58°.0	99°.6	183°.2
10000	–	9.60	–
9000	9.60	10.35	–
8000	10.40	11.85	–
7000	11.55	13.05	–
6000	12.30	14.70	–
5000	13.15	16.70	–
4000	14.00	20.15	–
3500	14.40	23.00	–
3000	–	26.40	29.10
2500	–	30.15	33.25
2000	–	35.20	40.95
1500	–	39.60	55.20
1000	–	–	76.00
500	–	–	117.20

TABLE 91. — Ammonia.

Original volume 100000 under one atmosphere of pressure and the temperature of the experiments as indicated at the top of the different columns.

Pressure in Atmos.	Corresponding Volume for Experiments at Temperature —		
	46°.6	99°.6	183°.6
10	9500	–	–
12.5	7245	7635	–
15	5880	6305	–
20	–	4645	4875
25	–	3560	3835
30	–	2875	3185
35	–	2440	2680
40	–	2080	2345
45	–	1795	2035
50	–	1490	1775
55	–	1250	1590
60	–	975	1450
70	–	–	1245
80	–	–	1125
90	–	–	1035
100	–	–	950

Volume.	Pressure in Atmospheres for Experiments at Temperature —			
	30°.2	46°.6	99°.6	183°.0
10000	8.85	9.50		–
9000	9.60	10.45		–
8000	10.40	11.50	12.00	–
7000	11.05	13.00	13.60	–
6000	11.80	14.75	15.55	–
5000	12.00	16.60	18.60	19.50
4000	–	18.35	22.70	24.00
3500	–	18.30	25.40	27.20
3000	–	–	29.20	31.50
2500	–	–	34.25	37.35
2000	–	–	41.45	45.50
1500	–	–	49.70	58.00
1000	–	–	59.65	93.60

* From the experiments of Roth, "Wied. Ann." vol. 11, 1880.

TABLE 92.

COMPRESSIBILITY OF LIQUIDS.

At the constant temperature t, the compressibility $\beta = (1/V_0)(dV/dP)$. In general as P increases, β decreases rapidly at first and then slowly; the change of β with t is large at low pressures but very small at pressures above 1000 to 2000 megabars. 1 megabar = 0.987 atmosphere = 10^6 dyne/cm².

Substance.	Temp. °C	Pressure, megabars.	Compressibility per megabars. $\beta \times 10^6$.	Reference.	Substance.	Temp. °C	Pressure, megabars.	Compressibility per megabars. $\beta \times 10^6$.	Reference.
Acetone	14	23	111	9	Ethyl ether, ct'd	20	1,000	61	1
"	20	500	61	1	" " "	20	12,000	10	1
"	20	1,000	52	1	Ethyl iodide	20	200	81	16
"	40	12,000	9	1	" "	20	400	69	16
Amyl alcohol	14	23	88	10	" "	20	500	64	1
" " iso	20	200	84	16	" "	20	1,000	50	1
" " iso	20	400	70	16	" "	20	12,000	8	1
" "	20	500	61	1	Gallium	30	300	3.97	6
" "	20	1,000	46	1	Glycerine	15	5	22	12
" "	20	12,000	8	1	Hexane	20	200	117	16
" "	40	12,000	8	1	"	20	400	91	16
Benzene	17	5	89	2, 3	Kerosene	20	500	55	1
"	20	200	77	16	"	20	1,000	45	1
"	20	400	67	16	"	20	12,000	8	1
Bromine	20	200	56	16	"	20	12,000	8	13
"	20	400	51	16	Mercury	20	300	3.95	7
Butyl alcohol, iso	18	8	97	2	"	22	500	3.97	8
" " iso	20	200	81	16	"	22	1,000	3.91	8
" " iso	20	400	64	16	"	22	12,000	2.37	8
" " iso	20	500	56	1	Methyl alcohol	15	23	103	10
" " iso	20	1,000	46	1	" "	20	200	95	16
" " iso	20	12,000	8	1	" "	20	400	80	16
Carbon bisulphide	16	21	86	10	" "	20	500	65	1
" "	20	500	57	1	" "	20	1,000	54	1
" "	20	1,000	48	1	" "	20	12,000	8	1
" "	20	12,000	6	1	Nitric acid	0	17	32	14
Carb. tetrachloride	20	200	86	16	Oils: Almond	15	5	53	12
" "	20	400	73	16	Castor	15	5	46	12
Chloroform	20	200	83	16	Linseed	15	5	51	12
"	20	400	70	16	Olive	15	5	55	12
Dichlorethylsulfide	32	1,000	34	5	Rape-seed	20	—	59	15
"	32	2,000	24	5	Phosph. trichloride	10	250	71	11
Ethyl acetate	13	23	103	10	" "	20	500	63	1
" "	20	200	90	16	" "	20	1,000	47	1
" "	20	400	75	16	" "	20	12,000	8	1
Ethyl alcohol	14	23	100	10	Propyl alcohol, n	20	200	77	16
" "	20	500	63	1	" " n	20	400	67	16
" "	20	1,000	54	1	" " (n?)	20	500	65	1
" "	20	12,000	8	1	" " (n?)	20	1,000	47	1
Ethyl bromide	20	200	100	16	" " (n?)	20	12,000	7	1
" "	20	400	82	16	Toluene	20	200	74	16
" "	20	500	70	1	"	20	400	64	16
" "	20	1,000	54	1	Turpentine	20	—	74	15
" "	20	12,000	8	1	Water	20	13	49	11
Ethyl chloride	15	23	151	10	"	20	200	43	16
" "	20	500	102	1	"	20	400	41	16
" "	20	1,000	66	1	"	20	500	39	4
" "	20	12,000	8	1	"	40	500	38	4
Ethyl ether	25	23	188	10	"	40	1000	33	4
" "	20	500	84	1	"	40	12,000	9	4
					Xylene, meta	20	200	69	16
					" "	20	400	60	16

For references, see page 108.

COMPRESSIBILITY OF SOLIDS.

If V is the volume of the material under a pressure P megabars and V_0 is the volume at atmospheric pressure, then the compressibility $\beta = -(1/V_0)\ (dV/dP)$. Its unit is cm^2/megadynes (reciprocal megabars). $10^6/\beta$ is the bulk modulus in absolute units (dynes/cm^2). The following values of β, arranged in order of increasing compressibility, are for $P = 0$ and room temperature. 1 megabar = 10^6 dynes = 1.013 kg/cm^2 = 0.987 atmosphere.

Substance.	Compression per unit vol. per megabar $\times 10^6$	Bulk modulus. dynes/cm^2 $\times 10^{12}$	Reference.	Substance.	Compression per unit vol. per megabar $\times 10^6$	Bulk modulus. dynes/cm^2 $\times 10^{12}$	Reference.
Tungsten	0.27	3.7	2	Plate glass	2.23	0.45	4
Boron	0.3	3.0	2	Lead	2.27	0.44	1, 2
Silicon	0.32	3.1	2	Thallium	2.3	0.43	2
Platinum	0.38	2.6	2	Antimony	2.4	0.42	2
Nickel	0.43	2.3	2	Quartz	2.7	0.37	1
Molybdenum	0.46	2.2	2	Magnesium	2.9	0.34	2
Tantalum	0.53	1.9	2	Bismuth	3.0	0.33	1
Palladium	0.54	1.9	2	Graphite	3.0	0.33	2
Iron	0.60	1.67	3	Silica glass	3.1	0.32	1
Gold	0.60	1.67	1, 2	Sodium chloride	4.12	0.24	1
Pyrite	0.7	1.4	4	Arsenic	4.5	0.22	2
Copper	0.75	1.33	1	Calcium	5.7	0.175	2
Manganese	0.84	1.19	2	Potassium chloride	7.4	0.135	6
Brass	0.89	1.12	1	Lithium	9.0	0.111	2
Chromium	0.9	1.12	1	Phosphorus (red)	9.2	0.109	2
Silver	0.99	1.01	1, 2	Selenium	12.0	0.083	2
Mg. silicate, crys.	1.03	0.97	4	Sulphur	12.9	0.078	2
Aluminum	1.33	0.75	1–3	Iodine	13.0	0.077	2
Calcite	1.39	0.72	1	Sodium	15.6	0.064	2
Zinc	1.74	0.57	1	Phosphorus (white)	20.5	0.049	2
Tin	1.89	0.53	1	Potassium	31.7	0.032	2
Gallium	2.09	0.48	5	Rubidium	40.0	0.025	2
Cadmium	2.17	0.46	1, 2	Calcium	61.0	0.016	2

NOTE. — Winklemann, Schott, and Straulel (Wied Ann. 61, 63, 1897, 68, 1899) give the following coefficients (among others) for various Jena glasses in terms of the volume decrease divided by the increase of pressure expressed in kilograms per square millimeter:

No.	Glass.	Compressibility.	No.	Glass.	Compressibility.
665	....	7520	2154	Kalibleisilicat	3660
1299	Barytborosilicat	5800	S 208	Heaviest Bleisilicat	3550
16	Natronkalkzinksilicat	4530	500	Very Heavy Bleisilicat	3510
278	....	3790	S 196	Tonerdborat with sodium, baryte	3470

The following values in cm^2/kg of $10^6 \times$ Compressibility are given for the corresponding temperatures by Grüneisen, Ann. der Phys. 33, p. 65, 1910.

Al — 191°, 1.32; 17°, 1.46; 125°, 1.70.
Cu — 191°, 0.72; 17°, 0.77; 165°, 0.83.
Pt — 189°, 0.37; 17°, 0.39; 164°, 0.40.
Fe — 190°, 0.61; 18°, 0.63; 165°, 0.67.
Ag — 191°, 0.71; 16°, 0.76; 166°, 0.86.
Pb — 191°, (2.5); 14°, (3.2).

References to Table 92, p. 107:

(1) Bridgman, Pr. Am. Acad. 49, 1, 1913;
(2) Roentgen, Ann. Phys. 44, 1, 1891;
(3) Pagliani-Palazzo, Mem. Acad. Lin. 3, 18, 1883;
(4) Bridgman, Pr. Am. Acad. 48, 341, 1912;
(5) Adams, Williamson, J. Wash. Acad. Sc. 9, Jan. 19, 1919;
(6) Richards, Boyer, Pr. Nat. Acad. Sc. 4, 389, 1918;
(7) Richards, J. Am. Ch. Soc. 37, 1646, 1915;
(8) Bridgman, Pr. Am. Acad. 47, 381, 1911;
(9) Amagat, C. R. 73, 143, 1872;
(10) Amagat, C. R. 68, 1170, 1869;
(11) Amagat, Ann. chim. phys. 29, 68, 505, 1893;
(12) de Metz, Ann. Phys. 41, 663, 1890;
(13) Adams, Williamson, Johnston, J. Am. Chem. Soc. 41, 27, 1919;
(14) Colladon, Sturm, Ann. Phys. 12, 39, 1828;
(15) Quincke, Ann. Phys. 19, 401, 1883;
(16) Richards *et al.* J. Am. Ch. Soc. 34, 988, 1912.

References to Table 93, p. 108:

(1) Adams, Williamson, Johnston, J. Am. Ch. Soc. 41, 39, 1919;
(2) Richards, *ibid.* 37, 1646, 1915;
(3) Bridgman, Pr. Am. Acad. 44, 279, 1909; 47, 366, 1911;
(4) Adams, Williamson, unpublished;
(5) Richards, Boyer, Pr. Nat. Acad. Sc. 4, 388, 1918;
(6) Voigt, Ann. Phys. 31, 1887; 36, 1888.

SPECIFIC GRAVITIES CORRESPONDING TO THE BAUMÉ SCALE.

The specific gravities are for 15.56°C (60°F) referred to water at the same temperature as unity For specific gravities less than unity the values are calculated from the formula:

$$\text{Degrees Baumé} = \frac{140}{\text{Specific Gravity}} - 130.$$

For specific gravities greater than unity from:

$$\text{Degrees Baumé} = 145 - \frac{145}{\text{Specific Gravity}}.$$

Specific Gravities less than 1.

Specific Gravity.	0.00	0.01	0.02	0.03	0.04	0.05	0.06	0.07	0.08	0.09
	Degrees Baumé.									
0.60	103.33	99.51	95.81	92.22	88.75	85.38	82.12	78.95	75.88	72.90
.70	70.00	67.18	64.44	61.78	59.19	56.67	54.21	51.82	49.49	47.22
.80	45.00	42.84	40.73	38.68	36.67	34.71	32.79	30.92	29.09	27.30
.90	25.56	23.85	22.17	20.54	18.94	17.37	15.83	14.33	12.86	11.41
1.00	10.00									

Specific Gravities greater than 1.

Specific Gravity.	0.00	0.01	0.02	0.03	0.04	0.05	0.06	0.07	0.08	0.09
	Degrees Baumé.									
1.00	0.00	1.44	2.84	4.22	5.58	6.91	8.21	9.49	10.74	11.97
1.10	13.18	14.37	15.54	16.68	17.81	18.91	20.00	21.07	22.12	23.15
1.20	24.17	25.16	26.15	27.11	28.06	29.00	29.92	30.83	31.72	32.60
1.30	33.46	34.31	35.15	35.98	36.79	37.59	38.38	39.16	39.93	40.68
1.40	41.43	42.16	42.89	43.60	44.31	45.00	45.68	46.36	47.03	47.68
1.50	48.33	48.97	49.60	50.23	50.84	51.45	52.05	52.64	53.23	53.80
1.60	54.38	54.94	55.49	56.04	56.58	57.12	57.65	58.17	58.69	59.20
1.70	59.71	60.20	60.70	61.18	61.67	62.14	62.61	63.08	63.54	63.99
1.80	64.44	64.89	65.33	65.76	66.20	66.62				

TABLE 94 (a). Degrees A. P. I. Corresponding to Specific Gravities at 60°/60° F.

(15.56°/15.56° C) for petroleum oils.

In order to avoid confusion and misunderstanding the American Petroleum Institute, the Bureau of Mines, and the Bureau of Standards have agreed that a scale based on the modulus 141.5 shall be used in the United States Petroleum Industry and shall be known as the A. P. I. scale. The United States Baumé scale based on the modulus 140 will continue to be used for other liquids lighter than water.

$$\text{Calculated from the formula, degrees A. P. I.} = \frac{141.5}{\text{Sp. Gr. } 60°/60° \text{ F}} - 131.5$$

Degrees A. P. I. 60°/60°F	0	1	2	3	4	5	6	7	8	9
0.6	104.33	100.47	96.73	93.10	89.59	86.19	82.89	79.69	76.59	73.57
.7	70.64	67.80	65.03	62.34	59.72	57.17	54.68	52.27	49.91	47.61
.8	45.38	43.19	44.06	38.98	36.95	34.97	33.03	31.14	29.30	27.49
.9	25.72	23.99	22.30	20.65	19.03	17.45	15.90	14.38	12.89	11.43
1.0	10.00									

TABLE 95.

DENSITY IN GRAMS PER CUBIC CENTIMETER OF THE ELEMENTS, LIQUID OR SOLID.

N. B. The density of a specimen may depend considerably on its state and previous treatment.

Element.	Physical State.	Grams per cu. cm.*	Temperature °C.†	Authority.
Aluminum	commercial h'd d'n	2.70	20°	Wolf, Dellinger, 1910
"	wrought	2.65–2.80		
Antimony	vacuo-distilled	6.618	20	Kahlbaum, 1902.
"	ditto-compressed	6.691	20	"
"	amorphous	6.22		Hérard.
Argon	liquid	1.3845	—183	Baly-Donnan.
"	"	1.4233	—189	" "
Arsenic	crystallized	5.73	14	
"	amorph. br.-black	3.70		Geuther.
"	yellow	3.88		Linck.
Barium		3.78		Guntz.
Bismuth	solid	9.70–9.90		
"	electrolytic	9.747		Classen, 1890.
"	vacuo-distilled	9.781	20	Kahlbaum, 1902.
"	liquid	10.00	271	Vincentini-Omodei.
"	solid	9.67	271	" "
Boron	crystal	2.535		Wigand.
"	amorph. pure	2.45		Moissan.
Bromine	liquid	3.12		Richards-Stull.
Cadmium	cast	8.54–8.57		
"	wrought	8.67		
"	vacuo-distilled	8.648	20	Kahlbaum, 1902.
"	solid	8.37	318	Vincentini-Omodei.
"	liquid	7.99	318	" "
Cæsium		1.873	20	Richards-Brink.
Calcium		1.54		Brink.
Carbon	diamond	3.52		Wigand.
"	graphite	2.25		"
Cerium	electrolytic	6.79		Muthmann-Weiss.
"	pure	7.02		" "
Chlorine	liquid	1.507	—33.6	Drugman-Ramsay.
Chromium		6.52–6.73		
"	pure	6.92	20	Moissan.
Cobalt		8.71	21	Tilden, Ch. C. 1898.
Columbium		8.4	15	Muthmann-Weiss.
Copper	cast	8.30–8.95		
"	annealed	8.89	20	Dellinger, 1911
"	wrought	8.85–8.95		
"	hard drawn	8.89	20	" "
"	vacuo-distilled	8.9326	20	Kahlbaum, 1902.
"	ditto-compressed	8.9376	20	" "
"	liquid	8.217		Roberts-Wrightson.
Erbium		4.77		St. Meyer, Z. Ph. Ch. 37.
Fluorine	liquid	1.14	—200	Moissan-Dewar.
Gallium		5.93	23	de Boisbaudran.
Germanium		5.46	20	Winkler.
Glucinum		1.85		Humpidge.
Gold	cast	19.3		
"	wrought	19.33		
"	vacuo-distilled	18.88	20	Kahlbaum, 1902.
"	ditto-compressed	19.27	20	"
Helium	liquid	0.15	—269	Onnes, 1908.
Hydrogen	liquid	0.070	—252	Dewar, Ch. News, 1904.
Indium		7.28		Richards.

*To reduce to pounds per cu. ft. multiply by 62.4.

† Where the temperature is not given, ordinary atmospheric temperature is understood.

Compiled from Clarke's Constants of Nature, Landolt-Börnstein-Meyerhoffer's Tables, and other sources. Where no authority is stated, the values are mostly means from various sources.

DENSITY IN GRAMS PER CUBIC CENTIMETER OF THE ELEMENTS, LIQUID OR SOLID.

Element.	Physical State	Grams per cu. cm.*	Temperature °C.†	Authority.
Iridium		22.42	17	Deville-Debray
Iodine		4.940	20	Richards-Stull
Iron	pure	7.85–7.88		
"	gray cast	7.03–7.13		
"	white cast	7.58–7.73		
"	wrought	7.80–7.90		
"	liquid	6.88		Roberts-Austen
"	steel	7.60–7.80		
Krypton	liquid	2.16	—146	Ramsay-Travers
Lanthanum		6.15		Muthmann-Weiss
Lead	vacuo-distilled	11.342	20	Kahlbaum, 1902
"	ditto-compressed	11.347	20	" "
"	solid	11.005	325	Vincentini-Omodei
"	liquid	10.645	325	" "
"	"	10.597	400°	Day, Sosman, Hostetter,
"	"	10.078	850°	1914
Lithium		0.534	20	Richards-Brink, '07
Magnesium		1.741		Voigt
Manganese		7.42		Prelinger
Mercury	liquid	13.596	0	Regnault, Volkmann
"	"	13.546	20	
"	"	13.690	—38.8	Vincentini-Omodei
"	solid	14.193	—38.8	Mallet
"	"	14.383	—188	Dewar, 1902
Molybdenum		9.01		Moissan
Neodymium		6.96		Muthmann-Weiss
Nickel		8.60–8.90		
Nitrogen	liquid	0.810	—195	Baly-Donnan, 1902
"	"	0.854	—205	" " "
Osmium		22.5		Deville-Debray
Oxygen	liquid	1.14	—184	
Palladium		12.16		Richards-Stull
Phosphorus ‡	white	1.83		
"	red	2.20		
"	metallic	2.34	15	Hittorf
Platinum		21.37	20	Richards-Stull
Potassium		0.870	20	Richards-Brink, '07
"	solid	0.851	62.1	Vincentini-Omodei
"	liquid	0.830	62.1	" "
Præsodymium		6.475		Muthmann-Weiss
Rhodium		12.44		Holborn Henning
Rubidium		1.532	20	Richards-Brink, '07
Ruthenium		12.06	0	Toby
Samarium		7.7–7.8		Muthmann-Weiss
Selenium		4.3–4.8		
Silicon	cryst.	2.42	20	Richards-Stull-Brink
"	amorph.	2.35	15	Vigoroux
Silver	cast	10.42–10.53		
"	wrought	10.6		
"	vacuo-distilled	10.492	20	Kahlbaum, 1902
"	ditto-compressed	10.503	20	" "
"	liquid	9.51		Wrightson
Sodium		0.9712	20	Richards-Brink, '07
"	solid	0.9519	97.6	Vincentini-Omodei
"	liquid	0.9287	97.6	" "
"		1.0066	—188	Dewar
Strontium		2.50–2.58		Matthiessen
Sulphur		2.0–2.1		
"	liquid	1.811	113	Vincentini-Omodei

* To reduce to pounds per cubic ft. multiply by 62.4.
† Where the temperature is not given, ordinary atmosphere temperature is understood.
‡ Black phosphorus, 2.69, Bridgman, 1918.

TABLE 95 (*continued*). — **Density in grams per cubic centimeter of the elements, liquid or solid.**

Element.	Physical State.	Grams per cu. cm.	Temperature °C.	Authority.
Tantalum		16.6		
Tellurium	crystallized	6.25		
"	amorphous	6.02	20	Beljankin.
Thallium		11.86		Richards-Stull.
Thorium		12.16	17	Bolton.
Tin	white, cast	7.29		Matthiessen.
"	" wrought	7.30		
"	" crystallized	6.97–7.18		
"	" solid	7.184	226	Vincentini-Omodei
"	liquid	6.99	226	" See Table 65
"	gray	5.8		
Titanium		4.5	18	Mixter.
Tungsten		18.6–19.1		
Uranium		18.7	13	Zimmermann.
Vanadium		5.69		Ruff-Martin.
Xenon	liquid	3.52	109	Ramsay-Travers.
Yttrium		3.80		St. Meyer.
Zinc	cast	7.04–7.16		
"	wrought	7.19		
"	vacuo-distilled	6.92	20	Kahlbaum, 1902.
"	ditto-compressed	7.13	20	" "
"	liquid	6.48		Roberts-Wrightson.
Zirconium		6.44		

TABLE 96. — **Density in grams per cubic centimeter and in pounds per cubic foot of different kinds of wood.**

Wood is to be seasoned and of average dryness. See also pages 96 to 99 and 114.

Wood.	Grams per cubic centimeter.	Pounds per cubic foot.	Wood.	Grams per cubic centimeter.	Pounds per cubic foot.
Alder	0.42–0.68	26–42	Hazel	0.60–0.80	37–49
Apple	0.66–0.84	41–52	Hickory	0.60–0.93	37–58
Ash	0.65–0.85	40–53	Holly	0.76	47
Bamboo	0.31–0.40	19–25	Iron-bark	1.03	64
Basswood. See Linden.			Juniper	0.56	35
Beech	0.70–0.90	43–56	Laburnum	0.92	57
Blue gum	1.00	62	Lancewood	0.68–1.00	42–62
Birch	0.51–0.77	32–48	Lignum vitæ	1.17–1.33	73–83
Box	0.95–1.16	59–72	Linden or Lime-tree	0.32–0.59	20–37
Bullet-tree	1.05	65	Locust	0.67–0.71	42–44
Butternut	0.38	24	Logwood	.91	57
Cedar	0.49–0.57	30–35	Mahogany, Honduras	0.66	41
Cherry	0.70–0.90	43–56	" Spanish	0.85	53
Cork	0.22–0.26	14–16	Maple	0.62–0.75	39–47
Dogwood	0.76	47	Oak	0.60–0.90	37–56
Ebony	1.11–1.33	69–83	Pear-tree	0.61–0.73	38–45
Elm	0.54–0.60	34–37	Plum-tree	0.66–0.78	41–49
Fir or Pine, American			Poplar	0.35–0.5	22–31
White	0.35–0.50	22–31	Satinwood	0.95	59
" Larch	0.50–0.56	31–35	Sycamore	0.40–0.60	24–37
" Pitch	0.83–0.85	52–53	Teak, Indian	0.66–0.88	41–55
" Red	0.48–0.70	30–44	" African	0.98	61
" Scotch	0.43–0.53	27–33	Walnut	0.64–0.70	40–43
" Spruce	0.48–0.70	30–44	Water gum	1.00	62
" Yellow	0.37–0.60	23–37	Willow	0.40–0.60	24–37
Greenheart	0.93–1.04	58–65			

* Where the temperature is not given, ordinary atmospheric temperature is understood.

DENSITY IN GRAMS PER CUBIC CENTIMETER AND POUNDS PER CUBIC FOOT OF VARIOUS SOLIDS.

N. B. The density of a specimen depends considerably on its state and previous treatment; especially is this the case with porous materials.

Material.	Grams per cu. cm.	Pounds per cu. foot.
Agate	2.5–2.7	156–168
Alabaster:		
Carbonate	2.69–2.78	168–173
Sulphate	2.26–2.32	141–145
Albite	2.62–2.65	163–165
Amber	1.06–1.11	66– 69
Amphiboles	2.9–3.2	180–200
Anorthite	2.74–2.76	171–172
Anthracite	1.4–1.8	87–112
Asbestos	2.0–2.8	125–175
Asphalt	1.1–1.5	69– 94
Basalt	2.4–3.1	150–190
Beeswax	0.96–0.97	60– 61
Beryl	2.69–2.7	168–168
Biotite	2.7–3.1	170–190
Bone	1.7–2.0	106–125
Brick	1.4–2.2	87–137
Butter	0.86–0.87	53– 54
Calamine	4.1–4.5	255–280
Caoutchouc	0.92–0.99	57– 62
Celluloid	1.4	87
Cement, set	2.7–3.0	170–190
Chalk	1.9–2.8	118–175
Charcoal: oak	0.57	35
pine	0.28–0.44	18– 28
Chrome yellow	6.00	374
Chromite	4.32–4.57	270–285
Cinnabar	8.12	507
Clay	1.8–2.6	122–162
Coal, soft	1.2–1.5	75– 94
Cocoa butter	0.89–0.91	56– 57
Coke	1.0–1.7	62–105
Copal	1.04–1.14	65– 71
Corundum	3.9–4.0	245–250
Diamond:		
Anthracitic	1.66	104
Carbonado	3.01–3.25	188–203
Diorite	2.52	157
Dolomite	2.84	177
Ebonite	1.15	72
Emery	4.0	250
Epidote	3.25–3.5	203–218
Feldspar	2.55–2.75	159–172
Flint	2.63	164
Fluorite	3.18	198
Gamboge	1.2	75
Garnet	3.15–4.3	197–268
Gas carbon	1.88	117
Gelatine	1.27	180
Glass: common	2.4–2.8	150–175
flint	2.9–5.9	180–370
Glue	1.27	80
Granite	2.64–2.76	165–172
Graphite	2.30–2.72	144–170

Material.	Grams per cu. cm.	Pounds per cu. foot.
Gum arabic	1.3–1.4	80– 85
Gypsum	2.31–2.33	144–145
Hematite	4.9–5.3	306–330
Hornblende	3.0	187
Ice	0.917	57.2
Ilmenite	4.5–5.	280–310
Ivory	1.83–1.92	114–120
Labradorite	2.7–2.72	168–170
Lava: basaltic	2.8–3.0	175–185
trachytic	2.0–2.7	125–168
Leather: dry	0.86	54
greased	1.02	64
Lime: mortar	1.65–1.78	103–111
slaked	1.3–1.4	81– 87
Limestone	2.68–2.76	167–171
Litharge:		
Artificial	9.3–9.4	580–585
Natural	7.8–8.0	490–500
Magnetite	4.9–5.2	306–324
Malachite	3.7–4.1	231–256
Marble	2.6–2.84	160–177
Meerschaum	0.99–1.28	62– 80
Mica	2.6–3.2	165–200
Muscovite	2.76–3.00	172–225
Ochre	3.5	218
Oligoclase	2.65–2.67	165–167
Olivine	3.27–3.37	204–210
Opal	2.2	137
Orthoclase	2.58–2.61	161–163
Paper	0.7–1.15	44– 72
Paraffin	0.87–0.91	54– 57
Peat	0.84	52
Pitch	1.07	67
Porcelain	2.3–2.5	143–156
Porphyry	2.6–2.9	162–181
Pyrite	4.95–5.1	309–318
Quartz	2.65	165
Quartzite	2.73	170
Resin	1.07	67
Rock salt	2.18	136
Rutile	4.2	260
Sandstone	2.14–2.36	134–147
Serpentine	2.50–2.65	156–165
Slag, furnace	2.0–3.9	125–240
Slate	2.6–3.3	162–205
Soapstone	2.6–2.8	162–175
Starch	1.53	95
Sugar	1.61	100
Talc	2.7–2.8	168–174
Tallow	0.91–0.97	57– 60
Topaz	3.5–3.6	219–223
Tourmaline	3.0–3.2	190–200
Zircon	4.68–4.70	292–293

DENSITY IN GRAMS PER CUBIC CENTIMETER AND POUNDS PER CUBIC FOOT OF VARIOUS ALLOYS.

Alloy.	Grams per cubic centimeter.	Pounds per cubic foot.
Brasses: Yellow, 70Cu + 30Zn, cast	8.44	527
" " " rolled	8.56	534
" " " drawn	8.70	542
" Red, 90Cu + 10Zn	8.60	536
" White, 50Cu + 50Zn	8.20	511
Bronzes: 90Cu + 10Sn	8.78	548
" 85Cu + 15Sn	8.89	555
" 80Cu + 20Sn	8.74	545
" 75Cu + 25Sn	8.83	551
German Silver: Chinese, 26.3Cu + 36.6Zn + 36.8Ni	8.30	518
" " Berlin (1) 52Cu + 26Zn + 22Ni	8.45	527
" " " (2) 59Cu + 30Zn + 11Ni	8.34	520
" " " (3) 63Cu + 30Zn + 6Ni	8.30	518
" " Nickelin	8.77	547
Lead and Tin: 87.5Pb + 12.5Sn	10.60	661
" " " 84Pb + 16Sn	10.33	644
" " " 77.8Pb + 22.2Sn	10.05	627
" " " 63.7Pb + 36.3Sn	9.43	588
" " " 46.7Pb + 53.3Sn	8.73	545
" " " 30.5Pb + 69.5Sn	8.24	514
Bismuth, Lead, and Cadmium: 53Bi + 40Pb + 7Cd	10.56	659
Wood's Metal: 50Bi + 25Pb + 12.5Cd + 12.5Sn	9.70	605
Cadmium and Tin: 32Cd + 68Sn	7.70	480
Gold and Copper: 98Au + 2Cu	18.84	1176
" " " 96Au + 4Cu	18.36	1145
" " " 94Au + 6Cu	17.95	1120
" " " 92Au + 8Cu	17.52	1093
" " " 90Au + 10Cu	17.16	1071
" " " 88Au + 12Cu	16.81	1049
" " " 86Au + 14Cu	16.47	1027
Aluminum and Copper: 10Al + 90Cu	7.69	480
" " " 5Al + 95Cu	8.37	522
" " " 3Al + 97Cu	8.69	542
Aluminum and Zinc: 91Al + 9Zn	2.80	175
Platinum and Iridium: 90Pt + 10Ir	21.62	1348
" " " 85Pt + 15Ir	21.62	1348
" " " 66.67Pt + 33.33Ir	21.87	1364
Constantan: 60Cu + 40Ni	8.88	554
Magnalium: 70Al + 30Mg	2.0	125
Manganin: 84Cu + 12Mn + 4Ni	8.5	530
Platinoid: German silver + little Tungsten	9.0	560
Stellite: Co 59.5; Mo 22.5; Cr 10.8; Fe 3.1; Mn 2.0; Co 0.9; Si 0.8	8.3	518

TABLE 96 (a).—Densities (g/cm³) of some foreign woods on the American market.

(See also Tables 72–75 and 96.)

Almon	0.464	Gardner	Olive	0.94	Boulger
Bullet-wood, Guiana	1.03 – 1.23	Boulger	Orange Wood	.70	"
Boxwood, West India	.83 – .88	"	Padouk	.89 – 1.29	"
Balsa	.11	Carpenter	Prima Vera	.58	Howard
Carreto	.84	USFPL	Purple-heart	.72 – .97	Boulger
Cedar, Spanish	.38	"	Quebracho	1.25	"
Cocobola	1.20	Boulger	Rosewood, Brazil	.77 – .84	"
Cocus	1.25	Stone	Rosewood, Honduras	1.09 – 1.23	"
Fustic	.68	Boulger	Sabicu	.90 – .96	"
Koa	.83	Howard	Snakewood	1.05 – 1.33	"
Lauaan Red	.41	Gardner	Tamarind	1.32	"
Mahogany, African	.55	Boulger	Tanguile	.47 – .51	Gardner
Mahogany, E. Indian	.38	"	Wallaba	.93 – .94	Boulger
Mora	1.07 – 1.09	"	Zebra Wood	1.03	"
Oak, English	.60 – .78	"			

Table prepared by W. M. N. Watkins, U. S. National Museum.

TABLE 99.—DENSITIES OF VARIOUS NATURAL AND ARTIFICIAL MINERALS.

(See also Table 97.)

Name and Formula.	Density grams per cc.	Sp. Vol. cc. per gram.	Reference.
Pure compounds, all at 25°C			
Magnesia, MgO	3.603	.2775	1
Lime, CaO	3.306	.3025	2
Forms of SiO_2:			
Quartz, natural	2.646	.3779	"
" artificial	2.642	.3785	"
Cristobalite, artificial	2.319	.4312	"
Silica glass	2.206	.4533	"
Forms of Al_2SiO_5:			
Sillimanite glass	2.53	.395	3
Sillimanite cryst.	3.022	.3309	"
Forms of $MgSiO_3$:			
β Monoclinic pyroxene	3.183	.3142	5
α' Orthorhombic pyroxene	3.166	.3159	"
β' Monoclinic amphibole			"
γ' Orthorhombic amphibole	2.849	.3510	"
Glass	2.735	.3656	"
Forms of $CaSiO_3$:			
α (Pseudo-wollastonite)	2.904	.3444	2
β (Wollastonite)	2.906	.3441	"
Glass	2.895	.3454	"
Forms of Ca_2SiO_4:			
α — calcium-orthosilicate	3.26	.307	"
β — " "	3.27	.306	"
γ — " "	2.965	.337	"
β' — " "			
Lime-alumina compounds:			
$3CaO \cdot Al_2O_3$	3.029	.3301	3
$5CaO \cdot 3Al_2O_3$	2.820	.3546	"
$CaO \cdot Al_2O_3$	2.972	.3365	"
$3CaO \cdot 5Al_2O_3$			
$3CaO \cdot 5Al_2O_3$, unstable form	3.04	.329	"
Forms of $MgSiO_3 \cdot CaSiO_3$:			
Diopside, natural, cryst.	3.258	.3069	4
" artificial, "	3.265	.3063	"
" glass	2.846	.3514	1

Name and Formula.	Density grams per cc.	Sp. Vol. cc. per gram.	Reference.
Feldspars:			
Albite glass, $NaAlSi_3O_8$, art.	2.375	.4210	6
Albite cryst., $NaAlSi_3O_8$, art.	2.597	.3851	"
Anorthite glass, $CaAl_2Si_2O_8$, art.	2.692	.3715	"
Anorthite cryst., $CaAl_2Si_2O_8$, art.	2.757	.3627	"
Soda anorthite, $NaAlSiO_4$, art.	2.563	.3902	7
Borax, glass, $Na_2B_4O_7$	2.36	.423	6
" cryst. "	2.27	.440	"
Fluorite, natural, CaF_2 (20°)	3.180	.3145	8
$(NH_4)_2SO_4$ (30°)	1.765	.5666	9
K_2SO_4 (30°)	2.657	.3764	"
KCl, fine powder (30°)	1.984	.5040	"
Forms of ZnS:			
Sphalerite, natural*	4.090	.2444	10
Wurtzite, artificial†	4.087	.2447	"
Greenockite, artificial	4.820	.2075	"
Forms of HgS:			
Cinnabar, artificial	8.176	.1223	"
Metacinnabar, artificial	7.58	.132	"
Minerals:			
Gehlenite, from Velardena	3.03	.330	11
Spurrite, from Velardena, $2Ca_2SiO_4 \cdot CaCO_3$	3.005	.3328	"
Hillebrandite, from Velardena, $CaSiO_3 \cdot Ca(OH)_2$	2.684	.3726	"
Pyrite, natural, FeS_2	5.012	.1995	10
Marcasite, natural, FeS_2	4.873	.2052	"

* Only 0.15% Fe total impurity.
† Same composition as Sphalerite.

References: 1, Larsen 1909; 2, Day and Shepherd; 3, Shepherd and Rankin, 1909; 4, Allen and White, 1909; 5, Allen, Wright and Clement, 1906; 6, Day and Allen, 1905; 7, Washington and Wright, 1910; 8, Merwin, 1911; 9, Johnston and Adams, 1911; 10, Allen and Crenshaw, 1912; 11, Wright, 1908.

All the data of this table are from the Geophysical Laboratory, Washington.

TABLE 100.—DENSITIES OF MOLTEN TIN AND TIN-LEAD EUTECTIC.

Temperature	250°C.	300°	400°	500°	600°	900°	1200°	1400°	1600°
Molten tin	6.982	6.943	6.875	6.814	6.755	6.578	6.399	6.280	6.162
37 pts. Pb, 63, Sn.*	8.011	7.965	7.879	7.800	7.731	–	–	–	–

* Melts at 181. Day and Sosman, Geophysical Laboratory, unpublished.

For further densities inorganic substances see table 219.
" " " organic " " " 220.

WEIGHT OF SHEET METAL.

TABLE 101.—Weight of Sheet Metal. (Metric Measure.)

This table gives the weight in grams of a plate one meter square and of the thickness stated in the first column.

Thickness in thousandths of a cm.	Iron.	Copper.	Brass.	Aluminum.	Platinum.	Gold.	Silver.
1	78.0	89.0	85.6	26.7	215.0	193.0	105.0
2	156.0	178.0	171.2	53.4	430.0	386.0	210.0
3	234.0	267.0	256.8	80.1	645.0	579.0	315.0
4	312.0	356.0	342.4	106.8	860.0	772.0	420.0
5	390.0	445.0	428.0	133.5	1075.0	965.0	525.0
6	468.0	534.0	513.6	160.2	1290.0	1158.0	630.0
7	546.0	623.0	599.2	186.9	1505.0	1351.0	735.0
8	624.0	712.0	684.8	213.6	1720.0	1544.0	840.0
9	702.0	801.0	770.4	240.3	1935.0	1737.0	945.0
10	780.0	890.0	856.0	267.0	2150.0	1930.0	1050.0

TABLE 102.—Weight of Sheet Metal. (British Measure.)

Thickness in Mils.	Iron.	Copper.	Brass.	Aluminum.		Platinum.	
	Pounds per Sq. Foot.	Pounds per Sq. Foot.	Pounds per Sq. Foot.	Pounds per Sq. Foot.	Ounces per Sq. Foot.	Pounds per Sq. Foot.	Ounces per Sq. Foot.
1	.04058	.04630	.04454	.01389	.2222	.1119	1.790
2	.08116	.09260	.08908	.02778	.4445	.2237	3.579
3	.12173	.13890	.13363	.04167	.6667	.3356	5.369
4	.16231	.18520	.17817	.05556	.8890	.4474	7.158
5	.20289	.23150	.22271	.06945	1.1112	.5593	8.948
6	.24347	.27780	.26725	.08334	1.3335	.6711	10.738
7	.28405	.32411	.31179	.09723	1.5557	.7830	12.527
8	.32463	.37041	.35634	.11112	1.7780	.8948	14.317
9	.36520	.41671	.40088	.12501	2.0002	1.0067	16.106
10	.40578	.46301	.44542	.13890	2.2224	1.1185	17.896

Thickness in Mils.	Gold.		Silver.	
	Troy Ounces per Sq. Foot.	Grains per Sq. Foot.	Troy Ounces per Sq. Foot.	Grains per Sq. Foot.
1	1.4642	702.8	0.7967	382.4
2	2.9285	1405.7	1.5933	764.8
3	4.3927	2108.5	2.3900	1147.2
4	5.8570	2811.3	3.1867	1529.6
5	7.3212	3514.2	3.9833	1912.0
6	8.7854	4217.0	4.7800	2294.4
7	10.2497	4919.8	5.5767	2676.8
8	11.7139	5622.7	6.3734	3059.2
9	13.1782	6325.5	7.1700	3441.6
10	14.6424	7028.3	7.9667	3824.0

DENSITY OF LIQUIDS.

Density or mass in grams per cubic centimeter and in pounds per cubic foot of various liquids.

Liquid.	Grams per cubic centimeter.	Pounds per cubic foot.	Temp. C.
Acetone	0.792	49.4	20°
Alcohol, ethyl	0.807	50.4	0
" methyl	0.810	50.5	0
Aniline	1.035	64.5	0
Benzene	0.899	56.1	0
Bromine	3.187	199.0	0
Carbolic acid (crude)	0.950–0.965	59.2–60.2	15
Carbon disulphide	1.293	80.6	0
Chloroform	1.4827	92.5	21.5
Cocoa-butter	0.857	53.5	100
Ether	0.736	45.9	0
Gasoline	0.66–0.69	41.0–43.0	–
Glycerine	1.260	78.6	0
Japan wax	0.875	54.6	100
Milk	1.028–1.035	64.2–64.6	–
Naphtha (wood)	0.848–0.810	52.9–50.5	0
Naphtha (petroleum ether)	0.665	41.5	15
Oils: Amber	0.800	49.9	15
Anise-seed	0.996	62.1	16
Camphor	0.910	56.8	–
Castor	0.969	60.5	15
Clove	1.04–1.06	65.–66.	25
Cocoanut	0.925	57.7	15
Cotton Seed	0.926	57.8	16
Creosote	1.040–1.100	64.9–68.6	15
Lard	0.920	57.4	15
Lavender	0.877	54.7	16
Lemon	0.844	52.7	16
Linseed (boiled)	0.942	58.8	15
Neat's foot	0.913–.917	57.0–57.2	–
Olive	0.918	57.3	15
Palm	0.905	56.5	15
Pentane	0.650	40.6	0
"	0.623	38.9	25
Peppermint	0.90–.92	56–57	25
Petroleum	0.878	54.8	0
" (light)	0.795–0.805	49.6–50.2	15
Pine	0.850–0.860	53.0–54.0	15
Poppy	0.924	57.7	–
Rapeseed (crude)	0.915	57.1	15
" (refined)	0.913	57.0	15
Resin	0.955	59.6	15
Sperm	0.88	55.	25
Soya-bean	0.919	57.3	30
" "	0.906	56.5	90
Train or Whale	0.918–0.925	57.3–57.7	15
Turpentine	0.873	54.2	16
Valerian	0.965	60.2	16
Wintergreen	1.18	74.	25
Pyroligneous acid	0.800	49.9	0
Water	1.000	62.4	4

Table 104.

DENSITY OF PURE WATER FREE FROM AIR. 0° TO 41° C.

[Under standard pressure (76 cm), at every tenth part of a degree of the international hydrogen scale from 0° to 41° C, in grams per milliliter [1]]

Degrees Centigrade.	Tenths of Degrees. 0	1	2	3	4	5	6	7	8	9	Mean Differences.
0	0.999 8681	8747	8812	8875	8936	8996	9053	9109	9163	9216	+ 59
1	9267	9315	9363	9408	9452	9494	9534	9573	9610	9645	+ 41
2	9679	9711	9741	9769	9796	9821	9844	9866	9887	9905	+ 24
3	9922	9937	9951	9962	9973	9981	9988	9994	9998	*0000	+ 8
4	1.000 0000	*9999	*9996	*9992	*9986	*9979	*9970	*9960	*9947	*9934	— 8
5	0.999 9919	9902	9884	9864	9842	9819	9795	9769	9742	9713	— 24
6	9682	9650	9617	9582	9545	9507	9468	9427	9385	9341	— 39
7	9296	9249	9201	9151	9100	9048	8994	8938	8881	8823	— 53
8	8764	8703	8641	8577	8512	8445	8377	8308	8237	8165	— 67
9	8091	8017	7940	7863	7784	7704	7622	7539	7455	7369	— 81
10	7282	7194	7105	7014	6921	6826	6729	6632	6533	6432	— 95
11	6331	6228	6124	6020	5913	5805	5696	5586	5474	5362	—108
12	5248	5132	5016	4898	4780	4660	4538	4415	4291	4166	—121
13	4040	3912	3784	3654	3523	3391	3257	3122	2986	2850	—133
14	2712	2572	2431	2289	2147	2003	1858	1711	1564	1416	—145
15	1266	1114	0962	0809	0655	0499	0343	0185	0026	*9865	—156
16	0.998 9705	9542	9378	9214	9048	8881	8713	8544	8373	8202	—168
17	8029	7856	7681	7505	7328	7150	6971	6791	6610	6427	—178
18	6244	6058	5873	5686	5498	5309	5119	4927	4735	4541	—190
19	4347	4152	3955	3757	3558	3358	3158	2955	2752	2549	—200
20	2343	2137	1930	1722	1511	1301	1090	0878	0663	0449	—211
21	0233	0016	*9799	*9580	*9359	*9139	*8917	*8694	*8470	*8245	—221
22	0.997 8019	7792	7564	7335	7104	6873	6641	6408	6173	5938	—232
23	5702	5466	5227	4988	4747	4506	4264	4021	3777	3531	—242
24	3286	3039	2790	2541	2291	2040	1788	1535	1280	1026	—252
25	0770	0513	0255	*9997	*9736	*9476	*9214	*8951	*8688	*8423	—261
26	0.996 8158	7892	7624	7356	7087	6817	6545	6273	6000	5726	—271
27	5451	5176	4898	4620	4342	4062	3782	3500	3218	2935	—280
28	2652	2366	2080	1793	1505	1217	0928	0637	0346	0053	—289
29	0.995 9761	9466	9171	8876	8579	8282	7983	7684	7383	7083	—298
30	6780	6478	6174	5869	5564	5258	4950	4642	4334	4024	—307
31	3714	3401	3089	2776	2462	2147	1832	1515	1198	0880	—315
32	0561	0241	*9920	*9599	*9276	*8954	*8630	*8304	*7979	*7653	—324
33	0.994 7325	6997	6668	6338	6007	5676	5345	5011	4678	4343	—332
34	4007	3671	3335	2997	2659	2318	1978	1638	1296	0953	—340
35	0610	0267	*9922	*9576	*9230	*8883	*8534	*8186	*7837	*7486	—347
36	0.993 7136	6784	6432	6078	5725	5369	5014	4658	4301	3943	—355
37	3585	3226	2866	2505	2144	1782	1419	1055	0691	0326	—362
38	0.992 9960	9593	9227	8859	8490	8120	7751	7380	7008	6636	—370
39	6263	5890	5516	5140	4765	4389	4011	3634	3255	2876	—377
40	2497	2116	1734	1352	0971	0587	0203	*9818	*9433	*9047	—384
41	0.991 8661										

[1] According to P. Chappuis, Bureau international des Poids et Mesures, Travaux et Mémoires, 13; 1907.

VOLUME IN CUBIC CENTIMETERS AT VARIOUS TEMPERATURES OF A CUBIC CENTIMETER OF WATER FREE FROM AIR AT THE TEMPERATURE OF MAXIMUM DENSITY. 0° TO 36° C.

Hydrogen Thermometer Scale.

Temp. C.	.0	.1	.2	.3	.4	.5	.6	.7	.8	.9
0	1.000132	125	118	112	106	100	095	089	084	079
1	073	069	064	059	055	051	047	043	039	035
2	032	029	026	023	020	018	016	013	011	009
3	008	006	005	004	003	002	001	001	000	000
4	000	000	000	001	001	002	003	004	005	007
5	008	010	012	014	016	018	021	023	026	029
6	032	035	039	042	046	050	054	058	062	066
7	070	075	080	085	090	095	101	106	112	118
8	124	130	137	142	149	156	162	169	176	184
9	191	198	206	214	222	230	238	246	254	263
10	272	281	290	299	308	317	327	337	347	357
11	367	377	388	398	409	420	430	441	453	464
12	476	487	499	511	522	534	547	559	571	584
13	596	609	623	636	649	661	675	688	702	715
14	729	743	757	772	786	800	815	830	844	859
15	873	890	905	920	935	951	967	983	998	015*
16	1.001031	047	063	080	097	113	130	147	164	182
17	198	216	233	252	269	287	305	323	341	358
18	378	396	415	433	452	471	490	510	529	548
19	568	588	606	626	646	667	687	707	728	748
20	769	790	811	832	853	874	895	916	938	960
21	981	002*	024*	046*	068*	091*	113*	135*	158*	181*
22	1.002203	226	249	271	295	319	342	364	389	412
23	436	459	483	507	532	556	581	605	629	654
24	679	704	729	754	779	804	829	854	879	905
25	932	958	983	010*	036*	061*	088*	115*	141*	168*
26	1.003195	221	248	275	302	330	357	384	412	439
27	467	495	523	550	579	607	635	663	692	720
28	749	776	806	836	865	893	922	951	981	011*
29	1.004041	069	100	129	160	189	220	250	280	310
30	341	371	403	432	464	494	526	557	588	619
31	651	682	713	744	777	808	840	872	904	936
32	968	001*	033*	066*	098*	132*	163*	197*	229*	263*
33	1.005296	328	361	395	427	461	496	530	562	597
34	631	665	698	732	768	802	836	871	904	940
35	975	009*	044*	078*	115*	150*	185*	219*	255*	290*

Reciprocals of the preceding table.

Influence of Pressure.*

kg/cm²	0° C	20° C	40° C	kg/cm²	20° C	40° C
1	1.0000	1.0016	1.0076	7,000	0.8404	0.8485
500	.9771	.9808	.9873	8,000	.8275	.8360
1,000	.9578	.9630	.9700	9,000	.8160	.8249
2,000	.9260	.9327	.9403	10,000	—	.8149
3,000	.9015	.9087	.9164	11,000	—	.8056
5,000	.8632	.8702	.8778	12,000	—	.7966
6,000	.8480	.8545	.8623	12,500	—	.7922

* Williamson, Change of Physical Properties with Pressure, J. Frank. Inst. 193, p. 491, 1922.

TABLE 106.

DENSITY AND VOLUME OF WATER.

—10° TO +250° C.

The mass of one cubic centimeter at 4° C. is taken as unity.

Temp. C.	Density.	Volume.	Temp. C.	Density.	Volume.
—10°	0.99815	1.00186	**+35°**	0.99406	1.00598
—9	843	157	36	371	633
—8	869	131	37	336	669
—7	892	108	38	300	706
—6	912	088	39	263	743
—5	0.99930	1.00070	**40**	0.99225	1.00782
—4	945	055	41	187	821
—3	958	042	42	147	861
—2	970	031	43	107	901
—1	979	021	44	066	943
+0	0.99987	1.00013	**45**	0.99025	1.00985
1	993	007	46	0.98982	1.01028
2	997	003	47	940	072
3	999	001	48	896	116
4	1.00000	1.00000	49	852	162
5	0.99999	1.00001	**50**	0.98807	1.01207
6	997	003	51	762	254
7	993	007	52	715	301
8	988	012	53	669	349
9	981	019	54	621	398
10	0.99973	1.00027	**55**	0.98573	1.01448
11	963	037	60	324	705
12	952	048	65	059	979
13	940	060	70	0.97781	1.02270
14	927	073	75	489	576
15	0.99913	1.00087	**80**	0.97183	1.02899
16	897	103	85	0.96865	1.03237
17	880	120	90	534	590
18	862	138	95	192	959
19	843	157	100	0.95838	1.04343
20	0.99823	1.00177	**110**	0.9510	1.0515
21	802	198	120	.9434	1.0601
22	780	220	130	.9352	1.0693
23	757	244	140	.9264	1.0794
24	733	268	150	.9173	1.0902
25	0.99708	1.00293	**160**	0.9075	1.1019
26	682	320	170	.8973	1.1145
27	655	347	180	.8866	1.1279
28	627	375	190	.8750	1.1429
29	598	404	200	.8628	1.1590
30	0.99568	1.00434	**210**	0.850	1.177
31	537	465	220	.837	1.195
32	506	497	230	.823	1.215
33	473	530	240	.809	1.236
34	440	563	250	.794	1.259

* From — 10° to 0° the values are due to means from Pierre, Weidner, and Rosetti; from 0° to 41°, to Chappuis, 42° to 100°, to Thiesen; 110° to 250°, to means from the works of Ramsey, Young, Waterston, and Hirn.

DENSITY OF MERCURY

Density or mass in grams per cubic centimeter, and the volume in cubic centimeters of one gram of mercury.

Temp. C.	Mass in grams per cu. cm.	Volume of 1 gram in cu. cms.	Temp. C	Mass in grams per cu. cm.	Volume of 1 gram in cu. cms.
—10°	13.6198	0.0734225	30°	13.5213	0.0739572
—9	6173	4358	31	5189	9705
—8	6148	4492	32	5164	9839
—7	6124	4626	33	5140	9973
—6	6099	4759	34	5116	40107
—5	13.6074	0.0734893	35	13.5091	0.0740241
—4	6050	5026	36	5066	0374
—3	6025	5160	37	5042	0508
—2	6000	5293	38	5018	0642
—1	5976	5427	39	4994	0776
—0	13.5951	0.0735560	40	13.4969	0.0740910
1	5926	5694	50	4725	2250
2	5901	5828	60	4482	3592
3	5877	5961	70	4240	4936
4	5852	6095	80	3998	6282
5	13.5827	0.0736228	90	13.3723	0.0747631
6	5803	6362	100	3515	8981
7	5778	6496	110	3279	50305
8	5754	6629	120	3040	1653
9	5729	6763	130	2801	3002
10	13.5704	0.0736893	140	13.2563	0.0754[illegible]54
11	5680	7030	150	2326	5708
12	5655	7164	160	2090	7064
13	5630	7298	170	1853	8422
14	5606	7431	180	1617	9784
15	13.5581	0.0737565	190	13.1381	0.0761149
16	5557	7699	200	1145	2516
17	5532	7832	210	0910	3886
18	5507	7966	220	0677	5260
19	5483	8100	230	0440	6637
20	13.5458	0.0738233	240	13.0206	0.0768017
21	5434	8367	250	12.9972	9402
22	5409	8501	260	9738	7090
23	5385	8635	270	9504	2182
24	5360	8768	280	9270	3579
25	13.5336	0.0738902	290	12.9036	0.0774979
26	5311	9036	300	8803	6385
27	5287	9170	310	8569	7795
28	5262	9304	320	8336	9210
29	5238	9437	330	8102	80630
30	13.5213	0.0739571	340	12.7869	0.0782054
			350	7635	3485
			360	7402	4921

Based upon Thiesen und Scheel, Tätigkeitber. Phys.-Techn. Reichsanstalt, 1897-1898; Chappuis, Trav. Bur. Int. 13, 1903. Thiesen, Scheel, Sell; Wiss. Abh. Phys.-Techn. Reichsanstalt 2, p. 184, 1895, and 1 liter =1.000027 cu. dm.

TABLE 108.

DENSITY OF AQUEOUS SOLUTIONS.*

The following table gives the density of solutions of various salts in water. The numbers give the weight in grams per cubic centimeter. For brevity the substance is indicated by formula only.

Substance.	Weight of the dissolved substance in 100 parts by weight of the solution.									Temp. C.	Authority.
	5	10	15	20	25	30	40	50	60		
K_2O	1.047	1.098	1.153	1.214	1.284	1.354	1.503	1.659	1.809	15.	Schiff.
KOH . . .	1.040	1.082	1.127	1.176	1.229	1.286	1.410	1.538	1.666	15.	"
Na_2O . . .	1.073	1.144	1.218	1.284	1.354	1.421	1.557	1.689	1.829	15.	"
$NaOH$ . . .	1.058	1.114	1.169	1.224	1.279	1.331	1.436	1.539	1.642	15.	"
NH_3	0.978	0.959	0.940	0.924	0.909	0.896	–	–	–	16.	Carius.
NH_4Cl . . .	1.015	1.030	1.044	1.058	1.072	–	–	–	–	15.	Gerlach.
KCl	1.031	1.065	1.099	1.135	–	–	–	–	–	15.	"
$NaCl$	1.035	1.072	1.110	1.150	1.191	–	–	–	–	15.	"
$LiCl$	1.029	1.057	1.085	1.116	1.147	1.181	1.255	–	–	15.	"
$CaCl_2$ . . .	1.041	1.086	1.132	1.181	1.232	1.286	1.402	–	–	15.	"
$CaCl_2 + 6H_2O$	1.019	1.040	1.061	1.083	1.105	1.128	1.176	1.225	1.276	18.	Schiff.
$AlCl_3$ . . .	1.030	1.072	1.111	1.153	1.196	1.241	1.340	–	–	15.	Gerlach.
$MgCl_2$ . . .	1.041	1.085	1.130	1.177	1.226	1.278	–	–	–	15.	"
$MgCl_2 + 6H_2O$	1.014	1.032	1.049	1.067	1.085	1.103	1.141	1.183	1.222	24.	Schiff.
$ZnCl_2$ . . .	1.043	1.089	1.135	1.184	1.236	1.289	1.417	1.563	1.737	19.5	Kremers.
$CdCl_2$ . . .	1.043	1.087	1.138	1.193	1.254	1.319	1.469	1.653	1.887	19.5	"
$SrCl_2$	1.044	1.092	1.143	1.198	1.257	1.321	–	–	–	15.	Gerlach.
$SrCl_2 + 6H_2O$	1.027	1.053	1.082	1.111	1.042	1.174	1.242	1.317	–	15.	"
$BaCl_2$	1.045	1.094	1.147	1.205	1.269	–	–	–	–	15.	"
$BaCl_2 + 2H_2O$	1.035	1.075	1.119	1.166	1.217	1.273	–	–	–	21.	Schiff.
$CuCl_2$ . . .	1.044	1.091	1.155	1.221	1.291	1.360	1.527	–	–	17.5	Franz.
$NiCl_2$ . . .	1.048	1.098	1.157	1.223	1.299	–	–	–	–	17.5	"
$HgCl_2$ . . .	1.041	1.092	–	–	–	–	–	–	–	20.	Mendelejeff.
Fe_2Cl_6 . . .	1.041	1.086	1.130	1.179	1.232	1.290	1.413	1.545	1.668	17.5	Hager.
$PtCl_4$	1.046	1.097	1.153	1.214	1.285	1.362	1.546	1.785	–	–	Precht.
$SnCl_2 + 2H_2O$	1.032	1.067	1.104	1.143	1.185	1.229	1.329	1.444	1.580	15.	Gerlach.
$SnCl_4 + 5H_2O$	1.029	1.058	1.089	1.122	1.157	1.193	1.274	1.365	1.467	15.	"
$LiBr$	1.033	1.070	1.111	1.154	1.202	1.252	1.366	1.498	–	19.5	Kremers.
KBr	1.035	1.073	1.114	1.157	1.205	1.254	1.364	–	–	19.5	"
$NaBr$ . . .	1.038	1.078	1.123	1.172	1.224	1.279	1.408	1.563	–	19.5	"
$MgBr_2$ . . .	1.041	1.085	1.135	1.189	1.245	1.308	1.449	1.623	–	19.5	"
$ZnBr_2$ . . .	1.043	1.091	1.144	1.202	1.263	1.328	1.473	1.648	1.873	19.5	"
$CdBr_2$ . . .	1.041	1.088	1.139	1.197	1.258	1.324	1.479	1 678	–	19.5	"
$CaBr_2$ . . .	1.042	1.087	1.137	1.192	2.250	1.313	1.459	1.639	–	19.5	"
$BaBr_2$ . . .	1.043	1.090	1.142	1.199	1.260	1.327	1.483	1.683	–	19.5	"
$SrBr_2$ . . .	1.043	1.089	1.140	1.198	1.260	1.328	1.489	1.693	1.953	19.5	"
KI	1.036	1.076	1.118	1.164	1.216	1.269	1.394	1.544	1.732	19.5	"
LiI	1.036	1.077	1.122	1.170	1.222	1.278	1.412	1.573	1.775	19.5	"
NaI	1.038	1.080	1.126	1.177	1.232	1.292	1.430	1.598	1.808	19.5	"
ZnI_2 . . .	1.043	1.089	1.138	1.194	1.253	1.316	1.467	1.648	1.873	19.5	"
CdI_2	1.042	1.086	1.136	1.192	1.251	1.317	1.474	1.678	–	19.5	"
MgI_2	1.041	1.086	1.137	1.192	1.252	1.318	1.472	1.666	1.913	19.5	"
CaI_2	1.042	1.088	1.138	1.196	1.258	1.319	1.475	1.663	1.908	19.5	"
SrI_2	1.043	1.089	1.140	1.198	1.260	1.328	1.489	1.693	1.953	19.5	"
BaI_2	1.043	1.089	1.141	1.199	1.263	1.331	1.493	1.702	1.968	19.5	"
$NaClO_3$ . . .	1.035	1.068	1.106	1.145	1.188	1.233	1.329	–	–	19.5	"
$NaBrO_3$ . . .	1.039	1.081	1.127	1.176	1.229	1.287	–	–	–	19.5	"
KNO_3 . . .	1.031	1.064	1.099	1.135	–	–	–	–	–	15.	Gerlach.
$NaNO_3$ . . .	1.031	1.065	1.101	1.140	1.180	1.222	1.313	1.416	–	20.2	Schiff.
$AgNO_3$ . . .	1.044	1.090	1.140	1.195	1.255	1.322	1.479	1.675	1.918	15.	Kohlrausch.

* Compiled from two papers on the subject by Gerlach in the "Zeit. für Anal. Chim.," vols. 8 and 27.

DENSITY OF AQUEOUS SOLUTIONS.

Substance.	Weight of the dissolved substance in 100 parts by weight of the solution.									Temp. C.	Authority.
	5	10	15	20	25	30	40	50	60		
NH_4NO_3	1.020	1.041	1.063	1.085	1.107	1.131	1.178	1.229	1.282	17.5	Gerlach.
$Zn(NO_3)_2$	1.048	1.095	1.146	1.201	1.263	1.325	1.456	1.597	–	17.5	Franz.
$Zn(NO_3)_2+6H_2O$	–	1.054	–	1.113	–	1.178	1.250	1.329	–	14.	Oudemans.
$Ca(NO_3)_2$	1.037	1.075	1.118	1.162	1.211	1.260	1.367	1.482	1.604	17.5	Gerlach.
$Cu(NO_3)_2$	1.044	1.093	1.143	1.203	1.263	1.328	1.471	–	–	17.5	Franz.
$Sr(NO_3)_2$	1.039	1.083	1.129	1.179	–	–	–	–	–	19.5	Kremers.
$Pb(NO_3)_2$	1.043	1.091	1.143	1.199	1.262	1.332	–	–	–	17.5	Gerlach.
$Cd(NO_3)_2$	1.052	1.097	1.150	1.212	1.283	1.355	1.536	1.759	–	17.5	Franz.
$Co(NO_3)_2$	1.045	1.090	1.137	1.192	1.252	1.318	1.465	–	–	17.5	"
$Ni(NO_3)_2$	1.045	1.090	1.137	1.192	1.252	1.318	1.465	–	–	17.5	"
$Fe_2(NO_3)_6$	1.039	1.076	1.117	1.160	1.210	1.261	1.373	1.496	1.657	17.5	"
$Mg(NO_3)_2+6H_2O$	1.018	1.038	1.060	1.082	1.105	1.129	1.179	1.232	–	21	Schiff.
$Mn(NO_3)_2+6H_2O$	1.025	1.052	1.079	1.108	1.138	1.169	1.235	1.307	1.386	8	Oudemans.
K_2CO_3	1.044	1.092	1.141	1.192	1.245	1.300	1.417	1.543	–	15	Gerlach.
$K_2CO_3 + 2H_2O$	1.037	1.072	1.110	1.150	1.191	1.233	1.320	1.415	1.511	15.	"
$Na_2CO_3 10H_2O$	1.019	1.038	1.057	1.077	1.098	1.118	–	–	–	15.	"
$(NH_4)_2SO_4$	1.027	1.055	1.084	1.113	1.142	1.170	1.226	1.287	–	19.	Schiff.
$Fe_2(SO_4)_3$	1.045	1.096	1.150	1.207	1.270	1.336	1.489	–	–	18.	Hager.
$FeSO_4 + 7H_2O$	1.025	1.053	1.081	1.111	1.141	1.173	1.238	–	–	17.2	Schiff.
$MgSO_4$	1.051	1.104	1.161	1.221	1.284	–	–	–	–	15	Gerlach.
$MgSO_4 + 7H_2O$	1.025	1.050	1.075	1.101	1.129	1.155	1.215	1.278	–	15.	"
$Na_2SO_4 + 10H_2O$	1.019	1.039	1.059	1.081	1.102	1.124	–	–	–	15.	"
$CuSO_4 + 5H_2O$	1.031	1.064	1.098	1.134	1.173	1.213	–	–	–	18.	Schiff.
$MnSO_4 + 4H_2O$	1.031	1.064	1.099	1.135	1.174	1.214	1.303	1.398	–	15.	Gerlach.
$ZnSO_4 + 7H_2O$	1.027	1.057	1.089	1.122	1.156	1.191	1.269	1.351	1.443	20.5	Schiff.
$Fe_2(SO_4)_3 \cdot K_2SO_4 + 24H_2O$	1.026	1.045	1.066	1.088	1.112	1.141	–	–	–	17.5	Franz.
$Cr_2(SO_4)_3 \cdot K_2SO_4 + 24H_2O$	1.016	1.033	1.051	1.073	1.099	1.126	1.188	1.287	1.454	17.5	"
$MgSO_4 + K_2SO_4 + 6H_2O$	1.032	1.066	1.101	1.138	–	–	–	–	–	15.	Schiff.
$(NH_4)_2SO_4 + FeSO_4 + 6H_2O$	1.028	1.058	1.090	1.122	1.154	1.191	–	–	–	19.	"
K_2CrO_4	1.039	1.082	1.127	1.174	1.225	1.279	1.397	–	–	19.5	"
$K_2Cr_2O_7$	1.035	1.071	1.108	–	–	–	–	–	–	19.5	Kremers.
$Fe(Cy)_6K_4$	1.028	1.059	1.092	1.126	–	–	–	–	–	15.	Schiff.
$Fe(Cy)_6K_3$	1.025	1.053	1.070	1.113	–	–	–	–	–	13	"
$Pb(C_2H_3O_2)_2 + 3H_2O$	1.031	1.064	1.100	1.137	1.177	1.220	1.315	1.426	–	15.	Gerlach.
$2NaOH + As_2O_5 + 24H_2O$	1.020	1.042	1.066	1.089	1.114	1.140	1.194	–	–	14.	Schiff.

Substance.	5	10	15	20	30	40	60	80	100	Temp. C.	Authority.
SO_3	1.040	1.084	1.132	1.179	1.277	1.389	1.564	1.840	–	15.	Brineau.
SO_2	1.013	1.028	1.045	1.063	–	–	–	–	–	4.	Schiff.
N_2O_5	1.033	1.069	1.104	1.141	1.217	1.294	1.422	1.506	–	15.	Kolb.
$C_4H_6O_6$	1.021	1.047	1.070	1.096	1.150	1.207	–	–	–	15.	Gerlach.
$C_6H_8O_7$	1.018	1.038	1.058	1.079	1.123	1.170	1.273	–	–	15.	"
Cane sugar	1.019	1.039	1.060	1.082	1.129	1.178	1.289	–	–	17.5	"
HCl	1.025	1.050	1.075	1.101	1.151	1.200	–	–	–	15.	Kolb.
HBr	1.035	1.073	1.114	1.158	1.257	1.376	–	–	–	14.	Topsöe.
HI	1.037	1.077	1.118	1.165	1.271	1.400	–	–	–	13.	"
H_2SO_4	1.032	1.069	1.106	1.145	1.223	1.307	1.501	1.732	1.838	15.	Kolb.
H_2SiF_6	1.040	1.082	1.127	1.174	1.273	–	–	–	–	17.5	Stolba.
P_2O_5	1.035	1.077	1.119	1.167	1.271	1.385	1.676	–	–	17.5	Hager.
$P_2O_5 + 3H_2O$	1.027	1.057	1.086	1.119	1.188	1.264	1.438	–	–	15.	Schiff.
HNO_3	1.028	1.056	1.088	1.119	1.184	1.250	1.373	1.459	1.528	15.	Kolb.
$C_2H_4O_2$	1.007	1.014	1.021	1.028	1.041	1.052	1.068	1.075	1.055	15.	Oudemans

TABLE 109.

DENSITIES OF MIXTURES OF ETHYL ALCOHOL AND WATER IN GRAMS PER MILLILITER.

The densities in this table are numerically the same as specific gravities at the various temperatures in terms of water at 4° C. as unity. Based upon work done at U. S. Bureau of Standards. See Bulletin Bur. Stds. vol. 9, no. 3; contains extensive bibliography; also Circular 19, 1913.

Per cent C_2H_5OH by weight	Temperatures.						
	10° C.	15° C.	20° C.	25° C.	30° C.	35° C.	40° C.
0	0.99973	0.99913	0.99823	0.99708	0.99568	0.99406	0.99225
1	785	725	636	520	379	217	034
2	602	542	453	336	194	031	.98846
3	426	365	275	157	014	.98849	663
4	258	195	103	.98984	.98839	672	485
5	098	032	.98938	817	670	501	311
6	.98946	.98877	780	656	507	335	142
7	801	729	627	500	347	172	.97975
8	660	584	478	346	189	009	808
9	524	442	331	193	031	.97846	641
10	393	304	187	043	.97875	685	475
11	267	171	047	.97897	723	527	312
12	145	041	.97910	753	573	371	150
13	026	.97914	775	611	424	216	.96989
14	.97911	790	643	472	278	063	829
15	800	669	514	334	133	.96911	670
16	692	552	387	199	.96990	760	512
17	583	433	259	062	844	607	352
18	473	313	129	.96923	697	452	189
19	363	191	.96997	782	547	294	023
20	252	068	864	639	395	134	.95856
21	139	.96944	729	495	242	.95973	687
22	024	818	592	348	087	809	516
23	.96907	689	453	199	.95929	643	343
24	787	558	312	048	769	476	168
25	665	424	168	.95895	607	306	.94991
26	539	287	020	738	442	133	810
27	406	144	.95867	576	272	.94955	625
28	268	.95996	710	410	098	774	438
29	125	844	548	241	.94922	590	248
30	.95977	686	382	067	741	403	055
31	823	524	212	.94890	557	214	.93860
32	665	357	038	709	370	021	662
33	502	186	.94860	525	180	.93825	461
34	334	011	679	337	.93986	626	257
35	162	.94832	494	146	790	425	051
36	.94986	650	306	.93952	591	221	.92843
37	805	464	114	756	390	016	634
38	620	273	.93919	556	186	.92808	422
39	431	079	720	353	.92979	597	208
40	238	.93882	518	148	770	385	.91992
41	042	682	314	.92940	558	170	774
42	.93842	478	107	729	344	.91952	554
43	639	271	.92897	516	128	733	332
44	433	062	685	301	.91910	513	108
45	226	.92852	472	085	692	291	.90884
46	017	640	257	.91868	472	069	660
47	.92806	426	041	649	250	.90845	434
48	593	211	.91823	429	028	621	207
49	379	.91995	604	208	.90805	396	.89979
50	162	776	384	.90985	580	168	750

DENSITY OF MIXTURES OF ETHYL ALCOHOL AND WATER IN GRAMS PER MILLILITER.

Per cent C_2H_5OH by weight	Temperature.						
	10° C.	15° C.	20° C.	25° C.	30° C.	35° C.	40° C.
50	0.92162	0.91776	0.91384	0.90985	0.90580	0.90168	0.89750
51	.91943	555	160	760	353	.89940	519
52	723	333	.90936	534	125	710	288
53	502	110	711	307	.89896	479	056
54	279	.90885	485	079	667	248	.88823
55	055	659	258	.89850	437	016	589
56	.90831	433	031	621	206	.88784	356
57	607	207	.89803	392	.88975	552	122
58	381	.89980	574	162	744	319	.87888
59	154	752	344	.88931	512	085	653
60	.89927	523	113	699	278	.87851	417
61	698	293	.88882	466	044	615	180
62	468	062	650	233	.87809	379	.86943
63	237	.88830	417	.87998	574	142	705
64	006	597	183	763	337	.86905	466
65	.88774	364	.87948	527	100	667	227
66	541	130	713	291	.86863	429	.85987
67	308	.87895	477	054	625	190	747
68	074	660	241	.86817	387	.85950	507
69	.87839	424	004	579	148	710	266
70	602	187	.86766	340	.85908	470	025
71	365	.86949	527	100	667	228	.84783
72	127	710	287	.85859	426	.84986	540
73	.86888	470	047	618	184	743	297
74	648	229	.85806	376	.84941	500	053
75	408	.85988	564	134	698	257	.83809
76	168	747	322	.84891	455	013	564
77	.85927	505	079	647	211	.83768	319
78	685	262	.84835	403	.83966	523	074
79	442	018	590	158	720	277	.82827
80	197	.84772	344	.83911	473	029	578
81	.84950	525	096	664	224	.82780	329
82	702	277	.83848	415	.82974	530	079
83	453	028	599	164	724	279	.81828
84	203	.83777	348	.82913	473	027	576
85	.83951	525	095	660	220	.81774	322
86	697	271	.82840	405	.81965	519	067
87	441	014	583	148	708	262	.80811
88	181	.82754	323	.81888	448	003	552
89	.82919	492	062	626	186	.80742	291
90	654	227	.81797	362	.80922	478	028
91	386	.81959	529	094	655	211	.79761
92	114	688	257	.80823	384	.79941	491
93	.81839	413	.80983	549	111	669	220
94	561	134	705	272	.79835	393	.78947
95	278	.80852	424	.79991	555	114	670
96	.80991	566	138	706	271	.78831	388
97	698	274	.79846	415	.78981	542	100
98	399	.79975	547	117	684	247	.77806
99	094	670	243	.78814	382	.77946	507
100	.79784	360	.78934	506	075	641	203

TABLE 110.

DENSITIES OF AQUEOUS MIXTURES OF METHYL ALCOHOL, CANE SUGAR, OR SULPHURIC ACID.

Per cent by weight of substance.	Methyl Alcohol. $D\frac{15^\circ}{4^\circ}$ C.	Cane Sugar. 20° See p. 444	Sulphuric Acid. $D\frac{20^\circ}{4^\circ}$ C.	Per cent by weight of substance.	Methyl Alcohol. $D\frac{15^\circ}{4^\circ}$ C.	Cane Sugar. 20° See p. 444	Sulphuric Acid. $D\frac{20^\circ}{4^\circ}$ C.
0	0.99913	0.998234	0.99823	50	0.91852	1.229567	1.39505
1	.99727	1.002120	1.00506	51	.91653	1.235085	1.40487
2	.99543	1.006015	1.01178	52	.91451	1.240641	1.41481
3	.99370	1.009934	1.01839	53	.91248	1.246234	1.42487
4	.99198	1.013881	1.02500	54	.91044	1.251866	1.43503
5	.99029	1.017854	1.03168	55	.90839	1.257535	1.44530
6	.98864	1.021855	1.03843	56	.90631	1.263243	1.45568
7	.98701	1.025885	1.04527	57	.90421	1.268989	1.46615
8	.98547	1.029942	1.05216	58	.90210	1.274774	1.47673
9	.98394	1.034029	1.05909	59	.89996	1.280595	1.48740
10	.98241	1.038143	1.06609	60	.89781	1.286456	1.49818
11	.98093	1.042288	1.07314	61	.89563	1.292354	1.50904
12	.97945	1.046462	1.08026	62	.89341	1.298291	1.51999
13	.97802	1.050665	1.08744	63	.89117	1.304267	1.53102
14	.97660	1.054900	1.09468	64	.88890	1.310282	1.54213
15	.97518	1.059165	1.10199	65	.88662	1.316334	1.55333
16	.97377	1.063460	1.10936	66	.88433	1.322425	1.56460
17	.97237	1.067789	1.11679	67	.88203	1.328554	1.57595
18	.97096	1.072147	1.12428	68	.87971	1.334722	1.58739
19	.96955	1.076537	1.13183	69	.87739	1.340928	1.59890
20	.96814	1.080959	1.13943	70	.87507	1.347174	1.61048
21	.96673	1.085414	1.14709	71	.87271	1.353456	1.62213
22	.96533	1.089900	1.15480	72	.87033	1.359778	1.63384
23	.96392	1.094420	1.16258	73	.86792	1.366139	1.64560
24	.96251	1.098971	1.17041	74	.86546	1.372536	1.65738
25	.96108	1.103557	1.17830	75	.86300	1.378971	1.66917
26	.95963	1.108175	1.18624	76	.86051	1.385446	1.68095
27	.95817	1.112828	1.19423	77	.85801	1.391956	1.69268
28	.95668	1.117512	1.20227	78	.85551	1.398505	1.70433
29	.95518	1.122231	1.21036	79	.85300	1.405091	1.71585
30	.95366	1.126984	1.21850	80	.85048	1.411715	1.72717
31	.95213	1.131773	1.22669	81	.84794	1.418374	1.73827
32	.95056	1.136596	1.23492	82	.84536	1.425072	1.74904
33	.94896	1.141453	1.24320	83	.84274	1.431807	1.75943
34	.94734	1.146345	1.25154	84	.84009	1.438579	1.76932
35	.94570	1.151275	1.25992	85	.83742	1.445388	1.77860
36	.94404	1.156238	1.26836	86	.83475	1.452232	1.78721
37	.94237	1.161236	1.27685	87	.83207	1.459114	1.79509
38	.94067	1.166269	1.28543	88	.82937	1.466032	1.80223
39	.93894	1.171340	1.29407	89	.82667	1.472986	1.80864
40	.93720	1.176447	1.30278	90	.82396	1.479976	1.81438
41	.93543	1.181592	1.31157	91	.82124	1.487002	1.81950
42	.93365	1.186773	1.32043	92	.81849	1.494063	1.82401
43	.93185	1.191993	1.32938	93	.81568	1.501158	1.82790
44	.93001	1.197247	1.33843	94	.81285	1.508289	1.83115
45	.92815	1.202540	1.34759	95	.80999	1.515455	1.83368
46	.92627	1.207870	1.35686	96	.80713	1.522656	1.83548
47	.92436	1.213238	1.36625	97	.80428	1.529891	1.83637
48	.92242	1.218643	1.37574	98	.80143	1.537161	1.83605
49	.92048	1.224086	1.38533	99	.79859	1.544462	
50	.91852	1.229567	1.39505	100	.79577	1.551800	

(1) Calculated from the specific gravity determinations of Doroschevski and Rozhdestvenski at 15°/15° C.; J. Russ., Phys. Chem. Soc., 41, p. 977, 1909.
(2) According to Dr. F. Plato; Wiss. Abh. der K. Normal-Eichungs-Kommission, 2, p. 153, 1900.
(3) Calculated from Dr. Domke's table; Wiss. Abh. der K. Normal-Eichungs-Kommission, 5, p. 131, 1900.

All reprinted from Circular 19, U.S. Bureau of Standards, 1913.

TABLE 111.
DENSITY OF GASES.

The following table gives the density as the weight in grams of a liter (normal liter) of the gas at 0° C, 76 cm pressure and standard gravity, 980.665 cm/sec², (sea-level, 45° latitude), the specific gravity referred to dry, carbon-dioxide-free air and to pure oxygen, and the weight in pounds per cubic foot. Dry, carbon-dioxide-free air is of remarkably uniform density; Guye, Kovacs and Wourtzel found maximum variations in the density of only 7 to 8 parts in 10,000. For highest accuracy pure oxygen should be used as the standard gas for specific gravities. Observed densities are closely proportional to the molecular weights. The following data are derived from advance sheets of "A Review of the Densities of Various Gases," to be published by the U. S. Bureau of Standards.

Gas	Formula	Weight in grams. Normal liter	Specific gravity		Pounds per cubic foot	
			Air = 1	O_2 = 1		
Acetylene	C_2H_2	1.1791	0.9120	0.8251	0.07361	1
Air	—	1.2929	1.	0.9048	0.08072	2
Ammonia	NH_3	0.7708	0.5962	0.5394	0.04812	3
Argon	A	1.7824	1.3786	1.2473	0.11127	4
Carbon dioxide	CO_2	1.9768	1.5290	1.3834	0.12341	5
Carbon monoxide	CO	1.2504	0.9671	0.8750	0.07806	6
Chlorine	Cl_2	3.214	2.4859	2.249	0.2006	7
Ethane	C_2H_6	1.3565	1.0492	0.9493	0.08468	1
Ethylene	C_2H_4	1.2603	0.9748	0.8820	0.07868	8
Fluorine	F_2	1.695	1.313	1.186	0.1058	9
Helium	He	0.1785	0.1381	0.1249	0.01114	10
Hydrogen	H_2	0.08987	0.06951	0.06289	0.005610	11
Hydrogen bromide	HBr	3.6443	2.8188	2.5503	0.22751	12
Hydrogen chloride	HCl	1.6392	1.2679	1.1471	0.10233	13
Hydrogen sulphide	H_2S	1.5392	1.1905	1.0771	0.09609	14
Krypton	Kr	3.708	2.867	2.595	0.2315	15
Methane	CH_4	0.7168	0.5545	0.5016	0.04475	14
Methyl chloride	$(CH_3)Cl$	2.3045	1.7824	1.6127	0.14387	16
Methyl ether	$(CH_3)O$	2.1096	1.6317	1.4763	0.13170	16
Methyl fluoride	$(CH_3)F$	1.5454	1.1953	1.0815	0.09648	17
Neon	Ne	0.9002	0.6962	0.6300	0.05620	18
Nitric oxide	NO	1.3402	1.0366	0.9379	0.08367	19
Nitrogen	N_2	1.2506	0.9673	0.8752	0.07807	20
Nitrogen, atmospheric	—	1.2568	0.9721	0.8795	0.07846	2
Nitrous oxide	N_2O	1.9777	1.5297	1.3840	0.12347	5
Oxygen	O_2	1.4290	1.1053	1.0000	0.08921	2
Phosphine	PH_3	1.5293	1.1828	1.0702	0.09547	21
Propane	C_3H_8	2.0200	1.5624	1.4136	0.12611	22
Silicon tetrachloride	SiF_4	4.6840	3.6229	3.2779	0.29242	23
Sulphur dioxide	SO_2	2.9267	2.2636	2.0481	0.18271	24
Xenon	X	5.851	4.514	4.094	0.3653	15

(1) Stahrfoss; (2) average; (3) Guye; (4) Ramsay, Leduc; (5) Guye, Puitza; (6) Rayleigh; (7) Jaquerod, Tourapian; (8) Batuecas; (9) Moissan; (10) Taylor; (11) Morley; (12) Moles, Reemair; (13) Gray, Burt; (14) Baume, Perrot; (15) Moore; (16) Baume; (17) Batuecas, Moles; (18) Watson; (19) Gray, Guye, Davila; (20) Gray, Moles; (21) Ter Gazarian; (22) Timmeraus; (23) Germann, Booth; (24) Scheuer.

The weight of the normal liter of the following gases has been determined with less accuracy: *Arsine,* 3.484, Dumas; *Carbon oxysulphide,* 2.7208, Von Than; *Butane,* 2.594, Frankland; *Cyanogen,* 2.335, Gay-Lussac; *Hydrogen fluoride,* 0.9212, Thorpe, Hambly; *Hydrogen iodide,* 5.657, Thomson; *Hydrogen selenide,* 3.6702, Bruylante; *Hydrogen telluride,* 5.805, Ernyei; *Methylamine,* 1.396, Leduc; *Nitrosyl chloride,* 2.9919, Wourtzel.

TABLE 112.

VOLUME OF GASES.

Values of 1 + .00367 *t*.

The quantity 1 + .00367 *t* gives for a gas the volume at t° when the pressure is kept constant, or the pressure at t° when the volume is kept constant, in terms of the volume or the pressure at 0°.

(a) This part of the table gives the values of 1 + .00367 *t* for values of *t* between 0° and 10° C. by tenths of a degree.

(b) This part gives the values of 1 + .00367 *t* for values of *t* between — 90° and + 1990° C. by 10° steps.

These two parts serve to give any intermediate value to one tenth of a degree by a simple computation as follows: — In the (*b*) table find the number corresponding to the nearest lower temperature, and to this number add the decimal part of the number in the (*a*) table which corresponds to the difference between the nearest temperature in the (*b*) table and the actual temperature. For example, let the temperature be 682°.2:

We have for 680 in table (*b*) the number	3.49560
And for 2.2 in table (*a*) the decimal	.00807
Hence the number for 682.2 is	3.50367

(c) This part gives the logarithms of 1 + .00367 *t* for values of *t* between — 49° and + 399° C. by degrees.

(d) This part gives the logarithms of 1 + .00367 *t* for values of *t* between 400° and 1990° C. by 10° steps.

(a) Values of 1 + .00367 *t* for Values of *t* between 0° and 10° C. by Tenths of a Degree.

t	0.0	0.1	0.2	0.3	0.4
0	1.00000	1.00037	1.00073	1.00110	1.00147
1	.00367	.00404	.00440	.00477	.00514
2	.00734	.00771	.00807	.00844	.00881
3	.01101	.01138	.01174	.01211	.01248
4	.01468	.01505	.01541	.01578	.01615
5	1.01835	1.01872	1.01908	1.01945	1.01982
6	.02202	.02239	.02275	.02312	.02349
7	.02569	.02606	.02642	.02679	.02716
8	.02936	.02973	.03009	.03046	.03083
9	.03303	.03340	.03376	.03413	.03450
t	0.5	0.6	0.7	0.8	0.9
0	1.00184	1.00220	1.00257	1.00294	1.00330
1	.00550	.00587	.00624	.00661	.00697
2	.00918	.00954	.00991	.01028	.01064
3	.01284	.01321	.01358	.01395	.01431
4	.01652	.01688	.01725	.01762	.01798
5	1.02018	1.02055	1.02092	1.02129	1.02165
6	.02386	.02422	.02459	.02496	.02532
7	.02752	.02789	.02826	.02863	.02899
8	.03120	.03156	.03193	.03290	.03266
9	.03486	.03523	.03560	.03597	.03633

VOLUME OF GASES.

(b) Values of $1 + .00367t$ for Values of t between $-90°$ and $+1990°$ C. by 10° Steps.

t	00	10	20	30	40
—000	1.00000	0.96330	0.92660	0.88990	0.85320
+000	1.00000	1.03670	1.07340	1.11010	1.14680
100	1.36700	1.40370	1.44040	1.47710	1.51380
200	1.73400	1.77070	1.80740	1.84410	1.88080
300	2.10100	2.13770	2.17440	2.21110	2.24780
400	2.46800	2.50470	2.54140	2.57810	2 61480
500	2.83500	2.87170	2.90840	2.94510	2.98180
600	3.20200	3.23870	3.27540	3.31210	3.34880
700	3.56900	3.60570	3.64240	3.67910	3.71580
800	3.93600	3.97270	4.00940	4.04610	4.08280
900	4.30300	4.33970	4.37640	4.41310	4.44980
1000	4.67000	4.70670	4.74340	4.78010	4.81680
1100	5.03700	5.07370	5.11040	5.14710	5.18380
1200	5.40400	5.44070	5.47740	5.51410	5.55080
1300	5.77100	5.80770	5.84440	5.88110	5.91780
1400	6.13800	6.17470	6.21140	6.24810	6.28480
1500	6.50500	6.54170	6.57840	6.61510	6.65180
1600	6.87200	6.90870	6.94540	6.98210	7.01880
1700	7.23900	7.27570	7.31240	7.34910	7.38580
1800	7.60600	7.64270	7.67940	7.71610	7.75280
1900	7.97300	8.00970	8.04640	8.08310	8.11980
2000	8.34000	8.37670	8.41340	8.45010	8.48680

t	50	60	70	80	90
—000	0.81650	0.77980	0.74310	0.70640	0.66970
+000	1.18350	1.22020	1.25690	1.29360	1.33030
100	1.55050	1.58720	1.62390	1.66060	1.69730
200	1.91750	1.95420	1.99090	2.02760	2.06430
300	2.28450	2.32120	2.35790	2.39460	2.43130
400	2.65150	2.68820	2.72490	2.76160	2.79830
500	3.01850	3.05520	3.09190	3.12860	3.16530
600	3.38550	3.42220	3.45890	3.49560	3.53230
700	3.75250	3.78920	3.82590	3.86260	3.89930
800	4.11950	4.15620	4.19290	4.22960	4.26630
900	4.48650	4.52320	4.55990	4.59660	4.63330
1000	4.85350	4.89020	4.92690	4.96360	5.00030
1100	5.22050	5.25720	5.29390	5.33060	5.36730
1200	5.58750	5.62420	5.66090	5.69760	5.73430
1300	5.95450	5.99120	6.02790	6.06460	6.10130
1400	6.32150	6.35820	6.39490	6.43160	6.46830
1500	6.68850	6.72520	6.76190	6.79860	6.83530
1600	7.05550	7.09220	7.12890	7.16560	7.20230
1700	7.42250	7.45920	7.49590	7.53260	7.56930
1800	7.78950	7.82620	7.86290	7.89960	7.93630
1900	8.15650	8.19320	8.22990	8.26660	8.30330
2000	8.52350	8.56020	8.59690	8.63360	8.67030

VOLUME OF

(c) Logarithms of 1 + .00367 t for Values

t	0	1	2	3	4	Mean diff. per degree.
−40	1̄.931051	1̄.929179	1̄.927299	1̄.925410	1̄.923513	1884
−30	.949341	.947546	.945744	.943934	.942117	1805
−20	.966892	.965169	.963438	.961701	.959957	1733
−10	.983762	.982104	.980440	.978769	.977092	1667
−0	0.000000	.998403	.996801	.995192	.993577	1605
+0	0.000000	0.001591	0.003176	0.004755	0.006329	1582
10	.015653	.017188	.018717	.020241	.021760	1526
20	.030762	.032244	.033721	.035193	.036661	1474
30	.045362	.046796	.048224	.049648	.051068	1426
40	.059488	.060875	.062259	.063637	.065012	1381
50	0.073168	0.074513	0.075853	0.077190	0.078522	1335
60	.086431	.087735	.089036	.090332	.091624	1299
70	.099301	.100567	.101829	.103088	.104344	1259
80	.111800	.113030	.114257	.115481	.116701	1226
90	.123950	.125146	.126339	.127529	.128716	1191
100	0.135768	0.136933	0.138094	0.139252	0.140408	1158
110	.147274	.248408	.149539	.150667	.151793	1129
120	.158483	.159588	.160691	.161790	.162887	1101
130	.169410	.170488	.171563	.172635	.173705	1074
140	.180068	.181120	.182169	.183216	.184260	1048
150	0.190472	0.191498	0.192523	0.193545	0.194564	1023
160	.200632	.201635	.202635	.203634	.204630	1000
170	.210559	.211540	.212518	.213494	.214468	976
180	.220265	.221224	.222180	.223135	.224087	956
190	.229759	.230697	.231633	.232567	.233499	935
200	0.239049	0.239967	0.240884	0.241798	0.242710	916
210	.248145	.249044	.249942	.250837	.251731	897
220	.257054	.257935	.258814	.259692	.260567	878
230	.265784	.266648	.267510	.268370	.269228	861
240	.274343	.275189	.276034	.276877	.277719	844
250	0.282735	0.283566	0.284395	0.285222	0.286048	828
260	.290969	.291784	.292597	.293409	.294219	813
270	.299049	.299849	.300648	.301445	.302240	798
280	.306982	.307768	.308552	.309334	.310115	784
290	.314773	.315544	.316314	.317083	.317850	769
300	0.322426	0.323184	0.323941	0.324696	0.325450	756
310	.329947	.330692	.331435	.332178	.332919	743
320	.337339	.338072	.338803	.339533	.340262	730
330	.344608	.345329	.346048	.346766	.347482	719
340	.351758	.352466	.353174	.353880	.354585	707
350	0.358791	0.359488	0.360184	0.360879	0.361573	696
360	.365713	.366399	.367084	.367768	.368451	684
370	.372525	.373201	.373875	.374549	.375221	674
380	.379233	.379898	.380562	.381225	.381887	664
390	.385439	.386494	.387148	.387801	.388453	654

GASES.

of t between −49° and +399° C. by Degrees.

t	5	6	7	8	9	Mean diff. per degree.
−40	$\bar{1}$.921608	$\bar{1}$.919695	$\bar{1}$.917773	$\bar{1}$.915843	$\bar{1}$.913904	**1926**
−30	.940292	.938460	.936619	.934771	.932915	1845
−20	.958205	.956447	.954681	.952909	.951129	1771
−10	.975409	.973719	.972022	.970319	.968609	1699
−0	.991957	.990330	.988697	.987058	.985413	1636
+0	0.007897	0.009459	0.011016	0.012567	0.014113	**1554**
10	.023273	.024781	.026284	.027782	.029274	1500
20	.038123	.039581	.041034	.042481	.043924	1450
30	.052482	.053893	.055298	.056699	.058096	1402
40	.066382	.067748	.069109	.070466	.071819	1359
50	0.079847	0.081174	0.082495	0.083811	0.085123	**1315**
60	.092914	.094198	.095486	.096765	.098031	1281
70	.105595	.106843	.108088	.109329	.110566	1243
80	.117917	.119130	.120340	.121547	.122750	1210
90	.129899	.131079	.132256	.133430	.134601	1175
100	0.141559	0.142708	0.143854	0.144997	0.146137	**1144**
110	.152915	.154034	.155151	.156264	.157375	1115
120	.163981	.164072	.166161	.167246	.168330	1087
130	.174772	.175836	.176898	.177958	.179014	1060
140	.185301	.186340	.187377	.188411	.189443	1035
150	0.195581	0.196596	0.197608	0.198619	0.199626	**1011**
160	.205624	.206615	.207605	.208592	.209577	988
170	.215439	.216409	.217376	.218341	.219304	966
180	.225038	.225986	.226932	.227876	.228819	946
190	.234429	.235357	.236283	.237207	.238129	925
200	0.243621	0.244529	0.245436	0.246341	0.247244	**906**
210	.252623	.253512	.254400	.255287	.256172	887
220	.261441	.262313	.263184	.264052	.264919	870
230	.270085	.270940	.271793	.272644	.273494	853
240	.278559	.279398	.280234	.281070	.281903	836
250	0.286872	0.287694	0.288515	0.289326	0.290153	**820**
260	.295028	.295835	.296640	.297445	.298248	805
270	.303034	.303827	.304618	.305407	.306196	790
280	.310895	.311673	.312450	.313226	.314000	776
290	.318616	.319381	.320144	.320906	.321667	763
300	0.326203	0.326954	0.327704	0.328453	0.329201	**750**
310	.333659	.334397	.335135	.335871	.336606	737
320	.340989	.341715	.342441	.343164	.343887	724
330	.348198	.348912	.349624	.350337	.351048	713
340	.355289	.355991	.356693	.357394	.358093	701
350	0.362266	0.362957	0.363648	0.364337	0.365025	**690**
360	.369132	.369813	.370493	.371171	.371849	678
370	.375892	.376562	.377232	.377900	.378567	668
380	.382548	.383208	.383868	.384525	.385183	658
390	.389104	.389754	.390403	.391052	.391699	648

VOLUME OF GASES.

(d) Logarithms of $1 + .00367t$ for Values of t between 400° and 1990° C. by 10° Steps.

t	00	10	20	30	40
400	0.392345	0.398756	0.405073	0.411300	0.417439
500	0.452553	0.458139	0.463654	0.469100	0.474479
600	.505421	.510371	.515264	.520103	.524889
700	.552547	.556990	.561388	.565742	.570052
800	.595055	.599086	.603079	.607037	.610958
900	.633771	.637460	.641117	.644744	.648341
1000	0.669317	0.672717	0.676090	0.679437	0.682759
1100	.702172	.705325	.708455	.711563	.714648
1200	.732715	.735655	.738575	.741475	.744356
1300	.761251	.764004	.766740	.769459	.772160
1400	.788027	.790616	.793190	.795748	.798292
1500	0.813247	0.815691	0.818120	0.820536	0.822939
1600	.837083	.839396	.841697	.843986	.846263
1700	.859679	.861875	.864060	.866234	.868398
1800	.881156	.883247	.885327	.887398	.889459
1900	.901622	.903616	.905602	.907578	.909545

t	50	60	70	80	90
400	0.423492	0.429462	0.435351	0.441161	0.446894
500	0.479791	0.485040	0.490225	0.495350	0.500415
600	.529623	.534305	.538938	.543522	.548058
700	.574321	.578548	.582734	.586880	.590987
800	.614845	.618696	.622515	.626299	.630051
900	.651908	.655446	.658955	.662437	.665890
1000	0.686055	0.689327	0.692574	0.695797	0.698996
1100	.717712	.720755	.723776	.726776	.729756
1200	.747218	.750061	.752886	.755692	.758480
1300	.774845	.777514	.780166	.782802	.785422
1400	.800820	.803334	.805834	.808319	.810790
1500	0.825329	0.827705	0.830069	0.832420	0.834758
1600	.848528	.850781	.853023	.855253	.857471
1700	.870550	.872692	.874824	.876945	.879056
1800	.891510	.893551	.895583	.897605	.899618
1900	.911504	.913454	.915395	.917327	.919251

RELATIVE DENSITY OF MOIST AIR FOR DIFFERENT PRESSURES AND HUMIDITIES.

TABLE 113.—Values of $\frac{h}{760}$, from $h=1$ to $h=9$, for the Computation of Different Values of the Ratio of Actual to Normal Barometric Pressure.

This gives the density of moist air at pressure h in terms of the same air at normal atmosphere pressure. When air contains moisture, as is usually the case with the atmosphere, we have the following equation for pressure term: $h=B-0.378e$, where e is the vapor pressure, and B the corrected barometric pressure. When the necessary psychrometric observations are made the value of e may be taken from Table 189 and then $0.378e$ from Table 115, or the dew-point may be found and the value of $0.378e$ taken from Table 115.

h	$\frac{h}{760}$
1	0.0013158
2	.0026316
3	.0039474
4	0.0052632
5	.0065789
6	.0078947
7	0.0092105
8	.0105263
9	.0118421

EXAMPLES OF USE OF THE TABLE.

To find the value of $\frac{h}{760}$ when $h = 754.3$

$h = 700$	gives	.92105
50	"	.065789
4	"	.005263
.3	"	.000395
754.3		.992497

To find the value of $\frac{h}{760}$ when $h = 5.73$

$h = 5$	gives	.0065789
.7	"	.0009210
.03	"	.0000395
5.73		.0075394

TABLE 114.—Values of the logarithms of $\frac{h}{760}$ for values of h between 80 and 340.

Values from 8 to 80 may be got by subtracting 1 from the characteristic, and from 0.8 to 8 by subtracting 2 from the characteristic, and so on.

h	Values of $\log \frac{h}{760}$.									
	0	1	2	3	4	5	6	7	8	9
80	$\bar{1}$.02228	$\bar{1}$.02767	$\bar{1}$.03300	$\bar{1}$.03826	$\bar{1}$.04347	$\bar{1}$.04861	$\bar{1}$.05368	$\bar{1}$.05871	$\bar{1}$.06367	$\bar{1}$.06858
90	.07343	.07823	.08297	.08767	.09231	.09691	.10146	.10596	.11041	.11482
100	$\bar{1}$.11919	$\bar{1}$.12351	$\bar{1}$.12779	$\bar{1}$.13202	$\bar{1}$.13622	$\bar{1}$.14038	$\bar{1}$.14449	$\bar{1}$.14857	$\bar{1}$.15261	$\bar{1}$.15661
110	.16058	.16451	.16840	.17226	.17609	.17988	.18364	.18737	.19107	.19473
120	.19837	.20197	.20555	.20909	.21261	.21611	.21956	.22299	.22640	.22978
130	.23313	.23646	.23976	.24304	.24629	.24952	.25273	.25591	.25907	.26220
140	.26531	.26841	.27147	.27452	.27755	.28055	.28354	.28650	.28945	.29237
150	$\bar{1}$.29528	$\bar{1}$.29816	$\bar{1}$.30103	$\bar{1}$.30388	$\bar{1}$.30671	$\bar{1}$.30952	$\bar{1}$.31231	$\bar{1}$.31509	$\bar{1}$.31784	$\bar{1}$.32058
160	.32331	.32601	.32870	.33137	.33403	.33667	.33929	.34190	.34450	.34707
170	.34964	.35218	.35471	.35723	.35974	.36222	.36470	.36716	.36961	.37204
180	.37446	.37686	.37926	.38164	.38400	.38636	.38870	.39128	.39334	.39565
190	.39794	.40022	.40249	.40474	.40699	.40922	.41144	.41365	.41585	.41804
200	$\bar{1}$.42022	$\bar{1}$.42238	$\bar{1}$.42454	$\bar{1}$.42668	$\bar{1}$.42882	$\bar{1}$.43094	$\bar{1}$.43305	$\bar{1}$.43516	$\bar{1}$.43725	$\bar{1}$.43933
210	.44141	.44347	.44552	.44757	.44960	.45162	.45364	.45565	.45764	.45963
220	.46161	.46358	.46554	.46749	.46943	.47137	.47329	.47521	.47712	.47902
230	.48091	.48280	.48467	.48654	.48840	.49025	.49210	.49393	.49576	.49758
240	.49940	.50120	.50300	.50479	.50658	.50835	.51012	.51188	.51364	.51539
250	$\bar{1}$.51713	$\bar{1}$.51886	$\bar{1}$.52059	$\bar{1}$.52231	$\bar{1}$.52402	$\bar{1}$.52573	$\bar{1}$.52743	$\bar{1}$.52912	$\bar{1}$.53081	$\bar{1}$.53249
260	.53416	.53583	.53749	.53914	.54079	.54243	.54407	.54570	.54732	.54894
270	.55055	.55216	.55376	.55535	.55694	.55852	.56010	.56167	.56323	.56479
280	.56634	.56789	.56944	.57097	.57250	.57403	.57555	.57707	.57858	.58008
290	.58158	.58308	.58457	.58605	.58753	.58901	.59048	.59194	.59340	.59486
300	$\bar{1}$.59631	$\bar{1}$.59775	$\bar{1}$.59919	$\bar{1}$.60063	$\bar{1}$.60206	$\bar{1}$.60349	$\bar{1}$.60491	$\bar{1}$.60632	$\bar{1}$.60774	$\bar{1}$.60914
310	.61055	.61195	.61334	.61473	.61611	.61750	.61887	.62025	.62161	.62298
320	.62434	.62569	.62704	.62839	.62973	.63107	.63240	.63373	.63506	.63638
330	.63770	.63901	.64032	.64163	.64293	.64423	.64553	.64682	.64810	.64939
340	.65067	.65194	.65321	.65448	.65574	.65701	.65826	.65952	.66077	.66201

DENSITY OF AIR.

Values of logarithms of $\frac{h}{760}$ for values of h between 350 and 800.

h	Values of log $\frac{h}{760}$.									
	0	1	2	3	4	5	6	7	8	9
350	$\bar{1}$.66325	$\bar{1}$.66449	$\bar{1}$.66573	$\bar{1}$.66696	$\bar{1}$.66819	$\bar{1}$.66941	$\bar{1}$.67064	$\bar{1}$.67185	$\bar{1}$.67307	$\bar{1}$.67428
360	.67549	.67669	.67790	.67909	.68029	.68148	.68267	.68385	.68503	.68621
370	.68739	.68856	.68973	.69090	.69206	.69322	.69437	.69553	.69668	.69783
380	.69897	.70011	.70125	.70239	.70352	.70465	.70577	.70690	.70802	.70914
390	.71025	.71136	.71247	.71358	.71468	.71578	.71688	.71798	.71907	.72016
400	$\bar{1}$.72125	$\bar{1}$.72233	$\bar{1}$.72341	$\bar{1}$.72449	$\bar{1}$.72557	$\bar{1}$.72664	$\bar{1}$.72771	$\bar{1}$.72878	$\bar{1}$.72985	$\bar{1}$.73091
410	.73197	.73303	.73408	.73514	.73619	.73723	.73828	.73932	.74036	.74140
420	.74244	.74347	.74450	.74553	.74655	.74758	.74860	.74961	.75063	.75164
430	.75265	.75366	.75467	.75567	.75668	.75768	.75867	.75967	.76066	.76165
440	.76264	.76362	.76461	.76559	.76657	.76755	.76852	.76949	.77046	.77143
450	$\bar{1}$.77240	$\bar{1}$.77336	$\bar{1}$.77432	$\bar{1}$.77528	$\bar{1}$.77624	$\bar{1}$.77720	$\bar{1}$.77815	$\bar{1}$.77910	$\bar{1}$.78005	$\bar{1}$.78100
460	.78194	.78289	.78383	.78477	.78570	.78664	.78757	.78850	.78943	.79036
470	.79128	.79221	.79313	.79405	.79496	.79588	.79679	.79770	.79861	.79952
480	.80043	.80133	.80223	.80313	.80403	.80493	.80582	.80672	.80761	.80850
490	.80938	.81027	.81115	.81203	.81291	.81379	.81467	.81554	.81642	.81729
500	$\bar{1}$.81816	$\bar{1}$.81902	$\bar{1}$.81989	$\bar{1}$.82075	$\bar{1}$.82162	$\bar{1}$.82248	$\bar{1}$.82334	$\bar{1}$.82419	$\bar{1}$.82505	$\bar{1}$.82590
510	.82676	.82761	.82846	.82930	.83015	.83099	.83184	.83268	.83352	.83435
520	.83519	.83602	.83686	.83769	.83852	.83935	.84017	.84100	.84182	.84264
530	.84346	.84428	.84510	.84591	.84673	.84754	.84835	.84916	.84997	.85076
540	.85158	.85238	.85319	.85399	.85479	.85558	.85638	.85717	.85797	.85876
550	$\bar{1}$.85955	$\bar{1}$.86034	$\bar{1}$.86113	$\bar{1}$.86191	$\bar{1}$.86270	$\bar{1}$.86348	$\bar{1}$.86426	$\bar{1}$.86504	$\bar{1}$.86582	$\bar{1}$.86660
560	.86737	.86815	.86892	.86969	.87047	.87123	.87200	.87277	.87353	.87430
570	.87506	.87582	.87658	.87734	.87810	.87885	.87961	.88036	.88111	.88186
580	.88261	.88336	.88411	.88486	.88560	.88634	.88708	.88782	.88856	.88930
590	.89004	.89077	.89151	.89224	.89297	.89370	.89443	.89516	.89589	.89661
600	$\bar{1}$.89734	$\bar{1}$.89806	$\bar{1}$.89878	$\bar{1}$.89950	$\bar{1}$.90022	$\bar{1}$.90094	$\bar{1}$.90166	$\bar{1}$.90238	$\bar{1}$.90309	$\bar{1}$.90380
610	.90452	.90523	.90594	.90665	.90735	.90806	.90877	.90947	.91017	.91088
620	.91158	.91228	.91298	.91367	.91437	.91507	.91576	.91645	.91715	.91784
630	.91853	.91922	.91990	.92059	.92128	.92196	.92264	.92333	.92401	.92469
640	.92537	.92604	.92672	.92740	.92807	.92875	.92942	.93009	.93076	.93143
650	$\bar{1}$.93210	$\bar{1}$.93277	$\bar{1}$.93343	$\bar{1}$.93410	$\bar{1}$.93476	$\bar{1}$.93543	$\bar{1}$.93609	$\bar{1}$.93675	$\bar{1}$.93741	$\bar{1}$.93807
660	.93873	.93939	.94004	.94070	.94135	.94201	.94266	.94331	.94396	.94461
670	.94526	.94591	.94656	.94720	.94785	.94849	.94913	.94978	.95042	.95106
680	.95170	.95233	.95297	.95361	.95424	.95488	.95551	.95614	.95677	.95741
690	.95804	.95866	.95929	.95992	.96055	.96117	.96180	.96242	.96304	.96366
700	$\bar{1}$.96428	$\bar{1}$.96490	$\bar{1}$.96552	$\bar{1}$.96614	$\bar{1}$.96676	$\bar{1}$.96738	$\bar{1}$.96799	$\bar{1}$.96861	$\bar{1}$.96922	$\bar{1}$.96983
710	.97044	.97106	.97167	.97228	.97288	.97349	.97410	.97471	.97531	.97592
720	.97652	.97712	.97772	.97832	.97892	.97951	.98012	.98072	.98132	.98191
730	.98251	.98310	.98370	.98429	.98488	.98547	.98606	.98665	.98724	.98783
740	.98842	.98900	.98959	.99018	.99076	.99134	.99193	.99251	.99309	.99367
750	$\bar{1}$.99425	$\bar{1}$.99483	$\bar{1}$.99540	$\bar{1}$.99598	$\bar{1}$.99656	$\bar{1}$.99713	$\bar{1}$.99771	$\bar{1}$.99828	$\bar{1}$.99886	$\bar{1}$.99942
760	0.00000	0.00057	0.00114	0.00171	0.00228	0.00285	0.00342	0.00398	0.00455	0.00511
770	.00568	.00624	.00680	.00737	.00793	.00849	.00905	.00961	.01017	.01072
780	.01128	.01184	.01239	.01295	.01350	.01406	.01461	.01516	.01571	.01626
790	.01681	.01736	.01791	.01846	.01901	.01955	.02010	.02064	.02119	.02173

TABLE 115. — Values of 0.378*e*.*

This table gives the humidity term 0.378*e*, which occurs in the equation $\delta = \delta_0 \frac{h}{760} = \delta_0 \frac{B - 0.378e}{760}$ for the calculation of the density of air containing aqueous vapor at pressure *e*; δ_0 is the density of dry air at normal temperature and barometric pressure, *B* the observed barometric pressure, and $h = B - 0.378e$, the pressure corrected for humidity. For values of $\frac{h}{760}$, see Table 113. Temperatures are in degrees Centigrade, and pressures in millimeters of mercury.

Dew point.	*e* Vapor pressure (ice).	0.378*e*	Dew point.	*e* Vapor pressure (water).	0.378*e*	Dew point.	*e* Vapor pressure (water).	0.378*e*
C	mm	mm	C	mm	mm	C	mm	mm
−50°	0.029	0.01	**0°**	4.58	1.73	**30°**	31.86	12.0
−45	0.054	0.02	1	4.92	1.86	31	33.74	12.8
−40	0.096	0.04	2	5.29	2.00	32	35.70	13.5
−35	0.169	0.06	3	5.68	2.15	33	37.78	14.3
−30	0.288	0.11	4	6.10	2.31	34	39.95	15.1
−25	0.480	0.18	**5**	6.54	2.47	**35**	42.23	16.0
24	0.530	0.20	6	7.01	2.66	36	44.62	16.9
23	0.585	0.22	7	7.51	2.84	37	47.13	17.8
22	0.646	0.24	8	8.04	3.04	38	49.76	18.8
21	0.712	0.27	9	8.61	3.25	39	52.51	19.8
−20	0.783	0.30	**10**	9.21	3.48	**40**	55.40	20.9
19	0.862	0.33	11	9.85	3.72	41	58.42	22.1
18	0.947	0.36	12	10.52	3.98	42	61.58	23.3
17	1.041	0.39	13	11.24	4.25	43	64.89	24.5
16	1.142	0.43	14	11.99	4.53	44	68.35	25.8
−15	1.252	0.47	**15**	12.79	4.84	**45**	71.97	27.2
14	1.373	0.52	16	13.64	5.16	46	75.75	28.6
13	1.503	0.57	17	14.54	5.50	47	79.70	30.1
12	1.644	0.62	18	15.49	5.85	48	83.83	31.7
11	1.798	0.68	19	16.49	6.23	49	88.14	33.3
−10	1.964	0.74	**20**	17.55	6.63	**50**	92.6	35.0
9	2.144	0.81	21	18.66	7.06	51	97.3	36.8
8	2.340	0.88	22	19.84	7.50	52	102.2	38.6
7	2.550	0.96	23	21.09	7.97	53	107.3	40.6
6	2.778	1.05	24	22.40	8.47	54	112.7	42.6
−5	3.025	1.14	**25**	23.78	8.99	**55**	118.2	44.7
4	3.291	1.24	26	25.24	9.54	56	124.0	46.9
3	3.578	1.35	27	26.77	10.12	57	130.0	49.1
2	3.887	1.47	28	28.38	10.73	58	136.3	51.5
1	4.220	1.60	29	30.08	11.37	59	142.8	54.0
0	4.580	1.73	**30**	31.86	12.04	**60**	149.6	56.5

* Table quoted from Smithsonian Meteorological Tables.

TABLE 116. — Maintenance of Air at Definite Humidities.

Taken from Stevens, Phytopathology, 6, 428, 1916; see also Curtis, Bul. Bur. Standards, 11, 359, 1914; Dieterici, Ann. d. Phys. u. Chem., 50, 47, 1893. The relative humidity and vapor pressure of aqueous vapor of moist air in equilibrium conditions above aqueous solutions of sulphuric acid are given below.

Density of acid sol.	Relative humidity.	Vapor pressure.		Density of acid sol.	Relative humidity.	Vapor pressure.	
		20° C	30° C			20° C	30° C
		mm	mm			mm	mm
1.00	100.0	17.4	31.6	1.30	58.3	10.1	18.4
1.05	97.5	17.0	30.7	1.35	47.2	8.3	15.0
1.10	93.9	16.3	29.6	1.40	37.1	6.5	11.9
1.15	88.8	15.4	28.0	1.50	18.8	3.3	6.0
1.20	80.5	14.0	25.4	1.60	8.5	1.5	2.7
1.25	70.4	12.2	22.2	1.70	3.2	0.6	1.0

TABLE 117.

PRESSURE OF COLUMNS OF MERCURY AND WATER.

British and metric measures. Correct at 0° C. for mercury and at 4° C. for water.

Metric Measure.			British Measure.		
Cms. of Hg.	Pressure in grams per sq. cm.	Pressure in pounds per sq. inch.	Inches of Hg.	Pressure in grams per sq. cm.	Pressure in pounds per sq. inch.
1	13.5956	0.193376	1	34.533	0.491174
2	27.1912	0.386752	2	69.066	0.982348
3	40.7868	0.580128	3	103.598	1.473522
4	54.3824	0.773504	4	138.131	1.964696
5	67.9780	0.966880	5	172.664	2.455870
6	81.5736	1.160256	6	207.197	2.947044
7	95.1692	1.353632	7	241.730	3.438218
8	108.7648	1.547008	8	276.262	3.929392
9	122.3604	1.740384	9	310.795	4.420566
10	135.9560	1.933760	10	345.328	4.911740
Cms. of H_2O.	Pressure in grams per sq. cm.	Pressure in pounds per sq. inch.	Inches of H_2O.	Pressure in grams per sq. cm.	Pressure in pounds per sq. inch.
1	1	0.0142234	1	2.54	0.036127
2	2	0.0284468	2	5.08	0.072255
3	3	0.0426702	3	7.62	0.108382
4	4	0.0568936	4	10.16	0.144510
5	5	0.0711170	5	12.70	0.180637
6	6	0.0853404	6	15.24	0.216764
7	7	0.0995638	7	17.78	0.252892
8	8	0.1137872	8	20.32	0.289019
9	9	0.1280106	9	22.86	0.325147
10	10	0.1422340	10	25.40	0.361274

REDUCTION OF BAROMETRIC HEIGHT TO STANDARD TEMPERATURE.*

Corrections for brass scale and English measure.	
Height of barometer in inches.	α in inches for temp. F.
15.0	0.00135
16.0	.00145
17.0	.00154
17.5	.00158
18.0	.00163
18.5	.00167
19.0	.00172
19.5	.00176
20.0	0.00181
20.5	.00185
21.0	.00190
21.5	.00194
22.0	.00199
22.5	.00203
23.0	.00208
23.5	.00212
24.0	0.00217
24.5	.00221
25.0	.00226
25.5	.00231
26.0	.00236
26.5	.00240
27.0	.00245
27.5	.00249
28.0	0.00254
28.5	.00258
29.0	.00263
29.2	.00265
29.4	.00267
29.6	.00268
29.8	.00270
30.0	.00272
30.2	0.00274
30.4	.00276
30.6	.00277
30.8	.00279
31.0	.00281
31.2	.00283
31.4	.00285
31.6	.00287

Corrections for brass scale and metric measure.	
Height of barometer in mm.	α in mm. for temp. C.
400	0.0651
410	.0668
420	.0684
430	.0700
440	.0716
450	.0732
460	.0749
470	.0765
480	.0781
490	.0797
500	0.0813
510	.0830
520	.0846
530	.0862
540	.0878
550	.0894
560	.0911
570	.0927
580	.0943
590	.0959
600	0.0975
610	.0992
620	.1008
630	.1024
640	.1040
650	.1056
660	.1073
670	.1089
680	.1105
690	.1121
700	0.1137
710	.1154
720	.1170
730	.1186
740	.1202
750	.1218
760	.1235
770	.1251
780	.1267
790	.1283
800	.1299

Corrections for glass scale and metric measure.	
Height of barometer in mm.	α in mm. for temp. C.
50	0.0086
100	.0172
150	.0258
200	.0345
250	.0431
300	.0517
350	.0603
400	0.0689
450	.0775
500	.0861
520	.0895
540	.0930
560	.0965
580	.0999
600	0.1034
610	.1051
620	.1068
630	.1085
640	.1103
650	.1120
660	.1137
670	0.1154
680	.1172
690	.1189
700	.1206
710	.1223
720	.1240
730	.1258
740	0.1275
750	.1292
760	.1309
770	.1327
780	.1344
790	.1361
800	.1378
850	0.1464
900	.1551
950	.1639
1000	.1723

* The height of the barometer is affected by the relative thermal expansion of the mercury and the glass, in the case of instruments graduated on the glass tube, and by the relative expansion of the mercury and the metallic inclosing case, usually of brass, in the case of instruments graduated on the brass case. This relative expansion is practically proportional to the first power of the temperature. The above tables of values of the coefficient of relative expansion will be found to give corrections almost identical with those given in the International Meteorological Tables. The numbers tabulated under α are the values of α in the equation $H_t = H_{t'} - \alpha(t' - t)$ where H_t is the height at the standard temperature, $H_{t'}$ the observed height at the temperature t', and $\alpha(t' - t)$ the correction for temperature. The standard temperature is 0° C. for the metric system and 28°.5 F. for the English system. The English barometer is correct for the temperature of melting ice at a temperature of approximately 28°.5 F., because of the fact that the brass scale is graduated so as to be standard at 62° F., while mercury has the standard density at 32° F.

EXAMPLE.—A barometer having a brass scale gave $H = 765$ mm. at 25° C.; required, the corresponding reading at 0° C. Here the value of α is the mean of .1235 and .1251, or .1243; $\therefore \alpha(t' - t) = .1243 \times 25 = 3.11$. Hence $H_0 = 765 - 3.11 = 761.89$.

N. B.—Although α is here given to three and sometimes to four significant figures, it is seldom worth while to use more than the nearest two-figure number. In fact, all barometers have not the same values for α, and when great accuracy is wanted the proper coefficients have to be determined by experiment.

TABLE 119.

REDUCTION OF BAROMETER TO STANDARD GRAVITY.

Free-air Altitude Term. Correction to be subtracted.

The correction to reduce the barometer to sea-level is $(g_1 - g)/g \times B$ where B is the barometer reading and g and g_1 the value of gravity at sea-level and the place of observation respectively. The following values were computed for free-air values of gravity g_1 (Table 565). It has been customary to assume for mountain stations that the value of g_1 = say about $\frac{2}{3}$ the free-air value, but a comparison of modern determinations of g_1 in this country shows that little reliance can be placed on such an assumption. Where g_1 is known its value should be used in the above correction term. (See Tables 566 and 567. Similarly for the latitude term, see succeeding tables, the true value of g should be used if known; the succeeding tables are based on the theoretical values, Table 565.)

Height above sea-level.	$g_1 - g$	Observed height of barometer in millimeters.										
		400	450	500	550	600	650	700	750	800		
meters.												
100	0.031	Correction in mm to be subtracted for height above sea-level in first column and barometer reading in the top line.						.02	.02	.02	—	—
200	0.062							.04	.05	.05	—	—
300	0.093							.07	.07	.07	—	—
400	0.123							.09	.10	.10	—	—
500	0.154	—	—	—	—	—	—	.11	.12	.13	—	—
600	0.185	—	—	—	—	—	.12	.13	.14	—	—	—
700	0.216	—	—	—	—	—	.14	.15	.16	—	—	—
800	0.247	—	—	—	—	—	.16	.18	.19	—	—	—
900	0.278	—	—	—	—	—	.18	.20	.22	—	—	—
1000	0.309	—	—	—	.18	.19	.20	.22	.24	—	—	—
1100	0.339	—	—	—	.19	.21	.22	.24	—	—	—	—
1200	0.370	—	—	—	.21	.23	.24	.26	—	—	—	—
1300	0.401	—	—	—	.22	.24	.26	.29	—	—	—	—
1400	0.432	—	—	—	.24	.26	.28	.31	—	—	—	—
1500	0.463	—	—	.24	.26	.28	.30	.33	—	—	—	—
1600	0.494	—	—	.25	.28	.30	.32	—	—	—	—	—
1700	0.525	—	—	.27	.30	.32	.34	—	—	—	—	—
1800	0.555	—	—	.28	.31	.34	.36	—	—	.020	.0463	15000
1900	0.586	—	—	.30	.33	.36	.39	—	—	.019	.0447	14500
2000	0.617	—	.28	.31	.34	.38	.41	—	.021	.019	.0432	14000
2100	0.648	—	.30	.33	.36	.40	—	—	.021	.018	.0416	13500
2200	0.679	—	.31	.35	.38	.41	—	—	.020	.017	.0401	13000
2300	0.710	—	.32	.36	.40	.43	—	.021	.019	.017	.0386	12500
2400	0.740	—	.34	.38	.42	.45	—	.021	.018	.016	.0370	12000
2500	0.771	.31	.35	.39	.43	.47	—	.020	.018	.015	.0355	11500
2600	0.802	.33	.37	.41	—	—	.021	.019	.017	.015	.0339	11000
2700	0.833	.34	.38	.42	—	—	.020	.018	.016	.014	.0324	10500
2800	0.864	.35	.40	.44	—	—	.019	.017	.015	.013	.0308	10000
2900	0.895	.36	.41	.46	—	.020	.018	.016	.015	.013	.0293	9500
3000	0.926	.38	.42	.47	—	.019	.017	.016	.014	.012	.0278	9000
3100	0.957	.39	.44	—	—	.018	.016	.015	.013	—	.0262	8500
3200	0.988	.40	.46	—	—	.017	.015	.014	.012	—	.0247	8000
3300	1.019	.42	.47	—	.017	.016	.014	.013	—	—	.0231	7500
3400	1.049	.43	.48	—	.016	.015	.013	.012	—	—	.0216	7000
3500	1.080	.44	.49	—	.015	.014	.012	.011	—	—	.0200	6500
3600	1.111	.45	—	—	.014	.013	.011	—	—	—	.0185	6000
3700	1.142	.46	—	—	.013	.012	.011	—	—	—	.0170	5500
3800	1.173	.48	—	.012	.011	.011	.010	—	—	—	.0154	5000
3900	1.204	.49	—	.011	.010	.010	—	—	—	—	.0139	4500
4000	1.235	.50	—	.010	.009	.009	—	—	—	—	.0123	4000
—	—	—	.008	.008	.007	.007	Corrections in in. to be subtracted for height above sea-level in last column and barometer reading in bottom line.				.0092	3000
—	—	.006	.005	.005	.004	—					.0062	2000
—	—	.003	.003	.003	—	—					.0031	1000
												feet.
		30	28	26	24	22	20	18	16	14	$g_1 - g$	Height above sea-level.
		Observed height of barometer in inches.										

TABLE 120.

REDUCTION OF BAROMETER TO STANDARD GRAVITY.*

METRIC MEASURES.

From Latitude 0° to 45°, the Correction is to be Subtracted.

Latitude.	520	540	560	580	600	620	640	660	680	700	720	740	760	780
	mm.	mm.	mm.	mm.	mm.	mm.	mm.	mm.	mm.	mm.	mm.	mm.	mm.	mm.
0	−1.39	−1.45	−1.50	−1.55	−1.61	−1.66	−1.71	−1.77	−1.82	−1.87	−1.93	−1.98	−2.04	−2.09
5	−1.37	−1.42	−1.48	−1.53	−1.58	−1.64	−1.69	−1.74	−1.79	−1.85	−1.90	−1.95	−2.00	−2.06
6	1.36	1.42	1.47	1.52	1.57	1.63	1.68	1.73	1.78	1.83	1.89	1.94	1.99	2.04
7	1.35	1.40	1.46	1.51	1.56	1.61	1.66	1.72	1.77	1.82	1.87	1.92	1.98	2.03
8	1.34	1.39	1.44	1.49	1.55	1.60	1.65	1.70	1.75	1.80	1.85	1.91	1.96	2.01
9	1.33	1.38	1.43	1.48	1.53	1.58	1.63	1.68	1.73	1.78	1.84	1.89	1.94	1.99
10	−1.31	−1.36	−1.41	−1.46	−1.51	−1.56	−1.61	−1.66	−1.71	−1.76	−1.81	−1.86	−1.92	−1.97
11	1.29	1.34	1.39	1.44	1.49	1.54	1.59	1.64	1.69	1.74	1.79	1.84	1.89	1.94
12	1.27	1.32	1.37	1.42	1.47	1.52	1.57	1.62	1.67	1.72	1.76	1.81	1.86	1.91
13	1.25	1.30	1.35	1.40	1.45	1.50	1.54	1.59	1.64	1.69	1.74	1.78	1.83	1.88
14	1.23	1.28	1.33	1.38	1.42	1.47	1.52	1.56	1.61	1.66	1.71	1.75	1.80	1.85
15	−1.21	−1.26	−1.30	−1.35	−1.40	−1.44	−1.49	−1.54	−1.58	−1.63	−1.67	−1.72	−1.77	−1.81
16	1.19	1.23	1.28	1.32	1.37	1.41	1.46	1.50	1.55	1.60	1.64	1.69	1.73	1.78
17	1.16	1.20	1.25	1.29	1.34	1.38	1.43	1.47	1.52	1.56	1.60	1.65	1.69	1.74
18	1.13	1.18	1.22	1.26	1.31	1.35	1.39	1.44	1.48	1.52	1.57	1.61	1.65	1.70
19	1.10	1.15	1.19	1.23	1.27	1.32	1.36	1.40	1.44	1.48	1.53	1.57	1.61	1.65
20	−1.07	−1.11	−1.16	−1.20	−1.24	−1.28	−1.32	−1.36	−1.40	−1.44	−1.49	−1.53	−1.57	−1.61
21	1.04	1.08	1.12	1.16	1.20	1.24	1.28	1.32	1.36	1.40	1.44	1.48	1.52	1.56
22	1.01	1.05	1.09	1.13	1.16	1.20	1.24	1.28	1.32	1.36	1.40	1.44	1.48	1.51
23	0.98	1.01	1.05	1.09	1.13	1.16	1.20	1.24	1.28	1.31	1.35	1.39	1.43	1.46
24	0.94	0.98	1.01	1.05	1.08	1.12	1.16	1.19	1.23	1.27	1.30	1.34	1.37	1.41
25	−0.90	−0.94	−0.97	−1.01	−1.04	−1.08	−1.11	−1.15	−1.18	−1.22	−1.25	−1.29	−1.32	−1.36
26	0.87	0.90	0.93	0.97	1.00	1.03	1.07	1.10	1.13	1.17	1.20	1.23	1.27	1.30
27	0.83	0.86	0.89	0.92	0.96	0.99	1.02	1.05	1.08	1.12	1.15	1.18	1.21	1.24
28	0.79	0.82	0.85	0.88	0.91	0.94	0.97	1.00	1.03	1.06	1.09	1.12	1.15	1.18
29	0.75	0.78	0.81	0.84	0.86	0.89	0.92	0.95	0.98	1.01	1.04	1.07	1.10	1.12
30	−0.71	−0.74	−0.76	−0.79	−0.82	−0.85	−0.87	−0.90	−0.93	−0.95	−0.98	−1.01	−1.04	−1.06
31	0.67	0.69	0.72	0.74	0.77	0.80	0.82	0.85	0.87	0.90	0.92	0.95	0.98	1.00
32	0.62	0.65	0.67	0.70	0.72	9.74	0.77	0.79	0.82	0.84	0.86	0.89	0.91	0.94
33	0.58	0.60	0.63	0.65	0.67	0.69	0.72	0.74	0.76	0.78	0.80	0.83	0.85	0.87
34	0.54	0.56	0.58	0.60	0.62	0.64	0.66	0.68	0.70	0.72	0.74	0.76	0.79	0.81
35	−0.49	−0.51	−0.53	−0.55	−0.57	−0.59	−0.61	−0.63	−0.64	−0.66	−0.68	−0.70	−0.72	−0.74
36	0.45	0.46	0.48	0.50	0.52	0.53	0.55	0.57	0.58	0.60	0.62	0.64	0.65	0.67
37	0.40	0.42	0.43	0.45	0.46	0.48	0.49	0.51	0.52	0.54	0.56	0.57	0.59	0.60
38	0.36	0.37	0.38	0.40	0.41	0.42	0.44	0.45	0.46	0.48	0.49	0.51	0.52	0.53
39	0.31	0.32	0.33	0.34	0.36	0.37	0.38	0.39	0.40	0.42	0.43	0.44	0.45	0.46
40	−0.26	−0.27	−0.28	−0.29	−0.30	−0.31	−0.32	−0.33	−0.34	−0.35	−0.36	−0.37	−0.38	−0.39
41	0.21	0.22	0.23	0.24	0.25	0.26	0.26	0.27	0.28	0.29	0.30	0.30	0.31	0.32
42	0.17	0.17	0.18	0.19	0.19	0.20	0.21	0.21	0.22	0.22	0.23	0.24	0.24	0.25
43	0.12	0.12	0.13	0.13	0.14	0.14	0.15	0.15	0.16	0.16	0.16	0.17	0.17	0.18
44	0.07	0.07	0.08	0.08	0.08	0.08	0.09	0.09	0.09	0.10	0.10	0.10	0.10	0.11
45	−0.02	−0.02	−0.03	−0.03	−0.03	−0.03	−0.03	−0.03	−0.03	−0.03	−0.03	−0.03	−0.03	−0.04

* "Smithsonian Meteorological Tables."

TABLE 121.

REDUCTION OF BAROMETER TO STANDARD GRAVITY.*

METRIC MEASURES.

From Latitude 46° to 90°, the Correction is to be Added.

Latitude.	520	540	560	580	600	620	640	660	680	700	720	740	760	780
	mm.	mm.	mm.	mm.	mm.	mm.	mm.	mm.	mm.	mm.	mm.	mm.	mm.	mm.
45	−0.02	−0.02	−0.03	−0.03	−0.03	−0.03	−0.03	−0.03	−0.03	−0.03	−0.03	−0.03	−0.03	−0.04
46	+0.02	+0.03	+0.03	+0.03	+0.03	+0.03	+0.03	+0.03	+0.03	+0.03	+0.03	+0.03	+0.04	+0.04
47	0.07	0.08	0.08	0.08	0.08	0.09	0.09	0.09	0.09	0.10	0.10	0.10	0.10	0.11
48	0.12	0.12	0.13	0.13	0.14	0.14	0.15	0.15	0.16	0.16	0.17	0.17	0.18	0.18
49	0.17	0.17	0.18	0.19	0.19	0.20	0.21	0.21	0.22	0.23	0.23	0.24	0.25	0.25
50	0.22	0.22	0.23	0.24	0.25	0.26	0.26	0.27	0.28	0.29	0.30	0.31	0.31	0.32
51	+0.26	+0.27	+0.28	+0.29	+0.30	+0.31	+0.32	+0.33	+0.34	+0.35	+0.36	+0.37	+0.38	+0.39
52	0.31	0.32	0.33	0.34	0.36	0.37	0.38	0.39	0.40	0.42	0.43	0.44	0.45	0.46
53	0.36	0.37	0.38	0.40	0.41	0.42	0.44	0.45	0.46	0.48	0.49	0.51	0.52	0.53
54	0.40	0.42	0.43	0.45	0.46	0.48	0.49	0.51	0.52	0.54	0.56	.057	0.59	0.60
55	0.45	0.46	0.48	0.50	0.52	0.53	0.55	0.57	0.58	0.60	0.62	0.64	0.65	0.67
56	+0.49	+0.51	+0.53	+0.55	+0.57	+0.59	+0.60	+0.62	+0.64	+0.66	+0.68	+0.70	+0.72	+0.74
57	0.54	0.56	0.58	0.60	0.62	0.64	0.66	0.68	0.70	0.72	0.74	0.76	0.78	0.80
58	0.58	0.60	0.62	0.65	0.67	0.69	0.71	0.74	0.76	0.78	0.80	0.82	0.85	0.87
59	0.62	0.65	0.67	0.69	0.72	0.74	0.77	0.79	0.81	0.84	0.86	0.89	0.91	0.93
60	0.66	0.69	0.72	0.74	0.77	0.79	0.82	0.84	0.87	0.89	0.92	0.94	0.97	1.00
61	+0.71	+0.73	+0.76	+0.79	+0.81	+0.84	+0.87	+0.89	+0.92	+0.95	+0.98	+1.00	+1.03	+1.06
62	0.74	0.77	0.80	0.83	0.85	0.88	0.91	0.94	0.97	1.00	1.02	1.05	1.08	1.11
63	0.78	0.81	0.85	0.88	0.91	0.94	0.97	1.00	1.03	1.06	1.09	1.12	1.15	1.18
64	0.82	0.85	0.89	0.92	0.95	0.98	1.01	1.04	1.08	1.11	1.14	1.17	1.20	1.23
65	0.86	0.89	0.93	0.96	0.99	1.03	1.06	1.09	1.13	1.16	1.19	1.22	1.26	1.29
66	+0.90	+0.93	+0.97	+1.00	+1.04	+1.07	+1.10	+1.14	+1.17	+1.21	+1.24	+1.28	+1.31	+1.35
67	0.93	0.97	1.00	1.04	1.08	1.11	1.15	1.18	1.22	1.25	1.29	1.33	1.36	1.40
68	0.97	1.00	1.04	1.08	1.11	1.15	1.19	1.23	1.26	1.30	1.34	1.37	1.41	1.45
69	1.00	1.04	1.08	1.11	1.15	1.19	1.23	1.27	1.31	1.34	1.38	1.42	1.46	1.50
70	1.03	1.07	1.11	1.15	1.19	1.23	1.27	1.31	1.35	1.39	1.43	1.47	1.51	1.55
71	+1.06	+1.10	+1.14	+1.18	+1.22	+1.26	+1.31	+1.35	+1.39	+1.43	+1.47	+1.51	+1.55	+1.59
72	1.09	1.13	1.17	1.22	1.26	1.30	1.34	1.38	1.42	1.47	1.51	1.55	1.59	1.63
73	1.12	1.16	1.20	1.25	1.29	1.33	1.37	1.42	1.46	1.50	1.55	1.59	1.63	1.67
74	1.14	1.19	1.23	1.28	1.32	1.36	1.41	1.45	1.50	1.54	1.58	1.63	1.67	1.72
75	1.17	1.21	1.26	1.30	1.35	1.39	1.44	1.48	1.53	1.57	1.62	1.66	1.71	1.75
76	+1.19	+1.24	+1.28	+1.33	+1.37	+1.42	+1.47	+1.51	+1.56	+1.60	+1.65	+1.70	+1.74	+1.79
77	1.21	1.26	1.31	1.35	1.40	1.45	1.49	1.54	1.59	1.63	1.68	1.73	1.77	1.82
78	1.23	1.28	1.33	1.38	1.42	1.47	1.52	1.57	1.61	1.66	1.71	1.76	1.80	1.85
79	1.25	1.30	1.35	1.40	1.45	1.49	1.54	1.59	1.64	1.69	1.73	1.78	1.83	1.88
80	1.27	1.32	1.37	1.42	1.47	1.51	1.56	1.61	1.66	1.71	1.76	1.81	1.86	1.90
81	+1.29	+1.33	+1.38	+1.43	+1.48	+1.53	+1.58	+1.63	+1.68	+1.73	+1.78	+1.83	+1.88	+1.93
82	1.30	1.35	1.40	1.45	1.50	1.55	1.60	1.65	1.70	1.75	1.80	1.85	1.90	1.95
83	1.31	1.36	1.41	1.46	1.51	1.56	1.61	1.67	1.72	1.77	1.82	1.87	1.92	1.97
84	1.32	1.37	1.42	1.48	1.53	1.58	1.63	1.68	1.73	1.78	1.83	1.88	1.93	1.98
85	1.33	1.38	1.43	1.49	1.54	1.59	1.64	1.69	1.74	1.79	1.84	1.90	1.95	2.00
90	+1.35	+1.41	+1.46	+1.51	+1.56	+1.61	+1.67	+1.72	+1.77	+1.82	+1.87	+1.93	+1.98	+2.03

* "Smithsonian Meteorological Tables."

REDUCTION OF BAROMETER TO STANDARD GRAVITY.*

ENGLISH MEASURES.

From Latitude 0° to 45°, the Correction is to be Subtracted.

Latitude.	19	20	21	22	23	24	25	26	27	28	29	30
	Inch.	Inch.	Inch.	Inch.	Inch.	Inch.	Inch.	Inch.	Inch.	Inch.	Inch.	Inch.
0	-0.051	-0.054	-0.056	-0.059	-0.062	-0.064	-0.067	-0.070	-0.072	-0.075	-0.078	-0.080
5	-0.050	-0.053	-0.055	-0.058	-0.061	-0.063	-0.066	-0.069	-0.071	-0.074	-0.077	-0.079
6	0.050	0.052	0.055	0.058	0.060	0.063	0.066	0.068	0.071	0.073	0.076	0.079
7	0.049	0.052	0.055	0.057	0.060	0.062	0.065	0.068	0.070	0.073	0.075	0.078
8	0.049	0.052	0.054	0.057	0.059	0.062	0.064	0.067	0.070	0.072	0.075	0.077
9	0.048	0.051	0.054	0.056	0.059	0.061	0.064	0.066	0.069	0.071	0.074	0.076
10	-0.048	-0.050	-0.053	-0.055	-0.058	-0.060	-0.063	-0.066	-0.068	-0.071	-0.073	-0.076
11	0.047	0.050	0.052	0.055	0.057	0.060	0.062	0.065	0.067	0.070	0.072	0.075
12	0.047	0.049	0.051	0.054	0.056	0.059	0.061	0.064	0.066	0.069	0.071	0.074
13	0.046	0.048	0.051	0.053	0.055	0.058	0.060	0.063	0.065	0.068	0.070	0.072
14	0.045	0.047	0.050	0.052	0.055	0.057	0.059	0.062	0.064	0.066	0.069	0.071
15	-0.044	-0.047	-0.049	-0.051	-0.053	-0.056	-0.058	-0.060	-0.063	-0.065	-0.067	-0.070
16	0.043	0.046	0.048	0.050	0.052	0.055	0.057	0.059	0.062	0.064	0.066	0.068
17	0.042	0.045	0.047	0.049	0.051	0.053	0.056	0.058	0.060	0.062	0.065	0.067
18	0.041	0.044	0.046	0.048	0.050	0.052	0.054	0.057	0.059	0.061	0.063	0.065
19	0.040	0.042	0.045	0.047	0.049	0.051	0.053	0.055	0.057	0.059	0.062	0.064
20	-0.039	-0.041	-0.043	-0.045	-0.047	-0.050	-0.052	-0.054	-0.056	-0.058	-0.060	-0.062
21	0.038	0.040	0.042	0.044	0.046	0.048	0.050	0.052	0.054	0.056	0.058	0.060
22	0.037	0.039	0.041	0.043	0.045	0.047	0.049	0.050	0.052	0.054	0.056	0.058
23	0.036	0.038	0.039	0.041	0.043	0.045	0.047	0.049	0.051	0.053	0.054	0.056
24	0.034	0.036	0.038	0.040	0.042	0.043	0.045	0.047	0.049	0.051	0.052	0.054
25	-0.033	-0.035	-0.037	-0.038	-0.040	-0.042	-0.043	-0.045	-0.047	-0.049	-0.050	-0.052
26	0.032	0.033	0.035	0.037	0.038	0.040	0.042	0.043	0.045	0.047	0.048	0.050
27	0.030	0.032	0.033	0.035	0.037	0.038	0.040	0.041	0.043	0.045	0.046	0.048
28	0.029	0.030	0.032	0.033	0.035	0.036	0.038	0.039	0.041	0.043	0.044	0.046
29	0.027	0.029	0.030	0.032	0.033	0.035	0.036	0.037	0.039	0.040	0.042	0.043
30	-0.026	-0.027	-0.029	-0.030	-0.031	-0.033	-0.034	-0.035	-0.037	-0.038	-0.040	-0.041
31	0.024	0.026	0.027	0.028	0.030	0.031	0.032	0.033	0.035	0.036	0.037	0.038
32	0.023	0.024	0.025	0.026	0.028	0.029	0.030	0.031	0.032	0.034	0.035	0.036
33	0.021	0.022	0.023	0.025	0.026	0.027	0.028	0.029	0.030	0.031	0.032	0.034
34	0.020	0.021	0.022	0.023	0.024	0.025	0.026	0.027	0.028	0.029	0.030	0.031
35	-0.018	-0.019	-0.020	-0.021	-0.022	-0.023	-0.024	-0.025	-0.026	-0.027	-0.027	-0.028
36	0.016	0.017	0.018	0.019	0.020	0.021	0.022	0.022	0.023	0.024	0.025	0.026
37	0.015	0.015	0.016	0.017	0.018	0.019	0.019	0.020	0.021	0.022	0.022	0.023
38	0.013	0.014	0.014	0.015	0.016	0.016	0.017	0.018	0.018	0.019	0.020	0.020
39	0.011	0.012	0.012	0.013	0.014	0.014	0.015	0.015	0.016	0.017	0.017	0.018
40	-0.010	-0.010	-0.011	-0.011	-0.012	-0.012	-0.013	-0.013	-0.014	-0.014	-0.015	-0.015
41	0.008	0.008	0.009	0.009	0.009	0.010	0.010	0.011	0.011	0.012	0.012	0.012
42	0.006	0.006	0.007	0.007	0.007	0.008	0.008	0.008	0.009	0.009	0.009	0.010
43	0.004	0.005	0.005	0.005	0.005	0.005	0.006	0.006	0.006	0.006	0.007	0.007
44	0.003	0.003	0.003	0.003	0.003	0.003	0.003	0.004	0.004	0.004	0.004	0.004
45	-0.001	-0.001	-0.001	-0.001	-0.001	-0.001	-0.001	-0.001	-0.001	-0.001	-0.001	-0.001

* "Smithsonian Meteorological Tables."

REDUCTION OF BAROMETER TO STANDARD GRAVITY.*

ENGLISH MEASURES.

From Latitude 46° to 90° the Correction is to be Added.

Latitude.	19	20	21	22	23	24	25	26	27	28	29	30
	Inch.	Inch.	Inch.	Inch.	Inch.	Inch.	Inch.	Inch.	Inch.	Inch.	Inch.	Inch.
45	−0.001	−0.001	−0.001	−0.001	−0.001	−0.001	−0.001	−0.001	−0.001	−0.001	−0.001	−0.001
46	+0.001	+0.001	+0.001	+0.001	+0.001	+0.001	+0.001	+0.001	+0.001	+0.001	+0.001	+0.001
47	0.003	0.003	0.003	0.003	0.003	0.003	0.003	0.004	0.004	0.004	0.004	0.004
48	0.004	0.005	0.005	0.005	0.005	0.006	0.006	0.006	0.006	0.006	0.007	0.007
49	0.006	0.006	0.007	0.007	0.007	0.008	0.008	0.008	0.009	0.009	0.009	0.010
50	0.008	0.008	0.009	0.009	0.010	0.010	0.010	0.011	0.011	0.012	0.012	0.012
51	+0.010	+0.010	+0.011	+0.011	+0.012	+0.012	+0.013	+0.013	+0.014	+0.014	+0.015	+0.015
52	0.011	0.012	0.012	0.013	0.014	0.014	0.015	0.015	0.016	0.016	0.017	0.018
53	0.013	0.014	0.014	0.015	0.016	0.016	0.017	0.018	0.018	0.019	0.020	0.020
54	0.015	0.015	0.016	0.017	0.018	0.019	0.019	0.020	0.021	0.022	0.022	0.023
55	0.016	0.017	0.018	0.019	0.020	0.021	0.021	0.022	0.023	0.024	0.025	0.026
56	+0.018	+0.019	+0.020	+0.021	+0.022	+0.023	+0.024	+0.024	+0.026	+0.026	+0.027	+0.028
57	0.020	0.021	0.022	0.023	0.024	0.025	0.026	0.027	0.028	0.029	0.030	0.031
58	0.021	0.022	0.023	0.025	0.026	0.027	0.028	0.029	0.030	0.031	0.032	0.033
59	0.023	0.024	0.025	0.026	0.028	0.029	0.030	0.031	0.032	0.033	0.035	0.036
60	0.024	0.026	0.027	0.028	0.029	0.031	0.032	0.033	0.034	0.036	0.037	0.038
61	+0.026	+0.027	+0.028	+0.030	+0.031	+0.033	+0.034	+0.035	+0.037	+0.038	+0.039	+0.041
62	0.027	0.029	0.030	0.032	0.033	0.034	0.036	0.037	0.039	0.040	0.042	0.043
63	0.029	0.030	0.032	0.033	0.035	0.036	0.038	0.039	0.041	0.042	0.044	0.045
64	0.030	0.032	0.033	0.035	0.036	0.038	0.040	0.041	0.043	0.044	0.046	0.047
65	0.031	0.033	0.035	0.036	0.038	0.040	0.041	0.043	0.045	0.046	0.048	0.050
66	+0.033	+0.034	+0.036	+0.038	+0.040	+0.041	+0.043	+0.045	+0.047	+0.048	+0.050	+0.052
67	0.034	0.036	0.038	0.039	0.041	0.043	0.045	0.047	0.048	0.050	0.052	0.054
68	0.035	0.037	0.039	0.041	0.043	0.045	0.046	0.048	0.050	0.052	0.054	0.056
69	0.036	0.038	0.040	0.042	0.044	0.046	0.048	0.050	0.052	0.054	0.056	0.058
70	0.038	0.040	0.042	0.044	0.046	0.048	0.050	0.052	0.053	0.055	0.057	0.059
71	+0.039	+0.041	+0.043	+0.045	+0.047	+0.049	+0.051	+0.053	+0.055	+0.057	+0.059	+0.061
72	0.040	0.042	0.044	0.046	0.048	0.050	0.052	0.054	0.057	0.059	0.061	0.063
73	0.041	9.043	0.045	0.047	0.049	0.052	0.054	0.056	0.058	0.060	0.062	0.064
74	0.042	0.044	0.046	0.048	0.051	0.053	0.055	0.057	0.059	0.062	0.064	0.066
75	0.043	0.045	0.047	0.049	0.052	0.054	0.056	0.058	0.061	0.063	0.065	0.067
76	+0.044	+0.046	+0.048	+0.050	+0.053	+0.055	+0.057	+0.060	+0.062	+0.064	0.066	0.069
77	0.044	0.047	0.049	0.051	0.054	0.056	0.058	0.061	0.063	0.065	0.068	0.070
78	0.045	0.047	0.050	0.052	0.055	0.057	0.059	0.062	0.064	0.066	0.069	0.071
79	0.046	0.048	0.051	0.053	0.055	0.058	0.060	0.063	0.065	0.067	0.070	0.072
80	0.046	0.049	0.051	0.054	0.056	0.059	0.061	0.063	0.066	0.068	0.071	0.073
81	+0.047	+0.049	+0.052	+0.054	+0.057	+0.059	+0.062	+0.064	+0.067	+0.069	+0.072	+0.074
82	0.047	0.050	0.052	0.055	0.057	0.060	0.062	0.065	0.067	0.070	0.072	0.075
83	0.048	0.050	0.053	0.056	0.058	0.061	0.063	0.066	0.068	0.071	0.073	0.076
84	0.048	0.051	0.053	0.056	0.059	0.061	0.064	0.066	0.069	0.071	0.074	0.076
85	0.049	0.051	0.054	0.056	0.059	0.061	0.064	0.067	0.069	0.072	0.074	0.077
90	+0.049	+0.052	+0.055	+0.057	+0.060	+0.062	+0.065	+0.068	+0.070	+0.073	+0.075	+0.078

* "Smithsonian Meteorological Tables."

TABLE 124. — Correction of the Barometer for Capillarity.*

1. Metric Measure.

Diameter of tube in mm.	Height of Meniscus in Millimeters.							
	0.4	0.6	0.8	1.0	1.2	1.4	1.6	1.8
	Correction to be added in millimeters.							
4	0.83	1.22	1.54	1.98	2.37	–	–	–
5	.47	0.65	0.86	1.19	1.45	1.80	–	–
6	.27	.41	.56	0.78	0.98	1.21	1.43	–
7	.18	.28	.40	.53	.67	0.82	0.97	1.13
8	–	.20	.29	.38	.46	.56	.65	0.77
9	–	.15	.21	.28	.33	.40	.46	.52
10	–	–	.15	.20	.25	.29	.33	.37
11	–	–	.10	.14	.18	.21	.24	.27
12	–	–	.07	.10	.13	.15	.18	.19
13	–	–	.04	.07	.10	.12	.13	.14

2. British Measure.

Diameter of tube in inches.	Height of Meniscus in Inches.							
	.01	.02	.03	.04	.05	.06	.07	.08
	Correction to be added in inches.							
.15	0.024	0.047	0.069	0.092	0.116	–	–	–
.20	.011	.022	.033	.045	.059	0.078	–	–
.25	.006	.012	.019	.028	.037	.047	0.059	–
.30	.004	.008	.013	.018	.023	.029	.035	0.042
.35	–	.005	.008	.012	.015	.018	.022	.026
.40	–	.004	.006	.008	.010	.012	.014	.016
.45	–	–	.003	.005	.007	.008	.010	.012
.50	–	–	.002	.004	.005	.006	.006	.007
.55	–	–	.001	.002	.003	.004	.005	.005

* The first table is from Kohlrausch (Experimental Physics), and is based on the experiments of Mendelejeff and Gutkowski (Jour. de Phys. Chem. Geo. Petersburg, 1877, or Wied. Beib. 1877). The second table has been calculated from the same data by conversion into inches and graphic interpolation.

TABLE 125. — Volume of Mercury Meniscus in Cu. Mm.

Height of meniscus.	Diameter of tube in mm.										
	14	15	16	17	18	19	20	21	22	23	24
mm.											
1.6	157	185	214	245	280	318	356	398	444	492	541
1.8	181	211	244	281	320	362	407	455	507	560	616
2.0	206	240	278	319	362	409	460	513	571	631	694
2.2	233	271	313	358	406	459	515	574	637	704	776
2.4	262	303	350	400	454	511	573	639	708	781	859
2.6	291	338	388	444	503	565	633	706	782	862	948

Scheel und Heuse, Annalen der Physik, 33, p. 291, 1910.

BAROMETRIC PRESSURES CORRESPONDING TO THE TEMPERATURE OF THE BOILING POINT OF WATER.

Useful when a boiling-point apparatus is used in the determination of heights. Copied from the Smithsonian Meteorological Tables, 4th revised edition.

(A) METRIC UNITS.

Temperature.	.0	.1	.2	.3	.4	.5	.6	.7	.8	.9
C	mm.	mm.	mm.	mm.	mm.	mm.	mm.	mm.	mm.	mm.
80°	355.40	356.84	358.28	359.73	361.19	362.65	364.11	365.58	367.06	368.54
81	370.03	371.52	373.01	374.51	376.02	377.53	379.05	380.57	382.09	383.62
82	385.16	386.70	388.25	389.80	391.36	392.92	394.49	396.06	397.64	399.22
83	400.81	402.40	404.00	405.61	407.22	408.83	410.45	412.08	413.71	415.35
84	416.99	418.64	420.29	421.95	423.61	425.28	426.95	428.64	430.32	432.01
85	433.71	435.41	437.12	438.83	440.55	442.28	444.01	445.75	447.49	449.24
86	450.99	452.75	454.51	456.28	458.06	459.84	461.63	463.42	465.22	467.03
87	468.84	470.66	472.48	474.31	476.14	477.99	479.83	481.68	483.54	485.41
88	487.28	489.16	491.04	492.93	494.82	496.72	498.63	500.54	502.46	504.39
89	506.32	508.26	510.20	512.15	514.11	516.07	518.04	520.01	521.99	523.98
90	525.97	527.97	529.98	531.99	534.01	536.04	538.07	540.11	542.15	544.21
91	546.26	548.33	550.40	552.48	554.56	556.65	558.75	560.85	562.96	565.08
92	567.20	569.33	571.47	573.61	575.76	577.92	580.08	582.25	584.43	585.61
93	588.80	591.00	593.20	595.41	597.63	599.86	602.09	604.33	606.57	608.82
94	611.08	613.35	615.62	617.90	620.19	622.48	624.79	627.09	629.41	631.73
95	634.06	636.40	638.74	641.09	643.45	645.82	648.19	650.57	652.96	655.35
96	657.75	660.16	662.58	665.00	667.43	669.87	672.32	674.77	677.23	679.70
97	682.18	684.66	687.15	689.65	692.15	694.67	697.19	699.71	702.25	704.79
98	707.35	709.90	712.47	715.04	717.63	720.22	722.81	725.42	728.03	730.65
99	733.28	635.92	738.56	741.21	743.87	740.54	749.22	751.90	754.59	757.29
100	760.00	762.72	765.44	768.17	770.91	773.66	776.42	779.18	781.95	784.73

(B) ENGLISH UNITS.

Temperature.	.0	.1	.2	.3	.4	.5	.6	.7	.8	.9
F.	Inches.	Inches.	Inches.	Inches.	Inches.	Inches.	Inches.	Inches.	Inches.	Inches.
185°	17.075	17.112	17.150	17.187	17.224	17.262	17.300	17.337	17.375	17.413
186	17.450	17.488	17.526	17.564	17.602	17.641	17.679	17.717	17.756	17.794
187	17.832	17.871	17.910	17.948	17.987	18.026	18.065	18.104	18.143	18.182
188	18.221	18.261	18.300	18.340	18.379	18.419	18.458	18.498	18.538	18.578
189	18.618	18.658	18.698	18.738	18.778	18.818	18.859	18.899	18.940	18.980
190	19.021	19.062	19.102	19.143	19.184	19.225	19.266	19.308	19.349	19.390
191	19.431	19.473	19.514	19.556	19.598	19.639	19.681	19.723	19.765	19.807
192	19.849	19.892	19.934	19.976	20.019	20.061	20.104	20.146	20.189	20.232
193	20.275	20.318	20.361	20.404	20.447	20.490	20.533	20.577	20.620	20.664
194	20.707	20.751	20.795	20.839	20.883	20.927	20.971	21.015	21.059	21.103
195	21.148	21.192	21.237	21.282	21.326	21.371	21.416	21.461	21.506	21.551
196	21.597	21.642	21.687	21.733	21.778	21.824	21.870	21.915	21.961	22.007
197	22.053	22.099	22.145	22.192	22.238	22.284	22.331	22.377	22.424	22.471
198	22.517	22.564	22.611	22.658	22.706	22.752	22.800	22.847	22.895	22.942
199	22.990	23.038	23.085	23.133	23.181	23.229	23.277	23.325	23.374	23.422
200	23.470	23.519	23.568	23.616	23.665	23.714	23.763	23.812	23.861	23.910
201	23.959	24.009	24.058	24.108	24.157	24.207	24.257	24.307	24.357	24.407
202	24.457	24.507	24.557	24.608	24.658	24.709	24.759	24.810	24.861	24.912
203	24.963	25.014	25.065	25.116	25.168	25.219	25.271	25.322	25.374	25.426
204	25.478	25.530	25.582	25.634	25.686	25.738	25.791	25.843	25.896	25.948
205	26.001	26.054	26.107	26.160	26.213	26.266	26.319	26.373	26.426	26.480
206	26.534	26.587	26.641	26.695	26.749	26.803	26.857	26.912	26.966	27.021
207	27.075	27.130	27.184	27.239	27.294	27.349	27.404	27.460	27.515	27.570
208	27.626	27.681	27.737	27.793	27.848	27.904	27.960	28.016	28.073	28.129
209	28.185	28.242	28.298	28.355	28.412	28.469	28.526	28.583	28.640	28.697
210	28.754	28.812	28.869	28.927	28.985	29.042	29.100	29.158	29.216	29.275
211	29.333	29.391	29.450	29.508	29.567	29.626	29.685	29.744	29.803	29.862
212	29.921	29.981	30.040	30.100	30.159	30.219	30.279	30.339	30.399	30.459
213	30.519	30.580	30.640	30.701	30.761	30.822	30.883	30.944	31.005	31.066
214	31.127	31.199	31.250	31.311	31.373	31.435	31.497	31.559	31.621	31.683

DETERMINATION OF HEIGHTS BY THE BAROMETER.

Formula of Babinet: $Z = C \frac{B_0 - B}{B_0 + B}$.

C (in feet) $= 52494 \left[1 + \frac{t_0 + t - 64}{900}\right]$ English measures.

C (in meters) $= 16000 \left[1 + \frac{2(t_0 + t)}{1000}\right]$ metric measures.

In which $Z =$ difference of height of two stations in feet or meters.

B_0, $B =$ barometric readings at the lower and upper stations respectively, corrected for all sources of instrumental error.

t_0, $t =$ air temperatures at the lower and upper stations respectively.

Values of *C*.

English Measures.			Metric Measures.		
$\frac{1}{2}(t_0+t)$.	C	Log C	$\frac{1}{2}(t_0+t)$.	C	Log C
Fahr.	Feet.		Cent.	Meters.	
10°	49928	4.69834	−10°	15360	4.18639
15	50511	.70339	−8	15488	.19000
			−6	15616	.19357
20	51094	4.70837	−4	15744	.19712
25	51677	.71330	−2	15872	.20063
30	52261	4.71818	0	16000	4.20412
35	52844	.72300	+2	16128	.20758
			4	16256	.21101
40	53428	4.72777	6	16384	.21442
45	54011	.73248	8	16512	.21780
50	54595	4.73715	10	16640	4.22115
55	55178	.74177	12	16768	.22448
			14	16896	.22778
60	55761	4.74633	16	17024	.23106
65	56344	.75085	18	17152	.23431
70	56927	4.75532	20	17280	4.23754
75	57511	.75975	22	17408	.24075
			24	17536	.24393
80	58094	4.76413	26	17664	.24709
85	58677	.76847	28	17792	.25022
90	59260	4.77276	30	17920	4.25334
95	59844	.77702	32	18048	.25643
			34	18176	.25950
100	60427	4.78123	36	18304	.26255

Values only approximate. Not good for great altitudes. A more accurate formula with corresponding tables may be found in Smithsonian Meteorological Tables.

TABLE 128.

VELOCITY OF SOUND IN SOLIDS.

The velocity of sounds in solids varies as $\sqrt{E/\rho}$, where E is Young's Modulus of elasticity and ρ the density. These constants for most of the materials given in this table vary through a somewhat wide range, and hence the numbers can only be taken as rough approximations to the velocity which may be obtained in any particular case. When temperatures are not marked, between 10° and 20° is to be understood.

Substance.	Temp. C.	Velocity in meters per second.	Velocity in feet per second.	Authority.
Metals: Aluminum	° –	5104	16740	Masson.
Brass	–	3500	11480	Various.
Cadmium	–	2307	7570	Masson.
Cobalt	–	4724	15500	"
Copper	20	3560	11670	Wertheim.
"	100	3290	10800	"
"	200	2950	9690	"
Gold (soft)	20	1743	5717	"
" (hard)	–	2100	6890	Various.
Iron and soft steel	–	5000	16410	"
Iron	20	5130	16820	Wertheim.
"	100	5300	17390	"
"	200	4720	15480	"
" cast steel	20	4990	16360	"
" " "	200	4790	15710	"
Lead	20	1227	4026	"
Magnesium	–	4602	15100	Melde.
Nickel	–	4973	16320	Masson.
Palladium	–	3150	10340	Various.
Platinum	20	2690	8815	Wertheim.
"	100	2570	8437	"
"	200	2460	8079	"
Silver	20	2610	8553	"
"	100	2640	8658	"
Tin	–	2500	8200	Various.
Zinc	–	3700	12140	"
Various: Brick	–	3652	11980	Chladni.
Clay rock	–	3480	11420	Gray & Milne.
Cork	–	500	1640	Stefan.
Granite	–	3950	12960	Gray & Milne.
Marble	–	3810	12500	"
Paraffin	15	1304	4280	Warburg.
Slate	–	4510	14800	Gray & Milne.
Tallow	16	390	1280	Warburg.
Tuff	–	2850	9350	Gray & Milne.
Glass } from	–	5000	16410	Various.
Glass } to	–	6000	19690	"
Ivory	–	3013	9886	Ciccone & Campanile.
Vulcanized rubber (black)	0	54	177	Exner.
" " (black)	50	31	102	"
" " (red)	0	69	226	"
" " "	70	34	111	"
Wax	17	880	2890	Stefan.
"	28	441	1450	"
Woods: Ash, along the fibre	–	4670	15310	Wertheim.
" across the rings	–	1390	4570	"
" along the rings	–	1260	4140	"
Beech, along the fibre	–	3340	10960	"
" across the rings	–	1840	6030	"
" along the rings	–	1415	4640	"
Elm, along the fibre	–	4120	13516	"
" across the rings	–	1420	4665	"
" along the rings	–	1013	3324	"
Fir, along the fibre	–	4640	15220	"
Maple "	–	4110	13470	"
Oak "	–	3850	12620	"
Pine "	–	3320	10900	"
Poplar "	–	4280	14050	"
Sycamore "	–	4460	14640	"

TABLE 129.

VELOCITY OF SOUND IN LIQUIDS AND GASES.

For gases, the velocity of sound $=\sqrt{\gamma P/\rho}$, where P is the pressure, ρ the density, and γ the ratio of specific heat at constant pressure to that at constant volume (see Table 253). For moderate temperature changes $V_t = V_0(1 + \alpha t)$ where α=0.00367. The velocity of sound in tubes increases with the diameter up to the free-air value as a limit. The values from ammonia to methane inclusive are for closed tubes.

Substance.	Temp. C.	Velocity in meters per second.	Velocity in feet per second.	Authority.
Liquids: Alcohol, 95%	12.5°	1241.	4072.	Dorsing, 1908.
"	20.5	1213.	3980.	"
Ammonia, conc.	16.	1663.	5456.	"
Benzol	17.	1166.	3826.	"
Carbon bisulphide	15.	1161.	3809.	"
Chloroform	15.	983.	3225.	"
Ether	15.	1032.	3386.	"
NaCl, 10% sol.	15.	1470.	4823.	"
" 15% "	15.	1530.	5020.	"
" 20% "	15.	1650.	5414.	"
Turpentine oil.	15.	1326.	4351.	"
Water, air-free	13.	1441.	4728.	"
" " "	19.	1461.	4794.	"
" " "	31.	1505.	4938.	"
" Lake Geneva	9.	1435.	4708.	Colladon-Sturm.
" Seine river	15.	1437.	4714.	Wertheim.
" " "	30.	1528.	5013.	"
" " "	60.	1724.	5657.	"
Explosive waves in water:				
Guncotton, 9 ounces		1732.	5680.	Threlfall, Adair, 1889, see Barton's Sound, p. 518.
" 10 "		1775.	5820.	
" 18 "		1942.	6372.	
" 64 "		2013.	6600.	
Gases: Air, dry, CO_2-free	0.	331.78	1088.5	Rowland.
" "	0.	331.36	1087.1	Violle, 1900.
" " CO_2-free	0.	331.92	1089.0	Thiesen, 1908.
" 1 atmosphere	0.	331.7	1088.	Mean.
" 25 "	0.	332.0	1089.	" (Witkowski).
" 50 "	0.	334.7	1098.	" "
" 100 "	0.	350.6	1150.	" "
"	20.	344.	1129.	
"	100.	386.	1266.	Stevens.
"	500.	553.	1814.	"
"	1000.	700.	2297.	"
Explosive waves in air:				
Charge of powder, 0.24 gms.		336.	1102.	Violle, Cong. Intern. Phys. 1, 243, 1900.
" " " 3.80 "		500.	1640.	
" " " 17.40 "		931.	3060.	
" " " 45.60 "		1268.	4160.	
Ammonia	0.	415.	1361.	Masson.
Carbon monoxide	0.	337.1	1106.	Wullner.
" "	0.	337.4	1107.	Dulong.
" dioxide	0.	258.0	846.	Brockendahl, 1906.
" disulphide	0.	189.	620.	Masson.
Chlorine	0.	206.4	677.	Martini.
"	0.	205.3	674.	Strecker.
Ethylene	0.	314.	1030.	Dulong.
Hydrogen	0.	1269.5	4165.	"
"	0.	1286.4	4221.	Zoch.
Illuminating gas	0.	490.4	1609.	"
Methane	0.	432.	1417.	Masson.
Nitric oxide	0.	325.	1066.	"
Nitrous oxide	0.	261.8	859.	Dulong.
Oxygen	0.	317.2	1041.	"
Vapors: Alcohol	0.	230.6	756.	Masson.
Ether	0.	179.2	588.	"
Water	0.	401.	1315.	"
"	100.	404.8	1328.	Treitz, 1903.
"	130.	424.4	1392.	"

MUSICAL SCALES.

The pitch relations between two notes may be expressed precisely (1) by the ratio of their vibration frequencies; (2) by the number of equally-tempered semitones between them (E. S.); also, less conveniently, (3) by the common logarithm of the ratio in (1); (4) by the lengths of the two portions of the tense string which will furnish the notes; and (5) in terms of the octave as unity. The ratio in (4) is the reciprocal of that in (1); the number for (5) is 1/12 of that for (2); the number for (2) is nearly 40 times that for (3).

Table 130 gives data for the middle octave, including vibration frequencies for three standards of pitch; A_3=435 double vibrations per second, is the international standard and was adopted by the American Piano Manufacturers' Association. The "just-diatonic scale" of C-major is usually deduced, following Chladni, from the ratios of the three perfect major triads reduced to one octave, thus:

				4	:	5	:	6				
4	:	5	:	6				4	:	5	:	6
F		A		C		E		G		B		D
16		20		24		30		36		45		54
				24	27	30	32	36	40	45	48	

Other equivalent ratios and their values in E. S. are given in Table 131. By transferring D to the left and using the ratio 10 : 12 : 15 the scale of A-minor is obtained, which agrees with that of C-major except that D=26 2/3. Nearly the same ratios are obtained from a series of harmonics beginning with the eighth; also by taking 12 successive perfect or Pythagorean fifths or fourths and reducing to one octave. Such calculations are most easily made by adding and subtracting intervals expressed in E. S. The notes needed to furnish a just major scale in other keys may be found by successive transpositions by fifths or fourths as shown in Table 131. Disregarding the usually negligible difference of 0.02 E. S., the table gives the 24 notes to the octave required in the simplest enharmonic organ; the notes fall into pairs that differ by a comma, 0.22 E. S. The line "mean tone" is based on Dom Bedos' rule for tuning the organ (1746). **The tables have been checked by the data in** Ellis' Helmholtz's "Sensations of Tone."

TABLE 130.

Note.	Interval.		Ratios.		Logarithms.		Number of double Vibrations per second.					
	Just.	Tempered.	Just.	Tempered.	Just.	Tempered.	Just.	Just.	Just.	Tempered-	Tempered.	Tempered
	E. S.	E. S.										
C_3	0.	0	1.00	1.00000	.0000	.00000	*256*	*264*	*258.7*	258.7	261.6	271.1
		1		1.05926		.02509				274.0	277.2	287.3
D_3	2.04	2	1.125	1.12246	.05115	.05017	288	297	291.0	290.3	293.7	304.3
		3		1.18921		.07526				307.6	311.1	322.4
E_3	3.86	4	1.25	1.25992	.09691	.10034	320	330	323.4	325.9	329.6	341.6
F_3	4.98	5	1.33	1.33484	.12494	.12543	341.3	352	344.9	345.3	349.2	361.9
		6		1.41421		.15051				365.8	370.0	383.4
G_3	7.02	7	1.50	1.49831	.17609	.17560	384	396	388	387.5	392.0	406.2
		8		1.58740		20069				410.6	415.3	430.4
A_3	8.84	9	1.67	1.68179	.22185	.22577	426.7	*440*	431.1	*435.0*	*440.0*	*456.0*
		10		1.78180		.25086				460.9	446.2	483.1
B_3	10.88	11	1.875	1.88775	.27300	.27594	480	495	485.0	488.3	493.9	511.8
C_4	12.00	12	2.00	2.00000	.30103	.30103	512	528	517.3	517.3	523.2	542.3

TABLE 131.

Key of		C		D		E	F		G		A		B	C
7 #s	C#		1.14		3.18		5.00	6.12		8.16		9.98		12.02
			0.92		2.96		4.78	5.90		7.94		9.76		11.80
6 "	F#		1.14		2.96		5.00	6.12		8.16		9.98	11.10	
			0.92		2.74		4.78	5.90		7.94		9.76	10.88	
5 "	B		1.14		2.96	4.08		6.12		7.94		9.98	11.10	
			0.92		2.74	3.86		5.90		7.72		9.76	10.88	
4 "	E		0.92		2.96	4.08		6.12		7.94	9.06		11.10	
			0.70		2.74	3.86		5.90		7.72	8.84		10.88	
3 "	A		0.92	2.04		4.08		5.90		7.94	9.06		11.10	
			0.70	1.82		3.86		5.68		7.72	8.84		10.88	
2 "	D		0.92	2.04		4.08		5.90	7.02		9.06		10.88	
1 #	G	0.00		2.04		3.86		5.90	7.02		9.06		10.88	12.00
	C	0.00		2.04		3.86	4.98		7.02		8.84		10.88	12.00
1 ♭	F	0.00		1.82		3.86	4.98		7.02		8.84	9.96		12.00
2 ♭s	B♭	0.00		1.82	2.94		4.98		6.80		8.84	9.96		12.00
3 "	E♭	−.22		1.82	2.94		4.98		6.80	7.92		9.96		11.78
4 "	A♭	−.22	0.90		2.94		4.76		6.80	7.92		9.96		11.78
5 "	D♭	−.22	0.90		2.94		4.76	5.88		7.92		9.74		11.78
6 "	G♭		0.90		2.72		4.76	5.88		7.92		9.74	10.86	
7 "	C♭		0.90		2.72	3.84		5.88		7.70		9.74	10.86	
Harmonic Series		8 0.0	(17 1.05)	9 2.04	(19 2.98)	10 3.86	(21 4.70)	11 5.51	12 7.02	(25 7.73)	13 8.41	14 9.69	15 10.88	16 12.00
Cycle of fifths		0.0	1.14	2.04	3.18	4.08	5.22	6.12	7.02	8.16	9.06	10.20	11.10	12.24
Cycle of fourths		0.0	0.90	1.80	2.94	3.84	4.98	5.88	6.78	7.92	8.82	9.96	10.86	11.76
Mean tone		0.0	0.76	1.93	3.11	3.86	5.03	5.79	6.97	7.72	8.90	10.07	10.83	12.00
Equal 7 step		0.0		1.71	3.43		5.14		6.86		8.57	10.29		12.00

MISCELLANEOUS SOUND DATA.

TABLE 132. — A Fundamental Tone, Its Harmonics (Overtones) and the Nearest Tone of the Equal-tempered Scale.

No. of partial	1	2	3	4	5	6	7	8	9	10
Frequency	129	259	388	517	647	776	905	1035	1164	1293
Nearest tempered note	C	C	G	C	E	G	B♭	C	D	E
Corresponding frequency	129	259	388	517	652	775	922	1035	1164	1293
No. of partial	11	12	13	14	15	16	17	18	19	20
Frequency	1423	1552	1681	1811	1940	2069	2199	2328	2457	2586
Nearest tempered note	G♭	G	G#	B♭	B	C	C#	D	D#	E
Corresponding frequency	1463	1550	1642	1843	1953	2069	2192	2323	2461	2607

NOTE. — Overtones of frequencies not exact multiples of the fundamental are sometimes called inharmonic partials.

TABLE 133. — Relative Strength of the Partials in Various Musical Instruments.

The values given are for tones of medium loudness. Individual tones vary greatly in quality and, therefore, in loudness.

Instrument.	Strength of partials in per cent of total tone strength.											
	1	2	3	4	5	6	7	8	9	10	11	12
Tuning fork on box	100	—	—	—	—	—	—	—	—	—	—	—
Flute	66	24	4	6	—	—	—	—	—	—	—	—
Violin, A string	26	25	9	10	27	1	0	2	—	—	—	—
Oboe	2	2	4	29	35	14	4	2	3	4	1	0
Clarinet	12	0	10	3	5	0	8	18	15	18	5	6
Horn	36	26	17	7	4	3	2	1	1	1	1	1
Trombone	6	11	35	12	8	11	6	4	3	2	1	1

TABLE 134. — Characteristics of the Vowels.

The larynx generates a fundamental tone of a *chosen* pitch with some 20 partials, usually of low intensity. The particular partial, or partials, most nearly in unison with the mouth cavity is greatly strengthened by resonance. Each vowel, for a given mouth, is characterized by a particular *fixed* pitch, or pitches, of resonance corresponding to that vowel's definite form of mouth cavity. These pitches may be judged by whispering the vowels. It is difficult to sing vowels true above the corresponding pitches. The greater part of the energy or loudness of a vowel of a *chosen* pitch is in those partials reinforced by resonance. The vowels may be divided into two classes, — the first having one characteristic resonance region, the second, two. The representative pitches of maximum resonance of a mouth cavity for selected vowels in each group are given in the following table.

Vowel indicated by italics in the words.	Pitch of maximum resonance.	Vowel indicated by italics in the words.	Pitch of maximum resonance.
f*a*ther, f*a*r, gu*a*rd	910	m*a*t, *a*dd, c*a*t	800 and 1840
r*aw*, f*a*ll, h*au*l	732	p*e*t, f*ea*ther, bl*e*ss	691 and 1953
n*o*, r*o*de, g*oa*l	461	th*ey*, b*ai*t, h*a*te	488 and 2461
gl*oo*m, m*o*ve, gr*ou*p	326	b*ee*, p*i*que, mach*i*ne	308 and 3100

TABLE 135. — Miscellaneous Sound Data.

Koenig's temperature coefficient for the frequency (n) of forks is nearly the same for all pitches. $n_t = n_0(1 - 0.00011t° \text{ C})$, Ann. d. Phys. 9, p. 408, 1880.

Vibration frequencies for continuous sound sensations are practically the same as for continuous light sensation, 10 or more per second. Helmholtz' value of 32 per sec. may be taken as the flicker value for the ear. Moving pictures use 16 or more per sec. For light the number varies with the intensity.

Pitch limits of voice: 60 to 1200 vibrations per second. (See also page 440.)

Piano pitch limits: 27.2 to 4138.4 v. per sec. (over 7 octaves).

Organ pitch limits: 16 (32 ft. pipe), sometimes 8 (64 ft.) to 4138 ($1\frac{1}{2}$ in.) (9 octaves).

Ear can detect frequencies of 20,000 to 30,000 v. per sec. Koenig, by means of dust figures, measured sounds from steel forks with frequencies up to 90,000.

The quality of a musical tone depends solely on the number and relative strength of its partials (simple tones) and probably not at all on their phases.

The wave-lengths of sound issuing from a closed pipe of length L are $4L$, $4L/3$, $4L/5$, etc., and from an open pipe, $2L$, $2L/2$, $2L/3$, etc. The end correction for a pipe with a flange is such that the antinode is 0.82 × radius of pipe beyond the end; with no flange the correction is 0.57 × radius of pipe.

The energy of a pure sine wave is proportional to n^2A^2; the energy per cm³ is on the average $2\rho\pi^2U^2A^2/\lambda^2$; the energy passing per sec. through 1 cm² perpendicular to direction of propagation is $2\rho\pi^2U^3A^2/\lambda^2$; the pressure is $\frac{1}{2}(\gamma + 1)$ (average energy per cm³); where n is the vibration number per sec., λ the wave-length, A the amplitude, V the velocity of sound, ρ the density of the medium, γ the specific heat ratio. Altberg (Ann. d. Phys. 11, p. 405, 1903) measured sound-wave pressures of the order of 0.24 dynes/cm² = 0.00018 mm Hg.

TABLE 136. — Aerodynamics.

KINETICS OF BODIES IN RESISTING MEDIUM.

The differential equation of a body falling in a resisting medium is $du/dt = g - ku^2$. The velocity tends asymptotically to a certain terminal velocity, $V = \sqrt{g/k}$. Integration gives $u = V \cdot \tanh(gt/V)$, $x = \frac{V^2}{g} \log\cosh(gt/V)$ if $u = x = t = 0$.

When body is projected upwards, $du/dt = -g - ku^2$, and if u_0 is velocity of projection, then $\tan^{-1} u/V = \tan^{-1}(u_0/V) - gt/V$, $x = (V^2/2g) \log(V^2 + u_0^2)(V^2 + u^2)$. The particle comes to rest when $t = (V/g)\tan^{-1}(u_0/V)$ and $x = (V^2/2g)\log(1 - u_0^2/V^2)$.

For small velocities the resistance is more nearly proportional to the velocity.

Stokes' Law for the rate of fall of a spherical drop of radius a under gravity g gives for the velocity, v,

$$v = \frac{2ga^2}{9\eta}(\sigma - \rho),$$

where σ and ρ are the densities of the drop and the medium, η the viscosity of the medium. This depends on five assumptions: (1) that the sphere is large compared to the inhomogeneities of the medium; (2) that it falls as in a medium of unlimited extent; (3) that it is smooth and rigid; (4) that there is no slipping of the medium over its surface; (5) that its velocity is so small that the resistance is all due to the viscosity of the medium and not to the inertia of the latter. Because of 5, the law does not hold unless the radius of the sphere is small compared with $\eta/v\rho$ (critical radius). Arnold showed that a must be less than 0.6 this radius.

If the medium is contained in a circular cylinder of radius R and length L, Ladenburg showed that the following formula is applicable (Ann. d. Phys. 22, 287, 1907, 23, 447, 1908):

$$V = \frac{2}{9}\frac{ga^2(\sigma - \rho)}{\eta(1 + 2.4a/R)(1 + 3.1a/L)}.$$

As the spheres diminish in size the medium behaves as if inhomogeneous because of its molecular structure, and the velocity becomes a function of l/a, where l is the mean free path of the molecules. Stokes' formula should then be modified by the addition of a factor, viz.:

$$v_1 = \frac{2}{9}\frac{ga^2}{\eta}(\sigma - \rho)\left\{1 + (0.864 + 0.29e^{-1.25(a/l)})\frac{l}{a}\right\}$$

(See chapter V, Millikan, The Electron, 1917; also Physical Review 15, p. 545, 1920.)

TABLE 137. — Flow of Gases through Tubes.*

When the dimensions of a tube are comparable with the mean free path (L) of the molecules of a gas, Knudsen (Ann. der Phys. 28, 75, 199, 1908) derives the following equation correct to 5% even when $D/L = 0.4$: Q, the quantity of gas in terms of PV which flows in a second through a tube of diameter D, length l, connecting two vessels at low pressure, difference of pressure $P_2 - P_1$, equals $(P_2 - P_1)/W\sqrt{\rho}$ where ρ is the density of the gas at one bar (1 dyne/cm²) = (molecular weight)/(83.15 × $10^6 T$) and W, which is of the nature of a resistance, $= 2.3941l/D^3 + 3.184/D^2$. The following table gives the cm³ of air and H at 1 bar which would flow through different sized tubes, difference of pressure 1 bar, room temperature.

l	D	W	Q, cm³ of air	cm³ of H_2
l = 1cm.	D = 1cm.	W = 5.58	Q, cm³ of air, 5200.	cm³ of H_2, 19700.
10	1	27.1	1070.	4050.
1	0.1	2710.	10.7	40.5
10	0.1	24300.	1.20	3.60

Knudsen derives the following equation, equivalent to Poiseuille's at higher, and to the above at lower pressures:

$Q = (P_2 - P_1)\{aP + b(1 + c_1P)/(1 + c_2P)\}$ where $a = \pi D^4/128\eta l$ (Poiseuille's constant); $b = 1/W\sqrt{\rho}$, (coefficient of molecular flow); $c_1 = \sqrt{\rho}\,D/\eta$; and $c_2 = 1.24\sqrt{\rho}\,D/\eta$; η = viscosity coefficient. The following are the volumes in cm³ at 1 bar, 20°C, that flow through tube, D = 1 cm, l = 10cm, $P_2 - P_1$ = 1 bar, average pressure of P bars:

P	Q	P	Q	P	Q
P = 10^6	Q = 13,000,000.	P = 5.	Q = 1026.	P = 1.	Q = 1044. cm³
100.	2,227.	4.	1024.	0.1	1065.
10.	1,058.	3.	1025.	0.01	1070.

When the velocity of flow is below a critical value, F (density, viscosity, diameter of tube), the stream lines are parallel to the axis of the tube. Above this critical velocity, V_c, the flow is turbulent. $V_c = k\eta/\rho r$ for small pipes up to about 5 cm diameter, where K is a constant, and r the tube radius. When these are in cgs units, k is 10^3 in round numbers. Below V_c the pressure drop along the tube is proportional to the velocity of gas flow; above it to the square of the velocity.

* See Dushman, The Production and Measurement of High Vacua, General Elec. Rev. 23, p. 493, 1920.

TABLE 138. — Air Pressures upon Large Square Normal Planes at Different Speeds through the Air.

The resistance F of a body of fixed shape and presentation moving through a fluid may be written

$$F = \rho L^2 V^2 f(LV/\nu)$$

in which ρ denotes the fluid density, ν the kinematic viscosity, L a linear dimension of the body, V the speed of translation. In general f is not constant, even for constant conditions of the fluid, but is practically so for normal impact on a plane of fixed size. In the following, ρ is taken as 1.230 g/l (.0768 lbs./ft^3).

The mean pressure on thin square plates of 1.1 m^2 (12 ft^2), or over, moving normally through air of standard density at ordinary transportation speeds may be written $P = .0060v^2$ for P in kg per m^2 and v in km per hour, or $P = .0032v^2$ for P in lbs. per ft^2 and v in miles per hour. The following values are computed from this formula. For smaller areas the correction factors as given in the succeeding table (Table 139) derived from experiments made at the British National Physical Laboratory, may be applied.

Units: the first of each group of three columns gives the velocity; the second, the corresponding pressure in kg/m^2 when the first column is taken as km per hour; the third in pds/ft^2 when in miles per hour.

Velocity.	Pressure.		Velocity.	Pressure.		Velocity.	Pressure.		Velocity.	Pressure.	
	Metric.	English.		Metric.	English.		Metric.	English.		Metric.	English.
10	0.60	0.32	40	9.60	5.12	70	29.4	15.7	100	60.0	32.0
11	0.73	0.39	41	10.09	5.38	71	30.2	16.1	101	61.2	32.6
12	0.86	0.46	42	10.58	5.64	72	31.1	16.6	102	62.4	33.3
13	1.01	0.54	43	11.09	5.92	73	32.0	17.0	103	63.7	33.9
14	1.18	0.63	44	11.6	6.20	74	32.8	17.5	104	64.9	34.6
15	1.35	0.72	45	12.1	6.48	75	33.7	18.0	105	66.1	35.3
16	1.54	0.82	46	12.7	6.77	76	34.7	18.5	106	67.4	36.0
17	1.73	0.92	47	13.3	7.07	77	35.6	19.0	107	68.7	36.6
18	1.94	1.04	48	13.8	7.37	78	36.5	19.5	108	70.0	37.2
19	2.17	1.16	49	14.4	7.68	79	37.4	20.0	109	71.3	38.0
20	2.40	1.28	50	15.0	8.00	80	38.4	20.5	110	72.6	38.7
21	2.65	1.41	51	15.6	8.32	81	39.4	21.0	111	73.9	39.4
22	2.90	1.55	52	16.2	8.65	82	40.3	21.5	112	75.3	40.1
23	3.17	1.69	53	16.9	8.99	83	41.3	22.0	113	76.6	40.9
24	3.46	1.84	54	17.5	9.33	84	42.3	22.6	114	78.0	41.6
25	3.75	2.00	55	18.1	9.68	85	43.3	23.1	115	79.3	42.3
26	4.06	2.16	56	18.8	10.04	86	44.4	23.7	116	80.8	43.1
27	4.37	2.33	57	19.5	10.40	87	45.4	24.2	117	82.1	43.7
28	4.70	2.51	58	20.2	10.76	88	46.4	24.8	118	83.5	44.6
29	5.05	2.69	59	20.9	11.14	89	47.5	25.4	119	84.9	45.3
30	5.40	2.88	60	21.6	11.52	90	48.6	25.9	120	86.4	46.1
31	5.77	3.08	61	22.3	11.91	91	49.7	26.5	121	87.8	46.8
32	6.14	3.28	62	23.0	12.3	92	50.8	27.1	122	89.3	47.6
33	6.54	3.48	63	23.8	12.7	93	51.9	27.7	123	90.8	48.4
34	6.93	3.70	64	24.6	13.1	94	53.0	28.3	124	92.2	49.2
35	7.35	3.92	65	25.4	13.5	95	54.2	28.9	125	93.7	50.0
36	7.74	4.15	66	26.2	13.9	96	55.3	29.5	126	95.3	50.8
37	8.22	4.38	67	26.9	14.4	97	56.5	30.1	127	96.8	51.6
38	8.66	4.62	68	27.7	14.8	98	57.6	30.7	128	98.4	52.5
39	9.12	4.87	69	28.6	15.2	99	58.8	31.4	129	98.7	53.2

TABLE 139. — Correction Factor for Small Square Normal Planes.

The values of Table 138 are to be multiplied by the following factors when the area of the surface is less than about 1 m^2 (12 ft^2).

Metric.				English.			
Area. m^2	Factor.	Area. m^2	Factor.	Area. ft^2	Factor.	Area. ft^2	Factor
0.03	0.845	5.0	0.969	0.03	0.842	5.0	0.968
0.10	0.859	6.0	0.975	0.10	0.857	6.0	0.973
0.50	0.884	7.0	0.979	0.50	0.884	7.0	0.977
0.75	0.890	8.0	0.984	0.75	0.889	8.0	0.981
1.00	0.898	9.0	0.989	1.00	0.896	9.0	0.986
2.00	0.919	10.0	0.993	2.00	0.917	10.0	0.990
3.00	0.933	11.0	0.999	3.00	0.930	11.0	0.994
4.00	0.950	12.0	1.000	4.00	0.943	12.0	1.000

TABLE 140. — Effect of Aspect Ratio upon Normal Plane Pressure (Eiffel).

The mean pressure on a rectangular plane varies with the "aspect ratio," a name introduced by Langley to denote the ratio of the length of the leading edge to the chord length. The effect of aspect ratio on normally moving rectangular plates is given in the following table, derived from Eiffel's experiments.

Aspect ratio	1.00	1.5	3.00	6.00	10.000	14.60	20.00	30.00	41.500	50.00
Pressure on rectangle / Pressure on square	1.00	1.04	1.07	1.10	1.145	1.25	1.34	1.40	1.435	1.47

TABLE 141. — Ratio of Pressures on Inclined and Normal Planes.

The pressure on a slightly inclined plane is proportional to the angle of incidence a, and is given by the formula $P_a = c \cdot P_{90} \cdot a$. The value of c, which is constant for incidences up to about 12°, is given for various aspect ratios. The angle of incidence is taken in degrees.

Aspect ratio	1	2	3	4	5	6	7	8	9	10
Value of c	0.036	0.043	0.050	0.053	0.057	0.061	0.065	0.070	0.075	0.080

TABLE 142. — Skin Friction.

The skin friction on an even rectangular plate moving edgewise through ordinary air is given by Zahm's equation,

$$F(\text{kg/m}^2) = 0.00030\{A(\text{m}^2)\}^{0.93}\{V(\text{km/hr.})\}^{1.86} \text{ in metric units}$$

or

$$F(\text{pds./ft.}^2) = 0.0000082\{A(\text{ft.}^2)\}^{0.93}\{V(\text{ft./sec.})\}^{1.86},$$

where A is the surface area and V the speed of the plane. The following table gives the friction per unit area on one side of a plate.

Speed.	Skin friction. Kg per sq. m. Plane.		Speed.		Skin friction. Lbs. per sq. ft. Plane.	
km/hr.	1 m long.	32 m long.	miles/hr.	ft./sec.	1 ft. long.	32 ft. long.
5	0.0059	0.0047	5	7.3	0.00033	0.00026
10	0.0217	0.0171	10	14.7	0.00121	0.00095
15	0.0464	0.0364	15	22.0	0.00258	0.00202
20	0.079	0.062	20	29.3	0.00439	0.00345
25	0.122	0.095	25	36.7	0.0068	0.00530
30	0.169	0.133	30	44.0	0.0094	0.0074
40	0.288	0.225	40	58.7	0.0160	0.0125
50	0.439	0.346	50	73.3	0.0244	0.0192
60	0.616	0.482	60	88.0	0.0342	0.0268
70	0.82	0.64	70	102.7	0.0455	0.0357
80	1.06	0.83	80	117.3	0.0587	0.0461
90	1.31	1.03	90	132.0	0.073	0.0572
100	1.58	1.24	100	146.7	0.088	0.069
110	1.89	1.49	110	161.2	0.105	0.083
120	2.20	1.73	120	175.8	0.122	0.096
125	2.39	1.87	125	183.4	0.133	0.104
130	2.56	2.01	130	190.5	0.142	0.112
135	2.68	2.10	135	197.8	0.149	0.117
140	2.94	2.31	140	205.4	0.164	0.128
145	3.15	2.47	145	212.5	0.175	0.137
150	3.37	2.65	150	220.0	0.188	0.147

The following tables, based on Eiffel, show the variation of the resistance coefficient K, with the angle of impact i, the aspect (ratio of leading edge to chord length), shape and velocity V in the formula

$$R(\text{kg/m}^2) = KS(\text{m}^2)\{V(\text{m/sec.})\}^2$$

The value of K for km/hour would be 0.77 times greater.

TABLE 143. — Variation of Air Resistance with Aspect and Angle.

Size of plane.	Aspect.	Values of i. 6°	10°	20°	30°	40°	45°	60°	75°	Max. ratio. Value.	Max. ratio. i.
		Values of Ki/K_{90}.									
15 x 90 cm.....	$\frac{1}{6}$	.07	.13	.40	0.67	0.92	1.08	1.07	1.03	1.07	60
15 x 45 cm.....	$\frac{1}{3}$	.11	.21	.51	0.89	1.20	1.22	1.06	1.02	1.22	45
25 x 25 cm.....	1	.20	.36	.80	1.24	1.17	1.08	1.03	1.02	1.46	38
30 x 15 cm.....	2	.26	.43	.91	0.72	0.79	0.82	0.90	0.97	0.91	20
45 x 15 cm.....	3	.31	.50	.77	0.77	0.84	0.88	0.94	0.99	0.77	20
90 x 15 cm.....	6	.37	.58	.70	0.78	0.84	0.88	0.93	0.98	0.69	15
90 x 10 cm.....	9	.45	.62	.73	0.80	0.85	0.88	0.94	0.99	—	—

TABLE 144. — Variation of Air Resistance with Shape and Size.

Cylinder, base ⊥ to wind:	Length.	0 cm	1R*	2R*	4R*	6R*	8R*	14R*
Diameter of base, 30 cm	K =	.0675	.068	.055	.050	—	—	—
Diameter of base, 15 cm	K =	.066	.066	.055	.051	.051	.0515	.059

Cylinder, base ‖ to wind: diameter base, 15 cm, length, 60 cm	K = .040
Cylinder, base ‖ to wind: diameter base, 3 cm, length, 100 cm	K = .060
Cone, angle 60°, diam. base, 40 cm, point to wind, solid	K = .032
Cone, angle 30°, diam. base, 40 cm, point to wind, solid	K = .021
Sphere, 25 cm diam.	K = .011
Hemisphere, same diam., convex to wind	K = .021
Hemisphere, same diam., concave to wind	K = .083
Sphero-conic body, diam., 20 cm, cone 20°, point forward	K = .010
Sphero-conic body, diam., 20 cm, cone 20°, point to rear	K = .0055
Cylinder, 120 cm long, spherical ends to wind	K = .012

The wind velocity for the values of this table was 10 m/sec.

Tables 143 and 144 were taken from "The Resistance of the Air and Aviation," Eiffel, translated by Hunsaker, 1913.

* In the case of these cylinders the percentages due to skin friction are 2, 3, 6, 8, 11 and 16 per cent respectively, excluding the disk.

TABLE 145. — Variation of Air Resistance with Shape, Size and Speed.

This table shows the peculiar drop in air resistance for speeds greater than 4 to 12 meters per second. Another change occurs when the velocity approaches that of sound.

Shape.	Speed, m/sec.	Values of K. 4	6	8	10	12	14	16	20	32
Sphere, 16.2 cm diameter		.033	.030	.028	.027	.024	.009	.0095	.010	.011
Sphere, 24.4 cm diameter		.025	.025	.021	.013	.010	.010	.010	.010	.010
Sphere, 33 cm diameter		.023	.017	.012	.010	.010	.010	.011	.012	.012
Concave cup, 25 cm diameter		.090	.090	.089	.087	.087	.088	.089	.095	.100
Convex cup, 25 cm diameter		.027	.022	.021	.022	.022	.021	.020	.019	.018
Disk, 25 cm diameter		.071	.070	.070	.070	.070	.070	.070	.070	.068
Cylinder	cm									
element ⊥ to wind, d = 15 cm, l =	15.0	.043	.042	.037	.030	.025	.022	.021	.022	.022
element ⊥ to wind, 30	30.0	.045	.032	.027	.023	.024	.025	.025	.025	.023
element ⊥ to wind, 15	7.5	.035	.034	.032	.031	.031	.031	.030	.030	.030
element ⊥ to wind, 15	12.0	.038	.037	.036	.032	.030	.028	.027	.025	.025
element ⊥ to wind, 15	22.5	.042	.041	.038	.034	.031	.028	.025	.022	.020
element ‖ to wind, 15	105.0	.069	.061	.057	.055	.053	.052	.051	.051	.050
Spherical ends, 15	120.0	.024	.022	.019	.018	.018	.018	.017	.016	.015

Taken from "Nouvelles Recherches sur la résistance de l'air et l'aviation," Eiffel, 1914.

TABLE 146. — Friction.

The required force F necessary to just move an object along a horizontal plane $= fN$ where N is the normal pressure on the plane and f the "coefficient of friction." The angle of repose Φ ($\tan \Phi = F/N$) is the angle at which the plane must be tilted before the object will move from its own weight. The following table of coefficients was compiled by Rankine from the results of General Morin and other authorities and is sufficient for ordinary purposes.

Material.	f	$1/f$	ϕ
Wood on wood, dry	.25–.50	4.00–2.00	14.0–26.5
" " " soapy	.20	5.00	11.5
Metals on oak, dry	.50–.60	2.00–1.67	26.5–31.0
" " " wet	.24–.26	4.17–3.85	13.5–14.5
" " " soapy	.20	5.00	11.5
" " elm, dry	.20–.25	5.00–4.00	11.5–14.0
Hemp on oak, dry	.53	1.89	28.0
" " " wet	.33	3.00	18.5
Leather on oak	.27–.38	3.70–2.86	15.0–19.5
" " metals, dry	.56	1.79	29.5
" " " wet	.36	2.78	20.0
" " " greasy	.23	4.35	13.0
" " " oily	.15	6.67	8.5
Metals on metals, dry	.15–.20	6.67–5.00	8.5–11.5
" " " wet	.3	3.33	16.5
Smooth surfaces, occasionally greased	.07–.08	14.3–12.50	4.0–4.5
" " continually greased	.05	20.00	3.0
" " best results	.03–.036	33.3–27.6	1.75–2.0
Steel on agate, dry *	.20	5.00	11.5
" " " oiled *	.107	9.35	6.1
Iron on stone	.30–.70	3.33–1.43	16.7–35.0
Wood on stone	About .40	2.50	22.0
Masonry and brick work, dry	.60–.70	1.67–1.43	33.0–35.0
" " " " damp mortar	.74	1.35	36.5
" on dry clay	.51	1.96	27.0
" " moist clay	.33	3.00	18.25
Earth on earth	.25–1.00	4.00–1.00	14.0–45.0
" " " dry sand, clay, and mixed earth	.38–.75	2.63–1.33	21.0–37.0
" " " damp clay	1.00	1.00	45.0
" " " wet clay	.31	3.23	17.0
" " " shingle and gravel	.81–1.11	1.23–0.9	39.0–48.0

* Quoted from a paper by Jenkin and Ewing, "Phil. Trans. R. S." vol. 167. In this paper it is shown that in cases where "static friction" exceeds "kinetic friction" there is a gradual increase of the coefficient of friction as the speed is reduced towards zero.

TABLE 147. — Lubricants.

The best lubricants are in general the following: Low temperatures, light mineral lubricating oils. Very great pressures, slow speeds, graphite, soapstone and other solid lubricants. Heavy pressures, slow speeds, ditto and lard, tallow and other greases. Heavy pressures and high speeds, sperm oil, castor oil, heavy mineral oils. Light pressures, high speeds, sperm, refined petroleum olive, rape, cottonseed. Ordinary machinery, lard oil, tallow oil, heavy mineral oils and the heavier vegetable oils. Steam cylinders, heavy mineral oils, lard, tallow. Watches and delicate mechanisms, clarified sperm, neat's-foot, porpoise, olive and light mineral lubricating oils.

TABLE 148. — Lubricants For Cutting Tools.

Material.	Turning.	Chucking.	Drilling.	Tapping Milling.	Reaming.
Tool Steel,	dry or oil	oil or s. w.	oil	oil	lard oil
Soft Steel,	dry or soda water	soda water	oil or s. w.	oil	lard oil
Wrought iron	dry or soda water	soda water	oil or s. w.	oil	lard oil
Cast iron, brass	dry	dry	dry	dry	dry
Copper	dry	dry	dry	dry	mixture
Glass	turpentine or kerosene				

Mixture = ⅓ crude petroleum, ⅔ lard oil. Oil = sperm or lard.

Tables 147 and 148 quoted from "Friction and Lost Work in Machinery and Mill Work," Thurston, Wiley and Sons.

VISCOSITY.

TABLE 149. — Viscosity of Fluids and Solids.

The coefficient of viscosity of a substance is the tangential force required to move a unit area of a plane surface with unit speed relative to another parallel plane surface from which it is separated by a layer a unit thick of the substance. Viscosity measures the temporary rigidity it gives to the substance. The viscosity of fluids is generally measured by the rate of flow of the fluid through a capillary tube the length of which is great in comparison with its diameter. The equation generally used is

$$\mu\text{, the viscosity, } = \frac{\gamma \pi g d^4 t}{128 Q(l + \lambda)} \left(h - \frac{m v^2}{g}\right),$$

where γ is the density (g/cm³), d and l are the diameter and length in cm of the tube, Q the volume in cm³ discharged in t sec., λ the Couette correction which corrects the measured to the effective length of the tube, h the average head in cm, m the coefficient of kinetic energy correction, mv^2/g, necessary for the loss of energy due to turbulent in distinction from viscous flow, g being the acceleration of gravity (cm/sec/sec), v the mean velocity in cm per sec. (See Technologic Paper of the Bureau of Standards, 100 and 112, Herschel, 1917–1918, for discussion of this correction and λ.)

The fluidity is the reciprocal of the absolute viscosity. The kinetic viscosity is the absolute viscosity divided by the density. Specific viscosity is the viscosity relative to that of some standard substance, generally water, at some definite temperature. The dimensions of viscosity are $ML^{-1}T^{-1}$. It is generally expressed in cgs units as dyne-seconds per cm² or poises.

The viscosity of solids may be measured in relative terms by the damping of the oscillations of suspended wires (see Table 78). Ladenburg (1906) gives the viscosity of Venice turpentine at 18.3° as 1300 poises; Trouton and Andrews (1904) of pitch at 0°, 51×10^{10}, at 15°, 1.3×10^{10}; of shoemakers' wax at 8°, 4.7×10^{6}; of soda glass at 575°, 11×10^{12}; Deeley (1908) of glacier ice as 12×10^{13}.

TABLE 150. — Viscosity of Water in Centipoises. Temperature Variation.

Bingham and Jackson, Bulletin Bureau of Standards, 14, 75, 1917.

°C.	Viscosity. cp	°C.	Viscosity. cp	°C.	Viscosity. cp	°C.	Viscosity. cp	°C.	Viscosity. cp	°C.	Viscosity. cp	°C.	Viscosity. cp
0	1.7921	10	1.3077	20	1.0050	30	0.8007	40	0.6560	50	0.5494	60	0.4688
1	1.7313	11	1.2713	21	0.9810	31	0.7840	41	0.6439	51	0.5404	65	0.4355
2	1.6728	12	1.2363	22	0.9579	32	0.7679	42	0.6321	52	0.5315	70	0.4061
3	1.6191	13	1.2028	23	0.9358	33	0.7523	43	0.6207	53	0.5229	75	0.3799
4	1.5674	14	1.1709	24	0.9142	34	0.7371	44	0.6097	54	0.5146	80	0.3565
5	1.5188	15	1.1404	25	0.8937	35	0.7225	45	0.5988	55	0.5064	85	0.3355
6	1.4728	16	1.1111	26	0.8737	36	0.7085	46	0.5883	56	0.4985	90	0.3165
7	1.4284	17	1.0828	27	0.8545	37	0.6947	47	0.5782	57	0.4907	95	0.2994
8	1.3860	18	1.0559	28	0.8360	38	0.6814	48	0.5683	58	0.4832	100	0.2838
9	1.3462	19	1.0299	29	0.8180	39	0.6685	49	0.5588	59	0.4759	153	0.181 *

* de Haas, 1894. Undercooled water: −2.10°, 1.33 cp; −4.70°, 2.12 cp; −6.20°, 2.25 cp; −8.48°, 2.46 cp; −9.30°, 2.55 cp; White, Twining, J. Amer. Ch. Soc., 50, 380, 1913.

TABLE 151. — Viscosity of Alcohol-water Mixtures in Centipoises. Temperature Variation.

°C.	Percentage by weight of ethyl alcohol.												
	0	10	20	30	39	40	45	50	60	70	80	90	100
0	1.792	3.311	5.319	6.94	7.25	7.14	6.94	6.58	5.75	4.762	3.690	2.732	1.773
5	1.519	2.577	4.065	5.29	5.62	5.59	5.50	5.26	4.63	3.906	3.125	2.309	1.623
10	1.308	2.179	3.165	4.05	4.39	4.39	4.35	4.18	3.77	3.268	2.710	2.101	1.466
15	1.140	1.792	2.618	3.26	3.52	3.53	3.51	3.44	3.14	2.770	2.309	1.802	1.332
20	1.005	1.538	2.183	2.71	2.88	2.91	2.88	2.87	2.67	2.370	2.008	1.610	1.200
25	0.894	1.323	1.815	2.18	2.35	2.35	2.39	2.40	2.24	2.037	1.748	1.424	1.096
30	0.801	1.160	1.553	1.87	2.00	2.02	2.02	2.02	1.93	1.767	1.531	1.279	1.003
35	0.722	1.006	1.332	1.58	1.71	1.72	1.73	1.72	1.66	1.529	1.355	1.147	0.914
40	0.656	0.907	1.160	1.368	1.473	1.482	1.495	1.499	1.447	1.344	1.203	1.035	0.834
45	0.599	0.812	1.015	1.189	1.284	1.289	1.307	1.294	1.271	1.189	1.081	0.939	0.764
50	0.549	0.734	0.907	1.050	1.124	1.132	1.148	1.155	1.127	1.062	0.968	0.848	0.702
60	0.469	0.609	0.736	0.834	0.885	0.893	0.907	0.913	0.902	0.856	0.789	0.704	0.592
70	0.406	0.514	0.608	0.683	0.725	0.727	0.740	0.740	0.729	0.695	0.650	0.589	0.504
80	0.356	0.430	0.505	0.567	0.598	0.601	0.609	0.612	0.604	—	—	—	—

Same authority as preceding table.

TABLE 152. — Viscosity and Density of Sucrose in Aqueous Solution.

See Scientific Paper 298, Bingham and Jackson, Bureau of Standards, 1917, and Technologic Paper 100, Herschel, Bureau of Standards, 1917.

Temperature.	Viscosity in centipoises.				Density d_4^t.			
	Per cent sucrose by weight.				Per cent sucrose by weight.			
	0	20	40	60	0	20	40	60
0° C	1.7921	3.804	14.77	238.	0.99987	1.08546	1.18349	1.29560
5	1.5188	3.154	11.56	156.	0.99999	1.08460	1.18192	1.29341
10	1.3077	2.652	9.794	109.8	0.99973	1.08353	1.18020	1.29117
15	1.1404	2.267	7.468	74.6	0.99913	1.08233	1.17837	1.28884
20	1.0050	1.960	6.200	56.5	0.99823	1.08094	1.17648	1.28644
30	0.8007	1.504	4.382	33.78	0.99568	1.07767	1.17214	1.28144
40	0.6560	1.193	3.249	21.28	0.99225	1.07366	1.16759	1.27615
50	0.5494	0.970	2.497	14.01	0.98807	1.06898	1.16248	1.27058
60	0.4688	0.808	1.982	9.83	0.98330	1.06358	1.15693	1.26468
70	0.4061	0.685	1.608	7.15				
80	0.3565	0.590	1.334	5.40	Densities due to Plato.			

TABLE 153. — Viscosity and Density of Glycerol in Aqueous Solution (20° C).

% Glycerol.	Density. g/cm³	Viscosity in centipoises.	100 × Kinematic viscosity.	% Glycerol.	Density. g/cm³	Viscosity in centipoises.	100 × Kinematic viscosity.	% Glycerol.	Density. g/cm³	Viscosity in centipoises.	100 × Kinematic viscosity.
5	1.0098	1.181	1.170	35	1.0855	3.115	2.870	65	1.1662	14.51	12.44
10	1.0217	1.364	1.335	40	1.0989	3.791	3.450	70	1.1797	21.49	18.22
15	1.0337	1.580	1.529	45	1.1124	4.692	4.218	75	1.1932	33.71	28.25
20	1.0461	1.846	1.765	50	1.1258	5.908	5.248	80	1.2066	55.34	45.86
25	1.0590	2.176	2.055	55	1.1393	7.664	6.727	85	1.2201	102.5	84.01
30	1.0720	2.585	2.411	60	1.1528	10.31	8.943	90	1.2335	207.6	168.3

The kinematic viscosity is the ordinary viscosity in cgs units (poises) divided by the density.

TABLE 154. — Viscosity and Density of Castor Oil (Temperature Variation).

°C	Density, g/cm³	Viscosity in poises.	Kinematic viscosity.	°C	Density, g/cm³	Viscosity in poises.	Kinematic viscosity.	°C	Density, g/cm³	Viscosity in poises.	Kinematic viscosity.	°C	Density, g/cm³	Viscosity in poises.	Kinematic viscosity.
5	.9707	37.6	38.7	14	.9645	16.61	17.22	23	.9583	7.67	8.00	32	.9520	3.94	4.14
6	.9700	34.5	35.5	15	.9638	15.14	15.71	24	.9576	7.06	7.37	33	.9513	3.65	3.84
7	.9693	31.6	32.6	16	.9631	13.80	14.33	25	.9569	6.51	6.80	34	.9506	3.40	3.58
8	.9686	28.9	29.8	17	.9624	12.65	13.14	26	.9562	6.04	6.32	35	.9499	3.16	3.33
9	.9679	26.4	27.3	18	.9617	11.62	12.09	27	.9555	5.61	5.87	36	.9492	2.94	3.10
10	.9672	24.2	25.0	19	.9610	10.71	11.15	28	.9548	5.21	5.46	37	.9485	2.74	2.89
11	.9665	22.1	22.8	20	.9603	9.86	10.27	29	.9541	4.85	5.08	38	.9478	2.58	2.72
12	.9659	20.1	20.8	21	.9596	9.06	9.44	30	.9534	4.51	4.73	39	.9471	2.44	2.58
13	.9652	18.2	18.9	22	.9589	8.34	8.70	31	.9527	4.21	4.42	40	.9464	2.31	2.44

Tables 153 and 154, taken from Technologic Paper 112, Bureau of Standards, 1918. Glycerol data due to Archbutt, Deeley and Gerlach; Castor Oil to Kahlbaum and Räber. See preceding table for definition of kinematic viscosity. Archbutt and Deeley give for the density and viscosity of castor oil at 65.6° C, 0.9284 and 0.605, respectively; at 100° C, 0.9050 and 0.169.

TABLE 155.

VISCOSITY OF LIQUIDS.

Viscosities are given in cgs units, dyne-seconds per cm², or poises.

Liquid.	°C	Viscosity.	Reference.
Acetaldehyde	0.	0.00275	1
"	10.	0.00252	1
"	20.	0.00231	1
Air	−192.3	0.00172	2
Aniline	20.	0.04467	3
"	60.	0.0156	3
Bismuth	285.	0.0161	4
"	365.	0.0146	4
Copal lac	22.	4.80	5
Glycerine	2.8	42.2	6
"	14.3	13.87	6
"	20.3	8.30	6
"	26.5	4.94	6
" 80.31% H_2O	8.5	1.021	6
" 64.05% H_2O	8.5	0.222	6
" 49.79% H_2O	8.5	0.092	6
Hydrogen, liquid	—	0.00011	2
Menthol, solid	14.9	2×10^{12}	7
" liquid	56.9	0.069	7
Mercury	−20.	0.0184	8
"	0.	0.01661	4
"	20.	0.01547	4
"	34.	0.01476	4
"	98.	0.01263	4
"	193.	0.01079	4
"	299.	0.00975	4
Oils:			
Dogfish-liver .923 ‡	30.	0.414	9
" " .918	50.	0.211	9
" " .908	90.	0.080	9
Linseed .925	30.	0.331	9
" .922	50.	0.176	9
" .914	90.	0.071	9
* Spindle oil .885	15.6	0.453	10
" "	37.8	0.162	10
" "	100.0	0.033	10
* Light machinery .907 ‡	15.6	1.138	10
* Light machinery	37.8	0.342	10
" "	100.0	0.049	10
* "Solar red" engine	15.6	1.915	10
" " "	37.8	0.496	10
" " "	100.0	0.058	10
* "Bayonne" engine	15.6	2.172	10
" "	37.8	0.572	10
* " "	100.0	0.063	10
* "Queen's red" engine	15.6	2.995	10
" " "	37.8	0.711	10
" " "	100.0	0.070	10
* "Galena" axle oil	15.6	4.366	10
* " " "	37.8	0.909	10
* Heavy machinery	15.6	6.606	10
" "	37.8	1.274	10
* Filtered cylinder	37.8	2.406	10
" "	100.0	0.187	10
* Dark cylinder	37.8	4.224	10
" "	100.0	0.240	10
* Dark cylinder	37.8	7.324	10
" "	100.0	0.341	10
* "Extra L. L."	37.8	11.156	10
" " "	100.0	0.451	10
Linseed .925 ‡	30.	0.331	9
" .922	50.	0.176	9
" .914	90.	0.071	9
Olive .9195	10.	1.38	11
"	15.	1.075	11
" .9130	20.	0.840	11
" .9065	30.	0.540	11
" .9000	40.	0.363	11
" .8935	50.	0.258	11
" .8800	70.	0.124	11
† Rape	15.6	1.118	10
"	37.8	0.422	10
"	100.0	0.080	10
" (another)	15.6	1.176	10
" (another)	100.0	0.085	10
Soya bean .919 ‡	30.0	0.406	9
" " .915	50.0	0.206	9
" " .906	90.0	0.078	9
† Sperm	15.6	0.420	10
"	37.8	0.185	10
"	100.0	0.046	10
Paraffins:			
Pentane	21.0	0.0026	12
Hexane	23.7	0.0033	12
Heptane	24.0	0.0045	12
Octane	22.2	0.0053	12
Nonane	22.3	0.0062	12
Decane	22.3	0.0077	12
Undecane	22.7	0.0095	12
Dodecane	23.3	0.0126	12
Tridecane	23.3	0.0155	12
Tetradecane	21.9	0.0213	12
Pentadecane	22.0	0.0281	12
Hexadecane	22.2	0.0359	12
Phenol	18.3	0.1274	13
"	90.0	0.0126	13
Sulphur	170.	320.0	14
"	180.	550.0	14
"	187.	560.0	14
"	200.	500.0	14
"	250.	104.0	14
"	300.	24.0	14
"	340.	6.2	14
"	380.	2.5	14
"	420.	1.13	14
"	448.	0.80	14
† Tallow	66.	0.176	10
"	100.	0.078	10
Zinc	280.	0.0168	4
"	357.	0.0142	4
"	389.	0.0131	4

* American mineral oils; based on water as .01028 at 20° C. † Based on water as per 1st footnote. ‡ Densities.

References: (1) Thorpe and Rodger, 1894–7; (2) Verschaffelt, Sc. Ab. 1917; (3) Wijkander, 1879; (4) Plüss. Z. An. Ch. 93, 1915; (5) Metz, C. R. 1903; (6) Schöttner, Wien. Ber. 77, 1878, 79, 1879; (7) Heydweiller, W. Ann. 63, 1897; (8) Koch, W. Ann. 14, 1881; (9) White, Bul. Bur. Fish. 32, 1912; (10) Archbutt-Deeley, Lubrication and Lubricants, 1912; (11) Higgins, Nat. Phys. Lab. 11, 1914; (12) Bartolli, Stracciati, 1885–6; (13) Scarpa, 1903–4; (14) Rotinganz, Z. Ph. Ch. 62, 1908.

Ratio of Viscosity at High to that at Atmospheric Pressure.

Pressure tons/in²	Pressure Kg/cm²	Bayonne oil (mineral)	FFF cylinder (mineral)	Trotter (animal)	Rape (vegetable)	castor (vegetable)	Sperm (fish)
1	157.5	1.3	1.4	1.2	1.1	1.2	1.2
2	315.	2.0	2.0	1.6	1.4	1.6	1.5
4	630.	4.0	4.5	2.4	2.3	2.7	2.4
6	945.	7.8	8.9	3.5	3.5	4.2	3.5
8	1200.	16.1	—	5.0	—	5.8	—

Hyde, Pr. Roy. Soc. 97A, 240, 1920.

TABLE 156.

VISCOSITY OF LIQUIDS.

Compiled from Landolt and Börnstein, 1912. Based principally on work of Thorpe and Rogers, 1894–97. Viscosity given in centipoises. One centipoise = 0.01 dyne-second per cm².

Liquid.	Viscosity in centipoises.								
	Formula.	0° C	10° C	20° C	30° C	40° C	50° C	70° C	100° C
Acids: Formic	CH_2O_2	solid	2.247	1.784	1.460	1.219	1.036	.780	.549
Acetic	$C_2H_4O_2$	solid	solid	1.222	1.040	0.905	0.796	.631	.465
Propionic	$C_3H_6O_2$	1.521	1.289	1.102	0.960	0.845	0.752	.607	.459
Butyric	$C_4H_8O_2$	2.286	1.851	1.540	1.304	1.120	0.975	.760	.551
i-Butyric	$C_4H_8O_2$	1.887	1.568	1.318	1.129	0.980	0.862	.683	.501
Alcohols: Methyl	CH_4O	0.817	0.690	0.596	0.520	0.456	0.403	—	—
Ethyl *	C_2H_6O	1.772	1.466	1.200	1.003	0.834	0.702	.510	—
Allyl	C_3H_6O	2.145	1.705	1.363	1.168	0.914	0.763	.553	—
Propyl	C_3H_8O	3.883	2.918	2.256	1.779	1.405	1.130	.760	—
i-Propyl	C_3H_8O	4.565	3.246	2.370	1.757	1.331	1.029	.646	—
Butyric	$C_4H_{10}O$	5.186	3.873	2.948	2.267	1.782	1.411	.930	.540
i-Butyric	$C_4H_{10}O$	8.038	5.548	3.907	2.864	2.122	1.611	—	.527
Amyl, op. act.	$C_5H_{12}O$	11.129	7.425	5.092	3.594	2.607	1.937	—	.610
Amyl, op. inact.	$C_5H_{12}O$	8.532	6.000	4.342	3.207	2.415	1.851	—	.632
Aromatics: Benzene	C_6H_6	0.906	0.763	0.654	0.567	0.498	0.444	.359	—
Toluene	C_7H_8	0.772	0.671	0.590	0.525	0.471	0.426	.354	.278
Ethylbenzole	C_8H_{10}	0.877	0.761	0.669	0.594	0.531	0.479	.397	.310
Orthoxylene	C_8H_{10}	1.105	0.937	0.810	0.709	0.627	0.560	.458	.352
Metaxylene	C_8H_{10}	0.806	0.702	0.620	0.552	0.497	0.451	.375	.296
Paraxylene	C_8H_{10}	solid	0.738	0.648	0.574	0.513	0.463	.383	.300
Bromides: Ethyl	C_2H_5Br	0.487	0.441	0.402	0.368	—	—	—	—
Propyl	C_3H_7Br	0.651	0.582	0.524	0.475	0.433	0.397	.338	—
i-Propyl	C_3H_7Br	0.611	0.545	0.489	0.443	0.403	0.368	—	—
Allyl	C_3H_5Br	0.626	0.560	0.504	0.458	0.419	0.384	.328	—
Ethylene	C_2H_4Br	2.438	2.039	1.721	1.475	1.286	1.131	.903	.678
Bromine	Br	1.267	1.120	1.005	0.911	0.830	0.761	—	—
Chlorides: Propyl	C_3H_7Cl	0.442	0.396	0.359	0.326	0.299	—	—	—
Allyl	C_3H_5Cl	0.413	0.372	0.337	0.307	0.282	—	—	—
Ethylene	C_2H_4Cl	1.132	0.966	0.838	0.736	0.652	0.584	.479	—
Chloroform	$CHCl_3$	0.706	0.633	0.571	0.519	0.474	0.435	—	—
Carbon-tetra	CCl_4	1.351	1.138	0.975	0.848	0.746	0.662	.534	—
Ethers: Diethyl	$C_4H_{10}O$	0.294	0.268	0.245	0.223	—	—	—	—
Methyl-propyl	$C_4H_{10}O$	0.314	0.285	0.260	0.237	—	—	—	—
Ethyl-propyl	$C_5H_{12}O$	0.402	0.360	0.324	0.294	0.268	0.245	—	—
Dipropyl	$C_6H_{14}O$	0.544	0.479	0.425	0.381	0.344	0.311	—	—
Esters: Methylformate	$C_2H_4O_2$	0.436	0.391	0.355	0.325	—	—	—	—
Ethylformate	C_3H_6O	0.510	0.454	0.408	0.369	0.336	0.308	—	—
Methylacetate	$C_3H_6O_2$	0.484	0.431	0.388	0.352	0.320	0.293	—	—
Ethylacetate	$C_4H_8O_2$	0.582	0.512	0.455	0.407	0.367	0.333	.279	—
Iodides: Methyl	CH_3I	0.606	0.548	0.500	0.460	0.424	—	—	—
Ethyl	C_2H_5I	0.727	0.654	0.592	0.540	0.495	0.456	.391	—
Propyl	C_3H_7I	0.944	0.833	0.744	0.669	0.607	0.552	.466	.371
Allyl	C_3H_5I	0.936	0.826	0.734	0.660	0.597	0.544	.458	.365
Paraffines: Pentane	C_5H_{12}	0.289	0.262	0.240	0.220	—	—	—	—
i-Pentane	C_5H_{12}	0.284	0.256	0.234	—	—	—	—	—
Hexane	C_6H_{14}	0.401	0.360	0.326	0.296	0.271	0.248	—	—
i-Hexane	C_6H_{14}	0.376	0.338	0.306	0.279	0.254	0.233	—	—
Heptane	C_7H_{16}	0.524	0.465	0.416	0.375	0.341	0.310	.262	—
i-Heptane	C_7H_{16}	0.481	0.428	0.384	0.347	0.315	0.288	.243	—
Octane	C_8H_{18}	0.706	0.616	0.542	0.483	0.433	0.391	.324	.252
Sulphides: Carbon di-	CS_2	0.438	0.405	0.376	0.352	0.330	—	—	—
Ethyl	$C_4H_{10}S$	0.563	0.501	0.450	0.407	0.369	0.338	.287	—
Turpentine †		2.248	1.783	1.487	1.272	1.071	0.926	.728	—

* Bureau of Standards, see special table. † Glaser.

VISCOSITY OF SOLUTIONS.

This table is intended to show the effect of change of concentration and change of temperature on the viscosity of solutions of salts in water. The specific viscosity × 100 is given for two or more densities and for several temperatures in the case of each solution. μ stands for specific viscosity, and *t* for temperature Centigrade.

Salt.	Percentage by weight of salt in solution.	Density.	μ	t	μ	t	μ	t	μ	t	Authority.
$BaCl_2$	7.60	–	77.9	10	44.0	30	35.2	50	–	–	Sprung.
"	15.40	–	86.4	"	56.0	"	39.6	"	–	–	"
"	24.34	–	100.7	"	66.2	"	47.7	"	–	–	"
$Ba(NO_3)_2$	2.98	1.027	62.0	15	51.1	25	42.4	35	34.8	45	Wagner.
"	5.24	1.051	68.1	"	54.2	"	44.1	"	36.9	"	"
$CaCl_2$	15.17	–	110.9	10	71.3	30	50.3	50	–	–	Sprung.
"	31.60	–	272.5	"	177.0	"	124.0	"	–	–	"
"	39.75	–	670.0	"	379.0	"	245.5	"	–	–	"
"	44.09	–	–	–	593.1	"	363.2	"	–	–	"
$Ca(NO_3)_2$	17.55	1.171	93.8	15	74.6	25	60.0	35	49.9	45	Wagner.
"	30.10	1.274	144.1	"	112.7	"	90.7	"	75.1	"	"
"	40.13	1.386	242.6	"	217.1	"	156.5	"	128.1	"	"
$CdCl_2$	11.09	1.109	77.5	15	60.5	25	49.1	35	40.7	45	"
"	16.30	1.181	88.9	"	70.5	"	57.5	"	47.2	"	"
"	24.79	1.320	104.0	"	80.4	"	64.6	"	53.6	"	"
$Cd(NO_3)_2$	7.81	1.074	61.9	15	50.1	25	41.1	35	34.0	45	"
"	15.71	1.159	71.8	"	58.7	"	48.8	"	41.3	"	"
"	22.36	1.241	85.1	"	69.0	"	57.3	"	47.5	"	"
$CdSO_4$	7.14	1.068	78.9	15	61.8	25	49.9	35	41.3	45	"
"	14.66	1.159	96.2	"	72.4	"	58.1	"	48.8	"	"
"	22.01	1.268	120.8	"	91.8	"	73.5	"	60.1	"	"
$CoCl_2$	7.97	1.081	83.0	15	65.1	25	53.6	35	44.9	45	"
"	14.86	1.161	111.6	"	85.1	"	73.7	"	58.8	"	"
"	22.27	1.264	161.6	"	126.6	"	101.6	"	85.6	"	"
$Co(NO_3)_2$	8.28	1.073	74.7	15	57.9	25	48.7	35	39.8	45	"
"	15.96	1.144	87.0	"	69.2	"	55.4	"	44.9	"	"
"	24.53	1.229	110.4	"	88.0	"	71.5	"	59.1	"	"
$CoSO_4$	7.24	1.086	86.7	15	68.7	25	55.0	35	45.1	45	"
"	14.16	1.159	117.8	"	95.5	"	76.0	"	61.7	"	"
"	21.17	1.240	193.6	"	146.2	"	113.0	"	89.9	"	"
$CuCl_2$	12.01	1.104	87.2	15	67.8	25	55.1	35	45.6	45	"
"	21.35	1.215	121.5	"	95.8	"	77.0	"	63.2	"	"
"	33.03	1.331	178.4	"	137.2	"	107.6	"	87.1	"	"
$Cu(NO_3)_2$	18.99	1.177	97.3	15	76.0	25	61.5	35	51.3	45	"
"	26.68	1.264	126.2	"	98.8	"	80.9	"	68.6	"	"
"	46.71	1.536	382.9	"	283.8	"	215.3	"	172.2	"	"
$CuSO_4$	6.79	1.055	79.6	15	61.8	25	49.8	35	41.4	45	"
"	12.57	1.115	98.2	"	74.0	"	59.7	"	52.0	"	"
"	17.49	1.163	124.5	"	96.8	"	75.9	"	61.8	"	"
HCl	8.14	1.037	71.0	15	57.9	25	48.3	35	40.1	45	"
"	16.12	1.084	80.0	"	66.5	"	56.4	"	48.1	"	"
"	23.04	1.114	91.8	"	79.9	"	65.9	"	56.4	"	"
$HgCl_2$	0.23	1.002	–	–	58.5	20	46.8	30	38.3	40	"
"	3.55	1.033	76.75	10	59.2	"	46.6	"	38.3	"	"

VISCOSITY OF SOLUTIONS.

Salt.	Percentage by weight of salt in solution.	Density.	μ	t	μ	t	μ	t	μ	t	Authority.
HNO_3	8.37	1.067	66.4	15	54.8	25	45.4	35	37.6	45	Wagner.
"	12.20	1.116	69.5	"	57.3	"	47.9	"	40.7	"	"
"	28.31	1.178	80.3	"	65.5	"	54.9	"	46.2	"	"
H_2SO_4	7.87	1.065	77.8	15	61.0	25	50.0	35	41.7	45	"
"	15.50	1.130	95.1	"	75.0	"	60.5	"	49.8	"	"
"	23.43	1.200	122.7	"	95.5	"	77.5	"	64.3	"	"
KCl	10.23	–	70.0	10	46.1	30	33.1	50	–	–	Sprung.
"	22.21	–	70.0	"	48.6	"	36.4	"	–	–	"
KBr	14.02	–	67.6	10	44.8	30	32.1	50	–	–	"
"	23.16	–	66.2	"	44.7	"	33.2	"	–	–	"
"	34.64	–	66.6	"	47.0	"	35.7	"	–	–	"
KI	8.42	–	69.5	10	44.0	30	31.3	50	–	–	"
"	17.01	–	65.3	"	42.9	"	31.4	"	–	–	"
"	33.03	–	61.8	"	42.9	"	32.4	"	–	–	"
"	45.98	–	63.0	"	45.2	"	35.3	"	–	–	"
"	54.00	–	68.8	"	48.5	"	37.6	"	–	–	"
$KClO_3$	3.51	–	71.7	10	44.7	30	31.5	50	–	–	"
"	5.69	–	–	"	45.0	"	31.4	"	–	–	"
KNO_3	6.32	–	70.8	10	44.6	30	31.8	50	–	–	"
"	12.19	–	68.7	"	44.8	"	32.3	"	–	–	"
"	17.60	–	68.8	"	46.0	"	33.4	"	–	–	"
K_2SO_4	5.17	–	77.4	10	48.6	30	34.3	50	–	–	"
"	9.77	–	81.0	"	52.0	"	36.9	"	–	–	"
K_2CrO_4	11.93	–	75.8	10	62.5	30	41.0	40	–	–	"
"	19.61	–	85.3	"	68.7	"	47.9	"	–	–	"
"	24.26	1.233	97.8	"	74.5	"	54.5	"	–	–	Slotte.
"	32.78	–	109.5	"	88.9	"	62.6	"	–	–	Sprung.
$K_2Cr_2O_7$	4.71	1.032	72.6	10	55.9	20	45.3	30	37.5	40	Slotte.
"	6.97	1.049	73.1	"	56.4	"	45.5	"	37.7	"	"
LiCl	7.76	–	96.1	10	59.7	30	41.2	50	–	–	Sprung.
"	13.91	–	121.3	"	75.9	"	52.6	"	–	–	"
"	26.93	–	229.4	"	142.1	"	98.0	"	–	–	"
$Mg(NO_3)_2$	18.62	1.102	99.8	15	81.3	25	66.5	35	56.2	45	Wagner.
"	34.19	1.200	213.3	"	164.4	"	132.4	"	109.9	"	"
"	39.77	1.430	317.0	"	250.0	"	191.4	"	158.1	"	"
$MgSO_4$	4.98	–	96.2	10	59.0	30	40.9	50	–	–	Sprung.
"	9.50	–	130.9	"	77.7	"	53.0	"	–	–	"
"	19.32	–	302.2	"	166.4	"	106.0	"	–	–	"
$MgCrO_4$	12.31	1.089	111.3	10	84.8	20	67.4	30	55.0	40	Slotte.
"	21.86	1.164	167.1	"	125.3	"	99.0	"	79.4	"	"
"	27.71	1.217	232.2	"	172.6	"	133.9	"	106.6	"	"
$MnCl_2$	8.01	1.096	92.8	15	71.1	25	57.5	35	48.1	45	Wagner.
"	15.65	1.196	130.9	"	104.2	"	84.0	"	68.7	"	"
"	30.33	1.337	256.3	"	193.2	"	155.0	"	123.7	"	"
"	40.13	1.453	537.3	"	393.4	"	300.4	"	246.5	"	"

VISCOSITY OF SOLUTIONS.

Salt.	Percentage by weight of salt in solution.	Density.	μ	t	μ	t	μ	t	μ	t	Authority.
$Mn(NO_3)_2$	18.31	1.148	96.0	15	76.4	25	64.5	35	55.6	45	Wagner.
"	29.60	1.323	167.5	"	126.0	"	104.6	"	88.6	"	"
"	49.31	1.506	396.8	"	301.1	"	221.0	"	188.8	"	"
$MnSO_4$	11.45	1.147	129.4	15	98.6	25	78.3	35	63.4	45	"
"	18.80	1.251	228.6	"	172.2	"	137.1	"	107.4	"	"
"	22.08	1.306	661.8	"	474.3	"	347.9	"	266.8	"	"
NaCl	7.95	–	82.4	10	52.0	30	31.8	50	–	–	Sprung.
"	14.31	–	94.8	"	60.1	"	36.9	"	–	–	"
"	23.22	–	128.3	"	79.4	"	47.4	"	–	–	"
NaBr	9.77	–	75.6	10	48.7	30	34.4	50	–	–	"
"	18.58	–	82.6	"	53.5	"	38.2	"	–	–	"
"	27.27	–	95.9	"	61.7	"	43.8	"	–	–	"
NaI	8.83	–	73.1	10	46.0	30	32.4	50	–	–	"
"	17.15	–	73.8	"	47.4	"	33.7	"	–	–	"
"	35.69	–	86.0	"	55.7	"	40.6	"	–	–	"
"	55.47	–	157.2	"	96.4	"	66.9	"	–	–	"
$NaClO_3$	11.50	–	78.7	10	50.0	30	35.3	50	–	–	"
"	20.59	–	88.9	"	56.8	"	40.4	"	–	–	"
"	33.54	–	121.0	"	75.7	"	53.0	"	–	–	"
$NaNO_3$	7.25	–	75.6	10	47.9	30	33.8	50	–	–	"
"	12.35	–	81.2	"	51.0	"	36.1	"	–	–	"
"	18.20	–	87.0	"	55.9	"	39.3	"	–	–	"
"	31.55	–	121.2	"	76.2	"	53.4	"	–	–	"
Na_2SO_4	4.98	–	96.2	10	59.0	30	40.9	50	–	–	"
"	9.50	–	130.9	"	77.7	"	53.0	"	–	–	"
"	14.03	–	187.9	"	107.4	"	71.1	"	–	–	"
"	19.32	–	302.2	"	166.4	"	106.0	"	–	–	"
Na_2CrO_4	5.76	1.058	85.8	10	66.6	20	53.4	30	43.8	40	Slotte.
"	10.62	1.112	103.3	"	79.3	"	63.5	"	52.3	"	"
"	14.81	1.164	127.5	"	97.1	"	77.3	"	63.0	"	"
NH_4Cl	3.67	–	71.5	10	45.0	30	31.9	50	–	–	Sprung.
"	8.67	–	69.1	"	45.3	"	32.6	"	–	–	"
"	15.68	–	67.3	"	46.2	"	34.0	"	–	–	"
"	23.37	–	67.4	"	47.7	"	36.1	"	–	–	"
NH_4Br	15.97	–	65.2	10	43.2	30	31.5	50	–	–	"
"	25.33	–	62.6	"	43.3	"	32.2	"	–	–	"
"	36.88	–	62.4	"	44.6	"	34.3	"	–	–	"
NH_4NO_3	5.97	–	69.6	10	44.3	30	31.6	50	–	–	"
"	12.19	–	66.8	"	44.3	"	31.9	"	–	–	"
"	27.08	–	67.0	"	47.7	"	34.9	"	–	–	"
"	37.22	–	71.7	"	51.2	"	38.8	"	–	–	"
"	49.83	–	81.1	"	63.3	"	48.9	"	–	–	"
$(NH_4)_2SO_4$	8.10	–	107.9	10	52.3	30	37.0	50	–	–	"
"	15.94	–	120.2	"	60.4	"	43.2	"	–	–	"
"	25.51	–	148.4	"	74.8	"	54.1	"	–	–	"

VISCOSITY OF SOLUTIONS.

Salt.	Percentage by weight of salt in solution.	Density.	μ	t	μ	t	μ	t	μ	t	Authority.
$(NH_4)_2CrO_4$	10.52	1.063	79.3	10	62.4	20	–	–	42.4	40	Slotte.
"	19.75	1.120	88.2	"	70.0	"	57.8	30	48.4	–	"
"	28.04	1.173	101.1	"	80.7	"	60.8	"	56.4	–	"
$(NH_4)_2Cr_2O_7$	6.85	1.039	72.5	10	56.3	20	45.8	30	38.0	40	"
"	13.00	1.078	72.6	"	57.2	"	46.8	"	39.1	"	"
"	19.93	1.126	77.6	"	58.8	"	48.7	"	40.9	"	"
$NiCl_2$	11.45	1.109	90.4	15	70.0	25	57.5	35	48.2	45	Wagner.
"	22.69	1.226	140.2	"	109.7	"	87.8	"	72.7	"	"
"	30.40	1.337	229.5	"	171.8	"	139.2	"	111.9	"	"
$Ni(NO_3)_2$	16.49	1.136	90.7	15	70.1	25	57.4	35	48.9	45	"
"	30.01	1.278	135.6	"	105.9	"	85.5	"	70.7	"	"
"	40.95	1.388	222.6	"	169.7	"	128.2	"	152.4	"	"
$NiSO_4$	10.62	1.092	94.6	15	73.5	25	60.1	35	49.8	45	"
"	18.19	1.198	154.9	"	119.9	"	99.5	"	75.7	"	"
"	25.35	1.314	298.5	"	224.9	"	173.0	"	152.4	"	"
$Pb(NO_3)_2$	17.93	1.179	74.0	15	59.1	25	48.5	35	40.3	45	"
"	32.22	1.362	91.8	"	72.5	"	59.6	"	50.6	"	"
$Sr(NO_3)_2$	10.29	1.088	69.3	15	56.0	25	45.9	35	39.1	45	"
"	21.19	1.124	87.3	"	69.2	"	57.8	"	48.1	"	"
"	32.61	1.307	116.9	"	93.3	"	76.7	"	62.3	"	"
$ZnCl_2$	15.33	1.146	93.6	15	72.7	25	57.8	35	48.2	45	"
"	23.49	1.229	111.5	"	86.6	"	69.8	"	57.5	"	"
"	33.78	1.343	151.7	"	117.9	"	90.0	"	72.6	"	"
$Zn(NO_3)_2$	15.95	1.115	80.7	15	64.3	25	52.6	35	43.8	45	"
"	30.23	1.229	104.7	"	85.7	"	69.5	"	57.7	"	"
"	44.50	1.437	167.9	"	130.6	"	105.4	"	87.9	"	"
$ZnSO_4$	7.12	1.106	97.1	15	79.3	25	62.7	35	51.5	45	"
"	16.64	1.195	156.0	"	118.6	"	94.2	"	73.5	"	"
"	23.09	1.281	232.8	"	177.4	"	135.2	"	108.1	"	"

TABLE 158.

SPECIFIC VISCOSITY.*

Dissolved salt.	Normal solution. Density.	Normal solution. Specific viscosity.	½ normal. Density.	½ normal. Specific viscosity.	¼ normal. Density.	¼ normal. Specific viscosity.	⅛ normal. Density.	⅛ normal. Specific viscosity.	Authority.
Acids: Cl_2O_3 . .	1.0562	1.012	1.0283	1.003	1.0143	1.000	1.0074	0.999	Reyher.
HCl . . .	1.0177	1.067	1.0092	1.034	1.0045	1.017	1.0025	1.009	"
$HClO_3$ . .	1.0485	1.052	1.0244	1.025	1.0126	1.014	1.0064	1.006	"
HNO_3 . .	1.0332	1.027	1.0168	1.011	1.0086	1.005	1.0044	1.003	"
H_2SO_4 . .	1.0303	1.090	1.0154	1.043	1.0074	1.022	1.0035	1.008	Wagner.
Aluminium sulphate	1.0550	1.406	1.0278	1.178	1.0138	1.082	1.0068	1.038	"
Barium chloride . .	1.0884	1.123	1.0441	1.057	1.0226	1.026	1.0114	1.013	"
" nitrate . .	–	–	1.0518	1.044	1.0259	1.021	1.0130	1.008	"
Calcium chloride .	1.0446	1.156	1.0218	1.076	1.0105	1.036	1.0050	1.017	"
" nitrate . .	1.0596	1.117	1.0300	1.053	1.0151	1.022	1.0076	1.008	"
Cadmium chloride .	1.0779	1.134	1.0394	1.063	1.0197	1.031	1.0098	1.020	"
" nitrate .	1.0954	1.165	1.0479	1.074	1.0249	1.038	1.0119	1.018	"
" sulphate .	1.0973	1.348	1.0487	1.157	1.0244	1.078	1.0120	1.033	"
Cobalt chloride . .	1.0571	1.204	1.0286	1.097	1.0144	1.048	1.0058	1.023	"
" nitrate . .	1.0728	1.166	1.0369	1.075	1.0184	1.032	1.0094	1.018	"
" sulphate . .	1.0750	1.354	1.0383	1.160	1.0193	1.077	1.0110	1.040	"
Copper chloride . .	1.0624	1.205	1.0313	1.098	1.0158	1.047	1.0077	1.027	"
" nitrate . .	1.0755	1.179	1.0372	1.080	1.0185	1.040	1.0092	1.018	"
" sulphate .	1.0790	1.358	1.0402	1.160	1.0205	1.080	1.0103	1.038	"
Lead nitrate . . .	1.1380	1.101	0.0699	1.042	1.0351	1.017	1.0175	1.007	"
Lithium chloride .	1.0243	1.142	1.0129	1.066	1.0062	1.031	1.0030	1.012	"
" sulphate .	1.0453	1.290	1.0234	1.137	1.0115	1.065	1.0057	1.032	"
Magnesium chloride	1.1375	1.201	1.0188	1.094	1.0091	1.044	1.0043	1.021	"
" nitrate .	1.0512	1.171	1.0259	1.082	1.0130	1.040	1.0066	1.020	"
" sulphate	1.0584	1.367	1.0297	1.164	1.0152	1.078	1.0076	1.032	"
Manganese chloride	1.0513	1.209	1.0259	1.098	1.0125	1.048	1.0063	1.023	"
" nitrate .	1.0690	1.183	1.0349	1.087	1.0174	1.043	1.0093	1.023	"
" sulphate	1.0728	1.364	1.0365	1.169	1.0179	1.076	1.0087	1.037	"
Nickel chloride . .	1.0591	1.205	1.0308	1.097	1.0144	1.044	1.0067	1.021	"
" nitrate . . .	1.0755	1.180	1.0381	1.084	1.0192	1.042	1.0096	1.019	"
" sulphate . .	1.0773	1.361	1.0391	1.161	1.0198	1.075	1.0017	1.032	"
Potassium chloride .	1.0466	0.987	1.0235	0.987	1.0117	0.990	1.0059	0.993	"
" chromate	1.0935	1.113	1.0475	1.053	1.0241	1.022	1.0121	1.012	"
" nitrate .	1.0605	0.975	1.0305	0.982	1.0161	0.987	1.0075	0.992	"
" sulphate	1.0664	1.105	1.0338	1.049	1.0170	1.021	1.0084	1.008	"
Sodium chloride . .	1.0401	1.097	1.0208	1.047	1.0107	1.024	1.0056	1.013	Reyher.
" bromide . .	1.0786	1.064	1.0396	1.030	1.0190	1.015	1.0100	1.008	"
" chlorate .	1.0710	1.090	1.0359	1.042	1.0180	1.022	1.0092	1.012	"
" nitrate . .	1.0554	1.065	1.0281	1.026	1.0141	1.012	1.0071	1.007	"
Silver nitrate . . .	1.1386	1.058	1.0692	1.020	1.0348	1.006	1.0173	1.000	Wagner.
Strontium chloride .	1.0676	1.141	1.0336	1.067	1.0171	1.034	1.0084	1.014	"
" nitrate .	1.0822	1.115	1.0419	1.049	1.0208	1.024	1.0104	1.011	"
Zinc chloride . . .	1.0590	1.189	1.0302	1.096	1.0152	1.053	1.0077	1.024	"
" nitrate . . .	1.0758	1.164	1.0404	1.086	1.0191	1.039	1.0096	1.019	"
" sulphate . . .	1.0792	1.367	1.0402	1.173	1.0198	1.082	1.0094	1.036	"

* In the case of solutions of salts it has been found (*vide* Arrhennius, Zeits. für Phys. Chem. vol. 1, p. 285) that the specific viscosity can, in many cases, be nearly expressed by the equation $\mu = \mu_1^n$, where μ_1 is the specific viscosity for a normal solution referred to the solvent at the same temperature, and n the number of gramme molecules in the solution under consideration. The same rule may of course be applied to solutions stated in percentages instead of gramme molecules. The table here given has been compiled from the results of Reyher (Zeits. für Phys. Chem. vol. 2, p. 749) and of Wagner (Zeits. für Phys. Chem. vol. 5, p. 31) and illustrates this rule. The numbers are all for 25° C.

Table 159.

VISCOSITY OF GASES AND VAPORS.

The values of μ given in the table are 10^6 times the coefficients of viscosity in C. G. S. units.

Substance.	Temp. °C	μ	Reference.	Substance.	Temp. °C		Reference.
Acetone	18.0	78.	1	Ether	16.1	73.2	1
Air *	−21.4	163.9	2	"	36.5	79.3	1
"	0.0	173.3	2	Ethyl chloride	0.	93.5	4
"	15.0	180.7	2	Ethyl iodide	72.3	216.0	3
"	99.1	220.3	2	Ethylene	0.0	96.1	2
"	182.4	255.9	2	Helium	0.0	189.1	5
"	302.0	299.3	2	"	15.3	196.9	5
Alcohol, Methyl	66.8	135.	3	"	66.6	234.8	5
Alcohol, Ethyl	78.4	142.	3	"	184.6	269.9	5
Alcohol, Propyl,				Hydrogen	−20.6	81.9	2
norm.	97.4	142.	3	"	0.0	86.7	10
Alcohol, Isopropyl	82.8	162.	3	"	15.	88.9	2
Alcohol, Butyl, norm.	116.9	143.	3	"	99.2	105.9	2
Alcohol, Isobutyl	108.4	144.	3	"	182.4	121.5	2
Alcohol, Tert. butyl.	82.9	160.	3	"	302.0	139.2	2
Ammonia	0.0	96.	4	Krypton	15.0	246.	11
"	20.0	108.	4	Mercury	270.0	489.†	8
Argon	0.0	210.4	5	"	300.0	532.†	8
"	14.7	220.8	5	"	330.0	582.†	8
"	17.9	224.1	5	"	360.0	627.†	8
"	99.7	273.3	5	"	390.0	671.†	8
"	183.7	322.1	5	Methane	20.0	120.1	4
Benzene	0.	70.	10	Methyl chloride	0.0	98.8	2
"	19.0	79.	6	" "	15.0	105.2	2
"	100.0	118.	6	" "	302.0	213.9	2
Carbon bisulphide	16.9	92.4	1	Methyl iodide	44.0	232.	3
Carbon dioxide	−20.7	129.4	2	Nitrogen	−21.5	156.3	7
" "	0.	142.	10	"	0.	166.	10
" "	15.0	145.7	2	"	10.9	170.7	7
" "	99.1	186.1	2	"	53.5	189.4	7
" "	182.4	222.1	2	Nitric oxide	0.	179.	10
" "	302.0	268.2	2	Nitrous oxide	0.	138.	10
Carbon monoxide	0.0	163.0	10	Oxygen	0.	189.	10
" "	20.0	184.0	4	"	15.4	195.7	7
Chlorine	0.0	128.7	4	"	53.5	215.9	7
"	20.0	147.0	4	Water Vapor	0.0	90.4	1
Chloroform	0.0	95.9	1	" "	16.7	96.7	1
"	17.4	102.9	1	" "	100.0	132.0	9
"	61.2	189.0	3	Xenon	15.	222.	11
Ether	0.0	68.9	1				

1 Puluj, Wien. Ber. 69 (2), 1874.
2 Breitenbach, Ann. Phys. 5, 1901.
3 Steudel, Wied. Ann. 16, 1882.
4 Graham, Philos. Trans. Lond. 1846, III.
5 Schultze, Ann. Phys. (4), 5, 6, 1901.
6 Schumann, Wied. Ann. 23, 1884.
7 Obermayer, Wien. Ber. 71 (2a), 1875.
8 Koch, Wied. Ann. 14, 1881, 19, 1883.
9 Meyer-Schumann, Wied. Ann. 13, 1881.
10 Jeans, assumed mean, 1916.
11 Rankine, 1910.
12 Vogel (Eucken, Phys. Z. 14, 1913). For summaries see: Fisher, Phys. Rev. 24, 1904; Chapman, Phil. Tr. A. 211, 1911; Gilchrist, Phys. Rev. 1, 1913. Schmidt, Ann. d. Phys. 30, 1909.

* Gilchrist's value of the viscosity of air may be taken as the most accurate at present available. His value at 20.2° C is 1.812×10^{-4}. The temperature variation given by Holman (Phil. Mag. 1886) gives $\mu = 1715.50 \times 10^{-7}(1 + .00275t - .00000034t^2)$. See Phys. Rev. 1, 1913. Millikan (Ann. Phys. 41, 759, 1913) gives for the most accurate value $\mu_t = 0.00018240 - 0.000000493(23 - t)$ when $(23 > t > 12)$ whence $\mu_{20} = 0.0001809 \pm 0.1\%$. For μ_0 he gives 0.0001711.

† The values here given were calculated from Koch's table (Wied. Ann. 19, p. 869, 1883) by the formula $\mu = 489 [1 + 746(t - 270)]$.

VISCOSITY OF GASES.

Variation of Viscosity with Pressure and Temperature.

According to the kinetic theory of gases the coefficient of viscosity $\mu = \frac{1}{3}(\rho \bar{c} l)$, ρ being the density, $\bar{c}$ the average velocity of the molecules, l the average path. Since l varies inversely as the number of molecules per unit volume, ρl is a constant and μ should be independent of the density and pressure of a gas (Maxwell's law). This has been found true for ordinary pressures; below $\frac{1}{60}$ atmosphere it may fail, and for certain gases it has been proved untrue for high pressures, e.g., CO_2 at 33° and above 50 atm. See Jeans, "Dynamical Theory of Gases."

If B is the amount of momentum transferred from a plane moving with velocity U and parallel to a stationary plane distant d, and s is a quantity (coefficient of slip) to allow for the slipping of the gas molecules over the plane, then $\mu = (B/U)(d + 2s)$; s is of the same magnitude as l, probably between .7 (Timiriazeff) and .9 (Knudsen) of it; at low pressures d becomes negligible compared with $2s$ and the viscosity should vary inversely as the pressure.

$\bar{c}$ depends only on the temperature and the molecular weight; viscosity should, therefore, increase with the temperature for gases. $\bar{c}$ varies as the $\sqrt{T}$, but μ has been found to increase much more rapidly. Meyer's formula, $\mu_t = \mu_0(1 + at)$, where a is a constant and μ_0 the viscosity at 0° C, is a convenient approximate relation. Sutherland's formula (Phil. Mag. 31, 1893).

$$\mu_t = \mu_o \frac{273 + C}{T + C}\left(\frac{T}{273}\right)^{\frac{3}{2}},$$

is the most accurate formula in use, taking in account the effect of molecular forces. It holds for temperatures above the critical and for pressures following approximately Boyle's law. It may be thrown into the form $T = KT^{\frac{3}{2}}/\mu - C$ which is linear in terms of T and $T^{\frac{3}{2}}/\mu$, with a slope equal to K and the ordinate intercept equal to $-C$. See Fisher, Phys. Rev. 24, 1907, from which most of the following table is taken. Onnes (see Jeans) shows that this formula does not represent Helium at low temperatures with anything like the accuracy of the simpler formula $\mu = \mu_0(T/273.1)^n$.

The following table contains the constants for the above three formulae, T being always the absolute temperature, Centigrade scale.

Gas.	C	$K \times 10^7$	a	n*	Gas.	C	$K \times 10^7$	a	n*
Air	124	150	—	.754	Hydrogen	72	66	—	.69
Argon	172	206	—	.819	Krypton	188	—	—	—
Carbon monoxide	102	135	.00269	.74	Neon	252	—	—	—
Carbon dioxide	240	158	.00348	.98	Nitrogen	110	143	.00269	.74
Chloroform	454	159	—	—	Nitrous oxide, N_2O	313	172	.00345	.93
Ethylene	226	106	.00350	—	Oxygen	131	176	—	.79
Helium	80	148	—	.683	Xenon	252	—	—	—
Helium	—	—	—	.647					

* The authorities for n are: Air, Rayleigh; Ar, Mean, Rayleigh, Schultze; CO, CO_2, N_2, N_2O, von Obermayer; Helium, Mean, Rayleigh, Schultze; 2d value, low temperature work of Onnes; H_2, O_2, Mean, Rayleigh, von Obermayer.

TABLE 161.

DIFFUSION OF AN AQUEOUS SOLUTION INTO PURE WATER.

If k is the coefficient of diffusion, dS the amount of the substance which passes in the time dt, at the place x, through q sq. cm. of a diffusion cylinder under the influence of a drop of concentration dc/dx, then

$$dS = -kq \frac{dc}{dx} dt.$$

k depends on the temperature and the concentration. c gives the gram-molecules per liter. The unit of time is a day.

Substance.	c	$t°$	k	Reference	Substance.	c	$t°$	k	Reference.
Bromine . . .	0.1	12.	0.8	1	Calcium chloride .	0.864	8.5	0.70	4
Chlorine . . .	"	12.	1.22	"	" " . .	1.22	9.	0.72	"
Copper sulphate .	"	17.	0.39	2	" " . .	0.060	9.	0.64	"
Glycerine . .	"	10.14	0.357	3	" " . .	0.047	9.	0.68	"
Hydrochloric acid .	"	19.2	2.21	2	Copper sulphate .	1.95	17.	0.23	2
Iodine . . .	"	12.	(0.5)	1	" " . .	0.95	17.	0.26	"
Nitric acid . .	"	19.5	2.07	2	" " . .	0.30	17.	0.33	"
Potassium chloride .	"	17.5	1.38	2	" " . .	0.005	17.	0.47	"
" hydroxide .	"	13.5	1.72	2	Glycerine . .	2/8	10.14	0.354	3
Silver nitrate . .	"	12.	0.985	2	" . . .	6/8	10.14	0.345	"
Sodium chloride .	"	15.0	0.94	2	" . . .	10/8	10.14	0.329	"
Urea . . .	"	14.8	0.97	3	" . . .	14/8	10.14	0.300	"
Acetic acid . .	0.2	13.5	0.77	4	Hydrochloric acid .	4.52	11.5	2.93	4
Barium chloride .	"	8.	0.66	4	" " .	3.16	11.	2.67	"
Glycerine . .	"	10.1	3.55	3	" " ,	0.945	11.	2.12	"
Sodium actetate .	"	12.	0.67	5	" " .	0.387	11.	2.02	"
" chloride .	"	15.0	0.94	2	" " .	0.250	11.	1.84	"
Urea . . .	"	14.8	0.969	3	Magnesium sulphate	2.18	5.5	0.28	4
Acetic acid . .	1.0	12.	0.74	6	" " .	0.541	5.5	0.32	"
Ammonia . .	"	15.23	1.54	7	" " .	3.23	10.	0.27	"
Formic acid . .	"	12.	0.97	7	" " .	0.402	10.	0.34	"
Glycerine . .	"	10.14	0.339	3	Potassium hydroxide	0.75	12.	1.72	6
Hydrochloric acid .	"	12.	2.09	6	" " .	0.49	12.	1.70	"
Magnesium sulphate	"	7.	0.30	4	" " .	0.375	12.	1.70	"
Potassium bromide .	"	10.	1.13	8	" nitrate .	3.9	17.6	0.89	2
" hydroxide .	"	12.	1.72	6	" " .	1.4	17.6	1.10	"
Sodium chloride .	"	15.0	0.94	2	" " .	0.3	17.6	1.26	"
" " .	"	14.3	0.964	3	" " .	0.02	17.6	1.28	"
" hydroxide .	"	12.	1.11	2	" sulphate	0.95	19.6	0.79	"
" iodide .	"	10.	0.80	8	" " .	0.28	19.6	0.86	"
Sugar . . .	"	12.	0.254	6	" " .	0.05	19.6	0.97	"
Sulphuric acid .	"	12.	1.12	6	" " .	0.02	19.6	1.01	"
Zinc sulphate . .	"	14.8	0.236	9	Silver nitrate . .	3.9	12.	0.535	"
Acetic acid . .	2.0	12.	0.69	6	" " . .	0.9	12.	0.88	"
Calcium chloride .	"	10.	0.68	8	" " . .	0.02	12.	1.035	"
Cadmium sulphate .	"	19.04	0.246	9	Sodium chloride .	2/8	14.33	1.013	3
Hydrochloric acid .	"	12.	2.21	6	" " . .	4/8	14.33	0.996	"
Sodium iodide .	"	10.	0.90	8	" " . .	6/8	14.33	0.980	2
Sulphuric acid .	"	12.	1.16	6	" " . .	10/8	14.33	0.948	"
Zinc acetate . .	"	18.05	0.210	9	" " . .	14/8	14.33	0.917	"
" " . .	"	0.04	0.120	9	Sulphuric acid .	9.85	18.	2.36	2
Acetic acid . .	3.0	12.	0.68	–	" " . .	4.85	18.	1.90	"
Potassium carbonate	"	10.	0.60	8	" " . .	2.85	18.	1.60	"
" hydroxide	"	12.	1.89	6	" " . .	0.85	18.	1.34	"
Acetic acid . .	4.0	12.	0.66	6	" " . .	0.35	18.	1.32	"
Potassium chloride .	"	10.	1.27	8	" " . .	0.005	18.	1.30	"

1 Euler, Wied. Ann. 63, 1897.
2 Thovert, C. R. 133, 1901; 134, 1902.
3 Heimbrodt, Diss. Leipzig, 1903.
4 Scheffer, Chem. Ber. 15, 1882; 16, 1883; Zeitschr. Phys. Chem. 2, 1888.
5 Kawalki, Wied. Ann. 52, 1894; 59, 1896.
6 Arrhenius, Zeitschr. Phys. Chem. 10, 1892.
7 Abegg, Zeitschr. Phys. Chem. 11, 1893.
8 Schuhmeister, Wien. Ber. 79 (2), 1879.
9 Seitz, Wied. Ann. 64, 1898.

Compiled from Landolt-Börnstein-Meyerhoffer's Physikalisch-chemische Tabellen.

DIFFUSION OF VAPORS.

Coefficients of diffusion of vapors in C. G. S. units. The coefficients are for the temperatures given in the table and a pressure of 76 centimeters of mercury.*

Vapor.	Temp. C. °	k_t for vapor diffusing into hydrogen.	k_t for vapor diffusing into air.	k_t for vapor diffusing into carbon dioxide.
Acids: Formic	0.0	0.5131	0.1315	0.0879
"	65.4	0.7873	0.2035	0.1343
"	84.9	0.8830	0.2244	0.1519
Acetic	0.0	0.4040	0.1061	0.0713
"	65.5	0.6211	0.1578	0.1048
"	98.5	0.7481	0.1965	0.1321
Isovaleric	0.0	0.2118	0.0555	0.0375
"	98.0	0.3934	0.1031	0.0696
Alcohols: Methyl	0.0	0.5001	0.1325	0.0880
"	25.6	0.6015	0.1620	0.1046
"	49.6	0.6738	0.1809	0.1234
Ethyl	0.0	0.3806	0.0994	0.0693
"	40.4	0.5030	0.1372	0.0898
"	66.9	0.5430	0.1475	0.1026
Propyl	0.0	0.3153	0.0803	0.0577
"	66.9	0.4832	0.1237	0.0901
"	83.5	0.5434	0.1379	0.0976
Butyl	0.0	0.2716	0.0681	0.0476
"	99.0	0.5045	0.1265	0.0884
Amyl	0.0	0.2351	0.0589	0.0422
"	99.1	0.4362	0.1094	0.0784
Hexyl	0.0	0.1998	0.0499	0.0351
"	99.0	0.3712	0.0927	0.0651
Benzene	0.0	0.2940	0.0751	0.0527
"	19.9	0.3409	0.0877	0.0609
"	45.0	0.3993	0.1011	0.0715
Carbon disulphide	0.0	0.3690	0.0883	0.0629
" "	19.9	0.4255	0.1015	0.0726
" "	32.8	0.4626	0.1120	0.0789
Esters: Methyl acetate	0.0	0.3277	0.0840	0.0557
" "	20.3	0.3928	0.1013	0.0679
Ethyl "	0.0	0.2373	0.0630	0.0450
" "	46.1	0.3729	0.0970	0.0666
Methyl butyrate	0.0	0.2422	0.0640	0.0438
" "	92.1	0.4308	0.1139	0.0809
Ethyl "	0.0	0.2238	0.0573	0.0406
" "	96.5	0.4112	0.1064	0.0756
" valerate	0.0	0.2050	0.0505	0.0366
" "	97.6	0.3784	0.0932	0.0676
Ether	0.0	0.2960	0.0775	0.0552
"	19.9	0.3410	0.0893	0.0636
Water	0.0	0.6870	0.1980	0.1310
"	49.5	1.0000	0.2827	0.1811
"	92.4	1.1794	0.3451	0.2384

* Taken from Winkelmann's papers (Wied. Ann. vols. 22, 23, and 26). The coefficients for 0° were calculated by Winkelmann on the assumption that the rate of diffusion is proportional to the absolute temperature. According to the investigations of Loschmidt and of Obermeyer the coefficient of diffusion of a gas, or vapor, at 0° C. and a pressure of 76 centimetres of mercury may be calculated from the observed coefficient at another temperature and pressure by the formula $k_0 = k_T \left(\frac{T_0}{T}\right)^n \frac{76}{p}$, where T is temperature absolute and p the pressure of the gas. The exponent n is found to be about 1.75 for the permanent gases and about 2 for condensible gases. The following are examples: Air — CO_2, $n = 1.968$; CO_2 — N_2O, $n = 2.05$; CO_2 — H, $n = 1.742$; CO — O, $n = 1.785$; H — O, $n = 1.755$; O — N, $n = 1.792$. Winkelmann's results, as given in the above table, seem to give about 2 for vapors diffusing into air, hydrogen or carbon dioxide.

DIFFUSION OF GASES, VAPORS, AND METALS.

TABLE 163. — Coefficients of Diffusion for Various Gases and Vapors.*

Gas or Vapor diffusing.	Gas or Vapor diffused into.	Temp. °C.	Coefficient of Diffusion.	Authority.
Air	Hydrogen	0	0.661	Schulze.
"	Oxygen	0	0.1775	Obermayer.
Carbon dioxide	Air	0	0.1423	Loschmidt.
" "	"	0	0.1360	Waitz.
" "	Carbon monoxide	0	0.1405	Loschmidt.
" "	" "	0	0.1314	Obermayer.
" "	Hydrogen	0	0.5437	"
" "	Methane	0	0.1465	"
" "	Nitrous oxide	0	0.0983	Loschmidt.
" "	Oxygen	0	0.1802	"
Carbon disulphide	Air	0	0.0995	Stefan.
Carbon monoxide	Carbon dioxide	0	0.1314	Obermayer.
" "	Ethylene	0	0.101	"
" "	Hydrogen	0	0.6422	Loschmidt.
" "	Oxygen	0	0.1802	"
" "	"	0	0.1872	Obermayer.
Ether	Air	0	0.0827	Stefan.
"	Hydrogen	0	0.3054	"
Hydrogen	Air	0	0.6340	Obermayer.
"	Carbon dioxide	0	0.5384	"
"	" monoxide	0	0 6488	"
"	Ethane	0	0.4593	"
"	Ethylene	0	0.4863	"
"	Methane	0	0.6254	"
"	Nitrous oxide	0	0.5347	"
"	Oxygen	0	0.6788	"
Nitrogen	"	0	0.1787	"
Oxygen	Carbon dioxide	0	0.1357	"
"	Hydrogen	0	0.7217	Loschmidt.
"	Nitrogen	0	0.1710	Obermayer.
Sulphur dioxide	Hydrogen	0	0.4828	Loschmidt.
Water	Air	8	0.2390	Guglilemo.
"	"	18	0.2475	"
"	Hydrogen	18	0.8710	"

* Compiled for the most part from a similar table in Landolt & Börnstein's Phys. Chem. Tab.

TABLE 164.— Diffusion of Metals into Metals.

$\frac{dv}{dt} = k\frac{d^2v}{dx^2}$; where x is the distance in direction of diffusion; v, the degree of concentration of the diffusing metal; t, the time; k, the diffusion constant = the quantity of metal in grams diffusing through a sq. cm. in a day when unit difference of concentration (gr. per cu. cm.) is maintained between two sides of a layer one cm. thick.

Diffusing Metal.	Dissolving Metal.	Temperature °C.	k.	Diffusing Metal.	Dissolving Metal.	Temperature °C.	k.
Gold	Lead	555	3.19	Platinum	Lead	492	1.69
"	"	492	3.00	Lead	Tin	555	3.18
"	"	251	0.03	Rhodium	Lead	550	3.04
"	"	200	0.008	Tin	Mercury	15	1.22*
"	"	165	0.004	Lead	"	15	1.0*
"	"	100	0.00002	Zinc	"	15	1.0*
"	Bismuth	555	4.52	Sodium	"	15	0.45*
"	Tin	555	4.65	Potassium	"	15	0.40*
Silver	"	555	4.14	Gold	"	15	0.72*

From Roberts-Austen, Philosophical Transactions, 187A, p. 383, 1896.
* These values are from Guthrie.

Table 165.

SOLUBILITY OF INORGANIC SALTS IN WATER; VARIATION WITH THE TEMPERATURE.

The numbers give the number of grams of the *anhydrous* salt soluble in 1000 grams of water at the given temperatures.

Salt.	Temperature Centigrade.										
	0°	10°	20°	30°	40°	50°	60°	70°	80°	90°	100°
$AgNO_3$	1150	1600	2150	2700	3350	4000	4700	5500	6500	7600	9100
$Al_2(SO_4)_3$	313	335	362	404	457	521	591	662	731	808	891
$Al_2K_2(SO_4)_4$	30	–	–	84	–	–	248	–	–	–	1540
$Al_2(NH_4)_2(SO_4)_4$	26	45	66	91	124	159	211	270	352	–	–
B_2O_3	11	15	22	–	40	–	62	–	95	–	157
$BaCl_2$	316	333	357	382	408	436	464	494	524	556	588
$Ba(NO_3)_2$	50	70	92	116	142	171	203	236	270	306	342
$CaCl_2$	595	650	745	1010	1153	–	1368	1417	1470	1527	1590
$CoCl_2$	405	450	500	565	650	935	940	950	960	–	1030
$CsCl$	1614	1747	1865	1973	2080	2185	2290	2395	2500	2601	2705
$CsNO_3$	93	149	230	339	472	644	838	1070	1340	1630	1970
Cs_2SO_4	1671	1731	1787	1841	1899	1949	1999	2050	2103	2149	2203
$Cu(NO_3)_2$	818	–	1250	–	1598	–	1791	–	2078	–	–
$CuSO_4$	149	–	–	255	295	336	390	457	535	627	735
$FeCl_2$	–	–	685	–	–	820	–	–	1040	1050	1060
Fe_2Cl_6	744	819	918	–	–	3151	–	–	5258	–	5357
$FeSO_4$	156	208	264	330	402	486	550	560	506	430	–
$HgCl_2$	43	66	74	84	96	113	139	173	243	371	540
KBr	540	–	650	–	760	–	860	–	955	–	1050
K_2CO_3	1050	–	–	1140	1170	1210	1270	1330	1400	1470	1560
KCl	285	312	343	373	401	429	455	483	510	538	566
$KClO_3$	33	50	71	101	145	197	260	325	396	475	560
K_2CrO_4	589	609	629	650	670	690	710	730	751	771	791
$K_2Cr_2O_7$	50	85	131	–	292	–	505	–	730	–	1020
$KHCO_3$	225	277	332	390	453	522	600	–	–	–	–
KI	1279	1361	1442	1523	1600	1680	1760	1840	1920	2010	2090
KNO_3	133	209	316	458	639	855	1099	1380	1690	2040	2460
KOH	970	1030	1120	1260	1360	1400	1460	1510	1590	1680	1780
K_2PtCl_6	7	9	11	14	18	22	26	32	38	45	52
K_2SO_4	74	92	111	130	148	165	182	198	214	228	241
$LiOH$	127	127	128	129	130	133	138	144	153	–	175
$MgCl_2$	528	535	545	–	575	–	610	–	660	–	730
$MgSO_4$ (7aq)	260	309	356	409	456	–	–	–	–	–	–
" (6aq)	408	422	439	453	–	504	550	596	642	689	738
NH_4Cl	297	333	372	414	458	504	552	602	656	713	773
NH_4HCO_3	119	159	210	270	–	–	–	–	–	–	–
NH_4NO_3	1183	–	–	2418	2970	3540?	4300?	5130?	5800	7400	8710
$(NH_4)_2SO_4$	706	730	754	780	810	844	880	916	953	992	1033
$NaBr$	795	845	903	–	1058	1160	1170	–	1185	–	1205
$Na_2B_4O_7$	–	16	–	39	–	105	200	244	314	408	523
Na_2CO_3 (10aq)	71	126	214	409	–	–	–	–	–	–	–
" (7aq)	204	263	335	435	(1aq)	475	464	458	452	452	452
$NaCl$	356	357	358	360	363	367	371	375	380	385	391
$NaClO_3$	820	890	990	–	1235	–	1470	–	1750	–	2040
Na_2CrO_4	317	502	900	–	960	1050	1150	–	1240	–	1260
$Na_2Cr_2O_7$	1630	1700	1800	1970	2200	2480	2830	3230	3860	–	4330
$NaHCO_3$	69	82	96	111	127	145	164	–	–	–	–
Na_2HPO_4	25	39	93	241	639	–	–	949	–	–	988
NaI	1590	1690	1790	1900	2050	2280	2570	–	2950	–	3020
$NaNO_3$	730	805	880	962	1049	1140	1246	1360	1480	1610	1755

Compiled from Landolt-Börnstein-Meyerhoffer's Physikalisch-chemische Tabellen.

SOLUBILITY OF SALTS AND GASES IN WATER.

TABLE 165 (*concluded*) — Solubility of Inorganic Salts in Water; Variation with the Temperature.

The numbers give the number of grams of the *anhydrous* salt soluble in 1000 grams of water at the given temperatures.

Salt.	Temperature Centigrade.										
	0°	10°	20°	30°	40°	50°	60°	70°	80°	90°	100°
$NaOH$	420	515	1090	1190	1290	1450	1740	–	3130	–	–
$Na_4P_2O_7$	32	39	62	99	135	174	220	255	300	–	–
Na_2SO_3	141	–	287	–	495	–	–	–	–	–	330
Na_2SO_4 (10aq)	50	90	194	400	} 482	468	455	445	437	429	427
" (7aq)	196	305	447	–							
$Na_2S_2O_3$	525	610	700	847	1026	1697	2067	–	2488	2542	2660
$NiCl_2$	–	600	640	680	720	760	810	–	–	–	–
$NiSO_4$	272	–	–	425	–	502	548	594	632	688	776
$PbBr_2$	5	6	8	12	15	20	24	28	33	–	48
$Pb(NO_3)_2$	365	444	523	607	694	787	880	977	1076	1174	1270
$RbCl$	770	844	911	976	1035	1093	1155	1214	1272	1331	1389
$RbNO_3$	195	330	533	813	1167	1556	2000	2510	3090	3750	4520
Rb_2SO_4	364	426	482	535	585	631	674	714	750	787	818
$SrCl_2$	442	483	539	600	667	744	831	896	924	962	1019
SnI_2	–	–	10	12	14	17	21	25	30	34	40
$Sr(NO_3)_2$	395	549	708	876	913	926	940	956	972	990	1011
$Th(SO_4)_2$ (9aq)	7	10	14	20	30	51	–	–	–	–	–
" (4aq)	–	–	–	–	40	25	16	11	–	–	–
$TlCl$	2	2	3	5	6	8	10	13	16	20	–
$TlNO_3$	39	62	96	143	209	304	462	695	1110	2000	4140
Tl_2SO_4	27	37	49	62	76	92	109	127	146	165	–
$Yb_2(SO_4)_3$	442	–	–	–	–	–	104	72	69	58	47
$Zn(NO_3)_2$	948	–	–	–	2069	–	–	–	–	–	–
$ZnSO_4$	–	–	–	–	700	768	–	890	860	920	785

TABLE 166. — Solubility of a Few Organic Salts in Water; Variation with the Temperature.

Salt.	0°	10°	20°	30°	40°	50°	60°	70°	80°	90°	100°
$H_2(CO_2)_2$	36	53	102	159	228	321	445	635	978	1200	–
$H_2(CH_2.CO_2)_2$	28	45	69	106	162	244	358	511	708	–	1209
Tartaric acid	1150	1260	1390	1560	1760	1950	2180	2440	2730	3070	3430
Racemic "	92	140	206	291	433	595	783	999	1250	1530	1850
$K(HCO_2)$	2900	–	3350	–	3810	–	4550	–	5750	–	7900
$KH(C_4H_4O_6)$	3	4	6	9	13	18	24	32	45	57	69

TABLE 167. — Solubility of Gases in Water; Variation with the Temperature.

The table gives the weight in grams of the gas which will be absorbed in 1000 grams of water when the partial pressure of the gas plus the vapor pressure of the liquid at the given temperature equals 760 mm.

Gas.	0°	10°	20°	30°	40°	50°	60°	70°	80°
O_2	.0705	.0551	.0443	.0368	.0311	.0263	.0221	.0181	.0135
H_2	.00192	.00174	.00160	.00147	.00138	.00129	.00118	.00102	.00079
N_2	.0293	.0230	.0189	.0161	.0139	.0121	.0105	.0089	.0069
Br_2	431.	248.	148.	94.	62.	40.	28.	18.	11.
Cl_2	–	9.97	7.29	5.72	4.59	3.93	3.30	2.79	2.23
CO_2	3.35	2.32	1.69	1.26	0.97	0.76	0.58	–	–
H_2S	7.10	5.30	3.98	–	–	–	–	–	–
NH_3	987.	689.	535.	422.	–	–	–	–	–
SO_2	228.	162.	113.	78.	54.	–	–	–	–

Compiled from Landolt-Börnstein-Meyerhoffer's Physikalisch-chemische Tabellen.

TABLE 168.

CHANGE OF SOLUBILITY PRODUCED BY UNIFORM PRESSURE.*

Pressure in atmospheres.	$CdSO_4 8/3H_2O$ at 25°		$ZnSO_4.7H_2O$ at 25°		Mannite at 24.05°		NaCl at 24.05°	
	Conc. of satd. soln. gs. $CdSO_4$ per 100 gs. H_2O	Percentage change.	Conc. of satd. soln. gs. $ZnSO_4$ per 100 gs. H_2O.	Percentage change.	Conc. of satd. soln. gs. mannite per 100 gs. H_2O.	Percentage change.	Conc. of satd. soln. gs. NaCl. per 100 gs. H_2O.	Percentage change.
1	76.80	—	57.95	—	20.66	—	35.90	—
500	78.01	+1.57	57.87	−0.14	21.14	+2.32	36.55	+1.81
1000	78.84	+2.68	57.65	−0.52	21.40	+3.57	37.02	+3.12
1500	—	—	—	—	21.64	+4.72	37.36	+4.07

* E. Cohen and L. R. Sinnige, *Z. physik. Chem.* 67, p. 432, 1909; 69, p. 102, 1909. E. Cohen, K. Inouye and C. Euwen, *ibid.* 75, p. 257, 1911. These authors give a critical résumé of earlier work along this line.

ABSORPTION OF GASES BY LIQUIDS.*

Temperature Centigrade. t	Absorption Coefficients, a_t, for Gases in Water.						
	Carbon dioxide. CO_2	Carbon monoxide. CO	Hydrogen. H	Nitrogen. N	Nitric oxide. NO	Nitrous oxide. N_2O	Oxygen. O
0	1.797	0.0354	0.02110	0.02399	0.0738	1.048	0.04925
5	1.450	.0315	.02022	.02134	.0646	0.8778	.04335
10	1.185	.0282	.01944	.01918	.0571	0.7377	.03852
15	1.002	.0254	.01875	.01742	.0515	0.6294	.03456
20	0.901	.0232	.01809	.01599	.0471	0.5443	.03137
25	0.772	.0214	.01745	.01481	.0432	–	.02874
30	–	.0200	.01690	.01370	.0400	–	.02646
40	0.506	.0177	.01644	.01195	.0351	–	.02316
50	–	.0161	.01608	.01074	.0315	–	.02080
100	0.244	.0141	.01600	.01011	.0263	–	.01690

Temperature Centigrade. t	Air.	Ammonia. NH_3	Chlorine. Cl	Ethylene. C_2H_4	Methane. CH_4	Hydrogen sulphide. H_2S	Sulphur dioxide. SO_2
0	0.02471	1174.6	3.036	0.2563	0.05473	4.371	79.79
5	.02179	971.5	2.808	.2153	.04889	3.965	67.48
10	.01953	840.2	2.585	.1837	.04367	3.586	56.65
15	.01795	756.0	2.388	.1615	.03903	3.233	47.28
20	.01704	683.1	2.156	.1488	.03499	2.905	39.37
25	–	610.8	1.950	–	.02542	2.604	32.79

Temperature Centigrade. t	Absorption Coefficients, a_t, for Gases in Alcohol, C_2H_5OH.								
	Carbon dioxide. CO_2	Ethylene. C_2H_4	Methane. CH_4	Hydrogen. H	Nitrogen. N	Nitric oxide. NO	Nitrous oxide. N_2O	Hydrogen sulphide. H_2S	Sulphur dioxide. SO_2
0	4.329	3.595	0.5226	0.0692	0.1263	0.3161	4.190	17.89	328.6
5	3.891	3.323	.5086	.0685	.1241	.2998	3.838	14.78	251.7
10	3.514	3.086	.4953	.0679	.1228	.2861	3.525	11.99	190.3
15	3.199	2.882	.4828	.0673	.1214	.2748	3.215	9.54	144.5
20	2.946	2.713	.4710	.0667	.1204	.2659	3.015	7.41	114.5
25	2.756	2.578	.4598	.0662	.1196	.2595	2.819	5.62	99.8

* This table contains the volumes of different gases, supposed measured at 0° C. and 76 centimeters' pressure, which unit volume of the liquid named will absorb at atmospheric pressure and the temperature stated in the first column. The numbers tabulated are commonly called the absorption coefficients for the gases in water, or in alcohol, at the temperature t and under one atmosphere of pressure. The table has been compiled from data published by Bohr & Bock, Bunsen, Carius, Dittmar, Hamberg, Henrick, Pagliano & Emo, Raoult, Schönfeld, Setschenow, and Winkler. The numbers are in many cases averages from several of these authorities.

NOTE. — The effect of increase of pressure is generally to increase the absorption coefficient. The following is approximately the magnitude of the effect in the case of ammonia in alcohol at a temperature of 23° C.:

P =	45 cms.	50 cms.	55 cms.	60 cms.	65 cms.
a_{23} =	69	74	79	84	88

According to Setschenow the effect of varying the pressure from 45 to 85 centimeters in the case of carbonic acid in water is very small.

CAPILLARITY.—SURFACE TENSION OF LIQUIDS.*

TABLE 170.—Water and Alcohol in Contact with Air.

Temp. C.	Surface tension in dynes per centimeter.		Temp. C.	Surface tension in dynes per centimeter.		Temp. C.	Surface tension in dynes per centimeter.
	Water.	Ethyl alcohol.		Water.	Ethyl alcohol.		Water.
0°	75.6	23.5	40°	70.0	20.0	80°	64.3
5	74.9	23.1	45	69.3	19.5	85	63.6
10	74.2	22.6	50	68.6	19.1	90	62.9
15	73.5	22.2	55	67.8	18.6	95	62.2
20	72.8	21.7	60	67.1	18.2	100	61.5
25	72.1	21.3	65	66.4	17.8	–	–
30	71.4	20.8	70	65.7	17.3	–	–
35	70.7	20.4	75	65.0	16.9	–	–

TABLE 171.—Miscellaneous Liquids in Contact with Air.

Liquid.	Temp. C.°	Surface tension in dynes per centimeter.	Authority.
Aceton	16.8	23.3	Ramsay-Shields.
Acetic acid . . .	17.0	30.2	Average of various.
Amyl alcohol . .	15.0	24.8	"
Benzole	15.0	28.8	"
Butyric acid . .	15.0	28.7	"
Carbon disulphide	20.0	30.5	Quincke.
Chloroform . .	20.0	28.3	Average of various.
Ether	20.0	18.4	"
Glycerine . . .	17.0	63.14	Hall.
Hexane	0.0	21.2	Schiff.
"	68.0	14.2	"
Mercury	18.0	520.0	Average of various.
Methyl alcohol .	15.0	24.7	"
Olive oil	20.0	34.7	"
Petroleum . . .	20.0	25.9	Magie.
Propyl alcohol .	5.8	25.9	Schiff.
" " . .	97.1	18.0	"
Toluol	15.0	29.1	"
"	109.8	18.9	"
Turpentine . . .	21.0	28.5	Average of various.

TABLE 172.—Solutions of Salts in Water.†

Salt in solution.	Density.	Temp. C.°	Tension in dynes per cm.
$BaCl_2$	1.2820	15–16	81.8
"	1.0497	15–16	77.5
$CaCl_2$	1.3511	19	95.0
"	1.2773	19	90.2
HCl	1.1190	20	73.6
"	1.0887	20	74.5
"	1.0242	20	75.3
KCl	1.1699	15–16	82.8
"	1.1011	15–16	80.1
"	1.0463	15–16	78.2
$MgCl_2$	1.2338	15–16	90.1
"	1.1694	15–16	85.2
"	1.0362	15–16	78.0
NaCl	1.1932	20	85.8
"	1.1074	20	80.5
"	1.0360	20	77.6
NH_4Cl	1.0758	16	84.3
"	1.0535	16	81.7
"	1.0281	16	78.8
$SrCl_2$	1.3114	15–16	85.6
"	1.1204	15–16	79.4
"	1.0567	15–16	77.8
K_2CO_3	1.3575	15–16	90.9
"	1.1576	15–16	81.8
"	1.0400	15–16	77.5
Na_2CO_3	1.1329	14–15	79.3
"	1.0605	14–15	77.8
"	1.0283	14–15	77.2
KNO_3	1.1263	14	78.9
"	1.0466	14	77.6
$NaNO_3$	1.3022	12	83.5
"	1.1311	12	80.0
$CuSO_4$	1.1775	15–16	78.6
"	1.0276	15–16	77.0
H_2SO_4	1.8278	15	63.0 ?
"	1.4453	15	79.7
"	1.2636	15	79.7
K_2SO_4	1.0744	15–16	78.0
"	1.0360	15–16	77.4
$MgSO_4$	1.2744	15–16	83.2
"	1.0680	15–16	77.8
$MnSO_4$	1.1119	15–16	79.1
"	1.0329	15–16	77.3
$ZnSO_4$	1.3981	15–16	83.3
"	1.2830	15–16	80.7
"	1.1039	15–16	77.8

* This determination of the capillary constants of liquids has been the subject of many careful experiments, but the results of the different experimenters, and even of the same observer when the method of measurement is changed, do not agree well together. The values here quoted can only be taken as approximations to the actual values for the liquids in a state of purity in contact with pure air. In the case of water the values given by Lord Rayleigh from the wave length of ripples (Phil. Mag. 1890) and by Hall from direct measurement of the tension of a flat film (Phil. Mag. 1893) have been preferred, and the temperature correction has been taken as 0.141 dyne per degree centigrade. The values for alcohol were derived from the experiments of Hall above referred to and the experiments on the effect of temperature made by Timberg (Wied. Ann. vol. 30).

The authority for a few of the other values given is quoted, but they are for the most part average values derived from a large number of results published by different experimenters.

† From Volkmann (Wied. Ann. vol. 17, p. 353).

For more recent data see especially Harkins, J. Am. Ch. Soc., 39, p. 55, 1917 (336 liquids), and 42, p. 702, 2543, 1920.

TENSION OF LIQUIDS.

TABLE 173.—Surface Tension of Liquids.*

Liquid.	Specific gravity.	Surface tension in dynes per centimeter of liquid in contact with — Air.	Water.	Mercury.
Water	1.0	75.0	0.0	(392)
Mercury	13.543	513.0	392.0	0
Bisulphide of carbon	1.2687	30.5	41.7	(387)
Chloroform	1.4878	(31.8)	26.8	(415)
Ethyl alcohol	0.7906	(24.1)	–	364
Olive oil	0.9136	34.6	18.6	317
Turpentine	0.8867	28.8	11.5	241
Petroleum	.7977	29.7	(28.9)	271
Hydrochloric acid	1.10	(72.9)	–	(392)
Hyposulphite of soda solution	1.1248	69.9	–	429

TABLE 174.—Surface Tension of Liquids at Solidifying Point.†

Substance.	Temperature of solidification. Cent.°	Surface tension in dynes per centimeter.	Substance.	Temperature of solidification. Cent.°	Surface tension in dynes per centimeter.
Platinum	2000	1691	Antimony	432	249
Gold	1200	1003	Borax	1000	216
Zinc	360	877	Carbonate of soda	1000	210
Tin	230	599	Chloride of sodium	–	116
Mercury	—40	588	Water	0	87.9‡
Lead	330	457	Selenium	217	71.8
Silver	1000	427	Sulphur	111	42.1
Bismuth	265	1390	Phosphorus	43	42.0
Potassium	58	371	Wax	68	34.1
Sodium	90	258			

TABLE 175.—Tension of Soap Films.

Elaborate measurements of the thickness of soap films have been made by Reinold and Rucker.‖ They find that a film of oleate of soda solution containing 1 of soap to 70 of water, and having 3 per cent of KNO_3 added to increase electrical conductivity, breaks at a thickness varying between 7.2 and 14.5 micro-millimeters, the average being 12.1 micro-millimeters. The film becomes black and apparently of nearly uniform thickness round the point where fracture begins. Outside the black patch there is the usual display of colors, and the thickness at these parts may be estimated from the colors of thin plates and the refractive index of the solution.

When the percentage of KNO_3 is diminished, the thickness of the black patch increases. For example,

KNO_3 =	3	1	0.5	0.0
Thickness =	12.4	13.5	14.5	22.1 micro-mm.

A similar variation was found in the other soaps.

It was also found that diminishing the proportion of soap in the solution, there being no KNO_3 dissolved, increased the thickness of the film.

1 part soap to 30 of water gave thickness 21.6 micro-mm.
1 part soap to 40 of water gave thickness 22.1 micro-mm.
1 part soap to 60 of water gave thickness 27.7 micro-mm.
1 part soap to 80 of water gave thickness 29.3 micro-mm.

* This table of tensions at the surface separating the liquid named in the first column and air, water or mercury as stated at the head of the last three columns, is from Quincke's experiments (Pogg. Ann. vol. 139, and Phil. Mag. 1871). The numbers given are the equivalent in dynes per centimeter of those obtained by Worthington from Quincke's results (Phil. Mag. vol. 20, 1885) with the exception of those in brackets, which were not corrected by Worthington; they are probably somewhat too high, for the reason stated by Worthington. The temperature was about 20° C.

† Quincke, "Pogg. Ann." vol. 135, p. 661.

‡ It will be observed that the value here given on the authority of Quincke is much higher than his subsequent measurements, as quoted above, give.

‖ "Proc. Roy. Soc." 1877, and "Phil. Trans. Roy. Soc." 1881, 1883, and 1893.

NOTE.—Quincke points out that substances may be divided into groups in each of which the ratio of the surface tension to the density is nearly constant. Thus, if this ratio for mercury be taken as unit, the ratio for the bromides and iodides is about a half: that of the nitrates, chlorides, sugars, and fats, as well as the metals, lead, bismuth, and antimony, about 1; that of water, the carbonates, sulphates, and probably phosphates, and the metals platinum, gold, silver, cadmium, tin, and copper, 2; that of zinc, iron, and palladium, 3; and that of sodium, 6.

TABLE 176. — Vapor Pressure of Elements.

Hydrogen. H scale.	Hydrogen. mm	Oxygen. H scale.	Oxygen. mm	Nitrogen. T	Nitrogen. mm
20.41° K	800	90.60° K	800	77.33° K	760.
20.22	760	90.10	760	76.83	714.5
19.93	700	89.33	700	76.65	700.
19.41	600	87.91	600	75.44	600.
18.82	500	86.29	500	74.03	500.
18.15	400	84.39	400	72.39	400.
17.36	300	82.09	300	70.42	300.
16.37	200	79.07	200	67.80	200.
14.93	100			63.65	100.
Travers, Jaquerod, 1902–5.		Travers, Senter, Jaquerod, 1902–3.		Fischer, Alt, 1902.	

Argon. °K	Argon. mm	Xenon. °K	Xenon. mm	Krypton. °K	Krypton. mm
155.6	40200.	287.7	44112	210.5	41245
139.0	23251.	273.3	31501	206.4	37006
137.8	21334.	255.6	21967	204.1	34693
136.8	20700.	254.0	21512	201.5	31621
123.1	10313.	252.6	19984	201.0	30837
87.8	821.2	248.7	18153	197.9	28808
86.5	704.5	244.2	15868	170.9	11970
85.5	633.4	239.7	13971	112.7	387
83.8	524.3	237.4	13505	88.6	17.4
82.6	465.0	231.4	11134	84.2	9.
81.7	410.1	183.2	2020	—	—
77.3	215.0	—	—	—	—
Ramsay, Travers, Zs. phys. Ch. 38, 1901.					

Chlorine. °C	Chlorine. Pressure.	Bromine. °C	Bromine. mm	Iodine. °C	Iodine. mm
+146.	93.50 atm.	+58.75	760	+55	3.084
+100.	41.70 atm.	56.3	700	50	2.154
+50.	14.70 atm.	51.95	600	45	1.498
+20.	6.62 atm.	46.8	500	40	1.025
0.	3.66 atm.	40.45	400	35	0.699
−20.	1.84 atm.	33.05	300	30	0.469
−33.6	760. mm	23.45	200	25	0.305
−40.	560. mm	16.95	150	15	0.131
−50.	350. mm	8.20	100	0	0.030
−60.	210. mm	−5.05	50		
−70.	118. mm	−7.0	45		
−80.	62.5 mm	−8.4	40		
−85.	45. mm	−12.0	30		
−88.	37.5 mm	−16.65	20		
Knietsch, W. Ann. 1890. Cu to Sn, Greenwood, Pr. Roy. Soc. 83A, 1910; Zs. ph. Ch. 76, 1911.		Ramsay, Young J. Ch. Soc. 1886.		Baxter, Hickey, Holmes, J. Am. Ch. Soc. 1907.	

Copper. °C	Copper. Atme.	Silver. °C	Silver. Atme.
2310	1.0	1955	1.0
2180	0.338	1780	0.346
1980	0.1315	1660	0.1355
Lead. °C	**Lead. Atme.**	**Bismuth. °C**	**Bismuth. Atme.**
2100	11.7	2060	16.5
1870	6.3	1950	11.7
1525	1.0	1740	6.3
1420	0.350	1420	1.0
1320	0.138	1310	0.338
—	—	1200	0.134
Zinc. °C	**Zinc. Atme.**	**Tin. °C**	**Tin. Atme.**
1510	53.0	2270	1.0
1280	21.5	2100	0.345
1230	11.7	1970	0.133
1120	6.3	—	—

TABLE 177. — Vapor Pressure and Rate of Evaporization.

°K	Mo mm	W mm	Evaporation rate. g/cm²/sec. Mo	Evaporation rate. g/cm²/sec. W	Platinum. °K	Platinum. mm	Platinum. g/cm²/sec.
1800	0.0_8643	—	$0.0_{10}863$	—	1000	$0.0_{17}324$	$0.0_{19}832$
2000	0.0_6789	$0.0_{11}645$	0.0_7100	$0.0_{12}114$	1200	$0.0_{12}111$	$0.0_{14}260$
2200	0.0_4396	0.0_9849	0.0_6480	$0.0_{10}144$	1400	0.0_9188	$0.0_{11}401$
2400	0.0_21027	0.0_7492	0.0_4120	0.0_9798	1600	0.0_7484	0.0_9966
2600	0.0160	0.0_5151	0.0_3179	0.0_7236	1800	0.0_5350	0.0_7667
2800	0.1679	0.0_4286	0.0_2181	0.0_6429	2000	0.0_3107	0.0_5195
3000	—	0.0_3362	—	0.0_5523	4180	760 mm	—
3200	3890° }	0.0_2333	—	0.0_4467	Langmuir, MacKay, Phys. Rev. 2, 1913; 4, 1914. Order of vacuum, 0.001 mm.		
3500	760 mm }	0.0572	—	0.0_3769			

$p = K.T^{-\frac{1}{2}}e^{-\lambda_0/RT}$ dynes/cm². Egerton, Phil. Mag. 33, p. 33, 1917.

Zn, $\lambda_0 = 3.28 \times 10^4$; $K = 1.17 \times 10^{14}$ Cd, $\lambda_0 = 2.77 \times 10^4$; $K = 5.27 \times 10^{13}$

Hg, $\lambda_0 = 1.60 \times 10^4$; $= 3.72 \times 10^{13}$ (Knudsen)

Table 178.

VAPOR PRESSURES.

The vapor pressures here tabulated have been taken, with one exception, from Regnault's results. The vapor pressure of Pictet's fluid is given on his own authority. The pressures are in centimeters of mercury.

Temperature Cent.	Acetone. C_3H_6O	Benzol. C_6H_6	Carbon bisulphide. CS_2	Carbon tetrachloride. CCl_4	Chloroform. $CHCl_3$	Ethyl alcohol. C_2H_6O	Ethyl ether. $C_4H_{10}O$	Ethyl bromide. C_2H_5Br	Methyl alcohol. CH_4O	Turpentine. $C_{10}H_6$
—25°	–	–	–	–	–	–	–	4.41	.41	–
—20	–	.58	4.73	.98	–	.33	6.89	5.92	.63	–
—15	–	.88	6.16	1.35	–	.51	8.93	7.81	.93	–
—10	–	1.29	7.94	1.85	–	.65	11.47	10.15	1.35	–
—5	–	1.83	10.13	2.48	–	.91	14.61	13.06	1.92	–
0	–	2.53	12.79	3.29	5.97	1.27	18.44	16.56	2.68	.21
5	–	3.42	16.00	4.32	–	1.76	23.09	20.72	3.69	–
10	–	4.52	19.85	5.60	10.05	2.42	28.68	25.74	5.01	.29
15	–	5.89	24.41	7.17	–	3.30	35.36	31.69	6.71	–
20	17.96	7.56	29.80	9.10	16.05	4.45	43.28	38.70	8.87	.44
25	22.63	9.59	36.11	11.43	20.02	5.94	52.59	46.91	11.60	–
30	28.10	12.02	43.46	14.23	24.75	7.85	63.48	56.45	15.00	.69
35	34.52	14.93	51.97	17.55	30.35	10.29	76.12	67.49	19.20	–
40	42.01	18.36	61.75	21.48	36.93	13.37	90.70	80.19	24.35	1.08
45	50.75	22.41	72.95	26.08	44.60	17.22	107.42	94.73	30.61	–
50	62.29	27.14	85.71	31.44	53.50	21.99	126.48	111.28	38.17	1.70
55	72.59	32.64	100.16	37.63	63.77	27.86	148.11	130.03	47.22	–
60	86.05	39.01	116.45	44.74	75.54	35.02	172.50	151.19	57.99	2.65
65	101.43	46.34	134.75	52.87	88.97	43.69	199.89	174.95	70.73	–
70	118.94	54.74	155.21	62.11	104.21	54.11	230.49	201.51	85.71	4.06
75	138.76	64.32	177.99	72.57	121.42	66.55	264.54	231.07	103.21	–
80	161.10	75.19	203.25	84.33	140.76	81.29	302.28	263.86	123.85	6.13
85	186.18	87.46	231.17	97.51	162.41	98.64	343.95	300.06	147.09	–
90	214.17	101.27	261.91	112.23	186.52	118.93	389.83	339.89	174.17	9.06
95	245.28	116.75	296.63	128.69	213.28	142.51	440.18	383.55	205.17	–
100	279.73	134.01	332.51	146.71	242.85	169.75	495.33	431.23	240.51	13.11
105	317.70	153.18	372.72	166.72	275.40	201.04	555.62	483.12	280.63	–
110	359.40	174.44	416.41	188.74	311.10	236.76	621.46	539.40	325.96	18.60
115	405.00	197.82	463.74	212.91	350.10	277.34	693.33	600.24	376.98	–
120	454.69	223.54	514.88	239.37	392.57	323.17	771.92	665.80	434.18	25.70
125	508.62	251.71	569.97	268.24	438.66	374.69	–	736.22	498.05	–
130	566.97	282.43	629.16	299.69	488.51	432.30	–	811.65	569.13	34.90
135	629.87	315.85	692.59	333.86	542.25	496.42	–	892.19	647.93	–
140	697.44	352.07	760.40	370.90	600.02	567.46	–	977.96	733.71	46.40
145	–	391.21	832.69	411.00	661.92	645.80	–	–	830.89	–
150	–	433.37	909.59	454.31	728.06	731.84	–	–	936.13	60.50
155	–	478.65	–	501.02	798.53	825.92	–	–	–	68.60
160	–	527.14	–	551.31	873.42	–	–	–	–	77.50
165	–	568.30	–	605.38	952.78	–	–	–	–	–
170	–	634.07	–	663.44	–	–	–	–	–	–

VAPOR PRESSURES.

Temperature, Centigrade.	Ammonia. NH_3	Carbon dioxide. CO_2	Ethyl chloride. C_2H_5Cl	Ethyl iodide. C_2H_5I	Methyl chloride. CH_3Cl	Methylic ether. C_2H_6O	Nitrous oxide. N_2O	Pictet's fluid. $64SO_2 + 44CO_2$ by weight	Sulphur dioxide. SO_2	Hydrogen sulphide. H_2S
−30°	86.61	–	11.02	–	57.90	57.65	–	58.52	28.75	–
−25	110.43	1300.70	14.50	–	71.78	71.61	1569.49	67.64	37.38	374.93
−20	139.21	1514.24	18.75	–	88.32	88.20	1758.66	74.48	47.95	443.85
−15	173.65	1758.25	23.96	–	107.92	107.77	1968.43	89.68	60.79	519.65
−10	214.46	2034.02	30.21	–	130.96	130.66	2200.80	101.84	76.25	608.46
−5	264.42	2344.13	37.67	–	157.87	157.25	2457.92	121.60	94.69	706.60
0	318.33	2690.66	46.52	4.19	189.10	187.90	2742.10	139.08	116.51	820.63
5	383.03	3075.38	56.93	5.41	225.11	222.90	3055.86	167.20	142.11	949.08
10	457.40	3499.86	69.11	6.92	266.38	262.90	3401.91	193.80	171.95	1089.63
15	543.34	3964.69	83.26	8.76	313.41	307.98	3783.17	226.48	206.49	1244.79
20	638.78	4471.66	99.62	11.00	366.69	358.60	4202.79	258.40	246.20	1415.15
25	747.70	5020.73	118.42	13.69	426.74	415.10	4664.14	297.92	291.60	1601.24
30	870.10	5611.90	139.90	16.91	494.05	477.80	5170.85	338.20	343.18	1803.53
35	1007.02	6244.73	164.32	20.71	569.11	–	6335.98	383.80	401.48	2002.43
40	1159.53	6918.44	191.96	25.17	–	–	–	434.72	467.02	2258.25
45	1328.73	7631.46	223.07	30.38	–	–	–	478.80	540.35	2495.43
50	1515.83	–	257.94	36.40	–	–	–	521.36	622.00	2781.48
55	1721.98	–	266.84	43.32	–	–	–	–	712.50	3069.07
60	1948.21	–	340.05	51.22	–	–	–	–	812.38	3374.02
65	2196.51	–	387.85	–	–	–	–	–	922.14	3696.15
70	2467.55	–	440.50	–	–	–	–	–	–	4035.32
75	2763.00	–	498.27	–	–	–	–	–	–	–
80	3084.31	–	561.41	–	–	–	–	–	–	–
85	3433.09	–	630.16	–	–	–	–	–	–	–
90	3810.92	–	704.75	–	–	–	–	–	–	–
95	4219.57	–	785.39	–	–	–	–	–	–	–
100	4660.82	–	872.28	–	–	–	–	–	–	–

TABLE 178 (a). — Vapor Pressures at Low Temperatures.

Many of the following values are extrapolations made by Langmuir by means of plots of log. p against 1/T. Gen. Elect. Rev. 23, 681, 1920. 1 bar = 0.000000987 atm. = 0.000750 mm Hg.

Gas	°C	mm	Gas	°C	bars
O_2	−182.9	760.	CO_2	−148	100.
	−211.2	7.75		−168	1.
N_2	−195.8	760.		−182	.01
	−210.5	86.		−193	.0001
CO	−190.	863.	Ice	−60	9.6
	−200.	249.		−75	1.0
CH_4	−185.8	79.8		−89	.1
	−201.5	50.2		−100	.01
A	−186.2	760.		−110	.001
	−194.2	300.	Hg	+30	3.7
C_2H_4	−175.7	.76		+20	1.6
	−188.	.076		+10	.65
	−197.	.0076		0	.25
	−205.	.00076		−10	.087
C_2H_6	−150.	7.6		−20	.029
	−180.	.076		−40	.0023
	−190.	.0076		−78	4.3×10^{-6}
	−198.	.00076		−180	2.3×10^{-24}

VAPOR PRESSURE.

TABLE 179. — Vapor Pressure of Ethyl Alcohol.*

Temp. C.	0°	1°	2°	3°	4°	5°	6°	7°	8°	9°
	Vapor pressure in millimeters of mercury at 0° C.									
0°	12.24	13.18	14.15	15.16	16.21	17.31	18.46	19.68	20.98	22.34
10	23.78	25.31	27.94	28.67	30.50	32.44	34.49	36.67	38.97	41.40
20	44.00	46.66	49.47	52.44	55.56	58.86	62.33	65.97	69.80	73.83
30	78.06	82.50	87.17	92.07	97.21	102.60	108.24	114.15	120.35	126.86
40	133.70	140.75	148.10	155.80	163.80	172.20	181.00	190.10	199.65	209.60
50	220.00	230.80	242.50	253.80	265.90	278.60	291.85	305.65	319.95	334.85
60	350.30	366.40	383.10	400.40	418.35	437.00	456.35	476.45	497.25	518.85
70	541.20	564.35	588.35	613.20	638.95	665.55	693.10	721.55	751.00	781.45

From the formula $\log p = a + b\alpha^t + c\beta^t$ Ramsay and Young obtain the following numbers.†

Temp. C.	0°	10°	20°	30°	40°	50°	60°	70°	80°	90°
	Vapor pressure in millimeters of mercury at 0° C.									
0°	12.24	23.73	43.97	78.11	133.42	219.82	350.21	540.91	811.81	1186.5
100	1692.3	2359.8	3223.0	4318.7	5686.6	7368.7	9409.9	11858.	14764.	18185.
200	22182.	26825.	32196.	38389.	45519.					

TABLE 180. — Vapor Pressure of Methyl Alcohol.‡

Temp. C.	0°	1°	2°	3°	4°	5°	6°	7°	8°	9°
	Vapor pressure in millimeters of mercury at 0° C.									
0°	29.97	31.6	33.6	35.6	37.8	40.2	42.6	45.2	47.9	50.8
10	53.8	57.0	60.3	63.8	67.5	71.4	75.5	79.8	84.3	89.0
20	94.0	99.2	104.7	110.4	116.5	122.7	129.3	136.2	143.4	151.0
30	158.9	167.1	175.7	184.7	194.1	203.9	214.1	224.7	235.8	247.4
40	259.4	271.9	285.0	298.5	312.6	327.3	342.5	358.3	374.7	391.7
50	409.4	427.7	446.6	466.3	486.6	507.7	529.5	552.0	575.3	599.4
60	624.3	650.0	676.5	703.8	732.0	761.1	791.1	822.0	–	–

* This table has been compiled from results published by Ramsay and Young (Jour. Chem. Soc. vol. 47, and Phil. Trans. Roy. Soc., 1886).

† In this formula $a = 5.0720301$; $\log b = \bar{2}.6406131$; $\log c = 0.6050854$; $\log \alpha = 0.003377538$; $\log \beta = \bar{1}.99682424$ (c is negative).

‡ Taken from a paper by Dittmar and Fawsitt (Trans. Roy. Soc. Edin. vol. 33).

Table 181.

VAPOR PRESSURE.*

Carbon Disulphide, Chlorobenzene, Bromobenzene, and Aniline.

Temp.	0°	1°	2°	3°	4°	5°	6°	7°	8°	9°
					(a) Carbon Disulphide.					
0°	127.90	133.85	140.05	146.45	153.10	160.00	167.15	174.60	182.25	190.20
10	198.45	207.00	215.80	224.95	234.40	244.15	254.25	264.65	275.40	286.55
20	298.05	309.90	322.10	334.70	347.70	361.10	374.95	389.20	403.90	419.00
30	434.60	450.65	467.15	484.15	501.65	519.65	538.15	557.15	576.75	596.85
40	617.50	638.70	660.50	682.90	705.90	729.50	753.75	778.60	804.10	830.25
					(b) Chlorobenzene.					
20°	8.65	9.14	9.66	10.21	10.79	11.40	12.04	12.71	13.42	14.17
30	14.95	15.77	16.63	17.53	18.47	19.45	20.48	21.56	22.69	23.87
40	25.10	26.38	27.72	29.12	30.58	32.10	33.69	35.35	37.08	38.88
50	40.75	42.69	44.72	46.84	49.05	51.35	53.74	56.22	58.79	61.45
60	64.20	67.06	70.03	73.11	76.30	79.60	83.02	86.56	90.22	94.00
70	97.90	101.95	106.10	110.41	114.85	119.45	124.20	129.10	134.15	139.40
80	144.80	150.30	156.05	161.95	168.00	174.25	181.70	187.30	194.10	201.15
90	208.35	215.80	223.45	231.30	239.35	247.70	256.20	265.00	274.00	283.25
100	292.75	302.50	312.50	322.80	333.35	344.15	355.25	366.65	378.30	390.25
110	402.55	415.10	427.95	441.15	454.65	468.50	482.65	497.20	512.05	527.25
120	542.80	558.70	575.05	591.70	608.75	626.15	643.95	662.15	680.75	699.65
130	718.95	738.65	758.80	–	–	–	–	–	–	–
					(c) Bromobenzene.					
40°	–	–	–	–	–	12.40	13.06	13.75	14.47	15.22
50	16.00	16.82	17.68	18.58	19.52	20.50	21.52	22.59	23.71	24.88
60	26.10	27.36	28.68	30.06	31.50	33.00	34.56	36.18	37.86	39.60
70	41.40	43.28	45.24	47.28	49.40	51.60	53.88	56.25	58.71	61.26
80	63.90	66.64	69.48	72.42	75.46	78.60	81.84	85.20	88.68	92.28
90	96.00	99.84	103.80	107.88	112.08	116.40	120.86	125.46	130.20	135.08
100	140.10	145.26	150.57	156.03	161.64	167.40	173.32	179.41	185.67	192.10
110	198.70	205.48	212.44	219.58	226.90	234.40	242.10	250.00	258.10	266.40
120	274.90	283.65	292.60	301.75	311.15	320.80	330.70	340.80	351.15	361.80
130	372.65	383.75	395.10	406.70	418.60	430.75	443.20	455.90	468.90	482.20
140	495.80	509.70	523.90	538.40	553.20	568.35	583.85	599.65	615.75	632.25
150	649.05	666.25	683.80	701.65	719.95	738.55	757.55	776.95	796.70	816.90
					(d) Aniline.					
80°	18.80	19.78	20.79	21.83	22.90	24.00	25.14	26.32	27.54	28.80
90	30.10	31.44	32.83	34.27	35.76	37.30	38.90	40.56	42.28	44.06
100	45.90	47.80	49.78	51.84	53.98	56.20	58.50	60.88	63.34	65.88
110	68.50	71.22	74.04	76.96	79.98	83.10	86.32	89.66	93.12	96.70
120	100.40	104.22	108.17	112.25	116.46	120.80	125.28	129.91	134.69	139.62
130	144.70	149.94	155.34	160.90	166.62	172.50	178.56	184.80	191.22	197.82
140	204.60	211.58	218.76	226.14	233.72	241.50	249.50	257.72	266.16	274.82
150	283.70	292.80	302.15	311.75	321.60	331.70	342.05	352.65	363.50	374.60
160	386.00	397.65	409.60	421.80	434.30	447.10	460.20	473.60	487.25	501.25
170	515.60	530.20	545.20	560.45	576.10	592.05	608.35	625.05	642.05	659.45
180	677.15	695.30	713.75	732.65	751.90	771.50	–	–	–	–

* These tables of vapor pressures are quoted from results published by Ramsay and Young (Jour. Chem. Soc. vol. 47). The tables are intended to give a series suitable for hot-jacket purposes.

Table 181 (*continued*).

VAPOR PRESSURE.

Methyl Salicylate, Bromonaphthalene, and Mercury.

Temp. C.	0°	1°	2°	3°	4°	5°	6°	7°	8°	9°
					(e) Methyl Salicylate.					
70°	2.40	2.58	2.77	2.97	3.18	3.40	3.62	3.85	4.09	4.34
80	4.60	4.87	5.15	5.44	5.74	6.05	6.37	6.70	7.05	7.42
90	7.80	8.20	8.62	9.06	9.52	9.95	10.44	10.95	11.48	12.03
100	12.60	13.20	13.82	14.47	15.15	15.85	16.58	17.34	18.13	18.95
110	19.80	20.68	21.60	22.55	23.53	24.55	25.61	26.71	27.85	29.03
120	30.25	31.52	32.84	34.21	35.63	37.10	38.67	40.24	41.84	43.54
130	45.30	47.12	49.01	50.96	52.97	55.05	57.20	59.43	61.73	64.10
140	66.55	69.08	71.69	74.38	77.15	80.00	82.94	85.97	89.09	92.30
150	95.60	99.00	102.50	106.10	109.80	113.60	117.51	121.53	125.66	129.90
160	134.25	138.72	143.31	148.03	152.88	157.85	162.95	168.19	173.56	179.06
170	184.70	190.48	196.41	202.49	208.72	215.10	221.65	228.30	235.15	242.15
180	249.35	256.70	264.20	271.90	279.75	287.80	296.00	304.48	313.05	321.85
190	330.85	340.05	349.45	359.05	368.85	378.90	389.15	399.60	410.30	421.20
200	432.35	443.75	455.35	467.25	479.35	491.70	504.35	517.25	530.40	543.80
210	557.50	571.45	585.70	600.25	615.05	630.15	645.55	661.25	677.25	693.60
220	710.10	727.05	744.35	761.90	779.85	798.10				
					(f) Bromonaphthalene.					
110°	3.60	3.74	3.89	4.05	4.22	4.40	4.59	4.79	5.00	5.22
120	5.45	5.70	5.96	6.23	6.51	6.80	7.10	7.42	7.76	8.12
130	8.50	8.89	9.29	9.71	10.15	10.60	11.07	11.56	12.07	12.60
140	13.15	13.72	14.31	14.92	15.55	16.20	16.87	17.56	18.28	19.03
150	19.80	20.59	21.41	22.25	23.11	24.00	24.92	25.86	26.83	27.83
160	28.85	29.90	30.98	32.09	33.23	34.40	35.60	36.83	38.10	39.41
170	40.75	42.12	43.53	44.99	46.50	48.05	49.64	51.28	52.96	54.68
180	56.45	58.27	60.14	62.04	64.06	66.10	68.19	70.34	72.55	74.82
190	77.15	79.54	81.99	84.51	87.10	89.75	92.47	95.26	98.12	101.05
200	104.05	107.12	110.27	113.50	116.81	120.20	123.67	127.22	130.86	134.59
210	138.40	142.30	146.29	150.38	154.57	158.85	163.25	167.70	172.30	176.95
220	181.75	186.65	191.65	196.75	202.00	207.35	212.80	218.40	224.15	230.00
230	235.95	242.05	248.30	254.65	261.20	267.85	274.65	281.60	288.70	295.95
240	303.35	310.90	318.65	326.50	334.55	342.75	351.10	359.65	368.40	377.30
250	386.35	395.60	405.05	414.65	424.45	434.45	444.65	455.00	465.60	476.35
260	487.35	498.55	509.90	521.50	533.35	545.35	557.60	570.05	582.70	595.60
270	608.75	622.10	635.70	649.50	663.55	677.85	692.40	707.15	722.15	737.45
					(g) Mercury.					
270°	123.92	126.97	130.08	133.26	136.50	139.81	143.18	146.61	150.12	153.70
280	157.35	161.07	164.86	168.73	172.67	176.79	180.88	185.05	189.30	193.63
290	198.04	202.53	207.10	211.76	216.50	221.33	226.25	231.25	236.34	241.53
300	246.81	252.18	257.65	263.21	268.87	274.63	280.48	286.43	292.49	298.66
310	304.93	311.30	317.78	324.37	331.08	337.89	344.81	351.85	359.00	366.28
320	373.67	381.18	388.81	396.56	404.43	412.44	420.58	428.83	437.22	445.75
330	454.41	463.20	472.12	481.19	490.40	499.74	509.22	518.85	528.63	538.56
340	548.64	558.87	569.25	579.78	590.48	601.33	612.34	623.51	634.85	646.36
350	658.03	669.86	681.86	694.04	706.40	718.94	731.65	744.54	757.61	770.87
360	784.31									

TABLE 182.

VAPOR PRESSURE OF SOLUTIONS OF SALTS IN WATER.*

The first column gives the chemical formula of the salt. The headings of the other columns give the number of gram-molecules of the salt in a liter of water. The numbers in these columns give the lowering of the vapor pressure produced by the salt at the temperature of boiling water under 76 centimeters barometric pressure.

Substance.	0.5	1.0	2.0	3.0	4.0	5.0	6.0	8.0	10.0
$Al_2(SO_4)_3$	12.8	36.5							
$AlCl_3$	22.5	61.0	179.0	318.0					
BaS_2O_6	6.6	15.4	34.4						
$Ba(OH)_2$	12.3	22.5	39.0						
$Ba(NO_3)_2$	13.5	27.0							
$Ba(ClO_3)_2$	15.8	33.3	70.5	108.2					
$BaCl_2$	16.4	36.7	77.6						
$BaBr_2$	16.8	38.8	91.4	150.0	204.7				
CaS_2O_3	9.9	23.0	56.0	106.0					
$Ca(NO_3)_2$	16.4	34.8	74.6	139.3	161.7	205.4			
$CaCl_2$	17.0	39.8	95.3	166.6	241.5	319.5			
$CaBr_2$	17.7	44.2	105.8	191.0	283.3	368.5			
$CdSO_4$	4.1	8.9	18.1						
CdI_2	7.6	14.8	33.5	52.7					
$CdBr_2$	8.6	17.8	36.7	55.7	80.0				
$CdCl_2$	9.6	18.8	36.7	57.0	77.3	99.0			
$Cd(NO_3)_2$	15.9	36.1	78.0	122.2					
$Cd(ClO_3)_2$	17.5								
$CoSO_4$	5.5	10.7	22.9	45.5					
$CoCl_2$	15.0	34.8	83.0	136.0	186.4				
$Co(NO_3)_2$	17.3	39.2	89.0	152.0	218.7	282.0	332.0		
$FeSO_4$	5.8	10.7	24.0	42.4					
H_3BO_3	6.0	12.3	25.1	38.0	51.0				
H_3PO_4	6.6	14.0	28.6	45.2	62.0	81.5	103.0	146.9	189.5
H_3AsO_4	7.3	15.0	30.2	46.4	64.9				
H_2SO_4	12.9	26.5	62.8	104.0	148.0	198.4	247.0	343.2	
KH_2PO_4	10.2	19.5	33.3	47.8	60.5	73.1	85.2		
KNO_3	10.3	21.1	40.1	57.6	74.5	88.2	102.1	126.3	148.0
$KClO_3$	10.6	21.6	42.8	62.1	80.0				
$KBrO_3$	10.9	22.4	45.0						
$KHSO_4$	10.9	21.9	43.3	65.3	85.5	107.8	129.2	170.0	
KNO_2	11.1	22.8	44.8	67.0	90.0	110.5	130.7	167.0	198.8
$KClO_4$	11.5	22.3							
KCl	12.2	24.4	48.8	74.1	100.9	128.5	152.2		
$KHCO_3$	11.6	23.6	59.0	77.6	104.2	132.0	160.0	210.0	255.0
KI	12.5	25.3	52.2	82.6	112.2	141.5	171.8	225.5	278.5
$K_2C_2O_4$	13.9	28.3	59.8	94.2	131.0				
K_2WO_4	13.9	33.0	75.0	123.8	175.4	226.4			
K_2CO_3	14.4	31.0	68.3	105.5	152.0	209.0	258.5	350.0	
KOH	15.0	29.5	64.0	99.2	140.0	181.8	223.0	309.5	387.8
K_2CrO_4	16.2	29.5	60.0						
$LiNO_3$	12.2	25.9	55.7	88.9	122.2	155.1	188.0	253.4	309.2
$LiCl$	12.1	25.5	57.1	95.0	132.5	175.5	219.5	311.5	393.5
$LiBr$	12.2	26.2	60.0	97.0	140.0	186.3	241.5	341.5	438.0
Li_2SO_4	13.3	28.1	56.8	89.0					
$LiHSO_4$	12.8	27.0	57.0	93.0	130.0	168.0			
LiI	13.6	28.6	64.7	105.2	154.5	206.0	264.0	357.0	445.0
Li_2SiFl_6	15.4	34.0	70.0	106.0					
$LiOH$	15.9	37.4	78.1						
Li_2CrO_4	16.4	32.6	74.0	120.0	171.0				

* Compiled from a table by Tammann, "Mém. Ac. St. Petersb." 35, No. 9, 1887. See also Referate, "Zeit. f. Phys." ch. 2, 42, 1886.

VAPOR PRESSURE OF SOLUTIONS OF SALTS IN WATER.

Substance.	0.5	1.0	2.0	3.0	4.0	5.0	6.0	8.0	10.0
$MgSO_4$	6.5	12.0	24.5	47.5					
$MgCl_2$	16.8	39.0	100.5	183.3	277.0	377.0			
$Mg(NO_3)_2$	17.6	42.0	101.0	174.8					
$MgBr_2$	17.9	44.0	115.8	205.3	298.5				
$MgH_2(SO_4)_2$	18.3	46.0	116.0						
$MnSO_4$	6.0	10.5	21.0						
$MnCl_2$	15.0	34.0	76.0	122.3	167.0	209.0			
NaH_2PO_4	10.5	20.0	36.5	51.7	66.8	82.0	96.5	126.7	157.1
$NaHSO_4$	10.9	22.1	47.3	75.0	100.2	126.1	148.5	189.7	231.4
$NaNO_3$	10.6	22.5	46.2	68.1	90.3	111.5	131.7	167.8	198.8
$NaClO_3$	10.5	23.0	48.4	73.5	98.5	123.3	147.5	196.5	223.5
$(NaPO_3)_6$	11.6								
$NaOH$	11.8	22.8	48.2	77.3	107.5	139.1	172.5	243.3	314.0
$NaNO_2$	11.6	24.4	50.0	75.0	98.2	122.5	146.5	189.0	226.2
Na_2HPO_4	12.1	23.5	43.0	60.0	78.7	99.8	122.1		
$NaHCO_3$	12.9	24.1	48.2	77.6	102.2	127.8	152.0	198.0	239.4
Na_2SO_4	12.6	25.0	48.9	74.2					
$NaCl$	12.3	25.2	52.1	80.0	111.0	143.0	176.5		
$NaBrO_3$	12.1	25.0	54.1	81.3	108.8	136.0			
$NaBr$	12.6	25.9	57.0	89.2	124.2	159.5	197.5	268.0	
NaI	12.1	25.6	60.2	99.5	136.7	177.5	221.0	301.5	370.0
$Na_4P_2O_7$	13.2	22.0							
Na_2CO_3	14.3	27.3	53.5	80.2	111.0				
$Na_2C_2O_4$	14.5	30.0	65.8	105.8	146.0				
Na_2WO_4	14.8	33.6	71.6	115.7	162.6				
Na_3PO_4	16.5	30.0	52.5						
$(NaPO_3)_3$	17.1	36.5							
NH_4NO_3	12.8	22.0	42.1	62.7	82.9	103.8	121.0	152.2	180.0
$(NH_4)_2SiFl_6$	11.5	25.0	44.5						
NH_4Cl	12.0	23.7	45.1	69.3	94.2	118.5	138.2	179.0	213.8
NH_4HSO_4	11.5	22.0	46.8	71.0	94.5	118.	139.0	181.2	218.0
$(NH_4)_2SO_4$	11.0	24.0	46.5	69.5	93.0	117.0	141.8		
NH_4Br	11.9	23.9	48.8	74.1	99.4	121.5	145.5	190.2	228.5
NH_4I	12.9	25.1	49.8	78.5	104.5	132.3	156.0	200.0	243.5
$NiSO_4$	5.0	10.2	21.5						
$NiCl_2$	16.1	37.0	86.7	147.0	212.8				
$Ni(NO_3)_2$	16.1	37.3	91.3	156.2	235.0				
$Pb(NO_3)_2$	12.3	23.5	45.0	63.0					
$Sr(SO_3)_2$	7.2	20.3	47.0						
$Sr(NO_3)_2$	15.8	31.0	64.0	97.4	131.4				
$SrCl_2$	16.8	38.8	91.4	156.8	223.3	281.5			
$SrBr_2$	17.8	42.0	101.1	179.0	267.0				
$ZnSO_4$	4.9	10.4	21.5	42.1	66.2				
$ZnCl_2$	9.2	18.7	46.2	75.0	107.0	153.0	195.0		
$Zn(NO_3)_2$	16.6	39.0	93.5	157.5	223.8				

PRESSURE OF SATURATED AQUEOUS VAPOR.

The following tables for the pressure of saturated aqueous vapor are taken principally from the Fourth Revised Edition (1918) of the Smithsonian Meteorological Tables.

TABLE 183. — At Low Temperatures, – 69° to 0° C over Ice.

Temp.	0	1	2	3	4	5	6	7	8	9
	mm	mm	mm	mm	mm	mm	mm	mm	mm	mm
–60	0.008	0.007	0.006	0.005	0.004	0.004	0.003	0.003	0.003	0.002
–50	0.029	0.026	0.023	0.020	0.017	0.015	0.013	0.012	0.010	0.009
–40	0.096	0.086	0.076	0.068	0.060	0.054	0.048	0.042	0.037	0.033
–30	0.288	0.259	0.233	0.209	0.188	0.169	0.151	0.135	0.121	0.108
–20	0.783	0.712	0.646	0.585	0.530	0.480	0.434	0.392	0.354	0.319
–10	1.964	1.798	1.644	1.503	1.373	1.252	1.142	1.041	0.947	0.861
– 0	4.580	4.220	3.887	3.578	3.291	3.025	2.778	2.550	2.340	2.144

TABLE 184. — At Low Temperatures, – 16° to 0° C over Water.

Temp.	0	1	2	3	4	5	6	7	8	9
	mm	mm	mm	mm	mm	mm	mm	mm	mm	mm
–10°	2.144	1.979	1.826	1.684	1.551	1.429	1.315	—	—	—
– 0°	4.579	4.255	3.952	3.669	3.404	3.158	2.928	2.712	2.509	2.321

TABLE 185. — For Temperatures 0° to 374° C over Water.

Temp.	.0	.1	.2	.3	.4	.5	.6	.7	.8	.9
	mm	mm	mm	mm	mm	mm	mm	mm	mm	mm
0°	4.580	4.614	4.647	4.681	4.715	4.750	4.784	4.819	4.854	4.889
1	4.924	4.960	4.996	5.032	5.068	5.105	5.142	5.179	5.216	5.254
2	5.291	5.329	5.368	5.406	5.445	5.484	5.523	5.562	5.602	5.642
3	5.682	5.723	5.763	5.804	5.846	5.887	5.929	5.971	6.013	6.056
4	6.098	6.141	6.185	6.228	6.272	6.316	6.361	6.406	6.450	6.496
5	6.541	6.587	6.633	6.680	6.726	6.773	6.820	6.868	6.916	6.964
6	7.012	7.061	7.110	7.159	7.209	7.259	7.309	7.360	7.410	7.462
7	7.513	7.565	7.617	7.669	7.722	7.775	7.828	7.882	7.936	7.991
8	8.045	8.100	8.156	8.211	8.267	8.324	8.380	8.437	8.494	8.552
9	8.610	8.669	8.727	8.786	8.846	8.906	8.966	9.026	9.087	9.148
10	9.21	9.27	9.33	9.40	9.46	9.52	9.59	9.65	9.72	9.78
11	9.85	9.91	9.98	10.04	10.11	10.18	10.25	10.31	10.38	10.45
12	10.52	10.59	10.66	10.73	10.80	10.87	10.94	11.02	11.09	11.16
13	11.24	11.31	11.38	11.46	11.53	11.61	11.68	11.76	11.84	11.92
14	11.99	12.07	12.15	12.23	12.31	12.39	12.47	12.55	12.63	12.71
15	12.79	12.88	12.96	13.04	13.13	13.21	13.30	13.38	13.47	13.56
16	13.64	13.73	13.82	13.91	14.00	14.08	14.17	14.26	14.36	14.45
17	14.54	14.63	14.73	14.82	14.91	15.01	15.10	15.20	15.29	15.39
18	15.49	15.58	15.68	15.78	15.88	15.98	16.08	16.18	16.28	16.39
19	16.49	16.59	16.70	16.80	16.91	17.01	17.12	17.22	17.33	17.44
20	17.55	17.66	17.77	17.88	17.99	18.10	18.21	18.32	18.44	18.55
21	18.66	18.78	18.90	19.01	19.13	19.25	19.36	19.48	19.60	19.72
22	19.84	19.96	20.09	20.21	20.33	20.46	20.58	20.71	20.83	20.96
23	21.09	21.22	21.34	21.47	21.60	21.73	21.87	22.00	22.13	22.26
24	22.40	22.53	22.67	22.80	22.94	23.08	23.22	23.36	23.50	23.64
25	23.78	23.92	24.06	24.21	24.35	24.50	24.64	24.79	24.94	25.09

PRESSURE OF SATURATED AQUEOUS VAPOR.

TABLE 185. — For Temperatures 0° to 374° C over Water.

Temperature.	.0	.1	.2	.3	.4	.5	.6	.7	.8	.9
	mm	mm	mm	mm	mm	mm	mm	mm	mm	mm
25°	23.78	23.92	24.06	24.21	24.35	24.50	24.64	24.79	24.94	25.09
26	25.24	25.38	25.54	25.69	25.84	25.99	26.14	26.30	26.46	26.61
27	26.77	26.92	27.08	27.24	27.40	27.56	27.72	27.89	28.05	28.22
28	28.38	28.55	28.71	28.88	29.05	29.22	29.39	29.56	29.73	29.90
29	30.08	30.25	30.43	30.60	30.78	30.96	31.14	31.32	31.50	31.68
30	31.86	32.04	32.23	32.41	32.60	32.79	32.97	33.16	33.35	33.54
31	33.74	33.93	34.12	34.32	34.51	34.71	34.91	35.10	35.30	35.50
32	35.70	35.91	36.11	36.32	36.52	36.73	36.94	37.14	37.35	37.56
33	37.78	37.99	38.20	38.42	38.63	38.85	39.06	39.28	39.50	39.72
34	39.95	40.17	40.39	40.62	40.85	41.07	41.30	41.53	41.76	41.99
35	42.23	42.46	42.70	42.93	43.17	43.41	43.65	43.89	44.13	44.37
36	44.62	44.86	45.11	45.36	45.61	45.86	46.11	46.36	46.62	46.87
37	47.13	47.38	47.64	47.90	48.16	48.43	48.69	48.95	49.22	49.49
38	49.76	50.02	50.30	50.57	50.84	51.12	51.39	51.67	51.95	52.23
39	52.51	52.79	53.08	53.36	53.65	53.94	54.23	54.52	54.81	55.10
40	55.40	55.69	55.99	56.29	56.59	56.89	57.19	57.50	57.80	58.11
41	58.42	58.73	59.04	59.35	59.66	59.98	60.30	60.62	60.94	61.26
42	61.58	61.90	62.23	62.56	62.89	63.22	63.55	63.88	64.22	64.55
43	64.89	65.23	65.57	65.91	66.26	66.60	66.95	67.30	67.64	68.00
44	68.35	68.70	69.06	69.42	69.78	70.14	70.50	70.87	71.23	71.60
45	71.97	72.34	72.71	73.09	73.46	73.84	74.22	74.60	74.98	75.36
46	75.75	76.14	76.53	76.92	77.31	77.70	78.10	78.50	78.90	79.30
47	79.70	80.11	80.51	80.92	81.33	81.74	82.16	82.57	82.99	83.41
48	83.83	84.25	84.68	85.10	85.53	85.96	86.39	86.83	87.26	87.70
49	88.14	88.58	89.02	89.47	89.92	90.36	90.82	91.27	91.72	92.18
	0.	1.	2.	3.	4.	5.	6.	7.	8.	9.
50	92.6	97.3	102.2	107.3	112.7	118.2	124.0	130.0	136.3	142.8
60	149.6	156.6	164.0	171.6	179.5	187.8	196.3	205.2	214.4	224.0
70	233.9	244.2	254.9	266.0	277.4	289.3	301.6	314.4	327.6	341.2
80	355.4	370.0	385.2	400.8	417.0	433.7	451.0	468.8	487.3	506.3
90	526.0	546.3	567.2	588.8	611.1	634.1	657.8	682.2	707.4	733.3
100	760.0	787.5	815.9	845.0	875.1	906.0	937.8	970.5	1004.2	1038.8
110	1074	1111	1149	1187	1227	1268	1310	1353	1397	1442
120	1489	1536	1585	1636	1687	1740	1794	1850	1907	1965
130	2025	2086	2149	2214	2280	2347	2416	2487	2559	2633
140	2709	2786	2866	2947	3030	3115	3201	3290	3381	3473
150	3568	3665	3763	3864	3967	4072	4180	4290	4402	4516
160	4632	4751	4873	4997	5123	5252	5383	5518	5654	5794
170	5936	6080	6228	6378	6532	6688	6847	7009	7174	7342
180	7513	7688	7865	8046	8230	8417	8608	8802	8999	9200
190	9404	9612	9823	10040	10260	10480	10700	10940	11170	11410
200	11650	11890	12140	12400	12650	12920	13180	13450	13730	14010
210	14290	14580	14870	15160	15470	15770	16080	16400	16720	17040
220	17370	17710	18050	18390	18740	19100	19450	19820	20190	20560
230	20950	21330	21720	22120	22520	22930	23350	23770	24190	24620
240	25060	25500	25950	26410	26870	27340	27810	28290	28780	29270
250	29770	30280	30790	31310	31830	32360	32900	33450	34000	34560
260	35130	35700	36280	36870	37470	38070	38680	39300	39920	40560
270	41200	41840	42500	43160	43840	44520	45200	45900	46600	47320
280	48040	48760	49500	50250	51000	51770	52540	53320	54110	54910
290	55710	56530	57360	58190	59040	59890	60750	61620	62510	63400
300	64300	65210	66130	67060	68000	68960	69920	70890	71870	72860
310	73870	74880	75910	76940	77990	79050	80120	81200	82290	83390
320	84500	85630	86760	87910	89070	90250	91430	92630	93840	95060
330	96290	97530	98790	100060	101350	102640	103950	105280	106600	108000
340	109300	110700	112100	113500	114900	116300	117800	119200	120700	122200
350	123700	125200	126800	128300	129900	131400	133000	134600	136300	137900
360	139600	141200	142900	144600	146300	148100	149800	151600	153400	155200
370	157000	158800	160700	162600	164400	—	—	—	—	—

TABLE 186. — Weight in Grams of a Cubic Meter of Saturated Aqueous Vapor.

Temp. °C	0°	1°	2°	3°	4°	5°	6°	7°	8°	9°
−20°	0.894	0.816	0.743	0.677	0.615	0.559	0.508	0.461	0.418	0.378
−10	2.158	1.983	1.820	1.671	1.531	1.403	1.284	1.174	1.073	0.980
− 0	4.847	4.482	4.144	3.828	3.534	3.261	3.006	2.770	2.551	2.347
+ 0°	4.847	5.192	5.559	5.947	6.360	6.797	7.261	7.751	8.271	8.821
+10	9.401	10.015	10.664	11.348	12.070	12.832	13.635	14.482	15.373	16.311
+20	17.300	18.338	19.430	20.578	21.783	23.049	24.378	25.771	27.234	28.765
+30	30.371	32.052	33.812	35.656	37.583	39.599	41.706	43.908	46.208	48.609

For higher temperatures, see Table 259.

TABLE 187. — Weight in Grains of a Cubic Foot of Saturated Aqueous Vapor.

Temp. °F.	0°	1°	2°	3°	4°	5°	6°	7°	8°	9°
−20°	0.167	0.158	0.150	0.141	0.134	0.126	0.119	0.112	0.106	0.100
−10	0.286	0.272	0.258	0.244	0.232	0.220	0.208	0.197	0.187	0.176
− 0	0.479	0.455	0.433	0.411	0.391	0.371	0.353	0.335	0.318	0.302
+ 0°	0.479	0.503	0.529	0.556	0.584	0.613	0.644	0.676	0.709	0.744
+10	0.780	0.818	0.858	0.900	0.943	0.988	1.035	1.084	1.135	1.189
+20	1.244	1.301	1.362	1.425	1.490	1.558	1.629	1.703	1.779	1.859
+30	1.942	2.028	2.118	2.200	2.286	2.375	2.466	2.560	2.658	2.759
+40	2.863	2.970	3.082	3.196	3.315	3.436	3.563	3.693	3.828	3.965
+50	4.108	4.255	4.407	4.564	4.725	4.891	5.062	5.238	5.420	5.607
+60	5.800	5.999	6.203	6.413	6.630	6.852	7.082	7.317	7.560	7.809
+70	8.066	8.329	8.600	8.879	9.165	9.460	9.761	10.072	10.392	10.720
+80	11.056	11.401	11.756	12.121	12.494	12.878	13.272	13.676	14.090	14.515
+90	14.951	15.400	15.858	16.328	16.810	17.305	17.812	18.330	18.863	19.407
100°	19.966	20.538	21.123	21.723	22.337	22.966	23.611	24.271	24.946	25.636
110	26.343	27.066	27.807	28.563	29.338	30.130	30.940	31.768	32.616	33.482

Tables are abridged from Smithsonian Meteorological Tables, fourth revised edition.

TABLE 188. — Pressure of Aqueous Vapor in the Atmosphere.

For various altitudes (barometric readings).

The first column gives the depression of the wet-bulb temperature t_1 below the air temperature t. The value corresponding to the barometric height at the altitude of observation is to be subtracted from the vapor pressure corresponding to the wet-bulb temperature taken from Table 185. The temperature corresponding to this vapor pressure taken from Table 185 is the dew point. The wet bulb should be ventilated about 3 meters per second. For sea-level use Table 189. Example: $t = 35°$, $t_1 = 30°$, barometer 74 cm. Then $31.83 - 2.46 = 29.37$ mm = aqueous vapor pressure; the dew point is 28.6° C.

Abridged from Smithsonian Meteorological Tables, 1907.

$t - t_1$ °C	Barometric pressure in centimeters.													
	74	**72**	**70**	**63**	**66**	**64**	**62**	**60**	**58**	**56**	**54**	**52**	**50**	**48**
	mm	mm	mm	mm	mm	mm	mm	mm	mm	mm	mm	mm	mm	mm
1°	0.50	0.48	0.47	0.46	0.44	0.43	0.42	0.40	0.39	0.38	0.36	0.35	0.34	0.32
2	0.98	0.96	0.93	0.90	0.88	0.85	0.82	0.80	0.77	0.75	0.72	0.69	0.67	0.64
3	1.47	1.43	1.39	1.35	1.32	1.28	1.24	1.20	1.15	1.12	1.08	1.04	1.00	0.96
4	1.97	1.91	1.86	1.81	1.75	1.70	1.65	1.60	1.54	1.49	1.44	1.38	1.33	1.28
5	2.46	2.39	2.32	2.26	2.19	2.13	2.06	1.99	1.93	1.86	1.80	1.73	1.66	1.60
6	2.95	2.87	2.79	2.71	2.63	2.55	2.47	2.39	2.32	2.24	2.16	2.08	2.00	1.92
7	3.45	3.36	3.26	3.17	3.08	2.99	2.89	2.80	2.71	2.61	2.52	2.43	2.33	2.24
8	3.95	3.84	3.73	3.63	3.53	3.42	3.31	3.20	3.10	2.99	2.88	2.78	2.67	2.56
9	4.44	4.32	4.21	4.09	3.97	3.85	3.73	3.61	3.49	3.37	3.25	3.13	3.00	2.88
10	4.94	4.81	4.68	4.54	4.41	4.28	4.14	4.01	3.88	3.74	3.61	3.48	3.34	3.21
11	5.44	5.30	5.15	5.00	4.86	4.71	4.56	4.42	4.27	4.12	3.97	3.83	3.68	3.53
12	5.94	5.78	5.62	5.46	5.30	5.14	4.98	4.82	4.66	4.50	4.34	4.18	4.02	3.85
13	6.45	6.27	6.10	5.92	5.75	5.57	5.40	5.23	5.05	4.88	4.70	4.53	4.36	4.18
14	6.95	6.76	6.58	6.39	6.20	6.01	5.83	5.64	5.45	5.26	5.07	4.88	4.70	4.51
15	7.46	7.26	7.06	6.85	6.65	6.45	6.25	6.05	5.85	5.64	5.44	5.24	5.04	4.84
16	7.96	7.75	7.54	7.32	7.11	6.89	6.68	6.46	6.24	6.03	5.81	5.60	5.38	5.17
17	8.47	8.24	8.02	7.79	7.56	7.33	7.10	6.87	6.64	6.41	6.18	5.95	5.72	5.50

PRESSURE OF AQUEOUS VAPOR IN THE ATMOSPHERE.

This table gives the vapor pressure corresponding to various values of the difference $t - t_1$ between the readings of dry and wet bulb thermometers and the temperature t_1 of the wet bulb thermometer. The difference $t - t_1$ is given by two-degree steps in the top line, and t_1 by degrees in the first column. Temperatures in Centigrade degrees, vapor pressures in millimeters of mercury are used throughout the table. The table was calculated for barometric pressure B equal to 76 centimeters. A correction is given for each centimeter at the top of the columns. Ventilating velocity of wet thermometer about 3 meters per second.

t_1	$t - t_1 = 0°$	2°	4°	6°	8°	10°	12°	14°	16°	18°	20°	Difference for 0.1° in $t - t_1$
Corrections for B per cm		.013	.026	.040	.053	.066	.079	.092	.106	.119	.132	
−10	1.96	0.97	—	—	—							0.050
−9	2.14	1.15	0.16	—	—							0.050
−8	2.34	1.35	0.35	—	—							0.050
−7	2.55	1.56	0.66	—	—							0.050
−6	2.78	1.78	0.79	—	—							0.050
−5	3.02	2.03	1.03	0.03	—							0.050
−4	3.29	2.29	1.29	0.29	—							0.050
−3	3.58	2.58	1.58	0.58	—							0.050
−2	3.89	2.89	1.89	0.88	—	—	—	—	—	—	—	0.050
−1	4.22	3.22	2.22	1.21	0.21	—	—	—	—	—	—	0.050
0	4.58	3.58	2.57	1.57	0.57	—	—	—	—	—	—	0.050
1	4.92	3.92	2.92	1.91	0.91	—	—	—	—	—	—	0.050
2	5.29	4.29	3.28	2.27	1.27	0.26	—	—	—	—	—	0.050
3	5.68	4.68	3.67	2.66	1.66	0.65	—	—	—	—	—	0.050
4	6.10	5.09	4.08	3.07	2.07	1.06	0.05	—	—	—	—	0.050
5	6.54	5.53	4.52	3.51	2.51	1.50	0.49	—	—	—	—	0.050
6	7.01	6.00	4.99	3.98	2.97	1.96	0.95	—	—	—	—	0.050
7	7.51	6.50	5.49	4.48	3.47	2.46	1.45	0.43	—	—	—	0.050
8	8.04	7.03	6.02	5.01	4.00	2.98	1.97	0.96	—	—	—	0.050
9	8.61	7.60	6.58	5.57	4.56	3.54	2.53	1.52	0.50	—	—	0.050
10	9.21	8.20	7.18	6.17	5.15	4.14	3.12	2.11	1.09	0.08	—	0.050
11	9.85	8.83	7.81	6.80	5.78	4.77	3.75	2.73	1.72	0.70	—	0.051
12	10.52	9.50	8.49	7.47	6.45	5.44	4.42	3.40	2.38	1.37	0.35	0.051
13	11.24	10.22	9.20	8.18	7.16	6.14	5.13	4.11	3.09	2.07	1.05	0.051
14	11.99	10.97	9.95	8.93	7.91	6.90	5.88	4.86	3.84	2.82	1.80	0.051
15	12.79	11.77	10.75	9.73	8.71	7.69	6.67	5.65	4.63	3.61	2.59	0.051
16	13.64	12.62	11.60	10.58	9.96	8.53	7.51	6.49	5.47	4.45	3.43	0.051
17	14.54	13.52	12.49	11.47	10.45	9.42	8.40	7.38	6.36	5.33	4.31	0.051
18	15.49	14.46	13.44	12.42	11.39	10.37	9.34	8.32	7.30	6.27	5.25	0.051
19	16.49	15.46	14.44	13.41	12.39	11.36	10.34	9.31	8.29	7.26	6.24	0.051
20	17.55	16.52	15.50	14.47	13.44	12.42	11.39	10.36	9.34	8.31	7.29	0.051
21	18.66	17.64	16.61	15.58	14.56	13.53	12.50	11.47	10.45	9.42	8.39	0.051
22	19.84	18.82	17.79	16.76	15.73	14.70	13.67	12.64	11.62	10.59	10.57	0.051
23	21.09	20.06	19.03	18.00	16.97	15.94	14.91	13.88	12.85	11.82	10.79	0.051
24	22.40	21.37	20.34	19.31	18.27	17.24	16.21	15.18	14.15	13.12	12.09	0.051
25	23.78	22.75	21.71	20.68	19.65	18.62	17.59	16.56	15.52	14.49	13.46	0.052
26	25.24	24.20	23.17	22.14	21.10	20.07	19.04	18.00	16.97	15.94	14.90	0.052
27	26.77	25.73	24.70	23.66	22.63	21.60	20.56	19.53	18.49	17.46	16.42	0.052
28	28.38	27.34	26.31	25.27	24.24	23.20	22.17	21.13	20.10	19.06	18.02	0.052
29	30.08	29.04	28.00	26.97	25.93	24.89	23.86	22.82	21.78	20.75	19.71	0.052
30	31.86	30.82	29.78	28.75	27.71	26.67	25.63	24.60	23.56	22.52	21.48	0.052
31	33.74	32.70	31.66	30.62	29.58	28.54	27.50	26.46	25.42	24.38	23.34	0.052
32	35.70	34.66	33.62	32.58	31.54	30.50	29.46	28.42	27.38	26.34	25.30	0.052
33	37.78	36.73	35.69	34.65	33.61	32.57	31.53	30.49	29.44	28.40	27.36	0.052
34	39.95	38.90	37.86	36.82	35.78	34.73	33.69	32.65	31.61	30.57	29.52	0.052
35	42.23	41.18	40.14	39.10	38.05	37.01	35.97	34.92	33.88	32.83	31.79	0.052
36	44.62	43.57	42.53	41.48	40.44	39.40	38.35	37.31	36.26	35.22	34.17	0.052
37	47.13	46.08	45.04	43.99	42.94	41.90	40.85	39.81	38.76	37.71	36.67	0.052
38	49.76	48.71	47.66	46.61	45.57	44.52	43.47	42.43	41.38	40.33	39.29	0.052
39	52.51	51.46	50.41	49.37	48.32	47.27	46.22	45.17	44.12	43.08	42.03	0.052
40	55.40	54.35	53.30	52.25	51.20	50.15	49.10	48.05	47.00	45.95	44.00	0.052

Example.

$t = 17.2$; $t_1 = 10.0$; $B = 74.5$ cm
$t - t_1 = 7.2$
From table: $6.17 - 12 \times 0.050 = 5.57$
For B, $1.5 \times .048 \quad = .07$
Hence $p \quad = 5.64$

RELATIVE HUMIDITY.

Vertical argument is the observed vapor pressure which may be computed from the wet and dry-bulb readings through Table 188 or 189. The horizontal argument is the observed air temperature (dry-bulb reading). Based upon Table 43, p. 142, Smithsonian Meteorological Tables, 3d Revised Edition, 1907.

Vapor Pressure. mm.	Air Temperatures, dry bulb, ° Centigrade.																
	0°	−1°	−2°	−3°	−4°	−5°	−6°	−7°	−8°	−9°	−10°	−11°	−12°	−13°	−14°	−15°	−20°
0.25	6	6	6	7	8	8	9	10	11	12	13	14	15	17	18	20	32
0.50	11	12	13	14	15	17	18	20	21	23	25	28	30	34	37	40	64
0.75	17	18	19	21	23	25	27	30	32	35	38	42	46	50	55	60	96
1.00	22	24	26	28	30	33	36	40	42	47	51	56	61	67	74	80	
1.25	27	30	32	35	38	42	45	49	54	58	64	70	76	84	92	100	
1.50	33	36	39	42	46	50	54	59	64	70	76	84	92	100			
1.75	38	42	45	49	53	58	63	69	75	82	89	98					
2.00	44	48	52	56	61	66	72	79	86	93							
2.25	49	53	58	63	69	75	81	89	96	–							
2.50	55	59	65	70	76	83	90	99	–	–							
2.75	60	65	71	77	84	91	100	–	–	–							
3.00	66	71	78	84	92	100	–	–	–	–							
3.25	71	77	84	91	99	–	–	–	–	–							
3.50	77	83	90	98	–	–	–	–	–	–							

mm.	0°	−1°	−2°	−3°
3.50	77	83	90	98
3.75	82	89	97	–
4.00	88	95	–	–
4.25	93	100	–	–
4.50	99	–	–	–

Vapor Pressure. mm.	Air Temperatures, dry bulb, ° Centigrade.																				
	0°	1°	2°	3°	4°	5°	6°	7°	8°	9°	10°	11°	12°	13°	14°	15°	16°	17°	18°	19°	20°
0.5	11	10	9	9	8	8	7	7	6	6	5	5	5	4	4	4	4	3	3	3	3
1.0	22	20	19	18	16	15	14	13	13	12	11	10	10	9	8	8	7	7	7	6	6
1.5	33	31	28	27	25	23	22	20	19	18	16	15	14	13	13	12	11	10	10	9	9
2.0	44	41	38	35	33	31	29	27	25	23	22	20	19	18	17	16	15	14	13	12	12
2.5	55	51	47	44	41	38	36	33	31	29	27	26	24	22	21	20	18	17	16	15	14
3.0	66	61	57	53	49	46	43	40	38	35	33	31	29	27	25	24	22	21	20	18	17
3.5	77	71	66	62	58	54	50	47	44	41	38	36	34	31	29	28	26	24	23	21	20
4.0	88	81	76	71	66	61	57	54	50	47	44	41	38	36	34	32	30	28	26	25	23
4.5	99	92	85	80	74	69	65	60	56	53	49	46	43	40	38	36	33	31	29	28	26
5.0	–	–	95	88	83	77	72	67	63	58	55	51	48	45	42	39	37	35	33	31	29
5.5	–	–	–	97	91	85	79	74	69	64	60	56	53	49	46	43	41	38	36	34	32
6.0	–	–	–	–	99	92	86	80	75	70	66	61	58	54	51	47	44	42	39	37	34
6.5	–	–	–	–	–	100	93	87	81	76	71	67	62	58	55	51	48	45	42	40	37
7.0	–	–	–	–	–	–	100	94	85	82	77	72	67	63	59	55	52	49	46	43	40
7.5	–	–	–	–	–	–	–	100	94	88	82	77	72	67	63	59	55	52	49	46	43
8.0	–	–	–	–	–	–	–	–	100	94	88	82	77	72	67	63	59	56	52	49	46
8.5	–	–	–	–	–	–	–	–	–	99	93	87	82	76	72	67	63	59	55	52	49
9.0	–	–	–	–	–	–	–	–	–	–	98	92	86	81	76	71	67	62	59	55	52
9.5	–	–	–	–	–	–	–	–	–	–	–	97	91	85	80	75	70	66	62	58	55
10.0	–	–	–	–	–	–	–	–	–	–	–	–	96	90	84	79	74	69	65	61	57
11.0	–	–	–	–	–	–	–	–	–	–	–	–	–	94	93	87	81	76	72	67	63
12.0	–	–	–	–	–	–	–	–	–	–	–	–	–	–	–	94	89	83	78	74	69
13.0	–	–	–	–	–	–	–	–	–	–	–	–	–	–	–	–	96	90	85	80	75
14.0	–	–	–	–	–	–	–	–	–	–	–	–	–	–	–	–	–	97	91	86	80
15.0	–	–	–	–	–	–	–	–	–	–	–	–	–	–	–	–	–	–	97	92	86
16.0	–	–	–	–	–	–	–	–	–	–	–	–	–	–	–	–	–	–	–	98	92
17.0	–	–	–	–	–	–	–	–	–	–	–	–	–	–	–	–	–	–	–	–	98

Table 190 *(continued).*

RELATIVE HUMIDITY.

Vapor Pressure. mm.	Air Temperatures, dry bulb, ° Centigrade.																				
	20°	21°	22°	23°	24°	25°	26°	27°	28°	29°	30°	31°	32°	33°	34°	35°	36°	37°	38°	39°	40°
1	6	5	5	5	5	4	4	4	4	3	3	3	3	3	3	3	2	2	2	2	2
2	12	11	10	10	9	8	8	8	7	7	6	6	6	5	5	5	5	4	4	4	4
3	17	16	15	14	14	13	12	11	11	10	10	9	9	8	8	7	7	6	6	6	5
4	23	22	20	19	18	17	16	15	14	13	13	12	11	11	10	10	9	9	8	8	7
5	29	27	25	24	23	21	20	19	18	17	16	15	14	13	13	12	11	11	10	10	9
6	34	32	31	29	27	26	24	23	21	20	19	18	17	16	15	14	14	13	12	12	11
7	40	38	36	34	32	30	28	26	25	24	22	21	20	19	18	17	16	15	14	13	13
8	46	43	41	38	36	34	32	30	29	27	25	24	23	21	20	19	18	17	16	15	15
9	52	49	46	43	41	38	36	34	32	30	29	27	25	24	23	22	20	19	18	17	16
10	57	54	51	48	45	43	40	38	36	34	32	30	28	27	25	24	23	21	20	19	18
11	63	60	56	53	50	47	44	42	39	37	35	33	31	29	28	26	25	24	22	21	20
12	69	65	61	58	54	51	48	45	43	40	38	36	34	32	30	29	27	26	24	23	22
13	75	70	66	62	59	55	52	49	46	44	41	39	37	35	33	31	29	28	26	25	24
14	80	76	71	67	63	60	56	53	50	47	44	42	40	37	35	33	32	30	28	27	26
15	86	81	76	72	68	64	60	57	53	50	48	45	42	40	38	36	34	32	30	29	27
16	92	87	82	77	72	68	64	60	57	54	51	48	45	43	41	38	36	34	32	31	29
17	98	92	87	81	77	72	68	64	61	57	54	51	48	45	43	41	38	36	34	33	31
18	–	97	92	86	81	77	72	68	64	60	57	54	51	48	46	43	41	39	37	35	33
19	–	–	97	91	86	81	76	72	68	64	60	57	54	51	48	45	43	41	39	36	35
20	–	–	–	96	90	85	80	76	71	67	63	60	57	53	51	48	45	43	41	38	36
21	–	–	–	–	95	89	84	79	75	71	67	63	59	56	53	50	48	45	43	40	38
22	–	–	–	–	100	94	88	83	78	74	70	66	62	59	56	53	50	47	45	42	40
23	–	–	–	–	–	98	92	87	82	77	73	69	65	62	58	55	52	49	47	44	42
24	–	–	–	–	–	–	96	91	85	81	76	72	68	64	61	57	54	51	49	46	44
25	–	–	–	–	–	–	100	94	89	84	79	75	71	67	63	60	56	54	51	48	46
26	–	–	–	–	–	–	–	98	93	87	83	78	74	70	66	62	59	56	53	50	47
27	–	–	–	–	–	–	–	–	96	91	86	81	76	72	68	65	61	58	55	52	49
28	–	–	–	–	–	–	–	–	100	94	89	84	79	75	71	67	63	60	57	54	51
29	–	–	–	–	–	–	–	–	–	97	92	87	82	78	73	69	65	62	59	56	53
30	–	–	–	–	–	–	–	–	–	–	95	90	85	80	76	72	68	64	61	58	55
31	–	–	–	–	–	–	–	–	–	–	98	93	88	83	78	74	70	66	63	60	56
32	–	–	–	–	–	–	–	–	–	–	–	96	91	86	81	77	72	69	65	62	58
33	–	–	–	–	–	–	–	–	–	–	–	99	93	88	84	79	75	71	67	63	60
34	–	–	–	–	–	–	–	–	–	–	–	–	96	91	86	81	77	73	69	65	62
35	–	–	–	–	–	–	–	–	–	–	–	–	99	94	89	84	79	75	71	67	64
36	–	–	–	–	–	–	–	–	–	–	–	–	–	96	91	86	81	77	73	69	66
37	–	–	–	–	–	–	–	–	–	–	–	–	–	99	94	89	84	79	75	71	67
38	–	–	–	–	–	–	–	–	–	–	–	–	–	–	96	91	86	81	77	73	69
39	–	–	–	–	–	–	–	–	–	–	–	–	–	–	99	93	88	83	79	75	71
40	–	–	–	–	–	–	–	–	–	–	–	–	–	–	–	96	90	86	81	77	73
41	–	–	–	–	–	–	–	–	–	–	–	–	–	–	–	98	93	88	83	79	75
42	–	–	–	–	–	–	–	–	–	–	–	–	–	–	–	100	95	90	85	81	77
43	–	–	–	–	–	–	–	–	–	–	–	–	–	–	–	–	97	92	87	83	78
44	–	–	–	–	–	–	–	–	–	–	–	–	–	–	–	–	99	94	89	84	80
45	–	–	–	–	–	–	–	–	–	–	–	–	–	–	–	–	–	96	91	86	82
46	–	–	–	–	–	–	–	–	–	–	–	–	–	–	–	–	–	99	93	88	84
47	–	–	–	–	–	–	–	–	–	–	–	–	–	–	–	–	–	–	95	90	86
48	–	–	–	–	–	–	–	–	–	–	–	–	–	–	–	–	–	–	97	92	87
49	–	–	–	–	–	–	–	–	–	–	–	–	–	–	–	–	–	–	99	94	89
50	–	–	–	–	–	–	–	–	–	–	–	–	–	–	–	–	–	–	–	96	91
51	–	–	–	–	–	–	–	–	–	–	–	–	–	–	–	–	–	–	–	98	93
52	–	–	–	–	–	–	–	–	–	–	–	–	–	–	–	–	–	–	–	100	95
53	–	–	–	–	–	–	–	–	–	–	–	–	–	–	–	–	–	–	–	–	97
54	–	–	–	–	–	–	–	–	–	–	–	–	–	–	–	–	–	–	–	–	98
55	–	–	–	–	–	–	–	–	–	–	–	–	–	–	–	–	–	–	–	–	100

TABLE 190 (*concluded*).—Relative Humidity.

(Data from 20° to 60° C. based upon Table 185).

Vapor Pressure. mm.	Air Temperatures, dry bulb, ° Centigrade.																				
	40°	41°	42°	43°	44°	45°	46°	47°	48°	49°	50°	51°	52°	53°	54°	55°	56°	57°	58°	59°	60°
5	9	9	8	8	7	7	7	6	6	6	5	5	5	5	4	4	4	4	4	4	3
10	18	17	16	15	15	14	13	13	12	11	11	10	10	9	9	8	8	8	7	7	7
15	27	26	24	23	22	21	20	19	18	17	16	15	15	14	13	13	12	12	11	10	10
20	36	34	33	31	29	28	26	25	24	23	22	21	20	19	18	17	16	15	15	14	13
25	45	43	41	39	37	35	33	31	30	28	27	26	24	23	22	21	20	19	18	18	17
30	54	51	49	46	44	42	40	38	36	34	32	31	29	28	27	25	24	23	22	21	20
35	63	60	57	54	51	49	46	44	42	40	38	36	34	33	31	30	28	27	26	25	23
40	72	68	65	62	59	56	53	50	48	45	43	41	39	37	36	34	32	31	29	28	27
45	81	77	73	69	66	63	59	57	54	51	49	46	44	42	40	38	36	35	33	32	30
50	90	86	81	77	73	70	66	63	60	57	54	51	49	47	44	42	40	38	37	35	33
55	99	94	89	85	81	76	73	69	66	62	59	57	54	51	49	46	44	42	40	39	37
60	–	–	98	93	88	83	79	75	72	68	65	62	60	56	53	51	48	46	44	42	40
65	–	–	–	100	95	90	86	82	78	74	70	67	64	61	58	55	52	50	48	46	43
70	–	–	–	–	–	97	92	88	84	80	76	72	68	65	62	59	56	54	51	49	47
75	–	–	–	–	–	–	99	94	90	85	81	77	74	70	67	64	60	58	55	53	50
80	–	–	–	–	–	–	–	100	96	91	86	82	78	75	71	68	64	62	59	56	54
85	–	–	–	–	–	–	–	–	–	97	92	87	84	79	75	72	69	65	62	60	57
90	–	–	–	–	–	–	–	–	–	–	97	93	88	84	80	76	73	69	66	63	60
95	–							–	–	–	–	98	94	89	84	80	77	73	70	67	64
100	–							–	–	–	–	–	98	93	89	85	81	77	73	70	67
105	–							–	–	–	–	–	–	98	93	89	85	81	77	74	70
110	–							–	–	–	–	–	–	–	98	93	89	85	81	77	74
115	–							–	–	–	–	–	–	–	–	97	93	88	84	81	77
120	–							–	–	–	–	–	–	–	–	–	97	92	88	84	80
125	–							–	–	–	–	–	–	–	–	–	–	96	92	88	84

mm.	57°	58°	59°	60°
125	96	92	88	84
130	100	95	91	87
135	–	99	95	90
140	–	–	98	94
145	–	–	–	97
150	–	–	–	100

TABLE 191.—Relative Humidity.

This table gives the relative humidity direct from the difference between the reading of the dry (t° C.) and the wet (t_1° C.) thermometer. It is computed for a barometer reading of 76 cm. The wet thermometer should be ventilated about 3 meters per second. From manuscript tables computed at the U.S. Weather Bureau.

t°	Depression of wet-bulb thermometer, t°-t_1°.																
	0.2°	0.4°	0.6°	0.8°	1.0°	1.2°	1.4°	1.6°	1.8°	2.0°	2.5°	3.0°	3.5°	4.0°	4.5°	5.0°	5.5°
−15	90	91	72	62	53	44	35	25	16	7	–	–	–	–	–	–	–
−12	92	85	77	69	62	54	47	39	32	25	7	–	–	–	–	–	–
−9	94	88	81	75	70	62	56	50	44	39	23	9	–	–	–	–	–
−6	95	89	85	80	74	69	64	59	54	49	36	25	13	2	–	–	–
−3	96	91	87	82	78	74	69	66	61	57	46	36	26	17	7	–	–
0	96	92	89	85	81	78	74	71	67	64	55	46	38	29	21	13	6
+3	97	94	91	87	84	81	78	75	72	69	62	54	46	40	32	25	18
	0.5°	1.0°	1.5°	2.0°	2.5°	3.0°	3.5°	4.0°	4.5°	5.0°	6.0°	7.0°	8.0°	9.0°	10.0°	11.0°	12.0°
+3	92	84	76	69	62	54	46	40	32	25	12	–	–	–	–	–	–
+6	94	87	80	73	66	60	54	47	41	35	23	11	–	–	–	–	–
+9	94	88	82	76	70	65	59	53	48	42	32	22	12	3	–	–	–
+12	94	89	84	78	73	68	63	58	53	48	38	30	21	12	4	–	–
+15	95	90	85	80	76	71	66	62	58	53	44	36	28	20	13	4	–
+18	95	90	86	82	78	73	69	65	61	57	49	42	35	27	20	13	6
+21	96	91	87	83	79	75	71	67	64	60	53	46	39	32	26	19	13
+24	96	92	88	85	81	77	74	70	66	63	56	49	43	37	31	26	21
+27	96	93	90	86	82	79	76	72	68	65	59	53	47	41	36	31	26
+30	96	93	90	86	82	79	76	73	70	67	61	55	50	44	39	35	30
+33	96	93	90	86	83	80	77	74	71	68	63	57	52	47	42	37	33
+36	97	93	90	87	84	81	78	75	72	70	64	57	54	50	45	41	36
+39	97	94	91	88	85	82	79	76	74	71	66	61	56	52	47	43	39

CORRECTION FOR TEMPERATURE OF EMERGENT MERCURIAL THERMOMETER THREAD.

When the temperature of a portion of a thermometer stem with its mercury thread differs much from that of the bulb, a correction is necessary to the observed temperature unless the instrument has been calibrated for the experimental conditions. This stem correction is proportional to $n\beta(T - t)$, where n is the number of degrees in the exposed stem, β the apparent coefficient of expansion of mercury in the glass, T the measured temperature, and t the mean temperature of the exposed stem. For temperatures up to 100° C, the value of β is for Jena 16^{III} or Greiner and Friedrich resistance glass, 0.000159, for Jena 59^{III}, 0.000164, and when of unknown composition it is best to use a value of about 0.000155. The formula requires a knowledge of the temperature of the emergent stem. This may be approximated in one of three ways: (1) by a "fadenthermometer" (see Buckingham, Bulletin Bureau of Standards, 8, p. 239, 1912); (2) by exploring the temperature distribution of the stem and calculating its mean temperature; and (3) by suspending along the side of, or attaching to the stem, a single thermometer. Table 192 is taken from the Smithsonian Meteorological Tables, Tables 193–195 from Rimbach, Z. f. Instrumentenkunde, 10, p. 153, 1890, and apply to thermometers of Jena or resistance glass.

TABLE 192. — Stem Correction for Centigrade Thermometers.

Values of $0.000155n(T - t)$.

n	$(T-t)$.							
	10°	20°	30°	40°	50°	60°	70°	80°
10° C	0.02	0.03	0.05	0.06	0.08	0.09	0.11	0.12
20	0.03	0.06	0.09	0.12	0.16	0.19	0.22	0.25
30	0.05	0.09	0.14	0.19	0.23	0.28	0.33	0.37
40	0.06	0.12	0.19	0.25	0.31	0.37	0.43	0.50
50	0.08	0.16	0.23	0.31	0.39	0.46	0.54	0.62
60	0.09	0.19	0.28	0.37	0.46	0.56	0.65	0.74
70	0.11	0.22	0.33	0.43	0.54	0.65	0.76	0.87
80	0.12	0.25	0.37	0.50	0.62	0.74	0.87	0.99
90	0.14	0.28	0.42	0.56	0.70	0.84	0.98	1.12
100	0.16	0.31	0.46	0.62	0.78	0.93	1.08	1.24

TABLE 193. — Stem Correction for Thermometer of Jena Glass (0° to 360° C).

Degree length 0.9 to 1.1 mm; t = the observed temperature; t' = that of the surrounding air 1 dm. away; n = the length of the exposed thread.

Correction to be added to the reading t.										
n	$t - t'$									
	70°	80°	90°	100°	120°	140°	160°	180°	200°	220°
10	0.01	0.01	0.03	0.04	0.07	0.10	0.13	0.17	0.19	0.21
20	0.08	0.12	0.14	0.19	0.25	0.28	0.32	0.40	0.49	0.54
30	0.25	0.28	0.32	0.36	0.42	0.48	0.54	0.66	0.78	0.87
40	0.30	0.35	0.41	0.48	0.60	0.67	0.77	0.92	1.08	1.20
50	0.41	0.46	0.52	0.59	0.79	0.89	0.98	1.16	1.38	1.53
60	0.52	0.60	0.68	0.79	0.99	1.11	1.23	1.46	1.70	1.87
70	0.63	0.74	0.85	0.98	1.20	1.32	1.45	1.70	1.99	2.21
80	0.75	0.87	1.01	1.15	1.38	1.53	1.70	1.98	2.29	2.54
90	0.87	0.99	1.13	1.28	1.62	1.82	1.94	2.25	2.60	2.89
100	0.98	1.12	1.29	1.47	1.82	2.03	2.20	2.55	2.92	3.24
120	—	—	—	1.88	2.28	2.49	2.68	3.13	3.59	3.96
140	—	—	—	—	2.75	2.97	3.22	3.75	4.24	4.69
160	—	—	—	—	—	3.35	3.80	4.35	4.92	5.45
180	—	—	—	—	—	—	4.37	4.99	5.63	6.22
200	—	—	—	—	—	—	—	5.68	6.34	6.98
220	—	—	—	—	—	—	—	—	7.05	7.82

CORRECTION FOR TEMPERATURE OF MERCURY IN THERMOMETER STEM (*continued*).

TABLE 194. — Stem Correction for Thermometer of Jena Glass (0°–360° C).

Degree length 1 to 1.6 mm.; t = the observed temperature; t' = that of the surrounding air one dm. away; n = the length of the exposed thread.

CORRECTION TO BE ADDED TO THERMOMETER READING.*

n	$t-t'$										*n*
	70°	80°	90°	100°	120°	140°	160°	180°	200°	220°	
10°	0.02	0.03	0.05	0.07	0.11	0.17	0.21	0.27	0.33	0.38	**10°**
20	0.13	0.15	0.18	0.22	0.29	0.38	0.46	0.53	0.61	0.67	20
30	0.24	0.28	0.33	0.39	0.48	0.59	0.70	0.78	0.88	0.97	30
40	0.35	0.41	0.48	0.56	0.68	0.82	0.94	1.04	1.16	1.28	40
50	0.47	0.53	0.62	0.72	0.88	1.03	1.17	1.31	1.44	1.59	**50**
60	0.57	0.66	0.77	0.89	1.09	1.25	1.42	1.58	1.74	1.90	60
70	0.69	0.79	0.92	1.06	1.30	1.47	1.67	1.86	2.04	2.23	70
80	0.80	0.91	1.05	1.21	1.52	1.71	1.94	2.15	2.33	2.55	80
90	0.91	1.04	1.19	1.38	1.73	1.96	2.20	2.42	2.64	2.89	**90**
100	1.02	1.18	1.35	1.56	1.97	2.18	2.45	2.70	2.94	3.23	100
110	–	–	–	1.78	2.19	2.43	2.70	2.98	3.26	3.57	110
120	–	–	–	1.98	2.43	2.69	2.95	3.26	3.58	3.92	120
130	–	–	–	–	2.68	2.94	3.20	3.56	3.89	4.28	**130**
140	–	–	–	–	2.92	3 22	3.47	3.86	4.22	4.64	140
150	–	–	–	–	–	–	3.74	4.15	4.56	5.01	150
160	–	–	–	–	–	–	4.00	4.46	4.90	5.39	160
170	–	–	–	–	–	–	4.27	4.76	5.24	5.77	**170**
180	–	–	–	–	–	–	4.54	5.07	5.59	6.15	180
190	–	–	–	–	–	–	–	5.38	5 95	6.54	190
200	–	–	–	–	–	–	–	5.70	6.30	6.94	200
210	–	–	–	–	–	–	–	–	6.68	7.35	**210**
220	–	–	–	–	–	–	–	–	7.04	7.75	220

* See Hovestadt's "Jena Glass" (translated by J. D. and A. Everett) for data on changes of thermometer zeros.

TABLE 195. — Stem Correction for a so-called Normal Thermometer of Jena Glass (0°–100° C).

Divided into tenth degrees; degree length about 4 mm.

CORRECTION TO BE ADDED TO THE READING t.

n	$t-t'$											
	30°	35°	40°	45°	50°	55°	60°	65°	70°	75°	80°	85°
10	0.04	0.04	0.05	0.05	0.05	0.06	0.06	0.07	0.08	0.09	0.10	0.10
20	0.12	0.12	0.13	0.14	0.15	0.16	0.17	0.18	0.19	0.20	0.22	0.23
30	0.21	0.22	0.23	0.24	0.25	0.25	0.27	0.29	0.31	0.33	0.35	0.37
40	0.28	0.29	0.31	0.33	0.35	0.37	0.39	0.41	0.43	0.45	0.48	0.51
50	0.36	0.38	0.40	0.42	0.44	0.46	0.48	0.50	0.53	0.57	0.61	0.65
60	0.45	0.48	0.51	0.53	0.55	0.57	0.60	0.63	0.66	0.69	0.73	0.78
70	–	–	–	–	–	0.66	0.69	0.71	0.75	0.81	0.87	0.92
80	–	–	–	–	–	–	0.76	0.81	0.87	0.93	1.00	1.06
90	–	–	–	–	–	–	–	0.92	0.99	1.06	1.13	1.20
100	–	–	–	–	–	–	–	–	1.10	1.18	1.26	1.34

THERMOMETERS.

TABLE 196. — Gas and Mercury Thermometers.

If t_H, t_N, t_{CO2}, t_{16}, t_{59}, t_T, are temperatures measured with the hydrogen, nitrogen, carbonic acid, 16[III], 59[III], and "verre dur" (Tonnelot), respectively, then

$$t_H - t_T = \frac{(100-t)t}{100^2}[-0.61859 + 0.0047351.t - 0.000011577.t^2]*$$

$$t_N - t_T = \frac{(100-t)t}{100^2}[-0.55541 + 0.0048240.t - 0.000024807.t^2]*$$

$$t_{CO2} - t_T = \frac{(100-t)t}{100^2}[-0.33386 + 0.0039910.t - 0.000016678.t^2]*$$

$$t_H - t_{16} = \frac{(100-t)t}{100^2}[-0.67039 + 0.0047351.t - 0.000011577.t^2]\dagger$$

$$t_H - t_{59} = \frac{(100-t)t}{100^2}[-0.31089 + 0.0047351.t - 0.000011577.t^2]\dagger$$

* Chappuis; Trav. et Mém. du Bur. internat. des Poids et Mes. 6, 1888.
† Thiesen, Scheel, Sell; Wiss. Abh. d. Phys. Techn. Reichanstalt, 2, 1895; Scheel; Wied. Ann. 58, 1896; D. Mech. Ztg. 1897.

TABLE 197. $t_H - t_{16}$ (Hydrogen — 16[III]).

	0°	1°	2°	3°	4°	5°	6°	7°	8°	9°
0°	.000°	—.007°	—.013°	—.019°	—.025°	—.031°	—.036°	—.042°	—.047°	—.051°
10	—.056	—.061	—.065	—.069	—.073	—.077	—.080	—.084	—.087	—.090
20	—.093	—.096	—.098	—.101	—.103	—.105	—.107	—.109	—.110	—.112
30	—.113	—.114	—.115	—.116	—.117	—.118	—.119	—.119	—.119	—.120
40	—.120	—.120	—.120	—.120	—.119	—.119	—.118	—.118	—.117	—.116
50	—.116	—.115	—.114	—.113	—.111	—.110	—.109	—.107	—.106	—.104
60	—.103	—.101	—.099	—.097	—.096	—.094	—.092	—.090	—.087	—.085
70	—.083	—.081	—.078	—.076	—.074	—.071	—.069	—.066	—.064	—.061
80	—.058	—.056	—.053	—.050	—.048	—.045	—.042	—.039	—.036	—.033
90	—.030	—.027	—.024	—.021	—.018	—.015	—.012	—.009	—.006	—.003
100	.000									

TABLE 198. $t_H - t_{59}$ (Hydrogen — 59[III]).

	0°	1°	2°	3°	4°	5°	6°	7°	8°	9°
0°	.000°	—.003°	—.006°	—.009°	—.011°	—.014°	—.016°	—.018°	—.020°	—.022°
10	—.024	—.025	—.027	—.028	—.030	—.031	—.032	—.033	—.034	—.035
20	—.035	—.036	—.036	—.037	—.037	—.037	—.038	—.038	—.038	—.038
30	—.038	—.037	—.037	—.037	—.037	—.036	—.036	—.035	—.035	—.034
40	—.034	—.033	—.032	—.032	—.031	—.030	—.029	—.028	—.028	—.027
50	—.026	—.025	—.024	—.023	—.022	—.021	—.020	—.019	—.018	—.017
60	—.016	—.015	—.015	—.014	—.013	—.012	—.011	—.010	—.009	—.008
70	—.008	—.007	—.006	—.005	—.005	—.004	—.003	—.003	—.002	—.001
80	—.001	—.001	.000	.000	+.001	+.001	+.001	+.002	+.002	+.002
90	+.002	+.002	+.002	+.002	+.002	+.002	+.001	+.001	+.001	.000
100	.000									

TABLE 199. (Hydrogen — 16[III]), (Hydrogen — 59[III]).

	—5°	—10°	—15°	—20°	—25°	—30°	—35°
$t_H - t_{16}$	+0.04°	+0.08°	+0.13°	+0.19°	+0.25°	+0.32°	+0.40°
$t_H - t_{59}$	+0.02°	+0.04°	+0.07°	+0.10°	+0.14°	+0.18°	+0.23°

All compiled from Landolt-Börnstein-Meyerhoffer's Physikalisch-chemische Tabellen.

AIR AND MERCURY THERMOMETERS.

TABLE 200. $t_{AIR} - t_{16}$. (Air — 16^{III}.)

°C.	0°	1°	2°	3°	4°	5°	6°	7°	8°	9°
0	.000	—.006	—.012	—.017	—.022	—.027	—.032	—.037	—.041	—.045
10	—.049	—.053	—.057	—.061	—.065	—.068	—.071	—.074	—.077	—.080
20	—.083	—.086	—.089	—.091	—.093	—.095	—.097	—.099	—.101	—.102
30	—.103	—.104	—.105	—.106	—.107	—.108	—.109	—.110	—.110	—.110
40	—.110	—.110	—.111	—.111	—.110	—.110	—.110	—.109	—.109	—.108
50	—.107	—.107	—.106	—.105	—.104	—.103	—.102	—.101	—.100	—.098
60	—.096	—.095	—.093	—.092	—.090	—.088	—.086	—.084	—.082	—.080
70	—.078	—.076	—.074	—.072	—.070	—.067	—.065	—.062	—.060	—.057
80	—.054	—.052	—.049	—.047	—.044	—.041	—.039	—.036	—.034	—.031
90	—.028	—.025	—.023	—.020	—.017	—.014	—.011	—.009	—.006	—.003
100	.000	+.003	+.006	+.008	+.011	+.014	+.017	+.019	+.022	+.025
110	+.028	+.030	+.033	+.035	+.038	+.041	+.043	+.046	+.048	+.050
120	+.053	+.055	+.057	+.060	+.062	+.064	+.066	+.068	+.070	+.072
130	+.074	+.076	+.078	+.080	+.081	+.083	+.084	+.086	+.087	+.089
140	+.090	+.091	+.092	+.093	+.094	+.095	+.096	+.096	+.097	+.097
150	+.098	+.098	+.098	+.099	+.099	+.099	+.098	+.098	+.098	+.097
160	+.097	+.096	+.095	+.094	+.093	+.092	+.090	+.089	+.088	+.086
170	+.084	+.082	+.080	+.078	+.076	+.073	+.071	+.068	+.065	+.062
180	+.059	+.055	+.052	+.048	+.045	+.041	+.037	+.033	+.028	+.023
190	+.019	+.014	+.009	+.004	—.001	—.007	—.013	—.019	—.025	—.031
200	—.038	—.045	—.051	—.058	—.066	—.073	—.080	—.088	—.096	—.105
210	—.113	—.122	—.130	—.139	—.148	—.158	—.168	—.177	—.187	—.198
220	—.208	—.219	—.230	—.241	—.252	—.264	—.275	—.287	—.300	—.312
230	—.325	—.338	—.351	—.365	—.378	—.392	—.407	—.421	—.436	—.450
240	—.466	—.481	—.497	—.513	—.529	—.546	—.562	—.579	—.597	—.614
250	—.632	—.650	—.668	—.687	—.706	—.725	—.745	—.765	—.785	—.805
260	—.825	—.846	—.867	—.889	—.911	—.933	—.955	—.978	—1.001	—1.025
270	—1.048	—1.072	—1.096	—1.121	—1.146	—1.171	—1.196	—1.222	—1.248	—1.274
280	—1.301	—1.328	—1.356	—1.384	—1.412	—1.440	—1.469	—1.498	—1.528	—1.558
290	—1.588	—1.618	—1.649	—1.680	—1.711	—1.743	—1.776	—1.808	—1.841	—1.874
300	—1.908									

Note: See Circular 8, Bureau of Standards relative to use of thermometers and the various precautions and corrections.

TABLE 201. $t_{AIR} - t_{59}$. (Air — 59^{III}.)

°C.	0°	1°	2°	3°	4°	5°	6°	7°	8°	9°
100	.000	.000	.000	.000	.000	.000	.000	.000	.000	.000
110	.000	.000	.000	—.001	—.001	—.001	—.001	—.001	—.002	—.002
120	—.002	—.002	—.002	—.002	—.002	—.003	—.003	—.003	—.004	—.004
130	—.004	—.004	—.005	—.005	—.006	—.006	—.006	—.007	—.007	—.008
140	—.008	—.008	—.009	—.009	—.010	—.010	—.011	—.011	—.012	—.012
150	—.013	—.013	—.014	—.015	—.016	—.016	—.016	—.017	—.018	—.019
160	—.019	—.020	—.021	—.021	—.022	—.023	—.024	—.025	—.026	—.027
170	—.028	—.029	—.030	—.031	—.032	—.033	—.034	—.035	—.037	—.038
180	—.039	—.040	—.041	—.043	—.044	—.045	—.046	—.048	—.049	—.051
190	—.052	—.053	—.055	—.056	—.057	—.059	—.060	—.062	—.064	—.066
200	—.067									

GAS, MERCURY, ALCOHOL, TOLUOL, PETROLETHER, PENTANE, THERMOMETERS.

TABLE 202. — $t^H—t_M$ (Hydrogen-Mercury).

Temperature, C.	Thuringer Glass.*	Verre dur. Tonnelot.†	Resistance Glass.*	English Crystal Glass.*	Choisy-le-Roi.*	122^{III}.*	Nitrogen Thermometer. $T_H—T_N$.†	CO_2 Thermometer. $T_H—T_{CO_2}$.†
°	°	°	°	°	°	°	°	°
0	.000	.000	.000	.000	.000	.000	.000	.000
10	—.075	—.052	—.066	—.008	—.007	—.005	—.006	—.025
20	—.125	—.085	—.108	—.001	—.004	—.006	—.010	—.043
30	—.156	—.102	—.131	+.017	+.004	—.002	—.011	—.054
40	—.168	—.107	—.140	+.037	+.014	+.001	—.011	—.059
50	—.166	—.103	—.135	+.057	+.025	+.004	—.009	—.059
60	—.150	—.090	—.119	+.073	+.033	+.008	—.005	—.053
70	—.124	—.072	—.095	+.079	+.037	+.009	—.001	—.044
80	—.088	—.050	—.068	+.070	+.032	+.007	+.002	—.031
90	—.047	—.026	—.034	+.046	+.022	+.006	+.003	—.016
100	.000	.000	.000	.000	.000	.000	.000	.000

* Schlösser, Zt. Instrkde. 21, 1901. † Chappuis, Trav. et mém. du Bur. Intern. des Poids et Mes. 6, 1888.

TABLE 203. — Comparison of Air and High Temperature Mercury Thermometers.

Comparison of the air thermometer with the high temperature mercury thermometer, filled under pressure and made of 59^{III} glass.

Air.	59^{III}.	Air.	59^{III}.
°	°	°	°
0	0.	375	385.4
100	100.	400	412.3
200	200.4	425	440.7
300	304.1	450	469.1
325	330.9	475	498.0
350	358.1	500	527.8

Mahlke, Wied. Ann. 1894.

TABLE 204. — Comparison of Hydrogen and Other Thermometers.

Comparison of the hydrogen thermometer with the toluol, alcohol, petrolether, and pentane thermometers (verre dur).

Hydrogen.	Toluol.*	Alcohol I.*	Alcohol II.*	Petrolether.†	Pentane.‡
°	°	°	°	°	°
0	0.00	0.00	0.00	-	0.00
—10	—8.54	—9.31	—9.44	-	—9.03
—20	—16.90	—18.45	—18.71	-	—17.87
—30	—25.10	—27.44	—27.84	-	—26.55
—40	—33.15	—36.30	—36.84	-	—35.04
—50	—41.08	—45.05	—45.74	—42.6	—43.36
—60	—48.90	—53.71	—54.55	-	—51.50
—70	—56.63	—62.31	—63.31	-	—59.46
—100	-	-	-	—80.2	—82.28
—150	-	-	-	—113.0	—116.87
—200	-	-	-	—140.7	—146.84

* Chappuis, Arch. sc. phys. (3) 18, 1892. † Holborn, Ann. d. Phys. (4) 6, 1901. ‡ Rothe, unpublished.

All compiled from Landolt-Börnstein-Meyerhoffer's Physikalisch-chemische Tabellen.

TABLE 205.—Platinum Resistance Thermometers.

Callendar has shown that if we define the platinum temperature, pt, by $pt = 100 \{ (R - R_0) / (R_{100} - R_0) \}$, where R is the observed resistance at t° C., R_0 that at 0°, R_{100} at 100°, then the relation between the platinum temperature and the temperature t on the scale of the gas thermometer is represented by $t - pt = \delta \{ t/100 - 1 \} t/100$ where δ is a constant for any given sample of platinum and about 1.50 for pure platinum (impure platinum having higher values). This holds good between —23° and 450° when δ has been determined by the boiling point of sulphur (445°.)

See Waidner and Burgess, Bul. Bureau Standards, 6, p. 149, 1909. Also Bureau reprints 124, 143 and 149.

TABLE 206.—Thermodynamic Temperature of the Ice Point, and Reduction to Thermodynamic Scale.

Mean = 273.1° C. (ice point).

For a discussion of the various values and for the corrections of the various gas thermometers to the thermodynamic scale see Buckingham, Bull. Bureau Standards, 3, p. 237, 1907.

Scale Corrections for Gas Thermometers.

Temp. C°.	Constant pressure = 100 cm.			Constant vol., p_0 = 100 cm, t_0 = 0°C		
	He	H	N	He	H	N
— 240°	—	+1.0	—	+0.02	+0.18	—
— 200	+0.13	+ .26	—	+ .01	+ .06	—
— 100	+ .04	+ .03	+0.40	.000	+ .010	+0.06
— 50	+ .012	+ .02	+ .12	.000	+ .004	+ .02
+ 25	— .003	— .003	— .020	.000	.000	— .006
+ 50	— .003	— .003	— .025	.000	.000	— .006
+ 75	— .003	— .003	— .017	.000	.000	— .004
+ 150	+ .007	+ .01	+ .04	+ .000	+ .001	+ .01
+ 200	+ .01	+ .02	+ .11	.000	+ .002	+ .04
+ 450	+ .1	+0.04	+ .5	0.00	+0.01	+ .2
+1000	+0.3	—	+1.7	—	—	+ .7
+1500	—	—	+3.	—	—	+1.3

See also Appendix, p. 438.

TABLE 207.—Standard Points for the Calibration of Thermometers.

Substance.	Point.	Atmosphere.	Crucible.	Temperatures. Nitrogen Scale. °C.	Temperatures. Thermodynamic. °C.
Water	boiling, 760 mm.	air	–	100.00	100.00
Naphthalene	" " "	"	–	218.0	218.0
Benzophenone	" " "	–	–	305.85 ± 0.1	305.9
Cadmium	melting or solidify.	air	graphite	320.8 ± 0.2	320.9
Zinc	" " "	"	"	419.3 ± 0.3	419.4
Sulphur	boiling, 760 mm.	–	–	444.45 ± 0.1	444.55
Antimony	melting or solidify.	CO_2	graphite	629.8 ± 0.5	630.0
Aluminum	solidification	"	"	658.5 ± 0.6	658.7
Silver	melting or solidify.	"	"	960.0 ± 0.7	
Gold	" " "	"	"	1062.4 ± 0.8	
Copper	" " "	"	"	1082.6 ± 0.8	
Li_2SiO_3	melting	air	platinum	1201.0 ± 1.0	
Diopside, pure	"	"	"	1391.2 ± 1.5	
Nickel	melting or solidify.	H and N	magnesia and Mg. aluminate	1452.3 ± 2.0	
Cobalt	" " "	"	magnesia	1489.8 ± 2.0	
Palladium	" " "	air	"	1549.2 ± 2.0	
Anorthite, pure	melting	"	platinum	1549.5 ± 2.0	
Platinum	"	"		1752. ± 5.*	
				1755. ± 5.†	

* Thermoelectric extrapolation. † Optical extrapolation.

(Day and Sosman, Journal de Physique, 1912. Mesure des témperatures élevées.) A few additional points are: H, boils —252.6°; O, boils —182.7°; CO_2, sublimes —78.5°; Hg. freezes —38.87°; Alumina melts 2000°; Tungsten melts 3400°. Quartz, α to β change, 573.3° ± 1.

TABLE 208.—Standard Calibration Curve for Pt.—Pt. Rh. (10% Rh.) Thermo-Element.

Giving the temperature for every 100 microvolts. For use in conjunction with a deviation curve determined by calibration of the particular element at some of the following fixed points:

Water	boiling-pt.	100.0	643mv.	Silver	melting-pt.	960.2	9111mv.
Naphthalene	" "	217.95	1585	Gold	" "	1062.6	10296
Tin	melting-pt.	231.9	1706	Copper	" "	1082.8	10534
Benzophenone	boiling-pt.	305.9	2365	Li_2SiO_3	" "	1201.	11941
Cadmium	melting-pt.	320.9	2503	Diopside	" "	1391.5	14230
Zinc	" "	419.4	3430	Nickel	" "	1452.6	14973
Sulphur	boiling-pt.	444.55	3672				
Antimony	melting-pt.	630.0	5530	Palladium	" "	1549.5	16144
Aluminum	" "	658.7	5827	Platinum	" "	1755.	18608

E micro-volts.	0	1000.	2000.	3000.	4000.	5000.	6000.	7000.	8000.	9000.	E micro-volts.
	Temperatures, °C.										
0.	0.0	147.1	265.4	374.3	478.1	578.3	675.3	769.5	861.1	950.4	0.
100.	17.8	159.7	276.6	384.9	488.3	588.1	684.8	778.8	870.1	959.2	100.
200.	34.5	172.1	287.7	395.4	498.4	597.9	694.3	788.0	879.1	968.0	200.
300.	50.3	184.3	298.7	405.9	508.5	607.7	703.8	797.2	888.1	976.7	300.
400.	65.4	196.3	309.7	416.3	518.5	617.4	713.3	806.4	897.1	985.4	400.
500.	80.0	208.1	320.6	426.7	528.6	627.1	722.7	815.6	906.1	994.1	500.
600.	94.1	219.7	331.5	437.1	538.6	636.8	732.1	824.7	915.0	1002.8	600.
700.	107.8	231.2	342.3	447.4	548.6	646.5	741.5	833.8	923.9	1011.5	700.
800.	121.2	242.7	353.0	457.7	558.5	656.1	750.9	842.9	932.8	1020.1	800.
900.	134.3	254.1	363.7	467.9	568.4	665.7	760.2	852.0	941.6	1028.7	900.
1000.	147.1	265.4	374.3	478.1	578.3	675.3	769.5	861.1	950.4	1037.3	1000.

E micro-volts.	10000.	11000.	12000.	13000.	14000.	15000.	16000.	17000.	18000.	E micro-volts.
	Temperatures, °C.									
0.	1037.3	1122.2	1205.9	1289.3	1372.4	1454.8	1537.5	1620.9	1704.3	0.
100.	1045.9	1130.6	1214.2	1297.7	1380.7	1463.0	1545.8	1629.2	1712.6	100.
200.	1054.4	1139.0	1222.6	1306.0	1389.0	1471.2	1554.1	1637.6	1721.0	200.
300.	1062.9	1147.4	1230.9	1314.3	1397.3	1479.4	1562.4	1645.9	1729.3	300.
400.	1071.4	1155.8	1239.3	1322.6	1405.6	1487.7	1570.8	1654.3	1737.7	400.
500.	1079.9	1164.2	1247.6	1330.9	1413.8	1496.0	1579.1	1662.6	1746.0	500.
600.	1088.4	1172.5	1255.9	1339.2	1422,0	1504.3	1587.5	1670.9	1754.3	600.
700.	1096.9	1180.9	1264.3	1347.5	1430.2	1512.6	1595.8	1679.3		700.
800.	1105.4	1189.2	1272.6	1355.8	1438.4	1520.9	1604.2	1687.6		800.
900.	1113.8	1197.6	1281.0	1364.1	1446.6	1529.2	1612.5	1696.0		900.
1000.	1122.2	1205.9	1289.3	1372.4	1454.8	1537.5	1620.9	1704.3		1000.

TABLE 209.—Standard Calibration Curve for Copper—Constantan Thermo-Element.

For use in conjunction with a deviation curve determined by the calibration of the particular element at some of the following fixed points:

Water, boiling-point, 100°, 4276 microvolts; Naphthalene, boiling-point, 217.95, 10248 mv.; Tin, melting-point, 231.9, 11009 mv.; Benzophenone, boiling-point, 305.9, 15203 mv.; Cadmium, melting-point, 320.9, 16083 mv.

E. micro-volts.	0	1000.	2000.	3000.	4000.	5000.	6000.	7000.	8000.	9000.	E micro-volts.
	Temperatures, °C.										
0.	0.00	25.27	49.20	72.08	94.07	115.31	135.91	155.95	175.50	194.62	0.
100.	2.60	27.72	51.53	74.31	96.23	117.40	137.94	157.92	177.43	196.51	100.
200.	5.17	30.15	53.85	76.54	98.38	119.48	139.96	159.89	179.36	198.40	200.
300.	7.73	32.57	56.16	78.76	100.52	121.56	141.98	161.86	181.28	200.28	300.
400.	10.28	34.98	58.46	80.97	102.66	123.63	143.99	163.82	183.20	202.16	400.
500.	12.81	37.38	60.76	83.17	104.79	125.69	146.00	165.78	185.11	204.04	500.
600.	15.33	39.77	63.04	85.37	106.91	127.75	148.00	167.73	187.02	205.91	600.
700.	17.83	42.15	65.31	87.56	109.02	129.80	150.00	169.68	188.93	207.78	700.
800.	20.32	44.51	67.58	89.74	111.12	131.84	151.99	171.62	190.83	209.64	800.
900.	22.80	46.86	69.83	91.91	113.22	133.88	153.97	173.56	192.73	211.50	900.
1000.	25.27	49.20	72.08	94.07	115.31	135.91	155.95	175.50	194.62	213.36	1000.

E micro-volts.	10000.	11000.	12000.	13000.	14000.	15000.	16000.	17000.	18000.	E micro-volts.
	Temperatures, °C.									
0.	213.36	231.74	249.82	267.60	285.13	302.42	319.49	336.36	353.09	0.
100.	215.21	233.56	251.61	269.36	286.87	304.14	321.19	338.04		100.
200.	217.06	235.38	253.40	271.12	288.61	305.85	322.88	339.72		200.
300.	218.91	237.20	255.18	272.88	290.35	307.56	324.57	341.40		300.
400.	220.75	239.01	256.96	274.64	292.08	309.27	326.26	343.07		400.
500.	222.59	240.82	258.74	276.40	293.81	310.98	327.95	344.74		500.
600.	224.43	242.63	260.52	278.15	295.54	312.69	329.64	346.41		600.
700.	226.26	244.43	262.29	279.90	297.26	314.39	331.32	348.08		700.
800.	228.09	246.23	264.06	281.65	298.98	316.09	333.00	349.75		800.
900.	229.92	248.03	265.83	283.39	300.70	317.79	334.68	351.42		900.
1000.	231.74	249.82	267.60	285.13	302.42	319.49	336.36	353.09		1000.

Cf. Day and Sosman, Am. Jour. Sci. 29, p. 93, 32, p. 517; *ibid.* R. B. Sosman, 30, p. 1.

MECHANICAL EQUIVALENT OF HEAT.

TABLE 210.—Summary of Older Work.

Taken from J. S. Ames, L'équivalent mécanique de la chaleur, Rapports présentés au congrès international du physique, Paris, 1900.
Reduced to Gram-calorie at 20° C. (Nitrogen thermometer).

		*
Joule	4.169 × 10⁷ ergs.	4.169 × 10⁷ ergs.
Rowland . . .	4.181 " "	4.181 " "
Griffiths . . .	4.192 " "	4.184 " "
Schuster-Gannon	4.189 " "	4.181 " "
Callendar-Barnes	4.186 " "	4.178 " "

* Admitting an error of 1 part per 1000 in the electrical scale.

The mean of the last four then gives

1 gram (20° C) calorie = 4.181 × 10⁷ ergs. See next table.

1 gram (15° C.) calorie = 4.185 × 10⁷ ergs assuming sp. ht. of water at 20° = 0.9990.

TABLE 211.—(1923.) Best Value, Electrical and Mechanical Equivalents of Heat.

The following values have been adopted for the International Critical Tables, prepared under the auspices of the International Research Council, 1923.

g (20° C) calorie = 4.180 international electrical joules
= 4.181 × 10⁷ ergs
g (15° C) calorie = 4.185 × 10⁷ ergs

The equivalance, 1 20° calorie = 4.183 joules, is so widely used it has been thought best to retain the following table computed with it as a basis:

TABLE 212—Conversion Factors for Units of Work.

	Joules.	Foot-pounds.	Kilogram-meters.	20° Calories.	British thermal units.	Kilowatt-hours.
1 Joule . . . . =	1	0.7376†	0.1020†	0.2391	0.0009486	0.2778×10^{-6}
1 Foot-pound . . =	1.356*	1	0.1383	0.3241*	0.001286*	0.3767×10^{-6}*
1 Kilogram-meter =	9.807*	7.234	1	2.345*	0.009302*	2.724×10^{-6}*
1 20° Calorie . . =	4.183	3.085†	0.4267†	1	0.003965	1.162×10^{-6}
1 British thermal unit . . . . =	1054.	777.5†	107.5†	251.9	1	0.0002928
1 Kilowatt-hour . =	3 600 000.	2 655 000.†	367 200.†	860 800.	3415.	1

The value used for g is the standard value, 980.665 cm. per sec. per sec. = 32.174 feet per sec. per sec.
* The values thus marked vary directly with "g."
† The values thus marked vary inversely with "g." For values of "g" see Tables 565–567.

TABLE 213.—Value of the English and American Horsepower (746 watts) in Local Foot-pounds and Kilogram-meters per Second at Various Altitudes and Latitudes.

Altitude,	Kilogram-meters per second. Latitude.					Foot-pounds per second. Latitude.				
	0°	30°	45°	60°	90°	0°	30°	45°	60°	90°
0 km.	76.275	76.175	76.074	75.973	75.873	551.70	550.97	550.24	549.52	548.79
1.5 "	76.297	76.197	76.095	75.995	75.895	551.86	551.13	550.41	549.68	548.95
3.0 "	76.320	76.220	76.119	76.018	75.918	552.03	551.30	550.57	549.85	549.12

TABLE 214.

MELTING POINTS OF THE CHEMICAL ELEMENTS.

The metals in heavier type are often used as standards.

The melting points are reduced as far as possible to a common (thermodynamic) temperature scale. This scale is defined in terms of Wien's law, with C_2 taken as 14,350, and on which the melting point of platinum is 1755° C (Nernst and Wartenburg, 1751; Waidner and Burgess, 1753; Day and Sosman, 1755; Holborn and Valentiner, 1770; see C. R. 148, p. 1177, 1909). Above 1100° C, the temperatures are expressed to the nearest 5° C. Temperatures above the platinum point may be uncertain by over 50° C.

Element.	Melting point. °C	Remarks.
Aluminum	658.7	Most samples give 657 or less (Burgess).
Antimony	630.0	
Argon	−188	Ramsay-Travers.
Arsenic	850	
Barium	850	(Guntz.)
Beryllium	1280	
Bismuth	271	Adjusted.
Boron	2200–2500?	
Bromine	−7.3	
Cadmium	320.9	Range: 320.7–320.9
Cæsium	26	Range: 26.37–25.3
Calcium	810	Adjusted.
Carbon	(>3500)	Sublimes.
Cerium	640	
Chlorine	−101.5	(Olszewski.)
Chromium	1615	Burgess-Waltenberg.
Cobalt	1480	Burgess-Waltenberg.
Copper	1083 ± 3	Mean, Holborn-Day, Day-Clement.
Erbium		
Fluorine	−223	(Moissan-Dewar.)
Gallium	30.1	
Germanium	958	
Gold	1063.0	Adjusted.
Helium	<−271	
Hydrogen	−259	
Indium	155	(Thiel.)
Iodine	113.5	Range: 112–115.
Iridium	2350?	
Iron	1530	Burgess-Waltenberg.
Krypton	−169	(Ramsay.)
Lanthanum	810?	(Muthmann-Weiss.)
Lead	327 ± 0.5	
Lithium	186	(Kahlbaum.)
Magnesium	651	(Grube) in clay crucibles, 635.
Manganese	1230	Burgess-Waltenberg.
Mercury	−38.87	
Molybdenum	2535	Mendenhall-Forsythe
Neodymium	840?	(Muthmann-Weiss.)
Neon	−253?	
Nickel	1452	Day, Sosman, Burgess, Waltenberg.
Niobium	1700?	
Nitrogen	−211	(Fischer-Alt.)
Osmium	About 2700	(Waidner-Burgess, unpublished.)
Oxygen	−218	
Palladium	1549 ± 5	(Waidner-Burgess, Nernst-Wartenburg, Day and Sosman.)
Phosphorus	44.2	
Platinum	1755 ± 5	See Note.
Potassium	62.3	
Praseodymium	940	(Muthmann-Weiss.)
Radium	700	
Rhodium	1950	(Mendenhall-Ingersoll.)
Rubidium	38	
Ruthenium	2450?	
Samarium	1300–1400	(Muthmann-Weiss.)
Scandium	?	
Selenium	217–220	
Silicon	1420	Adjusted.
Silver	960.5	Adjusted.
Sodium	97.5	
Strontium		Between Ca and Ba?
Sulphur	S_i 112.8; S_{ii} 119.2; S_{iii} 106.8	Various Forms. See Landolt-Börnstein.
Tantalum	2900	Adjusted from Waidner-Burgess = 2910.
Tellurium	452	Adjusted.
Thallium	302	
Thorium	>1700 <Mo	v. Wartenburg.
Tin	231.9 ± .2	
Titanium	1795	Burgess-Waltenberg.
Tungsten	3400	Adjusted.
Uranium	<1850	Moissan.
Vanadium	1720	Burgess-Waltenberg.
Xenon	−140	Ramsay.
Ytterbium		
Yttrium	1490	
Zinc	419.4	
Zirconium	1700?	Troost.

BOILING-POINTS OF THE CHEMICAL ELEMENTS.

Element.	Range.	Boiling-point. °C	Observer; Remarks.
	°	°	
Aluminum	–	1800.	Greenwood, Ch. News, 100, 1909.
Antimony	–	1440.	" " " "
Argon	–	—186.1	Ramsay-Travers, Z. Phys. Ch. 38, 1901.
Arsenic	449-450	–	Gray, sublimes, Conechy.
"	–	>360.	Black, sublimes, Engel, C. R. 96. 1883.
"	280-310	–	Yellow, sublimes.
Barium	–	–	Boils in vacuo, Guntz, 1903.
Bismuth	1420-1435	1430.	Barus, 1894; Greenwood, l. c.
Boron	–	–	Volatilizes without melting in electric arc.
Bromine	59-63	61.1	Thorpe, 1880; van der Plaats, 1886.
Cadmium	–	778.	Berthelot, 1902.
Cæsium	–	670.	Ruff-Johannsen.
Carbon	–	3600.	Conputed, Violle, C. R. 120, 1895.
"	–	–	Volatilizes without melting in electric oven. Moisson.
Chlorine	–	—33.6	Regnault, 1863.
Chromium	–	2200.	Greenwood, Ch. News, 100, 1909.
Copper	2100-2310	2310.	" l. c.
Fluorine	–	—187.	Moisson-Dewar, C. R. 136, 1903.
Helium	–	—267.	Computed, Tracers Ch. News, 86, 1902.
Hydrogen	—252.5-252.8	—252.6	Mean.
Iodine	–	>200.	
Iron	–	2450.	Greenwood, l. c.
Krypton	–	—151.7	Ramsay, Ch. News, 87, 1903.
Lead	–	1525.	Greenwood, l. c.
Lithium	–	1400.	Ruff-Johannsen, Ch. Ber. 38, 1905.
Magnesium	–	1120.	Greenwood, l c.
Manganese	–	1900.	" "
Mercury	–	357.	Crafts; Regnault.
Molybdenum	–	3620.	Langmuir, Mackay, Phys. Rev. 1914.
Neon	–	—239.	Dewar, 1901.
Nitrogen	—195.7-194.4	—195.	Mean.
Oxygen	—182.5-182.9	—182.7	"
Ozone	–	—119.	Troost. C. R. 126, 1898.
Phosphorus	287-290	288.	
Platinum	–	3910.	Langmuir, Mackay, Phys. Rev. 1914.
Potassium	667-757	712.	Perman; Ruff-Johannsen.
Rubidium	–	696.	Ruff-Johannsen.
Selenium	664-694	690.	
Silver	–	1955.	Greenwood, l. c.
Sodium	742-757	750.	Perman; Ruff-Johannsen.
Sulphur	444.7-445	444.7	Mean.
Tellurium	–	1390.	Deville-Troost, C. R. 91, 1880.
Thallium	–	1280.	v. Wartenberg, 25 Anorg. Ch. 56, 1908.
Tin	–	2270.	Greenwood, l. c.
Tungsten	–	5830.	Langmuir Phys. Rev. 1913.
Xenon	–	—109.1	Ramsay, Z. Phys, Ch. 44, 1903.
Zinc	916-942	930.	

TABLE 216. — Effect of Pressure on Melting Point.

Substance.	Melting point at 1 kg/sq. cm	Highest experimental pressure: kg/sq. cm	dt/dp at 1 kg/sq. cm.	Δt (observed) for 1000 kg/sq. cm	Reference
Hg	−38.85	12,000	0.00511	5.1 *	1
K	59.7	2,800	0.0136	13.8	2
Na	97.62	12,000	0.00860	+12.3†	4
Bi	271.0	12,000	−0.00342	−3.5†.	4
Sn	231.9	2,000	0.00317	3.17	3
Bi	270.9	2,000	−0.00344	−3.44	3
Cd	320.9	2,000	0.00609	6.09	3
Pb	327.4	2,000	0.00777	7.77	3

* Δt (observed) for 10,000 kg/sq. cm is 50.8°.

† Na melts at 177.5° at 12,000 kg/cm²; K at 179.6°; Bi at 218.3°; Pb at 644°. Luckey obtains melting point for tungsten as follows: 1 atme, 3623° K; 8, 3594; 18, 3572; 28, 3564. Phys. Rev. 1917.

References: (1) P. W. Bridgman, Proc. Am. Acad. 47, pp. 391–96, 416–19, 1911; (2) G. Tammann, Kristallisieren und Schmelzen, Leipzig, 1903, pp. 98–99; (3) J. Johnston and L. H. Adams, Am. J. Sci. 31, p. 516, 1911; (4) P. W. Bridgman, Phys. Rev. 6, 1, 1915.

A large number of organic substances, selected on account of their low melting points, have also been investigated: by Tammann, *loc. cit.;* G. A. Hulett, Z. physik. Chem. 28, p. 629, 1899; F. Körber, *ibid.*, 82, p. 45, 1913; E. A. Block, *ibid.*, 82, p. 403, 1913; Bridgman, Phys. Rev. 3, 126, 1914; Pr. Am. Acad. 51, 55, 1915; 51, 581, 1916; 52, 57, 1916; 52, 91, 1916. The results for water are given in the following table.

TABLE 217. — Effect of Pressure on the Freezing Point of Water (Bridgman*).

Pressure: † kg/sq. cm	Freezing point.	Phases in Equilibrium.
1	0.0	Ice I — liquid.
1,000	−8.8	Ice I — liquid.
2,000	−20.15	Ice I — liquid.
2,115	−22.0	Ice I — ice III — liquid (triple point).
3,000	−18.40	Ice III — liquid.
3,530	−17.0	Ice III — ice V — liquid (triple point).
4,000	−13.7	Ice V — liquid.
6,000	− 1.6	Ice V — liquid.
6,380	+ 0.16	Ice V — ice VI — liquid (triple point).
8,000	12.8	Ice VI — liquid.
12,000	37.9	Ice VI — liquid.
16,000	57.2	Ice VI — liquid.
20,000	73.6	Ice VI — liquid.

* P. W. Bridgman, Proc. Am. Acad. 47, pp. 441–558, 1912.

† 1 atm. = 1.033 kg/sq. cm.

TABLE 218. — Effect of Pressure on Boiling Point.*

Metal.	Pressure.	°C	Metal.	Pressure.	°C	Metal.	Pressure.	°C
Bi	10.2 cm Hg.	1200	Ag	26.3 cm Hg.	1780	Pb	20.6 cm Hg.	1410
Bi	25.7 cm Hg.	1310	Cu	10.0 cm Hg.	1980	Pb	6.3 atme.	1870
Bi	6.3 atme.	1740	Cu	25.7 cm Hg.	2180	Pb	11.7 atme.	2100
Bi	11.7 atme.	1950	Sn	10.1 cm Hg.	1970	Zn	11.7 atme.	1230
Bi	16.5 atme.	2060	Sn	26.2 cm Hg.	2100	Zn	21.5 atme.	1280
Ag	10.3 cm Hg.	1660	Pb	10.5 cm Hg.	1315	Zn	53.0 atme.	1510

* Greenwood, Pr. Roy. Soc., p. 483, 1910.

TABLE 219.
DENSITIES AND MELTING AND BOILING POINTS OF INORGANIC COMPOUNDS.

Substance.	Chemical formula.	Density, about 20° C	Melting point C	Authority.	Boiling point C	Pressure mm	Authority.
Aluminum chloride	$AlCl_3$	—	190.	1	183.°	752	1
" nitrate	$Al(NO_3)_3 + 9H_2O$	—	72.8	2	134.*	—	—
" oxide	Al_2O_3	4.00	2050.	28	—	—	—
Ammonia	NH_3	—	−75.	3	−33.5	760	7
Ammonium nitrate	NH_4NO_3	1.72	165.	—	210.*	—	—
" sulphate	$(NH_4)_2SO_4$	1.77	140.	4	—	—	—
" phosphite	$NH_4H_2PO_3$	—	123.	5	150.*	—	—
Antimony trichloride	$SbCl_3$	3.06	73.	—	223.	760	—
" pentachloride	$SbCl_5$	2.35	3.	11	102.	68	14
Arsenic trichloride	$AsCl_3$	2.20	−18.	8	130.2	760	23
Arsenic hydride	AsH_3	—	−113.5	6	−54.8	760	6
Barium chloride	$BaCl_2$	3.86	960.	11	—	—	—
" nitrate	$Ba(NO_3)_2$	3.24	575.	24	—	—	—
" perchlorate	$Ba(ClO_4)_2$	—	505.	10	—	—	—
Bismuth trichloride	$BiCl_3$	4.56	232.5	—	440.	760	—
Boric acid	H_3BO_3	1.46	185.	—	—	—	—
" anhydride	B_2O_3	1.79	577.	—	—	—	—
Borax (sodium borate)	$Na_2B_4O_7$	2.36	741.	27	—	—	—
Cadmium chloride	$CdCl_2$	4.05	560.	25	900 ±	—	9
" nitrate	$Cd(NO_3)_2 + 4H_2O$	2.45	59.5	2	132.	760	4
Calcium chloride	$CaCl_2$	2.26	774.0	—	—	—	—
" chloride	$CaCl_2 + 6H_2O$	1.68	29.6	—	—	—	—
" nitrate	$Ca(NO_3)_2$	2.36	499.	24	—	—	—
" nitrate	$Ca(NO_3)_2 + 4H_2O$	1.82	42.3	26	132.*	—	—
" oxide	CaO	3.3	2570.	28	—	—	—
Carbon tetrachloride	CCl_4	1.59	−24.	22	76.7	760	23
" trichloride	C_2Cl_6	1.63	184.	—	—	—	—
" monoxide	CO	—	−207.	6	−190.	760	6
" dioxide	CO_2	1.56	−57.	3	−80.	subl.	—
" disulphide	CS_2	1.26	−110.	13	46.2	760	—
Chloric (per) acid	$HClO_4 + H_2O$	1.81	50.	15	—	—	—
Chlorine dioxide	ClO_2	—	−76.	3	9.9	731	21
Chrome alum	$KCr(SO_4)_2 + 12H_2O$	1.83	89.	16	—	—	—
" nitrate	$Cr_2(NO_3)_6 + 18H_2O$	—	37.	2	170.	760	2
Chromium oxide	Cr_2O_3	5.04	1990.	28	—	—	—
Cobalt sulphate	$CoSO_4$	3.53	97.	16	880.*	—	—
Cupric chloride	$CuCl_2$	3.05	498.	9	*	—	—
Cuprous chloride	Cu_2Cl_2	3.7	421.	—	1000 ±	760	9
Cupric nitrate	$Cu(NO_3)_2 + 3H_2O$	2.05	114.5	2	170.*	760	2
Hydrobromic acid	HBr	—	−86.7	3	−68.7	760	—
Hydrochloric acid	HCl	—	−111.3	17	−83.1	755	17
Hydrofluoric acid	HFl	0.99	−92.3	6	−36.7	755	17
Hydriodic acid	HI	—	−51.3	17	−35.7	760	—
Hydrogen peroxide	H_2O_2	1.5	−2.	18	80.2	47	20
" phosphide	PH_3	—	−132.5	6	—	—	—
" sulphide	H_2S	—	−86.	3	−62.	—	—
Iron chloride	$FeCl_3$	2.80	301.	—	—	—	—
" nitrate	$Fe(NO_3)_3 + 9H_2O$	1.68	47.2	2	—	—	—
" sulphate	$FeSO_4 + 7H_2O$	1.90	64.	16	—	—	—
Lead chloride	$PbCl_2$	5.8	500.	9	900 ±	760	—
" metaphosphate	$Pb(PO_3)_2$	—	800.	9	—	—	—
Magnesium chloride	$MgCl_2$	2.18	708.	9	—	—	—
" oxide	MgO	3.4	2800.	28	—	—	—
" nitrate	$Mg(NO_3)_2 + 6H_2O$	1.46	90.	2	143.	760	2
" sulphate	$MgSO_4 + 5H_2O$	1.68	150.	16	—	—	—
Manganese chloride	$MnCl_2 + 4H_2O$	2.01	87.5	19	106.	760	19
" nitrate	$Mn(NO_3)_2 + 6H_2O$	1.82	26.	2	129.	760	2
" sulphate	$MnSO_4 + 5H_2O$	2.09	54.	16	—	—	—
Mercurous chloride	Hg_2Cl_2	7.10	450 ±	—	—	—	—
Mercuric chloride	$HgCl_2$	5.42	282.	—	305.	—	—

(1) Friedel and Crafts; (2) Ordway; (3) Faraday; (4) Marchand; (5) Amat; (6) Olszweski; (7) Gibbs; (8) Baskerville; (9) Carnelly; (10) Carnelly and O'Shea; (11) Ruff; (13) Wroblewski and Olszewski; (14) Anschütz; (15) Roscoe; (16) Tilden; (17) Ladenburg; (18) Staedel; (19) Clarke, Const. of Nature; (20) Bruhl; (21) Schacherl; (22) Tammann; (23) Thorpe; (24) Ramsay; (25) Lorenz; (26) Morgan; (27) Day; (28) Kanolt. * Decomposes.

DENSITIES AND MELTING AND BOILING POINTS OF INORGANIC COMPOUNDS.

Substance.	Chemical formula.	Density, about 20° C	Melting point C	Authority.	Boiling point C	Pressure mm	Authority.
Nickel carbonyl	NiC_4O_4	1.32	−25.	1	43.°	760	—
" nitrate	$Ni(NO_3)_2 + 6H_2O$	2.05	56.7	2	136.7	760	2
" oxide	NiO	6.69	—	—	—	—	—
" sulphate	$NiSO_4 + 7H_2O$	1.98	99.	3	—	—	—
Nitric acid	HNO_3	1.52	−42.	4	86.	760	16
" anhydride	N_2O_5	1.64	30.	5	48.	760	9
" oxide *	NO	1.27	−167.	—	−153.	760	6
" peroxide	N_2O_4	1.49	−9.6	8	21.6	760	—
Nitrous anhydride	N_2O_3	1.45	−111.	7	3.5	760	—
" oxide	N_2O	—	−102.4	8	−89.8	760	8
Phosphoric acid (ortho)	H_3PO_4	1.88	40 ±	—	—	—	—
Phosphorous acid	H_3PO_3	1.65	72.	—	—	—	—
Phosphorus trichloride	PCl_3	1.61	−111.8	10	76.	760	19
" oxychloride	$POCl_3$	1.68	+1.3	—	108.	760	—
" disulphide	P_3S_6	—	297.	12	—	760	—
" pentasulphide	P_2S_5	—	275.	13	522.	760	—
" sesquisulphide	P_4S_3	2.00	168.	—	400.	760	—
" trisulphide	P_2S_3	—	290 ±	14	490.	760	25
Potassium carbonate	K_2CO_3	2.29	909.	—	—	—	—
" chlorate	$KClO_3$	2.34	357.	15	—	—	—
" chromate	K_2CrO_4	2.72	975.	17	—	—	—
" cyanide	KCN	1.52	red h't	—	—	—	—
" perchlorate	$KClO_4$	2.52	610.	15	410.†	760	—
" chloride	KCl	1.99	772.	—	1500.	760	—
" nitrate	KNO_3	2.10	341.	—	400.†	—	—
" acid phosphate	KH_2PO_4	2.34	96.	3	—	—	—
" acid sulphate	$KHSO_4$	2.35	205.	—	dec.	—	—
Silver chloride	$AgCl$	5.56	451.	15	—	—	—
" nitrate	$AgNO_3$	4.35	218.	—	dec.	—	—
" perchlorate	$AgClO_4$	—	486.	18	—	—	—
" phosphate	Ag_3PO_4	6.37	849.	15	—	—	—
" metaphosphate	$AgPO_3$	—	482.	15	—	—	—
" sulphate	Ag_2SO_4	5.45	655 ±	—	1085.†	—	—
Sodium chloride	$NaCl$	2.17	800.	11	1490.	760	—
" hydroxide	$NaOH$	2.1	318.	27	—	—	—
" nitrate	$NaNO_3$	2.26	315.	—	380.†	—	—
" chlorate	$NaClO_3$	2.48	248.	28	†	—	—
" perchlorate	$NaClO_4$	—	482.	18	—	—	—
" carbonate	Na_2CO_3	2.48	852.	—	†	—	—
" carbonate	$Na_2CO_3 + 10H_2O$	1.46	34.	3	—	—	—
" phosphate	$Na_2HPO_4 + 12H_2O$	1.54	38.	—	—	—	—
" metaphosphate	$NaPO_3$	2.48	617.	15	—	—	—
" pyrophosphate	$Na_4P_2O_7$	2.45	970.	30	—	—	—
" phosphite	$(H_2NaPO_3)_2 + 5H_2O$	—	42.	20	—	—	—
" sulphate	Na_2SO_4	2.67	884.	11	—	—	—
" sulphate	$Na_2SO_4 + 10H_2O$	1.46	32.38	17	—	—	—
" hyposulphite	$Na_2S_2O_3 + 5H_2O$	1.73	48.16	—	†	—	—
Sulphur dioxide	SO_2	—	−76.	—	−10.	760	—
Sulphuric acid	H_2SO_4	1.83	10.4	21	338.	760	22
" acid	$12H_2SO_4 + H_2O$	—	−0.5	22	—	—	—
" acid	$H_2SO_4 + H_2O$	—	8.5	—	—	—	—
" acid (pyro)	$H_2S_2O_7$	1.89	35.	22	†	—	—
Sulphur trioxide	SO_3	1.91	16.8	—	44.9	760	—
Tin, stannic chloride	$SnCl_4$	2.28	−33.	23	114.	760	19
" stannous chloride	$SnCl_2$	—	250.	24	605.	760	—
Zinc chloride	$ZnCl_2$	2.91	365.	29	710.	760	—
" chloride	$ZnCl_2 + 3H_2O$	—	6.5	26	—	—	—
" nitrate	$Zn(NO_3)_2 + 6H_2O$	2.06	36.4	3	131.	760	2
" sulphate	$ZnSO_4 + 7H_2O$	2.02	50.	3	—	—	—

References: (1) Mond, Langer, Quincke; (2) Ordway; (3) Tilden; (4) Erdmann; (5) R. Weber; (6) Olszewski; (7) Birhaus; (8) Ramsay; (9) Deville; (10) Wroblewski; (11) Day, Sosman, White; (12) Ramme; (13) Meyer; (14) Lemoine; (15) Carnelly; (16) Mitscherlich; (17) LeChatelier; (18) Carnelly, O'Shea; (19) Thorpe; (20) Amat; (21) Mendelejeff; (22) Marignac; (23) Besson; (24) Clarke, Const. of Nature; (25) Isambert; (26) Mylius; (27) Hevesy; (28) Retgers; (29) Grünauer; (30) Richards and others.

* Under pressure 138 mm mercury. † Decomposes.

DENSITIES, MELTING-POINTS, AND BOILING-POINTS OF SOME ORGANIC COMPOUNDS.

N.B. — The data in this table refer only to normal compounds.

Substance.	Formula	Temp. °C.	Density.	Melting-point	Boiling-point.	Authority.
(a) Paraffin Series: C_nH_{2n+2}.						
Methane*	CH_4	—164.	0.415	—184.	—165.	Olszewski, Young.
Ethane†	C_2H_6	0	.446	—171.4	—93.	Ladenburg, "
Propane	C_3H_8	0	.536	—195.	—45.	Young, Hainlen.
Butane	C_4H_{10}	0	.60	—135.	1.	Butlerow, Young.
Pentane	C_5H_{12}	0	.647	—131.	36.3	Thorpe, Young.
Hexane	C_6H_{14}	17.	.663	—94.	69.	Schorlemmer.
Heptane	C_7H_{16}	0	.701	—97.	98.4	Thorpe, Young.
Octane	C_8H_{18}	0	.719	—56.6	125.5	" "
Nonane	C_9H_{20}	0	.733	—51.	150.	Krafft.
Decane	$C_{10}H_{22}$	0	.745	—31.	173.	"
Undecane	$C_{11}H_{24}$	0	.756	—26.	195.	"
Dodecane	$C_{12}H_{26}$	0	.765	—12.	214.	"
Tridecane	$C_{13}H_{28}$	0	.771	—6.	234.	"
Tetradecane	$C_{14}H_{30}$	4.	.775	5.	252.	"
Pentadecane	$C_{15}H_{32}$	10.	.776	10.	270.	"
Hexadecane	$C_{16}H_{34}$	18.	.775	18.	287.	"
Heptadecane	$C_{17}H_{36}$	22.	.777	22.	303.	"
Octadecane	$C_{18}H_{38}$	28.	.777	28.	317.	"
Nonadecane	$C_{19}H_{40}$	32.	.777	32.	330.	"
Eicosane	$C_{20}H_{42}$	37.	.778	37.	121.§	"
Heneicosane	$C_{21}H_{44}$	40.	.778	40.	129.§	"
Docosane	$C_{22}H_{46}$	44.	.778	44.	136.5§	"
Tricosane	$C_{23}H_{48}$	48.	.779	48.	142.5§	"
Tetracosane	$C_{24}H_{50}$	51.	.779	51.	243.‡	"
Heptacosane	$C_{27}H_{56}$	60.	.780	60.	172.§	"
Pentriacontane	$C_{31}H_{64}$	68.	.781	68.	199.§	"
Dicetyl	$C_{32}H_{66}$	70.	.781	70.	205.§	"
Penta-tria-contane	$C_{35}H_{72}$	75.	.782	75.	331.‡	"
(b) Olefines, or the Ethylene Series: C_nH_{2n}.						
Ethylene	C_2H_4	–	0.610	—169.	—103.	Wroblewski or Olszewski.
Propylene	C_3H_6	–	–	—180.	—50.2	Ladenburg, Krügel.
Butylene	C_4H_8	—13.5	.635	–	1.	Sieben.
Amylene	C_5H_{10}	–	–	–	36.	Wagner or Saytzeff.
Hexylene	C_6H_{12}	0	.76	–	69.	Wreden or Znatowicz.
Heptylene	C_7H_{14}	19.5	.703	–	96.–99.	Morgan or Schorlemmer.
Octylene	C_8H_{16}	17.	.722	–	122.–123.	Möslinger.
Nonylene	C_9H_{18}	20.	.767	–	140.–142.	Beilstein, "Org. Chem."
Decylene	$C_{10}H_{20}$	–	–	–	175.	" " "
Undecylene	$C_{11}H_{22}$	20.	.773	–	196.–197.	" " "
Dodecylene	$C_{12}H_{24}$	—31.	.795	—31.	212.–214.	" " "
Tridecylene	$C_{13}H_{26}$	15.	.774	–	233.	Bernthsen.
Tetradecylene	$C_{14}H_{28}$	—12.	.794	—12.	127.‡	Krafft.
Pentadecylene	$C_{15}H_{30}$	–	.814	–	247.	Bernthsen.
Hexadecylene	$C_{16}H_{32}$	4.	.792	4.	155.‡	Krafft, Mendelejeff, etc.
Octadecylene	$C_{18}H_{36}$	18.	.791	18.	179.‡	Krafft.
Eicosylene	$C_{20}H_{40}$	0	.871	–	390.–400.	Beilstein, "Org. Chem."
Cerotene	$C_{27}H_{54}$	–	–	58.	–	Bernthsen.
Melene	$C_{30}H_{60}$	–	–	62.	–	"

* Liquid at —11.° C. and 180 atmospheres' pressure (Cailletet).
† " " + 4.° " " 46 " " "
‡ Boiling-point under 15 mm. pressure.
§ In vacuo.

DENSITIES, MELTING-POINTS, AND BOILING-POINTS OF SOME ORGANIC COMPOUNDS.

Substance.	Chemical formula.	Temp. C°.	Specific gravity.	Melting-point.	Boiling-point.	Authority.
		(c) Acetylene Series : C_nH_{2n-2}.				
Acetylene	C_2H_2	—80.	.613	—81.8	—83.6	Villard.
Allylene	C_3H_4	–	–	—110.	—23.5	
Ethylacetylene . . .	C_4H_6	–	–	—130.	+8.	Bruylants, Kutscheroff, and others.
Propylacetylene . . .	C_5H_8	–	–	–	48.–50.	Bruylants, Taworski.
Butylacetylene . . .	C_6H_{10}	–	–	–	68.–70.	Taworski.
Oenanthylidene . . .	C_7H_{12}	–	–	–	100.–101.	Beilstein, and others.
Caprylidene	C_8H_{14}	0.	0.771	–	133.–134.	Behal.
Undecylidene	$C_{11}H_{20}$	–	–	–	210.–215.	Bruylants.
Dodecylidene . . .	$C_{12}H_{22}$	—9.	.810	—9.	105.*	Krafft.
Tetradecylidene . . .	$C_{14}H_{26}$	+6.5	.806	+6.5	134.*	"
Hexadecylidene . . .	$C_{16}H_{30}$	20.	.804	20.	160.*	"
Octadecylidene . . .	$C_{18}H_{34}$	30.	.802	30.	184.*	"
		(d) Monatomic alcohols : $C_nH_{2n+1}OH$.				
Methyl alcohol . . .	CH_3OH	0.	0.812	—94.9	64.6	
Ethyl alcohol . . .	C_2H_5OH	0.	.806	—114.2	78.3	
Propyl alcohol . . .	C_3H_7OH	0.	.817	—127.	97.	From Zander, "Lieb. Ann." vol. 224, p. 85, and Krafft, "Ber." vol. 16, 1714, " 19, 2221, " 23, 2360, and also Wroblewski and Olszewski, "Monatshefte," vol. 4, p. 338.
Butyl alcohol . . .	C_4H_9OH	0.	.823	—80.	117.	
Amyl alcohol . . .	$C_5H_{11}OH$	0.	.829	–	138.	
Hexyl alcohol . . .	$C_6H_{13}OH$	0.	.833	–	155.	
Heptyl alcohol . . .	$C_7H_{15}OH$	0.	.836	—36.	176.	
Octyl alcohol . . .	$C_8H_{17}OH$	0.	.839	—18.	195.	
Nonyl alcohol . . .	$C_9H_{19}OH$	0.	.842	—5.	213.	
Decyl alcohol . . .	$C_{10}H_{21}OH$	+7.	.839	+7.	231.	
Dodecyl alcohol . . .	$C_{12}H_{25}OH$	24.	.831	24.	143.*	
Tetradecyl alcohol . .	$C_{14}H_{29}OH$	38.	.824	38.	167.*	
Hexadecyl alcohol . .	$C_{16}H_{33}OH$	50.	.818	50.	190.*	
Octadecyl alcohol . .	$C_{18}H_{37}OH$	59.	.813	59.	211.*	
		(e) Alcoholic ethers : $C_nH_{2n+2}O$.				
Dimethyl ether . . .	C_2H_6O	–	–	–	—23.6	Erlenmeyer, Kreichbaumer.
Diethyl ether	$C_4H_{10}O$	4.	0.731	—117	+34.6	Regnault, Olszewski.
Dipropyl ether . . .	$C_6H_{14}O$	0.	.763	–	90.7	Zander and others.
Di-iso-propyl ether . .	$C_6H_{14}O$	0.	.743	–	69.	"
Di-n-butyl ether . . .	$C_8H_{18}O$	0.	.784	–	141.	Lieben, Rossi, and others.
Di-sec-butyl ether . .	$C_8H_{18}O$	21.	.756	–	121.	Kessel.
Di-iso-butyl " . .	$C_8H_{18}O$	15.	.762	–	122.	Reboul.
Di-iso-amyl " . .	$C_{10}H_{22}O$	0.	.799	–	170.–175.	Wurtz.
Di-sec-hexyl " . .	$C_{12}H_{26}O$	–	–	–	203.–208.	Erlenmeyer and Wanklyn.
Di-norm-octyl " . .	$C_{16}H_{34}O$	17.	.805	–	280.–282.	Moslinger.
		(f) Ethyl ethers : $C_nH_{2n+2}O$.				
Ethyl-methyl ether . .	C_3H_8O	0.	0.725	–	11.	Wurtz, Williamson.
" propyl " . .	$C_5H_{12}O$	20.	0.739	–	63.–64.	Chancel, Brühl.
" iso-propyl ether .	$C_5H_{12}O$	0.	.745	–	54.	Markownikow.
" norm-butyl ether	$C_6H_{14}O$	0.	.769	–	92.	Lieben, Rossi.
" iso-butyl ether .	$C_6H_{14}O$	–	.751	–	78.–80.	Wurtz.
" iso-amyl ether .	$C_7H_{16}O$	18.	.764	–	112.	Williamson and others.
" norm-hexyl ether	$C_8H_{18}O$	–	–	–	134.–137.	Lieben, Janeczek.
" norm-heptyl ether	$C_9H_{20}O$	16.	.790	–	165.	Cross.
" norm-octyl ether	$C_{10}H_{22}O$	17.	.794	–	182.–184.	Moslinger.

* Boiling-point under 15 mm. pressure.
† Liquid at —11.° C. and 180 atmospheres' pressure (Cailletet).

DENSITIES AND MELTING AND BOILING POINTS OF SOME ORGANIC COMPOUNDS.

(*g*) MISCELLANEOUS.

Substance	Chemical formula.	Density and temperature.		Melting point C	Boiling point C	Authority.
Acetic acid	CH_3COOH	1.115	0°	16.7	118.5	Young, '09
Acetone	CH_3COCH_3	0.812	0	−94.6	56.1	
Aldehyde	C_2H_4O	0.806	0	−120.	+20.8	
Aniline	$C_6H_5NH_2$	1.038	0	−8.	183.9	
Beeswax		0.96 ±		62.		
Benzoic acid	$C_7H_6O_2$	1.293	4	121.	249.	
Benzene	C_6H_6	0.879	20	5.48	80.2	Richards
Benzophenone	$(C_6H_5)_2CO$	1.090	50	48.	305.9	Holborn-Henning
Butter		0.86–7		30 ±		
Camphor	$C_{10}H_{16}O$	0.99	10	176.	209.	
Carbolic acid	C_6H_5OH	1.060	21	43.	182.	
Carbon bisulphide	CS_2	1.292	0	−110.	46.2	
" tetrachloride	CCl_4	1.582	21	−30.	76.7	Young
Chlorbenzene	C_6H_5Cl	1.111	15	−40.	132.	
Chloroform	$CHCl_3$	1.4989	15	−63.3	61.2	
Cyanogen	C_2N_2	—		−35.	−21.	
Ethyl bromide	C_2H_5Br	1.45	15	−117.	38.4	
" chloride	C_2H_5Cl	0.918	8	−141.6	14.	
" ether	$C_4H_{10}O$	0.736	0	−118.	34.6	
" iodide	C_2H_5I	1.944	14	—	72.	
Formic acid	HCOOH	1.242	0	8.6	100.8	
Gasolene		0.68 ±		—	70–90	
Glucose	$CHO(HCOH)_4CH_2OH$	1.56		146.	—	
Glycerine	$C_3H_8O_3$	1.269	0	20.	290.	
Iodoform	CHI_3	4.01	25	119.	—	
Lard				29 ±	—	
Methyl chloride	CH_3Cl	0.992	−24	−103.6	−24.1	
Methyl iodide	CH_3I	2.285	15	−64.	42.3	
Naphthalene	$C_6H_4 \cdot C_4H_4$	1.152	15	80.	218.	Holborn-Henning
Nitrobenzene	$C_6H_5O_2N$	1.212	7.5	5.	211.	
Nitroglycerine	$C_3H_5N_3O_9$	1.60		—	—	
Olive oil		0.92		20 ±	300 ±	
Oxalic acid	$C_2H_2O_4 \cdot 2H_2O$	1.68		190.	—	
Paraffin wax, soft		—		38–52	350–390	
" " hard		—		52–56	390–430	
Pyrogallol	$C_6H_3(OH)_3$	1.46	40	133.	293.	
Spermaceti		0.95	15	45 ±	—	
Starch	$C_6H_{10}O_5$	1.56		none	—	
Sugar, cane	$C_{12}H_{22}O_{11}$	1.588	20	160.	—	
Stearine	$(C_{18}H_{35}O_2)_3C_3H_5$	0.925	65	71.	—	
Tallow, beef		0.94	15	27–38	—	
" mutton		0.94	15	32–41	—	
Tartaric acid	$C_4H_6O_6$	1.754		170.	—	
Toluene	$C_6H_5CH_3$	0.882	00	−92.	110.31	Richards
Xylene (o)	$C_6H_4(CH_3)_2$	0.863	20	−28.	142.	
" (m)	$C_6H_4(CH_3)_2$	0.864	20	54.	140.	
" (p)	$C_6H_4(CH_3)_2$	0.861	20	15.	138.	

TABLE 221. — Melting-point of Mixtures.

Metals.	Melting-points, C°. Percentage of metal in second column.											Reference.
	0%	10%	20%	30%	40%	50%	60%	70%	80%	90%	100%	
Pb. Sn.	326	295	276	262	240	220	190	185	200	216	232	1
Bi.	322	290	–	–	179	145	126	168	205	–	268	7
Te.	322	710	790	880	917	760	600	480	410	425	446	8
Ag.	328	460	545	590	620	650	705	775	840	905	959	9
Na.	–	360	420	400	370	330	290	250	200	130	96	13
Cu.	326	870	920	925	945	950	955	985	1005	1020	1084	2
Sb.	326	250	275	330	395	440	490	525	560	600	632	16
Al. Sb.	650	750	840	925	945	950	970	1000	1040	1010	632	17
Cu.	650	630	600	560	540	580	610	755	930	1055	1084	18
Au.	655	675	740	800	855	915	970	1025	1055	675	1062	10
Ag.	650	625	615	600	590	580	575	570	650	750	954	17
Zn.	654	640	620	600	580	560	530	510	475	425	419	11
Fe.	653	860	1015	1110	1145	1145	1220	1315	1425	1500	1515	3
Sn.	650	645	635	625	620	605	590	570	560	540	232	17
Sb. Bi.	632	610	590	575	555	540	520	470	405	330	268	16
Ag.	630	595	570	545	520	500	505	545	680	850	959	9
Sn.	622	600	570	525	480	430	395	350	310	255	232	19
Zn.	632	555	510	540	570	565	540	525	510	470	419	17
Ni. Sn.	1455	1380	1290	1200	1235	1290	1305	1230	1060	800	232	17
Na. Bi.	96	425	520	590	645	690	720	730	715	570	268	13
Cd.	96	125	185	245	285	325	330	340	360	390	322	13
Cd. Ag.	322	420	520	610	700	760	805	850	895	940	954	17
Tl.	321	300	285	270	262	258	245	230	210	235	302	14
Zn.	322	280	270	295	313	327	340	355	370	390	419	11
Au. Cu.	1063	910	890	895	905	925	975	1000	1025	1060	1084	4
Ag.	1064	1062	1061	1058	1054	1049	1039	1025	1006	982	963	5
Pt.	1075	1125	1190	1250	1320	1380	1455	1530	1610	1685	1775	20
K. Na.	62	17.5	—10	—3.5	5	11	26	41	58	77	97.5	15
Hg.	–	–	–	–	–	90	110	135	162	265	–	13
Tl.	62.5	133	165	188	205	215	220	240	280	305	301	14
Cu. Ni.	1080	1180	1240	1290	1320	1335	1380	1410	1430	1440	1455	17
Ag.	1082	1035	990	945	910	870	830	788	814	875	960	9
Sn.	1084	1005	890	755	725	680	630	580	530	440	232	12
Zn.	1084	1040	995	930	900	880	820	780	700	580	419	6
Ag. Zn.	959	850	755	705	690	660	630	610	570	505	419	11
Sn.	959	870	750	630	550	495	450	420	375	300	232	9
Na. Hg.	96.5	90	80	70	60	45	22	55	95	215	–	13

1 Means, Landolt-Börnstein-Roth Tabellen.
2 Friedrich-Leroux, Metal. 4, 1907.
3 Gwyer, Zs. Anorg. Ch. 57, 1908.
4 Means, L.-B.-R. Tabellen.
5 Roberts-Austen Chem. News, 87, 2, 1903.
6 Shepherd J. ph. ch. 8, 1904.
7 Kapp, Diss., Königsberg, 1901.
8 Fay and Gilson, Trans. Am. Inst. Min. Eng. Nov. 1901.
9 Heycock and Neville, Phil. Trans. 189A, 1897.
10 " " " " " 194A, 201, 1900.
11 Heycock and Neville, J. Chem. Soc. 71, 1897.
12 " " " Phil. Trans. 202A, 1, 1903.
13 Kurnakow, Z. Anorg. Chem. 23, 439, 1900.
14 " " " " 30, 86, 1902.
15 " " " " 30, 109, 1902.
16 Roland-Gosselin, Bul. Soc. d'Encour. (5) 1, 1896.
17 Gautier, " " " (5) 1, "
18 Le Chatelier, " " " (4) 10, 573, 1895.
19 Reinders, Z. Anorg. Chem. 25, 113, 1896.
20 Erhard and Schertel, Jahrb. Berg-u. Hüttenw. Sachsen. 1879, 17.

TABLE 222. — Alloy of Lead, Tin, and Bismuth.

	Per cent.									
Lead	32.0	25.8	25.0	43.0	33.3	10.7	50.0	35.8	20.0	70.9
Tin	15.5	19.8	15.0	14.0	33.3	23.1	33.0	52.1	60.0	9.1
Bismuth	52.5	54.4	60.0	43.0	33.3	66.2	17.0	12.1	20.0	20.0
Solidification at	96°	101°	125°	128°	145°	148°	161°	181°	182°	234°

Charpy, Soc. d'Encours, Paris, 1901.

TABLE 223. — Low Melting-point Alloy.

	Per cent.						
Cadmium	10.8	10.2	14.8	13.1	6.2	7.1	6.7
Tin	14.2	14.3	7.0	13.8	9.4	–	–
Lead	24.9	25.1	26.0	24.3	34.4	39.7	43.4
Bismuth	50.1	50.4	52.2	48.8	50.0	53.2	49.9
Solidification at	65.5°	67.5°	68.5°	68.5°	76.5°	89.5°	95°

Drewitz, Diss. Rostock, 1902.

All compiled from Landolt-Börnstein-Meyerhoffer's Physikalisch-chemische Tabellen.

TRANSFORMATION AND MELTING TEMPERATURES OF LIME-ALUMINA-SILICA COMPOUNDS AND EUTECTIC MIXTURES.

The majority of these determinations are by G. A. Rankin. (Part unpublished.)

Substance.	% CaO	Al_2O_3	SiO_2	Transformation.	Temp.
$CaSiO_3$ . . .	48.2	—	51.8	Melting	1540° ±2°
$CaSiO_3$ . . .	48.2	—	51.8	α to β and reverse	1200 ±2
Ca_2SiO_4 . . .	65.	—	35.	Melting	2130 ±10
" . . .	65.	—	35.	γ to β and reverse	675 ±5
" . . .	65.	—	35.	β to α and reverse	1420 ±2
$Ca_3Si_2O_7$ . . .	58.2	—	41.8	Dissociation into Ca_2SiO_4 and liquid	1475 ±5
Ca_3SiO_5 . . .	73.6	—	26.4	Dissociation into Ca_2SiO_4 and CaO	1900 ±5
$Ca_3Al_2O_6$ . . .	62.2	37.8	—	Dissociation into CaO and liquid	1535 ±5
$Ca_5Al_6O_{14}$ . .	47.8	52.2	—	Melting	1455 ±5
$CaAl_2O_4$ . . .	35.4	64.6	—	Melting	1600 ±5
$Ca_3Al_{10}O_{18}$ . .	24.8	75.2	—	Melting	1720 ±10
Al_2SiO_5 . . .	—	62.8	37.1	Melting	1816 ±10
$CaAl_2Si_2O_8$ . .	20.1	36.6	43.3	Melting	1550 ±2
$Ca_2Al_2SiO_7$ . .	40.8	37.2	22.0	Melting	1590 ±2
$Ca_3Al_2SiO_8$ . .	50.9	30.9	18.2	Dissociation into Ca_2SiO_4 + $Ca_2Al_2SiO_7$ and liquid . .	1335 ±5

EUTECTICS.

Crystalline Phases.	% CaO	Al_2O_3	SiO_2	Melting Temp.
$CaSiO_3$, SiO_2	37.	—	63.	1436°
Ca,SiO_3 / 3CaO,2SiO_2	54.5	—	45.5	1455±
Ca_2SiO_4 / CaO.	67.5	—	32.5	2065±
Al_2SiO_5, SiO_2	—	13.	87.	1610
Al_2SiO_5, Al_2O_3	—	64.	36.	1810
$CaAl_2Si_2O_8$ / $CaSiO_3$	34.1	18.6	47.3	1299
$CaAl_2Si_2O_8$ / SiO_2	10.5	19.5	70.	1359
$CaAl_2Si_2O_8$ / SiO_2, $CaSiO_3$	23.2	14.8	62.	1165
$Ca_2Al_2SiO_7$ / Ca_2SiO_4	49.6	23.7	26.7	1545
Al_2O_3 / $CaAl_2Si_2O_8$	19.3	39.3	41.4	1547
$CaAl_2Si_2O_8$ / Al_2SiO_5, SiO_2	9.8	19.8	70.4	1345
$Ca_2Al_2SiO_7$ / $Ca_3Al_{10}O_{18}$	35.	50.8	14.2	1552
$Ca_2Al_2SiO_7$ / $CaAl_2O_4$	37.8	52.9	9.3	1512
$Ca_2Al_2SiO_7$ / $CaAl_2O_4$ / $Ca_3Al_{10}O_{18}$	37.5	53.2	9.3	1505
$CaAl_2Si_2O_8$ / $Ca_2Al_2SiO_7$	30.2	36.8	33.	1385
$Ca_2Al_2SiO_7$ / $Ca_3Si_2O_7$ / $CaSiO_3$	47.2	11.8	41.	1310
$Ca_2Al_2SiO_7$ / $CaSiO_3$	45.7	13.2	41.1	1316

EUTECTICS.

Crystalline Phases.	% CaO	Al_2O_3	SiO_2	Melting Temp.
$CaAl_2Si_2O_8$ / $Ca_2Al_2SiO_7$ / $CaSiO_3$	38.	20.	42.	1265°
$CaAl_2Si_2O_8$ / $Ca_2Al_2SiO_7$ / Al_2O_3	29.2	39.	31.8	1380
Ca_2SiO_4 / $CaAl_2O_4$ / $Ca_5Al_6O_{14}$	49.5	43.7	6.8	1335

QUINTUPLE POINTS.

Crystalline Phases.	% CaO	Al_2O_3	SiO_2	Melting Temp.
$Ca_2Al_2SiO_7$ / $Ca_3Si_2O_7$ / Ca_2SiO_4	48.2	11.9	39.9	1335
$Ca_2Al_2SiO_7$ / Ca_2SiO_4 / $CaAl_2O_4$	48.3	42.	9.7	1380
$CaAl_2Si_2O_8$ / Al_2O_3 / Al_2SiO_5	15.6	36.5	47.9	1512
$Ca_3Al_{10}O_{18}$ / $Ca_2Al_2SiO_7$ / Al_2O_3	31.2	44.5	24.3	1475

QUADRUPLE POINTS.

Crystalline Phases.	% CaO	Al_2O_3	SiO_2	Melting Temp.
3CaO.2SiO_2 / 2CaO.SiO_2	55.5	—	44.5	1475

The accuracy of the melting-points is 5 to 10 units. Geophysical Laboratory. See also Day and Sosman, Am. J. of Sc. xxxi, p. 341, 1911.

Table 225.

LOWERING OF FREEZING-POINTS BY SALTS IN SOLUTION.

In the first column is given the number of gram-molecules (anhydrous) dissolved in 1000 grams of water; the second contains the molecular lowering of the freezing-point; the freezing-point is therefore the product of these two columns. After the chemical formula is given the molecular weight, then a reference number.

g. mol. / 1000 g. H_2O	Molecular Lowering.
$Pb(NO_3)_2$, 331.0: 1, 2.	
0.000362	5.5°
.001204	5.30
.002805	5.17
.005570	4.97
.01737	4.69
.5015	2.99
$Ba(NO_3)_2$, 261.5: 1.	
0.000383	5.6°
.001259	5.28
.002681	5.23
.005422	5.13
.008352	5.04
$Cd(NO_3)_2$, 236.5: 3.	
0.00298	5.4°
.00689	5.25
.01997	5.18
.04873	5.15
$AgNO_3$, 167.0: 4, 5.	
0.1506	3.32°
.5001	2.96
.8645	2.87
1.749	2.27
2.953	1.85
3.856	1.64
0.0560	3.82
.1401	3.58
.3490	3.28
KNO_3, 101.9: 6, 7.	
0.0100	3.5
.0200	3.5
.0500	3.41
.100	3.31
.200	3.19
.250	3.08
.500	2.94
.750	2.81
1.000	2.66
$NaNO_3$, 85.09: 2, 6, 7.	
0.0100	3.6°
.0250	3.46
.0500	3.44
.2000	3.345
.500	3.24
.5015	3.30
1.000	3.15
1.0030	3.03
NH_4NO_3, 80.11: 6, 8.	
0.0100	3.6°
.0250	3.50

g. mol. / 1000 g H_2O	Molecular Lowering.
0.0500	3.47°
.1000	3.42
.2000	3.32
.500	3.26
1.000	3.14
$LiNO_3$, 69.07: 9.	
0.0398	3.4°
.1671	3.35
.4728	3.35
1.0164	3.49
$Al_2(SO_4)_3$, 342.4: 10.	
0.0131	5.6°
.0261	4.9
.0543	4.5
.1086	4.03
.217	3.83
$CdSO_4$, 208.5: 1, 11.	
0.000704	3.35°
.002685	3.05
.01151	2.69
.03120	2.42
.1473	2.13
.4129	1.80
.7501	1.76
1.253	1.86
K_2SO_4, 174.4: 3, 5, 6, 10, 12.	
0.00200	5.4°
.00398	5.3
.00865	4.9
.0200	4.76
.0500	4.60
.1000	4.32
.200	4.07
.454	3.87
$CuSO_4$, 159.7: 1, 4, 11.	
0.000286	3.3°
.000843	3.15
.002279	3.03
.006670	2.79
.01463	2.59
.1051	2.28
.2074	1.95
.4043	1.84
.8898	1.76
$MgSO_4$, 120.4: 1, 4, 11.	
0.000675	3.29
.002381	3.10
.01263	2.72
.0580	2.65
.2104	2.23

g. mol. / 1000 g. H_2O	Molecular Lowering.
0.4978	2.02°
.8112	2.01
1.5233	2.28
$BaCl_2$, 208.3: 3, 6, 13.	
0.00200	5.5°
.00498	5.2
.0100	5.0
.0200	4.95
.04805	4.80
.100	4.69
.200	4.66
.500	4.82
.586	5.03
.750	5.21
$CdCl_2$, 183.3: 3, 14.	
0.00299	5.0°
.00690	4.8
.0200	4.64
.0541	4.11
.0818	3.93
.214	3.39
.429	3.03
.858	2.71
1.072	2.75
$CuCl_2$, 134.5: 9.	
0.0350	4.9°
.1337	4.81
.3380	4.92
.7149	5.32
$CoCl_2$, 129.9: 9.	
0.0276	5.0°
.1094	4.9
.2369	5.03
.4399	5.30
.538	5.5
$CaCl_2$, 111.0: 5, 13–16.	
0.0100	5.1°
.05028	4.85
.1006	4.79
.5077	5.33
.946	5.3
2.432	8.2
3.469	11.5
3.829	14.4
0.0478	5.2
.153	4.91
.331	5.15
.612	5.47
.998	6.34

g. mol. / 1000 g. H_2O	Molecular Lowering.
$MgCl_2$, 95.26: 6, 14.	
0.0100	5.1°
.0500	4.98
.1500	4.96
.3000	5.186
.6099	5.69
KCl, 74.60: 9, 17–19.	
0.02910	3.54°
.05845	3.46
.112	3.43
.3139	3.41
.476	3.37
1.000	3.286
1.989	3.25
3.269	3.25
$NaCl$, 58.50: 3, 20, 12, 16.	
0.00399	3.7°
.01000	3.67
.0221	3.55
.04949	3.51
.1081	3.48
.2325	3.42
.4293	3.37
.700	3.43
NH_4Cl, 53.52: 6, 15.	
0.0100	3.6°
.0200	3.56
.0350	3.50
.1000	3.43
.2000	3.396
.4000	3.393
.7000	3.41
$LiCl$, 42.48: 9, 15.	
0.00992	3.7°
.0455	3.5
.09952	3.53
.2474	3.50
.5012	3.61
.7939	3.71
$BaBr_2$, 297.3: 14.	
0.100	5.1°
.150	4.9
.200	5.00
.500	5.18
$AlBr_3$, 267.0: 9.	
0.0078	1.4°
.0559	1.2
.1971	1.07
.4355	1.07

1 Hausrath, Ann. Phys. 9, 1902.
2 Leblanc-Noyes, Z. Phys. Ch. 6, 1890.
3 Jones, Z. Phys. Ch. 11, 1893.
4 Raoult, Z. Phys. Ch. 2, 1888.
5 Arrhenius, Z. Phys. Ch. 2, 1888.
6 Loomis, Wied. Ann. 57, 1896.
7 Jones, Am. Chem. J. 27, 1902.
8 Jones-Caldwell, Am. Chem. J. 25, 1901.
9 Biltz, Z. Phys. Ch. 40, 1902.
10 Jones-Mackay, Am. Chem. J. 19, 1897.
11 Kahlenberg, J. Phys. Ch. 5, 1901.
12 Abegg, Z. Phys. Ch. 20, 1896.
13 Jones-Getman, Am. Ch. J. 27, 1902.
14 Jones-Chambers, Am. Ch. J. 23, 1900.
15 Loomis, Wied. Ann. 60, 1897.
16 Roozeboom, Z. Phys. Ch. 4, 1889.
17 Raoult, Z. Phys. Ch. 27, 1898.
18 Roloff, Z. Phys. Ch. 18, 1895.
19 Kistiakowsky, Z. Phys. Ch. 6, 1890.
20 Loomis, Wied. Ann. 51, 1894.

Compiled from Landolt-Börnstein-Meyerhoffer's Physikalisch-chemische Tabellen.

LOWERING OF FREEZING-POINTS BY SALTS IN SOLUTION (*continued*).

$\frac{\text{g. mol.}}{1000\ \text{g. } H_2O}$	Molecular Lowering.
$CdBr_2$, 272.3: 3, 14.	
0.00324	5.1°
.00718	4.6
.03627	3.84
.0719	3.39
.1122	3.18
.220	2.96
.440	2.76
.800	2.59
$CuBr_2$, 223.5: 9.	
0.0242	5.1°
.0817	5.1
.2255	5.27
.6003	5.89
$CaBr_2$, 200.0: 14.	
0.0871	5.1°
.1742	5.18
.3484	5.30
.5226	5.64
$MgBr_2$, 184.28: 14.	
0.0517	5.4°
.103	5.16
.207	5.26
.517	5.85
KBr, 119.1: 9, 21.	
0.0305	3.61°
.1850	3.49
.6801	3.30
.250	3.78
.500	3.56
CdI_2, 366.1: 3, 5, 22.	
0.00210	4.5°
.00626	4.0
.02062	3.52
.04857	2.70
.1360	2.35
.333	2.13
.684	2.23
.888	2.51
KI, 166.0: 9, 2.	
0.0651	3.5°
.2782	3.50
.6030	3.42
1.003	3.37
SrI_2, 341.3: 22.	
0.054	5.1°
.108	5.2
.216	5.35
.327	5.52
NaOH, 40.06: 15.	
0.02002	3.45°
.05005	3.45
.1001	3.41
.2000	3.407

$\frac{\text{g. mol.}}{1000\ \text{g. } H_2O}$	Molecular Lowering.
KOH, 56.16: 1, 15, 23.	
0.00352	3.60°
.00770	3.59
.02002	3.44
.05006	3.43
.1001	3.42
.2003	3.424
.230	3.50
.465	3.57
CH_3OH, 32.03: 24, 25.	
0.0100	1.8°
.0301	1.82
.2018	1.811
1.046	1.86
3.41	1.88
6.200	1.944
C_2H_5OH, 46.04: 1, 12, 17, 24–27	
0.000402	1.67°
.004993	1.67
.0100	1.81
.02892	1.707
.0705	1.85
.1292	1.829
.2024	1.832
.5252	1.834
1.0891	1.826
1.760	1.83
3.901	1.92
7.91	2.02
11.11	2.12
18.76	1.81
0.0173	1.80
.0778	1.79
K_2CO_3, 138.30: 6	
0.0100	5.1°
.0200	4.93
.0500	4.71
.100	4.54
.200	4.39
Na_2CO_3, 106.10: 6.	
0.0100	5.1°
.0200	4.93
.0500	4.64
.1000	4.42
.2000	4.17
Na_2SO_3, 126.2: 28.	
0.1044	4.51°
.3397	3.74
.7080	3.38
Na_2HPO_4, 142.1: 22, 29.	
0.01001	5.0°
.02003	4.84
.05008	4.60
.1002	4.34

$\frac{\text{g. mol}}{1000\ \text{g. } H_2O}$	Molecular Lowering.
Na_2SiO_3, 122.5: 15.	
0.01052	6.4°
.05239	5.86
.1048	5.28
.2099	4.66
.5233	3.99
HCl, 36.46: 1–3, 6, 13, 18, 22.	
0.00305	3.68°
.00695	3.66
.0100	3.6
.01703	3.59
.0500	3.59
.1025	3.56
.2000	3.57
.3000	3.612
.464	3.68
.516	3.79
1.003	3.95
1.032	4.10
1.500	4.42
2.000	4.97
2.115	4.52
3.000	6.03
3.053	4.90
4.065	5.67
4.657	6.19
HNO_3, 63.05: 3, 13, 15.	
0.02004	3.55°
.05015	3.50
.0510	3.71
.1004	3.48
.1059	3.53
.2015	3.45
.250	3.50
.500	3.62
1.000	3.80
2.000	4.17
3.000	4.64
H_3PO_2, 66.0: 29.	
0.1260	2.90°
.2542	2.75
.5171	2.59
1.071	2.45
H_3PO_3, 82.0: 4, 5.	
0.0745	3.0°
.1241	2.8
.2482	2.6
1.00	2.39
H_3PO_4, 98.0: 6, 22.	
0.0100	2.8°
.0200	2.68
.0500	2.49
.1000	2.36
.2000	2.25

$\frac{\text{g mol.}}{1000\ \text{g. } H_2O}$	Molecular Lowering.
0.472	2.20°
.944	2.27
1.620	2.60
$(COOH)_2$, 90.02: 4, 15.	
0.01002	3.3°
.02005	3.19
.05019	3.03
.1006	2.83
.2022	2.64
.366	2.56
.648	2.3
$C_3H_5(OH)_3$, 92.06: 24, 25.	
0.0200	1.86°
.1008	1.86
.2031	1.85
.535	1.91
2.40	1.98
5.24	2.13
$(C_2H_5)_2O$, 74.08: 24	
0.0100	1.6°
.0201	1.67
.1011	1.72
.2038	1.702
Dextrose, 180.1: 24, 30.	
0.0198	1.84°
.0470	1.85
.1326	1.87
.4076	1.894
1.102	1.921
Levulose, 180.1: 24, 25.	
0.0201	1.87°
.2050	1.871
.554	2.01
1.384	2.32
2.77	3.04
$C_{12}H_{22}O_{11}$, 342.2: 1, 24, 26.	
0.000332	1.90°
.001410	1.87
.009978	1.86
.0201	1.88
.1305	1.88
H_2SO_4, 98.08: 13, 20, 31–33.	
0.00461	4.8°
.0100	4.49
.0200	4.32
.0461	4.10
.100	3.96
.200	3.85
.400	3.98
1.000	4.19
1.500	4.96
2.000	5.65
2.500	6.53

1–20 *See* page 217.
21 Sherrill, Z. Phys. Ch. 43, 1903.
22 Chambers-Frazer, Am. Ch. J. 23, 1900.
23 Noyes-Whitney, Z. Phys. Ch. 15, 1894.
24 Loomis, Z. Phys. Ch. 32, 1900.
25 Abegg, Z. Phys. Ch. 15, 1894.
26 Nernst-Abegg, Z. Phys. Ch. 15, 1894.
27 Pictet-Altschul, Z. Phys. Ch. 16, 1895.
28 Barth, Z. Phys. Ch. 9, 1892.
29 Petersen, Z. Phys. Ch. 11, 1893.
30 Roth, Z. Phys. Ch. 43, 1903.
31 Wildermann, Z. Phys. Ch. 15, 1894.
32 Jones-Carroll, Am. Ch. J. 28, 1902.
33 Jones-Murray, Am. Ch. J. 30, 1903.

RISE OF BOILING-POINT PRODUCED BY SALTS DISSOLVED IN WATER.*

This table gives the number of grams of the salt which, when dissolved in 100 grams of water, will raise the boiling-point by the amount stated in the headings of the different columns. The pressure is supposed to be 76 centimeters.

Salt.	1° C.	2°	3°	4°	5°	7°	10°	15°	20°	25°
$BaCl_2 + 2H_2O$	15.0	31.1	47.3	63.5	(71.6 gives 4°.5 rise of temp.)					
$CaCl_2$	6.0	11.5	16.5	21.0	25.0	32.0	41.5	55.5	69.0	84.5
$Ca(NO_3)_2 + 2H_2O$	12.0	25.5	39.5	53.5	68.5	101.0	152.5	240.0	331.5	443.5
KOH	4.7	9.3	13.6	17.4	20.5	26.4	34.5	47.0	57.5	67.3
$KC_2H_3O_2$	6.0	12.0	18.0	24.5	31.0	44.0	63.5	98.0	134.0	171.5
KCl	9.2	16.7	23.4	29.9	36.2	48.4	(57.4 gives a rise of 8°.5)			
K_2CO_3	11.5	22.5	32.0	40.0	47.5	60.5	78.5	103.5	127.5	152.5
$KClO_3$	13.2	27.8	44.6	62.2						
KI	15.0	30.0	45.0	60.0	74.0	99.5	134.	185.0	(220 gives 18°.5)	
KNO_3	15.2	31.0	47.5	64.5	82.0	120.5	188.5	338.5		
$K_2C_4H_4O_6 + \frac{1}{2}H_2O$	18.0	36.0	54.0	72.0	90.0	126.5	182.0	284.0		
$KNaC_4H_4O_6$	17.3	34.5	51.3	68.1	84.8	119.0	171.0	272.5	390.0	510.0
$KNaC_4H_4O_6 + 4H_2O$	25.0	53.5	84.0	118.0	157.0	266.0	554.0	5510.0		
$LiCl$	3.5	7.0	10.0	12.5	15.0	20.0	26.0	35.0	42.5	50.0
$LiCl + 2H_2O$	6.5	13.0	19.5	26.0	32.0	44.0	62.0	92.0	123.0	160.5
$MgCl_2 + 6H_2O$	11.0	22.0	33.0	44.0	55.0	77.0	110.0	170.0	241.0	334.5
$MgSO_4 + 7H_2O$	41.5	87.5	138.0	196.0	262.0					
$NaOH$	4.3	8.0	11.3	14.3	17.0	22.4	30.0	41.0	51.0	60.1
$NaCl$	6.6	12.4	17.2	21.5	25.5	33.5	(40.7 gives 8°.8 rise)			
$NaNO_3$	9.0	18.5	28.0	38.0	48.0	68.0	99.5	156.0	222.0	
$NaC_2H_3O_2 + 3H_2O$	14.9	30.0	46.1	62.5	79.7	118.1	194.0	480.0	6250.0	
$Na_2S_2O_3$	14.0	27.0	39.0	49.5	59.0	77.0	104.0	152.0	214.5	311.0
Na_2HPO_4	17.2	34.4	51.4	68.4	85.3					
$Na_2C_4H_4O_6 + 2H_2O$	21.4	44.4	68.2	93.9	121.3	183.0	(237.3 gives 8°.4 rise)			
$Na_2S_2O_3 + 5H_2O$	23.8	50.0	78.6	108.1	139.3	216.0	400.0	1765.0		
$Na_2CO_3 + 10H_2O$	34.1	86.7	177.6	369.4	1052.9					
$Na_2B_4O_7 + 10H_2O$	39.	93.2	254.2	898.5	(5555.5 gives 4°.5 rise)					
NH_4Cl	6.5	12.8	19.0	24.7	29.7	39.6	56.2	88.5		
NH_4NO_3	10.0	20.0	30.0	41.0	52.0	74.0	108.0	172.0	248.0	337.0
$(NH_4)_2SO_4$	15.4	30.1	44.2	58.0	71.8	99.1	(115.3 gives 108.2)			
$SrCl_2 + 6H_2O$	20.0	40.0	60.0	81.0	103.0	150.0	234.0	524.0		
$Sr(NO_3)_2$	24.0	45.0	63.6	81.4	97.6					
$C_4H_6O_6$	17.0	34.4	52.0	70.0	87.0	123.0	177.0	272.0	374.0	484.0
$C_2H_2O_4 + 2H_2O$	19.0	40.0	62.0	86.0	112.0	169.0	262.0	540.0	1316.0	50000.0
$C_6H_8O_7 + H_2O$	29.0	58.0	87.0	116.0	145.0	208.0	320.0	553.0	952.0	

Salt.	40°	60°	80°	100°	120°	140°	160°	180°	200°	240°
$CaCl_2$	137.5	222.0	314.0							
KOH	92.5	121.7	152.6	185.0	219.8	263.1	312.5	375.0	444.4	623.0
$NaOH$	93.5	150.8	230.0	345.0	526.3	800.0	1333.0	2353.0	6452.0	–
NH_4NO_3	682.0	1370.0	2400.0	4099.0	8547.0	∞				
$C_4H_6O_6$	980.0	3774.0	(infinity gives 170)							

* Compiled from a paper by Gerlach, "Zeit. f. Anal. Chem." vol. 26.

FREEZING MIXTURES.*

Column 1 gives the name of the principal refrigerating substance, *A* the proportion of that substance, *B* the proportion of a second substance named in the column, *C* the proportion of a third substance, *D* the temperature of the substances before mixture, *E* the temperature of the mixture, *F* the lowering of temperature, *G* the temperature when all snow is melted, when snow is used, and *H* the amount of heat absorbed in heat units (small calories when *A* is grams). Temperatures are in Centigrade degrees.

Substance.	*A*	*B*	*C*	*D*	*E*	*F*	*G*	*H*
$NaC_2H_3O_2$ (cryst.)	85	H_2O–100	–	10.7	– 4.7	15.4	–	–
NH_4Cl . . .	30	" "	–	13.3	– 5.1	18.4	–	–
$NaNO_3$ . . .	75	" "	–	13.2	– 5.3	18.5	–	–
$Na_2S_2O_3$ (cryst.) .	110	" "	–	10.7	– 8.0	18.7	–	–
KI	140	" "	–	10.8	– 11.7	22.5	–	–
$CaCl_2$ (cryst.) .	250	" "	–	10.8	– 12.4	23.2	–	–
NH_4NO_3 . .	60	" "	–	13.6	– 13.6	27.2	–	–
$(NH_4)_2SO_4$ . .	25	" 50	NH_4NO_3–25	–	–	26.0	–	–
NH_4Cl . . .	25	" "	" "	–	–	22.0	–	–
$CaCl_2$ . . .	25	" "	" "	–	–	20.0	–	–
KNO_3 . . .	25	" "	NH_4Cl–25	–	–	20.0	–	–
Na_2SO_4 . .	25	" "	" "	–	–	19.0	–	–
$NaNO_3$ . . .	25	" "	" "	–	–	17.0	–	–
K_2SO_4 . . .	10	Snow 100	–	– 1	– 1.9	0.9	–	–
Na_2CO_3 (cryst.) .	20	" "	–	– 1	– 2.0	1.0	–	–
KNO_3 . . .	13	" "	–	– 1	– 2.85	1.85	–	–
$CaCl_2$ . . .	30	" "	–	– 1	– 10.9	9.9	–	–
NH_4Cl . . .	25	" "	–	– 1	– 15.4	14.4	–	–
NH_4NO_3	45	" "	–	– 1	– 16.75	15.75	–	–
$NaNO_3$ . . .	50	" "	–	– 1	– 17.75	16.75	–	–
NaCl . . .	33	" "	–	– 1	– 21.3	20.3	–	–
	1	" 1.097	–	– 1	– 37.0	36.0	– 37.0	0.0
	1	" 1.26	–	– 1	– 36.0	35.0	– 30.2	17.0
$H_2SO_4 + H_2O$ (66.1 % H_2SO_4)	1	" 1.38	–	– 1	– 35.0	34.0	– 25.0	27.0
	1	" 2.52	–	– 1	– 30.0	29.0	– 12.4	133.0
	1	" 4.32	–	– 1	– 25.0	24.0	– 7.0	273.0
	1	" 7.92	–	– 1	– 20.0	19.0	– 3.1	553.0
	1	" 13.08	–	– 1	– 16.0	15.0	– 2.1	967.0
	1	" 0.35	–	0	–	–	0.0	52.1
	1	" .49	–	0	–	–	– 19.7	49.5
	1	" .61	–	0	–	–	– 39.0	40.3
$CaCl_2 + 6H_2O$	1	" .70	–	0	–	–	– 54.9†	30.0
	1	" .81	–	0	–	–	– 40.3	46.8
	1	" 1.23	–	0	–	–	– 21.5	88.5
	1	" 2.46	–	0	–	–	– 9.0	192.3
	1	" 4.92	–	0	–	–	– 4.0	392.3
Alcohol at 4°	77	" 73	–	0	– 30.0	–	–	–
	–	CO_2 solid	–	–	– 72.0	–	–	–
Chloroform . .	–	" "	–	–	– 77.0	–	–	–
Ether . . .	–	" "	–	–	– 77.0	–	–	–
Liquid SO_2 . .	–	" "	–	–	– 82.0	–	–	–
	1	H_2O–.75	–	20	5.0	–	–	33.0
	1	" .94	–	20	– 4.0	–	–	21.0
	1	" "	–	10	– 4.0	–	–	34.0
	1	" "	–	5	– 4.0	–	–	40.5
	1	Snow "	–	0	– 4.0	–	–	122.2
NH_4NO_3 .	1	H_2O–1.20	–	10	– 14.0	–	–	17.9
	1	Snow "	–	0	– 14.0	–	–	129.5
	1	H_2O–1.31	–	10	– 17.5†	–	–	10.6
	1	Snow "	–	0	– 17.5†	–	–	131.9
	1	H_2O–3.61	–	10	– 8.0	–	–	0.4
	1	Snow "	–	0	– 8.0	–	–	327.0

* Compiled from the results of Cailletet and Colardeau, Hammerl, Hanamann, Moritz, Pfaundler, Rudorf, and Tollinger.

† Lowest temperature obtained.

CRITICAL TEMPERATURES, PRESSURES, VOLUMES, AND DENSITIES OF GASES.*

θ = Critical temperature.
P = Critical pressure in atmospheres.
ϕ = Critical volume referred to volume at 0° and 76 centimeters pressure.
d = Critical density in grams per cubic centimeter.

a, b, Van der Waals constants in $\left(p + \frac{a}{v^2}\right)\left(v - b\right) = 1 + \alpha t.$

Substance.	θ	P	ϕ	d	a × 10^5	b × 10^6	Observer
Air	—140.0	39.0	–	–	257	1560	1
Alcohol (C_2H_6O)	243.6	62.76	0.00713	0.288	2407	3769	2
" (CH_4O)	239.95	78.5	–	–	1898	2992	3
Ammonia	130.0	115.0	–	–	798	1606	4
Argon	—117.4	52.9	–	–	259	1348	5
Benzene	288.5	47.9	–	0.305	3726	5370	3
Bromine	302.2	–	0.00605	1.18	1434	2020	6
Carbon dioxide	31.2	73.	0.0044	0.46	717	1908	–
" monoxide	—141.1	35.9	–	–	275	1683	7
" disulphide	273.	72.9	0.0090	–	2316	3430	8
Chloroform	260.0	54.9	–	–	2930	4450	9
Chlorine	141.0	83.9	–	–	1157	2259	4
"	146.0	93.5	–	–	1063	2050	10
Ether	197.0	35.77	0.01584	0.208	3496	6016	11
"	194.4	35.61	0.01344	0.262	3464	6002	3
Ethane	32.1	49.0	–	–	1074	2848	12
Ethylene	9.9	51.1	–	–	886	2533	–
Helium	<—268.0	2.3	–	–	5	700	13
Hydrogen	—240.8	14.	–	–	42	880	14
" chloride	51.25	86.0	–	–	692	1726	15
" "	52.3	86.0	–	0.61	697	1731	4
" sulphide	100.0	88.7	–	–	888	1926	1
Krypton	—62.5	54.3	–	–	462	1776	5
Methane	—81.8	54.9	–	–	376	1557	1
"	—95.5	50.0	–	–	357	1625	4
Neon	<—205.0	29.	–	–	–	–	5,13
Nitric oxide (NO)	—93.5	71.2	–	–	257	1160	1
Nitrogen	—146.0	35.0	–	0.44	259	1650	1
" monoxide (N_2O)	35.4	75.0	0.0048	0.41	720	1888	4,17
Oxygen	—118.0	50.0	–	0.6044	273	1420	1
Sulphur dioxide	155.4	78.9	0.00587	0.49	1316	2486	9,17
Water	358.1	–	0.001874	0.429	–	–	6
"	374.	217.5	–	–	1089	1362	16

(1) Olszewski, C. R. 98, 1884; 99, 1884; 100, 1885; Beibl. 14, 1890; Z. Phys. Ch. 16, 1893.
(2) Ramsay-Young, Tr. Roy. Soc. 177, 1886.
(3) Young, Phil. Mag. 1900.
(4) Dewar, Phil. Mag. 18, 1884; Ch. News, 84, 1901.
(5) Ramsay, Travers, Phil. Trans. 16, 17, 1901.
(6) Nadejdine, Beibl. 9, 1885.
(7) Wroblewski, Wied. Ann. 20, 1883; Stz. Wien. Ak. 91, 1885.
(8) Batelli, 1890.
(9) Sajotschewsky, Beibl. 3, 1879.
(10) Knietsch, Lieb. Ann. 259, 1890.
(11) Batelli, Mem. Torino (2), 41, 1890.
(12) Cardozo, Arch. sc. phys. 30, 1910.
(13) Kamerlingh-Onnes, Comno. Phys. tab. Leiden, 1908, 1909, Proc. Amst. 11, 1908, C. R. 147, 1908.
(14) Olszewski, Ann. Phys. 17, 1905.
(15) Ansdell, Chem. News, 41, 1880.
(16) Holborn, Baumann Ann. Phys. 31, 1910.
(17) Cailletet, C. R. 102, 1886; 104, 1887.

*Abridged for the most part from Landolt and Börnstein's "Phys. Chem. Tab."

TABLE 229.

CONDUCTIVITY FOR HEAT. METALS AND ALLOYS.

The coefficient k is the quantity of heat in small calories which is transmitted per second through a plate one centimeter thick per square centimeter of its surface when the difference of temperature between the two faces of the plate is one degree Centigrade. The coefficient k is found to vary with the absolute temperature of the plate, and is expressed approximately by the equation $k_t = k_0[1 + \alpha(t - t_0)]$. k_0 is the conductivity at t_0, the lower temperature of the bracketed pairs in the table, k_t that at temperature t, and α is a constant. k_t in g-cal. per degree C per sec. across cm cube = $0.239 \times k_t$ in watts per degree C per sec. across cm cube.

Substance	t°C	k_t	α	Reference.
Aluminum	−160	0.514	—	1
"	18	0.480	+.0030	2
"	100	0.492		
"	200	0.545	+.0020	3
"	400	0.760		
"	500	0.885	+.0014	3
"	600	1.01		
Antimony	0	0.0442	−.00104	4
"	100	0.0396		
Bismuth	−186	0.025	—	5
"	18	0.0194	−.0021	2
"	100	0.0161		
Brass	−160	0.181	—	1
"	17	0.260	—	1
" , yellow	0	0.204	+.0024	4
" , red	0	0.246	+.0015	4
Cadmium, pure	−160	0.239	—	1
" "	18	0.222	−.00038	2
"	100	0.215		
Constantan	18	0.0540	+.00227	2
(60 Cu+40 Ni)	100	0.0640		
Copper,* pure	−160	1.079	—	1
" "	18	0.918	−.00013	2
" "	100	0.908		
German silver	0	0.070	+.0027	4
Gold	17	0.705	−.00007	6
Graphite	17	0.037	+.0003	6
Iridium	17	0.141	−.0005	8
Iron,† pure	18	0.161	−.0008	2
" "	100	0.151		
Iron, wrought	−160	0.152	—	1
" "	18	0.144	−.00008	2
" "	100	0.143		
" steel, 1%	18	0.108	−.0001	2
C	100	0.107		
Lead, pure	−160	0.092	—	1
" "	18	0.083	−.0001	2
" "	100	0.081		
Magnesium	0 to 100	0.376	—	4
Manganin	−160	0.035	—	1
" (84 CU+4	18	0.0519	+.0026	2
Ni 12 Mn)	100	0.0630		
Mercury	0	0.0148	+.0055	7
"	50	0.0189		
Molybdenum	17	0.346	−.0001	6
Nickel	−160	0.129	—	1
"	18	0.1420	—	2
"	0	0.1425	−.00032	3
"	100	0.1380		
"	200	0.1325	−.00095	3
"	700	0.069		
"	1000	0.064	−.00047	3
"	1200	0.058		
Palladium	18	0.1683	+.0010	2
"	100	0.182		
Platinum	18	0.1664	+.00051	2
"	100	0.1733		
Pt 10% Ir	17	0.074	+.0002	6
Pt 10% Rh	17	0.072	+.0002	6
Platinoid	18	0.060	—	1
Potassium	5.0	0.232	−.0013	8
"	57.4	0.216		
Rhodium	17	0.210	−.0010	6
Silver, pure	−160	0.998	—	1
"	18	1.006	−.00017	2
"	100	0.992		
Sodium	5.7	0.321	−.0012	8
"	88.1	0.288		
Tantalum	17	0.130	−.0001	6
"	1700	0.174	—	9
"	1900	0.186	+.00032	9
"	2100	0.198		
Tin	0	0.155	−.00069	4
"	100	0.145		
" , pure	−160	0.192	—	1
Tungsten	17	0.476	−.0001	6
Tungsten	1600	0.249	+.00023	10
"	2000	0.272		
"	2400	0.294	+.00016	10
"	2800	0.313		
Wood's alloy	—	0.319	—	7
Zinc, pure	−160	0.278	—	1
" "	18	0.2653	−.00016	2
" "	100	0.2619		

References: (1) Lees, Phil. Trans. 1908; (2) Jaeger and Diesselhorst, Wiss. Abh. Phys. Tech. Reich. 3, 1900; (3) Angell, Phys. Rev. 1911; (4) Lorenz; (5) Macchia, 1907; (6) Barratt, Pr. Phys. Soc. 1914; (7) H. F. Weber, 1879; (8) Hornbeck, Phys. Rev. 1913; (9) Worthing, Phys. Rev. 1914; (10) Worthing, Phys. Rev. 1917.

* Copper: 100–197° C, k_t = 1.043; 100–268°, 0.969; 100–370°, 0.931; 100–541°, 0.902 (Hering; for reference see next page).

† Iron: 100–727° C, k_t = 0.202; 100–912°, 0.184; 100–1245°, 0.191 (Hering).

CONDUCTIVITY FOR HEAT.

TABLE 230. — Thermal Conductivity at High Temperatures.

(See also Table 229 for metals; k in gram-calories per degree centigrade per second across a centimeter cube.)

Material.	Temperature, ° C	k	Reference.	Material.	Temperature, ° C	k	Reference.
Amorphous carbon...	37-163	.028-.003	1	Brick: Carborundum	150-1200	.0032-.027	3
	170-330	.027-.004	1	Building / Terra-cotta	15-1100	.0018-.0038	3
	240-523	.020-.003	1				
	283-597	.011-.004	1	Fire-clay....	125-1220	.0032-.0054	3
	100-360	.089	2	Gas-retort...	100-1125	.0038	3
	100-751	.124	2	Graphite....	300-700	.024	3
	100-842	.129	2	Magnesia....	50-1130	.0027-.0072	3
Graphite (artificial)...	100-390	.338	2	Silica.......	100-1000	.002 -.0033	3
	100-546	.324	2	Granite...........	100	.0045-.0050	4
	100-720	.306	2		200	.0043-.0097	4
	100-914	.291	2		500	.0040	4
	30-2830	.162	1	Limestone.........	40	.0046-.0057	4
	2800-3200	.002	1		100	.0039-.0049	4
	90-110	.55-.45	1		350	.0032-.0035	4
	180-120	.44-.34	1	Porcelain (Sèvres)..	165-1055	.0039-.0047	3
	500-700	.31-.22	1	Stoneware mixtures.	70-1000	.0029-.0053	3

References: (1) Hansen, Tr. Am. Electrochem. Soc. 16, 329, 1909; (2) Hering, Tr. Am. Inst. Elect. Eng. 1910; (3) Bul. Soc. Encouragement, 111, 879, 1909; Electroch. and Met. Ind. 7, 383, 433, 1909; (4) Poole, Phil. Mag. 24, 45, 1912; see also Clement, Egy, Eng. Exp. Univers. Ill. Bull. 36, 1909; Dewey, Progressive Age, 27, 772, 1909; Woolson, Eng. News, 58, 166, 1907, heat transmission by concretes; Richards, Met. and Chem. Eng. 11, 575, 1913. The ranges in values under 1 do not depend on variability in material but on possible errors in method; reduced from values expressed in other units.

TABLE 231. — Thermal Conductivity of Various Substances.

Substance, temperature.	kt	Reference.	Substance, temperature.	kt	Reference.
Aniline BP 183° C., −160	.000112	1	Naphthalene MP 79° C., −160	.0013	1
Carbon, gas	.010	–	Naphthalene MP 79° C., 0	.00081	1
Carbon, graphite	.012	–	Naphthol — β, MP 122° C., −160	.00068	1
Carborundum	.00050	2	Naphthol, 0	.00062	1
Concrete, cinder	.00081	–	Nitrophenol, MP 114° C., −160	.00106	1
stone	.0022	3	Nitrophenol, 0	.00065	1
Diatomaceous earth	.00013	4	Paraffin MP 54° C., −160	.00062	1
Earth's crust	.004	–	Paraffin, 0	.00059	1
Fire-brick	.00028	4	Porcelain	.0025	–
Fluorite, −190	.093	5	Quartz ⊥ to axis, −190	.0586	5
Fluorite, 0	.025	5	" , 0	.0173	5
Glass: window	.0025	–	" , 100	.0133	5
crown, O3572, −190	.00118	5	Quartz ‖ to axis, 0	.0325	5
crown, O3572, 0	.00280	5	Rock salt, 0	.0167	5
crown, O3572, 100	.00324	5	Rock salt, 30	.0150	5
h'vy flint O165, −190	.00081	5	Rubber, vulcanized, −160	.00033	5
h'vy flint O165, 0	.00170	5	Rubber, 0	.00037	5
h'vy flint O165, 100	.00181	5	Rubber, para	.00045	–
Glycerine, −160	.00077	1	Sand, white, dry	.00093	6
Granite	.0053	6	Sandstone, dry	.0055	6
Ice, −160	.0066	1	Sawdust	.00012	–
Ice, 0	.0050	1	Slate ⊥ to cleavage	.0034	6
Iceland spar, −190	.038	5	Slate ‖ to cleavage	.0060	6
Iceland spar, 0	.0103	5	Snow, fresh, dens. = 0.11	.00026	7
Lime	.00029	4	Snow, old	.0012	7
Limestones, calcite	.0047 to	6	Soil, average, sl't moist	.0037	–
Marbles, dolomite	.0056	6	Soil, very dry	.0037	–
Mica	.0018	–	Sulphur, rhombic, 0	.00070	5
Flagstone ⊥ to cleavage	.0063	6	Vaseline, 20	.00022	8
Micaceous ‖ to cleavage	.0044	6	Vulcanite	.00087	9

References: (1) Lees, Tr. R. S. 1905; (2) Lorenz; (3) Norton; (4) Hutton, Blard; (5) Eucken, Ann. d. Phys., 1911; (6) Herschel, Lebour, Dunn, B. A. Committee, 1879; (7) Jansson, 1904; (8) Melmer, 1911; (9) Stefan.

THERMAL CONDUCTIVITIES OF INSULATING MATERIALS.

Conductivity in g-cal. flowing in 1 sec. through plate 1 cm thick per cm² for 1° C difference of temperature.

Material.	Conductivity.	Density. g/cm³	Remarks.
Air	0.00006	—	Horizontal layer, heated from above.
Calorox	0.000076	0.064	Fluffy, finely divided mineral matter.
Hair felt	0.000085	0.27	
Keystone hair	0.000093	0.30	Felt between layers of bldg. paper.
Pure wool	0.000084	0.107	Firmly packed.
" "	0.000084	0.102	" "
" "	0.000090	0.061	Loosely packed.
" "	0.000101	0.039	Very loosely packed.
Cotton wool	0.00010	—	Firmly packed.
Insulite	0.000102	1.9	Pressed wood-pulp — rigid, fairly strong.
Linofelt	0.000103	0.18	Vegetable fibers between layers of paper — soft and flexible.
Corkboard (pure)	0.000106	0.18	
Eel grass	0.00011	0.25	Inclosed in burlap.
Flaxlinum	0.000113	0.18	Vegetable fibers — firm and flexible.
Fibrofelt	0.000113	0.18	
Rock cork	0.000119	0.33	Rock wool pressed with binder, rigid.
Balsa wood	0.00012	0.12	Very light and soft.
Waterproof lith	0.00014	0.27	Rock wool, vegetable fiber and binder, not flexible.
Pulp board	0.00015	—	Stiff pasteboard.
Air cell ½ in. thick	0.000154	0.14	Corr. asbestos paper with air space.
Air cell 1 in. thick	0.000165	0.14	" " " " " "
Asbestos paper	0.00017	0.50	Fairly firm, but easily broken.
Infusorial earth, block	0.00020	0.69	
Fire-felt, sheet	0.000205	0.42	Asbestos sheet coated with cement, rigid.
Fire-felt, roll	0.00022	0.68	Soft, flexible asbestos.
Three-ply regal roofing	0.00024	0.88	Flexible tar roofing.
Asbestos mill board	0.00029	0.97	Pressed asbestos, firm, easily broken.
Woods, kiln dried:			
Cypress	0.00023	0.46	
White pine	0.00027	0.50	
Mahogany	0.00031	0.55	
Virginia pine	0.00033	0.55	
Oak	0.00035	0.61	
Hard maple	0.00038	0.71	
Asbestos wood, sanded	0.00093	1.97	Asbestos and cement, very hard, rigid.

Dickinson and van Dusen, Am. Soc. Refrigerating Eng. J. 3, Sept. 1916.

CONDUCTIVITY FOR HEAT.

TABLE 233. — Various Substances.

k_t is the heat in gram-calories flowing in 1 sec. through a plate 1 cm. thick per sq. cm. for 1°C drop in temperature.

Substance.	Density.	°C.	k_t	Substance.	k_t	Authority.
Asbestos fiber	0.201	500	.00019	Asbestos paper	0.00043	Lees-Chorlton.
85% magnesia asbestos	.216	100	.00016	Blotting paper	.00015	
		500	.00017	Portland cement	.00071	
Cotton	.021	100	.000111	Cork, t, 0°C	.0007?	Forbes.
"	.101	"	.000071	Chalk	.0020	H, L, D, see p. 205.
Eiderdown	.0021	150	.00015	Ebonite, t, 49°	.00037	
"	.109	"	.000046	Glass, mean	.002	Various.
Lampblack, Cabot number 5	.193	100	.000074	Ice	.0057	Neumann.
		500	.000107	Leather, cow-hide	.00042	Lees-Chorlton.
Quartz, mesh 200	1.05	500	.00024	" chamois	.00015	
Poplox, popped Na_2SiO_3	0.093	200	.000091	Linen	.00021	
		500	.000160	Silk	.000095	
Wool fibers	.015	100	.000118	Caen stone, limestone	.0043	H, L, D.
" "	.054	"	.000085	Free stone, sandstone	.0021	
" "	.192	"	.000054			

Left-hand half of table from Randolph, Tr. Am. Electroch. Soc. XXI., p. 550, 1912; k_t (Randolph's values) is mean conductivity between given temperature and about 10°C. Note effect of compression (density). The following are from Barratt Proc. Phys. Soc., London, 27, 81, 1914.

Substance.	Density.	k_t at 20°C.	k_t at 100°C.	Substance.	Density.	k_t at 20°C.	k_t at 100°C.
Brick, fire	1.73	.00110	.00109	Boxwood	0.90	.00036	.00041
Carbon, gas	1.42	.0085	.0095	Greenheart	1.08	.00112	.00110
Ebonite	1.19	.00014	.00013	Lignumvitæ	1.16	.00060	.00072
Fiber, red	1.29	.00112	.00119	Mahogany	0.55	.00051	.00060
Glass, soda	2.59	.00172	.00182	Oak	0.65	.00058	.00061
Silica, fused	2.17	00237	.00255	Whitewood	0.58	.00041	.00045

The following values are from unpublished data furnished by C. E. Skinner of the Westinghouse Co., Pittsburgh, Penn. They give the mean conductivity in gram-calories per sec. per cm. cube per °C. when the mean temperature of the cube is that stated in the table. Resistance in thermal ohms (watts/inch²/inch/°C.) $= \frac{1}{10.6}$ conductivity.

Substance.	Grams. per cm³.	Conductivity. 100° C.	200° C.	300° C.	400° C.	500° C.	Safe temp.
Air-cell asbestos	0.232	0.00034	0.00043	0.00050	—	—	320
Cork, ground	.168	.00015	.00019	—	—	—	180
Diatomit	.326	.00028	.00032	.00037	0.00042	0.00046	600
Infusorial earth, natural	.506	.00034	.00032	.00040	—	—	
" " h'd pressed blocks	.321	.00030	.00029	.00033	.00036	—	400
Magnesium carbonate	.450	.00023	.00025	.00025	—	—	300
Vitribestos	.362	.00049	.00066	.00079	.00090	.00102	600

TABLE 234. — Water and Salt Solutions.

Substance.	°C.	k_t	Authority.	Solution in water.	Density.	°C.	k_t	Authority.
Water	0	0.00150	Goldschmidt, '11.	$CuSO_4$	1.160	4.4	0.00118	H. F. Weber.
	11	.00147	Lees, '98.	KCl	1.026	13.	.00116	Graetz.
	25	.00136		NaCl	1.178	4.4	.00115	H. F. Weber.
	20	.00143	Milner, Chattock, '98	"	—	26.3	.00135	
				H_2SO_4	1.054	20.5	.00126	Chree.
				"	1.180	21.	.00130	
				$ZnSO_4$	1.134	4.5	.00118	H. F. Weber.
				"	1.136	4.5	.00115	

TABLE 235. — Thermal Conductivity of Organic Liquids.

Substance.	°C	k_t	Refer.	Substance.	°C	k_t	Refer.	Substance.	°C	k_t	Refer.
Acetic acid........	9–15	$.0_3472$	1	Carbon disulphide.	0	$.0_3387$	3	Oils: olive........	—	$.0_3395$	4
Alcohols: methyl...	11	$.0_352$	2	Chloroform......	9–15	$.0_3288$	1	" castor.......	—	$.0_3425$	4
" ethyl.....	11	$.0_346$	2	Ether...........	9–15	$.0_3303$	1	Toluene..........	0	$.0_3349$	3
" amyl.....	0	$.0_3345$	3	Glycerine........	25	$.0_368$	2	Vaseline..........	25	$.0_344$	2
Aniline............	0	$.0_3434$	—	Oils: petroleum...	13	$.0_3355$	5	Xylene...........	0	$.0_3343$	3
Benzene	9–15	$.0_3333$	1	" turpentine..	13	$.0_3325$	5				

References: (1) H. F. Weber; (2) Lees; (3) Goldschmidt; (4) Wachsmuth; (5) Graetz.

TABLE 236. — Thermal Conductivity of Gases.

The conductivity of gases, $k_t = \frac{1}{4}(9\gamma - 5)\mu C_v$, where γ is the ratio of the specific heats, C_p/C_v, and μ is the viscosity coefficient (Jeans, Dynamical Theory of Gases, 1916). Theoretically k_t should be independent of the density and has been found to be so by Kundt and Warburg and others within a wide range of pressure below one atm. It increases with the temperature.

Gas.	t° C	k_t	Ref.	Gas.	t° C	k_t	Ref.	Gas.	t° C	k_t	Ref.
Air*...	−191	0.0000180	1	CO_2	100	0.0000496	1	Hg	203	0.0000185	3
"	0	0.0000566	1	C_2H_4	0	0.0000395	2	N_2	−191	0.0000183	1
"	100	0.0000719	1	He	−193	0.000146	1	"	0	0.0000568	1
Ar	−183	0.0000142	1	"	0	0.000344	4	"	100	0.0000718	1
"	0	0.0000388	1	"	100	0.000398	1	O_2	−191	0.0000172	1
"	100	0.0000509	1	H_2	−192	0.000133	1	"	0	0.0000570	1
CO	0	0.0000542	1	"	0	0.000416	4	"	100	0.0000743	1
CO_2	−78	0.0000219	1	"	100	0.000499	1	NO	8	0.000046	2
"	0	0.0000332	1	CH_4	0	0.0000720	4	N_2O	0	0.0000353	4

References: (1) Eucken, Phys. Z. 12, 1911; (2) Winkelmann, 1875; (3) Schwarze, 1903; (4) Weber, 1917.

* Air: $k_0 = 5.22 (10^{-5})$ cal. cm $^{-1}$ sec.$^{-1}$ deg. C^{-1}; 5.74 at 22°; temp. coef. = .0029; Hercus-Laby, Pr. R. Soc. A95, 190, 1919.

TABLE 237. — Diffusivities.

The diffusivity of a substance = $h^2 = k/c\rho$, where k is the conductivity for heat, c the specific heat and ρ the density (Kelvin). The values are mostly for room temperatures, about 18° C.

Material.	Diffusivity.	Material.	Diffusivity.
Aluminum........................	0.826	Coal........................	0.002
Antimony........................	0.139	Concrete (cinder)................	0.0032
Bismuth........................	0.0678	Concrete (stone)................	0.0058
Brass (yellow)..................	0.339	Concrete (light slag)............	0.006
Cadmium........................	0.467	Cork (ground)...................	0.0017
Copper........................	1.133	Ebonite........................	0.0010
Gold........................	1.182	Glass (ordinary)................	0.0057
Iron (wrought, also mild steel)........	0.173	Granite........................	0.0155
Iron (cast, also 1% carbon steel)........	0.121	Ice........................	0.0112
Lead........................	0.237	Limestone........................	0.0092
Magnesium........................	0.883	Marble (white)..................	0.0090
Mercury........................	0.0327	Paraffin........................	0.00098
Nickel........................	0.152	Rock material (earth aver.)..........	0.0118
Palladium........................	0.240	Rock material (crustal rocks).........	0.0064
Platinum........................	0.243	Sandstone........................	0.0133
Silver........................	1.737	Snow (fresh).....................	0.0033
Tin........................	0.407	Soil (clay or sand, slightly damp)......	0.005
Zinc........................	0.402	Soil (very dry)..................	0.0031
Air........................	0.179	Water........................	0.0014
Asbestos (loose)................	0.0035	Wood (pine, cross grain)..............	0.00068
Brick (average fire).............	0.0074	Wood (pine with grain)...............	0.0023
Brick (average building).............	0.0050		

Taken from An Introduction to the Mathematical Theory of Heat Conduction, Ingersoll and Zobel, 1913.

LINEAR EXPANSION OF THE ELEMENTS.

In the heading of the columns t is the temperature or range of temperature; C is the coefficient of linear expansion; A_1 is the authority for C; M is the mean coefficient of expansion between 0° and 100° C; α and β are the coefficients in the equation $l_t = l_0(1 + \alpha t + \beta t^2)$, where l_0 is the length at 0° C and l_t the length at t° C; A_2 is the authority for α, β, and M. See footnote for Molybdenum and Tungsten.

Substance.	t	$C \times 10^4$	A_1	$M \times 10^4$	$\alpha \times 10^4$	$\beta \times 10^6$	A_2
Aluminum	40	0.2313	1	0.2220	—	—	2
"	600	0.3150	3	—	—	—	—
"	−191 to +16	0.1835	4	—	.23536	.00707	5
Antimony: ‖ to axis	40	0.1692	1	—	—	—	—
⊥ to axis	40	0.0882	1	—	—	—	—
Mean	40	0.1152	1	0.1056	.0923	.0132	6
Arsenic	40	0.0559	1	—	—	—	—
Bismuth: ‖ to axis	40	0.1621	1	—	—	—	—
⊥ to axis	40	0.1208	1	—	—	—	—
Mean	40	0.1346	1	0.1316	.1167	.0149	6
Cadmium	40	0.3069	1	0.3159	.2693	.0466	6
Carbon: Diamond	40	0.0118	1	—	—	—	—
Gas carbon	40	0.0540	1	—	—	—	—
Graphite	40	0.0786	1	—	.0055	.0016	13
Anthracite	40	0.2078	1	—	—	—	—
Cobalt	40	0.1236	1	—	—	—	—
Copper	40	0.1678	1	0.1666	.1481	.0185	6
"	−191 to +16	0.1409	4	—	.16070	.00403	5
Gold	40	0.1443	1	0.1470	.1358	.0112	6
"	−170	0.117	15	—	—	—	—
Indium	40	0.4170	1	—	—	—	—
Iridium	18	0.088	16	0.090	—	—	16
Iron: Soft	40	0.1210	1	—	—	—	—
Cast	40	0.1061	1	—	—	—	—
Cast	−191 to +16	0.0850	4	—	—	—	—
Wrought	−18 to 100	0.1140	7	—	.11705	.005254	8
Steel	40	0.1322	1	—	.09173	.008336	8
Steel annealed	40	0.1095	1	0.1089	.1038	.0052	9
Lead	40	0.2924	1	0.2709	.273	.0074	6
Lead (cast)	−170	0.24	15	—	—	—	—
Magnesium	40	0.2694	1	0.261	—	—	16
Nickel	40	0.1279	1	—	.13460	.003315	8
"	−191 to +16	0.1012	4	—	—	—	—
Osmium	40	0.0657	1	—	—	—	—
Palladium	40	0.1176	1	—	.11670	.002187	8
Phosphorus	0–40	1.2530	10	—	—	—	—
Platinum	40	0.0899	1	—	.08868	.001324	8
Potassium	0–50	0.8300	11	—	—	—	—
Rhodium	40	0.0850	1	—	—	—	—
Ruthenium	40	0.0963	1	—	—	—	—
Selenium	40	0.3680	1	0.6604	—	—	12
Silicon	40	0.0763	1	—	—	—	—
Silver	40	0.1921	1	—	.18270	.004793	8
"	−191 to +16	0.1704	4	0.189	—	—	16
Sulphur: Cryst. mean	40	0.6413	1	1.180	—	—	12
Tellurium	40	0.1675	1	0.3687	—	—	12
Thallium	40	0.3021	1	—	—	—	—
Tin	40	0.2234	1	0.2296	.2033	.0263	6
Zinc	40	0.2918	1	0.2976	.2741	.0234	6
Zinc (cast)	−170	0.190	15	—	—	—	—

References: (1) Fizeau; (2) Calvert, Johnson and Lowe; (3) Chatelier; (4) Henning; (5) Dittenberger; (6) Matthiessen; (7) Andrews; (8) Holborn-Day; (9) Benoit; (10) Pisati and De Franchis; (11) Hagen; (12) Spring; (13) Day and Sosman; (14) Griffiths; (15) Dorsey; (16) Grüneisen.

Tungsten: $(L - L_0)/L_0 = 4.44 \times 10^{-6}(T - 300) + 45 \times 10^{-11}(T - 300)^2 + 2.20 \times 10^{-13}(T - 300)^3$. L_0 = length at 300° K. Coefficient at 300° K = 4.44×10^{-6}; 1300° K, 5.19×10^{-6}; 2300° K, 7.26×10^{-6}. Worthing, Phys. Rev. 1917.

Molybdenum: $L_t = L_0(1 + 5.15t \times 10^{-6} + 0.00570t^2 \times 10^{-6})$, for 19° to −142° C; $= L_0(1 + 5.01t \times 10^{-6} + 0.00138t^2 \times 10^{-6})$, for 19° to +305° C; Schad and Hidnert, Phys. Rev. 1919.

The Holborn-Day and Sosman data are for temperatures from 20° to 1000° C. The Dittenberger, 0° to 600° C.

TABLE 239.

LINEAR EXPANSION OF MISCELLANEOUS SUBSTANCES.

The coefficient of cubical expansion may be taken as three times the linear coefficient. *t* is the temperature or range of temperature, *C* the coefficient of expansion, and A. the authority.

Substance.	*t*	$C \times 10^4$	A.
Brass:			
Cast	0–100	0.1875	1
Wire	"	0.1930	1
"	"	.1783–.193	2
71.5 Cu + 27.7 Zn + 0.3 Sn + 0.5 Pb	40	0.1859	3
71 Cu + 29 Zn	0–100	0.1906	4
Bronze:			
3 Cu + 1 Sn	16.6–100	0.1844	5
" " " " "	16.6–350	0.2116	5
" " " " "	16.6–957	0.1737	5
86.3 Cu + 9.7 Sn + 4 Zn	40	0.1782	3
97.6 Cu + 2.2 Sn + 0.2 P (hard)	0–80	0.1713	6
97.6 Cu + 2.2 Sn + 0.2 P (soft)	"	0.1708	6
Caoutchouc	—	0.657–0.686	2
"	16.7–25.3	0.770	7
Constantan	4–29	0.1523	—
Ebonite	25.3–35.4	0.842	7
Fluor spar: CaF_2	0–100	0.1950	8
German silver	"	0.1836	8
Gold-platinum: 2 Au + 1 Pt	"	0.1523	4
Gold-copper: 2 Au + 1 Cu	"	0.1552	4
Glass: Tube	"	0.0833	1
"	"	0.0828	9
Plate	"	0.0891	10
Crown (mean)	"	0.0897	10
" "	50–60	0.0954	11
Flint	"	0.0788	11
Jena thermometer 16III normal	0–100	0.081	12
" 59III	"	0.058	12
" "	−191 to +16	0.424	13
Gutta percha	20	1.983	14
Ice	−20 to −1	0.51	15
Iceland spar: Parallel to axis	0–80	0.2631	6
Perpendicular to axis	"	0.0544	6
Lead-tin (solder) 2 Pb + 1 Sn	0–100	0.2508	1
Magnalium	12–39	0.238	16
Manganin	—	0.181	—
Marble	15–100	0.117	17
Paraffin	0–16	1.0662	18
"	16–38	1.3030	18
"	38–49	4.7707	18
Platinum-iridium 10 Pt + 1 Ir	40	0.0884	3

Substance.	*t*	$C \times 10^4$	A.
Platinum-silver: 1 Pt + 2Ag	0–100	0.1523	4
Porcelain	20–790	0.0413	19
" Bayeux	1000–1400	0.0553	20
Quartz: Parallel to axis	0–80	0.0797	6
" "	−190 to +16	0.0521	21
Perpend. to axis	0–80	0.1337	6
Quartz glass	−190 to +16	−0.0026	13
" "	16 to 500	0.0057	26
" "	16–1000	0.0058	26
Rock salt	40	0.4040	3
Rubber, hard	0°	0.691	27
" "	−160	0.300	27
Speculum metal	0–100	0.1933	1
Topaz: Parallel to lesser horizontal axis	"	0.0832	8
Parallel to greater horizontal axis	"	0.0836	8
Parallel to vertical axis	"	0.0472	8
Tourmaline: Parallel to longitudinal axis	"	0.0937	8
Parallel to horizontal axis	"	0.0773	8
Type metal	16.6–254	0.1952	5
Vulcanite	0–18	0.6360	22
Wedgwood ware	0–100	0.0890	5
Wood: Parallel to fiber: Ash	"	0.0951	23
Beech	2.34	0.0257	24
Chestnut	"	0.0649	24
Elm	"	0.0565	24
Mahogany	"	0.0361	24
Maple	"	0.0638	24
Oak	"	0.0492	24
Pine	"	0.0541	24
Walnut	"	0.0658	24
Across the fiber: Beech	"	0.614	24
Chestnut	"	0.325	24
Elm	"	0.443	24
Mahogany	"	0.404	24
Maple	"	0.484	24
Oak	"	0.544	24
Pine	"	0.341	24
Walnut	"	0.484	24
Wax: White	10–26	2.300	25
" "	26–31	3.120	25
" "	31–43	4.860	25
" "	43–57	15.227	25

References:

(1) Smeaton.
(2) Various.
(3) Fizeau.
(4) Matthiessen.
(5) Daniell.
(6) Benoit.
(7) Kohlrausch.
(8) Pfaff.
(9) Deluc.
(10) Lavoisier and Laplace.
(11) Pulfrich.
(12) Schott.
(13) Henning.
(14) Russner.
(15) Mean.
(16) Stadthagen.
(17) Fröhlich.
(18) Rodwell.
(19) Braun.
(20) Deville and Troost.
(21) Scheel.
(22) Mayer.
(23) Glatzel.
(24) Villari.
(25) Kopp.
(26) Randall.
(27) Dorsey.

CUBICAL EXPANSION OF SOLIDS.

If v_2 and v_1 are the volumes at t_2 and t_1 respectively, then $v_2 = v_1 (1 + C\Delta t)$, C being the coefficient of cubical expansion and Δt the temperature interval. Where only a single temperature is stated C represents the true coefficient of cubical expansion at that temperature.*

Substance.	t or Δt	$C \times 10^4$	Authority.
Antimony	0–100	0.3167	Matthiessen
Beryl	0–100	0.0105	Pfaff
Bismuth	0–100	0.3948	Matthiessen
Copper	0–100	0.4998	"
Diamond	40	0.0354	Fizeau
Emerald	40	0.0168	"
Galena	0–100	0.558	Pfaff
Glass, common tube	0–100	0.276	Regnault
" hard	0–100	0.214	"
" Jena, borosilicate 59 III	20–100	0.156	Scheel
" pure silica	0–80	0.0129	Chappuis
Gold	0–100	0.4411	Matthiessen
Ice	—20——1	1.1250	Brunner
Iron	0–100	0.3550	Dulong and Petit
Lead	0–100	0.8399	Matthiessen
Paraffin	20	5.88	Russner
Platinum	0–100	0.265	Dulong and Petit
Porcelain, Berlin	20	0.0814	Chappuis and Harker
Potassium chloride	0–100	1.094	Playfair and Joule
" nitrate	0–100	1.967	" " "
" sulphate	20	1.0754	Tutton
Quartz	0–100	0.3840	Pfaff
Rock salt	50–60	1.2120	Pulfrich
Rubber	20	4.87	Russner
Silver	0–100	0.5831	Matthiessen
Sodium	20	2.13	E. Hazen
Stearic acid	33.8–45.5	8.1	Kopp
Sulphur, native	13.2–50.3	2.23	"
Tin	0–100	0.6889	Matthiessen
Zinc	0–100	0.8928	"

* For tables of cubical expansion complete to 1876, see Clark's Constants of Nature, Smithsonian Collections, 289.

Table 241.

CUBICAL EXPANSION OF LIQUIDS.

If V_o is the volume at 0° then at t° the expansion formula is $V_t = V_o (1 + \alpha t + \beta t^2 + \gamma t^3)$. The table gives values of α, β and γ and of C, the true coefficient of cubical expansion, at 20° for some liquids and solutions. Δt is the temperature range of the observation and A the authority.

Liquid.	Δt	$\alpha\ 10^3$	$\beta\ 10^6$	$\gamma\ 10^8$	$C\ 10^3$ at 20°	A
Acetic acid	16–107	1.0630	0.12636	1.0876	1.071	3
Acetone	0–54	1.3240	3.8090	—0.87983	1.487	3
Alcohol:						
Amyl	—15–80	0.9001	0.6573	1.18458	0.902	4a
Ethyl, 30% by vol.	18–39	0.2928	10.790	—11.87	–	6
" 50% "	0–39	0.7450	1.85	0.730	–	6
" 99.3% "	27–46	1.012	2.20	–	1.12	6
" 500 atmo. press.	0–40	0.866	–	–	–	1
" 3000 " "	0–40	0 524	–	–	–	1
Methyl	0–61	1.1342	1.3635	0.8741	1.199	5a
Benzene	11–81	1.17626	1.27776	0.80648	1.237	5a
Bromine	0–59	1.06218	1.87714	—0.30854	1.132	2
Calcium chloride:						
5.8% solution	18–25	0.07878	4.2742	–	0.250	7
40.9% "	17–24	0.42383	0.8571	–	0.458	7
Carbon disulphide	—34–60	1.13980	1.37065	1.91225	1.218	4a
500 atmos. pressure	0–50	0.940	–	–	–	1
3000 " "	0–50	0.581	–	–	–	1
Carbon tetrachloride	0–76	1.18384	0.89881	1.35135	1.236	4b
Chloroform	0–63	1.10715	4.66473	—1.74328	1.273	4b
Ether	—15–38	1.51324	2.35918	4.00512	1.656	4a
Glycerine	–	0.4853	0.4895	–	0.505	8
Hydrochloric acid:						
33.2% solution	0–33	0.4460	0.215	–	0.455	9
Mercury	0–100	0.18182	0.0078	–	0.18186	13
Olive oil	–	0.6821	1.1405	—0.539	0.721	10
Pentane	0–33	1.4646	3.09319	1.6084	1.608	14
Potassium chloride:						
24.3% solution	16–25	0.2695	2.080	–	0.353	7
Phenol	36–157	0.8340	0.10732	0.4446	1.090	11
Petroleum:						
Density 0.8467	24–120	0.8994	1.396	–	0.955	12
Sodium chloride:						
20.6% solution	0–29	0 3640	1.237	–	0.414	9
Sodium sulphate:						
24% solution	11–40	0.3599	1.258	–	0.410	9
Sulphuric acid:						
10.9% solution	0–30	0.2835	2.580	–	0.387	9
100.0%	0–30	0.5758	—0.432	–	0.558	9
Turpentine	—9–106	0.9003	1.9595	—0.44998	0.973	5b
Water	0–33	—0.06427	8.5053	—6.7900	0.207	13

Authorities.

1. Amagat: C. R. 105, p. 1120; 1887.
2. Thorpe: Proc. Roy. Soc. 24, p. 283; 1876.
3. Zander: Lieb. Ann. 225, p. 109; 1884.
4. Pierre: a. Lieb. Ann. 56, p. 139; 1845.
   b. Lieb. Ann. 80, p. 125; 1851.
5. Kopp: a. Lieb. Ann. 94, p. 257; 1855.
   b. Lieb. Ann. 93, p. 129; 1855.
6. Recknagel: Sitzber. bayr. Ak. p. 327, 2 Abt.; 1866.
7. Drecker: Wied. Ann. 34, p. 952; 1888.
8. Emo: Ber. Chem. Ges. 16, 1857; 1883.
9. Marignac: Lieb. Ann., Supp. VIII, p. 335; 1872.
10. Spring: Bull. Brux. (3) 3, p. 331; 1882.
11. Pinette: Lieb. Ann. 243, p. 32; 1888.
12. Frankenheim: Pogg. Ann. 72, p. 422; 1847.
13. Scheel: Wiss. Abh. Reichsanstalt, 4, p. 1; 1903.
14. Thorpe and Jones: J. Chem. Soc. 63, p. 273; 1893.

TABLE 242.

COEFFICIENTS OF THERMAL EXPANSION.

Coefficients of Expansion of Gases.

Pressures are given in centimeters of mercury.

Coefficient at Constant Volume.

Substance.	Pressure cm.	Coefficient × 100.	Reference.
Air	.6	.37666	1
"	1.3	.37172	"
"	10.0	.36630	"
"	25.4	.36580	"
"	75.2	.36660	"
" 0°–100° . .	100.1	.36744	2
"	76.0	.36650	3
"	200.0	.36903	"
"	2000.	.38866	"
"	10000.	.4100	"
Argon . . .	51.7	.3668	4
Carbon dioxide .	76.0	.36856	3
" " . .	1.8	.36753	1
" " . .	5.6	.36641	"
" " . .	74.9	.37264	"
" " 0°–20°	51.8	.36985	2
" " 0°–40°	51.8	.36972	"
" " 0°–100°	51.8	.36981	"
" " 0°–20°	99.8	.37335	"
" " 0°–100°	99.8	.37262	"
" " 0°–100°	100.0	.37248	5
Carbon monoxide .	76.	.36667	3
Helium . . .	56.7	.3665	4
Hydrogen 16°–132°	.0077	.3328	6
" 15°–132°	.025	.3623	"
" 12°–185°	.47	.3656	"
" . . .	.93	.37002	1
" . . .	11.2	.36548	"
" . . .	76.4	.36504	"
" 0°–100°	100.0	.36626	2
Nitrogen 13°–132°	.06	.3021	6
" 9°–133°	.53	.3290	"
" 0°–20°	100.2	.36754	2
" 0°–100°	100.2	.36744	"
" . . .	76.	.36682	7
Oxygen 11°–132° .	.007	.4161	6
" 9°–132° .	.25	.3984	"
" 11°–132° .	.51	.3831	"
"	1.9	.36683	8
"	18.5	.36690	"
"	75.9	.36681	"
Nitrous oxide .	76.	.3676	3
Sulph'r dioxide SO_2	76.	.3845	"

Coefficient at Constant Pressure.

Substance.	Pressure cm.	Coefficient × 100.	Reference.
Air . . .	76.	.3671	3
"	257.	.3693	"
" 0°–100° . .	100.1	.36728	2
Hydrogen 0°–100°	100.0	.36600	"
" . . .	200 Atm.	.332	9
" . . .	400 "	.295	"
" . . .	600 "	.261	"
" . . .	800 "	.242	"
Carbon dioxide .	76.	.3710	3
" " 0°–20°	51.8	.37128	2
" " 0°–40°	51.8	.37100	"
" " 0°–100°	51.8	.37073	"
" " 0°–20°	99.8	.37602	"
" " 0°–100°	99.8	.37410	"
" " 0°–20°	137.7	.37972	"
" " 0°–100°	137.7	.37703	"
" " 0°–7.5°	2621.	.1097	6
" " 64°–100°	2621.	.6574	"
Carbon monoxide .	76.	.3669	3
Nitrous oxide .	76.	.3719	"
Sulphur dioxide .	76.	.3903	"
" " .	98.	.3980	"
Water-vapor 0°–119°	76.	.4187	10
Water-vapor 0°–141°	76.	.4189	"
Water-vapor 0°–162°	76.	.4071	"
Water-vapor 0°–200°	76.	.3938	"
Water-vapor 0°–247°	76.	.3799	"

Thomson has given, Encyc. Brit. "Heat," the following for the calculation of the expansion, E, between 0° and 100° C. Expansion is to be taken as the change of volume under constant pressure:

Hydrogen, $E = .3662(1 - .00049\,V/v)$,
Air, $E = .3662(1 - .0026\;V/v)$,
Oxygen, $E = .3662(1 - .0032\;V/v)$,
Nitrogen, $E = .3662(1 - .0031\;V/v)$,
CO_2 $E = .3662(1 - .0164\;V/v)$.

V/v is the ratio of the actual density of the gas at 0° C to what it would have at 0° C and 1 Atm. pressure.

1 Meleander, Wied. Beibl. 14, 1890; Wied. Ann. 47, 1892.
2 Chappuis, Trav. Mem. Bur. Intern. Wts. Meas. 13, 1903.
3 Regnault, Ann. chim. phys. (3) 5, 1842.
4 Keunen-Randall, Proc. R. Soc. 59, 1896.
5 Chappuis, Arch. sc. phys. (3), 18, 1892.
6 Baly-Ramsay, Phil. Mag. (5), 38, 1894.
7 Andrews, Proc. Roy. Soc. 24, 1876.
8 Meleander, Acta Soc. Fenn. 19, 1891.
9 Amagat, C. R. 111, 1890.
10 Hirn, Théorie méc. chaleur, 1862.

TABLE 243.

SPECIFIC HEAT OF THE CHEMICAL ELEMENTS.

Element.	Range * of temperature, ° C	Specific heat.	Reference.
Aluminum	−240.6	.0092	45
"	−190.0	.0889	45
"	−73.0	.190	46
"	−190 to −82	.1466	47
"	−76 to −1	.1962	47
"	+16 to +100	.2122	48
"	+16 to +304	.2250	48
"	−250	.1428	1
"	0	.2089	1
"	100	.2226	1
"	250	.2382	1
"	500	.2739	1
"	16–100	.2122	43
Antimony	15	.0489	2
"	100	.0503	2
"	200	.0520	2
Arsenic, gray	0–100	.0822	3
Arsenic, black	0–100	.0861	3
Barium	−185 to +20	.068	4
Bismuth	−186	.0284	5
"	0	.0301	6
"	75	.0309	6
"	20–100	.0302	7
" fluid	280–380	.0363	8
Boron	0–100	.307	9
"	−191 to −78	.0707	47
"	−76 to −0	.1677	47
Bromine, solid	−78 to −20	.0843	10
" solid	−192 to −80	.0702	49
" fluid	13–45	.107	11
Cadmium	−223	.0308	46
"	−173	.0478	46
"	−73	.0533	46
"	21	.0551	2
"	100	.0570	2
"	200	.0594	2
"	300	.0617	2
Cæsium	0–26	.0482	12
Calcium	−185 to +20	.157	4
"	0–181	.170	13
Carbon, graphite	−191 to −79	.0573	47
" "	−76 to −0	.1255	47
" "	−50	.114	14
" "	+11	.160	14
" "	977	.467	14
" "	1730	.50	52
Acheson	−244	.005	50
	−186	.027	50
Carbon, diamond	−50	.0635	47
" "	+11	.113	47
" "	985	.459	47
Cerium	0–100	.0448	15
Chlorine, liquid	0–24	.2262	16
Chromium	−200	.0666	17
"	0	.1039	17
"	100	.1121	17
"	600	.1872	17
"	−185 to +20	.086	4

Element.	Range * of temperature, ° C	Specific heat.	Reference.
Cobalt	500	.1452	18
"	1000	.204	18
"	−182 to +15	.0822	19
"	15–100	.1030	19
Copper †	−249.5	.0035	45
"	−223	.0208	46
"	−185	.0532	45
"	−63	.0865	46
"	+25	.0917	44
"	76	.0937	51
"	84	.0938	51
"	100	.0942	2
"	362	.0997	51
"	900	.1259	20
"	15–238	.0951	43
"	−181 to 13	.0868	21
"	23–100	.0940	21
Gallium, liquid	12 to 113	.080	22
" solid	12–23	.079	22
Germanium	0–100	.0737	23
Gold	−185 to +20	.033	4
"	0–100	.0316	24
Indium	0–100	.0570	13
Iodine	−90 to +17	.0485	49
"	−191 to −80	.0454	49
"	9–98	.0541	25
Iridium	−186 to +18	.0282	26
"	18–100	.0323	26
Iron	−223	.0176	46
"	−163	.0622	46
"	−63	.0961	46
"	+37	.1092	46
" cast	20–100	.1189	27
" wrought	15–100	.1152	28
" wrought	1000–1200	.1989	28
" wrought	500	.176	28
" hard-drawn	0–18	.0986	29
" hard-drawn	20–100	.1146	29
"	−185 to +20	.0958	4
"	0 to +200	.1175	53
"	0 to +300	.1233	53
"	0 to +400	.1282	53
"	0 to +500	.1338	53
"	0 to +600	.1396	53
"	0 to +700	.1487	53
"	0 to +800	.1597	53
"	0 to +900	.1644	53
"	0 to +1000	.1557	53
"	0 to +1100	.1534	53
Lanthanum	0–100	.0448	15
Lead	−250	.0143	46
"	−236	.0217	46
"	−193	.0276	46
"	−73	.0295	46
"	15	.0299	2
"	100	.0311	2
"	300	.0338	2
" fluid	310	.0356	30

* When one temperature is given, the "true" specific heat is indicated, otherwise the "mean" specific heat.

† $0.3834 + 0.00020(t - 25)$ intern. j per g degree $= 0.0917 + 0.000048(t - 25)$ cal_{20} per g degree. (Griffith, 1913.)

Table 243 (*continued*).

SPECIFIC HEAT OF THE CHEMICAL ELEMENTS.

Element.	Range* of temperature, °C	Specific heat.	Reference.
Lead	90	0.0312	51
"	210	0.0334	51
"	18–100	0.0310	43
"	16–256	0.0319	43
Lithium	−191 to −80	0.521	47
"	−78 to 0	0.595	47
"	−75 to +19	0.629	47
"	−100	0.5997	31
"	0	0.7951	31
"	50	0.9063	31
"	100	1.0407	31
"	190	1.3745	31
Magnesium	−185 to +20	0.222	4
"	60	0.2492	7
"	325	0.3235	7
"	625	0.4352	7
"	20–100	0.2492	7
Manganese	−188 to −79	0.0820	49
"	−79 to +15	0.1091	49
"	60	0.1211	49
"	325	0.1783	49
"	20–100	0.1211	49
"	−100	0.0979	31
"	0	0.1072	31
"	100	0.1143	31
Mercury, sol.	−77 to −42	0.0329	47
" liq.	−36 to −3	0.0334	47
"	−185 to +20	0.032	4
"	0	0.03346	32
"	85	0.0328	32
"	100	0.03284	2
"	250	0.03212	2
Molybdenum	−185 to +20	0.062	4
"	60	0.0647	7
"	475	0.0750	7
"	20 to 100	0.0647	7
Nickel	−185 to +20	0.092	4
"	100	0.1128	18
"	300	0.1403	18
"	500	0.1299	18
"	1000	0.1608	18
"	18–100	0.109	26
Osmium	19–98	0.0311	10
Palladium	−186 to +18	0.0528	26
"	0–100	0.0592	24
"	0–1265	0.0714	24
Phosphorus, red	0–51	0.1829	33
" yellow.	13–36	0.202	33
" yellow.	−186 to +20	0.178	4
Platinum	−186 to +18	0.0293	26
"	100	0.0275	34
"	200	0.0330	35
"	500	0.0349	35
"	750	0.0365	35
"	1000	0.0381	35
"	1300	0.0400	35
"	20–100	0.0319	35
"	20–500	0.0333	35
"	20–1000	0.0346	35
"	20–1300	0.0359	35
Potassium	−191 to −80	0.1568	47
"	−78 to 0	0.1666	47
"	−185 to +20	0.170	4
Rhodium	10–97	0.0580	25
Rubidium	0	0.0802	—
Ruthenium	0–100	0.0611	13
Selenium	−188 to +18	0.068	36
Silicon	−185 to +20	0.123	4
"	−39.8	0.1360	14
"	+57.1	0.1833	14
"	232	0.2029	14
Silver	−238	0.0146	46
"	−213	0.0307	46
"	−173	0.0447	46
"	−73	0.0540	46
"	+27	0.0560	46
"	0–100	0.0559	13
"	23	0.05498	2
"	100	0.05663	2
"	500	0.0581	34
"	17–507	0.05987	43
"	800	0.076	18
" fluid	907–1100	0.0748	18
Sodium	−185 to +20	0.253	4
"	−191 to −83	0.243	47
"	−77 to 0	0.276	47
"	−223	0.152	46
"	−183	0.219	46
Sulphur	−188 to +18	0.137	36
" rhombic.	0–54	0.1728	33
" monoclin.	0–52	0.1809	33
" liquid	119–147	0.235	2
Tantalum	−185 to +20	0.033	4
"	1400	0.043	—
Tellurium	−188 to +18	0.047	36
" crys.	15–100	0.0483	37
Thallium	−185 to +20	0.038	4
"	20–100	0.0326	27
Thorium	0–100	0.0276	38
Tin	−196 to −79	0.0486	26
"	−76 to +18	0.0518	26
" cast	21–109	0.0551	30
" fluid	250	0.05799	18
" fluid	1100	0.0758	18
Titanium	−185 to +20	0.082	4
"	0–100	0.1125	39
Tungsten	−185 to +20	0.036	4
"	0–100	0.0336	40
"	1000	0.0337	52
"	2000	0.042	52
"	2400	0.045	52
Uranium	0–98	0.028	41
Vanadium	0–100	0.1153	40
Zinc	−243	0.0144	46
"	−193	0.0625	46
"	−153	0.0788	46
"	20–100	0.0931	27
"	100	0.0951	2
"	300	0.1040	2
Zirconium	0–100	0.0660	42

* When one temperature is given, the "true" specific heat is indicated, otherwise the "mean" specific heat. See page 226 for references.

HEAT CAPACITIES, TRUE AND MEAN SPECIFIC HEATS, AND LATENT HEATS AT FUSION.

The following data are taken from a research and discussion entitled "Die Temperatur-Wärmeinhaltskurven der technisch wichtigen Metalle," Wüst, Meuthen und Durrer, Forschungsarbeiten herausgegeben vom Verein Deutscher Ingenieure, Springer, Heft 204, 1918.

(*a*) There follow the constants of the equation for the heat capacity: $W = a + bt + ct^2$; for the mean specific heat: $s = at^{-1} + b + ct$; and for the true specific heat: $s' = b + 2ct$; also the latent heats at fusion. (See also Table 243, pp. 223–224.)

Element.	Temperature range. °C	*a*	*b*	$c \times 10^6$	Latent heat. cal./g	Element.	Temperature range. °C	*a*	*b*	$c \times 10^6$	Latent heat cal./g.
Cr	0–1500	—	0.10233	33.47	—	Ag	0–961	—	0.05725	5.48	26.0
Mo	0–1500	—	0.06162	10.99	—		961–1300	53.17	0.00710	28.30	—
W	0–1500	—	0.03325	1.07	—	Au	0–1064	—	0.03171	1.30	15.9
Pt	0–1500	—	0.03121	3.54	—		1064–1300	26.35	0.01420	8.52	—
Sn	0–232	—	0.06829	—	13.8	Cu	0–1084	—	0.10079	3.05	41.0
	232–1000	14.33	0.07020	−18.30	—		1084–1300	130.74	−.04150	65.6	—
Bi	0–270	—	0.03141	5.22	10.2	Mn	0–1070	—	0.12037	25.41	36.6
	270–1000	10.31	0.03107	5.41	—		1130–1210	−7.41	0.17700	—	24.14*
Cd	0–321	—	0.05550	6.28	10.8		1230–1250	3.83	0.19800	—	—
	321–1000	6.30	0.06952	6.37	—	Ni	0–320	—	0.10950	52.40	56.1
Pb	0–327	—	0.03591	−11.47	5.47		330–1451	0.41	0.12931	0.11	1.33*
	327–1000	6.07	0.02920	3.30	—		1451–1520	50.21	0.13380	—	—
Zn	0–419	—	0.08777	43.48	23.0	Co	0–950	—	0.09119	40.77	58.2
	419–1000	14.34	0.13340	−16.10	—		1100–1478	22.00	0.11043	14.57	14.70*
Sb	0–630	—	0.05179	3.00	38.9		1478–1600	57.72	0.14720	—	—
	630–1000	39.42	0.05090	2.96	—	Fe	0–725	—	0.10545	56.84	49.4
Al	0–657	—	0.22200	38.57	94.0		785–919	−1.63	0.1592	—	6.56*
	657–1000	102.39	0.21870	24.00	—		919–1404	18.31	0.14472	0.05	6.67*
							1405–1528	−77.18	0.21416	—	1.94*
							1528–1600	70.03	0.15012	—	—

* Allotropic heat of transformation: Mn, 1070–1130°; Ni, 320–330°; Co, 950–1100°; Fe, 725–785°; 919° ± 1; 1404.5° ± 0.5.

(*b*) TRUE SPECIFIC HEATS.

°C	Pb	Zn	Al	Ag	Au	Cu	Ni	Fe	Co	Quartz.
0° C	0.0359	0.0878	0.2220	0.0573	0.0317	0.1008	0.1095	0.1055	0.0912	—
100	0.0336	0.0965	0.2297	0.0583	0.0320	0.1014	0.1200	0.1168	0.0993	0.2372
200	0.0313	0.1052	0.2374	0.0594	0.0322	0.1020	0.1305	0.1282	0.1073	0.2416
300	0.0290	0.1139	0.2451	0.0605	0.0325	0.1026	0.1409	0.1396	0.1154	0.2460
400	0.0266	0.1226	0.2529	0.0616	0.0328	0.1032	0.1294	0.1509	0.1235	0.2504
500	0.0259	0.1173	0.2606	0.0627	0.0330	0.1038	0.1294	0.1623	0.1316	0.2548
600	0.0252	0.1141	0.2683	0.0638	0.0333	0.1045	0.1294	0.1737	0.1396	0.2592
700	0.0246	0.1109	0.2523	0.0649	0.0335	0.1051	0.1295	0.1850	0.1477	0.2636
800	0.0239	0.1076	0.2571	0.0660	0.0338	0.1057	0.1295	0.1592	0.1558	0.2680
900	0.0233	0.1044	0.2619	0.0671	0.0341	0.1063	0.1295	0.1592	0.1639	0.2724
1000	0.0226	0.1012	0.2667	0.0637	0.0343	0.1069	0.1295	0.1448	—	0.2768
1100	—	—	—	0.0694	0.0329	0.1028	0.1296	0.1448	0.1424	0.2812
1200	—	—	—	0.0750	0.0346	0.1159	0.1296	0.1448	0.1454	0.2856
1300	—	—	—	0.0807	0.0364	0.1291	0.1296	0.1449	0.1483	0.2900
1400	—	—	—	—	—	—	0.1296	0.1449	0.1512	0.2944
1500	—	—	—	—	—	—	0.1338	0.2142	0.1472	0.2988
1600	—	—	—	—	—	—	—	0.1501	0.1472	—

For more elaborate tables and for all the elements in upper table, see original reference.

ATOMIC HEATS (50° K). SPECIFIC HEATS (50° K). ATOMIC VOLUMES OF THE ELEMENTS.

The atomic and specific heats are due to Dewar, Pr. Roy. Soc. 89A, 168, 1913.

Element.	Specific heat −223° C.	Atomic heat −223°C.	Atomic volume.	Element.	Specific heat −223° C.	Atomic heat −223°C.	Atomic volume.	Element.	Specific heat −223° C.	Atomic heat −223°C.	Atomic volume.
Li	0.1924	1.35	13.0	Cr	0.0142	0.70	7.6	Sn	0.0286	3.41	20.3
Gl	0.0137	0.125	4.9	Mn	0.0229	1.26	7.4	Sb	0.0240	2.89	18.2
B	0.0212	0.24	4.5	Fe	0.0175	0.98	7.1	I	0.0361	4.59	25.7
C *	0.0137	0.16	5.1	Ni	0.0208	1.22	6.7	Te	0.0288	3.68	21.2
C †	0.0028	0.03	3.4	Co	0.0207	1.22	6.8	Cs	0.0513	6.82	71.0
Na	0.1519	3.50	23.6	Cu	0.0245	1.56	7.1	Ba¶	0.0350	4.80	36.6
Mg	0.0713	1.74	14.1	Zn	0.0384	2.52	9.2	La	0.0322	4.60	22.6
Al	0.0413	1.12	10.0	As	0.0258	1.94	15.9	Ce	0.0330	4.64	20.3
Si ‡	0.0303	0.86	14.2	Se	0.0361	2.86	18.5	W	0.0095	1.75	9.8
Si §	0.0303	0.77	11.4	Br	0.0453	3.62	24.9	Os	0.0078	1.49	8.5
P				Rb	0.0711	6.05	55.8	Ir	0.0099	1.92	8.6
yel.	0.0774	2.40	17.0	Sr¶	0.0550	4.82	34.5	Pt	0.0135	2.63	9.2
P				Zr	0.0262	2.38	21.8	Au	0.0160	3.16	10.2
red	0.0431	1.34	13.5	Mo	0.0141	1.36	9.3	Hg	0.0232	4.65	14.8
S	0.0546	1.75	16.	Ru	0.0109	1.11	9.0	Tl	0.0235	4.80	17.2
Cl	0.0967	3.43	24.6	Rh	0.0134	1.38	8.5	Pb	0.0240	4.96	18.3
K	0.1280	5.01	44.7	Pd	0.0190	2.03	9.2	Bi	0.0218	4.54	21.3
Ca	0.0714	2.86	25.9	Ag	0.0242	2.62	10.2	Th	0.0197	4.58	21.1
Ti	0.0205	0.99	10.7	Cd	0.0308	3.46	13.0	U	0.0138	3.30	12.8

* Graphite. † Diamond. ‡ Fused. § Crystallized. ¶ Impure.

References to Table 243:

(1) Bontschew.
(2) Naccari, Atti Torino, 23, 1887–88.
(3) Wigand, Ann. d. Phys. (4) 22, 1907.
(4) Nordmeyer-Bernouli, Verh. d. phys. Ges. 9, 1907; 10, 1908.
(5) Giebe, Verh. d. phys. Ges. 5, 1903.
(6) Lorenz, Wied. Ann. 13, 1881.
(7) Stücker, Wien. Ber. 114, 1905.
(8) Person, C. R. 23, 1846; Ann. d. chim. (3) 21, 1847; 24, 1848.
(9) Moisson-Gautier, Ann. chim. phys. (7) 17, 1896.
(10) Regnault, Ann. d. chim. (3) 26, 1849; 63, 1861.
(11) Andrews, Pog. Ann. 75, 1848.
(12) Eckardt-Graefe, Z. Anorg. Ch. 23, 1900.
(13) Bunsen, Pogg. Ann. 141, 1870; Wied. Ann. 31, 1887.
(14) Weber, Phil. Mag. (4) 49, 1875.
(15) Hillebrand, Pog. Ann. 158, 1876.
(16) Knietsch.
(17) Adler, Beibl. 27, 1903.
(18) Pionchon, C. R. 102–103, 1886.
(19) Tilden, Phil. Trans. (A) 201, 1903.
(20) Richards, Ch. News, 68, 1893.
(21) Trowbridge, Science, 8, 1898.
(22) Berthelot, Ann. d. chim. (5) 15, 1878.
(23) Pettersson-Hedellius, J. Pract. Ch. 24, 1881.
(24) Violle, C. R. 85, 1877; 87, 1878.
(25) Regnault, Ann. d. chim (2) 73, 1840; (3) 63, 1861.
(26) Behn, Wied. Ann. 66, 1898; Ann. d. Phys. (4) 1, 1900.
(27) Schmitz, Pr. Roy. Soc. 72, 1903.
(28) Nichol, Phil. Mag. (5) 12, 1881.
(29) Hill, Verh. d. phys. Ges. 3, 1901.
(30) Spring, Bull. de Belg. (3) 11, 1886; 29, 1895.
(31) Laemmel, Ann. d. Phys. (4) 16, 1905.
(32) Barnes-Cooke, Phys. Rev. 16, 1903.
(33) Wiegand, Fort. d. Phys. 1906.
(34) Tilden, Pr. Roy. Soc. 66, 1900; 71, 1903; Phil. Trans. (A) 194, 1900; 201, 1903.
(35) White, Phys. Rev. 12, 436, 1918.
(36) Dewar, Ch. News, 92, 1905.
(37) Kopp, Phil. Trans. London, 155, 1865.
(38) Nilson, C. R. 96, 1883.
(39) Nilson-Pettersson, Zt. phys. Ch. 1, 1887.
(40) Mache, Wien, Ber. 106, 1897.
(41) Blümcke, Wied. Ann. 24, 1885.
(42) Mixter-Dana, Lieb Ann. 169, 1873.
(43) Magnus, Ann. d. Phys. 31, 1910.
(44) Harper, Bull. Bureau of Stds. 11, p. 259, 1914.
(45) Nernst, Lindemann, 1910, 1911.
(46) Nernst, Dewar.
(47) Kosef, Ann. d. Phys. 36, 1911.
(48) Magnus, Ann. d. Phys. 31, 1910.
(49) Estreicher, Straniewski, 1912.
(50) Nernst, Ann. d. Phys. 36, 395, 1911.
(51) King, Phys. Rev. 11, 1918.
(52) Worthing, Phys. Rev. 12, 1918.
(53) Harker, Pr. Phys. Soc. London, 19, 703, 1905; Fe .01; C .02; Si .03; S .04; P, Mn trace.

TABLE 246.—Specific Heat of Various Solids.

Solid.	Temperature °C.	Specific heat.	Authority.
Alloys:			
Bell metal	15–98	0.0858	R
Brass, red	0	.08991	L
" yellow	0	.08831	"
80 Cu + 20 Sn	14–98	.0862	R
88.7 Cu + 11.3 Al	20–100	.10432	Ln
German silver	0–100	.09464	T
Lipowitz alloy: 24.97 Pb + 10.13 Cd + 50.66 Bi + 14.24 Sn	5–50	.0345	M
" "	100–150	.0426	"
Rose's alloy: 27.5 Pb + 48.9 Bi + 23.6 Sn	—77–20	.0356	S
" "	20–89	.0552	"
Wood's alloy: 25.85 Pb + 6.99 Cd + 52.43 Bi + 14.73 Sn	5–50	.0352	M
" " (fluid)	100–150	.0426	"
Miscellaneous alloys:			
17.5 Sb + 29.9 Bi + 18.7 Zn + 33.9 Sn	20–99	.05657	R
37.1 Sb + 62.9 Pb	10–98	.03880	"
39.9 Pb + 60.1 Bi	16–99	.03165	P
" " (fluid)	144–358	.03500	"
63.7 Pb + 36.3 Sn	12–99	.04073	R
46.7 Pb + 53.3 Sn	10–99	.04507	"
63.8 Bi + 36.2 Sn	20–99	.04001	"
46.9 Bi + 53.1 Sn	20–99	.04504	"
Gas coal	20–1040	.3145	—
Glass, normal thermometer 16^{III}	19–100	.1988	W
" French hard thermometer	–	.1869	Z
" crown	10–50	.161	H M
" flint	10–50	.117	"
Ice	—188– —252	.146	D
"	—78– —188	.285	"
"	—18– —78	.463	"
India rubber (Para)	?–100	.481	G T
Mica	20	.10	—
Paraffin	—20– +3	.3768	R W
"	—19– +20	.5251	"
"	0–20	.6939	"
"	35–40	.622	B
" fluid	60–63	.712	"
Vulcanite	20–100	.3312	A M
Woods	20	.327	—

TABLE 247.—Specific Heat of Water and of Mercury.

Specific Heat of Water.							Specific Heat of Mercury.			
Temperature, °C.	Barnes.	Rowland.	Barnes-Regnault.	Temperature, °C.	Barnes.	Barnes-Regnault.	Temperature, °C.	Specific Heat.	Temperature, °C.	Specific Heat.
−5	1.0155	–	–	60	0.9988	0.9994	0	0.03346	90	0.03277
0	1.0091	1.0070	1.0094	65	.9994	1.0004	5	.03340	100	.03269
+5	1.0050	1.0039	1.0053	70	1.0001	1.0015	10	.03335	110	.03262
10	1.0020	1.0016	1.0023	80	1.0014	1.0042	15	.03330	120	.03255
15	1.0000	1.0000	1.0003	90	1.0028	1.0070	20	.03325	130	.03248
20	0.9987	.9991	0.9990	100	1.0043	1.0101	25	.03320	140	.03241
25	.9978	.9989	.9981	120	–	1.0162	30	.03316	150	.0324
30	.9973	.9990	.9976	140	–	1.0223	35	.03312	170	.0322
35	.9971	.9997	.9974	160	–	1.0285	40	.03308	190	.0320
40	.9971	1.0006	.9974	180	–	1.0348	50	.03300	210	.0319
45	.9973	1.0018	.9976	200	–	1.0410	60	.03294	–	–
50	.9977	1.0031	.9980	220	–	1.0476	70	.03289	–	–
55	.9982	1.0045	.9985	–	–	–	80	.03284	–	–

Barnes's results: Phil. Trans. (A) 199, 1902; Phys. Rev. 15, 1902; 16, 1903. (H thermometer.)
Bousfield, Phil. Trans. A 211, p. 199, 1911. Barnes-Regnault's as revised by Peabody; Steam Tables.
The mercury data from 0° C to 80, Barnes-Cooke (H thermometer); from 90° to 140, mean of Winklemann, Naccari and Milthaler (air thermometer); above 140°, mean of Naccari and Milthaler.

TABLE 248. — Specific Heat of Various Liquids.

Liquid.	Temp. °C.	Spec. heat.	Authority.
Alcohol, ethyl	−20	0.5053	R
" "	0	0.548	"
" "	40	0.648	"
Alcohol, methyl	5–10	0.590	"
" "	15–20	0.601	"
Anilin	15	0.514	G
"	30	0.520	"
"	50	0.529	"
Benzole, C_6H_6	10	0.340	H–D
" "	40	0.423	"
" C_6H_6	65	0.482	"
$CaCl_2$, sp. gr. 1.14	−15	0.764	DMG
" " " "	0	0.775	"
" " " "	+20	0.787	"
" " " 1.20	−20	0.695	"
" " " "	0	0.712	"
" " " "	+20	0.725	"
" " " 1.26	−20	0.651	"
" " " "	0	0.663	"
" " " "	+20	0.676	"
$CuSo_4 + 50\ H_2O$	12–15	0.848	Pa
" + 200 "	12–14	0.951	"
" + 400 "	13–17	0.975	"
Diphenylamine, $C_{12}H_{11}N$	53	0.464	B
"	65	0.482	"
Ethyl ether	0	0.529	R
Glycerine	15–50	0.576	E
$KOH + 30H_2O$	18	0.876	TH
" + 100 "	18	0.975	"
$NaOH + 50H_2O$	18	0.942	"
" + 100 "	18	0.983	"
$NaCl + 10H_2O$	18	0.791	"
" + 200 "	18	0.978	"
Naphthalene, $C_{10}H_8$	80–85	0.396	B
"	90–95	0.409	"
Nitrobenzole	14	0.350	A
"	28	0.362	"
Oils: castor	—	0.434	W
citron	5.4	0.438	HW
olive	6.6	0.471	"
sesame	—	0.387	W
turpentine	0	0.411	R
Petroleum	21–58	0.511	Pa
Sea water, sp. gr. 1.0043	17.5	0.980	"
" " " " 1.0235	17.5	0.938	"
" " " " 1.0463	17.5	0.903	"
Toluol, C_6H_8	10	0.364	H–D
"	65	0.490	"
"	85	0.534	"
$ZnSO_4 + 50\ H_2O$	20–52	0.842	Ma
" + 200 "	20–52	0.952	"

References: (A) Abbot; (B) Batelli; (E) Emo; (G) Griffiths; (DMG) Dickinson, Mueller, and George; (H–D) de Heen and Deruyts; (Ma) Marignac; (Pa) Pagliani; (R) Regnault; (Th) Thomsen; (W) Wachsmuth; (Z) Zouloff; (HW) H. F. Weber.

TABLE 249. — Specific Heat of Liquid Ammonia under Saturation Conditions.

Expressed in $Calories_{20}$ per Gram per Degree C. Osborne and van Dusen, Bul. Bureau of Standards, 1918.

Temp. °C.	0	1	2	3	4	5	6	7	8	9
−40	1.062	1.061	1.060	1.059	1.058	1.058	1.057	1.056	1.055	1.055
−30	1.070	1.069	1.068	1.067	1.066	1.065	1.064	1.064	1.063	1.062
−20	1.078	1.077	1.076	1.075	1.074	1.074	1.073	1.072	1.071	1.070
−10	1.088	1.087	1.086	1.085	1.084	1.083	1.082	1.081	1.080	1.079
− 0	1.099	1.098	1.097	1.096	1.094	1.093	1.092	1.091	1.090	1.089
+ 0	1.099	1.100	1.101	1.103	1.104	1.105	1.106	1.108	1.109	1.110
+10	1.112	1.113	1.114	1.116	1.117	1.118	1.120	1.122	1.123	1.125
+20	1.126	1.128	1.129	1.131	1.132	1.134	1.136	1.137	1.139	1.141
+30	1.142	1.144	1.146	1.148	1.150	1.152	1.154	1.156	1.158	1.160
+40	1.162	1.164	1.166	1.169	1.171	1.173	1.176	1.178	1.181	1.183

TABLE 250. — Heat Content of Saturated Liquid Ammonia.

Heat content = $H = \epsilon + pv$, where ϵ is the internal or intrinsic energy. Osborne and van Dusen, Bul. Bureau of Standards, 1918.

Temperature	−50°	−40°	−30°	−20°	−10°	0°	+10°	+20°	+30°	+40°	+50°
$H = \epsilon + pv$	−53.8	−43.3	−32.6	−21.8	−11.0	0.0	+11.1	+22.4	−33.9	−45.5	−57.4

SPECIFIC HEATS OF MINERALS AND ROCKS.

TABLE 251.—Specific Heat of Minerals and Rocks.

Substance.	Temperature ° C.	Specific Heat.	Reference.
Andalusite	0–100	0.1684	1
Anhydrite, $CaSO_4$	0–100	.1753	1
Apatite	15–99	.1903	2
Asbestos	20–98	.195	3
Augite	20–98	.1931	3
Barite, $BaSO_4$	10–98	.1128	4
Beryl	15–99	.1979	2
Borax, $Na_2B_4O_7$ fused	16–98	.2382	4
Calcite, $CaCO_3$	0–50	.1877	1
" "	0–100	.2005	1
" "	0–300	.2204	1
Cassiterite SnO_2	16–98	.0933	4
Chalcopyrite	15–99	.1291	2
Corundum	9–98	.1976	4
Cryolite, $Al_2F_6.6NaF$	16–99	.2522	2
Fluorite, CaF_2	15–99	.2154	4
Galena, PbS	0–100	.0466	5
Garnet	16–100	.1758	2
Hematite, Fe_2O_3	15–99	.1645	2
Hornblende	20–98	.1952	3
Hypersthene	20–98	.1914	3
Labradorite	20–98	.1949	3
Magnetite	18–45	.156	6
Malachite, $Cu_2CO_4H_2O$	15–99	.1763	2
Mica (Mg)	20–98	.2061	3
" (K)	20–98	.2080	3
Oligoclase	20–98	.2048	3
Orthoclase	15–99	.1877	2
Pyrolusite, MnO_2	17–48	.159	6
Quartz, SiO_2	12–100	.188	7
" "	0	.1737	8
" "	350	.2786	8
" "	400–1200	.305	8

Substance.	Temperature ° C.	Specific Heat.	Reference.
Rock-salt	13–45	0.219	6
Serpentine	16–98	.2586	2
Siderite	9–98	.1934	4
Spinel	15–47	.194	6
Talc	20–98	.2092	3
Topaz	0–100	.2097	1
Wollastonite	19–51	.178	6
Zinc blende, ZnS	0–100	.1146	1
Zircon	21–51	.132	6
Rocks:			
Basalt, fine, black	12–100	.1996	6
" " "	20–470	.199	9
" " "	470–750	.243	9
" " "	750–880	.626	9
" " "	880–1190	.323	9
Dolomite	20–98	.222	3
Gneiss	17–99	.196	10
"	17–213	.214	10
Granite	12–100	.192	7
Kaolin	20–98	.224	3
Lava, Aetna	23–100	.201	11
" "	31–776	.259	11
" Kilauea	25–100	.197	11
Limestone	15–100	.216	12
Marble	0–100	.21	–
Quartz sand	20–98	.191	3
Sandstone	–	.22	–

1 Lindner. 2 Oeberg. 3 Ulrich. 4 Regnault. 5 Tilden. 6 Kopp. 7 Joly. 8 Pionchon. 9 Roberts-Austen, Rücker. 10 R. Weber. 11 Bartoli. 12 Morano.

Compiled from Landolt-Börnstein-Meyerhoffer's Physikalisch-chemische Tabellen.

TABLE 252.—Specific Heats of Silicates.

Silicate.	Mean specific heats. 0° C to				True specific heats. at				
	100°	500°	900°	1400°	0°C	100°	500°	1000°	1300°
Albite	.1948	.2363	.2561	–	.178	.211	.269	.294	–
" glass	.1977	.2410	.2640	–	–	–	–	–	–
Amphibole, Mg. silicate	.2033	.2461	.2661	.2731*	.185	.219	.279	.304	–
" glass	.2040	.2474	–	–	–	–	–	–	–
Andesine	.1925	.2330	.2525	–	–	–	.265	–	–
" glass	.1934	–	.2615	–	–	–	–	–	–
Anorthite	.1901	.2296	02481	.2674	.174	.205	.260	.286	.318
" glass	.1883	.2305	–	–	–	–	–	–	–
Cristobalite	.1883	.2426	.2568	.2680	–	–	–	–	–
Diopside	.1924	.2314	.2500	.2604†	.176	.207	.262	.284	–
" glass	.1939	.2332	–	–	–	–	–	–	–
Microcline	.1871	.2262	.2450	–	.171	.201	.258	.279	–
" glass	.1919	.2321	.2514	.2598*	.176	.206	.264	.299	–
Pyroxene	.2039	.2484	–	–	–	–	–	–	–
Quartz	.1868	.2379	.2596	.2640*	.168	.204	.294	.285	–
Silica glass	.1845	.2302	.2512	–	.166	.202	.266	.29	–
Wollastonite	–	–	.2344	–	–	–	–	–	–
" glass	.1852	.2206	–	–	–	–	–	–	–
" pseudo	.1844	.2170	.2324	.2448	.171	.197	.243	.262	.272

*0°–1100°; †0°–1250°; Taken from White, Am. J. Sc. 47, 1, 1919.

TABLE 253.

SPECIFIC HEATS OF GASES AND VAPORS.

Substance.	Range of temp. °C	Sp. ht. constant pressure.	Authority.	Range of temp. °C	Mean ratio of specific heats. C_p/C_v.	Authority.
Acetone, C_3H_6O	26–110	0.3468	Wiedemann.			
Air	−30–+10	0.2377	Regnault.	20	1.4011	Moody.
“	0–200	0.2375	“	−79.3	1.405	Koch, 1907.
“	20–440	0.2366	Holborn and	−79.3	2.333	“ 200 atm
“	20–630	0.2429	Austin.	0	1.828	“ “ “
“	20–800	0.2430	“	500	1.399	Fürstenau.
Alcohol, C_2H_5OH	108–220	0.4534	Regnault.	53	1.133	Jaeger.
“ “	—	—	—	100	1.134	Stevens.
“ CH_3OH	101–223	0.4580	Regnault.	100	1.256	“
Ammonia	23–100	0.5202	Wiedemann.	0	1.3172	Wüllner.
“	27–200	0.5356	“	100	1.2770	“
Argon	20–90	0.1233	Dittenberger.	0	1.667	Niemeyer.
Benzene, C_6H_6	34–115	0.2990	Wiedemann.	20	1.403	Pagliani.
“ “	35–180	0.3325	“	60	1.403	“
“ “	116–218	0.3754	Regnault.	99.7	1.105	Stevens.
Bromine	83–228	0.0555	“	20–388	1.293	Strecker.
Carbon dioxide, CO_2	−28– +7	0.1843	“	4–11	1.2995	Lummer and Pringsheim.
“ “ “	15–100	0.2025	“			
“ “ “	11–214	0.2169	“	0	1.3003	Moody, 1912.
“ monoxide, CO	23–99	0.2425	Wiedemann.	0	1.403	Wüllner.
“ “ “	26–198	0.2426	“	100	1.395	“
“ disulphide, CS_2	86–190	0.1596	Regnault.	3–67	1.205	Beyme.
Chlorine	16–343	0.1125	Strecker.	0	1.336	Martini.
Chloroform, $CHCl_3$	27–118	0.1441	Wiedemann.	22–78	1.102	Beyme.
“ “	28–189	0.1489	“	99.8	1.150	Stevens.
Ether, $C_4H_{10}O$	69–224	0.4797	Regnault.	42–45	1.029	Müller.
“ “	25–111	0.4280	Wiedemann.	12–20	1.024	Low, 1894.
Helium	—	—	—	0	1.64	Mean, Jeans.
Hydrochloric acid, HCl	13–100	0.1940	Strecker.	20	1.389	Strecker.
“ “ “	22–214	0.1867	Regnault.	100	1.400	“
Hydrogen	−28– +9	3.3996	“	4–16	1.4080	Lummer and Pringsheim.
“	12–198	3.4090	“			
“	21–100	3.4100	Wiedemann.	—	1.419	Hartmann.
“ sulphide, H_2S	20–206	0.2451	Regnault.	—	1.324	Capstick.
Krypton	—	—	—	19	1.666	Ramsay, '12.
Mercury	—	—	—	310	1.666	Kundt and Warburg.
Methane, CH_4	18–208	0.5929	Regnault.	11–30	1.316	Müller.
Neon	—	—	—	19	1.642	Ramsay, '12
Nitrogen	0–200	0.2438	Regnault.	—	1.41	Cazin.
“	20–440	0.2419	Holborn and	—	1.405	Masson.
“	20–630	0.2464	Austin.			
“	20–800	0.2497	“			
Nitric oxide, NO	13–172	0.2317	Regnault.	—	1.394	“
Nitrogen tetroxide, NO_2	27–67	1.625	Berthelot and	—	1.31	Natanson.
“ “ “	27–150	1.115	Olger.			
“ “ “	27–280	0.65	“			
Nitrous oxide, N_2O	16–207	0.2262	Regnault.	0	1.311	Wüllner.
“ “ “	26–103	0.2126	Wiedemann.	100	1.272	“
“ “ “	27–206	0.2241	“	—	1.324	Leduc, '98.
Oxygen	13–207	0.2175	Regnault.	5–14	1.3977	Lummer and Pringsheim.
“	20–440	0.2240	Holborn and			
“	20–630	0.2300	Austin.			
Sulphur dioxide, SO_2	16–202	0.1544	Regnault.	16–34	1.256	Müller.
Water vapor, H_2O	0	0.4655	Thiesen.	78	1.274	Beyme.
“ “ “	100	0.421	“	94	1.33	Jaeger.
“ “ “	180	0.51	“	100	1.305	Makower.
Xenon	—	—	—	19	1.666	Ramsay, '12.

Table 254.

LATENT HEAT OF VAPORIZATION.

The temperature of vaporization in degrees Centigrade is indicated by *t*, the latent heat in large calories per kilogram or in small calories or therms per gram by *r*; the total heat from 0° C, in the same units by *H*. The pressure is that due to the vapor at the temperature *t*.

Substance.	Formula.	$t°$ C	*r*	*H*	Authority.
Acetic acid	$C_2H_4O_2$	118°	84.9	—	Ogier.
Air	—	—	50.97	—	Fenner-Richtmyer.
Alcohol: Amyl	$C_5H_{12}O$	131	120	—	Schall.
Ethyl	C_2H_6O	78.1	205	255	Wirtz.
"	"	0	236	236	Regnault.
"	"	50	—	264	"
"	"	100	—	267	"
"	"	150	—	285	"
Methyl	CH_4O	64.5	267	307	Wirtz.
"	"	0	289	—	Ramsay and Young.
"	"	50	274	—	" " "
"	"	100	246	—	" " "
"	"	150	206	—	" " "
"	"	200	152	—	" " "
"	"	238.5	44.2	—	" " "
Aniline	C_6H_7N	184	110	—	Mean.
Benzene	C_6H_6	80.1	92.9	127.9	Wirtz.
Bromine	Br	61	45.6	—	Andrews.
Carbon dioxide, solid	CO_2	—	—	138.7	Favre.
" " liquid	"	−25	72.23	—	Cailletet and Mathias.
" " "	"	0	57.48	—	" " "
" " "	"	12.35	44.97	—	Mathias.
" " "	"	22.04	31.8	—	"
" " "	"	29.85	14.4	—	"
" " "	"	30.82	3.72	—	"
" disulphide	CS_2	46.1	83.8	94.8	Wirtz.
" "	"	0	90	90	Regnault.
" "	"	100	—	100.5	"
" "	"	140	—	102.4	"
Chloroform	$CHCl_3$	60.9	58.5	72.8	Wirtz.
Ether	$C_4H_{10}O$	34.5	88.4	107	"
"	"	34.9	90.5	—	Andrews.
"	"	0	94	94	Regnault.
"	"	50	—	115.1	"
"	"	120	—	140	"
Ethyl bromide	C_2H_5Br	38.2	60.4	—	Wirtz.
" chloride	C_2H_5Cl	12.5	—	98	Regnault.
" iodide	C_2H_5I	71	47	—	Mean.
Heptane	C_7H_{16}	90	77.8	—	Young.
Hexane	C_6H_{14}	70	79.2	—	"
Iodine	I	—	23.95	—	Favre and Silbermann.
Mercury	Hg	357.	65	—	Mean.
Nitrogen	N_2	−195.6	47.65	—	Alt.
Octane	C_8H_{18}	130	70.0	—	Young.
Oxygen	O_2	−182.9	50.97	—	Alt.
Pentane	C_5H_{12}	30	85.8	—	Young.
Sulphur	S	316	362.0	—	Person.
Sulphur dioxide	SO_2	0	91.2	—	Cailletet and Mathias.
" "	"	30	80.5	—	" " "
" "	"	65	68.4	—	" " "
Toluene	C_7H_8	111	86.0	—	Mean.
Turpentine	$C_{10}H_{10}$	159.3	74.04	—	Brix.

LATENT HEAT OF VAPORIZATION.

TABLE 255. — Formulae for Latent and Total Heats of Vapors.

r = latent heat of vaporization at *t*° C; *H* = total heat from fluid at 0° to vapor at *t*° C. *T*° refers to Kelvin scale. Same units as preceding table.

Substance	Formula	Range	Authority
Acetone, C_3H_6O	$H = 140.5 + 0.36644t - 0.000516t^2$	−3° to 147°	R
	$= 139.9 + 0.23356t + 0.00055358t^2$	−3 to 147	W
	$r = 139.9 - 0.27287t + 0.0001571t^2$	−3 to 147	W
Benzene C_6H_6	$H = 109.0 + 0.24429t - 0.0001315t^2$	7 to 215	R
Carbon dioxide	$r^2 = 118.485(31 - t) - 0.4707(31 - t)^2$	−25 to 31	C
Carbon bisulphide, CS_2	$H = 90.0 + 0.14601t - 0.0004123t^2$	−6 to 143	R
	$H = 89.5 + 0.16993t - 0.0010161t^2 + 0.0_5342t^3$	−6 to 143	W
	$r = 89.5 - 0.06530t - 0.0010976t^2 + 0.0_5342t^3$	−6 to 143	W
Carbon tetrachloride, CCl_4	$H = 52.0 + 0.14625t - 0.000172t^2$	8 to 163	R
	$H = 51.9 + 0.17867t - 0.0009599t^2 + 0.0_53733t^3$	8 to 163	W
	$r = 51.9 - 0.01931t - 0.0010505t^2 + 0.0_53733t^3$	8 to 163	W
Chloroform, $CHCl_3$	$H = 67.0 + 0.1375t$	−5 to 159	R
	$H = 67.0 + 0.14716t - 0.0000937t^2$	−5 to 159	W
	$r = 67.0 - 0.08519t - 0.0001444t^2$	−5 to 159	W
Ether, $C_4H_{10}O$	$H = 94.0 + 0.45000t - 0.0005556t^2$	−4 to 121	R
	$r = 94.0 - 0.07900t - 0.0008514t^2$	−4 to 121	R
Molybdenum	$r = 177000 - 2.5T$ (cal/g-atom)	—	L
Nitrogen, N_2	$r = 68.85 - 0.2736T$	—	A
Nitrous oxide, N_2O	$r^2 = 131.75(36.4 - t) - 0.928(36.4 - t)^2$	−20 to 36	C
Oxygen, O_2	$r = 69.67 - 0.2080T$	—	A
Platinum	$r = 128000 - 2.5T$ (cal/g-atom)	—	L
Sulphur dioxide	$r = 91.87 - 0.3842t - 0.000340t^2$	0 to 20	M
Tungsten	$r = 217800 - 1.8T$ (cal/g-atom)	—	L
Water, H_2O	$H = 638.9 + 0.3745(t - 100) - 0.00099(t - 100)^2$	—	D
	$r = 94.210(365 - t)^{0.31249}$ (See Table 259)	0 to 100	H

R, Regnault; W, Winkelmann; C, Cailletet and Mathias; A, Alt.; D, Davis; H, Henning; L, Langmuir.

TABLE 256.—Latent Heat of Vaporization of Ammonia.

CALORIES PER GRAM.

°C	0	1	2	3	4	5	6	7	8	9
−40	331.7	332.3	333.0	333.6	334.3	334.9	335.5	336.2	336.8	337.5
−30	324.8	325.5	326.2	326.9	327.6	328.3	329.0	329.7	330.3	331.0
−20	317.6	318.3	319.1	319.8	320.6	321.3	322.0	322.7	323.4	324.1
−10	309.9	310.7	311.5	312.2	313.0	313.8	314.6	315.3	316.1	316.8
− 0	301.8	302.6	303.4	304.3	305.1	305.9	306.7	307.5	308.3	309.1
+ 0	301.8	300.9	300.1	299.2	298.4	297.5	296.6	295.7	294.9	294.0
+10	293.1	292.2	291.3	290.4	289.5	288.6	287.6	286.7	285.7	284.8
+20	283.8	282.8	281.8	280.9	279.9	278.9	277.9	276.9	275.9	274.9
+30	273.9	272.8	271.8	270.7	269.7	268.6	267.5	266.4	265.3	264.2
+40	263.1	262.0	260.8	259.7	258.5	257.4	256.2	255.0	253.8	252.6

Osborne and van Dusen, Bul. Bureau Standards, 14, p. 439, 1918.

TABLE 257. — "Latent Heat of Pressure Variation" of Liquid Ammonia.

When a fluid undergoes a change of pressure, there occurs a transformation of energy into heat or vice versa, which results in a change of temperature of the substance unless a like amount of heat is abstracted or added. This change expressed as the heat so transformed per unit change of pressure is the "latent heat of pressure variation." It is expressed below as Joules per gram per kg/cm². Osborne and van Dusen, *loc. cit.*, p. 433, 1918.

Temperature ° C	−44.1	−39.0	−24.2	−0.2	+16.5	+26.5	+35.4	+40.3
Latent heat	−.055	−.057	−.068	−.088	−.107	−.123	−.140	−.150

TABLE 258.

LATENT AND TOTAL HEATS OF VAPORIZATION OF THE ELEMENTS.

The following table of theoretical values is taken from J. W. Richards, Tr. Amer. Electrch. Soc. 13, p. 447, 1908. They are computed as follows: $8T_m$ (8 = mean value atomic specific heat, Dulong-Petit constant, 0° to T° K, T_m = melting point, Kelvin scale) plus $2T_m$ (latent heat of fusion is approximately $2T_m$, J. Franklin Inst. 1897) plus $10(T_b - T_m)$ (specific heat of liquid metals is nearly constant and equal to that of the solid at T_m, T_b = boiling point, Kelvin scale) plus $23T_b$ (23 = Trouton constant; latent heat of vaporization of molecular weight in grams is approximately 23 times T_b) = $33T_b$. Total heat of vapor when raised from 273° K (0° C) equals $33T_b - 1700$ (mean value of Dulong-Petit constant between 0° and 273° K is 1700). Heats given in small calories per gram.

Element.	T_b °K	$23T_b$	Latent heat of vaporization.	$33T_b - 1700$	Total heat vapor from 273° K	Element.	T_b °K	$23T_b$	Latent heat of vaporization.	$33T_b - 1700$	Total heat of vapor from 273° K
Hg	630	14,500	72	19,100	96	Rh	2773	63,800	620	90,000	870
K	993	22,800	590	31,100	800	Ru	2790	64,100	630	90,000	880
Cd	1050	24,200	230	33,000	310	Au	2800	64,500	330	91,000	460
Na	1170	27,000	1170	37,000	1610	Pd	2810	64,600	610	91,000	850
Zn	1200	27,700	430	38,000	580	Ir	2820	64,800	340	91,300	470
In	1270	29,300	—	40,300	—	Os	2870	66,000	350	93,000	490
Mg	1370	31,600	1320	43,600	1820	U	3170	73,000	305	103,000	430
Te	1660	38,200	300	54,900	430	Mo	3470	80,000	830	113,000	1180
Bi	1710	39,300	190	56,400	270	W	3970	91,400	500	129,000	700
Sb	1870	43,100	360	60,000	510	H_2	20	460	230	—	—
Tl	1970	45,400	220	63,400	310	N_2	77	1,770	63	—	—
Pb	2070	47,700	230	66,700	320	O_2	85	1,960	61	—	—
Ag	2310	53,000	490	74,600	690	Cl_2	251	5,780	81	—	—
Cu	2370	54,500	860	76,600	1210	Br_2	331	7,600	48	—	—
Sn	2440	56,100	480	78,800	670	I_3	447	10,300	27	—	—
Mn	2470	56,500	1030	79,500	1440	P_3	560	13,000	138	—	—
Ni	2690	59,800	1010	84,000	1420	As_3	723	16,600	74	—	—
Cr	2640	60,700	1170	85,400	1640	Se_3	963	22,100	94	—	—
Fe	2690	62,000	1110	87,200	1560	B_2	3970	91,000	4200	—	—
Pt	2720	62,600	320	88,000	450	C_2	3970	91,000	3800	—	—
Ti	2750	63,200	1320	89,000	1850						

TABLE 250.

PROPERTIES OF SATURATED STEAM.

Metric and Common Units.

Reprinted by permission of the author and publishers from "Tables of the Properties of Steam," Cecil H. Peabody, 8th edition, rewritten in 1909. Calorie used is heat required to raise 1 Kg. water from 15° to 16° C. B. T. U. is heat required to raise 1 pd. water from 62° to 63° F. Mechanical Equiv. of heat used, 778 ft. pds. or 427 m. Kg. Specific heats, see Barnes-Regnault-Peabody results, p. 227. Heat of Liquid, q. heat required to raise 1 Kg. (1 lb.) to corresponding temperature from 0° C. Heat of vaporization, r. heat required to vaporize 1 Kg. (1 lb.) at corresponding temperature to dry saturated vapor against corresponding pressure; see Henning, Ann. der Phys., 21, p. 849, 1906. Total Heat, $H = r + q$, see Davis, Tr. Am. Soc. Mech. Eng., 1908.

Temperature Degrees Centigrade.	Pressure.			Heat of the Liquid.		Heat of Vaporization.		Heat Equivalent of Internal Work.		Temperature Degrees Fahrenheit.
	Mm. of Mercury.	Kg. per sq. cm.	Pds. per sq. in.	Calories.	B. T. U.	Calories.	B. T. U.	Calories.	B. T. U.	
t.	p.	p.	p.	q.	q.	r.	r.	ρ.	ρ.	t.
0	4.579	0.00623	0.0886	0.00	0.0	595.4	1071.7	565.3	1017.5	32.0
5	6.541	.00889	.1265	5.04	9.1	592.8	1067.1	562.2	1011.9	41.0
10	9.205	.01252	.1780	10.06	18.1	590.2	1062.3	559.0	1006.2	50.0
15	12.779	.01737	.2471	15.06	27.1	587.6	1057.6	555.9	1000.5	59.0
20	17.51	.02381	.3386	20.06	36.1	584.9	1052.8	552.7	994.8	68.0
25	23.69	.03221	.4581	25.05	45.1	582.3	1048.1	549.5	989.1	77.0
30	31.71	.04311	.6132	30.04	54.1	579.6	1043.3	546.3	983.4	86.0
35	42.02	.05713	.8126	35.03	63.1	576.9	1038.5	543.1	977.6	95.0
40	55.13	.07495	1.0661	40.02	72.0	574.2	1033.5	539.9	971.7	104.0
45	71.66	.09743	1.3858	45.00	81.0	571.3	1028.4	536.5	965.7	113.0
50	92.30	.12549	1.7849	49.99	90.0	568.4	1023.2	533.0	959.6	122.0
55	117.85	.16023	2.279	54.98	99.0	565.6	1018.1	529.7	953.5	131.0
60	149.19	.20284	2.885	59.97	108.0	562.8	1013.1	526.4	947.5	140.0
65	187.36	.2547	3.623	64.98	117.0	559.9	1007.8	523.0	941.3	149.0
70	233.53	.3175	4.516	69.98	126.0	556.9	1002.5	519.5	935.0	158.0
75	289.0	.3929	5.589	74.99	135.0	554.0	997.3	516.0	928.8	167.0
80	355.1	.4828	6.867	80.01	144.0	551.1	991.9	512.6	922.6	176.0
85	433.5	.5894	8.383	85.04	153.1	548.1	986.5	509.1	916.3	185.0
90	525.8	.7149	10.167	90.07	162.1	544.9	980.9	505.4	909.9	194.0
91	546.1	.7425	10.560	91.08	163.9	544.3	979.8	504.7	908.5	195.8
92	567.1	.7710	10.966	92.08	165.7	543.7	978.7	504.0	907.2	197.6
93	588.7	.8004	11.384	93.09	167.5	543.1	977.6	503.3	906.0	199.4
94	611.0	.8307	11.815	94.10	169.3	542.5	976.5	502.6	904.7	201.2
95	634.0	.8620	12.260	95.11	171.2	541.9	975.4	501.9	903.4	203.0
96	657.7	.8942	12.718	96.12	173.0	541.2	974.2	501.1	902.1	204.8
97	682.1	.9274	13.190	97.12	174.8	540.6	973.1	500.4	900.8	206.6
98	707.3	.9616	13.678	98.13	176.6	539.9	971.9	499.6	899.4	208.4
99	733.3	.9970	14.180	99.14	178.5	539.3	970.8	498.9	898.2	210.2
100	760.0	1.0333	14.697	100.2	180.3	538.7	969.7	498.2	896.9	212.0
101	787.5	1.0707	15.229	101.2	182.1	538.1	968.5	497.5	895.5	213.8
102	815.9	1.1093	15.778	102.2	183.9	537.4	967.3	496.8	894.1	215.6
103	845.1	1.1490	16.342	103.2	185.7	536.8	966.2	496.1	892.9	217.4
104	875.1	1.1898	16.923	104.2	187.6	536.2	965.1	495.4	891.6	219.2
105	906.1	1.2319	17.522	105.2	189.4	535.6	964.0	494.7	890.3	221.0
106	937.9	1.2752	18.137	106.2	191.2	534.9	962.8	493.9	889.0	222.8
107	970.6	1.3196	18.769	107.2	193.0	534.2	961.6	493.1	887.6	224.6
108	1004.3	1.3653	19.420	108.2	194.8	533.6	960.5	492.4	886.3	226.4
109	1038.8	1.4123	20.089	109.3	196.7	532.9	959.3	491.6	885.0	228.2
110	1074.5	1.4608	20.777	110.3	198.5	532.3	958.1	490.9	883.6	230.0
111	1111.1	1.5106	21.486	111.3	200.3	531.6	956.9	490.2	882.3	231.8
112	1148.7	1.5617	22.214	112.3	202.1	530.9	955.7	489.4	880.9	233.6
113	1187.4	1.6144	22.962	113.3	203.9	530.3	954.5	488.7	879.5	235.4
114	1227.1	1.6684	23.729	114.3	205.8	529.6	953.3	487.9	878.2	237.2
115	1267.9	1.7238	24.518	115.3	207.6	528.9	952.1	487.1	876.8	239.0
116	1309.8	1.7808	25.328	116.4	209.4	528.2	950.8	486.3	875.4	240.8
117	1352.8	1.8393	26.160	117.4	211.2	527.5	949.5	485.5	873.9	242.6
118	1397.0	1.8993	27.015	118.4	213.0	526.9	948.4	484.8	872.6	244.4
119	1442.4	1.9611	27.893	119.4	214.9	526.2	947.2	484.0	871.3	246.2

TABLE 259 (*continued*).

PROPERTIES OF SATURATED STEAM.

Metric and Common Units.

If a is the reciprocal of the Mechanical Equivalent of Heat, p the pressure, s and σ the specific volumes of the liquid and the saturated vapor, s —.σ, the change of volume, then the heat equivalent of the external work is Apu = Ap(s — σ). Heat equivalent of internal work, ρ = r — Apu. For experimental sp. vols. see Knoblauch, Linde and Klebe, Mitt. über Forschungarbeiten, 21, p. 33, 1905. Entropy = S dQ/T, where dQ = amount of heat added at absolute temperature T. For pressures of saturated steam see Holborn and Henning, Ann. der Phys. 26, p. 833, 1908; for temperatures above 205° C. corrected from Regnault.

Temperature Degrees Centigrade.	Heat Equivalent of External Work.		Entropy of the Liquid.	Entropy of Evaporation.	Specific Volume.		Density.		Temperature Degrees Fahrenheit.
	Calories.	B.T.U.			Cubic Meters per Kilogram.	Cubic Feet per Pound.	Kilograms per Cubic Meter.	Pounds per Cubic Foot.	
t	Apu.	Apu.	θ	$\frac{r}{T}$	s	s	$\frac{1}{s}$	$\frac{1}{s}$	t
0	30.1	54.2	0.0000	2.1804	206.3	3304.	0.00485	0.000303	32.0
5	30.6	55.2	.0183	2.1320	147.1	2356.	.00680	.000424	41.0
10	31.2	56.1	.0361	2.0850	106.3	1703.	.00941	.000587	50.0
15	31.7	57.1	.0537	2.0396	77.9	1248.	.01283	.000801	59.0
20	32.2	58.0	.0709	1.9959	57.8	926.	.01730	.001080	68.0
25	32.8	59.0	.0878	1.9536	43.40	695.	.02304	.001439	77.0
30	33.3	59.9	.1044	1.9126	32.95	528.	.03035	.001894	86.0
35	33.8	60.9	.1207	1.8728	25.25	404.7	.03960	.002471	95.0
40	34.3	61.8	.1368	1.8341	19.57	313.5	.0511	.003190	104.0
45	34.8	62.7	.1526	1.7963	15.25	244.4	.0656	.004092	113.0
50	35.4	63.6	.1682	1.7597	12.02	192.6	.0832	.00519	122.0
55	35.9	64.6	.1835	1.7242	9.56	153.2	.1046	.00653	131.0
60	36.4	65.6	.1986	1.6899	7.66	122.8	.1305	.00814	140.0
65	36.9	66.5	.2135	1.6563	6.19	99.2	.1615	.01008	149.0
70	37.4	67.4	.2282	1.6235	5.04	80.7	.1984	.01239	158.0
75	38.0	68.5	.2427	1.5918	4.130	66.2	.2421	.01510	167.0
80	38.5	69.3	.2570	1.5609	3.404	54.5	.2938	.01835	176.0
85	39.0	70.2	.2711	1.5307	2.824	45.23	.3541	.02211	185.0
90	39.5	71.0	.2851	1.5010	2.358	37.77	.4241	.02648	194.0
91	39.6	71.3	.2879	1.4952	2.275	36.45	.4395	.02743	195.8
92	39.7	71.5	.2906	1.4894	2.197	35.19	.4552	.02842	197.6
93	39.8	71.6	.2934	1.4836	2.122	34.00	.4713	.02941	199.4
94	39.9	71.8	.2961	1.4779	2.050	32.86	.4878	.03043	201.2
95	40.0	72.0	.2989	1.4723	1.980	31.75	.505	.03149	203.0
96	40.1	72.1	.3016	1.4666	1.913	30.67	.523	.03260	204.8
97	40.2	72.3	.3043	1.4609	1.849	29.63	.541	.03375	206.6
98	40.3	72.5	.3070	1.4552	1.787	28.64	.560	.03492	208.4
99	40.4	72.6	.3097	1.4496	1.728	27.69	.579	.03611	210.2
100	40.5	72.8	.3125	1.4441	1.671	26.78	.598	.03734	212.0
101	40.6	73.0	.3152	1.4386	1.617	25.90	.618	.03861	213.8
102	40.6	73.2	.3179	1.4330	1.564	25.06	.639	.03990	215.6
103	40.7	73.3	.3205	1.4275	1.514	24.25	.661	.04124	217.4
104	40.8	73.5	.3232	1.4220	1.465	23.47	.683	.04261	219.2
105	40.9	73.7	.3259	1.4165	1.419	22.73	.705	.04400	221.0
106	41.0	73.8	.3286	1.4111	1.374	22.01	.728	.04543	222.8
107	41.1	74.0	.3312	1.4057	1.331	21.31	.751	.04692	224.6
108	41.2	74.2	.3339	1.4003	1.289	20.64	.776	.04845	226.4
109	41.3	74.3	.3365	1.3949	1.248	19.99	.801	.0500	228.2
110	41.4	74.5	.3392	1.3895	1.209	19.37	.827	.0516	230.0
111	41.4	74.6	.3418	1.3842	1.172	18.77	.853	.0533	231.8
112	41.5	74.8	.3445	1.3789	1.136	18.20	.880	.0550	233.6
113	41.6	75.0	.3471	1.3736	1.101	17.64	.908	.0567	235.4
114	41.7	75.1	.3498	1.3683	1.068	17.10	.936	.0585	237.2
115	41.8	75.3	.3524	1.3631	1.036	16.59	.965	.0603	239.0
116	41.9	75.4	.3550	1.3579	1.005	16.09	.995	.0622	240.8
117	42.0	75.6	.3576	1.3527	0 9746	15.61	1.026	.0641	242.6
118	42.1	75.8	.3602	1.3475	0.9460	15.16	1.057	.0659	244.4
119	42.2	75.9	.3628	1.3423	0.9183	14.72	1.089	.0679	246.2

TABLE 259 (*continued*).

PROPERTIES OF SATURATED STEAM.

Metric and Common Units.

Temperature Degrees Centigrade.	Pressure.			Heat of the Liquid.		Heat of Vaporization.		Heat Equivalent of Internal Work.		Temperature Degrees Fahrenheit.
	Mm. of Mercury.	Kg. per sq. cm.	Pds. per sq. in.	Calories.	B. T. U.	Calories.	B. T. U.	Calories.	B. T. U.	
t.	p.	p.	p.	q.	q	r.	r.	ρ.	ρ.	t.
120	1489	2.024	28.79	120.4	216.7	525.6	946.0	483.4	870.0	248.0
121	1537	2.089	29.72	121.4	218.5	524.9	944.8	482.6	868.6	249.8
122	1586	2.156	30.66	122.5	220.4	524.2	943.5	481.8	867.1	251.6
123	1636	2.224	31.64	123.5	222.2	523.5	942.3	481.0	865.8	253.4
124	1688	2.294	32.64	124.5	224.1	522.8	941.0	480.2	864.3	255.2
125	1740	2.366	33.66	125.5	225.9	522.1	939.9	479.4	863.0	257.0
126	1795	2.440	34.71	126.5	227.7	521.4	938.6	478.6	861.6	258.8
127	1850	2.516	35.78	127.5	229.5	520.7	937.3	477.8	860.2	260.6
128	1907	2.593	36.88	128.6	231.4	520.0	936.1	477.0	858.8	262.4
129	1966	2.673	38.01	129.6	233.3	519.3	934.8	476.3	857.4	264.2
130	2026	2.754	39.17	130.6	235.1	518.6	933.6	475.5	856.0	266.0
131	2087	2.837	40.36	131.6	236.9	517.9	932.3	474.7	854.6	267.8
132	2150	2.923	41.57	132.6	238.7	517.3	931.1	474.0	853.2	269.6
133	2214	3.010	42.81	133.7	240.6	516.6	929.8	473.3	851.8	271.4
134	2280	3.100	44.09	134.7	242.4	515.9	928.5	472.5	850.4	273.2
135	2348	3.192	45.39	135.7	244.2	515.1	927.2	471.6	848.9	275.0
136	2416	3.285	46.73	136.7	246.0	514.4	925.9	470.8	847.5	276.8
137	2487	3.382	48.10	137.7	247.9	513.7	924.6	470.1	846.1	278.6
138	2560	3.480	49.50	138.8	249.7	513.0	923.3	469.3	844.6	280.4
139	2634	3.581	50.93	139.8	251.6	512.3	922.1	468.5	843.3	282.2
140	2710	3.684	52.39	140.8	253.4	511.5	920.7	467.6	841.8	284.0
141	2787	3.789	53.89	141.8	255.3	510.7	919.3	466.8	840.2	285.8
142	2866	3.897	55.43	142.8	257.1	510.1	918.1	466.1	838.9	287.6
143	2948	4.008	57.00	143.9	259.0	509.3	916.7	465.3	837.4	289.4
144	3030	4.121	58.60	144.9	260.8	508.6	915.4	464.4	835.9	291.2
145	3115	4.236	60.24	145.9	262.7	507.8	914.1	463.6	834.5	293.0
146	3202	4.354	61.92	146.9	264.5	507.1	912.8	462.8	833.1	294.8
147	3291	4.474	63.64	148.0	266.4	506.4	911.5	462.0	831.6	296.6
148	3381	4.597	65.39	149.0	268.2	505.6	910.1	461.2	830.1	298.4
149	3474	4.723	67.18	150.0	270.1	504.9	908.8	460.4	828.7	300.2
150	3569	4.852	69.01	151.0	271.9	504.1	907.4	459.5	827.2	302.0
151	3665	4.984	70.88	152.1	273.8	503.4	906.1	458.7	825.7	303.8
152	3764	5.118	72.79	153.1	275.6	502.6	904.7	457.9	824.2	305.6
153	3865	5.255	74.74	154.1	277.4	501.9	903.3	457.1	822.7	307.4
154	3968	5.395	76.73	155.1	279.2	501.1	901.9	456.3	821.2	309.2
155	4073	5.538	78.76	156.2	281.1	500.3	900.5	455.4	819.6	311.0
156	4181	5.684	80.84	157.2	283.0	499.6	899.2	454.6	818.2	312.8
157	4290	5.833	82.96	158.2	284.8	498.8	897.8	453.8	816.7	314.6
158	4402	5.985	85.12	159.3	286.7	498.1	896.5	453.0	815.3	316.4
159	4517	6.141	87.33	160.3	288.5	497.3	895.1	452.1	813.7	318.2
160	4633	6.300	89.59	161.3	290.4	496.5	893.7	451.2	812.2	320.0
161	4752	6.462	91.89	162.3	292.2	495.7	892.3	450.4	810.7	321.8
162	4874	6.628	94.25	163.4	294.1	494.9	890.9	449.5	809.2	323.6
163	4998	6.796	96.65	164.4	295.9	494.2	889.5	448.7	807.7	325.4
164	5124	6.967	99.09	165.4	297.7	493.4	888.1	447.9	806.2	327.2
165	5253	7.142	101.6	166.5	299.6	492.6	886.7	447.0	804.7	329.0
166	5384	7.320	104.1	167.5	301.5	491.9	885.4	446.3	803.3	330.8
167	5518	7.502	106.7	168.5	303.3	491.1	883.9	445.4	801.7	332.6
168	5655	7.688	109.4	169.5	305.1	490.3	882.5	444.6	800.1	334.4
169	5794	7.877	112.0	170.6	307.0	489.5	881.0	443.7	798.5	336.2

TABLE 239 (*continued*).

PROPERTIES OF SATURATED STEAM.

Metric and Common Units.

Temperature Degrees Centigrade.	Heat Equivalent of External Work.		Entropy of the Liquid.	Entropy of Evaporation.	Specific Volume.		Density.		Temperature Degrees Fahrenheit.
	Calories.	B. T. U.			Cubic Meters per Kilogram.	Cubic Feet per Pound.	Kilograms per Cubic Meter.	Pounds per Cubic Foot.	
t.	Apu.	Apu.	θ.	$\frac{r}{T}$.	s.	s.	$\frac{1}{s}$	$\frac{1}{s}$	t.
120	42.2	76.0	0.3654	1.3372	0.8914	14.28	1.122	0.0700	248.0
121	42.3	76.2	.3680	1.3321	.8653	13.86	1.156	.0721	249.8
122	42.4	76.4	.3705	1.3269	.8401	13.46	1.190	.0743	251.6
123	42.5	76.5	.3731	1.3218	.8158	13.07	1.226	.0765	253.4
124	42.6	76.7	.3756	1.3167	.7924	12.69	1.262	.0788	255.2
125	42.7	76.8	.3782	1.3117	.7698	12.33	1.299	.0811	257.0
126	42.8	77.0	.3807	1.3067	.7479	11.98	1.337	.0835	258.8
127	42.9	77.1	.3833	1.3017	.7267	11.64	1.376	.0859	260.6
128	43.0	77.3	.3858	1.2967	.7063	11.32	1.416	.0883	262.4
129	43.0	77.4	.3884	1.2917	.6867	11.00	1.456	.0909	264.2
130	43.1	77.6	.3909	1.2868	.6677	10.70	1.498	.0935	266.0
131	43.2	77.7	.3934	1.2818	.6493	10.40	1.540	.0961	267.8
132	43.3	77.9	.3959	1.2769	.6315	10.12	1.583	.0988	269.6
133	43.3	78.0	.3985	1.2720	.6142	9.839	1.628	.1016	271.4
134	43.4	78.1	.4010	1.2672	.5974	9.569	1.674	.1045	273.2
135	43.5	78.3	.4035	1.2623	.5812	9.309	1.721	.1074	275.0
136	43.6	78.4	.4060	1.2574	.5656	9.060	1.768	.1104	276.8
137	43.6	78.5	.4085	1.2526	.5506	8.820	1.816	.1134	278.6
138	43.7	78.7	.4110	1.2479	.5361	8.587	1.865	.1165	280.4
139	43.8	78.8	.4135	1.2431	.5219	8.360	1.916	.1196	282.2
140	43.9	78.9	.4160	1.2383	.5081	8.140	1.968	.1229	284.0
141	43.9	79.1	.4185	1.2335	.4948	7.926	2.021	.1262	285.8
142	44.0	79.2	.4209	1.2288	.4819	7.719	2.075	.1296	287.6
143	44.0	79.3	.4234	1.2241	.4694	7.519	2.130	.1330	289.4
144	44.2	79.5	.4259	1.2194	.4574	7.326	2.186	.1365	291.2
145	44.2	79.6	.4283	1.2147	.4457	7.139	2.244	.1401	293.0
146	44.3	79.7	.4307	1.2100	.4343	6.957	2.303	.1437	294.8
147	44.4	79.9	.4332	1.2054	.4232	6.780	2.363	.1475	296.6
148	44.4	80.0	.4356	1.2008	.4125	6.609	2.424	.1513	298.4
149	44.5	80.1	.4380	1.1962	.4022	6.443	2.486	.1552	300.2
150	44.6	80.2	.4405	1.1916	.3921	6.282	2.550	.1592	302.0
151	44.6	80.4	.4429	1.1870	.3824	6.126	2.615	.1632	303.8
152	44.7	80.5	.4453	1.1824	.3729	5.974	2.682	.1674	305.6
153	44.8	80.6	.4477	1.1778	.3637	5.826	2.750	.1716	307.4
154	44.8	80.7	.4501	1.1733	.3548	5.683	2.818	.1759	309.2
155	44.9	80.9	.4525	1.1688	.3463	5.546	2.888	.1803	311.0
156	45.0	81.0	.4549	1.1644	.3380	5.413	2.959	.1847	312.8
157	45.0	81.1	.4573	1.1599	.3298	5.282	3.032	.1893	314.6
158	45.1	81.2	.4596	1.1554	.3218	5.154	3.108	.1940	316.4
159	45.2	81.4	.4620	1.1509	.3140	5.029	3.185	.1988	318.2
160	45.3	81.5	.4644	1.1465	.3063	4.906	3.265	.2038	320.0
161	45.3	81.6	.4668	1.1421	.2989	4.789	3.345	.2088	321.8
162	45.4	81.7	.4692	1.1377	.2920	4.677	3.425	.2138	323.6
163	45.5	81.8	.4715	1.1333	.2855	4.571	3.503	.2188	325.4
164	45.5	81.9	.4739	1.1289	.2792	4.469	3.582	.2238	327.2
165	45.6	82.0	.4763	1.1245	.2729	4.368	3.664	.2289	329.0
166	45.6	82.1	.4786	1.1202	.2666	4.268	3.751	.2343	330.8
167	45.7	82.2	.4810	1.1159	.2603	4.168	3.842	.2399	332.6
168	45.7	82.4	.4833	1.1115	.2540	4.070	3.937	.2457	334.4
169	45.8	82.5	.4857	1.1072	.2480	3.975	4.032	.2516	336.2

TABLE 259 (*continued*).

PROPERTIES OF SATURATED STEAM.

Metric and Common Units.

Temperature Degrees Centigrade.	Pressure.			Heat of the Liquid.		Heat of Vaporization.		Heat Equivalent of Internal Work.		Temperature Degrees Fahrenheit.
	Mm. of Mercury.	Kg. per sq. cm.	Pds. per sq. in.	Calories.	B. T. U.	Calories.	B. T. U.	Calories.	B. T. U.	
t.	p.	p.	p.	q.	q.	r.	r.	ρ.	ρ.	t.
170	5937	8.071	114 8	171.6	308.9	488.7	879.6	442.8	797.0	338.0
171	6081	8.268	117.6	172.6	310.7	487.9	878.3	441.9	795.6	339.8
172	6229	8.469	120 4	173.7	312.6	487.1	876.9	441.1	794.1	341.6
173	6379	8.673	123.4	174.7	314.5	486.3	875.4	440.2	792.5	343.4
174	6533	8.882	126.3	175.7	316.3	485.5	873.9	439.4	790.9	345.2
175	6689	9.094	129.4	176.8	318.2	484.7	872.4	438.5	789.3	347.0
176	6848	9.310	132.4	177.8	320.0	483.9	871.0	437.7	787.8	348.8
177	7010	9.531	135.6	178.8	321.8	483.1	869.5	436.8	786.2	350.6
178	7175	9.755	138.8	179.9	323.7	482.3	868.1	436.0	784.7	352.4
179	7343	9.983	142.0	180.9	325.6	481.4	866.6	435.0	783.1	354.2
180	7514	10.216	145.3	181.9	327.5	480.6	865.1	434.2	781.5	356.0
181	7688	10.453	148.7	183.0	329.3	479.8	863.6	433.3	779.9	357.8
182	7866	10.695	152.1	184.0	331.2	479.0	862.2	432.5	778.4	359.6
183	8046	10.940	155.6	185.0	333.0	478.2	860.7	431.6	776.9	361.4
184	8230	11.189	159.2	186.1	334.9	477.4	859.2	430.8	775.3	363.2
185	8417	11.44	162.8	187.1	336.8	476.6	857.7	429.9	773.7	365.0
186	8608	11.70	166.5	188.1	338.6	475 7	856.3	429.0	772.2	366.8
187	8802	11.97	170.2	189.2	340.5	474.8	854.7	428.0	770.5	368.6
188	8999	12.24	174.0	190.2	342.4	474.0	853.2	427.2	768.9	370.4
189	9200	12.51	177.9	191.2	344.2	473.2	851.7	426.3	767.4	372.2
190	9404	12.79	181.8	192.3	346.1	472.3	850.2	425.4	765.8	374.0
191	9612	13.07	185.9	193.3	347.9	471.5	848.7	424.5	764.2	375.8
192	9823	13.36	190.0	194.4	349.8	470.6	847.1	423.6	762.5	377.6
193	10038	13.65	194.1	195.4	351.7	469.8	845.6	422.8	761.0	379.4
194	10256	13.94	198.3	196.4	353.5	468.9	844.1	421.9	759.4	381.2
195	10480	14.25	202.6	197.5	355.4	468.1	842.5	421.0	757.7	383.0
196	10700	14.55	207.0	198.5	357.3	467.2	841.0	420.1	756.1	384.8
197	10930	14.87	211.4	199.5	359.2	466.4	839.5	419.2	754.6	386.6
198	11170	15.18	216.0	200.6	361.1	465.6	838.0	418.4	753.0	388.4
199	11410	15.51	220.6	201.6	362.9	464.7	836.4	417.4	751.3	390.2
200	11650	15.84	225.2	202.7	364.8	463.8	834.8	416.5	749.7	392.0
201	11890	16.17	223.0	203.7	366.7	462.9	833.3	415.6	748.1	393.8
202	12140	16.51	234.8	204.7	368.5	462.1	831.8	414.8	746.6	395.6
203	12400	16.85	239.7	205.8	370.4	461.2	830.2	413.8	744.9	397.4
204	12650	17.20	244.7	206.8	372.3	460.3	828.6	412.9	743.3	399.2
205	12920	17.56	249.8	207.9	374.1	459.4	827.0	412.0	741.6	401.0
206	13180	17.92	254.9	208.9	376.0	458.6	825.4	411.1	740.0	402.8
207	13450	18.29	260.1	210.0	377.9	457.7	823.8	410.2	738.3	404.6
208	13730	18.66	265.4	211.0	379.8	456.8	822.2	409.3	736.7	406.4
209	14010	19.04	270.8	212.0	381.6	455.9	820.6	408.4	735.1	408.2
210	14290	19.43	276.3	213.1	383.5	455.0	819.1	407.5	733.6	410.0
211	14580	19.82	281.9	214.1	385.4	454.1	817.4	406.6	731.9	411.8
212	14870	20.22	287.6	215.2	387.3	453.2	815.8	405.7	730.2	413.6
213	15170	20.62	293.3	216.2	389.2	452.4	814.3	404.9	728.7	415.4
214	15470	21.03	299.2	217.3	391.1	451.5	812.7	404.0	727.1	417.2
215	15780	21.45	305.1	218.3	392.9	450.6	811.0	403.1	725.4	419.0
216	16090	21.88	311.1	219.3	394.8	449.6	809.3	402.1	723.7	420.8
217	16410	22.31	317.3	220.4	396.7	448.7	807.7	401.2	722.1	422.6
218	16730	22.74	323.5	221.4	398.5	447.8	806.1	400.3	720.5	424.4
219	17060	23.19	329.8	222.5	400.4	446.9	804.5	399.4	718.9	426 2
220	17390	23.64	336.2	223.5	402.3	446.0	802.9	398.5	717.3	428.0

TABLE 259 (*continued*).

PROPERTIES OF SATURATED STEAM.

Metric and Common Units.

Temperature Degrees Centigrade.	Heat Equivalent of External Work.		Entropy of the Liquid.	Entropy of Evaporation.	Specific Volume.		Density.		Temperature Degrees Fahrenheit.
	Calories.	B. T. U.			Cubic Meters per Kilogram.	Cubic Feet per Pound.	Kilograms per Cubic Meter.	Pounds per Cubic Foot.	
t.	Apu.	Apu.	θ.	$\frac{r}{T}$.	s.	s.	$\frac{1}{s}$.	$\frac{1}{s}$.	t.
170	45.9	82.6	0.4880	1.1029	0.2423	3.883	4.127	0.2575	338.0
171	46.0	82.7	.4903	1.0987	.2368	3.794	4.223	.2636	339.8
172	46.0	82.8	.4926	1.0944	.2314	3.709	4.322	.2696	341.6
173	46 1	82.9	.4949	1.0901	.2262	3.626	4.421	.2758	343.4
174	46.1	83.0	.4972	1.0859	.2212	3.545	4.521	.2821	345.2
175	46.2	83.1	.4995	1.0817	.2164	3.467	4.621	.2884	347.0
176	46.2	83.2	.5018	1.0775	.2117	3.391	4.724	.2949	348.8
177	46.3	83.3	.5041	1.0733	.2072	3.318	4.826	.3014	350.6
178	46.3	83.4	.5064	1.0691	.2027	3.247	4.933	.3080	352.4
179	46.4	83.5	.5087	1.0649	.1983	3.177	5.04	.3148	354.2
180	46.4	83.6	.5110	1.0608	.1941	3.109	5.15	.3217	356.0
181	46.5	83.7	.5133	1.0567	.1899	3.041	5.27	.3288	357.8
182	46.5	83.8	.5156	1.0525	.1857	2.974	5.38	.3362	359.6
183	46.6	83.8	.5178	1.0484	.1817	2.911	5.50	.3435	361.4
184	46.6	83.9	.5201	1.0443	.1778	2.849	5.62	.3510	363.2
185	46.7	84.0	.5224	1.0403	.1740	2.787	5.75	.3588	365.0
186	46.7	84.1	.5246	1.0362	.1702	2.727	5.88	.3667	366.8
187	46.8	84.2	.5269	1.0321	.1666	2.669	6.00	.3746	368.6
188	46.8	84.3	.5291	1.0280	.1632	2.614	6.13	.3826	370.4
189	46.9	84.3	.5314	1.0240	.1598	2.560	6.26	.3906	372.2
190	46.9	84.4	.5336	1.0200	.1565	2.507	6.39	.3989	374 0
191	47.0	84.5	.5358	1.0160	.1533	2.456	6.52	.4072	375.8
192	47.0	84.6	.5381	1.0120	.1501	2.405	6.66	.4158	377.6
193	47.0	84.6	.5403	1.0080	.1470	2.355	6.80	.4246	379.4
194	47.0	84.7	.5426	1.0040	.1440	2.306	6.94	.4336	381.2
195	47.1	84.8	.5448	1.0000	.1411	2.259	7.09	.4426	383.0
196	47.1	84.9	.5470	0.9961	.1382	2.214	7.23	.4516	384.8
197	47.2	84.9	.5492	.9922	.1354	2.169	7.38	.4610	386.6
198	47.2	85.0	.5514	.9882	.1327	2.126	7.53	.4704	388.4
199	47.3	85.1	.5536	.9843	.1300	2.083	7.69	.4801	390.2
200	47 3	85.1	.5558	.9804	.1274	2.041	7.84	.4900	392.0
201	47.3	85.2	.5580	.9765	.1249	2.001	8.00	.4998	393.8
202	47.3	85.2	.5602	.9727	.1225	1.962	8.16	.510	395.6
203	47.4	85.3	.5624	.9688	.1201	1.923	8.33	.520	397.4
204	47.4	85.3	.5646	.9650	.1177	1.885	8.50	.531	399.2
205	47.4	85.4	.5668	.9611	.1153	1.847	8.67	.541	401.0
206	47.5	85.4	.5690	.9572	.1130	1.810	8.85	.552	402.8
207	47.5	85.5	.5712	.9534	.1108	1.774	9.03	.564	404.6
208	47.5	85.5	.5733	.9496	.1086	1.739	9.21	.575	406.4
209	47.5	85.5	.5755	.9458	.1065	1.705	9.39	.587	408.2
210	47.5	85.5	.5777	.9420	.1044	1.673	9.58	.598	410.0
211	47.5	85.5	.5799	.9382	.1024	1.640	9.77	.610	411.8
212	47.5	85.6	.5820	.9344	.1004	1.608	9.96	.622	413.6
213	47.5	85.6	.5842	.9307	.0984	1.577	10.16	.634	415.4
214	47.5	85.6	.5863	.9269	.0965	1.546	10.36	.647	417.2
215	47.5	85.6	.5885	.9232	.0947	1.516	10.56	.660	419.0
216	47.5	85.6	.5906	.9195	.0928	1.486	10.78	.673	420.8
217	47.5	85.6	.5927	.9157	.0910	1.458	10.99	.686	422.6
218	47.5	85.6	.5948	.9120	.0893	1.430	11.20	.699	424.4
219	47.5	85.6	.5969	.9084	.0876	1.403	11.41	.713	426.2
220	47.5	85.6	.5991	.9047	.0860	1.376	11.62	.727	428.0

TABLE 260.

LATENT HEAT OF FUSION.

This table contains the latent heat of fusion of a number of solid substances in large calories per kilogram or small calories or therms per gram. It has been compiled principally from Landolt and Börnstein's tables. *C* indicates the composition, *T* the temperature Centigrade, and *H* the latent heat.

Substance.	*C*	*T*	*H*	Authority.
Alloys: $30.5Pb + 69.5Sn$	$PbSn_4$	183	17.	Spring.
$36.9Pb + 63.1Sn$	$PbSn_3$	179	15.5	"
$63.7Pb + 36.3Sn$	$PbSn$	177.5	11.6	"
$77.8Pb + 22.2Sn$	Pb_2Sn	176.5	9.54	"
Britannia metal, $9Sn + 1Pb$	–	236	28.0*	Ledebur.
Rose's alloy, $24Pb + 27.3Sn + 48.7Bi$	–	98.8	6.85	Mazzotto.
Wood's alloy $\{25.8Pb + 14.7Sn + 52.4Bi + 7Cd\}$	–	75.5	8.40	"
Aluminum	Al	658.	76.8	Glaser.
Ammonia	NH_3	—75.	108.	Massol.
Benzene	C_6H_6	5.4	30.6	Mean.
Bromine	Br	—7.3	16.2	Regnault.
Bismuth	Bi	268	12.64	Person.
Cadmium	Cd	320.7	13.66	"
Calcium chloride	$CaCl_2 + 6H_2O$	28.5	40.7	"
Copper	Cu	1083	42.	Mean.
Iron, Gray cast	–	–	23.	Gruner.
" White "	–	–	33.	"
" Slag	–	–	50.	"
Iodine	I	–	11.71	Favre and Silbermann.
Ice	H_2O	0	79.63	Dickinson, Harper, Osborne.†
"	"	0	79.59	Smith.‡
" (from sea-water)	H_2O + 3.535 of solids	—8.7	54.0	Petterson.
Lead	Pb	327	5.36	Mean.
Mercury	Hg	—39	2.82	Person.
Naphthalene	$C_{10}H_8$	79.87	35.62	Pickering.
Nickel	Ni	1435	4.64	Pionchon.
Palladium	Pd	1545	36.3	Violle.
Phosphorus	P	44.2	4.97	Petterson.
Platinum	Pt	1755	27.2	Violle.
Potassium	K	62	15.7	Joannis.
Potassium nitrate	KNO_3	333.5	48.9	Person.
Phenol	C_6H_6O	25.37	24.93	Petterson.
Paraffin	–	52.40	35.10	Batelli.
Silver	Ag	961	21.07	Person.
Sodium	Na	97	31.7	Joannis.
" nitrate	$NaNO_3$	305.8	64.87	"
" phosphate	$Na_2HPO_4 + 12H_2O$	36.1	66.8	"
Spermaceti	–	43.9	36.98	Batelli.
Sulphur	S	115	9.37	Person.
Tin	Sn	232	14.0	Mean.
Wax (bees)	–	61.8	42.3	"
Zinc	Zn	419	28.13	"

* Total heat from 0° C.
† U. S. Bureau of Standards, 1913, in terms of 15° calorie.
‡ 1903, based on electrical measurements, assuming mechanical equivalent = 4.187, and in terms of the value of the international volt in use after 1911.

TABLE 261. — Heat of Combustion of Some Carbon Compounds.

Compound.	Formula.	Kg. cal. per g-mol.	Kg. cal. per g
Paraffins:			
Methane, g	CH_4	214*p*	13.3*v*
Ethane, g	C_2H_6	371*p*	12.4*v*
Propane, g	C_3H_8	528*p*	12.0*p*
i-Butane, g	C_4H_{10}	687*p*	11.8*p*
n-Hexane, l	C_6H_{14}	995*p*	11.6*v*
n-Heptane, l	C_7H_{16}	1139*p*	11.4*p*
n-Octane, l	C_8H_{18}	1315*p*	11.5*v*
Dekane, l	$C_{10}H_{22}$	1626*p*	11.4*v*
Olefines:			
Ethylene, g	C_2H_4	343*p*	12.2*p*
Propylene, g	C_3H_6	496*p*	11.8*v*
i-Butylene, g	C_4H_8	651*p*	11.6*p*
Amylene, l	C_5H_{10}	804*p*	11.5*p*
Hexylene, l	C_6H_{12}	962*p*	11.4*v*
Acetylene, g	C_2H_2	313*p*	12.0*p*
Trimethylene, g	C_3H_6	503*p*	11.9*p*
Benzene, l	C_6H_6	781*p*	10.0*p*
Benzene, g	C_6H_6	788*p*	10.1*p*
Naphthalene, l	$C_{10}H_8$	1235*p*	9.6*v*
Toluene, l	C_7H_8	937*p*	10.2*v*
Chloroform, v	$CHCl_3$	70	—
Carbon disulphide, l	CS_2	253*p*	3.28*v*
Methyl-chloride, g	CH_3Cl	169*p*	3.26*p*
Ethyl-chloride, v	C_2H_5Cl	332*p*	5.10*p*
Alcohols:			
Methyl, l	CH_4O	170*p*	5.31*p*
Ethyl, l	C_2H_6O	327*p*	7.10*p*
n-propyl, l	C_3H_8O	483*p*	8.00*p*
n-butyl, l	$C_4H_{10}O$	644*p*	8.68*v*
Amyl, l	$C_5H_{12}O$	788*p*	8.96*p*
Ethers:			
Dimethyl, g	C_2H_6O	346*p*	7.60*p*
Diethyl, v	$C_4H_{10}O$	660*p*	8.92*p*
Ethyl-methyl, v	C_3H_8O	506*p*	8.43*p*
Acids:			
Formic, l	CH_2O_2	62*p*	1.357*v*
Acetic, l	$C_2H_4O_2$	210*p*	3.49*v*
Propionic, l	$C_3H_6O_2$	368*p*	4.96*v*
n-butyric, l	$C_4H_8O_2$	525*p*	5.95*v*
Lactic, l	$C_3H_6O_3$	330*p*	3.66*v*
Cellulose, s	$C_6H_{10}O_5$	680	4.18*v*
Dextrine, s	$C_{12}H_{20}O_{10}$	414	—
Glycerine, l	$C_3H_8O_3$	397	4.32
Phenol, l	C_6H_6O	735	7.81
Sugar, cane, s	$C_{12}H_{22}O_{11}$	1353	3.95*v*
Starch, s	$C_6H_{10}O_5$	685	4.23
Thymol, l	$C_{10}H_{14}O$	1353	9.02*p*
Urea, l	$CO(NH_2)_2$	152	2.53

v, *p*, following the heats of combustion, signify at constant volume and pressure respectively. When referred to constant pressure, the values are 0.58 Kg-cal. greater (at about 18° C) for each condensed gaseous molecule. The values are means from various observers. The combustion products are gaseous CO_2, liquid water, etc.

TABLE 262. — Heat of Combustion — Miscellaneous.

Substance.	Small calories per g substance.	Reference.
Asphalt	9530	1
Butter	9200	–
Carbon: amorphous	8080	2
charcoal	8100	2
diamond	7860	3
graphite	7900	3
Copper (to CuO)	590	5
Dynamite, 75%	1290	4
Egg, white of	5700	–
Egg, yolk of	8100	–
Fats, animal	9500	2
Hemoglobin	5900	–
Hydrogen	33900	2
Iron (to Fe_2O_3)	1582	–
Magnesium (to MgO)	6080	–
Oils: cotton-seed	9500	–
lard	9300	2
olive	9400	2
Oils: petroleum:		
crude	11500	2
light	10000	2
heavy	10200	2
rape	9500	6
sperm	10000	7
Paraffin (to CO_2, H_2O l)	11140	6
Paraffin (to CO_2, H_2O g)	10340	6
Pitch	8400	–
Sulphur, rhombic	2200	2
Sulphur, monoclinic	2240	5
Tallow	9500	6
Woods: beech, 13% H_2O	4170	8
birch, 12% H_2O	4210	8
oak, 13% H_2O	3990	8
pine, 12% H_2O	4420	8

References: (1) Slossen, Colburn; (2) Mean; (3) Berthellot; (4) Roux, Sarran; (5) Thomsen; (6) Stohmann; (7) Gibson; (8) Gottlieb.

TABLE 263.

HEAT VALUES AND ANALYSES OF VARIOUS TYPES OF FUEL.

(a) COALS.

Coal.		Moisture.	Volatile matter.	Fixed Carbon.	Ash.	Sulphur.	Hydrogen.	Carbon.	Nitrogen.	Oxygen.	Calories per gram.	B. T. U.'s per pound.
Lignite	Low grade.	38.81	25.48	27.29	8.42	0.97	7.09	37.45	0.50	45.57	3526	6347
	High grade	33.38	27.44	29.62	9.56	0.94	6.77	41.31	0.67	40.75	3994	7189
Sub-bituminous	Low grade.	22.71	34.78	36.60	5.91	0.29	6.14	52.54	1.03	34.09	5115	9207
	High grade	15.54	33.03	46.06	5.37	0.58	5.89	60.08	1.05	27.03	5865	10557
Bituminous	Low grade.	11.44	33.93	43.92	10.71	4.94	5.39	60.06	1.02	17.88	6088	10958
	High grade	3.42	34.36	58.83	3.39	0.58	5.25	77.98	1.29	11.51	7852	14134
Semi-bituminous	Low grade	2.7	14.5	75.5	7.3	0.99	4.58	80.65	1.82	4.66	7845	14121
	High grade	3.26	14.57	78.20	3.97	0.54	4.76	84.62	1.02	5.09	8166	14699
Semi-anthracite		2.07	9.81	78.82	9.30	1.74	3.62	80.28	1.47	3.59	7612	13702
Anthracite	Low grade.	2.76	2.48	82.07	12.69	0.54	2.23	79.22	0.68	4.64	6987	12577
	High grade	3.33	3.27	84.28	9.12	0.60	3.08	81.35	0.79	5.06	7417	13351
Oven coke	Low grade.	1.92	1.58	88.87	8.99	1.18	—	—	—	—	7946	14300
	High grade	1.14	0.04	94.66	3.57	0.69	—	—	—	—	8006	14410

(b) PEATS AND WOOD (air dried).

	Vol. hydro-carbon.	Fixed carbon.	Ash.	Sul-phur.	Hydro-gen.	Carbon.	Nitro-gen.	Oxygen.	Calories per gram.	B.T.U.'s per pound.
Peats:										
Franklin Co., N. Y...	67.10	28.99	3.91	0.15	5.93	57.17	1.48	31.36	5726	10307
Sawyer Co., Wis.....	56.54	27.92	15.54	0.29	4.71	51.00	1.92	26.54	4867	8761
Woods:										
Oak, dry............	—	—	0.37	—	6.02	50.16	0.09	43.36	4620	8316
Birch, dry...........	—	—	0.29	—	6.06	48.88	0.10	44.67	4771	8588
Pine, dry...........	—	—	0.37	—	6.20	50.31	0.04	43.08	5085	9153

(c) LIQUID FUELS.

Fuel.	Specific gravity at 15° C.	Calories per gram.	British thermal units per pound.
Petroleum ether.........................	.684–.694	12210–12220	21978–21996
Gasoline................................	.710–.730	11100–11400	19980–20520
Kerosene................................	.790–.800	11000–11200	19800–20160
Fuel oils, heavy petroleum or refinery residue	.960–.970	10200–10500	18360–18900
Alcohol, fuel or denatured with 7 to 9 per cent water and denaturing material.....	.8196–.8202	6440–6470	11592–11646

(d) GASES.

Gas.	H_2	CH_4	C_2H_2	Illuminants.	CO_2	CO	O_2	N_2	Cal. per m^3	B.T.U. per cu. ft.
Natural gas, Cal.........	—	88.0	—	—	11.10	—	—	0.90	8339	937
Natural gas, Pa..........	—	53.3	45.8*	—	—	—	—	0.90	12635	1420
Natural gas, France......	—	98.81	—	—	0.58	—	0.1	0.48	9364	1052
Coal gas, high grade	34.80	28.80	9.50	1.70	0.20	10.40	0.40	14.20	6151	657
Coal gas, low grade	57.2	18.8	—	0.8	2.00	3.20	—	18.0	3736	399
Water gas, low grade.....	52.88	2.16	—	3.47	—	36.8	—	4.69	2642	283
Water gas, high grade....	36.4	23.2	—	14.05	3.02	19.1	1.15	3.08	6140	657

* C_2H_6. Data from the Geological Survey, Poole's The Calorific Power of Fuels, and for natural gas from Snelling (Van Nostrand's Chemical Annual).

Table 264.

CHEMICAL AND PHYSICAL PROPERTIES OF FIVE DIFFERENT CLASSES OF EXPLOSIVES.

Explosive.	Specific gravity.	Number of large calories developed by 1 kilogram of the explosive.	Pressure developed in own volume after elimination of surface influence.	Unit disruptive charge by ballistic pendulum.	Rate of detonation. Cartridges 1¼ in. diam.	Duration of flame from 100 grams of explosive.	Length of flame from 100 grams.	Cartridge 1¼ in. transmitted explosion at a distance of	Products of combustion from 200 grams; gaseous, solid, and liquid, respectively.	Ignition occurred in 4% fire damp & coal dust mixture with
			Kg. per sq. cm.	Grams.	Meters per second.	Millisec-onds.	Inches.	Inches.	Grams.	Grams.
(A) Forty-per-cent nitroglycerin dynamite	1.22	1221.4	8235	227*	4688	.358	24.63	12	88.4 79.7 14.5	25
(B) FFF black blasting powder	1.25	789.4	4817	374† 458*	469.4‡	925.	54.32	–	154.4 126.9 4.1‖	25
(C) Permissible explosive; nitroglycerin class	1.10	760.5	5912	301*	3008	.471	27.79	4	103.9 65.1 15.4	1000
(D) Permissible explosive; ammonium nitrate class	0.97	992.8	7300	279*	3438§	.483	25.68	1	89.8 27.5 75.5	800
(E) Permissible explosive; hydrated class	1.54	610.6	6597	434*	2479	.338	17.49	3	86.1 56.0 33.0	Over 1000

Chemical Analyses.

(A)	Moisture	0.91
	Nitroglycerin	39.68
	Sodium nitrate	42.46
	Wood pulp	13.58
	Calcium carbonate	3.37
(B)	Moisture	0.80
	Sodium nitrate	70.57
	Charcoal	17.74
	Sulphur	10.89
(C)	Moisture	7.89
	Nitroglycerin	24.02
	Sodium nitrate	36.25
	Wood pulp and crude fibre from grains	9.20
	Starch	21.31
	Calcium carbonate	0.97
	Magnesium "	0.36
(D)	Moisture	0.23
	Ammonium nitrate	83.10
	Sulphur	0.46
	Starch	2.61
	Wood pulp	1.89
	Poisonous matter	2.54
	Manganese peroxide	2.64
	Sand	6.53
(E)	Moisture	2.34
	Nitroglycerin	30.85
	Ammonium nitrate	9.94
	Sand	1.75
	Coal	11.98
	Clay	7.64
	Ammonium sulphate	8.96
	Zinc sulphate (7HO)	6.89
	Potassium sulphate	19.65

* One pound of clay tamping used. † Two pounds of clay tamping used. ‡ Rate of burning.
§ Cartridges 1¾ in. diam. ‖ For 300 grammes.

Compiled from U. S. Geological Survey Results, — "Investigation of Explosives for use in Coal Mines, 1909."

TABLE 265. — Additional Data on Explosives.

Explosive. (Ref. Young, Nature, 102, 216, 1918.)	Vol. gas per g in cc = V	Calories per g = Q	Coefficient = QV ÷ 1000	Coefficient GP = 1	Calculated Temperature Q/C C, sp. ht. gases = 0.24
	cc				
Gunpowder	280	738	207	1	2240° C
Nitroglycerine	741	1652	1224	6	6880
Nitrocellulose, 13% N_2	923	931	859	4.3	3876
Cordite, Mk. I. (NG, 57; NC, 38; Vaseline, 5)	871	1242	1082	5.2	5175
Cordite, MD (NG, 30; NC, 65; Vaseline, 5)	888	1031	915	4.4	4225
Ballistite (NG, 50; NC, 50; Stabilizer, 5)	817	1349	1102	5.3	5621
Picric acid (Lyddite)	877	810	710	3.4	3375

Shattering power of explosive = vol. gas per g × cals./g × V_d × density where V_d is the velocity of detonation.
Trinitrotoluene: V_d = 7000 m/sec. Shattering effect = .87 picric acid.
Amatol (Ammonium nitrate + trinitrotoluene, TNT): V_d = 4500 m/sec.
Ammonal (Ammonium nitrate, TNT, Al): 1578 cal/g; 682 cc gas; V_d = 4000 m/sec.
Sabulite (Ammonium nitrate, 78, TNT 8, Ca silicide 14): about same as ammonal.

TABLE 266. — Ignition Temperatures Gaseous Mixtures.

Ignition temperature taken as temperature necessary for hot body immersed in gas to cause ignition; slow combination may take place at lower temperatures. McDavid, J. Ch. Soc. Trans. 111, 1003, 1917. Gases were mixed with air. Practically same temperatures as with O_2 (Dixon, Conrad, *loc. cit.* 95, 1909).

Benzene and air	1062° C	Ether and air	1033° C
Coal gas and air	878	Ethylene and air	1000
CO and air	931	Hydrogen and air	747

TABLE 267. — Time of Heating for Explosive Decomposition.

Temperature ° C.	170	180	190	200	220	Ignition temperature.	
Time.	sec.	sec.	sec.	sec.	sec.	° C †	° C ‡
Black powder	*n*	*n*	*n*	*n*	*n*	440	—
Smokeless powder A	600	195	130	45	23	{ 300	—
Smokeless powder B	190	130	—	90	25		—
Celluloid Pyroxylin	170	60	—	21	9	—	—
Collodion cotton	870	165	67	56	18	300	—
Celluloid *	160	100	60	50	30	590	450
Safety matches	*n*	340	240	150	60	—	—
Parlor matches	*n*	*n*	*n*	590	480	—	—
Cotton wool	—	—	—	—	—	900	—

n, failure to explode in twenty minutes. * The decomposition of nitrocellulose in celluloid commences at about 100° C; above that the heat of decomposition may raise the mass to the ignition point if loss of heat is prevented. Above 170°, decomposition occurs with explosive violence as with nitrocellulose. Rate of combustion is 5 to 10 times that of poplar, pine, or paper of the same size and conditions.
† Measured by contact with porcelain tube of given temperature. Average.
‡ Measured by contact with molten lead. Average.
Taken from Technologic Paper of Bureau of Standards, No. 98, 1917.

TABLE 268. — Flame Temperatures.

Measures made with optical pyrometer by Féry, J. de Phys. (4) 6, 1907.

Alcohol, with NaCl	1705° C	Hydrogen flame	1900° C
Bunsen flame, no air	1712	Hydrogen-oxygen	2420
Bunsen flame, ½ air	1812	Acetylene burner	2458
Bunsen flame, full air	1871	Acetylene-oxygen	3000
Illuminating gas-oxygen	2200	Cooper-Hewitt Hg	3500

THERMO-CHEMISTRY. CHEMICAL ENERGY DATA.

The total heat generated in a chemical reaction is independent of the steps from initial to final state. Heats of formation may therefore be calculated from steps chemically impracticable. Chemical symbols now represent the chemical energy in a gram-molecule or mol(*e*); treat reaction equations like algebraic equations: $CO + O = CO_2 + 68$ Kg-cal; subtract $C + 2\ O = CO_2 + 97$ Kg-cal, then $C + O = CO + 29$ Kg-cal. We may substitute the negative values of the formation heats in an energy equation and solve $MgCl_2 + 2\ Na = 2\ NaCl + Mg + x$ Kg-cal; $-151 = -196 + x$; $x = 45$ Kg-cal. Heats of formation of organic compounds can be found from the heats of combustion since burned to H_2O and CO_2. When changes are at constant volume, energy of external work is negligible; also generally for solid or liquid changes in volume. When a gas forms a solid or liquid at constant pressure, or vice versa, it must be allowed for. For N mols of gas formed (disappearing) at $T_K°$ the energy of the substance is decreased (increased) by $0.002 \cdot N \cdot T_K$ Kg-cal. $H_2 + O = H_2O + 67.5$ Kg-cal. at 18°C. at constant volume; $\frac{1}{2}(2\ H_2 + O_2 - 2\ H_2O = 135.0 + 0.002 \times 3 \times 291 = 136.7) = 68.4$ Kg-cal.

The heat of solution is the heat, + or —, liberated by the solution of 1 mol of substance in so much water that the addition of more water will produce no additional heat effects. Aq. signifies this amount of water; H_2O, one mol.; $NH_3 + Aq = NH_4OH \cdot Aq. + 8$ Kg-cal.

TABLE 269. (a). Heats of Formation from Elements in Kilogram Calories.

At ordinary temperatures.

Compound.	Heat of Formation.	Compound.	Heat of Formation.	Compound.	Heat of Formation.	Compound.	Heat of Formation.
Al_2O_3	380.	HgO	21.4	KCl	105.7	Li_2SO_4	334.2
Ag_2O	6.5	Na_2O	100.	$LiCl$	93.8	$(NH_4)_2SO_4$	283.
BaO	126.	Nd_2O_3	435.	$MgCl_2$	151.0	Na_2SO_4	328.3
BaO_2	142.	NiO	57.9	$MnCl_2$	112.3	$MgSO_4$	301.6
Bi_2O_3	138.	P_2O_5 sgs	370.	$NaCl$	97.8	$PbSO_4$	216.2
CO am	29.0	PbO	50.3	$NdCl_3$	250.	Tl_2SO_4	221.0
CO di	26.1	PbO_2	62.4	NH_4Cl	76.3	$ZnSO_4$	229.6
CO_2 am	97.0	Pr_2O_3	412.	$NiCl_2$	74.5	$CaCO_3$	270.
CO_2 gr	94.8	Rb_2O	89.2	$PbCl_2$	83.4	$CuCO_3$	143.
CO_2 di	94.3	SO_2 rh sgg	70.	$PdCl_2$	40.5	$FeCO_3$	179.
CaO	152.	SiO_2	191.0	$PtCl_4$	60.4	K_2CO_3	280.
CeO_2	225.	SnO	66.9	$SnCl_2$	80.8	$MgCO_3$	267.
Cl_2O g	-16.5	SnO_2 cr	137.5	$SnCl_4$	128.	Na_2CO_3	272.
CoO am	50.5	SrO_2	135.	$SrCl_2$	185.	$ZnCO_3$	194.
CoO cr	57.5	ThO_2	326.	$ThCl_4$	300.	$AgNO_3$	28.7
Co_3O_4	193.4	TiO_2 am	215.6	$TlCl$	48.6	$Ca(NO_3)_2$	209.
CrO_3	140.	TiO_2 cr	218.4	$RbCl$	105.9	$Cu(NO_3)_2 6\ H_2O$	92.9
Cs_2O	91.3	TlO_2	42.2	$ZnCl_2$	97.3	HNO_3 gggl	41.6
Cu_2O	42.3	WO_2	131.	HBr glg	8.6	KNO_3	119.2
CuO	37.2	WO_3	194.	NH_4Br	66.	$LiNO_3$	112.
FeO	65.7	ZnO	85.2	HI gsg	-6.2	NH_4NO_3	88.3
Fe_2O_3	196.5	$AgCl$	29.2	HF ggg	38.	$NaNO_3$	111.0
Fe_3O_4	270.8	Ag_2Cl	29.5	Ag_2S	3.3	$TlNO_3$	58.2
H_2O ggl	68.4	$AlCl_3$	161.4	CS_2 sgg	-26.0	CH_4 sgg	20.
H_2O_2 ggl	46.8	$AuCl$ y	5.81	CaS	90.8	C_2H_6 sgg	25.
Hg_2O	22.2	$AuCl_3$ y	22.8	$(NH_4)_2S$	66.2	C_2H_2 sgg	-53.
HgO	21.4	$BaCl_2$	197.	Cu_2S	18.3	HCN di gsgg	-30.5
K_2O	91.	$BiCl_3$	90.6	CuS	11.6	NH_3 ggg	12.0
La_2O_3	447.	CCl_4 am	21.0	H_2S gsg	2.73	$Ca(OH)_2$	230.
LiO_2	141.6	$CaCl_2$	187.	K_2S	103.4	NH_4OH	88.8
MgO	143.6	$CdCl_2$	93.2	MgS	79.4	$NaOH$	102.
MnO	90.8	$CoCl_2$	76.5	Na_2S	89.3	$Na \cdot H_2O \cdot Aq - H$	44.*
MnO_2	123.	$CuCl_2$	51.5	PbS	19.3	$\frac{1}{2}(2\ Na \cdot O \cdot H_2O)$	68.*
Mn_3O_4	325.	$CuCl$	34.1	$CaSO_4$	262.	$\frac{1}{2}(Na_2O \cdot H_2O \cdot Aq)$	30.*
MoO_2	143.	$FeCl_2$	82.1	$CuSO_4$	111.5	KOH	103.5
MoO_3	174.	$FeCl_3$	96.0	H_2SO_4 sggg	193.	$K \cdot H_2O \cdot Aq - H$	45.*
N_2O ggg	-18.2	$GlCl_2$	155.	$-SO_3 \cdot H_2O$*	21.3	$\frac{1}{2}(2\ K \cdot O \cdot H_2O)$	69.*
NO ggg	-21.6	HCl ggl	22.	Hg_2SO_4	175.	$\frac{1}{2}(K_2O \cdot H_2O \cdot Aq)$	35.5*
NO_2	- 8.1	$HgCl$	31.3	$HgSO_4$	165.		
Na_2O_4	- 2.6	$HgCl_2$	53.3	K_2SO_4	344.3		

am = amorphous; di = diamond; gr = graphite; cr = crystal; g = gas; l = liquid; s = solid; y = yellow (gold); rh = rhombic (sulphur). * Heats of formation not from elements but as indicated.

HEATS OF FORMATION OF IONS IN KILOGRAM-CALORIES.

+ and — signs indicate signs of ions and the number of these signs the valency. For the ionisation of each gram-molecule of an element divide the numbers in the table by the valency, e. g., 9.03 gr. Al = 9.03 gr. Al^{+} + 40.3 Kg. cal. When a solution is of such dilution that further dilution does not increase its conductivity, then the heats of formation of substances in such solutions may be found as follows: $FeCl_2Aq = +22.2 + 2 \times 39.1 = 100.4$ Kg. cal. $CuSO_4Aq = -15.8 + 214.0 = 198.2$ Kg. cal.

Ag^{+}	−25.3	NH_4^{+}	+32.7	AsO_4^{---}	+215.0	IO_3^{-}	+55.8
Al^{+++}	+121.0	NH_4O^{+}	+37.5	Br^{-}	+28.2	IO_4^{-}	+46.5
Co^{++}	+170.0	Na^{+}	+57.3	BrO_3^{-}	+11.2	OH^{-}	+54.4
Ca^{++}	+133.?	Ni^{++}	+16.0	CO_3^{--}	+160.8	PO_4^{---}	+298.0
Cd^{++}	+18.4	Mg^{++}	+108.8	Cl^{-}	+39.1	$S_2O_3^{--}$	+138.6
Cu^{++}	−16.0	Mn^{++}	+50.2	ClO^{-}	+26.0	$S_2O_6^{--}$	+278.2
Cu^{+}	−15.8?	Pb^{++}	+4.0	ClO_3^{-}	+23.4	$S_4O_6^{--}$	+260.8
Fe^{++}	+22.2	Rb^{+}	+625.0	ClO_4^{-}	−38.7	SO_3^{--}	+151.0
Fe^{+++}	−9.3	Sn^{+++}	+3.3	HCO_3^{-}	+163.0	SO_4^{--}	+214.0
H^{+}	0.0	Sr^{++}	+119.6	HPO_2^{-}	+143.9	Se^{--}	−35.6
Hg^{+}	−19.8	Tl^{+}	+1.7	HPO_3^{--}	+229.6	SeO_3^{--}	+119.6
K^{+}	+61.8	Zn^{++}	+35.0	HPO_4^{--}	+304.8	SeO_4^{--}	+144.8
Li^{+}	+62.8			HS^{-}	+1.2	Te^{--}	−34.8
				NO_2^{-}	+27.0	TeO_3^{--}	+77.0
				NO_3^{-}	+48.9	TeO_4^{--}	+98.4
				I^{-}	+13.1	S^{--}	−12.6

TABLE 271.—Heats of Neutralization in Kilogram-Calories.

The heat generated by the neutralization of an acid by a base is equal, for each gram-molecule of water formed, to 13.7 Kg. cal. plus the heat produced by the amount of un-ionized salt formed, plus the sum of the heats produced in the completion of the ionizations of the acid and the base. (See also p. 209).

Base.	HCl·aq	HNO_3·aq	H_2SO_4·aq	HCN·aq	CH_3COOH·aq	H_2CO_3·aq
KOH · aq	13.7	13.8	15.7	2.9	13.3	10.1
NaOH · aq	13.7	13.7	15.7	2.9	13.3	10.2
NH_4OH · aq	12.4	12.5	14.5	1.3	12.0	8.
½ $Ca(OH)_2$ · aq	14.0	13.9	15.6	3.2	13.4	9.5
½ $Zn(OH)_2$ · aq	9.9	9.9	11.7	8.1	8.9	5.5
½ $Cu(OH)_2$ · aq	7.5	7.5	9.2	—	6.2	—

TABLE 272.—Heat of Dilution, H_2SO_4.

In Kilogram-calories by the dilution of one gram-molecule of sulphuric acid by m gram-molecules of water.

m	1	2	3	5	19	49	99	199	399	1599
Kg. Cal. . .	6.38	9.42	11.14	13.11	16.26	16.68	16.86	17.06	17.31	17.86

RADIATION CONSTANTS.

TABLE 273.—Radiation Formulæ and Constants for Perfect Radiator.

The radiation per sq. cm. from a "black body" (exclusive of convection losses) at the temperature $T°$ (absolute, C) to one at $t°$ is equal to

$$J = \sigma(T^4 - t^4) \quad \text{(Stefan–Boltzmann);}$$

where $\sigma = 1.364 \times 10^{-12}$ gram-calories per second per sq. centimeter.
$= 8.20 \times 10^{-11}$ " " " minute " " "
$= 5.71 \times 10^{-12}$ watts per sq. centimeter.

The distribution of this energy in the spectrum is represented by Planck's formula:

$$J_\lambda = C_1 \lambda^{-5} \left[e^{\frac{C_2}{\lambda T}} - 1 \right]^{-1}$$

where J_λ is the intensity of the energy at the wave-length λ (λ expressed in microns, μ) and e is the base of the Napierian logarithms.

$C_1 = 8.86 \times 10^3$ for J in $\frac{gram.\ cal.}{sec.\ cm.^2} = 3.70 \times 10^4$ for J in $\frac{watts}{cm.^2}$

$C_2 = 14325$ for λ in μ

$J_{max} = 3.11 \times 10^{-16}\ T^5$ for J in $\frac{gram.\ cal.}{sec.\ cm.^2} = 1.30 \times 10^{-15}\ T^5$ for J in $\frac{watts}{cm.^2}$

$\lambda_{max}\ T = 2885$ for λ in μ

h = Planck's unit = elementary "Wirkungs quantum" $= 6.554 \times 10^{-27}$ ergs. sec.

k = constant of entropy equation $= 1.37 \times 10^{-16}$ ergs./degrees.

TABLE 274.—Radiation in Gram-Calories per 24 Hours per sq. cm. from a Perfect Radiator at $t°$ C to an absolutely Cold Space (−273° C).

Computed from the Stefan-Boltzmann formula.

$t°$ C	J	$t°$ C	J	$t°$ C	J	$t°$ C	J	$t°$ C	J	$t°$ C	J
−273	0	−120	65	−10	565	+12	778	+34	1048	+56	1380
−220	1	−110	83	−8	582	+14	801	+36	1076	+58	1420
−210	2	−100	106	−6	600	+16	823	+38	1104	+60	1450
−200	3	−90	132	−4	618	+18	846	+40	1133	+70	1630
−190	5	−80	164	−2	636	+20	870	+42	1162	+80	1830
−180	9	−70	201	0	655	+22	893	+44	1192	+90	2050
−170	13	−60	243	+2	675	+24	918	+46	1222	+100	2280
−160	19	−50	292	+4	695	+26	943	+48	1253	+200	5905
−150	27	−40	348	+6	715	+28	969	+50	1277	+1000	310×10^3
−140	38	−30	410	+8	736	+30	995	+52	1316	+2000	315×10^4
−130	50	−20	483	+10	757	+32	1021	+54	1348	+5000	912×10^5

TABLE 275.—Values of J_λ for Various Temperatures Centigrade.

Ekholm, Met. Z. 1902, used $C_1 = 8346$ and $C_2 = 14349$, and for the unit of time the day.
For 100°, the values for $J\lambda$ have been multiplied by 10, for the other temperatures by 100.

λ	T = 100° C	30° C	15° C	0° C	−30° C	−80° C	λ	100° C	30° C	15° C	0° C	−30° C	−80° C
μ							μ						
2	1	0	0	0	0	0	18	511	2961	2557	2175	1491	623
3	80	41	18	7	1	0	19	443	2626	2281	1954	1363	594
4	469	508	272	138	27	1	20	386	2329	2034	1754	1242	561
5	1047	1777	1085	628	172	8	21	337	2068	1816	1574	1129	527
6	1526	3464	2296	1454	493	39	22	295	1840	1622	1413	1026	494
7	1768	4954	3481	2353	931	105	23	259	1639	1448	1270	931	460
8	1810	5928	4352	3088	1372	203	24	228	1462	1298	1141	846	428
9	1724	6382	4834	3646	1730	316	25	202	1307	1165	1028	768	398
10	1573	6386	4979	3781	1971	426	26	179	1170	1047	926	698	369
11	1398	6127	4833	3798	2098	520	28	142	947	850	757	579	317
12	1225	5712	4633	3676	2114	592	30	114	771	696	623	482	272
13	1063	5222	4300	3467	2090	640	40	44	311	285	259	209	130
14	918	4713	3930	3215	2004	666	50	20	146	135	124	102	67
15	792	4220	3556	2944	1889	673	60	10	77	72	66	55	38
16	683	3759	3198	2674	1760	663	80	4	27	25	24	20	14
17	590	3340	2862	2417	1626	649	100	2	12	11	10	9	7

TABLE 276.

BLACK-BODY SPECTRUM INTENSITIES (J_λ).

Values of J_λ using for C_1, 9.23×10^3, C_2, 14350., λ in μ. If the figures given for J_λ are plotted in cms as ordinates to a scale of abscissae of 1 cm to 1 μ, then the area in cm² between the smooth curve through the resulting points and the axis of abscissae is equivalent to the radiation in calories per sec. from 1 cm² of a black body at the corresponding temperature, radiating to absolute zero. The intensities when radiating to a body at a lower temperature may be obtained by subtracting the intensities corresponding to the lower temperature from those of the higher. The nature of the black-body formula is such that when λT is small, a small change in C_2 produces a great change in J_λ; e.g., when $C_2/\lambda T$ is 100 or 10, the change is 100 and 10 fold respectively; as λT increases, the change becomes proportional; e.g., when $C_2/\lambda T$ is less than 0.05, the change in J_λ is proportional to the change in C_2.

λ	50° K.	100° K.	150° K.	200° K.	250° K.	273° K.	300° K.	373° K.	400° K.	500° K.	600° K.
μ											
1.0	—	.0583	.0372	.0276	.0201	.0181	.0161	.0122	.01124	.0831	.0638
1.5	—	.0383	.0242	.0172	0133	.0127	.0102	.088	.0749	.0558	.03143
2.0	.0591	.0282	.0185	.0137	.091	.0911	.0712	.0513	.0546	.03168	.00184
2.5	.0471	.0221	.0142	.0103	.0710	.077	.0646	.0419	.0450	.0397	.0066
3.0	.0409	.0196	.0125	.082	.0618	.069	.0545	.03102	.03242	.00265	.0131
3.5	.0344	.0163	.0102	.072	.0513	.055	.0420	.0329	.03620	.00482	.0189
4.0	.0306	.0142	.094	.0614	.0552	.0418	.0457	.0360	.00115	.00690	.0229
5.0	.0243	.0111	.0714	.0517	.0430	.048	.0321	.00134	.00226	.00952	.0249
6.0	.02019	.0105	.0614	.058	.048	.0318	.0341	.00195	.00301	.01001	.0224
7.0	.01883	.096	.066	.0419	.0315	.0330	.0359	.00225	.00328	.00925	.0186
8.0	.01672	.085	.0518	.0436	.0322	.0339	.0371	.00232	.00321	.00801	.0149
9.0	.01422	.0718	.0538	.0454	.0327	.0345	.0377	.00220	.00295	.00672	.0118
10.0	.01331	.0754	.0565	.0471	.0330	.0348	.0378	.00201	.00262	.00554	.00929
12.0	.01115	.0624	.0413	.0494	.0331	.0347	.0370	.00157	.00196	.00374	.00585
14.0	.01021	.0661	.0418	.04102	.0329	.0341	.0358	.00117	.00144	.00254	.00380
16.0	.0914	.0511	.0422	.04100	.0325	.0334	.0346	.0387	.00105	.00176	.00254
18.0	.0957	.0517	.0424	.0492	.0321	.0328	.03368	.03653	.03760	.00124	.00176
20.0	.0816	.0522	.0424	.0482	.0317	.03224	.03290	.03493	.03575	.03902	.00125
25.0	.0897	.0530	.0421	.0457	.03122	.03131	.03164	.03258	.03295	.03439	.03589
30.0	.0726	.0532	.0416	.0438	.0466	.0479	.0497	.03146	.03164	.03237	.03311
40.0	.0769	.0526	.059	.0418	.04282	.0433	.04391	.04558	.04620	.04858	.03110
50.0	.0795	.0518	.0551	.0592	.04150	.04158	.04184	.04255	.04281	.04381	.04482
75.0	.0787	.0667	.0515	.0524	.05338	.05383	.05436	.05580	.05634	.05834	.04103
100.0	.0755	.0629	.0657	.0688	.05119	.05134	.05150	.05197	.05214	.05277	.05342

λ	800° K.	1000° K.	1500° K.	2000° K.	3000° K.	4000° K.	5000° K.	6000° K.	8000° K.	10000° K.	20000° K.
μ											
0.1	—	—	—	0.0226	0.01115	0.0624	0.0331	0.038	15.	540.	710000.
0.2	—	—	—	0.087	0.0012	0.46	15.4	184.	3660.	22100.	820000.
0.3	—	—	—	0.0315	0.44	24.2	263.	1310.	9640.	31000.	3820000.
0.4	—	—	—	0.0145	5.75	115.	690.	2280.	10300.	25600.	180000.
0.5	—	—	—	0.172	20.6	226.	952.	2490.	8400.	17800.	92300.
0.6	—	.0548	0.014	0.757	40.8	301.	1000.	2240.	6290.	11950.	51460.
0.7	.0640	.0468	0.064	1.93	59.2	328.	925.	1860.	4590.	8110.	30700.
0.8	.0551	.00045	0.180	3.58	71.5	321.	800.	1490.	3350.	5620.	19400.
0.9	.0434	.00183	0.378	5.35	77.3	295.	671.	1177.	2470.	3980.	12820.
1.0	.00015	.00538	0.645	7.06	77.8	262.	554.	928.	1842.	2880.	8800.
1.5	.0775	.0848	2.07	10.25	52.2	122.	210.	309.	527.	758.	1980.
2.0	.0367	.221	2.43	8.19	29.0	57.6	90.2	125.	198.	275.	668.
2.5	.0719	.305	2.10	5.68	16.4	29.5	43.9	58.9	90.1	121.9	284.
3.0	.0964	.320	1.64	3.82	9.66	16.4	23.7	31.1	46.4	61.9	140.7
3.5	.1050	.296	1.22	2.60	6.02	9.84	13.8	17.9	26.3	34.7	77.3
4.0	.1027	.256	0.907	1.80	3.90	6.20	8.59	11.0	15.9	20.9	45.9
5.0	.0839	.178	0.511	0.923	1.84	2.81	3.81	4.81	6.84	8.89	19.15
6.0	.0629	.119	0.302	0.514	0.973	1.45	1.935	2.42	3.40	4.39	9.34
7.0	.0459	.0811	0.188	0.307	0.560	0.820	1.165	1.348	1.88	2.41	5.09
8.0	.0335	.0562	0.122	0.194	0.344	0.498	0.653	0.808	1.20	1.43	3.00
9.0	.0247	.0398	0.0824	0.128	0.223	0.319	0.416	0.513	0.709	0.90	1.87
10.0	.0184	.0288	0.0575	0.0880	0.151	0.214	0.278	0.342	0.470	0.598	1.24
12.0	.01072	.0160	0.0304	0.0553	0.0757	0.107	0.1373	0.168	0.230	0.292	0.602
14.0	.00660	.0096	0.0175	0.0256	0.0421	0.0587	0.0754	0.0921	0.125	0.159	0.326
16.0	.00425	.00606	0.0108	0.0155	0.0253	0.0350	0.0448	0.0546	0.0742	0.0938	0.192
18.0	.00285	.00400	0.00697	0.00997	0.0160	0.0221	0.0282	0.0344	0.0466	0.0585	0.120
20.0	.00198	.00275	0.00470	0.00668	0.01068	0.0147	0.01868	0.0227	0.0307	0.0388	0.0789
25.0	.00090	.00122	0.00203	0.00284	0.00448	0.00612	0.00777	0.00941	0.0127	0.0160	0.0325
30.0	.03464	.03619	0.00101	0.00141	0.00220	0.00299	0.00378	0.00455	0.00616	0.00775	0.0157
40.0	.03159	.03209	0.03334	0.03459	0.03710	0.03960	0.00121	0.00146	0.00197	0.00247	0.00498
50.0	.04684	.04888	0.03140	0.03191	0.03294	0.03397	0.03500	0.03603	0.03808	0.00101	0.00204
75.0	.04144	.04184	0.04286	0.04387	0.04591	0.04794	0.04997	0.03120	0.03161	0.03201	0.03406
100.0	.05470	.05598	0.05919	0.04124	0.04188	0.04252	0.04317	0.04381	0.04510	0.04639	0.03128

See Forsythe, J. Opt. Soc., 4,331, 1920, relative values, 0.4 to 0.76 μ (steps 0.01 μ), 12 temperatures, 1000 to 5000° K.

RADIATION EMISSIVITIES.

TABLE 277. — Relative Emissive Powers for Total Radiation.

Emissive power of black body = 1. Receiving surface platinum black at 25° C; oxidized surfaces oxidized at 600 + ° C. Randolph and Overholzer, Phys. Review, 2, p. 144, 1913.

	Temperature, Deg. C. 200	400	600
Silver	0.020	0.030	0.038
Platinum (1)	0.060	0.086	0.110
Oxidized zinc	—	0.110	—
Oxidized aluminum	0.113	0.153	0.192
Calorized copper, oxidized	0.180	0.185	0.190
Cast iron	0.210	—	—
Oxidized nickel	0.369	0.424	0.478
Oxidized monel	0.411	0.439	0.463
Calorized steel, oxidized	0.521	0.547	0.570
Oxidized copper	0.568	0.568	0.568
Oxidized brass	0.610	0.600	0.589
Oxidized lead	0.631	—	—
Oxidized cast iron	0.643	0.710	0.777
Oxidized steel	0.790	0.788	0.787
Black body	1.00	1.00	1.00

Remark: For radiation properties of bodies at temperatures so low that the radiations of wave-length greater than 20 μ or thereabouts are important, doubt must exist because of the possible and perhaps probable lack of blackness of the receiving body to radiations of those wave-lengths or greater. For instance, see Table 379 for the transparency of soot.

TABLE 278. — Emissivities of Metals and Oxides.

Emissivities for radiation of wave-length 0.55 and 0.65 μ. Burgess and Waltenberg, Bul. Bureau of Standards, 11, 591, 1914.

In the solid state practically all the metals examined appear to have a negligible or very small temperature coefficient of emission for λ = 0.55 and 0.65 μ within the temperature range 20° C to melting point. Nickel oxide has a well-defined negative coefficient, at least to the melting point. There is a discontinuity in emissivity, for λ = 0.65 μ at the melting point for some but not all the metals and oxides. This effect is most marked for gold, copper, and silver, and is appreciable for platinum and palladium. Palladium, in addition, possesses for radiation a property analogous to suffusion, in that the value of emissivity (λ = 0.65 μ) natural to the liquid state may persist for a time after solidification of the metal. The Violle unit of light does not appear to define a constant standard. Article contains bibliography.

Metals.	Cu	Ag	Au	Pd	Pt	Ir	Rh	Ni	Co	Fe	Mn	Ti
eλ, 0.55 μ solid	0.38	0.35	0.38	0.38	0.38	—	0.29	0.44	—	—	—	0.75
0.55 μ liquid	0.36	0.35	0.38	—	—	—	—	0.46	—	—	—	0.75
0.65 μ solid	0.10	0.04	0.14	0.33	0.33	0.30	0.29	0.36	0.36	0.37	0.59	0.63
liquid	0.15	0.07	0.22	0.37	0.38	—	0.30	0.37	0.37	0.37	0.59	0.65
Metals	Zr	Th	Y	Er	Be	Cb	V	Cr	Mo	W	U	
eλ, 0.55 μ solid	—	0.36	—	—	0.61	0.61	0.29	0.53	—	—	0.77	
liquid	—	—	—	0.30	0.81	—	—	—	—	—	—	
0.65 μ solid	0.32	0.36	0.35	0.55	0.61	0.49	0.35	0.39	0.43	0.39	0.54	
liquid	0.30	0.40	0.35	0.38	0.61	0.40	0.32	0.39	0.40	—	0.34	
Oxides: 0.65 μ	NiO	Co_3O_4	Fe_3O_4	Mn_3O_4	TiO_2	ThO_2	Y_2O_3	BeO	CbO_x	V_2O_3	Cr_2O_3	U_3C_8
eλ, solid	0.89	0.77	0.63	—	0.52	0.57	0.61	0.37	0.71	0.69	0.60	0.30
liquid	0.68	0.63	0.53	0.47	0.51	0.69	—	—	—	—	—	0.31

RADIATION EMISSIVITIES.

TABLE 279. — Relative Emissivities of Metals and Oxides.

Emissivity of black body taken as 100.

True temperature C.		500°	600°	700°	800°	900°	1000°	1100°	1200°	Ref.
60 FeO.40 Fe_2O_3 = Fe heated in air	Total	85	85	86	87	87	88	88	89	1
	λ = 0.65 μ	—	—	—	98	97	95	93	92	1
NiO	Total	—	54	62	68	72	75	81	86	2
	λ = 0.65 μ	—	—	98	96	94	92	88	87	2

Platinum:													
True temp. C.	0	100	200	300	400	500	750	1000	1200	1400	1600	1700	3
App.* temp. C.	—	—	—	—	—	—	—	486	630	780	930	1005	3
Total emiss. Pt.	3.1	4.0	5.1	6.1	7.0	8.0	10.3	12.4	14.0	15.5	16.9	17.5	3

Tungsten:											
True temp. K (abs.)	200	600	1000	1400	1800	2200	2600	3000	3400	3800	4
λ = 0.467	51.8	50.8	49.8	48.9	47.9	47.0	46.0	45.0	44.1	—	4
λ = 0.665	48.2	47.2	46.3	45.3	44.3	43.3	42.4	41.4	40.4	39.5	4

* As observed with total radiation pyrometer sighted on the platinum.

References: (1) Burgess and Foote, Bul. Bureau of Standards, 12, 83, 1915; (2) Burgess and Foote, *loc. cit.* 11, 41, 1914; (3) Foote, *loc. cit.* 11, 607, 1914; (4) Worthing, Phys. Rev. 10, 377, 1917.

TABLE 280. — Temperature Scale for Tungsten.

Hyde, Cady, Forsythe, J. Franklin Inst. 181, 418, 1916. See also Phys. Rev. 10, 395, 1917. The color temperature = temperature of black body at which its color matches the given radiation.

Lumens/watt	Color temperature.	Black-body temperature.	True temperature.	True temperature.	True — color.	True — brightness.
1	1763° K.	1627° K.	1729° K.	1700°	12°	100°
2	1917	1753	1875	1800	20	115
3	2025	1840	1976	1900	26	128
4	2109	1909	2056	2000	31	142
5	2179	1967	2125	2100	36	158
6	2237	2017	2184	2200	39	175
7	2290	2062	2238	2300	41	191
8	2338	2102	2286	2400	43	208
9	2383	2140	2332			
10	2425	2174	2373			

TABLE 281. — Color minus Brightness Temperatures for Carbon.

Hyde, Cady, Forsythe, Phys. Rev. 10, 395, 1917.

Brightness temp. ° K.	1600°	1700°	1800°	1900°	2000°	2100°	2200°
Color — brightness	2	7	12	16	22	28	33

COOLING BY RADIATION AND CONVECTION.

TABLE 282. — At Ordinary Pressures.

According to McFarlane * the rate of loss of heat by a sphere placed in the centre of a spherical enclosure which has a blackened surface, and is kept at a constant temperature of about 14° C, can be expressed by the equations

$$e = .000238 + 3.06 \times 10^{-6}t - 2.6 \times 10^{-8}t^2,$$

when the surface of the sphere is blackened, or

$$e = .000168 + 1.98 \times 10^{-6}t - 1.7 \times 10^{-8}t^2,$$

when the surface is that of polished copper. In these equations, e is the amount of heat lost in c. g. s. units, that is, the quantity of heat, small calories, radiated per second per square centimeter of surface of the sphere, per degree difference of temperature t, and t is the difference of temperature between the sphere and the enclosure. The medium through which the heat passed was moist air. The following table gives the results.

Difference of temperature t	Value of e. Polished surface.	Value of e. Blackened surface.	Ratio.
5	.000178	.000252	.707
10	.000186	.000266	.699
15	.000193	.000279	.692
20	.000201	.000289	.695
25	.000207	.000298	.694
30	.000212	.000306	.693
35	.000217	.000313	.693
40	.000220	.000319	.693
45	.000223	.000323	.690
50	.000225	.000326	.690
55	.000226	.000328	.690
60	.000226	.000328	.690

TABLE 283. — At Different Pressures.

Experiments made by J. P. Nicol in Tait's Laboratory show the effect of pressure of the enclosed air on the rate of loss of heat. In this case the air was dry and the enclosure kept at about 8° C.

Polished surface. t	Polished surface. et	Blackened surface. t	Blackened surface. et
PRESSURE 76 CMS. OF MERCURY.			
63.8	.00987	61.2	.01746
57.1	.00862	50.2	.01360
50.5	.00736	41.6	.01078
44.8	.00628	34.4	.00860
40.5	.00562	27.3	.00640
34.2	.00438	20.5	.00455
29.6	.00378	–	–
23.3	.00278	–	–
18.6	.00210	–	–
PRESSURE 10.2 CMS. OF MERCURY.			
67.8	.00492	62.5	.01298
61.1	.00433	57.5	.01158
55	.00383	53.2	.01048
49.7	.00340	47.5	.00898
44.9	.00302	43.0	.00791
40.8	.00268	28.5	.00490
PRESSURE 1 CM. OF MERCURY.			
65	.00388	62.5	.01182
60	.00355	57.5	.01074
50	.00286	54.2	.01003
40	.00219	41.7	.00726
30	.00157	37.5	.00639
23.5	.00124	34.0	.00569
–	–	27.5	.00446
–	–	24.2	.00391

* "Proc. Roy. Soc." 1872.
† "Proc. Roy. Soc." Edinb. 1869.
See also Compan, Annal. de chi. et phys. 26, p. 526.

COOLING BY RADIATION AND CONVECTION.

TABLE 284. — Cooling of Platinum Wire in Copper Envelope.

Bottomley gives for the radiation of a bright platinum wire to a copper envelope when the space between is at the highest vacuum attainable the following numbers: —

$t = 408°$ C., $et = 378.8 \times 10^{-4}$, temperature of enclosure 16° C.
$t = 505°$ C., $et = 726.1 \times 10^{-4}$, " " 17° C.

It was found at this degree of exhaustion that considerable relative change of the vacuum produced very small change of the radiating power. The curve of relation between degree of vacuum and radiation becomes asymptotic for high exhaustions. The following table illustrates the variation of radiation with pressure of air in enclosure.

Temp. of enclosure 16° C., $t=408°$ C.		Temp. of enclosure 17° C., $t=505°$ C.	
Pressure in mm.	*et*	Pressure in mm.	*et*
740.	8137.0×10^{-4}	0.094	1688.0×10^{-4}
440.	7971.0 "	.053	1255.0 "
140.	7875.0 "	.034	1126.0 "
42.	7591.0 "	.013	920.4 "
4.	6036.0 "	.0046	831.4 "
0.444	2683.0 "	.00052	767.4 "
.070	1045.0 "	.00019	746.4 "
.034	727.3 "	Lowest reached but not measured	726.1 "
.012	539.2 "		
.0051	436.4 "		
.00007	378.8 "		

TABLE 285.—Effect of Pressure on Loss of Heat at Different Temperatures.

The temperature of the enclosure was about 15° C. The numbers give the total radiation in therms per square centimeter per second.

Temp. of wire in C°.	Pressure in mm.				
	10.0	1.0	0.25	0.025	About 0.1 M.
100°	0.14	0.11	0.05	0.01	0.005
200	.31	.24	.11	.02	.0055
300	.50	.38	.18	.04	.0105
400	.75	.53	.25	.07	.025
500	–	.69	.33	.13	.055
600	–	.85	.45	.23	.13
700	–	–	–	.37	.24
800	–	–	–	.56	.40
900	–	–	–	–	.61

NOTE. — An interesting example (because of its practical importance in electric lighting) of the effect of difference of surface condition on the radiation of heat is given on the authority of Mr. Evans and himself in Bottomley's paper. The energy required to keep up a certain degree of incandescence in a lamp when the filament is dull black and when it is "flashed" with coating of hard bright carbon, was found to be as follows: —

Dull black filament, 57.9 watts.
Bright " " 39.8 watts.

TABLE 286. — Conduction of Heat across Air Spaces (Ordinary Temperatures).

Loss of heat by air from surfaces takes place by radiation (dependent upon radiating power of surface; for small temperature differences proportional to temperature difference; follows Stefan-Boltzmann formula, see p. 247), conduction, and convection. The two latter are generally inextricably mixed. For horizontal air spaces, upper surface warm, the loss is all radiation and conduction; with warm lower surface the loss is greater than for similar vertical space.

Vertical spaces: The following table shows that for spaces of less than 1 cm width the loss is nearly proportional to the space width, when the radiation is allowed for; for greater widths the increase is less rapid, then reaches a maximum, and for yet greater widths is slightly less. The following table is from Dickinson and van Dusen, A. S. Refrigerating Engineers J. 3, 1916.

HEAT CONDUCTION AND THERMAL RESISTANCES, RADIATION ELIMINATED, AIR SPACE 20 CM HIGH.

Air space, cm.	Heat conduction. Cal./hour/cm²/° C. Temperature difference.				Thermal resistance. Same units. Temperature difference.			
	10°	15°	20°	25°	10°	15°	20°	25°
0.5	0.46	0.46	0.46	0.46	2.17	2.17	2.17	2.17
1.0	0.24	0.24	0.24	0.24	4.25	4.20	4.15	4.10
1.5	0.160	0.172	0.182	0.192	6.25	5.80	5.50	5.20
2.0	0.161	0.178	0.200	0.217	6.20	5.60	5.00	4.60
3.0	0.172	0.196	0.208	0.217	5.80	5.10	4.80	4.60

Variation with height of air space: Max. thermal resistance = 4.0 at 1.4 cm air space, 10 cm high; 6.0 at 1.6 cm, 20 cm high; 8.9 at 2.5 cm, 60 cm high.

TABLE 287. — Heat Convection in Air at Ordinary Temperatures.

In very narrow layers of air between vertical surfaces at different temperatures the convection currents, in the main, flow up one side and down the other, with eddyless (stream-line) motion. It follows that these currents transport heat to or from the surfaces only when they turn and flow horizontally, from which fact it follows, in turn, that the convective heat transfer is independent of the height of the surface. It is, according to the laws of eddyless flow, proportional to the square of the temperature difference, and to the cube of the distance between the surfaces. As the flow becomes more rapid (e.g., for a 20° difference and a distance of 1.2 cm) turbulence enters, and the above relations begin to change. For the dimensions tested, convection in horizontal layers was a little over twice that in vertical.

Taken from White, Physical Review, 10, 743, 1917.

Heat Transfer, in the Usual C.G.S. Unit, i.e., Calories per Second per Degree of Thermal Head per Square Cm of Flat Surface, at 22.8° Mean Temperature.

Where two values are given, they show the range among determinations with different methods of getting the temperature of the outer plate. It will be seen that the value of the convection is practically unaffected by this difference of method.

Thermal head.	8 mm gap.		12 mm gap.		24 mm gap.	
	Total.	Convection.	Total.	Convection.	Total.	Convection.
0.99°	—	—	.000 083 9 .000 084 8	—	.000 065	—
1.98°	.000 109 110	—	.000 084 0 .000 085 2	.000 000 1 000 4	—	—
4.95°	.000 111	.000 001	.000 086 6 88 1	.000 002 8 003 7	.000 090	over .000 025
9.89°	.000 112 113	.000 003 003	.000 093 7 95 2	.000 010 .000 011	.000 106	over .000 040
19.76°	.000 116	.000 007	.000 107 7 109 4	.000 024 026	.000 126	over .000 060

TABLE 288.

CONVECTION AND CONDUCTION OF HEAT BY GASES AT HIGH TEMPERATURES.*

The loss of heat from wires at high temperatures occurs as if by conduction across a thin film of stationary gas adhering to the wire (vertical and horizontal losses very similar). Thickness of film is apparently independent of temperature of wire, but probably increases with the temperature of the gas and varies with the diameter of the wire according to the formula $b \cdot \log b/a = 2B$, where B = constant for any gas, b = diameter of film, a, of wire. The rate of convection (conduction) of heat is the product of two factors, one the shape factor, s, involving only a and B, the other a function ϕ of the heat conductivity of the gas. If W = the energy loss in watts/cm, then $W = s(\phi_2 - \phi_1)$. s may be found from the relation

$$\frac{s}{\pi} e^{-\frac{2\pi}{s}} = \frac{a}{B}; \quad \phi = 4.19 \int_0^T k\,dt,$$

where k is the heat conductivity of the gas at temperature T in calories/cm °C. ϕ_2 is taken at the temperature T_2 of the wire, ϕ_1 at that of the atmosphere. The following may be taken as the conductivities of the corresponding gases at high temperatures:

For hydrogen............. $k = 28 \times 10^{-6}\sqrt{T}\{(1 + .0002T)/(1 + 77T^{-1})\}$
air.................... $k = 4.6 \times 10^{-6}\sqrt{T}\{(1 + .0002T)/(1 + 124T^{-1})\}$
mercury vapor........ $k = 2.4 \times 10^{-6}\sqrt{T}\{1/(1 + 960T^{-1})\}$.

To obtain the heat loss: B may be assumed proportional to the viscosity of the gas and inversely proportional to the density. For air (see Table 289(*b*)) B may be taken as 0.43 cm; for H_2, 3.05 cm; for Hg vapor as 0.078. Obtain s from section (*a*) below from a/B; then from section (*b*) obtain ϕ_2 and ϕ_1 for the proper temperatures; the loss will be $s(\phi_2 - \phi_1)$ in watts/cm.

(*a*) s AS FUNCTION OF a/B.

s	a/B	s	a/B	s	a/B	s	a/B
0.0	0.0	5.0	0.453	10	1.696	30	7.738
0.5	0.735×10^{-6}	5.5	0.558	12	2.263	32	8.370
1.0	0.594×10^{-3}	6.0	0.671	14	2.844	34	8.995
1.5	0.725×10^{-2}	6.5	0.788	16	3.438	36	9.622
2.0	2.75×10^{-2}	7.0	0.908	18	4.040	38	10.25
2.5	0.0644	7.5	1.032	20	4.645	40	10.87
3.0	0.1176	8.0	1.160	22	5.263	42	11.50
3.5	0.185	8.5	1.291	24	5.877	44	12.14
4.0	0.265	9.0	1.424	26	6.505	46	12.77
4.5	0.354	9.5	1.561	28	7.122	48	13.14
5.0	0.453	10.0	1.696	30	7.738	50	14.03

(*b*) TABLE OF ϕ IN WATTS PER CM AS FUNCTION OF ABSOLUTE TEMP. (°K.).

T° K.	H_2	Air	Hg	T° K.	H_2	Air	Hg
0°	0.0000	0.0000	—	1500°	4.787	0.744	0.1783
100	0.0329	0.0041	—	1700	5.945	0.931	0.228
200	0.1294	0.0168	—	1900	7.255	1.138	0.284
300	0.278	0.0387	—	2100	8.655	1.363	0.345
400	0.470	0.0669	—	2300	10.18	1.608	0.411
500	0.700	0.1017	0.0165	2500	11.82	1.871	0.481
700	1.261	0.189	0.0356	2700	13.56	—	0.556
900	1.961	0.297	0.0621	2900	15.54	—	0.636
1100	2.787	0.426	0.0941	3100	17.42	—	0.719
1300	3.726	0.576	0.1333	3300	19.50	—	0.807
1500	4.787	0.744	0.1783	3500	21.79	—	0.898

* Langmuir Physical Review, 34, p. 401, 1912.

TABLE 289.

HEAT LOSSES FROM INCANDESCENT FILAMENTS.

(a) WIRES OF PLATINUM SPONGE SERVED AS RADIATORS (TO ROOM-TEMPERATURE SURROUNDINGS). HARTMAN, PHYSICAL REVIEW, 7, p. 431, 1916.

Diameter wire, cm.	(A) Observed heat losses in watts per cm. Absolute temperatures.											
	900°	1000°	1100°	1200°	1300°	1400°	1500°	1600°	1700°	1800°	1900°	2000°
0.0690	1.70	2.26	3.01	3.88	4.92	6.18	7.70	9.63	12.15	15.33	19.25	23.75
0.0420	1.35	1.75	2.26	2.84	3.53	4.29	5.33	6.60	8.25	10.20	12.45	14.75
0.0275	1.12	1.40	1.76	2.23	2.73	3.23	3.91	4.67	5.72	7.00	8.64	10.45
0.0194	0.92	1.15	1.39	1.74	2.12	2.54	3.04	3.64	4.32	5.10	6.10	7.35
	(B) Heat losses corrected for radiation, watts per cm (A–C).											
0.0690	0.91	1.05	1.23	1.36	1.45	1.51	1.54	1.66	2.00	2.56	3.40	4.30
0.0420	0.87	1.02	1.17	1.31	1.42	1.45	1.57	1.76	2.08	2.43	2.80	3.26
0.0275	0.80	0.92	1.05	1.22	1.35	1.37	1.46	1.50	1.67	1.91	2.32	2.70
0.0194	0.70	0.81	0.89	1.03	1.15	1.23	1.31	1.40	1.47	1.51	1.64	1.88
	(C) Computed radiation, watts per cm, $\sigma = 5.61 \times 10^{-12}$.*											
0.0690	0.79	1.21	1.78	2.52	3.47	4.67	6.16	7.97	10.15	12.77	15.85	19.45
0.0420	0.48	0.73	1.09	1.53	2.11	2.84	3.74	4.84	6.17	7.77	9.65	11.85
0.0275	0.32	0.48	0.71	1.01	1.38	1.86	2.45	3.17	4.05	5.09	6.32	7.75
0.0195	0.22	0.34	0.50	0.71	0.97	1.31	1.73	2.24	2.85	3.59	4.46	5.47
	(D) Conduction loss by silver leads, watts per cm.											
0.0420	0.42	0.46	0.49	0.61	0.75	0.88	1.00	1.07	1.13	1.22	—	—
0.0275	0.18	0.21	0.28	0.35	0.43	0.48	0.55	0.57	0.60	0.67	—	—
0.0195	0.06	0.08	0.08	0.09	0.11	0.12	0.14	0.15	0.22	0.23	—	—
	(E) Convection loss by air, watts per cm.											
0.0420	0.45	0.56	0.68	0.70	0.67	0.57	0.59	0.69	0.95	1.21	—	—
0.0275	0.62	0.71	0.77	0.87	0.92	0.89	0.91	0.93	1.07	1.24	—	—
0.0195	0.64	0.73	0.81	0.94	1.04	1.11	1.17	1.25	1.29	1.30	—	—

* This value is lower than the presently (1919) accepted value of 5.72.

(b) WIRES OF BRIGHT PLATINUM 40–50 CM LONG SERVED AS RADIATORS TO SURROUNDINGS AT 300° K. LANGMUIR, PHYSICAL REVIEW, 34, p. 401, 1912.

Diameter wire, cm.	Observed energy losses in watts per cm. Absolute temperatures.							
	500°	700°	900°	1100°	1300°	1500°	1700°	1900°
0.0510	0.22	0.52	0.90	1.42	2.03	2.89	4.10	5.65
0.02508	0.17	0.39	0.68	1.02	1.45	2.00	2.68	3.55
0.01262	0.13	0.31	0.53	0.79	1.11	1.46	1.95	2.71
0.00691	0.12	0.29	0.48	0.72	0.99	1.33	1.79	2.48
0.00404	0.11	0.24	0.41	0.61	0.84	1.14	1.54	2.13
	Energy radiated in watts per cm.*							
0.0510	0.002	0.013	0.049	0.137	0.323	0.67	1.25	2.15
0.02508	0.001	0.007	0.024	0.067	0.159	0.33	0.62	1.06
0.01262	0.001	0.003	0.012	0.034	0.080	0.17	0.31	0.53
0.00691	0.000	0.002	0.007	0.019	0.044	0.09	0.17	0.29
0.00404	0.000	0.001	0.004	0.011	0.026	0.05	0.10	0.17
	"Convection" losses in watts per cm.							
0.0510	0.22	0.51	0.85	1.28	1.71	2.22	2.85	3.50
0.02508	0.17	0.38	0.66	0.95	1.29	1.67	2.06	2.49
0.01262	0.13	0.31	0.52	0.75	1.03	1.29	1.64	2.18
0.00691	0.12	0.29	0.47	0.70	0.95	1.24	1.62	2.19
0.00404	0.11	0.24	0.41	0.60	0.81	1.09	1.44	1.96

	Thickness of theoretical conducting air film.							
								Means.
0.0510	0.28	0.30	0.33	0.33	0.36	0.37	0.35	0.36	0.34
0.02508	0.30	0.37	0.37	0.41	0.45	0.45	0.51	0.56	0.43
0.01262	0.42	0.42	0.44	0.49	0.56	0.69	0.69	0.47	0.54
0.00691	0.31	0.32	0.38	0.40	0.43	0.47	0.38	0.26	0.37
0.00404	0.27	0.43	0.43	0.47	0.56	0.47	0.40	0.25	0.41
Means.	0.31	0.37	0.39	0.42	0.49	0.49	0.47	0.38	†0.43

* Computed with $\sigma = 5.32$, black-body efficiency of platinum as follows (Lummer and Kurlbaum): 492° K. 0.039; 654°, 0.060; 795°, 0.075; 1108°, 0.112; 1481°, 0.154; 1761° K., 0.180. For significance of last group of data, see next page. † Weighted mean.

THE EYE AND RADIATION.

Definitions: A meter-candle is the intensity of illumination due to a standard candle at a meter distance. The millilambert (0.001 lambert) measures the brightness of a perfectly diffusing (according to Lambert's cosine law) surface diffusing .001 lumen/cm². A brightness of 10 meter-candles equals 1 millilambert. 0.001 ml corresponds roughly to night exteriors, 0.1, to night interiors, 10 ml to daylight interiors and 1000, to daylight exteriors. A brightness of 100,000 meter-candles is about that of a horizontal plane for summer day with sun in zenith, 500, on a cloudy day, 4, 1st magnitude stars just visible, 0.2, full moon in zenith, .001, by starlight; in winter the intensity at noon may drop about ⅓.

TABLE 290. — Spectral Variation of Sensitiveness as a Function of Intensity.

Radiation is easily visible to most eyes from 0.330 μ (violet) to 0.770 μ (red). At low intensities near threshold values (gray, rod vision) the maximum of spectral sensibility lies near 0.503 μ (green) for 90% of all persons. At higher intensities, after the establishment of cone vision, the max. shifts as far as 0.560 μ. See Table 297 for more accurate values of sensitiveness after this shift has been accomplished. The ratio of optical sensation to the intensity of energy increases with increasing energy more rapidly for the red than for the shorter wave-lengths (Purkinje phenomenon); i.e., a red light of equal intensity to the eye with a green one will appear darker as the intensities are equally lowered. This phenomenon disappears above a certain intensity (above 10 millilamberts). Table due to Nutting, Bulletin Bureau of Standards.

The intensity is given for the spectrum at 0.535μ (green).

Intensity (meter-candles) = Ratio to preceding step =	.00024 —	.00225 9.38	.0360 16	.575 16	2.30 4	9.22 4	36.9 4	147.6 4	590.4 4
Wave-length, λ.	Sensitiveness.								
0.430 μ	0.081	0.093	0.127	0.128	0.114	0.114	—	—	—
0.450	0.33	0.30	0.29	0.31	0.23	0.175	0.16	—	—
0.470	0.63	0.59	0.54	0.58	0.51	0.29	0.26	0.23	—
0.490	0.96	(0.89)	(0.76)	(0.89)	(0.83)	0.50	0.45	0.38	0.35
0.505	1.00	1.00	1.00	1.00	0.99	(0.76)	0.66	0.61	0.54
0.520	0.88	0.86	0.86	0.94	0.99	(0.85)	0.85	0.85	0.82
0.535	0.61	0.62	0.63	0.72	0.91	(0.98)	0.98	0.99	0.98
0.555	0.26	0.30	0.34	0.41	0.62	0.84	0.93	0.97	0.98
0.575	0.074	0.102	0.122	0.168	(0.39)	(0.63)	(0.76)	(0.82)	(0.84)
0.590	0.025	0.034	0.054	0.091	0.27	0.49	0.61	0.68	0.69
0.605	0.008	0.012	0.024	0.056	0.173	0.35	(0.45)	0.54	0.55
0.625	0.004	0.004	0.011	0.027	0.098	0.20	0.27	0.35	0.35
0.650	0.000	0.000	0.003	0.007	0.025	0.060	0.085	0.122	0.133
0.670	0.000	0.000	0.001	0.002	0.007	0.017	0.025	0.030	0.030
λ, maximum sensitiveness	0.503	0.504	0.504	0.508	0.513	0.530	0.541	0.543	0.544

TABLE 291. — Threshold Sensibility as Related to Field Brightness.

The eye perceives with ease and comfort a billion-fold range of intensities. The following data were obtained with the eye fully adapted to the sensitizing field, *B*, the field flashed off, and immediately the intensity, *T*, of a test spot (angular size at eye about 5°) adjusted to be just visible. This table gives a measure of the brightness, *T*, necessary to just pick up objects when the eye is adapted to a brightness, *B*. Intensities are indicated log intensities in millilamberts. Blanchard, Physical Review, 11, p. 81, 1918.

Log *B*	−7.0	−6.0	−5.0	−4.0	−3.0	−2.0	−1.0	0.0	+1.0	+2.0	+3.0
Log *T*, white	—	−5.81	−5.42	−4.87	−4.17	−3.30	−2.59	−2.02	−1.42	−0.75	+0.28
T/*B*	—	1.5	0.38	.13	.068	.050	.026	.0096	.0038	.0018	.0019
Log *T*, blue	−6.70	−6.38	−5.82	−5.12	−4.23	−3.46	−2.70	−2.18	−1.62	—	—
Log *T*, green	−6.42	−6.20	−5.62	−5.00	−4.23	−3.39	−2.60	−2.08	−1.62	−0.90	—
Log *T*, yellow	—	−5.47	−5.17	−4.61	−4.03	−3.33	−2.57	−1.97	−1.62	—	—
Log *T*, red	—	—	−4.27	−4.00	−3.47	−2.96	−2.43	−1.92	−1.37	−0.90	—

THE EYE AND RADIATION.

TABLE 292. — Heterochromatic Threshold Sensibility.

The following table shows the decrease in sensitiveness of the eye for comparing intensities of different colors. The numbers in the body of the table correspond to the line marked *T/B* of Table 291. The intensity of the field was probably between 10 and 100 millilamberts (25 photons).

Comparison color.		0.693 μ	0.640 μ	0.575 μ	0.505 μ	0.475 μ	0.430 μ
Standard color: red	0.693 μ	0.044	0.088	0.165	0.180	0.197	0.150
yellow	0.575 μ	0.174	0.160	0.032	0.166	0.174	0.134
green	0.505 μ	0.211	0.180	0.138	0.030	0.116	0.126
blue	0.475 μ	0.168	0.180	0.130	0.130	0.068	0.142

TABLE 293. — Contrast or Photometric Sensibility.

For the following table the eye was adapted to a field of 0.1 millilambert and the sensitizing field flashed off. A neutral gray test spot (angular size at eye, 5 × 2.5°) the two halves of which had the contrast indicated (½ transparent, ½ covered with neutral screen of transparency = contrast indicated) was then observed and the brightness of the transparent part measured necessary to just perceive the contrast after the lapse of the various times. One eye only used, natural pupil. Blanchard, Physical Review, 11, p. 88, 1918. Values are log brightness of brighter field in millilamberts.

Time in seconds.	0	1	2	5	10	20	40	60
Contrast: 0.00	−2.80	−3.47	−3.82	−4.30	−4.49	−4.60	−4.89	−5.03
0.39	−2.63	−3.36	−3.58	−3.74	−3.85	−3.97	−4.06	−4.23
0.67	−2.40	−3.00	−3.13	−3.22	−3.21	−3.33	−3.46	−3.48
0.87	−2.10	−2.46	−2.49	−2.48	−2.55	−2.54	−2.67	−2.73
0.97	−1.20	−1.57	−1.67	−1.69	−1.59	−1.63	−1.73	−1.78

TABLE 294. — Glare Sensibility.

When an eye is adapted to a certain brightness and is then exposed suddenly to a much greater brightness, the latter may be called glaring if uncomfortable and instinctively avoided. Observers naturally differ widely. The data are the means of three observers, and are log brightnesses in millilamberts. The glare intensity may be taken as roughly 1700 times the cube root of the field intensity in millilamberts. Angle of glare spot, 4°. Blanchard, Physical Review, *loc. cit.*

Log. field...	−6.0	−4.0	−2.0	−1.0	0.0	+1.0	2.0	3.0	4.0
Log. glare...	1.35	1.90	2.60	2.90	3.28	3.60	3.90	4.18	4.48

TABLE 295. — Rate of Adaptation of Sensibility.

This table furnishes a measure of the rate of increase of sensibility after going from light into darkness, and the values were obtained immediately from the instant of turning off the sensitizing field. Both eyes were used, natural pupil, angular size of test spot, 4.9°, viewed at 35 cm. Blanchard, *loc. cit.* Retinal light persists only 10 to 20 m when one has been recently in darkness, then in a dimly lighted room; it persists fully an hour when a subject has been in bright sunlight for some time. A person who has worked much in the dark "gets his eyes" quicker than one who has not, but his final sensitiveness may be no greater.

Sensitizing field.	Logarithmic thresholds in millilamberts after										
	0 sec.	1 sec.	2 sec.	5 sec.	10 sec.	20 sec.	40 sec.	60 sec.	5 min.	30 min.	60 min.
White, 0.1 ml	−2.79	−3.82	−4.13	−4.50	−4.75	−4.96	−5.16	−5.32	−5.68	−5.91	−6.06
1.0 ml	−2.20	−2.99	−3.27	−3.79	−4.15	−4.51	−4.82	−5.06	−5.52	−5.86	−6.04
10.0 ml	−1.60	−2.30	−2.53	−3.08	−3.54	−3.94	−4.31	−4.61	−5.22	−5.83	−6.01
100.0 ml	−0.90	−1.66	−2.00	−2.46	−2.64	−2.88	−3.20	−3.84	−4.76	−5.77	−5.97
Blue 0.1 ml	−2.82	−3.92	−4.36	−4.91	−5.27	−5.53	−5.68	−5.81	−6.23	—	—
Green 0.1 ml	−2.69	−4.08	−4.39	−4.82	−5.11	−5.26	−5.43	−5.56	−5.80	—	—
Yellow 0.1 ml	−2.61	−3.84	−4.17	−4.41	−4.65	−4.78	−5.02	−5.09	−5.39	—	—
Red 0.1 ml	−2.32	−2.69	−2.98	−3.37	−3.57	−3.65	−3.73	−3.80	−4.02	—	—

THE EYE AND RADIATION.

TABLE 296. — Apparent Diameter of Pupil and Flux Density at Retina.

Flashlight measures of the pupil (both eyes open) viewed through the eye lens and adapted to various field intensities. For eye accommodated to 25 cm, ratio apparent to true pupil, 1.02, for the unaccommodated eye, 1.14. The pupil size varies considerably with the individual. It is greater with one eye closed; e.g., it was found to be for 0.01 millilambert, 6.7 and 7.2 mm; for 0.6 ml, 5.3 and 6.5; for 6.3 ml, 4.1 and 5.7; for 12.6 ml, 4.1 and 5.7 mm for both and one eye open respectively for a certain individual. At the extreme intensities the two values approach each other. The ratio of the extreme pupil openings is about $\frac{1}{18}$, whereas the light intensities investigated vary over 1,000,000-fold. (Blanchard and Reeves, partly unpublished data.)

Field millilamberts.	Diameter, mm		Effective area, mm²	Flux at retina, lumens per mm²
	Observed.	(1.14/1.02) × Obs.		
0.00001	8	8.96	64	8.4×10^{-12}
0.001	7.6	8.51	57	7.6×10^{-10}
0.1	6.5	7.28	42	5.6×10^{-8}
10	4.0	4.48	16	2.1×10^{-6}
1000	2.07	2.35	4.3	5.8×10^{-5}

TABLE 297. — Relative Visibility of Radiation.

This table gives the relation between luminous sensation (light) and radiant energy. The results of two methods are given: one from measures of the direct equality of brightness, which some consider the true method, as more direct, but criticized because of the difficulty of judging heterochromatic light (Hyde, Forsythe, Cady, A. J. 48, 87, 1918, 29 observers); the other (Coblentz, Emerson, Bul. Bureau of Standards, 14, 219, 1917, 130 observers) depends on the disappearance of flicker when two lights of different color and intensity are alternated rapidly. Color has a lower critical frequency than brightness and disappears first. Data determined for intensities above Purkinje effect. See Table 290. Ratio of light unit (lumen) to energy unit (watt) at 0.55μ, 0.00162 (Ives, Coblentz, Kingsbury).

λ μ	Visibility.		λ μ	Visibility.		λ μ	Visibility.		λ μ	Visibility.		λ μ	Visibility.	
	HFC	CE		HFC	CE		HFC	CE		HFC	CE		HFC	CE
.40	.049	.010	.48	.138	.125	.56	.995	.998	.64	.154	.194	.72	$.0_{3}74$	$.0_{3}97$
.41	$.0_{3}62$	.017	.49	.216	.194	.57	.944	.968	.65	.094	.115	.73	$.0_{3}36$	$.0_{3}48$
.42	.0041	.024	.50	.328	.316	.58	.855	.898	.66	.051	.0645	.74	$.0_{3}18$	$.0_{3}28$
.43	.0115	.029	.51	.515	.503	.59	.735	.800	.67	.026	.0338	.75	.049	$.0_{3}20$
.44	.022	.033	.52	.698	.710	.60	.600	.687	.68	.0125	.0178	.76	.045	—
.45	.036	.041	.53	.847	.862	.61	.464	.557	.69	.0062	.0085	—	—	—
.46	.055	.056	.54	.968	.954	.62	.341	.427	.70	.0031	.0040	—	—	—
.47	.087	.083	.55	.996	.994	.63	.238	.302	.71	.0015	.00203	—	—	—

TABLE 298. — Miscellaneous Eye Data.

Light passing to the retina traverses in succession (*a*) front surface of the cornea (curvature, 7.9 mm); (*b*) cornea (equivalent water path for energy absorption, .06 cm); (c.) back surface cornea (curv., 7.9 mm); (*d*) aqueous humour (equiv. H_2O, .34 cm, $n = 1.337$); (*e*) front surface lens (c, 10 mm); (*f*) lens (equiv. H_2O, .42 cm, $n = 1.445$); (*g*) back surface lens (c., 6 mm); (*h*) vitreous humour (equiv. H_2O, 1.46 cm, $n = 1.337$). An equivalent simple lens has its principal point 2.34 mm behind (*a*), nodal point 0.48 mm in front of (*g*), posterior principal focus 22.73 mm behind (*a*), anterior principal focus 12.83 mm. in front of (*a*), curvature, 5.125 mm. At the rear surface of the retina (.15 mm thick) are the rods (30 × 2μ) and cones (10 (6 outside fovea) μ long). Rods are more numerous, 2 to 3 between 2 cones, over 3,000,000 cones in eye. Macula lutea, yellow spot, on temporal side, 4 mm from center of retina, long axis 2 mm. Central depression, fovea centralis, .3 mm diameter, 7000 cones alone present, 6 × 2 or 3μ. In region of distinct vision (fovea centralis) smallest angle at which two objects are seen separate is 50″ to 70″ = 3.65 to 5.14μ at retina; 50 cones in 100μ here; 4μ between centers, 3μ to cone, 1μ to interval. Distance apart for separation greater as depart from fovea. No vision in blind spot, nasal side, 2.5 mm from center of eye, 15 mm in diam.

Persistence of vision as related to color (Allen, Phys. Rev. 11, 257, 1900) and intensity (Porter, Pr. Roy. Soc. 70, 313, 1912) is measured by increasing speed of rotating sector until flicker disappears: for color, .4μ, .031 sec.; .45μ, .020 sec.; .5μ, .015 sec.; .57μ, .012 sec.; .68μ, .014 sec.; .76μ, .018 sec.; for intensity, .06 meter-candle, .028 sec.; 1 mc, .020 sec.; 6 mc, .014 sec.; 100 mc, .010 sec; 142 mc., .007 sec.

Sensibility to small differences in color has two pronounced maxima (in yellow and green) and two slight ones (extreme blue, extreme red). The sensibility to small differences in intensity is nearly independent of the intensity (Fechner's law) as indicated by the following data due to König:

I/I_0	1,000,000	100,000	10,000	1000	100	50	10	5	1	0.1	I_0 in mc
dI/I, white.......	.036	.019	.018	.018	.030	.032	.048	.059	.123	.377	.00072
.60 μ.......	—	.024	.016	.020	.028	.038	.061	.103	.212	—	.0056
.50 μ.......	—	—	.018	.018	.024	.025	.036	.049	.080	.133	.00017
.43 μ.......	—	—	—	.018	.025	.027	.040	.049	.074	.137	.00012

TABLE 299.

PHOTOMETRIC DEFINITIONS AND UNITS.

Radiant flux = Φ = rate of flow of radiation as energy, measured as ergs per second or watts.

Luminous flux = F or Ψ = rate of flow of radiation measured according to power to produce visual sensation. Although strictly thus defined, for photometric purposes it may be regarded as an entity, since the rate of flow for such purposes is invariable. Unit is the *lumen*, the flux emitted in a unit solid angle (steradian) by a point source of unit candle power.

Visibility of radiation of wave-length $\lambda = K_\lambda$ = ratio of luminous to radiant flux for that λ, $= F_\lambda/\Phi_\lambda$.

Mechanical equivalent of light = ratio of Φ/F for the λ of max. visibility expressed in ergs/sec /lumen or watts/lumen; it is the reciprocal of max. visibility. See p. 261.

Luminosity at wave-length $\lambda = (K_\lambda)(\Phi_\lambda)$. Spectral luminosity curve expresses this as a function of λ and is different for various sources.

Luminous efficiency = F/Φ expressed in lumens/watt.

Luminous intensity of (approximate) point source = I = solid-angle (ω) density of luminous flux in direction considered = $dF/d\omega$, or F/ω when the intensity is uniform. Unit, the *candle*.

Illumination on surface = E = flux density on surface = dF/dS (S is surface area) = F/S when uniform. Units, meter-candle, foot-candle, phot, lux.

Lux = one lumen per m^2; phot one lumen per cm^2.

Brightness of a luminous surface may be expressed in two ways:

(1) $b_I = dI/dS. \cos\theta$ where θ is the angle between normal to surface and the line of sight; normal brightness when θ is zero.

(2) $b_F = dF/dS'$ assuming that the surface is a perfect diffuser, obeying cos. law of emission or reflection. Unit, the lambert.

Specific luminous radiation, E' = luminous flux density emitted by a surface, or the flux emitted per unit of emissive area, expressed in lumens per cm^2. For surfaces obeying Lambert's cosine law, $E' = \pi b_0$.

The lambert, the cgs unit of brightness, is the brightness of a perfectly diffusing surface radiating or reflecting one lumen per cm^2. Equivalent to a perfectly diffusing surface with illumination of one phot. A perfectly diffusing surface emitting one lumen per ft^2 has a brightness of 1.076 millilamberts. Brightness in candles per cm^2 is reduced to lamberts by multiplying by π.

A uniform point source of one candle emits 4π lumens.
One lumen is emitted by .07958 spherical candle power.
One lumen emitted per ft^2 = 1.076 millilamberts (perfect diffusion).
One spherical candle power emits 12.57 lumens.
One lux = 1 lumen incident per m^2 = .0001 phot = .1 milliphot.
One phot = 1 lumen incident per cm^2 = 10,000 lux = 1000 milliphots.
One milliphot = .001 phot = .929 foot-candle.
One foot-candle = 1 lumen incident per ft^2 = 1.076 milliphots = 10.76 lux.
One lambert = 1 lumen emitted per cm^2 of a perfectly diffusing surface.
One millilambert = .929 lumen emitted per ft^2 (perfect diffusion).
One lambert = .3183 candle per cm^2 = 2.054 candles per in^2.
One candle per cm^2 = 3.1416 lamberts.
One candle per in^2 = .4868 lambert = 486.8 millilamberts.

Adapted from Reports of Committee on Nomenclature and Standards of Illuminating Engineering Society. 1916 to 1918.

TABLE 300. — Photometric Standards.

No primary photometric standard has been generally adopted by the various governments. In Germany the Hefner lamp is most used; in England the Pentane lamp and sperm candles are used; in France the Carcel lamp is preferred; in America the Pentane and Hefner lamps are used to some extent, but candles are more largely employed in gas photometry. For the photometry of electric lamps, and generally in accurate photometric work, electric lamps, standardized at a national standardizing institution, are commonly employed.

The "International candle" is the name recently employed to designate the value of the candle as maintained by coöperative effort between the national laboratories of England, France, and America; and the value of various photometric units in terms of this international candle is given in the following table (taken from Circular No. 15 of the Bureau of Standards).

1 International Candle = 1 Pentane Candle.
1 International Candle = 1 Bougie Decimale.
1 International Candle = 1 American Candle.
1 International Candle = 1.11 Hefner Unit.
1 International Candle = 0.104 Carcel Unit.

Therefore 1 Hefner Unit = 0.90 International Candle.

The values of the flame standards most commonly used are as follows:

1. Standard Pentane Lamp, burning pentane . . . . . . 10.0 candles.
2. Standard Hefner Lamp, burning amyl acetate . . . . . 0.9 candles.
3. Standard Carcel Lamp, burning colza oil . . . . 9.6 candles.
4. Standard English Sperm Candle, approximately . . . . 1.0 candles.

TABLE 301. — Intrinsic Brightness of Various Light Sources.

	Barrows.	Ives & Luckiesh.		National Electric Lamp Association.
	C. P. per Sq. In. of surface of light.	C. P. per Sq. In. of surface of light.	C. P. per Sq. Mm. of surface of light.	C. P. per Sq. In. of surface of light.
Sun at Zenith	600,000	–	–	600,000
Crater, carbon arc	200,000	84,000	130.	200,000
Open carbon arc	10,000–50,000	–	–	10,000–50,000
Flaming arc	5,000	–	–	5,000
Magnetite arc	–	4,000	6.2	–
Nernst Glower	800–1,000	(115v.6 amp. d.c.) 3,010	4.7	(1.5 w.p.c.) 2,200
Tungsten incandescent, 1.15 w. p. c.	–	–	–	1,000
Tungsten incandescent, 1.25 w. p. c.	1,000	1,000	1.64	875
Tantalum incandescent, 2.0 w. p. c.	750	580	0.9	750
Graphitized carbon filament, 2.5 w. p. c.	625	750	1.2	625
Carbon incandescent, 3.1 w. p. c.	480	485	0.75	480
Carbon incandescent, 3.5 w. p. c.	375	400	0.63	375
Carbon incandescent, 4.0 w. p. c.	300	325	0.50	–
Inclosed carbon arc (d. c.)	100–500	–	–	100–500
Inclosed carbon arc (a. c.)	–	–	–	75–200
Acetylene flame (1 ft. burner)	75–100	53.0	0.082	75–100
Acetylene flame (¼ ft. burner)	–	33.0	0.057	–
Welsbach mantle	20–25	31.9	0.048	20–50
Welsbach (mesh)	–	56.0	0.067	–
Cooper Hewitt mercury vapor lamp	16.7	14.9	0.023	17
Kerosene flame	4–8	9.0	0.014	3–8
Candle flame	3–4	–	–	3–4
Gas flame (fish tail)	3–8	2.7	0.004	3–8
Frosted incandescent lamp	4–8	–	–	2–5
Moore carbon-dioxide tube lamp	0.6	–	–	0.3–1.75

Taken from *Data*, 1911.

TABLE 302. — Visibility of White Lights.

Range.	Candle Power.	
	1	2
1 sea-mile = 1855 meters	0.47	0.41
2 " "	1.9	1.6
5 " "	11.8	10.

[1] Paterson and Dudding. [2] Deutsche Seewarte.

1 micro-calorie through 1 cm. at 1 m. = 0.034 sperm candle = 0.0385 Hefner unit (no diaphragm) = 0.043 Hefner unit (diap. 14 × 50 mm.). Coblentz Bul. B. of S., 11, p. 87, 1914.

BRIGHTNESS OF BLACK BODY, CROVA WAVE-LENGTH, MECHANICAL EQUIVALENT OF LIGHT, LUMINOUS INTENSITY AND EFFICIENCY OF BLACK BODY.

The values of L, the luminous intensity, are given in light watts/steroradian/cm² of radiating surface $= (1/\pi) \int_0^\infty V_\lambda E_\lambda d\lambda$, where V_λ is the visibility of radiation function.

Mechanical equivalent. The unit of power is the watt; of lumininous flux, the lumen. The ratio of these two quantities for light of maximum visibility, $\lambda = 0.556\ \mu$, is the stimulus coefficient V_m; its reciprocal is the (least) mechanical equivalent of light, i.e., least since applicable to radiation of maximum visibility. A better term is "luminous equivalent of radiation of maximum visibility." One lumen = 0.001496 watts (Hyde, Forsythe, Cady); or 1 watt of radiation of maximum visibility ($\lambda = 0.556\ \mu$) = 668 lumens.

White light has sometimes been defined as that emitted by a black body at 6000° K.

The Crova wave-length for a black body is that wave-length, λ, at which the luminous intensity varies by the same fractional part that the total luminous intensity varies for the same change in temperature.

TABLE 303. — Brightness, Crova Wave-length of Black Body, Mechanical Equivalent of Light.*

Temp. °K.	Brightness, candles per cm²	Crova wave-length, μ	Mech. equiv. watts per l.
1700°	5.1	0.584	0.001478
1750	7.6	0.583	—
1800	11.3	0.582	0.001491
1850	16.3	0.581	—
1900	23.1	0.580	0.001498
1950	32.2	0.579	—
2000	44.3	0.578	0.001498
2050	60.0	0.577	—
2100	80.1	0.576	0.001497
2150	105.7	0.576	—
2200	137.6	0.575	0.001496
2250	177.	0.574	—
2300	226.	0.574	0.001497
2350	284.	0.573	—
2400	354.	0.572	0.001497
2450	438.	0.572	—
2500	537.	0.571	0.001502
2550	651.	0.570	—
2600	785.	0.570	0.001511
2650	939.	0.569	—
Mean			0.001496

* Hyde, Forsythe, Cady, Phys. Rev. 13, p. 45, 1919.

TABLE 304. — Luminous, Total Intensity and Radiant Luminous Efficiency of Black Body.*

T, degrees absolute.	Luminous intensity L watt/cm²	Total intensity $\sigma_0 T^4$ watt/cm²	Radiant luminous efficiency.
1,200	2.34×10^{-5}	3.762	.000006
1,600	3.45×10^{-3}	1.189	.000290
1,700	8.46×10^{-3}	1.515×10	.000558
1,800	1.88×10^{-2}	1.905×10	.000987
1,900	3.85×10^{-2}	2.365×10	.00163
2,000	7.34×10^{-2}	2.903×10	.00253
2,100	1.32×10^{-1}	3.529×10	.00374
2,200	2.26×10^{-1}	4.250×10	.00532
2,300	3.69×10^{-1}	5.077×10	.00727
2,400	5.79×10^{-1}	6.020×10	.00962
2,500	8.77×10^{-1}	7.087×10	.0124
2,600	1.29	8.291×10	.0156
3,000	4.66	1.470×10^2	.0317
4,000	3.85×10	4.645×10^2	.0829
5,000	1.36×10^2	1.134×10^3	.1201
6,000	3.26×10^2	2.351×10^3	.1386
7,000	6.03×10^2	4.356×10^3	.1385
8,000	9.59×10^2	7.432×10^3	.1290
10,000	1.84×10^3	1.814×10^4	.1014

* Coblentz, Emerson, Bul. Bureau of Standards, 14, p. 255, 1917.

NOTE. — Minimum energy necessary to produce the sensation of light: Ives, 38×10^{-10}; Russell, 7.7×10^{-10}; Reeves, 19.5×10^{-10}; Buisson, 12.6×10^{-10} erg. sec. (Buisson, J. de Phys. 7, 68, 1917.)

Color temperature (temp. black-body same color) 500 w. gas-filled lamp (22 l/w) 3082°k; 900 w. gas-filled movie lamp, 22.7 l/w, 3086°k; crater 65v. 10 amp. arc, solid carbon, 3780°k; cored carbon 3420°k. Priest, 1922.

TABLE 305. — Color of Light Emitted by Various Sources.*

Source.	Color, per cent white.	Hue.	Source.	Color, per cent white.	Hue.
Sunlight	100	—	N-filled tungsten, 0.50 wpc	45	584
Average clear sky	60	472	N-filled tungsten, 0.35 wpc	53	584
Standard candle	13	593	Mercury vapor arc	70	490
Hefner lamp	14	593	Helium tube	32	598
Pentane lamp	15	592	Neon tube	6	605
Tungsten glow lamp, 1.25 wpc	35	588	Crater of carbon arc, 1.8 amp	59	585
Carbon glow lamp, 3.8 wpc	25	592	Crater of carbon arc, 3.2 amp	62	585
Nernst glower, 1.50 wpc	31	587	Crater of carbon arc, 5.0 amp	67	583
N-filled tungsten, 1.00 wpc	34	586	Acetylene flame (flat)	36	586

* Jones, L. A., Trans. Ill. Eng. Soc., Vol. 9 (1914).

TABLE 306.

EFFICIENCY OF VARIOUS ELECTRIC LIGHTS.

Bryant and Hake, Eng. Exp. Station, Univ. of Ill.	Amperes.	Terminal Watts.	Lumens.	Kw-hours for 100,000 Lumen-hours.	Total cost per 100,000 Lumen-hours at 10 cts. per Kw-hour.
Regenerative d.-c., series arc	5.5	385	11,670	3.3	0.339
Regenerative d.-c., multiple arc	5.5	605	11,670	5.18	0.527
Magnetite d.-c., series arc	6.6	528	7,370	7.16	0.729
Flame arc, d.-c., inclined electrodes	10.0	550	8,640	6.37	0.837
Mercury arc, d.-c., multiple	3.5	385	4,400	15.92	0.89
Flame arc, d.-c., inclined electrodes	8.0	440	6,140	7.16	0.966
Flame arc, d.-c., vertical electrodes	8.0	440	6,140	7.16	0.966
Luminous arc, d.-c., multiple	6.6	726	7,370	9.85	0.988
Open arc, d.-c., series	9.6	480	5,025	9.55	1.079
Magnetite arc, d.-c., series	4.0	320	2,870	11.15	1.13
Flame arc, a.-c., vertical electrodes	10.0	467	5,340	8.75	1.275
Flame arc, a.-c., inclined electrodes	10.0	467	5,340	8.75	1.275
Open arc, d.-c., series	6.6	325	2,920	11.15	1.305
Tungsten series	6.6	75	626	12.0	1.384
Flame arc, a.-c., inclined electrodes	8.0	374	3,910	9.55	1.405
Inclosed arc, d.-c., series	6.6	475	3,315	14.32	1.459
Luminous arc, d.-c., multiple	4.0	440	2,870	15.32	1.547
Tungsten, multiple	0.545	60	475	12.6	1.55
Nernst, a.-c., 3-glower	1.87	414	2,160	19.2	1.88
Nernst, d.-c., 3-glower	1.87	414	2,160	19.2	1.90
Inclosed arc, a.-c., series	7.5	480	2,410	19.9	2.05
Inclosed arc, a.-c., series	6.6	425	2,020	21.3	2.193
Tantalum, d.-c., multiple	—	40	199	21.1	2.31
Tantalum, a.-c., multiple	—	40	199	21.1	2.504
Carbon, 3.1 w. p. c., multiple	—	49.6	166	29.9	3.24
Carbon, 3.5 w. p. c., series	6.6	210	626	33.6	3.47
Carbon, 3.5 w. p. c., multiple	—	56	166	33.7	3.50
Inclosed arc, d.-c., multiple	5.0	550	1,535	35.8	3.66
Inclosed arc, d.-c., multiple	3.5	385	1,030	37.4	3.84
Inclosed arc, a.-c., multiple	6.0	430	1,124	38.3	3.94
Inclosed arc, a.-c., multiple	4.0	285	688	41.4	4.265

Ives, Phys. Rev., V, p. 390, 1915 (see also VI, p. 332, 1915); computed assuming 1 lumen = 0.00159 watt.	Commercial Rating	Lumens per Watt.	Luminous Watts Flux ÷ Watts Input or True Efficiency.
Open flame gas burner	Bray 6′ high pressure	0.22	0.00035
Petroleum lamp		.26	.0004
Acetylene	1.0 liters per hour	.67	.0011
Incandescent gas (low pressure)	.350 lumens per B. t. u. per hr.	1.2	.0019
Incandescent gas (high pressure)	.578 lumens per B. t. u. per hr.	2.0	.0031
Nernst lamp		4.8	.0076
Moore nitrogen vacuum tube	220-v. 60-cycle, 113 ft.	5.21	.0083
Carbon incandescent (treated filament)	4-watts per mean hor. C. P.	2.6	.0041
Tungsten incandescent (vacuum)	1.25 watts per hor. C. P.	8.	.013
Carbon arc, open arc	9.6 amp. clear globe	11.8	.019
Mazda, type C	500-watt multiple .7 w. p. c.	15.	.024
Mazda, type C	600 C. P. -20 amp. .5 w. p. c.	19.6	.031
Magnetite arc, series	6.6 amp. direct current	21.6	.034
Glass mercury arc	40-70 volt; 3.5 amperes	23.	.036
Quartz mercury arc	174-197 volt; 4.2 amperes	42.	.067
Enclosed white flame carbon arc	10 ampere, A. C.	26.7	.042
" " " " "	6.5 ampere, D. C.	35.5	.057
Open arc " " inclined	10 ampere, A. C.	29.	.046
" " " " "	10 ampere, D. C.	27.7	.044
Enclosed yellow flame carbon arc	10 ampere, A. C.	31.4	.050
" " " " "	6.5 ampere, D. C.	34.2	.054
Open arc, " " , inclined	10 ampere, A. C.	41.5	.066
" " " " "	10 ampere, D. C.	44.7	.071

PHOTOGRAPHIC DATA.

TABLE 307. — Numerical Constants Characteristic of Photographic Plates.

Abscissae of figure are log E = log It (meter-candles-seconds);
Ordinates are densities, $D = 1/T$;
E = exposure = I (illumination in meter-candles) × t seconds;
D, the density of deposit = $1/T$, where T is the ratio of the transmitted to incident intensity on developed plate.
i = inertia = intercept straight line portion of curve on log E axis.
S = speed = (some constant)/i; γ = gamma = tangent of angle α.
L = latitude = projected straight line portion of characteristic curve on log E axis, expressed in exposure units = Anti log ($b - a$).
The curve illustrates the characteristic curve of a photographic plate.

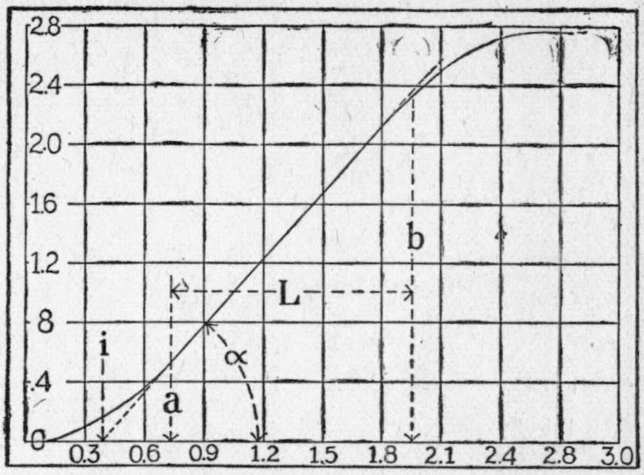

TYPICAL CHARACTERISTIC CURVE OF PHOTOGRAPHIC PLATE.

TABLE 308. — Relative Speeds of Photographic Materials.

The approximate exposure may be obtained when the intensity of the image on the plate is known. Let L be the intensity in meter-candles; E, the exposure in seconds; P, the speed number from the following table; then $E = 1{,}350{,}000/(L \times P)$ approximately.

Plate.	Relative speed.	Paper.	Relative speed.
Extremely high speed	100,000	Fast bromide	1000.0
High speed	75,000	Slow enlarging	60.0
Medium speed	60,000		
Rapid high contrast	50,000	Rapid gas-light, soft grade	6.5
Medium speed high contrast	25,000	Rapid gas-light, medium contrasty	3.5
Process, slow contrast	10,000	Rapid gas-light, contrasty	1.0
Lantern plate	3,000	Professiona	1.25

TABLE 309. — Variation of Resolving Power with Plate and Developer.

The resolving power is expressed as the number of lines per millimeter which is just resolvable, the lines being opaque and separated by spaces of the same width. The developer used for the comparison of plates was Pyro-soda; the plate for the comparison of developers, Seed Lantern. The numbers are all in the same units. Huse, J. Opt. Soc. America, July, 1917.

Plate.	Albumen.	Resolution.	Process.	Lantern.	Medium speed.	High speed.
Resolving power	125	81	67	62	35	27

Developer.	Resolving power.	Developer.	Resolving power.	Developer.	Resolving power.
Pyro-caustic	77	Pyrocatechin	62	Amidol	51
Glycin	69	Pyro-metol	62	Process hydroquinone	50
Hydroquinone	64	Eikon.-hydroquinone	61	Ortol	49
Pyro	64	Ferrous oxalate	61	Rodinal	49
MQ_{25}	64	Caustic hydroquinone	57	X-ray powders	49
Metol	63	Eikonogen	57	Edinol	47
Nepera	62	Kachin	54		

PHOTOGRAPHIC DATA.

TABLE 310. — Photographic Efficiencies of Various Lights.

Source.	Visual efficiency. Lumens per watt.	Photographic efficiency. (a) Ordinary plate.	(a) Orthochromatic plate.	(a) Panchromatic plate.	(b) Ordinary plate.	(b) Orthochromatic plate.	(b) Panchromatic plate.
Sun	150	100	100	100	100	100	100
Sky	—	181	155	130	—	—	—
Acetylene	0.7	30	44	52	0.14	0.21	0.24
" (screened)	0.07	81	85	89	0.037	0.040	0.042
Pentane	0.045	18	28	42	0.053	0.086	0.13
Mercury arc, quartz	40	600	500	367	158	132	99
" " "Nultra" glass	35	218	195	165	50	46	39
" " crown glass	37	324	275	249	79	68	62
Carbon arc, ordinary	12	126	112	104	10	10	8.5
" " white flame	29	257	234	215	52	45	2.0
" " enclosed	9	175	177	165	11	11	10
Carbon arc, "Artisto"	12	796	1070	744	62	86	60
Magnetite arc	18	106	115	82	12	14	10
Carbon glow-lamp	2.44	23	32	42	0.37	0.52	0.68
Carbon glow-lamp	3.16	25	35	45	0.51	0.74	0.95
Tungsten vacuum lamp	8	33	41	50	1.74	2.2	2.7
" vacuum lamp	9.9	37	45	53	2.41	3.0	3.5
" nitrogen lamp	16.6	56	62	70	6.1	6.8	7.7
" nitrogen lamp	21.6	64	68	76	8.9	9.8	11.0
" blue bulb	8.9	—	—	—	5.5	5.2	5.6
" blue bulb	11	108	99	106	7.8	7.3	7.9
Mercury arc (Cooper Hewitt)	23	316	354	273	47	54.2	42

(a) Relative efficiencies based on equal illumination.
(b) Relative efficiencies based on equal energy density.
Taken from Jones, Hodgson, Huse, Tr. Ill. Eng. Soc. 10, p. 963, 1915.

TABLE 311. — Relative Intensification of Various Intensifiers.

Bleaching solution.	Blackening solution.	Reference	Intensification.
Mercuric bromide	Amidol developer	$HgBr_2$ solution (Monckhoven sol. A).*	1.15
Mercuric chloride	Ammonia	Bleach according to Bennett; blackener.*	1.15
Potassium bichromate + hydrochloric acid	Amidol developer	Piper.*	1.45
Mercuric iodide	Schlippe's salt	Debenham, B. J.,† p. 186, '17.	2.50
Lead ferricyanide	Sodium sulphide	B. J. Almanac.*	2.28
Uranium formula	—	B. J. Almanac.*	3.50
Potassium permanganate + hydrochloric acid	Sodium stannate		2.05
Cupric chloride	Sodium stannate	Desalme, B. J.,† p. 215, '12.	1.93
Potassium ferricyanide + potassium bromide	Sodium sulphide	Ordinary sepia developer.	1.33
Mercuric iodide	Paraminophenol developer	HgI_2 according to Bennett.	1.23

See Nietz and Huse, J. Franklin Inst. March 3, 1918.
* B. J. Almanac, see annual Almanac of British Journal of Photography.
† B. J. refers to British Journal of Photography.

Table 312.

WAVE-LENGTHS OF FRAUNHOFER LINES.

For convenience of reference the values of the wave-lengths corresponding to the Fraunhofer lines usually designated by the letters in the column headed "index letters," are here tabulated separately. The values are in ten millionths of a millimeter, on the supposition that the D line value is 5896.155. The table is for the most part taken from Rowland's table of standard wave-lengths.

Index Letter.	Line due to —	Wave-length in centimeters × 10^8.	Index Letter.	Line due to —	Wave-length in centimeters × 10^8.
A	O	7621.28*	G	Fe	4308.081
	O	7594.06*		Ca	4307.907
a	–	7164.725	g	Ca	4226.904
B	O	6870.182 †	h or H_δ	H	4102.000
C or H_α	H	6563.045	H	Ca	3968.625
α	O	6278.303 ‡	K	Ca	3933.825
D_1	Na	5896.155	L	Fe	3820.586
D_2	Na	5890.186	M	Fe	3727.778
D_3	He	5875.985	N	Fe	3581.349
E_1	Fe	5270.558	O	Fe	3441.155
	Ca	5270.438	P	Fe	3361.327
E_2	Fe	5269.723	Q	Fe	3286.898
b_1	Mg	5183.791	R	Ca	3181.387
b_2	Mg	5172.856		Ca	3179.453
b_3	Fe	5169.220	S_1, S_2	Fe	3100.787
	Fe	5169.069		Fe	3100.430
b_4	Fe	5167.678		Fe	3100.046
	Mg	5167.497	s	Fe	3047.725
F or H_β	H	4861.527	T	Fe	3020.76
d	Fe	4383.721	t	Fe	2994.53
G′ or H_γ	H	4340.634	U	Fe	2947.99
f	Fe	4325.939			

* The two lines here given for A are stated by Rowland to be: the first, a line "beginning at the head of A, outside edge"; the second, a "single line beginning at the tail of A."

† The principal line in the head of B.

‡ Chief line in the α group.

See Table 321, Rowland's Solar Wave-lengths (foot of page) for correction to reduce these values to standard system of wave-lengths, Table 314.

STANDARD WAVE-LENGTHS.

TABLE 313.—Absolute Wave-length * of Red Cadmium Line in Air, 760 mm. Pressure, 15° C.

6438.4722 Michelson, Travaux et Mém. du Bur. intern. des Poids et Mesures, 11, 1895.
6438.4700 Michelson, corrected by Benoit, Fabry, Perot, C. R. 144, 1082, 1907.
6438.4696 (accepted primary standard) Benoit, Fabry, Perot, C. R. 144, 1082, 1907.

* In Ångströms. 10 Ångströms = 1 mμ = 10^{-6} mm.

TABLE 314.—International Secondary Standards. Iron Arc Lines in Ångströms.

Adopted as secondary standards at the International Union for Coöperation in Solar Research (transactions, 1910). Means of measures of Fabry-Buisson (1), Pfund (2), and Eversheim (3). Referred to primary standard = Cd. line, λ = 6438.4696 Ångströms (serving to define an Ångström). 760 mm., 15° C. Iron rods, 7 mm. diam. length of arc, 6 mm.; 6 amp. for λ greater than 4000 Ångströms, 4 amp. for lesser wave-lengths; continuous current, + pole above the —, 220 volts; source of light, 2 mm. at arc's center. Lines adopted in 1910.

Wave-length.	Wave-length.	Wave-length.	Wave-length.	Wave-length.	Wave-length.	Wave-length.
4282.408	4547.853	4789.657	5083.344	5405.780	5615.661	6230.734
4315.089	4592.658	4878.225	5110.415	5434.527	5658.836	6265.145
4375.934	4602.947	4903.325	5167.492	5455.614	5763.013	6318.028
4427.314	4647.439	4919.007	5192.363	5497.522	6027.059	6335.341
4466.556	4691.417	5001.881	5232.957	5506.784	6065.492	6393.612
4494.572	4707.288	5012.073	5266.569	5569.633	6137.701	6430.859
4531.155	4736.786	5049.827	5371.495	5586.772	6191.568	6494.993

TABLE 315.—International Secondary Standards. Iron Arc Lines in Ångströms.

Adopted in 1913. (4) Means of measures of Fabry-Buisson, Pfund, Burns and Eversheim.

Wave-length.	Wave-length.	Wave-length.	Wave-length.	Wave-length.	Wave-length.	Wave-length.
3370.789	3606.682	3753.615	3906.482	4076.642	4233.615	6750.250
3399.337	3640.392	3805.346	3907.937	4118.552	5709.396	5857.759 Ni
3485.345	3676.313	3843.261	3935.818	4134.685	6546.250	5892.882 Ni
3513.821	3677.629	3850.820	3977.746	4147.676	6592.928	
3556.881	3724.380	3865.527	4021.872	4191.443	6678.004	

(1) Astrophysical Journal, 28, p. 169, 1908; (2) Ditto, 28, p. 197, 1908; (3) Annalen der Physik, 30, p. 815, 1909. See also Eversheim, *ibid.* 36, p. 1071, 1911; Buisson et Fabry, *ibid.* 38, p. 245, 1912; (4) Astrophysical Journal, 39, p. 93, 1914.

TABLE 316.—Neon Wave-Lengths.

Intensity.	Wave length.	Intensity.	Wave length.	Intensity.	Wave length.	Intensity.	Wave length.	Intensity.	Wave length.
5	3369.904	5	3515.192	2	5820.155	4	6217.280	5	6717.043
6	3417.906	8	3520.474	10	5852.488	7	6266.495	8	6929.468
6	3447.705	4	3593.526	6	5881.895	4	6304.789	3	7024.049
6	3454.197	4	3593.634	8	5944.834	8	6334.428	9	7032.413
5	3460.526	5	3600.170	4	5975.534	8	6382.991	3	7059.111
4	3464.340	5	3633.664	4	6029.997	10	6402.245	5	7173.939
5	3466.581	8	5330.779	7	6074.338	9	6506.528	8	7245.167
6	3472.578	7	5341.096	8	6096.163	4	6532.883	6	7438.902
4	3498.067	6	5400.562	9	6143.062	5	6598.953	5	7488.885
4	3501.218	4	5764.419	5	6163.594	8	6678.276	5	7535.784

International Units (Ångströms). Burns, Meggers, Merrill, Bull. Bur. Stds. 14, 765, 1918.

TERTIARY STANDARD WAVE-LENGTHS. IRON ARC LINES.

For arc conditions see Table 314, p. 266. For lines of group *c* class 5 for best results the slit should be at right angles to the arc at its middle point and the current should be reversed several times during the exposure.

Wave-lengths.	Class.	Intensity.	Wave-lengths.	Class.	Intensity.	Wave-lengths.	Class.	Intensity.
*2781.840		4	4337.052	b3	5	5332.909	a4	2
*2806.985		7	4369.777	b3	3	5341.032	a4	5
*2831.559		3	4415.128	b1	8r	5365.404	a1	2
*2858.341		3	4443.198	b3	3	5405.780	a	6
*2901.382		4	4461.658	a3	4	5434.528	a	6
*2926.584		5	4489.746	a3	3	5473.913	a	4
*2986.460		3	4528.620	c4	7	5497.521	a	4
*3000.453		4	4619.297	c4	4	5501.471	a	4
*3053.070		4	4786.811	c4	3	5506.784	a	3
*3100.838		2	4871.331	c5	8	‡5535.419	a	2
*3154.202		4	4890.769	c5	7	5563.612	b	3
*3217.389		4	4924.773	a	3	5975.352	b	2
*3257.603		4	4939.685	a	3	6027.059	b	3
*3307.238		4	4973.113	a	2	6065.495	b	4
*3347.932		4	4994.133	a	3	6136.624	b	5
*3389.748		3	5041.076	a	3	6157.734	b	4
*3476.705		5	5041.760	a	4	6165.370	b	3
*3506.502		5	5051.641	a	4	6173.345	b	4
*3553.741		5	5079.227	a	3	6200.323	b	4
*3617.789		6	5079.743	a	3	6213.441	b	5
*3659.521		5	5098.702	a	4	6219.290	b	5
*3705.567		6R	5123.729	a	4	6252.567	b	6
*3749.487		8R	5127.366	a	3	6254.269	b	4
*3820.430		8R	5150.846	a	4	6265.145	b	5
*3859.913		7R	5151.917	a	3	6297.802	b	4
*3922.917		6R	5194.950	a	5	6335.342	b	6
*3956.682		6	5202.341	a	5	6430.859	b	5
*4009.718		5	5216.279	a	5	6494.992	b	6
*4062.451		4	5227.191	a4	8			
†4132.063	b1	7	5242.495	a	3			
†4175.639	b	4	5270.356	a4	8			
†4202.031	b1	7r	5328.043	a1	7			
†4250.791	b2	7	5328.537	a4	4			

* Measures of Burns. † Means of St. John and Burns.
‡ Means of St. John and Goos. Others are means of measures by all three. References: St. John and Ware, Astrophysical Journal, 36, 1912; 38, 1913; Burns, Z. f. wissen. Photog. 12, p. 207, 1913, J. de Phys. 1913, and unpublished data; Goos, Astrophysical Journal, 35, 1912; 37, 1913. The lines in the table have been selected from the many given in these references with a view to equal distribution and where possible of classes *a* and *b*.

For class and pressure shifts see Gale and Adams, Astrophysical Journal, 35, p. 10, 1912. Class *a:* "This involves the well-known flame lines (de Watteville, Phil. Trans. A 204, p. 139. 1904), i.e. the lines relatively strengthened in low-temperature sources, such as the flame of the arc, the low-current arc, and the electric furnace. (Astrophysical Journal, 24, p. 185, 1906, 30, p. 86, 1909, 34, p. 37, 1911, 35, p. 185, 1912.) The lines of this group in the yellow-green show small but definite pressure displacements, the mean being 0.0036 Ångström per atmosphere in the arc." Class *b:* "To this group many lines belong; in fact all the lines of moderate displacement under pressure are assigned to it for the present. These are bright and symmetrically widened under pressure, and show mean pressure displacements of 0.009 Ångström per atmosphere for the lines in the region λ 5975–6678 according to Gale and Adams. Group *c* contains lines showing much larger displacements. The numbers in the class column have the following meaning: 1, symmetrically reversed; 2, unsymmetrically reversed; 3, remain bright and fairly narrow under pressure; 4, remain bright and symmetrical under pressure but become wide and diffuse; 5, remain bright and are widened very unsymmetrically toward the red under pressure."

For further measures in International units see Kayser, Bericht über den gegenwärtigen Stand der Wellenlängenmessungen, International Union for Coöperation in Solar Research, 1913. For further spectroscopic data see Kayser's Handbuch der Spectroscopie.

REDUCTION OF WAVE-LENGTH MEASURES TO STANDARD CONDITIONS.

The international wave-length standards are measured in dry air at 15° C, 76 cm pressure. Density variations of the air appreciably affect the absolute wave-lengths when obtained at other temperatures and pressures. The following tables give the corrections for reducing measures to standard conditions, viz.: $\delta = \lambda_0(n_0 - n_0')(d - d_0)/d_0$ in ten-thousandths of an Angstrom, when the temperature t° C, the pressure B in cm of Hg, and the wave-length λ in Angstroms are given; n and d are the indices of refraction and densities, respectively; the subscript $_0$ refers to standard conditions, none, to the observed; the prime ′ to the standard wave-length, none, to the new wave-length. The tables were constructed for the correction of wave-length measures in terms of the fundamental standard 6438.4696 A of the cadmium red radiation in dry air, 15° C, 76 cm pressure. The density factor is, therefore, zero for 15° C and 76 cm, and the correction always zero for λ = 6438 A. As an example, find the correction required for λ when measured as 3000.0000 A in air at 25° C and 72 cm. Section (*a*) of table gives $(d - d_0)/d_0 = -.085$ and for this value of the density factor section (*b*) gives the correction to λ of −.0038 A. Again, if λ, under the same atmospheric conditions, is measured as 8000.0000 A in terms of a standard λ' of wave-length 4000.000 A, say, the measurement will require a correction of (0.0020 + 0.0008) = +.0028 A. Taken from Meggers and Peters, Bulletin Bureau of Standards, 14, p. 728, 1918.

TABLE 318 (*a*). — $1000 \times (d - d_0)/d_0$.

B cm	60.0	62.5	65.0	67.5	70	71	72	73	74	75	76	77	78
9° C	−192	−160	−126	−92	−59	−46	−32	−19	−5	+8	+22	+35	+48
11	−200	−167	−133	−100	−67	−53	−40	−27	−13	0	+13	+27	+40
13	−206	−172	−139	−106	−73	−60	−46	−33	−20	−7	+6	+20	+33
15	−211	−178	−145	−112	−79	−66	−53	−39	−26	−13	0	+13	+26
17	−216	−184	−151	−118	−86	−73	−60	−47	−34	−21	−8	+5	+19
19	−222	−189	−156	−124	−92	−79	−66	−53	−40	−27	−14	−1	+12
21	−227	−195	−163	−130	−98	−85	−72	−59	−46	−33	−21	−8	+5
23	−232	−200	−168	−136	−104	−91	−78	−65	−52	−40	−27	−14	−1
25	−238	−206	−174	−143	−111	−98	−85	−72	−60	−47	−34	−22	−9
27	−243	−211	−179	−148	−116	−104	−91	−78	−66	−53	−40	−28	−15
29	−248	−216	−185	−154	−122	−109	−97	−84	−72	−59	−46	−34	−21
31	−253	−222	−190	−159	−128	−116	−103	−91	−78	−66	−54	−41	−29
33	−258	−227	−196	−165	−134	−121	−109	−97	−84	−72	−59	−47	−34
35	−262	−231	−200	−170	−139	−127	−114	−102	−90	−77	−65	−53	−41

TABLE 318 (*b*). — $\delta = \lambda_0(n_0 - n_0')(d - d_0)/d_0$, in Ten-thousandth Angstroms.

$1000 \times \frac{d - d_0}{d_0}$	Wave-lengths in Angstroms.														
	2000	2500	3000	3500	4000	4500	5000	5500	6000	6500	7000	7500	8000	9000	10000
	Corrections in ten-thousandth Angstroms.														
−260	−259	−166	−116	−84	−61	−44	−30	−18	−8	+1	+9	+17	+24	+37	+50
−240	−239	−154	−107	−78	−57	−41	−28	−17	−7	+1	+9	+16	+22	+35	+46
−220	−219	−141	−98	−71	−52	−37	−26	−15	−7	+1	+8	+14	+20	+32	+42
−200	−199	−128	−89	−65	−47	−34	−23	−14	−6	+1	+7	+13	+19	+29	+38
−180	−179	−115	−80	−58	−42	−30	−21	−13	−6	+1	+6	+12	+17	+26	+34
−160	−159	−102	−71	−52	−38	−27	−19	−11	−5	+1	+6	+10	+15	+23	+31
−140	−139	−90	−62	−45	−33	−24	−16	−10	−4	+1	+5	+9	+13	+20	+27
−120	−119	−77	−54	−39	−28	−20	−14	−8	−4	+0	+4	+8	+11	+17	+23
−100	−100	−64	−45	−32	−24	−17	−12	−7	−3	+0	+4	+7	+9	+14	+19
−80	−80	−51	−36	−26	−19	−14	−9	−6	−2	+0	+3	+5	+7	+12	+15
−60	−60	−38	−27	−19	−14	−10	−7	−4	−2	+0	+2	+4	+6	+9	+11
−40	−40	−26	−18	−13	−9	−7	−5	−3	−1	+0	+1	+3	+4	+6	+8
−20	−20	−13	−9	−6	−5	−3	−2	−1	−1	0	+1	+1	+2	+3	+4
0	0	0	0	0	0	0	0	0	0	−0	0	0	0	0	0
+20	+20	+13	+9	+6	+5	+3	+2	+1	+1	−0	−1	−2	−2	−3	−4
+40	+40	+26	+18	+13	+9	+7	+5	+3	+1	−0	−1	−3	−4	−6	−8

SPECTRA OF THE ELEMENTS.

The following figure gives graphically the positions of some of the more prominent lines in the spectra of some of the elements. Flame spectra are indicated by lines in the lower parts of the panels, arc spectra in the upper parts, and spark spectra by dotted lines.

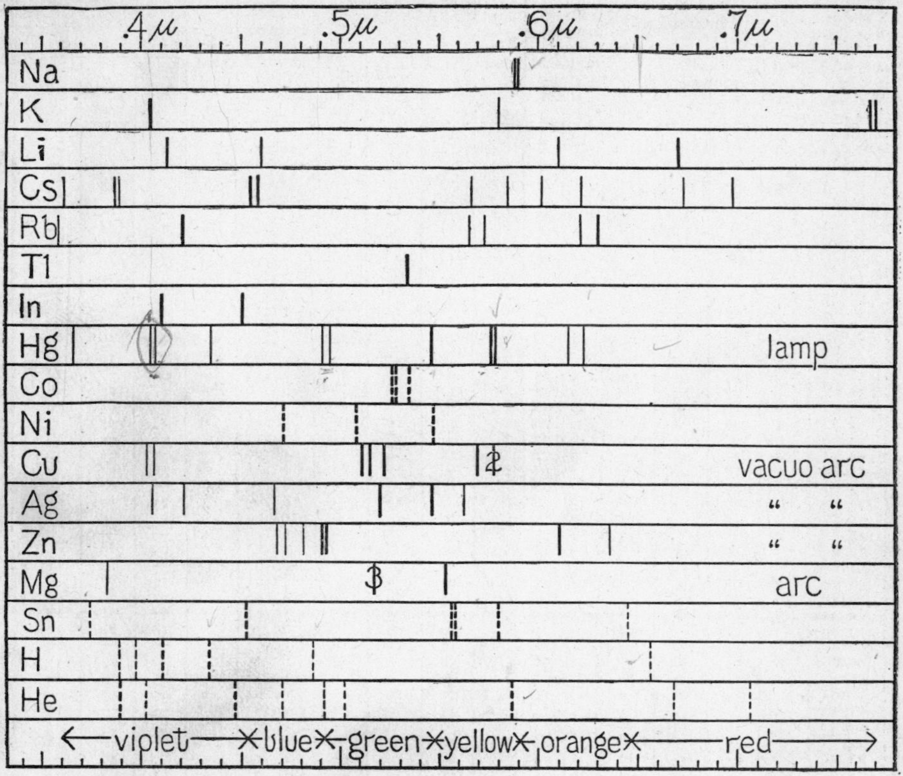

The following wave-lengths are in Angstroms.

Na	5889.965	Rb	4202	Cu	4023	Mg	5168
	5895.932		4216		4063		5173
K	4044		5648		5105.543*		5184
	4047		5724		5153.251*		5529
	5802		6207		5218.202*	Sn	4525
	7668		6299		5700		5563
	7702	Tl	5351		5782.090*		5589
Li	4132	In	4102		5782.159*		5799
	4602		4511	Ag	4055		6453
	6104	Hg	4046.8		4212	H	3970
	6707.846*		4078.1		4669		4102
Cs	4555		4358.3		5209.081*		4340
	4593		4916.4		5465.489*		4861
	5664		4959.7		5472		6563
	5945		5460.742*		5623	He	3187.743†
	6011		5769.598*	Zn	4680.138*		3888.646†
	6213		5790.659*		4722.164*		4026.189†
	6724		6152		4810.535*		4471.477†
	6974		6232		4912		4713.143†
					4925		4921.929†
					6103		5015.675†
					6362.345*		5875.618†
							6678.149†
							7065.188†

For other elements, see Kayser's Handbuch der Spectroscopie.
* Fabry and Perot. † Merrill.

TABLE 320.

SPECTRUM LINES OF THE ELEMENTS.

Table of brighter lines only abridged from more extensive table compiled from Kayser and containing 10,000 lines (Kayser's Handbuch der Spectroscopie, Vol. 6, 1912).

Wave-lengths, inter-national Ang-stroms.	Ele-ment.	Intensities.			Wave-lengths, inter-national Ang-stroms.	Ele-ment.	Intensities.			Wave-lengths, inter-national Ang-stroms.	Ele-ment.	Intensities		
		Arc.	Spark.	Tube.			Arc.	Spark.	Tube.			Arc.	Spark.	Tube.
3802.98	Nb	15	4	—	3968.48	Ca	30	40	—	4116.50	V	15	5	—
08.21	I	—	—	10	72.01	Eu	20	20	—	18.48	Pr	15	10	—
10.73	Nh	10	20	—	74.71	Er	15	5	—	23.24	La	10	15	—
14.45	Ra	20	20	—	76.85	Tb	20	10	—	28.3	Y	15	8	—
19.65	Eu	20	20	—	80.43	Br	—	—	10	28.70	I	—	—	10
22.15	Rh	12	15	—	81.68	Em	—	—	15	28.91	Rh	15	10	—
28.47	Rh	12	10	—	81.89	Tb	15	10	—	29.75	Eu	50	50	—
29.35	Mg	15	8	—	82.60	Y	12	12	—	30.42	Gd	15	10	—
32.30	Mg	20	10	—	88.00	Ny	50	20	—	35.29	Rh	12	10	—
36.83	Zr	—	15	—	88.52	La	10	15	—	35.80	Os	15	5	—
38.29	S	—	8	10	91.13	Zr	8	12	—	37.13	Nb	12	4	—
38.29	Mg	20	10	—	98.96	Zr	8	12	—	39.74	Nb	15	4	—
45.45	Co	10	15	—	4000.47	Dy	15	12	—	42.86	Y	15	8	—
47.98	Tm	15	10	—	05.50	Tb	15	10	—	43.14	Pr	15	10	—
48.75	Tb	15	15	—	05.73	V	—	20	—	45.12	S	—	—	10
51.02	Cl	—	—	10	08.73	Pr	12	8	—	49.20	Zr	10	15	—
56.50	Rh	10	12	—	19.62	Pb	12	10	—	51.12	Er	15	4	—
58.29	Ni	20	8	—	22.70	Cu	15	10	—	52.63	Nb	15	5	—
60.86	Cl	—	5	10	23.35	V	—	20	—	53.11	S	—	—	10
64.11	Mo	20	10	—	23.71	Se	12	8	—	58.62	A	—	—	10
71.65	La	8	15	—	25.1	F	—	—	10	61.83	Ar	10	20	—
73.07	Co	10	12	—	30.80	Mn	18	8	—	62.70	S	—	—	10
74.16	Tb	15	15	—	31.70	La	8	15	—	63.64	Nb	15	10	—
76.66	Lu	15	10	—	33.03	Ga	10	30	—	64.66	Nb	12	5	—
88.64	He	—	—	10	33.06	Mn	15	8	—	66.43	Em	—	—	20
88.96	Nh	15	10	—	34.48	Mn	15	8	—	68.14	Nb	15	5	—
91.01	Nh	20	15	—	35.62	V	—	20	—	69.0	Se	—	10	10
94.09	Co	10	15	—	41.43	Mn	12	8	—	72.05	Ga	15	20	—
94.22	Pd	15	15	—	42.92	La	8	15	—	77.53	Y	12	20	—
96.36	Er	15	6	—	44.15	K	20	10	—	79.04	Ge	—	20	—
97.63	I	—	—	10	45.45	Nh	20	10	—	79.43	Pr	15	10	—
3900.53	Ti	15	10	—	45.82	Fe	6	15	—	80.04	X	—	—	20
02.95	Mo	15	8	—	46.00	Dy	12	4	—	84.25	Lu	20	15	—
05.5	Si	15	4	—	46.6	Se	—	4	10	89.52	Pr	15	10	—
06.34	Er	15	10	—	47.21	K	20	10	—	90.91	Nb	15	9	—
07.14	Eu	30	20	—	48.73	Mn	11	6	—	4200.65	A	—	—	10
07.52	Sc	12	6	—	55.53	Ag	50	6	—	01.82	Rb	20	15	—
11.85	Sc	15	6	—	57.84	Pb	30	20	—	03.23	Em	—	—	10
14.26	Br	—	—	10	58.97	Nb	15	10	—	05.04	Eu	50	30	—
14.94	Sc	12	—	—	62.75	Cu	15	10	—	05.32	Nb	15	4	—
22.52	X	—	—	10	62.83	Pr	12	8	—	06.72	Pr	15	12	—
25.43	Tb	15	10	—	63.47	Gd	20	—	—	08.96	Zr	0	12	—
30.51	Eu	50	50	—	77.34	La	10	12	—	11.14	Rh	15	10	—
31.10	I	—	—	10	77.37	Y	15	5	—	11.69	Dy	12	5	—
33.67	Ca	40	50	—	77.75	Sr	50	50	—	14.74	Nb	12	—	—
39.55	Tb	15	10	—	77.97	Dy	12	11	—	15.52	Sr	30	30	—
40.07	I	—	—	10	78.79	X	—	—	10	15.56	Rb	20	10	—
40.47	Rb	—	15	—	79.73	Nb	15	6	—	17.95	Nb	15	—	—
44.68	Dy	12	10	—	80.62	Ra	12	10	—	21.08	I	—	—	10
45.33	O	—	—	10	86.70	La	10	15	—	23.00	Pr	15	12	—
49.10	La	12	20	—	92.68	V	15	—	—	25 34	Pr	15	12	—
50.35	Y	12	12	—	99.80	V	20	—	—	26.56	Ge	7	50	—
51.01	X	—	—	10	4100.74	Pr	15	12	—	26.72	Ca	20	10	—
51.95	V	—	15	—	00.97	Nb	20	6	—	38.21	X	—	—	10
58.22	Zr	8	15	—	01.82	In	20	12	—	41.04	Pr	12	10	—
58.66	Pd	15	10	—	02.40	Y	12	8	—	44.34	Rb	—	15	—
58.85	Rh	15	12	—	03.4	F	—	—	10	45.2	Pb	—	20	—
66.23	Nb	12	—	—	09.78	V	15	10	—	45.38	X	—	—	10
67.59	X	—	—	10	11.80	V	20	—	—	46.3	F	—	—	30
3968.40	Dy	15	12	—	4112.03	Os	12	4	—	4246.85	Sc	15	20	—

SPECTRUM LINES OF THE ELEMENTS.

Wave-lengths, international Angstroms.	Element.	Intensity.			Wave-lengths, international Angstroms.	Element.	Intensity.			Wave-lengths, international Angstroms.	Element.	Intensity.		
		Arc.	Spark.	Tube.			Arc	Spark.	Tube.			Arc.	Spark.	Tube.
4253.61	S	—	—	10	4477.77	Br	—	—	10	4994.13	Lu	20	—	—
54.34	Cr	12	12	—	81.17	Mg	—	20	—	5035.36	Ni	12	10	—
54.42	Nb	15	8	—	96.43	Pr	15	10	—	53.30	W	12	12	—
59.69	Bi	—	20	—	98.76	Pt	12	10	—	5135.08	Lu	15	—	—
60.84	Os	15	5	—	4510.15	Pr	12	10	—	56.20	Sr	20	—	—
73.96	Kr	—	—	10	22.59	Eu	20	20	—	61.19	I	—	—	10
74.80	Cr	12	10	—	24.74	Sn	10	20	—	63.78	Pd	15	—	—
86.97	La	10	12	—	54.97	Cr	15	—	—	72.68	Mg	15	15	—
4301.11	Nb	12	5	—	55.52	Ru	10	12	—	83.60	Mg	20	20	—
02.12	Bi	—	15	—	72.74	H	—	—	10	64.51	Cr	12	8	—
02.28	Y	12	—	—	73.09	Nb	12	5	—	5206.05	Cr	12	9	—
03.61	Nd	20	10	—	74.26	I	—	—	10	08.42	Cr	12	10	—
05.49	Sr	10	20	—	85.47	X	—	—	10	09.08	Ag	30	20	—
05.78	Pr	15	10	—	89.35	Dy	15	5	—	24.70	W	12	12	—
07.92	Fe	6	15	—	94.09	Eu	30	20	—	56.95	Sr	20	6	—
08.1	Em	—	—	10	4603.03	X	—	—	10	92.23	X	—	—	10
09.63	Y	12	12	—	06.77	Nb	12	10	—	95.62	Pd	15	—	—
14.11	Sc	12	12	—	07.34	Sr	30	20	—	5330.65	O	—	—	10
19.60	Kr	—	—	10	09.22	Em	—	—	10	32.01	Br	—	—	10
25.77	Nd	15	5	—	24.28	X	—	—	15	32.8	Sn	—	20	—
25.78	Fe	6	15	—	25.40	Em	—	—	15	35.14	Ny	5	20	—
26.36	Nb	12	—	—	27.29	Eu	20	15	—	50.49	Tl	20	10	—
30.47	X	—	—	15	27.98	H	—	—	10	52.86	Ny	—	20	—
33.77	La	12	12	—	33.86	H	—	—	10	60.59	Mo	15	12	—
40.67	Ra	15	10	—	34.02	H	—	—	10	69.85	Se	—	—	10
43.69	Cl	—	5	10	44.11	Sm	—	—	15	74.08	Se	—	—	10
48.01	A	—	—	10	46.16	Cr	12	10	—	95.27	Pd	12	—	—
49.65	Em	—	—	15	48.66	Ni	15	—	—	5419.19	X	—	—	10
55.47	Kr	—	—	10	61.92	Eu	20	15	—	64.5	I	—	—	10
65.58	Br	—	4	10	66.65	I	—	—	10	65.49	Ag	30	20	—
68.30	O	—	—	10	71.24	X	—	—	10	76.69	Lu	20	10	—
74.51	Sc	10	12	—	72.12	Nb	12	10	—	76.91	Ni	12	10	—
74.81	Rh	15	12	—	75.36	Nb	12	8	—	80.95	Sr	20	10	—
74.94	Y	15	20	—	80.138	Zn	10	20	—	96.78	I	—	—	10
79.24	V	30	30	—	80.74	Em	—	—	10	5504.26	Sr	20	—	—
79.77	Zr	10	12	—	82.18	Ra	20	15	—	06.51	Mo	20	15	—
81.66	Mo	12	6	—	87.80	Zr	7	12	—	14.71	W	20	20	—
82.8	Se	—	8	10	4704.93	Br	—	15	10	21.80	Sr	20	—	—
83.55	Fe	10	20	—	08.26	I	—	—	10	33.01	Mo	15	12	—
84.73	V	20	30	—	14.42	Ni	15	8	—	42.78	Pd	12	—	—
86.9	Pb	—	20	—	22.164	Zn	10	20	—	56.49	Ny	15	—	—
89.98	V	20	20	—	22.54	Bi	10	20	—	62.5	Sn	—	30	—
93.17	X	—	—	10	30.86	Se	—	—	10	70.46	Mo	15	10	—
95.24	V	15	10	—	38.12	Tl	—	15	—	89.2	Sa	—	20	—
95.74	X	—	—	10	85.49	Br	—	10	10	5608.9	Pb	—	12	—
98.03	Y	10	15	—	94.48	Cl	—	20	10	20.64	As	—	—	10
4401.54	Ni	15	15	—	4806.68	I	—	—	10	25.64	I	—	—	10
04.75	Fe	8	15	—	08.23	I	—	—	10	51.34	As	—	—	10
08.50	V	15	20	—	09.97	Cl	—	9	10	62.93	Y	—	12	—
08.83	Pr	12	10	—	10.534	Zn	10	20	—	70.05	Pd	15	—	—
10.09	I	—	—	10	11.83	Sr	15	8	—	98.54	V	15	15	—
11.71	Mo	12	6	—	19.28	Mo	12	—	—	5751.40	Mo	15	—	—
20.46	Os	15	10	—	25.93	Ra	15	10	—	99.4	Sa	—	20	—
24.36	Sm	20	10	—	32.07	Sr	15	6	—	5813.63	Ra	—	15	—
29.23	Pr	15	12	—	40.6	Se	—	4	10	52.49	Ne	—	—	15
34.26	I	—	—	10	44.32	X	—	—	10	57.76	Ni	15	—	—
35.58	Eu	20	20	—	44.8	Se	—	6	10	58.27	Mo	12	8	—
37.23	Nb	12	8	—	50.49	I	—	—	10	75.64	He	—	—	10
42.56	Pt	12	5	—	54.89	Y	10	15	—	88.33	Mo	15	10	—
46.6	F	—	—	20	83.71	Y	12	20	—	89.96	Na	20	20	—
48.11	X	—	—	10	4900.13	Y	10	20	—	95.93	Mo	15	10	—
51.56	Nd	10	15	—	11.7	Zn	10	20	—	95.93	Na	20	20	—
53.00	I	—	—	10	24.0	Zn	10	20	—	5928.82	Mo	15	15	—
59.8	Em	—	—	10	57.41	Dy	15	—	—	6090.22	V	15	15	—
4462.21	X	—	—	20	4962.27	Sr	15	—	—	6121.80	H	—	—	10

NOTE. — This table, somewhat unsatisfactory in its abridged form, is included with the hope to occupy its space later with a better table; e.g., no mercury lines appear since the scale of intensity used in the original table results in the intensity of all mercury lines falling below the critical value used in this table.

STANDARD SOLAR WAVE-LENGTHS. ROWLAND'S VALUES.

Wave-lengths are in Ångström units (10^{-7} mm.), in air at 20° C and 76 cm. of mercury pressure. The intensities run from 1, just clearly visible on the map, to 1000 for the H and K lines; below 1 in order of faintness to 0000 as the lines are more and more difficult to see. This table contains only the lines above 5.

N indicates a line not clearly defined, probably an undissolved multiple line; **s**, a faded appearing line; **d**, a double. In the "substance" column, where two or more elements are given, the line is compound; the order in which they are given indicates the portion of the line due to each element; when the solar line is too strong to be due wholly to the element given, it is represented, **–Fe**, for example; when commas separate the elements instead of a dash, the metallic lines coincide with the same part of the solar line, **Fe, Cr**, for example.

Capital letters next the wave-length numbers are the ordinary designations of the lines. **A** indicates atmospheric lines, (**wv**), due to water vapor, (**O**), due to Oxygen.

Wave-length.	Substance.	Intensity.	Wave-length.	Substance.	Intensity.	Wave-length.	Substance.	Intensity.
3037.510s	Fe	10 N	3372.947	Ti–Pd	10 d?	3533.345	Fe	6
3047.725s	Fe	20 N	3380.722	Ni	6 N	3536.709	Fe	7
3053.530s	–	7 d?	3414.911	Ni	15	3541.237	Fe	7
3054.429	Mn, Ni	10	3423.848	Ni	7	3542.232	Fe	6
3057.552s	Ti, Fe	20	3433.715	Ni, Cr	8 d?	3555.079	Fe	9
3059.212s	Fe	20	3440.762s } O	Fe	20	3558.672s	Fe	8
3067.369s	Fe	8	3441.155s } O	Fe	15	3565.535s	Fe	20
3073.091	Ti, –	6 Nd?	3442.118	Mn	6	3566.522	Ni	10
3078.769s	Ti, –	8 d?	3444.020s	Fe	8 N	3570.273s	Fe	20
3088.145s	Ti	7 d?	3446.406	Ni	15	3572.014	Ni	6
3134.230s	Ni, Fe	8	3449.583	Co	6 d?	3572.712	Se, –	6
3188.656	–, Fe	6 d?	3453.039	Ni	6 d?	3578.832	Cr	10
3236.703s	Ti	7 N	3458.601	Ni	8	3581.349s	Fe	30
3239.170	Ti	7	3461.801	Ni	8	3584.800	Fe	6
3242.125	Ti, –	8	3462.950	Co	6	3585.105	Fe	6
3243.189	–, Ni	6	3466.015s	Fe	6	3585.479	Fe	7
3247.688s	Cu	10	3475.594s	Fe	10	3585.859	Fe	6
3256.021	Fe?	6	3476.849s	Fe	8	3587.130	Fe	8
3267.834s	V	6	3483.923	Ni	6 d?	3587.370	Co	7
3271.129	Fe	6	3485.493	Fe Co	6	3588.084	Ni	6
3271.791	Ti, Fe	6 d?	3490.733s	Fe	10 N	3593.636	Cr	9
3274.096s	Cu	10	3493.114	Ni	10 N	3594.784	Fe	6
3277.482	Co–Fe	7 d?	3497.982s	Fe	8	3597.854	Ni	8
3286.898	Fe	7 N	3500.996s	Ni	6 d?	3605.479s	Cr	7
3295.951s	Fe, Mn	6	3510.466	Ni	8	3606.838s	Fe	6
3302.510s	Na	6	3512.785	Co	6	3609.008s	Fe	20
3315.807	Ni	7 d?	3513.965s	Fe	7	3612.882	Ni	6 d?
3318.160s	Ti	6	3515.206	Ni	12	3617.934s	Fe	6
3320.391	Ni	7	3519.904	N	7	3618.919s	Fe	20
3336.820	Mg	8 N	3521.410s	Fe	8	3619.539	Ni	8
3349.597	Ti	7	3524.677	Ni	20	3621.612s	Fe	6
3361.327	Ti	8	3526.183	Fe	6	3622.147s	Fe	6
3365.908	Ni	6	3526.988	Co	6	3631.605s	Fe	15
3366.311	Ti, Ni	6 d?	3529.964	Fe–Co	6	3640.535s	Cr–Fe	6
3369.713	Fe, Ni	6	3533.156	Fe	6	3642.820	Ti	7

Corrections to reduce Rowland's wave-lengths to standards of Table 314 (the accepted standards, 1913). Temperature 15° C, pressure 760 mm.

The differences "(Fabry-Buisson-arc-iron) — (Rowland-solar-iron)" lines were plotted, a smooth curve drawn, and the following values obtained:

Wave-length	3000.	3100.	3200.	3300.	3400.	3500.	3600.	3700.
Correction	—.106	—.115	—.124	—.137	—.148	—.154	—.155	—.140

H. A. Rowland, "A preliminary table of solar-spectrum wave-lengths," Astrophysical Journal, 1–6, 1895–1897.

STANDARD SOLAR WAVE-LENGTHS. ROWLAND'S VALUES.

Wave-length.	Substance.	Intensity.	Wave-length.	Substance.	Intensity.	Wave-length.	Substance.	Intensity.
3647.988s	Fe	12	3826.027s	Fe	20	4045.975s	Fe	30
3651.247	Fe,–	6	3827.980	Fe	8	4055.701s	Mn	6
3651.614	Fe	7	3829.501s	Mg	10	4057.668	–	7
3676.457	Fe, Cr	6	3831.837	Ni	6	4063.759s	Fe	20
3680.069s	Fe	9	3832.450s	Mg	15	4068.137	Fe–Mn	6
3684.258s	Fe	7d?	3834.364	Fe	10	4071.908s	Fe	15
3685.339	Ti	10d?	3838.435s	Mg–C	25	4077.885s	Sr	8
3686.141	Ti–Fe	6	3840.580s	Fe–C	8	4102.000Hδ	H, In	40N
3687.610s	Fe	6	3841.195	Fe–Mn	10	4121.477s	Cr–Co	6d?
3689.614	Fe	6	3845.606	C–Co	8d?	4128.251	Ce–V,–	6d
3701.234	Fe	8	3850.118	Fe–Cr	10	4132.235	Fe–Co	10
3705.708s	Fe	9	3856.524s	Fe	8	4137.156	Fe	6
3706.175	Ca, Mn	6d?	3857.805	Cr–C	6d?	4140.089	Fe	6
3709.389s	Fe	8	3858.442	Ni	7	4144.038	Fe	15
3716.591s	Fe	7	3860.055s	Fe–C	20	4167.438	–	8
3720.084s	Fe	40	3865.674	Fe–C	7	4187.204	Fe	6
3722.692s	Ni	10	3872.639	Fe	6	4191.595	Fe	6
3724.526	Fe	6	3878.152	Fe–C	8	4202.198s	Fe	8
3732.545s	Co–Fe	6	3878.720	Fe	7Nd?	4226.904sg	Ca	20d?
3733.469s	Fe–	7d?	3886.434s	Fe	15	4233.772	Fe	6
3735.014s	Fe	40	3887.196	Fe	7	4236.112	Fe	8
3737.281s	Fe	30	3894.211	–	8d	4250.287s	Fe	8
3738.466	–	6	3895.803	Fe	7	4250.945s	Fe	8
3743.508	Fe–Ti	6	3899.850	Fe	8	4254.505s	Cr	8
3745.717s	Fe	8	3903.090	Cr, Fe, Mo	10	4260.640s	Fe	10
3746.058s	Fe	6	3904.023	–	8d	4271.934s	Fe	15
3748.408s	Fe	10	3905.660s	Si	12	4274.958s	Cr	7d?
3749.631s	Fe	20	3906.628	Fe	10	4308.081sG	Fe	6
3753.732	Fe–Ti	6d?	3920.410	Fe	10	4325.939s	Fe	8
3758.375s	Fe	15	3923.054	Fe	12d?	4340.634Hγ	H	20N
3759.447	Ti	12d?	3928.075s	Fe	8	4376.107s	Fe	6
3760.196	Fe	5	3930.450	Fe	8	4383.720s	Fe	15
3761.464	Ti	7	3933.523	–	8N	4404.927s	Fe	10
3763.945s	Fe	10	3933.825sK	Ca	1000	4415.293s	Fe	8
3765.689	Fe	6	3934.108	Co, V–Cr	8N	4442.510	Fe	6
3767.341s	Fe	8	3944.160s	Al	15	4447.892s	Fe	6
3775.717	Ni	7	3956.819	Fe	6	4494.738s	Fe	6
3783.674s	Ni	6	3957.177s	Fe–Ca	7d?	4528.798	Fe	8
3788.046s	Fe	9	3961.674s	Al	20	4534.139	Ti–Co	6
3795.147s	Fe	8	3968.350	–, Zr	6N	4549.808	Ti–Co	6d?
3798.655s	Fe	6	3968.625sH	Ca	700	4554.211s	Ba	8
3799.693s	Fe	7	3968.886	–	6N	4572.156s	Ti–	6
3805.486s	Fe	6	3969.413	Fe	10	4603.126	Fe	6
3806.865	Mn–Fe	8d?	3974.904	Co–Fe	6d?	4629.521s	Ti–Co	6
3807.293	Ni	6	3977.891s	Fe	6	4679.027s	Fe	6
3807.681	V–Fe	6	3986.903s	–	6	4703.177s	Mg	10
3814.698	–	8	4005.408	Fe	7	4714.599s	Ni	6
3815.987s	Fe	15	4030.918s	Mn	10d?	4736.963	Fe	6
3820.586sL	Fe–C	25	4033.224s	Mn	8d?	4754.225s	Mn	7
3824.591	Fe	6	4034.644s	Mn	6d	4783.613s	Mn	6

Corrections to reduce Rowland's wave-lengths to standards of Table 314 (the accepted standards, 1913). Temperature 15° C, pressure 760 mm.:

Wave-length	3600.	3700.	3800.	3900.	4000.	4100.	4200.	4300.	4400.	4500.	4600.	4700.	4800.
Correction	–.155	–.140	–.141	–.144	–.148	–.152	–.156	–.161	–.167	–.172	–.176	–.179	–.179

STANDARD SOLAR WAVE-LENGTHS. ROWLAND'S VALUES.

Wave-length.	Substance.	Intensity.	Wave-length.	Substance.	Intensity.	Wave-length.	Substance.	Intensity.
4861.527sF	H	30	5948.765s	Si	6	6563.045sC	H	40
4890.948s	Fe	6	5985.040s	Fe	6	6593.161s	Fe	6
4891.683	Fe	8	6003.239s	Fe	6	6867.457sB	A(O)	6d?
4919.174s	Fe	6	6008.785s	Fe	6	6868.336 } s	A(O)	6
4920.685	Fe	10	6013.715s	Mn	6	6868.478 }	A(O)	6
4957.785s	Fe	8	6016.861s	Mn	6	6869.142s	A(O)	7
5050.008s	Fe	6	6022.016s	Mn	6	6869.353s	A(O)	6
5167.497sb$_4$	Mg	15	6024.281s	Fe	7	6870.116 } s	A(O)	7 } d
5171.778s	Fe	6	6065.709s	Fe	7	6870.249 }	A(O)	7 }
5172.856sb$_2$	Mg	20	6102.392s	Fe	6	6871.180s	A(O)	8
5183.791sb$_1$	Mg	30	6102.937s	Ca	9	6871.532s	A(O)	10
5233.122s	Fe	7	6108.334s	Ni	6	6872.486s	A(O)	11
5266.738s	Fe	6	6122.434s	Ca	10	6873.080s	A(O)	12
5269.723sE	Fe	8d?	6136.829s	Fe	8	6874.037s	A(O)	12
5283.802s	Fe	6	6137.915	Fe	7	6874.899s	A(O)	13
5324.373s	Fe	7	6141.938s	Fe, Ba	7	6875.830s	A(O)	13
5328.236	Fe	8d?	6155.350	–	7	6876.958s	A(O)	13
5340.121	Fe	6	6162.390s	Ca	15	6877.882s	A(O)	12
5341.213	Fe	7	6169.249s	Ca	6	6879.288s	A(O)	12
5367.669s	Fe	6	6169.778s	Ca	7	6880.172s	A(O)	6
5370.166s	Fe	6	6170.730	Fe–Ni	6	6884.076s	A(O)	10
5383.578s	Fe	6	6191.393s	Ni	6	6886.000s	A(O)	11
5397.344s	Fe	7d?	6191.779s	Fe	9	6886.990s	A(O)	12
5405.989s	Fe	6	6200.527s	Fe	6	6889.192s	A(O)	13
5424.290s	Fe	6	6213.644s	Fe	6	6890.151s	A(O)	14
5429.911	Fe	6d?	6219.494s	Fe	6	6892.618s	A(O)	14
5447.130s	Fe	6d?	6230.943s	V–Fe	8	6893.560s	A(O)	15
5528.641s	Mg	8	6246.535s	Fe	8	6896.289s	A(O)	14
5569.848	Fe	6	6252.773s	–Fe	7	6897.208s	A(O)	15
5573.075	Fe	6	6256.572s	Ni–Fe	6	6900.199s	A(O)	14
5586.991	Fe	7	6301.718	Fe	7	6901.117s	A(O)	15
5588.985s	Ca	6	6318.239	Fe	6	6904.362s	A(O)	14
5615.877s	Fe	6	6335.554	Fe	6	6905.271s	A(O)	14
5688.436s	Na	6	6337.048	Fe	7	6908.783s	A(O)	13
5711.313s	Mg	6	6358.898	Fe	6	6909.676s	A(O)	13
5763.218s	Fe	6	6393.820s	Fe	7	6913.448s	A(O)	11
5857.674s	Ca	8	6400.217s	Fe	8	6914.337s	A(O)	11
5862.582s	Fe	6	6411.865s	Fe	7	6918.370s	A(O)	9
5890.186sD$_2$	Na	30	6421.570s	Fe	7	6919.250s	A(O)	9
5896.155 D$_1$	Na	20	6439.293s	Ca	8	6923.553s	A(O)	9
5901.682s	A(wv)	6	6450.033s	Ca	6	6924.427s	A(O)	9
5914.430s	–, A(wv)	6	6494.004s	Ca	6	7191.755	A, –	6N
5919.860s	A(wv)	7	6495.213	Fe	8	7206.692	–, A	6
5930.406s	Fe	6	6546.479s	Ti–Fe	6			

Corrections to reduce Rowland's wave-lengths to standards of Table 314 (the accepted standards, 1913). Temperature 15° C, pressure 760 mm.:

Wave-length	4800.	4900.	5000.	5100.	5200.	5300.	5400.	5500.	5600.	5700.	5800.
Correction	–.179	–.176	–.173	–.170	–.166	–.172	–.212	–.217	–.218	–.213	–.209

Wave-length	5800.	5900.	6000.	6100.	6200.	6300.	6400.	6500.	6600.	6700.	6800.
Correction	–.209	–.209	–.213	–.214	–.213	–.210	–.209	–.210.			

Table 322.

SPECTRUM SERIES.

The flame spectrum lines of the elements are comparatively few. These remain prominent in the arc with the appearance of many more. In the spark the typical arc lines tend to disappear with the appearance of entirely new lines. Those thus intensified or only appearing in the violent action of the condensed spark have been called "enhanced" lines. This order of development may be taken as one due to increasing temperatures. The spectra of compounds are invariably banded; different sets of bands indicate oxides, chlorides, etc. Banded spectra of the elements are generally assumed to be due to molecules, the simpler spectra to atoms, and the enhanced spectra to atoms with one electron lost (ionized.) The enhanced spectrum of He is similar to that of H except that the wave-lengths are shortened because of the increased attraction of the heavier nucleus, 2e in place of e.

In the spectra of many elements and compounds certain lines or groups of lines (doublets, triplets, etc.) occur in orderly sequence, each series with definite order of intensity (generally decreasing with decreasing wave-length), pressure effect, Zeeman effect, etc. Such series generally obey approximately a law of the form

$$\nu = \frac{1}{\lambda} = L - \frac{N}{(m+R)^2},$$

where ν is the wave-number in vacuo (reciprocal of the wave-length λ) generally expressed in waves per cm; m is a variable integer, each integer giving a line of the series; L is the wave number of the limit of the series ($m = \infty$); N, the "Universal Series Constant"; and R is a function of m, or a constant in some simple cases.

Balmer's formula (1885) results if $L = N/n^2$, where n is another variable integer and $R = 0$. Rydberg's formula (1889) makes R a constant, and L is not known to be connected with N. Other formulae have been used with more success. Mogendorff (1906) requires R = constant/m, while Ritz (1903) has R = constant/m^2. Often no simple formula fits the case; either R must be a more complex function of m, or the shape of the formula is incorrect.

Bohr's theory (see also Table 515) gives for Hydrogen.

$$N = \left\{2\pi^2 me^4(M+m)\right\}/Mh^3,$$

where e and m are the charge and mass of an electron, M the atomic weight, and h, Planck's constant. The best value for N is 109678.7 international units (Curtis, Birge, Astrophys. J. 32, 1910). The theory has been elaborated by Sommerfeld (Ann. der Phys. 1916), and the present indications are that N is a complex function varying somewhat from element to element.

Among the series (of singles, doublets, etc.), there is apt to be one more prominent, its lines easily reversible, called the principal series, $P(m)$. With certain relationships to this there may be two subordinate series, the first generally diffuse, $D(m)$, and another, $S(m)$. Related to these there is at times another, the Bergmann or fundamental series, $B(m)$. or $F(m)$. m is the variable integer first used above and indicates the order of the line.

The following laws are in general true among these series: (1) In the $P(m)$ the components of the lines, if double, triple, etc., are closer with increasing order; in the subordinate series the distance of the components (in vibration number) remains constant. (2) Further, in two related $D(m)$ and $S(m)$, $\Delta\nu$ (vibration number difference) remains the same. (3) The limits (L) of the subordinate series, $D(m)$ and $S(m)$, are the same. (4) $\Delta\nu$ of the subordinate series is the same $\Delta\nu$ as for the first pair of the corresponding $P(m)$. (5) The limits (L) of the components of the doublets (triplets, etc.) of the $P(m)$ are the same. (6) The difference between the vibration numbers of the end of the $P(m)$ and of the two corresponding subordinate series gives the vibration number of the first term of the $P(m)$. The first line of the $S(m)$ coincides with the first line of the $P(m)$ (Rydberg-Schuster law). The limit of the Bergmann or fundamental series is the first term of the diffuse series (Runge law).

In the spectrum of an element several of these families of series $P(m)$, $D(m)$, $S(m)$, $B(m)$ may be found. For further information see Baly's Spectroscopy and Konen's Das Leuchten der Gasen, 1916, from the latter of which is taken the following tables, based greatly upon Dunz's Die Seriengesetze der Linienspektra, Diss., Tubingen, 1911, which has also appeared in book form, Hirzel, Leipzig.

The "complexity" of the lines of a series is constant throughout a column in the periodic table, varying from one column to another. The displacement law of Kossel and Sommerfeld states that when an element is ionized (losing an electron) the enhanced spectrum takes on the same type of "complexity" as the arc spectrum belonging to the element of the preceding column (to the left) but with lines shifted to a higher frequency. If two electrons are lost, the displacement is two columns to the left. If the outer ring has an odd number of electrons the spectrum will consist of doublets; if even, of triplets and singlets.

(discussion continued on page 441)

Series Spectra of the Elements. — The ordinary spectrum of H contains 3 series of the same kind: one in the ultra-violet region; Schumann region, $\nu = N(1/1^2 - 1/n^2)$, n, 2, 3 . . .; one in the visible, $\nu = N(1/2^2 - 1/n^2)$, n, 3, 4, 5 . . .; and one in the infra-red, $\nu = N(1/3^2 - 1/n^2)$, n, 4, 5, 6 . . . He has three systems of series, one 'enhanced," including the Pickering series formerly supposed to be due to H. The next two tables give some of the data for other elements.

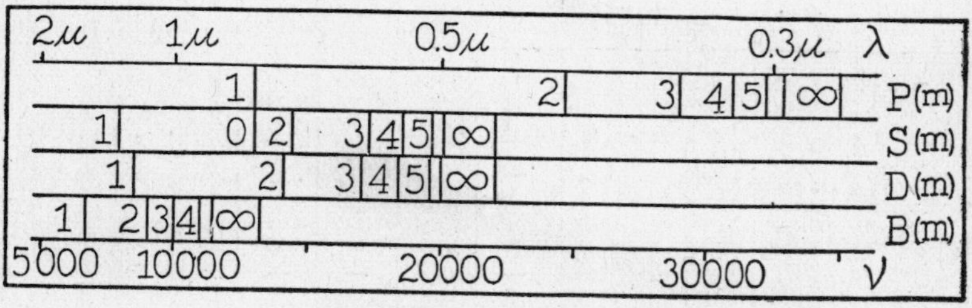

Series System of Potassium.

SPECTRUM SERIES.

TABLE 323. — Limits of Some of the Series.

	$P_1(\infty)$	$D_1(\infty) = S_1(\infty)$	$B_1(\infty)$	$P_2(\infty)$	$D_2(\infty) = S_2(\infty)$	$B_2(\infty)$	$P_3(\infty)$	$D_3(\infty) = S_3(\infty)$	$B_3(\infty)$	$R(\infty)$
H	48,764	27,429	12,186	48,764	27,419	12,186	48,744	27,429	12,186	—
He	32,031	27,173	12,204	38,453	{ 29,221 / 29,222	12,208	—	—	—	—
Li	—	—	—	43,484	28,581	12,202	—	—	—	—
Na	—	—	—	*41,445	{ 24,472 / 24,489	12,274	—	—	—	—
K	—	—	—	35,006	{ 21,963 / 22,020	13,471	—	—	—	—
Rb	—	—	—	33,685	{ 20,868 / 21,106	14,330	—	—	—	—
Cs	—	—	—	31,407	{ 19,674 / 20,228	16,809 / 16,907	—	—	—	—
Cu	—	—	—	62,306	{ 31,523 / 31,771	12,372 / 12,366	—	—	—	—
Ag	—	—	—	61,093	{ 30,621 / 31,542	12,351	—	—	—	—
Mg	—	26,613	—	?	?	?	20,467	{ 39,752 / 39,793 / 39,813	13,707	—
Ca	—	27,510	—	?	60,423 / 60,646	28,929	17,761	{ 33,983 / 34,089 / 34,142	28,929 / 28,950 / 28,964	49,353
Sr	--	25,745	—	—	55,029 / 55,830	—	—	31,026 / 31,420 / 31,607	27,605 / 27,705 / 27,766	45,895
Ba	—	—	—	—	{ 49,926 / 51,616	—	?	?	?	48,318

For the series of Zn, Cd, Hg, Al, Sn, Tl, O, S, Sn, see original reference.
* 48 lines have been measured in this series from 16,956 to 41,417.

TABLE 324. — First Terms of Some of the Series. Vibration Number Differences of Pairs $\Delta\nu$, and Triplets $\Delta\nu_1$, $\Delta\nu_2$.

For the $P(m)$ and the $S(m)$ is given only the first or second term, since the term with index o may be omitted as coinciding with the first term of the $S(m)$ or $P(m)$ respectively. Consequently the numbers always proceed from greater to smaller wave-lengths. Which is the common line can always be recognized from the vibration numbers. See figure on the preceding page. The vibration differences can be obtained from Table 323.

	$P(1)$	$D(1)$	$S(1)$	$B(1)$
H	21,334	15,233	9,871	5332
He	4,857	14,970	13,729	5348
	9,231	{ 17,014 / 17,015	14,149 / 14,148	5351
Li	14,903	16,379	12,301	5347
Na	{ 16,973 / 16,956	12,215 / 12,198	8,782 / 8,766	5416
K	{ 13,043 / 12,985	8,552 / 8,493	8,040 / 7,983	6592
Rb	{ 12,817 / 12,579	6,776 / 6,538	7,552 / 7,315	7437
Cs	{ 11,733 / 11,178	3,321 / 2,767	7,357 / 6,803	9972 / 9875
Cu	{ 30,783 / 30,535	19,158 / 19,151	12,601 / 12,352	5495
Ag	{ 30,472 / 30,551	19,191 / 18,271	13,003 / 12,083	5439
	—	11,352	—	—
Mg	35,760 / 35,668	35,831 / 35,739	34,135 / 34,043	—

	$P(1)$	$D(1)$	$S(1)$	$B(1)$
Mg	{ 6654 / 6650 / 6650	26,106 / 26,086 / 26,045	19,346 / 19,326 / 19,285	6,720
Ca	—	20,495	19,828	—
	—	11,763 / 11,541	25,414 / 25,191	22,153 / —
	{ 5036 / 5020 / 5012	5,019 / 5,125 / 5,177	16,381 / 16,329 / 16,223	21,834 / 21,820 / 21,799
Sr	—	19,390	—	—
	—	9,959 / 9,159	23,715 / 23,518	—
	—	3,842 / 3,655 / 3,260	14,721 / 14,533 / 14,139	20,591 / 20,533 / 20,435
Ba	—	12,176 / 10,493	21,952 / 20,261	—
	—	—	—	13,894 / 13,523 / 12,645

	$\Delta\nu$	$\Delta\nu_1$	$\Delta\nu_2$
He	1	—	—
Na	17	—	—
K	58	—	—
Rb	237	—	—
Cs	552	—	—
Cu	249	—	—
Ag	921	—	—
Mg	91	41	20
Ca	223	106	52
Sr	801	394	187
Ba	1690	878	370
Zn	872?	389	190
Cd	2484?	1171	542
Hg	—	4632	1769
Al	112	—	—
In	2213	—	—
Tl	7793	—	—
O	—	3.7	2.1
S	—	18.2	11.1
Se	—	104	45

TABLE 325. — Index of Refraction of Glass.

Indices of refraction of optical glass made at the Bureau of Standards. Correct probably to 0.00001. The composition given refers to the raw material which went into the melts and does not therefore refer to the composition of the finished glass.

Melt.	123	241	135	116	188	151	163	76
Wave-length.	Ordinary crown.	Borosilicate crown.	Barium flint.	Light barium crown.	Light flint.	Dense barium crown.	Medium flint.	Dense flint.
Hg 4046.8	1.53189	1.53817	1.58851	1.59137	1.60507	1.63675	1.65788	1.69005
Hg 4078.1	1.53147	1.53775	1.58791	1.59084	1.60430	1.63619	1.65692	1.68894
H 4340.7	1.52818	1.53468	1.58327	1.58698	1.59860	1.63189	1.64973	1.68079
Hg 4358.6	1.52798	1.53450	1.58299	1.58674	1.59826	1.63163	1.64931	1.68030
H 4861.5	1.52326	1.53008	1.57646	1.58121	1.59029	1.62548	1.63941	1.66911
Hg 4916.4	1.52283	1.52967	1.57587	1.58071	1.58958	1.62492	1.63854	1.66814
Hg 5461.0	1.51929	1.52633	1.57105	1.57657	1.58380	1.62033	1.63143	1.66016
Hg 5769.6	1.51771	1.52484	1.56894	1.57473	1.58128	1.61829	1.62834	1.65671
Hg 5790.5	1.51760	1.52475	1.56881	1.57460	1.58112	1.61817	1.62815	1.65650
Na 5893.2	1.51714	1.52430	1.56819	1.57406	1.58038	1.61756	1.62725	1.65548
Hg 6234.6	1.51573	1.52297	1.56634	1.57242	1.57818	1.61576	1.62458	1.65250
H 6563.0	1.51458	1.52188	1.56482	1.57107	1.57638	1.61427	1.62241	1.65007
Li 6708.2	1.51412	1.52145	1.56423	1.57054	1.57567	1.61369	1.62157	1.64913
K 7682.0	1.51160	1.51908	1.56100	1.56762	1.57183	1.61047	1.61701	1.64405
(Percentage composition)								
SiO_2	67.0	64.2	53.7	48.0	53.9	37.0	45.6	39.0
Na_2O	12.0	9.4	1.7	2.0	1.0	—	3.4	3.0
K_2O	5.0	8.3	8.3	6.1	7.6	2.7	4.1	4.0
B_2O_3	3.5	11.0	2.7	4.0	—	5.0	—	—
BaO	10.6	6.1	14.3	29.5	—	47.0	—	—
ZnO	1.5	—	2.5	10.0	—	7.7	—	—
As_2O_3	0.4	0.4	—	1.4	0.3	—	—	—
CaO	—	1.0	—	—	2.0	—	3.0	4.0
PbO	—	—	16.7	—	35.2	—	44.0	49.0
Sb_2O_3	—	—	—	—	—	—	—	1.0

TABLE 326. — Dispersion of Glasses of Table 325.

Melt.	123	241	135	116	188	151	163	76
n_D	1.51714	1.52430	1.56819	1.57406	1.58038	1.61756	1.62725	1.65548
$n_F - n_C$	0.00868	0.00820	0.01164	0.01014	0.01391	0.01121	0.01700	0.01904
$\frac{n_D - 1}{n_F - n_C} = \nu$	59.6	63.9	48.8	56.6	41.7	55.1	36.9	34.4
$n_D - n_F$	0.00612	0.00578	0.00827	0.00715	0.00991	0.00792	0.01216	0.01363
$n_F - n_{G'}$	0.00492	0.00460	0.00681	0.00577	0.00831	0.00641	0.01032	0.01168
$n_D - n_C$	0.00256	0.00242	0.00337	0.00299	0.00400	0.00329	0.00484	0.00541

TABLE 327. — Glasses Made by Schott and Gen, Jena.

The following constants are for glasses made by Schott and Gen, Jena: n_A, n_C, n_D, n_F, n_G, are the indices of refraction in air for A=0.7682μ, C=0.6563μ, D=0.5893, F=0.4861, G′=0.4341. $v=(n_D-1)/(n_F-n_C)$. Ultra-violet indices: Simon, Wied. Ann. 53, 1894. Infra-red: Rubens, Wied. Ann. 45, 1892. Table is revised from Landolt, Börnstein and Meyerhoffer, Kayser, Handbuch der Spectroscopie, and Schott and Gen's list No. 751, 1909. See also Hovestadt's "Jena Glass."

		O 546	O 381	O 184	O 102	O 165	S 57
Catalogue Type =		O 546	O 381	O 184	O 102	O 165	S 57
Designation =		Zinc-Crown.	Higher Dispersion Crown.	Light Silicate Flint.	Heavy Silicate Flint.	Heavy Silicate Flint.	Heaviest Silicate Flint.
Melting Number =		1092	1151	451	469	500	163
v =		60.7	51.8	41.1	33.7	27.6	22.2
Kind of Light and Wave-length.	Cd 0.2763μ	1.56759	–	–	–	–	–
	Cd .2837	1.56372	–	–	–	–	–
	Cd .2980	1.55723	1.57093	1.65397	–	–	–
	Cd .3403	1.54369	1.55262	1.63320	1.71968	1.85487	–
	Cd .3610	1.53897	1.54664	1.61388	1.70536	1.83263	–
	H .4340μ	1.52788	1.53312	1.59355	1.67561	1.78800	1.94493
	H .4861	1.52299	1.52715	1.58515	1.66367	1.77091	1.91890
	Na .5893	1.51698	1.52002	1.57524	1.64985	1.75130	1.88995
	H .6563	1.51446	1.51712	1.57119	1.64440	1.74368	1.87893
	K .7682	1.51143	1.51368	1.56669	1.63820	1.73530	1.86702
	.800μ	1.5103	1.5131	1.5659	1.6373	1.7339	1.8650
	1.200	1.5048	1.5069	1.5585	1.6277	1.7215	1.8481
	1.600	1.5008	1.5024	1.5535	1.6217	1.7151	1.8396
	2.000	1.4967	1.4973	1.5487	1.6171	1.7104	1.8316
	2.400	–	–	1.5440	1.6131	–	1.8286

Percentage composition of the above glasses:

O 546, SiO_2, 65.4; K_2O, 15.0; Na_2O, 5.0; BaO, 9.6; ZnO, 2.0; Mn_2O_3, 0.1; As_2O_3, 0.4; B_2O_3, 2.5.
O 381, SiO_2, 68.7; PbO, 13.3; Na_2O, 15.7; ZnO, 2.0; MnO_2, 0.1; As_2O_5, 0.2.
O 184, SiO_2, 53.7; PbO, 36.0; K_2O, 8.3; Na_2O, 1.0; Mn_2O_3, 0.06; As_2O_3, 0.3.
O 102, SiO_2, 40.0; PbO, 52.6; K_2O, 6.5; Na_2O, 0.5; Mn_2O_3, 0.09; As_2O_5, 0.3.
O 165, SiO_2, 29.26; PbO, 67.5; K_2O, 3.0; Mn_2O_3, 0.04; As_2O_3, 0.2.
S 57, SiO_2, 21.9; PbO, 78.0; As_2O_5, 0.1.

TABLE 328. — Jena Glasses.

No. and Type of Jena Glass.	n_D for D	n_F-n_C	$v=\frac{n_D-1}{n_F-n_C}$	n_D-n_A	n_F-n_D	$n_{G'}-n_F$	Specific Weight.
O 225 Light phosphate crown	1.5159	.00737	70.0	.00485	.00515	.00407	2.58
O 802 Boro-silicate crown	1.4967	0765	64.9	0504	0534	0423	2.38
UV 3199 Ultra-violet crown	1.5035	0781	64.4	0514	0546	0432	2.41
O 227 Barium-silicate crown	1.5399	0909	59.4	0582	0639	0514	2.73
O 114 Soft-silicate crown	1.5151	0910	56.6	0577	0642	0521	2.55
O 608 High-dispersion crown	1.5149	0943	54.6	0595	0666	0543	2.60
UV 3248 Ultra-violet flint	1.5332	0964	55.4	0611	0680	0553	2.75
O 381 High-dispersion crown	1.5262	1026	51.3	0644	0727	0596	2.70
O 602 Baryt light flint	1.5676	1072	53.0	0675	0759	0618	3.12
S 389 Borate flint	1.5686	1102	51.6	0712	0775	0629	2.83
O 726 Extra light flint	1.5398	1142	47.3	0711	0810	0669	2.87
O 154 Ordinary light flint	1.5710	1327	43.0	0819	0943	0791	3.16
O 184 " " "	1.5900	1438	41.1	0882	1022	0861	3.28
O 748 Baryt flint	1.6235	1599	39.1	9965	1142	0965	3.67
O 102 Heavy flint	1.6489	1919	33.8	1152	1372	1180	3.87
O 41 " "	1.7174	2434	29.5	1439	1749	1521	4.49
O 165 " "	1.7541	2743	27.5	1607	1974	1730	4.78
S 386 Heavy flint	1.9170	4289	21.4	2451	3109	2808	6.01
S 57 Heaviest flint	1.9626	4882	19.7	2767	3547	3252	6.33

TABLE 329. — Change of Indices of Refraction for 1° C in Units of the Fifth Decimal Place.

No. and Designation.	Mean Temp.	C	D	F	G′	$\frac{-\Delta n}{n}100$
S 57 Heavy silicate flint	58.8°	1.204	1.447	2.090	2.810	0.0166
O 154 Light silicate flint	58.4	0.225	0.261	0.334	0.407	0.0078
O 327 Baryt flint light	58.3	−0.008	0.014	0.080	0.137	0.0079
O 225 Light phosphate crown	58.1	−0.202	−0.190	−0.168	−0.142	0.0049

Pulfrich, Wied. Ann. 45, p. 609, 1892.

TABLE 330. — Index of Refraction of Rock Salt in Air.

$\lambda(\mu)$.	n.	Observer.	$\lambda(\mu)$.	n.	Observer.	$\lambda(\mu)$.	n.	Observer.
0.185409	1.89348	M	0.88396	1.534011	L	5.8932	1.516014	P
.204470	1.76964	"	.972298	1.532532	"	"	1.515553	L
.291368	1.61325	"	.98220	1.532435	P	6.4825	1.513628	P
.358702	1.57932	"	1.036758	1.531762	L	"	1.513467	L
.441587	1.55962	"	1.1786	1.530372	P	7.0718	1.511062	P
.486149	1.55338	"	"	1.530374	L	7.6611	1.508318	"
"	1.553406	L	1.555137	1.528211	"	7.9558	1.506804	"
"	1.553399	P	1.7680	1.527440	P	8.8398	1.502035	"
.58902	1.544340	L	"	1.527441	L	10.0184	1.494722	"
.58932	1.544313	P	2.073516	1.526554	"	11.7864	1.481816	"
.656304	1.540672	P	2.35728	1.525863	P	12.9650	1.471720	"
"	1.540702	L	"	1.525849	L	14.1436	1.460547	"
.706548	1.538633	P	2.9466	1.524534	P	14.7330	1.454404	"
.766529	1.536712	P	3.5359	1.523173	"	15.3223	1.447494	"
.76824	1.53666	M	4.1252	1.521648	P	15.9116	1.441032	"
.78576	1.536138	P	"	1.521625	L	20.57	1.3735	RN
.88396	1.534011	P	5.0092	1.518978	P	22.3	1.340	"

$$n^2=a^2+\frac{M_1}{\lambda^2-\lambda_1{}^2}+\frac{M_2}{\lambda^2-\lambda_2{}^2}-k\lambda^2-h\lambda^4 \text{ or } =b^2+\frac{M_1}{\lambda^2-\lambda_1{}^2}+\frac{M_2}{\lambda^2-\lambda_2{}^2}-\frac{M_3}{\lambda_3{}^2-\lambda^2}$$

where $a^2=2.330165$, $M_1=0.01278685$, $\lambda_1{}^2=0.0148500$, $M_2=0.005343924$;
$\lambda_2{}^2=0.02547414$, $k=0.0009285837$, $h=0.000000286086$;
$b^2=5.680137$, $M_3=12059.95$, $\lambda_3{}^2=3600.$ (P)

TABLE 331. — Change of Index of Refraction for 1° C in Units of the 5th Decimal Place.

0.202μ	+3.134	Mi	0.441μ	−3.425	Mi	C line	−3.749	Pl	0.760μ	−3.73	L
.210	+1.570	"	.508	−3.517	"	D "	−3.739	"	1.368	−3.88	L
.224	−0.187	"	.643	−3.636	"	F "	−3.648	"	1.88	−3.85	L
.298	−2.727	"				G′ "	−3.585	"	4.3	−3.82	L

L Annals of the Astrophysical Observatory of the Smithsonian Institution, Vol. I, 1900.
M Martens, Ann. d. Phys. 6, 1901, 8, 1902.
Mi Micheli, Ann. d. Phys. 7, 1902.
P Paschen, Wied. Ann. 26, 1908.
Pl Pulfrich, Wied. Ann. 45, 1892.
RN Rubens and Nichols, Wied. Ann. 60, 1897.

TABLE 332. — Index of Refraction of Sylvite (Potassium Chloride) in Air.

$\lambda(\mu)$.	n	Observer.	$\lambda(\mu)$.	n.	Observer.	$\lambda(\mu)$.	n.	Observer.
0.185409	1.82710	M	1.1786	1.478311	P	8.2505	1.462726	P
.200090	1.71870	"	"	1.47824	W	"	1.46276	W
.21946	1.64745	"	1.7680	1.475890	P	8.8398	1.460858	P
.257317	1.58125	"	"	1.47589	W	"	1.46092	W
.281640	1.55836	"	2.35728	1.474751	P	10.0184	1.45672	P
.308227	1.54136	"	2.9466	1.473834	"	"	1.45673	W
.358702	1.52115	"	"	1.47394	W	11.786	1.44919	P
.394415	1.51219	"	3.5359	1.473049	P	"	1.44941	W
.467832	1.50044	"	"	1.47304	W	12.965	1.44346	P
.508606	1.49620	"	4.7146	1.471122	P	"	1.44385	W
.58933	1.49044	P	"	1.47129	W	14.144	1.43722	P
.67082	1.48669	M	5.3039	1.470013	P	15.912	1.42617	"
.78576	1.483282	P	"	1.47001	W	17.680	1.41403	"
.88398	1.481422	P	5.8932	1.468804	P	20.60	1.3882	RN
.98220	1.480084	"	"	1.46880	W	22.5	1.369	"

$$n^2=a^2+\frac{M_1}{\lambda^2-\lambda_1{}^2}+\frac{M_2}{\lambda^2-\lambda_2{}^2}-k\lambda^2-h\lambda^4 \text{ or } =b^2+\frac{M_1}{\lambda^2-\lambda_1{}^2}+\frac{M_2}{\lambda^2-\lambda_2{}^2}+\frac{M_3}{\lambda_3{}^2-\lambda^2}$$

$a^2=2.174967$, $M_1=0.008344206$, $\lambda_1{}^2=0.0119082$, $M_2=0.00698382$;
$\lambda_2{}^2=0.0255550$, $k=0.000513495$, $h=0.000000167587$;
$b^2=3.866619$, $M_3=5569.715$, $\lambda_3{}^2=3292.47$ (P)

W Weller, see Paschen's article. Other references as under Table 331, above.

INDEX OF REFRACTION.

TABLE 333.—Index of Refraction of Fluorite in Air.

λ (μ)	n	Observer	λ (μ)	n	Observer	λ (μ)	n	Obser-ver.
0.1856	1.50940	S	1.4733	1.42641	P	4.1252	1.40855	P
.19881	1.49629	"	1.5715	1.42596	"	4.4199	1.40559	"
.21441	1.48462	"	1.6206	1.42582	"	4.7146	1.40238	"
.22645	1.47762	"	1.7680	1.42507	"	5.0092	1.39898	"
.25713	1.46476	"	1.9153	1.42437	"	5.3036	1.39529	"
.32525	1.44987	"	1.9644	1.42413	"	5.5985	1.39142	"
.34555	1.44697	"	2.0626	1.42359	"	5.8932	1.38719	"
.39681	1.44214	"	2.1608	1.42308	"	6.4825	1.37819	"
.48607	1.43713	P	2.2100	1.42288	"	7.0718	1.36805	"
.58930	1.43393	P	2.3573	1.42199	"	7.6612	1.35680	"
.65618	1.43257	S	2.5537	1.42088	"	8.2505	1.34444	"
.68671	1.43200	"	2.6519	1.42016	"	8.8398	1.33079	"
.71836	1.43157	"	2.7502	1.41971	"	9.4291	1.31612	"
.76040	1.43101	"	2.9466	1.41826	"	51.2	3.47	RA
.8840	1.42982	P	3.1430	1.41707	"	61.1	2.66	"
1.1786	1.42787	"	3.2413	1.41612	"	∞	2.63	S
1.3756	1.42690	"	3.5359	1.41379	"			
1.4733	1.42641	"	3.8306	1.41120	"	References under Table 331.		

$$n^2 = a^2 + \frac{M_1}{\lambda^2 - \lambda_1^2} - e\lambda^2 - f\lambda^4 \text{ or } = b^2 + \frac{M_2}{\lambda^2 - \lambda_v^2} + \frac{M_3}{\lambda^2 - \lambda_r^2}$$

where $a^2 = 2.03882$ $f = 0.000002916$ $M_3 = 5114.65$
$M_1 = 0.0062183$ $b^2 = 6.09651$ $\lambda_r^2 = 1260.56$
$\lambda_1^2 = 0.007706$ $M_2 = 0.0061386$ $\lambda_v = 0.0940\mu$
$e = 0.0031999$ $\lambda_v^2 = 0.00884$ $\lambda_r = 35.5\mu$ (P)

TABLE 334.—Change of Index of Refraction for 1° C in Units of the 5th Decimal Place.

C line, —1.220; D, —1.206; F, —1.170; G, —1.142. (Pl)

TABLE 335.—Index of Refraction of Iceland Spar ($CaCO_3$) in Air.

λ (μ)	n_o	n_e	Obser-ver.	λ (μ)	n_o	n_e	Obser-ver.	λ (μ)	n_o	n_e	Obser-ver.
0.198	–	1.5780	M	0.508	1.6653	1.4896	M	0.991	1.6438	1.4802	C
.200	1.9028	1.5765	"	.533	1.6628	1.4884	"	1.229	1.6393	1.4787	"
.208	1.8673	1.5664	"	.589	1.6584	1.4864	"	1.307	1.6379	1.4783	"
.226	1.8130	1.5492	–	.643	1.6550	1.4849	"	1.497	1.6346	1.4774	"
.298	1.7230	1.5151	C	.656	1.6544	1.4846	"	1.682	1.6313	–	"
.340	1.7008	1.5056	M	.670	1.6537	1.4843	"	1.749	–	1.4764	"
.361	1.6932	1.5022	C	.760	1.6500	1.4826	–	1.849	1.6280	–	"
.410	1.6802	1.4964	–	.768	1.6497	1.4826	M	1.908	–	1.4757	"
.434	1.6755	1.4943	M	.801	1.6487	1.4822	C	2.172	1.6210	–	"
.486	1.6678	1.4907	"	.905	1.6458	1.4810	"	2.324	–	1.4739	"

C Carvallo, J. de Phys. (3), 9, 1900.
M Martens, Ann. der Phys. (4) 6, 1901, 8, 1902.
P Paschen, Wied. Ann. 56, 1895.
Pl Pulfrich, Wied. Ann 45, 1892.
RA Rubens-Aschkinass, Wied. Ann. 67, 1899.
S Starke, Wied. Ann. 60, 1897.

TABLE 336.—Index of Refraction of Nitroso-dimethyl-aniline. (Wood.)

λ	n	λ	n	λ	n	λ	n	λ	n
0.497	2.140	0.525	1.945	0.584	1.815	0.636	1.647	0.713	1.718
.500	2.114	.536	1.909	.602	1.796	.647	1.758	.730	1.713
.506	2.074	.546	1.879	.611	1.783	.659	1.750	.749	1.709
.508	2.025	.557	1.857	.620	1.778	.669	1.743	.763	1.697
.516	1.985	.569	1.834	.627	1.769	.696	1.723		

Nitroso-dimethyl-aniline has enormous dispersion in yellow and green, metallic absorption in violet. See Wood, Phil. Mag. 1903.

INDEX OF REFRACTION.

TABLE 337. — Index of Refraction of Quartz (SiO_2).

Wave-length.	Index Ordinary Ray.	Index Extraordinary Ray.	Temperature ° C.	Wave-length.	Index Ordinary Ray.	Index Extraordinary Ray.	Temperature ° C.
μ				μ			
0.185	1.67582	1.68999	18	0.656	1.54189	1.55091	18
.193	.65997	.67343	"	.686	.54099	.54998	"
.198	.65090	.66397	"	.760	.53917	.54811	"
.206	.64038	.65300	"	1.160	.5329	Rubens.	–
.214	.63041	.64264	"	.969	.5216	"	–
.219	.62494	.63698	"	2.327	.5156	"	–
.231	.61399	.62560	"	.84	.5039	"	–
.257	.59622	.60712	"	3.18	.4944	"	–
.274	.58752	.59811	"	.63	.4799	"	–
.340	.56748	.57738	"	.96	.4679	"	–
.396	.55815	.56771	"	4.20	.4569	"	–
.410	.55650	.56600	"	5.0	.417	"	–
.486	.54968	.55896	"	6.45	.274	"	–
0.589	1.54424	1.55334	"	7.0	1.167	"	–

Except Rubens' values, — means from various authorities.

TABLE 338. — Indices of Refraction for various Alums.*

R	Density.	Temp. C°	Index of refraction for the Fraunhofer lines: a	B	C	D	E	b	F	G
			Aluminium Alums. $R\text{Al}(SO_4)_2+12H_2O$.†							
Na	1.667	17–28	1.43492	1.43563	1.43653	1.43884	1.44185	1.44231	1.44412	1.44804
$NH_3(CH_3)$	1.568	7–17	.45013	.45062	.45177	.45410	.45691	.45749	.45941	.46363
K	1.735	14–15	.45226	.45303	.45398	.45645	.45934	.45996	.46181	.46609
Rb	1.852	7–21	.45232	.45328	.45417	.45660	.45955	.45999	.46192	.46618
Cs	1.961	15–25	.45437	.45517	.45618	.45856	.46141	.46203	.46386	.46821
NH_4	1.631	15–20	.45509	.45599	.45693	.45939	.46234	.46288	.46481	.46923
Tl	2.329	10–23	.49226	.49317	.49443	.49748	.50128	.50209	.50463	.51076
			Chrome Alums. $R\text{Cr}(SO_4)_2+12H_2O$.†							
Cs	2.043	6–12	1.47627	1.47732	1.47836	1.48100	1.48434	1.48491	1.48723	1.49280
K	1.817	6–17	.47642	.47738	.47865	.48137	.48459	.48513	.48753	.49309
Rb	1.946	12–17	.47660	.47756	.47868	.48151	.48486	.48522	.48775	.49323
NH_4	1.719	7–18	.47911	.48014	.48125	.48418	.48744	.48794	.49040	.49594
Tl	2.386	9–25	.51692	.51798	.51923	.52280	.52704	.52787	.53082	.53808
			Iron Alums. $R\text{Fe}(SO_4)_2+12H_2O$.†							
K	1.806	7–11	1.47639	1.47706	1.47837	1.48169	1.48580	1.48670	1.48939	1.49605
Rb	1.916	7–20	.47700	.47770	.47894	.48234	.48654	.48712	.49003	.49700
Cs	2.061	20–24	.47825	.47921	.48042	.48378	.48797	.48867	.49136	.49838
NH_4	1.713	7–20	.47927	.48029	.48150	.48482	.48921	.48993	.49286	.49980
Tl	2.385	15–17	.51674	.51790	.51943	.52365	.52859	.52946	.53284	.54112

* According to the experiments of Soret (Arch. d. Sc. Phys. Nat. Genève, 1884, 1888, and Comptes Rendus, 1885).
† *R* stands for the different bases given in the first column.
For other alums see reference on Landolt-Börnstein-Roth Tabellen.

TABLE 339.

INDEX OF REFRACTION.

Selected Monorefringent or Isotropic Minerals.

The values are for the sodium D line unless otherwise stated and are arranged in the order of increasing indices. Selected by Dr. Edgar T. Wherry from a private compilation of Dr. E. S. Larsen of the U. S. Geological Survey.

Mineral.	Formula.	Index of refraction, $\lambda = 0.589\mu$.
Villiaumite	NaF	1.328
Cryolithionite	$3NaF.3LiF.2AlF_3$	1.339
Opal	$SiO_2.nH_2O$	1.406–1.440
Fluorite	CaF_2	1.434
Alum	$K_2O.Al_2O_3.4SO_3.24H_2O$	1.456
Sodalite	$3Na_2O.3Al_2O_3.6SiO_2.2NaCl$	1.483
Cristobalite	SiO_2	1.486
Analcite	$Na_2O.Al_2O_3.4SiO_2.2H_2O$	1.487
Sylvite	KCl	1.490
Noselite	$5Na_2O.3Al_2O_3.6SiO_2.2SO_3$	1.495
Hauynite	Like preceding + CaO	1.496
Lazurite	$4Na_2O.3Al_2O_3.6SiO_2.Na_2S_6$	1.500 ±
Leucite	$K_2O.Al_2O_3.4SiO_2$	1.509
Pollucite	$2Cs_2O.2Al_2O_3.9SiO_2.H_2O$	1.525
Halite	$NaCl$	1.544
Bauxite	$Al_2O_3.nH_2O$	1.570 ±
Pharmacosiderite	$3Fe_2O_3.2As_2O_5.3K_2O.5H_2O$	1.676
Spinel	$MgO.Al_2O_3$	1.723 ±
Berzeliite	$3(Ca, Mg, Mn)O.As_2O_5$	1.727
Periclasite	MgO	1.736
Grossularite	$3CaO.Al_2O_3.3SiO_2$	1.736
Helvite	$3(Mn, Fe)O.3BeO.3SiO_2.MnS$	1.739
Pyrope	$3MgO.Al_2O_3.3SiO_2$	1.745
Arsenolite	As_2O_3	1.755
Hessonite	$3CaO.(Al, Fe)_2O_3.3SiO_2$	1.763
Pleonaste	$(Mg, Fe)O.Al_2O_3$	1.770 ±
Almandite	$3FeO.Al_2O_3.3SiO_2$	1.778
Hercynite	$FeO.Al_2O_3$	1.800 ±
Gahnite	$ZnO.Al_2O_3$	1.800 ±
Spessartite	$3MnO.Al_2O_3.3SiO_2$	1.811
Lime	CaO	1.830
Uvarovite	$3CaO.Cr_2O_3.3SiO_2$	1.838
Andradite	$3CaO.Fe_2O_3.3SiO_2$	1.857
Microlite	$6CaO.3Ta_2O_5.CbOF_3$	1.925
Nantokite	$CuCl$	1.930
Pyrochlore	Contains CaO, Ce_2O_3, TiO_2, etc.	1.960–2.000
Schorlomite	$3CaO.(Fe, Ti)_2O_3.3(Si, Ti)O_2$	1.980
Percylite	$PbO.CuCl_2.H_2O$	2.050
Picotite	$(Mg, Fe)O.(Al, Cr)_2O_3$	2.050 ±
Eulytite	$2Bi_2O_3.3SiO_2$	2.050
Cerargyrite	$AgCl$	2.061
Mosesite	Contains Hg, NH_4, Cl, etc.	2.065
Chromite	$FeO.Cr_2O_3$	2.070
Senarmontite	Sb_2O_3	2.087
Embolite	$Ag(Br, Cl)$	2.150 ±
Manganosite	MnO	2.160
Bunsenite	NiO	2.18 (Li light)
Lewisite	$5CaO.2TiO_2.3Sb_2O_5$	2.200
Miersite	$CuI.4AgI$	2.200
Bromyrite	$AgBr$	2.253
Dysanalite	Contains CaO, FeO, TiO_2, etc.	2.330
Marshite	CuI	2.346
Franklinite	$(Zn, Fe, Mn)O.(Fe, Mn)_2O_3$	2.360 (Li light)
Sphalerite	$(Zn, Fe)S$	2.370–2.470
Perovskite	$CaO.TiO_2$	2.380
Diamond	C	2.419
Eglestonite	$HgO.2HgCl$	2.490 (Li light)
Hauerite	MnS_2	2.690 (Li light)
Alabandite	MnS	2.700 (Li light)
Cuprite	Cu_2O	2.849

TABLE 340.

INDEX OF REFRACTION.

Miscellaneous Monorefringent or Isotropic Solids.

Substance.	Spectrum line.	Index of refraction.	Authority.
Albite glass	D	1.4890	Larsen, 1909
Amber	D	1.546	Mühlheim
Ammonium chloride	D	1.6422	Grailich
Anorthite glass	D	1.5755	Larsen, 1909
Asphalt	D	1.635	E. L. Nichols
"	0.670μ	1.621	" " "
Bell metal	D	1.0052	Beer
Boric Acid, melted	C	1.4623	Bedson and Williams
" " "	D	1.4637	" " "
" " "	F	1.4694	" " "
Borax, melted	C	1.4624	" " "
" "	D	1.4630	" " "
" "	F	1.4702	" " "
Camphor	D	1.532	Kohlrausch
"	D	1.5462	Mühlheim
Canada balsam	D	1.530	Mean
Ebonite	red	1.66	Ayrton, Perry
Fuchsin	A	2.03	Mean
"	B	2.19	"
"	C	2.33	"
"	G	1.97	"
"	H	1.32	"
Gelatin, Nelson no. 1	D	1.530	Jones, 1911
" various	D	1.516–1.534	" "
Gum Arabic	red	1.480	Jamin
" "	red	1.514	Wollaston
Obsidian	D	1.482–1.496	Various
Phosphorus	D	2.1442	Gladstone, Dale
Pitch	red	1.531	Wollaston
Potassium bromide	D	1.5593	Topsöe, Christiansen
" chlorstannate	D	1.6574	" "
" iodide	D	1.6666	" "
Resins: Aloes	red	1.619	Jamin
Canada balsam	red	1.528	Wollaston
Colophony	red	1.548	Jamin
Copal	red	1.528	"
Mastic	red	1.535	Wollaston
Peru balsam	D	1.593	Baden Powell
Selenium	A	2.61	Wood
"	B	2.68	"
"	C	2.73	"
"	D	2.93	"
Sodium chlorate	D	1.5150	Dussaud
Strontium nitrate	D	1.5667	Fock

TABLE 341.

INDEX OF REFRACTION.

Selected Uniaxial Minerals.

The values are arranged in the order of increasing indices for the ordinary ray and are for the sodium D line unless otherwise indicated. Selected by Dr. Edgar T. Wherry from a private compilation of Dr. Esper S. Larsen of the U. S. Geological Survey.

Mineral.	Formula.	Index of refraction. Ordinary ray.	Index of refraction. Extraordinary ray.
(a) UNIAXIAL POSITIVE MINERALS.			
Ice	H_2O	1.309	1.313
Sellaite	MgF_2	1.378	1.390
Chrysocolla	$CuO.SiO_2.2H_2O$	1.460 ±	1.570 ±
Laubanite	$2CaO.Al_2O_3.5SiO_2.6H_2O$	1.475	1.486
Chabazite	$(Ca, Na_2)O.Al_2O_3.4SiO_2.6H_2O$	1.480 ±	1.482 ±
Douglasite	$2KCl.FeCl_2.2H_2O$	1.488	1.500
Hydronephelite	$2Na_2O.3Al_2O_3.6SiO_2.7H_2O$	1.490	1.502
Apophyllite	$K_2O.8CaO.16SiO_2.16H_2O$	1.535 ±	1.537 ±
Quartz	SiO_2	1.544	1.553
Coquimbite	$Fe_2O_3.3SO_3.9H_2O$	1.550	1.556
Brucite	$MgO.H_2O$	1.559	1.580
Alunite	$K_2O.3Al_2O_3.4SO_3.6H_2O$	1.572	1.592
Penninite	$5(Mg, Fe)O.Al_2O_3.3SiO_2.4H_2O$	1.576	1.579
Cacoxenite	$2Fe_2O_3.P_2O_5.12H_2O$	1.582	1.645
Eudialite	$6Na_2O.6(Ca, Fe)O.20(Si, Zr)O_2.NaCl$	1.606	1.611
Dioptasite	$CuO.SiO_2.H_2O$	1.654	1.707
Phenacite	$2BeO.SiO_2$	1.654	1.670
Parisite	$2CeOF.CaO.3CO_2$	1.676	1.757
Willemite	$2ZnO.SiO_2$	1.694	1.723
Vesuvianite	$2(Ca, Mn, Fe)O.(Al, Fe)(OH, F)O.2SiO_2$	1.716 ±	1.718 ±
Xenotime	$Y_2O_3.P_2O_5$	1.721	1.816
Connellite	$20CuO.SO_3.2CuCl_2.20H_2O$	1.724	1.746
Benitoite	$BaO.TiO_2.3SiO_2$	1.757	1.804
Ganomalite	$6PbO.4(Ca, Mn)O.6SiO_2.H_2O$	1.910	1.945
Scheelite	$CaO.WO_3$	1.918	1.934
Zircon	$ZrO_2.SiO_2$	1.923 ±	1.968 ±
Powellite	$CaO.MoO_3$	1.967	1.978
Calomel	$HgCl$	1.973	2.650
Cassiterite	SnO_2	1.997	2.093
Zincite	ZnO	2.008	2.029
Phosgenite	$PbO.PbCl_2.CO_2$	2.114	2.140
Penfieldite	$PbO.2PbCl_2$	2.130	2.210
Iodyrite	AgI	2.210	2.220
Tapiolite	$FeO.(Ta, Cb)_2O_5$	2.270	2.420 (Li light)
Wurtzite	ZnS	2.356	2.378
Derbylite	$6FeO.Sb_2O_3.5TiO_2$	2.450	2.510 (Li light)
Greenockite	CdS	2.506	2.529
Rutile	TiO_2	2.616	2.903
Moissanite	CSi	2.654	2.697
Cinnabarite	HgS	2.854	3.201
(b) UNIAXIAL NEGATIVE MINERALS.			
Chiolite	$2NaF.AlF_3$	1.349	1.342
Hanksite	$11Na_2O.9SO_3.2CO_2.KCl$	1.481	1.461
Thaumasite	$3CaO.CO_2.SiO_2.SO_3.15H_2O$	1.507	1.468
Hydrotalcite	$6MgO.Al_2O_3.CO_2.15H_2O$	1.512	1.498
Cancrinite	$4Na_2O.CaO.4Al_2O_3.2CO_2.9SiO_2.3H_2O$	1.524	1.496
Milarite	$K_2O.4CaO.2Al_2O_3.24SiO_2.H_2O$	1.532	1.529
Kaliophilite	$K_2O.Al_2O_3.2SiO_2$	1.537	1.533
Mellite	$Al_2O_3.C_{12}O_9.18H_2O$	1.539	1.511
Marialite	"Ma" = $3Na_2O.3Al_2O_3.18SiO_2.2NaCl$	1.539	1.537
Nephelite	$Na_2O.Al_2O_3.2SiO_2$	1.542	1.538

INDEX OF REFRACTION.

TABLE 341 (*Continued*). — Selected Uniaxial Minerals.

Mineral.	Formula.	Index of refraction. Ordinary ray.	Extraordinary ray.
	(*b*) UNIAXIAL NEGATIVE MINERALS (*continued*).		
Wernerite	Me_1Ma_1 ±	1.578	1.551
Beryl	$3BeO.Al_2O_3.6SiO_2$	1.581 ±	1.575 ±
Torbernite	$CuO.2UO_3.P_2O_5.8H_2O$	1.592	1.582
Meionite	"Me" = $4CaO.3Al_2O_3.6SiO_2$	1.597	1.560
Melilite	Contains Na_2O, CaO, Al_2O_3, SiO_2, etc.	1.634	1.629
Apatite	$9CaO.3P_2O_5.Ca(F, Cl)_2$	1.634	1.631
Calcite	$CaO.CO_2$	1.658	1.486
Gehlenite	$2CaO.Al_2O_3.SiO_2$	1.669	1.658
Tourmaline	Contains Na_2O, FeO, Al_2O_3, B_2O_3, SiO_2, etc.	1.669 ±	1.638 ±
Dolomite	$CaO.MgO.2CO_2$	1.682	1.503
Magnesite	$MgO.CO_2$	1.700	1.509
Pyrochroite	$MnO.H_2O$	1.723	1.681
Corundum	Al_2O_3	1.768	1.760
Smithsonite	$ZnO.CO_2$	1.818	1.618
Rhodochrosite	$MnO.CO_2$	1.818	1.595
Jarosite	$K_2O.3Fe_2O_3.4SO_3.6H_2O$	1.820	1.715
Siderite	$FeO.CO_2$	1.875	1.635
Pyromorphite	$9PbO.3P_2O_5.PbCl_2$	2.050	2.042
Barysilite	$3PbO.2SiO_2$	2.070	2.050
Mimetite	$9PbO.3As_2O_5.PbCl_2$	2.135	2.118
Matlockite	$PbO.PbCl_2$	2.150	2.040
Stolzite	$PbO.WO_3$	2.269	2.182
Geikielite	$(Mg, Fe)O.TiO_2$	2.310	1.950
Vanadinite	$9PbO.3V_2O_5.PbCl_2$	2.354	2.299
Wulfenite	$PbO.MoO_3$	2.402	2.304 (Li light)
Octahedrite	TiO_2	2.554	2.493
Massicotite	PbO	2.665	2.535 (Li light)
Proustite	$3Ag_2S.As_2S_3$	2.979	2.711 " "
Pyrargyrite	$3Ag_2S.Sb_2S_3$	3.084	2.881 " "
Hematite	Fe_2O_3	3.220	2.940 " "

TABLE 342. — Miscellaneous Uniaxial Crystals.

Crystal.	Spectrum line.	Index of refraction. Ordinary ray.	Extraordinary ray.	Authority.
Ammonium arseniate $NH_4H_2AsO_4$	D	1.5766	1.5217	T. and C.*
Benzil $(C_6H_5CO)_2$	D	1.6588	1.6784	Mean
Corundum, Al_2O_3, sapphire, ruby	D	1.769	1.760	Osann
Ice at −8° C.	D	1.308	1.313	Meyer
" " " "	Li	1.297	1.304	"
Ivory	D	1.539	1.541	Kohlrausch
Potassium arseniate KH_2AsO_4	F	1.5762	1.5252	T. and C.
" " "	D	1.5674	1.5179	" " "
" " "	C	1.5632	1.5146	" " "
Sodium arseniate $Na_3AsO_4.12H_2O$	D	1.457	1.466	Mean
" nitrate $NaNO_3$	D	1.586	1.336	"
" phosphate $Na_3PO_4.12H_2O$	D	1.447	1.453	"
Nickel sulphate $NiSO_4.6H_2O$	F	1.5173	1.4930	T. and C.
" " "	D	1.5109	1.4873	" " "
" " "	C	1.5078	1.4844	" " "
Strychnine sulphate	D	1.614	1.599	Martin

* Topsöe and Christiansen.

TABLE 343.

INDEX OF REFRACTION.

Selected Biaxial Minerals.

The values are arranged in the order of increasing β index of refraction and are for the sodium D line except where noted. Selected by Dr. Edgar T. Wherry from private compilation of Dr. Esper S. Larsen of the U. S. Geological Survey.

Mineral.	Formula.	Index of refraction.		
		n_α	n_β	n_γ
(a) BIAXIAL POSITIVE MINERALS.				
Stercorite	$Na_2O.(NH_4)_2O.P_2O_5.9H_2O$	1.439	1.441	1.469
Aluminite	$Al_2O_3.SO_3.9H_2O$	1.459	1.464	1.470
Tridymite	SiO_2	1.469	1.470	1.473
Thenardite	$Na_2O.SO_3$	1.464	1.474	1.485
Carnallite	$KCl.MgCl_2.6H_2O$	1.466	1.475	1.494
Alunogenite	$Al_2O_3.3SO_3.16H_2O$	1.474	1.476	1.483
Melanterite	$FeO.SO_3.7H_2O$	1.471	1.478	1.486
Natrolite	$Na_2O.Al_2O_3.3SiO_2.2H_2O$	1.480	1.482	1.493
Arcanite	$K_2O.SO_3$	1.494	1.495	1.497
Struvite	$(NH_4)_2O.2MgO.P_2O_5.12H_2O$	1.495	1.496	1.504
Heulandite	$CaO.Al_2O_3.6SiO_2.3H_2O$	1.498	1.499	1.505
Thomsonite	$(Na_2, Ca)O.Al_2O_3.2SiO_2.3H_2O$	1.497	1.503	1.525
Harmotomite	$(K_2, Ba)O.Al_2O_3.5SiO_2.5H_2O$	1.503	1.505	1.508
Petalite	$Li_2O.Al_2O_3.8SiO_2$	1.504	1.510	1.516
Monetite	$2CaO.P_2O_5.H_2O$	1.515	1.518	1.525
Newberyite	$2MgO.P_2O_5.7H_2O$	1.514	1.519	1.533
Gypsum	$CaO.SO_3.2H_2O$	1.520	1.523	1.530
Mascagnite	$(NH_4)_2O.SO_3$	1.521	1.523	1.533
Albite	"Ab" = $Na_2O.Al_2O_3.6SiO_2$	1.525	1.529	1.536
Hydromagnesite	$4MgO.3CO_2.4H_2O$	1.527	1.530	1.540
Wavellite	$3Al_2O_3.2P_2O_5.12(H_2O, 2HF)$	1.525	1.534	1.552
Kieserite	$MgO.SO_3.H_2O$	1.523	1.535	1.586
Copiapite	$2Fe_2O_3.5SO_3.18H_2O$	1.530	1.543	1.595
Whewellite	$CaO.C_2O_3.H_2O$	1.491	1.555	1.650
Variscite	$Al_2O_3.P_2O_5.4H_2O$	1.551	1.558	1.582
Labradorite	Ab_2An_3	1.559	1.563	1.568
Gibbsite	$Al_2O_3.3H_2O$	1.566	1.566	1.587
Wagnerite	$3MgO.P_2O_5.MgF_2$	1.569	1.570	1.582
Anhydrite	$CaO.SO_3$	1.571	1.576	1.614
Colemanite	$2CaO.3B_2O_3.5H_2O$	1.586	1.592	1.614
Fremontite	$Na_2O.Al_2O_3.P_2O_5.(H_2O, 2HF)$	1.594	1.603	1.615
Vivianite	$3FeO.P_2O_5.8H_2O$	1.579	1.603	1.633
Pectolite	$Na_2O.4CaO.6SiO_2.H_2O$	1.595	1.606	1.634
Calamine	$2ZnO.SiO_2.H_2O$	1.614	1.617	1.636
Chondrodite	$4MgO.2SiO_2.Mg(F, OH)_2$	1.609	1.619	1.639
Turquois	$CuO.3Al_2O_3.2P_2O_5.9H_2O$	1.610	1.620	1.650
Topaz	$2AlOF.SiO_2$	1.619	1.620	1.627
Celestite	$SrO.SO_3$	1.622	1.624	1.631
Prehnite	$2CaO.Al_2O_3.3SiO_2.H_2O$	1.616	1.626	1.649
Barite	$BaO.SO_3$	1.636	1.637	1.648
Anthophyllite	$MgO.SiO_2$	1.633	1.642	1.657
Sillimanite	$Al_2O_3.SiO_2$	1.638	1.642	1.653
Forsterite	$2MgO.SiO_2$	1.635	1.651	1.670
Enstatite	$MgO.SiO_2$	1.650	1.653	1.658
Euclasite	$2BeO.Al_2O_3.2SiO_2.H_2O$	1.652	1.655	1.671
Triplite	$3MnO.P_2O_5.MnF_2$	1.650	1.660	1.672
Spodumenite	$Li_2O.Al_2O_3.4SiO_2$	1.660	1.666	1.676
Diopside	$CaO.MgO.2SiO_2$	1.664	1.671	1.694
Olivine	$2(Mg, Fe)O.SiO_2$	1.662	1.680	1.699
Triphylite	$Li_2O.2(Fe, Mn)O.P_2O_5$	1.688	1.688	1.692

INDEX OF REFRACTION.

Selected Biaxial Minerals.

Mineral.	Formula.	Index of refraction.		
		n_α	n_β	n_γ
(*a*) BIAXIAL POSITIVE MINERALS (*continued*).				
Zoisite	$4CaO.3Al_2O_3.6SiO_2.H_2O$	1.700	1.702	1.706
Strengite	$Fe_2O_3.P_2O_5.4H_2O$	1.710	1.710	1.745
Diaspcrite	$Al_2O_3.H_2O$	1.702	1.722	1.750
Staurolite	$2FeO.5Al_2O_3.4SiO_2.H_2O$	1.736	1.741	1.746
Chrysoberyl	$BeO.Al_2O_3$	1.747	1.748	1.757
Azurite	$3CuO.2CO_2.H_2O$	1.730	1.758	1.838
Scorodite	$Fe_2O_3.As_2O_5.4H_2O$	1.765	1.774	1.797
Olivenite	$4CuO.As_2O_5.H_2O$	1.772	1.810	1.863
Anglesite	$PbO.SO_3$	1.877	1.882	1.894
Titanite	$CaO.TiO_2.SiO_2$	1.900	1.907	2.034
Claudetite	As_2O_3	1.871	1.920	2.010
Sulfur	S	1.950	2.043	2.240
Cotunnite	$PbCl_2$	2.200	2.217	2.260
Huebnerite	$MnO.WO_3$	2.170	2.220	2.320
Manganite	$Mn_2O_3.H_2O$	2.240	2.240	2.530 (Li)
Raspite	$PbO.WO_3$	2.270	2.270	2.300
Mendipite	$2PbO.PbCl_2$	2.240	2.270	2.310
Tantalite	$(Fe, Mn)O.Ta_2O_5$	2.260	2.320	2.430 (Li)
Wolframite	$(Fe, Mn)O.WO_3$	2.310	2.360	2.460 (Li)
Crocoite	$PbO.CrO_3$	2.310	2.370	2.660 (Li)
Pseudobrookite	$2Fe_2O_3.3TiO_2$	2.380	2.390	2.420 (Li)
Stibiotantalite	$Sb_2O_3.Ta_2O_5$	2.374	2.404	2,457
Montroydite	HgO	2.370	2.500	2.650 (Li)
Brookite	TiO_2	2.583	2.586	2.741
Lithargite	PbO	2.510	2.610	2.710
(*b*) BIAXIAL NEGATIVE MINERALS.				
Mirabilite	$Na_2O.SO_3.10H_2O$	1.394	1.396	1.398
Thomsenolite	$NaF.CaF_2.AlF_3.H_2O$	1.407	1.414	1.415
Natron	$Na_2O.CO_2.10H_2O$	1.405	1.425	1.440
Kalinite	$K_2O.Al_2O_3.4SO_3.24H_2O$	1.430	1.452	1.458
Epsomite	$MgO.SO_3.7H_2O$	1.433	1.455	1.461
Sassolite	$B_2O_3.H_2O$	1.340	1.456	1.459
Borax	$Na_2O.2B_2O_3.10H_2O$	1.447	1.470	1.472
Goslarite	$ZnO.SO_3.7H_2O$	1.457	1.480	1.484
Pickeringite	$MgO.Al_2O_3.4SO_3.22H_2O$	1.476	1.480	1.483
Bloedite	$Na_2O.MgO.2SO_3.4H_2O$	1.486	1.488	1.489
Trona	$3Na_2O.4CO_2.5H_2O$	1.410	1.492	1.542
Thermonatrite	$Na_2O.CO_2.H_2O$	1.420	1.495	1.518
Stilbite	$(Ca, Na_2)O.Al_2O_3.6SiO_2.5H_2O$	1.494	1.498	1.500
Niter	$K_2O.N_2O_5$	1.334	1.505	1.506
Kainite	$MgO.SO_3.KCl.3H_2O$	1.494	1.505	1.516
Gaylussite	$Na_2O.CaO.2CO_2.5H_2O$	1.444	1.516	1.523
Scolecite	$CaO.Al_2O_3.3SiO_2.3H_2O$	1.512	1.519	1.519
Laumontite	$CaO.Al_2O_3.4SiO_2.4H_2O$	1.513	1.524	1.525
Orthoclase	$K_2O.Al_2O_3.6SiO_2$	1.518	1.524	1.526
Microcline	Same as preceding	1.522	1.526	1.530
Anorthoclase	$(Na, K)_2O.Al_2O_3.6SiO_2$	1.523	1.529	1.531
Glauberite	$Na_2O.CaO.2SO_3$	1.515	1.532	1.536
Cordierite	$4(Mg, Fe)O.4Al_2O_3.10SiO_2.H_2O$	1.534	1.538	1.540
Chalcanthite	$CuO.SO_3.5H_2O$	1.516	1.539	1.546
Oligoclase	Ab_4An	1.539	1.543	1.547

INDEX OF REFRACTION.

Selected Biaxial Minerals.

Mineral.	Formula.	Index of refraction.		
		n_α	n_β	n_γ
(b) BIAXIAL NEGATIVE CRYSTALS (continued).				
Beryllonite	$Na_2O.2BeO.P_2O_5$	1.552	1.558	1.561
Kaolinite	$Al_2O_3.2SiO_2.2H_2O$	1.561	1.563	1.565
Biotite	$K_2O.4(Mg, Fe)O.2Al_2O_3.6SiO_2.H_2O$	1.541	1.574	1.574
Autunite	$CaO.2UO_3.P_2O_5.8H_2O$	1.553	1.575	1.577
Anorthite	"An" = $CaO.Al_2O_3.2SiO_2$	1.576	1.584	1.588
Lanthanite	$La_2O_3.3CO_2.9H_2O$	1.520	1.587	1.613
Pyrophyllite	$Al_2O_3.4SiO_2.H_2O$	1.552	1.588	1.600
Talc	$3MgO.4SiO_2.H_2O$	1.539	1.589	1.589
Hopeite	$3ZnO.P_2O_5.4H_2O$	1.572	1.590	1.590
Muscovite	$K_2O.3Al_2O_3.6SiO_2.2H_2O$	1.561	1.590	1.594
Amblygonite	$Al_2O_3.P_2O_5.2LiF$	1.579	1.593	1.597
Lepidolite	$Al_2O_3.3SiO_2.2(K, Li)F$	1.560	1.598	1.605
Phlogopite	$K_2O.6MgO.Al_2O_3.6SiO_2.2H_2O$	1.562	1.606	1.606
Tremolite	$CaO.3MgO.4SiO_2$	1.609	1.623	1.635
Actinolite	$CaO.3(Mg, Fe)O.4SiO_2$	1.611	1.627	1.636
Wollastonite	$CaO.SiO_2$	1.616	1.629	1.631
Lazulite	$(Fe, Mg)O.Al_2O_3.P_2O_5.H_2O$	1.603	1.632	1.639
Danburite	$CaO.B_2O_3.2SiO_2$	1.632	1.634	1.636
Glaucophanite	$Na_2O.2FeO.Al_2O_3.6SiO_2$	1.621	1.638	1.638
Andalusite	$Al_2O_3.SiO_2$	1.632	1.638	1.643
Hornblende	Contains Na_2O, MgO, FeO, SiO_2, etc.	1.629	1.642	1.653
Datolite	$2CaO.2SiO_2.B_2O_3.H_2O$	1.625	1.653	1.669
Erythrite	$3CoO.As_2O_5.8H_2O$	1.626	1.661	1.699
Monticellite	$CaO.MgO.SiO_2$	1.651	1.662	1.668
Strontianite	$SrO.CO_2$	1.520	1.667	1.667
Witherite	$BaO.CO_2$	1.529	1.676	1.677
Aragonite	$CaO.CO_2$	1.531	1.682	1.686
Axinite	$6(Ca, Mn)O.2Al_2O_3.B_2O_3.8SiO_2.H_2O$	1.678	1.685	1.688
Dumortierite	$8Al_2O_3.B_2O_3.6SiO_2.H_2O$	1.678	1.686	1.689
Cyanite	$Al_2O_3.SiO_2$	1.712	1.720	1.728
Epidote	$4CaO.3(Al, Fe)_2O_3.6SiO_2.H_2O$	1.729	1.754	1.768
Atacamite	$3CuO.CuCl_2.3H_2O$	1.831	1.861	1.880
Fayalite	$2FeO.SiO_2$	1.824	1.864	1.874
Caledonite	$2(Pb, Cu)O.SO_3.H_2O$	1.818	1.866	1.909
Malachite	$2CuO.CO_2.H_2O$	1.655	1.875	1.909
Lanarkite	$2PbO.SO_3$	1.930	1.990	2.020
Leadhillite	$4PbO.SO_3.2CO_2.H_2O$	1.870	2.000	2.010
Cerussite	$PbO.CO_2$	1.804	2.076	2.078
Laurionite	$PbCl_2.PbO.H_2O$	2.077	2.116	2.158
Matlockite	$PbO.PbCl_2$	2.040	2.150	2.150
Baddeleyite	ZrO_2	2.130	2.190	2.200
Lepidocrocite	$Fe_2O_3.H_2O$	1.930	2.210	2.510
Limonite	$2Fe_2O_3.3H_2O$ in part	2.170	2.290	2.310
Goethite	$Fe_2O_3.H_2O$	2.210	2.350	2.350 (Li)
Valentinite	Sb_2O_3	2.180	2.350	2.350
Turgite	$2Fe_2O_3.H_2O$ in part	2.450	2.550	2.550 (Li)
Realgar	AsS	2.460	2.590	2.610 (Li)
Terlinguaite	Hg_2OCl	2.350	2.640	2.670 (Li)
Hutchinsonite	$(Tl, Ag)_2S.PbS.2As_2S_3$	3.078	3.176	3.188
Stibnite	Sb_2S_3	3.194	4.303	4.460

INDEX OF REFRACTION.

TABLE 344. — Miscellaneous Biaxial Crystals.

Crystal.	Spectrum line.	Index of refraction.			Authority.
		n_α	n_β	n_γ	
Ammonium oxalate, $(NH_4)_2C_2O_4.H_2O$	D	1.4381	1.5475	1.5950	Brio
Ammonium acid tartrate, $(NH_4)H(C_4H_4O_6)$	D	1.5188	1.5614	1.5910	T. and C.*
Ammonium tartrate, $(NH_4)_2C_4H_4O_6$	D	—	1.581	—	Cloisaux
Antipyrin, $C_{11}H_{12}NO_2$	D	1.5697	1.6935	1.7324	Liweh
Citric acid, $C_6H_8O_7.H_2O$	D	1.4932	1.4977	1.5089	Schrauf
Codein, $C_{18}H_{21}NO_3.H_2O$	D	1.5390	1.5435	—	Grailich
Magnesium carbonate, $MgCO_3.3H_2O$	D	1.495	1.501	1.526	Genth
" sulphate, $MgSO_4.7H_2O$	D	1.432	1.455	1.461	Means
" "	Cd, 0.226μ	1.4990	1.5266	1.5326	Borel
" "	H, 0.656μ	1.4307	1.4532	1.4584	"
Potassium bichromate, $K_2Cr_2O_7$	D	1.7202	1.7380	1.8197	Dufet
" chromate, K_2CrO_4	D	—	1.7254	—	T. and C.
" "	red	1.6873	1.722	1.7305	Mallard
" nitrate, KNO_3	D	1.3346	1.5056	1.5064	Schrauf
" sulphate, K_2SO_4	F	1.4976	1.4992	1.5029	T. and C.
" "	D	1.4932	1.4946	1.4980	" " "
" "	C	1.4911	1.4928	1.4959	" " "
Racemic acid, $C_4H_6O_6.H_2O$	yellow	—	1.526	—	Groth
Resorcin, $C_6H_6O_2$	D	—	1.555	—	"
Sodium bichromate, $Na_2Cr_2O_7.2H_2O$	D	1.6610	1.6994	1.7510	Dufet
" acid tartrate, $NaH(C_4H_4O_6).2H_2O$	red	—	1.5332	—	Brio
Sugar (cane), $C_{12}H_{22}O_{11}$	Tl	1.5422	1.5685	1.5734	Calderon
" "	D	1.5397	1.5667	1.5716	"
" "	Li	1.5379	1.5639	1.5693	"
Tartaric acid, $C_4H_6O_6$ (right-)	D	1.4953	1.5353	1.6046	Means
Zinc sulphate, $ZnSO_4.7H_2O$	F	1.4620	1.4860	1.4897	T. and C.
" "	D	1.4568	1.4801	1.4836	" " "
" "	C	1.4544	1.4776	1.4812	" " "

* Topsöe and Christiansen.

TABLE 345. — Miscellaneous Liquids (see also Table 346), Liquefied Gases, Oils, Fats and Waxes.

Substance.	Temp. °C	Index for D 0.589μ.	Reference.	Substance.	Temp. °C	Index for D 0.589μ.	Reference.
Liquefied gases:				Oils:			
Br_2	15	1.659	a	Lavendar	20	1.464–1.466	e
Cl_2	14	1.367	b	Linseed	15	1.4820–1.4852	e
CO_2	15	1.195	b	Maize	15.5	1.4757–1.4768	d
C_2N_2	18	1.325	b	Mustard seed	15.5	1.4750–1.4762	d
C_2H_4	6	1.180	b	Neat's foot	15	1.4695–1.4708	e
H_2S	18.5	1.384	b	Olive	15.5	1.4703–1.4718	d
N_2	−190	1.205	c	Palm	60	1.4510	d
NH_3	16.5	1.325	b	Peanut	15.5	1.4723–1.4731	d
NO	−90	1.330	c	Peppermint	20	1.464–1.468	e
N_2O	15	1.194	b	Poppy	15.5	1.4770	d
O_2	−181	1.221	c	Porpoise	25	1.4677	e
SO_2	15.	1.350	b	Rape (Colza)	15.5	1.4748–1.4752	d
HCl	16.5	1.252	b	Seal	25	1.4741	e
HBr	10	1.325	b	Sesame	15.5	1.4742	d
HI	16.5	1.466	b	Soja bean	15.5	1.4760–1.4775	e
Oils:				Sperm	15.5	1.4665–1.4672	e
Almond	15.5	1.4728–1.4753	d	Sunflower	15.5	1.4739	d
Castor	15	1.4799–1.4803	e	Tung	19	1.503	e
Citronella	20	1.47–1.48	e	Whale	40	1.4649	e
Clove	20	1.5301–1.5360	e	Fats and Waxes:			
Cocoanut	15.5	1.4587	d	Beef tallow	40	1.4552–1.4587	e
Cod liver	15	1.4790–1.4833	e	Beeswax	75	1.4398–1.4451	e
Cotton seed	15.5	1.4737–1.4757	d	Carnauba wax	84	1.4520–1.4541	e
Croton	27	1.4757–1.4768	e	Cocoa butter	40	1.4560–1.4518	e
Eucalyptus	20	1.460–1.467	e	Lard	40	1.4584–1.4601	e
Lard	15.5	1.4702–1.4720	d	Mutton tallow	60	1.4510	e

References: (a) Martens; (b) Bleekrode, Pr. Roy. Soc. 37, 339, 1884; (c) Liveing, Dewar, Phil. Mag., 1892–3; (d) Tolman, Munson, Bul. 77, B. of C., Dept. Agriculture, 1905; (e) Seeker, Van Nostrand's Chemical Annual. For the oils of reference d, the average temperature coefficient is 0.000365 per ° C.

TABLE 346.

INDEX OF REFRACTION.

Indices of Refraction of Liquids Relative to Air.

Substance.	Density.	Temp. °C	Indices of refraction. 0.397μ H	0.434μ G′	0.486μ F	0.589μ D	0.656μ C	Authority.
Acetaldehyde, CH_3CHO	0.780	20	—	1.3394	1.3359	1.3316	1.3298	1a
Acetone, CH_3COCH_3	0.791	20	—	1.3678	1.3639	1.3593	1.3573	Means
Aniline, $C_6H_5.NH_2$	1.022	20	—	1.6204	1.6041	1.5863	1.5793	"
Alcohol, methyl, $CH_3.OH$	0.794	20	1.3399	1.3362	1.3331	1.3290	1.3277	"
" ethyl $C_2H_5.OH$	0.808	0	—	1.3773	1.3739	1.3695	1.3677	1b
" "	0.800	20	—	1.3700	1.3666	1.3618	1.3605	Means
" " *dn/dt*	—	20	—	−.0004	−.0004	−.0004	−.0004	2
" n-propyl $C_3H_7.OH$	0.804	20	—	1.3938	1.3901	1.3854	1.3834	1
Benzene, C_6H_6	0.880	20	—	1.5236	1.5132	1.5012	1.4965	Means
" C_6H_6 *dn/dt*	—	20	—	−.0007	−.0006	−.0006	−.0006	"
Bromnaphthalene, $C_{10}H_7Br$	1.487	20	1.7289	1.7041	1.6819	1.6582	1.6495	3
Carbon disulphide, CS_2	1.293	0	1.7175	1.6920	1.6688	1.6433	1.6336	4
" "	1.263	20	1.6994	1.6748	1.6523	1.6276	1.6182	"
" tetrachloride, CCl_4	1.591	20	—	1.4729	1.4676	1.4607	1.4579	1d
Chinolin, C_9H_7N	1.090	20	—	1.6679	1.6470	1.6245	1.6161	1c
Chloral, $CCl_3.CHO$	1.512	20	—	1.4679	1.4624	1.4557	1.4530	1c
Chloroform, $CHCl_3$	1.489	20	1.4[illegible]3	1.458	1.4530	1.4467	1.4443	Means
Decane, $C_{10}H_{22}$	0.728	14.9	—	1.4200	1.4160	1.4108	1.4088	1e
Ether, ethyl, $C_2H_5.O.C_2H_5$	0.715	20	—	1.3607	1.3576	1.3538	1.3515	Means
" " *dn/dt*	—	20	—	−.0006	−.0006	−.0006	−.0006	"
Ethyl nitrate, $C_2H_5.O.NO_3$	1.109	20	—	1.395	1.392	1.3853	1.3830	"
Formic acid, $H.CO_2H$	1.219	20	—	1.3804	1.3764	1.3714	1.3693	1a
Glycerine, $C_3H_8O_3$	1.260	20	—	1.4828	1.4784	1.4730	1.4706	5
Hexane, $CH_3(CH_2)_4CH_3$	0.660	20	—	1.3836	1.3799	1.3754	1.3734	1c
Hexylene, $CH_3(CH_2)_3CH.CH_2$	0.679	23.3	—	1.4059	1.4007	1.3945	1.3920	1c
Methylene iodide CH_2I_2	3.318	20	1.8027	—	1.7692	1.7417	1.7320	Means
" " *dn/dt*	—	20	—	—	−.0007	−.0007	−.0006	"
Naphthalene, $C_{10}H_8$	0.962	98.4	—	—	1.6031	1.5823	1.5746	1f
Nicotine, $C_{10}H_{14}N_2$	1.012	22.4	—	1.5439	—	1.5239	1.5198	1c
Octane, $CH_3(CH_2)_6CH_3$	0.707	15.1	—	1.4097	1.4046	1.4007	1.3987	1e
Oil, almond	0.92	0	—	—	1.4847	1.4782	1.4755	6
anise seed	0.99	15.1	1.6084	—	1.5743	1.5572	1.5508	7
"	0.99	21.4	—	—	1.5647	1.5475	1.5410	8
bitter almond	1.06	20	—	1.5775	1.5623	—	1.5391	5
cassia	—	10	1.7039	—	1.6389	1.6104	1.6007	7
"	—	22.5	1.6985	—	1.6314	1.6026	1.5930	7
cinnamon	1.05	23.5	—	—	1.6508	1.6188	1.6077	8
olive	0.92	0	—	—	1.4825	1.4763	1.4738	6
rock	—	0	—	—	1.4644	1.4573	1.4545	6
turpentine	0.87	10.6	1.4939	—	1.4817	1.4744	1.4715	9
"	0.87	20.7	1.4913	—	1.4793	1.4721	1.4692	8
Pentane, $CH_3(CH_2)_3CH_3$	0.625	15.7	—	1.3645	1.3610	1.3581	1.3570	1e
Phenol, C_6H_5OH	1.060	40.6	—	1.5684	1.5558	1.5425	1.5369	1g
"	1.021	82.7	—	—	1.5356	—	1.5174	1h
Styrene, $C_6H_5CH.CH_2$	0.910	16.6	—	1.5816	1.5659	1.5485	1.5419	1i
Thymol, $C_{10}H_{14}O$	0.982	—	—	—	1.5386	—	1.5228	1h
Toluene, $CH_3.C_6H_5$	0.86	20	—	1.5170	1.5070	1.4955	1.4911	10
Water, H_2O	—	20	1.3435	1.3404	1.3372	1.3330	1.3312	Means
"	—	0	1.3444	1.3413	1.3380	1.3338	1.3319	"
"	—	40	1.3411	1.3380	1.3349	1.3307	1.3290	"
"	—	80	1.3332	1.3302	1.3270	1.3230	1.3313	"

References: 1, Landolt and Börnstein (a, Landolt; b, Korten; c, Brühl; d, Haagen; e, Landolt, Jahn; f, Nasini, Bernheimer; g, Eisenlohr; h, Eykman; i, Auwers, Eisenlohr); 2, Korten; 3, Walter; 4, Ketteler; 5, Landolt; 6, Olds; 7, Baden Powell; 8, Willigen; 9, Fraunhofer; 10, Brühl.

INDEX OF REFRACTION.

Indices of Refraction relative to Air for Solutions of Salts and Acids.

Substance.	Density.	Temp. C.	Indices of refraction for spectrum lines. C	D	F	H_γ	H	Authority.
(a) Solutions in Water.								
Ammonium chloride	1.067	27°.05	1.37703	1.37936	1.38473	–	1.39336	Willigen.
" "	.025	29.75	.34850	.35050	.35515	–	.36243	"
Calcium chloride .	.398	25.65	.44000	.44279	.44938	–	.46001	"
" "	.215	22.9	.39411	.39652	.40206	–	.41078	"
" "	.143	25.8	.37152	.37369	.37876	–	.38666	"
Hydrochloric acid .	1.166	20.75	1.40817	1.41109	1.41774	–	1.42816	"
Nitric acid	.359	18.75	.39893	.40181	.40857	–	.41961	"
Potash (caustic) . .	.416	11.0	.40052	.40281	.40808	–	.41637	Fraunhofer.
Potassium chloride .	normal	solution	.34087	.34278	.34719	1.35049	–	Bender.
" "	double	normal	.34982	.35179	.35645	.35994	–	"
" "	triple	normal	.35831	.36029	.36512	.36890	–	"
Soda (caustic) . .	1.376	21.6	1.41071	1.41334	1.41936	–	1.42872	Willigen.
Sodium chloride . .	.189	18.07	.37562	.37789	.38322	1.38746	–	Schutt.
" "	.109	18.07	.35751	.35959	.36442	.36823	–	"
" "	.035	18.07	.34000	.34191	.34628	.34969	–	"
Sodium nitrate . .	1.358	22.8	1.38283	1.38535	1.39134	–	1.40121	Willigen.
Sulphuric acid . .	.811	18.3	.43444	.43669	.44168	–	.44883	"
" "	.632	18.3	.42227	.42466	.42967	–	.43694	"
" "	.221	18.3	.36793	.37009	.37468	–	.38158	"
" "	.028	18.3	.33663	.33862	.34285	–	.34938	"
Zinc chloride . . .	1.359	26.6	1.39977	1.40222	1.40797	–	1.41738	"
" " . . .	.209	26.4	.37292	.37515	.38026	–	.38845	"
(b) Solutions in Ethyl Alcohol.								
Ethyl alcohol . . .	0.789	25.5	1.35791	1.35971	1.36395	–	1.37094	Willigen.
" "	.932	27.6	.35372	.35556	.35986	–	.36662	"
Fuchsin (nearly saturated)	–	16.0	.3918	.398	.361	–	.3759	Kundt.
Cyanin (saturated) .	–	16.0	.3831	–	.3705	–	.3821	"

NOTE. — Cyanin in chloroform also acts anomalously; for example, Sieben gives for a 4.5 per cent. solution $\mu_A = 1.4593$, $\mu_B = 1.4695$, μ_F (green) $= 1.4514$, μ_G (blue) $= 1.4554$. For a 9.9 per cent. solution he gives $\mu_A = 1.4902$, μ_F (green) $= 1.4497$, μ_G (blue) $= 1.4597$.

(c) Solutions of Potassium Permanganate in Water.*

Wave-length in cms. × 10⁶.	Spectrum line.	Index for 1 % sol.	Index for 2 % sol.	Index for 3 % sol.	Index for 4 % sol.	Wave-length in cms. × 10⁶.	Spectrum line.	Index for 1 % sol.	Index for 2 % sol.	Index for 3 % sol.	Index for 4 % sol.
68.7	B	1.3328	1.3342	–	1.3382	51.6	–	1.3368	1.3385	–	–
65.6	C	.3335	.3348	1.3365	.3391	50.0	–	.3374	.3383	1.3386	1.3404
61.7	–	.3343	.3365	.3381	.3410	48.6	F	.3377	–	–	.3408
59.4	–	.3354	.3373	.3393	.3426	48.0	–	.3381	.3395	.3398	.3413
58.9	D	.3353	.3372	–	.3426	46.4	–	.3397	.3402	.3414	.3423
56.8	–	.3362	.3387	.3412	.3445	44.7	–	.3407	.3421	.3426	.3439
55.3	–	.3366	.3395	.3417	.3438	43.4	–	.3417	–	–	.3452
52.7	E	.3363	–	–	–	42.3	–	.3431	.3442	.3457	.3468
52.2	–	.3362	.3377	.3388	–	–	–	–	–	–	–

* According to Christiansen.

TABLE 348.

INDEX OF REFRACTION.

Indices of Refraction of Gases and Vapors.

A formula was given by Biot and Arago expressing the dependence of the index of refraction of a gas on pressure and temperature. More recent experiments confirm their conclusions. The formula is $n_t - 1 = \frac{n_0 - 1}{1 + at} \cdot \frac{p}{760}$, where n_t is the index of refraction for temperature t, n_0 for temperature zero, a the coefficient of expansion of the gas with temperature, and p the pressure of the gas in millimeters of mercury. For air see Table 349.

(a) Indices of refraction.

Spectrum line.	10^3 (n–1) Air.	Spectrum line.	10^3 (n–1) Air.	Wave-length.	(n–1) 10^3. Air.	O.	N.	H.
				μ				
A	.2905	M	.2993	.4861	.2951	.2734	.3012	.1406
B	.2911	N	.3003	.5461	.2936	.2717	.2998	.1397
C	.2914	O	.3015	.5790	.2930	.2710	—	.1393
D	.2922	P	3023	.6563	.2919	.2698	.2982	.1387
E	.2933	Q	.3031	.4360	.2971	.2743	CO_2	.1418
F	.2943	R	.3043	.5462	.2937	.2704	.4506	.1397
G	.2962	S	.3053	.6709	.2918	.2683	.4471	.1385
H	.2978	T	.3064	6.709	.2881	.2643	.4804	.1361
K	.2980	U	.3075	8.678	.2888	.2650	.4579	.1361
L	.2987			First 4, Cuthbertsons; the rest, Koch, 1909.				

(b) The following are compiled mostly from a table published by Brühl (Zeits. für Phys. Chem. vol. 7, pp. 25–27). The numbers are from the results of experiments by Biot and Arago, Dulong, Jamin, Ketteler, Lorenz, Mascart, Chappius, Rayleigh, and Rivière and Prytz. When the number given rests on the authority of one observer the name of that observer is given. The values are for 0° Centigrade and 760 mm. pressure.

Substance.	Kind of light.	Indices of refraction and authority.	Substance.	Kind of light.	Indices of refraction and authority.
Acetone	D	1.001079–1.001100	Hydrogen	white	1.000138–1.000143
Ammonia	white	1.000381–1.000385	"	D	1.000132 Burton.
"	D	1.000373–1.000379	Hydrogen sulphide	D	1.000644 Dulong.
Argon	D	1.000281 Rayleigh.		D	1.000623 Mascart.
Benzene	D	1.001700–1.001823	Methane	white	1.000443 Dulong.
Bromine	D	1.001132 Mascart.	"	D	1.000444 Mascart.
Carbon dioxide	white	1.000449–1.000450	Methyl alcohol	D	1.000549–1.000623
" "	D	1.000448–1.000454	Methyl ether	D	1.000891 Mascart.
Carbon disulphide	white	1.001500 Dulong.	Nitric oxide	white	1.000303 Dulong.
	D	1.001478–1.001485	" "	D	1.000297 Mascart.
Carbon monoxide	white	1.000340 Dulong.	Nitrogen	white	1.000295–1.000300
	white	1.000335 Mascart.	"	D	1.000296–1.000298
Chlorine	white	1.000772 Dulong.	Nitrous oxide	white	1.000503–1.000507
"	D	1.000773 Mascart.	" "	D	1.000516 Mascart.
Chloroform	D	1.001436–1.001464	Oxygen	white	1.000272–1.000280
Cyanogen	white	1.000834 Dulong.	"	D	1.000271–1.000272
"	D	1.000784–1.000825	Pentane	D	1.001711 Mascart.
Ethyl alcohol	D	1.000871–1.000885	Sulphur dioxide	white	1.000665 Dulong.
Ethyl ether	D	1.001521–1.001544	" "	D	1.000686 Ketteler.
Helium	D	1.000036 Ramsay.	Water	white	1.000261 Jamin.
Hydrochloric acid	white	1.000449 Mascart.	"	D	1.000249–1.000259
	D	1.000447 "			

TABLE 349. — Index of Refraction of Air (15°C, 76 cm).

Corrections for reducing wave-lengths and frequencies in air (15° C, 76 cm) to vacuo.

The indices were computed from the Cauchy formula $(n - 1)10^7 = 2726.43 + 12.288/(\lambda^2 \times 10^{-8}) + 0.3555/(\lambda^4 \times 10^{-16})$. For 0° C and 76 cm the constants of the equation become 2875.66, 13.412 and 0.3777 respectively, and for 30° C and 76 cm, 2589.72, 12.259 and 0.2576. Sellmeier's formula for but one absorption band closely fits the observations: $n^2 = 1 + 0.0005737 8\lambda^2/(\lambda^2 - 595260)$. If $n - 1$ were strictly proportional to the density, then $(n-1)_0/(n - 1)_t$ would equal $1 + \alpha t$ where α should be 0.00367. The following values of α were found to hold:

λ	0.85μ	0.75μ	0.65μ	0.55μ	0.45μ	0.35μ	0.25μ
α	0.003672	0.003674	0.003678	0.003685	0.003700	0.003738	0.003872

The indices are for dry air (0.05 ± % CO_2). Corrections to the indices for water vapor may be made for any wave-length by Lorenz's formula, $+ 0.000041(m/760)$, where m is the vapor pressure in mm. The corresponding frequencies in waves per cm and the corrections to reduce wave-lengths and frequencies in air at 15° C and 76 cm pressure to vacuo are given. E.g., a light wave of 5000 Angstroms in dry air at 15° C, 76 cm becomes 5001.391 Å in vacuo; a frequency of 20,000 waves per cm correspondingly becomes 19994.44. Meggers and Peters, Bul. Bureau of Standards, 14, p. 731, 1918.

Wave-length, λ Angstroms.	Dry air $(n-1) \times 10^7$ 15° C 76 cm	Vacuo correction for λ in air $(n\lambda - \lambda)$. Add.	Frequency waves per cm $\frac{1}{\lambda}$ in air.	Vacuo correction for $\frac{1}{\lambda}$ in air $\left(\frac{1}{n\lambda} - \frac{1}{\lambda}\right)$. Subtract.	Wave-length, λ Angstroms.	Dry air $(n-1) \times 10^7$ 15° C 76 cm	Vacuo correction for λ in air $(n\lambda - \lambda)$ Add.	Frequency waves per cm $\frac{1}{\lambda}$ in air.	Vacuo correction for $\frac{1}{\lambda}$ in air $\left(\frac{1}{n\lambda} - \frac{1}{\lambda}\right)$. Subtract.
2000	3256	0.651	50,000	16.27	5500	2771	1.524	18,181	5.04
2100	3188	0.670	47,619	15.18	5600	2769	1.551	17,857	4.94
2200	3132	0.689	45,454	14.23	5700	2768	1.578	17,543	4.85
2300	3086	0.710	43,478	13.41	5800	2766	1.604	17,241	4.77
2400	3047	0.731	41,666	12.69	5900	2765	1.631	16,949	4.68
2500	3014	0.754	40,000	12.05	6000	2763	1.658	16,666	4.60
2600	2986	0.776	38,461	11.48	6100	2762	1.685	16,393	4.53
2700	2962	0.800	37,037	10.97	6200	2761	1.712	16,129	4.45
2800	2941	0.824	35,714	10.50	6300	2760	1.739	15,873	4.38
2900	2923	0.848	34,482	10.08	6400	2759	1.766	15,625	4.31
3000	2907	0.872	33,333	9.69	6500	2758	1.792	15,384	4.24
3100	2893	0.897	32,258	9.33	6600	2757	1.819	15,151	4.18
3200	2880	0.922	31,250	9.00	6700	2756	1.846	14,925	4.11
3300	2869	0.947	30,303	8.69	6800	2755	1.873	14,705	4.05
3400	2859	0.972	29,411	8.41	6900	2754	1.900	14,492	3.99
3500	2850	0.998	28,571	8.14	7000	2753	1.927	14,285	3.93
3600	2842	1.023	27,777	7.89	7100	2752	1.954	14,084	3.88
3700	2835	1.049	27,027	7.66	7200	2751	1.981	13,888	3.82
3800	2829	1.075	26,315	7.44	7300	2751	2.008	13,698	3.77
3900	2823	1.101	25,641	7.24	7400	2750	2.035	13,513	3.72
4000	2817	1.127	25,000	7.04	7500	2749	2.062	13,333	3.66
4100	2812	1.153	24,390	6.86	7600	2749	2.089	13,157	3.62
4200	2808	1.179	23,809	6.68	7700	2748	2.116	12,987	3.57
4300	2803	1.205	23,255	6.52	7800	2748	2.143	12,820	3.52
4400	2799	1.232	22,727	6.36	7900	2747	2.170	12,658	3.48
4500	2796	1.258	22,222	6.21	8000	2746	2.197	12,500	3.43
4600	2792	1.284	21,739	6.07	8100	2746	2.224	12,345	3.39
4700	2789	1.311	21,276	5.93					
4800	2786	1.338	20,833	5.80	8250	2745	2.265	12,121	3.33
4900	2784	1.364	20,406	5.68	8500	2744	2.332	11,764	3.23
					8750	2743	2.400	11,428	3.13
5000	2781	1.391	20,000	5.56	9000	2742	2.468	11,111	3.05
5100	2779	1.417	19,607	5.45	9250	2741	2.536	10,810	2.96
5200	2777	1.444	19,230	5.34	9500	2740	2.604	10,526	2.88
5300	2775	1.471	18,867	5.23	9750	2740	2.671	10,256	2.81
5400	2773	1.497	18,518	5.13	10000	2739	2.739	10,000	2.74

MEDIA FOR DETERMINATIONS OF REFRACTIVE INDICES WITH THE MICROSCOPE.

TABLE 350. — Liquids, n_D (0.589μ) = 1.74 to 1.87.

In 100 parts of methylene iodide at 20° C. the number of parts of the various substances indicated in the following table can be dissolved, forming saturated solutions having the permanent refractive indices specified. When ready for use the liquids can be mixed by means of a dropper to give intermediate refractions. Commercial iodoform (CHI_3) powder is not suitable, but crystals from a solution of the powder in ether may be used, or the crystalized product may be bought. A fragment of tin in the liquids containing the SnI_4 will prevent discoloration.

CHI_3.	SnI_4.	AsI_3.	SbI_3.	S.	n_{na} at 20°.
			12		1.764
	25				1.783
	25		12		1.806
	30			6	1.820
	27	13	7		1.826
40	27	16			1.842
	31	14	8	10	1.853
35	31	16	8	10	1.868

TABLE 351. — Resin-like Substances, n_D (0.589μ) = 1.68 to 2.10.

Piperine, one of the least expensive of the alkaloids, can be obtained very pure in straw-colored crystals. When melted it dissolves the tri-iodides of arsenic and antimony very freely. The solutions are fluid at slightly above 100° and when cold, resin-like. A solution containing 3 parts antimony iodide to one part of arsenic iodide with varying proportions of piperine is easier to manipulate than one containing either iodide alone. The following table gives the necessary data concerning the composition and refractive indices for sodium light. In preparing, the constituents, in powder of about 1 mm. grain, should be weighed out and then fused *over*, not *in*, a low flame. Three-inch test tubes are suitable.

Per cent Iodides.	00.	10.	20.	30.	40.	50.	60.	70.	80.
Index of refraction	1.683	1.700	1.725	1.756	1.794	1.840	1.897	1.968	2.050

TABLE 352. — Permanent Standard Resinous Media, n_D (0.589μ) = 1.546 to 1.682.

Any proportions of piperine and rosin form a homogeneous fusion which cools to a transparent resinous mass. The following table shows the refractive indices of various mixtures. On account of the strong dispersion of piperine the refractive indices of minerals apparently matched with those of mixtures rich in this constituent are 0.005 to 0.01 too low. To correct this error a screen made of a thin film of 7 per cent antimony iodide and 93 per cent piperine should be used over the eye-piece. Any amber-colored rosin in lumps is suitable.

Per cent Rosin.	00.	10.	20.	30.	40.	50.	60.	70.	80.	90.	100.
Index of refraction	1.683	1.670	1.657	1.643	1.631	1.618	1.604	1.590	1.575	1.560	1.544

All taken from Merwin, Jour. Wash. Acad. of Sc. 3, p. 35, 1913.

OPTICAL CONSTANTS OF METALS.

TABLE 353.

Two constants are required to characterize a metal optically, the refractive index, *n*, and the absorption index, *k*, the latter of which has the following significance: the amplitude of a wave after travelling one wave-length, λ^1 measured in the metal, is reduced in the ratio[1] $1 : e^{-2\pi k}$ or for any distance *d*, $1 : e^{-\frac{2\pi dk}{\lambda^1}}$; for the same wave-length measured in air this ratio becomes $1 : e^{-\frac{2\pi dnk}{\lambda^1}}$. *nk* is sometimes called the extinction coefficient. Plane polarized light reflected from a polished metal surface is in general elliptically polarized because of the relative change in phase between the two rectangular components vibrating in and perpendicular to the plane of incidence. For a certain angle, $\bar{\phi}$ (principal incidence) the change is 90° and if the plane polarized incident beam has a certain azimuth $\bar{\psi}$ (Principal azimuth) circularly polarized light results. Approximately, (Drude, Annalen der Physik, 36, p. 546, 1889),

$$k = \tan 2\bar{\psi}\,(1 - \cot^2\bar{\phi}) \text{ and } n = \frac{\sin\bar{\phi}\tan\bar{\phi}}{(1+k^2)^{\frac{1}{2}}}\,(1 + \tfrac{1}{2}\cot^2\bar{\phi}).$$

For rougher approximations the factor in parentheses may be omitted. R = computed percentage reflection.

(The points have been so selected that a smooth curve drawn through them very closely indicates the characteristics of the metal.)

Metal.	λ	$\bar{\phi}$	$\bar{\psi}$	Computed.				Authority.
				n	k	nk	R	
	μ						%	
Cobalt	0.231	64°31′	29°39	1.10	1.30	1.43	32.	Minor.
	.275	70 22	29 59	1.41	1.52	2.14	46.	"
	.500	77 5	31 53	1.93	1.93	3.72	66.	"
	.650	79 0	31 25	2.35	1.87	4.40	69.	Ingersoll.
	1.00	81 45	29 6	3.63	1.58	5.73	73.	"
	1.50	83 21	26 18	5.22	1.29	6.73	75.	"
	2.25	83 48	26 5	5.65	1.27	7.18	76.	"
Copper	.231	65 57	26 14	1.39	1.05	1.45	29.	Minor.
	.347	65 6	28 16	1.19	1.23	1.47	32.	"
	.500	70 44	33 46	1.10	2.13	2.34	56.	"
	.650	74 16	41 30	0.44	7.4	3.26	86.	Ingersoll.
	.870	78 40	42 30	0.35	11.0	3.85	91.	"
	1.75	84 4	42 30	0.83	11.4	9.46	96.	"
	2.25	85 13	42 30	1.03	11.4	11.7	97.	"
	4.00	87 20	42 30	1.87	11.4	21.3		Först.-Fréed.
	5.50	88 00	41 50	3.16	9.0	28.4		" "
Gold	1.00	81 45	44 00	0.24	28.0	6.7		" "
	2.00	85 30	43 56	0.47	26.7	12.5		" "
	3.00	87 05	43 50	0.80	24.5	19.6		" "
	5.00	88 15	43 25	1.81	18.1	33.		" "
Iridium	1.00	82 10	29 15	3.85	1.60	6.2		" "
	2.00	83 10	29 40	4.30	1.66	7.1		" "
	3.00	81 40	30 40	3.33	1.79	6.0		" "
	5.00	79 00	32 20	2 27	2.03	4.6		" "
Nickel	0.420	72 20	31 42	1.41	1.79	2.53	54.	Tool.
	0.589	76 1	31 41	1.79	1.86	3.33	62.	Drude.
	0.750	78 45	32 6	2.19	1.99	4.36	70.	Ingersoll.
	1.00	80 33	32 2	2.63	2.00	5.26	74.	"
	2.25	84 21	33 30	3.95	2.33	9.20	85.	"
Platinum	1.00	75 30	37 00	1.14	3.25	3.7		Först.-Fréed.
	2.00	74 30	39 50	0.70	5.06	3.5		" "
	3.00	73 50	41 00	0.52	6.52	3.4		" "
	5.00	72 00	42 10	0.34	9.01	3.1		" "
Silver	0.226	62 41	22 16	1.41	0.75	1.11	18.	Minor.
	.293	63 14	18 56	1.57	0.62	0.97	17.	"
	.316	52 28	15 38	1.13	0.38	0.43	4.	"
	.332	52 1	37 2	0.41	1.61	0.65	32.	"
	.395	66 36	43 6	0.16	12.32	1.91	87.	"
	.500	72 31	43 29	0.17	17.1	2.94	93.	"
	.589	75 35	43 47	0.18	20.6	3.64	95.	"
	.750	79 26	44 6	0.17	30.7	5.16	97.	Ingersoll.
	1.00	82 0	44 2	0.24	29.0	6.96	98.	"
	1.50	84 42	43 48	0.45	23.7	10.7	98.	"
	2.25	86 18	43 34	0.77	19.9	15.4	99.	"
	3.00	87 10	42 40	1.65	12.2	20.1		Först.-Fréed.
	4.50	88 20	41 10	4.49	7.42	33.3		" "
Steel	0.226	66 51	28 17	1.30	1.26	1.64	35.	Minor.
	.257	68 35	28 45	1.38	1.35	1.86	40.	"
	.325	69 57	30 9	1.37	1.53	2.09	45.	"
	.500	75 47	29 2	2.09	1.50	3.14	57.	"
	.650	77 48	27 9	2.70	1.33	3.59	59.	Ingersoll.
	1.50	81 48	28 51	3.71	1.55	5.75	73.	"
	2.25	83 22	30 36	4.14	1.79	7.41	80.	"

Drude, Annalen der Physik und Chemie, 39, p. 481, 1890; 42, p. 186, 1891; 64, p. 159, 1898. Minor, Annalen der Physik, 10, p. 581, 1903. Tool, Physical Review, 31, p. 1, 1910. Ingersoll, Astrophysical Journal, 32, p. 265, 1910; Försterling and Fréedericksz, Annalen der Physik, 40, p. 201, 1913.

OPTICAL CONSTANTS OF METALS.

TABLE 354.

Metal.	λ.	n.	k.	R.	Ref.
	μ				
Al.*	0.589	1.44	5.32	83	1
Sb.*	.589	3.04	4.94	70	1
Bi.†‡	white	2.26	–	–	2
Cd.*	.589	1.13	5.01	85	1
Cr.*	.579	2.97	4.85	70	3
Cb.*	.579	1.80	2.11	41	3
Au.†	.257	0.92	1.14	28	4
	.441	1.18	1.85	42	4
	.589	0.47	2.83	82	4
I. crys.	.589	3.34	0.57	30	4
Ir.*	.579	2.13	4.87	75	3
Fe.§	.257	1.01	0.88	16	4
	.441	1.28	1.37	28	4
	.589	1.51	1.63	33	4
Pb.*	.589	2.01	3.48	62	1
Mg.*	.589	0.37	4.42	93	1
Mn.*	.579	2.49	3.89	64	3
Hg. (liq.)	.326	0.68	2.26	66	4
	.441	1.01	3.42	74	4
	.589	1.62	4.41	75	4
	.668	1.72	4.70	77	4
Pd.*	.579	1.62	3.41	65	3
Pt.†	.257	1.17	1.65	37	4
	.441	1.94	3.16	58	4
	.589	2.63	3.54	59	4
	.668	2.91	3.66	59	4
Ni.*	.275	1.09	1.16	24	4
	.441	1.16	1.23	25	4
	.589	1.30	1.97	43	4

Metal.	λ.	n.	k.	R.	Ref.
	μ				
Rh.*	0.579	1.54	4.67	78	3
Se.‡	.400	2.94	2.31	44	5
	.490	3.12	1.49	35	5
	.589	2.93	0.45	25	5
	.760	2.60	0.06	20	5
Si.*	.589	4.18	0.09	38	6
	1.25	3.67	0.08	33	6
	2.25	3.53	0.08	31	6
Na. (liq.)	.589	.004	2.61	99	1
Ta.*	.579	2.05	2.31	44	3
Sn.*	.589	1.48	5.25	82	1
W.*	.579	2.76	2.71	49	3
V.*	.579	3.03	3.51	58	3
Zn.*	.257	0.55	0.61	20	4
	.441	0.93	3.19	73	4
	.589	1.93	4.66	74	4
	.668	2.62	5.08	73	4

λ = wave-length, n = refraction index.
k = absorption index, R = reflection.

(1) Drude, see Table 205; (2) Kundt, prism used, Ann. der Physik und Chemie, 34, p. 477, 36, p. 824, 1889; (3) v. Wartenberg, Verh. deutsch. Physik. Ges. 12, p. 105, 1910; (4) Meier, Annales der Physik, 10, p. 581, 1903; (5) Wood, Phil. Mag. (6), 3, 607, 1902; (6) Ingersoll, *see* Table 205.

* solid, † electrolytic, ‡ prism, § deposited as film in vacuo.

TABLE 355.—Reflecting Power of Metals. (See page 298.)

Wave-length	Al.	Sb.	Cd.	Co.	Graphite.	Ir.	Mg.	Mo.	Pd.	Rh.	Si.	Ta.	Te.	Sn.	W	Va.	Zn.
μ	Per cents.																
.5	–	–	–	–	22	–	72	46	–	76	34	38	–	–	49	57	–
.6	–	53	–	–	24	–	73	48	–	77	32	45	49	–	51	58	–
.8	–	54	–	–	25	–	74	52	–	81	29	64	48	–	56	60	–
1.0	71	55	72	67	27	78	74	58	72	84	28	78	50	54	62	61	80
2.0	82	60	87	72	35	87	77	82	81	91	28	90	52	61	85	69	92
4.0	92	68	96	81	48	94	84	90	88	92	28	93	57	72	93	79	97
7.0	96	71	98	93	54	95	91	93	94	94	28	94	68	81	95	88	98
10.0	98	72	98	97	59	96	–	94	97	95	28	–	–	84	96	–	98
12.0	98	–	99	97	–	96	–	95	97	–	–	95	–	85	96	–	99

Coblentz, Bulletin Bureau of Standards, 2, p. 457, 1906, 7, p. 197, 1911. The surfaces of some of the samples were not perfect so that the corresponding values have less weight. The methods for polishing the various metals are described in the original articles. The following more recent values are given by Coblentz and Emerson, Bul. Bur. Stds. 14, p. 207, 1917; Stellite, an exceedingly hard and untarnishable alloy of Co, Cr, Mo, Mn, and Fe (C, Si, S, P) was obtained from the Haynes Stellite Co, Kokomo, Indiana.

Wave-length, μ,	.15	.20	.30	.50	.75	1.00	2.00	3.00	4.00	5.00	9.00
Tungsten,	—	—	—	.50	.52	.576	.900	.943	.948	.953	—
Stellite,	.32	.42	.50	.64	.67	.689	.747	.792	.825	.848	.880

According to Fresnel, the amount of light reflected by the surface of a transparent medium $= \frac{1}{2}(A+B) = \frac{1}{2}\left\{\frac{\sin^2(i-r)}{\sin^2(i+r)} + \frac{\tan^2(i-r)}{\tan^2(i+r)}\right\}$; A is the amount polarized in the plane of incidence; B is that polarized perpendicular to this; i and r are the angles of incidence and refraction.

TABLE 356. — Light reflected when $i = 0°$ or Incident Light is Normal to Surface $(n-1)^2/(n+1)^2$.

n.	$\frac{1}{2}(A+B)$.	n.	$\frac{1}{2}(A+B)$.	n.	$\frac{1}{2}(A+B)$.	n.	$\frac{1}{2}(A+B)$.
1.00	0.00	1.4	2.78	2.0	11.11	5.	44.44
1.02	0.01	1.5	4.00	2.25	14.06	5.83	50.00
1.05	0.06	1.6	5.33	2.5	18.37	10.	66.67
1.1	0.23	1.7	6.72	2.75	22.89	100.	96.08
1.2	0.83	1.8	8.16	3.	25.00	∞	100.00
1.3	1.70	1.9	9.63	4.	36.00		

TABLE 357. — Light reflected when n is near Unity or equals $1 + dn$.

i.	A.	B.	$\frac{1}{2}(A+B)$.	$\frac{A-B}{A+B}$.*
0°	1.000	1.000	1.000	0.0
5	1.015	.985	1.000	1.5
10	1.063	.939	1.001	6.2
15	1.149	.862	1.005	14.3
20	1.282	.752	1.017	26.0
25	1.482	.612	1.047	41.5
30	1.778	.444	1.111	60.0
35	2.221	.260	1.240	79.1
40	2.904	.088	1.496	94.5
45	4.000	.000	2.000	100.0
50	5.857	.176	3.016	94.5
55	9.239	1.081	5.160	79.1
60	16.000	4.000	10.000	60.0
65	31.346	12.952	22.149	41.5
70	73.079	42.884	57.981	26.0
75	222.85	167.16	195.00	14.3
80	1099.85	971.21	1035.53	6.2
85	17330.64	16808.08	17069.36	1.5
90	∞	∞	∞	0.0

The values for A and B are strictly $(dn^2/4)\sec^4 i$ and $(dn^2/4)(1-\tan^2 i)$; In columns 2, 3, and 4 $dn^2/4$ is omitted.

TABLE 358. — Light reflected when $n = 1.55$.

i.	r.	A.	B.	dA.†	dB.†	$\frac{1}{2}(A+B)$.	$\frac{A-B}{A+B}$.*
° ′	° ′						
0	0 0.0	4.65	4.65	0.130	0.130	4.65	0.0
5	3 13.4	4.70	4.61	.131	.129	4.65	1.0
10	6 25.9	4.84	4.47	.135	.126	4.66	4.0
15	9 36.7	5.09	4.24	.141	.121	4.66	9.1
20	12 44.8	5.45	3.92	.150	.114	4.68	16.4
25	15 49.3	5.95	3.50	.161	.105	4.73	25.9
30	18 49.1	6.64	3.00	.175	.094	4.82	37.8
35	21 43.1	7.55	2.40	.191	.081	4.98	51.7
40	24 30.0	8.77	1.75	.210	.066	5.26	66.7
45	27 8.5	10.38	1.08	.233	.049	5.73	81.2
50	29 37.1	12.54	0.46	.263	.027	6.50	92.9
55	31 54.2	15.43	0.05	.303	.007	7.74	99.3
60	33 58.1	19.35	0.12	.342	−.013	9.73	98.8
65	35 47.0	24.69	1.13	.375	−.032	12.91	91.2
70	37 19.1	31.99	4.00	.400	−.050	18.00	77.7
75	38 32.9	42.00	10.38	.410	−.060	26.19	61.8
80	39 26.8	55.74	23.34	.370	−.069	39 54	41.0
82 30	39 45.9	64.41	34.04	.320	−.067	49.22	30.8
85 0	39 59.6	74.52	49.03	.250	−.061	61.77	20.6
86 0	40 3.6	79.02	56.62	.209	−.055	67.82	16.5
87 0	40 6.7	83.80	65.32	.163	−.046	74.56	12.4
88 0	40 8.9	88.88	75.31	.118	−.036	82.10	8.3
89 0	40 10.2	94.28	86.79	.063	−.022	90.54	4.1
90 0	40 10.7	100.00	100.00	.000	−.000	100.00	0.0

Angle of total polarization $= 57° 10'.3$, $A = 16.99$.

* This column gives the degree of polarization. † Columns 5 and 6 furnish a means of determining A and B for other values of n. They represent the change in these quantities for a change of n of 0.01.

Taken from E. C. Pickering's "Applications of Fresnel's Formula for the Reflection of Light."

REFLECTING POWER OF METALS.

TABLE 359. — Perpendicular Incidence and Reflection. (See also Tables 352-355.)

The numbers give the per cents of the incident radiation reflected.

Wave-length, μ.	Silver-backed Glass.	Mercury-backed Glass.	Mach's Magnalium. $69Al+31Mg$.	Brandes-Schünemann Alloy. $32Cu+34Sn+29Ni+5Fe$.	Ross' Speculum Metal. $68.2Cu+31.8Sn$.	Nickel. *Electrolytically Deposited.*	Copper. *Electrolytically Deposited.*	Steel. *Untempered.*	Copper. *Commercially Pure.*	Platinum. *Electrolytically Deposited.*	Gold. *Electrolytically Deposited.*	Brass. *(Trowbridge).*	Silver. *Chemically Deposited.*
.251	–	–	67.0	35.8	29.9	37.8	–	32.9	25.9	33.8	38.8	–	34.1
.288	–	–	70.6	37.1	37.7	42.7	–	35.0	24.3	38.8	34.0	–	21.2
.305	–	–	72.2	37.2	41.7	44.2	–	37.2	25.3	39.8	31.8	–	9.1
.316	–	–	–	–	–	–	–	–	–	–	–	–	4.2
.326	–	–	75.5	39.3	–	45.2	–	40.3	24.9	41.4	28.6	–	14.6
.338	–	–	–	–	–	46.5	–	–	–	–	–	–	55.5
.357	–	–	81.2	43.3	51.0	48.8	–	45.0	27.3	43.4	27.9	–	74.5
.385	–	–	83.9	44.3	53.1	49.6	–	47.8	28.6	45.4	27.1	–	81.4
.420	–	–	83.3	47.2	56.4	56.6	–	51.9	32.7	51.8	29.3	–	86.6
.450	85.7	72.8	83.4	49.2	60.0	59.4	48.8	54.4	37.0	54.7	33.1	–	90.5
.500	86.6	70.9	83.3	49.3	63.2	60.8	53.3	54.8	43.7	58.4	47.0	–	91.3
.550	88.2	71.2	82.7	48.3	64.0	62.6	59.5	54.9	47.7	61.1	74.0	–	92.7
.600	88.1	69.9	83.0	47.5	64.3	64.9	83.5	55.4	71.8	64.2	84.4	–	92.6
.650	89.1	71.5	82.7	51.5	65.4	66.6	89.0	56.4	80.0	66.5	88.9	–	94.7
.700	89.6	72.8	83.3	54.9	66.8	68.8	90.7	57.6	83.1	69.0	92.3	–	95.4
.800	–	–	84.3	63.1	–	69.6	–	58.0	88.6	70.3	94.9	–	96.8
1.0	–	–	84.1	69.8	70.5	72.0	–	63.1	90.1	72.9	–	–	97.0
1.5	–	–	85.1	79.1	75.0	78.6	–	70.8	93.8	77.7	97.3	–	98.2
2.0	–	–	86.7	82.3	80.4	83.5	–	76.7	95.5	80.6	96.8	91.0	97.8
3.0	–	–	87.4	85.4	86.2	88.7	–	83.0	97.1	88.8	–	93.7	98.1
4.0	–	–	88.7	87.1	88.5	91.1	–	87.8	97.3	91.5	96.9	95.7	98.5
5.0	–	–	89.0	87.3	89.1	94.4	–	89.0	97.9	93.5	97.0	95.9	98.1
7.0	–	–	90.0	88.6	90.1	94.3	–	92.9	98.3	95.5	98.3	97.0	98.5
9.0	–	–	90.6	90.3	92.2	95.6	–	92.9	98.4	95.4	98.0	97.8	98.7
11.0	–	–	90.7	90.2	92.9	95.9	–	94.0	98.4	95.6	98.3	96.6	98.8
14.0	–	–	92.2	90.3	93.6	97.2	–	96.0	97.9	96.4	97.9	–	98.3

Based upon the work of Hagen and Rubens, Ann. der Phys (1) 352, 1900; (8) 1, 1902; (11) 873, 1903.
Taken partly from Landolt-Börnstein-Meyerhoffer's Physikalisch-chemische Tabellen.

TABLE 360. — Percentage Diffuse Reflection from Miscellaneous Substances.

Wave-length μ	Lamp-blacks.					Pt. black electrol.	Green leaves.	Lead oxide.	Al. oxide.	Zinc oxide.	White Paper.	Lead carbonate.	Asphalt.	Black velvet.	Black felt.	Red brick.
	Paint.	Rosin.	Sperm candle.	Acetylene	Camphor.											
*.60	3.2						25.	52.	84.	82.		89.	15.	1.8	14.	30.
*.95	3.4	1.3	1.1	0.6	1.3	1.1			88.	86.	75.	93.			21.	
4.4	3.2	1.3	.9	.8	1.2	1.4		51.	21.	8.	18.	29.		3.7		
8.8	3.8		1.3	1.2	1.6	2.1		26.	2.	3.	5.	11.		2.7		12.
24.0	4.4	3.0	4.0	2.1	5.7	4.2		10.	6.	5.		7.				

*Not monochromatic (max.) means from Coblentz, J. Franklin Inst. 1912. Bulletin Bureau of Standards, 9, p. 283, 1912, contains many other materials.

REFLECTING POWER OF PIGMENTS.

TABLE 361. — Percentage Reflecting Power of Dry Powdered Pigments.

Taken from "The Physical Basis of Color Technology," Luckiesh, J. Franklin Inst., 1917. The total reflecting power depends on the distribution of energy in the illuminant and is given in the last three columns for noon sun, blue sky, and for a 7.9 lumens/watt tungsten filament.

Spectrum color.	Violet.	Blue.		Green.			Yellow.		Orange.			Red.			Noon sun.	Sky light.	Tungsten lamp.
Wave-length in μ	0.44	0.46	0.48	0.50	0.52	0.54	0.56	0.58	0.60	0.62	0.64	0.66	0.68	0.70			
American vermilion	8	6	5	5	6	6	9	11	24	39	53	61	66	65	14	12	12
Venetian red	5	5	5	5	5	6	7	12	19	24	28	30	32	32	11	10	13
Tuscan red	7	7	7	8	8	8	8	12	16	18	20	22	23	24	11	10	12
Indian red	8	7	7	7	7	7	7	11	15	18	20	22	23	24	10	9	11
Burnt sienna	4	4	4	4	5	6	9	14	18	20	21	23	24	25	11	9	13
Raw sienna	12	13	13	13	18	26	35	43	46	46	45	44	45	43	33	30	37
Golden ochre	22	22	23	27	40	53	63	71	75	74	73	73	73	72	58	55	63
Chrome yellow ochre	8	9	7	7	10	19	30	46	60	62	66	82	81	80	33	29	40
Yellow ochre	20	20	21	24	32	42	53	63	64	61	60	59	59	59	49	46	53
Chrome yellow medium	5	5	6	8	18	48	66	75	78	79	81	81	81	81	54	50	63
Chrome yellow light	13	13	18	30	56	82	88	89	90	89	88	87	85	84	76	70	82
Chrome green light	10	10	14	23	26	23	20	17	14	11	9	8	7	6	19	19	18
Chrome green medium	7	7	10	21	21	17	13	11	9	7	6	6	6	5	14	14	12
Cobalt blue	59	58	49	35	23	15	11	10	10	10	11	15	20	25	16	18	13
Ultramarine blue	67	54	38	21	10	6	4	3	3	4	5	7	10	17	7	10	6

TABLE 362. — Infra-red Diffuse Percentage Reflecting Powers of Dry Pigments.

Wave-length in μ	Co_2O_3	CuO	Cr_2O_3	PbO	Fe_2O_3	Y_2O_3	$PbCrO_4$	Al_2O_3	ThO_2	ZnO	MgO	CaO	ZrO_2	$PbCO_3$	$MgCO_3$	White lead paint.	Zn oxide paint.
0.60 *	3	—	27	52	26	74	70	84	86	82	86	85	86	88	85	76	68
0.95 *	4	24	45	—	41	—	—	88	—	86	—	—	84	93	89	79	72
4.4	14	15	33	51	30	34	41	21	47	8	16	22	23	29	11	—	—
8.8	13	—	5	26	4	11	5	20	7	3	2	4	5	10	4	—	—
24.0	6	4	8	10	9	10	7	6	10	5	9	6	5	7	9	—	—

* Non-monochromatic means from Coblentz, Bul. Bureau Standards 9, p. 283, 1912.

For the REFLECTING (and transmissive) power of ROUGHENED SURFACES at various angles of incidence, see Gorton, Physical Review, 7, p. 66, 1916. A surface of plate glass, ground uniformly with the finest emery and then silvered, used at an angle of 75°, reflected 90 per cent at 4μ, approached 100 for longer waves, only 10 at 1μ, less than 5 in the visible red and approached 0 for shorter waves. Similar results were obtained with a plate of rock salt for transmitted energy when roughened merely by breathing on it. In both cases the finer the surface, the more suddenly it cuts off the short waves.

REFLECTING POWER.

TABLE 363.—Reflecting Power of Powders (White Light).

Various pure chemicals, very finely powdered and surface formed by pressing down with glass plate. White (noon sunlight) light. Reflection in per cent. Nutting, Jones, Elliott, Tr. Ill. Eng. Soc. 9, 593, 1914.

Aluminum oxide	83.6	Magnesium carbonate	86.6	Sodium chloride	78.1
Barium sulphate	81.1	" " (block)	97.5	Sodium sulphate	77.9
Borax	81.6	Magnesium oxide	85.7	Starch	80.3
Boric acid	83.2	Rochelle salt	79.3	Sugar	87.8
Calcium carbonate	83.8	Salicylic acid	81.1	Tartaric acid	79.1
Citric acid	81.5	Sodium carbonate	81.8		

TABLE 364.—Variation of Reflecting Power of Surfaces with Angle.

Illumination at normal incidence, 1½ watt tungsten lamp, reflection at angles indicated with normal. Ill. Eng. Soc., Glare Committee, Tr. Ill. Eng. Soc. 11, p. 92, 1916.

Angle of observation.	0°	1°	3°	5°	10°	15°	30°	45°	60°
Magnesium carbonate block	0.88	—	—	0.88	0.88	0.87	0.83	0.72	0.68
Magnesium oxide	0.80	—	—	0.80	0.80	0.80	0.77	0.75	0.66
Matt photographic paper	0.78	—	—	0.78	0.78	0.78	0.78	0.76	0.72
White blotter	0.76	—	—	0.76	0.76	0.76	0.73	0.70	0.67
Pot opal, ground	0.69	0.69	0.69	0.69	0.69	0.69	0.68	0.66	0.64
Flashed opal, not ground	11.3	11.3	11.3	0.31	0.22	0.21	0.20	0.20	0.18
Glass, fine ground	0.29	0.29	0.29	0.29	0.27	0.20	0.14	0.13	0.12
Glass, course ground	0.23	0.22	0.21	0.20	0.19	0.16	0.11	0.11	0.12
Matt varnish on foil	0.83	—	0.78	0.72	0.62	0.49	0.28	0.21	0.16
Mirror with ground face	4.9	—	—	4.55	3.86	3.03	0.78	0.42	0.35

The following figures, taken from Fowle, Smithsonian Misc. Col. 58, No. 8, indicate the amount of energy scattered on each side of the directly reflected beam from a silvered mirror; the energy at the center of the reflected beam was taken as 100,000, and the angle of incidence was about 3°.

Angle of reflection, 3° ±	0′	8′	10′	15′	20′	30′	45′	60′	100′
Energy	100,000	600	244	146	107	66	33	22	11

Wave-length of max. energy of Nernst lamp used as source about 2μ.

TABLE 365.—Infra-red Reflectivity of Tungsten (Temperature Variation).

Three tungsten mirrors were used,—a polished Coolidge X-ray target and two polished flattened wires mounted in evacuated soft-glass bulbs with terminals for heating electrically. Weniger and Pfund, J. Franklin Inst.

Wave-length in μ.	Absolute reflectivity at room temperature in per cent.	Per cent increase in reflectivity in going from room temperature to			
		1377° K	1628° K	1853° K	2056° K
0.67	51	+6.0	+7.4	+8.7	+9.8
0.80	55	—	—	—	+8.2
1.27	70	0.0	0.0	0.0	0.0
1.90	83	−6.6	−8.2	−9.6	−11.0
2.00	85	−7.5	−9.3	−10.9	−12.3
2.90	92	−7.7	−9.4	−11.1	−12.5
4.00	93	—	—	—	−12.5

See also Weniger and Pfund, Phys. Rev. 15, p. 427, 1919.

TABLE 366.

TRANSMISSIBILITY OF RADIATION BY DYES.

Percentage transmissions of aqueous solutions taken from The Physical Basis of Color-Technology, Luckiesh, J. Franklin Inst. 184, 1917.

Spectrum color →	Violet.	Blue.		Green.			Yellow.		Orange.			Red.		
Wave-length in μ →	.44	.46	.48	.50	.52	.54	.56	.58	.60	.62	.64	.66	.68	.70
Carmen ruby opt.	—	—	—	—	—	—	—	—	—	4	18	37	49	60
Amido naphthol red	—	—	—	—	—	—	—	—	4	38	75	92	96	96
Coccinine	—	—	—	—	—	—	—	4	56	96	98	98	98	98
Erythrosine	6	—	—	—	—	—	1	53	90	95	96	96	96	96
Hematoxyline	1	3	7	13	14	12	13	25	44	54	63	73	78	82
Alizarinered	1	1	2	3	4	6	11	22	39	54	65	72	77	79
Acid rosolic (pure)	4	3	1	—	—	—	2	38	78	88	90	91	92	92
Rapid filter red	—	—	—	—	—	1	10	47	86	95	96	96	96	96
Aniline red fast extra A	—	—	—	—	—	2	12	34	55	72	84	88	90	92
Pinatype red fast	—	—	—	—	—	—	—	—	11	35	55	65	68	69
Eosine	—	—	—	—	—	—	1	54	87	93	92	92	92	92
Rose bengal	80	70	34	6	1	—	14	82	96	97	98	98	98	98
Cobalt nitrate	69	51	40	31	32	48	67	82	87	90	90	90	90	90
Tartrazine	—	—	—	—	7	52	75	86	91	95	96	97	98	98
Chrysoidin	—	—	—	—	—	—	—	—	2	23	50	71	79	79
Aurantia	—	—	—	—	—	3	23	53	82	92	96	96	96	96
Aniline yellow phosphine	—	—	—	—	2	20	43	60	67	75	81	85	86	87
Fluorescein	15	1	—	—	48	91	97	98	98	98	98	98	98	98
Aniline yellow fast S	—	—	1	7	43	84	96	96	96	96	96	96	96	96
Methyl orange indicator	—	—	—	—	—	—	1	31	70	79	80	81	81	81
Uranine	15	1	—	1	58	96	97	97	97	97	97	97	97	97
Uranine naphthaline	—	—	—	4	53	77	82	83	84	85	86	86	87	87
Orange B naphthol	—	—	—	—	—	1	43	88	95	96	97	97	97	97
Safranine	—	—	—	—	—	—	—	—	3	27	64	85	93	93
Martius gelb	—	—	—	1	43	84	91	94	95	95	95	95	95	95
Naphthol yellow	—	—	1	18	74	91	96	97	98	98	98	98	98	98
Potassium bichromate, sat.	—	—	—	—	—	10	60	84	88	89	89	89	89	88
Cobalt chromate	17	36	62	82	88	90	92	93	95	96	96	96	96	95
Naphthol green	2	4	7	21	30	36	29	16	7	2	1	—	—	—
Brilliant green	4	39	69	52	23	4	—	—	—	—	—	2	23	64
Filter blue green	35	49	64	70	60	37	13	2	—	—	—	—	—	—
Malachite green	—	12	20	8	1	—	—	—	—	—	—	—	12	50
Saurgrün	3	29	57	57	39	19	4	1	—	—	—	—	4	30
Methylengrün	28	31	32	26	17	7	2	1	—	—	—	—	3	28
Aniline green naphthol B	2	6	14	24	34	40	32	14	4	1	—	—	—	—
Neptune green	—	40	63	41	13	1	—	—	—	—	—	—	—	5
Cupric chloride	77	84	89	92	92	89	80	67	52	36	19	6	2	—
Turnbull's blue	58	60	56	51	38	28	18	9	5	3	1	—	—	—
Victoria blau	52	23	9	1	—	—	—	—	—	1	4	21	49	73
Prussian blue (soluble)	66	71	76	69	60	46	32	20	12	7	5	3	3	—
Wasser blau	89	75	51	26	7	1	—	—	1	2	6	18	37	60
Resorcine blue	25	18	6	2	1	—	—	—	1	2	14	41	64	72
Toluidin blau	66	31	13	3	1	—	—	—	—	—	1	4	16	40
Patent blue	83	91	84	76	65	46	24	8	2	—	—	6	42	78
Dianil blue	77	69	59	48	35	24	15	9	5	5	7	14	29	53
Filter blue	84	79	66	44	27	17	14	19	36	56	74	81	88	92
Aniline blue, methyl	92	88	78	52	27	9	3	2	2	4	8	16	25	45
Magenta	21	8	2	1	—	—	1	22	73	93	97	97	97	97
Gentiana violet	89	83	64	44	26	19	15	10	13	42	75	92	93	94
Rosazeine	50	28	2	—	—	—	—	6	55	90	98	98	98	98
Iodine (dense)	—	—	—	—	—	—	—	—	—	—	1	93	11	23
Rhodamine B	81	71	45	13	2	—	—	23	83	96	96	96	95	94
Acid violet	84	76	68	50	33	26	27	34	49	70	84	96	96	96
Cyonine in alcohol	7	1	—	—	—	—	—	—	—	—	—	1	13	23
Xylene red	39	23	1	—	—	—	—	1	27	79	97	97	97	96
Methyl violet B	25	4	—	—	—	—	—	—	—	—	3	26	63	89

For the infra-red transmission (to 12μ) and reflection powers of a number of aniline dyes, see Johnson and Spence, Phys. Rev. 5, p. 349, 1915.

Scientific Paper 440 of the Bureau of Standards, 1922, gives spectrum transmission curves (0.24 to 1.36 μ) for the following dyes: Napthol Yellow S, Orange I, Amaranth, Erythrosine, Indigo Disulpho Acid, Ponceau 3R, and Light Green S F Yellowish.

TRANSMISSIBILITY OF RADIATION BY JENA GLASSES.

TABLE 367.

Coefficients, a, in the formula $I_t = I_0 a^t$, where I_0 is the Intensity before, and I_t after, transmission through the thickness t. Deduced from observations by Müller, Vogel, and Rubens as quoted in Hovestadt's Jena Glass (English translation).

Unit t=1 dm.	Coefficient of transmission, a.									
	.375 μ	390 μ	.400 μ	.434 μ	.436 μ	.455 μ	.477 μ	.503 μ	.580 μ	.677 μ
O 340, Ord. light flint	.388	.456	.614	.569	.680	.834	.880	.880	.878	.939
O 102, H'vy silicate flint	–	.025	.463	.502	.566	.663	.700	.782	.828	.794
O 93, Ord. " "	–	–	–	–	.714	.807	.899	.871	.903	.943
O 203, " " crown	.583	.583	.695	.667	.806	.822	.860	.872	.872	.903
O 598, (Crown)	–	–	–	–	.797	.770	.771	.776	.818	.860

Unit t=1 cm.	0.7 μ	0.95 μ	1.1 μ	1.4 μ	1.7 μ	2.0 μ	2.3 μ	2.5 μ	2.7 μ	2.9 μ	3.1 μ
S 204, Borate crown	1.00	.99	.94	.90	.85	.81	.69	.43	.29	.18	–
S 179, Med. phosp. cr.	–	.98	.95	.90	.84	.67	.49	.87	.18	–	–
O 1143, Dense, bor. sil. cr.	.98	–	.97	–	.95	.93	.90	.84	.71	.47	.27
O 1092, Crown	.99	.96	.95	.99	.99	.91	.82	.71	.60	.48	.29
O 1151, "	.98	–	.99	.99	.98	.94	.90	.79	.75	.45	.32
O 451, Light flint	1.00	–	.99	–	.98	.95	.92	.84	.78	.54	.34
O 469, Heavy "	1.00	–	.98	–	.99	.98	.98	.97	.90	.66	.50
O 500, " "	1.00	–	1.00	–	1.00	–	1.00	.99	.92	.74	.53
S 163, " "	1.00	–	.98	–	.99	–	.99	–	.94	.78	.60

TABLE 368.

Note: With the following data, t must be expressed in millimeters; i. e. the figures as given give the transmissions for thickness of 1 mm.

No. and Type of Glass.	Wave-length in μ.												
	Visible Spectrum.							Ultra-violet Spectrum.					
	.644 μ	.578 μ	.546 μ	.509 μ	.480 μ	.436 μ	.405 μ	.384 μ	.361 μ	.340 μ	.332 μ	.309 μ	.280 μ
F 3815 Dark neutral	.35	.35	.37	.35	.34	.30	.15	.06					
F 4512 Red filter	.94	.05											
F 2745 Copper ruby	.72	.39	.47	.47	.45	.43	.43						
F 4313 Dark yellow	.98	.97	.93	.83	.09								
F 4351 Yellow	.98	.97	.96	.93	.44	.15							
F 4937 Bright yellow	1.0	1.0	1.0	.99	.74	.40	.31	.28	.22	.18	.14	.06	
F 4930 Green filter	.17	.50	.64	.62	.44								
F 3873 Blue filter	–	–	–	.18	.50	.73	.69	.59	.36	.10			
F 3654 Cobalt glass, transparent for outer red	–	–	–	.15	.44	.85	1.0	1.0	1.0	1.0	1.0	.58	
F 3653 Blue, ultraviolet	–	–	–	–	.11	.65	1.0	1.0	1.0	1.0	1.0	.81	.18
F 3728 Didymium, str'g bands	.99	.72	.99	.96	.95	.96	.99	.99	.89	.89	.77	.54	

This and the following table are taken from Jenaer Glas für die Optik, Liste 751, 1909

TABLE 369. — Transmissibility by Jena Ultra-violet Glasses.

No. and Type of Glass.	Thickness.	0.397 μ	0.383 μ	0.361 μ	0.346 μ	0.325 μ	0.309 μ	0.280 μ
UV 3199 Ultra-violet	1 mm.	1.00	1.00	1.00	1.00	1.00	0.95	0.56
" "	2 mm.	0.99	0.99	0.99	0.97	0.90	0.57	
" "	1 dm.	0.95	0.95	0.89	0.70	0.36		
UV 3248 "	1 mm.	1.00	1.00	1.00	1.00	0.98	0.91	0.35
" "	2 mm.	0.98	0.98	0.98	0.92	0.78	0.38	
" "	1 dm.	0.96	0.87	0.79	0.45	0.08		

Table 370.

TRANSMISSIBILITY OF RADIATION BY GLASSES.

The following data giving the percentage transmission of radiation of various substances, mostly glasses, are selected from Spectroradiometric Investigation of the Transmission of Various substances, Coblentz, Emerson and Long, Bul. Bureau Standards, 14, p. 653, 1918.

Glass or substance, manufacturer.	Thickness, mm	Transmission per cents. Wave-lengths in μ. 0.5	1.0	1.5	2.0	2.5	3.0	3.5	4.0	4.5	5.0
Purple fluorite	4.98	—	—	—	47	48	48	57	60	62	62
Gold film on Crooke's glass	—	22	3	2	1	1	1	0	0	0	0
" " " crown glass	—	34	8	3	2	1	1	0	0	0	0
Molybdenite	.007	0	41	43	44	46	46	47	48	48	48
$Cr_2(SO_4)_3.18\,H_2O$	.24	0	83	63	37	11	0	0	0	0	0
Chrome alum, 10 g to 100 g H_2O	10	—	73	0	0	—	—	—	—	—	—
$CoCl_2$, 10 g to 100 g H_2O	10	—	50	0	0	—	—	—	—	—	—
GLASSES:											
Copper ruby, flashed	1.95	—	50	64	72	76	40	33	36	7	0
G24, Corning, red	5.90	—	60	70	72	65	2	1	0	0	0
Schott's red, No. 2745	3.18	—	83	89	89	75	10	10	0	0	0
G34, Corning, orange	3.55	—	50	62	67	68	15	3	1	0	0
Pyrex, Corning	1.55	90	90	90	91	87	35	13	7	2	0
Noviol, B, Corning, yellow	2.88	80	75	60	82	75	23	4	4	0	0
Novieweld 3, Corning, dk-yellow	2.2	12	1	2	6	13	6	7	7	1	0
Schott's 43111, green	3.43	50	4	53	79	83	25	9	0	0	0
G170N, green, Corning	5.11	—	1	23	53	68	20	9	8	0	0
G174J, Corning, heat abs'b'g	2.6	—	2	4	12	19	11	4	6	0	0
G124JA, Corning	1.5	52	0	1	5	10	3	5	6	0	0
Cobalt blue	2.43	—	74	43	63	79	36	27	28	0	0
Schott's F3086, blue	2.58	—	0	1	2	31	11	5	4	0	0
G4013, Corning, blue	6.36	—	0	15	50	61	11	1	2	0	0
G584, Corning, blue	3.70	—	0	24	60	75	45	20	20	1	0
G1711Z, Corning, blue	3.23	—	23	60	74	78	45	13	12	1	0
Amethyst, C, Corning	2.11	55	91	91	91	88	42	20	25	7	0
G172BW5, Corning, red-purple	4.43	—	0	0	2	5	6	8	12	2	0
Crookes' A, A. O. Co.	1.96	90	92	91	90	83	38	23	27	5	0
" sage green 30, A. O. Co	1.98	50	0	0	4	11	8	8	11	3	0
Lab. 58, A. O. Co.	2.04	72	86	91	91	89	51	35	38	7	0
Fieurzal B, A. O. Co.	2.04	59	76	80	82	81	30	20	25	2	0
Akopos green, J. K. O. Co.	1.58	76	91	91	91	90	70	52	51	10	0

Manufacturers: **Corning** Glass Works, **Corning**, N. Y.; A. O. Co., American Optical Co., Southbridge, Mass.; J. K. O. Co., Julius King Optical Co., New York City. For other glasses see original reference. See also succeeding table, which contains data for many of the same glasses.

TABLE 371. — Transmission of the Radiations from a Gas-filled Tungsten Lamp, the Sun, a Magnetite Arc, and from a Quartz Mercury Vapor Lamp (no Globe) through Various Substances, especially Colored Glasses.

Color.	Trade name.	Source.*	Thickness in mm	Transmission, per cent.			
				Gas-filled tungsten.	Quartz mercury vapor.†	Magnetite arc.†	Solar radiation.
Greenish-yellow.....	Fieuzal, B	A. O. C.	2.04	71.6	26.9	46.0	63
" "	Fieuzal, 63	F. H. E.	1.80	75.5	34.3	55.0	72
" "	Fieuzal, 64	F. H. E.	1.65	50.7	22.0	—	—
" "	Euphos	B. S.	3.27	78.9	25.0	—	—
" "	Euphos, B	B. & L.	3.12	78.8	24.7	53.0	64
" "	Akopos green	J. K.	1.58	84.6	29.5	59.0	74
" "	Hallauer, 65	B. S.	2.36	70.3	17.7	—	—
" "	Hallauer, 64	F. H. E.	1.35	58.7	25.9	—	55
Smoky green.......	G 124, IP	C. G. W.	2.81	0.4	0.2	—	—
Yellow-green.......	Noviweld, 30%	C. G. W.	2.14	5.1	7.8	—	9
" "	Noviweld, shade 3	C. G. W.	2.20	3.4	4.2	2.7	—
" "	Noviweld, shade 4½	C. G. W.	2.20	1.6	1.2	0.8	—
" "	Noviweld, shade 6	C. G. W.	2.17	0.9	0.4	0.2	0.9
" "	Noviweld, shade 7	C. G. W.	2.17	0.8	0.2	—	—
Amber............		B. S.	3.12	51.6	15.2	—	—
"	Saniweld, dark	J. K.	1.32	78.1	10.6	43.0	50
Orange...........	G 34	C. G. W.	3.57	56.9	17.0	—	47
Yellow...........	Noviol, shade A	C. G. W.	2.00	—	—	—	81
"	Noviol, shade B	C. G. W.	2.88	74.1	32.2	56.0	75
"	Noviol, shade C	C. G. W.	2.00	—	—	—	72
Sage green.........	Ferrous No. 30	A. O. C.	1.95	5.3	17.5	—	17
Yellow-green.......	No. 61	A. O. C.	2.10	82.7	28.6	—	72
Blue-green.........	Lab. No. 59	A. O. C.	1.93	3.7	17.3	11.5	—
" "	G 124 JA	C. G. W.	1.53	5.3	21.5	12.5	19
Black............	Smoke, C	B. & L.	2.26	65.3	31.2	52.0	60
"	Smoke, D	B. & L.	2.45	50.9	16.0	39.0	43
Neutral tint........	Crookes, A	A. O. C.	1.97	85.3	46.1	—	89
" "	Crookes, B	A. O. C.	2.00	75.7	32.0	64.0	69
Gold plate.........	Pfund	A. O. C.	—	2.6	7.2	1.2	12
" " (darker)..	Pfund	A. O. C.	—	—	1.3	—	—
Colorless..........	Lab. No. 58	A. O. C.	1.58	83.3	40.0	66	88
"	Lab. No. 57	A. O. C.	2.00	—	51.9	—	—
Amethyst..........	Shade C	A. O. C.	2.11	82.8	44.3	—	79
Purple............	Electric smoke	A. O. C.	1.89	36.6	2.2	—	11
"	G 55 A 62	C. G. W.	2.85	17.4	17.0	—	16
Blue.............	Shade D	B. & L.	2.09	37.6	20.7	39	—
Blue, dark.........	G 53	C. G. W.	2.51	2.9	3.9	—	—
Blue-green.........	G 171-IZ	C. G. W.	3.21	46.6	41.7	—	—
Blue-green, pale....	G 584	C. G. W.	3.75	24.9	25.2	—	—
Red-purple.........	G 172 BW 5	C. G. W.	4.93	72.4	26.5	—	—
Blue-purple........	G 585	C. G. W.	3.13	35.8	34.0	—	41
Red..............	Selenium	C. G. W.	2.90	67.8	7.9	48	48
"		Schotts	3.22	69.4	—	—	46
"	Flashed	B. S.	—	—	4.8	—	—
Colorless..........	Window	B. S.	1.85	—	59.5	—	82
"	Crown	B. S.	1.56	—	64.9	—	92
Brown............	Mica	B. S.	1.30	—	35.4	—	—
Colorless..........	Mica	B. S.	0.09	—	43.1	—	—
Clear.............	Water	B. S.	10.0	34.2	‡54.0	—	—

* A. O. C., Amer. Optical Co., Southbridge, Mass.; C. G. W., Corning Glass Works, Corning, N. Y.; B. & L., Bausch & Lomb, Rochester, N. Y.; J. K., Julius King Optical Co., New York City; F. H. E., F. H. Edmonds, optician, Washington, D. C.; B. S., Bureau of Standards; scrap material, source unknown.

† Infra-red radiation absorbed by quartz cell containing 1 cm layer of water. Taken from Coblentz-Emerson & Long, Bul. Bureau Standards, 14, 653, 1918.

‡ Transmission of 1 cm cell having glass windows.

Table 372.

TRANSMISSIBILITY OF RADIATION.

Transmissibility of the Various Substances of Tables 330 to 338.

Alum: Ordinary alum (crystal) absorbs the infra-red.
Metallic reflection at 9.05μ and 30 to 40μ.

Rock-salt: Rubens and Trowbridge (Wied. Ann. 65, 1898) give the following transparencies for a 1 cm. thick plate in %:

λ	9	10	12	13	14	15	16	17	18	19	20.7	23.7μ
%	99.5	99.5	99.3	97.6	93.1	84.6	66.1	51.6	27.5	9.6	0.6	0.

Pflüger (Phys. Zt. 5. 1904) gives the following for the ultra-violet, same thickness: 280$\mu\mu$, 95.5%; 231, 86%; 210, 77%; 186, 70%.
Metallic reflection at 0.110μ, 0.156, 51.2, and 87μ.

Sylvite: Transparency of a 1 cm. thick plate (Trowbridge, Wied. Ann. 60, 1897).

λ	9	10	11	12	13	14	15	16	17	18	19	20.7	23.7μ
%	100.	98.8	99.0	99.5	99.5	97.5	95.4	93.6	92.	86.	76.	58.	15.

Metallic reflection at 0.114μ, 0.161, 61.1, 100.

Fluorite: Very transparent for the ultra-violet nearly to 0.1μ.
Rubens and Trowbridge give the following for a 1 cm. plate (Wied. Ann. 60, 1897):

λ	8μ	9	10	11	12μ
%	84.4	54.3	16.4	1.0	0

Metallic reflection at 24μ, 31.6, 40μ.

Iceland Spar: Merritt (Wied. Ann. 55, 1895) gives the following values of k in the formula $i = i_0 e^{-kd}$ (d in cm.):

For the ordinary ray:

λ	1.02	1.45	1.72	2.07	2.11	2.30	2.44	2.53	2.60	2.65	2.74μ
k	0.0	0.0	0.03	0.13	0.74	1.92	3.00	1.92	1.21	1.74	2.36

λ	2.83	2.90	2.95	3.04	3.30	3.47	3.62	3.80	3.98	4.35	4.52	4.83μ
k	1.32	0.70	1.80	4.71	22.7	19.4	9.6	18.6	∞	6.6	14.3	6.1

For the extraordinary ray:

λ	2.49	2.87	3.00	3.28	3.38	3.59	3.76	3.90	4.02	4.41	4.67μ
k	0.14	0.08	0.43	1.32	0.89	1.79	2.04	1.17	0.89	1.07	2.40

λ	4.91	5.04	5.34	5.50μ
k	1.25	2.13	4.41	12.8

Quartz: Very transparent to the ultra-violet; Pflüger gets the following transmission values for a plate 1 cm. thick: at 0.222μ, 94.2%; 0.214, 92; 0.203, 83.6; 0.186, 67.2%.
Merritt (Wied. Ann. 55, 1895) gives the following values for k (see formula under Iceland Spar):

For the ordinary ray:

λ	2.72	2.83	2.95	3.07	3.17	3.38	3.67	3.82	3.96	4.12	4.50μ
k	0.20	0.47	0.57	0.31	0.20	0.15	1.26	1.61	2.04	3.41	7.30

For the extraordinary ray:

λ	2.74	2.89	3.00	3.08	3.26	3.43	3.52	3.59	3.64	3.74	3.91	4.19	4.36μ
k	0.0	0.11	0.33	0.26	0.11	0.51	0.76	1.88	1.83	1.62	2.22	3.35	8.0

For $\lambda > 7\ \mu$, becomes opaque, metallic reflection at 8.50μ, 9.02, 20.75–24.4μ, then transparent again.

The above are taken from Kayser's "Handbuch der Spectroscopie," vol. iii.

TRANSMISSIBILITY OF RADIATION.

TABLE 373. — Color Screens.

The following light-filters are quoted from Landolt's "Das optische Drehungsvermögen, etc." 1898. Although only the potassium salt does not keep well it is perhaps safer to use freshly prepared solutions.

Color.	Thickness. mm.	Water solutions of	Grammes of substance in 100 c.cm.	Optical centre of band. μ	Transmission.
Red	20	Crystal-violet, 5BO	0.005	0.6659	{ begins about 0.718μ. ends sharp at 0.639μ.
"	20	Potassium monochromate	10.		
Yellow	20	Nickel-sulphate, $NiSO_4.7aq.$	30.	0.5919	0.614–0.574μ,
"	15	Potassium monochromate	10.		
"	15	Potassium permanganate	0.025		
Green	20	Copper chloride, $CuCl_2.2aq.$	60.	0.5330	0.540–0.505μ
"	20	Potassium monochromate	10.		
Bright { blue	20	Double-green, SF	0.02	0.4885	{ 0.526–0.494 and 0.494–0.458μ
	20	Copper-sulphate, $CuSO_4.5aq.$	15.		
Dark { blue	20	Crystal-violet, 5BO	0.005	0.4482	0.478–0.410μ
	20	Copper sulphate, $CuSO_4.5aq.$	15.		

TABLE 374. — Color Screens.

The following list is condensed from Wood's Physical Optics :

Methyl violet, 4R· (Berlin Anilin Fabrik) very dilute, and nitroso-dimethyl-aniline transmits 0.365μ. Methyl violet + chinin-sulphate (separate solutions), the violet solution made strong enough to blot out 0.4359μ, transmits 0.4047 and 0.4048, also faintly 0.3984.

Cobalt glass + aesculin solution transmits 0.4359μ.

Guinea green B extra (Berlin) + chinin sulphate transmits 0.4916μ.

Neptune green (Bayer, Elberfeld) + chrysoidine. Dilute the latter enough to just transmit 0.5790 and 0.5461; then add the Neptune green until the yellow lines disappear.

Chrysoidine + eosine transmits 0.5790μ. The former should be dilute and the eosine added until the green line disappears.

Silver chemically deposited on a quartz plate is practically opaque except to the ultra-violet region 0.3160–0.3260 where 90% of the energy passes through. The film should be of such thickness that a window backed by a brilliantly lighted sky is barely visible.

In the following those marked with a * are transparent to a more or less degree to the ultra-violet :

* Cobalt chloride: solution in water, — absorbs 0.50–.53μ; addition of $CaCl_2$ widens the band to 0.47–.50. It is exceedingly transparent to the ultra-violet down to 0.20. If dissolved in methyl alcohol + water, absorbs 0.50–.53 and everything below 0.35. In methyl alcohol alone 0.485–0.555 and below 0.40μ.

Copper chloride: in ethyl alcohol absorbs above 0.585 and below 0.535; in alcohol + 50% water, above 0.595 and below 0.37μ.

Neodymium salts are useful combined with other media, sharpening the edges of the absorption bands. In solution with bichromate of potash, transmits 0.535–.565 and above 0.60μ, the bands very sharp (a useful screen for photographing with a visually corrected objective).

Praseodymium salts: three strong bands at 0.482, .468, .444. In strong solutions they fuse into a sharp band at 0.435–.485μ. Absorption below 0.34.

Picric acid absorbs 0.36–.42μ, depending on the concentration.

Potassium chromate absorbs 0.40–.35, 0.30–.24, transmits 0.23μ.

* Potassium permanganate: absorbs 0.555–.50, transmits all the ultra-violet.

Chromium chloride: absorbs above 0.57, between 0.50 and .39, and below 0.33μ. These limits vary with the concentration.

Aesculin: absorbs below 0.363μ, very useful for removing the ultra-violet.

* Nitroso-dimethyl-aniline: very dilute aqueous solution absorbs 0.49–.37 and transmits all the ultra-violet.

Very dense cobalt glass + dense ruby glass or a strong potassium bichromate solution cuts off everything below 0.70 and transmits freely the red.

Iodine: saturated solution in CS_2 is opaque to the visible and transparent to the infra-red.

TRANSMISSIBILITY OF RADIATION.

TABLE 375. — Color Screens. Jena Glasses.

	Kind of Glass.	Maker's No.	Color.	Region Transmitted.	Thickness. mm.
1	Copper-ruby . .	2728	Deep red	Only red to 0.6μ	1.7
1a	Gold-ruby . . .	459^{III}	Red	Red, yellow; in thin layers also blue and violet.	
2	Uranium . . .	454^{III}	Bright yellow . . .	Red, yellow, green to E_b; in thin layer also blue	16.
2a	"	455^{III}	Bright yellow, fluoresces.		
3	Nickel	440^{III}	Bright yellow-brown	Red, yellow, green (weakened), blue (very weakened)	11.
4	Chromium . .	414^{III}	Yellow-green . . .	Yellowish-green	10.
4a	"	433^{III}	Greenish-yellow . .	Red, green; from 0.65–.50μ . . .	5.
4b	Green copper . .	431^{III}	Green	Green, yellow, some red and blue .	2–3
5	Chromium . . .	432^{III}	Yellow-green . . .	Yellowish-green, some red . . .	2.5
6	Copper chromium	436^{III}	Grass-green . . .	Green	5.
7	Green-filter . .	437^{III}	Dark green	Green (in thin sheets some blue) .	5.
8	" " . . .	438^{III}	" "	Green	
10	Copper	2742	Blue, as $CuSO_4$ . .	Green, blue, violet	5–12
11	Blue-violet . .	447^{III}	Blue, as cobalt glass	Blue, violet	5.
"	" " . . .	"	" " " " .	Blue, violet, blue-green (weakened), *no red*	2–5
12	Cobalt	424^{III}	Blue	Blue, violet, extreme red	4–5
13	Nickel	450^{III}	Dark violet	Violet (G–H), extreme red . . .	6.
14	Violet	452^{III}	" "	Violet (G–H), some weakened . .	7.
15	Gray	444^{III}	Gray, no recognizable color	All parts of the spectrum weakened	0.1–8
16	"	445^{III}			0.1–3

See "Über Farbgläser für wissenschaftliche und technische Zwecke," by Zsigmondy, Z. für Instrumentenkunde, 21, 1901 (from which the above table is taken), and "Über Jenenser Lichtfilter," by Grebe, same volume.

(The following notes are quoted from Everett's translation of the above in the English edition of Hovestadt's "Jena Glass.")

Division of the spectrum into complementary colors:

1st by 2728 (deep red) and 2742 (blue, like copper sulphate).
2nd by 454^{III} (bright yellow) and 447^{III} (blue, like cobalt glass).
3rd by 433^{III} (greenish-yellow) and 424^{III} (blue).

Thicknesses necessary in above: 2728, 1.6–1.7 mm.; 2742, 5; 454^{III}, 16; 447^{III}, 1.5–2.0; 433^{III}, 2.5–3.5; 424^{III}, 3 mm.

Three-fold division into red, green and blue (with violet):

2728, 1.7 mm.; 414^{III}, 10 mm.; 447^{III}, 1.5 mm., or by
2728, 1.7 mm.; 436^{III}, 2.6 mm.; 447^{III}, 1.8 mm.

Grebe found the three following glasses specially suited for the additive methods of three-color projection:

2745, red; 438^{III}, green; 447^{III}, blue violet;

corresponding closely to Young's three elementary color sensations.

Most of the Jena glasses can be supplied to order, but the absorption bands vary somewhat in different meltings.

See also "Atlas of Absorption Spectra," Uhler and Wood, Carnegie Institution Publications, 1907.

TABLE 376.—Water.

Values of a in $I = I_0 e^{ad}$, d in c. m. I_0; I, intensity before and after transmission.

Wave-length μ,	.186	.193	.200	.210	.220	.230	.240	.260	.300	.415
a	.0688	.0165	.009	.0061	.0057	.0034	.0032	.0025	.0015	.00035
Wave-length μ,	.430	.450	.487	.500	.550	.600	.650	.779	.865	.945
a	.00023	.0002	.0001	.0002	.0003	.0016	.0025	.272	.296	.538

First 9; Kreusler, Drud. Ann. 6, 1901; next Ewan, Proc. R. Soc. 57, 1894, Aschkinass, Wied Ann. 55, 1895; last 3, Nichols, Phys. Rev. 1, 1.

See Rubens, Ladenburg, Verh. D. Phys. Ges., p. 19, 1909, for extinction coefs., reflective power and index of refraction, 1 μ to 18 μ.

TABLE 377.

TRANSMISSION PERCENTAGES OF RADIATION THROUGH MOIST AIR.

(For bodies at laboratory temperatures; for transmission of shorter-wave energy, see Table 553.)

The values of this table will be of use for finding the transmission of energy through air containing a known amount of water vapor. An approximate value for the transmission may be had if the amount of energy from the source between the wave-lengths of the first column is multiplied by the corresponding transmission coefficients of the subsequent columns. The values for the wave-lengths greater than 18μ are tentative and doubtful. Fowle, Water-vapor Transparency, Smithsonian Misc. Collections, 68, No. 8, 1917; Fowle, The Transparency of Aqueous Vapor, Astrophysical J. 42, p. 394, 1915.

Range of wave-lengths.		Precipitable water in centimeters.												
μ	μ	.001	.003	.006	.01	.03	.06	.10	.25	.50	1.0	2.0	6.0	10.0
0.75 to	1.0	—	—	—	*100*	99	99	98	97	95	93	90	83	*78*
1.0	1.25	—	—	—	*99*	99	98	97	95	92	89	85	74	*69*
1.25	1.5	—	—	—	*96*	*92*	*84*	*80*	*66*	57	51	44	31	*28*
1.5	2.0	—	—	—	*98*	*97*	*94*	*88*	*79*	73	70	66	60	*57*
* 2	3	*96*	*92*	*87*	*84*	*77*	*70*	*64*	—	—	—	—	—	—
3	4	*95*	*88*	*84*	*78*	*72*	*66*	*63*	—	—	—	—	—	—
* 4	5	*92*	*83*	*76*	*71*	*65*	*60*	*53*	—	—	—	—	—	—
5	6	95	82	75	68	56	51	47	35	—	—	—	—	—
6	7	*85*	54	50	31	24	8	4	3	*2*	0	0	0	0
7	8	*94*	84	76	68	57	46	35	16	*10*	*2*	0	0	0
8	9	100	100	100	99	98	96	94	*65*	—	—	—	—	—
† 9	10	100	100	100	100	100	100	100	100	100	100	100	—	—
† 10	11	100	100	100	100	100	100	100	100	100	100	100	—	—
11	12	100	100	100	100	100	99	98	96	95	*93*	—	—	—
12	13	100	100	100	100	99	99	97	*86*	*82*	—	—	—	—
* 13	14	100	100	100	99	97	94	90	*80*	*60*	—	—	—	—
* 14	15	—	—	*96*	*93*	*80*	*75*	*50*	*15*	*0*	*0*	*0*	*0*	*0*
* 15	16	—	—	—	—	*70*	*55*	*40*	*0*	*0*	*0*	*0*	*0*	*0*
16	17	—	—	—	—	—	*50*	*20*	*0*	0	0	0	0	0
17	18	—	—	—	—	—	*25*	*10*	*0*	0	0	0	0	0
18	∞	98	94	89	82	45	0	0	0	0	0	0	0	0

* These places require multiplication by the following factors to allow for losses in CO_2 gas. Under average sea-level outdoor conditions the CO_2 (partial pressure = 0.0003 atmos.) amounts to about 0.6 gram per cu. m. Paschen gives 3 times as much for indoor conditions.

2μ to 3μ, for 2 grams in m^2 path (95); for 140 grams in m^2 path (93);
4 " 5 " " " " " " (93); " " " " " " (70); more CO_2 no further effect;
13 " 14, slight allowance to be made;
14 " 15, 80 grams in m^2 path reduces energy to zero;
15 " 16, " " " " " " " " "

† These places require multiplication by 0.90 and 0.70 respectively for one air mass and 0.85 and 0.65 for two air masses to allow for ozone absorption when the radiation comes from a celestial body.

In the above table italicized figures indicate extrapolated values.

F. Paschen gives (Annalen d. Physik u. Chemie, 51, p. 14, 1894) the absorption of the radiation from a blackened strip at 500° C by a layer 33 centimeters thick of water vapor at 100° C and atmospheric pressure as follows:

Wave-length..................	2.20–3.10μ	5.33–7.67μ	7.67–10(?)μ
Percentage absorption.........	80	94	94–13

The following table, due to Rubens and Aschkinass (Annalen d. Physik u. Chemie, 64, p. 598, 1898), gives the absorption of radiation from a zircon burner by a layer 75 centimeters thick of water vapor saturated at 100° C. This amount of vapor is about equivalent to a layer of water 0.45 millimeter thick or to 1.5% of the water in a total vertical atmospheric column whose dew point at sea-level is 10° C. The region of spectrum examined includes most of the region of terrestrial radiation.

Wave-length.............	7.0μ	8.0μ	9.0–12.0μ	12.4μ	12.8μ	13.4μ	14.0μ
Percentage absorption.....	75	40	6	20	13	28	22
Wave-length.............	14.3μ	15.0μ	15.7μ	16.0μ	17.5μ	18.3μ	20.0μ
Percentage absorption.....	43	35	65	52	88	80	100

REFLECTION AND ABSORPTION OF LONG-WAVE RADIATIONS.

TABLE 378. — Long-wave Absorption by Gases.

Unless otherwise noted, gases were contained in a 20 cm long tube. Rubens, Wartenberg, Verh. d. Phys. Ges. 13, p. 796, 1911.

Gas.	Pressure, cm	Percentage absorption. 23μ	52μ	110μ	Long λ, Hg lamp.	Long λ, Hg lamp. Filtered, 314μ	Gas.	Pressure, cm	Percentage absorption. 23μ	52μ	110μ	Long λ. Hg lamp.	Long λ. Hg lamp. Filtered, 314μ
H_2	76	100	100	100	100	100	NH_3	76	83.1	0.5	99.2	43.3	66.7
Cl_2	76	100	99.6	99.5	98.5	97.6	CH_4	76	91	94.3	99.2	100	100
Br_2	20	100	100	100	100	100	C_2H_2	76	99.5	87.4	97.3	97.9	100
SO_2	76	22.6	76.9	12.7	6	4.8	C_2H_4	76	99	96.4	92.8	100	100
CO_2	76	100	100	100	100	100	CS_2	26	97.8	100	100	99.5	100
CO	76	100	100	94.1	92.1	91.6	C_2H_6O	6	85.4	5.4	58	52.4	49.9
H_2S	76	99.6	11.6	5.4	10.3	21.4	$C_4H_{10}O$	51	26.8	46	34	21.8	10.7
N_2O	76	100	96.8	98.4	93.3	90.8	C_5H_{12}	46	66 †	44.5	88.8	87	84.2
NO	76	—	94	99	87.3	85.5	CH_3Cl	14	98	100	100	95.4	94.7
$(CN)_2$	76	100	97.8	100	99.3	—	H_2O *	76	39.6	0.7	19.6	33.6	49.2

* Tube 40 cm long. † Pentane vapor, pressure 36 cm.

TABLE 379. — Properties with Wave-lengths 108 ± μ.

Rubens and Woods, Verh. d. Phys. Ges. 13, p. 88, 1911.

With quartz, 1.7 cm thick: 60 to 80μ, absorption very great; 63μ, 99%; 82μ, 97.5; 97μ, 83.

(a) PERCENTAGE REFLECTION.

Wave-length.	Iceland spar.	Marble.	Rock salt.	Sylvite	KBr	Kl	Fluorite.	Glass.	Water.	Alcohol.
λ = 82μ *	—	—	25.8	36.0	82.6	29.6	19.7	—	9.6	—
λ = 108μ †	47.1	43.8	20.3	19.3	31.1	35.5	20.2	19.2	11.6	1.6

* Restrahlung from KBr. † Isolated with quartz lens.

(b) PERCENTAGE TRANSPARENCY.
Uncorrected for reflections.

Solid.	Thickness.	Transparency.	Liquid.	Thickness.	Thickness, precipitable liquid.	Transparency.
Paraffin	3.03	57.0	Benzene	1.00	—	56.8
Mica	0.055	16.6	Ethyl alcohol	0.158	—	7.9
Hard rubber	0.40	39.0	Ethyl ether	0.158	—	37.1
Quartz ‖ axis	2.00	62.6	Water	0.029	—	25.8
Quartz, amorph.	3.85	0	Water	0.044	—	13.6
Rock salt	0.21	21.5				
Fluorite	0.59	5.3	Vapors:			
Diamond	1.26	45.3	Alcohol	2.00	0.023	88
Quartz ⊥ axis	2.00	81.3	Ether	2.00	0.350	33.5
" "	4.03	66.4	Benzene	2.00	0.063	100
" "	7.26	49.8	Water	4.00	0.21	19.6
" "	11.74	35.5	CO_2	2.00	—	100
" "	14.66	29.0				

(c) TRANSPARENCY OF BLACK ABSORBERS.

Method and wave-length.		Black silk paper, .025 mm thick.	Opaque black paper, 0.11 mm thick.	Black card-board, 0.4 mm thick.	Candle lamp-black, 10 cm^2 = 1.8 mg
Spectrometer	2μ	0	0	0	0.5
	4	0.9	0	0	8.6
	6	1.7	0	0	16.0
	12	8.2	1.4	0	37.6
Fluorite "reststrahlen"	26	24.2	3.2	0	76.7
Rock salt "reststrahlen"	52	46.0	15.1	0	91.3
Quartz lens isolation	108	61.5	33.5	1.6	91.5

TABLE 380.—Tartaric Acid; Camphor; Santonin; Santonic Acid; Cane Sugar.

A few examples are here given showing the effect of wave-length on the rotation of the plane of polarization. The rotations are for a thickness of one decimeter of the solution. The examples are quoted from Landolt & Börnstein's "Phys. Chem. Tab." The following symbols are used:—

p = number grams of the active substance in 100 grams of the solution.
c = " " solvent " " "
q = " " active " " cubic centimeter "

Right-handed rotation is marked +, left-handed —.

Line of spectrum.	Wave-length according to Ångström in cms. × 10⁶.	Tartaric acid,* $C_4H_6O_6$, dissolved in water. q = 50 to 95, temp. = 24° C.	Camphor,* $C_{10}H_{16}O$, dissolved in alcohol. q = 50 to 95, temp. = 22.9° C.	Santonin,† $C_{15}H_{18}O_3$, dissolved in chloroform. q = 75 to 96.5, temp. = 20° C.
B	68.67			— 140°.1 + 0.2085 q
C	65.62	+ 2°.748 + 0.09446 q	38°.549 — 0.0852 q	— 149.3 + 0.1555 q
D	58.92	+ 1.950 + 0.13030 q	51.945 — 0.0964 q	— 202.7 + 0.3086 q
E	52.69	+ 0.153 + 0.17514 q	74.331 — 0.1343 q	— 285.6 + 0.5820 q
b_1	51.83	— —	— —	— 302.38 + 0.6557 q
b_2	51.72	— 0.832 + 0.19147 q	79.348 — 0.1451 q	— —
F	48.61	— 3.598 + 0.23977 q	99.601 — 0.1912 q	— 365.55 + 0.8284 q
e	43.83	— 9.657 + 0.31437 q	149.696 — 0.2346 q	— 534.98 + 1.5240 q

Line of spectrum.	Wave-length	Santonin,† $C_{15}H_{18}O_3$,* dissolved in alcohol. c = 1.782. temp. = 20° C.	Santonin,† $C_{15}H_{18}O_3$, dissolved in alcohol. c = 4.046. temp. = 20° C.	Santonin,† $C_{15}H_{18}O_3$, dissolved in chloroform c = 3.1–30.5. temp. = 20° C.	Santonic acid,† $C_{15}H_{20}O_4$, dissolved in chloroform. c = 27.192. temp. = 20° C.	Cane sugar,‡ $C_{12}H_{22}O_{11}$, dissolved in water. p = 10 to 30.
B	68.67	— 110.4°	442°	484°	— 49°	
C	65.62	— 118.8	504	549	— 57	
D	58.92	— 161.0	693	754	— 74	
E	52.69	— 222.6	991	1088	— 105	
b_1	51.83	— 237.1	1053	1148	— 112	See page 444
b_2	51.72	—	—	—	—	
F	48.61	— 261.7	1323	1444	— 137	
e	43.83	— 380.0	2011	2201	— 197	
G	43.07	—	—	—	—	
g	42.26	—	2381	2610	— 230	

* Arndtsen, "Ann. Chim. Phys." (3) 54, 1858.
† Narini, "R. Acc. dei Lincei," (3) 13, 1882.
‡ Stefan, "Sitzb. d. Wien. Akad." 52, 1865.

TABLE 381.—Sodium Chlorate; Quartz.

Sodium chlorate (Guye, C. R. 108, 1889).

Spectrum line.	Wave-length.	Temp. C.	Rotation per mm.
a	71.769	15°.0	2°.068
B	67.889	17.4	2.318
C	65.073	20.6	2.599
D	59.085	18.3	3.104
E	53.233	16.0	3.841
F	48.912	11.9	4.587
G	45.532	10.1	5.331
G	42.834	14.5	6.005
H	40.714	13.3	6.754
L	38.412	14.0	7.654
M	37.352	10.7	8.100
N	35.818	12.9	8.861
P	33.931	12.1	9.801
Q	32.341	11.9	10.787
R	30.645	13.1	11.921
T	29.918	12.8	12.424
Cd_{17}	28.270	12.2	13.426
Cd_{18}	25.038	11.6	14.965

Quartz (Soret & Sarasin, Arch. de Gen. 1882, or C. R. 95, 1882).*

Spectrum line.	Wave-length.	Rotation per mm.	Spectrum line.	Wave-length.	Rotation per mm.
A	76.04	12°.668	Cd_9	36.090	63°.628
a	71.836	14.304	N	35.818	64.459
B	68.671	15.746	Cd_{10}	34.655	69.454
			O	34.406	70.587
C	65.621	17.318			
D_1	58.951	21.684	Cd_{11}	34.015	72.448
D_2	58.891	21.727	P	33.600	74.571
			Q	32.858	78.579
E	52.691	27.543	Cd_{12}	32.470	80.459
F	48.607	32.773			
G	43.072	42.604	R	31.798	84.972
			Cd_{17}	27.467	121.052
h	41.012	47.481	Cd_{18}	25.713	143.266
H	39.681	51.193	Cd_{23}	23.125	190.426
K	39.333	52.155			
			Cd_{24}	22.645	201.824
L	38.196	55.625	Cd_{25}	21.935	220.731
M	37.262	58.894	Cd_{26}	21.431	235.972

* The paper is quoted from a paper by Ketteler in "Wied. Ann." vol. 21, p. 444. The wave-lengths are for the Fraunhofer lines, Ångström's values for the ultra violet sun, and Cornu's values for the cadmium lines.

Abbreviations: int'n'l, international; emu, electromagnetic units; esu, electrostatic units; cgs, centimeter-gram-second units. (Taken from Circular 60 of U. S. Bureau of Standards, 1916, Electric Units and Standards.)

RESISTANCE:

1 international ohm =
- 1.00052 absolute ohms
- 1.0001 int'n'l ohms (France, before 1911)
- 1.00016 Board of Trade units (England, 1903)
- 1.01358 B. A. units
- 1.00283 "legal ohms" of 1884
- 1.06300 Siemens units

1 absolute ohm =
- 0.99948 int'n'l ohms
- 1 "practical" emu
- 10^9 cgs emu
- 1.1124×10^{-12} cgs esu

CURRENT:

1 international ampere =
- 0.99991 absolute ampere
- 1.00084 int'n'l amperes (U. S. before 1911)
- 1.00130 int'n'l amperes (England, before 1906)
- 1.00106 int'n'l amperes (England, 1906–08)
- 1.00010 int'n'l amperes (England, 1909–10)
- 1.00032 int'n'l amperes (Germany, before 1911)
- 1.0002 int'n'l amperes (France, before 1911)

1 absolute ampere =
- 1 00009 int'n'l amperes
- 1 "practical" emu
- 0.1 cgs emu
- 2.9982×10^9 cgs esu

ELECTROMOTIVE FORCE:

1 international volt =
- 1.00043 absolute volts
- 1.00084 int'n'l volts (U. S. before 1911)
- 1.00130 int'n'l volts (England, before 1906)
- 1.00106 int'n'l volts (England, 1906–08)
- 1.00010 int'n'l volts (England, 1909–10)
- 1.00032 int'n'l volts (Germany, before 1911)
- 1.00032 int'n'l volts (France, before 1911)

1 absolute volt =
- 0.99957 int'n'l volt
- 1 "practical" emu
- 10^8 cgs emu
- 0.0033353 cgs esu

QUANTITY OF ELECTRICITY:

(Same as current equivalents.)

1 international coulomb =
- 1/3600 ampere-hour
- 1/96500 faraday

CAPACITY:

1 international farad =
- 0.99948 absolute farad

1 absolute farad =
- 1.00052 int'n'l farads
- 1 "practical" emu
- 10^{-9} cgs emu
- 8.9892×10^{11} cgs esu

INDUCTANCE:

1 international henry =
- 1.00052 absolute henries

1 absolute henry =
- 0.99948 int'n'l henry
- 1 "practical" emu
- 10^9 emu
- 1.1124×10^{-12} cgs esu

ENERGY AND POWER:

(standard gravity = 980.665 cm/sec/sec.)

1 international joule =
- 1.00034 absolute joules

1 absolute joule =
- 0.99966 int'n'l joule
- 10^7 ergs
- 0.737560 standard foot-pound
- 0.101972 standard kilogram-meter
- 0.277778×10^{-6} kilowatt-hour

RESISTIVITY:

1 ohm-cm = 0.393700 ohm-inch
- = 10,000 ohm (meter, mm^2)
- = 12,732.4 ohm (meter, mm)
- = 393,700 microhm-inch
- = 1,000,000 microhm-cm
- = 6,015,290 ohm (mil, foot)

1 ohm (meter, gram) = 5710.0 ohm (mile, pound)

MAGNETIC QUANTITIES:

- 1 int'n'l gilbert = 0.99991 absolute gilbert
- 1 absolute gilbert = 1.00009 int'n'l gilberts
- 1 int'n'l maxwell = 1.00043 absolute maxwells
- 1 absolute maxwell = 0.99957 int'n'l maxwell
- 1 gilbert = 0.7958 ampere-turn
- 1 gilbert per cm = 0.7958 ampere-turn per cm
  - = 2.021 ampere-turns per inch
- 1 maxwell = 1 line
  - = 10^{-8} volt-second
- 1 maxwell per cm^2 = 6.452 maxwells per in^2

COMPOSITION AND ELECTROMOTIVE FORCE OF VOLTAIC CELLS.

The electromotive forces given in this table approximately represent what may be expected from a cell in good working order, but with the exception of the standard cells all of them are subject to considerable variation.

(a) DOUBLE FLUID CELLS.

Name of cell.	Negative pole.	Solution.	Positive pole.	Solution.	E.M.F. in volts.
Bunsen . .	Amalgamated zinc	1 part H_2SO_4 to 12 parts H_2O .	Carbon	Fuming HNO_3 .	1.94
" . .	" "	"	"	HNO_3, density 1.38	1.86
Chromate .	" "	12 parts $K_2Cr_2O_7$ to 25 parts of H_2SO_4 and 100 parts H_2O . .	"	1 part H_2SO_4 to 12 parts H_2O .	2.00
" . .	" "	1 part H_2SO_4 to 12 parts H_2O .	"	12 parts $K_2Cr_2O_7$ to 100 parts H_2O	2.03
Daniell* .	" "	1 part H_2SO_4 to 4 parts H_2O .	Copper	Saturated solution of $CuSO_4+5H_2O$	1.06
" .	" "	1 part H_2SO_4 to 12 parts H_2O .	"	"	1.09
" .	" "	5% solution of $ZnSO_4+6H_2O$	"	"	1.08
" .	" "	1 part NaCl to 4 parts H_2O .	"	"	1.05
Grove . .	" "	1 part H_2SO_4 to 12 parts H_2O .	Platinum	Fuming HNO_3 . .	1.93
" . .	" "	Solution of $ZnSO_4$	"	HNO_3, density 1.33	1.66
" . .	" "	H_2SO_4 solution, density 1.136 .	"	Concentrated HNO_3	1.93
" . .	" "	H_2SO_4 solution, density 1.136 .	"	HNO_3, density 1.33	1.79
" . .	" "	H_2SO_4 solution, density 1.06 .	"	"	1.71
" . .	" "	H_2SO_4 solution, density 1.14 .	"	HNO_3, density 1.19	1.66
" . .	" "	H_2SO_4 solution, density 1.06 .	"	" " "	1.61
" . .	" "	NaCl solution . .	"	" density 1.33	1.88
Marié Davy	" "	1 part H_2SO_4 to 12 parts H_2O	Carbon	Paste of protosulphate of mercury and water . . .	1.50
Partz . .	" "	Solution of $MgSO_4$	"	Solution of $K_2Cr_2O_7$	2.06

* The Minotto or Sawdust, the Meidinger, the Callaud, and the Lockwood cells are modifications of the Daniell, and hence have about the same electromotive force.

COMPOSITION AND ELECTROMOTIVE FORCE OF VOLTAIC CELLS.

Name of cell.	Negative pole.	Solution.	Positive pole.	E. M. F. in volts.
(b) SINGLE FLUID CELLS.				
Leclanche . . .	Amal. zinc	Solution of sal-ammoniac	Carbon. Depolarizer: manganese peroxide with powdered carbon	1.46
Chaperon . . .	" "	Solution of caustic potash	Copper. Depolarizer: CuO . . .	0.98
Edison-Lelande .	" "	"	"	0.70
Chloride of silver	Zinc . .	23 % solution of sal-ammoniac	Silver. Depolarizer: silver chl'ride	1.02
Law	" . .	15 % " "	Carbon	1.37
Dry cell (Gassner)	" . .	1 pt. ZnO, 1 pt. NH_4Cl, 3 pts. plaster of paris, 2 pts. $ZnCl_2$, and water to make a paste . .	"	1.3
Poggendorff . .	Amal. zinc	Solution of chromate of potash	"	1.08
" . . .	" "	12 parts $K_2Cr_2O_7$ + 25 parts H_2SO_4 + 100 parts H_2O . .	"	2.01
J. Regnault . . .	" "	1 part H_2SO_4 + 12 parts H_2O + 1 part $CaSO_4$ . .	Cadmium . . .	0.34
Volta couple . .	Zinc . .	H_2O	Copper	0.98
(c) STANDARD CELLS.				
Weston normal .	Cadmi'm am'lgam	Saturated solution of $CdSO_4$	Mercury. Depolarizer: paste of Hg_2SO_4 and $CdSO_4$	1.0183* at 20° C
Clark standard .	Zinc am'lgam	Saturated solution of $ZnSO_4$	Mercury. Depolarizer: paste of Hg_2SO_4 and $ZnSO_4$	1.434‡ at 15° C
(d) SECONDARY CELLS.				
Lead accumulator	Lead . .	H_2SO_4 solution of density 1.1 . . .	PbO_2	2.2†
Regnier (1) . . .	Copper .	$CuSO_4 + H_2SO_4$ . .	"	1.68 to 0.85, average 1.3.
" (2) . . .	Amal. zinc	$ZnSO_4$ solution . . .	" in H_2SO_4 .	2.36
Main	Amal. zinc	H_2SO_4 density ab't 1.1	"	2.50
Edison	Iron . .	KOH 20 % solution .	A nickel oxide .	1.1, mean of full discharge.

* The temperature formula is $E_t = E_{20} - 0.0000406\,(t-20) - 0.00000095\,(t-20)^2 + 0.00000001\,(t-20)^3$.

‡ The value given for the Clark cell is the old one adopted by the Chicago International Electrical Congress in 1893. The temperature formula is $E_t = E_{15} - 0.00119\,(t-15) - 0.000007\,(t-15)^2$.

† F. Streintz gives the following value of the temperature variation $\frac{dE}{dt}$ at different stages of charge:

E. M. F.	1.9223	1.9828	2.0031	2.0084	2.0105	2.0779	2.2070
$dE/dt \times 10^6$	140	228	335	285	255	130	73

Dolezalek gives the following relation between E. M. F. and acid concentration:

Per cent H_2SO_4	64.5	52.2	35.3	21.4	5.2
E.M.F., 0° C	2.37	2.25	2.10	2.00	1.89

CONTACT DIFFERENCE OF

Solids with Liquids and

Temperature of substances

	Carbon.	Copper.	Iron.	Lead.	Platinum.	Tin.	Zinc.
Distilled water	.01 to .17	.269 to .100	.148	.171	.285 to .345	.177	—.105 to +.156
Alum solution: saturated at 16°.5 C.	-	—.127	—.653	—.139	.246	—.225	—.536
Copper sulphate solution: sp. gr. 1.087 at 16°.6 C.	-	.103	-	-	-	-	-
Copper sulphate solution: saturated at 15° C. . .	-	.070	-	-	-	-	-
Sea salt solution: sp. gr. 1.18 at 20°.5 C. . . .	-	—.475	—.605	-	—.856	—.334	—.565
Sal-ammoniac solution: saturated at 15°.5 C. .	-	—.396	—.652	—.189	.059	—.364	—.637
Zinc sulphate solution: sp. gr. 1.125 at 16°.9 C. . .	-	-	-	-	-	-	—.238
Zinc sulphate solution: saturated at 15°.3 C. .	-	-	-	-	-	-	—.430
One part distilled water + 3 parts saturated zinc sulphate solution . . .	-	-	-	-	-	-	—.444
Strong sulphuric acid in distilled water:							
1 to 20 by weight . . .	-	-	-	-	-	-	—.344
1 to 10 by volume . . .	about —.035	-	-	-	-	-	-
1 to 5 by weight	-	-	-	-	-	-	-
5 to 1 by weight	.01 to 3.0	-	-	—.120	-	—.25	-
Concentrated sulphuric acid	.55 to .85	1.113	-	.72 to 1.252	1.3 to 1.6	-	-
Concentrated nitric acid .	-	-	-	-	.672	-	-
Mercurous sulphate paste .	-	-	-	-	-	-	-
Distilled water containing trace of sulphuric acid	-	-	-	-	-	-	—.241

* Everett's "Units and Physical Constants:" Table of

POTENTIAL IN VOLTS.

Liquids with Liquids in Air.*

during experiment about 16° C.

	Amalgamated zinc.	Brass.	Mercury.	Distilled water.	Alum solution: saturated at 16°.5 C.	Copper sulphate solution: saturated at 15° C.	Zinc sulphate solution: sp. gr. 1.25 at 16°.9 C.	Zinc sulphate solution: saturated at 15°.3 C.	One part distilled water + 3 pts. zinc sulphate.	Strong nitric acid.
Distilled water	.100	.231	–	–	–	—.043	–	.164	–	–
Alum solution: saturated at 16°.5 C.	–	—.014	–	–	–	–	–	–	–	–
Copper sulphate solution: sp. gr. 1.087 at 16°.6 C.	–	–	–	–	–	–	.090	–	–	–
Copper sulphate solution: saturated at 15° C. . .	–	–	–	—.043	–	–	–	.095	.102	–
Sea salt solution: sp. gr. 1.18 at 20°.5 C. . . .	–	—.435	–	–	–	–	–	–	–	–
Sal-ammoniac solution: saturated at 15°.5 C. .	–	—.348	–	–	–	–	–	–	–	–
Zinc sulphate solution: sp. gr. 1.125 at 16°.9 C.	–	–	–	–	–	–	–	–	–	–
Zinc sulphate solution: saturated at 15°.3 C. .	—.284	–	–	—.200	–	—.095	–	–	–	–
One part distilled water + 3 parts saturated zinc sulphate solution . .	–	–	–	–	–	—.102	–	–	–	–
Strong sulphuric acid in distilled water:										
1 to 20 by weight . . .	–	–	–	–	–	–	–	–	–	–
1 to 10 by volume . . .	—.358	–	–	–	–	–	–	–	–	–
1 to 5 by weight	.429	–	–	–	–	–	–	–	–	–
5 to 1 by weight	–	—.016	–	–	–	–	–	–	–	–
Concentrated sulphuric acid	.848	–	–	1.298	1.456	1.269	–	1.699	–	–
Concentrated nitric acid .			–	–	–	–	–	–	–	–
Mercurous sulphate paste .	–	–	.475	–	–	–	–	–	–	–
Distilled water containing trace of sulphuric acid .	–	–	–	–	–	–	–	–	–	.078

Ayrton and Perry's results, prepared by Ayrton.

TABLE 385.

DIFFERENCE OF POTENTIAL BETWEEN METALS IN SOLUTIONS OF SALTS.

The following numbers are given by G. Magnanini * for the difference of potential in hundredths of a volt between zinc in a normal solution of sulphuric acid and the metals named at the head of the different columns when placed in the solution named in the first column. The solutions were contained in a U-tube, and the sign of the difference of potential is such that the current will flow from the more positive to the less positive through the external circuit.

Strength of the solution in gram molecules per liter.		Zinc.†	Cadmium.†	Lead.	Tin.	Copper.	Silver.
No. of molecules.	Salt.	Difference of potential in centivolts.					
0.5	H_2SO_4	0.0	36.6	51.3	51.3	100.7	121.3
1.0	NaOH	—32.1	19.5	31.8	0.2	80.2	95.8
1.0	KOH	—42.5	15.5	32.0	—1.2	77.0	104.0
0.5	Na_2SO_4	1.4	35.6	50.8	51.4	101.3	120.9
1.0	$Na_2S_2O_8$	—5.9	24.1	45.3	45.7	38.8	64.8
1.0	KNO_3	11.8‡	31.9	42.6	31.1	81.2	105.7
1.0	$NaNO_3$	11.5	32.3	51.0	40.9	95.7	114.8
0.5	K_2CrO_4	23.9‡	42.8	41.2	40.9	94.6	121.0
0.5	$K_2Cr_2O_7$	72.8	61.1	78.4	68.1	123.6	132.4
0.5	K_2SO_4	1.8	34.7	51.0	40.9	95.7	114.8
0.5	$(NH_4)_2SO_4$	—0.5	37.1	53.2	57.6‡	101.5	125.7
0.25	$K_4FeC_6N_6$	—6.1	33.6	50.7	41.2	—‡	87.8
0.167	$K_6Fe_2(CN)_{12}$	41.0§	80.8	81.2	130.9	110.7	124.9
1.0	KCNS	—1.2	32.5	52.8	52.7	52.5	72.5
1.0	$NaNO_3$	4.5	35.2	50.2	49.0	103.6	104.6?
0.5	$Sr(NO_3)_2$	14.8	38.3	50.6	48.7	103.0	119.3
0.125	$Ba(NO_3)_2$	21.9	39.3	51.7	52.8	109.6	121.5
1.0	KNO_3	—‡	35.6	47.5	49.9	104.8	115.0
0.2	$KClO_3$	15–10‡	39.9	53.8	57.7	105.3	120.9
0.167	$KBrO_3$	13–20‡	40.7	51.3	50.9	111.3	120.8
1.0	NH_4Cl	2.9	32.4	51.3	50.9	81.2	101.7
1.0	KF	2.8	22.5	41.1	50.8	61.3	61.5
1.0	NaCl	—	31.9	51.2	50.3	80.9	101.3
1.0	KBr	2.3	31.7	47.2	52.5	73.6	82.4
1.0	KCl	—	32.1	51.6	52.6	81.6	107.6
0.5	Na_2SO_3	—8.2	28.7	41.0	31.0	68.7	103.7
–‖	NaOBr	18.4	41.6	73.1	70.6 ‡	89.9	99.7
1.0	$C_4H_6O_6$	5.5	39.7	61.3	54.4§	104.6	123.4
0.5	$C_4H_6O_6$	4.1	41.3	61.6	57.6	110.9	125.7
0.5	$C_4H_4KNaO_6$	—7.9	31.5	51.5	42–47	100.8	119.7

* "Rend. della R. Acc. di Roma," 1890.
† Amalgamated.
‡ Not constant.
§ After some time.
‖ A quantity of bromine was used corresponding to NaOH = 1.

TABLE 386.

THERMOELECTRIC POWER.

The thermoelectric power of a circuit of two metals is the electromotive force produced by one degree C difference of temperature between the junctions. The thermoelectric power varies with the temperature, thus: thermoelectric power $= Q = dE/dt = A + Bt$, where A is the thermoelectric power at 0° C, B is a constant, and t is the mean temperature of the junctions. The neutral point is the temperature at which $dE/dt = 0$, and its value is $-A/B$. When a current is caused to flow in a circuit of two metals originally at a uniform temperature, heat is liberated at one of the junctions and absorbed at the other. The rate of production or liberation of heat at each junction, or Peltier effect, is given in calories per second, by multiplying the current by the coefficient of the Peltier effect. This coefficient in calories per coulomb $= QT/\mathcal{J}$, in which Q is in volts per degree C, T is the absolute temperature of the junction, and $\mathcal{J} = 4.19$. Heat is also liberated or absorbed in each of the metals as the current flows through portions of varying temperature. The rate of production or liberation of heat in each metal, or the Thomson effect, is given in calories per second by multiplying the current by the coefficient of the Thomson effect. This coefficient, in calories per coulomb $= BT\theta/\mathcal{J}$, in which B is in volts per degree C, T is the mean absolute temperature of the junctions, and θ is the difference of temperature of the junctions. (BT) is Sir W. Thomson's "Specific Heat of Electricity." The algebraic signs are so chosen in the following table that when A is positive, the current flows in the metal considered from the hot junction to the cold. When B is positive, Q increases (algebraically) with the temperature. The values of A, B, and thermoelectric power in the following table are *with respect to lead* as the other metal of the thermoelectric circuit. The thermoelectric power of a couple composed of two metals, 1 and 2, is given by subtracting the value for 2 from that for 1; when this difference is positive, the current flows from the hot junction to the cold in 1. In the following table, A is given in microvolts per degree, B in microvolts per degree per degree, and the neutral point in degrees.

The table has been compiled from the results of Becquerel, Matthiessen and Tait; in reducing the results, the electromotive force of the Grove and Daniell cells has been taken as 1.95 and 1.07 volts. The value for constantan was reduced from results given in Landolt-Börnstein's tables. The thermoelectric powers of antimony and bismuth alloys are given by Becquerel in the reference given below.

Substance.	A Microvolts.	B Microvolts.	Thermoelectric power at mean temp. of junctions (microvolts).		Neutral point $-\frac{A}{B}$	Authority.
			20° C	50° C		
Aluminum	−0.76	+0.0039	−0.68	−0.56	+195	T
Antimony, comm'l pressed wire	—	—	+6.0	—	—	M
" axial	—	—	+22.6	—	—	"
" equatorial	—	—	+26.4	—	—	"
Argentan	−11.94	−0.0506	−12.95	−14.47	−236	T
"	—	—	—	−12.7	—	B
Arsenic	—	—	−13.56	—	—	M
Bismuth, comm'l pressed wire	—	—	−97.0	—	—	"
" pure " "	—	—	−89.0	—	—	"
" crystal, axial	—	—	−65.0	—	—	"
" " equatorial	—	—	−45.0	—	—	"
Cadmium	+2.63	+0.0424	+3.48	+4.75	−62	T
" fused	—	—	—	+2.45	—	B
Calcium	—	—	—	+8.9	—	S'
Cobalt	—	—	−22	—	—	M
Constantan	—	—	—	−19.3	—	—
Copper	+1.34	+0.0094	+1.52	+1.81	−143	T
" commercial	—	—	+0.10	—	—	M
" galvanoplastic	—	—	+3.8	—	—	"
Gallium	—	—	−0.2	—	—	S
Gold	+2.80	+0.0101	+3.0	+3.30	[−277]	T
Iron	+17.15	−0.0482	+16.2	+14.74	+356	T
" pianoforte wire	—	—	+17.5	—	—	M
" commercial	—	—	—	+12.10	—	B
" "	—	—	—	+9.10	—	"
Lead	—	0.0000	−0.00	0.00	—	—
Magnesium	+2.22	−0.0094	+2.03	+1.75	+236	T
Molybdenum	—	—	+5.9	—	—	S
Mercury	—	—	−0.413	−3.30	—	MB
Nickel	—	—	—	−15.50	—	B
" (−18° to 175°)	−21.8	−0.0506	−22.8	−24.33	[−431]	T
" (250°–300°)	−83.57	+0.2384	—	—	—	"
" (above 340°)	−3.04	−0.0506	—	—	—	"

TABLE 386.—Thermoelectric Power (continued).

Substance.	*A* Microvolts.	*B* Microvolts.	Thermoelectric power at mean temp. of junctions (microvolts). 20° C.	50° C.	Neutral point $-\frac{A}{B}$.	Authority.
Palladium	−6.18	−0.0355	−6.9	−7.96	−174	T
Phosphorus (red)	–	–	+29.9	–	–	M
Platinum	–	–	+0.9	–	–	"
" (hardened)	+2.57	−0.0074	+2.42	+2.20	347	T
" (malleable)	−0.60	−0.0109	−.818	−1.15	−55	"
" wire	–	–	–	+0.94	–	B
" another specimen	–	–	–	−2.14	–	"
Platinum-iridium alloys:						
85% Pt + 15% Ir	+7.90	+0.0062	+8.03	+8.21	[−1274]	T
90% Pt + 10% Ir	+5.90	−0.0133	+5.63	+5.23	444	"
95% Pt + 5% Ir	+6.15	+0.0055	+6.26	+6.42	[−1118]	"
Selenium	–	–	+807.	–	–	M
Silver	+2.12	+0.0147	+2.41	+2.86	−144	T
" (pure hard)	–	–	+3.00	–	–	M
" wire	–	–	–	+2.18	–	B
Steel	+11.27	−0.0325	+10.62	+9.65	347	T
Tantalum	–	–	−2.6	–	–	–
Tellurium *β*	–	–	+500.	–	–	H
" *α*	–	–	+160.	–	–	H
Thallium	–	–	+0.8	–	–	–
Tin (commercial)	–	–	–	+0.33	–	H
"	–	–	+0.1	–	–	M
"	−0.43	+0.0055	−0.33	−0.16	78	T
Tungsten	–	–	−2.0	–	–	–
Zinc	+2.32	+0.0238	+2.79	+3.51	−98	T
" pure pressed	–	–	+3.7	–	–	M

B Ed. Becquerel, "Ann. de Chim. et de Phys." [4] vol. 8. S. Bureau of Standards.
M Matthiesen, "Pogg. Ann." vol. 103, reduced by Fleming Jenkin.
T Tait, "Trans. R. S. E." vol. 27, reduced by Mascart.
H Haken, Ann. der Phys. 32, p. 291, 1910. (Electrical conductivity of Teβ = 0.04, Teα 1.7 e. m. units.) Swisher, 1917.

TABLE 387.—Thermoelectric Power of Alloys.

The thermoelectric powers of a number of alloys are given in this table, the authority being Ed. Becquerel. They are relative to lead, and for a mean temperature of 50° C. In reducing the results from copper as a reference metal, the thermoelectric power of lead to copper was taken as — 1.9.

Substance.	Relative quantity.	Thermoelectric power in microvolts.
Antimony	806	227
Cadmium	696	
Antimony	4	146
Cadmium	2	
Zinc	1	
Antimony	806	137
Cadmium	696	
Bismuth	121	
Antimony	806	95
Zinc	406	
Antimony	806	8.1
Zinc	406	
Bismuth	121	
Antimony	4	76
Cadmium	2	
Lead	1	
Zinc	1	
Antimony	4	46
Cadmium	2	
Zinc	1	
Tin	1	

Substance.	Relative quantity.	Thermoelectric power in microvolts.
Antimony	2	43
Zinc	1	
Tin	1	
Antimony	12	35
Cadmium	10	
Zinc	3	
Antimony	10	10.2
Tellurium	1	
Antimony	10	8.8
Bismuth	1	
Antimony	4	2.5
Iron	1	
Antimony	8	1.4
Magnesium	1	
Antimony	8	−0.4
Lead	1	
Bismuth	–	−43.8
Bismuth	2	−33.4
Antimony	1	

Substance.	Relative quantity.	Thermoelectric power in microvolts.
Bismuth	4	−51.4
Antimony	1	
Bismuth	8	−63.2
Antimony	1	
Bismuth	10	−68.2
Antimony	1	
Bismuth	12	−66.9
Antimony	1	
Bismuth	2	60
Tin	1	
Bismuth	10	−24.5
Selenium	1	
Bismuth	12	−31.1
Zinc	1	
Bismuth	12	−46.0
Arsenic	1	
Bismuth	1	68.1
Bismuth sulphide	1	

TABLE 388. — Thermoelectric Power against Platinum.

One junction is supposed to be at 0° C; + indicates that the current flows from the 0° junction into the platinum. The rhodium and iridium were rolled, the other metals drawn.*

Temperature, °C.	Au.	Ag.	90%Pt+10%Pd.	10%Pt+90%Pd.	Pd.	90%Pt+10%Rh.	90%Pt+10%Ru.	Ir.	Rh.
−185	−0.15	−0.16	−0.11	+0.24	+0.77	-	−0.53	−0.28	−0.24
−80	−0.31	−0.30	−0.09	+0.15	+0.39	-	−0.39	−0.32	−0.31
+100	+0.74	+0.72	+0.26	−0.19	−0.56	-	+0.73	+0.65	+0.65
+200	+1.8	+1.7	+0.62	−0.31	−1.20	-	+1.6	+1.5	+1.5
+300	+3.0	+3.0	+1.0	−0.37	−2.0	+2.3	+2.6	+2.5	+2.6
+400	+4.5	+4.5	+1.5	−0.35	−2.8	+3.2	+3.6	+3.6	+3.7
+500	+6.1	+6.2	+1.9	−0.18	−3.8	+4.1	+4.6	+4.8	+5.1
+600	+7.9	+8.2	+2.4	+0.12	−4.9	+5.1	+5.7	+6.1	+6.5
+700	+9.9	+10.6	+2.9	+0.61	−6.3	+6.2	+6.9	+7.6	+8.1
+800	+12.0	+13.2	+3.4	+1.2	−7.9	+7.2	+8.0	+9.1	+9.9
+900	+14.3	+16.0	+3.8	+2.1	−9.6	+8.3	+9.2	+10.8	+11.7
+1000	+16.8	-	+4.3	+3.1	−11.5	+9.5	+10.4	+12.6	+13.7
+1100	-	-	+4.8	+4.2	−13.5	+10.6	+11.6	+14.5	+15.8
+(1300)	-	-	-	-	-	+13.1	+14.2	+18.6	+20.4
+(1500)	-	-	-	-	-	+15.6	+16.9	+23.1	+25.6

* Holborn and Day.

TABLE 389. — Thermal E. M. F. of Platinum-Rhodium Alloys Against Pure Platinum, in Millivolts.*

t	1 p. ct.	5 p. ct.	10 p. ct.			15 p. ct.	20 p. ct.	30 p. ct.†	40 p. ct.†	100 p. ct.‡
			Low.	High.	Standard.					
100°	0.21	0.55	0.63	0.64	0.64	0.65	....	....	....	0.65
200	0.42	1.18	1.41	1.43	1.43	1.50	....	....	....	1.51
300	0.63	1.85	2.28	2.32	2.32	2.41	....	2.34	2.45	2.57
400	0.84	2.53	3.21	3.26	3.25	3.45	3.50	3.50	3.64	3.76
500	1.05	3.22	4.17	4.23	4.23	4.55	4.60	4.74	4.93	5.08
600	1.25	3.92	5.16	5.24	5.23	5.71	5.83	6.06	6.31	6.55
700	1.45	4.62	6.19	6.28	6.27	6.94	7.18	7.49	7.80	8.14
800	1.65	5.33	7.25	7.35	7.33	8.23	8.60	9.01	9.37	9.87
900	1.85	6.05	8.35	8.46	8.43	9.57	10.09	10.67	11.09	11.74
1000	2.05	6.79	9.47	9.60	9.57	10.96	11.65	12.42	12.94	13.74
1100	2.25	7.53	10.64	10.77	10.74	12.40	13.29	14.33	14.99	15.87
1200	2.45	8.29	11.82	11.97	11.93	13.87	14.96	16.39	17.13	18.10
1300	2.65	9.06	13.02	13.18	13.13	15.38	16.65	18.51	19.51	20.46
1400	2.86	9.82	14.22	14.39	14.34	16.98	18.39	20.67	21.73	....
1500	3.06	10.56	15.43	15.61	15.55	18.41	20.15	....	....	....
1600	3.26	11.31	16.63	16.82	16.75	19.94	21.90	....	....	....
1700	3.46	12.05	17.83	18.03	17.95	21.47	23.65	....	....	....
1755	3.56	12.44	18.49	18.70	18.61	22.31	24.55	....	....	....

* Carnegie Institution, Pub. 157, 1911.
† Holborn and Wien, 1892.
‡ Holborn and Day, mean value, 1899.

THERMOELECTRIC PROPERTIES: PRESSURE EFFECTS.

TABLE 390. — Thermoelectric Power; Pressure Effects.

The following values of the thermoelectric powers under various pressures are taken from Bridgman, Pr. Am. Acad. Arts and Sc. 53, p. 269, 1918. A positive emf means that the current at the hot junction flows from the uncompressed to the compressed metal. The cold junction is always at 0° C. The last two columns give the constants in the equation E = thermoelectric force against lead (0° to 100° C) = $(At + Bt^2) \times 10^{-6}$ volts, at atmospheric pressure, a positive emf meaning that the current flows from lead to the metal under consideration at the hot junction.

Metal.	Thermo-electric force, volts $\times 10^9$									Formula coefficients.	
	Pressure, kg/cm²										
	2000		4000		8000		12,000				
	Temperature, ° C										
	50°	100°	50°	100°	50°	100°	20°	50°	100°	A	B
Bi †	53,000	85,000	110,000	185,000	255,000	425,000	185,000	452,000	710,000	−74.42	+.0160
Zn †	6,200	14,100	13,000	28,500	26,100	58,100	14,400	38,500	87,400	+3.047	−.00495
Tl ‡	4,930	10,870	9,380	20,290	17,170	37,630	8,780	23,750	52,460	+1.659	−.00134 [1]
Cd †	2,040	7,120	4,620	14,380	10,960	28,740	6,680	19,180	45,560	+12.002	+.1619
Constantan ‡	2,850	5,950	5,800	11,810	11,530	23,790	6,750	17,200	35,470	−34.76	−.0397
Pd *	2,190	4,380	4,400	8,800	8,630	17,690	5,090	12,970	26,520	−5.496	−.01760
Pt *	1,810	3,600	3,600	7,310	7,370	14,350	3,880	11,030	21,570	− 3.092	−.01334
W ‡	1,190	2,530	2,360	4,990	4,690	10,120	2,700	7,050	15,140	+1.594	+.01705
Ni *	700	1,680	1,500	3,400	3,230	7,190	1,880	5,140	11,440	−17.61	−.0178
Ag *	840	1,870	1,720	3,720	3,350	7,190	+1,900	4,950	10,560	+2.556	+.00432
§ Fe †	390	1,670	590	3,250	5,300	5,820	−990	220	7,680	+16.18	−.0089 [2]
Pb ‡	460	1,050	920	2,120	1,860	4,210	+880	281	6,330	—	—
Au *	456	1,052	905	2,051	1,791	3,974	+990	2,627	5,760	+2.899	+.00467 [3]
Cu †	+292	584	+580	1,216	1,124	2,420	+596	1,616	3,546	+2.777	+.00483
§ Al †	−70	101	−91	294	32	929	−68	312	1,962	−0.416	+.00008 [4]
§ Mo ‡	+93	140	+187	278	375	555	+146	562	833	+5.892	+.02167 [5]
§ Sn †	+38	+87	+58	+165	+70	+292	−182	+10	+390	+0.230	−.00067
Manganin †	−123	−232	−242	−452	−489	−894	−308	−719	−1,314	+1.366	+.000414 [6]
Mg †	−84	−167	−181	−362	−395	−791	−259	−648	−1,296	−0.095	+.00004
Co †	−156	−348	−316	−692	−630	−1,360	−352	−937	−2,061	−17.32	−.0390

* Identical wire of Table 398. † Another wire of same sample. ‡ Different sample.
§ Results too irregular for interpolation for values at other temperature and pressures; see original article.
(1) $-.0_556t^3$; (2) $-.0_486t^3$, annealed ingot iron; (3) $-.0_5166t^3$; (4) $-.0_41t^3$; (5) $-.0_425t^3$; (6) $-.0_4112t^3$.

TABLE 391. — Peltier and Thomson Heats; Pressure Effects.

The following data indicate the magnitude of the effect of pressure on the Peltier and Thomson heats. They refer to the same samples as for the last table. The Peltier heat is considered positive if heat is absorbed by the positive current from the surroundings on flowing from uncompressed to compressed metal. A positive d^2E/dt^2 means a larger Thomson heat in the compressed metal, and the Thomson heat is itself considered positive if heat is absorbed by the positive current in flowing from cold to hot metal. Same reference and notes as for preceding table.

Metal.	Peltier heat, $10^6 \times$ Joules/coulomb.						Thomson heat, $10^8 \times$ Joules/coulomb/° C					
	Pressure kg/cm²						Pressure kg/cm²					
	6000			12,000			6000			12,000		
	Temperature ° C						Temperature ° C					
	0°	50°	100°	0°	50°	100°	0°	50°	100°	0°	50°	100°
§ Bi †	+1070	+1210	—	+2580	+2810	—	+1150	+650	—	−520	−405	—
§ Zn †	+98	+140	+190	+190	+278	+412	+41	+48	+56	+63	+133	+220
§ Tl ‡	+66	+95	+124	+112	+171	+229	+38	+28	+26	+79	+63	+50
§ Cd †	+19	+71	+118	+81	+148	+221	+109	+74	+63	+105	+92	+93
Constantan ‡	+46	+57	+70	+90	+114	+140	+5	+6	+6	+13	+14	+17
§ Pd *	+35	+43	+52	+68	+86	+103	+3	+4	+4	+9	+9	+8
§ Pt *	+23	+37	+35	+45	+76	+65	+49	−6	−18	+96	+17	+59
§ W ‡	+17	+25	+32	+36	+49	+65	+8	+7	+6	+9	+14	+20
§ Ni *	+11	+17	+23	+24	+37	+50	+9	+7	+8	+16	+15	+10
Ag *	+13	+17	+23	+25	+34	+44	+4	+5	+6	+7	+8	+10
§ Fe †	−11	+18	+15	−38	+38	+36	+79	+58	−121	−347	+120	−194
Pb ‡	+7	+10	+16	+14	+20	+30	+2	+6	+10	+6	+8	+20
Au *	+6	+10	+14	+13	+18	+25	+4	+4	+5	+6	+6	+7
Cu †	+4	+6	+8	+8	+11	+16	+4	+1	+4	+6	+3	+8
§ Al †	−2	+2	+8	−3	+7	+17	+6	+9	+11	+21	+16	+20
§ Mo ‡	+1	+2	+0	+2	+4	+1	+1	−5	−1	+2	−11	−2
§ Sn †	−1	+1	+1	−5	+2	+2	+6	+0	−1	+29	+2	−5
§ Manganin †	−2	−2	−2	−4	−4	−4	+1	+1	+0	+2	+1	+1
Mg †	−16	−18	−21	−35	−42	−48	0	0	0	0	0	0
§ Co †	−23	−33	−44	−46	−67	−90	−14	−11	−10	−20	−24	−28

* † ‡ § Same significance as in preceding table.

TABLE 392. — Peltier Effect.

The coefficient of Peltier effect may be calculated from the constants A and B of Table 386, as there shown. With Q (see Table 386) in microvolts per ° C. and $T=$ absolute temperature (K), the coefficient of Peltier effect$=\frac{QT}{42}$ cal. per coulomb$=0.00086\ QT$ cal. per ampere-hour$=QT/1000$ millivolts (=millijoules per coulomb). Experimental results, expressed in slightly different units, are here given. The figures are for the heat production at a junction of copper and the metal named, in calories per ampere-hour. The current flowing from copper to the metal named, a positive sign indicates a warming of the junction. The temperature not being stated by either author, and Le Roux not giving the algebraic signs, these results are not of great value.

	Calories per ampere-hour.										
	Sb. ‡	Sb. commercial.	Bi. pure.	Bi. §	Cd.	German Silver.	Fe.	Ni.	Pt.	Ag.	Zn.
Jahn* . .	–	–	–	–	—.62	–	—3.61	4.36	0.32	—.41	—.58
Le Roux† .	13.02	4.8	19.1	25.8	0.46	2.47	2.5	–	–	–	.39

* "Wied. Ann." vol. 34, p. 767.
† "Ann. de Chim. et de Phys." (4) vol. 10, p. 201.
‡ Becquerel's antimony is 806 parts Sb + 406 parts Zn + 121 parts Bi.
§ Becquerel's bismuth is 10 parts Bi + 1 part Sb.

TABLE 393. — Peltier Effect, Fe-Constantan, Ni-Cu, 0 — 560° C.

Temperature.	0°	20°	130°	240°	320°	560°	
Fe-Constantan . . .	3.1	3.6	4.5	6.2	8.2	12.5	in Gram. Cal. $\times 10^{-8}$
Ni-Cu	1.92	2.15	2.45	2.06	1.91	2.38	per coulomb.

TABLE 394. — Peltier Electromotive Force in Millivolts.

Metal against Copper.	Sb.	Fe.	Cd.	Zn.	Ag.	An.	Pb.	Sn.	Al.	Pt.	Pd.	Ni.	Bi.
Le Roux .	—5.64	—2.93	—.53	—.45	–	–	–	–	–	–	–	–	+22.3
Jahn . . .	–	—3.68	—.72	—.68	—.48	–	–	–	–	+.37	–	+5.07	–
Edlund . .	–	—2.96	—.16	—.01	+.03	+.33	+.50	+.56	+.70	+1.02	+2.17	–	+17.7
Caswell . .	–	–	–	–	+.03	–	–	–	+.70	+.85	–	+6.0	+16.1

Le Roux, 1867; Jahn, 1888; Edlund, 1870-71; Caswell, Phys. Rev. 33, p. 381, 1911.

TABLE 395.

THE TRIBO-ELECTRIC SERIES.

In the following table it is so arranged that any material in the list becomes positively electrified when rubbed by one lower in the list. The phenomenon depends upon surface conditions and circumstances may alter the relative positions in the list.

1 Asbestos (sheet).
2 Rabbit's fur, hair, (Hg).
3 Glass (combn. tubing).
4 Vitreous silica, opossum's fur.
5 Glass (fusn.).
6 Mica.
7 Wool.
8 Glass (pol.), quartz (pol.), glazed porcelain.
9 Glass (broken edge), ivory.
10 Calcite.
11 Cat's fur.
12 Ca, Mg, Pb, fluor spar, borax.
13 Silk.
14 Al, Mn, Zn, Cd, Cr, felt, hand, wash-leather.
15 Filter paper.
16 Vulcanized fiber.
17 Cotton.
18 Magnalium.
19 K-alum, rock-salt, satin spar.
20 Woods, Fe.
21 Unglazed porcelain, sal-ammoniac.
22 K-bichromate, paraffin, tinned-Fe.
23 Cork, ebony.
24 Amber.
25 Slate, chrome-alum.
26 Shellac, resin, sealing-wax.
27 Ebonite.
28 Co, Ni, Sn, Cu, As, Bi, Sb, Ag, Pd, C, Te, Eureka, straw, copper sulphate, brass.
29 Para rubber, iron alum.
30 Guttapercha.
31 Sulphur.
32 Pt, Ag, Au.
33 Celluloid.
34 Indiarubber.

Shaw, Pr. Roy. Soc. 94, p. 16, 1917; the original article shows the alterations in the series sequence due to varied conditions.

TABLE 396.

AUXILIARY TABLE FOR COMPUTING WIRE RESISTANCES.

For computing resistance in ohms per meter from resistivity, ρ, in michroms per cm. cube (see Table 397, etc.). *e. g.* to compute for No. 23 copper wire when $\rho = 1.724$: 1 meter = 0.0387 + .0271 + .0008 + .0002 = 0.0668 ohms; for No. 11 lead wire when $\rho = 20.4$; 1 meter = 0.0479 + .0010 = 0.0489 ohms. The following relation allows computation for wires of other gage numbers: resistance in ohms per meter of No. N = $2(n-3)$ within 1 %: *e. g.* resistance of meter of No. 18 = 2 × No. 15.

Gage. No.	Diam. in mm.	Section mm².	ρ in micro-ohms per cm. cube.									
			1.	2.	3.	4.	5.	6.	7.	8.	9.	10.
			Resistance of wire 1 meter long in ohms.									
0000	11.7	107.2	$.0_4933$	$.0_3187$	$.0_3280$	$.0_3373$	$.0_3466$	$.0_3560$	$.0_3653$	$.0_3746$	$.0_3840$	$.0_3933$
00	9.27	67.43	$.0_3148$	$.0_3297$	$.0_3445$	$.0_3593$	$.0_3742$	$.0_3890$	$.0_2104$	$.0_2119$	$.0_2133$	$.0_2148$
1	7.35	42.41	$.0_3236$	$.0_3472$	$.0_3707$	$.0_3943$	$.0_2118$	$.0_2141$	$.0_2165$	$.0_2189$	$.0_2212$	$.0_2236$
3	5.83	26.67	$.0_3375$	$.0_3750$	$.0_2112$	$.0_2150$	$.0_2187$	$.0_2225$	$.0_2262$	$.0_2300$	$.0_2337$	$.0_2375$
5	4.62	16.77	$.0_3596$	$.0_2119$	$.0_2179$	$.0_2239$	$.0_2298$	$.0_2358$	$.0_2417$	$.0_2477$	$.0_2537$	$.0_2596$
7	3.66	10.55	$.0_3948$	$.0_2190$	$.0_2284$	$.0_2379$	$.0_2474$	$.0_2569$	$.0_2664$	$.0_2758$	$.0_2853$	$.0_2948$
9	2.91	6.634	$.0_2151$	$.0_2301$	$.0_2452$	$.0_2603$	$.0_2754$	$.0_2904$	.0106	.0121	.0136	.0151
11	2.30	4.172	$.0_2240$	$.0_2479$	$.0_2719$	$.0_2959$	.0120	.0144	.0168	.0192	.0216	.0240
13	1.83	2.624	$.0_2381$	$.0_2762$	.0114	.0152	.0191	.0229	.0267	.0305	.0343	.0381
15	1.45	1.650	$.0_2606$	.0121	.0182	.0242	.0303	.0364	.0424	.0485	.0545	.0606
17	1.15	1.038	$.0_2963$	.0193	.0289	.0385	.0482	.0578	.0674	.0771	.0867	.0963
19	.912	.6527	.0153	.0306	.0460	.0613	.0766	.0919	.1072	.1226	.1379	.1532
21	.723	.4105	.0244	.0487	.0731	.0974	.1218	.1462	.1705	.1949	.2192	.2436
23	.573	.2582	.0387	.0775	.1162	.1549	.1936	.2324	.2711	.3098	.3486	.3873
25	.455	.1624	.0616	.1232	.1847	.2463	.3079	.3695	.4310	.4926	.5542	.6158
27	.361	.1021	.0979	.1959	.2938	.3918	.4897	.5877	.6856	.7835	.8815	.9794
29	.286	.0642	.1557	.3114	.4671	.6228	.7786	.9343	1.090	1.246	1.401	1.557
31	.227	.0404	.2476	.4952	.7428	.9904	1.238	1.486	1.733	1.981	2.228	2.476
33	.180	.0254	.3937	.7874	1.181	1.575	1.968	2.362	2.756	3.150	3.543	3.937
35	.143	.0160	.6262	1.252	1.879	2.505	3.131	3.757	4.383	5.009	5.636	6.262
37	.113	.0100	.9950	1.990	2.985	3.980	4.975	5.970	6.965	7.960	8.955	9.950
39	.090	.0063	1.583	3.166	4.748	6.331	7.914	9.497	11.08	12.66	14.25	15.83
40	.080	.0050	1.996	3.992	5.988	7.984	9.980	11.98	13.97	15.97	17.96	19.96

TABLE 297.

RESISTIVITY OF METALS AND SOME ALLOYS.

The resistivities are the values of ρ in the equation $R = \rho l/s$, where R is the resistance in microhms of a length l cm of uniform cross section s cm². The temperature coefficient is a_s in the formula $R_t = R_s\{1 + a_s(t - t_s)\}$. The information of column 2 does not necessarily apply to the temperature coefficient. See also next table for temperature coefficients 0° to 100° C.

Substance.	Remarks.	Temperature, ° C	Microhm-cm	Reference.	Temperature coefficient. t_s	a_s	Reference.
Advance	see constantan	—	—	—	—	—	—
Aluminum	see p. 334	20.	2.828	1	18°	+.0039	2
"	c. p.	−189.	0.64	3	25	+.0034	4
"	"	−100.	1.53	3	100	+.0040	4
"	"	0.	2.62	3	500	+.0050	4
"	"	+100.	3.86	3	—	—	—
"	"	400.	8.6	3	—	—	—
Antimony	—	20.	41.7	5	20	+.0036	5
"	—	−190.	10.5	6	—	—	—
"	liquid	+860.	120.	7	—	—	—
Arsenic	—	0.	35.	8	—	—	—
Bismuth	—	18.	119.0	9	20	+.004	5
"	—	100.	160.2	9	—	—	—
Brass	—	20.	7.	5	20	+.002	5
Cadmium	drawn	−160.	2.72	10	20	+.0038	5
"	"	18.	7.54	9	—	—	—
"	"	100.	9.82	9	—	—	—
"	liquid	318.	34.1	11	—	—	—
Caesium	—	−187.	5.25	12	—	—	—
"	—	0.	19.	11	—	—	—
"	solid	27.	22.2	13	—	—	—
"	liquid	30.	36.6	13	—	—	—
Calcium	99.57 pure	20.	4.6	14	—	+.0036	14
Calido	see constantan	—	—	—	—	—	—
Chromium	—	0.	2.6	15	—	—	—
Climax	—	20.	87.	5	20	+.0007	5
Cobalt	99.8 pure	20.	9.7	16	—	—	—
Constantan	60% Cu, 40% Ni	20.	49.	5	12	+.000008	4
"	—	—	—	—	25	+.000002	4
"	—	—	—	—	100	−.000033	4
"	—	—	—	—	200	−.000020	4
"	—	—	—	—	500	+.000027	4
Copper	annealed	20.	1.724	1	20 see col. 2	+.00393	5
"	hard-drawn	20.	1.77	1	" " " "	+.00382	5
"	electrolytic	−206.	0.144	17	100	+.0038	4
"	"	+205.	2.92	17	400	+.0042	4
"	pure	400.	4.10	3	1000	+.0062	4
"	very pure, ann'ld	20.	1.692	18	—	—	—
Eureka	see constantan	—	—	—	—	—	—
Excello	—	20.	92.	5	20	+.00016	5
Gallium	—	0.	53.	12	—	—	—
German silver	18% Ni	20.	33.	5	20	+.0004	5
Gold	99.9 pure	−183.	0.68	17	20	+.0034	5
"	—	0.	2.22	11	100 ann'ld	+.0025	4
"	pure, drawn	20.	2.44	9	500 "	+.0035	4
"	99.9 pure	194.5	3.77	17	1000 "	+.0049	4
Ia Ia	see constantan	—	—	—	—	—	—
Ideal	" "	—	—	—	—	—	—
Indium	—	0.	8.37	19	—	—	—
Iridium	—	−186.	1.92	20	—	—	—
"	—	0.	6.10	20	—	—	—
"	—	+100.	8.3	20	—	—	—

Arranged in order of increasing resistivity (ohm-cm³ × 10⁻⁶, 20° C).

Ag	1.468	Mn	5.±	Pd	10.21	Ga	53.
Cu	1.59	Mo	(5.3)	Pt	10.96	Os	56.
Au	2.22	Zn	5.75	Rb	13.0	Hg	94.07
Al	2.6	Ir	6.10	Sn	13.0	Bi	110.
Cr	2.6	K	6.1	Ta	14.6	Graphite	8×10^2
Ti	3.2	Ni	6.93	Tl	17.6	Carbon	3×10^3
Na	4.3	Cd	7.04	Cs	19.	Te	2×10^5
Ca	4.3	In	8.37	Pb	20.4	P	10^{12}
Mg	4.35	Li	8.55	Sr	(23.5)	B	8×10^{12}
Rh	4.69	Fe	8.8	As	35.	Se	10^{13}
W	5.0	Co	9.0	Sb	39.	S	10^{17}

TABLE 397 (*continued*).

RESISTIVITY OF METALS AND SOME ALLOYS.

Substance.	Remarks.	Temperature, °C	Microhm-cm	Reference.	Temperature coefficient. t_s	a_s	Reference.
Iron	99.98% pure	20.	10.	5	20	+.0050	5
"	pure, soft	−205.3	0.652	17	0	+.0062	21
"	" "	−78.	5.32	17	25	+.0052	4
"	" "	0.	8.85	17	100	+.0068	4
"	" "	+98.5	17.8	17	500	+.0147	4
"	" "	196.1	21.5	17	1000	+.0050	4
"	" "	400.	43.3	3	—	—	—
steel	E. B. B.	20.	10.4	5	20 see col. 2	+.005	5
"	B. B.	20.	11.9	5	" " " "	+.004	5
"	Siemens-Martin	20.	18.	5	" " " "	+.003	5
"	manganese	20.	70.	5	" " " "	+.001	5
"	35% Ni, "invar."	20.	81.	22	—	—	—
"	piano wire	0.	11.8	23	0 see col. 2	+.0032	23
"	temp. glass, hard	0.	45.7	23	" " " "	+.0016	23
"	" , yellow	0.	27.	23	—	—	—
"	" , blue	0.	20.5	23	0 see col. 2	+.0033	23
"	" , soft	0.	15.9	23	—	—	—
Lead	—	20.	22.	5	20	+.0039	5
"	cold pressed	−183.	6.02	17	18	+.0043	2
"	" "	−78.	14.1	17	—	—	—
"	" "	0.	20.4	17	—	—	—
"	" "	+90.4	28.0	17	—	—	—
"	" "	196.1	36.9	17	—	—	—
"	—	318.	94.	24	—	—	—
Lithium	solid	−187.	1.34	12	—	—	—
"	"	0.	8.55	12	—	—	—
"	"	99.3	12.7	12	—	—	—
"	liquid	230.	45.2	25	—	—	—
Magnesium	—	20.	4.6	5	20	+.004	5
"	free from Zn	−183.	1.00	17	0	+.0038	24
"	" " "	−78.	2.97	17	25	+.0050	4
"	" " "	0.	4.35	17	100	+.0045	4
"	" " "	+98.5	5.99	17	500	+.0036	4
"	pure	400.	11.9	3	600	+.0100	4
Manganese	—	—	5.0±	15	-	—	—
Manganin	84 Cu, 12 Mn, 4 Ni	20.	44.	5	12	+.000006	4
"	—	—	—	—	25	.000000	4
"	—	—	—	—	100	−.000042	4
"	—	—	—	—	250	−.000052	4
"	—	—	—	—	475	−.000000	4
"	—	—	—	—	500	−.00011	4
Mercury	—	20.	95.783	5	20	+.00089	5
"	solid	−183.5	6.97	17	0	+.00088	26
"	"	−102.9	15.04	17	—	—	—
"	"	−50.3	21.3	17	$R_t = R_0(1 +$	—	—
"	"	−39.2	25.5	17	$.00089t +$	—	—
"	liquid	−36.1	80.6	17	$.000001t^2)$	—	—
"	"	0.0	94.07	17	—	—	—
"	"	50.	98.50	27	—	—	—
"	"	100.	103.25	24	—	—	—
"	"	200.	114.27	24	—	—	—
"	"	350.	135.5	24	—	—	—
Molybdenum	drawn	20.	5.7	5	25	+.0033	4
"	—	—	—	—	100	+.0034	4
"	—	—	—	—	1000	+.0048	4
Monel metal	—	20.	42.	5	20	+.0020	5
Nichrome	—	20.	100.	5	20	+.0004	5
Nickel	—	20.	7.8	5	20	+.006	5
"	pure	−182.5	1.44	28	0	+.0062	24
"	"	−78.2	4.31	28	25	+.0043	4
"	"	0.	6.93	28	100	+.0043	4
"	"	94.9	11.1	28	500	+.0030	4
"	—	400.	60.2	3	1000	+.0037	4

RESISTIVITY OF METALS AND SOME ALLOYS.

Substance	Remarks.	Temperature, °C	Microhm cm	Reference.	Temperature coefficient.		
					t_s	a_s	Reference.
Osmium	—	20.	60.2	3	—	—	—
Palladium	—	20.	11.	5	20	+.0033	5
"	very pure	−183.	2.78	17	0	+.0035	21
"	" "	−78.	7.17	17	—	—	—
"	" "	0.	10.21	17	—	—	—
"	" "	98.5	13.79	17	—	—	—
Platinum	—	20.	10.	5	20	+.003	5
"	wire	−203.1	2.44	17	0	+.0037	21
"	"	−97.5	6.87	17	—	—	—
"	"	0.	10.96	17	—	—	—
"	"	100.	14.85	17	—	—	—
"	—	400.	26.	3	—	—	—
Potassium	—	−75.	4.0	13	—	—	—
"	—	0.	6.1	13	—	—	—
"	—	55.	8.4	13	—	—	—
Rhodium	—	−186.	0.70	20	—	—	—
"	—	−78.3	3.09	20	—	—	—
"	—	0.	4.69	20	—	—	—
"	—	100.	6.60	20	—	—	—
Rubidium	solid	−190.	2.5	13	—	—	—
"	"	0.	11.6	13	—	—	—
"	"	35.	13.4	13	—	—	—
"	liquid	40.	19.6	13	—	—	—
Silicium	—	20.	58. ±	—	—	—	—
Silver	99.98 pure	18.	1.629	2	20	+.0038	5
"	electrolytic	−183.	0.390	17	25	+.0030	4
"	"	−78.	1.021	17	100	+.0036	4
"	"	0.	1.468	17	500	+.0044	4
"	"	98.15	2.062	17	—	—	—
"	"	192.1	2.608	17	—	—	—
"	"	400.	3.77	3	—	—	—
Sodium	solid	−180.	1.0	13	—	—	—
"	"	−75.	2.8	13	—	—	—
"	"	0.	4.3	13	—	—	—
"	"	55.	5.4	13	—	—	—
"	liquid	116.	10.2	13	—	—	—
Strontium	—	20.	24.8	8	—	—	—
Tantalum	—	20.	15.5	5	20	+.0031	5
Tellurium*	—	19.6	200,000	8	—	—	—
Thallium	pure	−183.	4.08	17	—	—	—
"	"	−78.	11.8	17	—	—	—
"	"	0.	17.60	17	—	—	—
"	"	98.5	24.7	17	—	—	—
Therlo	—	20.	47.	5	20	+.00001	5
Tin	—	20.	11.5	5	20	+.0042	5
"	—	−184.	3.40	17	—	—	—
"	—	−78.	8.8	17	—	—	—
"	—	0.	13.0	17	—	—	—
"	—	91.45	18.2	17	—	—	—
Titanium	—	—	3.2	15	—	—	—
Tungsten	—	20.	5.51	29	18	+.0045	2
"	1000° K	727.	25.3	29	500	+.0057	4
"	1500° K	1227.	41.4	29	1000	+.0089	4
"	2000° K	1727.	59.4	29	—	—	—
"	3000° K	2727.	98.9	29	—	—	—
"	3500° K	3227.	118.	29	—	—	—
Zinc	trace Fe	−183.	1.62	17	20	+.0037	5
"	" "	−78.	3.34	17	—	—	—
"	" "	0.	5.75	17	—	—	—
"	" "	92.45	8.00	17	—	—	—
"	" "	191.5	10.37	17	—	—	—
"	liquid	440.	37.2	7	—	—	—

References to Table 397: (1) See page 334; (2) Jäger, Diesselhorst, Wiss. Abh. D. Phys. Tech. Reich. 3, p. 269, 1900; (3) Nicolai, 1907; (4) Somerville, Phys. Rev. 31, p. 261, 1910; 33, p. 77, 1911; (5) Circular 74 of Bureau of Standards, 1918; (6) Eucken, Gelhoff; (7) de la Rive; (8) Matthiessen; (9) Jäger, Diesselhorst; (10) Lees, 1908; (11) Mean; (12) Guntz, Broniewski; (13) Hackspill; (14) Swisher, 1917; (15) Shukow; (16) Reichardt, 1901; (17) Dewar, Fleming, Dickson, 1898; (18) Wolff, Dellinger, 1910; (19) Erhardt, 1881; (20) Broniewski, Hackspill, 1911; (21) Dewar, Fleming, 1893, 1896; (22) Circular 58, Bureau of Standards, 1916; (23) Strouhal, Barus, 1883; (24) Vincentini, Omodei, 1890; (25) Bernini, 1905; (26) Glazebrook, Phil. Mag. 20, p. 343, 1885; (27) Grimaldi, 1888; (28) Fleming, 1900; (29) Langmuir, Gen. Elec. Rev. 19, 1916.

* See note to Table 386.

TABLE 398.—Resistance of Metals under Pressure (Bridgman).

The average temperature coefficients are per ° C between 0° and 100° C. The instantaneous pressure coefficients are the values of the derivative $(1/r)\{dr/dp\}_t$, where r is the observed resistance at the pressure p and temperature t. The average coefficient is the total change of resistance between 0 and 12,000 kg/cm² divided by 12,000 and the resistance at atmospheric pressure and the temperature in question. Table taken from Proc. Nat. Acad. 3, p. 11, 1917. For coefficients at intermediate temperatures and pressures, see more detailed account in Proc. Amer. Acad. 52, p. 573, 1917. Sn, Cd, Zn, Kahlbaum's "K" grade; Tl, Bi, electrolytic, high purity; Pb, Ag, Au, Cu, Fe, Pt, of exceptional purity. Al better than ordinary, others only of high grade commercial purity.

	Average temperature coefficient 0° to 100° C		Pressure coefficients. Instantaneous coefficient. At 0° C		Instantaneous coefficient. At 100° C		Average coefficient 0 to 12,000 kg/cm²	
	At 0 kg	At 12,000 kg	0 kg	12,000 kg	0 kg	12,000 kg	At 0°	At 100°
In	+.00406	+.00383	$-.0_41226$	$-.0_41016$	$-.0_41510$‡	$-.0_41072$‡	$-.0_41021$	$-.0_41131$ ‡
Sn	.00447	.00441	$.0_41044$	$.0_40936$	$.0_41062$	$.0_40973$	$.0_40920$	$.0_40951$
Tl	.00517	.00499	$.0_41319$	$.0_41180$	$.0_41456$	$.0_41200$	$.0_41151$	$.0_41226$
Cd	.00424	.00418	$.0_41063$	$.0_40837$	$.0_41106$	$.0_40887$	$.0_40894$	$.0_40927$
Pb	.00421	.00412	$.0_41442$	$.0_41220$	$.0_41483$	$.0_41237$	$.0_41212$	$.0_41253$
Zn	.00416	.00420	$.0_40540$	$.0_40425$	$.0_40524$	$.0_40407$	$.0_40470$	$.0_40454$
Al	.00434	.00435	$.0_40416$	$.0_40365$	$.0_40397$	$.0_40373$	$.0_40382$	$.0_40377$
Ag	.004074	.004069	$.0_40358$	$.0_40321$	$.0_40355$	$.0_40331$	$.0_40333$	$.0_40336$
Au	.003968	.003964	$.0_40312$	$.0_40286$	$.0_40304$	$.0_40292$	$.0_40287$	$.0_40292$
Cu	.004293	.004303	$.0_40201$	$.0_40179$	$.0_40184$	$.0_40175$	$.0_40183$	$.0_40177$
Ni	.004873	.004855	$.0_40158$	$.0_40142$	$.0_40163$	$.0_40156$	$.0_40147$	$.0_40158$
Co	.003657	.003676	$.0_40094$	$.0_40081$	$.0_40076$	$.0_40070$	$.0_40087$	$.0_40073$
Fe	.006206	.006184	$.0_40241$	$.0_40218$	$.0_40247$	$.0_40230$	$.0_40226$	$.0_40235$
Pd	.003178	.003185	$.0_40198$	$.0_40190$	$.0_40189$	$.0_40187$	$.0_40190$	$.0_40186$
Pt	.003868	.003873	$.0_40198$	$.0_40181$	$.0_40190$	$.0_40182$	$.0_40187$	$.0_40184$
Mo	.004336	.004340	$.0_40133$	$.0_40126$	$.0_40130$	$.0_40125$	$.0_40129$	$.0_40126$
Ta	.002973	.002967	$.0_40149$	$.0_40139$	$.0_40153$	$.0_40147$	$.0_40143$	$.0_40149$
W	.003219	.003216	$.0_40128$	$.0_40121$	$.0_40130$	$.0_40123$	$.0_40123$	$.0_40126$
Mg	.00390 *	—	$.0_4055$	—	—	—	$.0_4055$	—
Sb	.00473	.00403	$+.0_41220$	$+.0_41064$	$+.0_40768$	$+.0_40723$	$+.0_41220$	$+.0_40768$
Bi	+.00438	+.00395	$+.0_4154$	$+.0_40213$	$+.0_4152$ §	$+.0_41895$§	$+.0_42228$	$+.0_41980$ §
Te	−.0063 †	—	$-.0_3129$	—	—	—	—	—

* 0° to 20°. † 0° to 24°. ‡ Extrapolated from 50°. § Extrapolated from 75°.

Additional data from P. Nat. Acad. Sc., 6, 505, 1920. Data are 10,000 × mean pressure coefficient, 0 − 12,000 kg, and 10,000 × instantaneous pressure coefficient at 0 kg. l = liquid; s = solid.

Li, s, 0°	+.0772	+ .068	Ca, 0°	+.106	+.129	Ti, 0°	±.001?	
Li, l, 240°	+.093	+ .093	Sr, 0°	+.680	+.502	Zr, 0°	−.0040	− .004
Na, s, 0°	−.345	− .663	Hg, s, 0°	−.236b		Bi, l, 275°	−.101c	− .123
Na, l, 200°	−.436	− .922	Hg, l, 25°	−.219	−.334	W, 0°	−.0135	− .014
K, s, 25°	−.604	−1.86	Ga, s, 0°	−.0247		La, 0°	−.0331	− .039
K, l, 165°	−.809a	−1.68	Ga, l, 30°	−.0531	−.064	P, black, 0°	−.81	− 2.00

a, 0 − 9,000 kg; b, 7,640 − 12,000 kg; c, 0 − 7,000 kg. The Ga, Na, K, Mg, Hg, Bi, W, P, of exceptional purity.

TABLE 399.—Resistance of Mercury and Manganin under Pressure.

Mercury, pure and free from air and with proper precautions, makes a reliable secondary electric-resistance pressure gauge. For construction and manipulation see "The Measurement of High Hydrostatic Pressure; a Secondary Mercury Resistance Gauge," Bridgman, Pr. Am. Acad. 44, p. 221, 1919.

Pressure, kg/cm²	—	500	1000	1500	2000	2500	3000	4000	5000	6000	6500
$R(p, -75°)$	0.9186	0.9055	0.8930	0.8818	0.8714	0.8582	0.8478	0.8268	0.8076	0.7896	0.7807
$R(p, 25°)$	1.0000	0.9836	0.9682	0.9535	0.9394	0.9258	0.9128	0.8882	0.8652	0.8438	0.8335
*	1.0000	0.9854	0.9716	0.9588	0.9462	0.9342	0.9228	0.9010	0.8806	0.8616	0.8527
$R(p, 125°)$	1.0970	1.0770	1.0580	1.0400	1.0230	1.0070	0.9908	0.9614	0.9342	0.9086	0.8966

* This line gives the Specific Mass Resistance at 25°, the other lines the specific volume resistance.

The use of mercury as above has the advantage of being perfectly reproducible so that at any time a pressure can be measured without recourse to a fundamental standard. However, at 0° C mercury freezes at 7500 kg/cm². Manganin is suitable over a much wider range. Over a temperature range 0 to 50° C the pressure resistance relation is linear within 1/10 per cent of the change of resistance up to 13,000 kg/cm². The coefficient varies slightly with the sample. Bridgman's samples (German) had values of $(\Delta R/pR_0) \times 10^9$ from 2295 to 2325. These are + instead of −, as with most of the above metals. See "The Measurement of Hydrostatic Pressure up to 20,000 Kilograms per Square Centimeter," Bridgman, Pr. Am. Acad. 47, p. 321, 1911.

Table 400.

CONDUCTIVITY AND RESISTIVITY OF MISCELLANEOUS ALLOYS.

TEMPERATURE COEFFICIENTS.

Conductivity in mhos or $\frac{1}{\text{ohms per cm.}^3} = \gamma_t = \gamma_0(1 - at + bt^2)$ and resistivity in microhms–cm $= \rho_t = \rho_0(1 + at - bt^2)$.

Metals and alloys.	Composition by weight.	$\frac{\gamma_0}{10^4}$	$a \times 10^6$	ρ_0	Authority.
Gold-copper-silver	58.3 Au + 26.5 Cu + 15.2 Ag	7.58	574*	13.2	1
" " "	66.5 Au + 15.4 Cu + 18.1 Ag	6.83	529†	14.6	1
" " "	7.4 Au + 78.3 Cu + 14.3 Ag	28.06	1830‡	3.6	1
Nickel-copper-zinc	12.84 Ni + 30.59 Cu + 6.57 Zn by volume	4.92	444§	20.3	1
Brass	Various	12.2–15.6	$1-2 \times 10^3$	6.4–8.4	2
" hard drawn	70.2 Cu + 29.8 Zn	12.16	–	8.2	3
" annealed	" "	14.35	–	7.0	3
German silver	Various	3–5	–	20.–33.	2
" "	60.16 Cu + 25.37 Zn + 14.03 Ni + .30 Fe with trace of cobalt and manganese	3.33	360	30.	4
Aluminum bronze	– – –	7.5–8.5	$5-7 \times 10^2$	12–13	2
Phosphor bronze	– – –	10–20	–	5–10	2
Silicium bronze	– – –	41	–	2.4	5
Manganese-copper	30 Mn + 70 Cu	1.00	40	100.	4
Nickel-manganese-copper	3 Ni + 24 Mn + 73 Cu	2.10	—30	48.	4
Nickelin	18.46 Ni + 61.63 Cu + 19.67 Zn + 0.24 Fe + 0.19 Co + 0.18 Mn	3.01	300	33.	4
Patent nickel	25.1 Ni + 74.41 Cu + 0.42 Fe + 0.23 Zn + 0.13 Mn + trace of cobalt	2.92	190	34.	4
Rheotan	53.28 Cu + 25.31 Ni + 16.89 Zn + 4.46 Fe + 0.37 Mn	1.90	410	53.	4
Copper-manganese-iron	91 Cu + 7.1 Mn + 1.9 Fe	4.98	120	20.	5
Copper-manganese-iron	70.6 Cu + 23.2 Mn + 6.2 Fe.	1.30	22	77.	6
Copper-manganese-iron	69.7 Cu + 29.9 Ni + 0.3 Fe	2.60	120	38.	7
Manganin	84 Cu + 12 Mn + 4 Ni.	2.3	6	44.	2
Constantan	60 Cu + 40 Ni	2.04	8	49.	8

[1] Matthiessen. [2] Various. [3] W. Siemens. [4] Feussner and Lindeck. [5] Van der Ven. [6] Blood. [7] Feussner. [8] Jaeger-Diesselhorst.

*, †, ‡, §, $b \times 10^9 = 924, 93, 7280, 51$, respectively.

TABLE 401.

CONDUCTING POWER OF ALLOYS.

This table shows the conducting power of alloys and the variation of the conducting power with temperature.* The values of C_0 were obtained from the original results by assuming silver $= \frac{10^6}{1.585}$ mhos. The conductivity is taken as $C_t = C_0 (1 - at + bt^2)$, and the range of temperature was from 0° to 100° C.

The table is arranged in three groups to show (1) that certain metals when melted together produce a solution which has a conductivity equal to the mean of the conductivities of the components, (2) the behavior of those metals alloyed with others, and (3) the behavior of the other metals alloyed together.

It is pointed out that, with a few exceptions, the percentage variation between 0° and 100° can be calculated from the formula $P = P_c \frac{l}{l'}$, where l is the observed and l' the calculated conducting power of the mixture at 100° C., and P_c is the calculated mean variation of the metals mixed.

Alloys.	Weight %	Volume %	$\frac{C_0}{10^4}$	$a \times 10^6$	$b \times 10^9$	Variation per 100° C.	
	of first named.					Observed.	Calculated.
GROUP 1.							
Sn_6Pb	77.04	83.96	7.57	3890	8670	30.18	29.67
Sn_4Cd	82.41	83.10	9.18	4080	11870	28.89	30.03
SnZn	78.06	77.71	10.56	3880	8720	30.12	30.16
PbSn	64.13	53.41	6.40	3780	8420	29.41	29.10
$ZnCd_2$	24.76	26.06	16.16	3780	8000	29.86	29.67
$SnCd_4$	23.05	23.50	13.67	3850	9410	29.08	30.25
$CdPb_6$	7.37	10.57	5.78	3500	7270	27.74	27.60
GROUP 2.							
Lead-silver ($Pb_{20}Ag$)	95.05	94.64	5.60	3630	7960	28.24	19.96
Lead-silver (PbAg)	48.97	46.90	8.03	1960	3100	16.53	7.73
Lead-silver ($PbAg_2$)	32.44	30.64	13.80	1990	2600	17.36	10.42
Tin-gold ($Sn_{12}Au$)	77.94	90.32	5.20	3080	6640	24.20	14.83
" " (Sn_5Au)	59.54	79.54	3.03	2920	6300	22.90	5.95
Tin-copper	92.24	93.57	7.59	3680	8130	28.71	19.76
" " †	80.58	83.60	8.05	3330	6840	26.24	14.57
" " †	12.49	14.91	5.57	547	294	5.18	3.99
" " †	10.30	12.35	6.41	666	1185	5.48	4.46
" " †	9.67	11.61	7.64	691	304	6.60	5.22
" " †	4.96	6.02	12.44	995	705	9.25	7.83
" " †	1.15	1.41	39.41	2670	5070	21.74	20.53
Tin-silver	91.30	96.52	7.81	3820	8190	30.00	23.31
" "	53.85	75.51	8.65	3770	8550	29.18	11.89
Zinc-copper †	36.70	42.06	13.75	1370	1340	12.40	11.29
" " †	25.00	29.45	13.70	1270	1240	11.49	10.08
" " †	16.53	23.61	13.44	1880	1800	12.80	12.30
" " †	8.89	10.88	29.61	2040	3030	17.41	17.42
" " †	4.06	5.03	38.09	2470	4100	20.61	20.62

NOTE. — Barus, in the "Am. Jour. of Sci." vol. 36, has pointed out that the temperature variation of platinum alloys containing less than 10% of the other metal can be nearly expressed by an equation $y = \frac{n}{x} - m$, where y is the temperature coefficient and x the specific resistance, m and n being constants. If a be the temperature coefficient at 0° C. and s the corresponding specific resistance, $s(a + m) = n$.

For platinum alloys Barus's experiments gave $m = -.000194$ and $n = .0378$.
For steel $m = -.000303$ and $n = .0620$.

Matthiessen's experiments reduced by Barus gave for

Gold alloys $m = -.000045$, $n = .00721$.
Silver " $m = -.000112$, $n = .00538$.
Copper " $m = -.000386$, $n = .00055$.

* From the experiments of Matthiessen and Vogt, "Phil. Trans. R. S." v. 154.
† Hard-drawn.

TABLE 401. — Conducting Power of Alloys.

GROUP 3.

Alloys.	Weight % of first named.	Volume % of first named.	$\frac{C_0}{10^4}$	$a \times 10^6$	$b \times 10^9$	Variation per 100° C. Observed.	Variation per 100° C. Calculated.
Gold-copper †	99.23	98.36	35.42	2650	4650	21.87	23.22
" " †	90.55	81.66	10.16	749	81	7.41	7.53
Gold-silver †	87.95	79.86	13.46	1090	793	10.09	9.65
" " *	87.95	79.86	13.61	1140	1160	10.21	9.59
" " †	64.80	52.08	9.48	673	246	6.49	6.58
" " *	64.80	52.08	9.51	721	495	6.71	6.42
" " †	31.33	19.86	13.69	885	531	8.23	8.62
" " *	31.33	19.86	13.73	908	641	8.44	8.31
Gold-copper †	34.83	19.17	12.94	864	570	8.07	8.18
" " †	1.52	0.71	53.02	3320	7300	25.90	25.86
Platinum-silver †	33.33	19.65	4.22	330	208	3.10	3.21
" " †	9.81	5.05	11.38	774	656	7.08	7.25
" " †	5.00	2.51	19.96	1240	1150	11.29	11.88
Palladium-silver †	25.00	23.28	5.38	324	154	3.40	4.21
Copper-silver †	98.08	98.35	56.49	3450	7990	26.50	27.30
" " †	94.40	95.17	51.93	3250	6940	25.57	25.41
" " †	76.74	77.64	44.06	3030	6070	24.29	21.92
" " †	42.75	46.67	47.29	2870	5280	22.75	24.00
" " †	7.14	8.25	50.65	2750	4360	23.17	25.57
" " †	1.31	1.53	50.30	4120	8740	26.51	29.77
Iron-gold †	13.59	27.93	1.73	3490	7010	27.92	14.70
" " †	9.80	21.18	1.26	2970	1220	17.55	11.20
" " †	4.76	10.96	1.46	487	103	3.84	13.40
Iron-copper †	0.40	0.46	24.51	1550	2090	13.44	14.03
Phosphorus-copper †	2.50	–	4.62	476	145	–	–
" " †	0.95	–	14.91	1320	1640	–	–
Arsenic-copper †	5.40	–	3.97	516	989	–	–
" " †	2.80	–	8.12	736	446	–	–
" " †	trace	–	38.52	2640	4830	–	–

* Annealed. † Hard-drawn.

TABLE 402. — Allowable Carrying Capacity of Rubber-covered Copper Wires.

(For inside wiring — Nat. Board Fire Underwriters' Rules.)

B + S Gage	18	16	14	12	10	8	6	5	4	3	2	1	0	00	0000
Amperes	3	6	12	17	24	33	46	54	65	76	90	107	127	150	210

500,000 circ. mills, 390 amp.; 1,000,000 c. m., 650 amp.; 2,000,000 c. m., 1,050 amp. For insulated al. wire, capacity = 84% of cu. Preece gives as formula for fusion of bare wires $I = ad^{\frac{3}{2}}$, where d = diam. in inches, a for cu. is 10,244; al., 7585; pt., 5172; German silver, 5230; platinoid, 4750; Fe, 3148; Pb., 1379; alloy 2 pts. Pb., 1 of Sn., 1318.

TABLE 403.

RESISTIVITIES AT HIGH AND LOW TEMPERATURES.

The electrical resistivity (ρ, ohms per cm. cube) of good conductors depends greatly on chemical purity. Slight contamination even with metals of lower ρ may greatly increase ρ. Solid solutions of good conductors generally have higher ρ than components. Reverse is true of bad conductors. In solid state allotropic and crystalline forms greatly modify ρ. For liquid metals this last cause of variability disappears. The + temperature coefficients of pure metals is of the same order as the coefficients of expansion of gases. For temperature resistance (t, ρ) plot at low temperatures the graph is convex towards the axis of t and probably approaches tangency to it. However for extremely low temperatures Onnes finds very sudden and great drops in ρ. *e.g.* for Mercury, $\rho_{3.6K} < 4 \times 10^{-10} \rho_0$ and for Sn., $\rho_{3.8K} < 10^{-7}\rho_0$. The t, ρ graph for an alloy may be nearly parallel to the t axis, cf. constantan; for poor conductors ρ may decrease with increasing t. At the melting-points there are three types of behavior of good conductors: those about doubling ρ and then possessing nearly linear t, ρ graphs (Al., Cu., Sn., Au., Ag., Pb.); those where ρ suddenly increases and then the + temp. coefficient is only approximately constant; (Hg., Na., K.); those about doubling ρ then having a –, slowly changing to a + temp. coef. (Zn., Cd.); those where ρ suddenly decreases and thereafter steadily increases (Sb., Bi.). The values from different authorities do not necessarily fit because of different samples of metals. The Shimank values (t given to tenths of °) are for material of theoretical purity and are determined by the α rule (see his paper, also Nernst, Ann. d Phys. 36, p. 403, 1911 for temperature resistance thermometry). The Shimank and Pirrani values are originally given as ratios to ρ_0. (Ann. d. Phys. 45, p. 706, 1914, 46, p. 176, 1915.) Resistivities are in micro-ohms per cm. cube unless stated. Italicized figures indicate liquid state.

Gold.			Copper.			Silver.			Zinc.		
°C.	ρ_t	ρ_t/ρ_0	°C.	ρ_t	ρ_t/ρ_0	°C.	ρ_t	ρ_t/ρ_0	°C.	ρ_t	ρ_t/ρ_0
-252.8	0.018	.0081	-258.6	0.014	.0091	-258.6	0.009	.0057	-252.9	.0511	.0089
-200.	.601	.267	-252.8	.016	.0103	-252.8	.014	.0090	-200.	1.39	.242
-192.5	.520	.231	-251.1	.028	.0178	-189.5	.334	.222	-191.1	1.23	.214
-150.	.997	.444	-206.6	.163	.1035	-200.	.357	.237	-150.	2.00	.348
-100.	1.400	.623	-192.9	.249	.1580	-150.	.638	.424	-100.	2.90	.504
-77.6	1.564	.696	-150.	.567	.359	-100.	.916	.608	- 77.8	3.97	.691
-50.	1.813	.806	-100.	.904	.573	-76.8	1.040	.690	- 50.	4.04	.703
0.	2.247	1.00	-50.	1.240	.786	-50.	1.212	.805	0.	5.75	1.00
100.	2.97	1.32	0.	1.578	1.00	0.	1.506	1.00	100.	7.95	1.38
200.	3.83	1.70	100.	2.28	1.44	100.	2.15	1.43	300.	13.25	2.30
500.	6.62	2.94	200.	2.96	1.88	200.	2.80	1.86	415.	17.00	2.96
750.	9.35	4.16	500.	5.08	3.22	400.	3.46	2.30	427.	*37.30*	*6.49*
1000.	12.54	5.58	750.	7.03	4.46	750.	6.65	4.42	450.	*37.08*	*6.46*
1063.	13.50	6.01	1000.	9.42	5.97	960.	8.4	5.58	500.	*36.60*	*6.36*
1063.	*30.82*	*13.7*	1083.	10.20	6.47	960.	*16.6*	*11.0*	600.	*35.90*	*6.25*
1200.	*32.8*	*14.6*	1083.	*21.30*	*13.5*	1000.	*17.01*	*11.3*	700.	*35.60*	*6.19*
1400.	*35.6*	*15.8*	1200.	*22.30*	*14.1*	1200.	*19.36*	*12.9*	800.	*35.60*	*6.19*
1500.	*37.0*	*16.5*	1400.	*23.86*	*15.1*	1400.	*21.72*	*14.4*	850.	*35.74*	*6.21*
			1500.	*24.62*	*15.6*	1500.	*23.0*	*15.3*			

Mercury.			Potassium.			Sodium.			Iron.		
°C.	ρ_t	ρ_t/ρ_0	°C.	ρ_t	ρ_t/ρ_0	°C.	ρ_t	ρ_t/ρ_0	°C.	ρ_t	ρ_t/ρ_0
-200.	5.38	.057	-200.	1.720	.246	-200.	0.605	.137	-252.7	0.011	.0010
-150.	10.30	.109	-150.	2.654	.379	-150.	1.455	.330	-200.	2.27	.212
-100.	15.42	.164	-100.	3.724	.532	-100.	2.380	.541	-192.5	.844	.079
-50.	21.4	.227	-50.	5.124	.732	-50.	3.365	.764	-100.	5.92	.554
-30.	*91.7*	*.975*	0.	7.000	1.00	0.	4.40	1.000	- 75.1	6.43	.602
0.	*94.1*	*1.000*	20.	7.116	1.016	20.	4.873	1.107	- 50.	8.15	.763
50.	*98.3*	*1.045*	60.	8.790	1.256	93.5	6.290	1.429	- 0.	10.68	1.00
100.	*103.1*	*1.096*	65.	*13.40*	*1.914*	100.	*9.220*	*2.095*	100.	16.61	1.554
200.	*114.0*	*1.212*	100.	*15.31*	*2.187*	120.	*9.724*	*2.209*	200.	24.50	2.293
300.	*127.0*	*1.350*	120.	*16.70*	*2.386*	140.	*10.34*	*2.349*	400.	43.29	4.052

Manganin.			German Silver.			Constantan.			90 % Pt. 10 % Rh.		
°C.	ρ_t	ρ_t/ρ_0	°C.	ρ_t	ρ_t/ρ_0	°C.	ρ_t	ρ_t/ρ_0	°C.	ρ_t	ρ_t/ρ_0
-200.	37.8	.974	-200.	27.9	.930	-200.	42.4	.961	-200.	14.49	.685
-150.	38.2	.985	-150.	28.7	.957	-150.	43.0	.975	-150.	16.29	.770
-100.	38.5	.992	-100.	29.3	.977	-100.	43.5	.986	-100.	18.05	.854
-50.	38.7	.997	-50.	29.7	.990	-50.	43.9	.995	- 50.	19.66	.930
0.	38.8	1.000	0.	30.0	1.000	0.	44.1	1.000	0.	21.14	1.000
100.	38.9	1.003	100.	33.1	1.103	100.	44.6	1.012	100.	24.20	1.145
400.	38.3	.987				400.	44.8	1.016			

Au. below 0°, Niccolai, Lincei Rend. (5), 16, p. 757, 906, 1907; above, Northrup, Jour. Franklin Inst. 177, p. 85, 1914. Cu. below, Niccolai, l. c. above, Northrup, ditto, 177, p. 1, 1914. Ag. below, Niccolai, l. c. above Northrup, ditto, 178, p. 85, 1914. Zn. below, Dewar, Fleming, Phil. Mag. 36, p. 271, 1893; above, Northrup, 175, p. 153, 1913. Hg. below Dewar, Fleming, Proc. Roy. Soc. 66, p. 76, 1900; above, Northrup, see Cd. K. below Guntz, Broniewski, C. R. 147, p. 1474, 1908, 148, p. 204, 1909. Above, Northrup, Tr. Am. Electroch. Soc. p. 185, 1911. Na, below, means, above, see K. Fe., Manganin, Constantan. Niccolai, l.c. German Silver, 90% Pt. 90% Rh., Dewar and Fleming — Phil. Mag. 36, p. 271, 1893.

RESISTIVITIES AT HIGH AND LOW TEMPERATURES.

(Ohms per cm. cube unless stated otherwise.)

Platinum.			Lead.			Bismuth.			Cadmium.		
°C.	ρ_t	$\frac{\rho_t}{\rho_o}$	°C.	ρ_t	$\frac{\rho_t}{\rho_o}$	°C.	ρ_t	$\frac{\rho_t}{\rho_o}$	°C.	ρ_t	ρ_o
−265.	0.10	.0092	−252.9	0.59	.0298	−200.	34.8	.314	−252.9	0.17	.0218
−253.	.15	.014	−203.	4.42	.223	−150.	55.3	.499	−200.	1.66	.214
−233.	.54	.049	−192.8	5.22	.264	−100.	75.6	.683	−190.2	2.00	.258
−153.	4.18	.378	−103.	11.8	.598	− 50.	94.3	.852	−183.1	2.22	.286
− 73.	7.82	.708	− 75.8	13.95	.705	0.	110.7	1.00	−139.2	3.60	.464
0.	11.05	1.00	− 53.	15.7	.792	17.	120.0	1.083	−100.	4.80	.619
100.	14.1	1.28	0.	19.8	1.00	100.	156.5	1.413	0.	7.75	1.00
200.	17.9	1.62	100.	27.8	1.403	200.	214.5	1.937	300.	16.50	2.13
400.	25.4	2.30	200.	38.0	1.919	259.	267.0	2.411	325.	*33.76*	*4.35*
800.	40.3	3.65	319.	50.0	2.52	263.	*127.5*	*1.150*	350.	*33.60*	*4.33*
1000.	47.0	4.25	333.	*95.0*	*4.80*	300.	*128.9*	*1.164*	400.	*33.70*	*4.35*
1200.	52.7	4.77	400.	*98.3*	*4.96*	500.	*139.9*	*1.263*	500.	*35.12*	*4.40*
1400.	58.0	5.25	600.	*107.2*	*5.41*	700.	*150.8*	*1.361*	700.	*35.78*	*4.62*
1600.	63.0	5.70	800.	*116.2*	*5.86*	750.	*153.5*	*1.386*			

Tin.			Carbon, Graphite.*			Fused silica.		Alundum cement.	
°C.	ρ_t	$\frac{\rho_t}{\rho_o}$	°C.	ρ in ohms, cm. cube.		°C.	ρ = megohms cm.	°C.	ρ in ohms cm. cube.
				Carbon	Graphite				
−200.	2.60	.199				15.	>200,000,000.	20.	$>9 \times 10^6$
−100.	7.57	.580	0.	0.0035	0.00080	230.	20,000,000.	800.	30800.
0.	13.05	1.00	500.	.0027	.00083	300.	200,000.	900.	13600.
200.	20.30	1.55	1000.	.0021	.00087	350.	30,000.	1000.	7600.
225.	22.00	1 69	1500.	.0015	.00090	450.	800.	1100.	6500.
235.	*47.60*	*3.65*	2000.	.0011	.00100	700.	30.	1200.	2300.
750.	*61.22*	*4.69*	2500.	.0009	.0011	850.	about 20.	1600.	190.

Pt. low, Nernst, l. c. high, Pirrani, Ber. Deutsch. Phys. Ges. 12, p. 305, Pb. low, Schimank, Nernst, l. c. high. Northrup, see Zn. Bi. low, means, high, Northrup, see Zn. Cd. low, Euchen, Gehlhoff, Verh. Deutsch. Phys. Ges. 14, p. 169, 1912, high, Northrup, see Zn. Sn. low, Dewar, Fleming, high, Northrup, see Zn. Carbon, graphite, Metallurg. Ch. Eng. 13, p. 23, 1915. Silica, Campbell, Nat. Phys. Lab. 11, p. 207, 1914. Alundum, Metallurg. Ch. Eng. 12, p. 125, 1914.

* Diamond 1030° C, $\rho > 10^7$; 1380°, 7.5×10^5, v. Wartenberg, 1912.

TABLE 404.—Volume and Surface Resistivity of Solid Dielectrics.

The resistance between two conductors insulated by a solid dielectric depends both upon the surface resistance and the volume resistance of the insulator. The volume resistivity, ρ, is the resistance between two opposite faces of a centimeter cube. The surface resistivity, σ, is the resistance between two opposite edges of a centimeter square of the surface. The surface resistivity usually varies through a wide range with the humidity. (Curtis, Bul. Bur. Standards, 11, 359, 1915, which see for discussion and data for many additional materials.)

Material.	σ; megohms 50% humidity.	σ; megohms 70% humidity.	σ; megohms 90% humidity.	ρ Megohms-cms.
Amber	6×10^8	2×10^8	1×10^5	5×10^{10}
Beeswax, yellow	6×10^8	6×10^8	5×10^8	2×10^9
Celluloid	5×10^4	2×10^4	2×10^3	2×10^4
Fiber, red	2×10^4	3×10^3	2×10^2	5×10^3
Glass, plate	5×10^4	6×10	2×10	2×10^7
" Kavalier	4×10^6	4×10^3	1×10^3	8×10^9
Hard rubber, new	3×10^9	1×10^8	2×10^3	1×10^{12}
Ivory	5×10^3	1×10^3	3×10	2×10^2
Khotinsky cement	7×10^8	3×10^8	5×10^5	2×10^9
Marble, Italian	3×10^3	2×10^2	2×10	1×10^5
Mica, colorless	2×10^7	4×10^5	8×10^3	2×10^{11}
Paraffin (parowax)	9×10^9	7×10^9	6×10^9	1×10^{10}
Porcelain, unglazed	6×10^5	7×10^3	5×10	3×10^8
Quartz, fused	3×10^6	2×10^3	2×10^2	5×10^{12}
Rosin	6×10^8	3×10^8	2×10^8	5×10^{10}
Sealing wax	2×10^9	6×10^8	9×10^7	8×10^9
Shellac	6×10^7	3×10^6	7×10^3	1×10^{10}
Slate	9×10	3×10	1×10	1×10^2
Sulphur	7×10^9	4×10^9	1×10^8	1×10^{11}
Wood, parafined mahogany	4×10^6	5×10^5	7×10^3	4×10^7

TABLE 405.—Variation of Electrical Resistance of Glass and Porcelain with Temperature.

The following table gives the values of a, b, and c in the equation

$$\log R = a + bt + ct^2,$$

where R is the specific resistance expressed in ohms, that is, the resistance in ohms per centimeter of a rod one square centimeter in cross section.*

No.	Kind of glass.	Density.	a	b	c	Range of temp. Centigrade.
1	Test-tube glass	–	13.86	—.044	.000065	0°–250°
2	" " "	2.458	14.24	—.055	.0001	37–131
3	Bohemian glass	2.43	16.21	—.043	.0000394	60–174
4	Lime glass (Japanese manufacture) .	2.55	13.14	—.031	—.000021	10–85
5	" " " " .	2.499	14.002	—.025	—.00006	35–95
6	Soda-lime glass (French flask) .	2.533	14.58	—.049	.000075	45–120
7	Potash-soda lime glass . . .	2.58	16.34	—.0425	.0000364	66–193
8	Arsenic enamel flint glass . .	3.07	18.17	—.055	.000088	105–135
9	Flint glass (Thomson's electrometer jar)	3.172	18.021	—.036	—.0000091	100–200
10	Porcelain (white evaporating dish) .	–	15.65	—.042	.00005	68–290

COMPOSITION OF SOME OF THE ABOVE SPECIMENS OF GLASS.

Number of specimen =	3	4	5	7	8	9
Silica	61.3	57.2	70.05	75.65	54.2	55.18
Potash	22.9	21.1	1.44	7.92	10.5	13.28
Soda	Lime, etc.	Lime, etc.	14.32	6.92	7.0	–
Lead oxide	by diff.	by diff.	2.70	–	23.9	31.01
Lime	15.8	16.7	10.33	8.48	0.3	0.35
Magnesia	–	–	–	0.36	0.2	0.06
Arsenic oxide	–	–	–	–	3.5	–
Alumina, iron oxide, etc. .	–	–	1.45	0.70	0.4	0.67

* T. Gray, "Phil. Mag." 1880, and "Proc. Roy. Soc." 1882.

TABLE 405a.—Temperature Resistance Coefficients of Glass, Porcelain and Quartz dr/dt.

Temperature.	450°	500°	575°	600°	700°	750°	800°	900°	1000°
Glass . . .	—32.	—6.	—1.5	—.8	—0.17	—0.1	—0.06	–	–
Porcelain . .	–	–	—16.	—9.8	—2.8	—1.6	—.70	—0.30	—0.12
Quartz. . .	–	–	–	–	–	—10.	—6.40	—2.60	—1.00

Somerville, Physical Review, 31, p. 261, 1910.

Table 406.

TABULAR COMPARISON OF WIRE GAGES.

Gage No.	American wire gage (B. & S.) mils.†	American wire gage (B. & S.) mm.†	Steel wire gage* mils.	Steel wire gage* mm.	Stubs' steel wire gage mils.	(British) standard wire gage mils.	Birmingham wire gage (Stubs') mils.	Gage No.
7-0			490.0	12.4		500.		7-0
6-0			461.5	11.7		464.		6-0
5-0			430.5	10.9		432.		5-0
4-0	460.	11.7	393.8	10.0		400.	454.	4-0
3-0	410.	10.4	362.5	9.2		372.	425.	3-0
2-0	365.	9.3	331.0	8.4		348.	380.	2-0
0	325.	8.3	306.5	7.8		324.	340.	0
1	289.	7.3	283.0	7.2	227.	300.	300.	1
2	258.	6.5	262.5	6.7	219.	276.	284.	2
3	229.	5.8	243.7	6.2	212.	252.	259.	3
4	204.	5.2	225.3	5.7	207.	232.	238.	4
5	182.	4.6	207.0	5.3	204.	212.	220.	5
6	162.	4.1	192.0	4.9	201.	192.	203.	6
7	144.	3.7	177.0	4.5	199.	176.	180.	7
8	128.	3.3	162.0	4.1	197.	160.	165.	8
9	114.	2.91	148.3	3.77	194.	144.	148.	9
10	102.	2.59	135.0	3.43	191.	128.	134.	10
11	91.	2.30	120.5	3.06	188.	116.	120.	11
12	81.	2.05	105.5	2.68	185.	104.	109.	12
13	72.	1.83	91.5	2.32	182.	92.	95.	13
14	64.	1.63	80.0	2.03	180.	80.	83.	14
15	57.	1.45	72.0	1.83	178.	72.	72.	15
16	51.	1.29	62.5	1.59	175.	64.	65.	16
17	45.	1.15	54.0	1.37	172.	56.	58.	17
18	40.	1.02	47.5	1.21	168.	48.	49.	18
19	36.	0.91	41.0	1.04	164.	40.	42.	19
20	32.	.81	34.8	0.88	161.	36.	35.	20
21	28.5	.72	31.7	.81	157.	32.	32.	21
22	25.3	.62	28.6	.73	155.	28.	28.	22
23	22.6	.57	25.8	.66	153.	24.	25.	23
24	20.1	.51	23.0	.58	151.	22.	22.	24
25	17.9	.45	20.4	.52	148.	20.	20.	25
26	15.9	.40	18.1	.46	146.	18.	18.	26
27	14.2	.36	17.3	.439	143.	16.4	16.	27
28	12.6	.32	16.2	.411	139.	14.8	14.	28
29	11.3	.29	15.0	.381	134.	13.6	13.	29
30	10.0	.25	14.0	.356	127.	12.4	12.	30
31	8.9	.227	13.2	.335	120.	11.6	10.	31
32	8.0	.202	12.8	.325	115.	10.8	9.	32
33	7.1	.180	11.8	.300	112.	10.0	8.	33
34	6.3	.160	10.4	.264	110.	9.2	7.	34
35	5.6	.143	9.5	.241	108.	8.4	5.	35
36	5.0	.127	9.0	.229	106.	7.6	4.	36
37	4.5	.113	8.5	.216	103.	6.8		37
38	4.0	.101	8.0	.203	101.	6.0		38
39	3.5	.090	7.5	.191	99.	5.2		39
40	3.1	.080	7.0	.178	97.	4.8		40
41			6.6	.168	95.	4.4		41
42			6.2	.157	92.	4.0		42
43			6.0	.152	88.	3.6		43
44			5.8	.147	85.	3.2		44
45			5.5	.140	81.	2.8		45
46			5.2	.132	79.	2.4		46
47			5.0	.127	77.	2.0		47
48			4.8	.122	75.	1.6		48
49			4.6	.117	72.	1.2		49
50			4.4	.112	69.	1.0		50

* The Steel Wire Gage is the same gage which has been known by the various names: "Washburn and Moen," "Roebling," "American Steel and Wire Co.'s." Its abbreviation should be written "Stl. W. G.," to distinguish it from "S. W. G.," the usual abbreviation for the (British) Standard Wire Gage.

† The American Wire Gage sizes have been rounded off to the usual limits of commercial accuracy. They are given to four significant figures in Tables 410 to 413. They can be calculated with any desired accuracy, being based upon a simple mathematical law. The diameter of No. 0000 is defined as 0.4600 inch and of No. 36 as 0.0050 inch. The ratio of any diameter to the diameter of the next greater number $\sqrt[39]{\frac{.4600}{.0050}} = 1.1229322$.

Taken from Circular No. 31. Copper Wire Tables, U.S. Bureau of Standards which contains more complete tables.

WIRE TABLES.

TABLE 407. — Introduction. Mass and Volume Resistivity of Copper and Aluminum.

The following wire tables are abridged from those prepared by the Bureau of Standards at the request and with the coöperation of the Standards Committee of the American Institute of Electrical Engineers (Circular No. 31 of the Bureau of Standards). The standard of copper resistance used is "The International Annealed Copper Standard" as adopted Sept. 5, 1913, by the International Electrotechnical Commission and represents the average commercial high-conductivity copper for the purpose of electric conductors. This standard corresponds to a conductivity of $58. \times 10^{-5}$ cgs. units, and a density of 8.89, at 20° C.

In the various units of mass resistivity and volume resistivity this may be stated as

0.15328 ohm (meter, gram) at 20° C.
875.20 ohms (mile, pound) at 20° C.
1.7241 microhm-cm. at 20° C.
0.67879 microhm-inch at 20° C.
10.371 ohms (mil, foot) at 20° C.

The temperature coefficient for this particular resistivity is $\alpha_{20} = 0.00393$ or $\alpha_0 = 0.00427$. The temperature coefficient of copper is proportional to the conductivity, so that where the conductivity is known the temperature coefficient may be calculated, and vice-versa. Thus the next table shows the temperature coefficients of copper having various percentages of the standard conductivity. A consequence of this relation is that the change of *resistivity* per degree is constant, independent of the sample of copper and independent of the temperature of reference. This resistivity-temperature constant, for volume resistivity and Centigrade degrees, is 0.00681 michrom-cm., and for mass resistivity is 0.000597 ohm (meter, gram).

The density of 8.89 grams per cubic centimeter at 20° C., is equivalent to 0.32117 pounds per cubic inch.

The values in the following tables are for annealed copper of standard resistivity. The user of the tables must apply the proper correction for copper of other resistivity. Hard-drawn copper may be taken as about 2.7 per cent higher resistivity than annealed copper.

The following is a fair average of the chemical content of commercial high conductivity copper:

Copper	99.91%	Sulphur	0.002%
Silver	.03	Iron	.002
Oxygen	.052	Nickel	Trace
Arsenic	.002	Lead	"
Antimony	.002	Zinc	"

The following values are consistent with the data above:

Conductivity at 0° C., in c.g.s. electromagnetic units	62.969×10^{-5}
Resistivity at 0° C., in michroms-cms.	1.5881
Density at 0° C.	8.90
Coefficient of linear expansion per degree C.	0.000017
"Constant mass" temperature coefficient of resistance at 0° C.	0.00427

The aluminum tables are based on a figure for the conductivity published by the U.S. Bureau of Standards, which is the result of many thousands of determinations by the Aluminum Company of America. A volume resistivity of 2.828 michrom-cm., and a density of 2.70 may be considered to be good average values for commercial hard-drawn aluminum. These values give:

Mass resistivity, in ohms (meter, gram) at 20° C.	0.0764
" " " " (mile, pound) at 20° C.	436.
Mass per cent conductivity	200.7%
Volume resistivity, in michrom-cm. at 20° C.	2.828
" " in microhm-inch at 20° C.	1.113
Volume per cent conductivity	61.0%
Density, in grams per cubic centimeter	2.70
Density, in pounds per cubic inch	0.0975

The average chemical content of commercial aluminum wire is

Aluminum	99.57%
Silicon	0.29
Iron	0.14

COPPER WIRE TABLES.

TABLE 408. — Temperature Coefficients of Copper for Different Initial Temperatures (Centigrade) and Different Conductivities.

Ohms (meter, gram) at 20° C.	Per cent conductivity.	a_0	a_{15}	a_{20}	a_{25}	a_{30}	a_{50}
0.161 34	95%	0.004 03	0.003 80	0.003 73	0.003 67	0.003 60	0.003 36
.159 66	96%	.004 08	.003 85	.003 77	.003 70	.003 64	.003 39
.158 02	97%	.004 13	.003 89	.003 81	.003 74	.003 67	.003 42
.157 53	97.3%	.004 14	.003 90	.003 82	.003 75	.003 68	.003 43
.156 40	98%	.004 17	.003 93	.003 85	.003 78	.003 71	.003 45
.154 82	99%	.004 22	.003 97	.003 89	.003 82	.003 74	.003 48
.153 28	100%	.004 27	.004 01	**.003 93**	.003 85	.003 78	.003 52
.151 76	101%	.004 31	.004 05	.00 397	.003 89	.003 82	.003 55

NOTE. — The fundamental relation between resistance and temperature is the following:

$$R_t = R_{t_1}(1 + a_{t_1}[t - t_1]),$$

where a_{t_1} is the "temperature coefficient," and t_1 is the "initial temperature" or "temperature of reference."

The values of a in the above table exhibit the fact that the temperature coefficient of copper is proportional to the conductivity. The table was calculated by means of the following formula, which holds for any per cent conductivity, n, within commercial ranges, and for centigrade temperatures. (n is considered to be expressed decimally: e.g., if per cent conductivity = 99 per cent, $n = 0.99$.)

$$a_{t_1} = \frac{1}{\frac{1}{n(0.00393)} + (t_1 - 20)}.$$

TABLE 409. — Reduction of Observations to Standard Temperature. (Copper.)

Temperature C.	Corrections to reduce Resistivity to 20° C.				Factors to reduce Resistance to 20° C.			Temperature C.
	Ohm (meter, gram).	Microhm—cm.	Ohm (mile, pound).	Microhm—inch.	For 96 per cent conductivity.	For 98 per cent conductivity.	For 100 per cent conductivity.	
0	+0.011 94	+0.1361	+ 68.20	+0.053 58	1.0816	1.0834	1.0853	0
5	+ .008 96	+ .1021	+ 51.15	+ .040 18	1.0600	1.0613	1.0626	5
10	+ .005 97	+ .0681	+ 34.10	+ .026 79	1.0392	1.0401	1.0409	10
11	+ .005 37	+ .0612	+ 30.69	+ .024 11	1.0352	1.0359	1.0367	11
12	+ .004 78	+ .0544	+ 27.28	+ .021 43	1.0311	1.0318	1.0325	12
13	+ .004 18	+ .0476	+ 23.87	+ .018 75	1.0271	1.0277	1.0283	13
14	+ .003 58	+ .0408	+ 20.46	+ .016 07	1.0232	1.0237	1.0242	14
15	+ .002 99	+ .0340	+ 17.05	+ .013 40	1.0192	1.0196	1.0200	15
16	+ .002 39	+ .0272	+ 13.64	+ .010 72	1.0153	1.0156	1.0160	16
17	+ .001 79	+ .0204	+ 10.23	+ .008 04	1.0114	1.0117	1.0119	17
18	+ .001 19	+ .0136	+ 6.82	+ .005 36	1.0076	1.0078	1.0079	18
19	+ .000 60	+ .0068	+ 3.41	+ .002 68	1.0038	1.0039	1.0039	19
20	0	0	0	0	1.0000	1.0000	1.0000	20
21	− .000 60	− .0068	− 3.41	− .002 68	0.9962	0.9962	0.9961	21
22	− .001 19	− .0136	− 6.82	− .005 36	.9925	.9924	.9922	22
23	− .001 79	− .0204	− 10.23	− .008 04	.9888	.9886	.9883	23
24	− .002 39	− .0272	− 13.64	− .010 72	.9851	.9848	.9845	24
25	− .002 99	− .0340	− 17.05	− .013 40	.9815	.9811	.9807	25
26	− .003 58	− .0408	− 20.46	− .016 07	.9779	.9774	.9770	26
27	− .004 18	− .0476	− 23.87	− .018 75	.9743	.9737	.9732	27
28	− .004 78	− .0544	− 27.28	− .021 43	.9707	.9701	.9695	28
29	− .005 37	− .0612	− 30.69	− .024 11	.9672	.9665	.9658	29
30	− .005 97	− .0681	− 34.10	− .026 79	.9636	.9629	.9622	30
35	− .008 96	− .1021	− 51.15	− .040 18	.9464	.9454	.9443	35
40	− .011 94	− .1361	− 68.20	− .053 58	.9298	.9285	.9271	40
45	− .014 93	− .1701	− 85.25	− .066 98	.9138	.9122	.9105	45
50	− .017 92	− .2042	−102.30	− .080 37	.8983	.8964	.8945	50
55	− .020 90	− .2382	−119.35	− .093 76	.8833	.8812	.8791	55
60	− .023 89	− .2722	−136.40	− .107 16	.8689	.8665	.8642	60
65	− .026 87	− .3062	−153.45	− .120 56	.8549	.8523	.8497	65
70	− .029 86	− .3403	−170.50	− .133 95	.8413	.8385	.8358	70
75	− .032 85	− .3743	−187.55	− .147 34	.8281	.8252	.8223	75

WIRE TABLE, STANDARD ANNEALED COPPER.

American Wire Gage (B. & S.). English Units.

Gage No.	Diameter in Mils. at 20° C.	Cross-Section at 20° C.		Ohms per 1000 Feet.*			
		Circular Mils.	Square Inches.	0° C (= 32° F)	20° C (= 68° F)	50° C (= 122° F)	75° C (= 167° F)
0000	460.0	211 600.	0.1662	0.045 16	0.049 01	0.054 79	0.059 61
000	409.6	167 800.	.1318	.056 95	.061 80	.069 09	.075 16
00	364.8	133 100.	.1045	.071 81	.077 93	.087 12	.094 78
0	324.9	105 500.	.082 89	.090 55	.098 27	.1099	.1195
1	289.3	83 690.	.065 73	.1142	.1239	.1385	.1507
2	257.6	66 370.	.052 13	.1440	.1563	.1747	.1900
3	229.4	52 640.	.041 34	.1816	.1970	.2203	.2396
4	204.3	41 740.	.032 78	.2289	.2485	.2778	.3022
5	181.9	33 100.	.026 00	.2887	.3133	.3502	.3810
6	162.0	26 250.	.020 62	.3640	.3951	.4416	.4805
7	144.3	20 820.	.016 35	.4590	.4982	.5569	.6059
8	128.5	16 510.	.012 97	.5788	.6282	.7023	.7640
9	114.4	13 090.	.010 28	.7299	.7921	.8855	.9633
10	101.9	10 380.	.008 155	.9203	.9989	1.117	1.215
11	90.74	8234.	.006 467	1.161	1.260	1.408	1.532
12	80.81	6530.	.005 129	1.463	1.588	1.775	1.931
13	71.96	5178.	.004 067	1.845	2.003	2.239	2.436
14	64.08	4107.	.003 225	2.327	2.525	2.823	3.071
15	57.07	3257.	.002 558	2.934	3.184	3.560	3.873
16	50.82	2583.	.002 028	3.700	4.016	4.489	4.884
17	45.26	2048.	.001 609	4.666	5.064	5.660	6.158
18	40.30	1624.	.001 276	5.883	6.385	7.138	7.765
19	35.89	1288.	.001 012	7.418	8.051	9.001	9.792
20	31.96	1022.	.000 802 3	9.355	10.15	11.35	12.35
21	28.45	810.1	.000 636 3	11.80	12.80	14.31	15.57
22	25.35	642.4	.000 504 6	14.87	16.14	18.05	19.63
23	22.57	509.5	.000 400 2	18.76	20.36	22.76	24.76
24	20.10	404.0	.000 317 3	23.65	25.67	28.70	31.22
25	17.90	320.4	.000 251 7	29.82	32.37	36.18	39.36
26	15.94	254.1	.000 199 6	37.61	40.81	45.63	49.64
27	14.20	201.5	.000 158 3	47.42	51.47	57.53	62.59
28	12.64	159.8	.000 125 5	59.80	64.90	72.55	78.93
29	11.26	126.7	.000 099 53	75.40	81.83	91.48	99.52
30	10.03	100.5	.000 078 94	95.08	103.2	115.4	125.5
31	8.928	79.70	.000 062 60	119.9	130.1	145.5	158.2
32	7.950	63.21	.000 049 64	151.2	164.1	183.4	199.5
33	7.080	50.13	.000 039 37	190.6	206.9	231.3	251.6
34	6.305	39.75	.000 031 22	240.4	260.9	291.7	317.3
35	5.615	31.52	.000 024 76	303.1	329.0	367.8	400.1
36	5.000	25.00	.000 019 64	382.2	414.8	463.7	504.5
37	4.453	19.83	.000 015 57	482.0	523.1	584.8	636.2
38	3.965	15.72	.000 012 35	607.8	659.6	737.4	802.2
39	3.531	12.47	.000 009 793	766.4	831.8	929.8	1012.
40	3.145	9.888	.000 007 766	966.5	1049.	1173.	1276.

* Resistance at the stated temperatures of a wire whose length is 1000 feet at 20° C.

WIRE TABLE, STANDARD ANNEALED COPPER (*continued*).

American Wire Gage (B. & S.). English Units (*continued*).

Gage No.	Diameter in Mils. at 20° C.	Pounds per 1000 Feet.	Feet per Pound.	Feet per Ohm.*			
				0° C (=32° F)	20° C (=68° F)	50° C (=122° F)	75° C (=167° F)
0000	460.0	640.5	1.561	22 140.	20 400.	18 250.	16 780.
000	409.6	507.9	1.968	17 560.	16 180.	14 470.	13 300.
00	364.8	402.8	2.482	13 930.	12 830.	11 480.	10 550.
0	324.9	319.5	3.130	11 040.	10 180.	9103.	8367.
1	289.3	253.3	3.947	8758.	8070.	7219.	6636.
2	257.6	200.9	4.977	6946.	6400.	5725.	5262.
3	229.4	159.3	6.276	5508.	5075.	4540.	4173.
4	204.3	126.4	7.914	4368.	4025.	3600.	3309.
5	181.9	100.2	9.980	3464.	3192.	2855.	2625.
6	162.0	79.46	12.58	2747.	2531.	2264.	2081.
7	144.3	63.02	15.87	2179.	2007.	1796.	1651.
8	128.5	49.98	20.01	1728.	1592.	1424.	1309.
9	114.4	39.63	25.23	1370.	1262.	1129.	1038.
10	101.9	31.43	31.82	1087.	1001.	895.6	823.2
11	90.74	24.92	40.12	861.7	794.0	710.2	652.8
12	80.81	19.77	50.59	683.3	629.6	563.2	517.7
13	71.96	15.68	63.80	541.9	499.3	446.7	410.6
14	64.08	12.43	80.44	429.8	396.0	354.2	325.6
15	57.07	9.858	101.4	340.8	314.0	280.9	258.2
16	50.82	7.818	127.9	270.3	249.0	222.8	204.8
17	45.26	6.200	161.3	214.3	197.5	176.7	162.4
18	40.30	4.917	203.4	170.0	156.6	140.1	128.8
19	35.89	3.899	256.5	134.8	124.2	111.1	102.1
20	31.96	3.092	323.4	106.9	98.50	88.11	80.99
21	28.46	2.452	407.8	84.78	78.11	69.87	64.23
22	25.35	1.945	514.2	67.23	61.95	55.41	50.94
23	22.57	1.542	648.4	53.32	49.13	43.94	40.39
24	20.10	1.223	817.7	42.28	38.96	34.85	32.03
25	17.90	0.9699	1031.	33.53	30.90	27.64	25.40
26	15.94	.7692	1300.	26.59	24.50	21.92	20.15
27	14.20	.6100	1639.	21.09	19.43	17.38	15.98
28	12.64	.4837	2067.	16.72	15.41	13.78	12.67
29	11.26	.3836	2607.	13.26	12.22	10.93	10.05
30	10.03	.3042	3287.	10.52	9.691	8.669	7.968
31	8.928	.2413	4145.	8.341	7.685	6.875	6.319
32	7.950	.1913	5227.	6.614	6.095	5.452	5.011
33	7.080	.1517	6591.	5.245	4.833	4.323	3.974
34	6.305	.1203	8310.	4.160	3.833	3.429	3.152
35	5.615	.095 42	10 480.	3.299	3.040	2.719	2.499
36	5.000	.075 68	13 210.	2.616	2.411	2.156	1.982
37	4.453	.060 01	16 660.	2.075	1.912	1.710	1.572
38	3.965	.047 59	21 010.	1.645	1.516	1.356	1.247
39	3.531	.037 74	26 500.	1.305	1.202	1.075	0.9886
40	3.145	.029 93	33 410.	1.035	0.9534	0.8529	.7840

* Length at 20° C. of a wire whose resistance is 1 ohm at the stated temperatures.

WIRE TABLE, STANDARD ANNEALED COPPER (*continued*).

American Wire Gage (B. & S.). English Units (*continued*).

Gage No.	Diameter in Mils at 20° C.	Ohms per Pound.			Pounds per Ohm.
		0° C. (=32° F.)	20° C. (=68° F.)	50° C. (=122° F.)	20° C. (=68° F.)
0000	460.0	0.000 070 51	0.000 076 52	0.000 085 54	13 070.
000	409.6	.000 1121	.000 1217	.000 1360	8219.
00	364.8	.000 1783	.000 1935	.000 2163	5169.
0	324.9	.000 2835	.000 3076	.000 3439	3251.
1	289.3	.000 4507	.000 4891	.000 5468	2044.
2	257.6	.000 7166	.000 7778	.000 8695	1286.
3	229.4	.001 140	.001 237	.001 383	808.6
4	204.3	.001 812	.001 966	.002 198	508.5
5	181.9	.002 881	.003 127	.003 495	319.8
6	162.0	.004 581	.004 972	.005 558	201.1
7	144.3	.007 284	.007 905	.008 838	126.5
8	128.5	.011 58	.012 57	.014 05	79.55
9	114.4	.018 42	.019 99	.022 34	50.03
10	101.9	.029 28	.031 78	.035 53	31.47
11	90.74	.046 56	.050 53	.056 49	19.79
12	80.81	.074 04	.080 35	.089 83	12.45
13	71.96	.1177	.1278	.1428	7.827
14	64.08	.1872	.2032	.2271	4.922
15	57.07	.2976	.3230	.3611	3.096
16	50.82	.4733	.5136	.5742	1.947
17	45.26	.7525	.8167	.9130	1.224
18	40.30	1.197	1.299	1.452	0.7700
19	35.89	1.903	2.065	2.308	.4843
20	31.96	3.025	3.283	3.670	.3046
21	28.46	4.810	5.221	5.836	.1915
22	25.35	7.649	8.301	9.280	.1205
23	22.57	12.16	13.20	14.76	.075 76
24	20.10	19.34	20.99	23.46	.047 65
25	17.90	30.75	33.37	37.31	.029 97
26	15.94	48.89	53.06	59.32	.018 85
27	14.20	77.74	84.37	94.32	.011 85
28	12.64	123.6	134.2	150.0	.007 454
29	11.26	196.6	213.3	238.5	.004 688
30	10.03	312.5	339.2	379.2	.002 948
31	8.928	497.0	539.3	602.9	.001 854
32	7.950	790.2	857.6	958.7	.001 166
33	7.080	1256.	1364.	1524.	.000 7333
34	6.305	1998.	2168.	2424.	.000 4612
35	5.615	3177.	3448.	3854.	.000 2901
36	5.000	5051.	5482.	6128.	.000 1824
37	4.453	8032.	8717.	9744.	.000 1147
38	3.965	12 770.	13 860.	15 490.	.000 072 15
39	3.531	20 310.	22 040.	24 640.	.000 045 38
40	3.145	32 290.	35 040.	39 170.	.000 028 54

WIRE TABLE, STANDARD ANNEALED COPPER.

American Wire Gage (B. & S.) Metric Units.

Gage No.	Diameter in mm. at 20° C.	Cross Section in mm.² at 20° C.	Ohms per Kilometer.*			
			0° C.	20° C.	50° C.	75° C.
0000	11.68	107.2	0.1482	0.1608	0.1798	0.1956
000	10.40	85.03	.1868	.2028	.2267	.2466
00	9.266	67.43	.2356	.2557	.2858	.3110
0	8.252	53.48	.2971	.3224	.3604	.3921
1	7.348	42.41	.3746	.4066	.4545	.4944
2	6.544	33.63	.4724	.5127	.5731	.6235
3	5.827	26.67	.5956	.6465	.7227	.7862
4	5.189	21.15	.7511	.8152	.9113	.9914
5	4.621	16.77	.9471	1.028	1.149	1.250
6	4.115	13.30	1.194	1.296	1.449	1.576
7	3.665	10.55	1.506	1.634	1.827	1.988
8	3.264	8.366	1.899	2.061	2.304	2.506
9	2.906	6.634	2.395	2.599	2.905	3.161
10	2.588	5.261	3.020	3.277	3.663	3.985
11	2.305	4.172	3.807	4.132	4.619	5.025
12	2.053	3.309	4.801	5.211	5.825	6.337
13	1.828	2.624	6.054	6.571	7.345	7.991
14	1.628	2.081	7.634	8.285	9.262	10.08
15	1.450	1.650	9.627	10.45	11.68	12.71
16	1.291	1.309	12.14	13.17	14.73	16.02
17	1.150	1.038	15.31	16.61	18.57	20.20
18	1.024	0.8231	19.30	20.95	23.42	25.48
19	0.9116	.6527	24.34	26.42	29.53	32.12
20	.8118	.5176	30.69	33.31	37.24	40.51
21	.7230	.4105	38.70	42.00	46.95	51.08
22	.6438	.3255	48.80	52.96	59.21	64.41
23	.5733	.2582	61.54	66.79	74.66	81.22
24	.5106	.2047	77.60	84.21	94.14	102.4
25	.4547	.1624	97.85	106.2	118.7	129.1
26	.4049	.1288	123.4	133.9	149.7	162.9
27	.3606	.1021	155.6	168.9	188.8	205.4
28	.3211	.080 98	196.2	212.9	238.0	258.9
29	.2859	.064 22	247.4	268.5	300.1	326.5
30	.2546	.050 93	311.9	338.6	378.5	411.7
31	.2268	.040 39	393.4	426.9	477.2	519.2
32	.2019	.032 03	496.0	538.3	601.8	654.7
33	.1798	.025 40	625.5	678.8	758.8	825.5
34	.1601	.020 14	788.7	856.0	956.9	1041.
35	.1426	.015 97	994.5	1079.	1207.	1313.
36	.1270	.012 67	1254.	1361.	1522.	1655.
37	.1131	.010 05	1581.	1716.	1919.	2087.
38	.1007	.007 967	1994.	2164.	2419.	2632.
39	.089 69	.006 318	2514.	2729.	3051.	3319.
40	.079 87	.005 010	3171.	3441.	3847.	4185.

*Resistance at the stated temperatures of a wire whose length is 1 kilometer at 20° C.

WIRE TABLE, STANDARD ANNEALED COPPER (*continued*).

American Wire Gage (B. & S.) Metric Units (*continued*).

Gage No.	Diameter in mm. at 20° C.	Kilograms per Kilometer.	Meters per Gram.	Meters per Ohm.* 0° C.	20° C.	50° C.	75° C.
0000	11.68	953.2	0.001 049	6749.	6219.	5563.	5113.
000	10.40	755.9	.001 323	5352.	4932.	4412.	4055.
00	9 266	599.5	.001 668	4245.	3911.	3499.	3216.
0	8.252	475.4	.002 103	3366.	3102.	2774.	2550.
1	7.348	377.0	.002 652	2669.	2460.	2200.	2022.
2	6.544	299.0	.003 345	2117.	1951.	1745.	1604.
3	5.827	237.1	.004 217	1679.	1547.	1384.	1272.
4	5.189	188.0	.005 318	1331.	1227.	1097.	1009.
5	4.621	149.1	.006 706	1056.	972.9	870.2	799.9
6	4.115	118.2	.008 457	837.3	771.5	690.1	634.4
7	3.665	93.78	.010 66	664.0	611.8	547.3	503.1
8	3.264	74.37	.013 45	526.6	485.2	434.0	399.0
9	2.906	58.98	.016 96	417.6	384.8	344.2	316.4
10	2.588	46.77	.021 38	331.2	305.1	273.0	250.9
11	2.305	37.09	.026 96	262.6	242.0	216.5	199.0
12	2.053	29.42	.034 00	208.3	191.9	171.7	157.8
13	1.828	23.33	.042 87	165.2	152.2	136.1	125.1
14	1.628	18.50	.054 06	131.0	120.7	108.0	99.24
15	1.450	14.67	.068 16	103.9	95.71	85 62	78.70
16	1.291	11.63	.085 95	82.38	75.90	67.90	62.41
17	1.150	9.226	.1084	65.33	60.20	53.85	49.50
18	1.024	7.317	.1367	51.81	47.74	42.70	39.25
19	0.9116	5.803	.1723	41.09	37.86	33.86	31.13
20	.8118	4.602	.2173	32.58	30.02	26.86	24.69
21	.7230	3.649	.2740	25.84	23.81	21.30	19.58
22	.6438	2.894	.3455	20.49	18.88	16.89	15.53
23	.5733	2.295	.4357	16.25	14.97	13.39	12.31
24	.5106	1.820	.5494	12.89	11 87	10.62	9.764
25	.4547	1.443	.6928	10.22	9.417	8.424	7.743
26	.4049	1.145	.8736	8.105	7.468	6.680	6.141
27	.3606	0.9078	1.102	6.428	5.922	5.298	4.870
28	.3211	.7199	1.389	5.097	4.697	4.201	3.862
29	.2859	.5709	1.752	4 042	3.725	3.332	3.063
30	.2546	.4527	2.209	3.206	2.954	2.642	2.429
31	.2268	.3590	2.785	2.542	2.342	2.095	1.926
32	.2019	.2847	3.512	2.016	1.858	1.662	1.527
33	.1798	.2258	4.429	1.599	1.473	1.318	1.211
34	.1601	.1791	5.584	1.268	1.168	1.045	0.9606
35	.1426	.1420	7.042	1.006	0.9265	0.8288	.7618
36	.1270	.1126	8.879	0.7974	.7347	.6572	.6041
37	.1131	.089 31	11.20	.6324	.5827	.5212	.4791
38	.1007	.070 83	14.12	.5015	.4621	.4133	.3799
39	.089 69	.056 17	17.80	.3977	.3664	.3278	.3013
40	.079 87	.044 54	22.45	.3154	.2906	.2600	.2390

*Length at 20° C. of a wire whose resistance is 1 ohm at the stated temperatures.

WIRE TABLE, STANDARD ANNEALED COPPER (*continued*).

American Wire Gage (B. & S.). Metric Units (*continued*).

Gage No.	Diameter in mm. at 20° C.	Ohms per Kilogram.			Grams per Ohm.
		0° C.	20° C.	50° C.	20° C.
0000	11.68	0.000 155 4	0.000 168 7	0.000 188 6	5 928 000.
000	10.40	.000 247 2	.000 268 2	.000 299 9	3 728 000.
00	9.266	.000 393 0	.000 426 5	.000 476 8	2 344 000.
0	8.252	.000 624 9	000 678 2	.000 758 2	1 474 000.
1	7.348	.000 993 6	.001 078	.001 206	927 300.
2	6.544	.001 580	.001 715	.001 917	583 200.
3	5.827	.002 512	.002 726	.003 048	366 800.
4	5.189	.003 995	.004 335	.004 846	230 700.
5	4.621	.006 352	.006 893	.007 706	145 100.
6	4.115	.010 10	.010 96	.012 25	91 230.
7	3.665	.016 06	.017 43	.019 48	57 380.
8	3.264	.025 53	.027 71	.030 98	36 080.
9	2.906	.040 60	.044 06	.049 26	22 690.
10	2.588	.064 56	.070 07	.078 33	14 270.
11	2.305	.1026	.1114	.1245	8976.
12	2.053	.1632	.1771	.1980	5645
13	1.828	.2595	.2817	.3149	3550.
14	1.628	.4127	.4479	.5007	2233.
15	1.450	.6562	.7122	.7961	1404.
16	1.291	1.043	1.132	1.266	883.1
17	1.150	1.659	1.801	2.013	555.4
18	1.024	2.638	2.863	3.201	349.3
19	0.9116	4.194	4.552	5.089	219.7
20	.8118	6.670	7.238	8.092	138.2
21	.7230	10.60	11.51	12.87	86.88
22	.6438	16.86	18.30	20.46	54.64
23	.5733	26.81	29.10	32.53	34.36
24	.5106	42.63	46.27	51.73	21.61
25	.4547	67.79	73.57	82.25	13.59
26	.4049	107.8	117.0	130.8	8.548
27	.3606	171.4	186.0	207.9	5.376
28	.3211	272.5	295.8	330.6	3.381
29	.2859	433.3	470.3	525.7	2.126
30	.2546	689.0	747.8	836.0	1.337
31	.2268	1096.	1189.	1329.	0.8410
32	.2019	1742.	1891.	2114.	.5289
33	.1798	2770.	3006.	3361.	.3326
34	.1601	4404.	4780.	5344.	.2092
35	.1426	7003.	7601.	8497.	.1316
36	.1270	11140.	12090.	13510.	.082 74
37	.1131	17710.	19220.	21480.	.052 04
38	.1007	28150.	30560.	34160.	.032 73
39	.089 69	44770.	48590.	54310.	.020 58
40	.079 87	71180.	77260.	86360.	.012 94

Hard-Drawn Aluminum Wire at 20° C. (or, 68° F.).

American Wire Gage (B. & S.). English Units.

Gage No.	Diameter in Mils.	Cross Section. Circular Mils.	Cross Section. Square Inches.	Ohms per 1000 Feet.	Pounds per 1000 Feet.	Pounds per Ohm.	Feet per Ohm.
0000	460.	212 000.	0.166	0.0804	195.	2420.	12 400.
000	410.	168 000.	.132	.101	154.	1520.	9860.
00	365.	133 000.	.105	.128	122.	957.	7820.
0	325.	106 000.	.0829	.161	97.0	602.	6200.
1	289.	83 700.	.0657	.203	76.9	379.	4920.
2	258.	66 400.	.0521	.256	61.0	238.	3900.
3	229.	52 600.	.0413	.323	48.4	150.	3090.
4	204.	41 700.	.0328	.408	38.4	94.2	2450.
5	182.	33 100.	.0260	.514	30.4	59.2	1950.
6	162.	26 300.	.0206	.648	24.1	37.2	1540.
7	144.	20 800.	.0164	.817	19.1	23.4	1220.
8	128.	16 500.	.0130	1.03	15.2	14.7	970.
9	114.	13 100.	.0103	1.30	12.0	9.26	770.
10	102.	10 400.	.008 15	1.64	9.55	5.83	610.
11	91.	8230.	.006 47	2.07	7.57	3.66	484.
12	81.	6530.	.005 13	2.61	6.00	2.30	384.
13	72.	5180.	.004 07	3.29	4.76	1.45	304.
14	64.	4110.	.003 23	4.14	3.78	0.911	241.
15	57.	3260.	.002 56	5.22	2.99	.573	191.
16	51.	2580.	.002 03	6.59	2.37	.360	152.
17	45.	2050.	.001 61	8.31	1.88	.227	120.
18	40.	1620.	.001 28	10.5	1.49	.143	95.5
19	36.	1290.	.001 01	13.2	1.18	.0897	75.7
20	32.	1020.	.000 802	16.7	0.939	.0564	60.0
21	28.5	810.	.000 636	21.0	.745	.0355	47.6
22	25.3	642.	.000 505	26.5	.591	.0223	37.8
23	22.6	509.	.000 400	33.4	.468	.0140	29.9
24	20.1	404.	.000 317	42.1	.371	.008 82	23.7
25	17.9	320.	.000 252	53.1	.295	.005 55	18.8
26	15.9	254.	.000 200	67.0	.234	.003 49	14.9
27	14.2	202.	.000 158	84.4	.185	.002 19	11.8
28	12.6	160.	.000 126	106.	.147	.001 38	9.39
29	11.3	127.	.000 099 5	134.	.117	.000 868	7.45
30	10.0	101.	.000 078 9	169.	.0924	.000 546	5.91
31	8.9	79.7	.000 062 6	213.	.0733	.000 343	4.68
32	8.0	63.2	.000 049 6	269.	.0581	.000 216	3.72
33	7.1	50.1	.000 039 4	339.	.0461	.000 136	2.95
34	6.3	39.8	.000 031 2	428.	.0365	.000 085 4	2.34
35	5.6	31.5	.000 024 8	540.	.0290	.000 053 7	1.85
36	5.0	25.0	.000 019 6	681.	.0230	.000 033 8	1.47
37	4.5	19.8	.000 015 6	858.	.0182	.000 021 2	1.17
38	4.0	15.7	.000 012 3	1080.	.0145	.000 013 4	0.924
39	3.5	12.5	.000 009 79	1360.	.0115	.000 008 40	.733
40	3.1	9.9	.000 007 77	1720.	.0091	.000 005 28	.581

TABLE 413.—ALUMINUM WIRE TABLE.

Hard-Drawn Aluminum Wire at 20° C.

American Wire Gage (B. & S.) Metric Units.

Gage No.	Diameter in mm.	Cross Section in mm.²	Ohms per Kilometer.	Kilograms per Kilometer.	Grams per Ohm.	Meters per Ohm.
0000	11.7	107.	0.264	289.	1 100 000.	3790.
000	10.4	85.0	.333	230.	690 000.	3010.
00	9.3	67.4	.419	182.	434 000.	2380.
0	8.3	53.5	.529	144.	273 000.	1890.
1	7.3	42.4	.667	114.	172 000.	1500.
2	6.5	33.6	.841	90.8	108 000.	1190.
3	5.8	26.7	1.06	72.0	67 900.	943.
4	5.2	21.2	1.34	57.1	42 700.	748.
5	4.6	16.8	1.69	45.3	26 900.	593.
6	4.1	13.3	2.13	35.9	16 900.	470.
7	3.7	10.5	2.68	28.5	10 600.	373.
8	3.3	8.37	3.38	22.6	6680.	296.
9	2.91	6.63	4.26	17.9	4200.	235.
10	2.59	5.26	5.38	14.2	2640.	186.
11	2.30	4.17	6.78	11.3	1660.	148.
12	2.05	3.31	8.55	8.93	1050.	117.
13	1.83	2.62	10.8	7.08	657.	92.8
14	1.63	2.08	13.6	5.62	413.	73.6
15	1.45	1.65	17.1	4.46	260.	58.4
16	1.29	1.31	21.6	3.53	164.	46.3
17	1.15	1.04	27.3	2.80	103.	36.7
18	1.02	0.823	34.4	2.22	64.7	29.1
19	0.91	.653	43.3	1.76	40.7	23.1
20	.81	.518	54.6	1.40	25.6	18.3
21	.72	.411	68.9	1.11	16.1	14.5
22	.64	.326	86.9	0.879	10.1	11.5
23	.57	.258	110.	.697	6.36	9.13
24	.51	.205	138.	.553	4.00	7.24
25	.45	.162	174.	.438	2.52	5.74
26	.40	.129	220.	.348	1.58	4.55
27	.36	.102	277.	.276	0.995	3.61
28	.32	.0810	349.	.219	.626	2.86
29	.29	.0642	440.	.173	.394	2.27
30	.25	.0509	555.	.138	.248	1.80
31	.227	.0404	700.	.109	.156	1.43
32	.202	.0320	883.	.0865	.0979	1.13
33	.180	.0254	1110.	.0686	.0616	0.899
34	.160	.0201	1400.	.0544	.0387	.712
35	.143	.0160	1770.	.0431	.0244	.565
36	.127	.0127	2230.	.0342	.0153	.448
37	.113	.0100	2820.	.0271	.009 63	.355
38	.101	.0080	3550.	.0215	.006 06	.282
39	.090	.0063	4480.	.0171	.003 81	.223
40	.080	.0050	5640.	.0135	.002 40	.177

TABLE 414.—Ratio of Alternating to Direct Current Resistances for Copper Wires.

This table gives the ratio of the resistance of straight copper wires with alternating currents of different frequencies to the value of the resistance with direct currents.

Diameter of wire in millimeters.	Frequency f =					
	60	100	1000	10,000	100,000	1,000,000
0.05	—	—	—	—	—	*1.001
0.1	—	—	—	—	*1.001	1.008
0.25	—	—	—	—	1.003	1.247
0.5	—	—	—	*1.001	1.047	2.240
1.0	—	—	—	1.008	1.503	4.19
2.0	—	—	1.001	1.120	2.756	8.10
3.	—	—	1.006	1.437	4.00	12.0
4.	—	—	1.021	1.842	5.24	17.4
5.	—	*1.001	1.047	2.240	6.49	19.7
7.5	1.001	1.002	1.210	3.22	7.50	29.7
10.	1.003	1.008	1.503	4.19	12.7	39.1
15.	1.016	1.038	2.136	6.14	18.8	—
20.	1.044	1.120	2.756	8.10	25.2	—
25.	1.105	1.247	3.38	10.1	28.3	—
40.	1.474	1.842	5.24	17.4	—	—
100.	3.31	4.19	13.7	39.1	—	—

Values between 1.000 and 1.001 are indicated by *1.001.

The values are for wires having an assumed conductivity of 1.60 microhm-cms; for copper wires at room temperatures the values are slightly less than as given in table.

The change of resistance of wire other than copper (iron wires excepted) may be calculated from the above table by taking it as proportional to $d\sqrt{f/\rho}$ where d = diameter, f the frequency and ρ the resistivity.

If a given wire be wound into a solenoid, its resistance, at a given frequency, will be greater than the values in the table, which apply to straight wires only. The resistance in this case is a complicated function of the pitch and radius of the winding, the frequency, and the diameter of the wire, and is found by experiment to be sometimes as much as twice the value for a straight wire.

TABLE 415.—Maximum Diameter of Wires for High-frequency Alternating-to-direct-current Resistance Ratio of 1.01.

Frequency ÷ 10^6....	0.1	0.2	0.4	0.6	0.8	1.0	1.2	1.5	2.0	3.0
Wave-length, meters	3000	1500	750	500	375	300	250	200	150	100
Material.	Diameter in centimeters.									
Copper	0.0356	0.0251	0.0177	0.0145	0.0125	0.0112	0.0102	0.0092	0.0079	0.0065
Silver	0.0345	0.0244	0.0172	0.0141	0.0122	0.0109	0.0099	0.0089	0.0077	0.0063
Gold	0.0420	0.0297	0.0210	0.0172	0.0149	0.0133	0.0121	0.0108	0.0094	0.0077
Platinum	0.1120	0.0793	0.0560	0.0457	0.0396	0.0354	0.0323	0.0290	0.0250	0.0205
Mercury	0.264	0.187	0.132	0.1080	0.0936	0.0836	0.0763	0.0683	0.0591	0.0483
Manganin	0.1784	0.1261	0.0892	0.0729	0.0631	0.0564	0.0515	0.0461	0.0399	0.0325
Constantan	0.1892	0.1337	0.0946	0.0772	0.0664	0.0598	0.0546	0.0488	0.0423	0.0345
German silver	0.1942	0.1372	0.0970	0.0792	0.0692	0.0614	0.0560	0.0500	0.0434	0.0354
Graphite	0.765	0.541	0.383	0.312	0.271	0.242	0.221	0.197	0.171	0.140
Carbon	1.60	1.13	0.801	0.654	0.565	0.506	0.462	0.414	0.358	0.292
Iron μ = 1000	0.00263	0.00186	0.00131	0.00108	0.00094	0.00083	0.00076	0.00068	0.00059	0.00048
μ = 500	0.00373	0.00264	0.00187	0.00152	0.00132	0.00118	0.00108	0.00096	0.00084	0.00068
μ = 100	0.00838	0.00590	0.00418	0.00340	0.00295	0.00264	0.00241	0.00215	0.00186	0.00152

Bureau of Standards Circular 74, Radio Instruments and Measurements, 1918.

ELECTROCHEMICAL EQUIVALENTS.

Every gram-ion involved in an electrolytic change requires the same number of coulombs or ampere-hours of electricity per unit change of valency. This constant is 96.494 coulombs or 26.804 ampere-hours per gram-hour (a Faraday) corresponding to an electrochemical equivalent for silver of 0.00111800 gram sec^{-1} amp^{-1}. It is to be noted that the *change of valence* of the element from its state before to that after the electrolytic action should be considered. The valence of a free, uncombined element is to be considered as 0. The same current will electrolyze "chemically equivalent" quantities per unit time. The valence is then included in the "chemically equivalent" quantity. The following table is based on the atomic weights of 1917.

Element.	Change of valency.	Mg per coulomb.	Coulombs per mg	Grams per amp.-hour.	Element.	Change of valency.	Mg per coulomb.	Coulombs per mg	Grams per amp.-hour.
Aluminum....	3	0.0936	10.682	0.3370	Nickel......	1	0.6081	1.6444	2.1892
Chlorine.....	1	0.3675	2.721	1.3229	"	2	0.3041	3.289	1.0946
"	3	0.1225	8.164	0.4410	"	3	0.2027	4.933	0.7298
"	5	0.0735	13.606	0.2646	Oxygen.....	2	0.08291	12.062	0.2985
"	7	0.0525	19.05	0.1890	"	4	0.04145	24.123	0.1492
Copper......	1	0.6588	1.518	2.3717	Platinum....	2	1.0115	0.9887	3.641
"	2	0.3294	3.036	1.1858	"	4	0.5057	1.9773	1.821
Gold........	1	2.044	0.4893	7.357	"	6	0.3372	2.966	1.214
"	3	0.6812	1.468	2.452	Potassium...	1	0.4052	2.468	1.459
Hydrogen....	1	0.010459	5.728	0.037607	Silver.......	1	1.1180	0.89445	4.0248
Lead........	1	2.1473	0.4657	7.7302	Sodium......	1	0.2384	4.195	0.8581
"	2	1.0736	0.9314	3.8651	Tin.........	2	0.6151	1.626	2.214
"	4	0.5368	1.8628	1.9326	"	4	0.3075	3.252	1.107
Mercury.....	1	2.0789	0.4810	7.484	Zinc........	2	0.3387	2.952	1.2194
"	2	1.0394	0.9620	3.742					

The electrochemical equivalent for silver is 0.00111800 g sec^{-1} amp^{-1}. (See p. xxxvii.)

For other elements the electrochemical equivalent = (atomic weight divided by change of valency) times 1/96494 g/sec/amp. or g/coulomb. The equivalent for iodine has been determined at the Bureau of Standards as 0.0013150 (1913).

For a unit change of valency for the diatomic gases Br_2, Cl_2, F_2, H_2, N_2 and O_2 there are required

8.619 coulombs/cm^3 0° C, 76 cm (0.1160 cm^3/coulomb)
2.394 ampere-hours/*l*, 0° C, 76 cm (0.4177 *l*/ampere-hour).

NOTE. — The change of valency for O_2 is usually 2, etc.

CONDUCTIVITY OF ELECTROLYTIC SOLUTIONS.

This subject has occupied the attention of a considerable number of eminent workers in molecular physics, and a few results are here tabulated. It has seemed better to confine the examples to the work of one experimenter, and the tables are quoted from a paper by F. Kohlrausch,* who has been one of the most reliable and successful workers in this field.

The study of electrolytic conductivity, especially in the case of very dilute solutions, has furnished material for generalizations, which may to some extent help in the formation of a sound theory of the mechanism of such conduction. If the solutions are made such that per unit volume of the solvent medium there are contained amounts of the salt proportional to its electrochemical equivalent, some simple relations become apparent. The solutions used by Kohlrausch were therefore made by taking numbers of grams of the pure salts proportional to their electrochemical equivalent, and using a liter of water as the standard of quantity of the solvent. Taking the electrochemical equivalent number as the chemical equivalent or atomic weight divided by the valence, and using this number of grams to the liter of water, we get what is called the normal or gram molecule per liter solution. In the table, m is used to represent the number of gram molecules to the liter of water in the solution for which the conductivities are tabulated. The conductivities were obtained by measuring the resistance of a cell filled with the solution by means of a Wheatstone bridge alternating current and telephone arrangement. The results are for 18° C., and relative to mercury at 0° C., the cell having been standardized by filling with mercury and measuring the resistance. They are supposed to be accurate to within one per cent of the true value.

The tabular numbers were obtained from the measurements in the following manner: —

Let K_{18} = conductivity of the solution at 18° C. relative to mercury at 0° C.

K^w_{18} = conductivity of the solvent water at 18° C. relative to mercury at 0° C.

Then $K_{18} - K^w_{18} = k_{18}$ = conductivity of the electrolyte in the solution measured.

$\frac{k_{18}}{m} = \mu$ = conductivity of the electrolyte in the solution per molecule, or the "specific molecular conductivity."

TABLE 417. — Value of k_{18} for a few Electrolytes.

This short table illustrates the apparent law that the conductivity in very dilute solutions is proportional to the amount of salt dissolved.

m	KCl	NaCl	$AgNO_3$	$KC_2H_3O_2$	K_2SO_4	$MgSO_4$
0.00001	1.216	1.024	1.080	0.939	1.275	1.056
0.00002	2.434	2.056	2.146	1.886	2.532	2.104
0.00006	7.272	6.162	6.462	5.610	7.524	6.216
0.0001	12.09	10.29	10.78	9.34	12.49	10.34

TABLE 418. — Electro-Chemical Equivalents and Normal Solutions.

The following table of the electro-chemical equivalent numbers and the densities of approximately normal solutions of the salts quoted in Table 419 may be convenient. They represent grams per cubic centimeter of the solution at the temperature given.

Salt dissolved.	Grams per liter.	m	Temp. C.	Density.	Salt dissolved.	Grams per liter.	m	Temp. C.	Density.
KCl . . .	74.59	1.0	15.2	1.0457	$\frac{1}{2}K_2SO_4$.	87.16	1.0	18.9	1.0658
NH_4Cl . .	53.55	1.0009	18.6	1.0152	$\frac{1}{2}Na_2SO_4$.	71.09	1.0003	18.6	1.0602
NaCl . . .	58.50	1.0	18.4	1.0391	$\frac{1}{2}Li_2SO_4$.	55.09	1.0007	18.6	1.0445
LiCl . . .	42.48	1.0	18.4	1.0227	$\frac{1}{2}MgSO_4$.	60.17	1.0023	18.6	1.0573
$\frac{1}{2}BaCl_2$ . .	104.0	1.0	18.6	1.0888	$\frac{1}{2}ZnSO_4$.	80.58	1.0	5.3	1.0794
$\frac{1}{2}ZnCl_2$ . .	68.0	1.012	15.0	1.0592	$\frac{1}{2}CuSO_4$.	79.9	1.001	18.2	1.0776
KI	165.9	1.0	18.6	1.1183	$\frac{1}{2}K_2CO_3$.	69.17	1.0006	18.3	1.0576
KNO_3 . .	101.17	1.0	18.6	1.0601	$\frac{1}{2}Na_2CO_3$.	53.04	1.0	17.9	1.0517
$NaNO_3$ . .	85.08	1.0	18.7	1.0542	KOH . .	56.27	1.0025	18.8	1.0477
$AgNO_3$ . .	169.9	1.0	–	–	HCl . .	36.51	1.0041	18.6	1.0161
$\frac{1}{2}Ba(NO_3)_2$.	65.28	0.5	–	–	HNO_3 . .	63.13	1.0014	18.6	1.0318
$KClO_3$ . .	61.29	0.5	18.3	1.0367	$\frac{1}{2}H_2SO_4$.	49.06	1.0006	18.9	1.0300
$KC_2H_3O_2$.	98.18	1.0005	18.6	1.0467					

* "Wied. Ann." vol. 26, pp. 161–226, 1885.

Table 419.

SPECIFIC MOLECULAR CONDUCTIVITY μ : MERCURY $= 10^8$.

Salt dissolved.	$m =$ 10	5	3	1	0.5	0.1	.05	.03	.01
$\frac{1}{2}K_2SO_4$	–	–	–	–	672	736	897	959	1098
KCl	–	–	827	919	958	1047	1083	1107	1147
KI	–	770	900	968	997	1069	1102	1123	1161
NH_4Cl	–	752	825	907	948	1035	1078	1101	1142
KNO_3	–	–	572	752	839	983	1037	1067	1122
$\frac{1}{2}BaCl_2$	–	–	487	658	725	861	904	939	1006
$KClO_3$	–	–	–	–	799	927	(976)	1006	1053
$\frac{1}{2}BaN_2O_6$	–	–	–	–	531	755	828	(870)	951
$\frac{1}{2}CuSO_4$	–	–	150	241	288	424	479	537	675
$AgNO_3$	–	351	448	635	728	886	936	(966)	1017
$\frac{1}{2}ZnSO_4$	–	82	146	249	302	431	500	556	685
$\frac{1}{2}MgSO_4$	–	82	151	270	330	474	532	587	715
$\frac{1}{2}Na_2SO_4$	–	–	–	475	559	734	784	828	906
$\frac{1}{2}ZnCl_2$	60	180	280	514	601	768	817	851	915
$NaCl$	–	398	528	695	757	865	897	(920)	962
$NaNO_3$	–	–	430	617	694	817	855	877	907
$KC_2H_3O_2$	30	240	381	594	671	784	820	841	879
$\frac{1}{2}Na_2CO_3$	–	–	254	427	510	682	751	799	899
$\frac{1}{2}H_2SO_4$	660	1270	1560	1820	1899	2084	2343	2515	2855
C_2H_4O	0.5	2.6	5.2	12	19	43	62	79	132
HCl	600	1420	2010	2780	3017	3244	3330	3369	3416
HNO_3	610	1470	2070	2770	2991	3225	3289	3328	3395
$\frac{1}{3}H_3PO_4$	148	160	170	200	250	430	540	620	790
KOH	423	990	1314	1718	1841	1986	2045	2078	2124
NH_3	0.5	2.4	3.3	8.4	12	31	43	50	92
Salt dissolved.	**.006**	**.002**	**.001**	**.0006**	**.0002**	**.0001**	**.00006**	**.00002**	**.00001**
$\frac{1}{2}K_2SO_4$	1130	1181	1207	1220	1241	1249	1254	1266	1275
KCl	1162	1185	1193	1199	1209	1209	1212	1217	1216
KI	1176	1197	1203	1209	1214	1216	1216	1216	1207
NH_4Cl	1157	1180	1190	1197	1204	1209	1215	1209	1205
KNO_3	1140	1173	1180	1190	1199	1207	1220	1198	1215
$\frac{1}{2}BaCl_2$	1031	1074	1092	1102	1118	1126	1133	1144	1142
$KClO_3$	1068	1091	1101	1109	1119	1122	1126	1135	1141
$\frac{1}{2}BaN_2O_6$	982	1033	1054	1066	1084	1096	1100	1114	1114
$\frac{1}{2}CuSO_4$	740	873	950	987	1039	1062	1074	1084	1086
$AgNO_3$	1033	1057	1068	1069	1077	1078	1077	1073	1080
$\frac{1}{2}ZnSO_4$	744	861	919	953	1001	1023	1032	1047	1060
$\frac{1}{2}MgSO_4$	773	881	935	967	1015	1034	1036	1052	1056
$\frac{1}{2}Na_2SO_4$	933	980	998	1009	1026	1034	1038	1056	1054
$\frac{1}{2}ZnCl_2$	939	979	994	1004	1020	1029	1031	1035	1036
$NaCl$	976	998	1008	1014	1018	1029	1027	1028	1024
$NaNO_3$	921	942	952	956	966	975	970	972	975
$KC_2H_3O_2$	891	913	919	923	933	934	935	943	939
$\frac{1}{2}Na_2CO_3$	956	1010	1037	1046	988	874	790	715	697*
$\frac{1}{2}H_2SO_4$	3001	3240	3316	3342	3280	3118	2927	2077	1413*
C_2H_4O	170	283	380	470	796	995	1133	1328	1304*
HCl	3438	3455	3455	3440	3340	3170	2968	2057	1254*
HNO_3	3421	3448	3427	3408	3285	3088	2863	1904	1144*
$\frac{1}{3}H_3PO_4$	858	945	968	977	920	837	746	497	402*
KOH	2141	2140	2110	2074	1892	1689	1474	845	747*
NH_3	116	190	260	330	500	610	690	700	560*

* Acids and alkaline salts show peculiar irregularities.

LIMITING VALUES OF μ. TEMPERATURE COEFFICIENTS.

TABLE 420.—Limiting Values of μ.

This table shows limiting values of $\mu = \frac{k}{m} \cdot 10^8$ for infinite dilution for neutral salts, calculated from Table **271**.

Salt.	μ	Salt.	μ	Salt.	μ	Salt.	μ
$\frac{1}{2}K_2SO_4$.	1280	$\frac{1}{2}BaCl_2$.	1150	$\frac{1}{2}MgSO_4$.	1080	$\frac{1}{2}H_2SO_4$.	3700
KCl . . .	1220	$\frac{1}{2}KClO_3$.	1150	$\frac{1}{2}Na_2SO_4$.	1060	HCl . .	3500
KI . . .	1220	$\frac{1}{2}BaN_2O_6$.	1120	$\frac{1}{2}ZnCl_2$ . .	1040	HNO_3 . .	3500
NH_4Cl . .	1210	$\frac{1}{2}CuSO_4$.	1100	NaCl . .	1030	$\frac{1}{3}H_3PO_4$.	1100
KNO_3 . .	1210	$AgNO_3$.	1090	$NaNO_3$.	980	KOH . .	2200
–	–	$\frac{1}{2}ZnSO_4$.	1080	$K_2C_2H_3O_2$	940	$\frac{1}{2}Na_2CO_3$.	1400

If the quantities in Table 420 be represented by curves, it appears that the values of the specific molecular conductivities tend toward a limiting value as the solution is made more and more dilute. Although these values are of the same order of magnitude, they are not equal, but depend on the nature of both the ions forming the electrolyte.

When the numbers in Table 421 are multiplied by Hittorf's constant, or 0.00011, quantities ranging between 0.14 and 0.10 are obtained which represent the velocities in millimetres per second of the ions when the electromotive force gradient is one volt per millimetre.

Specific molecular conductivities in general become less as the concentration is increased, which may be due to mutual interference. The decrease is not the same for different salts, but becomes much more rapid in salts of high valence.

Salts having acid or alkaline reactions show marked differences. They have small specific molecular conductivity in very dilute solutions, but as the concentration is increased the conductivity rises, reaches a maximum and again falls off. Kohlrausch does not believe that this can be explained by impurities. H_3PO_4 in dilute solution seems to approach a monobasic acid, while H_2SO_4 shows two maxima, and like H_3PO_4 approaches in very weak solution to a monobasic acid.

Kohlrausch concludes that the law of independent migration of the ions in media like water is sustained.

TABLE 421.—Temperature Coefficients.

The temperature coefficient in general diminishes with dilution, and for very dilute solutions appears to approach a common value. The following table gives the temperature coefficient for solutions containing 0.01 gram molecule of the salt.

Salt.	Temp. Coeff.	Salt.	Temp. Coeff.	Salt.	Temp. Coeff.	Salt.	Temp. Coeff.
KCl . . .	0.0221	KI . . .	0.0219	$\frac{1}{2}K_2SO_4$.	0.0223	$\frac{1}{2}K_2CO_3$ . .	0.0249
NH_4Cl . .	0.0226	KNO_3 . .	0.0216	$\frac{1}{2}Na_2SO_4$.	0.0240	$\frac{1}{2}Na_2CO_3$ . .	0.0265
NaCl . .	0.0238	$NaNO_3$ . .	0.0226	$\frac{1}{2}Li_2SO_4$.	0.0242	KOH . . .	0.0194
LiCl . . .	0.0232	$AgNO_3$ . .	0.0221	$\frac{1}{2}MgSO_4$.	0.0236	HCl . . .	0.0159
$\frac{1}{2}BaCl_2$ . .	0.0234	$\frac{1}{2}Ba(NO_3)_2$	0.0224	$\frac{1}{2}ZnSO_4$.	0.0234	HNO_3 . . .	0.0162
$\frac{1}{2}ZnCl_2$ . .	0.0239	$KClO_3$ . .	0.0219	$\frac{1}{2}CuSO_4$.	0.0229	$\frac{1}{2}H_2SO_4$ . .	0.0125
$\frac{1}{2}MgCl_2$.	0.0241	$KC_2H_3O_2$.	0.0229	–	–	$\frac{1}{2}H_2SO_4$ for $m = .001$	0.0159

THE EQUIVALENT CONDUCTIVITY OF SALTS, ACIDS AND BASES IN AQUEOUS SOLUTIONS.

In the following table the equivalent conductance is expressed in reciprocal ohms. The concentration is expressed in milli-equivalents of solute per litre of solution at the temperature to which the conductance refers. (In the cases of potassium hydrogen sulphate and phosphoric acid the concentration is expressed in milli-formula-weights of solute, $KHSO_4$ or H_3PO_4, per liter of solution, and the values are correspondingly the modal, or "formal," conductances.) Except in the cases of the strong acids the conductance of the water was subtracted, and for sodium acetate, ammonium acetate and ammonium chloride the values have been corrected for the hydrolysis of the salts. The atomic weights used were those of the International Commission for 1905, referred to oxygen as 16.00. Temperatures are on the hydrogen gas scale.

$$\text{Concentration in } \frac{\text{gram equivalents}}{\text{1000 liter}}.$$

$$\text{Equivalent conductance in } \frac{\text{reciprocal ohms per centimeter cube}}{\text{gram equivalents per cubic centimeter}}.$$

Substance.	Concentration.	Equivalent conductance at the following °C temperatures.									
		18°	25°	50°	75°	100°	128°	156°	218°	281°	306°
Potassium chloride .	0	130.1	(152.1)	(232.5)	(321.5)	414	(519)	625	825	1005	1120
" " .	2	126.3	146.4	–	–	393	–	588	779	930	1008
" " .	10	122.4	141.5	215.2	295.2	377	470	560	741	874	910
" " .	80	113.5	–	–	–	342	–	498	638	723	720
" " .	100	112.0	129.0	194.5	264.6	336	415	490			
Sodium chloride . .	0	109.0	–	–	–	362	–	555	760	970	1080
" " . .	2	105.6	–	–	–	349	–	534	722	895	955
" " . .	10	102.0	–	–	–	336	–	511	685	820	860
" " . .	80	93 5	–	–	–	301	–	450	500	674	680
" " . .	100	92.0	–	–	–	296	–	442			
Silver nitrate . . .	0	115.8	–	–	–	367	–	570	780	965	1065
" " . . .	2	112.2	–	–	–	353	–	539	727	877	935
" " . . .	10	108.0	–	–	–	337	–	507	673	790	818
" " . . .	20	105.1	–	–	–	326	–	488	639		
" " . . .	40	101.3	–	–	–	312	–	462	599	680	680
" " . . .	80	96.5	–	–	–	294	–	432	552	614	604
" " . . .	100	94.6	–	–	–	289					
Sodium acetate . .	0	78.1	–	–	–	285	–	450	660	–	924
" " . .	2	74.5	–	–	–	268	–	421	578	–	801
" " . .	10	71.2	–	–	–	253	–	396	542	–	702
" " . .	80	63.4	–	–	–	221	–	340	452		
Magnesium sulphate	0	114.1	–	–	–	426	–	690	1080		
" " .	2	94.3	–	–	–	302	–	377	260		
" " .	10	76.1	–	–	–	234	–	241	143		
" " .	20	67.5	–	–	–	190	–	195	110		
" " .	40	59.3	–	–	–	160	–	158	88		
" " .	80	52.0	–	–	–	136	–	133	75		
" " .	100	49.8	–	–	–	130	–	126			
" " .	200	43.1	–	–	–	110	–	109			
Ammonium chloride	0	131.1	152.0	–	–	(415)	–	(628)	(841)	–	(1176)
" " .	2	126.5	146.5	–	–	399	–	601	801	–	1031
" " .	10	122.5	141.7	–	–	382	–	573	758	–	925
" " .	30	118.1	–	–	–	–	–	–	–	–	828
Ammonium acetate .	0	(99.8)	–	–	–	(338)	–	(523)			
" " .	10	91.7	–	–	–	300	–	456			
" " .	25	88.2	–	–	–	286	–	426			

From the investigations of Noyes, Melcher, Cooper, Eastman and Kato; Journal of the American Chemical Society, 30, p. 335, 1908.

THE EQUIVALENT CONDUCTIVITY OF SALTS, ACIDS AND BASES IN AQUEOUS SOLUTIONS.

Substance.	Concentration.	Equivalent conductance at the following ° C temperatures.									
		18°	25°	50°	75°	100°	128°	156°	218°	281°	306°
Barium nitrate . . .	0	116.9	–	–	–	385	–	600	840	1120	1300
" " . . .	2	109.7	–	–	–	352	–	536	715	828	824
" " . . .	10	101.0	–	–	–	322	–	481	618	658	615
" " . . .	40	88.7	–	–	–	280	–	412	507	503	448
" " . . .	80	81.6	–	–	–	258	–	372	449	430	
" " . . .	100	79.1	–	–	–	249					
Potassium sulphate .	0	132.8	–	–	–	455	–	715	1065	1460	1725
" " . .	2	124.8	–	–	–	402	–	605	806	893	867
" " . .	10	115.7	–	–	–	365	–	537	672	687	637
" " . .	40	104.2	–	–	–	320	–	455	545	519	466
" " . .	80	97.2	–	–	–	294	–	415	482	448	396
" " . .	100	95.0	–	–	–	286					
Hydrochloric acid .	0	379.0	–	–	–	850	–	1085	1265	1380	1424
" " . .	2	373.6	–	–	–	826	–	1048	1217	1332	1337
" " . .	10	368.1	–	–	–	807	–	1016	1168	1226	1162
" " . .	80	353.0	–	–	–	762	–	946	1044	1046	862
" " . .	100	350.6	–	–	–	754	–	929	1006		
Nitric acid	0	377.0	421.0	570	706	826	945	1047	(1230)	–	(1380)
" "	2	371.2	413.7	559	690	806	919	1012	1166	–	1156
" "	10	365.0	406.0	548	676	786	893	978			
" "	50	353.7	393.3	528	649	750	845	917			
" "	100	346.4	385.0	516	632	728	817	880	–	–	454*
Sulphuric acid . . .	0	383.0	(429)	(591)	(746)	891	(1041)	1176	1505	–	(2030)
" " . . .	2	353.9	390.8	501	561	571	551	536	563	–	637
" " . . .	10	309.0	337.0	406	435	446	460	481	533		
" " . . .	50	253.5	273.0	323	356	384	417	448	502		
" " . . .	100	233.3	251.2	300	336	369	404	435	483	–	474*
Potassium hydrogen sulphate . . .	2	455.3	506.0	661.0	754	784	773	754			
	50	295.5	318.3	374.4	403	422	446	477			
	100	263.7	283.1	329.1	354	375	402	435			
Phosphoric acid . .	0	338.3	376	510	631	730	839	930			
" " . .	2	283.1	311.9	401	464	498	508	489			
" " . .	10	203.0	222.0	273	300	308	298	274			
" " . .	50	122.7	132.6	157.8	168.6	168	158	142			
" " . .	100	96.5	104.0	122.7	129.9	128	120	108			
Acetic acid	0	(347.0)	–	–	–	(773)	–	(980)	(1165)	–	(1268)
" "	10	14.50	–	–	–	25.1	–	22.2	14.7		
" "	30	8.50	–	–	–	14.7	–	13.0	8.65		
" "	80	5.22	–	–	–	9.05	–	8.00	5.34		
" "	100	4.67	–	–	–	8.10	–	–	4.82	–	1.57
Sodium hydroxide .	0	216.5	–	–	–	594	–	835	1060		
" " . .	2	212.1	–	–	–	582	–	814			
" " . .	20	205.8	–	–	–	559	–	771	930		
" " . .	50	200.6	–	–	–	540	–	738	873		
Barium hydroxide .	0	222	256	389	(520)	645	(760)	847			
" " . . .	2	215	–	359	4	591					
" " . . .	10	207	235	342	449	548	664	722			
" " . . .	50	191.1	215.1	308	399	478	549	593			
" " . . .	100	180.1	204.2	291	373	443	503	531			
Ammonium hydroxide	0	(238)	(271)	(404)	(526)	(647)	(764)	(908)	(1141)	–	(1406)
	10	9.66	–	–	–	23.2	–	22.3	15.6		
	30	5.66	–	–	–	13.6	–	13.0			
	100	3.10	3.62	5.35	6.70	7.47	–	7.17	4.82	–	1.33

* These values are at the concentration 80.0.

TABLE 423.

THE EQUIVALENT CONDUCTIVITY OF SOME ADDITIONAL SALTS IN AQUEOUS SOLUTION.

Conditions similar to those of the preceding table except that the atomic weights for 1908 were used.

Substance.	Concentration.	Equivalent conductance at the following ° C temperature.							
		0°	18°	25°	50°	75°	100°	128°	156°
Potassium nitrate . . .	0	80.8	126.3	145.1	219	299	384	485	580
" " . . .	2	78.6	122.5	140.7	212.7	289.9	370.3	460.7	551
" " . . .	12.5	75.3	117.2	134.9	202.9	276.4	351.5	435.4	520.4
" " . . .	50	70.7	109.7	126.3	189.5	257.4	326.1	402.9	476.1
" " . . .	100	67.2	104.5	120.3	180.2	244.1	308.5	379.5	447.3
Potassium oxalate . . .	0	79.4	127.6	147.5	230	322	419	538	653
" " . . .	2	74.9	119.9	139.2	215.9	300.2	389.3	489.1	587
" " . . .	12.5	69.3	111.1	129.2	199.1	275.1	354.1	438.8	524.3
" " . . .	50	63	101	116.5	178.6	244.9	312.2	383.8	449.5
" " . . .	100	59.3	94.6	109.5	167	227.5	288.9	353.2	409.7
" " . . .	200	55.8	88.4	102.3	155	210.9	265.1	321.9	372.1
Calcium nitrate . . .	0	70.4	112.7	130.6	202	282	369	474	575
" " . . .	2	66.5	107.1	123.7	191.9	266.7	346.5	438.4	529.8
" " . . .	12.5	61.6	98.6	114.5	176.2	244	314.6	394.5	473.7
" " . . .	50	55.6	88.6	102.6	157.2	216.2	276.8	343	405.1
" " . . .	100	51.9	82.6	95.8	146.1	199.9	255.5	315.1	369.1
" " . . .	200	48.3	76.7	88.8	135.4	184.7	234.4	288	334.7
Potassium ferrocyanide .	0	98.4	159.6	185.5	288	403	527		
" " .	0.5	91.6	–	171.1					
" " .	2.	84.8	137	158.9	243.8	335.2	427.6		
" " .	12.5	71	113.4	131.6	200.3	271	340		
" " .	50	58.2	93.7	108.6	163.3	219.5	272.4		
" " .	100	53	84.9	98.4	148.1	198.1	245		
" " .	200	48.8	77.8	90.1	135.7	180.6	222.3		
" " .	400	45.4	72.1	83.3	124.8	165.7	203.1		
Barium ferrocyanide . .	0	91	150	176	277	393	521		
" " . .	2	46.9	75	86.2	127.5	166.2	202.3		
" " . .	12.5	30.4	48.8	56.5	83.1	107	129.8		
Calcium ferrocyanide .	0	88	146	171	271	386	512		
" " . .	2	47.1	75.5	86.2	130				
" " . .	12.5	31.2	49.9	57.4					
" " . .	50	24.1	38.5	44.4	64.6	81.9			
" " . .	100	21.9	35.1	40.2	58.4	73.7	84.3		
" " . .	200	20.6	32.9	37.8	55	68.7	77.5		
" " . .	400	20.2	32.2	37.1	54	67.5	76.2		
Potassium citrate . . .	0	76.4	124.6	144.5	228	320	420		
" " . . .	0.5	–	120.1	139.4					
" " . . .	2	71	115.4	134.5	210.1	293.8	381.2		
" " . . .	5	67.6	109.9	128.2	198.7	276.5	357.2		
" " . . .	12.5	62.9	101.8	118.7	183.6	254.2	326		
" " . . .	50	54.4	87.8	102.1	157.5	215.5	273		
" " . . .	100	50.2	80.8	93.9	143.7	196.5	247.5		
" " . . .	300	43.5	69.8	81	123.5	167	209.5		
Lanthanum nitrate . .	0	75.4	122.7	142.6	223	313	413	534	651
" " . .	2	68.9	110.8	128.9	200.5	279.8	363.5	457.5	549
" " . .	12.5	61.4	98.5	114.4	176.7	243.4	311.2	383.4	447.8
" " . .	50	54	86.1	99.7	152.5	207.6	261.4	315.8	357.7
" " . .	100	49.9	79.4	91.8	139.5	189.1	236.7	282.5	316.3
" " . .	200	46	72.1	83.5	126.4	170.2	210.8	249.6	276.2

From the investigations of Noyes and Johnston, Journal of the American Chemical Society, 31, p. 287, 1909.

CONDUCTANCE OF IONS. — HYDROLYSIS OF AMMONIUM ACETATE.

TABLE 424. — The Equivalent Conductance of the Separate Ions.

Ion.	0°	18°	25°	50°	75°	100°	128°	156°
K	40.4	64.6	74.5	115	159	206	263	317
Na	26	43.5	50.9	82	116	155	203	249
NH_4	40.2	64.5	74.5	115	159	207	264	319
Ag	32.9	54.3	63.5	101	143	188	245	299
$\frac{1}{2}$Ba	33	55[2]	65	104	149	200	262	322
$\frac{1}{2}$Ca	30	51[2]	60	98	142	191	252	312
$\frac{1}{3}$La	35	61	72	119	173	235	312	388
Cl	41.1	65.5	75.5	116	160	207	264	318
NO_3	40.4	61.7	70.6	104	140	178	222	263
$C_2H_3O_2$	20.3	34.6	40.8	67	96	130	171	211
$\frac{1}{2}SO_4$	41	68[2]	79	125	177	234	303	370
$\frac{1}{2}C_2O_4$	39	63[2]	73	115	163	213	275	336
$\frac{1}{3}C_6H_5O_7$	36	60	70	113	161	214		
$\frac{1}{4}Fe(CN)_6$	58	95	111	173	244	321		
H	240	314	350	465	565	644	722	777
OH	105	172	192	284	360	439	525	592

From Johnson, Journ. Amer. Chem. Soc., 31, p. 1010, 1909.

TABLE 425. — Hydrolysis of Ammonium Acetate and Ionization of Water.

Temperature.	Percentage hydrolysis.	Ionization constant of water.	Hydrogen-ion concentration in pure water. Equivalents per liter.
t	100_h	$K_W \times 10^{14}$	$C_H \times 10^7$
0	–	0.089	0.30
18	(0.35)	0.46	0.68
25	–	0.82	0.91
100	4.8	48.	6.9
156	18.6	223.	14.9
218	52.7	461.	21.5
306	91.5	168.	13.0

Noyes, Kato, Kanolt, Sosman, No. 63 Publ. Carnegie Inst., Washington.

DIELECTRIC STRENGTH.

TABLE 426. — Steady Potential Difference in Volts required to produce a Spark in Air with Ball Electrodes.

Spark length. cm.	*R* = 0. Points.	*R* = 0.25 cm.	*R* = 0.5 cm.	*R* = 1 cm.	*R* = 2 cm.	*R* = 3 cm.	*R* = ∞. Plates.
0.02	–	–	1560	1530			
0.04	–	–	2460	2430	2340		
0.06	–	–	3300	3240	3060		
0.08	–	–	4050	3990	3810		
0.1	3720	5010	4740	4560	4560	4500	4350
0.2	4680	8610	8490	8490	8370	7770	7590
0.3	5310	11140	11460	11340	11190	10560	10650
0.4	5970	14040	14310	14340	14250	13140	13560
0.5	6300	15990	16950	17220	16650	16470	16320
0.6	6840	17130	19740	20070	20070	19380	19110
0.8	8070	18960	23790	24780	25830	26220	24960
1.0	8670	20670	26190	27810	29850	32760	30840
1.5	9960	22770	29970	37260			
2.0	10140	24570	33060	45480			
3.0	11250	28380					
4.0	12210	29580					
5.0	13050						

Based on the results of Baille, Bichat-Blondot, Freyburg, Liebig, Macfarlane, Orgler, Paschen, Quincke, de la Rue, Wolff. For spark lengths from 1 to 200 wave-lengths of sodium light, see Earhart, Phys. Rev. 15, p. 163; Hobbs, Phil. Mag. 10, p. 607, 1905.

TABLE 427. — Alternating Current Potentials required to produce a Spark in Air with various Ball Electrodes.

The potentials given are the maxima of the alternating waves used. Frequency, 33 cycles per second.

Spark length. cm.	*R* = 1 cm.	*R* = 1.92	*R* = 5	*R* = 7.5	*R* = 10	*R* = 15
0.08	3770					
.10	4400	4380	4330	4290	4245	4230
.15	5990	5940	5830	5790	5800	5780
.20	7510	7440	7340	7250	7320	7330
.25	9045	8970	8850	8710	8760	8760
0.30	10480	10400	10270	10130	10180	10150
.35	11980	11890	11670	11570	11610	11590
.40	13360	13300	13100	12930	12980	12970
.45	14770	14700	14400	14290	14330	14320
.50	16140	16070	15890	15640	15690	15690
0.6	18700	18730	18550	18300	18350	18400
.7	21350	21380	21140	20980	20990	21000
.8	23820	24070	23740	23490	23540	23550
0.9	26190	26640	26400	26130	26110	26090
1.0	28380	29170	28950	28770	28680	28610
1.2	32400	34100	33790	33660	33640	33620
1.4	35850	38850	38850	38580	38620	38580
1.6	38750	43400	43570	43250	43520	
1.8	40900	–	48300	47900		
2.0	42950	–	–	52400		

Based upon the results of Kawalski, Phil. Mag. 18, p. 699, 1909.

DIELECTRIC STRENGTH.

TABLE 428. — Potential Necessary to produce a Spark in Air between more widely Separated Electrodes.

Spark length, cm.	Dull points. Alternating current.	Steady potentials. Ball electrodes. R=1 cm.	Steady potentials. Ball electrodes. R=2.5 cm.	Steady potentials. Cup electrodes. Projection. 4.5 mm.	Steady potentials. Cup electrodes. Projection. 1.5 mm.	Spark length, cm.	Dull points. Alternating current.	Steady potentials. Ball electrodes. R=1 cm.	Steady potentials. Ball electrodes. R=2.5 cm.
0.3	–	–	–	–	11280	6.0	61000	–	86830
0.5	–	17610	17620	–	17420	7.0	–	52000	–
0.7	–	–	23050	–	22950	8.0	67000	52400	90200
1.0	12000	30240	31390	31400	31260	10.0	73000	74300	91930
1.2	–	33800	36810	–	36700	12.0	82600	–	93300
1.5	–	37930	44310	–	44510	14.0	92000	–	94400
2.0	29200	42320	56000	56500	56530	15.0	–	–	94700
2.5	–	45000	65180	–	68720	16.0	101000	–	101000
3.0	40000	46710	71200	80400	81140	20.0	119000		
3.5	–	–	75300	–	92400	25.0	140600		
4.0	48500	49100	78600	101700	103800	30.0	165700		
4.5	–	–	81540	–	114600	35.0	190900		
5.0	56500	50310	83800	–	126500				
5.5	–	–	–	–	135700				

This table for longer spark lengths contains the results of Voege, Ann. der Phys. 14, 1904, using alternating current and "dull point" electrodes, and the results with steady potential found in the recent very careful work of C. Müller, Ann. d. Phys. 28, p. 585, 1909.

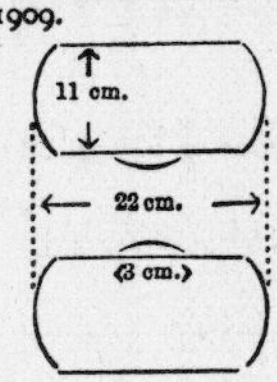

The specially constructed electrodes for the columns headed "cup electrodes" had the form of a projecting knob 3 cm. in diameter and having a height of 4.5 mm. and 1.5 mm. respectively, attached to the plane face of the electrodes. These electrodes give a very satisfactory linear relation between the spark lengths and the voltage throughout the range studied.

TABLE 429. — Effect of the Pressure of the Gas on the Dielectric Strength.

Voltages are given for different spark lengths *l*.

Pressure. cm. Hg.	l=0.04	l=0.06	l=0.08	l=0.10	l=0.20	l=0.30	l=0.40	l=0.50
2	–	–	–	–	744	939	1110	1266
4	–	483	567	648	1015	1350	1645	1915
6	–	582	690	795	1290	1740	2140	2505
10	–	771	933	1090	1840	2450	3015	3580
15	–	1060	1280	1490	2460	3300	4080	4850
25	1110	1420	1725	2040	3500	4800	6000	7120
35	1375	1820	2220	2615	4505	6270	7870	9340
45	1640	2150	2660	3120	5475	7650	9620	11420
55	1820	2420	3025	3610	6375	8950	11290	13455
65	2040	2720	3400	4060	7245	10210	12950	15470
75	2255	3035	3805	4565	8200	11570	14650	17450

This table is based upon the results of Orgler, 1899. See this paper for work on other gases (or Landolt-Börnstein-Meyerhoffer).

For long spark lengths in various gases see Voege, Electrotechn. Z. 28, 1907. For dielectric strength of air and CO_2 in cylindrical air condensers, see Wien, Ann. d. Phys. 29, p. 679, 1909.

DIELECTRIC STRENGTH.

TABLE 430. — Dielectric Strength of Materials.

Potential necessary for puncture expressed in kilovolts per centimeter thickness of the dielectric.

Substance.	Kilovolts per cm	Substance.	Kilovolts per cm.	Substance.	Kilovolts per cm.
Ebonite	300–1100	Oils: Thickness		Papers:	
Empire cloth . .	80–300	Castor 0.2 mm.	190	Beeswaxed . .	770
" paper . .	450	" 1.0 "	130	Blotting . . .	150
Fibre	20	Cottonseed	70	Manilla . . .	25
Fuller board . .	200–300	Lard 0.2 "	140	Paraffined . .	500
Glass	300–1500	" 1.0 "	40	Varnished . .	100–250
Granite (fused) .	90	Linseed, raw 0.2 "	185	Paraffine:	
Guttapercha . . .	80–200	" " 1.0 "	90	Melted . . .	75
Impregnated jute .	20	" boiled 0.2 "	190	Melt point.	
Leatheroid . . .	30–60	" " 1.0 "	80	Solid 43°	350
Linen, varnished .	100–200	Lubricating	50	" 47°	400
Liquid air . . .	40–90	Neatsfoot 0.2 "	200	" 52°	230
Mica: Thickness.		" 1.0 "	90	" 70°	450
Madras 0.1 mm.	1600	Olive 0.2 "	170	Presspaper . . .	45–75
" 1.0 "	300	" 1.0 "	75	Rubber	160–500
Bengal 0.1 "	2200	Paraffin 0.2 "	215	Vaseline	90–130
" 1.0 "	700	" 1.0 "	160	Thickness.	
Canada 0.1 "	1500	Sperm, mineral 0.2 "	180	Xylene 0.2 mm.	140
" 1.0 "	500	" " 1.0 "	85	" 1.0 "	80
South America .	1500	" natural 0.2 "	195		
Micanite . . .	400	" " 1.0 "	90		
		Turpentine 0.2 "	160		
		" 1.0 "	110		

TABLE 431. — Potentials in Volts to Produce a Spark in Kerosene.

Spark length. mm.	Electrodes Balls of Diam. *d*.			
	0.5 cm.	1 cm.	2 cm.	3 cm.
0.1	3800	3400	2750	2200
.2	7500	6450	4800	3500
.3	10250	9450	7450	4600
.4	11750	10750	9100	5600
.5	13050	12400	11000	6900
.6	14000	13550	12250	8250
.8	15500	15100	13850	10450
1.0	16750	16400	15250	12350

Determinations of the dielectric strength of the same substance by different observers do not agree well For a discussion of the sources of error see Mościcki, Electrotechn. Z. 25, 1904.

For more detailed information on the dependence of the sparking distance in oils as a function of the nature of the electrodes, see Edmondson, Phys. Review 6, p. 65, 1898.

DIELECTRIC CONSTANTS.

TABLE 432. — Dielectric Constant (Specific Inductive Capacity) of Gases. Atmospheric Pressure.

Wave-lengths of the measuring current greater than 10000 cm.

Gas.	Temp. ° C.	Dielectric constant referred to Vacuum=1	Dielectric constant referred to Air=1	Authority.
Air	0	1.000590	1.000000	Boltzmann, 1875.
"	–	1.000586	1.000000	Klemenčič, 1885.
Ammonia	20	1.00718	1.00659	Bädeker, 1901.
Carbon bisulphide . . .	0	1.00290	1.00231	Klemenčič.
" "	100	1.00239	1.00180	Bädeker.
Carbon dioxide	0	1.000946	1.000356	Boltzmann.
" "	0	1.000985	1.000399	Klemenčič.
Carbon monoxide	0	1.000690	1.000100	Boltzmann.
" "	0	1.000695	1.000109	Klemenčič.
Ethylene	0	1.00131	1.00072	Boltzmann.
"	0	1.00146	1.00087	Klemenčič.
Hydrochloric acid . . .	100	1.00258	1.00199	Bädeker.
Hydrogen	0	1.000264	0.999674	Boltzmann.
"	0	1.000264	0.999678	Klemenčič.
Methane	0	1.000944	1.000354	Boltzmann.
"	0	1.000953	1.000367	Klemenčič.
Nitrous oxide (N_2O) . .	0	1.00116	1.00057	Boltzmann.
" " " . . .	0	1.00099	1.00041	Klemenčič.
Sulphur dioxide	0	1.00993	1.00934	Bädeker.
" "	0	1.00905	1.00846	Klemenčič.
Water vapor, 4 atmospheres	145	1.00705	1.00646	Bädeker.

TABLE 433. — Variation of the Dielectric Constant with the Temperature.

For variation with the pressure see next table.

If D_θ = the dielectric constant at the temperature θ° C., D_t at the temperature t° C., and α and β are quantities given in the following table, then

$$D_\theta = D_t[1 - \alpha(t-\theta) + \beta(t-\theta)^2].$$

The temperature coefficients are due to Bädeker.

Gas.	α	β	Range of temp. ° C.
Ammonia . .	5.45×10^{-5}	2.59×10^{-7}	10 — 110
Sulphur dioxide	6.19×10^{-5}	1.86×10^{-7}	0 — 110
Water vapor .	1.4×10^{-4}	–	145

The dielectric constant of air at atmospheric pressure but with varying temperature may also be calculated from the fact that $D - 1$ is approximately proportional to the density.

TABLE 434. — Change of the Dielectric Constant of Gases with the Pressure.

Gas.	Temperature, ° C.	Pressure atmos.	Dielectric constant.	Authority.
Air	19	20	1.0108	Tangl, 1907.
"	–	40	1.0218	" "
"	–	60	1.0330	" "
"	–	80	1.0439	" "
"	–	100	1.0548	" "
"	11	20	1.0101	Occhialini, 1905.
"	–	40	1.0196	" "
"	–	60	1.0294	" "
"	–	80	1.0387	" "
"	–	100	1.0482	" "
"	–	120	1.0579	" "
"	–	140	1.0674	" "
"	–	160	1.0760	" "
"	–	180	1.0845	" "
Carbon dioxide . .	15	10	1.008	Linde, 1895.
" " . .	–	20	1.020	" "
" " . .	–	40	1.060	" "
Nitrous oxide, N_2O	15	10	1.010	"
" " "	–	20	1.025	"
" " "	–	40	1.070	"

TABLE 435. — Dielectric Constants of Liquids.

A wave-length greater than 10000 centimeters is denoted by ∞.

Substance.	Temp. ° C.	Wave-length, cm.	Dielectric constant.	Authority.	Substance.	Temp. ° C.	Wave-length, cm.	Dielectric constant.	Authority.
Alcohol:					Alcohol:				
Amyl . . .	frozen	∞	2.4	1	Methyl . .	−50	∞	45.3	1
" . . .	−100	"	30.1	1	" . . .	0	"	35.0	1
" . . .	−50	"	23.0	1	" . . .	+20	"	31.2	1
" . . .	0	"	17.4	1	" . . .	17	75	33.2	2
" . . .	+20	"	16.0	1	Propyl . .	−120	∞	46.2	1
" . . .	18	200	10.8	2	" . . .	−60	"	33.7	1
" . . .	18	73	4.7	2	" . . .	0	"	24.8	1
Ethyl . . .	frozen	∞	2.7	1	" . . .	+20	"	22.2	1
" . . .	−120	"	54.6	1	" . . .	15	75	12.3	2
" . . .	−80	"	44.3	1	Acetone . . .	−80	∞	33.8	5
" . . .	−40	"	35.3	1	" . . .	0	"	26.6	5
" . . .	0	"	28.4	1	" . . .	15	1200	21.85	6
" . . .	+20	"	25.8	1	" . . .	17	73	20.7	7
" . . .	17	200	24.4	2	Acetic acid .	18	∞	9.7	8
" . . .	"	75	23.0	2	" " . .	15	1200	10.3	6
" . . .	"	53	20.6	3	" " . .	17	200	7.07	2
" . . .	"	4	8.8	3	" " . .	19	75	6.29	2
" . . .	"	0.4	5.0	4	Amyl acetate .	19	∞	4.81	9
Methyl . .	frozen	∞	3.07	1	Amylene . .	16	"	2.20	10
" . . .	−100	"	58.0	1					

References on page 358.

DIELECTRIC CONSTANTS OF LIQUIDS.

A wave-length greater than 10000 centimeters is desiguated by ∞.

Substance.	Temp. °C.	Wave-length cm.	Diel. const.	Authority.
Aniline	18	∞	7.316	11
Benzol (benzene)	18	"	2.288	"
" "	19	73	2.26	2
Bromine	23	84	3.18	12
Carbon bisulphide	20	∞	2.626	13
" "	17	73	2.64	2
Chloroform	18	∞	5.2	11
"	17	73	4.95	2
Decane	14	∞	1.97	10
Decylene	17	"	2.24	"
Ethyl ether	−80	∞	7.05	5
" "	−40	"	5.67	"
" "	0	"	4.68	"
" "	18	"	4.368	11
" "	20	"	4.30	13
" "	60	"	3.65	"
" "	100	"	3.12	"
" "	140	"	2.66	"
" "	180	"	2.12	"
" "	Crit. temp. 192	"	1.53	"
" "	18	83	4.35	14
Formic acid	+2 (frozen)	73	19.0	2
" "	15	1200	62.0	6
" "	16	73	58.5	2
Glycerine	15	1200	56.2	6
"	15	200	39.1	2
"	15	75	25.4	"
"	–	8.5	4.4	15
"	–	0.4	2.6	4
Hexane	17	∞	1.880	16
Hydrogen peroxide 46% in H_2O	18	75	84.7	17

Substance.	Temp. °C.	Wave-length cm.	Diel. const.	Authority.
Nitrobenzol	(frozen) −10	∞	9.9	1
"	−5	"	42.0	"
"	0	"	41.0	"
"	+15	"	37.8	"
"	30	"	35.1	"
"	18	"	36.45	11
"	17	73	34.0	2
Octane	17	∞	1.949	16
Oils:				
Almond	20	∞	2.83	18
Castor	11	"	4.67	19
Colza	20	"	3.11	20
Cottonseed	14	"	3.10	21
Lemon	21	"	2.25	22
Linseed	13	"	3.35	21
Neatsfoot	–	"	3.02	20
Olive	20	"	3.11	23
Peanut	11.4	"	3.03	21
Petroleum	–	2000	2.13	24
Petroleum ether	20	∞	1.92	20
Rape seed	16	"	2.85	21
Sesame	13.4	"	3.02	"
Sperm	20	"	3.17	20
Turpentine	20	"	2.23	"
Vaseline	–	"	2.17	25
Phenol	48	73	9.68	2
Toluene	−83	∞	2.51	5
"	+16	"	2.33	"
"	19	73	2.31	2
Meta-xylene	18	∞	2.376	11
" "	17	73	2.37	2
Water	18	∞	81.07	11
for temp. coeff.	17	200	80.6	2
see Table 344.	17	74	81.7	"
	17	38	83.6	"

1 Abegg-Seitz, 1899.
2 Drude, 1896.
3 Marx, 1898.
4 Lampa, 1896.
5 Abegg, 1897.
6 Thwing, 1894.
7 Drude, 1898.
8 Francke, 1893.
9 Löwe, 1898.
10 Landolt-Jahn, 1892.
11 Turner, 1900.
12 Schlundt.
13 Tangl, 1903.
14 Coolidge, 1899.
15 v. Lang, 1896.
16 Nernst, 1894.
17 Calvert, 1900.
18 Hasenöhrl, 1896.
19 Arons-Rubens, 1892.
20 Hopkinson, 1881.
21 Salvioni, 1888.
22 Tomaszewski, 1888.
23 Heinke, 1896.
24 Marx.
25 Fuchs.

Addenda to Table 440, p. 361, Dielectric Constant of Rochelle Salt:

The polarization of the Rochelle salt dielectric in an electric field is somewhat analagous to the behavior of the magnetization of iron in a magnetic field, showing both saturation and hysteresis. The dielectric constant D depends on the initial and final fields and the hysteresis.

Initial field, v/cm.	Final field, v/cm.	Average D (23° C)
765	690	40
765	−153	205
765	−765	157
0	880	86

The last value may be fair value for ordinary purposes. The electrodes were tinfoil attached with shellac. The field was applied perpendicular to the a axis. Like piezoelectric properties, the dielectric constant varies with different crystals. It depends on the temperature as follows: (field 0 to 880 v/cm)

−70° C, D = 12; −40°, 14; −20°, 48; 0°, 174; +20°, 88; +30°, 52.

(Data from Valesek, University of Minnesota, 1921.)

DIELECTRIC CONSTANTS OF LIQUIDS (*continued*).

TABLE 436. — Temperature Coefficients of the Formula:

$$D_\theta = D_t[1 - \alpha(t-\theta) + \beta(t-\theta)^2].$$

Substance.	α	β	Temp. range, ° C.	Authority.
Amyl acetate	0.0024	–	–	Löwe.
Aniline	0.00351	–	–	Ratz.
Benzene	0.00106	0.0000087	10–40	Hasenöhrl.
Carbon bisulphide	0.000966	–	–	Ratz.
" "	0.000922	0.00000060	20–181	Tangl.
Chloroform	0.00410	0.000015	22–181	"
Ethyl ether	0.00459	–	–	Ratz.
Methyl alcohol	0.0057	–	–	Drude.
Oils: Almond	0.00163	0.000026	–	Hasenöhrl.
Castor	0.01067	–	–	Heinke, 1896.
Olive	0.00364	–	–	" "
Paraffine	0.000738	0.0000072	–	Hasenöhrl.
Toluene	0.000921	–	0–13	Ratz.
"	0.000977	0.00000046	20–181	Tangl.
Water	0.004474	–	5–20	Heerwagen.
"	0.004583	0.0000117	0–76	Drude.
"	0.00436	–	4–25	Coolidge.
Meta-xylene	0.000817	–	20–181	Tangl.

(See Table 433 for the signification of the letters.)

TABLE 437.—Dielectric Constants of Liquefied Gases.

A wave-length greater than 10000 centimeters is designated by ∞.

Substance.	Temp. ° C.	Wave-length cm.	Dial. constant.	Authority.	Substance.	Temp. ° C.	Wave-length cm.	Dial. constant.	Authority.
Air	−191	∞	1.432	1	Nitrous oxide N_2O	−88	∞	1.938	8
"	"	75	1.47–1.50	2	" "	−5	"	1.630	5
Ammonia	−34	75	21–23	3	" "	+5	"	1.578	"
"	14	130	16.2	4	" "	+15	"	1.520	"
Carbon dioxide	−5	∞	1.608	5	Oxygen	−182	"	1.491	9
" "	0	"	1.583	"	"	"	"	1.465	8
" "	+10	"	1.540	"	Sulphur dioxide	14.5	120	13.75	4
" "	+15	"	1.526	"	" "	20	∞	14.0	6
Chlorine	−60	"	2.150	"	" "	40	"	12.5	"
"	−20	"	2.030	"	" "	60	"	10.8	"
"	0	"	1.970	"	" "	80	"	9.2	"
"	+10	"	1.940	"	" "	100	"	7.8	"
"	0	"	2.08	6	" "	120	"	6.4	"
"	+14	100	1.88	4	" "	140	"	4.8	"
Cyanogen	23	84	2.52	7	Critical	154.2	"	2.1	"
Hydrocyanic acid	21	"	about 95	"					
Hydrogen sulph.	10	∞	5.93	6					
" "	50	"	4.92	"					
" "	90	"	3.76	"					

1 v. Pirani, 1903.
2 Bahn-Kiebitz, 1904.
3 Goodwin-Thompson, 1899.
4 Coolidge, 1899.
5 Linde, 1895.
6 Eversheim, 1904.
7 Schlundt. 1901.
8 Hasenöhrl, 1900.
9 Fleming-Dewar, 1896.

TABLE 438. — Standard Solutions for the Calibration of Apparatus for the Measuring of Dielectric Constants.

Turner.

Substance.	Diel. const. at 18°. $\lambda = \infty$.
Benzene	2.288
Meta-xylene	2.376
Ethyl ether	4.36[7]
Aniline	7.29[8]
Ethyl chloride . . .	10.90
O-nitro toluene . .	27.71
Nitrobenzene . . .	36.45
Water (conduct. 10^{-6})	81.07

Drude.

Per cent by weight.	Density 16°.	Dielectric constant.	Temp. coefficient.
Acetone in benzene at 19°. $\lambda = 75$ cm.			
0	0.885	2.26	0.1%
20	0.866	5.10	0.3
40	0.847	8.43	0.4
60	0.830	12.1	0.5
80	0.813	16.2	0.5
100	0.797	20.5	0.6
Water in acetone at 19°. $\lambda = 75$ cm.			
0	0.797	20.5	0.6%
20	0.856	31.5	0.5
40	0.903	43.5	0.5
60	0.940	57.0	0.5
80	0.973	70.6	0.5
100	0.999	80.9	0.4

Nernst.

Ethyl alcohol in water at 19.5°. $\lambda = \infty$.

Per cent by weight.	Dielectric constant.
100	26.0
90	29.3
80	33.5
70	38.0
60	43.1

TABLE 439. — Dielectric Constants of Solids.

Substance.	Condition.	Wave-length, cm.	Dielectric constant.	Authority.
Asphalt . .	–	∞	2.68	1
Barium sulphate . .	–	75	10.2	2
Caoutchouc	–	∞	2.22	3
Diamond . .	–	"	16.5	1
" . .	–	75	5.50	2
Ebonite . .	–	∞	2.72	4
" . .	–	"	2.86	5
" . .	–	1000	2.55	6
Glass *	Density.			
Flint (extra heavy) .	4.5	∞	9.90	7
Flint (very light) . .	2.87	"	6.61	7
Hard crown	2.48	"	6.96	7
Mirror . .	–	"	6.44–7.46	5
" . .	–	"	5.37–5.90	8
" . .	–	600	5.42–6.20	8
Lead (Powell) . . .	3.0–3.5	∞	5.4–8.0	9
Jena				
Boron .	–	"	5.5–8.1	10
Barium .	–	"	7.8–8.5	10
Borosilicate .	–	"	6.4–7.7	1
Gutta percha .	–	–	3.3–4.9	11
	Temp.			
Ice	—5	1200	2.85	12
"	—18	5000	3.16	13
"	—190	75	1.76–1.88	14

Substance.	Condition.	Wave-length, cm.	Dielectric constant.	Authority.
	Temp.			
Iodine (cryst.) .	23	75	4.00	2
Lead chloride . (powder)	–	"	42	2
" nitrate .	–	"	16	2
" sulphate .	–	"	28	2
" molybdenate . .	–	"	24	2
Marble (Carrara)	–	"	8.3	2
Mica	–	∞	5.66–5.97	5
"	–	"	5.80–6.62	15
Madras, brown	–	"	2.5–3.4	16
" green	–	"	3.9–5.5	16
" ruby .	–	"	4.4	16
Bengal, yellow	–	"	2.8	16
" white .	–	"	4.2	16
" ruby .	–	"	4.2–4.7	16
Canadian amber . .	–	"	3.0	16
South America	–	"	5.9	16
Ozokerite (raw)	–	"	2.21	1
Paper (telephone)	–	"	2.0	17
" (cable) .	–	"	2.0–2.5	1
Paraffine . . .		"	2.46	18
	Melting point.			
" . . .		"	2.32	19
" . . .	44–46	"	2.10	20
" . . .	54–56	"	2.14	20
" . . .	74–76	"	2.16	20

References on p. 361.

* For the effect of temperature, see Gray-Dobbie, Pr. Roy. Soc. 63, 1898; 67, 1900.
" " " " wave-length, see K. F. Löwe, Wied. Ann. 66, 1898.

DIELECTRIC CONSTANTS (*continued*).

TABLE 439. — Dielectric Constants of Solids (*continued*).

Substance.	Condition.	Wave-length, cm.	Diel. constant.	Authority.	Substance.	Condition.	Wave-length, cm.	Diel. constant.	Authority.
Paraffine . .	47.°6	61	2.16	21	Sulphur				
" . .	56.°2	61	2.25	21	Amorphous	–	∞	3.98	1
Phosphorus:					"	–	75	3.80	2
Yellow . .	–	75	3.60	2	Cast, fresh	–	∞	4.22	1
Solid . . .	–	80	4.1	22	" "	–	"	4.05	18
Liquid . .	–	80	3.85	22	" "	–	75	3.95	2
Porcelain:					Cast, old .	–	∞	3.60	18
Hard					" "	–	75	3.90	2
(Royal B'l'n)	–	∞	5.73	15					
Seger " " .	–	"	6.61	15	Liquid .	near melting-point	∞	3.42	1
Figure " " .	–	"	6.84	15					
Selenium . .	–	"	7.44	1	Strontium				
" . .	–	75	6.60	2	sulphate	–	75	11.3	2
" . .	–	∞	6.13	23	Thallium				
" . .	–	1000	6.14	23	carbonate	–	75	17	2
Shellac . . .	–	∞	3.10	4	" nitrate .	–	75	16.5	2
" . . .	–	"	2.95–3.73	24	Wood			dried	
" . . .	–	"	3.67	25	Red beech .	∥ fibres	∞	4.83–2.51	–
					" " .	⊥ "	"	7.73–3.63	–
Amber . . .	–	–	**2.86**	18	Oak . . .	∥ "	"	4.22–2.46	–
					" . . .	⊥ "	"	6.84–3.64	–

1 v. Pirani, 1903.
2 Schmidt, 1903.
3 Gordon, 1879.
4 Winklemann, 1889.
5 Elsas, 1891.
6 Ferry, 1897.
7 Hopkinson, 1891.
8 Arons-Rubens, 1891.
9 Gray-Dobbie, 1898.
10 Löwe, 1898.
11 (submarine-data).
12 Thwing, 1894.
13 Abegg, 1897.
14 Behn-Kiebitz, 1904.
15 Starke, 1897.
16 E. Wilson.
17 Campbell, 1906.
18 Fallinger, 1902.
19 Boltzmann, 1875.
20 Zietkowski, 1900.
21 Hormell, 1902.
22 Schlundt, 1904.
23 Vonwiller-Mason, 1907.
24 Wüllner, 1887.
25 Donle.

TABLE 440. — Dielectric Constants of Crystals.

Dα, Dβ, Dγ are the dielectric constants along the brachy, macro and vertical axes respectively.

Substance.	Wave-length, cm.	Diel. const.		Authority.	Substance.	Wave-length, cm.	Diel. const.			Authority.
		⊥ Axis.	∥ Axis.				Dα	Dβ	Dγ	
UNIAXIAL:					RHOMBIC:					
Apatite	75	9.50	7.40	1	Aragonite . . .	∞	9.14	–	7.13	4
Beryl	∞	7.85	7.44	2	" . . .	75	9.80	7.68	6.55	1
"	"	7.10	6.05	3	Barite	∞	6.97	10.09	7.00	4
"	75	6.05	5.52	1	"	75	7.65	12.20	7.70	1
Calcite	∞	8.49	7.56	4	Celestite . . .	75	7.70	18.5	8.30	1
"	"	8.78	8.29	5	Cerussite . . .	75	25.4	23.2	19.2	1
Dolomite . . .	75	7.80	6.80	1	$MgSO_4+7H_2O$.	∞	5.26	6.05	8.28	7
Iceland spar . .	75	8.50	8.00	1	K_2SO_4	"	6.09	5.08	4.48	7
Quartz	∞	4.69	5.06	4	Rochelle salt* .	"	6.70	6.92	8.89	7
"	"	4.38	4.46	6	Sulphur . . .	"	3.81	3.97	4.77	8
"	1000	4.27	4.34	6	"	"	3.65	3.85	4.66	7
Ruby (Siam) . .	–	13.3	11.3	4	"	75	3.62	3.85	4.66	1
Rutile (TiO_2) . .	75	89	1.73	1	Topaz	75	6.65	6.70	6.30	1
Tourmaline . . .	∞	7.13	6.54	4	" colorless .	–	6.25	6.54	6.44	4
" . . .	75	6.75	5.65	1						
Zircon	75	12.8	12.6	1						

* See page 358.

1 Schmidt, 1903.
2 Starke, 1897.
3 Curie, 1889.
4 Fallinger, 1902, 1919.
5 v. Pirani, 1903.
6 Ferry, 1897.
7 Borel, 1893.
8 Bolztmann, 1875.

TABLE 441.

WIRELESS TELEGRAPHY.

Wave-Length in Meters, Frequency in periods per second, and Oscillation Constant LC in Microhenries and Microfarads.

The relation between the free wave-length in meters, the frequency in cycles per second, and the capacity-inductance product in microfarads and microhenries are given for circuits between 1000 and 10,000 meters. For values between 100 and 1000 meters, multiply the columns for n by 10 and move the decimal point of the corresponding LC column two places to the left (dividing by 100); for values between 10,000 and 100,000, divide the n column by 10 and multiply the LC column by 100. The relation between wave-length and capacity-inductance may be relied upon throughout the table to within one part in 200.

Example 1: What is the natural wave-length of a circuit containing a capacity of 0.001 microfarad, and an inductance of 454 microhenries? The product of the inductance and capacity is 454 × 0.001 = 0.454. Find 0.454 under LC; opposite under meters is 1270 meters, the natural wave-length of the circuit.

Example 2: What capacity must be associated with an inductance of 880 microhenries in order to tune the circuit to 3500 meters? Find opposite 3500 meters the LC value 3.45; divide this by 880, and the quotient, 0.00397, is the desired capacity in microfarads.

Example 3: A condenser has the capacity of 0.004 microfarad. What inductance must be placed in series with this condenser in order that the circuit shall have a wave-length of 600 meters? From the table, the LC value corresponding to 600 meters is 0.101. Divide this by 0.004, the capacity of the condenser, and the desired inductance is 25.2 microhenries.

Meters.	n	LC	Meters.	n	LC	Meters.	n	LC
1000	300,000	0.281	1300	230,800	0.476	1600	187,500	0.721
1010	297,000	0.287	1310	229,000	0.483	1610	186,300	0.730
1020	294,100	0.293	1320	227,300	0.490	1620	185,200	0.739
1030	291,300	0.299	1330	225,600	0.498	1630	184,100	0.748
1040	288,400	0.305	1340	223,900	0.505	1640	182,900	0.757
1050	285,700	0.310	1350	222,200	0.513	1650	181,800	0.766
1060	283,600	0.316	1360	220,600	0.521	1660	180,700	0.776
1070	280,400	0.322	1370	218,900	0.529	1670	179,600	0.785
1080	277,800	0.328	1380	217,400	0.536	1680	178,600	0.794
1090	275,200	0.335	1390	215,800	0.544	1690	177,500	0.804
1100	272,700	0.341	1400	214,300	0.552	1700	176,500	0.813
1110	270,300	0.347	1410	212,800	0.559	1710	175,400	0.823
1120	267,900	0.353	1420	211,300	0.567	1720	174,400	0.833
1130	265,500	0.359	1430	209,800	0.576	1730	173,400	0.842
1140	263,100	0.366	1440	208,300	0.584	1740	172,400	0.852
1150	260,900	0.372	1450	206,900	0.592	1750	171,400	0.862
1160	258,600	0.379	1460	205,500	0.600	1760	170,500	0.872
1170	256,400	0.385	1470	204,100	0.608	1770	169,400	0.882
1180	254,200	0.392	1480	202,700	0.617	1780	168,500	0.892
1190	252,100	0.399	1490	201,300	0.625	1790	167,600	0.902
1200	250,000	0.405	1500	200,000	0.633	1800	166,700	0.912
1210	247,900	0.412	1510	198,700	0.642	1810	165,700	0.923
1220	245,900	0.419	1520	197,400	0.650	1820	164,800	0.933
1230	243,900	0.426	1530	196,100	0.659	1830	163,900	0.943
1240	241,900	0.433	1540	194,800	0.668	1840	163,000	0.953
1250	240,000	0.440	1550	193,600	0.676	1850	162,200	0.963
1260	238,100	0.447	1560	192,300	0.685	1860	161,300	0.974
1270	236,200	0.454	1570	191,100	0.694	1870	160,400	0.985
1280	234,400	0.461	1580	189,900	0.703	1880	159,600	0.995
1290	232,600	0.468	1590	188,700	0.712	1890	158,700	1.006

Adapted from table prepared by Greenleaf W. Picard; copyright by Wireless Specialty Apparatus Company, New York. Computed on basis of 300,000 kilometers per second for the velocity of propagation of electromagnetic waves.

WIRELESS TELEGRAPHY.

Wave-Length, Frequency and Oscillation Constant.

Meters.	n	LC	Meters.	n	LC	Meters.	n	LC
1900	157,900	1.016	2800	107,100	2.21	7000	42,860	13.8
1910	157,100	1.026	2820	106,400	2.24	7100	42,250	14.2
1920	156,300	1.037	2840	105,600	2.27	7200	41,670	14.6
1930	155,400	1.048	2860	104,900	2.30	7300	41,100	15.0
1940	154,600	1.059	2880	104,200	2.33	7400	40,540	15.4
1950	153,800	1.070	2900	103,400	2.37	7500	40,000	15.8
1960	153,100	1.081	2920	102,700	2.40	7600	39,470	16.3
1970	152,300	1.092	2940	102,000	2.43	7700	38,960	16.7
1980	151,500	1.103	2960	101,300	2.47	7800	38,460	17.1
1990	150,800	1.114	2980	100,700	2.50	7900	37,980	17.6
2000	150,000	1.126	3000	100,000	2.53	8000	37,500	18.0
2020	148,500	1.148	3100	96,770	2.70	8100	37,040	18.5
2040	147,100	1.171	3200	93,750	2.88	8200	36,590	18.9
2060	145,600	1.194	3300	90,910	3.07	8300	36,140	19.4
2080	144,200	1.218	3400	88,240	3.26	8400	35,710	19.9
2100	142,900	1.241	3500	85,910	3.45	8500	35,290	20.3
2120	141,500	1.265	3600	83,330	3.65	8600	34,880	20.8
2140	140,200	1.289	3700	81,080	3.85	8700	34,480	21.3
2160	138,900	1.313	3800	78,950	4.06	8800	34,090	21.8
2180	137,600	1.338	3900	76,920	4.28	8900	33,710	22.3
2200	136,400	1.362	4000	75,000	4.50	9000	33,330	22.8
2220	135,100	1.387	4100	73,170	4.73	9100	32,970	23.3
2240	133,900	1.412	4200	71,430	4.96	9200	32,610	23.8
2260	132,700	1.438	4300	69,770	5.20	9300	32,260	24.3
2280	131,600	1.463	4400	68,180	5.45	9400	31,910	24.9
2300	130,400	1.489	4500	66,670	5.70	9500	31,590	25.4
2320	129,300	1.515	4600	65,220	5.96	9600	31,250	25.9
2340	128,200	1.541	4700	63,830	6.22	9700	30,930	26.5
2360	127,100	1.568	4800	62,500	6.49	9800	30,610	27.0
2380	126,000	1.594	4900	61,220	6.76	9900	30,310	27.6
2400	125,000	1.621	5000	60,000	7.04	10000	30,000	28.1
2420	124,000	1.648	5100	58,820	7.32			
2440	129,000	1.676	5200	57,690	7.61			
2460	121,900	1.703	5300	56,600	7.91			
2480	121,000	1.731	5400	55,560	8.21			
2500	120,000	1.759	5500	54,550	8.51			
2520	119,000	1.787	5600	53,570	8.83			
2540	118,100	1.816	5700	52,630	9.15			
2560	117,200	1.845	5600	51,720	9.47			
2580	116,300	1.874	5900	50,850	9.81			
2600	115,400	1.903	6000	50,000	10.1			
2620	114,500	1.932	6100	49,180	10.5			
2640	113,600	1.962	6200	48,550	10.8			
2660	112,800	1.991	6300	47,620	11.1			
2680	111,900	2.02	6400	46,870	11.5			
2700	111,100	2.05	6500	46,150	11.9			
2720	110,300	2.08	6600	45,450	12.3			
2740	109,500	2.11	6700	44,780	12.6			
2760	108,700	2.14	6800	44,120	13.0			
2780	107,900	2.18	6900	43,480	13.4			
2800	107,100	2.21	7000	42,860	13.8			

TABLE 442.

WIRELESS TELEGRAPHY.

Radiation Resistances for Various Wave-Lengths and Antenna Heights.

The radiation theory of Hertz shows that the radiated energy of an oscillator may be represented by $E = \text{constant } (h^2/\lambda^2)\, I^2$, where h is the length of the oscillator, λ, the wave-length and I the current at its center. For a flat-top antenna $E = 1600\,(h^2/\lambda^2)\, I^2$ watts; $1600\, h^2/\lambda^2$ is called the radiation resistance.

(h = height to center of capacity of conducting system.)

h = / Wave-Length λ	40 Ft.	60 Ft.	80 Ft.	100 Ft.	120 Ft.	160 Ft.	200 Ft.	300 Ft.	450 Ft.	600 Ft.	1200 Ft.
m	*ohm*	*ohm*	*ohm*	*ohm*	*ohm*	*ohm*	*ohm*	*ohm*	*ohm*	*ohm*	*ohm*
200	6.0	13.4	24.0	37.0	54.0	95.0					
300	2.7	6.0	10.6	16.5	23.8	42.4					
400	1.5	3.4	6.0	9.3	13.4	23.8					
600	0.66	1.5	2.7	4.1	6.0	10.6	16.4	37.4	84.0	149.0	
800	0.37	0.84	1.5	2.3	3.4	6.0	9.2	21.0	47.0	84.0	
1000	0.24	0.54	0.95	1.5	2.1	3.8	6.0	13.5	30.0	54.0	215.0
1200	0.17	0.37	0.66	1.03	1.5	2.6	4.1	9.3	21.0	37.0	149.0
1500	0.11	0.24	0.42	0.66	0.95	1.7	2.6	6.0	13.4	24.0	95.0
2000		0.13	0.24	0.37	0.54	0.95	1.5	3.4	7.5	13.4	54.0
2500			0.15	0.24	0.34	0.61	0.95	2.2	4.8	8.6	34.0
3000			0.11	0.17	0.24	0.42	0.66	1.5	3.4	6.0	24.0
4000			0.06	0.09	0.13	0.24	0.37	0.84	1.9	3.4	13.4
5000							0.24	0.53	1.20	2.2	8.6
6000							0.16	0.37	0.84	1.5	6.0
7000							0.12	0.27	0.61	1.1	4.4

Austin, Jour. Wash. Acad. of Sci. 1, p. 190, 1911.

TABLE 443.

THE DIELECTRIC PROPERTIES OF NON-CONDUCTORS.

Phillips Thomas, J. Franklin Inst. 176, 283, 1913.

Results of tests at unit area and unit thickness of dielectric.

At 1000 cycles.	Mica.	Paper.	Celluloid.	Ice.
Max. breakdown volts per cm.	1.06×10^6	0.71×10^6	1.05×10^6	$.011 \times 10^6$
Specific induc. capacity	4.00	4.90	13.26	86.40
Max. absorbable energy, watts-sec/cm³	0.198	0.108	0.640	.00040
90°-angle of lead	0° 57′	2° 10′	3° 40′	13° 39′
Equiv. resistance ohms/cm³ × 10^{11}	3.91	9.84	48.3	1400
Conductivity per cm. cube × 10^{-10}	2.56	1.02	0.207	.00722
Percent change in cap. per cycle × 10^4	2.18	14.31	30.7	70.0
Percent change in resistance per cycle	0.258	0.146	0.106	0.127
At 15 cycles.				
Specific inductive capacity	4.09	5.77	18.60	429.0
Max. absorbable energy, watt-sec/cm³	0.203	0.126	0.90	0.002
Percent change in capacity per cycle	0.00	0.306	1.74	1.59
On direct current.				
Conductivity per cm³	2.42×10^{-17}	$2.27 \times 10^{-1?}$ [illegible]	71.5×10^{-14}	$163.\ 10^{-11}$

MAGNETIC PROPERTIES.

Unit pole is a quantity of magnetism repelling another unit pole with a force of one dyne; 4π lines of force radiate from it. M, pole strength; $4\pi M$ lines of force radiate from pole of strength M.

H, field strength, = no. of lines of force crossing unit area in normal direction; unit = gauss = one line per unit area.

M, magnetic moment, = Ml, where l is length between poles of magnet.

I, intensity of magnetization or pole strength per unit area, = $\mathbf{M}/V = M/A$ where A is cross section of uniformly magnetized pole face, and V is the volume of the magnet. $4\pi M/A = 4\pi I$ = no. lines of force leaving unit area of pole.

J, specific intensity of magnetism, = I/ρ where ρ = density, g/cm³.

ϕ, magnetic flux, = $4\pi M + HA$ for magnet placed in field of strength H (axis parallel to field). Unit, the maxwell.

B, flux density (magnetic) induction, = $\phi/A = 4\pi I + H$; unit the gauss, maxwell per cm.

μ, magnetic permeability, = B/H. Strength of field in air-filled solenoid = $H = (4\pi/10)\, ni$ in gausses, i in amperes, n, number of turns per cm length. If iron filled, induction increased, i.e., no. of lines of force per unit area, B, passing through coil is greater than H; $\mu = B/H$.

κ, susceptibility; permeability relates to effect of iron core on magnetic field strength of coil; if effect be considered on iron core, which becomes a magnet of pole strength M and intensity of magnetism I, then the ratio $I/H = (\mu - 1)/4\pi$ is the magnetic susceptibility per unit volume and is a measure of the magnetizing effect of a magnetic field on the material placed in the field. $\mu = 4\pi\kappa + 1$.

χ, specific susceptibility (per unit mass) = $\kappa/\rho = J/H$.

χ_A, atomic susceptibility, = $\chi \times$ (atomic weight); χ_M = molecular susceptibility.

J_A, J_M, similarly atomic and molecular intensity of magnetization.

Hysteresis is work done in taking a cm³ of the magnetic material through a magnetic cycle = $\int H\, dI = (1/4\pi)\int H\, dB$. Steinmetz's empirical formula gives a close approximation to the hysteresis loss; it is $aB^{1.6}$ where B is the max. induction and a is a constant (see Table 472). The retentivity (B_r) is the value of B when the magnetizing force is reduced to zero. The reversed field necessary to reduce the magnetism to zero is called the coercive force (H_c).

Ferromagnetic substances, μ very large, κ very large: Fe, Ni, Co, Heusler's alloy (Cu 62.5, Mn 23.5, Al 14. See Stephenson, Phys. Rev. 1910), magnetite and a few alloys of Mn. μ for Heusler's alloy, 90 to 100 for B = 2200; for Si sheet steel 350 to 5300.

Paramagnetic substances, $\mu > 1$, very small but positive, $\kappa = 10^{-3}$ to 10^{-6}: oxygen, especially at low temperatures, salts of Fe, Ni, Mn, many metallic elements. (See Table 474.)

Diamagnetic substances, $\mu < 1$, κ negative. Most diamagnetic substance known is Bi, -14×10^{-6}. Volume susceptibility (see Table 474).

Paramagnetic substances show no retentivity or hysteresis effect. Susceptibility independent of field strength. The specific susceptibility for both para- and diamagnetic substances is independent of field strength.

For Hall effect (galvanomagnetic difference of potential), Ettinghausen effect (galvanomagnetic difference of temperature), Nernst effect (thermomagnetic difference of potential) and the Leduc effect (thermomagnetic difference of temperature), see Tables 487 and 488.

Magneto-strictive phenomena:

Joule effect: Mechanical change in length when specimen is subjected to a magnetic field. With increasing field strength, iron and some iron alloys show first a small increment $\Delta l/l = (7 \text{ to } 35) \times 10^{-7}$, then a decrement, and for $H = 1600$, $\Delta l/l$ may amount to $-(6 \text{ to } 8) \times 10^{-6}$. Cast cobalt with increasing field first decreases, $\Delta l/l = -8 \times 10^{-6}$, $H = 150$, then increases in length, $\Delta l/l = +5 \times 10^{-6}$, $H = 2000$; annealed cobalt steadily contracts, $\Delta l/l = -25 \times 10^{-6}$, $H = 2000$. Ni rapidly then slowly contracts, $\Delta l/l = -30 \times 10^{-6}$, $H = 100$; -35×10^{-6}, $H = 300$; -36×10^{-6}, $H = 2000$ (Williams, Phys. Rev. 34, 44, 1912). A transverse field generally gives a reciprocal effect.

Wiedemann effect: The lower end of a vertical wire, magnetized longitudinally, when a current is passed through it, if free, twists in a certain direction, depending upon circumstances (see Williams, Phys. Rev. 32, 281, 1911). A reciprocal effect is observed in that when a rod of soft iron, exposed to longitudinal magnetizing force, is twisted, its magnetism is reduced.

Villari effect; really a reciprocal Joule effect. The susceptibility of an iron wire is increased by stretching when the magnetism is below a certain value, but diminished when above that value.

COMPOSITION AND MAGNETIC

This table and Table 456 below are taken from a paper by Dr. Hopkinson * on the magnetic properties of iron and steel. which is stated in the paper to have been 240. The maximum magnetization is not tabulated; but as stated in the by 4π. "Coercive force" is the magnetizing force required to reduce the magnetization to zero. The "demag- previous magnetization in the opposite direction to the "maximum induction" stated in the table. The "energy which, however, was only found to agree roughly with the results of experiment.

No. of Test.	Description of specimen.	Temper.	Chemical analysis.					
			Total Carbon.	Manganese.	Sulphur.	Silicon.	Phosphorus.	Other substances.
1	Wrought iron . . .	Annealed	–	–	–	–	–	–
2	Malleable cast iron . .	"	–	–	–	–	–	–
3	Gray cast iron . . .	–	–	–	–	–	–	–
4	Bessemer steel . . .	–	0.045	0.200	0.030	None.	0.040	–
5	Whitworth mild steel .	Annealed	0.090	0.153	0.016	"	0.042	–
6	" " .	"	0.320	0.438	0.017	0.042	0.035	–
7	" " .	Oil-hardened	"	"	"	"	"	–
8	" " .	Annealed	0.890	0.165	0.005	0.081	0.019	–
9	" " .	Oil-hardened	"	"	"	"	"	–
10	Hadfield's manganese steel .	–	1.005	12.360	0.038	0.204	0.070	–
11	Manganese steel . .	As forged	0.674	4.730	0.023	0.608	0.078	–
12	" " . . .	Annealed	"	"	"	"	"	–
13	" " . . .	Oil-hardened	"	"	"	"	"	–
14	" " . . .	As forged	1.298	8.740	0.024	0.094	0.072	–
15	" " . . .	Annealed	"	"	"	"	"	
16	" " . . .	Oil-hardened	"	"	"	"	"	–
17	Silicon steel . . .	As forged	0.685	0.694	"	3.438	0.123	–
18	" " . . .	Annealed	"	"	"	"	"	–
19	" " . . .	Oil-hardened	"	"	"	"	"	–
20	Chrome steel . . .	As forged	0.532	0.393	0.020	0.220	0.041	0.621 Cr.
21	" " . . .	Annealed	"	"	"	"	"	"
22	" " . . .	Oil-hardened	"	"	"	"	"	"
23	" " . . .	As forged	0.687	0.028	"	0.134	0.043	1.195 Cr.
24	" " . . .	Annealed	"	"	"	"	"	"
25	" " . . .	Oil-hardened	"	"	"	"	"	"
26	Tungsten steel . . .	As forged	1.357	0.036	None.	0.043	0.047	4.649 W.
27	" " . . .	Annealed	"	"	"	"	"	"
28	" " . . .	Hardened in cold water	"	"	"	"	"	"
29	" " . . .	Hardened in tepid water	"	"	"	"	"	"
30	" " (French) .	Oil-hardened	0.511	0.625	None.	0.021	0.028	3.444 W.
31	" " . . .	Very hard	0.855	0.312	–	0.151	0.089	2.353 W.
32	Gray cast iron . . .	–	3.455	0.173	0.042	2.044	0.151	2.064 C.†
33	Mottled cast iron . .	–	2.581	0.610	0.105	1.476	0.435	1.477 C.†
34	White " " . .	–	2.036	0.386	0.467	0.764	0.458	–
35	Spiegeleisen . . .	–	4.510	7.970	Trace.	0.502	0.128	–

* Phil. Trans. Roy. Soc. vol. 176.

† Graphitic carbon.

PROPERTIES OF IRON AND STEEL.

The numbers in the columns headed "magnetic properties" give the results for the highest magnetizing force used, paper, it may be obtained by subtracting the magnetizing force (240) from the maximum induction and then dividing netizing force" is the magnetizing force which had to be applied in order to leave no residual magnetization after dissipated" was calculated from the formula: — Energy dissipated = coercive force × maximum induction ÷ π

No. of Test.	Description of specimen.	Temper.	Specific electrical resistance.	Magnetic properties.				Energy dissipated per cycle.
				Maximum induction.	Residual induction.	Coercive force.	Demagnetizive force.	
1	Wrought iron . . .	Annealed	.01378	18251	7248	2.30	–	13356
2	Malleable cast iron . .	"	.03254	12408	7479	8.80	–	34742
3	Gray cast iron . . .	–	.10560	10783	3928	3.80	–	13037
4	Bessemer steel . . .	–	.01050	18196	7860	2.96	–	17137
5	Whitworth mild steel .	Annealed	.01080	19840	7080	1.63	–	10289
6	" " .	"	.01446	18736	9840	6.73	–	40120
7	" " .	Oil-hardened	.01390	18796	11040	11.00	–	65786
8	" " .	Annealed	.01559	16120	10740	8.26	–	42366
9	" " .	Oil-hardened	.01695	16120	8736	19.38	–	99401
10	Hadfield's manganese steel .	–	.06554	310	–	–		–
11	Manganese steel . .	As forged	.05368	4623	2202	23.50	37.13	34567
12	" " . .	Annealed	.03928	10578	5848	33.86	46.10	113963
13	" " . .	Oil-hardened	.05556	4769	2158	27.64	40.29	41941
14	" " . .	As forged	.06993	747	–	–	–	–
15	" " . .	Annealed	.06316	1985	540	24.50	50.39	15474
16	" " . .	Oil-hardened	.07066	733	–	–	–	–
17	Silicon steel . . .	As forged	.06163	15148	11073	9.49	12.60	45740
18	" " . . .	Annealed	.06185	14701	8149	7.80	10.74	36485
19	" " . . .	Oil-hardened	.06195	14696	8084	12.75	17.14	59619
20	Chrome steel . . .	As forged	.02016	15778	9318	12.24	13.87	61439
21	" " . . .	Annealed	.01942	14848	7570	8.98	12.24	42425
22	" " . . .	Oil-hardened	.02708	13960	8595	38.15	48.45	169455
23	" " . . .	As forged	.01791	14680	7568	18.40	22.03	85944
24	" " . . .	Annealed	.01849	13233	6489	15.40	19.79	64842
25	" " . . .	Oil-hardened	.03035	12868	7891	40.80	56.70	167050
26	Tungsten steel . . .	As forged	.02249	15718	10144	15.71	17.75	78568
27	" " . . .	Annealed	.02250	16498	11008	15.30	16.93	80315
28	" " . . .	Hardened in cold water	.02274	–	–	–	–	–
29	" " . . .	Hardened in tepid water	.02249	15610	9482	30.10	34.70	149500
30	" " (French) .	Oil hardened	.03604	14480	8643	47.07	64.46	216864
31	" " . . .	Very hard	.04427	12133	6818	51.20	70.69	197660
32	Gray cast iron . . .	–	.11400	9148	3161	13.67	17.03	39789
33	Mottled cast iron . .	–	.06286	10546	5108	12.24	–	41072
34	White " " . .	–	.05661	9342	5554	12.24	20.40	36383
35	Spiegeleisen . . .	–	.10520	385	77	–	–	–

TABLE 446. — Magnetic Properties of Iron and Steel.

		Electrolytic Iron.	Good Cast Steel.	Poor Cast Steel.	Steel.	Cast Iron.	Electrical Sheets. Ordinary.	Electrical Sheets. Silicon Steel.
Chemical composition in per cent	C	0.024	0.044	0.56	0.99	3.11	0.036	0.036
	Si	0.004	0.004	0.18	0.10	3.27	0.330	3.90
	Mn	0.008	0.40	0.29	0.40	0.56	0.260	0.090
	P	0.008	0.044	0.076	0.04	1.05	0.040	0.009
	S	0.001	0.027	0.035	0.07	0.06	0.068	0.006
Coercive force . . .		2.83 [0.36]	1.51 [0.37]	7.1 (44.3)	16.7 (52.4)	11.4 [4.6]	[1.30]	[0.77]
Residual B		11400 [10800]	10600 [11000]	10500 (10500)	13000 (7500)	5100 [5350]	[9400]	[9850]
Maximum permeability		1850 [14400]	3550 [14800]	700 (170)	375 (110)	240 [600]	[3270]	[6130]
B for H=150 . . .		19200 [18900]	18800 [19100]	17400 (15400)	16700 (11700)	10400 [11000]	[18200]	[17550]
4πI for saturation .		21620 [21630]	21420 [21420]	20600 (20200)	19800 (18000)	16400 [16800]	[20500]	[19260]

E. Gumlich, Zs für Electrochemie, 15, p. 599; 1909.

Brackets indicate annealing at 800° C in vacuum. *Parentheses* indicate hardening by quenching from cherry-red.

TABLE 447. — Cast Iron in Intense Fields.

Soft Cast Iron.				Hard Cast Iron.			
H	B	I	μ	H	B	I	μ
114	9950	782	87.3	142	7860	614	55.4
172	10800	846	62.8	254	9700	752	38.2
433	13900	1070	32.1	339	10850	836	30.6
744	15750	1200	21.2	684	13050	983	19.1
1234	17300	1280	14.0	915	14050	1044	15.4
1820	18170	1300	10.0	1570	15900	1138	10.1
12700	31100	1465	2.5	2020	16800	1176	8.3
13550	32100	1475	2.4	10900	26540	1245	2.4
13800	32500	1488	2.4	13200	28600	1226	2.2
15100	33650	1472	2.2	14800	30200	1226	2.0

B. O. Peirce, Proc. Am. Acad. 44, 1909.

TABLE 448. — Corrections for Ring Specimens.

In the case of ring specimens, the average magnetizing force is not the value at the mean radius, the ratio of the two being given in the table. The flux density consequently is not uniform, and the measured hysteresis is less than it would be for a uniform distribution. This ratio is also given for the case of constant permeability, the values being applicable for magnetizations in the neighborhood of the maximum permeability. For higher magnetizations the flux density is more uniform, for lower it is less, and the correction greater.

Ratio of Radial Width to Diameter of Ring.	Ratio of Average H to H at Mean Radius.		Ratio of Hysteresis for Uniform Distribution to Actual Hysteresis.	
	Rectangular Cross-section.	Circular Cross-section.	Rectangular Cross-section.	Circular Cross-section.
1/2	1.0986	1.0718	1.112	1.084
1/3	1.0397	1.0294	1.045	1.033
1/4	1.0216	1.0162	1.024	1.018
1/5	1.0137	1.0102	1.015	1.011
1/6	1.0094	1.0070	1.010	1.008
1/7	1.0069	1.0052	1.008	1.006
1/8	1.0052	1.0040	1.006	1.004
1/10	1.0033	1.0025	1.003	1.002
1/19	1.0009	1.0007	1.001	1.001

M. G. Lloyd, Bull. Bur. Standards, 5, p. 435; 1908.

MAGNETIC PROPERTIES OF IRONS AND STEELS.

TABLE 449. — Magnetic Properties of Various Types of Iron and Steel.

From tests made at the Bureau of Standards. *B* and *H* are measured in cgs units.

Values of *B*.		2000	4000	6000	8000	10,000	12,000	14,000	16,000	18,000	20,000
Annealed Norway iron	H	**.81**	**1.15**	**1.60**	**2.18**	**3.06**	**4.45**	**7.25**	**23.5**	**116.**	—
	μ	2470	3480	3750	3670	3270	2700	1930	680	150	—
Cast semi-steel......	H	**2.00**	**2.90**	**4.30**	**6.46**	**9.82**	**15.1**	**24.9**	**50.5**	**135.**	**325.**
	μ	1000	1380	1400	1240	1020	795	563	317	133	62.
Machinery steel......	H	**5.0**	**8.8**	**13.1**	**18.6**	**25.8**	**35.8**	**50.5**	**76.0**	**142.**	—
	μ	400	455	460	430	390	340	280	210	127	—

TABLE 450. — Magnetic Properties of a Specimen of Very Pure Iron (.017% C).

From tests at the Bureau of Standards. *B* and *H* are measured in cgs units.

Values of *B*		2000	4000	6000	8000	10,000	12,000	14,000	16,000	18,000	20,000
Very pure iron as received	H	**3.30**	**4.48**	**6.35**	**9.10**	**13.0**	**18.9**	**28.8**	**47.0**	**103.**	**240.**
	μ	606	893	945	880	770	635	486	340	175	83
Annealed in vacuo from 900° C	H	**.46**	**.60**	**.80**	**1.02**	**1.38**	**2.00**	**3.20**	**11.3**	**72.0**	**194.**
	μ	4350	6670	7500	7840	7250	6000	4380	1420	250	103

As received:		After annealing:	
H_{max}	150	H_{max}	150
B_{max}	18,900	B_{max}	19,500
B_r	7,650		—
H_c	2.8	H_c	0.53

TABLE 451. — Magnetic Properties of Electrical Sheets.

From tests at the Bureau of Standards. *B* and *H* are measured in cgs units.

Values of *B*		2000	4000	6000	8000	10,000	12,000	14,000	16,000	18,000	20,000
Dynamo steel.......	H	**1.00**	**1.10**	**1.43**	**2.00**	**3.10**	**4.95**	**9.20**	**34.0**	**114.**	—
	μ	2000	3640	4200	4000	3220	2420	1520	470	158	—
Ordinary transformer steel	H	**.60**	**.87**	**1.10**	**1.48**	**2.28**	**3.85**	**10.9**	**43.0**	**149.**	—
	μ	3340	4600	5450	5400	4380	3120	1280	372	121	—
High silicon transformer steel	H	**.50**	**.70**	**.90**	**1.28**	**1.99**	**3.60**	**9.80**	**47.4**	**165.**	—
	μ	4000	5720	6670	6250	5020	3340	1430	338	109	—

MAGNETIC PROPERTIES OF IRONS AND STEELS.

TABLE 452. — Magnetic Properties of Two Types of American Magnet Steel.

From tests at the Bureau of Standards. B and H are measured in cgs units.

Values of B		2000	4000	6000	8000	10,000	12,000	14,000	16,000	18,000	20,000
Tungsten steel.	H	35.0	53.3	63.3	72.0	83.4	109	200	—	—	—
	μ	57	75	95	111	120	110	70	—	—	—
Chrome steel...	H	34.5	49.0	63.5	88.4	143	270	—	—	—	—
	μ	58	82	95	91	70	45	—	—	—	—

Percentage composition: Tungsten steel, C 0.67 W 5.1 Mn 0.38 Si 0.26
Chrome steel, C 0.81 W 0.96 Cr 2.09 Si 0.25

Tungsten steel: Hmax 200 Bmax 14,000 Chrome steel: Hmax 200 Bmax 11,050
H_c 62.5 B_r 10,400 H_c 45.7 B_r 7,030

TABLE 453. — Magnetic Properties of a Ferro-Cobalt Alloy, Fe_2Co (35% Cobalt).

From tests at the Bureau of Standards. B and H are measured in cgs units.

Values of B		2000	4000	6000	8000	10,000	12,000	14,000	16,000	18,000	20,000
As received.....	H	3.10	4.28	5.50	7.17	9.65	13.4	19.1	27.3	40.0	65.0
	μ	645	935	1090	1115	1040	900	730	590	450	310
Annealed at 1000° C	H	3.00	4.11	5.05	6.45	8.40	11.3	15.4	21.9	31.7	50.6
	μ	670	970	1190	1240	1190	1060	910	730	570	400
Quenched from 1000° C	H	10.8	13.8	19.1	28.7	43.4	65.8	104	163	262	—
	μ	185	290	314	279	230	182	135	98	69	—

	Bmax	Hmax	B_r	H_c
As received	15,000	22.9	7750	3.79
Annealed at 1000° C	15,000	18.3	7460	3.95
Quenched from 1000° C	15,000	130.	8240	14.3

TABLE 454. — Magnetic Properties of a Ring Sample of Transformer Steel in Very Weak Fields.

From tests made at the Bureau of Standards. B and H are measured in cgs units.

Values of H...........	0.001	0.002	0.004	0.006	0.008	0.010	0.012	0.014	0.018	0.020
Values of B...........	0.45	0.91	1.85	2.87	3.94	5.05	6.30	7.51	10.19	11.64
Values of μ...........	450	455	462	478	492	505	525	536	566	582

TABLE 455. — Magnetic Properties of Iron in Very Weak Fields.

The effect of very small magnetizing forces has been studied by C. Baur and by Lord Rayleigh. The following short table is taken from Baur's paper, and is taken by him to indicate that the susceptibility is finite for zero values of H and for a finite range increases in simple proportion to H. He gives the formula $k = 15 + 100H$, or $I = 15H + 100H^2$ The experiments were made on an annealed ring of round bar 1.013 cms radius, the ring having a radius of 9.432 cms. Lord Rayleigh's results for an iron wire not annealed give $k = 6.4 + 5.1H$, or $I = 6.4H + 5.1H^2$. The forces were reduced as low as 0.00004 cgs, the relation of k to H remaining constant.

First experiment.			Second experiment.	
H	k	I	H	k
.01580	16.46	2.63	.0130	15.50
.03081	17.65	5.47	.0847	18.38
.07083	23.00	16.33	.0946	20.49
.13188	28.90	38.15	.1864	25.07
.23011	39.81	91.56	.2903	32.40
.38422	58.56	224.87	.3397	35.20

PERMEABILITY OF SOME OF THE SPECIMENS IN TABLE 445.

TABLE 456.

This table gives the induction and the permeability for different values of the magnetizing force of some of the specimens in Table 445. The specimen numbers refer to the same table. The numbers in this table have been taken from the curves given by Dr. Hopkinson, and may therefore be slightly in error; they are the mean values for rising and falling magnetizations.

Magnetizing force. H	Specimen 1 (iron).		Specimen 8 (annealed steel).		Specimen 9 (same as 8 tempered).		Specimen 3 (cast iron).	
	B	μ	B	μ	B	μ	B	μ
1	–	–	–	–	–	–	265	265
2	200	100	–	–	–	–	700	350
3	–	–	–	–	–	–	1625	542
5	10050	2010	1525	300	750	150	3000	600
10	12550	1255	9000	900	1650	165	5000	500
20	14550	727	11500	575	5875	294	6000	300
30	15200	507	12650	422	9875	329	6500	217
40	15800	395	13300	332	11600	290	7100	177
50	16000	320	13800	276	12000	240	7350	149
70	16360	234	14350	205	13400	191	7900	113
100	16800	168	14900	149	14500	145	8500	85
150	17400	116	15700	105	15800	105	9500	63
200	17950	90	16100	80	16100	80	10190	51

Tables 457-8, 463-5 give the results of some experiments by Du Bois,* on the magnetic properties of iron, nickel, and cobalt under strong magnetizing forces. The experiments were made on ovoids of the metals 18 centimeters long and 0.6 centimeters diameter. The specimens were as follows: (1) Soft Swedish iron carefully annealed and having a density 7.82. (2) Hard English cast steel yellow tempered at 230° C.; density 7.78. (3) Hard drawn best nickel containing 99 % Ni with some SiO_2 and traces of Fe and Cu; density 8.82. (4) Cast cobalt giving the following composition on analysis: Co = 93.1, Ni = 5.8, Fe = 0.8, Cu = 0.2, Si = 0.1, and C = 0.3. The specimen was very brittle and broke in the lathe, and hence contained a surfaced joint held together by clamps during the experiment. Referring to the columns, H, B, and μ have the same meaning as in the other tables, S is the magnetic moment per gram, and I the magnetic moment per cubic centimeter. H and S are taken from the curves published by Du Bois; the others have been calculated using the densities given.

MAGNETIC PROPERTIES OF SOFT IRON AT 0° AND 100° C.

TABLE 457.

Soft iron at 0° C.					Soft iron at 100° C.				
H	S	I	B	μ	H	S	I	B	μ
100	180.0	1408	17790	177.9	100	180.0	1402	17720	177.2
200	194.5	1521	19310	96.5	200	194.0	1511	19190	96.0
400	208.0	1627	20830	52.1	400	207.0	1613	20660	51.6
700	215.5	1685	21870	31.2	700	213.4	1663	21590	29.8
1000	218.0	1705	22420	22.4	1000	215.0	1674	22040	21.0
1200	218.5	1709	22670	18.9	1200	215.5	1679	22300	18.6

MAGNETIC PROPERTIES OF STEEL AT 0° AND 100° C.

TABLE 458.

Steel at 0° C.					Steel at 100° C.				
H	S	I	B	μ	H	S	I	B	μ
100	165.0	1283	16240	162.4	100	165.0	1278	16170	161.7
200	181.0	1408	17900	89.5	200	180.0	1395	17730	88.6
400	193.0	1500	19250	48.1	400	191.0	1480	19000	47.5
700	199.5	1552	20210	28.9	700	197.0	1527	19890	28.4
1000	203.5	1583	20900	20.9	1000	199.0	1543	20380	20.4
1200	205.0	1595	21240	17.7	1500	203.0	1573	21270	14.2
3750†	212.0	1650	24470	6.5	3000	205.5	1593	23020	7.7
					5000	208.0	1612	25260	5.1

* "Phil. Mag." 5 series, vol. xxix.

† The results in this and the other tables for forces above 1200 were not obtained from the ovoids above referred to, but from a small piece of the metal provided with a polished mirror surface and placed, with its polished face normal to the lines of force, between the poles of a powerful electromagnet. The induction was then inferred from the rotation of the plane of a polarized ray of red light reflected normally from the surface. (See Kerr's "Constants," p. 331.)

MAGNETISM AND TEMPERATURE.

TABLE 459. — Magnetism and Temperature, Critical Temperature.

The magnetic moment of a magnet diminishes with increasing temperature. Different specimens vary widely. In the formula $Mt/M_0 = (1 - at)$ the value of a may range from .0003 to .001 (see Tables 457–458). The effect on the permeability with weak fields may at first be an increase. There is a critical temperature (Curie point) above which the permeability is very small (paramagnetic?). Diamagnetic susceptibility does not change with the temperature. Paramagnetic susceptibility decreases with increase in temperature. This and the succeeding two tables are taken from Dushman, "Theories of Magnetism," General Electric Review, 1916.

Substance.	Critical temperature, Curie point.	Reference.	Substance.	Critical temperature, Curie point.	Reference.
Iron, α form	756° C	1	MnBi	360 to 380° C	4
" β form	920	1	MnSb	310 " 320	4
" γ form	1280	1	MnAs	45 " 50	4
Magnetite (Fe_3O_4)	536	1	MnP	18 " 25	4
"	589	2	Heusler alloy	310	5
"	555	3	Nickel	340	1
Cobalt-ferrite (Fe_2Co)	520	3	"	376	6
			Cobalt	1075	6

References: (1) P. Curie; (2) see Williams, Electron Theory of Magnetism, quoted from Weiss; (3) du Bois, Tr. Far. Soc. 8, 211, 1912; (4) Hilpert, Tr. Far. Soc. 8, 207, 1912; (5) Gumaer; (6) Stifler, Phys. Rev. 33, 268, 1911.

TABLE 460. — Temperature Variation for Paramagnetic Substances.

The relation deduced by Curie that $\chi = C/T$, where C is a constant and T the absolute temperature, holds for some paramagnetic substances over the ranges given in the following table. Many paramagnetic substances do not obey the law (Honda and Owen, Ann. d. Phys. 32, 1027, 1910; 37, 657, 1912). See the following table.

Substance.	$C \times 10^6$	Range ° C	Reference.	Substance.	$C \times 10^6$	Range ° C	Reference.
Oxygen	33,700	20° to 450° C	1	Gadolinium sulphate.	21,000	−259° to 17	2
Air	7,830	— — —	1	Ferrous sulphate	11,000	−259 " 17	2
Palladium	1,520	20 to 1370	1	Ferric sulphate	17,000	−208 " 17	3
Magnetite	28,000	850 " 1360	1	Manganese chloride.	30,000	−258 " 17	3
Cast iron	38,500	850 " 1267	1				

References: (1) P. Curie, London Electrician, 66, 500, 1912; see also Du Bois, Rap. du Cong. 2, 460, 1900; (2) Perrier, Onnes, Tables annuelles, 3, 288, 1914; (3) Oosterhuis, Onnes, *l.c.* 2, 389, 1913.

TABLE 461. — Temperature Effect on Susceptibility of Diamagnetic Elements.

No effect:

B Cryst. 400 to 1200°	P white	Se —	Sb −170 to 50°
C Diamond, +170 to 200°	S Cryst.; ppt.	Br −170 to 18°	Cs and Au
C "Sugar" carbon	Zn −170 to 300°	Zr Cryst. −170 to 500°	Hg −39 to +350°
Si Cryst.	As —	Cd −170 to 300°	Pb 327 to 600°

Increase with rise in Temperature:

Be —	C Diamond, 200 to 1200°	I −170 to 114°
B Cryst. +170 to 400°	Ag —	Hg −170 to −30°

Decrease with rise in Temperature:

C Amorphous	Gd −179 to 30°	In −170 to 150°	Tl —
C Ceylon graphite	Ge −170 to 900°	Sb +50 to +631°	Pb −170 to 327°
Cu —	Zr 500 to 1200°	Te —	Bi −170 to 268°
Zn +300 to 700°	Cd 300 to 700°	I +114 to +200°	

TABLE 462. — Temperature Effects on Susceptibility of Paramagnetic Elements.

No effect:

Li —	K −170 to 150°	Cr −170 to 500°	W —
Na −170 to 97°	Ca −170 to 18°	Mn −170 to 250°	Os —
Al 657 to 1100°	V −170 to 500°	Rb —	

Increase with rise in Temperature:

Ti −40 to 1100°	Cr 500 to 1100°	Ru +550 to 1200°	Ba −170 to 18°
V 500 to 1100°	Mo −170 to 1200°	Rh —	Ir and Th

Decrease with rise in Temperature:

(O) —	Ti −180 to −40°	Ni 350 to 800°	Pd and Ta
As −170 to 657°	Mn 250 to 1015°	Co above 1150°	Pt and U
Mg —	(Fe) —	Cb −170 to 400°	Rare earth metals

Tables 461 and 462 are due to Honda and Owen; for reference, see preceding table.

MAGNETIC PROPERTIES OF METALS.

TABLE 463. — Cobalt at 100° C.

H	*S*	*I*	*B*	μ
200	106	848	10850	54.2
300	116	928	11960	39.9
500	127	1016	13260	26.5
700	131	1048	13870	19.8
1000	134	1076	14520	14.5
1500	138	1104	15380	10.3
2500	143	1144	16870	6.7
4000	145	1164	18630	4.7
6000	147	1176	20780	3.5
9000	149	1192	23980	2.6
At 0° C. this specimen gave the following results:				
7900	154	1232	23380	3.0

TABLE 464. — Nickel at 100° C.

H	*S*	*I*	*B*	μ
100	35.0	309	3980	39.8
200	43.0	380	4966	24.8
300	46.0	406	5399	18.0
500	50.0	441	6043	12.1
700	51.5	454	6409	9.1
1000	53.0	468	6875	6.9
1500	56.0	494	7707	5.1
2500	58.4	515	8973	3.6
4000	59.0	520	10540	2.6
6000	59.2	522	12561	2.1
9000	59.4	524	15585	1.7
12000	59.6	526	18606	1.5
At 0° C. this specimen gave the following results:				
12300	67.5	595	19782	1.6

TABLE 465. — Magnetite.

The following results are given by Du Bois * for a specimen of magnetite.

H	*I*	*B*	μ
500	325	4580	9.16
1000	345	5340	5.34
2000	350	6400	3.20
12000	350	16400	1.37

Professor Ewing has investigated the effects of very intense fields on the induction in iron and other metals.† The results show that the intensity of magnetization does not increase much in iron after the field has reached an intensity of 1000 c. g. s. units, the increase of induction above this being almost the same as if the iron were not there, that is to say, dB/dH is practically unity. For hard steels, and particularly manganese steels, much higher forces are required to produce saturation. Hadfield's manganese steel seems to have nearly constant susceptibility up to a magnetizing force of 10,000. The following tables, taken from Ewing's papers, illustrate the effects of strong fields on iron and steel. The results for nickel and cobalt do not differ greatly from those given above.

TABLE 466. — Lowmoor Wrought Iron.

H	*I*	*B*	μ
3080	1680	24130	7.83
6450	1740	28300	4.39
10450	1730	32250	3.09
13600	1720	35200	2.59
16390	1630	36810	2.25
18760	1680	39900	2.13
18980	1730	40730	2.15

TABLE 467. — Vicker's Tool Steel.

H	*I*	*B*	μ
6210	1530	25480	4.10
9970	1570	29650	2.97
12120	1550	31620	2.60
14660	1580	34550	2.36
15530	1610	35820	2.31

TABLE 468. — Hadfield's Manganese Steel.

H	*I*	*B*	μ
1930	55	2620	1.36
2380	84	3430	1.44
3350	84	4400	1.31
5920	111	7310	1.24
6620	187	8970	1.35
7890	191	10290	1.30
8390	263	11690	1.39
9810	396	14790	1.51

TABLE 469. — Saturation Values for Steels of Different Kinds.

		H	*I*	*B*	μ
1	Bessemer steel containing about 0.4 per cent carbon . . .	17600	1770	39880	2.27
2	Siemens-Marten steel containing about 0.5 per cent carbon	18000	1660	38860	2.16
3	Crucible steel for making chisels, containing about 0.6 per cent carbon	19470	1480	38010	1.95
4	Finer quality of 3 containing about 0.8 per cent carbon . .	18330	1580	38190	2.08
5	Crucible steel containing 1 per cent carbon	19620	1440	37690	1.92
6	Whitworth's fluid-compressed steel	18700	1590	38710	2.07

* "Phil. Mag." 5 series, vol. xxix, 1890.

† "Phil. Trans. Roy. Soc." 1885 and 1889.

DEMAGNETIZING FACTORS FOR RODS.

TABLE 470.

H = true intensity o. magnetizing field, H' = intensity of applied field, I = intensity of magnetization, $H = H' - NI$.

Shuddemagen says: The demagnetizing factor is not a constant, falling for highest values of I to about 1/7 the value when unsaturated; for values of B $(= H + 4\pi I)$ less than 10000, N is approximately constant; using a solenoid wound on an insulating tube, or a tube of split brass, the reversal method gives values for N which are considerably lower than those given by the step-by-step method; if the solenoid is wound on a thick brass tube, the two methods practically agree.

Ratio of Length to Diameter.	Values of $N \times 10^4$.						
	Ellipsoid.	Cylinder.					
		Uniform Magnetization.	Magnetometric Method (Mann).	Ballistic Step Method.			
				Dubois.	Shuddemagen for Range of Practical Constancy.		
				Diameter.			
				0.158 cm.	0.3175 cm.	1.111 cm.	1.905 cm.
5	7015	–	6800				
10	2549	630	2550	2160	–	–	1960
15	1350	280	1400	1206	–	–	1075
20	848	160	898	775	–	–	671
30	432	70	460	393	388	350	343
40	266	39	274	238	234	212	209
50	181	25	182	162	160	145	149
60	132	18	131	118	116	106	106
70	101	13	99	89	88		
80	80	9.8	78	69	69	66	63
90	65	7.8	63	55	56		
100	54	6.3	51.8	45	46	41	41
150	26	2.8	25.1	20	23	21	21
200	16	1.57	15.2	11	12.5	11	11
300	7.5	0.70	7.5	5.0			
400	4.5	0.39	–	2.8			

C. R. Mann, Physical Review, 3, p. 359; 1896.
H. DuBois, Wied. Ann. 7, p. 942; 1902.
C. L. B. Shuddemagen, Proc. Am. Acad. Arts and Sci. 43, p. 185, 1907 (Bibliography).

TABLE 471.

Shuddemagen also gives the following, where B is determined by the step method and $H = H' - KB$.

Ratio of Length to Diameter.	Values of $K \times 10^4$.	
	Diameter 0.3175 cm.	Diameter 1.1 to 2.0 cm.
15	–	85.2
20	–	53.3
25	–	36.6
30	30.9	27.3
40	18.6	16.6
50	12.7	11.6
60	9.25	8.45
80	5.5	5.05
100	3.66	3.26
150	1.83	1.67

DISSIPATION OF ENERGY IN THE CYCLIC MAGNETIZATION OF VARIOUS SUBSTANCES.

C. P. Steinmetz concludes from his experiments * that the dissipation of energy due to hysteresis in magnetic metals can be expressed by the formula $e = aB^{1.6}$, where e is the energy dissipated and a a constant. He also concludes that the dissipation is the same for the same range of induction, no matter what the absolute value of the terminal inductions may be. His experiments show this to be nearly true when the induction does not exceed $\pm$ 15000 c. g. s. units per sq. cm. It is possible that, if metallic induction only be taken, this may be true up to saturation; but it is not likely to be found to hold for total inductions much above the saturation value of the metal. The law of variation of dissipation with induction range in the cycle, stated in the above formula, is also subject to verification.†

Values of Constant *a*.

The following table gives the values of the constant *a* as found by Steinmetz for a number of different specimens. The data are taken from his second paper.

Number of specimen.	Kind of material.	Description of specimen.	Value of *a*.
1	Iron	Norway iron	.00227
2	"	Wrought bar	.00326
3	"	Commercial ferrotype plate	.00548
4	"	Annealed " "	.00458
5	"	Thin tin plate	.00286
6	"	Medium thickness tin plate	.00425
7	Steel	Soft galvanized wire	.00349
8	"	Annealed cast steel	.00848
9	"	Soft annealed cast steel	.00457
10	"	Very soft annealed cast steel	.00318
11	"	Same as 8 tempered in cold water	.02792
12	"	Tool steel glass hard tempered in water	.07476
13	"	" " tempered in oil	.02670
14	"	" " annealed	.01899
15	"	Same as 12, 13, and 14, after having been subjected	.06130
16	"	to an alternating m. m. f. of from 4000 to 6000	.02700
17	"	ampere turns for demagnetization	.01445
18	Cast iron	Gray cast iron	.01300
19	" "	" " " containing $\frac{1}{8}$ % aluminium	.01365
20	" "	" " " " $\frac{1}{2}$ % "	.01459
21	Magnetite	A square rod 6 sq. cms. section and 6.5 cms. long, from the Tilly Foster mines, Brewsters, Putnam County, New York, stated to be a very pure sample	.02348
22	Nickel	Soft wire	.0122
23	"	Annealed wire, calculated by Steinmetz from Ewing's experiments	.0156
24	"	Hardened, also from Ewing's experiments	.0385
25	Cobalt	Rod containing about 2 % of iron, also calculated from Ewing's experiments by Steinmetz	.0120
26	Iron filings	Consisted of thin needle-like chips obtained by milling grooves about 8 mm. wide across a pile of thin sheets clamped together. About 30 % by volume of the specimen was iron.	
		1st experiment, continuous cyclic variation of m. m. f. 180 cycles per second	.0457
		2d experiment, 114 cycles per second	.0396
		3d " 79–91 cycles per second	.0373

* "Trans. Am. Inst. Elect. Eng." January and September, 1892.
† See T. Gray, "Proc. Roy. Soc." vol. lvi.

TABLE 473.

ENERGY LOSSES IN TRANSFORMER STEELS.

Determined by the wattmeter method.

Loss per cycle per $cc = AB^x + bnB^y$, where B = flux density in gausses and n = frequency in cycles per second. x shows the variation of hysteresis with B between 5000 and 10000 gausses, and y the same for eddy currents.

Designation.	Thickness. cm.	Ergs per Gramme per Cycle. 10000 Gausses. Hysteresis.	Ergs per Gramme per Cycle. 10000 Gausses. Eddy Currents at 60~	Ergs per Gramme per Cycle. 5000 Gausses. Hysteresis.	Ergs per Gramme per Cycle. 5000 Gausses. Eddy Currents at 60~	x	y	a	Watts per Pound at 60 Cycles and 10000 Gausses. Eddy Current Loss for Gage No. 29.‡	Watts per Pound at 60 Cycles and 10000 Gausses. Hysteresis.	Watts per Pound at 60 Cycles and 10000 Gausses. Total.
Unannealed											
A	0.0399	1599	186	562	46	1.51	2.02	0.00490	0.41	4.35	4.76
B	.0326	1156	134	384	36	1.59	1.89	.00358	0.44	3.14	3.58
C	.0422	1032	242	356	70	1.51	1.79	.00319	0.47	2.81	3.28
D	.0381	1009	184	353	48	1.52	1.94	.00312	0.44	2.74	3.18
Annealed											
E	.0476	735	236	246	58	1.58	2.02	.00227	0.36	2.00	2.36
F	.0280	666	100	220	27	1.60	1.88	.00206	0.44	1.81	2.25
G	.0394	563	210	193	54	1.54	1.96	.00174	0.47	1.53	2.00
H*	.0307	412	146	138.5	39	1.58	1.90	.00127	0.54	1.12	1.66
J	.0318	341	202	111.5	55	1.62	1.88	.00105	0.70	0.93	1.63
K*	.0282	394	124	130	32	1.61	1.90	.00122	0.54	1.07	1.61
L	.0346	381	184	125	50	1.61	1.88	.00118	0.535	1.035	1.57
B	.0338	354	200	116	57	1.61	1.81	.00110	0.61	0.96	1.57
M	.0335	372	178	127	46	1.55	1.95	.00115	0.55	1.01	1.56
N	.0340	321	210	105	56	1.62	1.90	.00099	0.63	0.87	1.50
P	.0437	334	184	107	50	1.64	1.88	.00103	0.34	0.91	1.25
Silicon steels											
Q†	.0361	303	54	98	15	1.63	–	.00094	0.14	0.825	0.965
R	.0315	288	42	93	11	1.64	–	.00089	0.15	0.78	0.93
S	.0452	278	72	90	18	1.63	–	.00086	0.12	0.755	0.875
T	.0338	250	60	78	18	1.68	–	.00077	0.18	0.68	0.86
U	.0346	270	42	86	12	1.66	–	.00084	0.12	0.735	0.855
V*	.0310	251.5	47	79	13	1.68	–	.00078	0.17	0.685	0.855
W*	.0305	197	43	62.3	12.4	1.67	–	.00061	0.16	0.535	0.695
X	.0430	200	65	64.2	16.6	1.65	–	.00062	0.12	0.545	0.665

* German. † English.

‡ In order to make a fair comparison, the eddy current loss has been computed for a thickness of 0.0357 cm. (Gage No. 29), assuming the loss proportional to the thickness.

Lloyd and Fisher, Bull. Bur. Standards, 5, p. 453; 1909.

Note. — **For formulæ and tables for the calculation of mutual and self inductance see Bulletin Bureau of Standards, vol. 8, p. 1–237, 1912.**

TABLE 474.

MAGNETIC SUSCEPTIBILITY.

If $\mathfrak{I}$ is the intensity of magnetization produced in a substance by a field strength $\mathfrak{H}$, then the magnetic susceptibility $H = \mathfrak{I}/\mathfrak{H}$. This is generally referred to the unit mass; italicized figures refer to the unit volume. The susceptibility depends greatly upon the purity of the substance, especially its freedom from iron. The mass susceptibility of a solution containing p per cent by weight of a water-free substance is, if H_0 is the susceptibility of water, $(p/100)\ H + (1 - p/100)\ H_0$.

Substance.	$H \times 10^6$	Temp. C.	Remarks	Substance.	$H \times 10^6$	Temp. C.	Remarks
Ag	−0.19	18°		K_2CO_3	−0.50	20°	Sol'n
AgCl	−0.28			Li	+0.38		
Air, 1 Atm.	*+0.024*	15		Mo	+0.04	18	
Al	+0.65	18		Mg	+0.55	18	
$Al_2K_2(SO_4)_4 24H_2O$	*−1.0*		Crys.	$MgSO_4$	−0.40		
A, 1 Atm	*−0.10*	0		Mn	+11.	18	
As	−0.3	18		$MnCl_2$	+122.	18	Sol'n
Au	−0.15	18		$MnSO_4$	+100.	18	"
B	−0.71	18		N_2, 1 Atm.	*0.001*	16	
$BaCl_2$	−0.36	20		NH_3	−1.1		
Be	+0.79	15	Powd.	Na	+0.51	18	
Bi	−1.4	18		NaCl	−0.50	20	
Br	−0.38	18		Na_2CO_3.	−0.19	17	Powd.
C, arc-carbon	−2.0	18		$Na_2CO_3 . 10\ H_2O$	−0.46	17	"
C, diamond	−0.49	18		Nb	+1.3	18	
CH_4, 1 Atm.	*+0.001*	16		$NiCl_2$	+40.	18	Sol'n
CO_2, 1 Atm.	*+0.002*	16		$NiSO_4$	+30.	20	"
CS_2	*−0.77*	18		O_2, 1 Atm.	+0.120	20	
CaO	−0.27	16	Powd.	Os	+0.04	20	
$CaCl_2$	−0.40	19	"	P, white	−0.90	20	
$CaCO_3$, marble	*−0.7*			P, red	−0.50	20	
Cd	−0.17	18		Pb	−0.12	20	
$CeBr_3$	+6.3	18		$PbCl_2$	−0.25	15	Powd.
Cl_2, 1 Atm.	−0.59	16		Pd	+5.8	18	
$CoCl_2$	+90.	18	Sol'n	$PrCl_3$	+13.	18	Sol'n
$CoBr_2$	+47.	18	"	Pt.	+1.1	18	
CoI_2	+33.	18	"	$PtCl_4$	0.0	22	Sol'n
$CoSO_4$	+57.	19	"	Rh	+1.1	18	
$Co(NO_3)_2$	+57.	18	"	S	−0.48	18	
Cr	+3.7	18		SO_2, 1 Atm.	−0.30	16	
CsCl	−0.28	17	Powd.	Sb	−0.94	18	
Cu	−0.09	18		Se	−0.32	18	
$CuCl_2$	+12.	20	Sol'n	Si	−0.12	18	Crys.
$CuSO_4$	+10.	20	Sol'n	SiO_2, Quartz	−0.44	20	
CuS	+0.16	17	Powd.	—Glass	−0.5±		
$FeCl_3$	+90.	18	Sol'n	Sn	+0.03	20	
$FeCl_2$	+90.	18	"	$SrCl_2$	−0.42	20	Sol'n
$FeSO_4$	+82.	20	"	Ta	+0.93	18	
$Fe_2(NO_3)_6$	+50.	18	"	Te	−0.32	20	
$FeCn_6K_4$	−0.44		Powd.	Th	+0.18	18	
$FeCn_6K_3$	+9.1		"	Ti	+3.1	18	
He, 1 Atm.	*−0.002*	0		Va	+1.5	18	
H_2, 1 Atm.	*0.000*	16		Wo	+0.33	20	
H_2, 40 Atm.	*0.000*	16		Zn	−0.15	18	
H_2O	−0.79	20		$ZnSO_4$	−0.40		
HCl	*−0.80*	20		Zr	−0.45	18	
H_2SO_4	*+0.78*	20		CH_3OH	−0.73		
HNO_3	*−0.70*	20		C_2H_5OH	−0.80		
Hg	−0.19	20		C_3H_7OH	−0.80		
I	−0.4	20		$C_2H_5OC_2H_5$	*−0.60*	20	
In	0.1±	18		$CHCl_3$	−0.58		
Ir	+0.15	18		C_6H_6	−0.78		
K	+0.40	20		Ebonite	*+1.1*		
KCl	−0.50	20		Glycerine	−0.64	22	
KBr	−0.40	20		Sugar	−0.57		
KI	−0.38	20		Paraffin	*−0.58*		
KOH	−0.35	22	Sol'n	Petroleum	−0.91		
K_2SO_4	−0.42	20		Toluene	−0.77		
$KMnO_4$	+2.0			Wood	−0.2–5		
KNO_3	−0.33	20		Xylene	−0.81		

Values are mostly means taken of values given in Landolt-Börnstein's Physikalisch-chemische Tabellen. See especially Honda, Annalen der Physik (4), 32, 1910.

MAGNETO-OPTIC ROTATION.

Faraday discovered that, when a piece of heavy glass is placed in magnetic field and a beam of plane polarized light passed through it in a direction parallel to the lines of magnetic force, the plane of polarization of the beam is rotated. This was subsequently found to be the case with a large number of substances, but the amount of the rotation was found to depend on the kind of matter and its physical condition, and on the strength of the magnetic field and the wave-length of the polarized light. Verdet's experiments agree fairly well with the formula—

$$\theta = clH\left(r - \lambda\frac{dr}{d\lambda}\right)\frac{r^2}{\lambda^2},$$

where c is a constant depending on the substance used, l the length of the path through the substance, H the intensity of the component of the magnetic field in the direction of the path of the beam, r the index of refraction, and λ the wave-length of the light in air. If H be different, at different parts of the path, lH is to be taken as the integral of the variation of magnetic potential between the two ends of the medium. Calling this difference of potential v, we may write $\theta = Av$, where A is constant for the same substance, kept under the same physical conditions, when the one kind of light is used. The constant A has been called "Verdet's constant," * and a number of values of it are given in Tables 476-480. For variation with temperature the following formula is given by Bichat:—

$$R = R_0\,(1 - 0.00104\,t - 0.000014\,t^2),$$

which has been used to reduce some of the results given in the table to the temperature corresponding to a given measured density. For change of wave-length the following approximate formula, given by Verdet and Becquerel, may be used:—

$$\frac{\theta_1}{\theta_2} = \frac{\mu_1^2(\mu_1^2 - 1)\lambda_2^2}{\mu_2^2(\mu_2^2 - 1)\lambda_1^2},$$

where μ is index of refraction and λ wave-length of light.

A large number of measurements of what has been called molecular rotation have been made, particularly for organic substances. These numbers are not given in the table, but numbers proportional to molecular rotation may be derived from Verdet's constant by multiplying in the ratio of the molecular weight to the density. The densities and chemical formulæ are given in the table. In the case of solutions, it has been usual to assume that the total rotation is simply the algebraic sum of the rotations which would be given by the solvent and dissolved substance, or substances, separately; and hence that determinations of the rotary power of the solvent medium and of the solution enable the rotary power of the dissolved substance to be calculated. Experiments by Quincke and others do not support this view, as very different results are obtained from different degrees of saturation and from different solvent media. No results thus calculated have been given in the table, but the qualitative result, as to the sign of the rotation produced by a salt, may be inferred from the table. For example, if a solution of a salt in water gives Verdet's constant less than 0.0130 at 20° C., Verdet's constant for the salt is negative.

The table has been for the most part compiled from the experiments of Verdet,† H. Becquerel,‡ Quincke,§ Koepsel,‖ Arons,¶ Kundt,** Jahn,†† Schönrock,‡‡ Gordon,§§ Rayleigh and Sidgewick,‖‖ Perkin,¶¶ Bichat.***

As a basis for calculation, Verdet's constant for carbon disulphide and the sodium line D has been taken as 0.0420 and for water as 0.0130 at 20° C.

* The constancy of this quantity has been verified through a wide range of variation of magnetic field by H. E. J. G. Du Bois (Wied. Ann. vol. 35), p. 137, 1888.
† "Ann. de Chim. et de Phys." [3] vol. 52, p. 129, 1858.
‡ "Ann. de Chim. et de Phys." [5] vol. 12; "C. R." vols. 90, p. 1407, 1880, and 100, p. 1374, 1885.
§ "Wied. Ann." vol. 24, p. 606, 1885.
‖ "Wied. Ann." vol. 26, p. 456, 1885.
¶ "Wied. Ann." vol. 24, p. 161, 1885.
** "Wied. Ann." vols. 23, p. 228, 1884, and 27, p. 191, 1886.
†† "Wied. Ann." vol. 43, p. 280, 1891.
‡‡ "Zeits. für Phys. Chem." vol. 11, p. 753, 1893.
§§ "Proc. Roy. Soc." 36, p. 4, 1883.
‖‖ "Phil. Trans. R. S." 176, p. 343, 1885.
¶¶ "Jour. Chem. Soc."
*** "Jour. de Phys." vols. 8, p. 204, 1879, and 9, p. 204 and p. 275, 1880.

TABLE 476.

MAGNETO-OPTIC ROTATION.

Solids.

Substance.	Formula.	Wave-length.	Verdet's Constant. Minutes.	Temp. C.	Authority.
		μ			
Amber		0.589	0.0095	18–20°	Quincke.
Blende	ZnS	"	0.2234	15	Becquerel.
Diamond	C	"	0.0127	15	"
Lead borate	PbB_2O_4	"	0.0600	15	"
Selenium	Se	0.687	0.4625	15	"
Sodium borate	$Na_2B_4O_7$	0.589	0.0170	15	"
Ziqueline (Cuprite)	Cu_2O	0.687	0.5908	15	"
Fluorite	$CaFl_2$	0.2534	0.05989	20	Meyer, Ann. der Physik, 30, 1909.
		.3655	.02526	"	
		.4358	.01717	"	
		.4916	.01329	"	
		.589	.00897	"	
		1.00	.00300	"	
		2.50	.00049	"	
		3.00	.00030	"	
Glass, Jena: Medium phosphate crn.		0.589	0.0161	18	DuBois, Wied. Ann. 51, 1894.
Heavy crown, O1143		"	0.0220	"	
Light flint, O451		"	0.0317	"	
Heavy flint O500		"	0.0608	"	
" " S163		"	0.0888	"	
Zeiss, Ultraviolet		0.313	0.0674	16	Landau, Phys. ZS. 9, 1908.
"		0.405	.0369	"	
"		0.436	.0311	"	
Quartz, along axis, i.e., plate cut ⊥ to axis	SiO_2	0.2194	0.1587	20	Borel, Arch. sc. phys. 16, 1903.
		.2573	.1079	"	
		.3609	.04617	"	
		.4800	.02574	"	
		.5892	.01664	"	
		.6439	.01368	"	
Rock salt	$NaCl$	0.2599	0.2708	20	Meyer, as above.
		.3100	.1561	"	
		.4046	.0775	"	
		.4916	.0483	"	
		.6708	.0245	"	
		1.00	.01050	"	
		2.00	.00262	"	
		4.00	.00069	"	
Sugar, cane: along axis IIA	$C_{12}H_{22}O_{11}$	0.451	.0122	20	Voigt, Phys. ZS. 9, 1908.
		.540	.0076	"	
		.626	.0066	"	
axis IIA[1]	–	0.451	0.0129	"	
		.540	.0084	"	
		.626	.0075	"	
Sylvite	KCl	0.4358	0.0534	20	Meyer, as above.
		.5461	.0316	"	
		.6708	.02012	"	
		.90	.01051	"	
		1.20	.00608	"	
		2.00	.00207	"	
		4.00	.00054	"	

TABLE 477.

MAGNETO-OPTIC ROTATION.

Liquids: Verdet's Constant for $\lambda = 0.589\mu$.

Substance.	Chemical formula.	Density in grams per c. c.	Verdet's constant in minutes.	Temp. C.	Authority.
Acetone	C_3H_6O	0.7947	0.0113	20°	Jahn.
Acids: Acetic	$C_2H_4O_2$	1.0561	.0105	21	Perkin.
" Butyric	$C_4H_8O_2$	0.9663	.0116	15	"
" Formic	CH_2O_2	1.2273	.0105	"	"
" Hydrochloric	HCl	1.2072	.0224	"	"
" Hydrobromic	HBr	1.7859	.0343	"	"
" Hydroiodic	HI	1.9473	.0515	"	"
" Nitric	HNO_3	1.5190	.0070	13	"
" Sulphuric	H_2SO_4	—	.0121	15	Becquerel.
Alcohols: Amyl	$C_5H_{11}OH$	0.8107	.0128	20	Jahn.
" Butyl	C_4H_9OH	0.8021	.0124	"	"
" Ethyl	C_2H_5OH	0.7900	.0112	"	"
" Methyl	CH_3OH	0.7920	.0093	"	"
" Propyl	C_3H_7OH	0.8042	.0120	"	"
Benzene	C_6H_6	0.8786	.0297	"	"
Bromides: Bromoform	$CHBr_3$	2.9021	.0317	15	Perkin.
" Ethyl	C_2H_5Br	1.4486	.0183	"	"
" Ethylene	$C_2H_4Br_2$	2.1871	.0268	"	"
" Methyl	CH_3Br	1.7331	.0205	0	"
" Methylene	CH_2Br_2	2.4971	.0276	15	"
Carbon bisulphide	CS_2	—	.0433	0	Gordon.
" "	"	—	.0420	18	Rayleigh.
Chlorides: Amyl	CHCl	0.8740	.0140	20	Jahn.
" Arsenic	$AsCl_3$	—	.0422	15	Becquerel.
" Carbon	CCl_4	—	.0321	"	"
" Chloroform	$CHCl_3$	1.4823	.0164	20	Jahn.
" Ethyl	C_2H_5Cl	0.9169	0.0138	6	Perkin.
" Ethylene	$C_2H_4Cl_2$	1.2589	.0166	15	"
" Methyl	CH_3Cl	—	.0170	"	Becquerel.
" Methylene	CH_2Cl_2	1.3361	.0162	"	Perkin.
" Sulphur bi-	S_2Cl_2	—	.0393	"	Becquerel.
" Tin tetra	$SnCl_4$	—	.0151	"	"
" Zinc bi-	$ZnCl_2$	—	.0437	"	"
Iodides: Ethyl	C_2H_5I	1.9417	.0296	"	Perkin.
" Methyl	CH_3I	2.2832	.0336	"	"
" Propyl	C_3H_7I	1.7658	.0271	"	"
Nitrates: Ethyl	$C_2H_5O.NO_2$	1.1149	.0091	"	"
" Methyl	$CH_3O.NO_2$	1.2157	.0078	"	"
" Propyl	$C_3H_7O.NO_2$	1.0622	.0100	"	"
Paraffins: Heptane	C_7H_{16}	0.6880	.0125	"	"
" Hexane	C_6H_{14}	0.6743	.0125	"	"
" Pentane	C_5H_{12}	0.6332	.0118	"	"
Phosphorus, melted	P	—	.1316	33	Becquerel.
Sulphur, melted	S	—	.0803	114	"
Toluene	C_7H_8	0.8581	.0269	28	Schönrock.
Water, $\lambda = 0.2496\ \mu$	H_2O		.1042		See Meyer,
0.275			.0776		Ann. der
0.3609			.0384		Physik, 30,
0.4046			.0293		1909. Meas-
0.500			.0184		ures by
0.589			.0131		Landau,
0.700			.0091		Siertsema,
1.000			.00410		Ingersoll.
1.300			.00264		
Xylene	C_8H_{10}	0.8746	.0263	27	Schönrock.

Table 478.

MAGNETO-OPTIC ROTATION.

Solutions of acids and salts in water. Verdet's constant for $\lambda = 0.589\mu$.

Chemical formula.	Density, grams per c. c.	Verdet's constant in minutes.	Temp. C.	*	Chemical formula.	Density, grams per c. c.	Verdet's constant in minutes.	Temp. C.	*
C_3H_6O	0.9715	0.0129	20°	J	LiCl	1.0619	0.0145	20°	J
HBr	1.3775	0.0244	"	P	"	1.0316	0.0143	"	"
"	1.1163	0.0168	"	"	$MnCl_2$	1.1966	0.0167	15	B
HCl	1.1573	0.0204	"	"	"	1.0876	0.0150	"	"
"	1.0762	0.0168	"	"	$HgCl_2$	1.0381	0.0137	16	S
"	1.0158	0.0140	"	J	"	1.0349	0.0137	"	"
HI	1.9057	0.0499	"	P	$NiCl_2$	1.4685	0.0270	15	B
"	1.4495	0.0323	"	"	"	1.2432	0.0196	"	"
"	1.1760	0.0205	"	"	"	1.1233	0.0162	"	"
HNO_3	1.3560	0.0105	"	"	KCl	1.6000	0.0163	"	"
NH_3	0.8918	0.0153	15	"	"	1.0732	0.0148	20	J
NH_4Br	1.2805	0.0226	"	"	NaCl	1.2051	0.0180	15	B
"	1.1576	0.0186	"	"	"	1.0546	0.0144	"	"
$BaBr_2$	1.5399	0.0215	20	J	"	1.0418	0.0144	"	J
"	1.2855	0.0176	"	"	$SrCl_2$	1.1921	0.0162	"	"
$CdBr_2$	1.3291	0.0192	"	"	"	1.0877	0.0146	"	"
"	1.1608	0.0162	"	"	$SnCl_2$	1.3280	0.0266	15	V
$CaBr_2$	1.2491	0.0189	"	"	"	1.1112	0.0175	"	"
"	1.1337	0.0164	"	"	$ZnCl_2$	1.2851	0.0196	"	"
KBr	1.1424	0.0163	"	"	"	1.1595	0.0161	"	"
"	1.0876	0.0151	"	"	K_2CrO_4	1.3598	0.0098	"	"
NaBr	1.1351	0.0165	"	"	$K_2Cr_2O_7$	1.0786	0.0126	"	"
"	1.0824	0.0152	"	"	$Hg(CN)_2$	1.0638	0.0136	16	S
$SrBr_2$	1.2901	0.0186	"	"	"	1.0605	0.0135	"	"
"	1.1416	0.0159	"	"	NH_4I	1.5948	0.0396	15	P
K_2CO_3	1.1906	0.0140	20	"	"	1.5109	0.0358	"	"
Na_2CO_3	1.1006	0.0140	"	"	"	1.2341	0.0235	"	"
"	1.0564	0.0137	"	"	CdI_2	1.5156	0.0291	20	J
NH_4Cl	1.0718	0.0178	15	V	"	1.1521	0.0177	"	"
$BaCl_2$	1.2897	0.0168	20	J	KI	1.6743	0.0338	15	B
"	1.1338	0.0149	"	"	"	1.3398	0.0237	"	"
$CdCl_2$	1.3179	0.0185	"	"	"	1.1705	0.0182	"	"
"	1.2755	0.0179	"	"	NaI	1.1939	0.0200	"	J
"	1.1732	0.0160	"	"	"	1.1191	0.0175	"	"
"	1.1531	0.0157	"	"	NH_4NO_3	1.2803	0.0121	15	P
$CaCl_2$	1.1504	0.0165	"	"	KNO_3	1.0634	0.0130	20	J
"	1.0832	0.0152	"	"	$NaNO_3$	1.1112	0.0131	"	"
$CuCl_2$	1.5158	0.0221	15	B	$U_2O_3N_2O_5$	2.0267	0.0053	"	B
"	1.1330	0.0156	"	"	"	1.1963	0.0115	"	"
$FeCl_2$	1.4331	0.0025	15	"	$(NH_4)_2SO_4$	1.2286	0.0140	15	P
"	1.2141	0.0099	"	"	$NH_4.HSO_4$	1.4417	0.0085	"	"
"	1.1093	0.0118	"	"	$BaSO_4$	1.1788	0.0134	20	J
Fe_2Cl_6	1.6933	—0.2026	"	"	"	1.0938	0.0133	"	"
"	1.5315	—0.1140	"	"	$CdSO_4$	1.1762	0.0139	"	"
"	1.3230	—0.0348	"	"	"	1.0890	0.0136	"	"
"	1.1681	—0.0015	"	"	Li_2SO_4	1.1762	0.0137	"	"
"	1.0864	0.0081	"	"	$MnSO_4$	1.2441	0.0138	"	"
"	1.0445	0.0113	"	"	K_2SO_4	1.0475	0.0133	"	"
"	1.0232	0.0122	"	"	Na_2SO_4	1.0661	0.0135	"	"

* J, Jahn, P, Perkin, V, Verdet, B, Becquerel, S, Schönrock; see p. 378 for references.

TABLE 479. — Magneto-Optic Rotation.

Gases.

Substance.	Pressure.	Temp.	Verdet's constant in minutes.	Authority.
Atmospheric air	Atmospheric	Ordinary	6.83×10^{-6}	Becquerel.
Carbon dioxide	"	"	13.00 "	"
Carbon disulphide	74 cms.	70° C.	23.49 "	Bichat.
Ethylene	Atmospheric	Ordinary	34.48 "	Becquerel.
Nitrogen	"	"	6.92 "	"
Nitrous oxide	"	"	16.90 "	"
Oxygen	"	"	6.28 "	"
Sulphur dioxide	"	"	31.39 "	"
" "	246 cms.	20° C.	38.40 "	Bichat.

See also Siertsema, Ziting. Kon. Akad. Watt., Amsterdam, 7, 1899; 8, 1900.

Du Bois shows that in the case of substances like iron, nickel, and cobalt which have a variable magnetic susceptibility the expression in Verdet's equation, which is constant for substances of constant susceptibility, requires to be divided by the susceptibility to obtain a constant. For this expression he proposes the name "Kundt's constant." These experiments of Kundt and Du Bois show that it is not the difference of magnetic potential between the two ends of the medium, but the product of the length of the medium and the induction per unit area, which controls the amount of rotation of the beam.

TABLE 480. — Verdet's and Kundt's Constants.

The following short table is quoted from Du Bois' paper. The quantities are stated in c. g. s. measure, circular measure (radians) being used in the expression of "Verdet's constant" and "Kundt's constant."

Name of substance.	Magnetic susceptibility.	Verdet's constant.		Wave-length of light in cms.	Kundt's constant.
		Number.	Authority.		
Cobalt	–	–	–	6.44×10^{-5}	3.99
Nickel	–	–	–	"	3.15
Iron	–	–	–	6.56 "	2.63
Oxygen: 1 atmo.	$+0.0126 \times 10^{-5}$	0.000179×10^{-5}	Becquerel.	5.89 "	0.014
Sulphur dioxide	-0.0751 "	0.302 "	"	"	-4.00
Water	-0.0694 "	0.377 "	Arons	"	-5.4
Nitric acid	-0.0633 "	0.356 "	Becquerel.	"	-5.6
Alcohol	-0.0566 "	0.330 "	De la Rive.	"	-5.8
Ether	-0.0541 "	0.315 "	"	"	-5.8
Arsenic chloride	-0.0876 "	1.222 "	Becquerel.	"	-14.9
Carbon disulphide	-0.0716 "	1.222 "	Rayleigh.	"	-17.1
Faraday's glass	-0.0982 "	1.738 "	Becquerel.	"	-17.7

TABLE 481. — Values of Kerr's Constant.*

Du Bois has shown that the rotation of the major axis of vibration of radiations normally reflected from a magnet is algebraically equal to the normal component of magnetization multiplied into a constant K. He calls this constant K, Kerr's constant for the magnetized substance forming the magnet.

Color of light.	Spectrum line.	Wave-length in cms. × 10[6]	Kerr's constant in minutes per c. g. s. unit of magnetization.			
			Cobalt.	Nickel.	Iron.	Magnetite.
Red	Li *a*	67.7	−0.0208	−0.0173	−0.0154	+0.0096
Red	—	62.0	−0.0198	−0.0160	−0.0138	+0.0120
Yellow	D	58.9	−0.0193	−0.0154	−0.0130	+0.0133
Green	*b*	51.7	−0.0179	−0.0159	−0.0111	+0.0072
Blue	F	48.6	−0.0180	−0.0163	−0.0101	+0.0026
Violet	G	43.1	−0.0182	−0.0175	−0.0089	−

* H. E. J. G. Du Bois, "Phil. Mag." vol. 29.

TABLE 482. — Dispersion of Kerr Effect.

Wave-length.	0.5μ	1.0μ	1.5μ	2.0μ	2.5μ
Steel . . .	−11′.	−16′.	−14′.	−11′.	−9′.0
Cobalt . . .	− 9.5	−11.5	− 9.5	−11.	−6.5
Nickel . . .	− 5.5	− 4.0	0	+1.75	+3.0

Field Intensity = 10,000 C. G. S. units. (Intensity of Magnetization = about 800 in steel, 700 to 800 in cobalt, about 400 in nickel). Ingersoll, Phil. Mag. 11, p. 41, 1906.

TABLE 483. — Dispersion of Kerr Effect.

Mirror.	Field (C. G. S.)	.41μ	.44μ	.48μ	.52μ	.56μ	.60μ	.64μ	.66μ
Iron . .	21,500	−.25	−.26	−.28	−.31	−.36	−.42	−.44	−.45
Cobalt . .	20,000	−.36	−.35	−.34	−.35	−.35	−.35	−.35	−.36
Nickel . .	19,000	−.16	−.15	−.13	−.13	−.14	−.14	−.14	−.14
Steel . .	19,200	−.27	−.28	−.31	−.35	−.38	−.40	−.44	−.45
Invar . .	19,800	−.22	−.23	−.24	−.23	−.23	−.22	−.23	−.23
Magnetite	16,400	−.07	−.02	+.04	+.06	+.08	+.06	+.04	+.03

Foote, Phys. Rev. 34, p. 96, 1912.

See also Ingersoll, Phys. Rev. 35, p. 312, 1912, for "The Kerr Rotation for Transverse Magnetic Fields," and Snow, l. c. 2, p. 29, 1913, "Magneto-optical Parameters of Iron and Nickel."

RESISTANCE OF METALS. MAGNETIC EFFECTS.

TABLE 484.—Variation of Resistance of Bismuth, with Temperature, in a Transverse Magnetic Field.

	Proportional Values of Resistance.								
H	−192°	−135°	−100°	−37°	0°	+18°	+60°	+100°	+183°
0	0.40	0.60	0.70	0.88	1.00	1.08	1.25	1.42	1.79
2000	1.16	0.87	0.86	0.96	1.08	1.11	1.26	1.43	1.80
4000	2.32	1.35	1.20	1.10	1.18	1.21	1.31	1.46	1.82
6000	4.00	2.06	1.60	1.29	1.30	1.32	1.39	1.51	1.85
8000	5.90	2.88	2.00	1.50	1.43	1.42	1.46	1.57	1.87
10000	8.60	3.80	2.43	1.72	1.57	1.54	1.54	1.62	1.89
12000	10.8	4.76	2.93	1.94	1.71	1.67	1.62	1.67	1.92
14000	12.9	5.82	3.50	2.16	1.87	1.80	1.70	1.73	1.94
16000	15.2	6.95	4.11	2.38	2.02	1.93	1.79	1.80	1.96
18000	17.5	8.15	4.76	2.60	2.18	2.06	1.88	1.87	1.99
20000	19.8	9.50	5.40	2.81	2.33	2.20	1.97	1.95	2.03
25000	25.5	13.3	7.30	3.50	2.73	2.52	2.22	2.10	2.09
30000	30.7	18.2	9.8	4.20	3.17	2.86	2.46	2.28	2.17
35000	35.5	20.35	12.2	4.95	3.62	3.25	2.69	2.45	2.25

TABLE 485.—Increase of Resistance of Nickel due to a Transverse Magnetic Field, expressed as % of Resistance at 0° and H = 0.

H	−190°	−75°	0°	+18°	+100°	+182°
0	+0	0	0	0	0	0
1000	+0.20	+0.23	+0.07	+0.07	+0.96	+0.04
2000	+0.17	+0.16	+0.03	+0.03	+0.72	−0.07
3000	0.00	−0.05	−0.34	−0.36	−0.14	−0.60
4000	−0.17	−0.15	−0.60	−0.72	−0.70	−1.15
6000	−0.19	−0.20	−0.70	−0.83	−1.02	−1.53
8000	−0.19	−0.23	−0.76	−0.90	−1.15	−1.66
10000	−0.18	−0.27	−0.82	−0.95	−1.23	−1.76
12000	−0.18	−0.30	−0.87	−1.00	−1.30	−1.85
14000	−0.18	−0.32	−0.91	−1.04	−1.37	−1.95
16000	−0.17	−0.35	−0.94	−1.09	−1.44	−2.05
18000	−0.17	−0.38	−0.98	−1.13	−1.51	−2.15
20000	−0.16	−0.41	−1.03	−1.17	−1.59	−2.25
25000	−0.14	−0.49	−1.12	−1.29	−1.76	−2.50
30000	−0.12	−0.56	−1.22	−1.40	−1.95	−2.73
35000	−0.10	−0.63	−1.32	−1.50	−2.13	−2.98

F. C. Blake, Ann. der Physik, 28, p. 449; 1909.

TABLE 486.—Change of Resistance of Various Metals in a Transverse Magnetic Field. Room Temperature.

Metal.	Field Strength in Gausses.	Per cent Increase.	Authority.
Nickel	10000	−1.2	Williams, Phil. Mag. 9, 1905.
"	"	−1.4	Barlow, Pr. Roy. Soc. 71, 1903.
"	6000	−1.0	Dagostino, Atti Ac. Linc. 17, 1908.
"	10000	−1.4	Grummach, Ann. der Phys. 22, 1906.
Cobalt	"	−0.53	"
Cadmium	"	+0.03	"
Zinc	"	+0.01	"
Copper	"	+0.004	"
Silver	"	+0.004	"
Gold	"	+0.003	"
Tin	"	+0.002	"
Palladium	"	+0.001	"
Platinum	"	+0.0005	"
Lead	"	+0.0004	"
Tantalum	"	+0.0003	"
Magnesium	6000	+0.01	Dagostino, *l. c.*
Manganin	"	+0.01	"
Tellurium	?	+0.02 to 0.34	Goldhammer, Wied Ann. 31, 1887.
Antimony	?	+0.02 to 0.16	"
Iron	Different specimens show very diverse results, usually an increase in weak fields, a decrease in strong.		Grummach, *l. c.* Barlow, *l. c.* Williams, *l. c.*
Nickel steel	Alloys behave similarly to iron.		Williams, *l. c.*

TABLE 487. — Transverse Galvanomagnetic and Thermomagnetic Effects.

Effects are considered positive when, the magnetic field being directed away from the observer, and the primary current of heat or electricity directed from left to right, the upper edge of the specimen has the higher potential or higher temperature.

E = difference of potential produced; T = difference of temperature produced; I = primary current; $\frac{dt}{dx}$ = primary temperature gradient; B = breadth, and D = thickness, of specimen; H = intensity of field. C. G. S. units.

Hall effect (Galvanomagnetic difference of Potential), $E = R\frac{HI}{D}$

Ettingshausen effect (" " " Temperature), $T = P\frac{HI}{D}$

Nernst effect (Thermomagnetic " " Potential), $E = QHB\frac{dt}{dx}$

Leduc effect (" " " Temperature), $T = SHB\frac{dt}{dx}$

Substance.	Values of R.	$P \times 10^6$.	$Q \times 10^6$.	$S \times 10^8$.
Tellurium	+400 to 800	+200	+360000	+400
Antimony	+ 0.9 " 0.22	+2	+9000 to 18000	+200
Steel	+.012 " 0.033	—0.07	—700 " 1700	+69
Heusler alloy	+.010 " 0.026	-	+1600 " 7000	
Iron	+.007 " 0.011	—0.06	—1000 " 1500	+39
Cobalt	+.0016 " 0.0046	+0.01	+1800 " 2240	+13
Zinc	-	-	—54 " 240	+13
Cadmium	+.00055			
Iridium	+.00040	-	up to —5.0	+5
Lead	+.00009	-	—5.0 (?)	
Tin	—.00003	-	—4.0 (?)	
Platinum	—.0002	-	-	—2
Copper	—.00052	-	—90 to 270	—18
German silver	—.00054			
Gold	—.00057 to .00071			
Constantine	—.0009			
Manganese	—.00093			
Palladium	—0007 to .0012	-	+50 to 130	—3
Silver	—.0008 " .0015	-	—46 " 430	—41
Sodium	—.0023			
Magnesium	—.00094 to .0035			
Aluminum	—.00036 " .0037			
Nickel	—.0045 " .024	+0.04 to 0.19	+2000 " 9000	—45
Carbon	—.017	+5.	+100	
Bismuth	— up to 16.	+3 to 40	+ up to 132000	—200

TABLE 488. — Variation of Hall Constant with the Temperature.

Bismuth.[1]						Antimony.[2]				
H	—182°	—90°	—23°	+11.5°	+100°	H	—186°	—79°	+21.5°	+58°
1000	62.2	28.0	17.0	13.3	7.28	1750	0.263	0.249	0.217	
2000	55.0	25.0	16.0	12.7	7.17	3960	0.252	0.243	0.211	
3000	49.7	22.9	15.1	12.1	7.06	6160	0.245	02.35	0.209	0.203
4000	45.8	21.5	14.3	11.5	6.95					
5000	42.6	20.2	13.6	11.0	6.84					
6000	40.1	18.9	12.9	10.6	6.72					

Bismuth.[3]

H	+14.5°	+104°	125°	189°	212°	239°	259°	269°	270°
890	5.28	2.57	2.12	1.42	1.24	1.11	0.97	0.83	0.77*

[1] Barlow, Ann. der Phys. 12, 1903.
[2] Everdingen, Comm. Phys. Lab. Leiden, 58.
[3] Traubenberg, Ann. der Phys. 17, 1905.
* Melting-point.

Both tables taken from Jahn, Jahrbuch der Radioactivität und Electronik, 5, p. 166; 1908, who has collected data of all observers and gives extensive bibliography.

RÖNTGEN (X-RAYS) RAYS.

TABLE 489. — Cathode and Canal Rays.

Cathode (negative) rays consist of negatively charged particles (charge 4.77×10^{-10} esu, 1.591×10^{-20} emu, mass, 9×10^{-28} g or 1/1800 H atom, diam. 4×10^{-13} cm) emitted at low pressures in an electric discharge tube perpendicularly to the cathode (∴ can be focused) with velocities (10^9 to 10^{10} cm/sec.) depending on the acting potential difference. When stopped by suitable body they produce heat, ionization (inversely proportional to velocity squared), photographic action, X-rays, phosphorescence, pressure. The bulk of energy is transformed into heat (Pt, Ta, W may be fused). In an ordinary X-ray tube carrying 10^{-3} ampere the energy given up may be of the order of 100 cal/m. Maximum thickness of glass or Al for appreciable transmission of high speed particles is .0015 cm. Maximum velocity V_d with which a cathode ray of velocity V_0 may pass through a material of thickness d is given by $V_0^4 - V_d^4 = ad \times 10^{40}$; $a = 2$ for air, 732 for Al and 2540 for Au, cm-sec. units (Whiddington, 1912). Cathode rays have a range of only a few millimeters in air.

Canal (positive) rays move from the anode with velocities about 10^8 cm/sec. in opposite direction to the cathode rays, carry a positive charge, a mass of the order of magnitude of the H molecule, cause strong ionization, fluorescence (LiCl fluoresces blue under cathode, red under canal ray bombardment), photographic action, strong pulverizing or disintegrating power and by bombardment of the cathode liberate the cathode rays.

TABLE 490. — Speed of Cathode Rays.

The speed of the cathode particles in cm/sec. as dependent upon the drop of potential to which they owe the speed, is given by the formula $v = 5.95 \sqrt{E} \cdot 10^7$. The following table gives values of $5.95 \sqrt{E}$.

Voltage..........	10	20	30	40	50	60	70	80	90	100
Velocity × 10^{-7}...	18.8	26.6	32.6	37.6	42.1	46.1	49.8	53.3	56.5	59.5
Voltage..........	100	200	300	400	500	600	700	800	900	1000
Velocity × 10^{-7}...	59.5	84.2	103.1	119.1	133.1	145.8	157.5	168.3	178.6	188.

For voltages 1000 to 10,000 multiply 2d line by 10, etc.

TABLE 491. — Cathodic Sputtering.

The disintegration of the cathode in an electric discharge tube is not a simple phenomenon. The particles taking part in the sputtering must be either large or of high speed or both (2000+ gauss field required for their deviation). It depends upon the nature of the residual gas. H, N, CO_2 are not generally favorable; Ar is especially favorable, also He, Ne, Kr and Xe. Raised temperature favors it. The relative sputtering from various metals is shown in the following table (Crookes, Pr. R. S. 1891); the residual gas was air, pressure about .05 mm Hg.

Metal.........	Pd	Au	Ag	Pb	Sn	Pt	Cu	Cd	Ni	Ir	Fe	Al	Mg	Brass
Sputtering.....	100	92	76	69	52	40	37	31	10	10	5	0	0	47

For further data on cathode, canal and X-rays, see X-rays by G. W. C. Kaye, Longmans, 1917, upon which much of the above and the following data for X-rays is based. See also J. J. Thomson, Positive Rays, Longmans, 1913.

TABLE 492. — X-rays, General Properties.

X-rays are produced whenever and wherever a cathode ray hits matter. They are invisible, of the same nature as, and travel with the velocity of light, affect photographic plates, excite phosphorescence, ionize gases and suffer deviation neither by magnetic nor electric fields as do cathode rays. In an ordinary X-ray tube (vacuum order 0.001 to 0.01 mm Hg) the cathode (concave for focusing, generally of aluminum) rays are focused on an anticathode of high atomic weight (W, Pt, high atomic weight, high melting point, low vapor pressure, to avoid sputtering, high thermal conductivity to avoid heating). Depth to which cathode rays penetrate, order of 0.2×10^{-5} cm in Pb, 90,000 volts (Ham, 1910), 24×10^{-5} cm in Al, 22,000 volts (Warburg, 1915). Note: High speed H and He molecules (2×10^8 cm/sec.) can penetrate 0.001 to 0.006 mm mica; He α particles (2×10^9 cm/sec.), 0.04 mm glass.

The X-rays from an ordinary bulb consist of two main classes:

Heterogeneous ("general," "independent") radiation, which depends solely on the speed of the parent cathode rays. It is always present and its range of hardness (wave-lengths) depends on the range of speeds of the cathode rays. Its energy is proportional to the 4th power of these speeds.

Homogeneous ("characteristic," "monochromatic") radiation (*K*, *L*, *M*, etc. radiations, see Table 498 for wave-lengths), characteristic of the metal of the anticathode. Generated only when cathode rays are sufficiently fast. There is a critical velocity for each characteristic radiation from each material, proportional to the atomic weight of the anticathode. The critical velocity for the *K* radiation is $V_K = A \times 10^8$, when A is the atomic weight of the radiator (e.g. anticathode); $V_L = 1/2(A - 48)10^8$.

The following relation has been found to hold experimentally between the voltage V through which the cathode particles fall and the maximum frequency ν of the X-rays produced: $eV = h\nu$, where e is the electronic charge and h, Planck's constant. Blake and Duane (Phys. Rev. 10, 624, 1917) found for h, 6.555×10^{-27} erg second.

As the speed of the cathode rays is increased, shorter and shorter wave-lengthed "independent" X-rays are produced until the critical speed is reached for the "characteristic" rays; with faster speeds, the cathode rays become at first increasingly effective for the characteristic radiation, then less so as the independent radiation again predominates.

When cathode rays hit the anticathode some 75 per cent are reflected, the more the heavier its atomic weight. The chances of the remainder hitting an atom so as to generate an X-ray are slight; only 1/1000 or 1/2000 of the original energy goes into X-rays. If E_x and E_c are the energies of the X and the parent cathode rays, A the atomic weight of the anticathode, β the velocity of the cathode rays as fraction of the light value (3×10^{10} cm/sec.), Beatty showed (Pr. R. S. 1913) that $E_x = E_c\,(.51 \times 10^4 A\beta^2)$; this refers only to the independent radiations; when characteristic radiations are excited their energy must be added and the tube becomes considerably more efficient. No quantitative expression for the latter has been developed.

When an X-ray strikes a substance three types of radiation result: scattered (sometimes called secondary) X-rays, characteristic X-rays and corpuscular rays (negatively charged particles). The proportions of the rays depend on the substance and the quality of the primary rays. When the substance is of low atomic weight, by far the greater portion of the X-rays, if of a penetrating type, are scattered. With elements of the Cr-Zn group most of the resulting radiation is "characteristic." With the Cu group the scattered radiation (1/200) is negligible. Heavier elements, both scattered and characteristic X-rays. Corpuscular radiation greater, mass for mass, for elements of high atomic weight and may mask and swamp the characteristic radiation. Hence an X-ray tube beam, heterogeneous in quality, allowed to fall on different metals, — Cu, Ag, Fe, Pt, etc., — excites characteristic X-rays of wide range of qualities. Exciting ray must be harder than the characteristic radiation wished. The higher the atomic weight of the material struck (radiator), the more penetrating the quality of the resulting radiation as shown by the following table, which gives λ, the reciprocal of the distance in cm in Al, through which the rays must pass in order that their intensity will be reduced to 1/2.7 of their original intensity.

TABLE 493. — Röntgen Secondary Rays.

Radiator.	Cr	Fe	Co	Ni	Cu	Zn	As	Se	Sr	Ag	Sn
Atomic weight....	52.	55.8	59.0	58.7	63.6	65.4	75.0	79.2	87.6	108.	119.
λ...............	367.	239.	193.	160.	129.	106.	61.	51.	35.2	6.75	4.33

With the radiator at 45° to the primary X-rays at most only about 50 per cent of the energy goes to characteristic rays and only about 1/10 of the latter escape the surface of the radiator. The β radiations of radioactive elements may possibly be regarded (Rutherford) as a characteristic radiation produced by the expulsion of the α particles. The hardness of some corresponds to the *K* and *L* radiations.

For more complete data on X-rays, see X-rays, G. W. C. Kaye, Longmans, 1917, upon which these X-ray tables are greatly based.

RÖNTGEN (X-RAYS) RAYS.

TABLE 494. — Corpuscular Rays.

Corpuscular rays are given off in greatest abundance when radiator emits its characteristic radiation. Intensity increases with atomic weight (4th power, Moore, Pr. Phys. Soc.). Greater number emitted at right angles to incident rays. Velocity range (6 to 8.5)10^9 cm/sec. v_0 = velocity when leaving radiator = $10^8(A$ = Atomic weight) = critical velocity necessary to excite characteristic radiation, therefore corpuscular rays have practically the same velocity as the original generating cathode rays. Are of uniform quality when excited by characteristic rays and follow exponential law of absorption in gases. If λ is the absorption coefficient and A the atomic weight, $\lambda A^4 = \lambda v_0^4$ = constant (Whiddington, Beatty). λ is defined by $I = I_0 e^{-\lambda d}$ where I and I_0 are the intensities after and before absorption and d the thickness of the absorptive layer in cm. The following values for λ in air for characteristic radiations from various substances are due to Sadler. (At 0° C and 76 cm Hg.)

Metal emitting corpuscles.	Exciting characteristic radiation from									
	Ni	Cu	Zn	As	Se	Sr	Mo	Rh	Ag	Sn
Al	—	—	—	29.6	—	20.0	15.2	—	8.90	6.54
Fe	38.9	37.0	35.8	30.2	26.4	21.5	15.5	10.9	8.84	6.41
Cu	—	—	36.2	30.4	—	20.8	15.2	10.8	8.81	6.67

TABLE 495. — Intensity of X-Rays. Ionization.

The intensity of the radiation from an X-ray bulb is proportional to the current. Except at low voltages it equals $Ki(v^2 - v_0^2)$ where i is the current, v the applied voltage, v_0 the break-down voltage and K a constant for the tube (Krönke). The intensity of X-rays is most accurately measured by the ionization they produce. This may be referred to the International Radium Standard (see Table 508). It is proportional to the 4th power of the speed of the parent cathode rays (Thomson), (true only of independent rays, Beatty, 1913). The saturation current due to X-ray ionization is usually of the order of 10^{-10} to 10^{-15} ampere. When X-rays pass through a substance, only once in a while is an atom struck, only perhaps 1 in a billion, and ionized. The ionization is probably an indirect process through the mediation of corpuscular rays. In the absence of secondary radiations the ionization is proportional to the mass of the gas (that is, its pressure at constant temperature). It depends on the nature of the gas, but is little affected by the quality of the rays. The following results are due to Crowther, 1908.

Gas or vapor.	Ionization relative to air = 1.		
	Density, air = 1.	Soft X-rays 6 mm spark.	Hard X-rays 27 mm spark.
Hydrogen H_2	0.07	0.01	0.18
Carbon dioxide CO_2	1.53	1.57	1.49
Ethyl chloride C_2H_5Cl	2.24	18.0	17.3
Carbon tetrachloride CCl_4	5.35	67.	71.
Ethyl bromide C_2H_5Br	3.78	72.	118.
Methyl iodide CH_3I	4.96	145.	125.
Mercury methyl $Hg(CH_3)_2$	7.93	425.	—

RÖNTGEN (X-RAYS) RAYS.

TABLE 496. — Mass Absorption Coefficients, λ/d.

The quality by which X-rays have been generally classified is their "hardness" or penetrating power. It is greater the greater the exhaustion of the tube, but for a given tube depends solely upon the potential difference of the electrodes. With extreme exhaustion the X-rays have an appreciable effect after passing through several millimeters of brass or Al. The penetrability of the characteristic radiation is in general proportional to the 5th power of the atomic weight of the radiator. The absorption of any substance is equal to the sum of the absorptions of the individual atoms and is independent of the chemical combination, its physical state and probably of the temperature. Most of the following table is from the work of Barkla and Sadler, Phil. Mag. 17, 739, 1909. For starred radiators, *L* radiations used; for others the *K*.

If I_0 be the intensity of a parallel beam of homogeneous radiation incident normally on a plate of absorbing material of thickness *t*, then $I = I_0 e^{-\lambda x}$ gives the intensity *I* at the depth *x*. Because of the greater homogeneity of the secondary X-rays they were used in the determination of the following coefficients. The coefficients λ have been divided by the density *d*.

Radiator.	Absorber.										
	C	Mg	Al	Fe	Ni	Cu	Zn	Ag	Sn	Pt	Au
Cr.	15.3	126.	136.	104.	129.	143.	170.	580.	714.	(517.)	(507.)
Fe.	10.1	80.	88.	66.	84.	95.	112.	381.	472.	340.	367.
Co.	8.0	64.	72.	67.	67.	75.	92.	314.	392.	281.	306.
Ni.	6.6	52.	59.	314.	56.	62.	74.	262.	328.	236.	253.
Cu.	5.2	41.	48.	268.	63.	53.	61.	214.	272.	194.	210.
Zn.	4.3	35.	39.	221.	265.	56.	50.	175.	225.	162.	178.
As.	2.5	19.	22.	134.	166.	176.	204.	105.	132.	106.	106.
Se.	2.0	16.	19.	116.	141.	150.	175.	88.	112.	93.	100.
Ag.	.46	2.2	2.5	17.	23.	24.	27.	13.	16.	56.	61.
Sn.	.35	—	1.6	—	—	—	—	16.	—	47.	52.
Sb.	.31	—	1.2	—	—	—	—	56.	—	—	—
I.	.29	—	.9	—	—	—	—	46.	—	—	—
Ba.	.26	—	.8	—	—	—	—	35.	—	—	—
W*.	—	—	30.	—	—	127.	—	140.	—	133.	—
Pt*.	—	—	22.	—	—	177.	—	106.	—	113.	—
Pb*.	—	—	17.	—	—	139.	—	78.	—	128.	—
Bi*.	—	—	16.	—	—	127.	—	73.	—	125	—
Th*.	—	—	8.	—	—	77.	—	42.	—	134.	—
U*.	—	—	7.	—	—	70.	—	40.	—	132 .	—

TABLE 497. — Absorption Coefficients of Characteristic Radiations in Gases.

The penetrating power of X-rays ranges in normal air from 1 to 10,000 cm or more. The absorptive power of 1 cm air = 1/820 that of water. λ (see preceding table for definition) for air for soft bulb (1.5 to 5 cm spark gap, 4 to 10 m air) ranges from .0010 to .0018; for hard bulb (30 cm spark gap, 4 to 10 m air), .00029. (Eve and Day, Phil. Mag. 1912.) The absorption coefficient for gases for characteristic or monochromatic radiations varies directly with the pressure. For different characteristic radiations it is proportional to the coefficients in air. It varies with the 5th power of the atomic weight of the radiator. The following table is taken from Kaye's X-rays and is based on the work of Barkla and Collier (Phil. Mag. 1912) and Owen. All are for the gas at 0° C and 76 cm Hg.

	Air		CO_2		SO_2		C_2H_5Br		CH_3I	
	λ	λ/d	λ	λ/d	λ	λ/d	λ	λ/d	λ	λ/d
Fe.	.0202	15.6	.0456	23.1	.24	83.3	.512	105.	2.16	339.
Co.	.0165	12.7	—	—	.20	69.4	.407	83.2	—	—
Ni.	.0136	10.5	.0319	16.1	.166	57.6	.325	66.3	1.80	282.
Cu.	.0109	8.43	.0227	11.5	.134	46.5	.260	53.1	1.54	241.
Zn.	.0090	6.96	.0184	9.31	.112	38.9	.215	43.9	1.27	198.
As.	.0053	4.10	.00988	5.00	.066	22.9	.128	26.1	.743	116.
Se.	.0044	3.40	.00782	3.96	.0546	19.1	.110	22.4	.619	97.
Br.	.0039	3.02	—	—	.050	17.4	.096	19.6	.552	86.5
Sr.	.0023	1.78	.00420	2.12	.0281	9.76	.325	66.3	.338	53.0
Mo.	.00127	0.98	.00281	1.42	.0160	5.56	.210	42.9	.197	30.9
Ag.	.00077	0.59	—	—	.0079	2.75	.108	22.0	.113	17.7

TABLE 498.

X-RAY SPECTRA AND ATOMIC NUMBERS.

Kaye has shown that an element excited by sufficiently rapid cathode rays emits Röntgen rays characteristic of that substance. These were analyzed and the wave-lengths determined by Moseley (Phil. Mag. 27, 703, 1914), using a crystal of potassium ferrocyanide as a grating. He noted the K series, showing two lines, and the L series with several. He found that every element from Al to Au was characterized by integer N, which determines its X-ray spectrum; N is identified with the number of positive units associated with its atomic nucleus. The order of these atomic numbers (N) is that of the atomic weights, except where the latter disagrees with the order of the chemical properties. Known elements now correspond with all the numbers between 1 and 92 except 6. There are here six possible elements still to be discovered (atomic nos. 43, 61, 72, 75, 85).

The frequency of any line in an X-ray spectrum is approximately proportional to $A(N - b)^2$, where A and b are constants. All X-ray spectra of each series are similar in structure, differing only in wave-lengths. $Q_K = (v/\frac{3}{4}v_0)$; $Q_L = (v/\frac{5}{36}v_0)$ where v is the frequency of the a line and v_0 the fundamental Rydberg frequency. The atomic number for the K series $= Q_K + 1$ and for the L series, $Q_L + 7.4$ approximately. $v_0 = 3.29 \times 10^{15}$

Moseley's work has been extended, and the following tables indicate the present (1919) knowledge of the X-ray spectra.

(a) K SERIES (WAVE-LENGTHS, $\lambda \times 10^8$ CM).

Element, atomic number.		β_2	β_1	a_4	a_3a_4 (not separable)	a_3	a_1	a_1a_2 (not separable)	a_2
11	Na	—	—	—	—	—	—	11.951	—
12	Mg	—	9.477	9.845	—	9.856	—	9.915	—
13	Al	—	7.986	8.300	—	8.310	—	8.360	—
14	Si	—	6.759	7.080	—	7.088	—	7.131	—
15	P	—	5.808	6.122	—	6.129	—	6.168	—
16	S	—	5.018	5.314	—	5.317	—	5.360	—
17	Cl	—	4.394	—	4.692	—	—	4.712	—
18	Ar	—	—	—	—	—	—	—	—
19	K	—	3.449	—	3.724	—	3.735	—	3.738
20	Ca	3.074	3.086	—	3.328	—	3.355	—	3.359
21	Sc	—	2.778	—	3.011	—	3.028	—	3.032
22	Ti	2.492	2.509	—	2.729	—	2.742	—	2.746
23	Va	—	2.281	—	—	—	2.498	—	2.502

Element, atomic number.		β_2	β_1	a_1	a_2	Element, atomic number.		β_2	β_1	a_1	a_2
24	Cr	2.069	2.079	2.284	2.288	43		—	—	—	—
25	Mn	1.892	1.902	2.093	2.097	44	Ru	—	0.574	0.645	—
26	Fe	1.736	1.748	1.928	1.932	45	Rh	0.537	.547	.615	0.619
27	Co	1.602	1.613	1.781	1.785	46	Pd	—	.501	.562	.567
28	Ni	1.488	1.497	1.653	1.657	47	Ag	.491	.501	.562	.567
29	Cu	1.379	1.391	1.539	1.543	48	Cd	—	.479	.538	.543
30	Zn	1.281	1.294	1.433	1.437	49	In	.440	.453	.510	.515
31	Ga	—	1.206	1.338	1.342	50	Sn	—	.432	.487	.490
32	Ge	1.121	1.131	1.257	1.251	51	Sb	.408	.416	.468	.472
33	As	1.038	1.052	1.170	1.174	52	Te	—	.404	.456	—
34	Se	—	0.993	1.104	1.109	53	I	—	.388	.437	—
35	Br	0.914	.929	1.035	1.040	54	X	—	—	—	—
36	Kr	—	—	—	—	55	Cs	—	.352	.398	.402
37	Rb	.813	.825	0.922	0.926	56	Ba	—	.343	.388	.393
38	Sr	.767	.779	.871	.876	57	La	—	.329	.372	.376
39	Y	.733	.746	.835	.840	58	Ce	—	.314	.355	.360
40	Zr	—	.705	.788	.793	59	Pr	—	.301	.342	.347
41	Nb	.657	.669	.749	.754	60	Nd	—	.292	.330	.335
42	Mo	—	.633	.710	.714	74	W		.177	.203	.335

X-RAY SPECTRA AND ATOMIC NUMBERS.

(*b*) L SERIES (WAVE-LENGTHS, $\lambda \times 10^8$ CM).

Element, atomic number.	l	α_2	α_1	α_3	Element, atomic number.	l	α_2	α_1	η
30 Zn	—	—	12.346	—	60 Nd	—	2.379	2.369	—
33 As	—	—	9.701	—	62 Sa	—	2.210	2.200	—
35 Br	—	—	8.391	8.360	63 Eu	—	2.131	2.121	—
37 Rb	—	—	7.335	7.305	64 Gd	—	2.054	2.043	—
38 Sr	—	—	6.879	—	65 Tb	—	1.983	1.973	1.935
39 Y	—	—	6.464	6.440	66 Dy	—	1.916	1.907	—
40 Zr	—	—	6.083	6.057	67 Ho	—	1.854	1.843	—
41 Nb	—	5.731	5.724	5.709	68 Er	—	1.794	1.783	1.725
42 Mo	—	5.410	5.403	5.381	70 Ad	1.892	1.681	1.670	1.618
44 Ru	—	4.853	4.845	4.823	71 Cp	1.834	1.629	1.619	—
45 Rh	—	—	4.596	4.577	73 Ta	—	1.528	1.518	1.435
46 Pd	—	4.374	4.365	4.352	74 W	1.672	1.481	1.471	—
47 Ag	—	4.155	4.146	4.133	76 Os	—	1.398	1.388	—
48 Cd	—	3.959	3.949	—	77 Ir	1.840	1.360	1.350	—
49 In	—	3.774	3.766	—	78 Pt	1.499	1.323	1.313	1.242
50 Sn	—	3.604	3.594	—	79 Au	1.457	1.283	1.271	1.197
51 Sb	—	3.443	3.434	—	80 Hg	—	1.251	1.240	—
52 Te	—	3.299	3.290	—	81 Tl	1.385	1.215	1.205	1.124
53 I	—	3.155	3.146	—	82 Pb	1.348	1.186	1.175	1.091
55 Cs	—	2.899	2.891	—	83 Bi	1.317	1.153	1.144	1.059
56 Ba	—	2.786	2.776	—	84 Po	—	—	1.109	—
57 La	—	2.674	2.665	—	88 Ra	—	—	1.010	—
58 Ce	—	2.573	2.563	—	90 Th	1.117	0.969	0.957	—
59 Pr	—	2.472	2.462	—	92 U	1.066	0.922	0.911	—

Element, atomic number.	β_4	β_1	β_2	β_3	β_5	γ_1	γ_2	γ_3	γ_4
33 As	—	9.449	—	—	—	—	—	—	—
35 Br	—	8.141	—	—	—	—	—	—	—
37 Rb	—	7.091	—	—	—	—	—	—	—
38 Sr	—	6.639	—	—	—	—	—	—	—
39 Y	—	6.227	—	—	—	—	—	—	—
40 Zr	—	5.851	—	—	—	5.386	—	—	—
41 Nb	—	5.493	5.317	—	—	—	—	—	—
42 Mo	—	5.175	—	—	—	—	—	—	—
44 Ru	—	4.630	—	—	—	—	—	—	—
45 Rh	—	4.372	—	—	—	—	—	—	—
46 Pd	4.071	4.144	3.904	4.030	—	3.720	3.597		—
47 Ag	3.861	3.928	3.698	3.823	—	3.515	—	—	—
48 Cd	3.676	3.733	3.514	3.639	—	3.331	—	—	—
49 In	—	3.550	3.354	—	—	3.160	—	—	—
50 Sn	3.337	3.381	3.172	3.300	—	2.999	2.903	2.889	2.831
51 Sb	3.184	3.222	3.021	3.149	—	2.849	2.782		—
52 Te	3.044	3.074	2.881	3.007	—	2.712	—	—	—
53 I	2.911	2.934	2.750	2.873	—	2.583	—	—	—
55 Cs	2.668	2.684	2.514	2.629	—	2.350	2.234		—
56 Ba	2.558	2.569	2.407	2.520	—	2.245	—	—	—
57 La	2.453	2.461	2.307	2.414	—	2.146	—	—	—
58 Ce	2.357	2.359	2.212	2.307	—	2.052	2.003		—
59 Pr	—	2.259	2.210	2.217	—	1.958	1.937	1.933	—
60 Nd	2.167	2.167	2.036	2.128	—	1.875	1.803	1.775	—
62 Sa	—	2.000	1.884	1.965	—	1.725	1.659		—
63 Eu	1.923	1.918	1.810	1.888	—	1.662	1.599	1.590	—
64 Gd	1.851	1.844	1.744	1.811	—	1.597	(1.562)	(1.558)	—
65 Tb	1.784	1.775	1.682	1.745	1.659	1.531	1.477	1.470	1.437

X-RAY SPECTRA AND ATOMIC NUMBERS.

(*b*) L SERIES (WAVE-LENGTHS, $\lambda \times 10^8$ CM).

Element, atomic number.		β_4	β_1	β_2	β_3	β_5	γ_1	γ_2	γ_3	γ_4
66	Dy	1.721	1.709	1.622	1.683	—	1.470	1.422	1.418	—
67	Ho	1.657	1.646	1.568	1.620	—	1.415	1.369	1.365	—
68	Er	1.599	1.586	1.514	1.560	—	1.367	1.323	1.316	—
70	Ad	1.490	1.474	1.414	1.451	1.422	1.267	1.228	1.223	—
71	Cp	1.437	1.421	1.368	1.399	—	1.224	1.188	1.183	—
73	Ta	1.343	1.323	1.280	1.303	—	1.135	1.101	1.097	—
74	W	1.296	1.278	1.241	1.258	—	1.105	1.064	1.058	—
76	Os	1.214	1.194	1.167	1.176	—	1.021	—	—	—
77	Ir	1.176	1.154	1.133	1.138	1.101	0.989	0.962	0.956	0.917
78	Pt	1.142	1.120	1.101	1.098	1.072	0.958	0.933	0.929	0.900
79	Au	1.102	1.080	1.065	1.059	1.035	0.922	0.898	0.894	0.869
80	Hg	—	1.049	1.042	—	—	0.896	—	—	—
81	Tl	1.036	1.012	1.006	0.998	0.977	0.864	0.844	0.840	0.808
82	Pb	1.008	0.983	0.983	0.968	—	0.842	0.820	0.816	0.792
83	Bi	0.977	0.950	0.954	0.937	0.923	0.810	0.794	0.790	0.762
84	Po	—	0.920	—	—	—	—	—	—	—
88	Ra	—	—	—	—	—	—	—	—	—
90	Th	—	0.766	0.797	0.758	—	0.654	0.635		—
92	U	—	0.720	0.756	0.710	—	0.615	0.596		—

(*c*) M SERIES (WAVE-LENGTHS, $\lambda \times 10^8$ CM).

Element, atomic number.		α	β	γ_1	γ_2	δ_1	δ_2	ϵ
79	Au	5.838	5.623	5.348	5.284	5.146	5.102	—
81	Tl	5.479	5.256	—	—	—	4.826	4.735
82	Pb	5.303	5.095	4.910	—	—	4.695	—
83	Bi	5.117	4.903	4.726	—	4.561	4.532	4.456
90	Th	4.139	3.941	3.812	3.678	—	—	—
92	U	3.905	3.715	—	3.480	3.363	3.324	—

Reference: Jahrbuch der Radioaktivität und Elektronik, 13, 296, 1916.

(*d*) TUNGSTEN X-RAY SPECTRUM (WAVE-LENGTHS, $\lambda \times 10^8$ CM).

The wave-lengths of the tungsten X-ray spectrum have been measured more frequently than those of any other element. The following values are perhaps the most accurate that have hitherto been published. Compton, Physical Review, 7, 646, 1916 (errata, 8, 753, 1916).

Line.	λ	Line.	λ	Line.	λ
a	1.0249	*e*	1.2185	*j*	1.3363
b	1.0399	*f*	1.2420	*k*	1.4735
c'	1.0582	*g*	1.2601	*l*	1.4844
c''	1.0652	*h*	1.2787		
d	1.0959	*i*	1.2985		

Other references on the X-ray spectrum of tungsten: Gorton, Physical Review, 7, 203, 1916; Hull, Proc. Nat. Acad. Sci. 2, 265, 1916; Dershem, Physical Review, 11, 461, 1918; Overn, Physical Review, 14, 137, 1919.

The following values for tungsten are from Duane and Patterson, Phys. Rev. 16, p. 526, 1920:

Critical Absorption wave-lengths $\times 10^8$ cm.

$K\alpha$	.17806	$L\alpha_1$	1.2136	$L\alpha_2$	1.0726	$L\alpha_3$	1.024

Emission wave-length $\times 10^8$ cm.

$K\alpha_2$	.21341	$K\alpha_1$	.20860	$K\beta$	.18420	$K\lambda$	.17901		
Ll	1.6756	$L\alpha_2$	1.4839	$L\alpha_1$	1.47306	$L\eta$	1.4176		
$L\beta_4$	1.2985	$L\beta_1$	1.27892	$L\beta_3$	1.2601	$L\beta_2$	1.24193	$L\beta_5$	1.2040
$L\gamma_1$	1.09608	$L\gamma_2$	1.0655	$L\gamma_3$	1.0596	$L\gamma_4$	1.0261		

TABLE 499.

X-RAY ABSORPTION SPECTRA AND ATOMIC NUMBERS.

A marked increase in the absorption of X-rays by a chemical element occurs at frequencies close to those of the X-rays characteristic of that element. The absorption coefficient is much greater on the short wave-length side. In the K series the α lines are much stronger than the corresponding β and γ lines, but the wave-lengths of the α lines are greater. There is a marked increase in the absorption at wave-lengths considerably shorter than the α lines and near the β lines. Bragg came to the conclusion that the critical absorption frequency lay at or above the γ of the K series. The γ line has a frequency about 1 per cent higher than the corresponding β line. For the L series there are 3 characteristic marked absorption changes (de Broglie).

The critical absorption wave-lengths of the following table are due to Blake and Duane, Phys. Rev. 10, 697, 1917. The equation $\nu = \nu_0(N - 3.5)^2$ where ν is Rydberg's fundamental frequency (109,675 × the velocity of light) and N the atomic number, represents the data with considerable accuracy. The nuclear charge is obtained by $Q = 2e(N - 3.5)$.

Element.	Atomic number.	ÅU	Element.	Atomic number.	ÅU	Element.	Atomic number.	ÅU
Bromine.....	35	.9179	Ruthenium	44	.5584	Tellurium..	52	.3896
Krypton.....	36	—	Rhodium..	45	.5324	Iodine.....	53	.3727
Rubidium....	37	.8143	Palladium.	46	.5075	Xenon.....	54	—
Strontium....	38	.7696	Silver.....	47	.4850	Caesium...	55	.3444
Yttrium.....	39	.7255	Cadmium..	48	.4632	Barium....	56	.3307
Zirconium....	40	.6872	Indium....	49	.4434	Lanthanum	57	.3188
Columbium..	41	.6503	Tin......	50	.4242	Cerium....	58	.3073
Molybdenum.	42	.6180	Antimony.	51	.4065			

Radioactivity is a property of certain elements of high atomic weight. It is an additive property of the atom, dependent only on it and not on the chemical compound formed nor affected by physical conditions controlling ordinary reactions, viz: temperature, whether solid or liquid or gaseous, etc.

With the exception of actinium, radioactive bodies emit α, β, or γ rays. α rays are easily absorbed by thin metal foil or a few cms. of air and are positively charged atoms of helium emitted with about 1/15 the velocity of light. They are deflected but very slightly by intense electric or magnetic fields. The β rays are on the average more penetrating, are negatively charged particles projected with nearly the velocity of light, easily deflected by electric or magnetic fields and identical in type with the cathode rays of a vacuum tube. The γ rays are extremely penetrating and non-deviable, analogous in many respects to the very penetrating Röntgen rays. These rays produce ionization of gases, act on the photographic plate, excite phosphorescence, produce certain chemical reactions such as the formation of ozone or the decomposition of water. All radioactive compounds are luminous even at the temperature of liquid air.

Table 506 is based very greatly on Rutherford's Radioactive Substances and their radiations (Oct. 1912). To this and to Landolt-Börnstein Physikalisch-chemische Tabellen the reader is referred for references. In the three radioactive series each successive product (except Ur. Y, and Ra. C_2) results from the transformation of the preceding product and in turn produces the following. When the change is accompanied by the ejection of an α particle (helium, atomic weight = 4.0) the atomic weight decreases by 4. The italicized atomic weights are thus computed. Each product with its radiation decays by an exponential law; the product and its radiation consequently depend on the same law. $I = I_0 e^{-\lambda t}$ where I_0 = radioactivity when t = O, I that at the time t, and λ the transformation constant. Radioactive equilibrium of a body with its products exists when that body is of such long period that its radiation may be considered constant and the decay and growth of its products are balanced.

International radium standard: As many radioactivity measures depend upon the purity of the radium used, in 1912 a committee appointed by the Congress of Radioactivity and Electricity, Brussels, 1910, compared a standard of 21.99 mg. of pure Ra. chloride sealed in a thin glass tube and prepared by Mme. Curie with similar standards by Hönigschmid and belonging to The Academy of Sciences of Vienna. The comparison showed an agreement of 1 in 300. Mme. Curie's standard was accepted and is preserved in the Bureau international des poids et mesures at Sèvres, near Paris. Arrangements have been made for the preparation of duplicate standards for governments requiring them.

TABLE 500.—Relative Phosphorescence Excited by Radium.

(Becquerel, C. R. 129, p. 912, 1899.)

Without screen,	Hexagonal zinc blende	13.36	With screen	.04
" "	Pt. cyanide of barium	1.99	" "	.05
" "	Diamond	1.14	" "	.01
" "	Double sulphate Ur and K	1.00	" "	.31
" "	Calcium fluoride	.30	" "	.02

The screen of black paper absorbed most of the α rays to which the phosphorescence was greatly due. For the last column the intensity without screen was taken as unity. The γ rays have very little effect.

TABLE 501.—The Production of α Particles (Helium).

(Geiger and Rutherford, Philosophical Magazine, 20, p. 691, 1910.)

Radioactive substance (1 gram.)	α particles per sec.	Helium per year.
Uranium	2.37×10^4	2.75×10^{-5} cu. mm.
Uranium in equilibrium with products	9.7×10^4	11.0×10^{-5} " "
Thorium " " " "	2.7×10^4	3.1×10^{-5} " "
Radium	3.4×10^{10}	39 " "
Radium in equilibrium with products	13.6×10^{10}	158 " "

TABLE 502.—Heating Effect of Radium and its Emanation.

(Rutherford and Robinson, Philosophical Magazine, 25, p. 312, 1913.)

	Heating effect in gram-calories per hour per gram radium.			
	α rays.	β rays.	γ rays.	Total.
Radium	25.1	–	–	25.1
Emanation	28.6	–	–	28.6
Radium A	30.5	–	–	30.5
Radium B + C	39.4	4.7	6.4	50.5
Totals	123.6	4.7	6.4	134.7

Other determinations: Hess, Wien. Ber. 121, p. 1, 1912, Radium (alone) 25.2 cal. per hour per gram. Meyer and Hess, Wien. Ber. 121, p. 603, 1912, Radium in equilibrium, 132.3 gram. cal. per hour per gram. See also, Callendar, Phys. Soc. Proceed. 23, p. 1, 1910; Schweidler and Hess, Ion. 1, p. 161, 1909; Ångström, Phys. ZS. 6, 685, 1905, etc.

RADIOACTIVITY.

TABLE 503.—Stopping Powers of Various Substances for α Rays.

s, the stopping power of a substance for the α rays is approximately proportional to the square root of the atomic weight, w.

Substance	H_2	Air	O_2	C_2H_2	C_2H_4	Al	N_2O	CO_2	CH_3Br	CS_2	Fe
s . . .	.24	1.0	1.05	1.11	1.35	1.45	1.46	1.47	2.09	2.18	2.26
$\sqrt{w}$ . . .	.26	1.0	1.05	1.17	1.44	1.37	1.52	1.51	2.03	1.95	1.97
Substance	Cu	Ni	Ag	Sn	C_6H_6	C_5H_{12}	C_2H_5I	CCl_4	Pt	Au	Pb
s . . .	2.43	2.46	3.17	3.37	3.37	3.59	3.13	4.02	4.16	4.45	4.27
$\sqrt{w}$ . . .	2.10	2.20	2.74	2.88	3.53	3.86	3.06	3.59	3.68	3.70	3.78

Bragg, Philosophical Magazine, 11, p. 617, 1906.

TABLE 504.—Absorption of β Rays by Various Substances.

μ, the coefficient of absorption for β rays is approximately proportional to the density, D. See Table 506 for μ for Al.

Substance . .	B	C	Na	Mg	Al	Si	P	S	K	Ca
μ/D	4.65	4.4	4.95	5.1	5.26	5.5	6.1	6.6	6.53	6.47
Atomic Wt. .	11	12	23	24.4	27	28	31	32	39	40
Substance . .	Ti	Cr	Fe	Co	Cu	Zn	Ar	Se	Sr	Zr
μ/D	6.2	6.25	6.4	6.48	6.8	6.95	8.2	8.65	8.5	8.3
Atomic Wt. .	48	52	56	59	63.3	65.5	75	79	87.5	90.7
Substance . .	Pd	Ag	Sn	Sb	I	Ba	Pt	Au	Pb	U
μ/D	8.0	8.3	9.46	9.8	10.8	8.8	9.4	9.5	10.8	10.1
Atomic Wt. .	106	108	118	120	126	137	195	197	207	240

For the above data the β rays from Uranium were used.

Crowther, Philosophical Magazine, 12, p. 379, 1906.

TABLE 505.—Absorption of γ Rays by Various Substances.

Substance.	Density.	Radium rays.		Uranium rays.		Th. D. $\mu(cm)^{-1}$	Meso. Th_2 $\mu(cm)^{-1}$	Range of thickness cm.
		μ (cm)$^{-1}$	100μ/D	μ(cm)$^{-1}$	100μ/D			
Hg . .	13.59	.642	4.72	.832	6.12			.3 to 3.5
Pb . .	11.40	.495	4.34	.725	6.36	.462	.620	.0 " 7.9
Cu . .	8.81	.351	3.98	.416	4.72	.294	.373	.0 " 7.6
Brass .	8.35	.325	3.89	.392	4.70	.271	.355	.0 " 5.86
Fe . .	7.62	.304	3.99	.360	4.72	.250	.316	.0 " 7.6
Sn . .	7.24	.281	3.88	.341	4.70	.236	.305	.0 " 5.5
Zn . .	7.07	.228	3.93	.329	4.65	.233	.300	.0 " 6.0
Slate. .	2.85	.118	4.14	.134	4.69	.096	–	.0 " 9.4
Al . .	2.77	.111	4.06	.130	4.69	.092	.119	.0 " 11.2
Glass .	2.52	.105	4.16	.122	4.84	.089	.113	.0 " 11.3
S . . .	1.79	.078	4.38	.092	5.16	.066	.083	.0 " 11.6
Paraffin .	.86	.042	4.64	.043	5.02	.031	.050	.0 " 11.4

In determining the above values the rays were first passed through one cm. of lead.

Russell and Soddy, Philosophical Magazine, 21, p. 130, 1911.

TABLE 506.

RADIOACTIVITY.

T = ½ period = time when body is ½ transformed. λ = transformation constant = $\frac{.6939}{P}$. $\theta = 1/\lambda$ is average life of radioactive atoms. Parentheses indicate feeble radiation. V is the velocity of the α or β rays relative to that of light. To convert to cm/sec multiply by 3×10^{10}. a_0 is the range of α particles in air 76 cm pres. 0°C; at other temperatures and pressures, $a_t = \{a_0\,(273 + t)760\}\,/273\,p$. For α-rays, $V = 0.0342a^{1/3}$.

Radioactive Element	Symbol	Atomic Weight	Atomic No.	Isotope	T ½ period	$\theta = \frac{1}{\lambda}$ average life	λ (sec.$^{-1}$)	No.
Uranium Radium Group								
Uranium I	UI	238	92	U	4.67×10^9 y	6.75×10^9 y	4.7×10^{-18}	1
Uranium X_1	UX_1	234	90	Th	24.6 d	35.5 d	3.26×10^{-7}	2
Uranium X_2	UX_2	234	91	Po	1.15 m	1.65 m	0.010	3
Uranium II	UII	234	92	U	2×10^6 y	3×10^6 y	10^{-14} (?)	4
Ionium	Io	230	90	Th	6.9×10^4 y	10^5 y	3.2×10^{-13}	5
Radium	Ra	226	88	Ra	1690 y	2440 y	1.30×10^{-11}	6
Radon	Rn	222	86	Rn	3.85 d	5.55 d	2.08×10^{-6}	7
Radium A	RaA	218	84	Po	3.0 m	4.32 m	3.85×10^{-3}	8
Radium B	RaB	214	82	Pb	26.8 m	38.7 m	4.30×10^{-4}	9
Radium C	RaC	214	83	Bi	19.5 m	28.1 m	5.92×10^{-4}	10
Radium C′	RaC′	214	84	Po	10^{-6} s	10^{-6} s	10^6 (?)	11
Radium D	RaD	210	82	Pb	16.5 y	23.8 y	1.33×10^{-9}	12
Radium E	RaE	210	83	Bi	5.0 d	7.2 d	1.61×10^{-6}	13
Radium F	RaF	210	84	Po	136 d	196 d	5.90×10^{-8}	14
Radium Ω′	RaΩ′	206	82	Pb				15
Radium C	RaC	214	83	Bi			(1.8×10^{-7})	16
Radium C″	RaC″	210	81	Tl	1.4 m	2.0 m	8.3×10^{-3}	17
Radium Ω″	RaΩ″	210	82	Pb				18
Actinium Group								
Uranium ?		?	92	U				19
Uranium Y	UY	?	90	Th	1.04 d	1.5 d	7.8×10^{-6}	20
Protoactinium	Pa	?	91	Pa	1.2×10^4 y	1.7×10^4 y	1.9×10^{-12}	21
Actinium	Ac	?	89	Ac	20 y	28.8 y	1.1×10^{-9}	22
Radioactinium	RdAc	?	90	Th	19.5 d	28.1 d	4.11×10^{-7}	23
Actinium X	AcX	?	88	Ra	11.4 d	16.4 d	7.06×10^{-7}	24
Actinon	An	?	86	Rn	3.9 s	5.6 s	0.178	25
Actinium A	AcA	?	84	Po	2.0×10^{-3} s	2.9×10^{-3} s	345.	26
Actinium B	AcB	?	82	Pb	36.1 m	52.1 m	3.2×10^{-4}	27
Actinium C	AcC	?	83	Bi	2.15 m	3.10 m	5.37×10^{-3}	28
Actinium C″	AcC″	?	81	Tl	4.71 m	6.83 m	2.44×10^{-3}	29
Actinium Ω″	AcΩ″	?	82	Pb				30
Thorium Group								
Thorium	Th	232	90	Th	1.31×10^{10} y	1.89×10^{10} y	1.68×10^{-18}	31
Mesothorium 1	MsTh1	228	88	Ra	6.7 y	9.67 y	3.28×10^{-9}	32
Mesothorium 2	MsTh2	228	89	Ac	6.2 h	8.9 h	3.12×10^{-5}	33
Radiothorium	RaTh	228	90	Th	2.02 y	2.91 y	1.09×10^{-8}	34
Thorium X	ThX	224	88	Ra	3.64 d	5.25 d	2.20×10^{-6}	35
Thoron	Tn	220	86	Rn	54 s	78 s	0.0128	36
Thorium A	ThA	216	84	Po	0.14 s	0.20 s	5.0	37
Thorium B	ThB	212	82	Pb	10.6 h	15.3 h	1.82×10^{-5}	38
Thorium C	ThC	212	83	Bi	60 m	87 m	1.92×10^{-4} q	39
Thorium C′	ThC′	212	84	Po	10^{-11} s	10^{-11} s	10^{11} (?)	40
Thorium Ω′	ThΩ′	208	82	Pb				41
Thorium C	ThC	212	83	Bi			(6.7×10^{-5})	42
Thorium C″	ThC″	208	81	Tl	3.1 m	4.5 m	3.70×10^{-3}	43
Thorium Ω″	ThΩ″	208	82	Pb				44
Potassium	K	39.1	19	K				45
Rubidium	Rb	85.5	37	Rb				46

NOTES. — (1) 1 g U emits 2.37×10^4 α-particles per sec. (3) also called Brevium; (7) also called Radium Emanation and Niton; inert gas, dens. 111 H; boils −65° C, condenses low pressure −150° C. (10) has double disintegration; 99.97% emits β-rays and give RaC′; rest, α-rays and give RaC″; (12) radiolead; (14) Polonium; (15) lead; (17) also called radium C_2; (18) hypothetical; (21) also called ekatantalum; Uranium Z, isotopic with Pa accompanies Ur in minute quantities; period 6–7 h; β-radiation; (25) also called act. emanation, inert gas, condenses between −120 and −150° C; (29) also called act. D; (36) also called th. emanation, inert gas.

TABLE 506 (*concluded*).

RADIOACTIVITY.

$\mu_{\beta Al}$ is the absorption coefficient of the β-rays in Al, thickness measured in cm.; $\mu_{\gamma Al}$ and $\mu_{\gamma Pb}$ ditto for γ-rays in Al and Pb, unit the cm; the latter is given only for the most penetrating type of γ-rays. If I_0 is the initial intensity and I the intensity after the rays have traversed x cm of the absorbent then, $I = I_0 e^{-\mu x}$; $\log_{10} I_0/I = 0.4343\ \mu x$. If D is the thickness corresponding to the absorption of ½ the rays then $D\mu = 0.693$. (Adapted from report of International Committee on Chemical Elements, 1923; J. Am. Ch. Soc. 45, p. 867, 1923. Col. 4 is from Geiger, Z. fur Physik, 1921.)

No.	Radiation	a_0	No. of ion pairs × 10^5	V for α and β radiations	$\mu_{\beta Al}$	$\mu_{\gamma Al}$	$\mu_{\gamma Pb}$
				Uranium Radium Group			
1	α	2.37	1.33	0.056			
2	β				463		
3	β (γ)				14.4	24; 0.7; 0.14	0.72
4	α	2.75	1.43	0.0479			
5	α	2.85	1.46	0.0485			
6	α (β+γ)	3.13	1.52	α 0.0500; β 0.52; 0.65	312	354; 16; 0.27	
7	α	3.94	1.71	0.0540			
8	α	4.50	1.87	0.0565			
9	β (γ)			0.36; 0.41; 0.63; 0.70; 0.74;	13.1; 80	230; 40; 0.51	
10	99.97%β+γ			0.786; 0.862; 0.949; 0.957	13.2; 53	0.115	0.50
11	α	6.57	2.37	0.0641			
12	(β and γ)			0.33; 0.39	5500	45; 0.99	
13	β				43.3		
14	α (γ)	3.58	1.67	0.0523		585	
15	—						
16	.03% α	?					
17	β						
18	—						
				Actinium Group			
19	α						
20	β				about 300		
21	α	3.314	1.60	0.0510			
22	—						
23	α (β)	4.36	1.87	α .0559; β .38; .43; .49; .53; .60; .67; .73	about 170		
24	α	4.17	1.78	0.0550			
25	α	5.40	2.11	0.0600			
26	α	6.16	2.28	0.0627			
27	(β and γ)				Very large	120; 31; 0.45	
28	α	5.12	2.05	0.0589			
29	β and γ				28.5	0.198	1.2–1.8
30							
				Thorium Group			
31	α	2.58	1.37	0.0469			
32	—						
33	β and γ			.37; .39; .43; .50; .57; .60; .66; .70	20.2 to 38.5	26; 0.116	0.62
34	α (β)	3.67	1.69	α 0.0527; β 0.47; 0.51			
35	α	4.08	1.77	0.0546			
36	α	4.74	1.95	0.0574			
37	α	5.40	2.09	0.600			
38	β and γ			0.63; 0.72	110	160; 32; 0.36	
39	65% β			C + C′: 0.29; 0.36; 0.93 to 0.95	14.4		
40	α	8.16	2.74	0.0688			
41	—						
42	35% α	4.55 4.69?	1.89	0.0572			
43	β and γ			See Thorium C	21.6	0.096	0.46
44	—						
45	β β				22 to 38 308 to 347		

NOTES: (38) ThC has double disintegration; 65% of atoms emit β-rays and produce ThC′ having α-rays; 35% emit α-rays and give ThC″ having β rays; (41) 469 corresponds to V 0.0572 directly measured; (42) also called ThD. Parentheses indicate relatively feeble radiation.

RADIOACTIVITY.

TABLE 507. — Total Number of Ions produced by the α, β, and γ Rays.

The total number of ions per second due to the complete absorption in air of the β rays due to 1 gram of radium is 9×10^{14}, to the γ rays, 13×10^{14}.

The total number of ions due to the α rays from 1 gram of radium in equilibrium is 2.56×10^{16}. If it be assumed that the ionization is proportional to the energy of the radiation, then the total energy emitted by radium in equilibrium is divided as follows: 92.1 parts to the α, 3.2 to the β, 4.7 to the γ rays. (Rutherford, Moseley, Robinson.)

TABLE 508. — Amount of Radium Emanation. Curie.

At the Radiology Congress in Brussels in 1910, it was decided to call the amount of emanation in equilibrium with 1 gram of pure radium one Curie. [More convenient units are the millicurie (10^{-3}Curie) and the microcurie (10^{-6}Curie)]. The rate of production of this emanation is 1.24×10^{-9} cu. cm. per second. The volume in equilibrium is 0.59 cu. mm. (760 cm., 0°C.) assuming the emanation mon-atomic.

The Mache unit is the quantity of Radium emanation without disintegration products which produces a saturation current of 10^{-3} unit in a chamber of large dimensions. 1 curie $= 2.5\times10^{9}$ Mache units.

The amount of the radium emanation in the air varies from place to place; the amount per cubic centimeter of air expressed in terms of the number of grams of radium with which it would be in equilibrium varies from 24×10^{-12} to 350×10^{-12}.

TABLE 509. — Vapor Pressure of the Radium Emanation in cms. of Mercury.

(Rutherford and Ramsay, Phil. Mag. 17, p. 723, 1909, Gray and Ramsay, Trans. Chem. Soc. 95, p. 1073, 1909.)

Temperature C°.	−127°	−101°	−65°	−56°	−10°	+17°	+49°	+73°	+100°	+104° (crit)
Vapor Pressure.	0.9	5	76	100	500	1000	2000	3000	4500	4745

TABLE 510. — References to Spectra of Radioactive Substances.

Radium spectrum:	Demarçay, C. R. 131, p. 258, 1900.
Radium emanation spectrum:	Rutherford and Royds, Phil. Mag. 16, p. 313, 1908; Watson, Proc. Roy. Soc. A 83, p. 50, 1909.
Polonium spectrum:	Curie and Debierne, Rad. 7, p. 38, 1910, C. R. 150, p. 386, 1910.

TABLE 511. — Molecular Velocities.

The probability of a molecular velocity x is $(4/\sqrt{\pi})x^2e^{-x^2}$, the most probable velocity being taken as unity. The number of molecules at any instant of speed greater than c is $2N(hm/\pi)^{\frac{1}{2}}\left\{\int_c^\infty e^{-hmc^2}dc + ce^{-hmc^2}\right\}$ (see table), where N is the total number of molecules. The mean velocity G (sq. rt. of mean sq.) is proportional to the mean kinetic energy and the pressure which the molecules exert on the walls of the vessel and is equal to $15{,}800\sqrt{T/m}$ cm/sec, where T is the absolute temperature and m the molecular weight. The most probable velocity is denoted by W, the average arithmetical velocity by Ω.

$$G = W\sqrt{3/2} = 1.225W; \qquad \Omega = W\sqrt{4/\pi} = 1.128W; \qquad G = \Omega\sqrt{3\pi/8} = 1.086\Omega.$$

The number of molecules striking unit area of inclosing wall is $(1/4)N\Omega$ (Meyer's equation), where N is the number of molecules per unit volume; the mass of gas striking is $(1/4)\rho\Omega$ where ρ is the density of the gas. For air at normal pressure and room temperature (20° C) this is about 14 g/cm²/sec. See Langmuir, Phys. Rev. 2, 1913 (vapor pressure of W) and J. Amer. Ch. Soc. 37, 1915 (Chemical Reactions at Low Pressures), for fertile applications of these latter equations. The following table is based on Kinetic Theory of Gases, Dushman, Gen. Elec. Rev. 18, 1915, and Jeans, Dynamical Theory of Gases, 1916.

Gas.	Molecular weight.	Sq. rt. mean sq. $G \times 10^{-2}$ cm/sec.			Arithmetical average velocity, $\Omega \times 10^{-2}$ cm/sec.							
		273°	293°	373°	223°	273°	293°	373°	1000°	1500°	2000°	6000°
Air	28.96	485	502	567	404	447	463	522	855	1047	1209	2094
Ammonia	17.02	633	655	740	527	583	604	681	1115	1367	1577	2734
Argon	39.88	413	428	483	344	381	395	445	729	892	1030	1784
Carbon monoxide	28.00	493	511	576	410	454	471	531	870	1065	1230	2130
Carbon dioxide	44.00	393	408	459	327	362	376	434	694	850	981	1700
Helium	4.00	1311	1358	1533	1092	1208	1252	1412	2300	2840	3270	5680
Hydrogen	2.01	1838	1904	2149	1534	1696	1755	1980	3241	3970	4583	7940
Krypton	82.92	286	296	335	238	263	272	308	502	618	712	1236
Mercury	200.6	184	191	215	154	170	176	199	325	398	459	796
Molybdenum	96.0	—	—	—	—	—	—	—	469	575	664	1150
Neon	20.2	584	605	683	486	538	557	629	1030	1260	1460	2520
Nitrogen	28.02	493	511	577	410	454	471	531	869	1064	1229	2128
Oxygen	32.00	461	478	539	384	425	440	497	813	996	1150	1992
Tungsten	184.0	—	—	—	—	—	—	—	339	416	480	832
Water vapor	18.02	615	637	720	512	566	587	662	1084	1317	1533	2634
Xenon	130.2	228	236	267	190	210	218	246	400	493	570	986

Free electron, molecular weight = 1/1835 when $H = 1$; $G = 1.114 \times 10^7$ at 0° C and $\Omega = 1.026 \times 10^7$ at 0° C.

TABLE 512. — Molecular Free Paths, Collision Frequencies and Diameters.

The following table gives the average free path L derived from Boltzmann's formula μ $(.3502\rho\Omega)$, μ being the viscosity, ρ the density, and from Meyer's formula $\mu(.3097\rho\Omega)$. Experimental values (Verh. d. Phys. Ges. 14, 596, 1912; 15, 373, 1913) agree better with Meyer's values, although many prefer Boltzmann's formula. As the pressure decreases, the free path increases, at one bar (ordinary incandescent lamp) becoming 5 to 10 cm. The diameters may be determined from L by Sutherland's equation $\{1.402/\sqrt{2}\pi NL(1 + C/T)\}^{\frac{1}{2}}$, N being the number of molecules per unit vol. and C Sutherland's constant; from van der Waal's b, $\{3b/2NV\pi\}^{\frac{1}{3}}$; from the heat conductivity k, the specific heat at constant volume c_v, $\{.146\rho Gc_v/Nk\}^{\frac{1}{2}}$ (Laby and Kaye); a superior limit from the maximum density in solid and liquid states (Jeans, Sutherland, 1916) and an inferior limit from the dielectric constant D, $\{(D-1)2/\pi N\}^{\frac{1}{3}}$, or the index of refraction n, $\{(n^2-1)2/\pi N\}^{\frac{1}{3}}$. The table is derived principally from Dushman, *l.c.*

Gas.	$L \times 10^6$ (cm) Average free path.*			Collision frequency. $\Omega/L \times 10^{-6}$ 20° C*	$10^8 \times$ Molecular diameters (cm):				
	Boltzmann.		Meyer.		From L (viscosity) μ	From van der Waal's b	From heat conductivity k	Limiting	
	0° C	20° C	20° C					Max. density ρ	Min. D or n
Ammonia	5.92	6.60	5.83	9150	2.97	3.08	—	—	—
Argon	8.98	9.88	8.73	4000	2.88	2.94	2.86	2.87	2.66
Carbon monoxide	8.46	9.23	8.16	5100	3.19	3.12	—	3.27	2.74
" dioxide	5.56	6.15	5.44	6120	3.34	3.23	3.40	3.35	2.90
Helium	25.25	27.45	33.10	4540	1.90	2.65	2.30	1.98	1.92
Hydrogen	16.00	17.44	15.40	10060	2.40	2.34	2.32	2.40	2.17
Krypton	9.5	—	—	—	—	(3.69)	3.14	3.35	(2.70)
Mercury	—	(14.70)	(13.0)	—	—	3.01	—	—	—
Nitrogen	8.50	9.29	8.21	5070	3.15	3.15	3.53	3.23	2.95
Oxygen	9.05	9.93	8.78	4430	2.98	2.92	—	2.99	2.71
Xenon	5.6	—	—	—	—	4.02	3.42	3.55	(3.18)

* Pressure = 10^6 bars = 10^6 dynes ÷ cm² = 75 cm Hg.

TABLE 513. — Cross Sections and Lengths of Some Organic Molecules.

According to Langmuir (J. Am. Ch. Soc. 38, 2221, 1916) in solids and liquids every atom is chemically combined to adjacent atoms. In most inorganic substances the identity of the molecule is generally lost, but in organic compounds a more permanent existence of the molecule probably occurs. When oil spreads over water evidence points to a layer a molecule thick and that the molecules are not spheres. Were they spheres and an attraction existed between them and the water, they would be dissolved instead of spreading over the surface. The presence of the —COOH, —CO or —OH groups generally renders an organic substance soluble in water, whereas the hydrocarbon chain decreases the solubility. When an oil is placed on water the —COOH groups are attracted to the water and the hydrocarbon chains repelled but attracted to each other. The process leads the oil over the surface until all the —COOH groups are in contact if possible. Pure hydrocarbon oils will not spread over water. Benzene will not mix with water. When a limited amount of oil is present the spreading ceases when all the water-attracted groups are in contact with water. If weight w of oil spreads over water surface A, the area covered by each molecule is AM/wN where M is the molecular weight of the oil (O = 16), N, Avogadro's constant. The vertical length of a molecule $l = M/a\rho N = W/\rho A$ where ρ is the oil density and a the horizontal area of the molecule.

Substance.	Cross section in cm² × 10¹⁶	l in cm (length) × 10⁸	Substance.	Cross section in cm² × 10¹⁶	l in cm (length) × 10⁸
Palmitic acid $C_{15}H_{31}COOH$	24	19.6	Cetyl alcohol $C_{16}H_{33}OH$	21	21.9
Stearic acid $C_{17}H_{35}COOH$	24	21.8	Myricyl alcohol $C_{30}H_{61}OH$	29	35.2
Cerotic acid $C_{25}H_{51}COOH$	25	29.0	Cetyl palmitate $C_{15}H_{31}COOC_{16}H_{33}$	21	44.0
Oleic acid $C_{17}H_{33}COOH$	48	10.8	Tristearin $(C_{18}H_{35}O_2)_3C_3H_5$	69	23.7
Linoleic acid $C_{17}H_{31}COOH$	47	10.7	Trielaidin $(C_{18}H_{33}O_2)_3C_3H_5$	137	11.9
Linolenic acid $C_{17}H_{29}COOH$	66	7.6	Triolein $(C_{18}H_{33}O_2)_3C_3H_5$	145	11.2
Ricinoleic acid $C_{17}H_{32}(OH)COOH$	90	5.8	Castor oil $(C_{17}H_{32}(OH)COO)_3C_3H_5$	280	5.7
			Linseed oil $(C_{17}H_{31}COO)_3C_3H_5$	143	11.0

TABLE 514. — Size of Diffracting Units in Crystals.¶

The use of crystals for the analysis of X-rays leads to estimates of the relative sizes of molecular magnitudes. The diffraction phenomenon is here not a surface one, as with gratings, but one of interference of radiations reflected from the regularly spaced atomic units in the crystals, the units fitting into the lattice framework of the crystal. In cubical crystals {100} this framework is built of three mutually perpendicular equidistant planes whose distance apart in crystallographic parlance is d_{100}. This method of analysis from the nature of the diffraction pattern leads also to a knowledge of the structure of the various atoms of the crystal. See Bragg and Bragg, X-rays and Crystal Structure, 1918.

Crystal.	Elementary diffracting element.	Side of cube.	Molecules or atoms in unit cube.
		cm	
KCl	Face-centered cube *	6.30×10^{-8}	4 molecules
NaCl	" " "	5.56×10^{-8}	"
ZnS	" " " †	5.46×10^{-8}	"
CaF_2	" " " ‡	5.40×10^{-8}	"
FeS_2	" " " §	5.26×10^{-8}	"
Fe	Body-centered cube	2.86×10^{-8}	2 atoms
Al	Face-centered cube	4.05×10^{-8}	4 "
Na	Body-centered cube	4.30×10^{-8}	2 "
Ni	" " "	2.76×10^{-8}	2 "
"	Face-centered cube	3.52×10^{-8}	4 "

* Each atom is so nearly equal in diffracting power (atomic weight) in KCl that the apparent unit diffracting element is a cube (simple) of ⅛ this size. Elementary body-centered cube, — atom at each corner, one in center; e.g., Fe, Ni (in part), Na, Li? Elementary face-centered cube, — atom at each corner, one in center of each face; e.g., Cu, Ag, Au, Pb, Al, Ni (in part), etc. Simple cubic lattice, — atom in each corner. Double face-centered cubic or diamond lattice — C (diamond); Si, Sb, Bi, As?, Te?.

† Diamond lattice. ‡ Cubic-holohedral. § Cubic-pyritohedral.

Metals taken from Hull, Phys. Rev. 10, p. 661, 1917

¶ See Table 528 for best values of calcite and rock-salt grating spaces.

Note: — (Hull, Science 52, 227, 1920). **Ca**, face-centered cube, side 5.56 A, each atom 12 neighbors 3.93 A distant. **Ti**, centered cube, cf. Fe, side 3.14 A, 8 neighbors 2.72 A. **Zn**, 6 nearest neighbors in own plane. 2.67 A, 3 above, 3 below, 2.92 A. **Cd**, cf. Zn, 2.98 A, 3.30 A. **In**, face-centered tetragonal, 4 nearest 3.24 A, 4 above, 4 below, 3.33 A. **Ru**, cf. Zn, 2.69 A, 2.64 A. **Pd**, face-centered cube, side 3.92 A, 12 neighbors. 2.77 A. **Ta**, centered cube, side 3.27 A, 8 neighbors 2.83 A. **Ir**, face-centered cube, side 3.80 A, 12 neighbors, 2.69 A (A $= 10^{-8}$ cm).

Note: — (Bragg, Phil. Mag. 40, 169, 1920). Crystals empirically considered as tangent spheres of diameter in table, atom at center of sphere. When lattice known allows estimation of dimensions of crystal unit. Table foot of next page (atomic numbers, elements, diameter in Angstroms, 10^{-8} cm).

TABLE 515.

ELECTRONS, PROTONS, ATOMIC STRUCTURE, MAGNETIC FIELD OF ATOMS.

Free negative electron: (corpuscle, **J. J.** Thomson); mass = 8.99×10^{-28}g = 1/1848 H atom, probably all of electrical origin due to inertia of self-induction.

Theory shows that when speed of electron = 1/10 velocity of light its mass should be appreciably dependent upon that speed. If m_0 be mass for small velocity v, m be the transverse mass for v, v/ (velocity of light) = β, then $m = m_0 (1 - \beta^2)^{\frac{1}{2}}$, Lorentz, Einstein;

for β =	0.01	0.10	0.2	0.3	0.4	0.5	0.6	0.7	0.8	0.9
m/m_0 =	1.00005	1.005	1.02	1.048	1.091	1.155	1.250	1.400	1.667	2.294

(Confirmed by Bucherer, Ann. d. Phys. 1909, Wolz, Ann. d. Phys. Radium ejects electrons with 3/10 to 98/100 velocity of light.) m, due to charge = $2E^2/3a$, E = charge, a = radius, whence radius of electron = 2×10^{-13} cm = 1/50,000 atomic radius. Cf. (radius of earth)/(radius of Neptune's orbit) = 1/360,000.

Evidence from collisions of α-particles indicate that the diameter of the electron cannot be greater than 4×10^{-13}. (Chadwick and Bieler, Phil. Mag. 1921.)

Positive Electron or Proton : heavy, extraordinarily small, never found associated with mass less than that of the H atom; mass 1.65×10^{-24}g. If mass all electrical, radius must be 1/2000 that of the electron. No experimental evidence as with the latter since high enough speeds not available. Penetrability of atom by β-particle (may penetrate 10,000 atomic systems before it happens to detach an electron) and α-particles (8,000 times more massive than — electron, pass through 500,000 atoms without apparent deflection by nucleus more than 2 or 3 times) shows extreme minuteness. Upper limit: not larger than 10^{-12} cm for Au (heavy atom) and 10^{-13} for H (light atom) (Rutherford). Cf. (radius sun) /(radius Neptune's orbit) = 1/3000, but sun is larger than planets. (Hg atoms by billions may pass through thin-walled highly-evacuated glass tubes without impairing vacuum, therefore massive parts of atoms must be extremely small compared to volume of atom.)

Rutherford Atom : Atoms of all elements are somewhat similarly built. At the center a + charged nucleus of minute dimensions, responsible for most of the mass of the atom; this is surrounded by a distribution of electrons held in equilibrium by the force from the nucleus. Resultant nuclear charge = atomic or ordinal no., varies from 1 for H to 92 for U. These atomic nos. represent the number of planetary electrons which surround the nucleus. By the action of light, the electric charge, bombardment by α-particles, one or more of the planetary electrons may be driven away from the nucleus; by X-rays or the swift β-rays some of the more strongly bound may be removed. New electrons are generally soon captured to replace these. The nucleus is much more stable and when disrupted (radioactive changes, bombardment with α-particles) shows no tendency to revert to original state.

Moseley (Phil. Mag. 26, 1912; 27, 1914) photographed and analyzed X-ray spectra, showing their exact similarity in structure from element to element, differing only in frequencies, the square roots of these frequencies forming an arithmetical progression from element to element. Moseley's series of increasing X-ray frequencies is with one or two exceptions that of increasing atomic weights, and these exceptions are less anomalous for the X-ray series than for the atomic-weight series. It seems plausible then that there are 92 elements (from H to U) built up by the addition of some electrical element. Moseley assigned successive integers to this series (see Table 531) known now as atomic numbers.

Moseley's discovery may be expressed in the form

$$\frac{n_1}{n_2} = \frac{E_1}{E_2} \text{ or } \frac{\lambda_2}{\lambda_1} = \frac{E_1{}^2}{E_2{}^2}$$

where E is the nuclear charge and λ the wave-length. Substituting for the highest frequency line of W, $\lambda_2 = 0.167 \times 10^{-8}$ cm (Hull), $E_2 = 74 = N_w$, and $E_1 = 1$, then λ_1 = highest possible frequency by element which has one + electron; $\lambda_1 = 91.4\ m\mu$. Now the H ultra-violet series highest frequency line = 91.2 $m\mu$ (Lyman); i.e., this ultra-violet line of H is nothing but its K X-ray line. Similarly, it seems equally certain that the ordinary Balmer series of H (head at 365 $m\mu$) is its L X-ray series and Paschen's infra-red series its M X-ray series.

The application of Newton's law to Moseley's law leads to $E_1/E_2 = a_2/a_1$, where the a's are the radii of the inmost — electronic orbits, i.e., the radii of these orbits are inversely proportional to the central charges or atomic numbers.

There are other negative electrons on the nucleus with corresponding + charges to make the atom neutral electrically. The negative nuclear charges may serve to hold the positive ones together. He, atomic no. = 2, has two free + charges, on nucleus; the nucleus has 4 + protons held together by 2 — electrons with 2 — electrons outside nucleus. H has one + proton and one — electron.

If the — electron is designated as e (charge — 1, mass negligible) and the + proton as p (charge + 1, mass 1 except in H) then the formula for the nucleus of any element from He to U may be written as $(p_2e)_N(pe)_n$ where N is the atomic number and n has values from 0 to 54. If n be taken as —1, then H may be included. (Masson, Phil. Mag. 41, 1921.) If brackets are used to designate the nucleus, then the complete element becomes $[(p_2e)_N(pe)_n]\,e_N$. In the formation of ions only the part exterior to the brackets is affected. For the α-transformation (emission of + charged He nucleus) $2(p_2e) = (p_2e)_2\uparrow$, the sub-chemical equation may be written $[(p_2e)_N(pe)_n]\,e_N = [(p_2e)_{N-2}(pe)_n]\,e_N + (p_2e)_2\uparrow$ (He nucleus); the new elements upon discharge of its — charge becomes $[(p_2e)_{N-2}(pe)_n]\,e_{N-2}$ showing the characteristic α-ray change with the atomic weight lowered by 2 and the mass by 4. The β-ray $2(pe) = (p_2e) + e\uparrow$ gives the equation $[(p_2e)_N(pe)_n]e_N = [(p_2e)_{N+1}(pe)_{n-2}]\,e_N + e\uparrow$, mass uncharged and forms the singly — charged ion of an isobar.

(This Table supplements Table 514.)

3 Li	3.00	13 Al	2.70	25 Mn	2.95†	36 Kr	2.35*	54 Xe	2.70*
4 Gl	2.30	14 Si	2.35	26 Fe	2.80	37 Rb	4.50	55 Cs	4.75
6 C	1.54	16 S	2.05	27 Co	2.75	38 Sr	3.90	56 Ba	4.20
7 N	1.30	17 Cl	2.10	28 Ni	2.70	47 Ag	3.55	81 Tl	4.50
8 O	1.30	18 A	2.05*	29 Cu	2.75	48 Cd	3.20	82 Pb	3.80
9 F	1.35	19 K	4.15	30 Zn	2.65	50 Sn	2.80	83 Bi	2.96
10 Ne	1.30*	20 Ca	3.40	33 As	2.52	51 Sb	2.80		
11 Na	3.55	22 Ti	2.80	34 Se	2.35	52 Te	2.65		
12 Mg	2.85	24 Cr	2.80†	35 Br	2.38	53 I	2.80		

* Outer electron shell. † Cr, "electronegative," 2.35; Mn., ditto, 2.35.

Broughall (Phil. Mag. 41, p. 872, 1921) computes in the same units from Van der Waal's constant "b" the diameters of He, N, A, Kr, and X as 2.3, 2.6, 2.9, 3.1, and 3.4. These inert elements correspond to Langmuir's completely filled successive electron shells. The corresponding atomic numbers are 2, 10, 18, 36 and 54. For Langmuir's theory see J. Am. Ch. Soc., p. 868, 1919, Science 54, p. 59, 1921.

LANGMUIR ATOM. BOHR ATOM. ATOMIC MAGNETIC FIELD.

From the emission of nuclear α-particles, $2(p_2e) = p_4e_2$, it seems probable that the nuclei are compounds of He and H nuclei. By the bombardment of the nuclei of atoms up to atomic number 40 with α-particles Rutherford has obtained H but only where H and He nuclei should both occur in the nucleus (Bo, N, Fl, Na, Al, P). Harkins has developed this idea (J. Franklin. Inst. 194, 213 et seq., 1922) and shown the much greater frequency in nature of the even-atomic numbered elements (97.6 per cent in stony meteorites, 99.2 Fe meteorites, 85.6 lithosphere, 5 unknown elements all odd, even radio-active most stable). Elements below atomic number 30 make up 99.99 per cent of all meteorites, 99.85 igneous rocks, 99.95 shale, 99.95 sandstones, 99.85 lithosphere. The stability of the He nucleus may be judged by the energy set free in the formation of He from H. According to "relativity" 1 g-mass = 9×10^{20} ergs ($E = mc^2$). The change of mass involved in the formation of 1 g-atom of He (4.000 g) from 4 g-atoms of H_2 (4×1.0078 g) = 2.81×10^{19} ergs = 6.71×10^{11} calories. 1 lb. H_2 changed to He equals heat from 10,000 tons coal. The nuclei of light even numbered atoms (most abundant isotope) up to Fe (26) almost wholly of He nuclei. To a 1st approximation the α-particle behaves in collision like an elastic oblate spheroid, semi-axes, 8×10^{-13} and 4×10^{-13} cm. (Chadwick, Bieler, P. M. 1921).

The theory of the arrangement of the extra-nuclear electrons has followed two developments: The physicist, to explain radiation phenomena, desires a planetary type of atom (Bohr atom); the chemist, for stereochemical phenomena, one with electrons not in co-planar revolution, rather vibrating in 3-dimensional position (Langmuir atom).

Langmuir atom (J. Am., Ch. Soc. 41, 868, 1919) postulates a 3-dimensional more or less symmetrical arrangement of the extra-nuclear electrons in concentric shells containing successively, when complete 2, 8, 8, 18, 18, 32 ($N = 2(1^2 + 2^2 + 2^2 + 3^2 + 3^2 + 4^2 +)$). No outer layer has electrons until the inner have their quota. The electrons in the outer shell determine the valence and the chemical and electrochemical properties of the element. This device very satisfactorily accounts for many chemical structures and reactions. The following table shows the arrangement of the electrons from shell to shell. N gives the number in the inner completed shells; E, the number (valence electrons in the outer shell).

E =		0	1	2	3	4	5	6	7	8	9	10	11	12	13	14	15	16	17	18
I	N		H	He																
IIa	2	He	Li	Be	B	C	N	O	F	Ne										
IIb	10	Ne	Na	Mg	Al	Si	P	S	Cl	A										
IIIa	18	A	K	Ca	Sc	Ti	V	Cr	Mn	Fe	Co	Ni	Cu	Zn	Ga	Ge	As	Se	Br	Kr
IIIb	36	Kr	Rb	Sr	Y	Zr	Cb	Mo	43	Ru	Rh	Pd	Ag	Cd	In	Sn	Sb	Te	I	Xe
IVa	54	Xe	Cs	Ba	La	Ce	Pr	Nd	61	Sa	Eu	Gd	Tb	Ho	Dy	Er	Tm	Tm_2	Yb	Lu
							19	20	21	22	23	24	25	26	27	28	29	30	31	32
							Ta	W	75	Os	Ir	Pt	Au	Hg	Tl	Pb	Bi	RaF	85	Nt
IVb	86	Nt	87	Ra	Ac	Th	Ux_2	U												

Bohr Atom: (Phil. Mag. 26, 1, 476, 857, 1913; 29, 332, 1915; 30, 394, 1915). The experimental facts and the law of circular electronic orbits limit the electrons to orbits of particular radii. When an electron is disturbed from its orbit, e.g., struck out by a cathode ray, or returns from space to a particular orbit, energy must be radiated. It is suggestive that the emission of a β ray requires a series of γ ray radiations. H does not radiate unless ionized and then gives out a spectrum represented by Balmer's formula $\nu = N(1/n_1^2 - 1/n^2)$ where ν is the frequency, N, a constant, and n_1 for all the lines in the visible spectrum has the value 2, n, the successive integers, 3, 4, 5, . . .; if $n_1 = 1$ and n, 2, 3, 4, . . ., Lyman's ultra-violet series results; if $n_1 = 3$, n, 4, 5, 6, . . ., Paschen's infra-red series. These considerations led Bohr to his atom and he assumed: (a) a series of circular non-radiating orbits governed as above; (b) radiation taking place only when an electron jumps from one to another of these orbits, the amount radiated and its frequency being determined by $h\nu = A_1 - A_2$, h being Planck's constant and A_1 and A_2 the energies in the two orbits; (c) the various possible circular orbits, for the case of a single electron rotating around a single positive nucleus. to be determined by $T = (1/2)\tau hn$, in which τ is a whole number, n is the orbital frequency, and T is the kinetic energy of rotation.

The remarkable test of this theory is not its agreement with the H series, which it was constructed to fit, but in the value found for N. From (a), (b), and (c) it follows that $N = (2\pi^2e^2E^2m)/h^3 = 3.294 \times 10^{15}$, within 1/10 per cent of the observed value (Science, 45, p. 327).

The radii of the stable orbits $= \tau^2h^2/4\pi^2me^4$, or the radii bear the ratios 1, 4, 9, 16, 25. If normal H be assumed to be with its electron in the inmost orbit, then $2a = 1.1 \times 10^{-8}$; best determination gives 2.2×10^{-8}. The fact that H emits its characteristic radiations only when ionized favors the theory that the emission process is a settling down to normal condition through a series of possible intermediate states, i. e., a change of orbit is necessary for radiation. That in the stars there are 33 lines in the Balmer series, while in the laboratory we never get more than 12, is easily explicable from the Bohr theory.

Bohr's theory leads to the relationship $\nu_{K_\beta} - \nu_{K_\alpha} = \nu_{L_\alpha}$ (see X-ray tables), Rydberg-Schuster law.

For further development, see Sommerfeld, Ann. d. Phys. 51, 1, 1916, Paschen, Ann. d. Phys., October, 1916; Harkins, Recent work on the structure of the atom, J. Am. Ch. Soc. 37, p. 1396, 1915; 39, p. 856, 1916.

Magnetic field of atom: From the Zeeman effect due to the action of a magnetic field on the radiating electron the strength of the atomic magnetic field comes out about 10^8 gauss, 2000 times the most intense field yet obtained by an electromagnet. A similar result is given by the rotation of a number of electrons, $A 10^3$, where A is the atomic weight; for Fe this gives 10^8 gauss. For other determinations, see Weiss (J. de Phys. 6, p. 661, 1907; 7, p. 249, 1908), Ritz (Ann. d. Phys. 25, p. 660, 1908), Oxley (change of magnetic susceptibility on crystallization, Phil. Tr. Roy. Soc. 215, p. 95, 1915) and Merritt (fluorescence, 1915); Humphreys, "The Magnetic Field of an Atom," Science, 46, p. 276, 1917.

Note: The phenomena of Electron Emission, Photo-electric Effect and Contact (Volta) Potential treated in the subsequent tables are extremely sensitive to surface conditions of the metal. The most consistent observations have been made in high vacua with freshly cut metal surfaces.

TABLE 516. Electron Emission from Hot Metals.

Among the free electrons within a metal some may have velocities great enough to escape the surface attraction. The number n reaching the surface with velocities above this critical velocity $= N(RT/2\pi M)^{\frac{1}{2}}e^{-\frac{W}{RT}}$ where N = number of electrons in each cm^3 of metal, R the gas constant (83.15×10^6 erg-dyne), T the absolute temperature, M the atomic weight of electron (.000546, O = 16), w the work done when a "gram-molecule" of electrons (6.06×10^{23} electrons or 96,500 coulombs) escape. It seems very probable that this work is done against the attraction of the electron's own induced image in the surface of the conductor. When a sufficiently high + field is applied to escaping electrons so that none return to the conductor, then the saturation current has been found to follow the equation

$$i = a\sqrt{T}e^{-b/T},$$

assuming N and w constant with the temperature; this is equivalent to the equation for n just given and is known as Richardson's equation. In the following table due to Langmuir (Tr. Am. Electroch. Soc. 29, 125, 1916) i_{2000} = saturation current per cm^2 for $T = 2000$ K°; $\phi = w/F = Rb/F$ = work done when electrons escape from metal in terms of equivalent potential difference in volts; F = Faraday constant = 96,500 coulombs.

Metal.	a amp/cm²	b	i_{2000} amp/cm²	ϕ (volts).
Tungsten *	2.36×10^7	52500	0.0042	4.52
Thorium	2.0×10^8	39000	30.0	3.36
Tantalum	1.12×10^7	50000	0.007	4.31
Molybdenum	2.1×10^7	50000	.013	4.31
Carbon (untreated)	—	48000	—	4.14
Titanium	1300?	28000?	.048?	2.4?
Iron	2400?	37000?	.0010?	3.2?
Platinum †	1.25×10^7	51060	.0035	4.4
BaO-SrO, Pt-6 % Ir core	**1.6×10^4**	**20000**	**3.25**	**1.7**

* Best determined value of table, pressure less than 10^{-7} mm Hg. † Schlichter, 1915.

TABLE 517. Photo-electric Effect.

A negatively charged body loses its charge under the influence of ultra-violet light because of the escape of negative electrons freed by the absorption of the energy of the light. The light must have a wave-length shorter than some limiting value λ_0 characteristic of the metal. The emission of these electrons, unlike that from hot bodies, is independent of the temperature. The relation between the maximum velocity v of the expelled electron and the frequency ν of the light is $(1/2)mv^2 = h\nu - P$ (Einstein's equation) where h is Planck's constant (6.58×10^{-27} erg. sec.); $h\nu$ sometimes taken as the energy of a "quanta," P, the work which must be done by the electron in overcoming surface forces. $(1/2)mv^2$ is the maximum kinetic energy the electron may have after escape. Richardson identifies the P of Einstein's formula with the w of electron emission of the preceding table. The minimum frequency ν_0 (corresponding to maximum wave-length λ_0) at which the photo-electric effect can be observed is determined by $h\nu = P$. P applies to a single electron, whereas w applies to one coulomb (6.062×10^{23} electrons); therefore $w = NP = .00399\nu_0$ ergs. $\phi = (12.4 \times 10^{-5})\lambda_0$ volts. See Millikan, Pr. Nat. Acad. 2, 78, 1916; Phys. Rev. 7, 355, 1916; 4, 73, 1914; Hennings, Phys. Rev. 4, 228, 1914.

TABLE 518. Ionizing and Resonance Potentials of the Elements.

(Abridged by permission from "Origin of Spectra," Foote and Mohler, 1922)

When electrons are accelerated through gases or vapors (especially monatomic gases of small electron affinity and metallic vapors), at well-defined velocities a large transfer of energy takes place between the moving electrons and the gas atoms. Below the critical value the collision is elastic. In general two types of inelastic encounters occur: the first, accompanied by the emission of the radiation of a single spectrum line at a potential called the resonance potential, V_r; an outer electron of the atom then undergoes an interorbital transition; the relation $h\nu = eV_r$ holds where ν is the frequency of the radiation and h Planck's quantum. The energy absorbed at the resonance is not enough to completely eject an electron but only displaces it to an outer orbit. e.g., in the alkali group the electron is displaced from the 1s to the 2p orbit, the first energy level outside. In returning the energy emitted has the frequency 1s–2p; there may be more than one resonance potential due to other displacements. The second type of encounter completely removes an electron and ionizes the gas (ionization potential V_i). This potential in general satisfies a relation $h\nu = eV$ except that now ν corresponds to the highest convergence frequency in the arc spectrum of the material (monatomic vapor), to the limit of a series the first line of which corresponds to the resonance potential. In the case of the ionization potential the electron may return by a variety of interorbital transitions, each resulting in an emission of a quantum of wave-number ν, subject to the conservation of energy condition: $\Sigma hc^2\nu_k = eV_i10^8$. With numerous atoms and electrons returning to equilibrium in different manners, there results the composite result of the emission of the complete arc spectrum.

Atomic Number; Element	Line	ν	Resonance Volts		Line	ν	Ionization Volts	
			Computed	Observed			Computed	Observed
3 Li	1s–2p₁	14,903	1.840	—	1s	43,486	5.368	—
11 Na	"	16,973	2.095	2.12	"	41,449	5.116	5.13
19 K	"	13,043	1.610	1.55	"	35,006	4.321	4.1
29 Cu	"	30,784	3.800	—	"	62,308	7.692	—
37 Rb	"	12,817	1.582	1.6	"	33,689	4.158	4.1
47 Ag	"	30,473	3.762	—	"	61,096	7.542	—
55 Cs	"	11,732	1.448	1.48	"	31,405	3.877	3.9
79 Au	"	41,174	5.1	—	"	70,000?	8 to 9?	—

(For conclusion of Table, see page 442).

TABLE 519.

CONTACT (VOLTA) POTENTIALS.

There has been considerable controversy over the reality and nature of the contact differences of potential between two metals. At present, due to the studies of Langmuir, there is a decided tendency to believe that this Volta difference of potential is an intrinsic property of metals closely allied to the phenomena just given in Tables 516 to 518 and that the discrepancies among different observers have been caused by the same disturbing surface conditions. The following values of the contact potentials with silver and the relative photo-sensitiveness of a few of the metals are from Henning, Phys. Rev. 4, 228, 1914. The values are for freshly cut surfaces in vacuo. Freshly cut surfaces are more electro-positive and grow more electro-negative with age. That the observed initial velocities of emission of electrons from freshly cut surfaces are nearly the same for all metals suggests that the more electro-positive a metal is the greater the actual velocity of emission of electrons from its surface.

	Ag	Cu	Fe	Brass	Sn	Zn	Al	Mg
Contact potential with Ag	0	.05	.19	.21	.27	.59	.99	1.42
Relative photo-sensitiveness	50	60	65	45	70	80	500	1000

From the equation $w = RT \log(N_A/N_B)$, where w is the work necessary per gram-molecule when electrons pass through a surface barrier separating concentrations N_A and N_B of electrons, it can be shown (Langmuir, Tr. Am. Eletroch. Soc. 29, 142, 1916, *et seq.*) that the Volta potential difference between two metals should be

$$v_1 - v_2 = \frac{1}{F}\{w_2 - w_1 + RT \log(N_A/N_B)\} = \frac{w_2 - w_1}{F} = \phi_2 - \phi_1$$

(see Table 517 for significance of symbols), since the number of free electrons in different metals per unit volume is so nearly the same that $RT \log (N_A/N_B)$ may be neglected. The contact potentials may thus be calculated from photoelectric phenomena (see Table 517 for references). They are independent of the temperature. The following table gives a summary of values of ϕ in volts obtained from the various phenomena where an electron is torn from the attraction of some surface. In the case of ionization potentials the work necessary to take an electron from an atom of metal vapor is only approximately equal to that needed to separate it from a solid metal surface.

(*a*) THE ELECTRON AFFINITY OF THE ELEMENTS, IN VOLTS.

Metal.	Contact. (Henning.)	Thermionic. (Langmuir.)	Photo-electric and contact. (Millikan.)	Photo-electric. (Richardson)	Miscellaneous.	Single-line spectra.	Adjusted mean.
Tungsten	—	4.52	—	—	—	—	4.52
Platinum	—	—	—	4.3	4.45	—	4.4?
Tantalum	—	4.31	—	—	—	—	4.3
Molybdenum	—	4.31	—	—	—	—	4.3
Carbon	—	4.14		—	—	—	4.1
Silver	4.05	—	—	—	—	—	4.1
Copper	(4.0)	—	—	4.1	—	—	4.0
Bismuth	—	—	—	3.7	—	—	3.7
Tin	3.78	—	—	3.5	—	—	3.8
Iron	3.86	3.2?	—	—	—	—	3.7
Zinc	3.46	—	—	3.4	—	4.04	3.4
Thorium	—	3.36	—	—	—	—	3.4
Aluminum	3.06	—	—	2.8	—	—	3.0
Magnesium	2.63	—	—	3.2	—	4.35	2.7
Titanium	—	2.4?	—	—	—	—	2.4
Lithium	—	—	2.35	—	—	1.85	2.35
Sodium	—	—	1.82	2.1	—	2.11	1.82

(*b*) It should not be assumed that all the emf of an electrolytic cell is contact emf. Its emf varies with the electrolyte, whereas the contact emf is an intrinsic property of a metal. There must be an emf between the two electrodes of such a cell dependent upon the concentration of the electrolyte used. The following table gives in its first line the electrode potential e_h of the corresponding metals (in solutions of their salts containing normal ion concentration) on assumption of no contact emf at the junction of the metals. The second line, $\phi - e_h - 3.7$ volts, gives an idea of the electrode potentials (arbitrary zero) exclusive of contact emf.

Metal	Ag	Cu	Bi	Sn	Fe	Zn	Mg	Li	Na
e_h	+0.80	+0.34	+0.20	−0.10	−0.43	−0.76	−1.55	−3.03	−2.73
$\phi - e_h - 3.7$	−0.40	+0.04	+0.20	−0.20	−0.43	−0.46	−0.55	−1.65	−0.85

IONIC MOBILITIES AND DIFFUSIONS.

The process of ionization is the removal of an electron from a neutral molecule, the molecule thus acquiring a resultant + charge and becoming a + ion. The negative carriers in all gases at high pressures, except inert gases, consist for the most part of carriers with approximately the same mobilities as the + ions. The negative electrons must, therefore, change initially to ions by union with neutral molecules.

The mobility, U, of an ion is its velocity in cm/sec. for an electrical field of one volt per cm. The rates of diffusion, D, are given in cm³/sec. $U = DP/Ne$, where P is the pressure, N, the number of molecules per unit volume of a gas and e the electronic charge.

Nature of the gas and the mobilities: (1) The mobilities are approximately proportional to the inverse sq. rts. of the molecular weights of the permanent gases; better yet when the proportionality is divided by the 4th root of the dielectric constant minus unity; (2) The ratio $U+/U-$ seems to be greater than unity in all the more electro-negative gases.

Mobilities of Gaseous Mixtures: Three types: (1) Inert gases have high mobilities; small traces of electro-negative gases make values normal. (2) Mixed gases: lowering of mobilities is greater than would be expected from simple law of mixture. (3) Abnormal changes produced by addition of small quantities of electro-negative gases:

	$U+$	$U-$	
e.g.: normal mobility	1.37	1.80	Wellisch, Pr.
6 mm C_2H_5Br gave	1.37	1.80	Roy. Soc. 82A,
6 mm C_2H_3I "	1.37	1.80	p. 500, 1909.
10 mm C_2H_5OH "	0.91	1.10	
9 mm C_3H_6O "	1.15	1.37	

Temperature Coefficient of Mobility: There is no decided change with the temperature.

Pressure Coefficient of Mobility: Mobility varies inversely with the pressure in air from 100 to 1/10 atmosphere for − ion, to 1/1000, for + ion; below 1/10 atmosphere all observers agree that the negative ion in air increases abnormally rapidly.

Free Electrons: In pure He, Ar, and N, the negative carriers have a high mobility and are, in part at any rate, free electrons; electrons become appreciable in air at 10 cm pressure.

TABLE 520. — Ionic Mobilities.

Dry gas.	Mobilities. +	Mobilities. −	$K-1$	Observer.	Dry gas.	Mobilities. +	Mobilities. −	$K-1$	Observer.
H	6.70	7.95	.000273	Zeleny	Nitrous oxide	0.82	0.90	.00107	Wellisch
He	5.09	6.31	.000074	Franck	Ethyl alcohol	0.34	0.27	.00940	"
Ar	1.37	—	.000100	"	CCl_4	0.30	0.31	.00426	"
N	1.27	—	.000590	"	Ethyl chloride	0.33	0.31	.01550	"
O	1.36	1.80	.000540	Zeleny	Ethyl ether	0.29	0.31	.00742	"
CO_2	0.81	0.85	.000960	Wellisch	Methyl bromide	0.29	0.28	.01460	"
NH_3	0.74	0.80	.00770	"	Ethyl formate	0.30	0.31	.00870	"
Air	1.40	1.78	.000590	Mean	Ethyl iodide	0.17	0.16	—	"

Franck, Jahr. d. Rad. u. Elek. 9, p. 2, 1912; Wellisch, Pr. Roy. Soc. 82A, p. 500, 1909. The following values are from Yen, Pr. Nat. Acad. 4, 19 8.

	H_2	N_2	Air.	SO_2	C_5H_{12}	C_2H_6O	C_2H_4O	C_2H_5Cl	CH_3I	C_2H_5I
$U+$	5.54	1.30	1.37	.412	.385	.363	.307	.304	.216	1.81
$U-$	8.45	1.80	1.81	.414	.451	.373	.331	.317	.226	1.81
$U-/U+$	1.53	1.38	1.34	1.00	1.17	1.03	1.07	1.04	1.05	1.00

TABLE 521. — Diffusion Coefficients.

The following table gives the observed and computed ($D = 300UP/Ne$ = very nearly $0.0236U$) values of the diffusion coefficients. The diffusion coefficients are given for some neutral molecules as actually determined for some gases into gases of nearly equal molecular weight. Table taken from Loeb, "The Nature of the Gaseous Ion," J. Franklin Inst. 184, p. 775, 1917.

Gas, diffusing.	Gas diffused into	D molecules.	$U+$	$D+$ for ions. Computed.	$D+$ for ions. Observed.
Ar	He	0.706	5.09	1.20	—
H_2	N_2	.739	6.02	0.143	0.123
Air	O_2	.178	1.35	0.0319	0.028
O_2	N_2	.171	1.27	.0299	.025
CO_2	N_2O	1.5–1.0	.82	.0193	.023 *
CO_2	CO	1.31	.81	.0193	—
C_2H_5OH	CO_2	0.0693	.34	.00805	—
Air	Ethyl acetate	.093	.30 †	.0071	—
H_2O	Air	.246	1.35	.0319	—
NH_3	NH_3	.190 ‡	0.74	.0174	—

* CO_2 into CO_2. † Ethyl formate. ‡ Estimated.

COLLOIDS.

TABLE 522. — General Properties of Colloids.

For methods of preparing colloids, see The Physical Properties of Colloidal Solutions, Burton, 1916; for general properties, see Outlines of Colloidal Chemistry, J. Franklin Inst. 185, p. 1, 1918 (contains bibliography).

The colloidal phase is conditioned by sufficiently fine division (1×10^{-4} to 10^{-7} cm). Colloids are suspensions (in gas, liquid, solid) of masses of small size capable of indefinite suspension; suspensions in water, alcohol, benzole, glycerine, are called hydrosols, alcosols, benzosols, glycerosols, respectively. The suspended mass is called the disperse phase, the medium the dispersion medium.

Colloids fall into 3 quite definite classes: 1st, those consisting of extremely finely divided particles (Cu, Au, Ag, etc.) capable of more or less indefinite suspension against gravity, in equilibrium of somewhat the same aspect as the gases of the atmosphere, depending as in the Brownian movement upon the bombardment of the molecules of the medium; 2nd, those resisting precipitation (hæmoglobin, etc.) probably because of charged nuclei and which may be coagulated and precipitated by the neutralization of the charges; 3rd, colloidal as distinguished from the crystalloidal condition, the colloid being very slowly diffusible and incapable unlike crystalloids of penetrating membranes (gelatine, silicic acid, caramel, glue, white of egg, gum, etc.).

Smallest particle of Au observed by Zsigmody (ultramicroscope)	1.7×10^{-7} cm.
" " visible in ordinary microscope about	2.5×10^{-5} cm.
" " " " ultramicroscope, with electric arc	15×10^{-7} cm.
" " " " " with direct sunlight	1×10^{-7} cm.

TABLE 523. — Molecular Weights of Colloids.

Determined from diffusion.		Determined from freezing point	
Gum arabic	1750	Glycogen (162) *	1625
Tannic acid (322) *	2730	Tungstic acid (250) *	1750
Egg albumen	7420	Gum	1800
Caramel	13200	Albumose	2400
(Due to Graham)		Ferric hydrate (107) *	6000
		Egg albumen	14000
		Starch (162) *	25000

* Formula weight.

TABLE 524. — Brownian Movement.

The Brownian movement is a microscopically observed agitation of colloidal particles. It is caused by the bombardment of them by the molecules of the medium and may be used to determine the value of Avogadro's number. Perrin, Chaudesaignes, Ehrenhaft and De Broglie found, respectively, 70, 64, 63 and 64×10^{22} as the value of this constant. The following table indicates the size and the dependence of this movement on the magnitude of the particles.

Material.	Diameter $\times 10^5$ cm	Medium.	Temp. °C	Velocity $\times 10^5$ cm/sec.	Observer.
Dust particles	2.0	Water	—	none	Zsigmody
Gold	0.35	"	20?	200.	"
Gold	0.1	"	"	280.	"
Gold	0.06	"	"	700.	"
Platinum	.4 to .5	Acetone	18	3900.	Svedberg, 1906–9
Platinum	" " "	Water	20	3200.	"
Rubber emulsion	10.	"	17	124.	Henri, 1908
Mastic	10.	"	20?	1.55	Perrin, Dabrowski, 1909.
Gamboge	4.5	"	20	2.4	Chaudesaignes, 1908.
"	2.13	"	"	3.4	"

The movement varies inversely as the size of the particles; in water, particles of diameter greater than 4μ show no perceptible movement; when smaller than $.1\mu$, lively movement begins, while at 10 $m\mu$ the trajectories amount up to $20 m\mu$.

COLLOIDS.

TABLE 525. — Adsorption of Gas by Finely Divided Particles. See also p. 439.

Fine division means great surface per unit weight. All substances tend to adsorb gas at surface, the more the higher the pressure and the lower the temperature. Since different gases vary in this adsorption, fractional separation is possible. Pt black can absorb 100 vols. H_2, 800 vols. O_2, Pd 3000 vols. H_2. In gas analysis Pd, heated to 100°, is used to remove H_2 (higher temperature used for faster adsorption, will take more at lower temperature). Pt can dissolve several vols. of H_2, Pd, nearly 100 at ordinary temperatures; but it seems probable that the bulk of the 100 vols. of H_2 taken by Pt and the 3000 by Pd must be adsorbed. In 1848 Rose found the density 21 to 22 for Pt foil, but 26 for precipitated Pt.

The film of adsorbed air entirely changes the behavior of very small particles. They flow like a liquid (cf. fog). With substances like carbon black as little as 5 per cent of the bulk is C; a liter of C black may contain 2.5 liters of air. Mitscherlich calculated that when CO_2 at atmospheric pressure, 12° C, is adsorbed by boxwood charcoal, it occupies 1/56 original vol. Apparent densities of gases adsorbed at low temperatures by cocoanut charcoal are of the same order (sometimes greater) as liquids.

Cm³ of Gas Adsorbed by a Cm³ of Synthetic Charcoal (corrected to 0° C, 76 cm?) (Hemperl and Vater).

° C	H_2	Ar	N_2	O_2	CO	CO_2	NO	N_2O
+20°	7.3	12.6	21.0	25.4	26.8	83.8	103.6	109.4
−78	19.5	92.6	107.4	122.4	139.4	568.4	231.3	330.1
−185	284.7	—	632.2	—	697.0	—	—	—

° C	CH_4	C_2H_6	C_2H_4	C_2H_2	NH_3	H_2S	Cl_2	SO_2
+20°	41.7	119.1	139.2	135.8	197.0	213.0	304.5	337.8
−78	174.3	275.5	360.7	488.5	—	—	—	—

Cm³ of Gas Adsorbed by a Cm³ of Cocoanut Charcoal (corrected to 0° C, 76 cm) (Dewar).

° C	He	H_2	N_2	O_2	CO	Ar
0°	2	4	15	18	21	12
−185	15	135	155	230	190	175

See Langmuir, J. Am. Ch. Soc. 40, 1361, 1918; Richardson, 39, 1829, 1916.

TABLE 526. — Heats of Adsorption.

Adsorber.	Amylene.	Water.	Acetone.	Methyl alcohol.	Ethyl alcohol.	Aniline.	Amyl alcohol.	Ethyl ether.	Chloroform.	Benzene.	Carbon disulphide.	Carbon tetrachloride.	Hexane.
Fuller's earth *	57.1	30.2	27.3	21.8	17.2	13.4	10.9	10.5	8.4	4.6	4.6	4.2	3.9
Bone charcoal *	—	18.5	19.3	17.6	16.5	—	10.6	—	14.0	11.1	8.4	13.9	8.9
Kaolin *	78.8	—	—	27.6	24.5	—	20.4	—	15.7	9.9	9.9	9.4	7.2
Fuller's earth †	—	.683	.684	.679	—	—	—	—	.611	.610	.621	.625	—

* Small calories liberated when 1 g of the adsorbent is added to a relatively large quantity of the liquid.
† Volume adsorped from saturated vapor by 1 g of fuller's earth.
Gurvich, J. Russ. Phys. Ch. Soc. 47, 805, 1915.

TABLE 527. — Molecular Heats of Adsorption and Liquefaction (Favre).

Adsorber.	Gas.	Molecular heats of adsorption.	Molecular heats of liquefaction.	Adsorber.	Gas.	Molecular heats of adsorption.	Molecular heats of liquefaction.
Platinum	H_2	46200	—	Charcoal	SO_2	10000–10900	5600
Palladium	H_2	18000	—	"	HCl	9200–10200	(3600)
Charcoal	NH_3	5900–8500	(5000)	"	HBr	15200–15800	(4000)
"	CO_2	6800–7800	6250	"	HI	21000–23000	(4400)
"	N_2O	7100–10900	4400				

TABLE 528. — Miscellaneous Constants (Fundamental, Atomic, Molecular, etc.).

c	Velocity of light in vacuo	299860 Km/sec.
γ	Newtonian gravitational constant	6.658×10^{-8} $cm^3/g.sec.^2$
e	electronic charge	4.774×10^{-10} esu.
		1.591×10^{-20} emu.
		1.591×10^{-19} coulombs.
m_0	mass isolated electron at rest	8.9999×10^{-28} g
m_p	mass of proton in heavier elements	1.649×10^{-24} g
m_H	mass neutral H atom at rest, $1.008 m_p$	1.663×10^{-24} g
h	Planck's quantum of action	6.554×10^{-27} erg. sec.
e/m_0	small velocities or at rest, Bucherer's constant	1.769×10^{7} emu g^{-1}
		5.307×10^{17} esu g^{-1}
F	Faraday constant, 10Ne/c	96500 coulombs
N	Avogadro's constant, no. molecules per g molecular weight	6.061×10^{23}
n	Loschmidt's number, no. molecules per cm^3 0°C, 76 cm	2.705×10^{19}
	No. molecules per cm^3, 0°C, 10^6 bars	2.670×10^{19}
	Radius of electron	about 2×10^{-13} cm
	Radius of H molecule	about 10^{-8} cm
L	Mean free path H molecule 76 cm 0°C	about 1.6×10^{-5} cm
G	Sq. rt. mean sq. velocity ditto	about 1.84×10^{5} cm/sec.
Ω	Arithmetical average velocity ditto	about 1.70×10^{5} cm/sec.
	No. ditto striking 1 cm^2 per sec, 25° C, 1 bar	10.82×10^{17}
	Average distance apart of molecules 76 cm 0°C	3×10^{-6} cm
	Absolute zero, centigrade scale, 0° K	−273.1°
V_0	Vol. per mol(e) or g. molecular weight ideal gas:	
	76 cm. 0° C (1,013,230 dynes/cm^2)	22.412 cm^3
	10^6 dynes, 0° C	22.708 cm^3
p_0	atmosphere, 76 cm Hg, 0° C	1,013,230 dynes/cm^2
E_0	Kinetic energy of translation of molecule at 0° C, (3/2) $(p_0 v_0/N)$	5.620×10^{-14} erg
ϵ	Constant of molecular energy, change of translational energy per °C, E_0/T	2.058×10^{-16} erg/°K
k	Boltzman gas constant, constant of entropy equation, $R/N = (2/3)\epsilon$	1.372×10^{-16} erg/°K.
R	Gas Constant in $PV_m = RT$:	
	V_m = molecular weight in g; P in g/cm^2, V_m, cm^3	84.77 g. cm/°K
	atm., l.	0.08206 l. atm./°K
	dynes, cm^3	8.315×10^{7} erg/°K
	calorimetric units	1.9869 cal (15°)/g. mol.
	1 megabar (= Meteorological "bar") = 10^6 dynes/cm^2 = 1.013 Kg/cm^2	0.98694 atm.
	Mechanical equivalent of heat, g. (20° C) calorie	4.180 intern. elec. joules
	g. (20° C) "	4.181×10^{7} ergs
	g. (15° C) "	4.185×10^{7} ergs
C_v	Atomic heat solids (D and P), 3Nk	24.95 joules per mol.°K
C_{v0}	Molar heat perfect gas, constant volume (3/2)R	2.979 cal per mol.°K
C_{p0}	Molar heat perfect gas, constant pressure (5/2)R	4.965 cal per mol.°K
c_v	Atomic heat solids, D and P, 3R	5.958 cal per mol.°K
R_∞	Rydberg fundamental frequency	3.2775×10^{15} $sec.^{-1}$
	Rydberg's constant, R_∞/c (Bohr)	1.0930×10^{5} cm^{-1}
σ	Stefan-Boltzmann constant, total radiation	5.71×10^{-12} watt-cm^2
c_2	Wien's constant of spectrum radiation ch/k	1.4325 λ in cm.
c_1	Planck's constant of radiation, $2 c^2 h$	3.703×10^{-5} erg/cm^2 sec.
	Photo-electric quantum ch/e	4.117×10^{-7} esu erg/°K
	Thermo-electric quantum k/e	2.874×10^{-7} esu erg/°K
$\lambda_m T$	Wien's displacement constant $c_2/4.9651$	0.2885 cm°K
M	Mechanical equivalent of light	0.00150 watt/lumen
	Calcite grating space	3.028 Angstroms
	Standard gravity	980.665 cm/$sec.^2$
	Mean solar second	1.00273791 sidereal sec.
c	Velocity of light in vacuo (Michelson 1926)	299796 Km/sec.

TABLE 529. — Radiation Wave-length Limits.

Hertzian waves, longest	2 000 000.0 cm
" " shortest	0.2 cm
Infra-red, longest, reststrahlung, focal-isolation	0.03 cm
Infra-red, spectroscopically studied	0.002 cm
Visible, longest	0.000 08 cm
" shortest	0.000 04 cm
Ultra-violet, Lyman, shortest*	0.000 006 cm
X-rays, longest	0.000 000 12 cm
" shortest	0.000 000 001 cm
γ rays, longest	0.000 000 013 cm
" shortest	0.000 000 000 7 cm

* 0.000 0020 cm (Millikan-Sawyer, 1920)

TABLE 530. — Periodic System of the Elements.

O	I	II	III	IV	V	VI	VII	
—	R_2O	RO	R_2O_3	RO_2	R_2O_5	RO_3	R_2O_7	RO_4 ☞ Oxides.
—	—	—	—	RH_4	RH_3	RH	RH	— ☞ Hydrides.
He 4	Li 7	Gl 9	B 11	C 12	N 14	O 16	F 19	— —
Ne 20	Na 23	Mg 24	Al 27	Si 28	P 31	S 32	Cl 35	— —
A 40	K 39	Ca 40	Sc 44	Ti 48	V 51	Cr 52	Mn 55	Fe 56, Ni 59, Co 59
— —	Cu 64	Zn 65	Ga 70	Ge 72	As 75	Se 79	Br 80	— —
Kr 82	Rb 85	Sr 88	Yt 89	Zr 91	Cb 94	Mo 96	— —	Ru 102, Rh 103, Pd 107
— —	Ag 108	Cd 112	In 115	Sn 119	Sb 120	Te 128	I 127	— —
X 128	Cs 133	Ba 137	La 139	Ce 140	Pr 141	Nd 144	— —	— —
— —	Sa 150	**Eu** 152	**Gd** 157	Tb 159	**Ds** 162	Er 168	— —	— —
— —	Tm 168	Yb 174	Lu 175	— —	Ta 181	W 184	— —	Os 191, Ir 193, Pt 195
— —	Au 197	Hg 201	Tl 204	Pb 207	Bi 208	Po 210	— —	— —
Em (222)	— —	Ra 226	Ac (227)	Th 232	UrX_2 234	U 238	— —	— —

TABLE 531. — Atomic Numbers.*

1 Hydrogen
2 Helium
3 Lithium
4 Beryllium
5 Boron
6 Carbon
7 Nitrogen
8 Oxygen
9 Fluorine
10 Neon
11 Sodium
12 Magnesium
13 Aluminum
14 Silicon
15 Phosphorus
16 Sulphur
17 Chlorine
18 Argon
19 Potassium
20 Calcium
21 Scandium
22 Titanium
23 Vanadium
24 Chromium
25 Manganese
26 Iron
27 Cobalt
28 Nickel
29 Copper
30 Zinc
31 Gallium
32 Germanium
33 Arsenic
34 Selenium
35 Bromine
36 Krypton
37 Rubidium
38 Strontium
39 Yttrium
40 Zirconium
41 Niobium ‡
42 Molybdenum
43 Masurium
44 Ruthenium
45 Rhodium
46 Palladium
47 Silver
48 Cadmium
49 Indium
50 Tin
51 Antimony
52 Tellurium
53 Iodine
54 Xenon
55 Caesium
56 Barium
57 Lanthanum
58 Cerium
59 Praseodymium
60 Neodymium
61 Illinium
62 Samarium
63 Europium
64 Gadolinium
65 Terbium
66 Dysprosium
67 Holmium
68 Erbium
69 Thulium
70 Ytterbium
71 Lutecium
72 Hafnium
73 Tantalum
74 Tungsten
75 Rhenium
76 Osmium
77 Iridium
78 Platinum
79 Gold
80 Mercury
81 Thalium
82 Lead
83 Bismuth
84 Polonium
85
86 Emanation
87
88 Radium
89 Actinium
90 Thorium
91 Uranium X_2
92 Uranium

† Glucinium. ‡ Columbium.

PERODIC SYSTEM AND THE RADIOACTIVE ISOTOPES.*

	4	5A	6A	7A	O	1A	2A	3A	4	
		Non-metals.			Inert-gases.	Light-metals.				
Vb	82 Pb	83 Bi	84 Po	85 —	86 Nt	87 —	88 Ra	89 Ac	90 Th	VI
IVb	50 Sn	51 Sb	52 Te	53 I	54 Xe	55 Cs	56 Ba	57 La	58 Ce	Va
IIIb	32 Ge	33 As	34 Se	35 Br	36 Kr	37 Rb	38 Sr	39 Y	40 Zr	IVa
IIb	14 Si	15 P	16 S	17 Cl	18 Ar	19 K	20 Ca	21 Sc	22 Ti	IIIa
Ib	6 C	7 N	8 O	9 F	10 Ne	11 Na	12 Mg	13 Al	14 Si	IIa
	—		1 H		2 He	3 Li	4 Be	5 B	6 C	Ia
		Heavy metals.								
III′	22 Ti	23 V	24 Cr	25 Mn	26 Fe 27 Co 28 Ni	29 Cu	30 Zn	31 Ga	32 Ge	III′
IV′	40 Zr	41 Cb	42 Mo	43 —	44 Ru 45 Rh 46 Pd	47 Ag	48 Cd	49 In	50 Sn	IV′
V″	58 Ce 59 Pr	60 Nd 61 —	62 Sa 63 Eu	64 Gd 65 Tb	66 Dy 67 Ho	68 Er	69 Ad	70 Cp 71 Yb	72 Lu	V″
V′	72 Lu	73 Ta	74 W	75 —	76 Os 77 Ir 78 Pt	79 Au	80 Hg	81 Tl	82 Pb	V′
VI	90 Th	91 Bv	92 U	—						VI
	4	5B	6B	7B		1B	2B	3B	4	

Radioactive isotopes.

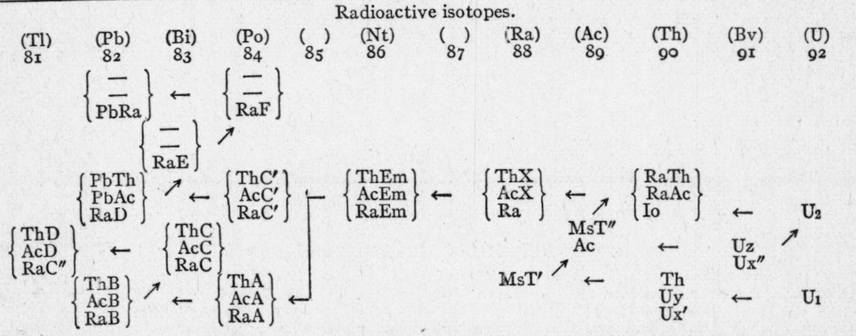

← Indicates the loss of an alpha particle (producing He); the element becomes more electro-positive and the atomic weight decreases by 4, position changing 2 columns to the left.

↗ Indicates beta radiation (loss of electron); the element becomes more electro-negative, atomic weight remains the same, position changes one column to the right and up.

Isotopes of an element have the same valency and the same chemical properties (solubility, reactivity, etc.), although their atomic weights may differ. The isotopes of Bi are, e.g., RaE, ThC, AcC, RaC.

In the *upper* half of the table are the elements possessing high electro-potential, simple spectra, colorless ions. The properties are analogous in the vertical direction (groups). In the *lower* half are the elements with low electro-potential, complex spectra, colored ions and tending to form complex double salts, the general properties of the elements being more pronounced in the horizontal direction (periods).

On the *left* side of the table are the electro-negative elements, those of the upper half forming strong acids, those of the lower half weak oxyacids.

On the *right* side of the table are the electro-positive elements, forming bases, oxysalts, sulfides, etc.

The *center* of the lower half is occupied by the amphoteric elements forming weak acids and bases, many complex compounds and double salts, many insoluble and mostly colored compounds.

A very striking point, however, is, as already mentioned, that the *similarity among the elements* in the upper half is in the vertical direction, and in the lower half in the horizontal direction. This justifies the use of the expressions group-relation and period-relation.

* Table adapted from Hackh, J. Am. Chem. Soc. 40, 1023, 1918, Phys. Rev. 13, 169, 1919.

For Isotopes see conclusion of this Table on page 445

ASTRONOMICAL DATA.

TABLE 533. — Stellar Spectra and Related Characteristics.

The spectra of almost all the stars can be arranged in a continuous sequence, the various types connected in a series of imperceptible gradations. With one unimportant exception, the sequence is linear, the transition between two given types always involving the same intermediate steps. According to the now generally adopted Harvard system of classification, certain principal types of spectrum are designated by letters, — O, B, A, F, G, K, M, R and N, — and the intermediate types by suffixed numbers. A spectrum halfway between classes B and A is denoted B5, while those differing slightly from Class A in the direction of Class B are called B8 or B9. In Classes M and O the notation Ma, Mb, Mc, etc., is employed. Classes R and N apparently form a side chain branching from the main series near Class K.

The colors of the stars, the degree to which they are concentrated into the region of the sky, including the Milky Way, and the average magnitudes of their peculiar velocities in space, referred to the center of gravity of the naked-eye stars as a whole, all show important correlations with the spectral type. In the case of colors, the correlation is so close as to indicate that both spectrum and color depend almost entirely on the surface temperature of the stars. The correlation in the other two cases, though statistically important, is by no means as close.

Examples of all classes from O to M are found among the bright stars. The brightest star of Class N is of magnitude 5.3; the brightest of Class R, 7.0.

TABLE 534. — The Harvard Spectral Classification.

Class.	Principal spectral lines (dark unless otherwise stated).	Example.	Number brighter than 6.25, mag.	Per cent in galactic region.	Color index.	Effective surface temperature, K	Mean peculiar velocity, km/sec.
O	Bright H lines, bright spark lines of He, N, O, C	γ Velorum	20	100	−0.3	—	—
B	H, He, spark lines of N and O, a few spark lines of metals	ε Orionis	696	82	−0.30	20,000°	6
A	H series very strong, spark lines of metals	Sirius	1885	66	0.00	11,000°	10
F	H lines fainter. Spark and arc lines of metals	Canopus	720	57	+0.33	7,500°	14
G	Arc lines of metals, spark lines very faint	The sun	609	58	+0.70	5,000°	15
K	Arc lines of metals, spectrum faint in violet	Arcturus	1719	56	+1.12	4,200°	17
M	Bands of TiO_2, flame and arc lines of metals	Antares	457	54	+1.00	3,100°	17
R	Bands of carbon, flame and arc lines of metals	B. D. −10° 5057	0	63	+1.7	3,000°	15
N	Bands of carbon, bright lines, very little violet light	19 Piscium	8	87	+2.5	2,300°	13

Compiled mainly from the Harvard Annals. Temperatures based on the work of Wilsing and Scheiner. Radial velocities from Campbell. Data for classes R and N from Curtis and Rufus. The color indices are the differences of the visual and photographic magnitudes. Negative values indicate bluish white stars; large positive values, red stars. The peculiar velocities are in the radial direction (towards or from the sun). The average velocities in space should be twice as great.

The "galactic region" here means the zone between galactic latitudes ± 30°, and including half the area of the heavens.

96% of the stars of known spectra belong to classes A, F, G, K, 99.7% including B and M (Innes, 1919).

TABLE 535. — Apex and Velocity of Solar Motion.

R. A. 1900.	Dec.	Velocity, km/sec.	Method.	No. of stars.	Authority.
18^h 02^m	+34.3	—	Proper motions	5413	Boss, Astron. J. 614, 1910
17 54	25.1	19.5	Radial velocities	1193	Campbell, Lick Bull. 196, 1911
18 00	29.2	21.4	" "	1405	Strömberg, Astrophys. J. 1918.

ASTRONOMICAL DATA.

TABLE 536. — Motions of the Stars.

The individual stars are moving in all directions, but, for the average of considerable groups, there is evidence of a drift away from the point in the heavens towards which the sun is moving (solar apex). The best determinations of the solar motion, relative to the stars as a whole, are given in Table 535. In round numbers this motion of the sun may be taken as 20 km/sec. towards the point R. A. 18 h. 0 m., Dec +30.0°.

After allowance is made for the solar motion, the motions of the stars in space, relative to the general mean, present marked peculiarities. If from an arbitrary origin a series of vectors are drawn, representing the velocities of the various stars, the ends of these vectors do not form a spherical cluster (as would occur if the motions of the stars were at random), but a decidedly elongated cluster, whose form can be approximately represented either by the superposition of two intermingling spherical clusters with different centers (Kapteyn's two-stream hypothesis) or by a single ellipsoidal cluster (Schwarzschild), the actual form, however, being more complicated than is indicated by either of these hypotheses. The direction of the longest axis of the cluster is known as that of preferential motion. The two opposite points in the heavens at the extremities of this axis are called the vertices. The components of velocity of the stars parallel to this axis average considerably larger than those parallel to any axis perpendicular to it.

The preferential motion varies greatly with spectral type, being practically absent in Class B, very strong in Class A, and somewhat less conspicuous in Classes F to M, on account of the greater mean velocities of these stars in all directions. The positions of the vertices are nearly the same for all.

Numerous investigators, from the more distant naked-eye stars, find substantially the same position for the vertex, the mean being R. A. 6 h, 6 m., Dec. +9°. The nearer stars, of large proper motion, give a mean of 6 h. 12m., +25°. (See Strömberg's discussion, cited above.)

In addition to these general phenomena, there are numerous clusters of stars whose members possess almost exactly equal and parallel motions, — for example, the Pleiades, the Hyades, and certain large groups in Ursa Major, Scorpius, and Orion. The vertices, and the directions toward which these clusters are moving, are all in the plane of the galaxy.

Several faint stars are known which have radial velocities between 300 and 350 km/sec. (e.g. A. G. Berlin 1366 R.A. 1900 = $4^h 8^m 6$, Dec. 1900 = +22.7°, mag. 8.9 velocity of recession 339 km/sec.), and it is probable that the actual velocity in space exceeds 500 km/sec. for some of these.

The 9th magnitude star A. G. Berlin 1366 has a radial velocity of 494 km/sec.

The greatest known proper motion is that of Barnard's star of the ninth magnitude in Ophiuchus, 10.3″ per year, position angle 356°. The parallax of this star is 0.52″, and its radial velocity about −100 km/sec.

The average radial velocity of the globular clusters is 100 km/sec. and that of the spiral nebulae 400 km. The globular clusters as a class are approaching the sun. The spiral nebulae, with a few exceptions, are receding. The greatest individual values are −410 km for the cluster N. G. C. 6934 and + 1800 km for the nebula N. G. C. 584.

Average velocities with regard to center of gravity of the stellar system, according to Campbell (Stellar Motion, 1913):

Type B Stars:	6.6 km per sec.	Type G Stars:	15.0 km. per sec.
" A "	10.9 " " "	" K "	16.8 " " "
" F "	14.4 " " "	" M "	17.1 " " "

TABLE 537. — Distances of the Stars.

Distances.	Parsecs.*	Light years.
Alpha Centauri (nearest star)	1.32	4.3
Barnard's Star	1.9	6.3
Sirius	2.7	8.7
Arcturus	13.0	43.0
The Hyades	40.	130.
Nebula of Orion (Kapteyn)	185.	600.
Globular Clusters (Shapley): omega Centauri (nearest)	6,500.	21,000.
N. G. C. 7006 (farthest)	67,000.	220,000.

* Parsec = 206,265 astronomical units = 3.08×10^{13} km = 3.26 light years. 1 astronomical unit = distance sun to earth.

Practically all the stars visible to the naked eye lie within 1000 parsecs of the sun, and most of them are more than 100 parsecs distant. In the vicinity of the sun, the majority of the stars lie within two or three hundred parsecs of the galactic plane; but along this plane the star-filled region extends far beyond 1000 parsecs in all directions, and may reach 30,000 parsecs in the great southern star clouds (Shapley).

Average parallax 6 planetary nebulae, 0.018″ (van Maanen, Pr. Nat. Acad. 4, p. 394, 1918).

ASTRONOMICAL DATA.

TABLE 538. — Brightness of the Stars.

Stellar magnitudes give the apparent brightness of the stars on a logarithmic scale, — a numerical increase of one magnitude corresponding to a *decrease* of the common logarithm of the light by 0.400, and a change of five magnitudes to a factor of 100. The brightest objects have negative stellar magnitudes. The visual magnitude of the Sun is −26.7; of the mean full Moon, −12.5; of Venus at her brightest, −4.3; of Jupiter, at opposition, −2.3; of Sirius, −1.6; of Vega, +0.2; of Polaris, +2.1. (The stellar magnitude of a standard candle 1 m distant is −14.18.) The faintest stars visible with the naked eye on a clear dark night are of about the sixth magnitude (though a single luminous point as faint as the eighth magnitude can be seen on a perfectly black background). The faintest stars visible with a telescope of aperture A in. are approximately of magnitude $9 + 5 \log_{10} A$. The faintest photographed with the 60-inch reflector at Mt. Wilson are of about the 21st magnitude. A standard candle, of the same color as the stars, would appear of magnitude +0.8 at a distance of one kilometer.

The actual luminosity (absolute magnitude) is the stellar magnitude which the star would have if placed at a distance of ten parsecs. The faintest star at present known (Innes), a distant companion to α Centauri, has the (visual) absolute magnitude +15.4, and a luminosity 0.00006 that of the sun. The brightest so far definitely measured, β Orionis, has (Kapteyn) the abs. mag. −5.5 and a luminosity 13,000 times the sun's. Canopus, and some other stars, may be still brighter.

The absolute magnitudes of 6 planetary nebulae average 9.1; average diameter, 4000 astronomical units (Solar system to Neptune = 60 astr. units), van Maanen, Pr. Nat. Acad. 4, p. 394, 1918.

Giant and Dwarf Stars.

The stars of Class B are all bright, and nearly all above the absolute magnitude zero. Stars of comparable brightness occur in all the other spectral classes, but the inferior limit of brightness diminishes steadily for the "later" or redder types. The distribution of absolute magnitudes conforms to the superposition of two series, in each of which the individual stars of each spectral class range through one or two magnitudes on each side of the mean absolute magnitude. Absolute magnitude giants roughly 0 to +1; dwarfs A, 1 to 2; F, 2 to 4; G, 4 to 6; K, 6 to 9; M, 9 to 11. The two series overlap in Classes A and F, are fairly well separated in Class K, and sharply so in Class M. Two very faint stars of Classes A and F fall into neither series.

The majority of the stars visible to the naked eye are giants, since these, being brighter, can be seen at much greater distances. The greatest percentage of dwarf stars among those visible to the eye is found in Classes F and G. The dwarf stars of Classes K and M are actually much more numerous per unit of volume, but are so faint that few of the former, and none of the latter, are visible to the naked eye.

Adams and Stromberg have shown that the mean peculiar velocities of the giant stars are all small, — increasing only from about 6 km/sec. for Class B to 12 for Class M, — while those of the dwarf stars are much greater, increasing within each spectral class by about 1.5 km per unit of absolute magnitude, and reaching fully 30 km for stars of Class M and abs. mag. 10. Both giant and dwarf stars show the phenomenon of preferential motion.

TABLE 539. — Masses and Densities.

Stars differ less in mass than in any other characteristic. The most massive star known is the brighter component of the spectroscopic binary B.D. 6°1309, 86 times the sun's mass, 113 times its luminosity, and spectrum Oe. The smallest known mass is that of the faint component of the visual binary Krueger 60, whose mass is 0.15, and luminosity 0.0004 of the sun's, and spectrum M.

The giant stars are in general more massive than the dwarfs. According to Russell (Publ. Astron. Soc. America, 3, 327, 1917) the mean values of Binary systems are:

Spectrum	B2	A0	F5 giant	K5 giant	F2 dwarf	G2 dwarf	K8 dwarf
Ratio of mass to Sun	12	6.5	8	10	3.0	1.2	0.9

The densities can be determined only for eclipsing variables. Stars of Classes B and A have densities averaging about one tenth that of the sun and a relatively small range; Classes F to K show a wide range in density, from 1.8 times that of the sun (W Urs. Maj.) to 0.000002 (W Crucis).

The surface brightness probably diminishes by at least one magnitude for each step along the Harvard scale from B to M. It follows that the dwarf stars are, in general, closely comparable with the sun in diameter, while the stars of Classes B and A, though larger, rarely exceed ten times the sun's diameter. The redder giant stars must be much larger, and a few, such as Antares, may have diameters exceeding that of the earth's orbit. The densities of these stars must be exceedingly low.

Arranged in order of increasing density, the stars form a single sequence starting with the giant stars of Class M, proceeding up that series to Class B, and then down the dwarf series to Class M. Russell and others believe this sequence indicates the order of evolution, — a star at first rising in temperature as it contracts and then cooling off again.

Star	Type	Mag.	Diam.	Parallax	Mass	Density	Brightness	Diameter (km)
Antares	Map	1.2	″.038	″.013	30.	.0000010	1600.	440,000,000
Betelgeuse	Ma	0.9	.044	.018	30.	.0000012	1450.	378,000,000
α Hercules	Mb	3.5	.015	.007	30.	.0000020	710.	320,000,000
Aldebaran	K5	1.5	.027	.075	10.	.00017	36.	53,000,000
Arcturus	K0	0.2	.023	.095	10.	.0007	78.	37,000,000
Rigel	B8	0.3	.0019	.007	30.	.0012	13500.	40,000,000
Capella	G0	0.2	.0082	.071	4.6	.006	78.	13,000,000
Vega	A0	0.1	.0026	.094	5.	.21	86.	4,200,000
Sirius	A0	− 1.6	.0057	.376	2.5	.62	26.	2,300,000
Procyon	F5	0.5	.0048	.328	2.	.60	5.	2,300,000
Our Sun	G	−26.5	960.	—	1.	1.	1.	1,391,000
Krueger 60	Mb	9.3	.0011	.260	.42	4.0	0.002	580,000
Prox. Cent.	N	11.0	.0017	.76	.055	4.0	.00006	333,000
Barnard's	Mb	9.7	.0009	.53	.023	4.0	.0004	249,000

Computed by Plaskett, Pub. Ast. Soc. Pac. 1922; Interferometer measurements, Antares, 0.024″, 30,600,000 km; Betelgeuse, 0.047″, 386, 000,000 km. (1921)

MISCELLANEOUS ASTRONOMICAL DATA.

Tropical (ordinary) year = {365.24219879 − 0.0000000614 (t − 1900)} days
Sidereal year = {365.25636042 + 0.0000000011 (t − 1900)} days
Anomalistic year = {365.25964134 + 0.0000000304 (t − 1900)} days
Eclipse year = {346.620000 + 0.00000036 (t − 1900)} days

Synodical (ordinary) month = {29.530588102 − 0.00000000294 (t − 1900)} days
Sidereal month = {27.321660890 − 0.00000000252 (t − 1900)} days

Sidereal day (ordinary, two successive transits of vernal equinox, might be called equinoctial day) = 86164.09054 mean solar seconds
= 23 h. 56 m. 4.09054 mean solar time

Two successive transits of same fixed star = 86164.09966 mean solar seconds

1920, Julian Period = 6633
January 1, 1920, Julian-day number = 2422325

Solar parallax = 8.7958″ ± 0.002″ (Weinberg)
8.807 ± 0.0027 (Hincks, Eros)
8.799 (Sampson, Jupiter satellites; Harvard observations)
8.80 Paris conference
Lunar parallax = 3422.63″ = 57′ 2.63″ (Newcomb)

Mean distance earth to sun = 149500000 kilometers = 92900000 miles
Mean distance earth to moon = 60.2678 terrestrial radii
= 384411 kilometers = 238862 miles
Light traverses mean radius of earth's orbit in 498.580 seconds
Velocity of light (mean value) in vacuo, 299860 kilometers/sec. (Michelson-Newcomb) = 186324 statute miles/sec.
Constant of aberration = 20.4874″ ± 0.005″
20.47 Paris conference (work of Doolittle and others indicates value not less than 20.51)
Light year = 9.5×10^{12} kilometers = 5.9×10^{12} miles
Parsec, distance star whose parallax is 1 sec. = 31×10^{12} km = 19.2×10^{12} m
General precession = 50.2564″ + 0.000222 (t − 1900)″ (Newcomb)
Obliquity of ecliptic = 23° 27′ 8.26″ − 0.4684 (t − 1900)″ (Newcomb)
Constant of nutation = 9.21″ (Paris conference)
Gravitation constant = 666.07×10^{-10} cm³/g sec² ± 0.16×10^{-10}
Eccentricity earth's orbit = e = 0.01675104 − 0.0000004180 (t − 1900) − 0.0000000000126 $(t - 1900)^2$
Eccentricity moon's orbit = e_2 = 0.05490056 (Brown)
Inclination moon's orbit = I = 5° 8′ 43.5″ (Brown)
Delaunay's $\gamma = \sin \frac{1}{2}I$ = 0.04488716 (Brown)
Lunar inequality of earth = L = 6.454″
Parallactic inequality moon = Q = 124.785″ (Brown)
Mean sidereal motion of moon's node in 365.25 days = − 19° 21′ 19.3838″ + 0.001294 (t − 1900)″
Pole of Milky Way = R. A., 12 h. 48 m.; Dec., +27°

ASTRONOMICAL DATA.

TABLE 541. — The First-magnitude Stars.

No.	Star.	Mag.	Spectrum.	R.A. 1900.	Dec. 1900.	Annual proper motion, μ	P.A. of μ	Parallax.	Abs. mag.	Radial velocity km.
1	Achernar......	0.6	B5	1^h 34.0^m	−57° 45′	0.094″	108°	+0.051″	−0.9	—
2	Aldebaran ‡....	1.1	K5	4 30.2	+16 18	0.203	160	+0.062	−0.0	+55.1
3	Capella † ‡.....	0.2	G	5 9.3	+45 54	0.437	168	+0.075	−0.5	+30.2
4	Rigel *†	0.3	B8	5 9.7	−8 19	0.001	135	+0.007	−5.5	+22.6
5	Betelgeuse † §..	0.6–1.2	Ma	5 49.8	+7 23	0.029	74	+0.019	−2.7	+21.3
6	Canopus.......	−0.9	F	6 21.7	−52 38	0.018	56	+0.007	−6.7	+20.8
7	Sirius *........	−1.6	A	6 40.7	−16 35	1.316	204	+0.376	+1.2	−7.4
8	Procyon *......	0.5	F5	7 34.1	+5 29	1.242	214	+0.309	+3.0	−3.5
9	Pollux §.......	1.2	K	7 39.2	+28 16	0.625	264	+0.064	+0.2	+3.9
10	Regulus ‡......	1.3	B8	10 3.0	+12 27	0.247	269	+0.033	−1.1	−9.1
11	α Crucis *......	1.1	B1	12 21.0	−62 33	0.048	240	+0.047	−0.5	+7.
12	β Crucis †.....	1.5	B1	12 41.9	−59 9	0.056	240	+0.008	−4.0	+13.
13	Spica †........	1.2	B2	13 19.9	−10 38	0.055	229	−0.012	—	+1.6
14	β Centauri †...	0.9	B1	13 56.8	−59 53	0.041	219	+0.037	−1.3	−7.
15	Arcturus.......	0.2	K	14 11.1	+19 42	2.282	209	+0.075	−0.5	−3.9
16	α Centauri *....	0.3	G	14 32.8	−60 25	3.680	281	+0.759	+4.7	−21.6
17	Antares ‡ †....	1.2	Ma	16 23.3	−26 13	0.034	192	+0.029	−1.5	−3.1
18	Vega §........	0.1	A	18 33.6	+38 41	0.346	36	+0.091	−0.1	−13.8
19	Altair §........	0.9	A5	19 45.9	+8 36	0.655	54	+0.214	+2.5	−33.
20	Deneb §.......	1.3	A2	20 38.0	+44 55	0.001	180	+0.002	−7.2	−4.
21	Fomalhaut.....	1.3	A3	22 52.1	−30 9	0.365	117	+0.138	+2.0	+6.7

* Visual binary. † Spectroscopic binary. ‡ Pair with common proper motion.
§ Wide pair probably optical.

Mass relative to sun of (7) is 3.1; of (8), 1.5; of (16), 2.0. For description of types, see Table 534 or Annals of Harvard College Observatory, 28, p. 146, or more concisely 56, p. 66, and 91, p. 5. The light ratio between successive stellar magnitudes is $\sqrt[5]{100}$ or the number whose logarithm is 0.4000, viz., 2.512. The absolute magnitude of a star is its magnitude reduced to a distance corresponding to 0.1″ parallax.

TABLE 542. — Wolf's Observed Sun-spot Numbers. Annual Means.

Sun-spot number = k(10 × number of groups and single spots observed + total number of spots in groups and single spots). k depends on condition of observation and telescope, equaling unity for Wolf with 3-in. telescope and power of 64. Wolf's numbers are closely proportional to spotted area on sun. 100 corresponds to about 1/500 of visible disk covered (umbras and penumbras). Periodicity: mean, 11.13, extremes, 7.3 and 17.1 years. Monthly Weather Review, 30, p. 171, 1902; monthly means, revised, 1749–1901; see A. Wolfer in Astronomische Mitteilungen and Zeitschrift für Meteorologie, daily and monthly values.

Year.	0	1	2	3	4	5	6	7	8	9
1750	83	48	48	31	12	*10*	10	32	48	54
1760	63	**86**	61	45	36	21	*11*	38	70	**106**
1770	101	82	66	35	31	*7*	20	92	**154**	126
1780	85	68	38	23	*10*	24	83	**132**	131	118
1790	90	67	60	47	41	21	16	6	*4*	7
1800	14	34	45	43	**48**	42	28	10	8	2
1810	*0*	1	5	12	14	35	**46**	41	30	24
1820	16	7	4	*2*	8	17	36	50	62	67
1830	**71**	48	28	*8*	13	57	122	**138**	103	86
1840	63	37	24	*11*	15	40	62	98	**124**	96
1850	66	64	54	39	21	7	*4*	23	55	94
1860	**96**	77	59	44	47	30	16	*7*	37	74
1870	**139**	111	102	66	45	17	11	12	*3*	6
1880	32	54	60	**64**	64	52	25	13	7	*6*
1890	7	36	73	**85**	78	64	42	26	27	12
1900	10	*3*	5	24	42	**63**	54	62	48	44
1910	19	6	4	*1*	10	46	**55**	**99**	78	63
1920	39	25	15	—	—	—	—	—	—	—

NOTE: The sun's apparent magnitude is −26.5, sending the earth 90,000,000,000 times as much light as the star Aldebaran. Its absolute magnitude is +4.8.

Ratio of total radiation of sun to that of moon about 100.000 to 1 } Langley
" " " light " " " " " " " 400,000 to 1 }

GEODETICAL AND ASTRONOMICAL TABLES.

TABLE 543.—Length of Degrees on the Earth's Surface.

At Lat.	Miles per degree		Km. per degree		At Lat.	Miles per degree		Km. per degree	
	of Long.	of Lat.	of Long.	of Lat.		of Long.	of Lat.	of Long.	of Lat.
0°	69.17	68.70	111.32	110.57	55°	39.77	69.17	64.00	111.33
10	68.13	68.72	109.64	110.60	60	34.67	69.23	55.80	111.42
20	65.03	68.79	104.65	110.70	65	29.32	69.28	47.18	111.50
30	59.96	68.88	96.49	110.85	70	23.73	69.32	38.19	111.57
40	53.06	68.99	85.40	111.03	75	17.96	69.36	28.90	111.62
45	49.00	69.05	78.85	111.13	80	12.05	69.39	19.39	111.67
50	44.55	69.11	71.70	111.23	90	0.00	69.41	0.00	111.70

For more complete table see "Smithsonian Geographical Tables."

TABLE 544.—Equation of Time.

The equation of time when + is to be added to the apparent solar time to give mean time. When the place is not on a standard meridian (75th, etc.) its difference in longitude in time from that meridian must be subtracted when east, added when west to get standard time (75th meridian time, etc.). The equation varies from year to year cyclically, and the figure following the ± sign gives a rough idea of this variation.

	M. S.		M. S.		M. S.		M. S.
Jan. 1	+3 26±14	Apr. 1	+4 2±7	July 1	+3 31±5	Oct. 1	−10 12±8
15	+9 25±9	15	+0 8±5	15	+5 42±3	15	−14 5±6
Feb. 1	+13 42±4	May 1	−2 54±3	Aug. 1	+6 9±3	Nov. 1	−16 19±2
15	+14 20±2	15	−3 49±1	15	+4 24±5	15	−15 22±4
Mar. 1	+12 34±4	June 1	−2 28±3	Sept. 1	+0 2±7	Dec. 1	−10 58±8
15	+9 9±6	15	+0 8±4	15	−4 41±9	15	−4 53±10

TABLE 545.—Planetary Data.

Body.	Reciprocals of masses.	Mean distance from the sun. Km.	Sidereal period. Mean days.	Equatorial diameter. Km.	Inclination of orbit.	Mean density. $H_2O=1$	Gravity at surface.
Sun	1.	—	—	1391107	—	1.42	28.0
Mercury	6000000.	58×10^6	87.97	4842	7°.003	5.61	0.4
Venus	408000.	108 "	244.70	12191	3.393	5.16	0.9
Earth*	329390.	149 "	365.26	12757	—	5.52	1.00
Mars	3093500.	228 "	686.98	6784	1.850	3.95	0.4
Jupiter	1047.35	778 "	4332.59	142745	1.308	1.34	2.7
Saturn	3501.6	1426 "	10759.20	120798	2.492	.69	1.2
Uranus	22869.	2869 "	30685.93	49693	0.773	1.36	1.0
Neptune	19700.	4495 "	60187.64	52999	1.778	1.30	1.0
Moon	†81.45	‡ 38×10^4	27.32	3476	5.145	3.36	0.17

*Earth and moon. † Relative to earth. Inclination of axes: Sun 7°.25; Earth 23°.45; Mars 24°.6; Jupiter 3°.1; Saturn 26°.8; Neptune 27°.2. Others doubtful. Approximate rates of rotation: Sun 25⅓d; Moon 27⅓d; Mercury 88d; Venus 225d; Mars $24^h\ 37^m$; Jupiter $9^h\ 55^m$; Saturn $10^h\ 14^m$. ‡ from earth

TABLE 546. — Numbers and Equivalent Light of the Stars.

The total of starlight is a sensible but very small amount. This table, taken from a paper by Chapman, shows that up to the 20th magnitude the total light emitted is equivalent to 687 1st-magnitude stars, equal to about the hundredth part of full moonlight. If all the remaining stars are included, following the formula, the equivalent addition would be only three more 1st-magnitude stars. The summation leaves off at a point where each additional magnitude is adding more stars than the last. But, according to the formula, between the 23d and 24th magnitudes there is a turning point, after which each new magnitude adds less than before. The actual counts have been carried so near this turning point that there is no reasonable doubt of its existence. Given its existence, the number of stars is probably finite, a conclusion open to very little doubt. All the indications of the earlier terms must be misleading if the margin between 1 and 2 thousand millions is not enough to cover the whole. (Census of the Sky, Sampson, Observatory, 1915.)

Magnitude, *m*	Number.	Equivalent number of 1st-magnitude stars.	Totals to magnitude, *m*	Magnitude, *m*	Number.	Equivalent number of 1st-magnitude stars.	Totals to magnitude, *m*
−1.6	Sirius	11	—	9.0–10.0	174,000	69	380
−0.9	α Carinæ	6	—	10.0–11.0	426,000	68	448
0.0	α Centauri	2	—	11.0–12.0	961,000	60	508
0.0–1.0	8	14	33	12.0–13.0	2,020,000	51	559
1.0–2.0	27	17	50	13.0–14.0	3,960,000	40	599
2.0–3.0	73	18	68	14.0–15.0	7,820,000	31	630
3.0–4.0	189	19	87	15.0–16.0	14,040,000	22	652
4.0–5.0	650	26	113	16.0–17.0	25,400,000	16	668
5.0–6.0	2,200	35	148	17.0–18.0	38,400,000	10	678
6.0–7.0	6,600	42	190	18.0–19.0	54,600,000	6	684
7.0–8.0	22,550	56	246	19.0–20.0	76,000,000	3	687
8.0–9.0	65,000	65	311	All stars fainter than 20.0	—	3	690

TABLE 547. — Albedos.

The albedo, according to Bond, is defined as follows: "Let a sphere *S* be exposed to parallel light. Then its Albedo is the ratio of the whole amount reflected from *S* to the whole amount of light incident on it." In the following table, m = the stellar magnitude at mean opposition; g = magnitude it would have at full phase and unit distance from earth and sun; σ = assumed mean semi-diameter at unit distance; p = ratio of observed brightness at full phase to that of a flat disk of same size and same position, illuminated and viewed normally and reflecting all the incident light according to Lambert's law; g depends on law of variation of light with phase; albedo = pq. Russell, Astrophysical Journal, 43, p. 173, 1916.

Albedo of the earth: A reduction of Very's observations by Russell gives 0.45 in close agreement with the recent value of Aldrich of 0.43 (see Aldrich, Smithsonian Misc. Collections, 69, 1919).

Object.	m	g	σ	p	q	Visual albedo.	Color index.	Photographic albedo.
Moon	−12.55	+0.40	2.40″	0.105	0.694	0.073	+1.18	0.051
Mercury	−2.94	−0.88	3.45	.164	0.42	.069	—	—
"	−2.12	−0.06	3.45	.077	0.72	.055	—	—
Venus	−4.77	−4.06	8.55	.492	1.20	.59	+0.78	.60
Mars	−1.85	−1.36	4.67	.139	1.11	.154	+1.38	.090
Jupiter	−2.29	−8.99	95.23	.375	1.5:	.56:	+0.50	.73:
Saturn	+0.89	−8.67	77.95	.420	1.5:	.63:	+1.12	0.47:
Uranus	+5.74	−6.98	36.0	.42	1.5:	.63:	—	—
Neptune	+7.65	−7.06	34.5	.49	1.5:	.73:	—	—

TABLE 548. — Duration of Sunshine.

Declination of sun:	−23° 27′	−15°	−10°	−5°	0°	+5°	+10°	+15°	+20°	+23° 27′
approx. date:	Dec. 22.	Feb. 9 Nov. 3.	Feb. 23 Oct. 19.	Mar. 8 Oct. 6.	Mar. 21 Sept. 23.	Sept. 10 Apr. 3.	Apr. 16 Aug. 28.	May 1 Aug. 13.	May 20 Jan. 24.	June 21
Latitude.	h m	h m	h m	h m	h m	h m	h m	h m	h m	h m
0°	12 07	12 07	12 07	12 07	12 07	12 07	12 07	12 07	12 07	12 07
10°	11 32	11 45	11 53	12 00	12 07	12 14	12 21	12 29	12 36	12 43
20°	10 55	11 22	11 38	11 53	12 07	12 22	12 37	12 52	13 08	13 20
30°	10 13	10 57	11 21	11 44	12 08	12 31	12 55	13 19	13 46	14 05
40°	9 19	10 25	11 01	11 35	12 09	12 43	13 17	13 53	14 32	15 01
50°	8 04	9 43	10 34	11 23	12 10	12 58	13 48	14 40	15 38	16 23
55°	7 09	9 12	10 15	11 14	12 12	13 09	14 09	15 13	16 26	17 23
60°	5 52	8 34	9 52	11 04	12 13	13 23	14 36	15 57	17 31	18 52
65°	3 34	7 39	9 19	10 50	12 16	13 43	15 15	17 01	19 19	22 03
70°	—	6 10	8 31	10 29	12 19	14 11	16 15	18 50	—	—
75°	—	2 37	7 04	9 55	12 26	15 00	18 05	—	—	—
80°	—	—	3 10	8 46	12 38	16 44	—	—	—	—

For more extensive table, see Smithsonian Meteorological Tables.

TABLE 549. — The Solar Constant.

Solar constant (amount of energy falling at normal incidence on one square centimeter per minute on body at earth's mean distance) = 1.932 calories = mean 696 determinations 1902—12. Apparently subject to variations, usually within the range of 7 per cent, and occurring irregularly in periods of a week or ten days.

Computed effective temperature of the sun: from form of black-body curves, 6000° to 7000° Absolute; from λmax. = 2930 and max. = 0.470μ, 6230°; from total radiation, $J = 76.8 \times 10^{-12} \times T^4$, 5830°.

TABLE 550. — Solar spectrum energy (arbitrary units) and its transmission by the earth's atmosphere.

Values computed from $e_m = e_o a^m$, where e_m is the intensity of solar energy after transmission through a mass of air m; m is unity when the sun is in the zenith, and approximately = sec. zenith distance for other positions (see table 556); e_o = the energy which would have been observed had there been no absorbing atmosphere; a is the fractional amount observed when the sun is in the zenith.

Wave-length. μ	Transmission coefficients, a. Washington.	Mount Wilson.	Mount Whitney.	One mile nearer earth.	Intensity Solar Energy. Arbitrary Units. m = 0	Mount Whitney. m = 1	Mount Wilson. m = 1	2	4	6	Washington. m = 1	2	3	4	6
0.30	—	(.460)	(.550)	—	54	30	25	11	2	1	—	—	—	—	—
.32	—	.520	.615	—	111	68	58	30	8	2	—	—	—	—	—
.34	—	.580	.692	—	232	160	135	78	26	9	—	—	—	—	—
.36	—	.635	.741	—	302	224	192	122	49	20	—	—	—	—	—
.38	(.380)	.676	.784	.562	354	278	239	162	74	34	134	51	19	7	3
.40	.560	.729	.809	.768	414	335	302	220	117	62	232	130	73	41	13
.46	.690	.832	.887	.829	618	548	514	428	296	205	426	294	203	140	67
.50	.733	.862	.919	.850	606	557	522	450	334	248	441	323	237	174	94
.60	.779	.900	.940	.866	504	474	454	409	331	268	393	306	238	185	112
.70	.858	.950	.964	.903	364	351	346	329	297	268	312	268	230	197	145
.80	.886	.970	.976	.915	266	260	258	250	235	221	236	209	185	164	145
1.00	.922	.980	.975	.941	166	162	163	160	154	147	153	141	130	120	102
1.50	.938	.976*	.965	.961	63	61	61*	60*	57*	55*	59	55	52	49	43
2.00	.912	.970*	.932	.940	25	23	24*	23*	21*	19*	23	21	19	17	14

Transmission coefficients are for period when there was apparently no volcanic dust in the air.
* Possibly too high because of increased humidity towards noon.

TABLE 551. — The intensity of Solar Radiation in different sections of the spectrum, ultra-violet, visual infra-red. Calories.

Wave-length. μ	μ	Mount Whitney. m = 0	m = 1	2	3	4	Mount Wilson. m = 1	2	3	4	Washington. m = 1	2	3	4
0.00 to	0.45	.31	.25	.19	.16	.13	.23	.16	.12	.09	.13	.06	.04	.02
0.45 to	0.70	.71	.67	.62	.58	.54	.65	.57	.51	.45	.53	.40	.30	.24
0.70 to	∞	.91	.87	.85	.82	.80	.69	.68	.66	.63	.69	.62	.57	.53
0.00 to	∞	1.93	1.78	1.66	1.56	1.47	1.57	1.42	1.28	1.17	1.35	1.08	.90	.79

TABLE 552. — Distribution of brightness (Radiation) over the Solar Disk.

(These observations extend over only a small portion of a sun-spot cycle.)

Wave-length.	μ 0.323	μ 0.386	μ 0.433	μ 0.456	μ 0.481	μ 0.501	μ 0.534	μ 0.604	μ 0.670	μ 0.699	μ 0.866	μ 1.031	μ 1.225	μ 1.655	μ 2.097
Fraction Radius. 0.00	144	338	456	515	511	489	463	399	333	307	174	111	77.6	39.5	14.0
0.40	128	312	423	486	483	463	440	382	320	295	169	108	75.7	38.9	13.8
0.55	120	289	395	455	456	437	417	365	308	284	163	105.5	73.8	38.2	13.6
0.65	112	267	368	428	430	414	396	348	295	273	159	103	72.2	37.6	13.4
0.75	99	240	333	390	394	380	366	326	281	258	152	99	69.8	36.7	13.1
0.825	86	214	296	351	358	347	337	304	262	243	145	94.5	67.1	35.7	12.8
0.875	76	188	266	317	324	323	312	284	247	229	138	90.5	64.7	34.7	12.5
0.92	64	163	233	277	290	286	281	259	227	212	130	86	61.6	33.6	12.2
0.95	49	141	205	242	255	254	254	237	210	195	122	81	58.7	32.3	11.7

Taken from vols. II and III and unpublished data of the Astrophysical Observatory of the Smithsonian Institution. Schwartzchild and Villiger: Astrophysical Journal, 23, 1906.

ATMOSPHERIC TRANSPARENCY AND SOLAR RADIATION.

TABLE 553. — Transmission of Radiation Through Moist and Dry Air.

This table gives the wave-length, λ; a the transmission of radiation by dry air above Mount Wilson (altitude = 1730 m. barometer, 620 mm.) for a body in the zenith; finally a correction factor, a_w, due to such a quantity of aqueous vapor in the air that if condensed it would form a layer 1 cm. thick. Except in the bands of selective absorption due to the air, a agrees very closely with what would be expected from purely molecular scattering. a_w is very much smaller than would be correspondingly expected, due possibly to the formation of ions by the ultra-violet light from the sun. The transmission varies from day to day. However, values for clear days computed as follows agree within a per cent or two of those observed when the altitude of the place is such that the effect due to dust may be neglected, e. g. for altitudes greater than 1000 meters. If B = the barometric pressure in mm., w, the amount of precipitable water in cm., then $a_B = a^{\frac{B}{620}} a_w^w$. w is best determined spectroscopically (Astrophysical Journal, 35, p. 149, 1912, 37, p. 359, 1913) otherwise by formula derived from Hann, $w = 2.3 e_w 10^{-\frac{h}{22000}}$, e_w being the vapor pressure in cm. at the station, h, the altitude in meters. See Table 377 for long-wave transmission.

λ (μ)	.360	.384	.413	.452	.503	.535	.574	.624	.653	.720	.986	1.74
a	(.660)	.713	.783	.840	.885	.898	.905	.929	.938	.970	.986	.990
a_w	.950	.960	.965	.967	.977	.980	.974	.978	.985	.988	.990	.990

Fowle, Astrophysical Journal, 38, 1913.

TABLE 554. — Brightness of (radiation from) Sky at Mt. Wilson (1730 m.) and Flint Island (sea level).

Zenith dist. of zone		0–15°	15–35°	35–50°	50–60°	60–70°	70–80°	80–90°	–	Sun.
10^8 × mean ratio sky/sun	Mt. Wilson	1500*	400	520	610	660	700	720	–	–
" "	Flint Island	115	122	128	150	185	210	460	–	–
Ditto × area of zone	Mt. Wilson	51.0	58.8	91.5	87.2	104.3	117.6	125.3	–	636
" " "	Flint Island	3.9	17.9	22.5	21.4	29.2	35.3	80.0	–	210
Altitude of sun		–	–	5°	15°	25°	35°	47½°	65°	82½°
Sun's brightness, cal. per cm.² per min.		–	–	.533	.900	1.233	1.358	1.413	1.496	1.521
Ditto on horizontal surface		–	–	.046	.233	.524	.780	1.041	1.355	1.507
Mean brightness on normal surface sky × 10^8/sun		–	–	423	403	.385	365	346	326	310
Total sky radiation on horizontal cal. per cm.² per m.		–	–	.056	.110	.162	.189	.205	.226	.240
Total sun + sky, ditto		–	–	.102	.343	.686	.969	1.246	1.581	1.747

* Includes allowance for bright region near sun. For the dates upon which the observation of the upper portion of table were taken, the mean ratios of total radiation sky/sun, for equal angular areas, at normal incidence, at the island and on the mountain, respectively, were 636×10^{-8} and 210×10^{-8}, on a horizontal surface, 305×10^{-8} and 77×10^{-8}; for the whole sky, at normal incidence, 0.57 and 0.20; on a horizontal surface 0.27 and 0.07. Annals of the Astrophysical Observatory of the Smithsonian Institution, vols. II and III, and unpublished researches (Abbot).

TABLE 555. — Relative Distribution in Normal Spectrum of Sunlight and Sky-light at Mount Wilson.

Zenith distance about 50°.

	μ	μ	μ	μ	μ	μ	C	D	b	F
Place in Spectrum	0.422	0.457	0.491	0.566	0.614	0.660				
Intensity Sunlight	186	232	227	211	191	166				
Intensity Sky-light	1194	986	701	395	231	174				
Ratio at Mt. Wilson	642	425	309	187	121	105	102	143	246	316
Ratio computed by Rayleigh	–	–	–	–	–	–	102	164	258	328
Ratio observed by Rayleigh	–	–	–	–	–	–	102	168	291	369

TABLE 556. — Air Masses.

See Table 174 for definition. Besides values derived from the pure secant formula, the table contains those derived from various other more complex formula, taking into account the curvature of the earth, refraction, etc. The most recent is that of Bemporad.

Zenith Dist.	0°	20°	40°	60°	70°	75°	80°	85°	88°
Secant	1.00	1.064	1.305	2.000	2.924	3.864	5.76	11.47	28.7
Forbes	1.00	1.065	1.306	1.995	2.902	3.809	5.57	10.22	18.9
Bouguer	1.00	1.064	1.305	1.990	2.900	3.805	5.56	10.20	19.0
Laplace	1.00	–	–	1.993	2.899	–	5.56	10.20	18.8
Bemporad	1.00	–	–	1.995	2.904	–	5.60	10.39	19.8

The Laplace and Bemporad values, Lindholm, Nova Acta R. Soc. Upsal. 3, 1913; the others, Radau's Actinometric, 1877.

RELATIVE INTENSITY OF SOLAR RADIATION.

TABLE 557. — Mean intensity J for 24 hours of solar radiation on a horizontal surface at the top of the atmosphere and the solar radiation A, in terms of the solar radiation, A_0, at earth's mean distance from the sun.

Date.	Motion of the sun in longitude.	Relative Mean Vertical Intensity $\left(\frac{J}{A_0}\right)$. Latitude North.										$\frac{A}{A_0}$.
		0°	10°	20°	30°	40°	50°	60°	70°	80°	90°	
Jan. 1	0.99°	0.303	0.265	0.220	0.169	0.117	0.066	0.018				1.0335
Feb. 1	31.54	.312	.282	.244	.200	.150	.100	.048	0.006			1.0288
Mar. 1	59.14	.320	.303	.279	.245	.204	.158	.108	.056	0.013		1.0173
Apr. 1	89.70	.317	.319	.312	.295	.269	.235	.195	.148	.101	0.082	1.0009
May 1	119.29	.303	.318	.330	.329	.320	.302	.278	.253	.255	.259	0.9841
June 1	149.82	.287	.315	.334	.345	.349	.345	.337	.344	.360	.366	0.9714
July 1	179.39	.283	.312	.333	.347	.352	.351	.345	.356	.373	.379	0.9666
Aug. 1	209.94	.294	.316	.330	.334	.330	.318	.300	.282	.295	.300	0.9709
Sept. 1	240.50	.310	.318	.316	.305	.285	.256	.220	.180	.139	.140	0.9828
Oct. 1	270.07	.317	.308	.289	.261	.225	.183	.135	.084	.065		0.9995
Nov. 1	300.63	.312	.286	.251	.211	.164	.114	.063	.018			1.0164
Dec. 1	330.19	.304	.267	.224	.175	.124	.072	.024				1.0288
Year....		0.305	0.301	0.289	0.268	0.241	0.209	0.173	0.144	0.133	0.126	

TABLE 558. — Mean Monthly and Yearly Temperatures.

Mean temperatures of a few selected American stations, also of a station of very high, two of very low temperature, and one of very great and one of very small range of temperature.

	Jan.	Feb.	Mar.	Apr.	May.	June.	July.	Aug.	Sept.	Oct.	Nov.	Dec.	Year.
1 Hebron-Rama (Labr.)	−20.7	−20.9	−15.6	− 6.9	+ 0.2	+ 4.5	+ 7.6	+ 8.0	+ 4.5	− 0.8	− 6.2	−16.2	− 5.2
2 Winnipeg (Canada) .	−21.6	−18.8	−11.0	+ 1.9	+10.9	+17.1	+18.9	+17.6	+11.6	+ 4.1	− 7.6	−15.7	+ 0.6
3 Montreal . . .	−10.9	− 9.1	− 4.3	+ 4.8	+12.6	+18.3	+20.5	+19.3	+14.7	+ 7.8	− 0.2	− 7.1	+ 5.5
4 Boston . . .	− 2.8	− 2.2	+ 1.2	+ 7.3	+13.6	+19.1	+21.8	+20.6	+16.9	+11.1	+ 4.8	− 0.5	+ 9.2
5 Chicago . . .	− 4.8	− 2.9	+ 1.2	+ 7.9	+13.4	+19.7	+22.2	+21.6	+17.9	+11.1	+ 3.6	− 1.5	+ 9.1
6 Denver . . .	− 2.1	+ 0.1	+ 3.8	+ 8.3	+13.6	+19.1	+22.1	+21.2	+16.6	+10.3	+ 3.3	0.0	+ 9.7
7 Washington . .	+ 0.7	+ 2.1	+ 5.2	+11.7	+17.7	+22.9	+24.9	+23.7	+19.9	+13.4	+ 6.9	+ 2.3	+12.6
8 Pikes Peak . . .	−16.4	−15.6	−13.4	−10.4	− 5.3	+ 0.4	+ 4.5	+ 3.6	− 0.3	− 5.8	−11.8	−14.4	− 7.1
9 St. Louis . . .	− 0.8	+ 1.7	+ 6.2	+13.4	+18.8	+24.0	+26.0	+24.9	+20.8	+14.2	+ 6.4	+ 2.0	+13.1
10 San Francisco . .	+10.1	+10.9	+12.0	+12.6	+13.7	+14.7	+14.6	+14.8	+15.8	+15.2	+13.5	+10.8	+13.2
11 Yuma	+12.3	+14.9	+18.1	+21.0	+25.1	+29.4	+33.1	+32.6	+29.1	+22.8	+16.6	+13.3	+22.3
12 New Orleans . .	+12.1	+14.5	+16.7	+20.6	+23.7	+26.8	+27.9	+27.5	+25.7	+21.0	+15.9	+13.1	+20.4
13 Massaua . . .	+25.6	+26.0	+27.1	+29.0	+31.1	+33.5	+34.8	+34.7	+33.3	+31.7	+29.0	+27.0	+30.3
14 Ft. Conger (Greenl'd)	−39.0	−40.1	−33.5	−25.3	−10.0	+ 0.4	+ 2.8	+ 1.0	− 9.0	−22.7	−30.9	−33.4	−20.0
15 Werchojansk . .	−51.0	−45.3	−32.5	−13.7	+ 2.0	+12.3	+15.5	+10.1	+ 2.5	−15.0	−37.8	−47.0	−16.7
16 Batavia . . .	+25.3	+25.4	+25.8	+26.3	+26.4	+26.0	+25.7	+25.9	+26.3	+26.4	+26.2	+25.6	+25.9

Lat., Long., Alt. respectively: (1) +58°.5, 63°.0 W, —; (2) +49.9, 97.1 W, 233m.; (3) +45.5, 73.6 W, 57m.; (4) +42.3, 71.1 W, 38m.; (5) +41.9, 87.6 W, 251m.; (6) +39.7, 105.0 W, 1613m.; (7) +38.9, 77.0 W, 34m.; (8) +38.8, 105.0 W, 4308m.; (9) +38.6, 90.2 W, 173m.; (10) +37.8, 122.5 W, 47m.; (11) +32.7, 114.6 W, 43m.; (12) +30.0, 90.1 W, 16m.; (13) +15.6, 37.5 E, 9m.; (14) +81.7, 64.7 W., —; (15) +67.6, 133.8 E, 140m.; (16) −6.2, 106.8 E, 7m.

Taken from Hann's Lehrbuch der Meteorologie, 2'nd edition, which see for further data.

Note: Highest recorded temperature in world = 57° C in Death Valley, California, July 10, 1913. Lowest recorded temperature in world = −68° C at Verkhoyansk, Feb. 1892.

THE EARTH'S ATMOSPHERE.

TABLE 559. — Miscellaneous Data. Variation with Latitude.

Optical evidence of atmosphere's extent: twilight 63 km, luminous clouds 83, meteors 200, aurora 44–360. Jeans computes a density at 170 km of 2×10^{13} molecules per cm^3, nearly all H (5% He); at 810 km, 3×10^{10} molecules per cm^3 almost all H. When in equilibrium, each gas forms an atmosphere whose density decrease with altitude is independent of the other components (Dalton's law, H_2O vapor does not). The lighter the gas, the smaller the decrease rate. A homogeneous atmosphere, 76 cm pressure at sea-level, of sea-level density, would be 7991 m high. Average sea-level barometer is 74 cm; corresponding homogeneous atmosphere (truncated cone) 7790 m, weighs (base, m^2) 10,120 kg; this times earth's area is 52×10^{14} metric tons or 10^{-6} of earth's mass. The percentage by vol. and the partial pressures of the dry-air components at sea-level are: N_2, 78.03, 593.02 mm; O_2, 20.99, 159.52; A, 0.94, 7.144; CO_2, 0.03, 0.228; H_2, 0.01, 0.076; Ne, 0.0012, 0.009; He, 0.0004, 0.003 (Hann). The following table gives the variation of the mean composition of moist air with the latitude (Hann).

Equator	N_2 75.99	O_2 20.44	A 0.92	H_2O 2.63	CO_2 0.02
50° N.	77.32	20.80	0.94	0.92	0.02
70° N.	77.87	20.94	0.94	0.22	0.03

TABLE 560. — Variation of Percentage Composition with Altitude (Humphreys).

Computed on assumptions: sea-level temperature 11° C; temperature uniformly decreasing 6° per km up to 11 km, from there constant with elevation at −55°. J. Franklin Inst. 184, p. 388, 1917.

Height, km	Argon.	Nitrogen.	Water vapor.	Oxygen.	Carbon dioxide.	Hydrogen.	Helium.	Total pressure, mm
140	—	0.01	—	—	—	99.15	0.84	0.0040
120	—	0.19	—	—	—	98.74	1.07	0.0052
100	—	2.95	0.05	0.11	—	95.58	1.31	0.0067
80	—	32.18	0.17	1.85	—	64.70	1.10	0.0123
60	0.03	81.22	0.15	7.69	—	10.68	0.23	0.0935
50	0.12	86.78	0.10	10.17	—	2.76	0.07	0.403
40	0.22	86.42	0.06	12.61	—	0.67	0.02	1.84
30	0.35	84.26	0.03	15.18	0.01	0.16	0.01	8.63
20	0.59	81.24	0.02	18.10	0.01	0.04	—	40.99
15	0.77	79.52	0.01	19.66	0.02	0.02	—	89.66
11	0.94	78.02	0.01	20.99	0.03	0.01	—	168.00
5	0.94	77.89	0.18	20.95	0.03	0.01	—	405.
0	0.93	77.08	1.20	20.75	0.03	0.01	—	760.

TABLE 561. — Variation of Temperature, Pressure and Density with Altitude.

Average data from sounding balloon flights (65 for summer, 52 for winter data) made at Trappes (near Paris), Uccle (near Brussels), Strassburg and Munich. Compiled by Humphreys, 16 to 20 m chiefly extrapolated.

Elevation, km	Summer.			Winter.		
	Temp. ° C	Pressure, mm of Hg.	Density, dry air, g/cm^3	Temp. ° C	Pressure, mm of Hg.	Density, dry air, g/cm^3
20.0	−51.0	44.1	0.000092	−57.0	39.5	0.000085
19.0	−51.0	51.5	.000108	−57.0	46.3	.000100
18.0	−51.0	60.0	.000126	−57.0	54.2	.000117
17.0	−51.0	70.0	.000146	−57.0	63.5	.000137
16.0	−51.0	81.7	.000171	−57.0	74.0	.000160
15.0	−51.0	95.3	.000199	−57.0	87.1	.000187
14.0	−51.0	111.1	.000232	−57.0	102.1	.000220
13.0	−51.0	129.6	.000270	−57.0	119.5	.000257
12.0	−51.0	151.2	.000316	−57.0	140.0	.000301
11.0	−49.5	176.2	.000366	−57.0	164.0	.000353
10.0	−45.5	205.1	.000419	−54.5	192.0	.000408
9.0	−37.8	237.8	.000470	−49.5	224.1	.000466
8.0	−29.7	274.3	.000524	−43.0	260.6	.000526
7.0	−22.1	314.9	.000583	−35.4	301.6	.000590
6.0	−15.1	360.2	.000649	−28.1	347.5	.000659
5.0	−8.9	410.6	.000722	−21.2	398.7	.000735
4.0	−3.0	466.6	.000803	−15.0	455.9	.000821
3.0	+2.4	528.9	.000892	−9.3	519.7	.000915
2.5	+5.0	562.5	.000942	−6.7	554.3	.000967
2.0	+7.5	598.0	.000990	−4.7	590.8	.001023
1.5	+10.0	635.4	.001043	−3.0	629.6	.001083
1.0	+12.0	674.8	.001100	−1.3	670.6	.001146
0.5	+14.5	716.3	.001157	0.0	714.0	.001215
0.0	+15.7	760.0	.001223	+0.7	760.0	.001290

760 mm = 29.921 in. = 1013.3 millibars. 1 mm = 1.33322387 millibars. 1 bar = 1,000,000 dynes; this value, sanctioned by International Meteorological Conferences, is 1,000,000 times that sometimes used by physicists.

TERRESTRIAL TEMPERATURES.

TABLE 562. — Temperature Variation over Earth's Surface (Hann).

Latitude.	Temperatures ° C						Mean ocean temp.	Land surface %
	Jan.	Apr.	July.	Oct.	Year.	Range.		
North pole	−41.0	−28.0	−1.0	−24.0	−22.7	40.0	−1.7	—
+80°	−32.2	−22.7	+2.0	−19.1	−17.1	34.2	−1.7	20
70	−26.3	−14.0	7.3	−9.3	−10.7	33.6	+0.7	53
60	−16.1	−2.8	14.1	+0.3	−1.1	30.2	4.8	61
50	−7.2	+5.2	17.9	6.9	+5.8	25.1	7.9	58
40	+5.5	13.1	24.0	15.7	14.1	18.5	14.1	45
30	14.7	20.1	27.3	21.8	20.4	12.6	21.3	43.5
20	21.9	25.2	*28.0*	26.4	25.3	6.1	25.4	31.5
+10	25.8	*27.2*	27.0	*26.9*	*26.8*	1.4	27.2	24
Equator	*26.5*	26.6	25.7	26.5	26.3	0.9	27.1	22
−10	26.4	25.9	23.0	25.7	25.5	3.4	25.8	20
20	25.3	24.0	19.8	22.8	23.0	5.5	24.0	24
30	21.6	18.7	14.5	18.0	18.4	7.1	19.5	20
40	15.4	12.5	8.8	11.7	11.9	6.6	13.3	4
50	8.4	5.4	3.0	4.8	5.4	5.4	+6.4	2
60	3.2	—	−9.3	—	−3.2	12.5	0.0	0
70	−1.2	—	−21.0	—	−12.0	19.8	−1.3	71
80	(−4.3)	—	(−28.7)	—	(−20.6)	(24.4)	—	100
South pole	(−6.0)	—	(−33.0)	—	(−25.0)	(27.0)	—	(100)

TABLE 563. — Temperature Variation with Depth (Land and Ocean).

Table illustrates temperature changes underground at moderate depths due to surface warming (read from plot for Tiflis, Lehrbuch der Meteorologie, Hann and Süring, 1915). Below 20–30 m (nearer the surface in tropics) there is no annual variation. Increase downwards at greater depths, 0.03 ± ° C per m (1° per 35 m) l. c. At Pittsburgh, 1524 m, 49.4°, .0294 per m; Oberschlesien, 2003 m, 70°, .0294 per m; or W. Virginia, 2200 m, 70°, .034° per m (Van Orstrand). Mean value outflow heat from earth's center, 0.00000172 g-cal/cm²/sec. or 54 g-cal/cm²/year (39 Laby). Open ocean temperatures: Greatest mean annual range (Schott) 40° N, 4.2° C; 30° S, 5.1°; but 10° N, only 2.2°; 50° S, 2.9°. Mean surface temp. whole ocean (Krümmel) 17.4°; all depths, 3.9°. Below 1 km nearly isothermal with depth. In tropics, surface 28°; at 183 m, 11°, 80% all water less than 4.4°. Deep-sea (bottom) temps. range −0.5° to +2.6°. Soundings in S. Atlantic: 0 km, 18.9°; .25 km, 15°; .5 km, 8.3°; 1 km, 3.3°; 3 km, 1.7°; 4.5 km, 0.6°.

Depth, m	Temperature, centigrade.											
	Jan.	Feb.	Mar.	Apr.	May.	June.	July.	Aug.	Sept.	Oct.	Nov.	Dec.
0	*1*	4	10	14	21	29	**32**	32	24	16	9	4
0.5	4	*4*	9	13	18	23	26	**28**	24	18	12	6
1.0	6	*6*	8	12	15	20	24	**26**	23	18	14	10
1.5	9	*8*	9	11	14	18	21	**23**	22	18	15	12
2.0	11	10	*10*	11	13	16	19	21	**21**	18	16	14
3.0	14	12	12	*11*	13	14	16	17	18	**18**	17	15
4.0	15	13	12	*12*	12	13	14	16	16	**17**	17	16
5.0	15	14	13	13	*13*	13	14	14	15	16	**16**	16
6.0	15	14	14	14	14	14	14	14	14	15	15	15

Table 564.

GEOCHEMICAL DATA.

Eighty-three chemical elements (86 including Po, Ac and UrX_2) **are found on the earth.** Besides the eight occurring uncombined as gases, 23 may be found native, Sb, As, Bi, C, Cu, Au, Ir, Fe, Pb?, Hg, Ni, Os, Pd, Pt, Rh, Ru, Se, Ag, S, Ta?, Te, Sn?, Zn?. Combined the elements form about 1000 known mineral species. Rocks are in general aggregates of these species. Some few (e. g., quartzite, limestone, etc.) consist of one specie. We have some knowledge of the earth to a depth of 10 miles. This portion may be divided into three parts: the innermost of crystalline or plutonic rocks, the middle, of sedimentary or fragmentary rocks, the outer of clays, gravels, etc. 93% of it is solid matter, 7% liquid, and the atmosphere amounts by weight to 0.03% of it. Besides the 9 major constituents of igneous rock (see 7th col. of table) 3 are notable by their almost universal occurrence, TiO_2, P_2O_5, and MnO. Bo, Gl, and Sc are also widely distributed.

The density of the earth as a whole is 5.52 (Burgess); continental surface, 2.67 and outer 10 miles of crust, 2.40 (Harkness). Computed from average chemical composition: outer ten miles as a whole, 2.77; northern continents 2.73; southern, 2.76; Atlantic basin, 2.83; Pacific basin, 2.88.

Data of Geochemistry, Clarke, Bul. 616, **U. S. Geological Survey,** 1916; **Washington, J. Franklin. Inst.** 190, p. 757, 1920.

Average Composition of Known Terrestrial Matter.

Atomic number and element.	Average composition.			Igneous rocks.	Average composition of lithosphere.					
	Lithosphere, 93%	Hydrosphere, 7%	Average including atmosphere.		Compound.	Igneous rocks, 95%	Shale, 4%	Sandstone, 0.75%	Limestone, 0.25%	Weighted average.
8 O	47.33	85.79	46.43	47.29	SiO_2	59.09	58.10	78.33	5.19	59.77
14 Si	27.74	—	27.77	28.02	Al_2O_3	15.35	15.40	4.77	0.81	14.89
13 Al	7.85	—	8.14	7.96	Fe_2O_3	3.08	4.02	1.07	.54	2.69
26 Fe	4.50	—	5.12	4.56	FeO	3.80	2.45	.30	—	3.39
20 Ca	3.47	0.05	3.63	3.47	MgO	3.49	2.44	1.16	7.89	3.74
12 Mg	2.24	0.14	2.09	2.29	CaO	5.08	3.11	5.50	42.57	4.86
11 Na	2.46	1.14	2.85	2.50	Na_2O	3.84	1.30	.45	.05	3.25
19 K	2.46	0.04	2.60	2.47	K_2O	3.13	3.24	1.31	.33	2.98
1 H	0.22	10.67	0.127	0.16	H_2O	1.14	5.00	1.63	.77	2.02
22 Ti	0.46	—	.629	.46	TiO_2	1.05	.65	.25	.06	.77
6 C	.19	0.002	.027	.13	ZrO_2	0.039	—	—	—	.02
17 Cl	.06	2.07	.055	.063	CO_2	.102	2.63	5.03	41.54	.70
35 Br	—	0.008	—	—	P_2O_5	.30	.17	.08	.04	.28
15 P	.12	—	.130	.13	S	.053	—	—	.09	.10
16 S	.12	.09	.052	.103	SO_3	—	.64	.07	.05	.03
56 Ba	.08	—	.048	.092	Cl	.056	—	—	.02	.06
25 Mn	.08	—	.096	.078	F	.078	—	—	—	.09
38 Sr	.02	—	.018	.033	BaO	.055	.05	.05	—	.09
7 N	—	—	—	—	SrO	.022	—	—	—	.04
9 Fl	.10	—	.077	.10	MnO	.125	—	—	.05	.09
etc.	.50	—	.111	.091	NiO	.025	—	—	—	.025
					Cr_2O_3	.056	—	—	—	.05
					V_2O_3	.032	—	—	—	.025
					Li_2O	.007	—	—	—	..01
					C	—	.80	—	—	.03

Average Composition of Meteorites: The following figures give in succession the element, atomic number (bracketed), and the percentage amount in stony meteorites (Merrill, Mem. Nat. Acad. Sc. 14, p. 28, 1916). The "iron" meteorites contain a much larger percentage of iron and nickel, but there is a tendency to believe that with such meteorites the composition is altered by the volatilization or burning up of the other material in passing through the air. Note the greater abundance of elements of even atomic number (97.2 per cent).

O (8)	36.53	Fe (26)	23.32	Si (14)	18.03	Mg (12)	13.60
S (16)	1.80	Ca (20)	1.72	Al (13)	1.53	Ni (28)	1.52
Na (11)	1.64	Cr (24)	0.32	Mn (25)	0.23	K (19)	0.17
C (6)	0.15	Co (27)	0.12	Ti (22)	0.11	P (15)	0.11
H (1)	0.09	Cu (29)	0.01	Cl (17)	0.09	V (23)	tr.
Ru (44)	tr.	Pd (46)	tr.	Pt (78)	tr.	Ir (77)	tr.

TABLE 565.

ACCELERATION OF GRAVITY.

For Sea Level and Different Altitudes.

Calculated from U. S. Coast and Geodetic Survey formula, p. 134 of Special Publication No. 40 of that Bureau.

$$g = 9.78039\ (1 + 0.005294 \sin^2 \phi - 0.000007 \sin^2 2\phi)\ \text{m}$$
$$g = 32.08783\ (1 + 0.005294 \sin^2 \phi - 0.000007 \sin^2 2\phi)\ \text{ft.}$$

Latitude ϕ	g cm/sec²	log g	g ft./sec²	Latitude ϕ	g cm/sec²	log g	g ft./sec²
0°	978.039	2.9903562	32.0878	50°	981.071	2.9917004	32.1873
5	.078	.9903735	.0891	51	.159	.9917394	.1902
10	.195	.9904254	.0929	52	.247	.9917784	.1931
12	.262	.9904552	.0951	53	.336	.9918177	.1960
14	.340	.9904898	.0977	54	.422	.9918558	.1988
15	978.384	2.9905094	32.0991	55	981.507	2.9918934	32.2016
16	.430	.9905298	.1007	56	.592	.9919310	.2044
17	.480	.9905520	.1023	57	.675	.9919677	.2071
18	.532	.9905750	.1040	58	.757	.9920040	.2098
19	.585	.9905985	.1057	59	.839	.9920403	.2125
20	978.641	2.9906234	32.1076	60	981.918	2.9920752	32.2151
21	.701	.9906500	.1095	61	.995	.9921073	.2176
22	.763	.9906775	.1116	62	982.070	.9921424	.2201
23	.825	.9907050	.1136	63	.145	.9921756	.2225
24	.892	.9907348	.1158	64	.218	.9922079	.2249
25	978.960	2.9907649	32.1180	65	982.288	2.9922388	32.2272
26	979.030	.9907960	.1203	66	.356	.9922689	.2295
27	.101	.9908275	.1227	67	.422	.9922981	.2316
28	.175	.9908603	.1251	68	.487	.9923268	.2338
29	.251	.9908940	.1276	69	.549	.9923542	.2358
30	979.329	2.9909286	32.1302	70	982.608	2.9923803	32.2377
31	.407	.9909632	.1327	71	.665	.9924055	.2396
32	.487	.9909987	.1353	72	.720	.9924298	.2414
33	.569	.9910350	.1380	73	.772	.9924528	.2431
34	.652	.9910718	.1407	74	.822	.9924749	.2448
35	979.737	2.9911095	32.1435	75	982.868	2.9924952	32.2463
36	.822	.9911472	.1463	76	.912	.9925147	.2477
37	.908	.9911853	.1491	77	.954	.9925332	.2491
38	.995	.9912238	.1520	78	.992	.9925500	.2503
39	980.083	.9912628	.1549	79	983.027	.9925655	.2515
40	980.171	2.9913018	32.1578	80	983.059	2.9925796	32.2525
41	.261	.9913417	.1607	81	.089	.9925929	.2535
42	.350	.9913812	.1636	82	.115	.9926043	.2544
43	.440	.9914210	.1666	83	.139	.9926149	.2552
44	.531	.9914613	.1696	84	.160	.9926242	.2558
45	980.621	2.9915011	32.1725	85	983.178	2.9926321	32.2564
46	.711	.9915410	.1755	86	.191	.9926379	.2569
47	.802	.9915814	.1785	87	.203	.9926432	.2572
48	.892	.9916212	.1814	88	.211	.9926467	.2575
49	.981	.9916606	.1844	90	983.217	.9926494	.2577

To reduce log g (cm. per sec. per sec.) to log g (ft. per sec. per sec.) add log 0.03280833 = 8.5159842 − 10.

The standard value of gravity, used in barometer reductions, etc., is 980.665. It was adopted by the International Committee on Weights and Measures in 1901. It corresponds nearly to latitude 45° and sea-level.

FREE-AIR CORRECTION FOR ALTITUDE.

−0.0003086 cm/sec²/m when altitude is in meters.
−0.000003086 ft/sec²/ft when altitude is in feet.

Altitude.	Correction.	Altitude.	Correction.
200 m.	−0.0617 cm/sec²	200 ft.	−0.000617 ft./sec²
300	.0926	300	.000926
400	.1234	400	.001234
500	.1543	500	.001543
600	.1852	600	.001852
700	.2160	700	.002160
800	.2469	800	.002469
900	.2777	900	.002777

GRAVITY.

The following more recent gravity determinations (Potsdam System) serve to show the accuracy which may be assumed for the values in Table 565, except for the three stations in the Arctic Ocean. The error in the observed gravity is probably not greater than 0.010 cm/sec^2, as the observations were made with the half-second invariable pendulum, using modern methods.

In recent years the Coast and Geodetic Survey has corrected the computed value of gravity for the effect of material above sea-level, the deficiency of matter in the oceans, the deficiency of density in the material below sea-level under the continents and the excess of density in the earth's crust under the ocean, in addition to the reduction for elevation. Such corrections make the computed values agree more closely with those observed. See special publication No. 40 of the U. S. Coast and Geodetic Survey entitled, "Investigations of Gravity and Isostasy," by William Bowie, 1917; also Special Publication No. 10 of same bureau entitled, "Effect of Topography and Isostatic Compensation upon the Intensity of Gravity," by J. F. Hayford and William Bowie, 1912.

Name.	Latitude.	Elevation, meters.	Gravity, cm/sec^2		Reference.
			Observed.	Reduced to sea-level.	
Kodaikanal, India	10° 14′	2336	977.645	978.366	1
Ootacamund, India	11 25	2254	977.735	978.427	2
Madras, India	13 4	6	978.279	978.281	2
Jamestown, St. Helena	−15 55	10	978.712	978.715	2
Cuttack, India	20 29	28	978.659	978.668	2
Amraoti, India	20 56	342	978.609	978.714	2
Jubbulpur, India	23 9	447	978.719	978.856	2
Gaya, India	24 48	110	978.884	978.918	2
Siliguri, India	26 42	118	978.887	978.923	2
Kuhrja, India	28 14	198	979.082	979.143	2
Galveston, Texas	29 18	3	979.272	979.273	2
Rajpur, India	30 24	1012	979.002	979.313	2
Alexandria, La	31 19	24	979.429	979.436	2
St. Georges, Bermuda	32 21	2	979.806	979.807	2
McCormick, S. C	33 55	163	979.624	979.674	2
Shamrock, Texas	35 13	708	979.577	979.795	2
Cloudland, Tenn	36 6	1890	979.383	979.966	2
Mount Hamilton, Cal	37 20	1282	979.660	980.056	2
Kala-i-Chumb, Turkestan	38 27	1345	979.462	979.877	2
Denver, Col	39 41	1638	979.609	980.114	2
Hachinohe, Japan	40 31	21	980.359	980.365	2
Chicago, Ill	41 47	182	980.278	980.334	2
Albany, N. Y	42 39	61	980.344	980.363	2
Florence, Italy	43 45	184	980.491	980.548	2
Minneapolis, Minn	44 59	256	980.597	980.676	2
Simplon Hospice, Switzerland	46 15	1998	980.202	980.819	2
Fort Kent, Me	47 15	160	980.765	980.814	2
Sandpoint, Idaho	48 16	637	980.680	980.877	2
Medicine Hat, Canada	50 2	664	980.865	981.070	2
Field, Canada	51 24	1239	980.745	981.127	2
Magleby, Denmark	54 47	14	981.502	981.506	1
Copenhagen, Denmark	55 41	14	981.559	981.563	1
St. Paul Island, Alaska	57 7	10	981.726	981.729	2
Fredericksvarn, Norway	59 0	10	981.874	981.877	1
Christiania, Norway	59 55	28	981.927	981.936	1
Ashe Inlet, Hudson Strait	62 33	15	982.105	982.110	3
St. Michael, Alaska	63 28	1	982.192	982.192	2
Hatnarfjordr, Iceland	64 3	4	982.266	982.267	1
Niantilik, Cumberland Sound	64 54	7	982.273	982.275	3
Glaesibaer, Iceland	65 46	10	982.342	982.345	1
Sorvagen, Norway	67 54	19	982.622	982.628	2
Umanak, Greenland	70 40	10	982.590	982.593	3
Danes Island, Spitzbergen	79 46	3	983.078	983.079	1
Arctic Sea	84 12	0	983.109	983.109	1
Arctic Sea	84 52	0	983.174	983.174	1
Arctic Sea	85 55	0	983.155	983.155	1

References: (1) Report 16th General Conference International Geodetic Association, London and Cambridge, 1909, 3d Vol. by Dr. E. Borrass, 1911; (2) U. S. Coast and Geodetic Survey, Special Publ. No. 40;* (3) U. S. Coast and Geodetic Survey, Report for 1897, Appendix 6.*

* For references (2) and (3), values were derived from comparative experiments with invariable pendulums, the value for Washington being taken as 980.112. For the latter, Appendix 5 of the Coast and Geodetic Survey Report for 1901, and pages 25 and 244 of the 3d vol. by Dr. E. Borrass in 1911 of the Report of the 16th General Conference of the Intern. Geodetic Association, London and Cambridge, 1909. As a result of the adjustment of the net of gravity base stations throughout the world by the Central Bureau of the Intern. Geodetic Association, the value of the Washington base station was changed to 980.112.

ACCELERATION OF GRAVITY (*g*) IN THE UNITED STATES.

The following table is abridged from one for 219 stations given on pp. 50 to 52, Special Publication No. 40, U. S. Coast and Geodetic Survey. The observed values depend on relative determinations and on adopted value of 980.112 for Washington (Coast and Geodetic Survey Office, see footnote, Table 566). There are also given terms necessary in reducing the theoretical value (Table 565) to the proper elevation (free-air) and to allow for topography and isostatic compensation by the Hayford method (see introductory note to Table 566).

To a certain extent, the greater the bulk of material below any station, the less its average density. This phenomenon is known as isostatic compensation. The depth below sea-level to which this compensation extends is about 96 km. Below this depth any mass element is subject to equal (fluid) pressure from all directions.

Station.	Latitude.	Longitude.	Elevation, meters.	Observed *g* cm/sec²	Correction.	
					Elevation, cm/sec²	Topography and compensation, cm/sec²
Key West, Fla.	24° 33.6′	81° 48.4′	1	978.970	0.000	+0.035
New Orleans, La.	29 57.0	90 4.2	2	979.324	−.001	+.013
Austin, Tex. university	30 17.2	97 44.2	189	979.283	−.058	−.001
El Paso, Tex.	31 46.3	106 29.0	1146	979.124	−.354	+.001
Yuma, Ariz.	32 43.3	114 37.0	54	979.529	−.017	−.010
Charleston, S. C.	32 47.2	79 56.0	6	979.546	−.002	+.016
Birmingham, Ala.	33 30.8	86 48.8	179	979.536	−.055	+.011
Arkansas City, Ark.	33 36.5	91 12.2	44	979.600	−.014	+.005
Atlanta, Ga. capitol	33 45.0	84 23.3	324	979.524	−.100	+.014
Beaufort, N. C.	34 43.1	76 39.8	1	979.729	−.000	+.036
Little Rock, Ark.	34 45.0	92 16.4	89	979.721	−.027	+.001
Memphis, Tenn.	35 8.8	90 3.3	80	979.740	−.025	+.002
Charlotte, N. C.	35 13.8	80 50.8	228	979.727	−.070	+.015
Las Vegas, N. Mex.	35 35.8	105 12.1	1960	979.204	−.605	+.017
Knoxville, Tenn.	35 57.7	83 55.	280	979.712	−.086	−.001
Grand Canyon, Ariz.	36 5.3	112 6.8	849	979.463	−.262	−.096
Cloudland, Tenn.	36 6.2	82 7.9	1890	979.383	−.583	+.130
Mount Hamilton, Cal., Obs'y.	37 20.4	121 38.6	1282	979.660	−.396	+.120
Richmond, Va.	37 32.2	77 26.1	30	979.960	−.009	+.010
San Francisco, Cal.	37 47.5	122 25.7	114	979.965	−.035	+.045
St. Louis, Mo., university	38 38.0	90 12.2	154	980.001	−.048	+.001
Pike's Peak, Col.	38 50.3	105 2.0	4293	978.954	−1.325	+.187
Colorado Springs, Col.	38 50.7	104 49.0	1841	979.490	−.568	−.007
Washington, D. C., Bur. St'ds.	38 56.3	77 4.0	103	980.095	−.032	+.012
Wallace, Kans.	38 54.7	101 35.4	1005	979.755	−.310	.000
Green River, Utah	38 59.4	110 9.9	1243	979.636	−.384	−.043
Cincinnati, Ohio, obs'y	39 8.3	84 25.3	245	980.004	−.076	+.002
Baltimore, Md., university	39 17.8	76 37.3	30	980.097	−.009	+.006
Terre Haute, Ind.	39 28.7	87 23.8	151	980.072	−.047	+.001
Denver, Col., university obs'y.	39 40.6	104 56.9	1638	979.609	−.505	−.015
Philadelphia, Pa., university	39 57.1	75 11.7	16	980.196	−.005	+.009
Wheeling, W. Va.	40 4.0	80 43.4	205	980.085	−.063	−.003
Princeton, N. J.	40 21.0	74 39.5	64	980.178	−.020	+.013
Pittsburg, Pa.	40 27.4	80 0.6	235	980.118	−.073	.000
Salt Lake City, Utah	40 46.1	111 53.8	1322	979.803	−.408	−.041
New York, N. Y., university	40 48.5	73 57.7	38	980.267	−.012	+.011
Winnemucca, Nev.	40 58.4	117 43.8	1311	979.844	−.404	−.004
Cleveland, Ohio	41 30.4	81 36.6	210	980.241	−.065	.000
Chicago, Ill., university	41 47.4	87 36.1	182	980.278	−.056	+.007
Worcester, Mass.	42 16.5	71 48.5	170	980.324	−.052	+.018
Cambridge, Mass. observatory	42 22.8	71 7.8	14	980.398	−.004	+.010
Ithaca, N. Y., university	42 27.1	76 29.0	247	980.300	−.076	+.005
Fort Dodge, Iowa	42 30.8	94 11.4	340	980.311	−.105	+.002
Grand Rapids, Mich.	42 58.0	85 40.8	236	980.372	−.073	+.003
Madison, Wis., university	43 4.6	89 24.0	270	980.365	−.083	+.003
Boise, Idaho	43 37.2	116 12.3	821	980.212	−.253	−.042
Mitchell, S. Dak. university	43 41.8	98 1.8	408	980.375	−.126	−.006
Lancaster, N. H.	44 29.5	71 34.3	261	980.486	−.081	+.007
Grand Canyon, Wyo.	44 43.3	110 29.7	2386	979.899	−.736	+.038
Minneapolis, Minn.	44 58.7	93 13.9	256	980.597	−.079	−.005
Calais, Me.	45 11.2	67 16.9	38	980.631	−.012	+.010
Miles City, Mont.	46 24.2	105 50.	718	980.539	−.222	−.020
Seattle, Wash. university	47 39.6	122 18.3	58	980.733	−.018	−.020
Pembina, N. Dak.	48 58.1	97 14.9	243	980.917	−.075	−.009

TABLE 568. — Length of Seconds Pendulum at Sea Level and for Different Latitudes.

Latitude	Length in cm	Log.	Length in inches.	Log.	Latitude	Length in cm	Log.	Length in inches.	Log.
0	99.0961	1.996056	39.0141	1.591222	50	99.4033	1.997401	39.1351	1.592566
5	.1000	.996074	.0157	.591239	55	.4475	.997594	.1525	.592760
10	.1119	.996126	.0204	.591292	60	.4891	.997776	.1689	.592941
15	.1310	.996210	.0279	.591375	65	.5266	.997939	.1836	.593104
20	.1571	.996324	.0382	.591490	70	.5590	.998081	.1964	.593246
25	99.1894	1.996465	39.0509	1.591631	75	99.5854	1.998196	39.2068	1.593361
30	.2268	.996629	.0656	.591794	80	.6047	.998280	.2144	.593446
35	.2681	.996810	.0819	.591976	85	.6168	.998332	.2191	.593498
40	.3121	.997002	.0992	.592168	90	.6207	.998350	.2207	.593515
45	.3577	.997201	.1171	.592367	—	—	—	—	—

Calculated from Table 565 by the formula $l = g/\pi^2$. For each 100 ft. of elevation subtract 0.000953 cm or 0.000375 in. or 0.0000313 ft. This table could also have been computed by either of the following formulae derived from the gravity formula at the top of Table 565.

$$l = 0.990961(1 + 0.005294 \sin^2\phi - 0.000007 \sin^2 2\phi) \text{ meters}$$
$$l = 0.990961 + 0.005246 \sin^2\phi - 0.000007 \sin^2 2\phi \text{ meters}$$
$$l = 39.014135(1 + 0.005294 \sin^2\phi - 0.000007 \sin^2 2\phi) \text{ inches.}$$
$$l = 39.014135 + 0.206535 \sin^2\phi - 0.000276 \sin^2 2\phi \text{ inches.}$$

TABLE 569. — Miscellaneous Geodetic Data.

	Clarke Spheroid.	U. S. C. & G. Survey.
Equatorial radius $= a$	$= 6378206$ meters; 3963.225 miles.	6378388 ± 18 meters; 3963.339 miles.
Polar semi-diameter $= b$	$= 6356584$ meters; 3949.790 miles.	6356909 meters; 3949.992 miles.
Reciprocal of flattening $= \dfrac{a}{a-b}$	$= 295.0$	297.0 ± 0.5
Square of eccentricity $= e^2 = \dfrac{a^2 - b^2}{a^2}$	$= 0.006768658$	0.0067237 ± 0.0000120

Difference between geographical and geocentric latitude $= \phi - \phi' =$
$688.2242'' \sin 2\phi - 1.1482'' \sin 4\phi + 0.0026'' \sin 6\phi$.

Mean density of the earth = 5.5247 ± 0.0013 (Burgess Phys. Rev. 1902).

Continental surface density of the earth = 2.67
Mean density outer ten miles of earth's crust = 2.40 } Harkness. See also page 423.

Constant of gravity, 6.66×10^{-8} c.g.s. units. Mass $= 5.997 \times 10^{27}$ c.g.s. units.

Rigidity $= n = 8.6 \times 10^{11}$ c.g.s. units.
Viscosity $= e = 10.0 \times 10^{16}$ c.g.s. units (comparable to steel). } A. A. Michelson, Astrophysical Journal, 39, p. 105, 1914.

Moments of inertia of the earth; the principal moments being taken as A, B, and C, and C the greatest:

$$\frac{C - A}{C} = 0.00326521 = \frac{1}{306.259};$$
$$C - A = 0.001064767\, Ea^2;$$
$$A = B = 0.325029\, Ea^2;$$
$$C = 0.326094\, Ea^2;$$

where E is the mass of the earth and a its equatorial semi-diameter.

Velocity of Compressional Earthquake Waves and Elastic Constants of Rocks.

Depth	Velocity*	Velocity†	Poisson's Ratio.†
0 km	7.4 km/sec.	7.17 km/sec.	0.258
100	7.76	7.59	.273
300	8.57	8.42	.272
600	—	9.62	.274
1000	11.24	11.07	.270
1400	12.06	12.40	—
2500	12.95	12.95	—

* Adams and Williamson. † Weichert.

TABLE 570.

TERRESTRIAL MAGNETISM.

Secular Change of Declination.

Changes in the magnetic declination between 1810, the date of the earliest available observations, and 1920. Based on tables in "Distribution of the Magnetic Declination in Alaska and Adjacent Regions in 1910" and "Distribution of the Magnetic Declination in the United States for January 1, 1915," published by the United States Coast and Geodetic Survey. For a somewhat different set of stations, see 6th Revised Edition of the Smithsonian Physical Tables.

State.	Station.	1810	1820	1830	1840	1850	1860	1870	1880	1890	1900	1910	1920
		°	°	°	°	°	°	°	°	°	°	°	°
Ala.	Ashland	6.0 E	6.2 E	6.1 E	5.9 E	5.6 E	5.2 E	4.7 E	4.1 E	3.4 E	3.0 E	2.9 E	3.0 E
	Tuscaloosa	7.1 E	7.3 E	7.3 E	7.2 E	6.9 E	6.6 E	6.1 E	5.5 E	4.8 E	4.4 E	4.4 E	4.6 E
Alas.	Sitka	—	—	—	—	—	28.7 E	29.0 E	29.3 E	29.5 E	29.7 E	30.2 E	30.4 E
	Kodiak	—	—	—	—	—	26.2 E	25.7 E	25.2 E	24.8 E	24.5 E	24.2 E	24.2 E
	Unalaska	—	—	—	—	—	20.4 E	20.1 E	19.6 E	19.0 E	18.3 E	17.5 E	17.2 E
	St. Michael	—	—	—	—	—	—	—	24.7 E	23.1 E	22.1 E	21.5 E	21.0 E
Ariz.	Holbrook	—	—	—	—	13.5 E	13.7 E	13.8 E	13.6 E	13.4 E	13.5 E	14.1 E	14.5 E
	Prescott	—	—	—	—	13.3 E	13.6 E	13.7 E	13.7 E	13.6 E	13.7 E	14.4 E	14.9 E
Ark.	Augusta	7.7 E	7.9 E	8.0 E	8.0 E	7.8 E	7.5 E	7.1 E	6.5 E	5.9 E	5.5 E	5.6 E	5.8 E
	Danville	—	—	9.3 E	9.3 E	9.2 E	9.0 E	8.6 E	8.1 E	7.6 E	7.2 E	7.4 E	7.7 E
Cal.	Bagdad	—	—	13.1 E	13.5 E	13.9 E	14.1 E	14.3 E	14.4 E	14.4 E	14.6 E	15.3 E	15.7 E
	Mojave	12.4 E	12.9 E	13.4 E	13.8 E	14.2 E	14.4 E	14.6 E	14.9 E	14.9 E	15.1 E	15.8 E	16.3 E
	Modesto	13.8 E	14.2 E	14.7 E	15.1 E	15.5 E	15.8 E	16.1 E	16.1 E	16.2 E	16.6 E	17.3 E	17.7 E
	Redding	15.6 E	16.1 E	16.6 E	17.0 E	17.4 E	17.8 E	18.1 E	18.2 E	18.3 E	18.7 E	19.4 E	19.7 E
Colo.	Pueblo	—	—	—	—	13.7 E	13.8 E	13.7 E	13.5 E	13.0 E	12.8 E	13.3 E	13.7 E
	Ouray	—	—	—	—	15.0 E	15.2 E	15.2 E	15.0 E	14.6 E	14.6 E	15.1 E	15.5 E
Conn.	Hartford	5.1 W	5.5 W	6.1 W	6.8 W	7.5 W	8.1 W	8.7 W	9.4 W	9.8 W	10.4 W	11.2 W	12.1 W
Del.	Dover	1.6 W	1.9 W	2.3 W	2.8 W	3.4 W	4.0 W	4.7 W	5.3 W	5.9 W	6.5 W	7.2 W	8.0 W
D. C.	Washington	0.5 E	0.3 E	0.0	0.5 W	1.0 W	1.7 W	2.4 W	3.0 W	3.6 W	4.2 W	4.9 W	5.6 W
Fla.	Miami	5.8 E	5.7 E	5.3 E	4.9 E	4.4 E	3.9 E	3.3 E	2.7 E	2.2 E	1.7 E	1.5 E	1.5 E
	Bartow	5.5 E	5.4 E	5.2 E	4.8 E	4.4 E	3.8 E	3.2 E	2.6 E	2.1 E	1.6 E	1.4 E	1.3 E
	Jacksonville	5.0 E	5.0 E	4.9 E	4.6 E	4.2 E	3.6 E	3.0 E	2.4 E	1.8 E	1.3 E	1.1 E	0.9 E
	Tallahassee	5.8 E	5.8 E	5.7 E	5.5 E	5.2 E	4.8 E	4.2 E	3.6 E	3.0 E	2.5 E	2.4 E	2.4 E
Ga.	Millen	4.9 E	4.8 E	4.6 E	4.3 E	3.9 E	3.4 E	2.7 E	2.1 E	1.5 E	0.9 E	0.7 E	0.5 E
	Americus	5.9 E	6.0 E	5.9 E	5.6 E	5.2 E	4.7 E	4.1 E	3.5 E	2.9 E	2.4 E	2.2 E	2.2 E
Haw.	Honolulu	—	—	—	—	9.4 E	9.4 E	9.5 E	9.8 E	10.1 E	10.4 E	10.7 E	11.1 E
Idaho	Pocatello	—	—	—	—	17.7 E	17.9 E	18.0 E	17.9 E	17.8 E	17.9 E	18.5 E	18.8 E
	Boise	—	—	—	—	18.0 E	18.5 E	18.8 E	18.8 E	18.6 E	18.8 E	19.5 E	19.8 E
	Pierce	—	—	—	20.2 E	20.6 E	21.0 E	21.2 E	21.1 E	21.2 E	21.4 E	22.0 E	22.2 E
Ill.	Kankakee	6.6 E	6.8 E	6.8 E	6.6 E	6.3 E	5.8 E	5.3 E	4.8 E	4.1 E	3.5 E	3.3 E	3.1 E
	Rushville	7.7 E	8.0 E	8.1 E	8.0 E	7.8 E	7.4 E	7.0 E	6.4 E	5.7 E	5.2 E	5.1 E	5.1 E
Ind.	Indianapolis	5.0 E	5.1 E	5.0 E	4.7 E	4.3 E	3.8 E	3.3 E	2.7 E	2.1 E	1.5 E	1.1 E	0.9 E
Iowa	Walker	—	8.9 E	9.1 E	9.1 E	8.9 E	8.6 E	8.2 E	7.5 E	6.8 E	6.2 E	6.2 E	6.2 E
	Sac City	—	10.4 E	10.7 E	10.8 E	10.8 E	10.5 E	10.2 E	9.6 E	8.8 E	8.4 E	8.6 E	8.6 E
Kans.	Emporia	—	—	—	—	11.5 E	11.4 E	11.2 E	10.8 E	10.2 E	9.9 E	10.1 E	10.3 E
	Ness City	—	—	—	—	12.4 E	12.4 E	12.2 E	11.9 E	11.3 E	11.2 E	11.4 E	11.7 E
Ky.	Manchester	3.5 E	3.6 E	3.4 E	3.1 E	2.8 E	2.2 E	1.6 E	1.0 E	0.3 E	0.3 W	0.6 W	0.8 W
	Louisville	4.8 E	4.9 E	4.8 E	4.6 E	4.3 E	3.8 E	3.2 E	2.5 E	1.9 E	1.5 E	1.3 E	1.2 E
	Princeton	6.8 E	6.9 E	6.9 E	6.8 E	6.5 E	6.0 E	5.5 E	4.8 E	4.2 E	3.9 E	3.7 E	3.8 E
La.	Winfield	8.6 E	8.9 E	9.0 E	9.0 E	8.9 E	8.6 E	8.2 E	7.6 E	7.1 E	6.8 E	7.0 E	7.4 E
Me.	Eastport	13.9 W	14.7 W	15.5 W	16.3 W	17.2 W	18.0 W	18.5 W	18.8 W	19.0 W	19.3 W	20.0 W	21.0 W
	Bangor	11.8 W	12.4 W	13.2 W	13.9 W	14.7 W	15.4 W	15.9 W	16.4 W	16.7 W	17.1 W	17.8 W	18.8 W
	Portland	9.3 W	9.9 W	10.6 W	11.2 W	11.9 W	12.6 W	13.1 W	13.6 W	14.1 W	14.5 W	15.3 W	16.3 W
Md.	Baltimore	0.9 W	1.1 W	1.4 W	1.9 W	2.4 W	3.1 W	3.8 W	4.4 W	5.0 W	5.6 W	6.3 W	7.0 W
Mass.	Boston	7.3 W	7.8 W	8.4 W	9.1 W	9.8 W	10.5 W	11.0 W	11.5 W	12.0 W	12.6 W	13.4 W	14.4 W
	Pittsfield	5.7 W	6.2 W	6.7 W	7.4 W	8.1 W	8.7 W	9.3 W	10.0 W	10.4 W	11.0 W	11.8 W	12.7 W
Mich.	Marquette	—	6.7 E	6.7 E	6.5 E	6.1 E	5.5 E	4.7 E	3.8 E	3.0 E	2.4 E	2.1 E	1.7 E
	Lapeer	—	2.6 E	2.4 E	2.1 E	1.6 E	1.0 E	0.3 E	0.5 W	1.2 W	1.8 W	2.3 W	2.8 W
	Grand Haven	—	5.1 E	5.0 E	4.8 E	4.4 E	3.8 E	3.1 E	2.4 E	1.6 E	1.1 E	0.7 E	0.3 E
Minn.	St. Paul	—	11.6 E	11.8 E	11.9 E	11.7 E	11.4 E	10.9 E	10.3 E	9.5 E	8.9 E	8.8 E	8.7 E
	Marshall	—	—	—	11.7 E	11.6 E	11.4 E	11.0 E	10.5 E	9.8 E	9.3 E	9.4 E	9.4 E
	Hibbing	—	10.5 E	10.7 E	10.8 E	10.6 E	10.3 E	9.7 E	9.0 E	8.2 E	7.6 E	7.7 E	7.5 E
	Bagley	—	—	13.0 E	13.1 E	13.1 E	12.8 E	12.3 E	11.7 E	11.0 E	10.4 E	10.6 E	10.5 E
Miss.	Meridian	7.3 E	7.4 E	7.5 E	7.4 E	7.2 E	6.9 E	6.5 E	5.9 E	5.2 E	4.8 E	4.9 E	5.1 E
	Vicksburg	8.2 E	8.4 E	8.5 E	8.4 E	8.2 E	8.0 E	7.6 E	7.1 E	6.4 E	6.0 E	6.1 E	6.4 E

TERRESTRIAL MAGNETISM (*continued*).

Secular Change of Declination (*concluded*).

State.	Station.	1810	1820	1830	1840	1850	1860	1870	1880	1890	1900	1910	1920
Mo.	Hermann.....	—	9.2 E	9.3 E	9.2 E	9.0 E	8.7 E	8.3 E	7.7 E	7.0 E	6.5 E	6.5 E	6.6 E
	Sedalia.......	—	9.9 E	10.0 E	10.0 E	9.9 E	9.6 E	9.3 E	8.7 E	8.0 E	7.6 E	7.8 E	8.0 E
Mont.	Miles City....	—	—	—	—	17.6 E	17.8 E	17.7 E	17.4 E	16.9 E	16.9 E	17.3 E	17.6 E
	Lewistown....	—	—	—	19.5 E	19.8 E	20.1 E	20.1 E	19.9 E	19.6 E	19.6 E	20.1 E	20.4 E
	Ovando......	—	—	—	20.4 E	20.8 E	21.1 E	21.2 E	21.1 E	20.9 E	21.1 E	21.6 E	22.0 E
Nebr.	Albion.......	—	12.4 E	12.7 E	12.9 E	12.9 E	12.8 E	12.5 E	12.0 E	11.4 E	11.0 E	11.2 E	11.5 E
	Valentine.....	—	—	—	—	14.1 E	14.1 E	13.9 E	13.4 E	12.8 E	12.6 E	12.8 E	13.1 E
	Alliance......	—	—	—	—	15.4 E	15.4 E	15.3 E	14.8 E	14.3 E	14.2 E	14.5 E	14.8 E
Nev.	Elko.........	—	—	—	—	17.3 E	17.6 E	17.7 E	17.7 E	17.6 E	17.8 E	18.4 E	18.9 E
	Hawthorne...	—	—	—	—	16.2 E	16.6 E	16.8 E	17.0 E	17.0 E	17.3 E	18.0 E	18.4 E
N. H.	Hanover.....	7.1 W	7.5 W	8.2 W	8.9 W	9.7 W	10.5 W	11.1 W	11.6 W	12.0 W	12.6 W	13.2 W	14.2 W
N. J.	Trenton......	2.8 W	3.1 W	3.5 W	4.1 W	4.7 W	5.4 W	6.0 W	6.7 W	7.2 W	7.8 W	8.6 W	9.4 W
N. M.	Santa Rosa...	—	—	—	—	12.7 E	12.8 E	12.7 E	12.4 E	12.0 E	11.9 E	12.5 E	12.9 E
	Laguna......	—	—	—	—	13.4 E	13.6 E	13.6 E	13.4 E	13.0 E	13.0 E	13.6 E	14.1 E
N. Y	Albany.......	5.7 W	5.9 W	6.4 W	7.0 W	7.8 W	8.5 W	9.2 W	10.0 W	10.3 W	10.9 W	11.6 W	12.5 W
	Elmira.......	2.2 W	2.4 W	2.8 W	3.3 W	4.0 W	4.8 W	5.4 W	6.3 W	7.0 W	7.5 W	8.2 W	9.0 W
	Buffalo.......	1.0 W	1.1 W	1.4 W	1.9 W	2.4 W	3.2 W	3.8 W	4.7 W	5.4 W	5.9 W	6.5 W	7.2 W
N. C.	Newbern.....	1.7 E	1.6 E	1.3 E	0.8 E	0.3 E	0.3 W	1.0 W	1.7 W	2.3 W	2.9 W	3.4 W	4.0 W
	Greensboro...	3.5 E	3.4 E	3.1 E	2.7 E	2.2 E	1.6 E	1.0 E	0.3 E	0.3 W	0.8 W	1.3 W	1.8 W
	Asheville.....	4.2 E	4.2 E	4.0 E	3.6 E	3.1 E	2.6 E	2.0 E	1.3 E	0.7 E	0.2 E	0.2 W	0.5 W
N. D.	Jamestown...	—	—	14.0 E	14.2 E	14.2 E	14.0 E	13.7 E	13.2 E	12.5 E	12.2 E	12.4 E	12.5 E
	Bismarck.....	—	—	—	—	16.4 E	16.3 E	16.1 E	15.6 E	15.0 E	14.7 E	15.0 E	15.2 E
	Dickinson....	—	—	—	—	17.7 E	17.7 E	17.5 E	17.1 E	16.5 E	16.3 E	16.7 E	16.9 E
Ohio	Canton.......	2.3 E	2.2 E	2.0 E	1.7 E	1.2 E	0.6 E	0.0	0.7 W	1.3 W	1.9 W	2.5 W	3.1 W
	Urbana.......	4.4 E	4.4 E	4.3 E	4.0 E	3.5 E	3.0 E	2.4 E	1.8 E	1.1 E	0.5 E	0.1 E	0.3 W
Okla.	Okmulgee....	—	—	—	—	10.2 E	10.1 E	9.8 E	9.5 E	9.1 E	8.7 E	8.9 E	9.2 E
	Enid.........	—	—	—	—	11.2 E	11.2 E	11.0 E	10.6 E	10.2 E	9.8 E	10.1 E	10.5 E
Ore.	Sumpter......	—	—	—	—	19.3 E	19.7 E	20.0 E	20.2 E	20.2 E	20.4 E	21.1 E	21.4 E
	Detroit.......	16.7 E	17.4 E	18.0 E	18.6 E	19.2 E	19.7 E	20.1 E	20.3 E	20.5 E	20.8 E	21.6 E	21.9 E
Pa.	Wilkes-Barre..	2.3 W	2.5 W	2.9 W	3.4 W	4.0 W	4.7 W	5.3 W	6.0 W	6.6 W	7.2 W	8.0 W	8.8 W
	Lockhaven....	1.4 W	1.5 W	1.9 W	2.4 W	3.0 W	3.6 W	4.3 W	5.0 W	5.6 W	6.3 W	7.0 W	7.7 W
	Indiana.......	0.6 E	0.5 E	0.3 E	0.1 W	0.7 W	1.3 W	2.0 W	2.6 W	3.3 W	3.9 W	4.6 W	5.2 W
P. R.	San Juan.....	—	—	—	—	—	—	—	—	—	1.0 W	2.0 W	3.4 W
R. I.	Newport.....	6.6 W	7.1 W	7.7 W	8.4 W	9.1 W	9.8 W	10.3 W	10.8 W	11.3 W	11.9 W	12.7 W	13.7 W
S. C.	Marion.......	3.4 E	3.3 E	3.0 E	2.6 E	2.1 E	1.6 E	0.9 E	0.3 E	0.4 W	1.0 W	1.4 W	1.8 W
	Aiken........	4.8 E	4.7 E	4.5 E	4.2 E	3.7 E	3.1 E	2.5 E	1.9 E	1.3 E	0.7 E	0.4 E	0.1 E
S. D.	Huron........	—	—	—	13.2 E	13.2 E	13.0 E	12.7 E	12.3 E	11.7 E	11.2 E	11.5 E	11.7 E
	Murdo.......	—	—	—	—	15.0 E	14.9 E	14.7 E	14.3 E	13.7 E	13.4 E	13.7 E	13.9 E
	Rapid City...	—	—	—	—	16.4 E	16.4 E	16.3 E	15.8 E	15.3 E	15.1 E	15.4 E	15.7 E
Tenn.	Knoxville.....	3.8 E	3.8 E	3.6 E	3.3 E	2.9 E	2.4 E	1.8 E	1.1 E	0.5 E	0.0	0.3 W	0.5 W
	Shelbyville....	6.4 E	6.5 E	6.4 E	6.2 E	5.9 E	5.5 E	4.9 E	4.3 E	3.7 E	3.2 E	3.0 E	2.9 E
	Huntingdon...	7.3 E	7.4 E	7.4 E	7.3 E	7.0 E	6.6 E	6.1 E	5.5 E	4.9 E	4.4 E	4.3 E	4.4 E
Tex.	Houston......	—	9.0 E	9.2 E	9.4 E	9.4 E	9.3 E	8.9 E	8.4 E	7.9 E	7.7 E	8.1 E	8.6 E
	San Antonio..	—	—	9.5 E	9.7 E	9.8 E	9.7 E	9.5 E	9.2 E	8.7 E	8.7 E	9.2 E	9.7 E
	Pecos........	—	—	10.7 E	11.0 E	11.1 E	11.1 E	11.0 E	10.8 E	10.4 E	10.3 E	10.8 E	11.3 E
	Wytheville....	2.9 E	2.9 E	2.7 E	2.4 E	2.0 E	1.4 E	0.8 E	0.1 E	0.5 W	1.1 W	1.5 W	1.9 W
Wash.	Wilson Creek..	—	—	—	—	21.2 E	21.6 E	21.8 E	21.9 E	22.1 E	22.4 E	23.0 E	23.3 E
	Seattle.......	18.9 E	19.5 E	20.1 E	20.7 E	21.2 E	21.6 E	22.0 E	22.2 E	22.4 E	22.8 E	23.5 E	23.8 E
W. Va.	Sutton.......	1.9 E	1.8 E	1.6 E	1.2 E	0.8 E	0.2 E	0.4 W	1.1 W	1.8 W	2.4 W	2.9 W	3.4 W
Wis.	Shawamo.....	—	7.4 E	7.4 E	7.3 E	7.0 E	6.5 E	5.9 E	5.0 E	4.3 E	3.7 E	3.4 E	3.1 E
	Floydada.....	—	—	—	—	11.2 E	11.3 E	11.2 E	10.9 E	10.4 E	10.3 E	10.7 E	11.1 E
Utah	Manti........	—	—	—	—	16.4 E	16.7 E	16.8 E	16.7 E	16.4 E	16.5 E	17.1 E	17.5 E
Vt.	Rutland......	6.6 W	7.1 W	7.6 W	8.3 W	9.1 W	9.8 W	10.5 W	11.2 W	11.6 W	12.1 W	12.8 W	13.8 W
Va.	Richmond.....	0.8 E	0.6 E	0.3 E	0.1 W	0.6 W	1.2 W	1.8 W	2.5 W	3.1 W	3.7 W	4.2 W	4.9 W
	Lynchburg....	1.6 E	1.5 E	1.3 E	0.9 E	0.5 E	0.1 W	0.7 W	1.4 W	2.0 W	2.6 W	3.1 W	3.7 W
	Stanley......	—	8.9 E	9.0 E	9.0 E	8.8 E	8.4 E	7.8 E	7.1 E	6.3 E	5.8 E	5.6 E	5.4 E
Wyo.	Douglas......	—	—	—	—	15.8 E	16.0 E	16.0 E	15.8 E	15.3 E	15.2 E	15.7 E	16.0 E
	Green River..	—	—	—	—	16.8 E	17.0 E	17.0 E	16.8 E	16.5 E	16.6 E	17.2 E	17.5 E

TERRESTRIAL MAGNETISM (*continued*).

TABLE 571. — Dip or Inclination.

This table gives for the epoch January 1, 1915, the values of the magnetic dip, *I*, corresponding to the longitudes west of Greenwich in the heading and the north latitudes in the first column.

λ / φ	65°	70°	75°	80°	85°	90°	95°	100°	105°	110°	115°	120°	125°
	°	°	°	°	°	°	°	°	°	°	°	°	°
19	—	—	50.4	49.4	48.5	47.2	46.1	45.1	44.1	—	—	—	—
21	—	—	52.7	51.9	51.1	50.1	48.9	47.9	46.9	—	—	—	—
23	—	—	55.1	54.2	53.7	52.8	51.7	50.4	49.7	48.7	—	—	—
25	—	—	57.6	56.8	56.1	55.2	54.2	53.1	52.2	51.2	50.1	—	—
27	—	—	59.8	59.3	58.3	57.6	56.6	55.6	54.6	53.6	52.4	—	—
29	—	—	61.9	61.3	60.5	59.7	58.9	57.9	56.8	55.8	54.6	53.8	—
31	—	63.6	63.8	63.4	62.8	62.0	61.1	60.1	59.0	58.1	57.0	55.8	—
33	—	65.4	65.6	65.3	64.7	64.0	63.1	62.4	61.2	60.2	59.1	58.0	—
35	—	67.2	67.3	67.2	66.6	66.1	65.3	64.3	63.2	62.2	61.0	60.1	—
37	—	69.1	69.2	69.0	68.9	68.1	67.3	66.4	65.2	64.2	63.1	62.1	—
39	—	70.6	70.8	70.6	70.6	70.0	69.2	68.3	67.3	66.2	64.9	63.9	62.5
41	—	72.2	72.3	72.5	72.2	71.7	71.0	70.1	69.0	68.0	66.6	65.5	64.3
43	—	73.6	74.0	74.1	74.0	73.5	72.6	71.8	70.7	69.7	68.4	67.2	65.9
45	74.3	74.9	75.4	75.5	75.5	75.2	74.5	73.5	72.4	71.3	70.2	69.0	67.8
47	75.6	76.3	76.8	76.9	76.9	77.0	76.1	75.1	74.2	72.9	71.7	70.5	69.5
49	76.5	77.4	78.2	78.5	78.5	78.3	77.7	76.7	75.7	74.5	73.2	72.1	71.2

TABLE 572. — Secular Change of Dip.

Values of the magnetic dip for places designated by the north latitudes and longitudes west of Greenwich in the first two columns for January 1 of the years in the heading. The degrees are given in the third column and the minutes in the suceeding columns.

Latitude.	Longitude.		1855	1860	1865	1870	1875	1880	1885	1890	1895	1900	1905	1910	1915
°	°	°	′	′	′	′	′	′	′	′	′	′	′	′	′
25	80	55+	32	32	31	29	26	23	18	18	22	31	43	73	108
25	110	49+	14	26	36	45	52	61	67	74	82	92	102	116	132
30	83	60+	66	70	73	74	73	67	57	51	53	63	78	101	126
30	100	57+	41	46	55	64	67	62	57	58	65	74	87	103	120
30	115	54+	47	56	63	65	64	66	69	73	79	85	90	96	102
35	80	66+	67	68	67	64	55	45	36	31	30	32	40	55	72
35	90	65+	67	61	53	46	39	34	28	27	27	29	38	51	66
35	105	62+	—	—	—	47	45	39	39	39	43	49	57	65	72
35	120	59+	56	59	61	61	60	59	61	64	66	66	66	66	66
40	75	71+	82	82	78	73	65	55	43	33	27	24	24	29	36
40	90	70+	30	31	34	37	36	32	29	26	25	26	30	38	48
40	105	67+	—	—	—	56	53	51	51	51	52	56	60	63	66
40	120	64+	—	—	—	51	52	54	57	58	58	54	50	45	42
45	65	74+	118	112	103	94	82	70	59	48	37	30	26	22	18
45	75	75+	91	87	83	78	73	61	50	41	31	26	24	24	24
45	90	74+	86	86	86	84	82	80	73	68	66	64	65	68	72
45	105	72+	—	—	—	—	—	30	28	27	26	26	25	25	24
45	122.5	68+	45	44	47	50	50	49	47	44	40	37	33	27	21
49	92	77+	80	79	78	76	74	74	69	66	65	63	60	58	60
49	120	72+	—	27	25	24	23	22	21	20	20	19	17	12	06

TERRESTRIAL MAGNETISM (*continued*).

TABLE 573. — Horizontal Intensity.

This table gives for the epoch January 1, 1915, the horizontal intensity, *H*, expressed in cgs units, corresponding to the longitudes in the heading and the latitudes in the first column.

λ / φ	65°	70°	75°	80°	85°	90°	95°	100°	105°	110°	115°	120°	125°
°													
19	—	—	.297	.303	.311	.316	.321	.325	.325	—	—	—	—
21	—	—	.290	.296	.303	.310	.315	.320	.320	—	—	—	—
23	—	—	.283	.288	.294	.301	.307	.311	.311	.311	—	—	—
25	—	—	.273	.281	.286	.292	.298	.302	.303	.303	.304	—	—
27	—	—	.264	.271	.276	.281	.288	.292	.295	.296	.297	—	—
29	—	—	.253	.258	.265	.272	.277	.283	.286	.287	.288	.288	—
31	—	.237	.242	.247	.254	.260	.266	.272	.276	.279	.280	.280	—
33	—	.225	.230	.236	.242	.248	.255	.259	.264	.270	.271	.272	—
35	—	.213	.217	.223	.232	.235	.241	.249	.251	.256	.260	.263	—
37	—	.202	.205	.210	.213	.222	.227	.234	.240	.244	.250	.253	—
39	—	.191	.193	.196	.200	.206	.212	.218	.226	.232	.237	.242	.245
41	—	.178	.178	.182	.185	.191	.197	.204	.212	.218	.226	.232	.236
43	—	.166	.166	.165	.171	.174	.182	.189	.198	.207	.214	.221	.227
45	.159	.154	.153	.153	.155	.160	.167	.174	.185	.192	.202	.210	.216
47	.146	.143	.139	.139	.141	.142	.150	.159	.168	.180	.187	.195	.202
49	.135	.130	.126	.123	.123	.129	.136	.144	.153	.164	.174	.182	.189

TABLE 574. — Secular Change of Horizontal Intensity.

Values of horizontal intensity, *H*, in cgs units for the places designated by the latitude and longitude in the first two columns for January 1 of the years in the heading.

Lat.	Long.	1860	1865	1870	1875	1880	1885	1890	1895	1900	1905	1910	1915
°	°												
25	80	.3086	.3073	.3057	.3042	.3025	.3008	.2990	.2970	.2949	.2917	.2870	.2810
25	110	.3216	.3202	.3187	.3168	.3153	.3141	.3128	.3115	.3102	.3088	.3063	.3030
30	83	.2775	.2768	.2760	.2752	.2743	.2732	.2720	.2705	.2686	.2658	.2614	.2560
30	100	—	.2978	.2959	.2941	.2924	.2908	.2894	.2882	.2867	.2847	.2817	.2780
30	115	.2996	.2981	.2966	.2949	.2934	.2922	.2910	.2899	.2890	.2880	.2863	.2840
35	80	.2367	.2362	.2357	.2355	.2351	.2347	.2340	.2335	.2325	.2306	.2272	.2230
35	90	—	—	.2460	.2460	.2459	.2456	.2453	.2445	.2435	.2418	.2387	.2350
35	105	—	—	—	.2619	.2607	.2598	.2589	.2582	.2572	.2559	.2537	.2510
35	120	—	—	.2727	.2714	.2702	.2690	.2679	.2670	.2663	.2657	.2645	.2630
40	75	.1876	.1884	.1895	.1904	.1912	.1918	.1923	.1924	.1921	.1911	.1889	.1860
40	90	.2080	.2076	.2073	.2070	.2069	.2068	.2066	.2062	.2054	.2042	.2019	.1990
40	105	—	—	.2269	.2263	.2258	.2254	.2250	.2245	.2237	.2227	.2210	.2190
40	120	—	—	.2439	.2430	.2422	.2416	.2409	.2402	.2396	.2390	.2381	.2370
45	65	.1504	.1515	.1527	.1543	.1557	.1568	.1579	.1590	.1598	.1600	.1596	.1590
45	75	.1487	.1490	.1497	.1508	.1518	.1529	.1540	.1548	.1552	.1552	.1543	.1530
45	90	.1648	.1646	.1644	.1641	.1639	.1637	.1636	.1637	.1636	.1633	.1620	.1600
45	105	—	—	.1895	.1894	.1893	.1891	.1888	.1885	.1881	.1875	.1864	.1850
45	122.5	.2183	.2175	.2166	.2158	.2148	.2140	.2134	.2130	.2128	.2128	.2125	.2120
49	92	.1336	.1334	.1330	.1327	.1325	.1324	.1324	.1327	.1330	.1336	.1330	.1320
49	120	.1846	.1845	.1844	.1841	.1836	.1831	.1826	.1824	.1825	.1825	.1823	.1820

TERRESTRIAL MAGNETISM (*continued*).

TABLE 575. — Total Intensity.

This table gives for the epoch January 1, 1915, the values of the total intensity, *F*, expressed in cgs units corresponding to the longitudes in the heading and the latitudes in the first column.

λ / φ	65°	70°	75°	80°	85°	90°	95°	100°	105°	110°	115°	120°	125°
°													
19	—	—	.466	.466	.469	.465	.463	.451	.453	—	—	—	—
21	—	—	.478	.480	.482	.483	.479	.477	.468	—	—	—	—
23	—	—	.495	.492	.497	.498	.495	.488	.481	.471	—	—	—
25	—	—	.509	.513	.513	.512	.510	.503	.494	.484	.474	—	—
27	—	—	.525	.531	.525	.524	.523	.517	.509	.499	.487	—	—
29	—	—	.537	.537	.538	.539	.536	.533	.522	.511	.497	.488	—
31	—	.533	.548	.552	.556	.554	.550	.546	.536	.528	.514	.498	—
33	—	.540	.557	.565	.566	.566	.564	.559	.548	.543	.528	.513	—
35	—	.550	.562	.576	.584	.580	.577	.574	.557	.549	.536	.528	—
37	—	.566	.577	.586	.592	.595	.588	.585	.572	.561	.552	.541	—
39	—	.575	.587	.590	.602	.602	.597	.590	.586	.575	.559	.550	.531
41	—	.582	.585	.605	.605	.608	.605	.599	.592	.582	.569	.559	.544
43	—	.588	.602	.602	.620	.613	.609	.605	.599	.597	.581	.570	.556
45	.588	.591	.607	.611	.619	.626	.625	.613	.612	.599	.596	.586	.572
47	.587	.604	.609	.613	.622	.631	.624	.618	.617	.612	.596	.584	.577
49	.578	.596	.616	.617	.617	.636	.638	.626	.619	.614	.602	.592	.587

TABLE 576. — Secular Change of Total Intensity.

Values of total intensity, *F*, in cgs units for places designated by the latitudes and longitudes in the first two columns for January 1 of the years in the heading.

Lat.	Long.	1855	1860	1865	1870	1875	1880	1885	1890	1895	1900	1905	1910	1915
°	°													
25	80	.5476	.5453	.5427	.5396	.5363	.5324	.5285	.5253	.5227	.5208	.5178	.5160	.5131
25	110	.4941	.4946	.4941	.4933	.4914	.4906	.4900	.4889	.4884	.4879	.4876	.4861	.4836
30	83	.5758	.5755	.5749	.5735	.5716	.5678	.5625	.5584	.5559	.5549	.5534	.5510	.5471
30	100	—	—	.5608	.5595	.5567	.5523	.5479	.5455	.5450	.5444	.5441	.5426	.5399
30	115	.5219	.5216	.5205	.5182	.5149	.5129	.5114	.5101	.5094	.5092	.5086	.5068	.5041
35	80	.6101	.6090	.6075	.6048	.6008	.5955	.5910	.5873	.5856	.5838	.5823	.5796	.5756
35	90	—	—	—	.5993	.5966	.5946	.5914	.5904	.5885	.5868	.5861	.5834	.5800
35	105	—	—	—	—	.5720	.5675	.5656	.5636	.5634	.5630	.5627	.5604	.5567
35	120	—	—	—	.5457	.5428	.5401	.5383	.5369	.5356	.5342	.5330	.5306	.5276
40	75	.6183	.6193	.6196	.6204	.6190	.6160	.6115	.6077	.6047	.6022	.5991	.5948	.5892
40	90	—	.6236	.6240	.6246	.6233	.6209	.6190	.6169	.6151	.6133	.6118	.6089	.6052
40	105	—	—	—	.6040	.6011	.5988	.5978	.5967	.5958	.5955	.5944	.5912	.5871
40	120	—	—	—	.5739	.5720	.5709	.5707	.5692	.5676	.5647	.5621	.5581	.5546
45	65	.6161	.6159	.6140	.6126	.6107	.6082	.6052	.6022	.5994	.5980	.5962	.5923	.5875
45	75	.6369	.6347	.6330	.6320	.6329	.6281	.6247	.6228	.6189	.6171	.6157	.6121	.6070
45	90	—	.6552	.6544	.6522	.6495	.6474	.6415	.6377	.6366	.6349	.6344	.6315	.6264
45	105	—	—	—	—	—	.6296	.6276	.6261	.6245	.6232	.6206	.6170	.6118
45	122.5	.6037	.6019	.6010	.6000	.5978	.5944	.5913	.5883	.5855	.5837	.5820	.5784	.5745
49	92	.6616	.6597	.6578	.6540	.6508	.6498	.6448	.6421	.6427	.6424	.6426	.6380	.6349
49	120	—	.6121	.6107	.6098	.6083	.6061	.6039	.6017	.6010	.6008	.5997	.5963	.5922

TERRESTRIAL MAGNETISM (*continued*).

TABLE 577. — Agonic Line.

The line of no declination appears to be still moving westward in the United States, but, as the line of no annual change is only a short distance to the west of it, it is probable that the extreme westerly position will soon be reached.

Lat. N.	Longitudes of the agonic line for the years					
	1800	1850	1875	1890	1905	1915
°	°	°	°	°	°	°
25	—	—	—	75.5	76.1	77.4
30	—	—	—	78.6	79.7	80.0
35	—	76.7	79.0	79.9	81.7	82.7
6	75.2	77.3	79.7	80.5	82.8	84.4
7	76.3	77.7	80.6	82.2	83.5	84.0
8	76.7	78.3	81.3	82.6	83.6	84.1
9	76.9	78.7	81.6	82.2	83.6	83.9
40	77.0	79.3	81.6	82.7	84.0	84.3
1	77.9	80.4	81.8	82.8	84.6	85.1
2	79.1	81.0	82.6	83.7	84.8	85.3
3	79.4	81.2	83.1	84.3	85.0	85.4
4	79.8	—	83.3	84.9	85.5	85.8
45	—	—	83.6	85.2	86.0	86.2
6	—	—	84.2	84.8	86.4	86.3
7	—	—	85.1	85.4	86.4	86.6
8	—	—	86.0	85.9	86.5	87.2
9	—	—	86.5	86.3	87.2	88.0

TABLE 578. — Mean Magnetic Character of Each Month in the Years 1906 to 1922.*

Means derived from daily magnetic characters based upon the following scale: 0, no disturbance; 1, moderate disturbance, and 2, large disturbance.

Year.	Jan.	Feb.	Mar.	Apr.	May.	June.	July.	Aug.	Sept.	Oct.	Nov.	Dec.	Year Mean.
1906	0.45	0.90	0.68	0.63	0.58	0.56	0.69	0.63	0.79	0.59	0.55	0.71	0.65
1907	0.69	0.83	0.58	0.55	0.72	0.67	0.67	0.66	0.68	0.71	0.61	0.53	0.66
1908	0.64	0.71	0.87	0.68	0.82	0.66	0.49	0.77	0.89	0.53	0.60	0.47	0.68
1909	0.76	0.63	0.79	0.49	0.59	0.54	0.53	0.65	0.70	0.69	0.49	0.58	0.62
1910	0.58	0.71	0.81	0.68	0.72	0.53	0.55	0.81	0.80	0.96	0.77	0.76	0.72
1911	0.78	0.89	0.78	0.76	0.70	0.53	0.61	0.53	0.50	0.59	0.49	0.45	0.63
1912	0.42	0.49	0.45	0.45	0.47	0.47	0.41	0.49	0.47	0.46	0.45	0.43	0.46
1913	0.51	0.53	0.53	0.54	0.45	0.45	0.42	0.46	0.58	0.57	0.42	0.36	0.48
1914	0.46	0.50	0.62	0.50	0.37	0.52	0.61	0.61	0.53	0.64	0.60	0.46	0.54
1915	0.53	0.64	0.68	0.61	0.58	0.61	0.47	0.60	0.59	0.77	0.82	0.54	0.62
1916	0.61	0.56	0.86	0.68	0.75	0.67	0.62	0.75	0.75	0.76	0.83	0.65	0.71
1917	0.81	0.69	0.59	0.63	0.66	0.55	0.61	0.85	0.61	0.74	0.53	0.72	0.67
1918	0.63	0.78	0.73	0.79	0.68	0.56	0.69	0.77	0.88	0.85	0.87	0.88	0.76
1919	0.78	0.81	0.89	0.70	0.82	0.55	0.54	0.70	0.83	0.91	0.52	0.56	0.72
1920	0.62	0.52	0.78	0.65	0.57	0.43	0.51	0.61	0.87	0.65	0.58	0.65	0.62
1921	0.54	0.51	0.68	0.67	0.83	0.55	0.54	0.58	0.50	0.63	0.62	0.65	0.76
1922	0.65	0.74	0.79	0.75	0.57	0.62	0.66	0.71	0.69	0.68	0.47	0.42	0.65
1923													
1924													

* Compiled from annual reviews of the "Caractère magnétique de chaque jour," prepared by the Royal Meteorological Institute of the Netherlands for the International Commission for Terrestrial Magnetism.

TABLE 579.

RECENT VALUES OF THE MAGNETIC ELEMENTS AT MAGNETIC OBSERVATORIES.

(Compiled by the Department of Terrestrial Magnetism, Carnegie Institution of Washington.)

Place.	Latitude.	Longitude.	Middle of year.	Magnetic elements.				
				Declination.	Inclination.	Intensity (cgs units).		
						Hor'l.	Ver'l.	Total.
	° ′	° ′		° ′	° ′			
Pavlovsk	59 41 N	30 29 E	1907	1 09.9 E	70 37.7 N	.1650	.4694	.4975
Sitka	57 03 N	135 20 W	1916	30 24.0 E	74 26.0 N	.1558	.5592	.5805
Katharinenburg	56 50 N	60 38 E	1907	10 35.5 E	70 52.2 N	.1762	.5081	.5378
Rude Skov	55 51 N	12 27 E	1915	8 44.3 W	68 50.6 N	.1726	.4459	.4781
Kasan	55 47 N	49 08 E	1912	8 09.1 E	69 17.3 N	.1802	.4765	.5094
Eskdalemuir	55 19 N	3 12 W	1913	17 54.9 W	69 37.3 N	.1682	.4528	.4831
Stonyhurst	53 51 N	2 28 W	1915	16 38.0 W	68 41.4 N	.1734	.4446	.4772
Wilhelmshaven	53 32 N	8 09 E	1911	11 28.2 W	67 30.7 N	.1811	.4375	.4735
Potsdam	52 23 N	13 04 E	1916	8 07.6 W	66 27.1 N	.1870	.4290	.4680
Seddin	52 17 N	13 01 E	1916	8 08.9 W	66 24.1 N	.1874	.4289	.4680
Irkutsk	52 16 N	104 16 E	1905	1 58.1 E	70 25.0 N	.2001	.5625	.5970
De Bilt	52 06 N	5 11 E	1914	12 22.6 W	66 46.5 N	.1851	.4314	.4694
Valencia	51 56 N	10 15 W	1913	20 19.6 W	68 09.2 N	.1789	.4463	.4808
Clausthal	51 48 N	10 20 E	1905	10 40.3 W	—	—	—	—
Bochum	51 29 N	7 14 E	1912	11 39.4 W	—	—	—	—
Kew	51 28 N	0 19 W	1915	15 18.4 W	66 56.6 N	.1846	.4338	.4714
Greenwich	51 28 N	0 00	1916	14 46.9 W	66 52.8 N	.1849	.4332	.4710
Uccle	50 48 N	4 21 E	1911	13 13.9 W	66 00.1 N	.1902	.4273	.4677
Hermsdorf	50 46 N	16 14 E	1913	6 58.2 W	—	—	—	—
Beuthen	50 21 N	18 55 E	1908	6 12.3 W	—	—	—	—
Falmouth	50 09 N	5 05 W	1912	17 24.2 W	66 26.6 N	.1880	.4312	.4704
Prague	50 05 N	14 25 E	1912	7 50.3 W	—	—	—	—
Cracow	50 04 N	19 58 E	1913	5 03.3 W	64 18.4 N	—	—	—
Val Joyeux	48 49 N	2 01 E	1913	13 59.2 W	64 38.9 N	.1974	.4167	.4611
Munich	48 09 N	11 37 E	1911	9 23.8 W	63 06.2 N	.2063	.4068	.4561
Kremsmünster	48 03 N	14 08 E	1904	9 02.4 W	—	—	—	—
O'Gyalla (Pesth)	47 53 N	18 12 E	1912	6 17.5 W	—	.2106	—	—
Odessa	46 26 N	30 46 E	1910	3 35.9 W	62 26.9 N	.2171	.4161	.4693
Pola	44 52 N	13 51 E	1915	7 39.0 W	60 05.1 N	.2217	.3853	.4445
Agincourt (Toronto)	43 47 N	79 16 W	1916	6 33.4 W	74 43.5 N	.1599	.5854	.6068
Perpignan	42 42 N	2 53 E	1910	12 44.8 W	—	—	—	—
Tiflis	41 43 N	44 48 E	1913	3 09.1 E	56 51.1 N	.2522	.3761	.4528
Capodimonte	40 52 N	14 15 E	1911	—	56 11.7 N	—	—	—
Ebro (Tortosa)	40 49 N	0 31 E	1914	12 51.6 W	57 47.5 N	.2330	.3698	.4371
Coimbra	40 12 N	8 25 W	1915	15 57.5 W	58 34.7 N	.2305	.3773	.4422
Baldwin *	38 47 N	95 10 W	1909	8 34.0 E	68 50.2 N	.2167	.5596	.6001
Cheltenham	38 44 N	76 50 W	1916	6 07.6 W	70 49.9 N	.1934	.5662	.5889
San Fernando	36 28 N	6 12 W	1913	14 51.7 W	54 26.6 N	.2494	.3489	.4289
Tokio	35 41 N	139 45 E	1912	5 03.4 W	48 53.7 N	.3000	.3438	.4563
Tucson	32 15 N	110 50 W	1916	13 44.4 E	59 26.1 N	.2706	.4582	.5322
Lukiapang **	31 19 N	121 02 E	1909	2 59.6 W	45 34.9 N	.3323	.3391	.4747
Dehra Dun	30 19 N	78 03 E	1914	2 18.8 E	44 22.9 N	.3316	.3246	.4641
Helwan	29 52 N	31 20 E	1913	2 17.0 W	40 47.6 N	.3003	.2592	.3967
Barrackpore †	22 46 N	88 22 E	1914	0 32.2 E	30 58.9 N	.3740	.2246	.4363
Hongkong	22 18 N	114 10 E	1916	0 13.8 W	30 51.8 N	.3716	.2220	.4328
Honolulu	21 19 N	158 04 W	1916	9 43.8 E	39 29.2 N	.2896	.2386	.3752
Toungoo	18 56 N	96 27 E	1914	0 02.6 E	23 06.1 N	.3898	.1663	.4238
Alibág	18 38 N	72 52 E	1915	0 40.6 E	24 21.1 N	.3687	.1669	.4047
Vieques	18 09 N	65 26 W	1916	3 19.4 W	50 56.7 N	.2815	.3470	.4468
Antipolo	14 36 N	121 10 E	1911	0 40.9 E	16 18.2 N	.3820	.1117	.3981
Kodaikánal	10 14 N	77 28 E	1914	1 17.1 W	4 11.2 N	.3757	.0275	.3767
Batavia-Buitenzorg	6 11 S	106 49 E	1912	0 47.3 E	31 19.4 S	.3668	.2232	.4229
St. Paul de Loanda	8 48 S	13 13 E	1910	16 12.3 W	35 32.2 S	.2012	.1437	.2473
Samoa (Apia)	13 48 S	171 46 W	1916	9 59.9 E	29 54.5 S	.3536	.2034	.4080
Tananarive	18 55 S	47 32 E	1907	9 29.7 W	54 05.7 S	.2533	.3499	.4319
Mauritius	20 06 S	57 33 E	1916	9 47.6 W	52 54.6 S	.2320	.3069	.3847
Pilar	31 40 S	63 53 W	1914	8 40.4 E	25 41.5 S	.2560	.1232	.2841
Santiago	33 27 S	70 42 W	1909	13 57.9 E	29 57.2 S	—	—	—
Christchurch	43 32 S	172 37 E	1914	16 44.8 E	67 59.8 S	.2241	.5546	.5982
New Year's Island	54 45 S ‡	64 03 W ‡	1906	15 41.6 E	50 03.6 S	.2717	.3244	.4231
Orcadas	60 45 S	42 32 W	1912	4 46.5 E	54 26.0 S	.2534	.3544	.4357

* Baldwin Obs'y replaced by Tucson Obs'y, Oct. 1909; mean given for Jan.–Oct. '09.
** Replaced Zi-ka-wei Obs'y, 1908. † Observations discontinued Apr. 26, 1915.
‡ Provisional values taken for position of Port Cork, p. 298, American Practical Navigator, 1914 edition.

APPENDIX.

DEFINITIONS OF UNITS.

ACTIVITY. Power or rate of doing work; unit, the watt.

AMPERE. Unit of electrical current. The international ampere, "which is one-tenth of the unit of current of the C. G. S. system of electro-magnetic units, and which is represented sufficiently well for practical use by the unvarying current which, when passed through a solution of nitrate of silver in water, and in accordance with accompanying specifications, deposits silver at the rate of 0.00111800 of a gram per second."

The ampere = 1 coulomb per second = 1 volt through 1 ohm = 10^{-1} E. M. U. = 3×10^{9} E. S. U.*

Amperes = volts/ohms = watts/volts = $(\text{watts/ohms})^{\frac{1}{2}}$.

Amperes $\times$ volts = amperes^2 $\times$ ohms = watts.

ANGSTROM. Unit of wave-length = 10^{-10} meter.

ATMOSPHERE. Unit of pressure.

English normal = 14.7 pounds per sq. in. = 29.929 in. = 760.18 mm Hg. 32° F.

French " = 760 mm of Hg. 0° C = 29.922 in. = 14.70 lbs. per sq. in.

BAR. A pressure of one dyne per cm.^{2} Meteorological "bar" = 10^6 dynes/cm^2.

BRITISH THERMAL UNIT. Heat required to raise one pound of water at its temperature of maximum density, 1° F. = 252 gram-calories.

CALORIE. Small calorie = gram-calorie = therm = quantity of heat required to raise one gram of water at its maximum density, one degree Centigrade.

Large Calorie = kilogram-calorie = 1000 small calories = one kilogram of water raised one degree Centigrade at the temperature of maximum density.

For conversion factors *see* page 197.

CANDLE, INTERNATIONAL. The international unit of candlepower maintained jointly by national laboratories of England, France and United States of America.

CARAT. The diamond carat standard in U. S. = 200 milligrams. Old standard = 205.3 milligrams = 3.168 grains.

The gold carat: pure gold is 24 carats; a carat is 1/24 part.

CIRCULAR AREA. The square of the diameter = 1.2733 $\times$ true area.

True area = 0.785398 $\times$ circular area.

COULOMB. Unit of quantity. The international coulomb is the quantity of electricity transferred by a current of one international ampere in one second. = 10^{-1} E. M. U. = 3×10^{9} E. S. U.

Coulombs = (volts-seconds)/ohms = amperes $\times$ seconds.

CUBIT = 18 inches.

DAY. Mean solar day = 1440 minutes = 86400 seconds = 1.0027379 sidereal day.

Sidereal day = 86164.10 mean solar seconds.

DIGIT. 3/4 inch; 1/12 the apparent diameter of the sun or moon.

DIOPTER. Unit of "power" of a lens. The number of diopters = the reciprocal of the focal length in meters.

DYNE. C. G. S. unit of force = that force which acting for one second on one gram produces a velocity of one cm per sec. = 1g $\div$ gravity acceleration in cm/sec./sec.

Dynes = wt. in g $\times$ acceleration of gravity in cm/sec./sec.

ELECTROCHEMICAL EQUIVALENT is the ratio of the mass in grams deposited in an electrolytic cell by an electrical current to the quantity of electricity.

ENERGY. *See* Erg.

ERG. C. G. S. unit of work and energy = one dyne acting through one centimeter.

For conversion factors *see* page 197.

FARAD. Unit of electrical capacity. The international farad is the capacity of a condenser charged to a potential of one international volt by one international coulomb of electricity = 10^{-9} E. M. U. = 9×10^{11} E. S. U.

The one-millionth part of a farad (microfarad) is more commonly used.

Farads = coulombs/volts.

* E. M. U. = C. G. S. electromagnetic units. E. S. U. = C. G. S. electrostatic units.

FOOT-POUND. The work which will raise one pound one foot high.
For conversion factors *see* page 197.
FOOT-POUNDALS. The English unit of work = foot-pounds/g.
For conversion factors *see* page 197.
g. The acceleration produced by gravity.
GAUSS. A unit of intensity of magnetic field = 1 E. M. U. = $\frac{1}{3} \times 10^{-10}$ E. S. U.
GRAM. *See* page 6.
GRAM-CENTIMETER. The gravitation unit of work = g. ergs.
GRAM-MOLECULE = x grams where x = molecular weight of substance.
GRAVITATION CONSTANT = G in formula $G\frac{m_1 m_2}{r^2} = 665.8 \times 10^{-10}$ cm.^{3}/gr. sec.^{2}
HEAT OF THE ELECTRIC CURRENT generated in a metallic circuit without self-induction is proportional to the quantity of electricity which has passed in coulombs multiplied by the fall of potential in volts, or is equal to (coulombs $\times$ volts) /4.181 in small calories.
The heat in small or gram-calories per second = (amperes^2 $\times$ ohms) /4.181 = volts^2/(ohms $\times$ 4.181) = (volts $\times$ amperes)/4.181 = watts/4.181.
HEAT. Absolute zero of heat = $-273.1°$ C.
HEFNER UNIT. Photometric standard; *see* page 260.
HENRY. Unit of induction. It is "the induction in a circuit when the electromotive force induced in this circuit is one international volt, while the inducing current varies at the rate of one ampere per second." = 10^9 E. M. U. = $1/9 \times 10^{-11}$ E. S. U.
HORSEPOWER. The English and American horsepower is defined by some authorities as 550 foot-pounds per second and by others as 746 watts. The continental horsepower is defined by some authorities as 75 kilogrammeters per second and by others as 736 watts. See page 197.
JOULE. Unit of work = 10^7 ergs. For electrical Joule see p. xxxvii.
Joules = (volts^2 $\times$ seconds) /ohms = watts $\times$ seconds = amperes^2 $\times$ ohms $\times$ sec.
For conversion factors *see* page 197.
JOULE'S EQUIVALENT. The mechanical equivalent of heat = 4.185×10^7 ergs. *See* page 197.
KILODYNE. 1000 dynes. About 1 gram.
KINETIC ENERGY in ergs = grams $\times$ (cm./sec.)^2/2.
LITER. *See* page 6.
LUMEN. Unit of flux of light-candles divided by solid angles.
MEGABAR. Unit of pressure = 1 000 000 bars = 0.987 atmospheres.
MEGADYNE. One million dynes. About one kilogram.
METER. *See* page 6.
METER CANDLE. The intensity of lumination due to standard candle distant one meter.
MHO. The unit of electrical conductivity. It is the reciprocal of the ohm.
MICRO. A prefix indicating the millionth part.
MICROFARAD. One-millionth of a farad, the ordinary measure of electrostatic capacity.
MICRON. (μ) = one-millionth of a meter.
MIL. One-thousandth of an inch.
MILE. *See* pages 5, 6.
MILE, NAUTICAL or GEOGRAPHICAL = 6080.204 feet.
MILLI-. A prefix denoting the thousandth part.
MONTH. The anomalistic month = time of revolution of the moon from one perigee to another = 27.55460 days.
The nodical month = draconitic month = time of revolution from a node to the same node again = 27.21222 days.
The sidereal month = the time of revolution referred to the stars = 27.32166 days (mean value), but varies by about three hours on account of the eccentricity of the orbit and "perturbations."
The synodic month = the revolution from one new moon to another = 29.5306 days (mean value) = the ordinary month. It varies by about 13 hours.

OHM. Unit of electrical resistance. The international ohm is based upon the ohm equal to 10^9 units of resistance of the C. G. S. system of electromagnetic units, and "is represented by the resistance offered to an unvarying electric current by a column of mercury, at the temperature of melting ice, 14.4521 grams in mass, of a constant cross section and of the length of 106.3 centimeters." $= 10^9$ E. M. U. $= 1/9 \times 10^{-11}$ E. S. U.

International ohm = 1.01367 B. A. ohms = 1.06292 Siemens' ohms.
B. A. ohm = 0.98651 international ohms.
Siemens' ohm = 0.94080 international ohms.

PENTANE CANDLE. Photometric standard. *See* page 260.

PI $= \pi =$ ratio of the circumference of a circle to the diameter = 3.14159265359.

POUNDAL. The British unit of force. The force which will in one second impart a velocity of one foot per second to a mass of one pound.

RADIAN $= 180°/\pi = 57.29578° = 57°\ 17'\ 45'' = 206265''$.

SECOHM. A unit of self-induction = 1 second $\times$ 1 ohm.

THERM = small calorie = (obsolete).

THERMAL UNIT, BRITISH = the quantity of heat required to warm one pound of water at its temperature of maximum density one degree Fahrenheit = 252 gram-calories.

VOLT. The unit of electromotive force (E. M. F.). The international volt is "the electromotive force that, steadily applied to a conductor whose resistance is one international ohm, will produce a current of one international ampere. The value of the E. M. F. of the Weston Normal cell is taken as 1.0183 international volts at 20° C. $= 10^8$ E. M. U. = 1/300 E. S. U. *See* page 197.

VOLT-AMPERE. Equivalent to Watt/Power factor.

WATT. The unit of electrical power $= 10^7$ units of power in the C. G. S. system. It is represented sufficiently well for practical use by the work done at the rate of one Joule per second.

Watts = volts $\times$ amperes = amperes^2 $\times$ ohms = volts^2/ohms (direct current or alternating current with no phase difference).
For conversion factors *see* page 197.
Watts $\times$ seconds = Joules.

WEBER. A name formerly given to the coulomb.

WORK in ergs = dynes $\times$ cm. Kinetic energy in ergs = grams $\times$ (cm./sec.)^2/2.

YEAR. *See* page 414.

Anomalistic	year	= 365	days,	6	hours,	13	minutes,	48	seconds.
Sidereal	"	= 365	"	6	"	9	"	9.314	"
Ordinary	"	= 365	"	5	"	48	"	46+	"
Tropical	"	same as the ordinary year.							

Table 580.

TEMPERATURE MEASUREMENTS.

The ideal standard temperature scale (Kelvin's thermodynamic scale, see introduction, p. xxxiv) is independent of the properties of any substance, and would be indicated by a gas thermometer using a perfect gas. The scale indicated by any actual gas can be corrected if the departure of that gas from a perfect gas be known (see Table 206, p. 195, — also Buckingham, Bull. Bur. Standards, 3, 237). The thermodynamic correction of the constant-pressure scale at any temperature is very nearly proportional to the constant pressure at which the gas is kept and that for the constant-volume scale is approximately proportional to the initial pressure at the ice-point. The gas thermometer has been carried up to the melting point of palladium, 1822° K (1549° C) (Day and Sosman, Am. J. Sc., 29, p. 93, 1910).

A proposed international agreement divides the temperature scale into three intervals. The first interval, −40° to 450° C, uses the platinum resistance thermometer calibrated at the melting point of ice, 0° C, at saturated steam, 100° C, and sulphur vapor, 444.6° C, all under standard atmospheric pressure. Points on the temperature scale are interpolated by the Callendar formulæ:

$$Pt = \frac{R_t - R_0}{R_{100} - R_0} 100 \quad \text{or} \quad t - Pt = \delta \left\{ \frac{t}{100} - 1 \right\} \frac{t}{100}$$

where t is the temperature, R, the resistance, Pt, the platinum temperature, and δ, a constant.

Temperatures in the second interval are measured by a standard platinum-platinum-rhodium couple calibrated say at the freezing points of zinc, 419.4° C, cadmium, 320.9° C, antimony, 630° C, and copper free from oxide, 1083° C. These points furnish constants for the formula, $e = a + bt + ct^2$ (see Sosman, Am. J. Sc., 30, p. 1, 1910).

For the region above 1100° C most experimenters base their results upon certain radiation laws. These laws all apply to a black body and the temperature of a non-black body cannot be determined directly without correction for its emissive power. For standard points the melting points of gold, 1336° K and palladium 1822° K, are convenient.

Above 1336° K the optical pyrometer is generally used with a calibration based upon Wien's equation

$$J_\lambda = c_1 \lambda^{-5} e^{-\frac{c_2}{\lambda T}}$$

By comparing the brightness of a black body at two temperatures and applying this equation, the following formula results:

$$\log R = \frac{c_2 \log e}{\lambda} \left\{ \frac{1}{T_2} - \frac{1}{T_1} \right\}$$

where R is the ratio of the brightnesses, λ, the wave-length used, T_1 and T_2, the two temperatures, and $c_2 = 14.250\ \mu$ deg. Thus if R is measured and one temperature known, the other can be calculated.

A table of the standard fixed points is given in Table 207, p. 195. With these determined there comes the difficulty of maintaining this temperature scale both from the standpoint of the standardizing laboratory and the man using the temperature scale in the practical field. In the region of the platinum-resistance thermometer and the thermocouple, standards of either can be obtained from the standardizing laboratories and used in checking up the secondary instruments. It is not very difficult to actually check up a resistance thermometer at any one of the standard points in the region −40° C to +450° C. It is a little more difficult to check the thermocouple in the region 450° C to 1100° C. Most of the standard fixed points in this region are given by melting points of metals that must be melted so as to avoid oxidation. This requires a neutral atmosphere, or that the sample be covered with some flux that will protect it.

Both the gold and the palladium, used to calibrate the scale above 1300° K, can be successfully melted in a platinum wound black-body furnace. The whole operation can be carried out in the open air, requiring neither a vacuum nor neutral atmosphere within the furnace. But because of the trouble necessitated by a black-body comparison, much time can be saved if a tungsten lamp with filament of suitable size is standardized so as to have the same brightness for a particular part of the filament, when observed with the optical pyrometer, as the standard black-body furnace for one or more definite temperatures. With such lamps properly calibrated, any one may maintain his own temperature scale for years, if the calibration does not extend higher than that of the palladium point and the standard lamp is not accidentally heated to a higher temperature.

(See 1919 Report of Standards Committee on Pyrometry, Forsythe, J. Opt. Soc. of America, 4, p. 205, 1920; The Measurement of High Temperatures, Burgess, Le Chatelier, 1912, The Disappearing Filament Type of Optical Pyrometer, Forsythe, Tr. Faraday Soc., 1919.)

The following additional adsorptio- tables (see page 407, Table 525) may be of use in the "cleaning-up of vacua." See Dushman, General Electric Review, 24, 58, 1921, Methods for the Production and Measurement of High Vacua.

TABLE 581. — Adsorption of H and He by Cocoanut Charcoal at the temperature of liquid air.

For the preparation of activated charcoal see Dushman, l. c. 5 g of charcoal at the temperature of liquid air will clean up the residual gases in a volume of 3000 cm^3 from an initial pressure of 1 bar (bar = 1 $dyne/cm^2$) to less than 0.0005 bars at the temperature of liquid air. 5 grams cleaned up 3000 cm^3 of H from an initial pressure at room temperature of 0.01 bar to a final pressure at liquid air temperature of less than 0.0004 bar. The clean-up is rapid at first but then slower taking about an hour to reach equilibrium. The figures of the following table are from Firth, Z. Phys. Ch. 74, 129, 1910; 86, 294, 1913. p is in mm of Hg; v = volume adsorbed per g of charcoal reduced to 0° C and 76 cm Hg.

Hydrogen				Helium	
p	v	p	v	p	v
9	21.5	90	59.3	120	0.337
17	32.1	126	63.1	171	.465
30	46.5	186	69.2	235	.81
51	53.3	245	76.0	428	1.17
59	56.0			705	1.84

TABLE 582. — Adsorption by Ch rcoal at Low Pressures and temperatures.

Extrapolated by Dushman from Claude, see l. c., and C.R. 158, 861, 1914. Amounts occluded in terms of volume measured at 1 bar, 0° C. e.g. at a pressure of 0.01 bar, 1 g charcoal would clean up 130 cm^3 hydrogen or 18,000 cm^3 nitrogen from a pressure of 1 bar down to 0.01 bar.

H, T = 77.6° K		N, T = 90.6° K	
p = 8.	v = 106,000.	p = 5.3	v = 9,500,000.
1.	13,250	1.	1,800,000
0.1	1,325	0.0	180,000
0.01	133	0.01	18,000
0.001	13	0.001	1,800

TABLE 583. — Adsorption of Hydrogen by Palladium Black.

Palladium, heated, allows hydrogen to pass through it freely; the gas is first adsorbed and then diffuses through. For the preparation of palladium black, see reference at top of page for Dushman. The following data are from Valentiner, Verh. Deutsch. Phys. Ges., 3, 1003, 1911. Different samples vary greatly. P gives the pressure in mm of Hg, and V the volume at standard pressure and temperature per g of palladium black.

−190° C:	P =	.0005	.001	.002	.005	.012	.025
	V =	2.05	3.06	33.0	40.0	47.2	63.0
+20° C:	P =	.001	.005	.037	.110	.315	.76
	V =	0.10	0.26	0.40	0.52	0.70	0.92

TABLE 584.

MISCELLANEOUS SOUND DATA.

TABLE 584. — Audibility as dependent on Sound Pressure and Frequency.

The auditory sense detects sounds varying over a range of pressure from about 0.001 to 1000 dynes/cm²; over much of this range it differentiates with accuracy between complex sounds so nearly alike that no existing physical device can distinguish them. Plot shows minimum audibility pressures from 72 normal ears from 60 to 4000 cycles (both scales logarithmic); standard deviation indicated by dotted curves. The maximum audibility curve was obtained from 48 normal ears. A louder sound becomes painful. The intensity of pressure necessary for the latter is about equal to that required to excite the tactile nerves in the finger tips. (Wegel, Pr. Nat. Acad. Sc., 8, p. 155, 1922.)

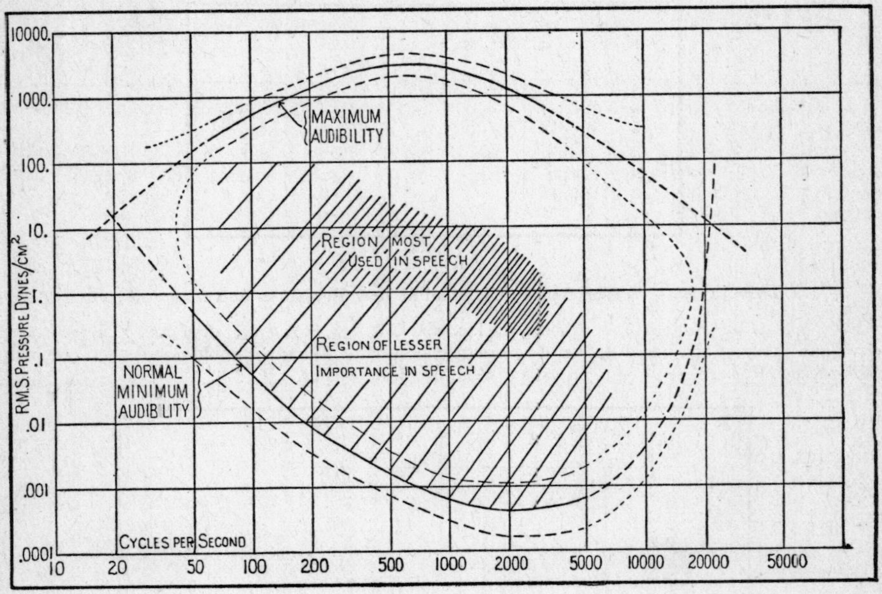

Table 322 (*concluded from page 275*).

SPECTRUM SERIES.

The following table gives the series types due to the neutral (arc) and ionized (spark) elements of the various groups. The question marks indicate merely that the series relations have not been disentangled from the maze of spectrum lines.

Series	VIII, O	I	II	III	IV	V	VI	VII
Arc	complex and triplet	doublet	triplet	doublet	triplet?	doublet?	triplet?	?
Spark	?	complex and triplet?	doublet	triplet	doublet?	triplet?	doublet?	triplet?

To a first approximation the equations leading to the numerical values of the various lines of a series may be expressed in the following form:

$\nu = N/(1+S)^2 - N/(m+P)^2$, for instance, for the principal series. In abbreviated notation these series take the following forms (Fowler). m generally takes for its lowest value, the first value which makes the expression positive. The abbreviated notation expresses the laws above stated.

Principal series = (1,S) − (m,P)
1st subordinate (diffuse) = (1,P) − (m,D)
2nd " (sharp) = (1,P) − (m,S)
Fundamental (Bergmann) = (1,D) − (m,F)

(Singlet, doublet and triplet series are distinguished by capital, Greek and small letters.)

Paschen notation is frequently used especially by writers on ionization and radiating potentials; the following table indicates in part the differences for corresponding formulae:

Series	Revised Paschen		Fowler	
Principal series of doublets of alkalis	1s−mp,	m=2, 3	$1\sigma - m\pi$,	m=1, 2
1st subordinate (diffuse) series of doublets of alkalis	2p−md,	3, 4	$1\pi - m\delta$,	2, 3
2nd subordinate (sharp) series of doublets of alkalis	2p−ms,	2, 3	$1\pi - m\sigma$,	2, 3
Bergmann (fundamental) series of doublets of alkalis	3d−mb,	4, 5	$2\delta - m\phi$,	3, 4
Principal series of triplets of alkali earths	1s−mp,	3, 4	1s −mp,	2, 3
1st subordinate series of triplets of alkali earths	2p−md,	3, 4	1p −md,	2, 3
2nd subordinate series of triplets of alkali earths	2p−ms,	1, 2	1p −ms,	1, 2
Principal series of singlets of alkali earths	1S−mP,	2, 3	1S −mP,	1, 2
Combination series of singlets of alkali earths	$1S - mp_2$,	2, 3	$1S - mp_2$,	1, 2
Principal series of doublets of ionized alkali earths	$1\mathfrak{S} - m\mathfrak{P}$,	2, 3	$1\sigma - m\pi$,	1, 2
1st subordinate series of doublets of ionized alkali earths	$2\mathfrak{P} - m\mathfrak{D}$,	3, 4	$1\pi - m\delta$,	2, 3
2nd subordinate series of doublets of ionized alkali earths	$2\mathfrak{P} - m\mathfrak{S}$,	2, 3	$1\pi - m\sigma$,	2, 3

Other minor differences exist; Paschen values fit the Ritz formula whereas Fowler uses the Hicks formula. The variable terms of the two series may take the following forms respectively: $N/\{m+a+\alpha N/m^2\}^2$ and $N/\{m+a'\alpha'/m\}^2$.

For further information see Foote and Mohler's Origin of Spectra, 1922 (From which considerable of the above is taken) and Fowler's Report of Series in Line Spectra, 1922.

TABLE 518 (*concluded from p.* 403).

IONIZING AND RESONANCE POTENTIALS OF THE ELEMENTS.

Atomic Number Element.	Line.	ν	Resonance volts.		Line.	ν	Ionization volts.	
			Comp.	Obs.			Comp.	Obs.
34 Be								
12 Mg	1S–2p2	21,871	2.700	2.65	1S	61,672	7.613	7.75
	1S–2P	35,051	4.327	4.42				
20 Ca	1S–2p2	15,210	1.878	1.90	1S	49,305	6.087	6.01
	1S–2P	23,652	2.920	2.85				
30 Zn	1S–2p2	32,502	4.012	4.18	1S	75,767	9.353	9.3
	1S–2P	46,745	5.771	5.65				
38 Sr	1S–2p2	14,504	1.791		1S	45,926	5.670	
	1S–2P	21,698	2.679					
48 Cd	1S–2p2	30,656	3.784	3.95	1S	72,539	8.955	8.92
	1S–2P	43,692	5.394	5.35				
56 Ba	1S–2p2	12,637	1.560		1S	42,029	5.188	
	1S–2P	18,060	2.230					
80 Hg	1S–2p2	39,413	4.866	4.76	1S	84,178	10.392	10.2
	1S–2P	54,066	6.674	6.45				
88 Ra	1S–2p2	12,500?	1.5?		1S	4–50000?	5–5.5?	
	1S–2P	20,700?	2.6?					

Atomic Number Element.	Resonance volts.	Ionization volts.	
7 N	8.18	16.9	Ionization at 17.75, 25.4, 30.7, Brandt
15 P	5.80	13.3	
33 As	4.7	11.5	
51 Sb			
83 Bi			
8 O	7.91	15.5	See Hughes for other data
16 S	4.78	12.2	
34 Se	3.0–3.5	12–13	12.7 observed by Udden
52 Te	2.3–2.9	?	
2 He	20.4–21.2	25.6	25.4, Franck and Knipping, later 25.3; 25.5, Compton
10 Ne	11.8–17.8	16.7 20.0 22.8	
18 A	11.5	15.1	
1 H	10.5–13.9	14.4 16.9	Horton and Davies
"	10.4–12.0	13.3 16.0	Mohler and Foote, revised

For further information and data see The Origin of Spectra, Foote and Mohler, 1922; Hughes, Bull. Nat. Research Council, 2, 127–169, 1921; Report on Series in Line Spectra, Fowler, 1922.

Table 585.

LIST OF STARS KNOWN TO BE WITHIN FIVE PARSECS.

The number of stars (doubles counted as singles) per cubic parsec in the neighborhood of the Sun has been estimated as .0451 (Kapteyn and Van Rhijn, Astr. J., 52, 32, 1920). This gives an expectation of 24 within 5 parsecs and 12 nearer than 4. The numbers actually known are 20 and 16. The agreement is good but it seems improbable that we should already know practically all within 4 or 5 parsecs. (Hertzsprung, Com. Observatory at Leiden, no. 5, 1922.)

Star		Right Ascension (1900)	Declination (1900)	vis. magn. (Harvard)	Parallax	abs. magnitude	Spectrum	Proper motion ″/year
α Canis majoris	A	6^h 40.7^m	−16° 34′	− 1.58	.376″	−3.70	A0	1.32
α Aquilae		19 45.9	+ 8 36	+ 0.89	.200	−2.60	A5	.65
α Canis minoris	A	7 34.1	+ 5 29	0.48	.312	−2.05	F3	1.24
Our Sun				−26.9		−0.33	G0	
α Centauri	A	14 32.8	−60 25	0.33	.759	−0.27	G6	3.66
α Centauri	B	14 32.8	−60 25	1.70	.759	+1.10	K4	3.66
τ Ceti		1 39.4	−16 28	3.65	.318	1.16	G7	1.92
ε Eridani		3 28.2	− 9 48	3.81	.311	1.27	K1	.97
ε Indi		21 55.7	−57 12	4.74	.284	2.01	K5	4.67
61 Cygni	A	21 2.4	+38 15	5.57	.306	3.00	K7	5.21
Lac 8760		21 11.4	−39 15	6.65	.251	3.65	Ma	3.53
61 Cygni	B	21 2.4	+38 15	6.28	.306	3.71	K8	5.21
Lac 9352		22 59.4	−36 26	7.44	.292	4.77	Ma	7.02
Gou 32416		23 59.5	−37 51	8.5	.203	5.04	K5	6.11
Groombr 34	A	0 12.5	+43 27	7.98	.281	5.22	Ma	2.85
Lal 21185		10 57.9	+36 38	7.60	.403	5.63	Ma	4.77
Zc 5^h243		5 7.7	−44 59	8.3	.319	5.82	K2	8.70
Oe Arg 17415–6		17 37.0	+68 26	9.2	.247	6.16	Mb	1.31
α Canis majoris	B	6 40.7	−16 34	8.44	.376	6.32	(A)	1.32
Σ 2398	A	18 41.8	+59 29	9.33	.287	6.62	Mb	2.28
Krüger 60	A	22 24.5	+57 12	9.64	.262	6.73	Mb	.94
Σ 2398	B	18 41.8	+59 29	10.01	.287	7.30	(M)	2.28
Groombr 34	B	0 12.5	+43 27	11.05	.281	8.29	(M)	2.85
(Barnard star)		17 52.9	+ 4 28	9.67	.533	8.30	Mb	10.30
Krüger 60	B	22 24.5	+57 12	11.34	.262	8.43	(M)	.94
(van Maanen)		0 43.9	+ 4 55	12.34	.246	9.29	F0	3.01
(Innes)		11 12.0	−57 2	(12.)	.339	(9.65)	—	2.69
α Centauri	C	14 22.9	−62 2	(10.5)	(.759)	(9.90)	(M)	(3.66)
α Canis minoris	B	7 34.1	+ 5 29	(12.5)	.312	(9.97	—	1.24

Degrees Brix, Specific Gravity, and Degrees Baumé of Sugar Solutions.
Degrees Brix = Per cent Sucrose by Weight.

Specific Gravities and Degrees Baumé corresponding to the Degrees Brix are for $\frac{20^\circ}{20^\circ}$ C.

The relation between the specific gravity and Degrees Baumé is given by Degrees Baumé $= 145 - \frac{145}{\text{specific gravity}}$.

Degrees Brix or per cent sucrose by weight	Specific gravity at 20°/20°C	Degrees Baumé (modulus 145)	Degrees Brix or per cent sucrose by weight	Specific gravity at 20°/20°C	Degrees Baumé (modulus 145)	Degrees Brix or per cent sucrose by weight	Specific gravity at 20°/20°C	Degrees Baumé (modulus 145)
0.0	1.00000	0.00	40.0	1.17853	21.97	80.0	1.41421	42.47
1.0	1.00389	0.56	41.0	1.18368	22.50	81.0	1.42088	42.95
2.0	1.00779	1.12	42.0	1.18887	23.04	82.0	1.42759	43.43
3.0	1.01172	1.68	43.0	1.19410	23.57	83.0	1.43434	43.91
4.0	1.01567	2.24	44.0	1.19936	24.10	84.0	1.44112	44.38
5.0	1.01965	2.79	45.0	1.20467	24.63	85.0	1.44794	44.86
6.0	1.02366	3.35	46.0	1.21001	25.17	86.0	1.45480	45.33
7.0	1.02770	3.91	47.0	1.21538	25.70	87.0	1.46170	45.80
8.0	1.03176	4.46	48.0	1.22080	26.23	88.0	1.46862	46.27
9.0	1.03586	5.02	49.0	1.22625	26.75	89.0	1.47559	46.73
10.0	1.03998	5.57	50.0	1.23174	27.28	90.0	1.48259	47.20
11.0	1.04413	6.13	51.0	1.23727	27.81	91.0	1.48963	47.66
12.0	1.04831	6.68	52.0	1.24284	28.33	92.0	1.49671	48.12
13.0	1.05252	7.24	53.0	1.24844	28.86	93.0	1.50381	48.58
14.0	1.05677	7.79	54.0	1.25408	29.38	94.0	1.51096	49.03
15.0	1.06104	8.34	55.0	1.25976	29.90	95.0	1.51814	49.49
16.0	1.06534	8.89	56.0	1.26548	30.42	96.0	1.52535	49.94
17.0	1.06968	9.45	57.0	1.27123	30.94	97.0	1.53260	50.39
18.0	1.07404	10.00	58.0	1.27703	31.46	98.0	1.53988	50.84
19.0	1.07844	10.55	59.0	1.28286	31.97	99.0	1.54719	51.28
20.0	1.08287	11.10	60.0	1.28873	32.49	100.0	1.55454	51.73
21.0	1.08733	11.65	61.0	1.29464	33.00			
22.0	1.09183	12.20	62.0	1.30059	33.51			
23.0	1.09636	12.74	63.0	1.30657	34.02			
24.0	1.10092	13.29	64.0	1.31260	34.53			
25.0	1.10551	13.84	65.0	1.31866	35.04			
26.0	1.11014	14.39	66.0	1.32476	35.55			
27.0	1.11480	14.93	67.0	1.33090	36.05			
28.0	1.11949	15.48	68.0	1.33708	36.55			
29.0	1.12422	16.02	69.0	1.34330	37.06			
30.0	1.12898	16.57	70.0	1.34956	37.56			
31.0	1.13378	17.11	71.0	1.35585	38.06			
32.0	1.13861	17.65	72.0	1.36218	38.55			
33.0	1.14347	18.19	73.0	1.36856	39.05			
34.0	1.14837	18.73	74.0	1.37496	39.54			
35.0	1.15331	19.28	75.0	1.38141	40.03			
36.0	1.15828	19.81	76.0	1.38790	40.53			
37.0	1.16329	20.35	77.0	1.39442	41.01			
38.0	1.16833	20.89	78.0	1.40098	41.50			
39.0	1.17341	21.43	79.0	1.40758	41.99			

The above table is abridged from Bureau of Standards Technologic Paper No. 115. The original table is given in steps of 0.1 Degrees Brix.

SUPPLEMENTARY TO TABLE 380, PAGE 310.

The values on page 310 given for the rotation of sucrose are from antiquated data and are incorrect by several whole degrees. Values obtained at the Bureau of Standards, but as yet published only in part, are given below.

Light Source	Rot. λ / Rot. λ = 5461	$[\alpha]^{20}_{\lambda}$	Light Source	Rot. λ / Rot. λ = 5461	$[\alpha]^{20}_{\lambda}$
Li 6708	.644	50.45	Cd 4678	1.403	109.9
Cd 6438	.711	55.70	Hg 4358	1.644	128.8
Na 5892.5	.84922	66.529	Ag 4208	1.786	139.9
Hg 5780	.8854	69.36	Hg 4047	1.95	152.8
Hg 5461	1.0000	78.342			
Ag 5209	1.108	86.80			
Cd 5086	1.167	91.43			
Cd 4800	1.323	103.65			

The above values are for a near normal solution, i.e. approximately 26 g of sucrose per 100 cc.

SUPPLEMENTARY TO TABLE 532, PAGE 410.

ISOTOPES.

(See J. Am. Ch. Soc. 45, p. 869, 1923.)

Element	Atomic number	Atomic weight	Minimum number of Isotopes	Masses of Isotopes	% Accuracy	Observer
H	1	1.008	1	1.008	0.2	A.
He	2	4.00	1	4		A.
Li	3	6.94	2	7; 6		A., T., D.
Gl	4	9.1	1	9		T.
B	5	10.9	2	11; 10	0.1	A.
C	6	12.005	1	12		A.
N	7	14.008	1	14	0.2	A.
O	8	16.000	1	16		A.
F	9	19.0	1	19	0.1	A.
Ne	10	20.2	2	20; 22	0.1	A.
Na	11	23.00	1	23		A.
Mg	12	24.32	3	24; 25; 26		D.
Al	13	27.0	1	27		A.
Si	14	28.1	2	28; 29; (30)	0.1	A.
P	15	31.04	1	31	0.2	A.
S	16	32.06	1	32	0.2	A.
Cl	17	35.46	2	35; 37	0.1	A.
A	18	39.9	2	40; 36	0.1	A.
K	19	39.10	2	39; 41		A.
Ca	20	40.07	(2)	40; (44)		D.
Fe	26	55.84	(1)	56; (54) ?		A.
Ni	28	58.68	2	58; 60	0.1	A.
Zn	30	65.37	4	64; 66; 68; 70		D.
As	33	74.96	1	75	0.1	A.
Se	34	79.2	6	80; 78; 76; 82; 77; 74	0.1	A.
Br	35	79.92	2	79; 81	0.1	A.
Kr	36	82.92	6	84; 86; 82; 83; 80; 78	0.1	A.
Rb	37	85.45	2	85; 87		A.
Sn	50	118.7	7 (8)	120; 118; 116; 124 119; 117; 122; (121)		A.
I	53	126.92	1	127	0.2	A.
Xe	54	130.2	7 (9)	129; 132; 131; 134; 136; 128; 130; (126); (124)	0.1	A.
Cs	55	132.81	1	133		A.
Hg	80	200.6	(6)	(197–200); 202; 204	0.1	A.

Observers: A = Aston, D = Dempster, T = Thompson (G. P.)

The masses are given in the order of the intensity of the mass spectrum bands. Table published by the International Committee on Chemical elements. The Phil. Mag. 45, 1923, gives also Sb, 2 isotopes, 121, 123.

INDEX.

(The new matter in the Appendix is not indexed.)

PAGE

α particles: energy of 396
helium 394
ions produced 396, 398
production of 394
range 396
stopping powers of substances 395
velocities, initial 396
Abbreviations 2
Aberration constant 414
Absolute units . . . . xxxvi, 311
Absolute zero of temperature 195, 408
Absorption coefficients: see transmission coef.
β-rays 395
γ-rays 395
X-rays 389
Acceleration of gravity 424–426
Accumulators, voltage 313
Actinium group of radioactive substances 396
Activity, definition 435
Adaptation, rate of eye 257
Adsorption, by charcoal 407
by fine particles 407
heat of 407
Aerodynamical tables 150–153
Agonic line 433
Air: composition of; variation with alt. and lat. . . . 421
density of moist 133–135
densities in air, reduction to vacuo 73
dielectric strength 353
humidity relative, via v. p. and dry 187
via wet and dry 189
index of refraction 292
masses 419
moist, transparency, short λ 419
long λ 308
transmission coefficients . . 308, 418, 419
resistance, fluid 151–153
sparkling potentials for 353–354
thermal conductivity, high temperatures . . . 254
thermometer, comparison with 59^{III} 193
vapor pressure of water in 185–186
viscosity 164
wave-lengths in air, reduction to vacuo . . . 293
Albedos 417
Altitudes by barometer 145
by boiling points of water 144
Alternating current resistances 344
Aluminum: mechanical properties 79–80
resistivity constants 334
wire-tables, English units 342
Metric units 343
Ammonia, latent and specific heats 228, 232
Ampere equivalents 311
Ampere turns xlvi
Angstrom, wave-length unit 266
Antennae, resistances 364
Antilogarithms, std. 4-place, p. 28; .9 to 1.0 . . 30
Apothecaries weights 7, 8
Arc, iron, lines 266–267
Astronomical data 411–420
Atmosphere: composition, alt. and lat. var. 421
density, altitude variation 421
height of 421
homogeneous, height of 421
pressure, altitude variation 421
"Atmosphere": value of pressure unit . . . 421, 435
Atmospheric water vapor 185–186
Atom: Bohr 401
Atom, hydrogen: mass, mean free path 408
radius, mean velocity 408
Rutherford 401

PAGE

Atomic heats, elements at 50° K 226
magnetic field 401
magnitudes 401, 408
numbers 409
volumes, elements 226
weights, international 71
β-rays, absorption coefficients 395
absorption of 397
ions produced by 398
velocity 397
Balmer series spectrum formula 275
Bar, definition 435
Barometer, altitude, variation with 421
heights, determination of by 145
reduction for capillarity 143
to std. gravity 138–143
to std. temperature 137
Batteries, composition, voltages 312–313
Baumé scale, conversion to densities 109
Bessel functions, 1st, 2d orders, roots 66–68
Biaxial crystals, formulae, + refractive indices . . 286
— refractive indices . . 287
Black absorbers, long-λ transparencies 309
Black-body: brightness (photometric) of 261
luminosity of 261
luminous efficiency of 261
Planck's constant 247
radiation, total, for various temp. . . 247
by wave-lenths, var. temp. . 248
Stefan-Boltzmann constant 247
-temperature for C, Pt and W 250
Bohr atom 275, 401
Boiling points: elements 199
pressure effect 200
inorganic compounds 201
organic compounds 203
rise of, salts in H_2O 210
water,- pressure variation 144
Boltzman gas-constant (entropy) 408
Bougie decimale 260
Break-down voltage, dielectrics 304, 355
Brightness of sky 419
of stars 413
of sun 260, 413
temperature of C, Pt and W 250
of various illuminations 256
of various light sources 260–262
Brinell hardness 74
British thermal unit 435
British weights and measures 8–11
Brownian movement 406
Buoyancy correction: of densities 73
weighings 73
γ-function 62
γ-rays: absorption coefficients 395
absorption of 397
ions produced by 398
Cadmium line, red, λ of intern. prim. std. 266
Calcite grating space 408
Calibration points for temperatures 195
for thermoelements 196
Calorie, definition 435
Canal rays 386
Candle, energy from 260
international standard 260
meter-candle, foot-candle 259
Capacity of wires for electrical current 329
Capacity, specific inductive: crystals 361
gases, f(t, p) . . 356–357
liquids 357
liq. gases 359
solids 360
standard solutions . 360

PAGE
Capillarity 173–174
correction to barometer 143
Carcel unit 260
Cathode rays: 386
energy of 386
penetration depths 387
X-rays, generative efficiency for . . 387
Cathodic sputtering 387
Cells: standard, voltages 313
Weston normal xli
Weston portable xliii
voltaic, composition, voltages 312–313
Centipoise 155
Characteristic X-rays 387–392
Charcoal, adsorption by 407
Charge, elementary electrical 408
Chemical energy data 241–246
Coals, heats of combustion 242
Collision frequencies, molecules 399
Colloids 406–407
Color: eye sensitiveness to 256–258
indices of various stars 411
lights, of various 261
screens 306–307
complimentary colors 307
temperature of C, Pt and W 250
Combination, heats of 245–246
Combustion, heats of: carbon and misc. cpds . . . 241
explosives 243–244
fuels 242
Compressibility: gases 104, 128–132
liquids 107
solids . . . . 74 seq, 108
Conductivity, electrical: see resistivity 322–332
alloys 327–328
electrolytic: 346–352
equivalent 349
ionic 352
sp. molecular 347
limiting values . . . 348
temp. coefs. 348
Conductivity, thermal: alloys, metals 213
building materials 215
earth 422
gases 217
high temp. 254
insulators 214–216
high temp. . . . 214
liquids 217
metals, high temp. 213
salt solutions 216
water 216
Cones, number and distance apart in eye 258
Constants: mathematical 14
miscellaneous, atomic, etc. 408
radiation, σ, C_1, C_2 247
Contact difference of potential 314, 316, 404
Contrast, eye sensitiveness to 257
Convection, cooling by 251–255
pressure effect 251–252
Conversion factors: general formulae, see introduction.
Baumé to densities 109
horse-power 197
work-units 197
Cooling of bodies 251–254
Copper: mechanical properties 82–83
wire, alternating current resistance 344
wire tables, English units 336
metric units 339
Corpuscle (Thomson) 401
Corpuscular radiation (X-rays) 387–388
Cosines, circular, natural, (° ′) 32–36
(radians) 37–40
logarithmic, (° ′) 32–36
(radians) . . . 37–40
hyperbolic, natural logarithmic 41–47
Cotangents, circular, natural, (° ′) 32–36
(radians) 37–40
logarithmic, (° ′) 32–36
(radians) . . .37–40
hyperbolic, natural, logarithmic . . 41–47
Coulomb, electrical equivalents 311
Critical data for gases 212
Crova wave-length 261

PAGE
Crystals: diffracting units, X-rays 400
elasticity 102–103
refraction indices: alums, quartz 281
fluorite, spar 280
rock-salt, silvine . . . 279
refr. indices: minerals, isotropic 282
uniaxial
(+ −) . 284, 285
biaxial
(+ −) . 286, 287
miscel. uniaxial 285
biaxial 289
Cubes of numbers 15
Cubical expansion coefficients: gases 222
liquids 221
solids 222, 227
Current measures: absolute units xli
equivalents 311
Curie, radium standard 398
point and constant, magnetic 372
Cutting-tool lubricants 154
Cylindrical harmonics (Bessel) 1st and 2nd deg. . 66
general formulæ . 68
roots 68

Day, length of sidereal 414
Declination, magnetic: secular change 428
Degree on earth's surface, length of 416
Demagnetizing factors for rods 374
Densities in air, reduction to vacuo 73
Densities: air moist, values of h/760 133–135
alcohol ethyl aqueous 124
methyl aqueous 126
alloys 114
aqueous solutions 122, 159–163
Baumé equivalents 109
cane sugar, aqueous 126
castor oil 156
earth 427
elements chemical 110
gases 127
glycerol, aqueous 156
inorganic compounds 201
insulators, thermal 215, 216
liquids 117
mercury, −10° to +360° C. 121
minerals 115
organic compounds 203
planets 416
solids various 113
stars 413
sucrose, aqueous 156
sulphuric acid, aqueous 126
tin, liquid; tin-lead eutectic 115
water, 0° to 41° C, −10° to 250° C. 118, 120
woods 95-98, 112
Developers and resolving power of photo. plate . . 263
Diamagnetic properties 365
Diamagnetic susceptibility, temperature effect . . . 372
Diameter molecules 399
some organic molecules 400
Dielectric constant: (sp. inductive capacity) . 356–360
crystals 361
gases f(t, p) 356–357
liquids 357
liquefied gases 359
solids 360
standard solutions 360
Dielectric strength (see sparking potentials) . 353–355
Dielectrics, volume and surface resistances 331
Differentials, formulae 12
Diffusion: aqueous solutions into water 166
gases 167, 168
integral 60
ionic 405
metals 168
vapors 167, 168
Diffusivities thermal 217
Digit 435
Dilution, heat of (H_2SO_4) 246
Dimensional formulae . . . . xxv, 3
reasoning xxv
Diopter 435
Dip, magnetic, 1915 value, secular variation . . . 430
Disk, distribution of brightness over sun's 418
Distance earth to moon 414
sun 414

PAGE

Distance of the stars, nebulae and clusters 412
Dyes, transparency of 301
Dynamical equivalent of thermal unit 197

e (base of natural logarithms) 14
e (elementary electrical charge) 408
e/m 408
e^x, e^{-x}, and their logarithms x=0 to 10 48
x fractional . . . 56
e^{x^2}, e^{-x^2} x=0.1 to 5.0 . . 54
$e^{\frac{\pi}{4}x}$, $e^{-\frac{\pi}{4}x}$ x=1 to 20 . . . 55
$e^{\frac{\sqrt{\pi}}{4}x}$, $e^{-\frac{\sqrt{\pi}}{4}x}$ x=1 to 20 . . . 55
$\frac{e^x+e^{-x}}{2}$, $\frac{e^x-e^{-x}}{2}$ and their logarithms 41
Earth: atmospheric data 421
conductivity, thermal 422
degrees on, length of 416
elements, percentage composition 423
geochemical data 423
geodetical 424–427
moments of inertia 427
moon, distance of 414
rigidity 427
size of, shape of 427
spheroid constants 427
sun, distance of 414
temperatures 420, 422
viscosity 427
Efficiency of various lights 262
Elastic limit (see mechanical properties) 74
Elastic modulus of rigidity, —temp. variation . . 100
Elasticity (see mechanical data) 74
crystals 102, 103
Young's modulus (see mechanical prop.) 74
Electrical charge, elementary 408
Electrical equivalents 311
Electrical units: international xxxvi
standards xxxviii
practical xxxvi
Electric lights, efficiency of 262
Electric, tribo-, series (frictional) 322
Electrochemical equivalents 345, 346
silver 345
Electrolytic conduction: ammonium acetate 352
equivalent conductance 345–346
hydrolysis 352
ionic 352
ionization water 352
solutions 345-346
spec. molecular 347
limiting value μ . . 348
temp. coef. 348
Electro-motive force: accumulators 313
contact 314, 316, 404
Peltier 321
standard cells 313
thermo-electric 317–320
pres. effect . 320
voltaic cells 312-313
Weston normal xli
Weston portable xliii
Electromagnetic system of units xxxi
Electromagnetic/electrostatic units=v . . . xxx, xxxvi
Electrons: —, + ... 401
affinity of elements 404
e/m 404
elementary charge 404
emission from hot bodies 403
ionization potentials 403
mass 408
photo-electric effect 403
radius 408
resonance potentials 403
work required to remove 403
Elements: atomic heats at 50° K 226
atomic numbers 409
atomic volumes 226
atomic weights (international) 71
boiling points 199
compressibility 108
conductivity electrical 323–326
thermal 213
densities 110
earth's crust, occurrence in 423
Elements: evaporation rate, Mo, W, Pt. 175
expansion, cubical (gaseous) 222
linear (solid) 218
hardness 73, 101
isotopes 410
latent heat of evaporation 233
melting points 198
meteorites, occurrence in 423
Peltier effect 317, 320–322
periodic system 409-410
resistance, electrical 323–326
specific heats 223, 225
spectra 266–270
thermal conductivities 213
expansion 128, 222
thermo-electric powers 317, 319
Thomson effect 317, 320
valencies 71
vapor pressures 175
Elementary electric charge 408
Elements, magnetic, at various observatories . . . 434
Elliptic integrals 69
Emanation (radioactive) 398
Emissivities, radiation 249, 250
Energy kinetic, definition 428
of molecule 408
Energy, minimum visible to eye 261
solar, data relating to 418–420
of candle radiation 260
of sound waves 149
Entropy constant (Boltzmann) 408
Entropy of steam 234–240
Equation of time 416
Equilibrium radioactive 394
Equivalent, electrochemical 345–346
mechanical, of heat 197
Erg 435
Erichson values 73
Errors, probable 57–59
Ettingshausen effect 385
Eutectic mixtures, melting points 206, 207
Evaporation rate of, Mo, Pt, W 175
Expansion, cubical: gases 222
liquids 221
solids 220
linear: elements 218
miscellaneous 219
Explosives: decomposition, ignition temp. 244
miscellaneous 243–244
Exponential functions: see index under e. 41–55
diffusion integral 60
gudermanians 41
hyperbolic functions (nat) . 41
hyperbolic functions (logs) . 41
probability integral . . . 56–57
Eye: adaptation rate 257
color sensitiveness 256–258
contrast sensibility 257
Fechner's law 258
glare sensibility 257
heterochromatic sensibility 257
minimum energy visible 261
miscellaneous data 258
persistence of vision 258
pupil size for various intensities 258
Purkinje phenomenon 256
sensitiveness to light 256-258
small dif. of color, sensitiveness to 258
threshold sensitiveness 256
visibility of radiation, relative 258

Factorials: γ function, n=1 to 2 62
$n!$, n=1 to 20 47
logs, n=1 to 100 40
Falling bodies (Stokes' law) 150
Farad 311
Faraday . . . . xliv, 311, 345
Faraday constant 408
Fechner's law 258
Ferromagnetism 365
Field, magnetic: earth's, components 428–434
metals, behavior of in . . . 365–377
resistance of metals in 384
rotation of plane of polarization . 378
thermo-galvanometric effects . . 385
Filaments, heat losses from 255
Flame temperatures 244
Fluidity 155

PAGE
Foot pound 436
Fork, tuning, temperature coefficient 149
Formation, heat of, for elements 245, 246
Formulæ: conversion 3
dimensional, see introduction xxv
least squares 59
Fraunhofer lines, solar, wave-lengths 265
Free path of molecules 399
Freezing mixtures 211
point, lowering of for salt solutions . . . 208
point of water, pressure effect 200
Frequencies, corresponding to wave-lengths 293
in air, reduction to vacuo 293
Friction, mechanical 154
internal of metals, temp. variation . . . 101
skin (air resistance) 152
Frictional electricity series 322
Functions: Bessel functions (roots, 68) . . . 66–68
cylindrical harmonics 66–68
elliptic 69
exponential 48–56
gamma 62
hyperbolic 41
probability 56–58
trigonometric, circular (° ′) 32
trigonometric, circular (radians) . . . 37
zonal harmonics 64
Fundamental frequency (Rydberg) 408
Fundamental standards xxxiii
units xxiii
Fusion current for wires 329
Fusion, latent heat of 240

Gages, wire 333
Galvano-magnetic effects 385
Gamma function 62
Gas constant 408
Gas thermometry 192–194
Gases: absorption of by liquids 172
absorption of by water 170–171
absorption coef: long-wave radiation . . . 309
X-rays 389
compressibility 104
conductivity, thermal 217
critical data 212
densities 127
dielectric constants 356–361
strength 353–355
diffusion 168
ionic 405
expansion coefficients 222
expansion of 128–132
flow in tubes 150
ignition temperatures of mixtures 244
magnetic susceptibility 377
magnetic-optical rotation 382
refractive indices 292
resistance (aerodynamical) 150–153
solubility in water 170–171
sound, velocity of, in 147
specific heats (also c_p/c_v) 230
viscosity 164–165
volume, f(t, p) 104–106
f(t), 1+0.00367, logs . . . 128–132
Gauss . . . . xlv, 365
Gaussian system of units xxxv
Geochemical data 423
Geodetic data 424–427
Gilbert 311
Glare sensibility of eye 257
Glasses: refraction indices, American 277
German, temp. var. . . 278
resistance electric, temp. var. 332
transparency of 302–304, 306–307
Glass vessels, volume of 72
Gram-molecule, definition 436
Grating space of calcite 408
Gravitation constant 427
Gravity, acceleration of, altitude variation 424
latitude variation 424
observed values . . . 425–426
Gravity specific, see densities.
Gudermanians 41
Gyration, radii of 70

Hall effect, temperature variation 385

PAGE
Hardness: (see mechanical properties) 74
Brinell test 74
elements 101
scleroscope test 74
Harmonics: cylindrical (Bessel) 66–68
roots, formulæ . . 68
zonal 64
Heat: adsorption heats 407
atomic heats of elements 226
combination 245–246
combustion: explosives 243–244
fuels 242
gases 242
organic compounds 241
conductivity: metals (also high temp.) . . . 213
gases 217, 254
liquids 217
diffusivities 217
dilution, heat of, H_2SO_4 246
formation 245–246
latent heat of fusion 240
vaporization, elements . . . 233
NH_3 232
steam 234
various . . 232–233
pressure variation, NH_3 liq. . . 232
losses from incandescent wires 255
mechanical equivalent of 197
neutralization, H_2SO_4 246
solution 246
specific: alloys 227
ammonia, liq. 228
electricity 317
elements 233
gases 230
liquids 227–228
mercury 227
minerals, rocks 229
silicates 229
solids 227
true [elements, f(t)] 225
vapors 230
water 227
total [elements, f(t)] 225
treatment of steels 76
Heating effect, radium and emanation 394
Hefner unit 260
Heights, barometric determination of 145
boiling point of water determination of . . 144
Helium, production, relation to radium 394
Henry . . . . xxxvii, xliv, 311
Heterochromatic sensibility of eye 257
Hertzen wave-lengths 408
High-frequency electric resistance of wires 344
Horizontal intensity earth's field, 1915 431
secular var. . . 431
Horse power 197
Humidity, relative: vapor-pressure and dry 187
wet and dry 189
Hydrogen: atomic data, mass, radius, etc. 408
series spectra 275, 401
thermometer 192–194
Hydrolysis of ammonium acetate 352
Hydrostatic pressures of Hg and H_2O columns . . 136
Hyperbolic functions, natural and logarithmic . . . 41
Hysteresis . . . . 365 et seq, 375–376

Ice, allotropic modifications 200
freezing point, pressure effects 200
Ice-point, thermodynamic scale 195
Ignition temperatures gaseous mixtures 244
Incandescent filaments, heat losses 255
Inclination (dip) of magnetic needle, 1915 430
secular var. . . 430
Index of refraction: air 293
alums 281
crystals, see minerals, etc. . 282–289
fats 289
fluorite, f(t) 280
gases and vapors 292
glass American 277
German f(t) 278
Iceland spar 280
liquefied gases 289
liquids 290
metals 296
minerals, isotropic 282
uniaxial 284
biaxial 286

PAGE

Index of refraction: miscellaneous, isotropic . . 283
uniaxial . . . 285
biaxial . . . 289
nitroso-dimthyl-aniline . . . 280
quartz 280
rock-salt, f(t) 279
salt solutions 291
silvine 279
solids, biaxial 286, 289
isotropic 282, 283
uniaxial 284, 285
standard media for microscope. 294
vapors 292
waxes 289
Induction, self 376
Inductive capacity, specific: crystals 361
gases, f(t, p) . . 356–357
liquids 357
liq. gases 359
solids 360
standard solutions . . 360
Inertia, moments of 70
Inorganic compounds: boiling points 201
densities 201
melting points 201
soluibilities 169
Insulators: break-down potentials 364
dielectric properties 364
resistance, thermal 214–216
electrical 331
Integral: diffusion 60
elliptic 69
formulæ 12
gamma function 62
probability 56–57, 60
Intensity, horizontal, earth field, 1915 431
secular var. . . . 431
Intensity, total, earth field, 1915 432
secular var. 432
International candle standard 260
electric units xxxvi
standards xxxviii
standard radium 394, 398
standard wave-lengths 266–267
Intrinsic brightness of various lights 260
Ionic charge 401, 408
diffusion 405
mobilities 405
Ionization potentials 403
Ions produced by α, β, γ rays 398
work required to detach 403–404
Ions, conductance of 352
heat of formation 246
Iron, magnetic properties (steels) 365–376
standard wave-lengths, international . . 266–267
Isostacy 426
Isotopes 410

Joule 197
Joule magnetic effect 365

K X-ray spectrum series 390
Kerr's constant, magneto-optic 383
Kinetic energy 436
molecular 408
Kundt's constant, magneto-optic 383

L X-ray series 391
Lambert, definition 256, 259
Latent heat of fusion 240
Latent heat of pressure variation liq. ammonia . . . 232
Latent heat of vaporization: ammonia 232
formulæ 232
steam tables 234
various 231
Latitude correction to barometer 139–142
of a few stations 420, 434
Least squares: formulæ 59
probability integral, arg, hx . . . 56
x/r . . . 57
inverse . . . 60
$0.6745\sqrt{1/(n-1)}$ 57
$0.6745\sqrt{1/n(n-1)}$ 58
$0.8453\sqrt{1/n(n-1)}$ 58
$0.8453[1/n\sqrt{n-1}]$ 58
Leduc thermomagnetic effect 385
Legal electrical units xxxvii

PAGE

Length, standards of . . . . xxxiv, 5
Light: eye, sensitiveness of to 256–258
flux, definition 259
intensities on various days 256
lambert, definition 256
least visible to eye 261
mechanical equivalent 261
photometric standards 260
units 259
polarized, reflection 295–297
rotation of plane by substances . . 310
rotation of plane, magnetic . 378–383
reflection of: formulæ 297
function of "n" 297
reflecting power: metals 295–298
pigments 299
powders 300
rough surfaces 299
scattered light 300
temperature variation . . . 300
sensitiveness of eye to 256–258
transparency to: crystals 305
dyes 301
glasses, American . . 303–304
Jena 302
water 307
velocity of 408, 414
wave-lengths: cadmium std. line 266
elements 269–271
Fraunhofer lines 265
solar, Rowland 272
Std. iron lines 266–267
Lights, brightness of various 260
color of various 261
efficiency of various electric 262
photographic efficiency of 264
visibility of white lights 260
Light-year 414
Limits of spectrum series 276
Linear expansion coefficients 218–219
Liquids: absorption of gases by 172
Baumé density scale 109
capillarity of 173–174
combustion heat, fuels 242
compressibilities 107
conductivity, thermal 217
contact emf 314–316
densities 115–117
mercury, f(t) 121
water, f(t) 118–120
dielectric constant 357–360
strength 355
diffusion, aqueous solutions 166
expansion coefficients 221
expansion coefficients 221
fuels, combustion heats 242
magnetic optic rotation 380
magnetic susceptibility 377
potential dif. with substances 314–316
refractive indices 289–291
sound velocity in 147
specific heats 228
surface tensions 173–174
thermal conductivity 217
expansion, cubical 221
vapor pressures 175–187
viscosity, absolute 157–159
specific, solutions 163
Logarithms: standard 4-place 26
1000 to 2000 24
anti-, standard 4-place 28
.9000 to 1.0000 30
Logarithmic functions 40
Longitudes of a few stations 420, 434
Long-wave transmissions 309
Loschmidt's number 408
Lowering of freezing points by salts 208
Lubricants for cutting tools 154
Lumen 259
Luminosity of black-body, f(t) 261
Lunar parallax 414
Lux 259

M X-ray spectrum 392
Mache radioactivity unit 398
Maclauren's theorem 13

PAGE

Magnetic field: atomic 402
bismuth, resistance in 384
Ettingshausen effect 385
galvanometric effects 385
Hall effect 385
Joule effect 365
Leduc effect 385
Nernst effect 385
nickel, resistance in 384
optical rotation polarization . 378–383
resistance of metals in 384
thermo-magnetic effects 385
Villari effect 365
Wiedemann effect 365
Magnetic observatories, magnetic elements 434
Magnetic properties: cobalt, 0° to 100° C. . . . 373
Curie constant 372
Curie point 372
definitions 365
demagnetizing factor for rods . 374
diamagnetism, f(t) . . 365, 372
ferro-cobalt alloy 370
ferromagnetism 365
hysteresis 375–376
iron: 367–376
cast, intense fields . . 368
pure 369
soft, 0 and 100° C. . . . 371
very weak fields 370
wrought 373
magnetite, 0°, 100° C. . . 373
magneto-strictive effects . . 365
magnet steel 370
maxwell 311, 365
nickel, 0°, 100° C. 373
paramagnetism, f(t) . . 365, 372
permeability 365, 371
saturation values for steels . .373
steel: 367–376
energy losses 376
magnet steel 370
manganese steel . . . 373
saturation values . . . 373
temperature effect . 371–372
tool steel 373
transformer steel . 371, 376
weak fields 370
Steinmetz constant 375
susceptibility . . 365, 372, 377
temperature effects . . 371–373
Magnetism terrestrial: agonic line 425
declination 420
dip 422
inclination 422
intensity, horizontal . . . 423
total 424
magnetic character yearly . 425
observatories, elements at . 426
Magneto-optic rotation, gases 382
Kerr constant 383
Kundt constant 383
liquids 380
solids 379
solutions 381
Verdet's constant . . 378–383
Magnitudes, absolute stellar 415, 412
stellar 415, 412
sun 415
Mass: electronic, f(velocity) 401, 408
fundamental standard xxxiv
hydrogen atom 408
Mass absorption coefficient for X-rays 389
Masses, stellar 404
Mathematical constants 14
physical 408
Maxwell . . . . .xlv, 311, 365
Mean free path, H molecule 408
velocity H molecule 408
Measures, weights: customary—metric 5–7
English—metric 8–11
Mechanical equivalent of heat 197
light 261
Mechanical properties: definitions 74
elastic limit . . 74
Erichson value . . 74
hardness 74
moduli 74

Mechanical properties: definitions: modulus of rupture 74
proportional limit. 74
scleroscope . . . 74
ultimate strength,
compr. 74
tension 74
yield point . . . 74
alloys: aluminum 81
brasses 83–85
bronzes 83–85
copper 83–85
iron 75–79
miscel. 88–89
steel 77
white metal 89
aluminum: 80–81
alloys 81
brick and brick piers 93
cement 90
cement mortars 90
clay products 93
concrete 91
copper: 82
brasses and bronzes 83–85
wire 82–83
heat treatment for steels . . 76
iron: 75
alloys 75
leather belting 94
ρ-ratio extension/contraction. 101
rigidity moduli f(t) 100
rope, manila 95
steel-wire 79
rubber, sheet 94
steel: 76
alloys 77
heat treatment for . . 76
semi- 78
wire 78
wire-rope 79
stone products 92
terra-cotta piers 93
tungsten 89
white metal 89
woods: conifers: English unit 99
metric unit . 97
hard: English unit . 98
metric unit . . 96
Melting points: alloys 206
elements 198
eutectics 207
inorganic compounds 201
lime-alumina-silica compounds . . 207
organic compounds 203
paraffins 203
pressure effect 200
water-ice, pressure effect 200
Meniscus, volume of mercury 143
Mercury: density and volume, —10° to 360° C. . 121
conductivity thermal, high temp. 254
electric resistance standard xxxviii
meniscus, volume of 143
pressure hydrostatic of columns 136
specific heat 227
thermometer 190–194
vapor pressure 180
Metals: conductivity, thermal 213
diffusion 168
potential differences, Volta 316, 404
reflection of light by 295–296, 298
refraction indices 295–296
optical constants 295-298
resistivity, temperature coefficient 323
pressure effect 326
Volta emf 316, 404
weight sheet metal 116
Metallic reflection 295–296, 298
Meteors, chemical composition 423
Meter-candle 256, 259
Metric weights and measures, equivalents . . . 5–10
Mho 436

PAGE

Micron, μ . . . 7, 436
Milky way, pole of . . . 414
Minerals: densities . . . 115
refractive indices: biaxial . . . 286
isotropic . . . 282
uniaxial . . . 284
specific heats . . . 229
Minimum energy for light sensation . . . 261
Mixtures freezing . . . 211
Mobilities, ionic . . . 405
Moduli, —see mechanical properties . . . 74–103
Mogendorf series formula . . . 275
Moist air, density of . . . 133–135
maintenance of . . . 135
transparency to radiation, .36 to 1.7μ. 411
to 20μ . . . 308
Molecular collision frequencies . . . 399
conductivities: equivalent . . . 349–352
specific . . . 346–348
crystal units . . . 400
diameters . . . 399, 400
free paths . . . 399
heats of adsorption . . . 407
liquefaction . . . 407
kinetic energy . . . 408
magnitudes . . . 399–400, 408
number in cm^3, 76 cm, 0° C. . . . 408
gram-molecule . . . 408
velocities . . . 399
weights of colloids . . . 406
Moments of inertia: earth . . . 427
formulæ . . . 70
Month . . . 414, 436
Moon: albedo . . . 417
distance from earth, parallax . . . 414
radiation compared with sun's . . . 415
Musical scale . . . 148
tone quality . . . 149
Mutual induction . . . 376

Nernst thermomagnetic potential difference . . . 385
Neutral points, thermoelectric . . . 317
Neutralization, heat of . . . 246
Nickel, Kerr's constants for . . . 383
magnetic properties, 0 to 100° . . . 373
resistance in magnetic field . . . 384
Nitrogen thermometer . . . 192
Nitroso-dimethyl-aniline, refractive index . . . 280
Nuclear charge, atomic . . . 393, 401
Number of stars . . . 417
Numbers atomic . . . 409
X-ray spectra and . . . 390–393
Numbers: magnetic character . . . 433
sun-spot . . . 415
Nutation . . . 414

Observatories, magnetic elements at . . . 434
Ohm: . . . xxxvii, xxxviii, 311
electrical equivalents . . . 311
Oersted . . . xlvi
Oils, viscosity of . . . 156, 157
Optical constants of metals . . . 295
Optical rotation magnetic . . . 378–383
Optical thermometry . . . 250
Organ pipes, pitch . . . 149
Organic compounds: boiling points . . . 203
densities . . . 203
melting points . . . 203
Organic salts, solubilities . . . 170
Oscillation constants wireless telegraphy . . . 362
times of wires, temperature variation . . . 101
Overtones . . . 149

π pi . . . 14, 436
P-limit (proportional limit) . . . 74 et seq.
Parsec . . . 414
Parallax, solar, lunar . . . 414
stellar . . . 412, 415
Paramagnetism . . . 365
Partials (sound) . . . 149
Particle, smallest visible . . . 406
Peltier effect: . . . 317, 321
pressure effect . . . 320

Pendulum, second: formula; latitude variation . . . 427
Penetration cathode rays . . . 387
high speed molecules . . . 387
Pentane candle . . . 260
thermometer . . . 194
Periodic system: Hackh . . . 410
Mendelejeff . . . 409
Permeability, magnetic . . . 365 et seq.
Persistence of vision . . . 258
Petrol-ether thermometer . . . 194
Phosphorescence (radio-active excitation) . . . 394
Phot . . . 259
Photoelectricity . . . 403
Photographic data: intensification . . . 264
lights, efficiencies . . . 264
plate characteristics . . . 263
resolving power . . . 263
speeds various materials . . . 263
Photometric definitions, units . . . 259
standards . . . 260
Physiological constants of the eye . . . 258
Pi (π) . . . 14, 436
Pigments, reflecting powers f(λ) . . . 299
Pipes, organ: pitch . . . 149
Pitch: . . . 148
organ pipes . . . 149
voice, limits . . . 149
Planck's "h" . . . 408
radiation formulæ, C_1 C_2 . . . 247
Plane, air resistance to . . . 150–152
Planetary data . . . 416
Platinum resistance thermometer . . . 195
thermoelectric thermometer . . . 196
thermoelectric powers against . . . 319
Poisson's ratio . . . 101
Polonium radioactive series . . . 398
Polarized light: reflection by . . . 295–296, 297
rotation of plane . . . 310
magnetic . . . 378–383
Porcelain, resistance, f(t) . . . 332
Positive rays . . . 386
Potential (emf): accumulators . . . 313
cells voltaic . . . 312–313
contact . . . 314, 316, 404
ionizing . . . 403
Peltier . . . 321
sparking, kerosene . . . 355
various . . . 353–355
resonance . . . 403
standard cells . . . 313
thermo-electric . . . 317–320
pressure effect . . . 320
Weston normal . . . xli
portable . . . xliii
Poundal . . . 436
Precession . . . 414
Pressure: air, on moving surfaces . . . 150–152
barometric, reductions, capillarity . . . 143
gravity . . . 138–143
temperature . . . 137
boiling water . . . 144
critical, gases . . . 212
mercury columns . . . 136
volume relations, gases . . . 104
water columns . . . 136
Pressure effect on boiling points . . . 200
melting points . . . 200
resistance electrical . . . 326
thermoelectric powers . . . 320
Pressure vapor: alcohol, methyl and ethyl . . . 178
aqueous (steam tables 234) . 183–186
elements . . . 175
mercury . . . 180
salt solutions . . . 181
water vapor (steam tables 234) 183–186
various . . . 176–181
Probable errors . . . 56–59
Probability integral . . . 56–57
inverse . . . 60
Proportional limit (P-limit) . . . 74 et seq.
Pupil diameter . . . 258
Purkinje phenomenon . . . 256

Quality, tone . . . 148
Quartz: refraction indices . . . 286
transmission of radiation by . . . 305

PAGE

R, gas constant 408
ρ, Poisson's ratio 101
Radian 436
  circular functions in terms of 37
Radiation: black-body, formulæ 247
      f(λ,T) 247, 248
      total, f(t) 247
  candle 260
  constants, σ, C_1, C_2 247
  cooling by, across airspaces 253
      high temperatures 254
      ordinary temperatures 253
      pressure effect 251–252
  emissivities 249, 250
  eye sensitiveness to 256-258
      f(λ) 256, 258
  moon's compared to sun's 415
  Planck's formula 247
  σ, Stefan's formula 247
  solar constant of 418
    variation with latitude 420
      month 420
  Stefan's formula 247
  sun's to earth 418
  sun's compared to moon 415
  temperature as function of 250
  transmissibility of by air, moist 308, 419
      alum 305
      atmosphere 308, 419
      crystals, f(λ) 305
      dyes, f( ) 301
      fluorite, f(λ) 305
      glass, f(λ) 302–304
      ice-land spar 305
      lamp-black 309
      long-wave, f(λ) 309
      quartz, f(λ) 305
      rock-salt, f(λ) 305
      water, f(λ) 307
  visibility of by eye 256, 258
Radii of gyration 70
Radio-activity 394–398
  α rays: helium 394
    ions produced 398
    kinetic energy 396
    number produced 396
    production of 394
    range 396
    stopping powers for 395
    velocity, initial 396
  actinium group 396
  β rays: absorption coefficients 395, 397
    ions produced 398
    velocities 397
  γ rays: absorption coefficients 395, 397
    ions produced 398
  constants, various 396–397
  Curie unit 398
  emanation 398
  general characteristics 394
  heating effects 394
  helium, Production of 394
  ions produced by α, β and γ rays 398
  isotopes 410
  Mache unit 398
  phosphorescence 394
  radium group 396
  spectra 398
  standard, international 394
  thorium group 396
  transformations 396, 410
  transformation constants 396
  vapor pressure of emanation 398
Radium emanation, vapor-pressure 398
  group 396
  spectra 398
Radius hydrogen atom 408
Reciprocals 15
Reflection of light: formulæ 297
  long-wave 309
  metals 295–298
  miscellaneous 298
  f(n,i) 297
  pigments dry, f(λ) 299
  polarization by 297
  rough surfaces 299
  stellite 296
  variation with angle 300
  temperature 300

PAGE

Refraction index: air, f(λ) 293
  alum 281
  crystals, see minerals 279–289
  fats 289
  fluorite, f(t) 280
  gases 292
  glasses, American, f(λ) 277
    Jena, f(λ, t) 278
  Iceland-spar 280
  liquids 290
  liquefied gases 289
  microscopic determination media 294
  minerals, biaxial, positive 286
      negative 287
    isotropic 282
    uniaxial, positive 284
      negative 284
  miscellaneous, biaxial 289
      isotropic 283
      uniaxial 285
  nitroso-dimethyl aniline 280
  oils 289
  quartz 280
  rock-salt, f(t) 279
  salt solutions 291
  silvine 279
  vapors 292
  waxes 289
Relative humidity, vapor pressure and dry 187
  wet and dry 189
Resistance, air (aerodynamical) 151–153
  planes 151
    angle factor 152
    aspect factor 152
  shape factor 153
  size factor 153
  skin friction 152
  speed factor 153
Resistance, resistivity electrical (see conductivity).
  alloys, f(t) 323, 327–328
  aluminum 334
    wire tables 342, 343
  alternating current values 344
  antennæ (wireless) 364
  copper, f(t) 334, 335
    reduction to standard temperature 335
    wire-tables, English units 336
      metric units 339
  dielectrics, volume, surface 331
  electrolytic, see conductivity 345–352
  equivalents 311
  glass, f(t) 332
  high frequency values 344
  high temperature values 330
  low temperature values 330
  magnetic field, effect of 384
  mercury resistance standards xxxviii
  metals, f(t) 323
  porcelain, f(t) 332
  pressure effect 326
  standards, mercury xxxviii
  surface, of dielectrics 331
  thermometer, platinum resistance 195
  volume, of dielectrics 331
  wire, auxiliary table for computing 322
  wire tables, aluminum, common units 342
      metric 343
    copper, common units 336
      metric 339
Resolving power photographic plate 263
Resonance potentials (spectra) 403
Retina, physiological data 258
  sensitiveness to light and colors 256–258
Rigidity of earth 427
Rigidity moduli, f(t) 100
Ritz spectrum series formula 275
Rocks, specific heats of 229
Rock-salt, index of refraction 279
Rods in retina of eye 258
Röntgen rays: 383–393
  absorption coefficients (mass) 389
  atomic numbers and spectra 390–393
  cathode efficiencies 387
  corpuscular radiation 387–388
  characteristic radiations 387
  energy relations 387
  general radiations 387

PAGE
Röntgen rays: heterogeneous radiations 387
homogeneous radiations 387
independent radiations 387
intensity 388
ionization 388
K series of radiations 390
L series of radiations 391
M series of radiations 392
monochromatic radiations 387
secondary radiations 387
spectra: absorption 393
K series 390
L series 391
M series 392
tungsten 392
wave-length and cathode fall 387
Roots of Bessel functions, 1st and 2nd orders 68
Roots square 15
Rope, manilla, mechanical properties 95
steel wire, mechanical properties 79
Rotation of polarized light 310
magnetic 378–382
Rough surfaces, reflecting power 299
Rowland solar wave-lengths 272
Rupture, moduli of 74 et seq
Rutherford atom 401
Rydberg constant 408
Rydberg series formula (spectrum) 275

σ, Stefan-Boltzmann 247
Salt solutions: boiling point raising 210
conductivity thermal 216
freezing point lowering 208
vapor pressure 181
Scales, musical 148
Scleroscope (hardness test) 74
Screens, color 306–307
Second pendulum, formula 419
sea-level values, f(ϕ) 419
Secohm xliv
Secondary batteries 313
X-rays 387
Self induction 376
Sensation, Minimum energy of light for 261
Sensitiveness of eye to light and radiation 256–258
Series, mathematical 13
Series spectra: Balmer formula 401, 275
first terms 276
limits of 276
Morgendorff formula 275
Ritz formula 275
Rydberg formula 275
vibration differences 276
Sheet metal, weight of 116
Silver, electrochemical equivalent 345
Silver voltameter xl
Silvine, refractive index 279
Sines, natural and logarithmic, circular 32
hyperbolic 41
Sky brightness 419
Skin friction, air resistance 152
Soap films 174
Solids: compressibility 108
contact potentials 314, 316, 404
densities 113
dielectric constants 360
expansion coefficients, cubical 227
linear 218–220
hardness 74 et seq, 101
magneto-optic rotation 379
refractive indices 277–289
resistance, electrical 323–344
velocity of sound 146
Verdet's constant 379
Solubility: gases in water 170
pressure effect 171
salts in water, inorganic, f(t) 169
organic, f(t) 170
Solutions: boiling point raise by salts in 210
conductivity electrolytic 346–352
thermal 216
densities of aqueous 122
diffusion of aqueous 166
dielectric constant, calibration stds. 360
freezing points lowering by salts in 208
magneto-optic rotation by 381

PAGE
Solutions: refractive indices 291
specific heats 228
surface tensions 173
vapor pressures 181
Verdet's constant 381
viscosities, specific 159, 163
Sound, velocity of: gases 147
liquids 147
solids 147
waves, energy of 149
Sparking potentials: air, alternating potentials 353
large spark-gaps, f(p) 354
steady potentials 353
dielectrics 355
kerosene 355
Specific heat of electricity 317
Specific heats: ammonia, sat. liq. 228
elements 223, 225, 226
gases, also c_p/c_v 230
liquids 228
mercury 227
minerals and rocks 229
rocks 229
silicates 229
solids 227
vapors 230
water 227
Specific gravities (see densities) 109–135
conversion of Baumé 109
Specific inductive capacities: crystals 361
gases, f(t,p) 356–357
liquids, f(t) 357–359
solids 360
std. solutions 360
Specific molecular conductivity 347–348
Specific resistance, see resistivity 323–326
Specific viscosity: 159, 163
Spectrum: black-body intensities 247, 248
elements, international units 267, 270
eye sensitiveness, f(λ) 256–258
iron standards, international units 266, 267
radium 398
series, limits, first terms, etc. 276
solar: intensities of energy 418
Rowland wave-lengths 272
cor. to intern. scale 272
standard wave-lengths, intern. units. 266–267
stellar 403
wave-lengths standards 266, 267
reduction to std. pressure. 268
X-ray: absorption 393
atomic numbers 390–393
K series 390
L series 391
M series 392
tungsten 392
Speed of corpuscles 401
Spherical harmonics 64
Sputtering, cathodic 386
Squares of numbers 15
Square roots of numbers 15
Squares, least, — formulæ and tables 56–59
Standard cells, emf of 313
radium, international 394
refractive media for microscope 294
resistance, mercury xxxviii
temperature calibration points 195
wave-lengths: primary (international) 266
secondary (international) 266
tertiary (international) 267
reduction to std. pressure 268
Standards: electrical, international xxxviii
fundamental xxxiii
photometric 260
Stars: brightness 413
densities 413
distances 412
equivalent 1st magnitude 417
first magnitude data (positions, etc.) 415
Harvard classification 411
light, total 417
magnitudes, apparent and absolute 413
masses 413
motions 412
number of 417
parallax 415

PAGE
Stars: size . . . 413
spectra . . . 411
temperatures, surface . . . 411
velocities . . . 412, 415
Steam tables . . . 234
Steel: magnetic properties . . . 367–376
mechanical properties . . . 76–79
Stefan-Boltzmann constant, and formula . . . 247
Steinmetz magnetic constant . . . 375
Stellite, reflecting powers . . . 150
Stem correction for thermometers . . . 190–191
Stokes law for falling bodies . . . 150
Stone, mechanical properties . . . 92
Storage batteries . . . 313
Strengths—see mechanical properties . . . 74–99
Sucrose, viscosities of solutions, f(t) . . . 156
Sugar cane, densities aqueous solutions . . . 126
Sulphuric acid, densities aqueous solutions . . . 126
Sun: apex of solar motion . . . 411
brightness . . . 260
disk brightness distribution . . . 418
distance to earth . . . 414
Fraunhofer lines . . . 265
magnitude, stellar . . . 413
motion . . . 411
numbers, Wolf's sun-spot . . . 415
parallax . . . 414
radiation compared to moon . . . 414
constant (solar constant) . . . 418
variation with month and latitude . . . 420
spectrum: energy intensities . . . 418
Fraunhofer lines . . . 265
Rowland's wave-lengths . . . 272
spot numbers, Wolf's . . . 415
temperature . . . 418
velocity . . . 411
wave-lengths: Fraunhofer lines . . . 265
Rowland's . . . 272
Sunshine, duration of f(month, latitude) . . . 417
Surface resistivities, solid dielectrics . . . 331
Surface tensions . . . 173, 174
Susceptibility magnetic, definition . . . 365
elements, etc . . . 377

Tangents circular, nat. and log., f(°, ′) . . . 32
f(radians) . . . 37
hyperbolic, nat. and log . . . 41
Taylor's series . . . 13
Telegraphy, wireless: . . . 362, 364
Temperature, black-body scale for W . . . 250
brightness black body as function of. 261
brightness scale for C . . . 250
color scale for W . . . 250
critical gas constants . . . 212
earth: f(altitude) . . . 421
f(latitude) . . . 422
monthly and yearly means . . . 420
variation below surface . . . 422
flame temperatures . . . 244
ignition, gaseous mixtures . . . 244
standards . . . xxxiv
stellar . . . 411
sun's . . . 418
thermodynamic . . . 195
zero absolute . . . 195
Tensile strengths, see mechanical properties . . . 74–99
Tension, surface . . . 173, 174
Tensions, vapor, see vapor pressures . . . 175–186
Terrestrial magnetism: agonic line . . . 425
declination . . . 420
dip . . . 422
inclination . . . 422
intensity, horizontal . . . 431
total . . . 424
magnetic character, yearly . 425
observatories, elements at . 426
Thermal unit, British . . . 435
standard calorie . . . 435
Thermal conductivity: alloys, metals . . . 213
building materials . 215
earth . . . 422
gases . . . 217
high temperature 254

PAGE
Thermal conductivity: insulators . . . 214–216
high temp. 214
liquids . . . 217
metals, high temperature . . . 213
salt solutions . . . 216
water . . . 216
Thermal diffusivities . . . 217
Thermal expansion: cubical: gases . . . 222
liquids . . . 221
solids . . . 220
linear: elements . . . 218
miscellaneous . . . 219
Thermal unit: British . . . 435
calorie . . . 435
dynamical equivalents . . . 197
Thermo-chemistry: heat of combustion: carbon cpds . 241
coals . . . 242
cokes . . . 242
gases . . . 242
liquid fuels . 242
miscellaneous . 241
peats . . . 242
heat of dilution, H_2SO_4 . . . 246
formation . . . 245
ions . . . 246
neutralization . . . 246
Thermodynamical scale of temperature, ice-point . . 195
Thermoelectrical properties: (emf) alloys . . . 318
platinum 319
elements . . . 317
Peltier . . 317, 321
pressure effects . 320
Thomson . 317, 320
Thermoelements, calibration of . . . 196
Thermogalvanometric effect . . . 385
Thermomagnetic effects . . . 385
Thermometry: absolute zero . . . 195
air — 16^{III} 0° to 300° C . . . 193
59^{III}, 100° to 200° C . . . 193
59^{III}, high temperature . . . 194
calibration points, standard . . . 195
gas-mercury, formulæ, comparisons 192–194
hydrogen — 16^{III}, 0° to 100° C . . 192
16^{III}, 59 —5° to —35° C . . . 192
various . . . 194
ice point . . . 195
Kelvin scale . . . 193
mercury cf with gas . . . 190–194
platinum resistance . . . 195
resistance electrical . . . 195
stem corrections . . . 190–191
thermodynamic scale, ice-point . . . 195
thermo-electric, Cu-Constantan . . . 196
Pt-PtRh . . . 196
Thomson thermo-electric effect . . . 317, 320
Thorium radio-active group, constants of . . . 396
Threshold sensitiveness of eye . . . 256
Timber, strength of . . . 96–99
Timbre (sound) . . . 149
Time, equation of . . . 416
Time, solar, sidereal . . . 414
Time standards . . . xxxiv
Transformation constants of radio-active substances. 396
Transformation points of minerals . . . 207
Transmissibility to radiation: air, moist . . . 308, 419
atmospheric . . . 418
crystals, various . . . 306
dyes . . . 301
glass, American . 303, 304
Jena . . . 302
water . . . 307
water-vapor . . . 308, 419
Trigonometric functions: circular, (° ′) nat., log. . 32
(radians), log . . 37
hyperbolic, nat., log. . . . 41
Tribo-electric series . . . 322
Tubes, flow of gas through . . . 150
Tuning forks, temperature coefficients . . . 149

Ultimate strengths of materials, see Mechanical. properties . . . 74–99

PAGE

Units of measurements, see Introduction xxiii
electrical, absolute xxxvi
international . . xxxvi
legal . . . xxxvii
practical . . . xxxvi
fundamental xxiii
photometric . . . 260
radioactive 394
work, transformation factors 197
Uranium group of radio-active substances 396

v, ratio electro-magnetic to — static units . . xxx, xxxvi
Vacuo, reduction of densities to 73
weighings to 73
Valencies of the elements 71
Van der Waal's constants 212
Vaporization, latent heat of: 231
ammonia 232
elements, theoretical . 233
formulæ 232
steam tables 234
Vapor pressure: alcohol ethyl 178
methyl 178
aqueous (see water below) . . 183, 234
carbon disulphide 179
elements 175
mercury 180
radium emanation 398
salt solutions 181
various 179–180
water: atmospheric via wet and dry
sea-level . . . 186
other altitudes . 185
saturated 183
steam tables 234
Vapor, water; weight per m^3 and ft^3 185
Vapors: densities 127, 234
diffusion of 167, 168
heats of vaporization 231–239
heats specific 230
pressures, see vapor pressures 175–186
refractive indices 292
specific heats 230
viscosities 164
water, transparency to radiation . . . 308, 419
Velocity of light 408, 414
molecules 399
sound, in gases 147
liquids 147
solids 146
stars 411, 412
sun 411
Verdet's constant (magneto-optic) 378–382
gases 382
liquids 380
solids 379
solutions, aqueous 381
Vessels, volume of glass, via Hg 72
Villari magnetic effect 365
Viscosity: air 164
alcohol ethyl, f(t, dilution) 155
castor oil, f(t) 156
centipoise, definition 155
definition 155
earth 427
gases, temperature and pressure var. 164–165
glycerol, dilution variation 156
liquids, f(t) 157–158
specific: solutions, f(dens., t) 159
atom. conc. 25° C. . 163
sucrose solutions, f(t, dilution) 156
vapors, f(p, t) 164–165
water, f(t) 155
Visibility of radiation 256–258
relative of various colors 256–258
white lights 260
Vision, distinct 258
persistence of 258
Voice, pitch limits of 149
Voltages: accumulators 313
contact 314, 316, 404
Peltier 321
standard cells 313
thermoelectric 317–320
pressure effect 320
voltaic cells 312–313
Weston normal xli
Weston portable xliii

PAGE

Voltaic cells; comp., emf: double fluid 312
secondary 313
single fluid 313
standard 313
storage 313
Voltameter, silver xl
Volts, electrical equivalents 311
Volts, legal, international . . . . xli, 311
Volume atomic, 50° K 226
critical for gases 212
glass vessels, determination of 72
gases, f(p) 104–106
f(t) 128–132
mercury, — 10° to + 360° C. 121
resistance of dielectrics 331
specific of elements 225
water, 0° to 40° C. 119
— 10° to 250° C. 120
Vowels, tone characteristics 149

Waal's (van der) constants 212
Water: boiling point, f(p) 144
density, f(t) 0° to 41°C, — 10° to 250°C 118–120
solutions, ethyl alcohol 124
glycerol 156
methyl alcohol 126
sucrose 156
sugar (cane) 126
sulphuric acid 126
various . . . 122, 159–163
freezing point, f(pressure) 200
ionization of 352
melting point, f(p) 200
pressure (hydrostatic) of columns of . . . 136
solutions: boiling points 210
densities, see water, density solutions . . 118–126, 156, 159–163
diffusion 166
electrolytic conduction . . 346–352
freezing points 208
viscosities 156, 159–163
specific heat 227
thermal conductivity 216
transparency to radiation 307
vapor pressure 234–239, 183–184
vapor pressure of in atmosphere . . . 185–189
viscosity, f(t) 155
volume, 0° to 40° C, — 10° to 250° C. 119, 120
Water-vapor, determination in atmosphere, via wet
and dry sea-level 186
various altitudes 185
relative humidity via wet and dry . . 189
dry and v.p. . . 187
in atmosphere, f(altitude) 421
transparency 308, 419
weight saturated per m^3 and ft^3 . . . 185
Watt . . . . xxxvii, xlv, 311
Wave-lengths: Angstrom, definition 266
cadmium red line 266
Crova 261
elements, international scale . . 267, 270
Fraunhofer solar lines 265
iron arc standards 266–267
neon, international scale 266
pipes (sound) 149
limits, Hertz, X, visible, etc. . . . 408
Röntgen 390–393
Rowland solar 272
corrections to Intern λ 272
solar, Fraunhofer lines 265
Rowland 272
standard pressure, correction to . . . 268
standards: international primary . . 266
secondary 266–267
tertiary . . 267
vacuo, reduction to 293
wireless 362–364
X-ray 390–393
Waves, energy of sound 149
Weighings, reduction to vacuo 73
Weight sheet metal 116
Weights, atomic 71
Weights and Measures: customary to metric . . . 5
metric to customary . . . 6
metric to imperial 8
imperial to metric 10
miscellaneous 7

PAGE
Weston normal cell xli
Weston portable cell xliii
Wiedemann magnetic effect 365
Wind pressures 150–153
Wire gages, comparison 333
Wire, mechanical properties: copper 82–83
steel 78
steel rope and cable . 79
Wire resistance, auxiliary table for computing . . . 322
Wire tables: aluminum, English measures 342
metric measures 343
see also 334
copper, English measures 336
metric measures 339
see also 334
temperature coefficients . 334, 335
reduc. to std. . . 335
Wires, alternating-current resistance 344
carrying capacity of 329
high-frequency resistance 344
Wires, heat losses from incandescent, bright Pt. . 255
Pt sponge . 255
Wireless telegraphy: antennæ resistances 364
wave-lengths, frequencies, oscillation constants 362
Wolf sun-spot numbers, 1750 to 1917 415
Woods: densities 96–99, 112
mechanical properties: conifers, Eng. units. 99
metric units 97
hard wds, Eng. units 98
metric units 96
Work, conversion factors 197

PAGE
X-rays: 383–393
absorption coefficients (mass) 389
atomic numbers and spectra 390–393
calcite grating space 408
cathode efficiencies 387
characteristic radiations 387
corpuscular radiation 387–388
crystals, diffraction with 401
energy relations 387
general radiations 387
heterogeneous radiations 387
homogeneous radiations 387
independent radiations 387
intensity 388
ionization 388
K series of radiations 390
L series of radiations 391
M series of radiations 392
monochromatic radiations 387
secondary radiations 387
spectra: absorption 393
K series 390
L series 391
M series 392
tungsten 392
wave-lengths 390–393
wave-lengths and cathode fall 387
Years 414, 437
Yearly temperature means 420
Young's modulus, definition 74
values 74–103
Yield point (mechanical property) 74–103
Zero, absolute, thermodynamic scale 195
Zonal harmonics 64

The Riverside Press
CAMBRIDGE · MASSACHUSETTS
PRINTED IN THE U.S.A.